COMPACT HEAT EXCHANGERS AND ENHANCEMENT TECHNOLOGY FOR THE PROCESS INDUSTRIES-2001

Proceedings of the Third International Conference on Compact Heat Exchangers and Enhancement Technology for the Process Industries held at the Davos Congress Centre, Davos, Switzerland, July 1-6, 2001

Editor
Ramesh K. Shah
Delphi Harrison Thermal Systems
Lockport, New York, USA

... ...

Associate Editors
Alan W. Deakin
BP Chemicals
Sunbury, Middlesex, United Kindgom

Hiroshi Honda
Kyushu University
Kasuga, Fukuoka, Japan

Thomas M. Rudy
ExxonMobil Research and Engineering Company
Florham Park, New Jersey, USA

Begell House, Inc.
New York, Wallingford (UK)

Compact Heat Exchangers and Enhancement Technology for the Process Industries 2001

EDITOR

Ramesh K. Shah

Library of Congress Cataloging-in-Publication Data

Catalog record is available from the Library of Congress.

This book represents information obtained from authentic and highly regarded sources. Reprinted material is quoted with permission, and sources are indicated. A wide variety of references are listed. Every reasonable effort has been made to give reliable data and information, but the authors and the publisher cannot assume responsibility for the validity of all materials or for the consequences of their use.

All rights reserved. This book, or any parts thereof, may not be reproduced in any form without written consent from the publisher.

Direct all inquiries to Begell House, Inc., 145 Madison Avenue, New York, NY 10016.

© 2001 by Begell House, Inc.

ISBN: 1-56700-164-5

Printed in the United States of America 1 2 3 4 5 6 7 8 9 0

AUTHOR INDEX

Afgan, N.
Agarwal, V.K.
Alabi, K.
Ando, K.
Augustyniak, J.D.
Auracher, H.

Bacquet, N.
Bae, K.Y.
Balzereit, F.
Bar-Cohen, A.
Baudoin, B.
Bernoux, P.
Bezgin, L.S.
Bhaumik, S.
Boccardi, G.
Bontemps, A.
Borisov, I.
Bougeard, D.
Boxma, B.

Cailloux, B.D.
Carvalho, M.G.
Celata, G.P.
Chatel, F.
Chopard, F.
Chun, S.H.
Chunangad, K.S.
Chung, H.S.
Chung, M.H.

Das, L.
Das, S.K.
Deng, B.
Ding, R.
Dreister, G.A.
Dubrovsky, E.V.
Dzyubenko, B.V.

Fiedler, S.
Filippov, Y.N.
Folomeev, E.A.
Fornalik, M.
Forster, R.N.G.
Fowles, P.
Fujita, H.

Galankin, V.M.
Galitseysky, B.M.
Gately, W.
Gitteau, J.
Gorenflo, D.
Grishaev, V.V.
Groll, M.

Gruszczynski, D.
Guo, Z.Y.
Gupta, S.C.
Hahne, E.
Hanamura, K.
Heidemann, W.
Hirota, M.
Holm, M.
Honda, H.
Hsu, K.C.
Huang, L.
Huang, S.H.
Hübner, P.
Huo, X.

Ichimiya, K.
Ilchenko, M.A.
Inaoka, K.
Ishizuka, M.
Iwasaki, H.

Jachuck, R.J.
Jeong, H.M.
Ji, M.K.
Jia, R.

Kakiyama, S.
Kalatov, A.
Kandlikar, S.G.
Kang, H.C.
Kapustenko, P.O.
Karayiannis, T.G.
Kawaguchi, K.
Kawano, K.
Kim, D.H.
Kim, E.
Kim, J.H.
Kim, J.K.
Kim, J.S.
Kim, J.W.
Kim, M.H.
Kong, T.W.
Kulenovic, R.
Kumada, M.

Ladeinde, F.
Lebouché, M.
Lee, H.S.
Lee, J.K.
Lee, J.M.G.
Lee, S.Y.
Lel, V.V.
Leznov, A.S.
Li, Z.X.

Lin, H.H.
Lin, Y.T.
Lobanov, I.E.
Lockett, M.J.
Luke, A.

Master, B.I.
Matsui, H.
Mercier, P.
Mertz, R.
Minakami, K.
Moon, C.G.
Morita, H.
Müller-Steinhagen, H.
Myacotchin, A.S.

Nagorna, O.G.
Nakayama, H.
Nazarov, A.D.
Niclout, N.
Nilsson, I.

Obot, N.T.
Ohara, T.
Onishi, H.

Panigrahi, A.
Park, D.S.
Pavlenko, A.N.
Perevertaylenko, O.
Polley, G.T.
Prstic, S.
Pua, L.M.
Puccini, C.

Rabas, T.J.
Ranganayakulu, Ch.
Rebholz, H.
Reneaume, J.M.
Rezai, A.
Roetzel, W.
Roques, J.F.
Roy, A.
Russeil, S.

Sadrameli, M.
Saito, M.
Sasaki, N.
Schäfer, P.
Sekimura, M.
Seol, W.S.
Serov, A.F.
Shah, R.K.
Shokouhmand, H.

Shyu, R.J.
Soudarev, A.V.
Soudarev, B.V.
Srinivasan, V.
Sundén, B.
Suzuki, K.
Svetlakov, A.L.

Tanaka, H.
Tao, W.Q.
Tabrizi, N.S.
Terao, T.
Thome, J.R.
Thonon, B.
Tian, Y.S.
Tolba, M.B.
Tolmachov, A.V.
Tourreuil, J.
Tovazhnyansky, L.L.
Tribes, C.

Ursenbacher, T.
Utriainen, E.

Vasiliev, V.Y.
Vijaya Vittala, C.B.
Volov, V.T.

Wadekar, V.V.
Wang, C.C.
Wang, H.S.
Wang, J.H.
Wang, L.
Wang, Q.W.
Wang, W.Q.
Werlen, E.
Wilcox, M.B.
Wilhelmsson, B.
Wojtan, L.

Xia, Z.Z.
Xuan, Y.M.

Yanagida, M.
Yang, B.C.
Yoo, S.Y.
Yoon, J.I.
Yoshida, H.

Zettler, H.U.
Zhang, H.L.
Zhu, X.X.
Zubair, S.M.

TABLE OF CONTENTS

Preface ... xi

SINGLE-PHASE FLOW AND HEAT TRANSFER FUNDAMENTAL STUDIES ... 1

Numerical Analysis of Laminar Flow and Heat Transfer in Corrugated Undulated Ducts for Recuperators ... 3
E. Utriainen and B. Sundén

A Three-Dimensional Unsteady Numerical Analysis for a Plate-Finned Heat Exchanger in the Middle Reynolds Number Range ... 9
H. Onishi, K. Inaoka and K. Suzuki

Numerical Heat Transfer in a Top Opening Rectangular with a Heating Source ... 17
K. Y. Bae, T. W. Kong, M. K. Ji, H. M. Jeong and H. S. Chung

Heat Transfer Characteristics in Two-Pass Rectangular Channels with Inclined Partition Wall ... 23
M. Hirota, H. Fujita, M. Yanagida and H. Nakayama

SINGLE-PHASE AUGMENTATION TECHNIQUES ... 31

Keynote Lecture: Some Enhancement Studies for Laminar and Turbulent Heat Transfer of Air in Tubes and Across Plates ... 33
W.Q. Tao and W.Q. Wang

Heat Transfer Characteristics of Fin-Tube Heat Exchanger with Vortex Generators ... 49
S.Y. Yoo, D.S. Park, M.H. Chung and J.K. Kim

Influence of Conjugate Heat Transfer in Compact Heat Exchangers with Vortex Generators ... 57
J. Tourreuil, D. Bougeard and B. Baudoin

Investigation of Heat Transfer Process in Channels with Cylindrical Intensifiers ... 65
B.M. Galitseysky

Enhancement of Heat and Mass Transfer in Heat Exchangers with Helical Tubes ... 73
B.V. Dzyubenko and G.A. Dreitser

Performance Evaluation of Heat Transfer Enhancement by Twisted Tapes ... 81
K. Ichimiya

Study on Convective Heat Transfer in Fiber-Finned Ducts Based on the Concept of Equivalent Thermal Conductivity ... 89
Z.Z. Xia, Z.X. Li and Z.Y. Guo

Combined Techniques of Heat Exchange Enhancement along the Paths of Gas-to-Gas Coolers for Small GTE ... 97
A.V. Soudarev, B.V. Soudarev, V.V. Grishaev and A.S. Leznov

Superlean Combustion Using Couterflow Heat Exchanger Filled with Porous Media ... 105
H. Yoshida, M. Saito and H. Matsui

Rules of Similarity for the Vortex Electro-Discharged Plasmotron ... 113
V.T. Volov

Heat-Hydraulic Performance of Rolled-up Thin-Wall Small-Diameter Tubes at Longitudinal Flow of Heat Carrier 117
A.L. Svetlakov, E.A. Folomeev, L.S. Bezgin, V.M. Galankin, M.A. Ilchenko, A.V. Tolmachov, and Y.N. Filippov

SINGLE-PHASE HEAT TRANSFER DESIGN DATA AND METHODS 125

Keynote Paper: From Fin Selection to Network Design – A Holistic Design Approach for Compact Heat Exchanger Networks 127
X.X. Zhu and L.M. Pua

Design and Optimization of Compact Heat Exchangers 135
R. Jia, B. Sundén and Y. Xuan

Optimal Design of Plate-Fin Heat Exchangers using Both Heuristic based Procedures and Mathematical Programming Techniques 143
J.M. Reneaume and N. Niclout

An Improved Approach to the Design of Plate Heat Exchangers 151
L. Wang and B. Sundén

A Method of Relative Comparison of Thermo-Hydraulic Efficiencies of Heat Transfer Surfaces and Heat Exchangers 159
E.V. Dubrovsky and V.Yu. Vasiliev

Numerical Investigation of Plate Heat Exchanger Design 169
H. Rebholz, W. Heidemann, E. Hahne, and H. Müller-Steinhagen

Two-Dimensional Numerical Method for Thermal and Hydraulic Design of Multistream Plate-Fin Heat Exchanger 179
B. Deng, Q.W. Wang and W.Q. Tao

Smooth- and Enhanced-Tube Heat Transfer and Pressure Drop: Part I. Effect of Prandtl Number with Air, Water, and Glycol/Water Mixtures 187
N.T. Obot, L. Das, and T.J. Rabas

Smooth- and Enhanced-Tube Heat Transfer and Pressure Drop: Part II. The Role of Transition to Turbulent Flow 199
N.T. Obot, L. Das, and T.J. Rabas

Theoretical and Experimental Study of Axial Dispersion in Packed-Bed Heat/Mass Transfer Equipment 205
S.K. Das, A. Roy, W. Roetzel and F. Balzereit

Heat Transfer Enhancement and Flow Visualization of Wavy-Perforated Plate-and-Fin Surface 215
Y. M. Xuan, H. L. Zhang and R. Ding

Positron Emission Particle Tracking (PEPT) – A New Technique to Investigate the Flow Pattern in a Plate and Frame Heat Exchanger with Corrugated Plates 223
H. U. Zettler, H. Müller-Steinhagen, R.N.G. Forster, P. Fowles

A Simple and Industrial Test Facility for Corrugated Fins Evaluation 231
F. Châtel and E. Werlen

SINGLE-PHASE HEAT EXCHANGER DEVELOPMENT AND APPLICATIONS 239

Keynote Lecture: Research and Development on the High Performance Ceramic Heat Exchangers ... 241
M. Kumada and K. Hanamura

Influence of Header Design on Pressure Drop and Thermal Performance of a Compact Heat Exchanger ... 251
Ch. Ranganayakulu and A. Panigrahi

Port Flow Distribution in Plate Heat Exchangers ... 259
L. Huang

The Spiral Heat Exchanger Concept and Manufacturing Technique ... 265
N. Bacquet

Reduced Total Life Cycle Costs Using Helixchanger™ Heat Exchangers ... 269
B.I. Master, K.S. Chunangad, B. Boxma, G.T. Polley, and M.B. Tolba

Micro Channel Heat Exchanger for Cooling Electrical Equipment ... 275
K. Kawano, M. Sekimura, K. Minakami, H. Iwasaki and M. Ishizuka

Sustainability Assessment of Aluminium Heat Sink Design ... 283
S. Prstic, A. Bar-Cohen, N. Afgan and M.G. Carvalho

A New Approach to the Thermal Analysis of Rotary Air Preheater ... 291
H. Shokouhmand and A. Rezai

Simulation of Regenerators Utilizing Encapsulated Phase Change Materials (PCM) ... 297
M. Sadrameli and N.S. Tabrizi

Performance of Material Saved Fin-Tube Heat Exchangers in Dehumidifying Conditions ... 303
H.C. Kang and M.H. Kim

Frictional Performance of Rectangular Fin in Partially Wet Conditions ... 311
Y.T. Lin, K.C. Hsu, R.J. Shyu and C.C. Wang

PHASE-CHANGE HEAT TRANSFER FUNDAMENTAL STUDIES ... 317

Keynote Lecture: Two-Phase Flow Patterns, Pressure Drop and Heat Transfer during Boiling in Minichannel and Microchannel Flow Passages of Compact Evaporators ... 319
S. G. Kandlikar

Review of Aspects of Two-Phase Flow and Boiling Heat Transfer in Small Diameter Tubes ... 335
X. Huo, Y.S. Tian, V.V. Wadekar and T.G. Karayiannis

Two-Phase Flow Distribution in a Compact Heat Exchanger ... 347
P. Bernoux, P. Mercier and M. Lebouché

Technique for Measurement of Void Fraction in Stratified Flows in Horizontal Tubes ... 353
L. Wojtan, T. Ursenbacher and J.R. Thome

Dividing Two-Phase Flow within a Small Vertical Rectangular Channel with a Horizontal Branch ... 361
J. K. Lee and S. Y. Lee

Heat Transfer during Reflux Condensation of R134a in a Small Diameter Inclined Tube ... 369
S. Fiedler and H. Auracher

Experimental Study of Condensation on Two Flat Vertical Parallel Plates ... 377
S. Russeil, C. Tribes and B. Baudoin

Circumferential Temperature Distributions on Plain and Finned Tubes in Pool Boiling — 383
P. Hübner, D. Gorenflo and A. Luke

Flow Patterns and Phenomena for Falling Liquid Films on Plain and Enhanced Tube Arrays — 391
J.F. Roques and J.R. Thome

Flow Dynamics and Intensification of the Heat Transfer in Precritical Regimes of Intensively Evaporating Wavy Liquid Film — 399
A.N. Pavlenko, V.V. Lel, A.F. Serov and A.D. Nazarov

VAPORIZATION, CONDENSATION AND ABSORPTION AUGMENTATION TECHNIQUES — 407

Pool Boiling Experiments with Liquid Nitrogen on Enhanced Boiling Surfaces — 409
V. Srinivasan, J.D. Augustyniak, and M.J. Lockett

Boiling of Hydrocarbons on Tubes with Subsurface Structures — 415
R. Mertz, R. Kulenovic, P. Schäfer and M. Groll

Enhancement of Boiling Heat Transfer of Ethanol by PTFE Coating on Heating Tube — 423
C.B. Vijaya Vittala, S. Bhaumik, S.C. Gupta, and V.K. Agarwal

Heat Transfer Enhancement with a Surfactant in Horizontal Bundle Tubes on an Absorber — 429
J.I. Yoon, E. Kim, W.S. Seol, C.G. Moon and D.H. Kim

VAPORIZATION AND CONDENSATION DATA AND METHODS — 435

A New Procedure for Two-Phase Thermal Analysis of Multi-Pass Industrial Plate-Fin Heat Exchangers — 437
F. Ladeinde and K. Alabi

A Simple Method for Evaluation of Heat Transfer Enhancement in Tubular Heat Exchangers under Single-Phase Flow, Boiling, Condensation and Fouling Conditions — 445
G.A. Dreitser, A.S. Myacotchin and I.E. Lobanov

Theoretical Study on the Effects of Tube Diameter and Tubeside Fin Geometry on the Heat Transfer Performance of Air-Cooled Condensers — 457
H.S. Wang and H. Honda

Effects of Several Factors in Tube on the Capacity of Air-Cooled Plate-Fin and Round-Tube Heat Exchangers — 467
N. Sasaki, S. Kakiyama and H. Morita

Alternative Refrigerants Performance in Evaporators of Plate Heat Exchanger Type — 471
G. Boccardi and G.P. Celata

Characteristics of Heat Transfer and Pressure Drop of R-22 Inside an Evaporating Tube with Small Diameter Helical Coil — 479
J.W. Kim, J.H. Kim and J.S. Kim

Condensation of Pure Hydrocarbons and Their Mixture in a Compact Welded Heat Exchanger — 487
J. Gitteau, B. Thonon and A. Bontemps

The Simulation of Multicomponent Mixtures Condensation in Plate Condensers — 494
L.L. Tovazhnyansky, P.O. Kapustenko, O.G. Nagorna, O. Perevertaylenko

PHASE-CHANGE HEAT EXCHANGER DEVELOPMENT AND APPLICATIONS — 501

Keynote Lecture: Use of Plate Heat Exchangers in Refinery and Petrochemical Plants — 503
J.H. Wang

Development and Application of Semiconductor Device Cooler Using High Performance Plate Fin Heat Exchanger — 509
K. Ando

Boiling Refrigerant Type Compact Cooling Unit for Computer Chip — 515
H. Tanaka, T. Ohara, K. Kawaguchi and T. Terao

Reduction of Maldistribution in Large Rising Film Plate Evaporators — 519
M. Holm, I. Nilsson and B. Wilhelmsson

Enhanced Reboilers for the Process Industry — 525
B. Thonon

Process Intensification: Performance Studies and Integration of Multifunctional Compact Condenser in Distillation Processes — 533
B.D. Cailloux, J.M.G. Lee and R.J. Jachuck

A Study on the Advanced Performance of an Absorption Heater/Chiller Using Waste Gas — 539
E. Kim, J.I. Yoon, H.S. Lee, C.G. Moon and S.H. Chun

The Experimental Analysis for the Application of High-Temperature Latent Heat Exchanger — 545
S.H. Huang, H.H. Lin and B.C. Yang

FOULING IN HEAT EXCHANGERS — 551

Fouling in Plate-and-Frame Heat Exchangers and Cleaning Strategies — 553
S.M. Zubair and R.K. Shah

Diagnosing, Characterizing, and Controlling a Liquid Transfer System Cleaning Problem — 567
M. Fornalik, D. Gruszczynski, C. Puccini, and M.B. Wilcox

Fouling Reduction in Gelatin-Based Photographic Emulsions Using Spiral Wound Membranes — 573
W. Gately and M.B. Wilcox

PREFACE

The drive to minimize capital investment and improve the energy efficiency of process industry plants has led to a reassessment of the desirability and practicality of incorporating compact heat exchangers (CHEs) and heat transfer enhancement technology into process plants. CHEs are characterized by high heat transfer areas per unit volume (above 400 m^2/m^3) and mass, usually achieved by construction techniques that result in a large number of small channels. Enhancement technology increases either the heat transfer area by extended surface elements ("fins") or the heat transfer coefficient by modifying the flow patterns near the surface. The resistance to introducing these technologies into process plants is driven by concerns about fouling and cleanability, ruggedness, safety (especially in high temperature, high pressure applications), and designability. However, continuing advances at both the fundamental and equipment development levels and a growing database of successful plant experiences are demonstrating that significant capital and operating cost savings can be achieved by the reasoned application of CHEs and enhanced heat transfer technology.

The first International Conference on Compact Heat Exchangers for the Process Industries was held during June 22-27, 1997 in Snowbird, Utah, USA. The Second International Conference on Compact Heat Exchangers and Enhancement Technology for the Process Industries was held in Banff, Canada during July 18-23, 1999. They were sponsored by The United Engineering Foundation, New York, USA. Each of these conferences was attended by about 80 specialists with more than 50% participants from process and other industries. The conference program included invited lectures, tutorial lectures/short course, panel discussions, and contributed technical papers. These conferences were highly successful and unique in that they brought together a good mix of technical specialists from industry, universities and government organizations on a very focused subject area. The objectives of this Conference were to exploit present applications of advanced heat exchanger technology, evaluating benefits and drawbacks; to identify further areas where advanced heat exchanger technology could be used; and to identify and discuss barriers and critical issues promoting the broader use of CHEs and enhancement technology for the process industry applications.

This book includes most of the papers presented at this Conference. A total of 72 papers from 18 countries are included in the book. The book is divided into the following sections.

Single-Phase Heat Transfer Fundamental Studies
Single-Phase Augmentation Techniques
Single-Phase Design Data and Methods
Single-Phase Heat Exchanger Development and Applications
Phase-Change Heat Transfer Fundamental Studies
Vaporization, Condensation and Absorption Augmentation Techniques
Vaporization and Condensation Design Data and Methods
Phase-Change Heat Exchanger Development and Applications
Fouling in Heat Exchangers

These papers represent a focused attention to the use of CHEs and Enhancement Technology in the process industries and indicate enormous opportunities in the process industries.

We appreciate the efforts and assistance provided by the following regional committee members who took the responsibility of encouraging appropriate experts from industry and academia to present papers and getting submitted papers reviewed.

B. Arman	R. Jachuck	W. Roetzel
M. Behnia	K. Kasano	R.J. Shyu
G.P. Celata	M. Kim	B. Sundén
Y. Chen	M. Kumada	K. Suzuki
G.A. Dreitser	F. Ladeinde	W.Q. Tao
B. J. Drazner	B. Ljubicic	J.R. Thome
H. Fujita	B.I. Master	B. Thonon
L. Huang	R.M. Manglik	V.V. Wadekar
M. Ishizuka	J. Meyer	

We also acknowledge the continued encouragement and assistance provided by Barbara Hickernell and Dr. Frank Schmidt of the Engineering Foundation in organizing this Conference. Our thanks to xxx xxx of Begell House, Inc. for his contribution toward publication of this book in a short lead time.

Editor
Ramesh K. Shah

Associate Editors
Alan W. Deakin
Hiroshi Honda
Thomas M. Rudy

LIST OF REVIEWERS

Andrews, M.J.
Arkharov, A.M.
Auracher, H.

Bang, K.H.
Bontemps, A.
Bourdakov, V.P.

Celata, G.P.
Chung, B.T.F.
Cornwell, K.
Curcio, L.A.

Deakin, A.W.
Drazner, B.J.
Dubinkin, Y.M.

Elsherbini, A.

Frennborn, P
Fujita, H.

Gibbard, I.

Hanamura, K
Heikal, M.
Hirota, M.
Honda, H
Huang, L-D.
Huang, S.Y
Hur, N.K.

Ichimiya, K.
Ishizuka, M.

Jachuck, R.
Jackson, J.D.
James, P.

Kang, H.C.
Kang, Y.T.
Kenning, D.
Khevesyuk, V.I.
Kim, B.J.
Kim, D.K.
Kim, D.S.
Kim, J.S.
Kim, M.S.
Kim, N.H.
Kim, Y.C.
Konukman, A.
Koyama, S.
Kwon, J.T.

Lee, J.H.
Leontiev, A.I.
Liu, W.
Ljubicic, B.

Mnatsakanyan, Y.

Ognev, V.V.

Palen, J.W.
Panchal, C.B.
Pechorkin, N.I.
Pribaturin, N.A.

Quarini, J.

Ramis, Y.A.
Ranasinghe, J.
Ravigururajan, T.
Roudakov, A.

Shah, R.K.
Sheipak, A.A.
Shevchenko, I.V.
Shi, M.H.
Sunden, B.
Suzuki, K.

Thome, J.R.
Thonon, B.
Tikhoplav, V.Y.
Tochon, P.
Tuzla, K.

Wadekar, V.
Watkinson, P.A.
Webb, R.L.
Winkelman, D.
Witte, L.

Xin, M.D.
Xing, L.

Yakimenko, R.I.
Yoshida, H.
You, S.M.
Youn, B.

Zhao, T.S

SINGLE-PHASE FLOW AND HEAT TRANSFER FUNDAMENTAL STUDIES

NUMERICAL ANALYSIS OF LAMINAR FLOW AND HEAT TRANSFER IN CORRUGATED UNDULATED DUCTS FOR RECUPERATORS

Esa Utriainen and Bengt Sundén
Division of Heat Transfer, Lund Institute of Technology, Box 118, 22100 Lund, Sweden
Esa.Utriainen@vok.lth.se, Bengt.Sunden@vok.lth.se

ABSTRACT

A numerical study was conducted to assess the hydraulic and thermal performance of a primary surface type heat exchanger surface, called the Corrugated Undulated (CU) duct aimed for modern gas turbine recuperators.

The governing equations, i.e., the mass conservation equation, Navier-Stokes equations and the energy equation, are solved numerically by a three dimensional finite volume method for boundary fitted coordinates. Partial periodic boundary conditions, i.e., pressure drop and temperature rise due to heating are taken into account, are imposed in two flow directions. In this particular case laminar convective flow and heat transfer prevail.

Details of the recuperator ducts and the numerical method as well as relevant results are presented.

INTRODUCTION

There has been a considerable interest in micro- and miniturbines (approximate output power ranges from 5 kW to 400 kW) following the deregulation of the electricity market in many countries around the world. The use of small power plants in the distribution network, known as embedded generation, allows distributors to relieve congested distribution networks and reduce transmission losses and costs. The small turbines make it possible for small energy consumers to generate their own electricity. The application of recuperators is mandatory to meet the low emission requirements of pollutants and to reduce the operating cost of the power plant by improving the cycle efficiency and thus cut down the fuel consumption. The requirements on recuperators may be summarized as: high effectiveness, low pressure losses, minimum volume and weight, high reliability and low cost etc., see Utriainen and Sundén (1998).

Present manufacturing techniques provide the means of designing very compact recuperator cores, having hydraulic diameters about one millimeter.

As the density and thermal conductivity of gases are both much lower than those of liquids, the heat transfer performance is poor. Passive enhancement techniques (Webb, 1994) may be used to reduce the physical size of the recuperator core. In the passive techniques secondary flow structures are created by means of curved and interrupted duct surfaces. The secondary gas and air flow structures in heat exchangers disturb the insulating near wall layers and thus improve the thermal properties of the duct (Jacobi and Shah, 1998).

Surface protrusion vortex generators like delta winglet pairs have been investigated by Tiggelbeck et al. (1994). Their study, for developing duct flow, shows that the average Nusselt number, for Re number about 2000, is increased by 46% with the accompanying increase of the friction factor by 177% compared to a channel without vortex generators.

An experimental study presented by Olsson and Sundén (1996) investigated thermal and hydraulic performance of flat, dimpled and rib-roughened tubes. Their results show that for Reynolds number about 1000, compared to a smooth flat tube, the average Nusselt number is enhanced by about 50% and the friction factor is increased by about 200%. The two examples above show that enhancing the heat transfer by ribs or other surface protrusions often results in unacceptable high pressure losses, at least for use in gas turbine recuperator cores.

The primary surface type of heat transfer surfaces offer heat transfer enhancement by accompanying acceptable increase of the pressure losses. For this type of surface, the ratio of heat transfer enhancement to pressure loss increase is often about unity.

A primary surface duct concept called the Trapezoidal Cross Wavy (TCW) duct has been investigated by Utriainen and Sundén (2000a). The TCW duct configuration shows enhancement of the Nusselt number up to 200% (at

Re=1000) with an increase of the friction factor f by a factor of the same order, compared to straight ducts. The geometrical shape, with relatively sharp corners, of the duct is not realistic for small size ducts due to limitations in the manufacturing process, i.e., pressing of metal sheets would give a duct shape which is more rounded than the TCW duct.

The Cross Wavy (CW), see Utriainen and Sundén (2000b), is similar to the TCW duct above, the difference being the cross sectional shape. The CW duct has a cross section with a rounded shape which is similar to the primary surface duct concept in a recuperator manufactured by Solar Turbines Inc., see Parsons (1985). For Re number about 600 the Nusselt number is enhanced by up to 600% and the friction factor f is increased by 400 – 2000%. The very high increase of the friction factor is due to a large ratio of amplitude of waviness to length of waviness.

Cross Corrugated (CC) ducts (also called herring bone pattern and chevron plate) have been investigated, both experimentally and numerically, by Ciofalo et al. (1996), Stasiek et al. (1996), Savostin and Tikhonov (1970). The thermal and hydraulic performance of these ducts are similar to the performance exhibited by the CW duct above but very much dependent of the angle between corrugations, see Stasiek et al. (1996).

A variant of the CC duct is the Corrugated Undulated (CU) duct type, see Fig. 1. It offers heat transfer enhancement with an acceptable increase of the pressure losses. Earlier experimental work, see Stasiek (1998), reported performance of the CU duct for turbulent flow (from Re=1500). These data will be completed, in this paper, with results of numerical simulations of laminar flow performance of the CU duct.

All simulations in this paper have been performed for laminar flow situations because an investigation of real recuperators showed that with the operating mass flow rates and the small hydraulic diameters, the Reynolds number is usually much less than 2000. Thus laminar flow prevails.

THE HEAT TRANSFER SURFACES

The Corrugated Undulated (CU) surface is projected to be used in air preheaters, often rotary regenerators, in fossil fueled power plants, see Ciofalo et al. (1996). The flue gases leaving the economizer are cooled and deliver warm air to the furnace. For this purpose hydraulic diameters of several millimeters (9-14mm) is used for the CU ducts to meet the requirements of the hydraulic and thermal performance of air preheaters.

For use in recuperators in gas turbine systems the hydraulic diameter has to be smaller to meet the requirements of small unit volume and weight (Utriainen and Sundén, 2001). A small diameter gives a small volume of the unit but may lead to bulky physical sizes in one or two dimensions relative to the other direction/directions.

The heat transfer surface configurations in the present study are formed from closely packed furrowed plates, alternately stacked in couples where plates with the deeper furrows (corrugated plates) alternate with plates having shallow furrows (undulated plates). The corrugations are aligned with the main flow direction ($\theta_c=0°$), while undulations form an angle ($\theta u=20°$, $30°$ and $50°$) to the corrugations, see also Ciofalo et al. (1996), Stasiek (1998) and Fig. 1. Three of the six CU configurations in Stasiek (1998), showing promising thermal and hydraulic performance, have been chosen for this study.

The cross sectional shape is composed of circle segments and straight connecting lines. Table 1 provides geometrical details of the different surface configurations.

GOVERNING EQUATIONS

The Reynolds number range considered in this study is Re=200-1600. A laminar flow assumption may not be correct for the upper limit as the complex flow pattern may result in transitional flow already at a moderate level of the Reynolds number, i.e., less than for straight ducts Re<2100.

The commercially available CFD code CFX 4 from AEA Technology, has been used for the numerical predictions of the TCW duct performance.

The methodology of periodic boundaries, placed at the inlet and outlet faces of the unit cell, see Fig. 1, for solving fully developed flows, is applied in the current investigation. Partial periodic boundary conditions, as presented by Patankar et al. (1977), are used in the two flow directions, i.e., at the inlet and outlet of furrows in the corrugated and the undulated plates respectively. By adding extra source terms in the momentum and the temperature equations the pressure drop due to wall friction and the temperature increase due to heating, are taken into account. In the current study the extra source terms, in the case of temperature rise compensation, are obtained by calculating a compensation power (total power entering the fluid from the walls) which then is compensated for in each cell of the computational domain (Ciofalo, 2000). The governing equations below show the details.

Table 1. Corrugated-Undulated geometries

Geometry	Corrugated plate	Undulated plate		
	CP	UCS	UP2	US
P [mm]	22.60	32.00	17.48	21.87
H_i [mm]	13.13	8.0	3.65	3.65
P /Hi	1.72	4.0	4.79	6.00
D_{eq} [mm]	-	15.57	12.53	12.70

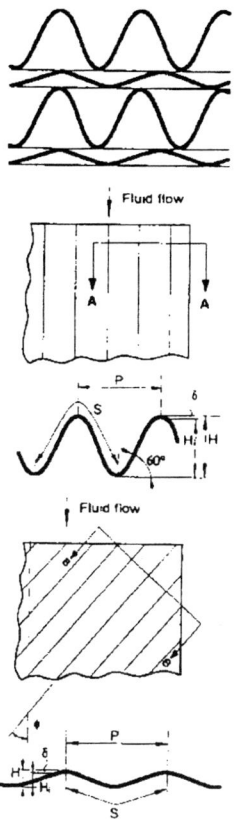

Figure 1. The Corrugated Undulated (CU) duct surface (from Stasiek, 1998)

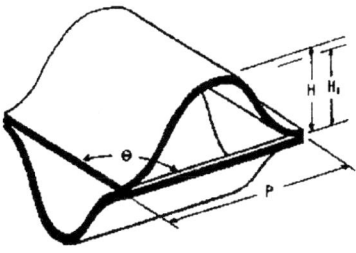

Figure 2. A representative cell of the CU duct surface

The following equations are valid for an incompressible fluid with constant properties at steady state simulation of periodically repeating flow.
The equation of continuity:
$$\frac{\partial U_i}{\partial x_i} = 0 \quad (1)$$
The Navier-Stokes equation of momentum conservation:
$$\frac{\partial}{\partial x_j}(\rho \cdot U_j \cdot U_i) = -\frac{\partial p^*}{\partial x_i} + \frac{\partial}{\partial x_j}\left[\mu\left(\frac{\partial U_i}{\partial x_j} + \frac{\partial U_j}{\partial x_i}\right)\right] + \beta \cdot \delta_{3i} \quad (2)$$
where the extra source term, assuming linear pressure drop characteristics in the flow direction is:

$$\beta = \frac{\partial p}{\partial x_i} \quad (3)$$

The pressure p is written as:
$$p = p^* - \beta \cdot z \quad (4)$$
The pressure p* behaves periodically and gives no contribution to the pressure drop over a repeated unit cell.
The temperature equation:
$$\frac{\partial}{\partial x_j}(\rho \cdot U_j \cdot T^*) = \frac{\partial}{\partial x_j}\left(\frac{\lambda}{c_p}\frac{\partial T^*}{\partial x_j}\right) - \rho \cdot U_j \cdot \sigma \cdot \delta_{3j} \quad (5)$$
where σ in the extra source term equals:
$$\sigma = \frac{\partial T_b}{\partial z} = \frac{P}{V_{tot}} \cdot \frac{U_1}{\overline{U_1}} \quad (6)$$
The power is obtained by:
$$P = \int_{wall} q_w \cdot dA_w \quad (7)$$
The temperature T is written as:
$$T = T^* + \sigma \cdot z \quad (8)$$
For prediction of the friction factor f and the Nusselt number, the following relations are used.
Fanning friction factor:
$$f = \frac{\beta \cdot D_h}{2 \cdot \rho \cdot \overline{W}^2} \quad (9)$$
The average Nusselt number for a unit cell is calculated from:
$$\overline{Nu} = \frac{\overline{q_w} \cdot D_h}{k \cdot \overline{(T_w - T_b)}} \quad (10)$$
where the mean wall to bulk temperature difference for a cross section is:
$$\overline{T_w - T_b} = \frac{\int (T_w - T_b) \cdot dA_c}{\int dA_c} \quad (11)$$
and the mass averaged bulk temperature is computed from:
$$T_b = \frac{\int T \cdot |W| \cdot dA_c}{\int |W| \cdot dA_c} \quad (12)$$
The equivalent hydraulic diameter is:
$$D_{eq} = 4 \cdot \frac{A_{wall}}{V_{tot}} \quad (13)$$

COMPUTATIONAL DETAILS

A finite volume and steady state method with a body fitted grid was employed in the computations. The commercially available CFD code CFX 4 from AEA Technology has been used for the numerical predictions of the CU duct performance.

The meshes used in this study are different for each CU configuration. Mesh independent solutions are ensured by increasing the number of computational cells to a level where further increase will not affect the solution significantly. The number of computational cells ranges from 28800 to 180000. A typical cell mesh used for, e.g., the CP/UCS-30 configuration has a mesh 40 x 33 x 32 = 42240 cells and mesh influence is tested by a mesh 60 x 60 x 50 = 180000 cells.

Patankar et al. (1977) presented the methodology of periodic boundaries for solving fully developed flows in one representative module of a periodically repeated duct. This methodology is adopted in the present investigation by applying partial periodic boundary conditions in two directions, i.e., in the directions of the furrows on the corrugated and the undulated plates, see Fig. 2. For all reported computations residuals, normalized with values at the periodic boundaries, less than 10^{-6} for velocities and less than 10^{-5} for enthalpy were reached. Using the first order Upwind Difference (UD) scheme for velocities usually forces the flow field calculations to converge to a steady state solution even for transitional and turbulent flow. Using a higher order scheme results in non-converged solutions for a larger range of Re number. These difficulties may indicate that transition from laminar to turbulent flow occurs in the Re range 500-1500. The lower number indicates transition of the flow for surface configurations having large angle (θ) between corrugations and undulations. All results in the present study are obtained by using the UD scheme as no change was detected using higher order schemes apart from the difficulties to reach converged solutions as mentioned above.

The fluid used in the CFD calculations is air with constant properties. For the wall boundary conditions both uniform heat flux as well as constant temperature has been implemented on all walls and results of these two cases are reported in the next section. This computational approach, predicted Nusselt numbers and f-values for straight ducts having rectangular and polygonal cross sections within 1% of the values given in the literature.

RESULTS

The complex geometry of the duct is advantageous for generation of secondary flow structures. The flow in the near wall region will be disturbed, by both the angled, relative to the walls, streamwise flow and the secondary flows, which results in an increase of velocity and temperature gradients and thus enhancement of the heat transfer, but also an increase in wall friction, i.e., the pressure drop increase. Hence a higher overall Nusselt number and friction factor f is obtained for the duct. It can be seen that the Nusselt number of a CU duct is not constant in the laminar flow regime, as is the case for laminar flow in straight ducts, see Fig. 3, where Nusselt numbers are plotted for the wall boundary condition, constant heat flux. The same applies to the product f*Re which also is a function of the Re number for the CW duct, but for a laminar flow in a straight duct this product f*Re is constant, see Fig. 4. In both Figs. 3 and 4 results for a straight duct having a square cross section is plotted for comparison. As expected large angle, θ, between the furrows of the respective plates gives an increase in both the Nusselt numbers and friction factors of all surface configurations. It appears that the internal height, H_i, also has great influence on the Nusselt number and the friction factor. The deep furrows of the CP/UCS configuration give high Nusselt numbers and friction factors compared to the other configurations having relatively shallow furrows. In Fig. 5 results of Nusselt numbers for the wall boundary condition, constant wall temperature, are plotted and the Nusselt numbers are greater for this wall boundary condition than for the constant wall heat flux case. These results were not expected because in straight ducts the Nusselt numbers are greater for the wall boundary condition, constant heat flux, see, e.g., Kakac et al. (1987). An explanation may be that the magnitude of secondary flow velocities are greater in ducts having complex non-straight duct geometries, than they are in straight ducts, and the thinning of the near wall layers is considerable at certain points in the duct which give great local temperature gradients. The great local temperature gradients give great local heat fluxes and therefore high local Nusselt numbers, see the numerator of Eq. (10). The corresponding mechanism of enhancement, for the wall boundary condition, constant heat flux, is that the temperature difference, between the wall and the bulk of the fluid, in the denominator of Eq. (10), is reduced giving higher local Nusselt numbers. From both Figs. 3 and 5 it is evident that the Nusselt number has not the same linear behaviour, with the Reynolds number as was the case in Utriainen and Sundén (2000a) and Utriainen and Sundén (2000b). The non-linear behaviour for the higher Re numbers could be due to the fact that the flow is in the transitional regime and the laminar flow assumption in the computations is not suitable.

In this study the volume goodness criterion is used to compare the different duct configurations, which like most other concepts of comparison, only consider the performance on one side of the recuperator, e.g., the gas side. The ducts presented in Fig. 6 have equal hydraulic diameter D_{eq}=1.54 mm. The axes represent the heat transfer coefficient (h) and pumping power per unit heat transfer area ($\Delta p \cdot \dot{V}/A_{ht}$), respectively. A high position in this plot signifies a high heat transfer coefficient, and hence a small volume of the heat exchanger core for a specified pressure drop. The heat transfer coefficients are obtained from the Nusselt numbers for constant heat flux as wall boundary conditions, i.e., values of Fig. 3. The CU ducts are plotted together with the reference straight duct having a square cross section. The volume goodness criterion shows that the CP/UCS plate configuration yields the smallest volume of the heat exchanger core for small pressure drop and relatively low Reynolds numbers (Re<600). For higher pressure drop and higher Reynolds numbers the CP/US configuration having large angle, θ=50°, shows best performance.

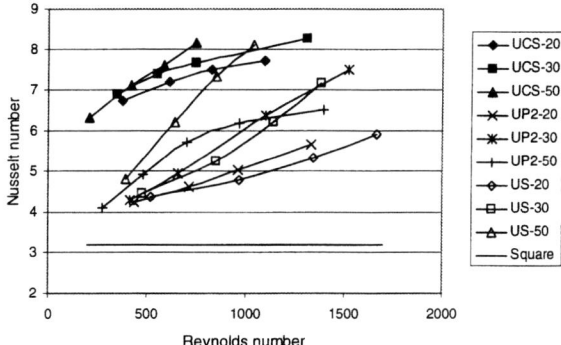

Figure 3. Nu number vs Re number for the boundary condition, constant wall heat flux, $\theta=(20°, 30° \text{ and } 50°)$

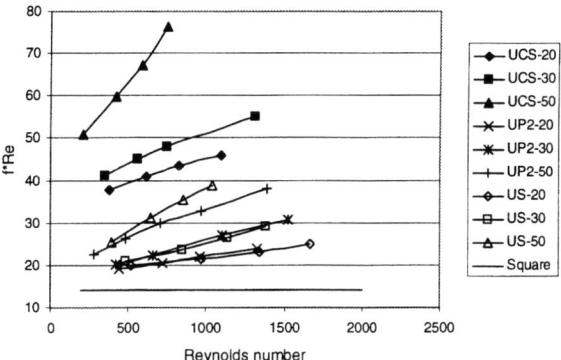

Figure 4. The product f*Re vs Re number, $\theta=(20°, 30° \text{ and } 50°)$

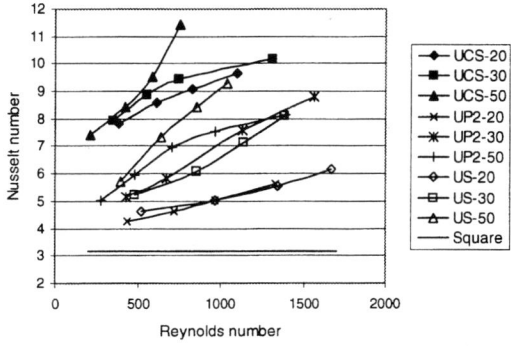

Figure 5. Nu number vs Re number for the boundary condition, constant wall temperature $\theta=(20°, 30° \text{ and } 50°)$

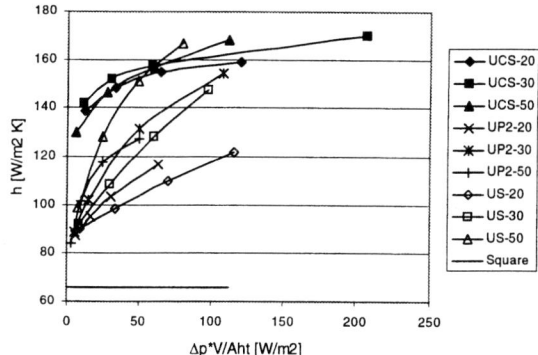

Figure 6. Volume goodness chart for the CU configurations $\theta=(20°, 30° \text{ and } 50°)$ and reference square duct

CONCLUSIONS

Numerical prediction of thermal and hydrodynamic performance, using CFD analysis, of a Corrugated Undulated (CU) heat transfer surface has been presented for laminar flow conditions suitable for compact heat exchangers such as gas turbine recuperators and furnace air preheaters. It is shown that the performance of the CU duct is superior to simple straight duct configurations. The Nusselt number of the CU duct is enhanced by up to 100% compared to straight ducts with an increase of pressure drop by 30%-400%.

Both the posed angle, θ, between the corrugations and the undulations and the internal height of the furrows (undulations) have a big influence on the performance of the CU duct concept.

The wall boundary condition, constant wall temperature, gives higher average Nusselt number for the CU surface than the constant heat flux condition.

It was also shown that the laminar flow characteristics of the CU duct show that the Nusselt number and the product f*Re are functions of the Re number, while in straight ducts these are constants, see Figs. 3 to 5.

NOMENCLATURE

A_w area of wall in computational cell
A_c area of computational cell face [m^2]
A_{ht} wall heat transfer area [m^2]
c_p specific heat of fluid [J kg^{-1} K^{-1}]
CU   Corrugated Undulated duct, see Fig. 1
D_h hydraulic diameter of duct [m]
f   Fanning friction factor
h   heat transfer coefficient [W m^{-2} K^{-1}]
k   thermal conductivity [W m^{-1} K^{-1}]
Nu   Nusselt number

P   compensation power [W]
p   static pressure [Pa]
p^* periodic pressure [Pa]
q_w heat flux from wall [W m^{-2}]
$\bar{q}_w$ wall area averaged heat flux [W m^{-2}]
Re   Reynolds number
T   temperature [K]
T_b Temperature of fluid bulk [K]
T^* periodic temperature [K]
U_i velocities using tensor notation [m s^{-1}]
V_{tot} Total volume of a computational unit cell
$\dot{V}$ volume flow rate [m^3 s^{-1}]
W   velocity in the flow direction (z-direction) [m s^{-1}]
$\overline{W}$ mass averaged velocity in the flow direction (z-direction) [m s^{-1}]
x_i coordinate directions using tensor notation [m]
x, y, z   coordinates in the Cartesian system

Greek Symbols

$\delta_{\alpha\beta}$ Kroneckers delta, equal to 1 when $\alpha = \beta$, otherwise 0
Δp pressure loss in duct [Pa]
θ angle between corrugations and undulations [°]
μ dynamic viscosity [kg m^{-1} s^{-1}]
ρ density of fluid [kg m^{-3}]

REFERENCES

Asako, Y. Faghri, M., and Sundén, B., 1996, Laminar Flow and Heat Transfer Characteristics of a Wavy Duct With a Trapezoidal Cross Section for Heat Exchanger Application, *2nd European Thermal Sciences and 14th UIT National Heat Transfer Conference 1996*, G. P. Celata, P. Di Marco and A. Mariani (Editors), pp. 1097-1104.

Ciofalo, M., Stasiek, J. and Collins, M. W., 1996, Investigation of Flow and Heat Transfer in Corrugated Passages – II. Numerical Simulations, *Int. J. Heat Mass Transfer*, Vol. 39, No. 1, pp. 165-192.

Ciofalo, M., Stasiek, J. A. and Collins, M. W., 1998, Flow and Heat Transfer Predictions in Flow Passages of Air Preheaters: Assesment of Alternative Modeling Approaches, *Computer Simulations in Compact Heat Exchangers*, eds. Sundén B. and Faghri M., Computational Mechanics Publications.

Ciofalo, M., 2000, Personal communication.

Jacobi A. M. and Shah R. K., Heat Transfer Surface Enhancement through the Use of Longitudinal Vortices: A Review of Recent Progress, *Exp. Thermal Fluid Science*, Vol 11, pp. 295-309, 1995.

Kakac, S., Shah, R. K. and Aung, W., 1987, *Handbook of Single-Phase Convective Heat Transfer*, John Wiley & Sons, Inc.

Olsson, C. O. and Sundén, B., 1996, Heat Transfer and Pressure Drop Characteristics of Ten Radiator Tubes, *Int. J. Heat Mass Transfer*, Vol. 39, pp. 3211-3220.

Parsons, E.L., 1985, Development, Fabrication and Application of a Primary Surface Gas Turbine Recuperator, *SAE Technical Paper Series No. 851254*.

Patankar, S. V., Liu, C. H., and Sparrow, E. M., 1977, Fully Developed Flow and Heat Transfer in Ducts Having Streamwise-periodic Variations of Cross Sectional Area, *ASME J. Heat Transfer*, Vol. 99, pp. 180-186.

Rokni, M., Sundén, B., 1998, 3D Numerical Investigation of Turbulent Forced Convection In Wavy Ducts With Trapezoidal Cross Section, *Int. J. Num. Heat & Fluid Flow*, Vol 8, No 1, pp. 118-141.

Savostin, A. F., Tikhonov, A. M. 1970, Investigation of the Characteristics of Plate-Type Heating Surfaces, *Teploenergetica*, Vol. 17, No. 9, pp. 75-78.

Stasiek, J., Ciofalo, M. and Collins, M.W., 1996, Investigation of Flow and Heat Transfer in Corrugated Passages-I. Experimental Results, *Int. J. Heat Mass Transfer*, Vol. 39, No. 1, pp. 149-164.

Stasiek, J. A., 1998, Experimental Studies of Heat Transfer and Fluid Flow across Corrugated-Undulated Heat Exchanger Surfaces, *Int. J. Heat Mass Transfer*, Vol. 41, Nos. 6-7, pp. 899-914.

Tiggelbeck, St., Mitra, N. K. and Fiebig, M., 1994, Comparison of Wing-Type Vortex Generators for Heat Transfer Enhancement in Channel Flows, *J. Heat Transfer*, Vol. 116, pp. 880-885.

Utriainen, E., Sundén, B., 1998, Recuperators in Gas Turbine Systems, *ASME paper 98-GT-165*.

Utriainen, E., Sundén, B., 2000a, Numerical Analysis of a Primary Surface Trapezoidal Cross Wavy Duct, *Int. J. Num. Meth. Heat & Fluid Flow*, Vol. 10, Nos. 5-6, pp. 634-648.

Utriainen, E., Sundén, B., 2000b, A Numerical investigation of a Primary Surface Rounded Cross Wavy Duct, *To be submitted*.

Utriainen, E., Sundén, B., 2001, A Comparison of Some Heat Transfer Surfaces for Small Gas Turbine Recuperators, *Submitted to the ASME TURBOEXPO 2001, 4-7 June, 2001, New Orleans, Louisiana, USA*.

Webb, R. L., 1994, *Principles of Enhanced Heat Transfer*, J. Wiley & Sons, Inc., USA.

A THREE-DIMENSIONAL UNSTEADY NUMERICAL ANALYSIS FOR A PLATE-FINNED TUBE HEAT EXCHANGER IN THE MIDDLE REYNOLDS NUMBER RANGE

Hajime ONISHI[1], Kyoji INAOKA[2] and Kenjiro SUZUKI[1]

[1] Department of Mechanical Engineering, Kyoto University, Kyoto 606-8501
[2] Department of Mechanical Engineering, Doshisha University, Kyoto 610-0321

ABSTRACT

In this study, a three-dimensional unsteady numerical analysis has been made for a one unit of single row plate-finned tube heat exchanger located in a uniform flow. The effects of Reynolds number and fin pitch on local and mean Nusselt number and pressure coefficient were examined parametrically. The unsteadiness of the flow revealed some important effects on heat transfer from the fin-and-tube surface. The flow unsteadiness generated downstream the fin depends not only on the tube diameter but also on the fin pitch. Karman vortex structure is generated in the instantaneous flow field and twin vortex structure can be seen in the time mean flow field. As Reynolds number or fin pitch increases, flow unsteadiness is intensified so that lower heat transfer area of the fin behind the tube becomes narrower. Local streamwise pressure coefficient shows a peak. It is related to the increase of the momentum transfer by the unsteady flow.

INTRODUCTION

Plate-finned tube arrangements are widely used as one of the most important fundamental elements in the practical heat exchange devices, for instance, indoor and outdoor heat exchangers of air conditioners, and refrigerators, etc. Achievement of high heat transfer performance is a key issue in the design of such heat exchangers. Therefore, various studies have been made for the plate-finned tube heat exchangers. In the previous studies (Onishi et al., 1998; Onishi et al., 1999), the present authors developed detailed discussions on the local flow and heat transfer characteristics for one unit of plate-finned tube heat exchangers based on the results of numerical computation, where the flow was assumed to be steady like as those in other studies (Bastani et al., 1990; Fiebig et al., 1995; Jang et al., 1996). However, as it was pointed out by Xi et al., (1995), flow can become unsteady in the studied Reynolds number range. Therefore, it is worth to know whether the flow is destabilized or not and it so, how such flow instability affects the heat transfer performance of the heat exchangers.

In the present study, a three-dimensional unsteady numerical analysis is made for a one unit of single row plate-finned tube heat exchanger located in a uniform flow and an effort has been poured into analyzing the unsteady structures of the flow and thermal fields and into seeing how they are affected by the flow unsteadiness. The influence of Reynolds number and the fin pitch on local and mean Nusselt number and pressure coefficient is examined parametrically.

COMPUTATIONAL PROCEDURE

Figure 1 shows the computational domain used in the present study. Studied geometrical parameters are also illustrated in the figure. Air, the working fluid, flows into the system with uniform velocity U_i and temperature T_i from the facet ABCD, and flows out from the system through the facet EFGH. The inlet boundary is placed one tube diameter D upstream the leading edge of the fin and the outlet boundary $9D$ downstream the fin trailing edge. In this study, relatively large downstream length is given than those in the previous studies so as to calculate the flow unsteadiness more precisely. Two kinds of coordinate systems are used in the computational domain; i.e. the main Cartesian coordinate system (x, y, z) covering the whole computational domain and the cylindrical sub coordinate system (r, θ, z) in the neighborhood of the circular tube. The origin of both main and sub coordinate systems is placed at the center of the tube on the bottom facet of the computational domain BCGF. At the boundaries except for the inlet and outlet; i.e., side, top, and bottom boundaries, both of the flow and thermal fields are assumed to be spatially periodic. Thus, computation is made within the region from $y = -P_T/2$ to $P_T/2$ in y-direction and from $z = 0$ to P_F in z-direction. Here, P_T and P_F denote the tube pitch and the fin pitch, respectively. At the outlet boundary, flow and thermal fields are treated to

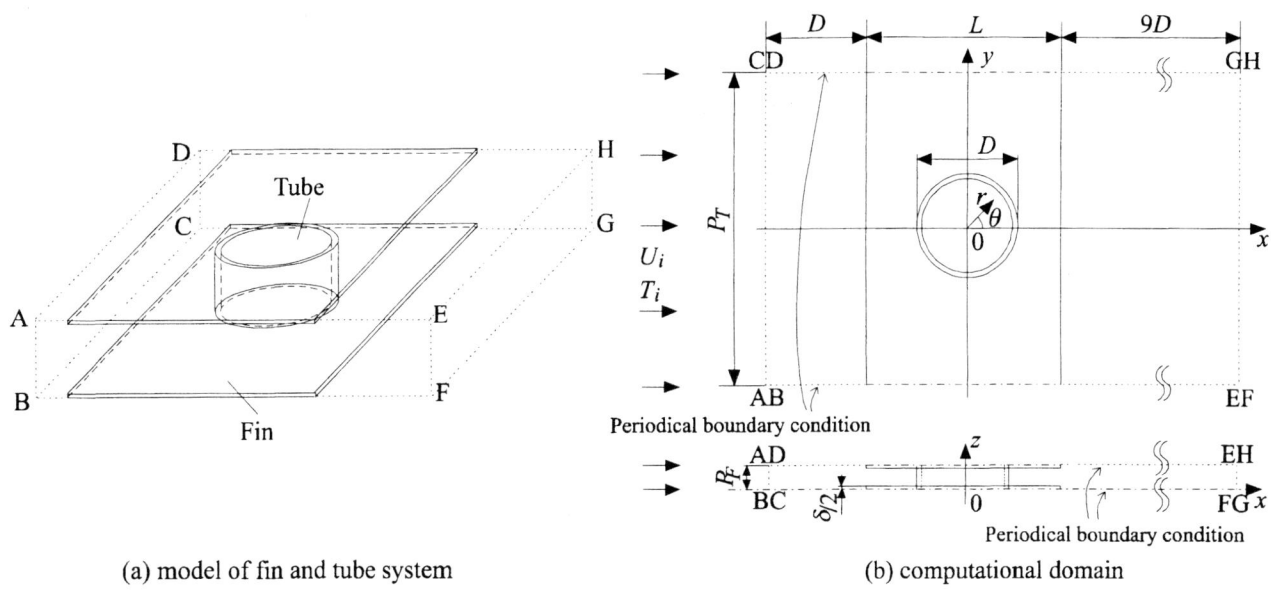

(a) model of fin and tube system (b) computational domain

Fig. 1 Studied fin and tube system.

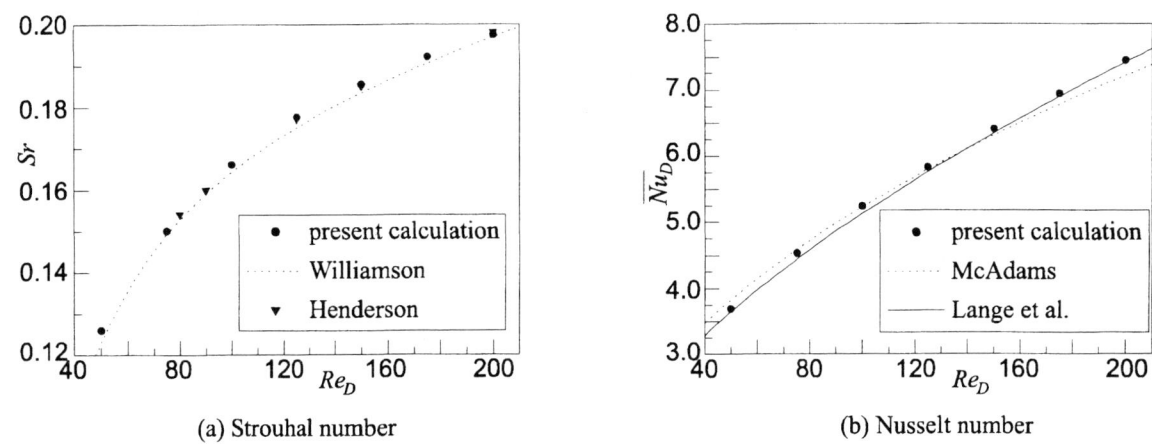

(a) Strouhal number (b) Nusselt number

Fig. 2 Comparison between the numerical results and experimental data for a circular tube.

obey the boundary layer approximations (Kieda and Suzuki, 1980).

Computational domain is covered by a composite grid system (Xi et al., 1995), which is composed of Cartesian coordinate system covering the main part of the whole domain except for the region around the tube. Another cylindrical coordinate system is provided to cover the region around the tube. This grid system is rather simple and is easily built up for the rectangular and circular shape body compared with the BFC (Boundary Fitted Curvilinear) grid system.

Finite difference equivalents of time-dependent momentum equations and energy equation are solved numerically in the air flow region. The governing equations are solved numerically using finite volume method. QUICK scheme (Leonard, 1979) is applied for finite-differencing of the convection terms and the central finite-differencing is used for the diffusion terms. Pressure field is solved by making use of SIMPLE algorithm (Patanker, 1980), the pressure correction method. Grid nodes are allocated non-uniformly. Staggered grids are used for the velocity components in all grids.

Analysis of the obtained results is made for the so-called middle Reynolds number range; i.e. the case where air flow becomes unsteady but remains laminar or regular in the studied system (Fiebig et al., 1995; Jang et al., 1996). At all parts of the finned tube surface, uniform surface temperature of T_W is assumed to prevail as the temperature boundary condition. In the present study, the fin pitch is basically changed in two steps as the main geometrical parameters, namely $P_F/D = 0.2, 0.4$, keeping other geometrical parameters, $L/D = 2.0$, $P_T/D = 3.0$ and $\delta/D = 0.04$. Under such fin arrangements, Reynolds number Re_D based on the tube diameter D and the inlet velocity U_i is varied in several

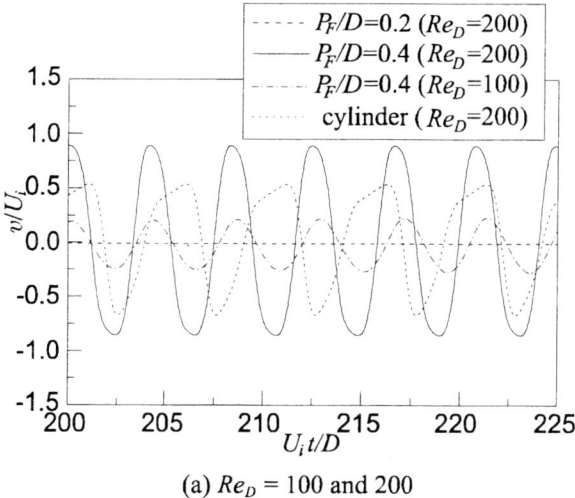
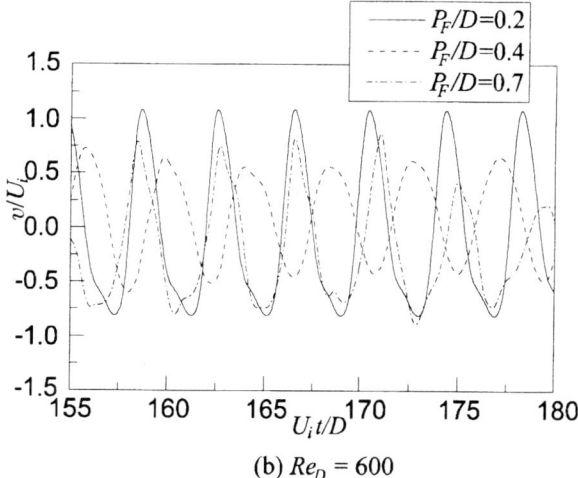

(a) $Re_D = 100$ and 200 (b) $Re_D = 600$

Fig.3 Time variations of velocity v for several cases.

steps from 200 to 2500. The definition of Reynolds number Re_D is

$$Re_D = \frac{\rho U_i D}{\mu} \quad (1)$$

where, ρ and μ denote air density and air viscosity, respectively and they are evaluated at the film temperature $T_f = (T_W + T_i)/2$. At $Re_D = 600$, in order to investigate the influence of the fin pitch more detail, the value of P_F/D is varied in several steps from 0.125 to 0.7.

The definitions of Nusselt number Nu_D, total pressure coefficient $C_{P(total)}$ and local pressure coefficient $C_{P(local)}$ to be used in the later discussions are as follows;

$$Nu_D = \frac{\alpha D}{\lambda} = \frac{q_W D}{(T_W - T_i)\lambda} \quad (2)$$

$$C_{P(total)} = \frac{2(P_i - P_o)}{\rho U_i^2} \quad (3)$$

$$C_{P(local)} = \frac{2(P_i - P_x)}{\rho U_i^2} \quad (4)$$

where, q_W, P_i, P_o and P_x are the heat flux of the finned tube surfaces, the time and spanwise spatial mean pressure at the inlet, that at the outlet and that at local streamwise positions.

RESULTS AND DISCUSSION

Flow around the cylinder

First of all, to confirm the validity of the present unsteady calculation, preliminary calculations were done for the case of a flow around a circular cylinder (with no fins). For this numerical calculation, a wider spanwise width is adopted ($P_T/D = 10$). It is well known that the flow behind the cylinder becomes three-dimensional if the value of Re_D is greater than 200 (Williamson, 1996), therefore, this test calculations were done for $Re_D \leq 200$.

Strouhal number Sr obtained by the frequency of the flow behind the cylinder for different Re_D is shown in Fig. 2 (a). Time and space mean Nusselt number $\overline{Nu_D}$ around the cylinder surface for different Re_D is shown in Fig. 2 (b). The reference values (Williamson, 1989; Henderson, 1995; McAdams, 1954; Lange et al., 1998) are also shown in each figure. As can be seen in Fig. 2 (a), the cylinder wake is steady for $Re_D < 40$, on the other hand, appearance of unsteady Karman vortex is observed in the cylinder wake for $Re_D \geq 50$. The value of this transition Reynolds number is almost the same as that of Williamson, (1996). Moreover, both values of Sr and $\overline{Nu_D}$ for the test results have good agreement with the reference values. Consequently, the unsteady calculation to be used in this study is confirmed to be valid both qualitatively and quantitatively.

Flow unsteadiness

Time variation of y-direction velocity v at the points just downstream the trailing edge of the fin is shown in Fig. 3 for the case of (a) $Re_D = 100$, 200 and of (b) $Re_D = 600$. The abscissa and the ordinate represent dimensionless time $U_i t/D$ and dimensionless velocity v/U_i, respectively. The corresponding result obtained in the preliminary calculation for the case around the cylinder without the fins for $Re_D = 200$ is also shown in Fig. 3 (a).

The periodic v signal can be confirmed in the case of the cylinder without the fins. Worth to note is that the flow remains steady in the case of $P_F/D = 0.2$ for $Re_D = 200$ while the flow behind the finned tube becomes unsteady in the case of $P_F/D = 0.4$. In the case of $P_F/D = 0.4$, the flow becomes unsteady even in the case of lower velocity at $Re_D = 100$. As Reynolds number based on the fin pitch in the case of $P_F/D = 0.4$ for $Re_D = 100$ and that in the case of $P_F/D = 0.2$ for $Re_D = 200$ are the same with each other, the fin pitch

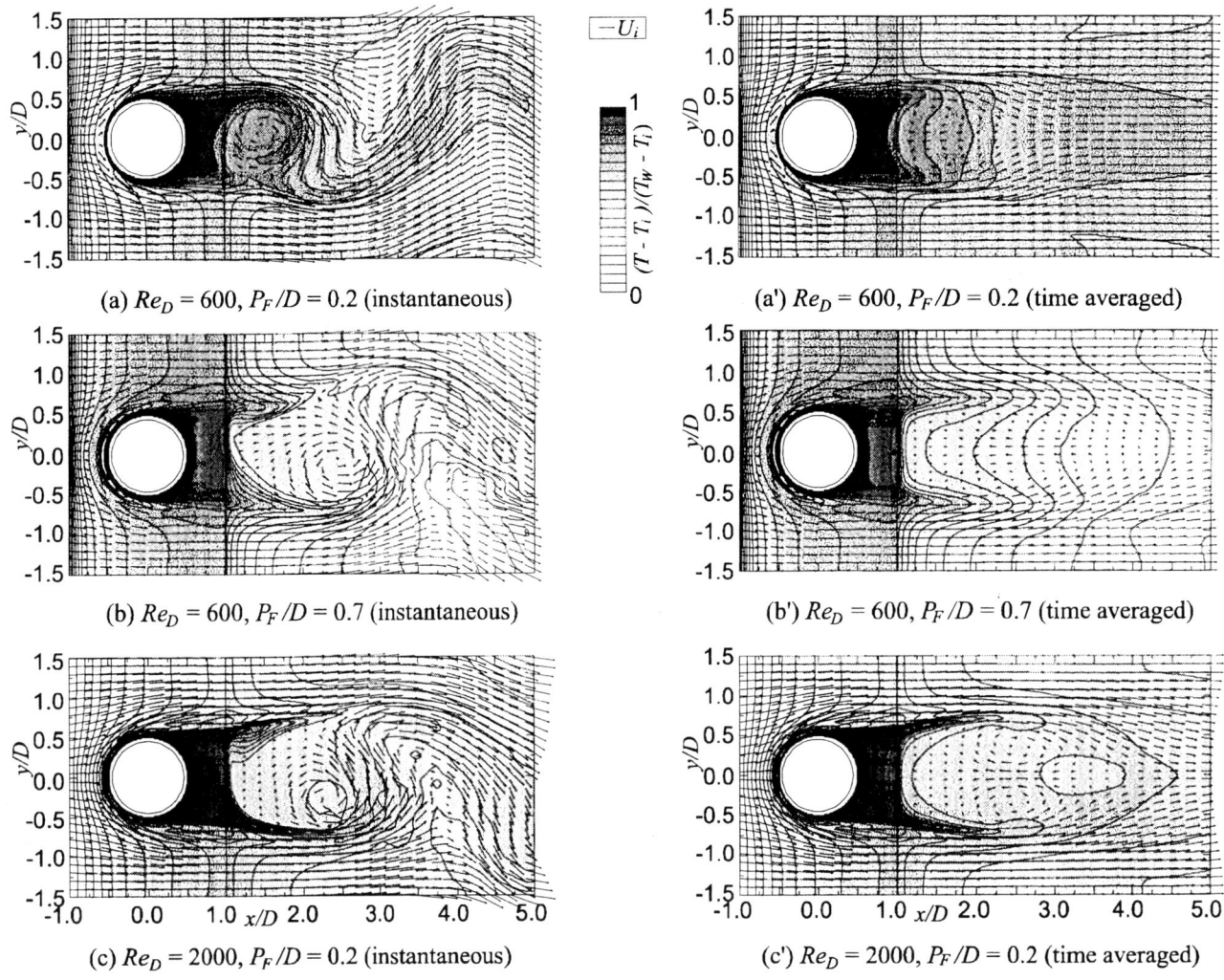

Fig.4 Velocity vector maps and temperature contours ($z/D = 0.05$)

might play an important role on the flow unsteadiness in this study. This velocity fluctuation has constant amplitude and a constant cycle and this behavior corresponds to Karman vortex shedding from a tube with a constant fluctuating intensity and period.

In the case of $Re_D = 600$, as is observed in Fig. 3 (b), flow becomes unsteady for all cases. The value of v varies periodically and shows constant amplitude in the case of smaller fin pitch, however its amplitude turns to vary with time in the case of larger fin pitch though its fluctuation period is not changed. This difference is related to the three-dimensionality of the unsteady flow as mentioned later.

Flow and thermal fields

Instantaneous velocity vector maps and temperature contours in x - y plane at $z/D = 0.05$ between the two plate fins for the cases of (a) $P_F/D = 0.2$ and (b) $P_F/D = 0.7$ for $Re_D = 600$ and (c) $P_F/D = 0.2$ for $Re_D = 2000$ are shown compared with their time-averaged counterparts (a') – (c') in Fig. 4. The instantaneous and time-averaged ones are presented in the left and right columns in Fig. 4, respectively. Deep gray area corresponds to the regions of high temperature.

Most noticeable point in this figure is that flow instability occurs in a very intensive manner downstream the fin while the flow remains almost steady in the space between the plate fins. Thus, the instantaneous temperature field looks quite similar to the time-averaged one in the flow space between the plate fins. Karman vortex motion is observed downstream area of the fin in the instantaneous map while a twin vortex structure is observed in the time-averaged map. Flow unsteadiness becomes more intensive as Reynolds number or the fin pitch is increased. Thus, it is found that the flow instability depends not only on the tube diameter but also on the fin pitch.

Velocity vector maps in y - z plane at $x/D = 2.5$ corresponding to the same instance of Fig. 4 (a), (b) and (c) are shown in Fig. 5 (a), (b) and (c), respectively. In order to see their distributions more clearly, these figures are presented

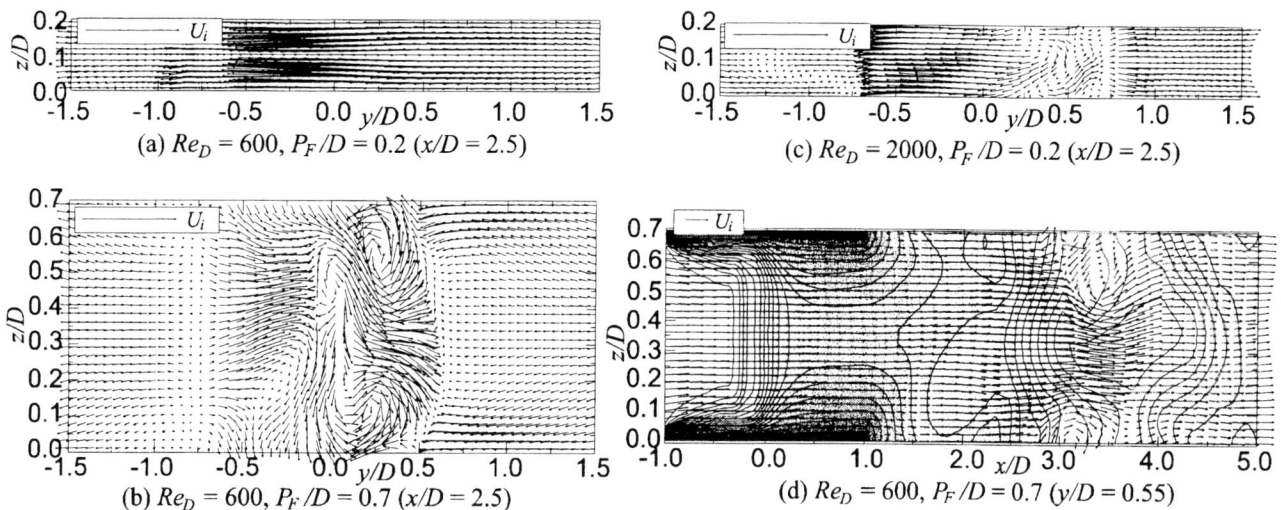

Fig.5 Instantaneous velocity vector maps and temperature contours (y - z and z - x plane)

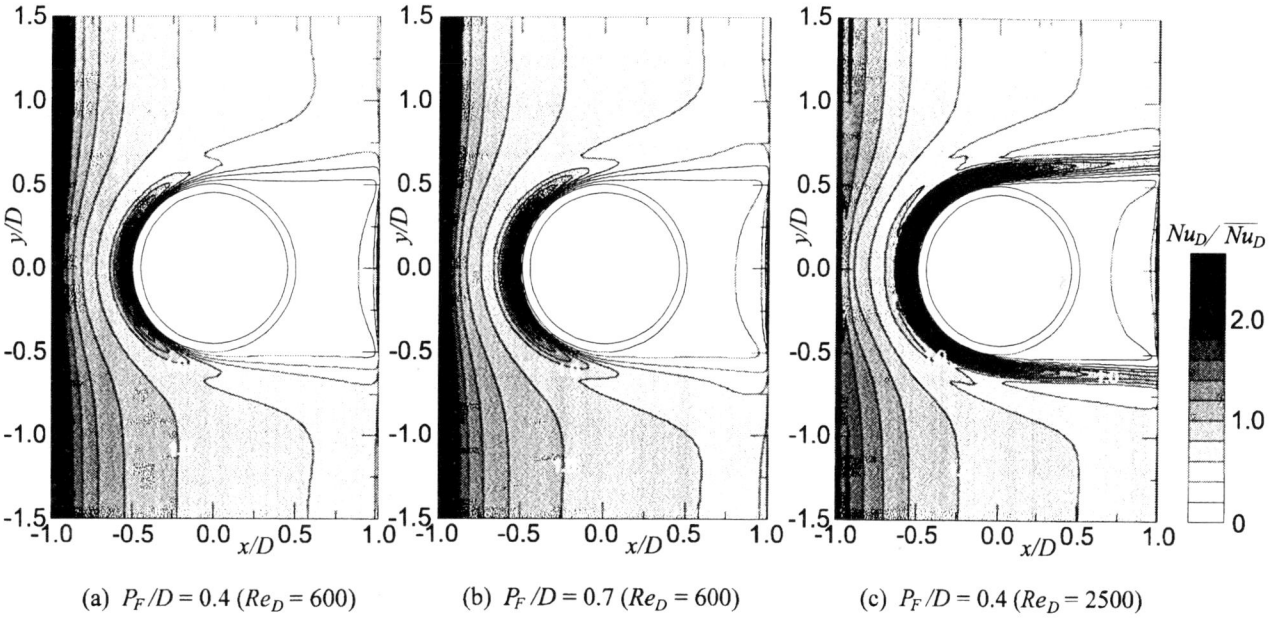

Fig.6 Time averaged local Nusselt number normalized by the time and space mean value.

in a way that the length in z-direction is magnified twice larger than that in y-direction. In addition to that, Fig. 5 (d) shows velocity vector and temperature contour in $x - z$ plane at $y/D = 0.55$ corresponding to the instantaneous map of Fig. 5 (b). In this figure, the length in z-direction is magnified three times larger than that in x-direction.

In the case of smaller fin pitch and of lower Reynolds number (Fig. 5 (a)), velocity vectors are almost in a parallel along y-axis and no vortex structure can be observed. In contrast to this, in the case of larger fin pitch and lower Reynolds number (Fig. 5 (b)), unsteady flow structure with vortices is apparently observed in the figure. Such flow structure with vortices can also be observed in the case of smaller fin pitch and of higher Reynolds number (Fig. 5 (c)). This structure can also be recognized in Fig. 5 (d). Thus, the present flow is found to be characterized by not only Karman vortex in the wake of the finned tube but also complex three-dimensional vortices as Reynolds number or the fin pitch is increased. The before mentioned amplitude variation of v signal in Fig. 3 (b) might be caused by the effect of such three dimensional vortex structure.

Distribution of local Nusselt number

In the next, spatial distributions of time averaged local Nusselt number Nu_D are illustrated in Fig. 6 for the cases of

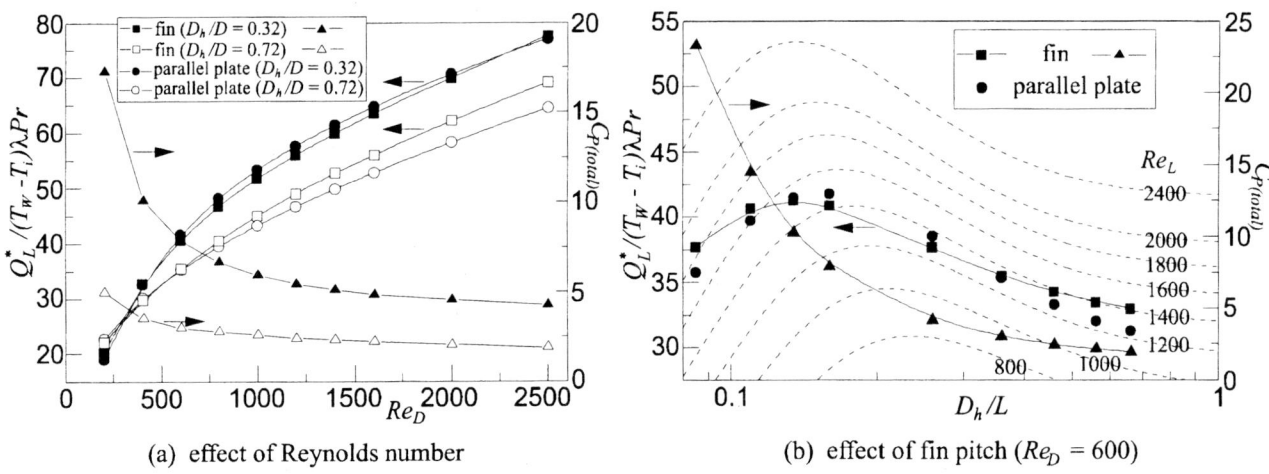

(a) effect of Reynolds number (b) effect of fin pitch ($Re_D = 600$)

Fig.7 Dimensionless heat transfer rate and total pressure coefficient.

(a) $P_F/D = 0.4$ and (b) $P_F/D = 0.7$ for $Re_D = 600$ and (c) $P_F/D = 0.4$ for $Re_D = 2500$. The value Nu_D is normalized by the time and space mean Nusselt number $\overline{Nu_D}$ on the fin surfaces for each case to investigate which part is contributed to the heat transfer enhancement. Deep gray area corresponds to the region of high local Nusselt number. Local Nusselt number shows a high value near the leading edge of the fin and decreases towards the downstream in a general trend but it shows a little higher value besides the tube. In contrary to this, it takes much smaller value in the wake region of the tube as is observed in the previous study (Onishi et al., 1998). In the case where the fin pitch or Reynolds number is relatively high (Fig. 6 (b), (c)), local Nusselt number takes large value in a region just upstream side of the tube. It is produced by the downwash fluid motion occurring related to the horseshoe vortex generated around the tube. Furthermore, as Reynolds number or the fin pitch is increased, flow unsteadiness is intensified so that lower heat transfer area of the fin behind the tube is decreased in size.

Heat transfer rate and total pressure coefficient

Total heat transfer rate from the fin surface is compared with that of the most basic value obtained from either surface of parallel plates in a channel where the flow is hydrodynamically and thermally developing (Shah and London, 1978). The total heat transfer rates Q^*_L normalized by $(T_W - T_i)\lambda Pr$ in the cases of (a) different Reynolds number Re_D and of (b) different hydraulic diameter D_h for $Re_D = 600$ are illustrated in Fig. 7. Here, hydraulic diameter $D_h = 2(P_F - \delta)$ is used instead of the fin pitch P_F so as to compare the fin heat transfer and the parallel plate heat transfer easily. In these figures, the circle symbols show the value of Q^*_L obtained from either surface of the parallel plates having the plate streamwise length L, the plate spanwise length 1 and the height between the plates $P_F - \delta$ for hydrodynamically and thermally developing flow. The plates are heated at constant wall temperature T_W and air flow mean velocity in the channel was assumed to be $U_i P_F/(P_F - \delta)$. On the other hand, the square symbols show the value of Q^*_L from an assumed fin surface having the fin streamwise length L and the fin spanwise length 1 where the mean heat flux of the predicted fin surface is distributed. Furthermore, total pressure coefficient $C_{P(total)}$ is also shown in the figures for finned tube cases by the triangle symbols.

At first, as is seen in Fig. 7 (a), the value of Q^*_L from the fin surface becomes larger in the case of small hydraulic diameter than in the case of large hydraulic diameter for $Re_D > 300$. Thus, as Reynolds number becomes higher, it is effective to make the hydraulic diameter small from the viewpoint of increase of the heat transfer rate. On the other hand, from the viewpoint of heat transfer enhancement, it is effective to make the hydraulic diameter large because the effect of the horseshoe vortices around the tube becomes intensified and it enhances the heat transfer. The dead water region behind the tube is got narrower by the unsteady vortex as mentioned at previous section. This can be clearly recognized by comparing the heat transfer rate from the fin surface and that from the parallel plate at same hydraulic condition; i.e. as Reynolds number becomes higher, heat transfer rate from the fin surface becomes rather larger than that from the parallel plate in the case of $D_h/D = 0.72$.

In the next, the dotted lines in Fig. 7 (b) show how the value of Q^*_L is changed with the value of D_h/L if the value of Re_L is kept constant in the case of the parallel plate heat transfer (Onishi et al., 1998). Here, Reynolds number Re_L is based on the plate length L and the mean velocity $U_i P_F/(P_F - \delta)$. Existence of maximum peak value of Q^*_L is observed at any value of Re_L. This feature having a maximum peak value is also recognized in the case of the fin heat transfer for $Re_D = 600$ and the optimum value of D_h/L appears at 0.14. This feature is caused by following four factors worth to note. (1) The development of thermal boundary layer is strongly affected by the acceleration of main flow caused by the thickening of the boundary layer. This flow acceleration

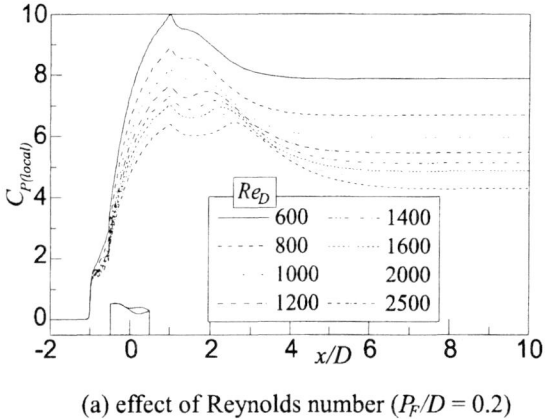 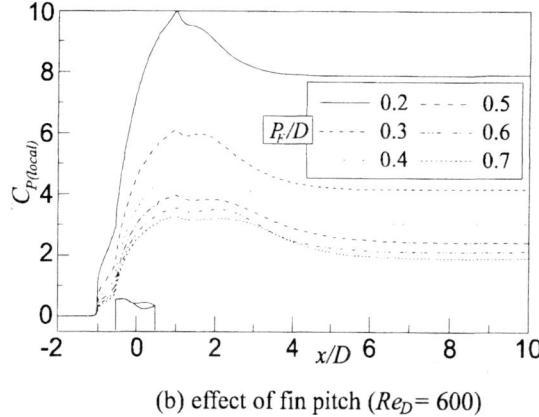

(a) effect of Reynolds number ($P_F/D = 0.2$) (b) effect of fin pitch ($Re_D = 600$)

Fig.8 Streamwise distribution of time averaged local pressure coefficient.

makes the boundary layer thinner and heat transfer coefficient higher and is intensified when the hydraulic diameter becomes smaller. (2) On the other hand, in a hydrodynamically and thermally developing region, heat transfer coefficient generally becomes smaller toward the downstream because of the thickening of thermal boundary layer and it approaches a constant value downstream in a fully developed region. Therefore, since the fractional length of developing region to the total channel length becomes smaller as the value of the hydraulic diameter is increased, total heat transfer rate over the total plate length becomes larger. (3) And then, as the hydraulic diameter becomes larger, heat transfer enhancement near the tube base caused by the horseshoe vortex is conspicuously obtained. (4) Moreover, as the hydraulic diameter becomes larger, the unsteady vortex downstream the fin trailing edge works so as to refresh the air near the fin trailing edge behind the tube. Thus, the dead water region becomes narrower and it leads to the increase of heat transfer coefficient. Consequently, as the value of D_h/L becomes larger, the value of Q^*_L for the finned tube case becomes much larger than that for the parallel plate case and this is caused by the combined effects of factors (3) and (4).

With respect to total pressure coefficient $C_{P(total)}$, its value decreases monotonically as Reynolds number or the hydraulic diameter is increased so that the increase of Reynolds number or the hydraulic diameter is preferable from the viewpoint of pressure loss. Thus, it can be said that when both the heat transfer rate and the pressure loss are considered, there exists an optimum value of D_h/L.

Local pressure coefficient

Streamwise local pressure coefficient $C_{P(local)}$ averaged in the spanwise direction for the case of (a) different Reynolds number and of (b) different fin pitch are shown in Fig. 8. The value of $C_{P(local)}$ increases discontinuously at the position where the leading edge of the fin is located and continuously increases toward the trailing edge of the fin, especially in the region where tube is located. In contrast, the value of $C_{P(local)}$ decreases downstream the trailing edge of the fin. In the cases where the fin pitch or the Reynolds number is small, the value of $C_{P(local)}$ monotonically decreases and then approaches asymptotically to the constant value. However, in the cases where the fin pitch or the Reynolds number is high, the distribution of $C_{P(local)}$ takes the second peak. As Reynolds number or the fin pitch is increased, the recovery of the pressure after the second peak becomes more noticeable. This is because when the unsteady flow is intensive, flow defect region in the tube wake recovers more quickly. Therefore, unsteady flow plays an important role in the determination of the pressure loss.

CONCLUSIONS

In this study, an unsteady numerical computation code of a plate-finned tube heat exchanger was developed for the convective heat transfer in middle Reynolds number range. This code was applied to a single row finned tube and the basic heat transfer effect of Reynolds number and the fin pitch were examined in order to supply useful information for the development of high performance heat exchanger in the practical application. Main results of this study are summarized as follows.

1. The flow unsteadiness generated downstream the fin trailing edge depends not only on the tube diameter but also on the fin pitch.
2. Karman vortex structure is generated in the instantaneous flow field and twin vortex structure is seen in the time mean flow field. As Reynolds number or fin pitch is increased, unsteadiness is intensified so that the wake of the finned tube becomes three-dimensional.
3. As Reynolds number or the fin pitch is increased, heat transfer enhancement caused by the horseshoe vortex is intensively generated near the tube base. Moreover, the dead water region near the fin trailing edge behind the

tube becomes narrower so that heat transfer enhancement occurs there.

4. Local pressure coefficient shows a peak. It is related to the increase of the momentum transfer caused by the unsteady flow. As Reynolds number or the fin pitch is increased, the pressure recovery after the second peak becomes more noticeable because flow defect region in the wake recovers more quickly as the unsteady flow is intensive.

NOMENCLATURE

A : area of surface [m^2]
C_p : specific heat of air [J/kgK]
$C_{P(local)}$: local pressure coefficient
$C_{P(total)}$: total pressure coefficient
D : tube diameter [m]
D_h : hydraulic diameter [m]
f : frequency of Karman vortex [Hz]
L : fin length [m]
Nu_D : local Nusselt number
$\overline{Nu_D}$: time and space mean Nusselt number
P : pressure [Pa]
P_F : fin pitch [m]
P_T : tube pitch [m]
Pr : Prandtl number
Q^*_L : heat transfer rate [W]
q_W : heat flux [w/m^2]
r : r-direction coordinate [m]
Re_D : Reynolds number based on D
Re_L : Reynolds number based on L, $= \rho U_i P_F L / (P_F - \delta) \mu$
Sr : Strouhal number $= fD/U_i$
t : time [s]
T_i : inlet temperature [K]
T_W : temperature of wall surface [K]
U_i : inlet velocity [m/s]
u, v, w : x, y, z -direction velocity [m/s]
x, y, z : streamwise, spanwise, normal coordinate [m]
α : local heat transfer coefficient [W/m^{2}K]
δ : fin thickness [m]
θ : radian coordinate [rad]
λ : thermal conductivity of air [W/mK]
μ : viscosity of air [Pa s]
ρ : density of air [kg/m^3]

subscript

i : inlet
o : outlet
x : local location of $x = x$
W : wall

REFERENCES

Bastani, A., Mitra, N.K. and Fiebig, M., Numerical Analysis of Three-Dimensional Flows and Heat Transfer in a Channel with a Staggered Array of Tubes, *Proc. Adv. Comp. Methods in Heat Transfer*, **3** (1990), 235, Comp. Mech. Publications.

Fiebig, M., Grosse-Gorgemann, A., Chen, Y. and Mitra, N.K., Conjugate Heat Transfer of a Finned Tube Part A: Heat Transfer Behavior and Occurrence of Heat Transfer Reversal, *Numer. Heat Transf. A.*, **28**-2 (1995), 133.

Henderson, R.D., Details of the Drag Curve Near the Onset of Vortex Shedding, *Phys. Fluids.*, **7** (1995), 2102.

Jang, J.Y., Wu, M.C. and Chang, W.J., Numerical and Experimental Studies of Three-Dimensional Plate-Fin and Tube Heat Exchangers, *Int. J. Heat Mass Transfer*, **39**-14 (1996), 3057.

Kieda, S. and Suzuki, K., 1980, Numerical Analysis on the Flow Pass the Finite Plate, *Trans. JSME*, **46**-409B, 1655.

Lange, C.F., Durst, F. and Breuer, M., Momentum and Heat Transfer from Cylinders in Laminar Crossflow at $10^{-4} \leq Re \leq 200$, *Int. J. Heat Mass Transfer*, **41** (1998), 3409.

Leonard, B.P., A Stable and Accurate Convective Modeling Procedure Based on Quadratic Upstream Interpolation, *Comp. Mech. Appl. Mech. Eng.*, **19** (1979), 59.

McAdams, W.H., *Heat Transmission*, (1954), 10., McGraw-Hill, New York.

Onishi, H., Inaoka, K., Suzuki, K. and Matsubara, K., Heat Transfer Performance of a Plate-Finned Tube Heat Exchanger - a Three-Dimensional Steady Numerical Analysis for a Single Row Tube -, *Proc. of 11th Int. Heat Transfer 1998*, Vol.6, 227, Taylor & Francis.

Onishi, H., Inaoka, K., Matsubara, K. and Suzuki, K., Numerical Analysis of Flow and Conjugate Heat Transfer of a Two-Row Plate-Finned Tube Heat Exchanger, *Proc. of Compact Heat Exchangers and Enhancement Technology for the Process Industries* 1999, 175.

Patankar, S.V., *Numerical Heat Transfer and Fluid Flow*, Hemisphere Publishing Corporation (1980).

Shah, R.K. and London, A.L., *Laminar Flow Forced Convection in Ducts (Advances in Heat Transfer)*, Supplement 1), (1978), Academic Press.

Williamson, C.H.K., Vortex Dynamics in the Cylinder Wake, *Annu. Rev. Fluid Mech.*, **28** (1996), 477.

Williamson, C.H.K., Oblique and Parallel Modes of Vortex Shedding in the Wake of a Circular Cylinder at Low Reynolds number, *J. Fluid Mech.*, **206** (1989), 579.

Xi, G.N., Ebisu, T. and Torikoshi, K., Differences in Simulation With Two-and Three-Dimensional Models for Finned Tube Heat Exchangers, *Proc. 10th Int. Symp. Transport Phenomena in Therm. Sci. and Proc. Eng. 1997*, **3**, 755.

Xi, G.N., Torikoshi, K., Kawabata, K. and Suzuki, K., Numerical Analysis of Unsteady Flow and Heat Transfer around Bodies Using a Compound Grid System, *Trans. JSME*, **61**-585B (1995), 1796.

NUMERICAL HEAT TRANSFER IN A TOP OPENING RECTANGULAR WITH A HEATING SOURCE

K. Y. Bae[1], T. W. kong[2], M. K. Ji[3], H. M. Jeong[4] and H. S. Chung[5]

[1]Gyeongsang National University(Graduate School), 445 Inpyeong-dong, Tongyoung, Kyongnam, 650-160, S. Korea;
E-mail: kybae7@yahoo.co.kr

[2]Gyeongsang National University(Graduate School), 445 Inpyeong-dong, Tongyoung, Kyongnam, 650-160, S. Korea;
E-mail: twkong@hanmail.net

[3]Gyeongsang National University(Graduate School), 445 Inpyeong-dong, Tongyoung, Kyongnam, 650-160, S. Korea;
E-mail: mk_ji73@hanmail.net

[4]Gyeongsang National University(Professor), 445 Inpyeong-dong, Tongyoung, Kyongnam, 650-160, S. Korea;
E-mail: hmjeong@gshp.gsnu.ac.kr

[5]Gyeongsang National University(Professor), 445 Inpyeong-dong, Tongyoung, Kyongnam, 650-160, S. Korea;
E-mail: hachung@gshp.gsnu.ac.kr

ABSTRACT

This study represents a numerical analysis in a top opening rectangular with a heating source. The governing equations were solved by a finite volume method, a SIMPLE algorithm was adopted to solve a pressure term. The top boundary with free surface was calculated by energy balance equation. As the results of simulations, the magnitudes of the velocity vectors were very small and the temperature was very low at the lower space of a heating source. But, at the center of the rectangular, the maximum velocity was appeared and the temperature was very high because of a uprising buoyant flow. The mean Nusselt numbers at the heating source surface are increased proportionally to the Rayleigh number, and the heat transfer at upper position was greater than the other heating source positions.

INTRODUCTION

The heat transfer in rectangular space with heating source has been applied to many of industrial problems. These researches are very important to the radiative heat control in PCB(printed Circuit Board), solar collector, accumulator and ship equipments, and they can change the machinery performances.

Thus, there are many kinds of calculation methods for basic or optimum design, and the natural convective heat transfer is very popular to understand the heat transfer mechanism. Also, the

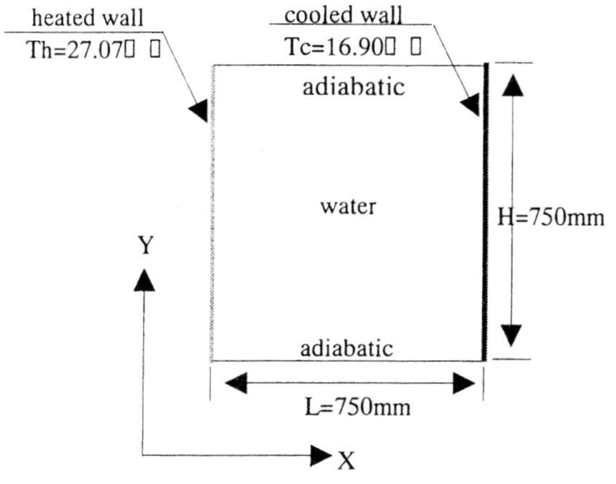

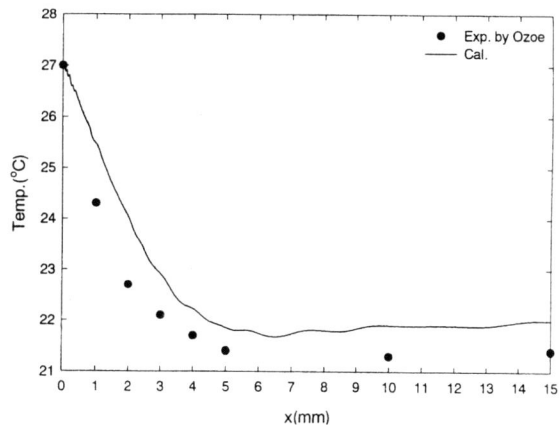

Fig. 2 Temperature distribution at Y=375mm

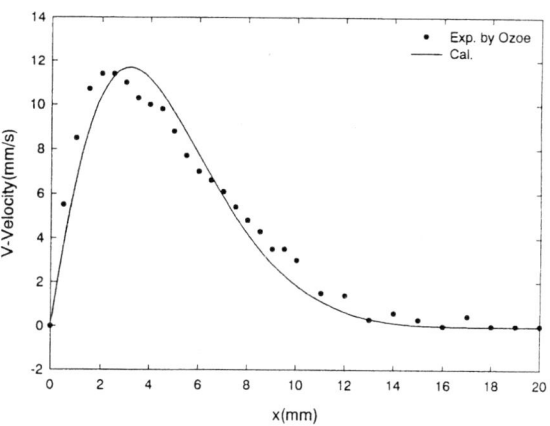

Fig. 3 Velocity distribution at Y=375mm

Fig. 1 Schematic diagram of comparison model(Ozoe)

thermal and fluid flow with air and water in rectangular space was carried out by the numerical analysis. But, most of this researches are about the rectangular space that is enclosed perfectly. In this paper, we performed the numerical simulation in case of a top opening rectangular and adopted the numerical model of the low Reynold number model by Launder and Sharma. This model was proved that had a good results at the near wall by a preceding researchers. For verification of the program with a water as a working fluid, we compared with the experimental results by Ozoe for selecting the most useful low Reynold number turbulence model.

Fig.1 represents a schematic diagram of the experiment by Ozoe. The working fluid in a rectangular enclosure is a water, the left wall is a heated wall, the right wall is a cooled wall, the top and bottom wall is adiabatic.

Fig.2 and Fig.3 show a temperature and velocity distribution, respectively. Generally, the boundary layer in water was very thin, and the temperature was a little large than the experimental value, but the velocity had a good agreements with a experimental values.

In this paper, we perform a numerical simulation of natural convection in a top opening rectangular with a heating source by using the low Reynolds number model.

NUMERICAL ANALYSIS

As a Numerical method, we introduced the low Reynolds number model by Launder and Sharma, and the governing equations are as follows:

Continuity

$$\frac{\partial (\rho U_i)}{\partial X_i} = 0 \qquad (1)$$

Momentum

$$\frac{\partial (\rho U_i U_j)}{\partial X_j} = -\frac{\partial P}{\partial X_i} + \frac{\partial}{\partial X_j}\left[\mu(\frac{\partial U_i}{\partial X_j}) + (\frac{\partial U_j}{\partial X_i})\right]$$
$$- \frac{\partial}{\partial X_j}\left[\rho \overline{u_i u_j}\right] + \delta_{i2} \rho g \beta \Delta T \qquad (2)$$

Energy

$$\frac{\partial (\rho U_j T)}{\partial X_j} = \frac{\partial}{\partial X_j}\left[(\frac{\mu}{P_r} + \frac{\mu_t}{\sigma_t})\frac{\partial T}{\partial X_j}\right] \qquad (3)$$

Turbulent kinetic energy

$$\frac{\partial (\rho U_j k)}{\partial X_j} = \frac{\partial}{\partial X_j}\left[(\frac{\mu_t}{\sigma_k} + \mu)\frac{\partial k}{\partial X_j}\right] + G - \rho \varepsilon + B - 2\mu(\frac{\partial \sqrt{k}}{\partial X_j})^2 \qquad (4)$$

Turbulent kinetic energy dissipation rate

$$\frac{\partial (\rho U_j \varepsilon)}{\partial X_j} = \frac{\partial}{\partial X_j}\left[(\frac{\mu_t}{\sigma_\varepsilon} + \mu)\frac{\partial \varepsilon}{\partial X_j}\right] + C_{1\varepsilon}\frac{\varepsilon}{k}(G + B) - C_{2\varepsilon} f_2 \frac{\rho \varepsilon^2}{k}$$
$$+ 2\frac{\mu \mu_t}{\rho}(\frac{\partial^2 U_i}{\partial X_j \partial X_k}) \qquad (5)$$

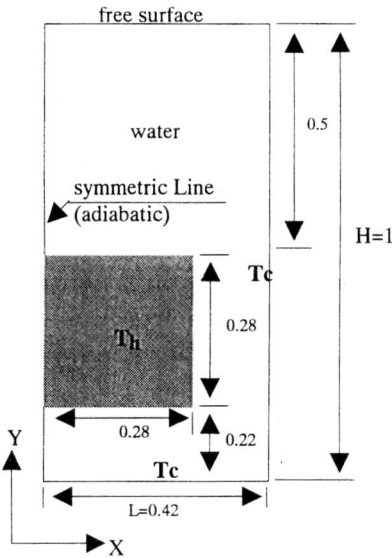

(a) Schematic diagram for calculation model

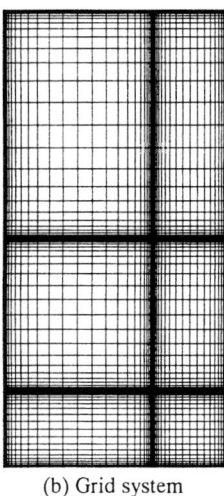

(b) Grid system

Fig. 4 Schematic diagram and grid system of calculation model

$$G = \mu_t (\frac{\partial U_i}{\partial X_j} + \frac{\partial U_j}{\partial X_i})\frac{\partial U_i}{\partial X_j}$$
$$B = -g\beta \frac{\mu_t}{\sigma_t}\frac{\partial T}{\partial y} \qquad (6)$$

where G and B are the turbulent generation term and the buoyancy term, respectively:

Here, the turbulence model constants and functions are given as follows:

(a) lower position

(b) center position

Ra. = 4.25☐ ☐0⁵   Ra. = 1.15☐ ☐0⁷   Ra. = 5.32☐ ☐0⁷

(c) upper position

Fig. 5 Isothermal distributions at Rayleigh Nu. = 4.25☐ ☐0⁵, 1.15☐ ☐0⁷ and 5.32☐ ☐0⁷

$$C_{1\varepsilon}=1.44, C_{2\varepsilon}=1.92, \sigma_\varepsilon=1.3, \sigma_k=1.0, C_\mu=0.09, \sigma_t=0.9$$
$$f_2=1-0.3\exp(-R_{et}^2), f_\mu=\exp\left\{\frac{-3.4}{(1+R_{et}/50)^2}\right\}$$
$$R_{et}=\frac{\rho k^2}{\mu\varepsilon}, \mu_t=\frac{f_\mu C_\mu \rho k^2}{\varepsilon} \quad (7)$$

Fig.4 shows the schematic diagram and the grid system of numerical analysis model. All of the scale have a dimensionless value by a height H. The model is rectangular with L☐ H=0.42☐ ☐ and the heating source have a size of 0.28☐ 0.28.

In this study, we adopted finite volume method for solving each values from given equations, SIMPLE algorithm was used to solve a pressure term. The calculation grid adopted a non-constant grid by using equation(8) for dense arrangement near a heat source and the wall side.

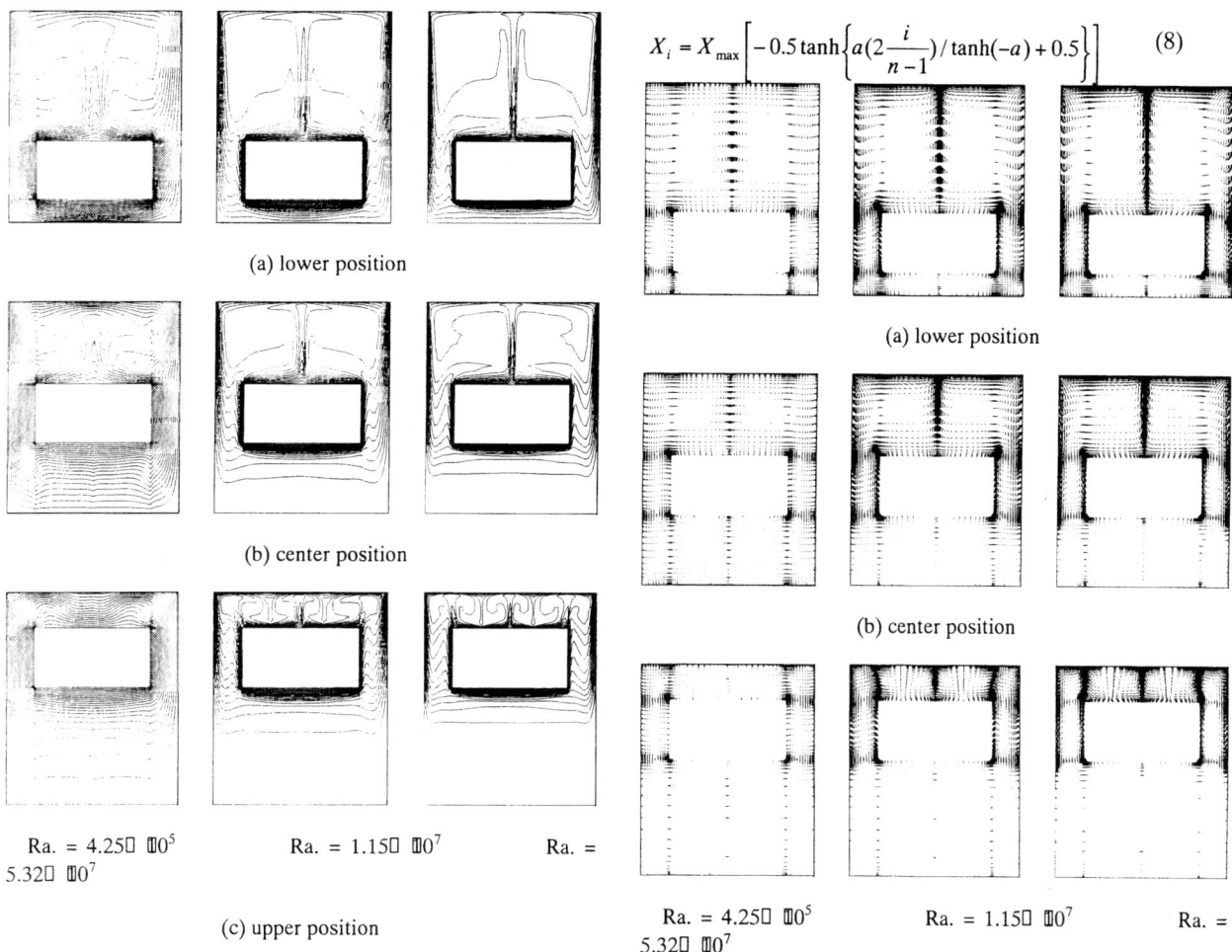

$$X_i = X_{max}\left[-0.5\tanh\left\{a(2\frac{i}{n-1})/\tanh(-a)+0.5\right\}\right] \quad (8)$$

(a) lower position

(b) center position

Ra. = 4.25☐ ☐0⁵   Ra. = 1.15☐ ☐0⁷   Ra. = 5.32☐ ☐0⁷

(c) upper position

Fig. 6 Velocity vector distributions at Rayleigh Nu. = 4.25☐ ☐0⁵, 1.15☐ ☐0⁷ and 5.32☐ ☐0⁷

where n is grid number of X and Y direction. The i is coordinate position and a is a coefficient for adjusting the grid interval.

The numerical analysis was performed for a half of right area. The bottom and right wall is cooling at constant temperature, T_c=15☐ ☐and the symmetric line has a adiabatic condition. The top boundary with free surface was calculated by a energy balance condition and a heat source is heating at constant temperature, T_h=55☐ ☐

The energy balance condition at the free surface is as follows:

$$(q_{cond}+q_{conv})_{water}=(q_{cond}+q_{conv})_{air} \quad (9)$$

where q_{cond} and q_{conv} are the heat flux per normal length by conduction and convection in the free surface, respectively.

The equation(9) yields the following equation(10).

$$-k_w \cdot A \cdot \frac{dT}{dy} + h_{c,w} \cdot A(T_{j-1,w} - T_{j,w}) = -k_a \cdot A \cdot \frac{dT}{dy} + h_{c,a} \cdot A(T_{j,w} - T_a)$$
(10)

where j is a water side grid position of a free surface, and $j-1$ is a one grid below of j point.

From the equation(10), we can calculate the temperature T_j of a surface, and introduced the following constants.

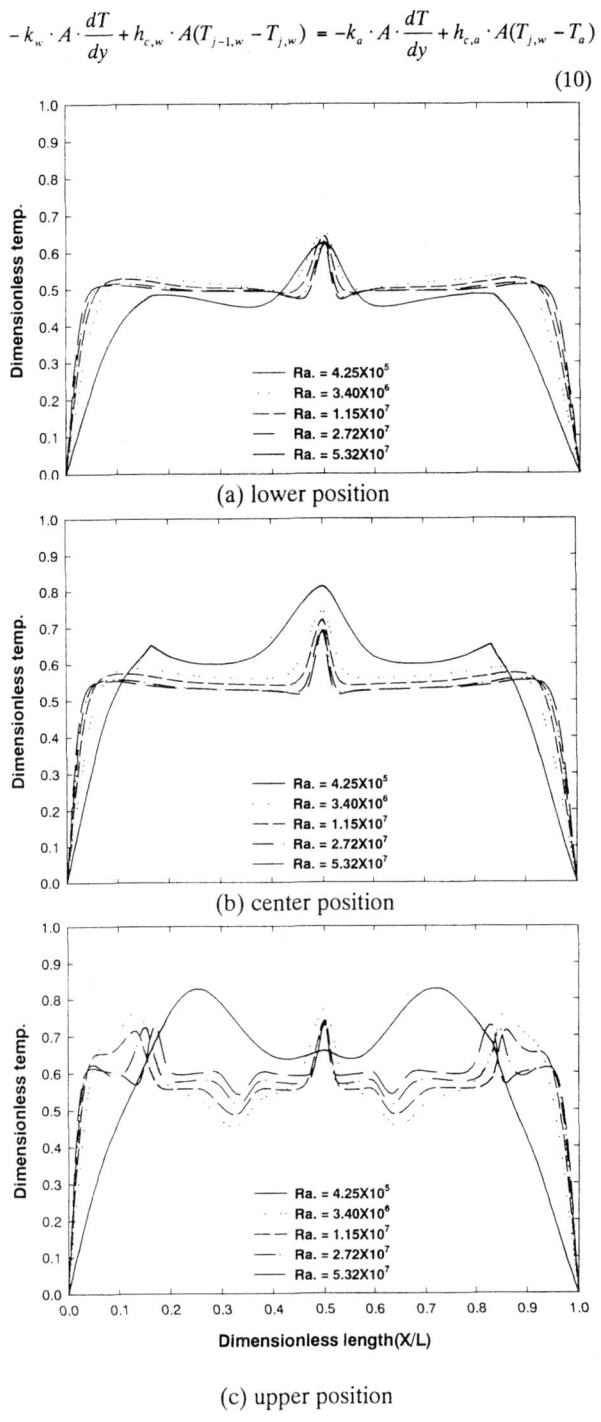

Fig.7 Dimensionless temperature distributions at middle section in the top space of the heat source

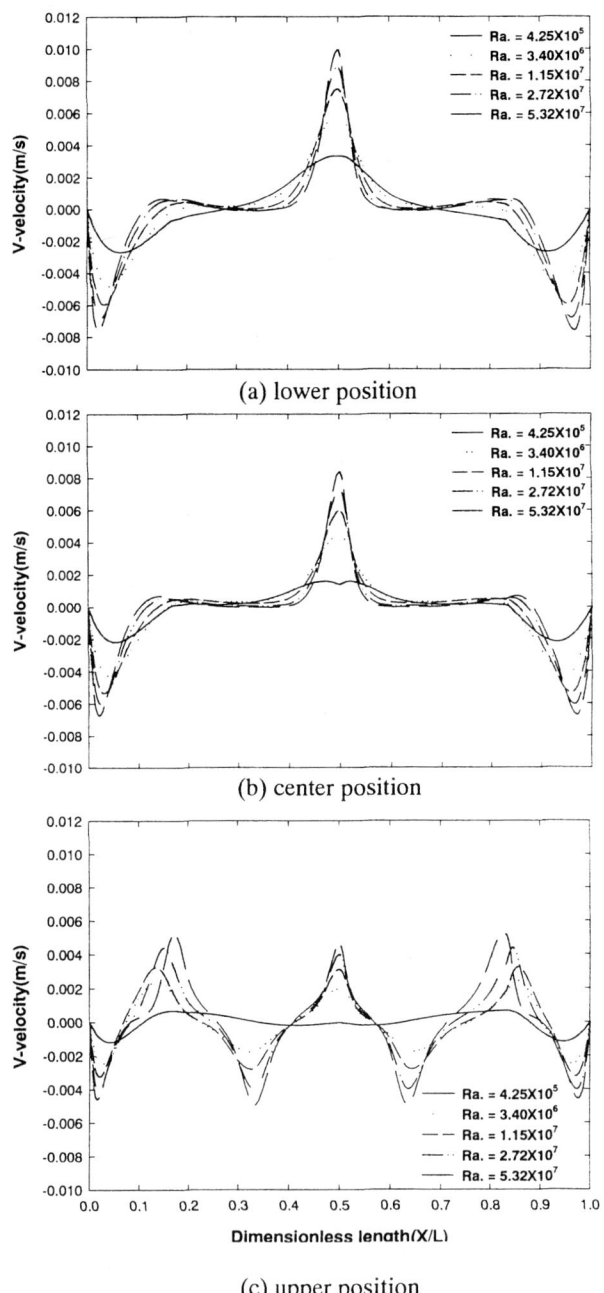

Fig.8 V-velocity distributions at middle section in the top space

$k_w = 0.6$, $k_a = 0.026$, $h_{c,w} = 50$, $h_{c,a} = 20$

RESULTS and DISCUSION

Fig.5 shows a distributions of isotherms according to the

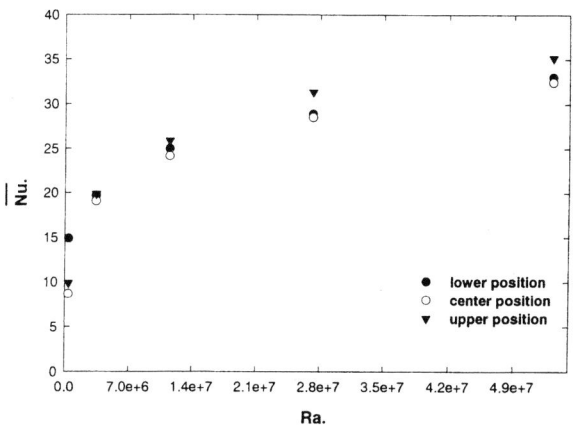

Fig. 9 Distributions of mean Nusselt vs. Rayleigh number for various positions of the heating source surface

position of a heating source and the variation of Rayleigh numbers.

The general distributions of isotherms has a uprising flow with a high temperature in the center of a heating source.

Especially, the low temperature region was appeared at the low space of a heating source point.

Fig.6 represents the velocity vectors according to the position of a heating source and the variation of Rayleigh numbers. In case of large Rayleigh number, the velocity vectors were very dense at near the wall of a heating source. Thus, the heat transfer was increased than small Rayleigh number.

Fig.7 shows the dimensionless temperature distribution of V-direction at middle section in the top space with several Rayleigh numbers. As the heating source moves to lower position, the maximum velocity vector and the high temperature was appeared in the center of heating source.

But, the heating source moves to upper position, the Bernard flow pattern was appeared in the upper space of the heating source.

Fig. 8 shows the velocity distribution of V-direction at middle section in the top space with several Rayleigh numbers. As Rayleigh numbers are increased, the magnitude of velocity vector was large and the heat transfer was increased. But at near the upper side of a heating source, the velocity is small and very complex by increasing Rayliegh numbers.

Fig.9 represented a distributions of mean Nusselt numbers versus Rayleigh number for various heating positions at the heating source surface by using equation(11).

$$\overline{Nu} = \frac{1}{H}\int Nu \cdot ds \qquad (11)$$

where, ds represents a heating source surfaces. The distributions of mean Nusselt number was high when the a heating source was located at the upper position but they was low when a heating source was located at the center position.

CONCLUSIONS

1. As the Rayleigh numbers are increased, the isotherms showed a uniformly distribution by a thermal convective flow in the upper space of a heating source.
2. The gradient of the heat transfer was very steep at the small Rayleigh number region.
3. The maximum velocity was appeared at the low space of a heating source but, at near the wall side of a heating source, the velocity was very small.
4. When the heating source was located at the upper position, the flow pattern was showed a Bernard flow pattern.

ACKNOWLENGEMENTS

This work was supported by Korea Maritime Institute and the Brain Korea 21 Projects, and the authors gratefully appreciate the support.

NOMENCLATURE

A   free surface area
a   air
g   gravity acceleration, m/s^2
Gr.   Grashof number ($Gr = \frac{g\beta(T_h - T_c)H^3}{\nu^2}$)
H   vertical wall length, m
h_c convective coefficient, w/m^{2}k
k   conductive coefficient, w/mk
k turbulent energy, m^2/s^2
L   horizontal wall length, m
Nu.   local Nusselt number ($Nu = \frac{hL}{k} = -\frac{\partial \theta}{\partial X}|_{X=0}$)
$\overline{Nu}$.   mean Nusselt number ($\overline{Nu} = \frac{1}{H}\int Nu \cdot dy$)
Pr   Prandtl number
T   temperature, °C
T_h heating source temperature, °C
T_c cooled wall temperature, °C
U   X direction velocity, m/s
V   Y direction velocity, m/s
w   water
â   thermal expansion coefficient, K^{-1}
ä$_{ij}$ Kronecker delta
å   turbulent energy dissipation rate
ì$_t$ turbulent eddy viscosity, kg/ms
ñ   density, kg/m^3

REFERENCES

H. M. Jeong, C. H. Lee and H. S. Chung, 2000, study on the Numerical Modeling of Turbulent Natural Convection in Rectangular Enclosure, Journal of SAREK, Vol. 10, No. 2, pp.33-39.

J. H. Lee, M. H. Kim and J. H. Moh, 1990, Natural convection Heat Transfer in Rectangular Air Enclosure with Adiabatic and Isothermal Horizontal Boundary Condition, Journal of KSME, Vol. 14, No. 1, pp.207-213.

D. K. Kang, W. S. Kim and K. S. Lee, 1995, A Numerical Study on the Two-Dimensional Turbulent Natural Convection Using a Low-Raynolds Number k-å Model, Journal of KSME, Vol. 19, No. 3, pp.741-750.

Suhas V. Patankar, 1980, Numerical Heat Transfer and Fluid Flow.

H. S, Chung, H. M. Jeong, K. K. Kim and S. T. Ro, 1999, The Turbulent Natural Convection in Membrane Type LNG Carrier Cofferdam, Journal of KSME, Part B, Vol. 23, No. 2, pp.75-81.

R. Cheesewright, K. H. King and S. Ziai, 1986, Experimental data of the Validation of Computer Codes for the Prediction of Two Dimensional Buoyancy Affected Enclosure or Cavity Flows, ASME-HTD, pp.75-81.

T. G. Kim and H. J. Sung, 1993, A Hybrid Turbulence Model for Prediction of Buoyancy-Driven Turbulent Thermal Convection Flow, Journal of KSME, Vol. 17, No. 8, pp.2069-2078.

H. T. Seo, D. S. Lee, S. H. Yoon and J. S. Boo, 1998, Numerical Analysis on Wall-Attaching Offset Jetwith Various Turbulent k-å Models and Upwind Scheme, Journal of KSME, Part B, pp.828-835.

Kumar. K, 1983, Mathematical Modeling of Natural Convection in Fire-a State of the Art Review of the Field Modeling of Variable Density Turbulent Flow, Fire and Materials, Vol. 103, pp.456-460.

C. J. Lee, H. M Jeong, H. S. Chung and K. K. Kim, 1998, Study on the Natural Convection of Rectangular Enclosure with Heating Source, Proceeding of KSME, part B, pp.417-421.

W. Shyy and M. M. Rao, 1993, Simulation of Transient Natural Convection Around an Enclosed Vertical Channel, Journal of Heat Transfer, the ASME, Vol. 115, pp.946-954.

Ozoe, The Natural Convective Measurement for a Heating and Cooling Walls, No. 5, NST-Symposium, Institute of Industrial Science, Tokyo Univ.

HEAT TRANSFER CHARACTERISTICS IN TWO-PASS RECTANGULAR CHANNELS WITH INCLINED PARTITION WALL

Masafumi Hirota[1], Hideomi Fujita[2], Makoto Yanagida[3], and Hiroshi Nakayama[4]

[1]Department of Mechanical Engineering, Nagoya University, Nagoya 464-8603, Japan; E-mail: hirota@mech.nagoya-u.ac.jp
[2]Department of Mechanical Engineering, Nagoya University, Nagoya 464-8603, Japan; E-mail: fujita@mech.nagoya-u.ac.jp
[3]Interior Design Department, ARACO Co. Ltd., Toyota 473-8512, Japan; E-mail: makoto-yanagida@arnotes.araco.co.jp
[4]Department of Mechanical Engineering, Nagoya University, Nagoya 464-8603, Japan; E-mail: naka@mech.nagoya-u.ac.jp

ABSTRACT

Serpentine channels with a sharp 180° turn are often used in compact heat exchangers. The local heat transfer characteristics in such channels are very complex, because the secondary flow arises in the turning flow, and the flow separation and reattachment occur around the sharp turn. In most studies conducted on this subject, the partition (inner) wall of the channel was settled parallel to the outer walls. If the partition wall is inclined with respect to the outer walls, the local heat transfer characteristics in the channel can be considerably altered from those in the parallel-walled channel. Moreover, it is expected that the heat transfer performance of the channel be improved than the case of the parallel partition wall.

We made an experimental study on the heat transfer characteristics in rectangular cross-sectioned two-pass serpentine channels with an inclined partition wall. The channel cross section at the entrance was 50 mm × 25 mm, and the inclination angle of the partition wall was changed from -6° (convergent channel) to +6° (divergent channel) at 2° intervals for three turn clearances of 30 mm, 50 mm, and 70 mm. Distributions of local heat (mass) transfer rates on the long-side wall were measured by the naphthalene sublimation method under a single Reynolds number of 3.5×10^4. In this paper, the influence of the inclination angle of the partition wall on heat transfer characteristics in the channels is discussed. Then, the optimum combination of the inclination angle and the turn clearance is examined from viewpoints of the trade-off between the mean heat transfer rate and pressure loss, and of the uniformity in the distribution of local heat transfer rates.

INTRODUCTION

Rectangular cross-sectioned two-pass channels with a sharp 180-degree turn are often used in heat exchangers. The flow structure in such a serpentine channel is very complex because, in addition to the secondary flow which presents in the turning flow, the flow separation and reattachment occur around the sharp turn (Metzger et al., 1986; Liou et al., 1999). As a result, the local heat transfer rates around the turn change steeply, and thus the non-uniformity of their distribution is very large there (Murata et al., 1994; Wang et al., 1994; Astarita et al., 1995; Mochizuki et al., 1998; Astarita et al., 2000). The non-uniformity in the distribution of the local heat transfer rates causes large temperature gradients over the heat transfer surface, and consequently increases the thermal stress. Thus, detailed data on the local heat transfer rates in the channels are indispensable for a design of thermal equipment used under a severe thermal condition. In response to such demand, many researches have been made on forced convection heat transfer in those channels with a sharp turn (Metzger et al., 1986; Fan et al., 1987; Chandra et al., 1988; Han et al., 1988; Chyu, 1991; Besserman et al., 1992; Murata et al., 1994; Wang et al., 1994; Astarita et al., 1995; Mochizuki et al., 1998; Astarita et al., 2000). The present authors also made clear the detailed characteristics of local heat transfer in two-pass rectangular channels with a sharp 180-degree turn by using the naphthalene sublimation technique. They presented 2-D maps of local Nusselt (Sherwood) number distributions on the channel walls, and examined the influences of the turn clearance and flow-inlet condition at the channel entrance on the local heat transfer characteristics (Hirota et al., 1997; Hirota et al., 1999a; 1999b).

In most studies conducted to date, the partition (inner) wall of the serpentine channel was settled parallel to the outer walls. If the partition wall is inclined with respect to the outer walls, as shown in Fig. 1, the primary flow in the convergent (or divergent) sections before and after the turn is accelerated (or decelerated) due to the streamwise change of the channel cross-sectional area. The influence of such acceleration or deceleration of the primary flow is then reflected on the secondary flow distribution and on the flow separation and reattachment around the turn. As a result, the local heat transfer characteristics in the channel can be considerably altered from those in the standard parallel-walled channel. Moreover, it is expected that the heat transfer performance of the channel be improved than the case of the parallel partition wall. In applying the serpentine channels to practical equipment used under severe thermal conditions, it is important not only to achieve higher heat transfer rates with lower pressure loss but also to increase the uniformity in the distributions of the local heat transfer rates for decreasing the thermal stress. By selecting some optimum inclination angle, both the enhancement of mean heat transfer and the improvement of local heat transfer uniformity may be achieved concurrently without a considerable increase of the pressure loss.

With these points as background, we have made an experimental study on forced convection heat transfer in the two-pass rectangular channels with an inclined partition wall. As the first step of the study, we have measured local heat transfer rates on the long-side wall of the channel (see Fig. 2) by the naphthalene sublimation method (Goldstein et al., 1995) to elucidate whether or not the inclined partition wall is effective to the enhancement of heat transfer and/or the improvement of the heat transfer uniformity. In this paper, detailed results of heat (mass) transfer obtained in the test channels are presented, and the influence of the inclined partition wall on the heat transfer characteristics is discussed in detail. Based on the experimental results, the optimum condition of the inclination angle and the turn clearance that can achieve both the enhancement of heat transfer and the improvement of heat transfer uniformity is examined.

EXPERIMENTS

Figure 1 shows a schematic diagram of the experimental apparatus. Air flows into the test channel through a settling chamber. A baffle plate is settled at the entrance of the test channel to set up a contracted-flow inlet condition, under which turbulent boundary layers cover the whole cross-section at the entrance (Sparrow et al., 1982; Hirota et al., 1999b). The test channel is a three-pass passage with a rectangular cross-section; the mass transfer

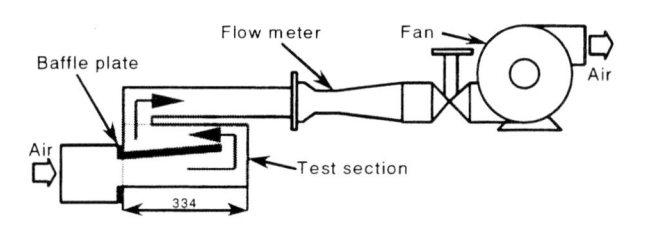

Fig. 1 Schematic diagram of the experimental apparatus

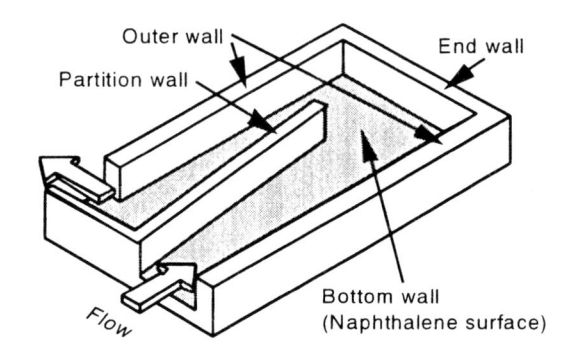

Fig. 2 Details of the test section

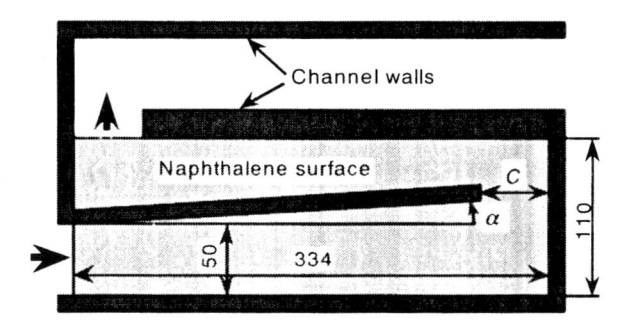

Fig. 3 Details of the mass transfer surface (long-side wall)

rate is measured in the darkened region in the figure, which is located just downstream of the settling chamber. Air contaminated by the naphthalene vapor is then exhausted from the building by a turbofan.

Figure 2 shows a schematic illustration of the test section. The test channel has a rectangular cross-section; the short-side length is 25 mm and is constant throughout the test section, while the long-side length is 50 mm at the channel entrance and varies in the flow direction depending on the inclination angle of the partition (inner) wall. The hydraulic diameter defined at the entrance of the channel, d_h, is 33.3 mm. The surface of one long-side wall was coated by naphthalene, and we measured the local mass transfer rates on it. The other long-side wall and all the short-side walls were made of smooth aluminum plates. From a preliminary experiment, it was confirmed that the Sh distribution obtained under the present situation agreed well with that obtained in the channel which had naphthalene-coated surfaces on the both long-side walls.

Figure 3 shows the details of the naphthalene surface in the mass transfer test section. The partition wall is made of a smooth aluminum plate of 10 mm thick. The inclination angle of this partition wall, denoted as α in the figure, was changed from -6° (clockwise direction in the figure) to +6° (counter-clockwise direction) at 2° intervals; totally seven inclination angles were tested in the present study. In this paper, the channels with α < 0° are called "convergent channels", and those with α > 0° are "divergent channels." The channel with α = 0° having a parallel partition wall is called "standard channel." The turn clearance, denoted C in Fig. 3, was varied from 30 mm to 70 mm at 20 mm pitch; in this paper, however, the results of medium turn clearance C = 50 mm are mainly presented due to the limited space. The edge of the partition-wall tip is parallel to the end wall. It should be noted that, in the channel of C = 50 mm and α = 0°, the turn clearance is equal to the spanwise length of the long-side wall. This channel geometry is the standard case that has been adopted in most of the preceding studies.

The measurement procedure of the mass transfer coefficients by the naphthalene sublimation method (Schmidt number = 2.5) is the same as that of the preceding experiment (Hirota et al., 1997; 1999a). We also measured the pressure distribution on the long-side wall to calculate the pressure loss of the channel. The pressure measurement was conducted under a non-sublimating condition by replacing the naphthalene surface with an acrylic resin plate on which the pressure taps of 0.4 mm diameter were distributed on the spanwise centerline (Hirota et al., 1999a). The mass transfer experiment and the pressure measurement were conducted under a single Reynolds number of 3.5×10^4, which was defined by the bulk velocity and hydraulic diameter at the channel entrance.

Here, it should be noted that, in a preliminary experiment, we conducted the measurements of the mass transfer rates and the pressure loss under the Reynolds number range of $(2.0 \sim 5.0) \times 10^4$ with a limited combination of α and C. It was found that the characteristics of heat (mass) transfer and pressure loss for $Re = 2.0 \times 10^4$ and 5.0×10^4 were consistent with those obtained under the condition of $Re = 3.5 \times 10^4$. That is, the mean and local Sherwood numbers in the channels normalized by the mean Sherwood numbers for fully developed turbulent flow in a straight pipe agreed qualitatively and quantitatively with one another irrespective of the values of Reynolds number. The coefficient of pressure loss for different Re was also in good agreement with one another. Therefore, it follows that the results presented in this paper can be reasonably applied to the estimation of the heat transfer characteristics of the channels for different Reynolds numbers.

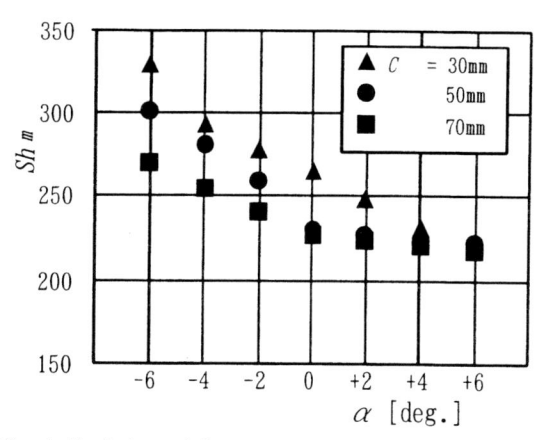

Fig. 4 Variation of the mean Sherwood number against the inclination angle of the partition wall

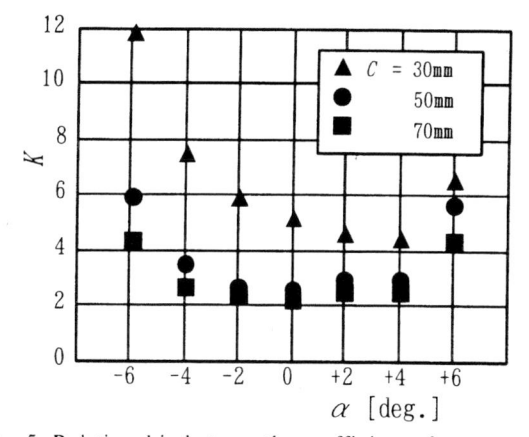

Fig. 5 Relationship between the coefficient of pressure loss and the inclination angle of the partition wall

RESULTS AND DISCUSSION

Mean Sherwood Number and Pressure Loss

Figure 4 shows the variation of the mean Sherwood number Sh_m against the inclination angle of the partition wall α obtained for all turn clearances. In the convergent channel of α < 0°, Sh_m increases in proportional to $|α|$, and Sh_m for α = -6° is about 20 - 30 % larger than that for the standard channel (α = 0°). On the other hand, in the divergent channel of α > 0°, Sh_m is almost constant (C = 50 mm and 70 mm) or decreased (C = 30 mm) against α. In general, Sh_m shows higher values as the turn clearance C is decreased. The dependence of Sh_m on the turn clearance is, however, rather weak in the divergent channels.

The coefficient of flow resistance of the channel K, which is defined by Eq. (1) (Metzger et al., 1984), was calculated from the pressure distribution, and the result is shown in Fig. 5.

$$K = \frac{2 \Delta P}{\rho U^2} \quad (1)$$

where ΔP is the pressure difference between two pressure taps; one is located 50 mm downstream from the channel entrance, and the other is 50 mm upstream from the end of the test section. In both the convergent and the divergent channels, K is increased as the inclination of the partition wall is increased or as the turn clearance is decreased. The values of K for $C = 30$ mm are significantly larger than those for wider turn clearances in all the test channels. However, the difference in K-values caused by C becomes smaller in the divergent channels; this tendency is quite similar to Sh_m.

Local Sherwood Number Distribution

In this section, the distributions of local Sherwood number Sh obtained in the channels with different inclination angles are presented in a form of 2-D map. The turn clearance is kept constant at a medium value, i. e., $C = 50$ mm, throughout the section. At first, Sh distribution in the standard channel with a parallel partition wall ($\alpha = 0°$) is reviewed briefly (Hirota et al., 1999a). Then, Sh maps obtained in the divergent channels and the convergent channels are presented, and the influence of the inclination angle of the partition wall on the local heat transfer characteristics is discussed in detail. In order to define the locations in the Sh maps, we divide the mass transfer surface into 14 blocks as shown in Fig. 6, and number the blocks as Block 1, 2, 3, etc. The boundary between Block 7 and Block 8 moves depending on α.

Standard channel ($\alpha = 0°$). Figure 7 shows a distribution of Sh obtained in the standard channel with a parallel partition wall. The variations in Sh-values are shown as contrasting light and dark areas. The map becomes whiter as the Sh-values are increased; the high Sh regions of $Sh > 400$ are shown in white. The arrow presents the flow direction. The broken lines in the map show the boundaries of the blocks defined in Fig. 6.

In the straight section before the turn (Blocks 1 to 6), Sh shows the local maximum at the channel entrance, then it decreases in the flow direction due to the development of the concentration boundary layer. In the turn section (Blocks 7 and 8), Sh increases gradually in the flow direction, and it rises steeply to attain the local maximums near the end wall in Block 7 and near the outer wall in Block 8. On the other hand, in the corner in Block 7, there appears a low Sh region that corresponds to the flow circulation caused by the abrupt change of the flow direction (Murata et al., 1994; Liou et al., 1999).

In the straight section after the turn (Blocks 9 to 14), a region of relatively low Sh appears along the partition wall near the partition-wall tip in Block 9. Sh then increases in the streamwise direction, and attains the local maximum in Block 10 (Liou et al., 1999). This means that the flow

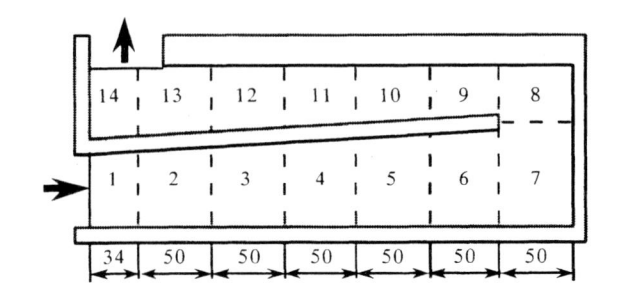

Fig. 6 Definition of Block on the mass transfer surface

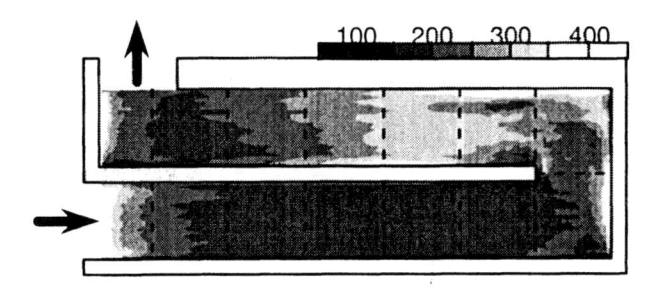

Fig. 7 Local Sherwood number distribution in the standard channel ($\alpha = 0°$)

is separated at the partition-wall tip, forms a separation bubble along the partition wall in Block 9, and then reattaches in Block 10. In Block 10, the region of high Sh-values occupies most part of the block and no spanwise deviations are observed in Sh distribution. From the measurement of Sh on all the channel walls, it was elucidated that this separated flow reattaches on the partition wall. After Block 11, Sh is decreased in the flow direction; if compared at the same streamwise location, Sh in the outer-wall side shows larger values than that in the partition-wall side.

Divergent channel ($\alpha > 0°$). Figures 8(a), 8(b), and 8(c) show the Sh maps for $\alpha = +2°, +4°$, and $+6°$, respectively, with a medium turn clearance ($C = 50$ mm). In the divergent section before the turn, there appear the regions of relatively low Sh near the short-side walls in Blocks 4 and 5, and these low Sh regions tend to extend toward the central area of the channel as α is increased.

In the upstream half of the turn section (Block 7), as similar to the standard channel shown in Fig. 7, Sh increases gradually in the flow direction and attains the local maximum near the end wall. (Here, it should be noted that the boundary between Blocks 7 and 8 moves downstream as α is increased.) Near the corner in Block 7, the low Sh region can be observed. As α is increased, Sh in Block 7 becomes smaller and the low Sh region in the corner occupies larger area. On the other hand, in the downstream half of the turn section (Block 8), Sh shows larger values as α is increased.

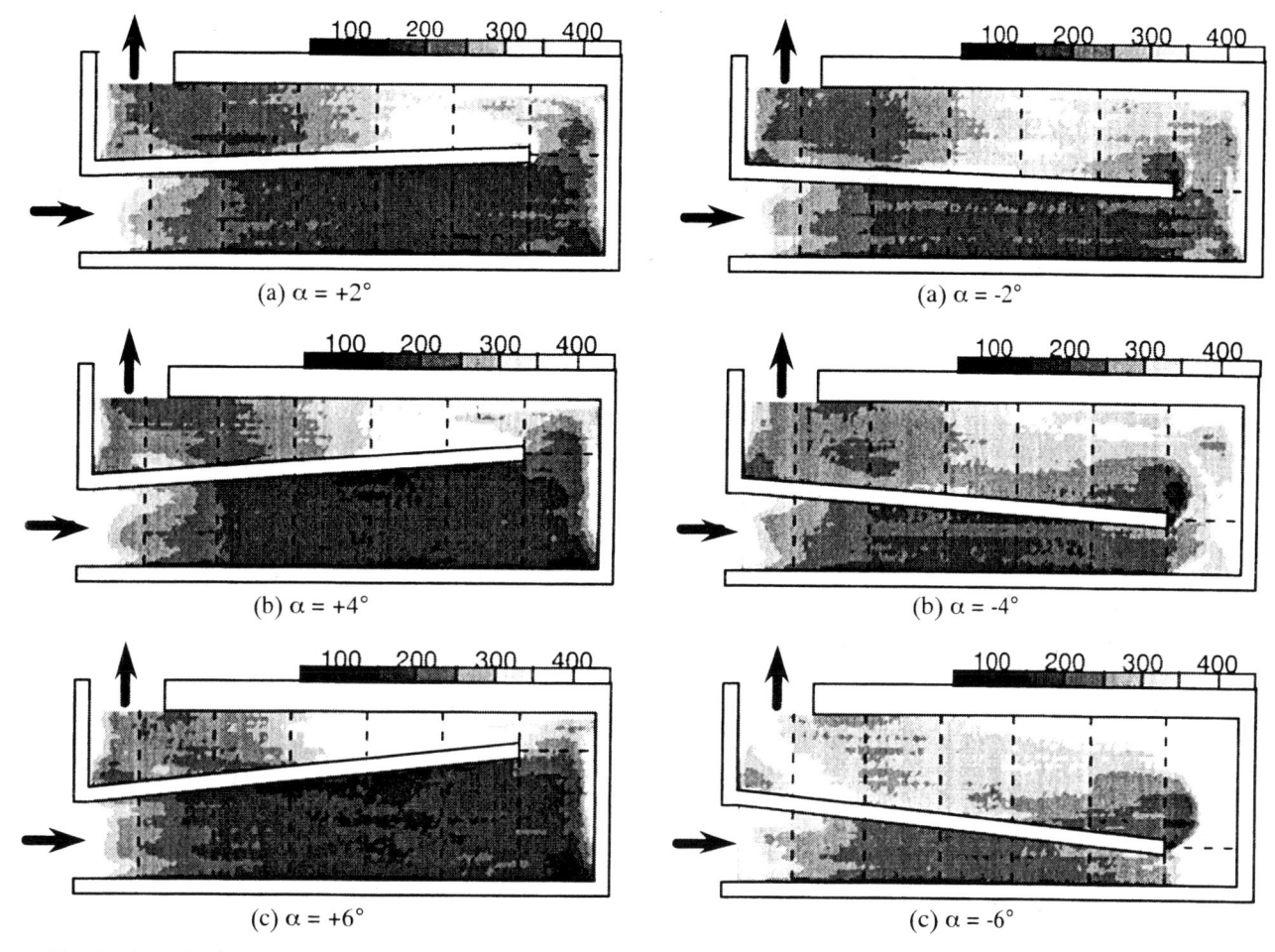

Fig. 8 Local Sherwood number distributions in the divergent channels ($\alpha > 0°$)

Fig. 9 Local Sherwood number distributions in the convergent channels ($\alpha < 0°$)

Characteristics of the Sh distribution after the turn are much influenced by the inclination angle of the partition wall. At first, the Sh distribution for $\alpha = +2°$ shown in Fig. 8(a) is examined in detail. Near the partition-wall tip, the region of relatively low Sh corresponding to a separation bubble is observed, but the scale of this low Sh region is considerably smaller than that observed in the standard channel. In Block 9, Sh shows large values ($Sh > 400$) outside this separation bubble and near the outer wall; the high Sh region in the partition-wall side (outside the separation bubble) extends into Block 10. Sh attains the maximum in Block 10, and it is much larger than that observed in the corresponding region of the standard channel. Such increase of Sh in Blocks 9 and 10 in the divergent channel is caused by the acceleration of the primary flow which passes through a narrower turn exit, and by high turbulence which is produced in the shear layer outside the separation bubble. After Block 11, Sh is decreased as the flow proceeds downstream. The distribution of Sh in this region is quite uniform in the spanwise direction. The spanwise non-uniformity of Sh distribution, i. e., deviation of high Sh region to the outer-wall side, such as observed in the standard channel does not appear in Fig. 8(a).

As the inclination angle of the partition wall is increased, as shown in Figs. 8(b) and 8(c), the values of Sh in the divergent section after the turn become larger because the flow velocity at the turn exit is increased. The low Sh region near the partition-wall tip corresponding to the separation bubble disappears and, in the channel of $\alpha = +6°$, the white region of $Sh > 400$ occupies almost whole area of Blocks 9 and 10. It has been found that, in the divergent channels of $\alpha = +4°$ and $+6°$, the maximum Sh appears not in Block 10 but in Block 9 located just downstream of the turn exit. After these high Sh-values in Blocks 9 and 10, Sh decreases quite rapidly in the flow direction. It can be found that Sh after Block 11 becomes larger as the inclination angle is increased.

Convergent channel ($\alpha < 0°$). The Sh maps obtained in the convergent channels with $\alpha = -2°$, $-4°$, and $-6°$ are shown in Figs. 9(a), 9(b), and 9(c), respectively. In the convergent section before the turn, the primary flow is accelerated due to the decrease of the cross-sectional area of the flow passage. Thus, Sh in this section shows larger values than those observed in the standard and the divergent channels, and it becomes larger as the inclination of the partition wall is increased.

In the turn section, Sh in the convergent channels is generally larger than that in the other channels. As similar to the case of the standard channel, Sh shows the local maximums near the end wall in Block 7 and near the outer wall in Block 8. These high Sh regions extend into the inside of the turn section as $|\alpha|$ is increased; in the channel of $\alpha = -6°$, the region of $Sh > 400$ occupies most part of the turn section. Near the corner in Block 7, the low Sh region corresponding to the flow circulation almost disappears.

In Block 9 located just downstream of the turn exit, the low Sh region corresponding to the separation bubble is observed near the partition-wall tip. This low Sh region occupies larger area than that in the standard channel, and it extends upstream into the turn section (Block 8) as the inclination angle is increased. On the other hand, near the outer wall in Blocks 9 and 10, Sh shows relatively large values and it becomes larger as $|\alpha|$ is increased. These characteristics of Sh distribution observed in Fig. 9 suggest that, after the turn section of the convergent channels, a separation bubble with a rather large scale is formed near the partition-wall tip, and the flow is accelerated in the outer-wall side due to the decrease of the substantial cross-sectional area of the flow passage.

After Block 11, Sh for a larger inclination angle shows higher values than that for smaller one. It should be also noted that Sh after Block 11 in the convergent channels is generally larger than that in the corresponding region of the standard and the divergent channels.

ASSESSMENT OF HEAT-TRANSFER PERFORMANCE

As described so far, the heat transfer characteristics and pressure loss in the channels are greatly altered by inclining the partition wall. In applying the serpentine channels with a sharp turn to practical equipment used under severe thermal conditions, it is important not only to achieve higher heat transfer rates with lower pressure loss, but also to increase the uniformity in the distributions of the local heat transfer rates for decreasing the thermal stress. Hence, in this chapter, the heat transfer performance of the channels is examined from viewpoints of (1) the trade-off between the mean heat (mass) transfer rates and the pressure loss, and (2) the uniformity in the distribution of the local heat (mass) transfer rates. Then, we determine the optimum condition of the inclination angle and the turn clearance that can achieve both the enhancement of mean heat transfer and the improvement of the local heat transfer

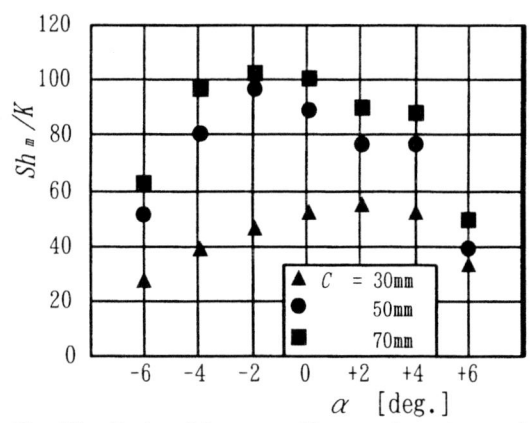

Fig. 10 Ratio of the mean Sherwood number to the coefficient of flow resistance of the channel

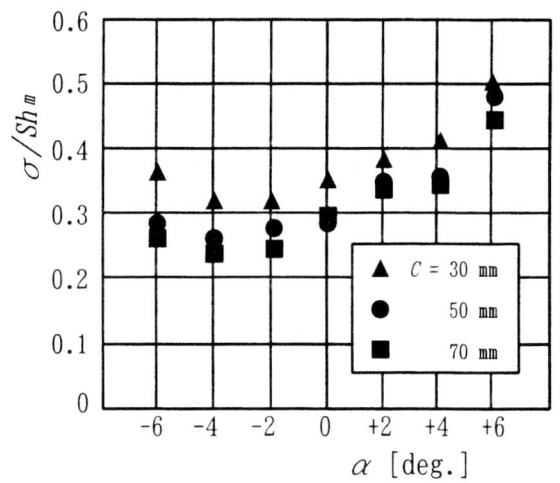

Fig. 11 Standard deviation of the local Sherwood numbers

uniformity concurrently without a considerable increase of pressure loss.

At first, in order to evaluate the increases of the mean heat transfer rate and of the pressure loss, we have calculated the ratio of mean Sherwood number to the coefficient of flow resistance Sh_m/K. The values of this parameter become larger when higher mean heat transfer rate can be achieved with lower pressure loss. Thus, the condition (α and C) under which this parameter becomes the maximum gives the optimum combination of α and C. Figure 10 shows the variation of Sh_m/K against α and C. In general, the value of Sh_m/K is increased as the turn clearance is widened, and the maximum of Sh_m/K is found under the condition of $C = 70$ mm and $\alpha = 0°$ or $-2°$. Therefore it follows that, from a viewpoint of the trade-off between the mean heat (mass) transfer rate and the pressure loss, the highest heat transfer performance can be achieved in the channels with $C = 70$ mm and $\alpha = 0°$ or $-2°$.

Next, the uniformity of the local Sherwood number distributions is examined quantitatively. In order to evaluate it, we have calculated the standard deviation of the local Sherwood number, denoted as σ and defined by the following equation, for all channels tested in this study.

$$\sigma = \sqrt{\frac{1}{N}\sum_{i=1}^{N}(Sh_i - Sh_m)^2} \qquad (2)$$

N denotes the total number of the measuring points of Sh, and it is about 65000. Lower value of σ corresponds to the higher uniformity of the Sh distribution. The variations of σ/Sh_m against α and C are shown in Fig. 11. It is clear that the values of σ/Sh_m for the convergent channels are generally smaller than those for the divergent channels; thus the convergent channels are more advantageous that the divergent channels from a viewpoint of uniformity in the distribution of the local heat transfer rates. The value of σ/Sh_m decreases as the turn clearance is increased, and σ can be minimized under the condition of $C = 70$ mm and $\alpha = -2°$ or $-4°$. Therefore, it follows that the combination of $C = 70$ mm and $\alpha = -2°$ or $-4°$, i. e., convergent channels with a wide turn clearance, is the optimum condition for realizing the most uniform distribution of local heat (mass) transfer rates.

The optimum conditions of C and α obtained based on Sh_m/K and σ/Sh_m partially overlap each other. From the above-mentioned results, it can be concluded that, under the present experimental condition, the highest heat-transfer performance can be achieved in the channel of $C = 70$ mm and $\alpha = -2°$, i. e., in the convergent channel with a relatively wide turn clearance. With this combination of C and α, both the enhancement of mean heat transfer and the improvement of local heat transfer uniformity can be realized concurrently without an serious increase of the pressure loss. Here, it must be noted that the present results are based on the mass transfer rates on the long-side wall of the channel only; the optimum condition of α and C may be slightly modified if it is evaluated based on Sh distributions on all the channel walls. At the present stage, however, it can be recommended that, in applying the two-pass serpentine channels with a sharp turn to the flow passages in heat exchangers, the convergent channel with a wide turn clearance should be used rather than the divergent channel with a narrow turn clearance. Higher heat transfer performance can be expected with the convergent channels having a wider turn clearance in the points of both mean and local heat transfer characteristics.

CONCLUSION

In this paper, the heat (mass) transfer characteristics in the rectangular cross-sectioned serpentine channels with an inclined partition wall have been elucidated; seven inclination angles (α) have been tested for three turn clearances (C) under a single Reynolds number. Based on the experimental results, the optimum combination of α and C has been examined from viewpoints of (i) trade-off between the mean heat transfer rate and the pressure loss, and (ii) increasing the uniformity in the distribution of the local heat transfer rates. Main results are summarized as follows.

(1) In the convergent channels ($\alpha < 0°$), the mean Sherwood number Sh_m increases as $|\alpha|$ is increased, whereas Sh_m in the divergent channels ($\alpha > 0°$) is nearly constant or decreased against α. Sh_m shows higher values as C is decreased, but the dependence of Sh_m on C is rather weak in the divergent channels. In both the convergent and the divergent channels, the pressure loss increases as $|\alpha|$ is increased or as C is decreased.

(2) In the divergent channels, the local Sherwood number Sh before the turn is smaller than that in the standard channel ($\alpha = 0°$); after the turn section, however, Sh shows rather large values. The low Sh region corresponding to the separation bubble near the partition-wall tip, which appears in the standard channel, is diminished in scale, and high Sh regions are formed outside the separation bubble and near the outer wall. The maximum Sh that appears just after the turn is much larger than that in the standard channel.

(3) In the convergent channels, Sh in the upstream half of the test section is larger than that in the standard channel. After the turn section, quite a large separation bubble and a corresponding low Sh region are formed near the partition-wall tip, while Sh shows quite high values in the outer-wall side. The separation bubble becomes larger as $|\alpha|$ is increased. After this separation bubble, Sh shows larger values with the increase of $|\alpha|$, and it is larger than that in the corresponding region of the divergent channels.

(4) From viewpoints of enhancing the mean heat transfer and improving the uniformity in the distribution of the local heat transfer rates concurrently without a serious increase of pressure loss, the heat transfer performance of the channel is improved by increasing the turn clearance and by inclining the partition wall to the convergent side. The best performance of heat transfer can be achieved in the convergent channel with $C = 70$ mm and $\alpha = -2°$.

ACKNOWLEDGEMENT

The authors express their thanks to Mr. N. Shiraki and Mr. K. Tachibana, Research Engineers in the School of Engineering of Nagoya University, for their assistance in producing the experimental apparatus. This study was partly supported by Japanese Ministry of Education through a Grant-in-Aid for Scientific Research (Grant No. 10450083).

NOMENCLATURE

C	turn clearance, m (see Fig. 3)
D	naphthalene-air molecular diffusion coefficient, m²/s
d_h	hydraulic diameter of the channel defined at the channel entrance, m (= 33.3 mm)
h_m	local mass transfer coefficient, m/s
$\overline{h_m}$	mean mass transfer coefficient averaged over the whole mass transfer surface, m/s
K	coefficient of pressure loss, defined by Eq. (1), dimensionless
Re	Reynolds number defined at the channel entrance, $U \cdot d_h/\nu$, dimensionless
Sh	local Sherwood number, $h_m \cdot d_h/D$, dimensionless
Sh_m	mean Sherwood number averaged over the whole test section, $\overline{h_m} \cdot d_h/D$, dimensionless
U	bulk velocity of air at the channel entrance, m/s
α	inclination angle of the partition wall, degree (see Fig. 3)
ν	kinematic viscosity of air, m²/s
ρ	density of air, kg/m³
σ	standard deviation of Sh, defined by Eq. (2), dimensionless

REFERENCES

Astarita, T., Cardone G. and Carlomagno, G. M., 1995, Heat transfer and surface flow visualization around a 180-deg turn in a rectangular channel, *Heat Transfer in Turbulent Flows, ASME HTD*-**318**, 161-168.

Astarita, T. and Cardone G., 2000, Thermodynamic analysis of the flow in a sharp 180° turn channel, *Exp. Thermal & Fluid Sci.*, **20**, 188-200.

Besserman, D. L. and Tanrikut, S., 1992, Comparison of heat transfer measurements with computations for turbulent flow around a 180 deg bend, *J. Turbomachinery*, **114**, 865-871.

Chandra, P. R., Lau, S. C and J. C. Han, 1988, Effect of rib angle on local heat/mass transfer distribution in a two-pass rib-roughened duct, *J. Turbomachinery*, **110**, 233-241.

Chyu, M. K., 1991, Regional heat transfer in two-pass and three-pass passages with 180-deg sharp turn, *J. Heat Transfer*, **113**, 63-70.

Fan, C. S. and Metzger, D. E., 1987, Effects of channel aspect ratio on heat transfer in rectangular passage with sharp 180-deg turns, *ASME Paper*, 87-GT-113.

Goldstein, R. J. and Cho, H. H., 1995, A review of mass transfer measurements using naphthalene sublimation, *Exp. Thermal & Fluid Sci.*, **10**, 416-434.

Han, J. C., Chandra, P. R. and Lau, S. C., 1988, Local heat/mass transfer distributions around sharp 180 deg turns in two-pass smooth and rib-roughened channels, *J. Heat Transfer*, **110**, 91-98.

Hirota, M., Fujita, H., Tanaka, A., Araki, S. and Tanaka, T., 1997, Local heat (mass) transfer characteristics in rectangular ducts with a sharp 180-degree turn, *Energy Conv. & Mgt.*, **38**, 1155-1168.

Hirota, M., Fujita, H., Syuhada, A., Araki, S., Yoshida, T. and Tanaka, T., 1999a, Heat/Mass Transfer Characteristics in Two-Pass Smooth Channels with a Sharp 180-deg Turn, *Int. J. Heat Mass Transfer*, **42**, 3757-3770.

Hirota, M., Fujita, H., Syuhada, A., Araki, S, Yanagida, N. and Tanaka, T., 1999b, Heat/Mass Transfer in Serpentine Flow Passage with a Sharp Turn (Influence of Entrance Configuration), *Compact Heat Exch. & Enhancement Tech. for Process Ind.*, 159-166.

Liou, T-M., Tzeng, Y-Y. and Chen, C-C., 1999, Fluid Flow in a 180-deg Sharp Turning Duct With Different Divider Thickness, *J. Turbomachinery*, **121**, 569-576.

Metzger, D. E., Plevich, C. W. and Fan, C. S., 1984, Pressure loss through sharp 180-deg turns in smooth rectangular channels, *J. Eng. for Gas Turbines Power*, **106**, 677-681.

Metzger, D. E. and Sahm, M. K., 1986, Heat transfer around sharp 180-deg turns in smooth rectangular channels, *J. Heat Transfer*, **108**, 500-506.

Mochizuki, S., Murata, A., Shibata, R. and Yang, W-J., 1998, Detailed measurements of local heat transfer coefficients of turbulent flow in smooth and rib-roughened serpentine passages with a 180-deg sharp bend, *Trans. Japan Soc. Mech. Engrs.*, **64**, 2216-2223 (in Japanese).

Murata, A., Mochizuki, S. and Fukunaga, M., 1994, Detailed measurement of local heat transfer in a square-cross-sectioned duct with a sharp 180-deg turn, *Proc. 10th Int. Heat Transfer Conf.*, **4**, 291-296.

Sparrow, E. M., and Cur, N., 1982, Turbulent heat transfer in a symmetrically or asymmetrically heated flat rectangular duct with flow separation at inlet, *J. Heat Transfer*, **104**, 82-89.

Wang, T. S. and Chyu, M. K., 1994, Heat convection in a 180-deg turning duct with different turn configurations, *J. Thermophysics & Heat Transfer*, **8**, 595-601.

SINGLE-PHASE AUGMENTATION TECHNIQUES

SOME ENHANCEMENT STUDIES FOR LAMINAR AND TUEBULENT HEAT TRANSFER OF AIR IN TUBES AND ACROSS PLATES

W.Q.Tao[1] and Q.W.Wang[2]

[1] Xi'an Jiaotong University, Xi'an,Shaanxi, 710049,China;E-mail;wqtao@xjtu.edu.cn
[2] Xi'an Jiaotong University, Xi'an,Shaanxi 710049,China;E-mail: qwwang@xjtu.edu.cn

ABSTACT

Presented in this invited lecture are systematic investigation results of the heat transfer and pressure drop characteristics for three classes of heat transfer enhancement surfaces. These are plate arrays with an oblique angle to the flow direction, plate fin-and-tube surfaces and the circular tubes with wave-like longitudinal fins and a blocked-center. For the plate arrays, it is found that from the identical pumping power and identical pressure drop constraints the best oblique angle is around 30 degrees; the non-uniformity in the successive plate length may enhance heat transfer by about 20 %, and at the very low Reynolds number region, both heat transfer and friction factor exhibit characteristics of flow and heat transfer in a straight duct. General correlations of heat transfer and friction factor are presented for four types of plate fin-and-tube exchanger surfaces, these are plain plate fin, slotted fin, plate fin with triangular cross section and plate fin with sinusoidal cross section. Results show that heat transfer is a weak function of both fin spacing and tube row number, while the friction factor is nearly independent of the tube row number. Experiments also show that the interwall cylinder between the plates can enhance the heat transfer and promote the transition from laminar to turbulent flow. For the circular tube with wave-like longitudinal fins and a blocked center-part, a general heat transfer correlation obtained from six tubes with different wave numbers is provided. Because of the small hydraulic diameter, and the extended surface, the tube side heat transfer of air can be greatly enhanced. Moreover, the blockage of the center part may further enhance the heat transfer..

INTRODUCTION

Plate-fin and tube-fin are probably the most widely used compact heat exchangers for gas flow on the fin-side in the process industries. A large amount of investigations have been performed to reveal the related heat transfer and pressure drop characteristics for complex flow passages in compact heat exchangers, and many results are documented in, for example, Kays and London(1993), Webb(1994), Rohsenow et. al.(1999) and Shah and Heikal (2000). Because of the complexity of fluid flow and heat transfer processes and the variety of the engineering applications, the matter is far from being resolved and research in this field has received increasing attention in recent years. The primary objective of this invited paper is to present systematically the results of both experimental and numerical investigations for complex flow passages conducted in the CFD/NHT & Enhancement Heat Transfer Group of Xi'an Jiaotong University in the past decade. The major part of the results were published in international journals, some were published in Chinese journals, and a few of them are recently-obtained new results. The presentation will be divided into three parts: (1) forced convective heat transfer of air across plate arrays; (2) forced convective heat transfer of air over plate-fin surfaces; and (3) forced convective heat transfer of air in enhanced tubes. Finally some conclusions will be made.

FORCED CONVECTIVE HEAT TRANSFER OF AIR ACROSS PLATE ARRAYS

Flow interruption in a flow passage, at periodical intervals, is a well-known technique for enhancing convective heat transfer, and the slit fins used in compact heat exchangers and locomotive radiators are examples of this technique. Due to the repeated interruption of thermal boundary layer, the slit fin has higher local heat transfer coefficients than that without slits. In the case of louvered fins, the fin segments are positioned obliquely to the flow direction, and additional heat transfer enhancement can be obtained because of the vorticity and turbulence created in the flow and the impinging effect. The heat transfer and pressure drop characteristics of such units are controlled mainly by the geometric factors. These include the positioning of the fin segments (i.e., plates), the transverse pitch between two adjacent plates, the angle of the plates, and the length ratio of two adjacent plates. In order to reveal the basic characteristics of flow and heat transfer characteristics of such units, two-dimensional models were often used. Based on the results of Zhang and Lang (1989) and Lee(1989), systematic investigations, both experimental and numerical, were conducted in our group to examine the effects of these factors on the flow and heat transfer characteristics (Pang, 1989;Pang et al., 1990; Huang and Tao, 1993; Lue et al, 1993;Wei and Tao,1994; Wang and Tao, 1995; Wang and Tao 1997; Wang et al., 1998). The major results are summarized as follows.

Configurations Studied

The configurations studied are presented in Fig. 1. Figure 1(a) is an interrupted plate array with uniform plate length which serves a two-dimensional model for the conventional louvered fins. The array in Fig. 1(b) has non-uniform plate length, while the plates in Fig. 1(c) are positioned convergently-divergently along the flow direction. The latter two configurations were proposed to further enhance the heat transfer with the conventional uniform plate length array as a reference.

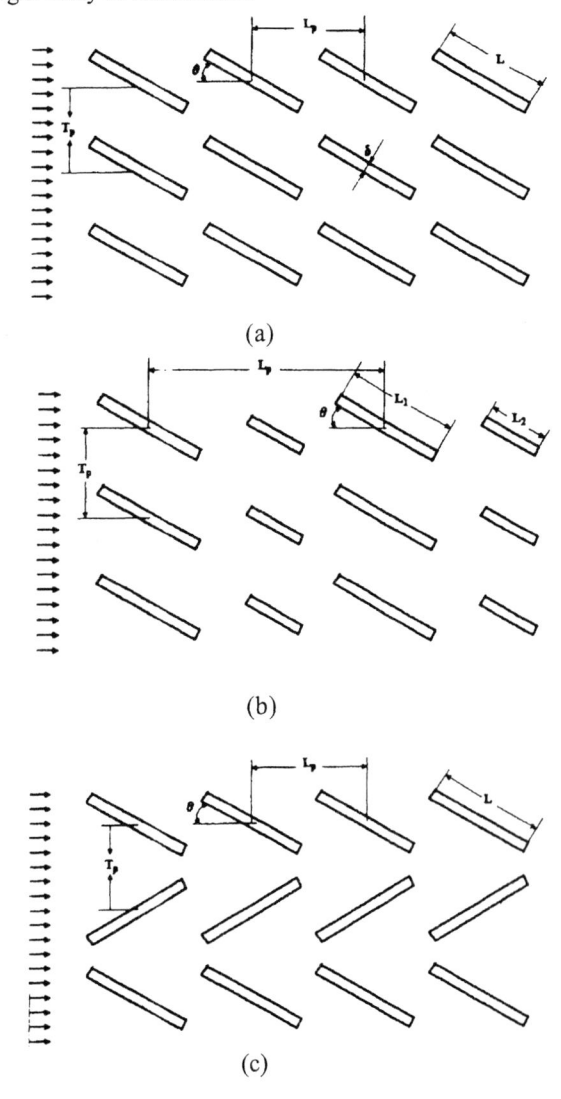

Figure 1 Plate Array Configurations studied

Experimental Method and Test Sections

The naphthalene sublimation technique was used to determine the local heat transfer coefficients. This technique is especially useful for the measurement of heat transfer coefficients in complicated geometries, and a comprehensive review was made by Goldstein and Cho(1993). The side view of the test section for the case of non-uniform plate length(Fig. 1(b)) is shown in Fig. 2, and a perspective view of test plate in Fig. 3. The substrate is made of 1.5 mm thick mild steel with a rectangular recessed area in the center of each surface. Thin naphthalene layers are coated onto the two recessed areas. The naphthalene coated plate is then mounted on a specially designed fixture and machined to the required thickness. Details of the test process may be found in Lue et al.(1993). For the other two configurations, the construction of the test sections are of the same type and will not be shown here. The geometric parameters investigated for the three configurations are listed in Tables 1, 2 and 3 respectively. The measurements were performed for both developing and developed fluid flow and mass/heat transfer.. The characteristic length of the Reynolds number and Nusselt number is the plate length , L or L_2. The test Reynolds number ranges of the three configurations are basically below 2000 - 3000, and will be indicated later.

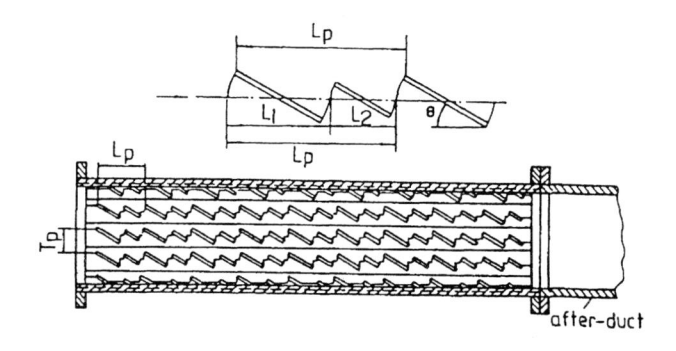

Figure 2 Test section of naphthalene sublimation technique

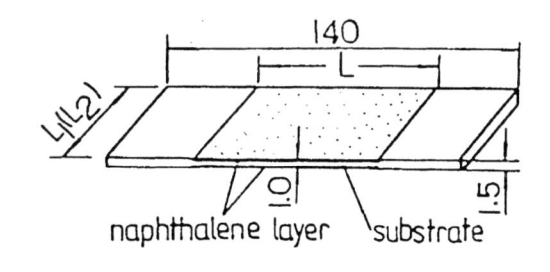

Figure 3 Perspective view of test plate

Numerical Methods

The finite volume method was used to discretize the governing

Table 1 Geometric parameters of configuration (a)

Case	ϑ (degree)	$L_p \times 10^3$	$L \times 10^3$	$T_p \times 10^3$
1	10	30	30	30
2	15	30	30	30
3	20	30	30	30
4	25	30	30	30
5	30	30	30	30
6	35	30	30	30
7	25	37.5	37.5	30
8	25	22.5	22.5	30

Table 2 Geometric parameters of configuration (b)

Case	$L_p \times 10^3$	$L_1 \times 10^3$	$T_p \times 10^3$	L_1/L_2	T_p/L_p
1	52.5	37.5	30.0	2.5	0.571
2	45.0	30.0	30.0	2.0	0.667
3	37.5	22.5	30.0	1.5	0.800
4	52.5	37.5	25.0	2.5	0.476
5	45.0	30.0	25.0	2.0	0.556
6	37.5	22.5	25.0	1.5	0.667
7	52.5	37.5	20.0	2.5	0.381
8	45.0	30.0	20.0	2.0	0.444
9	37.5	22.5	20.0	1.5	0.533

($\vartheta = 25$ degrees)

Table 3 Geometric parameters of configuration (c)

Case	ϑ (degree)	$L_p \times 10^3$	$L \times 10^3$	$T_p \times 10^3$
1	10	30	30	30
2	15	30	30	30
3	20	30	30	30
4	25	30	30	30
5	30	30	30	30
6	35	30	30	30
7	25	22.5	22.5	30
8	25	37.5	37.5	30

(a)

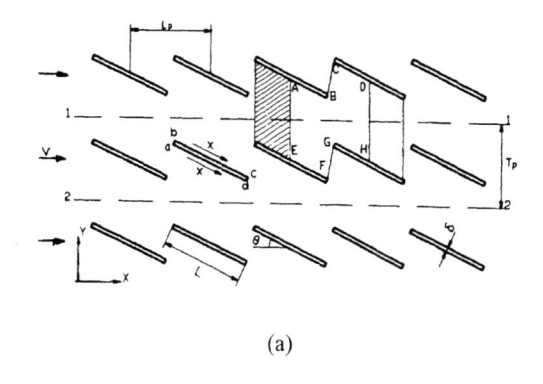

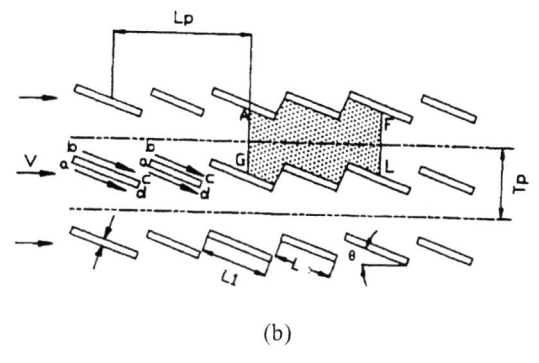

(b)

Figure 4 Computation domain

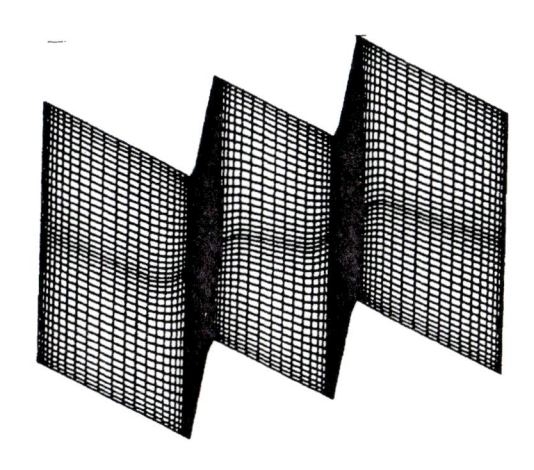

Figure 5 Grid system for the nonuniform plate length case

equations and the SIMPLE-like algorithms were adopted to deal with the linkage between the pressure and velocity (Patankar, 1980: Tao, 1988). Computations were conducted for the periodically fully developed laminar flow and heat transfer cases, and only one cycle region was taken as the computations domain. Figure 4 shows the computational domains for the uniform plate length case and the non-uniform one. To accommodate the inclined solid surface, stepwise approximation was once used in Pang et al.(1990). Later, the multisurface method (Eiseman, 1982, 1985) was adopted to generate a body-fitted coordinates. Figure 5 shows such a grid system for the non-uniform plate length case. Laminar flow model was used and computations were conducted in the Reynolds number range of 10-2300.(Wang and Tao, 1995, Wang and Tao, 1997, Wang et. al. 1998).

Major Findings of Experimental and Numerical Studies

Length of entrance region Figures 6 and 7 present the variation of the per-cycle average Sherwood number with the cycle number along the flow direction. As can be seen from Fig.6, for the uniform plate length, starting from the fifth cycle, the per-cycle average Sherwood number ceases change and the transport process reaches the fully developed region. For the non-uniform plate length cases, Figure 7 shows that after the fourth cycle, the fluid flow and mass transfer may be regarded as fully developed. Since in most applications, the cycle number in the main flow direction is much larger than 4 or 5, our focus is then put on the characteristics in the fully developed region.

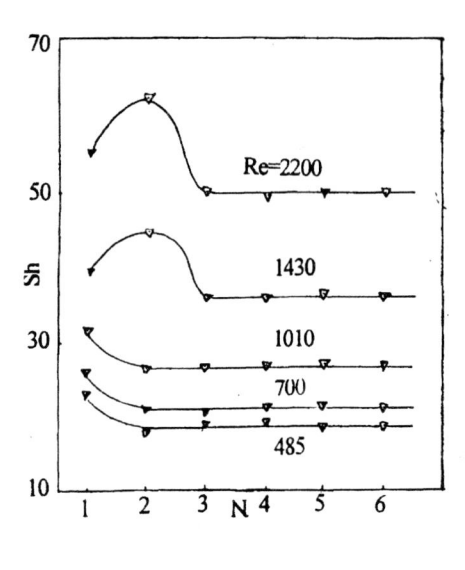

(a) Case 1 of Table 1

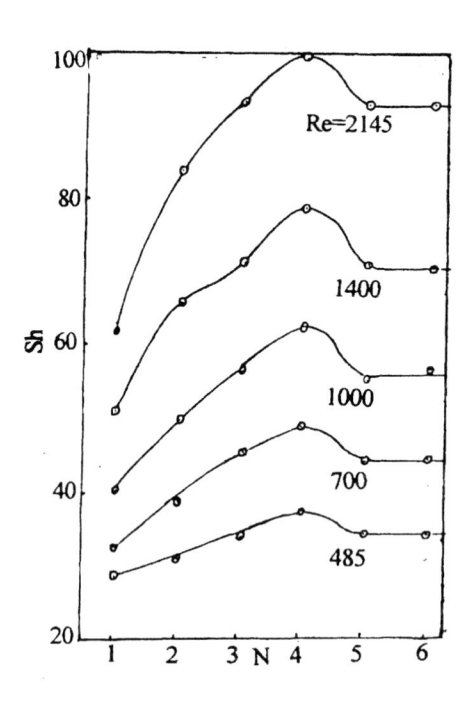

(b) Case 6 of Table 1

Figure 6 Variation of Sh with cycle number(uniform array)

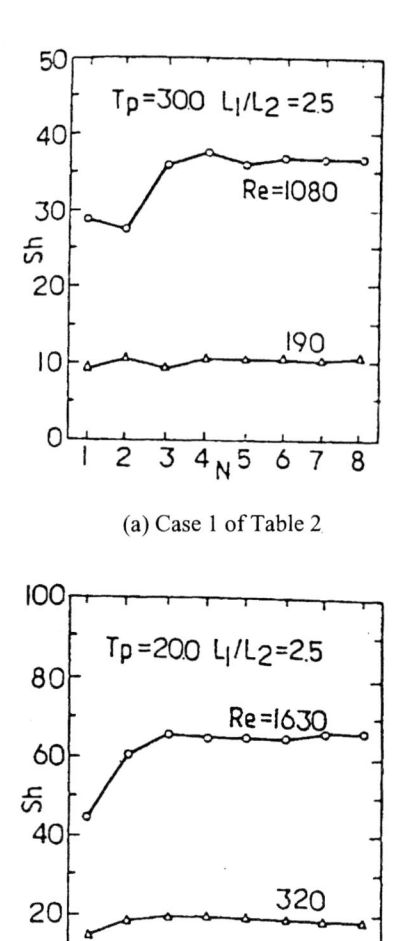

(a) Case 1 of Table 2

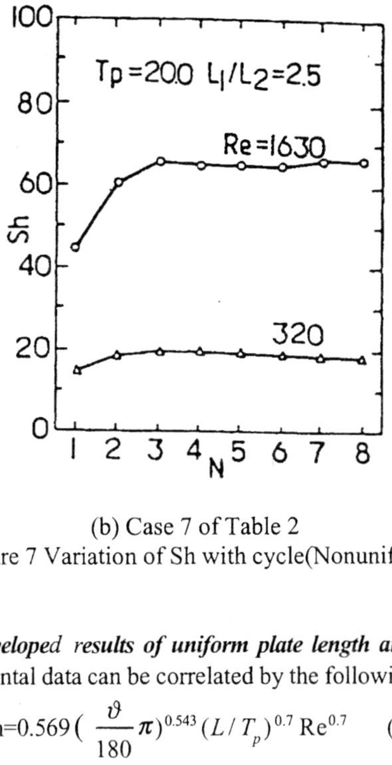

(b) Case 7 of Table 2

Figure 7 Variation of Sh with cycle(Nonuniform array)

Fully developed results of uniform plate length array. The experimental data can be correlated by the following equation

$$Sh = 0.569 \left(\frac{\vartheta}{180} \pi \right)^{0.543} (L/T_p)^{0.7} Re^{0.7} \qquad (1)$$

The test parameter ranges are:
$0.75 \leq L/T_p \leq 1.25$; $10° \leq \vartheta \leq 35°$; $480 \leq Re \leq 2\,300$: $Sc=2.5$

The maximum deviation of Eq.(1) from 25 test data is 4.3%, and the average one is 1.6%.

The per-cycle average friction factors are presented in Figure 8. It can be observed from Eq.(1) and Fig. 8, that the increase in the oblique angle leads to the increase in both heat transfer and friction factor. And when the Reynolds number reaches a certain value, the per-cycle average friction factor approaches a constant. The tested lower Reynolds number was limited by the measurement difficulty, and this limitation was partially overcome by the numerical simulation. Figure 9 provides the numerical results for Nusslet number using laminar fluid flow model with Reynolds number ranging from 50 to 2300. The dashed lines are the experimental results, while the solid curves are predicted ones. From the figures, it can be seen that in the low Reynolds number range, the plate-average Nusselt number approaches a constant when Reynolds number decreases, which characterizing the fully developed laminar heat transfer in a duct with constant cross section. In addition, the per-cycle friction factor is nearly inversely proportional to the Reynolds number. Therefore, in the very low Reynolds number region, the fluid flow and heat transfer across the plate array behave as an internal laminar flow in a straight duct

Fully developed results of non-uniform plate length array
The following correlation is obtained

$$Sh = 0.252 (T_p/L_p)^{-0.52} (L_1/L_2)^{0.107} Re^{0.681} \quad (2)$$

L1/L2=1.5-2.5, Tp/Lp=0.38-0.80, $\vartheta = 25°$, Re=198-1660
This equation agree with the test data to within 2.7% (average) for 72 test data, with the maximum deviation of 9.9%. According to the mass heat transfer analogy theory, Eq.(2) can be transformed into heat transfer correlation with Sc=2.5:

$$Nu = 0.175 (T_p/L_p)^{-0.52} (L_1/L_2)^{0.107} Re^{0.681} Pr^{0.4} \quad (3)$$

where $Nu = Sh(Pr/Sc)^{0.4}$ and Sc=2.5 have been introduced. The same transformation can be performed for Eq.(1). The friction data are presented in Fig. 10. Numerical simulation in the low Reynolds number region also reveals the same characteristics as for the uniform plate length case.

Fully developed results of converging-diverging array
Following relation is obtained:

$$Sh = 0.245 Re^{0.838} (/180)^{0.673} (L/T_p)^{0.806} \quad (4)$$

The corresponding tested friction factors are shown in Fig. 11.

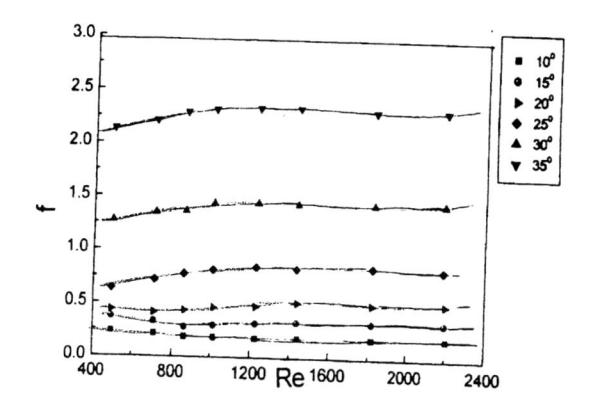

Figure 8 Friction factor data of uniform plate array

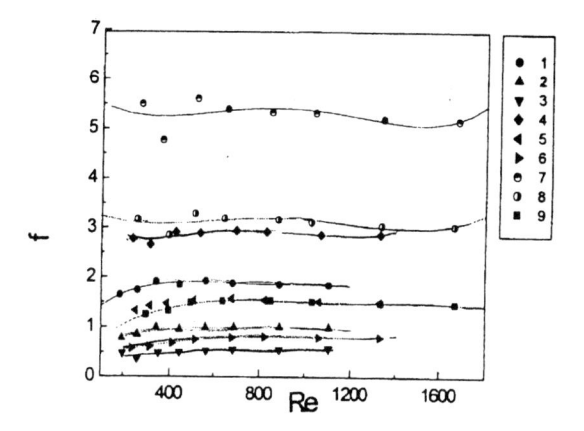

Figure 10 Friction factor data of nonuniform plate array

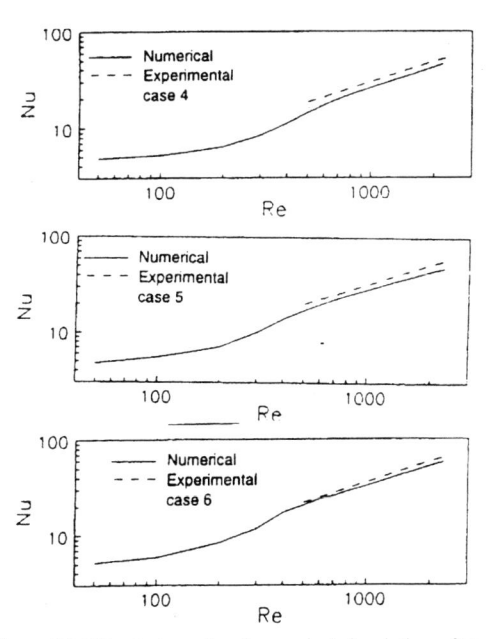

Figure 9(a) Typical results of numerical simulation of Nu~Re for uniform plate

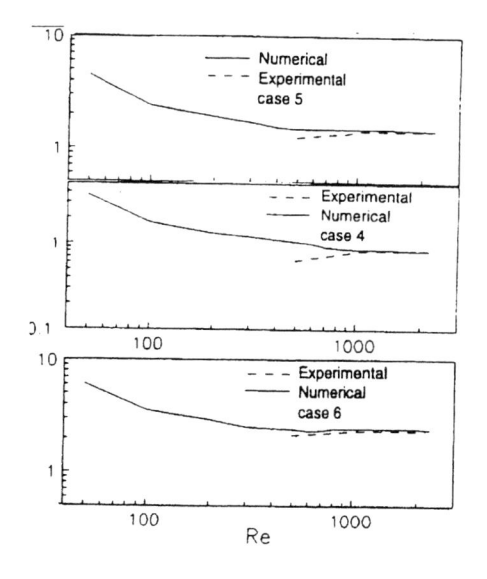

Figure 9(b) Typical results of numerical simulation of f~Re for of uniform plate array

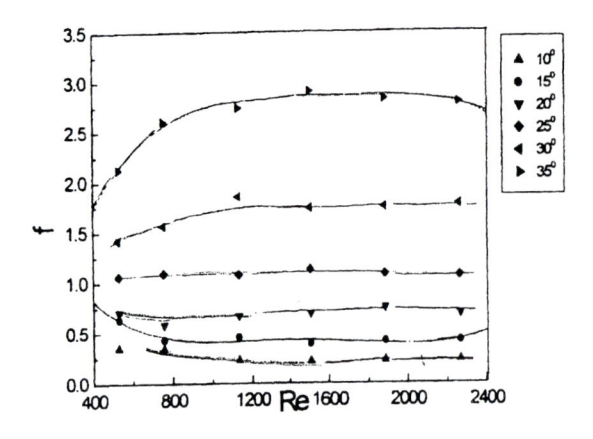

Figure 11 Friction data of convergently-divergently positioned plate array

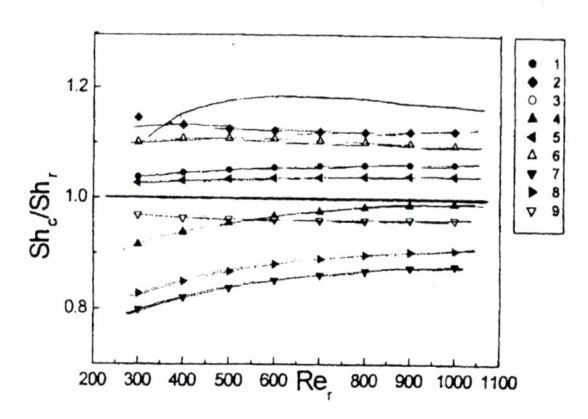

Figure 13 Uniform plate array vs. nonuniform plate array(IPPC)

Performance Comparisons

Performance comparisons were conducted for the following cases. (1) The effect of oblique angle of the uniform plate array on the heat transfer and pressure drop characteristics; (2) The effect of the non-uniformity of the successive plate length, and (3) The effect of the position pattern of the plates. Comparisons were conducted under two constraints: identical pumping power and identical pressure drop. The comparisons show that for the uniform plate length case, the best oblique angle under either identical pumping power or identical pressure drop is around 30 degrees(Fig.12). For the tested non-uniform plate length situation, taking the uniform plate length as the reference, case 3 may enhance heat transfer by 20 to 25 % (Figs. 13). While for the convergently-divergently positioned plate array, comparisons show that for most cases this position pattern can not provide heat transfer enhancement (Fig. 14).

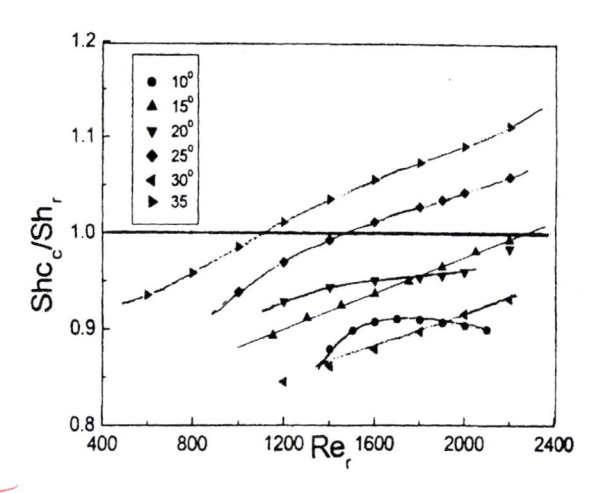

Figure 14 Convergently/divergently positioned array vs parallel aligned array (IPPC)

FORCED CONVECTIVE HEAT TRANSFER OF AIR ACROSS FOUR TYPES OF PLATE-FIN-AND-TUBES SURFACES

Of the various fin configurations for enhancing air-liquid heat transfer, the plate-fin-and-tube exchanger, consisting of round tube mechanically or hydraulically expanded into a block of parallel continuous fins of constant thickness, is probably one of the most important structures. In the past two-decades, different types of plate fins have been developed, of which the plain plate fin, wavy plate fin with streamwise triangular cross-section, wavy plate fin with sinusoidal cross-section and slotted plain fin are most often encountered. Although it is generally known that many experimental investigations have been conducted for different types of plate fin-and-tube configurations (Wang, et. al., 1998;. Wang, et. al., 1999), systematic investigations and comparisons of the heat transfer and pressure drop performances between them is

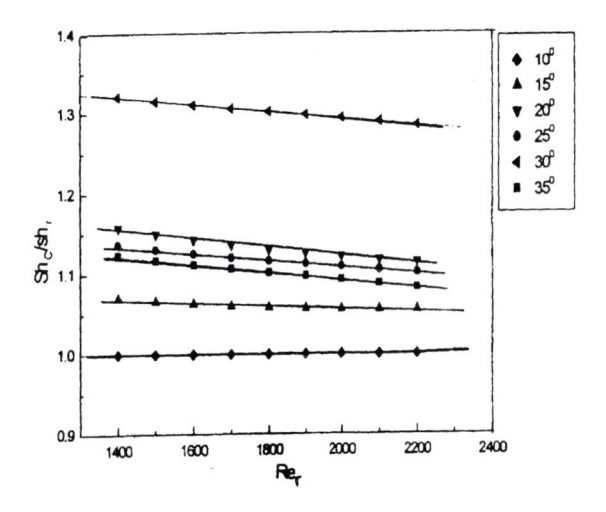

Figure 12 Performance comparison for different oblique angle under identical pumping power constraint(IPPC)

still limited. A long-term research project was executed in our group in the last nineties to reveal the basic characteristics of the above mentioned four types of plate-fin surfaces in full dry condition (H.Z.Li, et.al.,1992a; H.Z.Li,et.al.,1992b;Xin, et.al., 1992; Kang, et.al., 1994a; Kang, et.al.,1994b; Xin, et. al., 1992; Xin, et.al.,1994). The major results are summarized below.

Heat Transfer and Friction Factor Results of Four Types of Surfaces

Experimental methods and specifications of test sections The test apparatus consists of two loops, an air loop and a vapor loop. The air loop is operated in a suction mode. The steam is generated in an electrically-heated boiler. It is condensed in the vertical tubes of the test section, giving its latent heat to the air. The condensate flows to a volumetric flow meter and then returns to the boiler. With the exception of fin configuration and fin spacing, physical dimensions of the test cores are virtually identical. All the cores have the frontal area of 250×300 mm^2 and two, three or four-row deep in the air direction. The specific geometries of the four types of plate fins are presented in Fig. 15. Other data common to each type of fin are listed in Table 4.

Table 4 Some data of the test core

Tube row arrangement	staggered
Tube material	copper
Fin material	aluminum
Number of tube per row	10
Outer diameter of tube (after expansion)	10.15 mm
Inner diameter of tube (after expansion)	9.33 mm
Fin thickness	0.2 mm
Fin spacing	2.0, 2.6, 3.2 mm
Effective length of tube	0.3 m

With the vapor condensation heat transfer coefficient being determined by the Nusselt equation, the air side average heat transfer coefficient of the test core can be determined from the measured overall heat transfer coefficient based on air-side total heat transfer area by the separate-thermal-resistance method. The fin efficiency is iteratively determined and taken into account in the computation of fin-side heat transfer area. The total heat transfer rate used to calculate the overall heat transfer coefficient was an arithmetic mean of the vapor-side and air-side heat transfer rates. In nearly all cases the energy unbalance between the two sides was less than 5%, although slightly larger deviations were allowed at very low air velocities.

The maximum velocity in the test core was used to determine the Reynolds number, and the tube outer diameter (after expansion)was taken as the characteristic length. The friction factor was calculated according to the average pressure drop per each row.

Experimental correlations of heat transfer and friction factor Tables 5 and 6 provide the experimental correlations for the test

(a) Plain plate (PP)   (b) Slotted plate (SP)

c) Wavy fin with triangular cross section(WT)   (d) Wavy fin with sinusoidal cross section(WS)

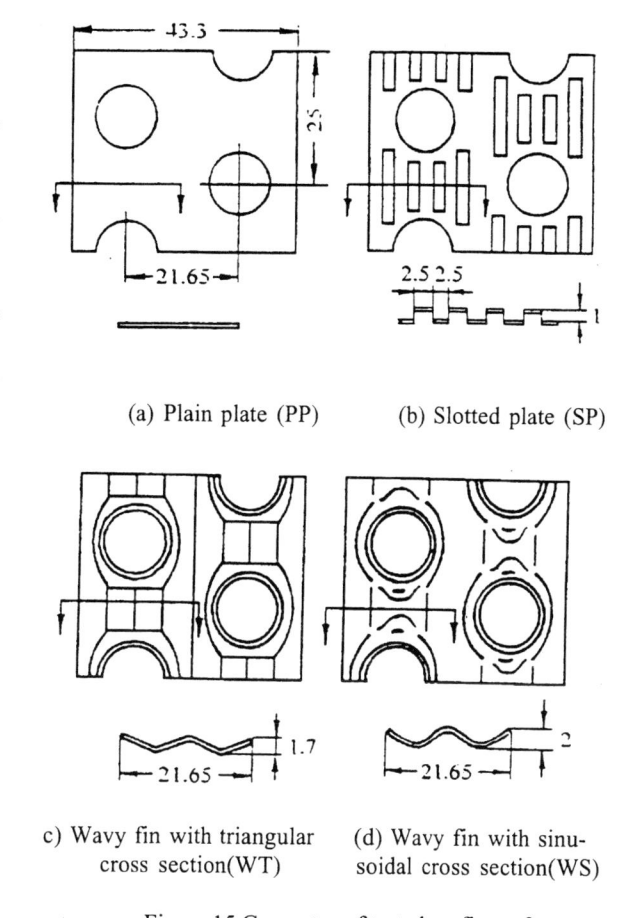

Figure 15 Geometry of test plate-fin surfaces

data with Reynolds number below 5000. In order to cover the whole tested Reynolds number range(from 500 to about 16000), more complicated formulas have to be used and the data of each test core should be correlated individually. In most applications, however, the frontal velocity is usually within 2–3 m/s, and the corresponding Reynolds number is well below 5000. Thus, only the results of Re $\leq$ 5000 are presented.

Analysis and discussion From Tables 5 and 6 following features may be noted. The exponent in the Reynolds number varies from 0.424 to 0.556, which is quite consistent with the variation range in Grimson(1937) correlation for heat transfer across tube banks. In the test range of fin spacing(s=2.0 to 3.2 mm), the decrease in it leads to some enhancement of heat transfer., meanwhile the friction factor is very sensitive to the fin spacing, with its exponent as high as 0.74 to 1.1. This implies that the mild increase in heat transfer is obtained at great cost of pressure drop. The production of N and s_2 in the Nusselt number correlation is the plate-fin length in stream direction, and the term (Ns_2/d) actually represents the effect of the streamwise tube-row number. The heat transfer correlations show that with the increase in the tube-row number, the heat transfer deteriorates to some extent. This is in contrast to the results for smooth tube banks. The per-tube average friction factor, however, is nearly independent of

Table 5 Heat transfer correlations

Fin type	Heat transfer correlations	Reynolds number Range	Maximum fitting deviation, %	RMS deviation %
Plain plate fin	$Nu = 0.982 Re^{0.424}(\frac{s}{d})^{-0.0887}(\frac{Ns_2}{d})^{-0.159}$	700 – 5 000	9.9 ~ −8.5	2.34
Slotted fin	$Nu = 0.772 Re^{0.477}(\frac{s}{d})^{-0.363}(\frac{Ns_2}{d})^{-0.217}$	500 ~ 5 000	11.3 ~ −11.2	4.45
Triangular fin	$Nu = 0.687 Re^{0.518}(\frac{s}{d})^{-0.0935}(\frac{Ns_2}{d})^{-0.199}$	580 ~ 5 000	11.9 ~ −10.0	3.88
Sinusoidal fin	$Nu = 0.274 Re^{0.556}(\frac{s}{d})^{-0.202}(\frac{Ns_2}{d})^{-0.0372}$	700 ~ 5 000	10.9 ~ −10.0	2.40

Table 6 Friction correlations

Fin type	Friction factor correlation	Reynolds number range	Maximum fitting deviation, %	RMS deviation %
Plain plate fin	$f = 5.504 Re^{-0.454}(\frac{s}{d})^{-0.940}$	700~ 5 000	10.3 ~ - 13.1	3.33
Slotted fin	$f = 5.541 Re^{-0.426}(\frac{s}{d})^{-1.10}$	500 ~ 5 000	10.8 ~ -10.8	5.12
Triangular fin	$f = 5.440 Re^{-0.392}(\frac{s}{d})^{-0.736}$	580 ~ 5 000	13.1 ~ 12.8	4.30
Sinusoidal fin	$f = 3.400 Re^{-0.353}(\frac{s}{d})^{-0.900}$	700 ~ - 5 000	10.9 ~ - 9.98	2.24

the tube-row number, hence, the term (Ns_2 / d) is not included in the friction factor correlations.

Thermal performance comparison was conducted based on the identical pumping power constraints. Typical results are shown in Fig. 16 for the four-row test cores with fin spacing of 2.6 mm. As can be seen there, based on the identical pumping power constraint, the slotted fin ranks first, then come the sinusoidal and triangular fins. Taking the plain plate fin surface as a reference, the above mentioned three types of plate-fin surfaces may provide by about 40%, 34% and 22% enhancement of heat transfer, respectively.

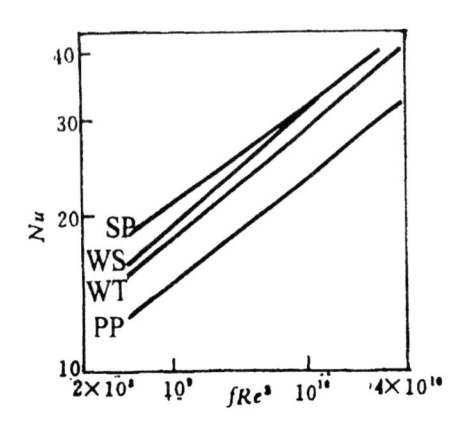

Figure 16 Performance comparison of four types of plate-fins (N=4, 2=2.6 mm)

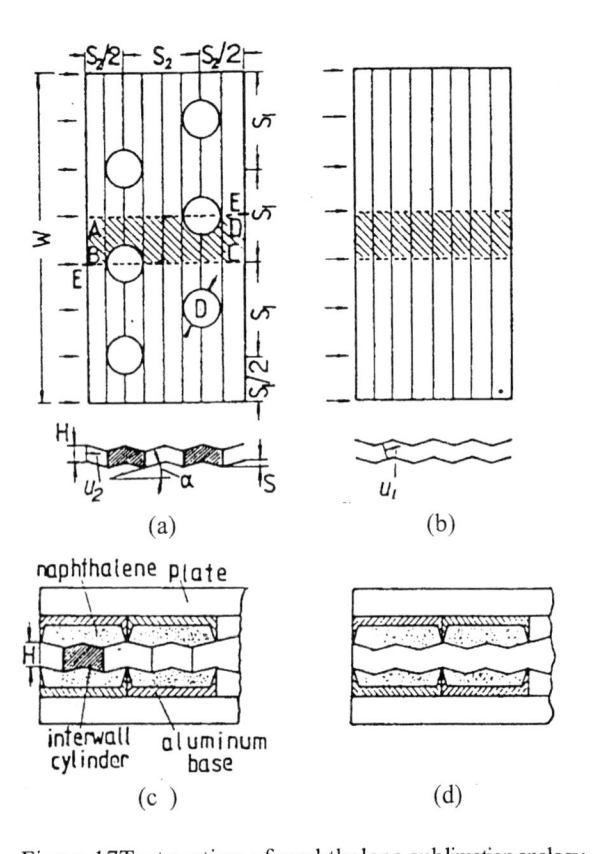

Figure 17 Test section of naphthalene sublimation analogy

Mass Transfer Study for the Effect of Interwall Cylinder

Because of the complexity of heat transfer experiments for the industrial test core described above, it is quite inconvenience to reveal many details by such experiments. The naphthalene sublimation technique provide a useful way to obtain some detail information for such configurations. This technique was used to reveal the effect of the interwall tube cylinder on the heat /mass transfer of corrugated configuration. The schematic diagram of the test section is shown in Fig. 17 When the configuration is incorporated with interwall tube cylinder, as shown in Fig. 17 it can be regarded as a model for the plate-fin-and-tube exchanger surface with triangular cross section, while that without the interwall cylinders is just a corrugated duct. The test results for four ratios of s/d are presented in Fig. 18 where the square symbols are the data with interwall tube cylinders while the circles are those for the corresponding corrugated duct without interwall cylinders. For the comparison purpose, the characteristic length of the Reynolds number and the Nusselt number is the hydraulic diameter of the duct ,i.e., 2s., and the velocity in the Reynolds number is based on the corresponding channel cross section area without the cylinder. In such a way the data for the two cases(corrugated ducts with or without interwall cylinders) can be directly compared. Figure 18reveals following important facts. First, for any Reynolds number tested, the Sherwood number of the corrugated plate-fin-and-tube exchanger is always larger than that of the duct. This implies that the interwall cylinders act as heat transfer enhancement turbulators, and the enhancement effect is more significant in the low Reynolds number region. For example, at Re=1100, the interwall cylinder can enhance the mass transfer about 70-90 percent, and at Re=4800 by 11-30 %. Second, for the four spacings investigated, the curves of Sh ~ Re for the corrugated ducts continuously change their slopes in the low Reynolds number range(from about 1000 to 2000), and then gradually approach straight lines. The continuous change in slope may be indicative of the transition from laminar to turbulent flow. Thus, these curves suggest that from a Reynolds number of about 2500 ~ 3000, the flow in a corrugated duct may be regarded as turbulent. This observation agrees quite well with the experimental results of Goldstein and Sparrow(1977).

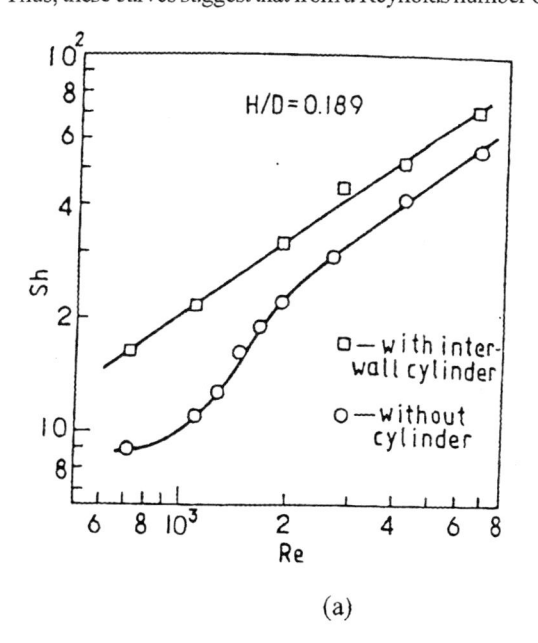

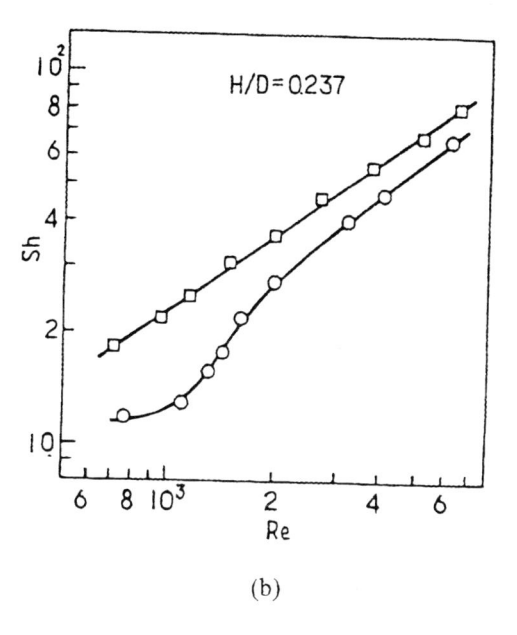

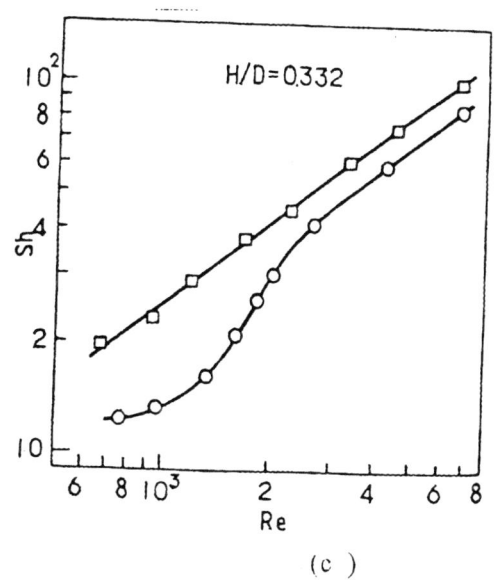

Figure 18 Effect of interwall cylinder on heat transfer

FORCED CONVECTIVE HEAT TRANSFER OF AIR IN ENHANCED CIRCULAR TUBES

The circular tube is one of the most widely used heat transfer surface in heat exchangers, and the problem of enhancing the heat transfer rate in circular tubes has long been an important research project. As an effective method for heat transfer enhancement, internally finned tube has been often employed. Many investigations, both experimental and numerical, have been conducted for different kinds of internally finned tubes. Webb (1994) made a comprehensive summary about this subject. As far as the geometry of the internal fins are concerned, most internal fins are usually strips with flat surfaces positioned longitudinally along the tube axis. In some gas/liquid heat exchangers used in the processing industry gases flow inside the tube and liquid, such as water, flows across the tubes. In such cases, the traditional longitudinal fins can not provide enough enhancement for the design of a compact air/liquid heat exchanger, and some new type of finned surface must be sought. In order to develop such enhanced surfaces, a systematic investigation was conducted in our group in the past decade, and finally a new type of enhanced tube surface was developed. The cross section view of the developed enhanced tube is shown in Fig. 19, which is a center-blocked tube with wave-like longitudinal fins in the annular space. In the developing process, serious studies, both experimental and numerical were conducted. These include: the study of laminar flow and heat transfer in the developing and developed region for annular-sector duct (Lin, et. al. 1995;Lin et. al. 2000); the turbulent fluid flow and heat transfer characteristics in the annular sector ducts (Tao, et.al., 2000); the study of laminar and turbulent fluid flow and heat transfer in the annular space with longitudinal wave-like fins (Yu, et. al., 1998; Yu, et. al.1999a; Yu, et. al. 1999b; Yu, et. al. 2000). Because of the space limitation, only the results of turbulent flow situations are presented here.

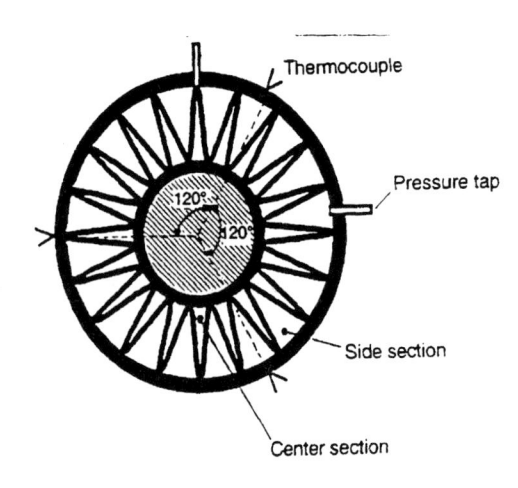

Figure 19 Centered-blocked internally finned tube

entire inner surface of the model wetted by the fluid was used as the heat transfer area. The hydraulic diameter was used as the characteristic length.

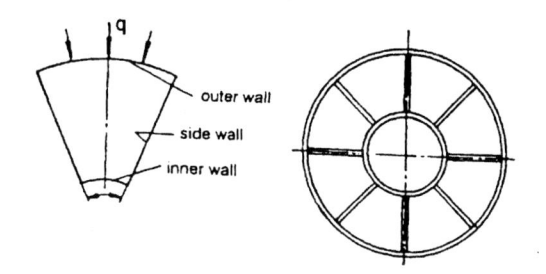

(a) Annular sector duct   (b) Integral fin tube

Figure 20 Schematic diagram of annular-sector duct

Table 7 Geometric parameters of annular sector duct

Model No	1	2	3	4	5
ϑ, degree	40	30	24	20	18
$d_o \times 10^3$	72.5	87.8	103.1	118.3	128.6
$d_i \times 10^3$	18.1	22.2	25.8	29.6	32.2
$A \times 10^4$, m^2	4.3	473	5.22	5.73	6.09
S, m	0.086	0.095	0.104	0.115	0.122
d_h, m	0.02	0.02	0.02	0.02	0.02

The test results of the friction factor in the fully developed region are presented in Fig. 21. These curves show that the test data can be correlated by the Blasius-type equation:
$$f = CRe^{-0.25} \qquad (5)$$
The values of C for the five test models and the related deviation between the correlation and the test data are listed in Table 8.

Table 8 Constants in Eq. (5)

Model No.	1	2	3	4	5
C	0.2685	0.2636	0.2559	0.2252	0.2192
δ, %	-14.4 ~ 13.4	-15.7 ~ 13.4	-9.6 ~ 10.8	-7.4 ~ 6.4	-5.8 ~ 5.9

Results of Developing and Developed Turbulent Fluid Flow and Heat Transfer in Annular-Sector Ducts

Our first suggestion for enhancing tube-side air heat transfer is to increase the inner surface in such a way that the circular tube becomes an annular-sector duct. Test models were manufactured with different apex angles of the sector. Experimental studies were conducted for the annular-sector ducts shown in Fig. 20(a), which may be regarded as one unit of the integral fin tube shown in Fig. 20(b). Five models were tested, whose geometric parameters are listed in Table 7. To simulate the working condition of the annular-sector tube, the outside surface of the test model was electrically heated, while all other surfaces were adiabatic. The

The streamwise local heat transfer coefficient distributions for the apex angle of 18° and 40° are shown in Fig. 22. It can be seen that the duct with an apex angle of 18 degree has an entrance length larger than that of circular tube. This observation is consistent with the well-known finding in the literature that the entrance region of non-circular duct is larger than that of circular cross section.. The mean Nusselt numbers in the fully developed region are presented in Fig. 23. All the data can be correlated by following equation:
$$Nu = 0.0281 Re^{0.75} \quad (5000 \leq Re \leq 50000) \qquad (6)$$
with a maximum deviation of 10%. Figure 29 also shows that the Dittus-Boelter equation overpredicts the heat transfer rate for the

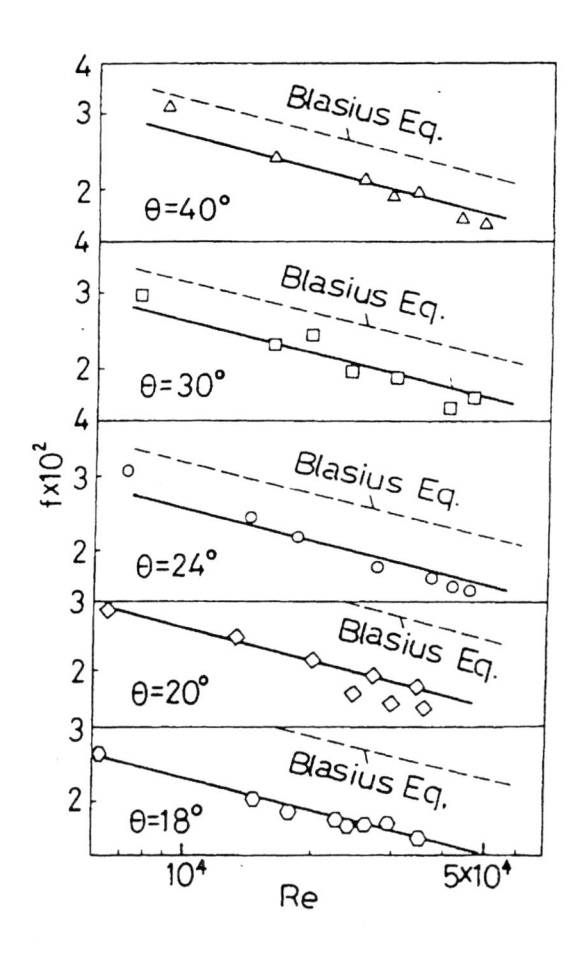

Figure 21 f~Re of the annular-sector duct

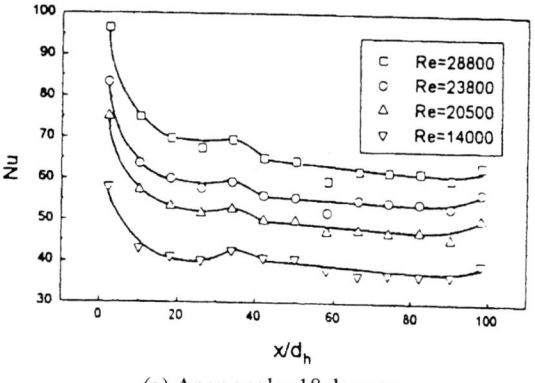

(a) Apex angle=18 degrees

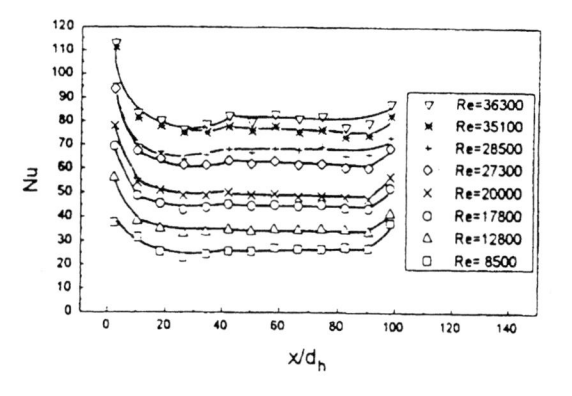

(b) Apex angle =40 degrees
Figure 22 Local Nusselt number distributions

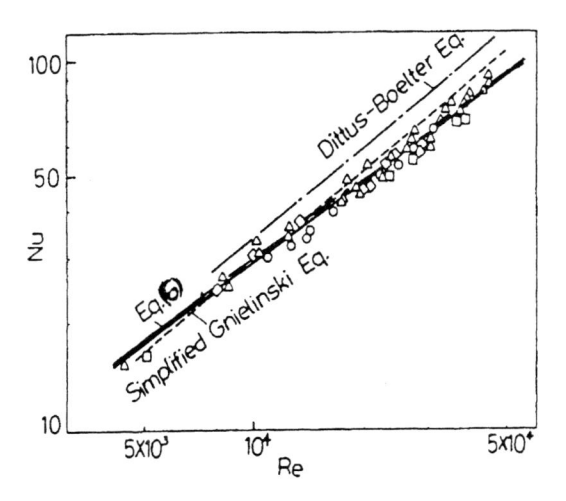

annular-sector ducts tested by about 20-25%, while the simplified Gnielinski equation suggested by Shah and Johnson(1981)
$$Nu = 0.0214(Re^{0.8} - 1000) \quad (7)$$
can fit the data quite well with most deviations being in the positive side but less than 13%. The above annular-sector tube, though can greatly enhance air-side heat transfer, still can not satisfy the space requirement for some compressor intercooler. Thus further improvement was made to meet the industrial needs.

Pressure Drop and Heat Transfer Characteristics in Tubes with Wave-like Longitudinal Fins

To further enhance the heat transfer, the annular-sector configuration was reformed into a dense wave-like longitudinal fins with the central tube as a supporter. Two types of such tubes were made, one with the center tube blocked, and the other is unblocked. The geometric dimensions for the two tubes are listed in Table 9. According to Shah and Heikal(2000), if the hydraulic diameter is less than 6 mm for operating in a gas steam, the heat exchanger is also referred to as a compact one. Thus the tested tubes are also belong to the compact category.

Detailed measurement of the local heat transfer and pressure drop characteristics were made(Yu, et.al., 1999a). Only the results for the fully developed region are presented here. In Fig. 24 the

Figure 23 Nu~Re of the annular sector duct

Table 9 Test tube dimensions

Case	L	c	D_o	D_i	d_o	d_i	d_h
(each of the seven terms $\times 10^3$, unit: m)							
Unblocked	1000	390	35	33	11.5	10.5	3.10
Blocked	1000	390	35	33	11.5	10.5	2.84

(Fin thickness = 0.25×10^{-3} m)

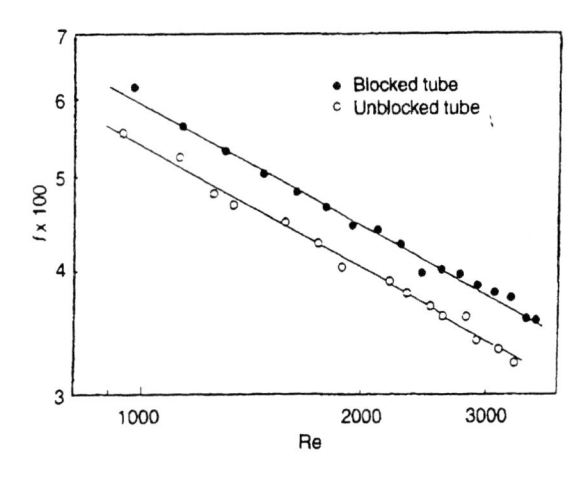

(a) Friction factor

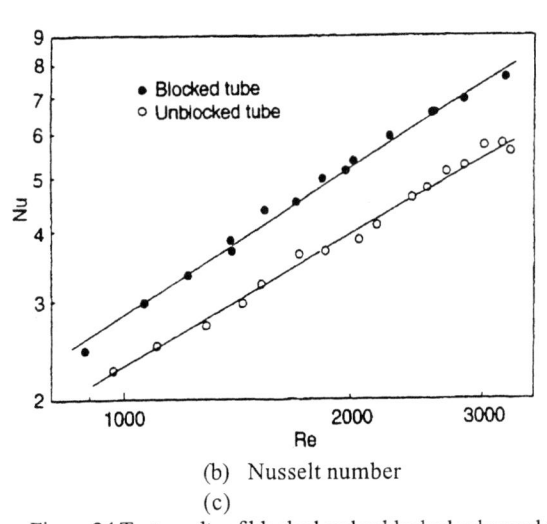

(b) Nusselt number

(c)
Figure 24 Test results of blocked and unblocked enhanced tubes

test data of the friction factor and the Nusselt number are provided, where the hydraulic diameter was taken as the characteristic length.. It can be seen that the blockage can enhance heat transfer appreciably, with a mild increase in friction factor. The four curves can be well correlated by the following equations:
For unblocked tube:

$$f = 0.971/Re^{0.419} \quad (Re=930 \sim 3300) \quad (8)$$

$$Nu = 0.00981 Re^{0.789} \quad (Re=930 \sim 3300) \quad (9)$$

For blocked tube

$$f = 0.991/Re^{0.407} \quad (Re=970 \sim 3500) \quad (10)$$

$$Nu = 0.00668 Re^{0.876} \quad (Re=880 \sim 3300) \quad (11)$$

Thermal performance comparisons were made between the two tubes under the identical pumping power and identical pressure drop. The results are shown in Fig. 25, where hF denotes the production of heat transfer coefficient and the surface area.

It can be seen that for the two constraints adopted, the performance of the two longitudinally finned tubes are much better than that of the plain tube, and the performance of the blocked tube is always superior to that of the unblocked one. The significant heat transfer enhancement obtained from the blocked longitudinally finned tube over the plain circular tube may be attributed to following three aspects. (1) The increase in heat transfer surface area. The total heat transfer surface area of the blocked case is about 7.1 times of the smooth one with a fin efficiency of 0.9. (2) The reduction of the hydraulic diameter, and (3) The insertion of the small central blocked tube, which makes the velocity in the annular space larger than that in the unblocked one at the same mass flow rate.

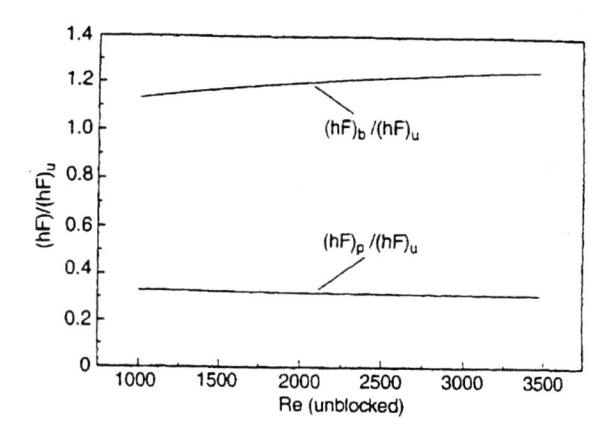

(a) Identical pumping power

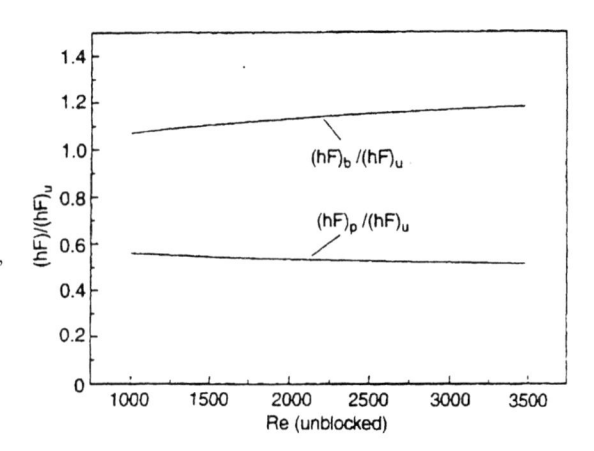

(b) Identical pressure drop
Figure 25 Performance comparison between two tubes

Above experimental investigation reveals the high heat transfer performance of the blocked tube with wave-like longitudinal fins. To further confirm the test results and to extend the parameter range, systematic experiments were conducted for five blocked tubes with different number of longitudinal wave-like fins. The test tube dimensions are listed in Table 10, and its cross section view is shown in Fig. 26. Presented in Fig. 27 are the fully developed heat transfer data which span the Reynolds number range from 850 to 14700. It turns out that Eq.(11) can be used to predict Nusselt number of the fully developed region with quite satisfactory accuracy: Of totally fifty data, the deviation of 86% data is within 8%, the average deviation is 9.3% and the maximum deviation is 22.5%. Therefore Eq(11) can be recommended to predict the fully developed Nusselt number for the circular enhanced configuration shown above in the Reynolds number range from 850 to 14700.

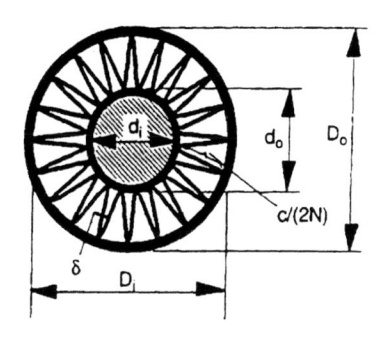

Figure 26 Cross section view of test enhanced tubes

Table 10 Dimensions of the five test tubes

Tube No.	Wave number	L	c	D_o	D_i	d_o	d_i	d_h
(each of the seven term $\times 10^3$, unit: m)
b-1	4	1000	87	30	26	12	10	5.4
b-2	8	650	143	30	26	12	10	3.77
b-3	12	550	203	30	26	12	10	2.79
b-4	16	550	315	35	32	15	13	2.82
b-5	20	550	389	35	32	15	13	2.29

Results of Numerical Simulations

Numerical simulation for the tube b-1 was performed by using the standard $k - \varepsilon$ turbulence model. To deal with the curved surface of the wave-like fins, unstructured grid generated by the Delaunay triangulation was used(Yu, et.al.,1999b). The predicted average Nusselt number are presented in Fig. 29. It can be seen that the predicted variation pattern of the Nusselt number vs Reynolds number is agreeable with the test data quite well. The qualitative discrepancy may be attributed to the turbulence model adopted and the set up of the computational conditions. The improvement of the numerical model is underway in the authors' group.

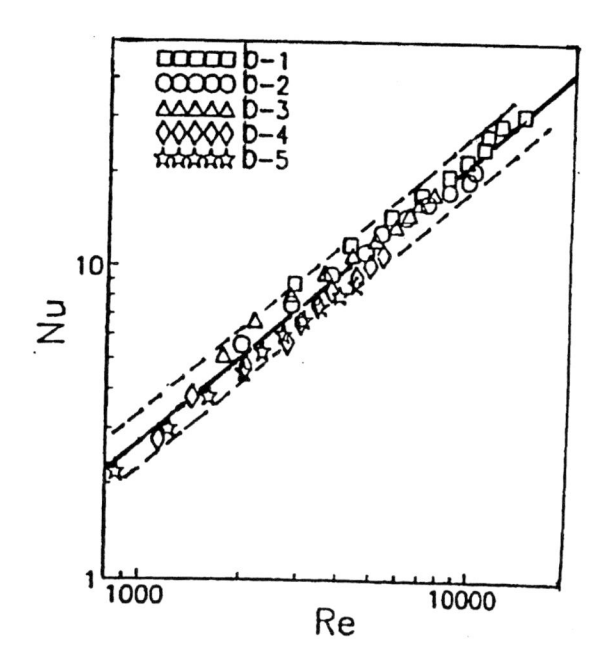

Figure 27 Average Nusselt number correlation for tubes with longitudinal wave-like fins

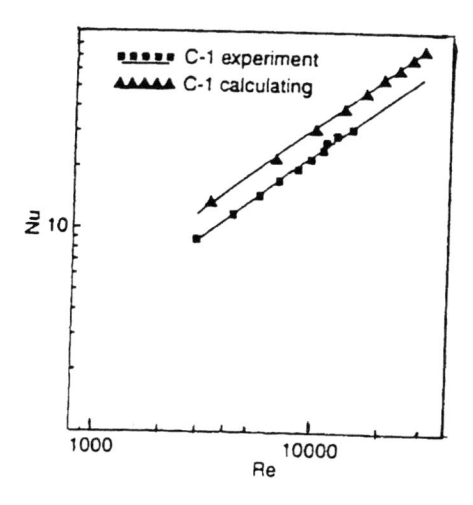

Figure 28 Predicted and measured Nusselt numbers for tubes with longitudinal wave-like fins

CONCLUSIONS

From the above presentation following conclusions may be obtained:
1. For the laminar heat transfer and fluid flow of air across different types of plate arrays, the fully developed fluid flow and heat transfer may be reached after 5 cycles. The plate oblique angle has significant effect on the heat transfer

and pressure drop. Under the identical pumping power and pressure drop constraints, the appropriate oblique angle is around 30 degree. The non-uniformity of successive plate length may enhance heat transfer. For the oblique angle of 25 degree, this enhancement may reach about 25. At very low Reynolds number region the heat transfer and friction factor of the air flow across the plates exhibit the basic characteristics of internal flow and heat transfer in straight duct. The convergently/divergently positioned arrays provide no benefit to enhance the heat transfer..

2. Four types of plate fin-and-tube exchanger surfaces are experimentally tested for their heat transfer and pressure drop characteristics. The Nusselt number is a weak function of the fin spacing and the tube row number, while the per tube friction factor is nearly independent on the tube row number. Correlations are obtained for the range of Reynolds number below 5000. Thermal performance comparison at the identical pumping power constraint shows that the slotted fin surface ranks first, then come the sinusoidal and triangular fins. Taking the plain plate-fin as a reference, the enhancement provided by the slotted fin, sinusoidal fin and triangular fin may reach around 40, 30 and 20%,respectively. It is found that the interwall cylinder between plates serves as a turbulator and a turbulence promotor. It may significantly enhance heat transfer in the low Reynolds number region.

3. When air flows in a tube and exchanges energy with outside liquid, the wave-like longitudinal fins can significantly enhance air-side heat transfer. Experiments were conducted for six model tubes with wave-like internal fins as well as the annular-sector ducts. Heat transfer and friction factor correlations for the fully developed region were provided for the both cases. It is revealed that (1) The simplified Gnielinski equation can be used to predict the fully developed turbulent heat transfer in the annular-sector duct with a reasonable accuracy;(2) The tube with wave-like longitudinal internal fins can significantly enhance the air-side heat transfer., because of its small hydraulic diameter and the extended heat transfer surfaces.; and (3)The blockage of the center part of the tube with wave-like longitudinal fins can further enhance the heat transfer..

ACKNOWLEDGMENTS

The most results presented this paper were supported by following funds: National Natural Science Foundation of China(Grant No. 59676019, 59806011, 50076034)); The Research Fund for the Doctorial Program of Higher Education of China (Grant No. 8969821,9269805, 98069835); Research and Development Fund of Ministry of Mechanical Industry. This paper was also supported by the National Key Project of Fundamental R&D of China (Grant No, 200026303). We also gratefully acknowledge Mr. Z.G. Qu for his assistance in preparing the comparison between three plate arrays and the related drawing.

NOMENCLATURE

A	cross sectional area, m^2
c	unfolded length of fin, m
d, D	diameter, m
d_h	hydraulic diameter, m
f	friction factor
H	spacing between two corrugated walls
h	heat transfer coefficient, $W/m^2\ °C$
k	turbulent kinetic energy, m^2/s^2
L	plate length, m
L_1	length of long plate, m
L_2	length of short plate, m
L_p	longitudinal pitch, m
N	tube row number, plate cycle number
s	fin spacing, m
S	wetted perimeter, m
S_1	transverse pitch, m
S_2	longitudinal pitch, m
T	transverse pitch, m
δ	plate thickness, m; percentage deviation
ε	dissipation rate of turbulent kinetic energy, m^2/s^2
ϑ	oblique or apex angle, degree

Subscripts

b	blocked tube
c	compared
i	inner
o	outer
p	plain tube
u	unblocked tube

REFERENCES

Eisemann, P.R.,1982. Coordinate Generation with Precise Controls over Mesh Peoperties. J.Comput Physics, col. 47, pp. 330-351

Eiseman, P.R., 1985. Grid Generation for Fluid mechanics Computations. Annu. Rev. Fluid mech, vol. 17, pp.487-522

Goldstein, R.J. and Cho, H.H., 1993. A review of Mass(Heat) Transfer Measurements Using Naphthalene Sublimation, in *Experimental Heat Transfer, Fluid Mechanics and Thermodynamics* 1993, M.C. Kelleher et. al.(eds), Elsevier Science Publishers, pp. 21-40

Goldstein, L. .Jr..and Sparrow, E. M., 1977. Heat/Mass Transfer Characteristics for Flow in a Corrugated Wall Channel. *ASME J. Heat Transfer*, 99:187-196

Grimson,E.D., 1937. Correlationa and Utilization of New Data on Flow Resistance and Heat Transfer for Cross flow of Gases over Tube Banks. *Trans. ASME*, 59:583-594

Huang,H.Z., and Tao,W.Q., 1993, An Experimental Study on Heat Transfer and Pressure Drop Characteristics for Arrays of Non-uniform Length Positioned Obliquely to the Flow Direction. *ASME J Heat Transfer*, vol. 115, pp. 568-575

Kang,H.J., Li,W., Li,H.Z., Xin,R.C., and Tao,W.Q., 1994a. Experimental Study on Heat Transfer and Pressure Drop

Characteristics of Four Types of Plate Fin-and-Tube Heat Exchanger Surfaces. *J Thermal Science*, vol.3,pp.34-42

Kang,H.J., Xin,R.C., Li,H.Z.,Li,W., and Tao,W.Q., 1994b, Experimental Study on Heat Transfer and Pressure Drop for Plain Plate-Fin and-Tube Heat Exchanger, *J. Xi'an Jiaotong University*(in Chinese), vol. 28, no. 1, pp.91-98

Kays,W.M., and London, A.L., 1993, *Compact Heat Exchanger*, McGraw-Hill., New York

Li,H.Z., Kang,H.J., Xin,R.C., Li,W., and Tao,W.Q., 1992a. Experimental Study on Heat Transfer and Flow Resistance of Slotted-Plate-Fin-and-Tube Heat Exchanger Surface. in *Proceedings of 4th Symposium on Engineering Thermophysics of Chinese Universities and Institutes*(in Chinese), pp.277-280

Li, H.Z., Li, W., Zin, R.X., Kang,H.J., Tao,W.Q.1992b. Experimental Study on Heat Transfer and Pressure Drop Characteristics of Plate-Fin-and –Tube Heat Exchanger Surfaces with Sinusoid Cross Section, *Fluids Engineering*(in Chinese), vol. 20, no. 12, pp..55-59.

Lin,M.J., Tao,W.Q., and Lue,S.S., 1995. Study on Friction Factor of Developing and Developed Laminar Flow in Annular-Sector Duct, *J. Thermal Science*, vol. 4, pp. 180-1184

Lin,M.J., Wang,Q.W., and Tao,W.Q. 2000. Developing Laminar Flow and Heat Transfer in Annular-Sector Ducts, *Heat Transfer Engineering*, 2000, vol. 21, No. 2 ,pp. 53 - 61.

Li,W., Tao,W.Q., Kang,H.J., Li,H.Z., Xin,R.C., 1997. Experimental Study on Heat Transfer and Pressure Drop Characteristics for Fin-and-Tube Heat Exchangers, *Chinese Journal of Mechanical Engineering*, vol. 33(1), pp.81-86

Lue, S.S., Huang,H.Z., and Tao,W.Q., 1993. Experimental Study on Heat Transfer and Pressure Drop Characteristics in Developing Region for Arrays of Obliquely-positioned Plates of Non-uniform Length, *Experimental Thermal Fluid Science*, vol. 7, pp.568-575

Pang, K., 1989. Experimental and Numerical studies on Heat Transfer and pressure Drop Characteristics for convergently-Divergently Positioned Arrays of Interrupted Plates. Thesis. Department of Power Machinery Engineering, Xi'an Jiaotong University

Pang, K., Tao, W. Q.,and Zhang, H. H.,1990. Numerical Analysis of Fully Developed Flow and Heat Transfer for Arrays of Interrupted Plates Positioned Convergently-divergently along the Flow Direction. *Numer Heat Transfer*, Part A,vol.18, pp.309-324

Patankar, S.V., 1980. Numerical Heat Transfer and Fluid Flow. McGraw-Hill, New York

Rohsenow, W. M., Hartnett,J. P., and Cho, Y. I.(eds), 1998. Handbook oh Heat Transfer, 3rd edition. McGraw-Hill,New York

Shah, R. K. and Johnson, R.S., 1981. Correlations for fully developed flow trough circular and non-circular channels. In: Proceedings of 6th National Heat and Mass Transfer Conference, Madras, India, pp. D75-D76

Shah, R. K., and Heikal M.R., Progrss in the numerical analysis of compact heat exchanger surfaces. In: Advances in Heat Transfer, vol. 34, 2001, pp.363-442

Tao,W.Q., 1988. Numerical Heat Transfer (in Chinese). Xi'an Jiaotong University Press, Xi'an

Tao,W.Q., Kang,H.J., Xin,R.C., Li,H.Z., and Li,W., 1997. Turbulent Heat Transfer measurement in Air-cooled tube Sets Using the Heat Resistance-Separating Method, *Heating, Ventilating, and air-Conditioning* (in Chinese), vol. 27 (Supplement), pp. 64-67

Tao,W.Q., Lu,S.S., Kang,H.J., and Lin,M.M., 2000. Experimental Study on Developing and Fully Developed Fluid Flow and Heat Transfer in Annular-sector Ducts, *Enhanced Heat Transfer*, vol. 7,pp. 51-60

Wang, L.B., and Tao,W.Q., 1995. Heat Transfer and Fluid Flow Characteristics of Plate Array Aligned at Angles to the Flow Direction. *Int J Heat Mass Transfer*, vol. 38:pp. 3053-3063

Wang,L.B., Jiang,G.D., Tao,W.Q., and Ozoe,H., 1998. Numerical Simulation on Heat Transfer and Fluid Flow Characteristics of Arrays with Non-uniform Plate Length Positioned obliquely to the Flow Direction. *ASME J Heat Transfer*, vol. 120, pp. 991-998

Wang,L.B., and Tao,W.Q., 1997. Numerical Analysis on Heat Transfer and Fluid Flow for Arrays of non-uniform Plate Length Aligned at Angles to the Flow Direction, *Int J Numer Methods Heat Fluid Flow*, vol. 7, pp. 479-496

Webb,R. L., Principles of Enhanced Heat Transfer. John Wiley & sons, New York, 1994

Wei J G, and Tao W Q. 1994. Experimental study on the flow resistance characteristics of plate arrays positioned convergently-divergently to the flow direction, in *Proceedings of the Fifth Conference of Engineering Thermophysics of Chinese Universities and Institutes*(in Chinese), vol. 1, pp. 213-216

Xiao,Q., and Tao,W.Q., 1990. Effect of Fin Spacing on Heat Transfer and Pressure Drop of Two-row Corrugated Fin-and-Tube Heat Exchangers. *Int Comm Heat Mass Transfer.*, vol. 17, pp. 577-588

Xiao,Q., Chen B., and Tao,W.Q., 1992. Experimental Study on Effect of Interwall Tube Cylinder on Heat Transfer of Corrugated Plate and Tube Heat Exchanger. *ASME J Heat Transfer*, vol. 114, pp. 755-758

Xin,R.C., Li,H.Z., Kang,H.J., Li,W., and Tao,W.Q., 1992. Heat Transfer and Pressure Drop Measurements on Four Types of Plate Fin-and-Tube Heat Exchangers, in *Heat Transfer Science and Technology*, ed..B.X.Wang, Higher Education Press, Beijing, pp. 942-947

Xin, R.C., Li,H.Z., Kang,H.J., Li,W., and Tao,W.Q., 1994. An Experimental Investigation on Heat Transfer and Pressure Drop Characteristics of Triangular Wavy Fin-and-Tube Heat Exchanger Surfaces, *J. Xi'an Jiaotong University* (in Chinese), vol.28,no. 2, pp.77-83

Yu,Bo., Nie,J.H., Wang,Q.W., and Tao,W.Q., 1999a. Experimental Study on the Pressure Drop and Heat Transfer Characteristics of Tubes with Internal Wave-like Longitudinal Fins. *Heat Mass Transfer*, vol. 35, pp. 65-73

Yu,B., Lin,M.J., and Tao,W.Q., 1999b. Automatic Generation of Unstructured Grids with Delaunay Triangulation and Its Applications. *Heat Mass Transfer*, vol. 36, pp. 361-370

Yu,B., Lin, M.J., Xy, J.Y. and Tao,W.Q., 1998. Numerical Simulation on Convective Heat Transfer of Fully Developed Laminar Flow in Wave-like Fin Tube. *J. of Xi'an Jiaotong University*, vol. 32, No. 9, pp. 51-55

Yu,B., Wang,Q.W., and Tao,W.Q., 2000. Pressure Drop and Heat Transfer Characteristics of Turbulent Flow in Tubes with Internal Wave-like Longitudinal Fins, submitted for publication

Zhang, H.H., and Lang,X.S., 1989. The Experimental Investigation of Oblique Angles and Interrupted Plate lengths for Louvered Fins in Compact Heat Exchanger.. *Experimental Thermal Fluid Science*, vol. 2,, pp. 100-106

HEAT TRANSFER CHARACTERISTICS OF FIN-TUBE HEAT EXCHANGER WITH VORTEX GENERATORS

S.Y. Yoo, D.S. Park, M.H. Chung[1] and J.K. Kim[2]

[1]Chungnam National University, Taejon, 305-764, Korea; E-mail: syyooh@cnu.ac.kr
[2]Dubal Gas Engineering, Incheon, 405-310, Korea; E-mail: dubalgas@unitel.co.kr

ABSTRACT

In the present work, vortex generators are fabricated on the fin surface of a fin-tube heat exchanger to augment the convective heat transfer. In addition to horseshoe vortices formed naturally around the tube of the fin-tube heat exchanger, longitudinal vortices are artificially created on the fin surface by vortex generators. The purpose of this study is to investigate the local heat transfer phenomena in the fin-tube heat exchangers with and without vortex generators, and to evaluate the effect of vortices on the heat transfer enhancement. Experiments were performed for the model of fin-circular tube heat exchangers with and without vortex generators, and of fin-flat tube heat exchangers with and without vortex generators. In addition, the prototype of heat exchangers with and without vortex generators used for Air Handling Unit were manufactured, performance tests were done, and results are compared each other. Average heat transfer coefficients of fin-flat tube heat exchanger without vortex generator were much lower than those of fin-circular tube heat exchanger. On the other hand, fin-flat tube heat exchanger with vortex generators was shown to have much higher heat transfer value than fin-circular tube heat exchanger.

INTRODUCTION

For energy conservation and the protection of environment, it becomes increasingly important to adopt heat transfer enhancement technique in the design of heat exchangers used in process industries for heating, cooling, air-conditioning and refrigeration. There are two enhancement technologies of convective heat transfer for compact heat exchangers. One is to extend heat transfer surface area like a fin, the other is to increase heat transfer coefficients between solid surface and fluid. In the present work, vortex generators are fabricated on the fin surface of a fin-tube heat exchanger to augment the convective heat transfer. In addition to horseshoe vortices formed naturally around the tube of the fin-tube heat exchanger, longitudinal vortices are artificially created on the fin surface by vortex generators. The purpose of this study is to investigate the local heat transfer phenomena in the fin-tube heat exchangers with and without vortex generators, and to evaluate the effect of vortices on the heat transfer enhancement.

Fiebig and Chen (1998) have summarized their systematic study on the characteristics of heat transfer surface with vortex generators. According to their research, the fin heat exchanger surface area may be reduced with vortex generators by more than 50 % compared to a plain fin for identical heat duty and pressure loss. Zhu et al. (1995) have studied numerically the effect of vortex generator on the flow and heat transfer in a rib-roughened channel. Yoo (1997) have developed fin-flat tube heat exchanger with vortex generators, and showed that vortex generators increase heat transfer rates on the fin surface by almost double.

When the heat is transferred by forced convection from the fin-tube heat exchanger with the vortex generators, complex flow phenomena - such as stagnation, separation, vortex formation or wake - affect the heat transfer characteristics. In a complicated flow situation, it is very difficult to measure local heat transfer coefficients by conventional methods of heat transfer measurement. Goldstein et al. (1990) showed that mass transfer experiments using naphthalene sublimation technique is an effective way for measuring local heat transfer distribution in a complex, three-dimensional flow region. In the present study, the naphthalene sublimation technique is employed to measure local mass transfer from fin-tube heat exchangers. Then, mass transfer data are converted to their counterpart of heat transfer processes using a heat/mass transfer analogy.

Experiments were performed for the model of fin-circular tube heat exchangers with and without vortex generators, and of fin-flat tube heat exchangers with and without vortex generators. At the same time, friction losses for four types of heat exchangers were measured and compared. In addition, the prototype of heat exchangers with and without vortex generators used for AHU(Air Handling Unit) were manufactured, performance tests were done at various conditions, and test results are compared each other.

EXPERIMENTAL APPARATUS AND PROCEDURE

Experimental Apparatus

The experimental apparatus for model test comprises a wind tunnel, a naphthalene casting facility and a sublimation depth measurement system. The open-circuit blowing type wind tunnel, which has a square test section of 300 mm × 46 mm, is used. The air speed is controlled by an inverter and the freestream turbulence intensity is less than 1.0 % over the entire range of speed.

The naphthalene casting facility has a mold, heating plate, hot air gun, mold separating device and suction hood. Automated sublimation depth measurement system is used to measure local mass transfer coefficients. The depth measurement system consists of a depth gage, a signal conditioner, two stepper motor-driven traversing table, a data acquisition board and a personal computer. The depth gage used to measure the naphthalene surface profile is a linear variable differential transformer(LVDT) which has a ±0.254 mm(0.01 in) linear range and a resolution of ±25.4 nm(1 μ-in). It is connected to a signal conditioner, which supplies excitation voltage to LVDT and amplifies the output signal from LVDT.

The experimental apparatus for prototype test, shown in Fig. 1, comprises a wind tunnel, hot water supply system, heat exchanger, and measurement and control system.

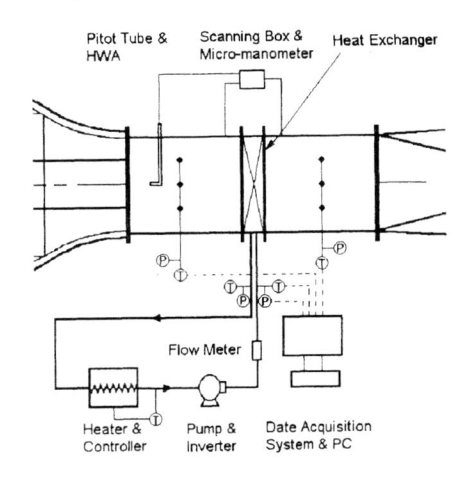

Fig. 1 Schematic of prototype test apparatus on heat exchanger

The open-circuit suction type wind tunnel, which has a square test section of 400 mm × 300 mm is used. Air speed and temperature are controlled by inverter and electric heater, respectively. Water supply system consists of constant temperature bath, pump and flowmeter. The temperature of bath is controlled with RTD sensor PID controller, and circulation pump is operated by an inverter to control flowrate.

Experimental Procedure

A new naphthalene casting is made for each test run. The testpiece is clamped to a highly polished mold, and molten naphthalene is then poured into the mold. After the naphthalene solidifies, the mold is separated from the testpiece by applying a shear force. The casted testpiece is placed and then clamped on the sublimation depth measurement table. Initial readings of the naphthalene surface elevation are taken at predetermined locations using the automated sublimation depth measurement system. The testpiece is then installed in the wind tunnel and exposed to the air stream for about one hour. During a test run, the naphthalene surface temperature, tunnel air temperature and pressure, and freestream velocity are measured. The testpiece is then removed and a second set of surface elevation is obtained at the same locations as before. Finally, data reduction program calculates Sherwood numbers and Nusselt numbers from the measured sublimation depth and other related data.

Data Reduction

The mass transfer coefficient is defined by

$$\dot{m}/A = h_m(\rho_{v,w} - \rho_\infty) \qquad (1)$$

where $\dot{m}$ is mass transfer rate, A is naphthalene sublimation area, h_m is mass transfer coefficient, $\rho_{v,w}$ is naphthalene vapor density on the surface, and ρ_∞ is naphthalene vapor density in the freestream, which is ignored in this study. The empirical equation of Ambrose et al. (1975) is used to determine the naphthalene vapor pressure from the measured surface temperature and then, using the ideal gas law, naphthalene vapor density on the surface is evaluated.

The local mass transfer rate can be determined from

$$\dot{m}/A = \rho_s \Delta t / \Delta \tau \qquad (2)$$

where ρ_s is the density of the solid naphthalene, Δt is the net sublimation depth, and $\Delta \tau$ is the total exposure time in the wind tunnel. Total naphthalene sublimation depth is calculated from the variation in measured surface elevations before and after the exposure, and the excess sublimation due to natural convection during the sublimation depth measurement period is subtracted from the total sublimation to calculate net sublimation.

Combining equations (1) and (2) gives

$$h_m = \frac{\rho_s \Delta t / \Delta \tau}{\rho_{v,w}} \qquad (3)$$

The Sherwood number can be expressed as

$$Sh = h_m H / D_{iff} \qquad (4)$$

where H is the characteristic length, fin spacing in this study. The mass diffusion coefficient of naphthalene in air, D_{iff} is calculated from Cho's (1989) correlation. The estimated errors of Sherwood number are found to be within 6 % in the entire range of our measurements. The measured mass transfer coefficients or Sherwood numbers are converted to their counterpart of heat transfer using the following analogy relation,

$$Nu / Sh = (hH / k)/(h_m H / D_{iff}) = (Pr / Sc)^n \qquad (5)$$

where 1/3 is used for exponent n in the present study.

RESULTS AND DISCUSSION

Model Test

Heat transfer for fin-circular tube. Figure 2 shows the schematic of the fin-tube heat exchanger model with circular tubes in a staggered arrangement. The heat exchanger models consist of five parallel plate representing the fins and three rows of tubes. Fins are made of 0.8 mm thick stainless-steel and tubes are made of acryl. Vortex generators, 7 mm × 7 mm, made of acryl are mounted vertically on the fins, and fin spacing is the same as the height of vortex generator. The bottom plate of the wind tunnel is casted with naphthalene and heat exchanger model is mounted on it. Results obtained with the inlet air velocity of 7 m/s are presented

Distribution of the local heat transfer coefficients for fin-circular tube heat exchanger without vortex generator is shown in Fig. 3 (a). When the boundary layer flow meet the protruding object, tube in the study, flow is slowdown and dynamic pressure converted to static pressure. A pressure gradient across the boundary layer is, therefore, created. This pressure gradient causes the flow to move toward the fin and reverse flow at the region nearest to the fin. This reverse flow of the boundary layer produces the horseshoe vortex, which is then carried around the tube by the main flow. This horseshoe vortex enhances the convective heat transfer dramatically. Horseshoe-like peak, starting in front of the tube and continuing around the tube, corresponds to the trail of horseshoe vortex. Meanwhile, the heat transfer coefficient behind the circular tube is found to be very low. The reason is that wide wake zone is formed behind the tube, so the fluid in this region is scarcely mixed with the main flow. Figure 3 (b) shows the distribution of the local heat transfer coefficients for fin-circular tube heat exchanger with vortex generators. Heat transfer enhancement by the horseshoe vortices is similar to that without vortex generator. In addition to this enhancement, another peak trail is found in the downstream of each vortex generator. Longitudinal vortices are generated by flow separation due to the pressure

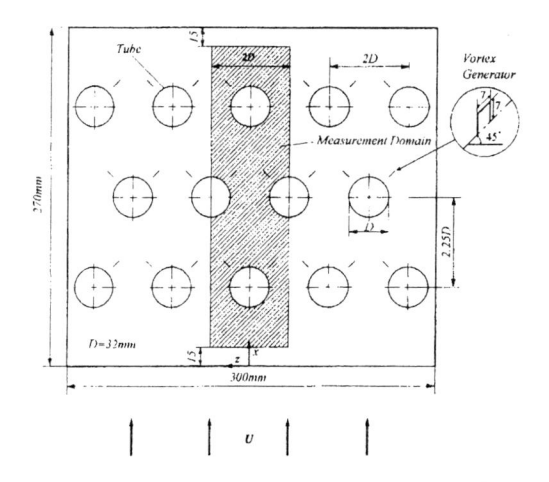

Fig. 2 Schematic of fin-circular tube heat exchanger model

(a) without vortex generator

(b) with vortex generator

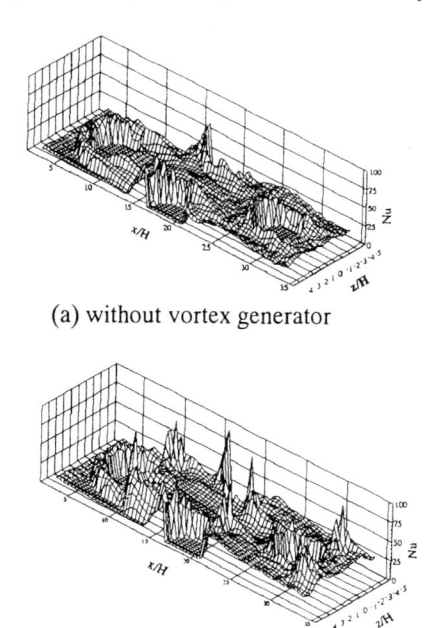

Fig. 3 Distribution of local Nusselt number for fin-circular tube heat exchanger

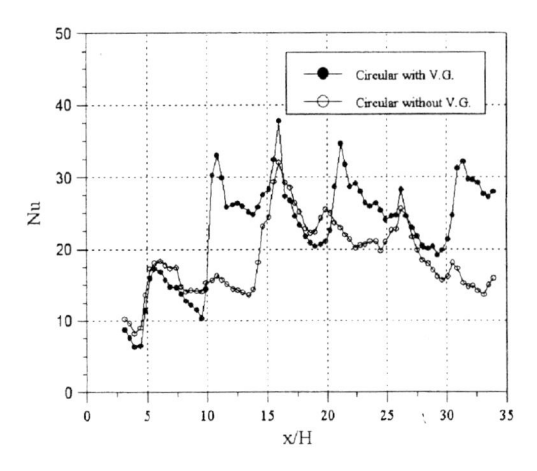

Fig. 4 Comparison of spanwise-averaged heat transfer for fin-circular tube heat exchanger

difference between upstream and downstream side of vortex generator. These vortices turn the flow field perpendicular to the main flow direction. They enhance the mixing of the fluids close to and far from the fin and, thereby, the fin heat transfer.

In the Fig. 4, average Nusselt numbers are compared on the fin surface of the fin-circular tube heat exchanger with and without vortex generators. Average Nusselt number is calculated by integrating the local Nusselt numbers along the spanwise-direction. In case of a heat exchanger without vortex generator, three local maximum is found at x/H≈6, x/H≈16 and x/H≈26, which is caused by the horseshoe vortices formed in front of tubes located in the first, second and third row, respectively. Second and third peaks are relatively higher than first peak, because blockage due to the first tube increases the velocity and produces stronger horseshoe vortices in front of second tube. In case of a heat exchanger with vortex generators on the other hand, three additional peaks appear at x/H≈11, x/H≈21 and x/H≈31, which correspond to the location of vortex generators. The magnitude of enhancement by vortex generators are almost same regardless of location, and longitudinal vortices does not influence horseshoe vortices. The average heat transfer coefficient on the whole fin surface of the fin-circular tube heat exchanger with vortex generators found to be 20 % higher than that of the fin-circular tube heat exchanger without vortex generator

Heat transfer for fin-flat tube. Figure 5 shows the schematic of the fin-tube heat exchanger of model with flat tube. The cross sectional area and tube spacing of flat tubes are equal to those of circular tubes. The same vortex generators as is used in the fin-circular tube heat exchanger, are mounted in the upstream side of flat tube. Fin-flat tube heat exchangers are often adopted in air-conditioning and gas-boiler system instead of fin-circular tube heat exchangers, to reduce the pressure loss, consequently pumping cost. But it is well known that heat transfer

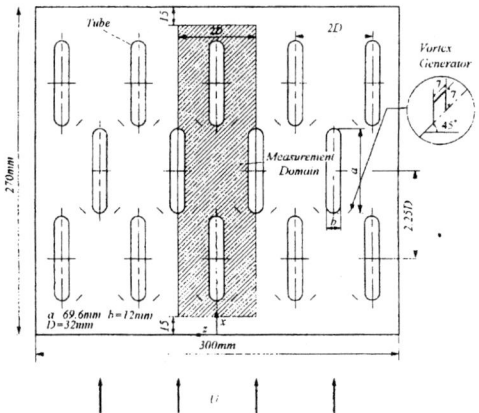

Fig. 5 Schematic of fin-flat tube heat exchanger model

efficiency of the fin-flat tube heat exchanger is much lower than that of circular type. Yoo (1997) have reported that dual effects, the enhancement of heat transfer and the reduction of pressure loss, could be accomplished by implementing vortex generators on the fin surface.

Figure 6 (a) shows the distributions of local heat transfer coefficients in the fin-flat tube heat exchanger without vortex generator. Heat transfer enhancement by horseshoe vortices found in front of tubes, as in fin-circular tube heat exchanger, but their magnitude is much less than that of fin-circular tube. Blockage area of flat tube is smaller than circular tube, so intensity of horseshoe vortices formed around flat tubes are weaker. This is the reason why fin-flat tube heat exchanger has lower heat transfer efficiency than fin-circular tube. Distribution of local heat transfer coefficients in the fin-flat

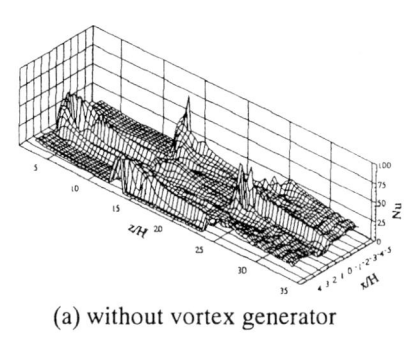

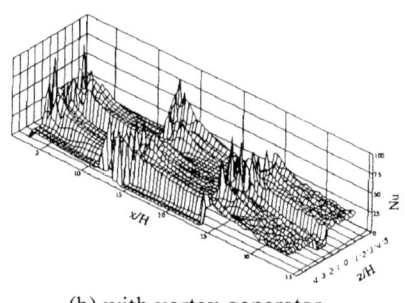

(a) without vortex generator

(b) with vortex generator

Fig. 6 Distribution of local Nusselt number for fin-flat tube heat exchanger

tube heat exchanger with vortex generators is shown in Fig. 6 (b). In addition to heat transfer enhancement by horseshoe vortices, longitudinal vortices created by vortex generators are found to augment heat transfer drastically.

Effect of vortex generators on the heat transfer enhancement in the fin-flat tube heat exchanger is shown in Fig. 7. Because vortex generators are installed nearby upstream side of flat tube, heat transfer enhancement caused by horseshoe vortices and by longitudinal vortices are coupled. The increase in Nusselt number by vortex generators is found to be much higher in the fin-flat tube heat exchanger than fin-circular tube. Overall average heat transfer coefficient of fin-flat tube heat exchanger with vortex generators is increased by 60 % compared to that of the fin-flat tube without vortex generator, and is increased by 30 % compared to that of the fin-circular tube heat exchanger without vortex generator.

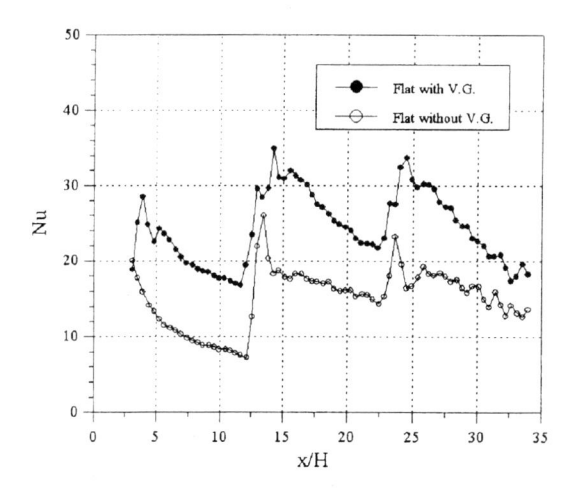

Fig. 7 Comparison of spanwise-averaged heat transfer for fin-flat tube heat exchanger

Pressure drop. Pressure drop in the fin-tube heat exchangers are measured using a micro-manometer, and the apparent friction factor f is calculated from the following expression:

$$f = \frac{\Delta P}{\rho V^2 / 2} \cdot \frac{H}{L} \qquad (6)$$

where H is fin spacing, the distance between fins and L is the distance between pressure measuring points.

Figure 8 shows the apparent friction factor against Reynolds number, based on fin spacing, for circular and flat tube heat exchangers with and without vortex generators. Friction factors decrease as the Reynolds number increase, and vortex generators increase pressure drop in both cases.

Friction factors for fin-circular tube heat exchanger are much higher than that of flat tube because of relatively larger blockage area. Pressure drop of the fin-flat tube heat exchanger with vortex generators is increased by 60 % compared to that of the fin-flat tube without vortex generator, but is decreased by 45 % compared to that of the fin-circular tube heat exchanger without vortex generator. Consequently, fin-flat tube heat exchanger with vortex generators has higher heat transfer efficiency and lower flow loss than conventional fin-circular tube heat exchanger.

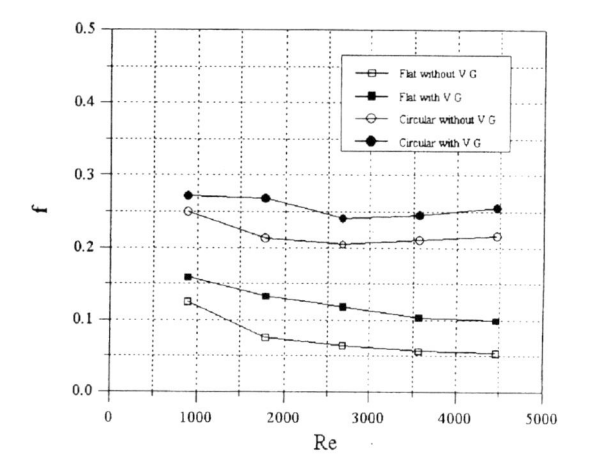

Fig. 8 Comparison of apparent friction factors for four Type of heat exchanger

Prototype Test

Figure 9 shows the schematic of wavy fin of fin-circular tube heat exchanger which is used for AHU(Air Handling Unit). It is difficult to fabricate the vortex generator on the wavy fin surface, so fin-circular tube heat exchanger with vortex generator is excluded in the prototype test. Prototype heat exchanger has a dimension of 400(W) × 304(H) × 252(D) mm, which is manufactured to fit wind tunnel test section. It consists of 32(4 rows × 8 columns) copper tubes in a staggered arrangement, and 124 copper fins of 0.15 mm thickness and 3.2 mm pitch. In a water flow line, header isattached to the first row of tubes so that 8 tubes in the equivalent row has same temperature condition, and air vent and drain are mounted on the inlet and outlet of header

The prototype of fin-flat tube heat exchangers with and without vortex generators are manufactured. Square wing type vortex generator(3.2 mm × 3.2 mm) is used, and two

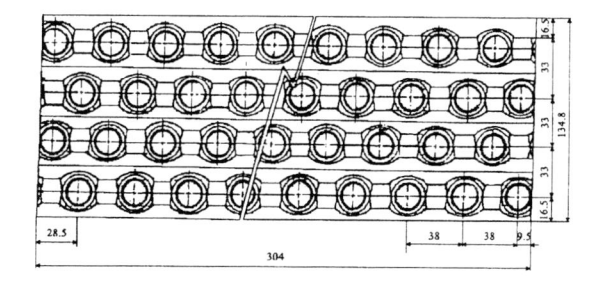

Fig. 9 Schematic of wavy fin of fin-circular tube heat exchanger

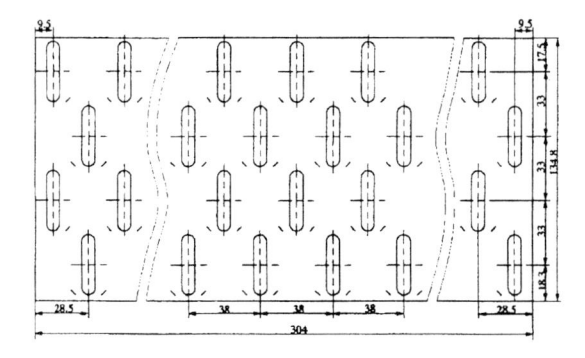

Fig. 10 Schematic of fin of fin-flat tube heat exchanger with vortex generator

vortex generators for each tube is punched from the fin surface with an angle of 45° to the main flow direction, as shown in Fig. 10. The overall size, material, number of fins and tubes, cross-sectional area of tube, tube location of fin-flat tube heat exchangers are the same as those of wavy fin-circular tube heat exchanger.

Thermal performance in the fin-tube heat exchangers is expressed in terms of UAF, which is defined by the following equation:

$$Q_a = (UAF)_a \Delta T_m \qquad (7)$$

where Q_a is heat transfer rate from hot water to air, U is overall heat transfer coefficient between them, A is heat transfer area, F is correction factor, and ΔT_m is LMTD (Log Mean Temperature Difference), which is calculated from the inlet and outlet temperature of hot water and cooling air. Figure 11 shows the UAF against air velocity at water inlet temperature of 40℃, water flow rate of 4.9 m³/h. The UAF increases as the air velocity increases. The UAF of fin-flat tube heat exchanger without vortex generators is lower than those of fin-circular tube heat exchanger. But the UAF of fin-flat tube heat exchanger with vortex generators is almost same as those of the fin-circular tube heat exchanger. Heat transfer rates of fin-circular tube heat exchanger is shown to be higher than those obtained in the model test, because wavy fin is used in the prototype test instead of plain fin, which is used in the model test.

To evaluate pressure drop, apparent friction factor in the air side, defined by eq. (6), is plotted in Fig. 12. Friction factors decrease as the Reynolds number, based on fin spacing, increases and are linearly dependent on Reynolds number in the log-log scale except low velocity region. Friction factor of fin-circular tube heat exchanger is almost three times that of fin-flat tube heat exchanger without vortex generators, and two times that of fin-flat tube heat exchanger with vortex generators.

Figure 13 shows thermal performance against equivalent fan power, which is defined by the following equation:

$$P = \Delta P \cdot Q \qquad (8)$$

where ΔP is the pressure loss in air side, and Q is the air

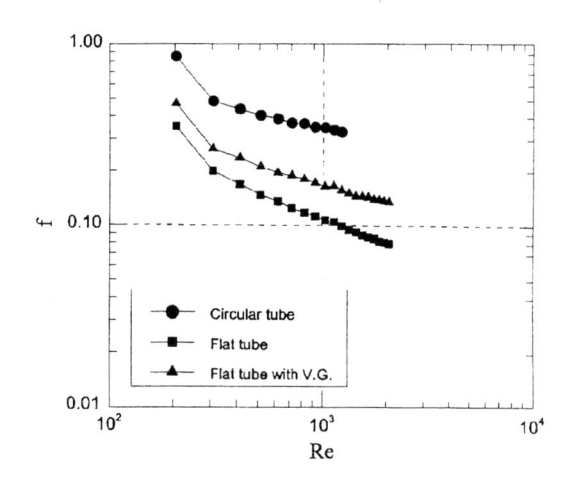

Fig. 12 Variation of apparent friction factor of air side

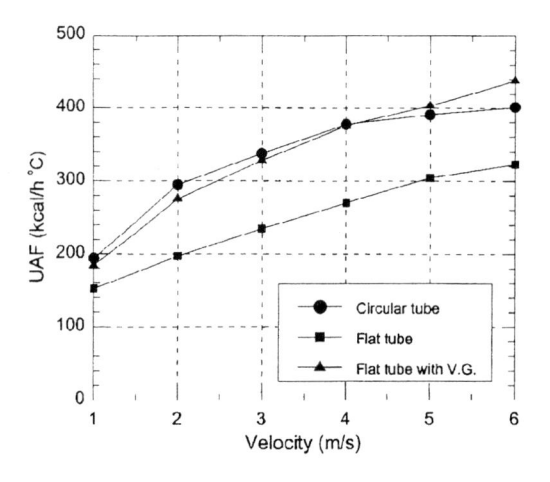

Fig. 11 Variation of UAF against air velocity

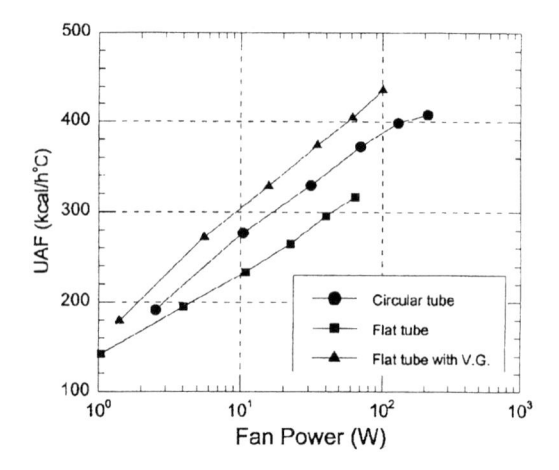

Fig. 13 Variation of UAF against equivalent fan power

flowrate. Thermal performance of fin-flat tube heat exchanger with vortex generators is increased by 12% compared to that of wavy fin-circular tube heat exchanger. Much higher heat transfer enhancement is expected, if comparison is made with plain fin-circular tube heat exchanger. And thermal performance of fin-flat tube heat exchanger with vortex generators is increased by 29 % compared to that of fin-flat tube heat exchanger without vortex generators.

CONCLUSIONS

Heat transfer and pressure loss for the model and the prototype of fin-tube heat exchangers with and without vortex generators are measured and compared. Summary of major results are as follows.

1. Horseshoe vortices formed around the tube enhance the heat transfer dramatically, and their effect is relatively weaker in the flat tube.
2. In case of fin-tube heat exchanger with vortex generators, longitudinal vortices augment heat transfer in addition to horseshoe vortices.
3. Average heat transfer coefficient on the fin surface of fin-flat tube heat exchanger with vortex generators is increased by 60 % compared to that of the fin-flat tube without vortex generator, and 30 % compared to that of the fin-circular tube without vortex generator.
4. Fin-flat tube heat exchanger with vortex generators has much lower pressure loss than conventional fin-circular tube heat exchanger without vortex generator.
5. Thermal performance of fin-flat tube heat exchanger with vortex generators is increased by 12 % compared to that of wavy fin-circular tube, and 29 % compared to that of fin-flat tube without vortex generators.

ACKNOWLEDGMENT

This study is financially supported by the Ministry of Commerce, Industry and Energy through Energy Conservation Program. Authors gratefully appreciate the support.

NOMENCLATURE

A naphthalene sublimation area, m^2
D_{iff} mass diffusion coefficient, m^2/s
F correction factor, dimensionless
f apparent friction factor, dimensionless
H fin spacing, m
h heat transfer coefficient, W/m^2·K
h_m mass transfer coefficient, m/s
k thermal conductivity, W/m·K
L distance between pressure measuring points, m
$\dot{m}$ mass transfer rate, kg/s
Nu Nusselt number, dimensionless
P fan power, W
Pr Prandtl number, dimensionless
Q air flow rate, m^3/s
Q_a heat transfer rate, W
Sc Schmidt number, dimensionless
Sh Sherwood number, dimensionless
U overall heat transfer coefficient, W/m^2·K
V air velocity, m/s
ρ_s naphthalene solid density, kg/m^3
$\rho_{v,w}$ naphthalene vapor density on the surface, kg/m^3
ρ_∞ naphthalene vapor density in the freestream, kg/m^3
ΔP pressure loss, Pa
ΔT_m log mean temperature difference, K
Δt net sublimation depth, m
$\Delta \tau$ total exposure time in the wind tunnel, s

REFERENCE

M. Fiebig and Y. Chen, 1998, Heat Transfer Enhancement by Wing-type Longitudinal Vortex Generators and Their Application to Finned Oval Tube Heat Exchanger Elements, *Proc. of the NATO Advanced Study Institute on Heat Transfer Enhancement of Heat Exchangers*, pp. 79-105.

J. X. Zhu, M. Fiebig and N. K. Mitra, 1995, Numerical Investigation of Turbulent Flows and Heat Transfer in a Rib-Roughened Channel with Longitudinal Vortex Generators, *Int. J. Heat Mass Transfer*, Vol. **38**, No. 3, pp. 495-501.

S. Y. Yoo, 1997, Development of High Efficiency Fin-Flat Tube Heat Exchanger Using Vortex Generators, *MOTI Report*, **95-P-10-2**.

R. J. Goldstein, S. Y. Yoo, and M. K. Chung, 1990, Convective Mass Transfer from a Square Cylinder and Its Base Plate, *Int. J. of Heat and Mass Transfer*, Vol. **33**, pp. 9-18.

D. Ambrose, I. J. Lawrenson and C. H. S. Sparke, 1975, The Vapor Pressure of Naphthalene, *J. of Chemical Thermodynamics*, Vol. **7**, pp. 1173-1176.

K. Cho, 1989, Measurement of Diffusion Coefficient of Naphthalene into Air, *Ph. D Thesis, State University of New York*.

INFLUENCE OF CONJUGATE HEAT TRANSFER IN COMPACT HEAT EXCHANGERS WITH VORTEX GENERATORS

J. Tourreuil, D. Bougeard and B. Baudoin

Département Energétique Industrielle, Ecole des Mines de Douai
941, rue Charles Bourseul - B.P. 838 - 59508 DOUAI Cedex
daniel.bougeard@ensm-douai.fr

ABSTRACT : An experimental study was carried out to highlight conjugate heat transfer effect in compact heat exchangers ($400 < Re_{2H} < 2600$). In order to enhance local heat transfer coefficient, Vortex Generators (VG) are punched out of the fin. In a first time, the heat transfer augmentation due to the addition of VG was measured on a constant temperature fin. An unsteady method with infrared thermography measure the local heat transfer coefficient. In addition, to show the influence of the heat diffusion in the fin, a steady method was developed to determine the local heat flux density. This work showed that the fin efficiency has a strong influence on the Vortex Generator performance. VG can increase the heat transfer by up to 35% on a fin with an efficiency equal to 1 (constant temperature), whereas this enhancement is about 23% for a brass fin and 7% for a steel fin. Finally, we show that the theory of the fin efficiency (with separate effect of convective and conductive transfer) is valid in the majority of industrial applications ($\eta > 0.75$).

INTRODUCTION AND LITTERATURE REVIEW

Heat exchangers are used in many industrial applications. They are used in air-conditioning, automotive, chemical and agroalimentary industry...More particularly, our study relates to the automotive radiators. These finned tube heat exchangers remove excess heat generated by the engine (fig. 1).

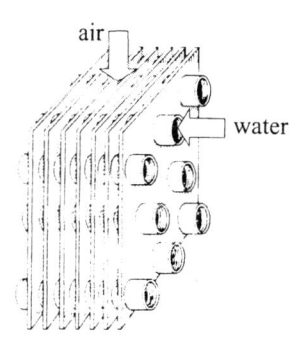

Fig. 1: Compact finned tube heat exchanger

According to the industrial feature of our work, research objectives are as follows :

- to reduce the size of the exchangers without decreasing the dissipation (improvement of the heat transfer coefficient)
- to reduce the pressure losses (a lower pumping power).

Because the heat transfer coefficient between the air and the wall can be 100 times weaker than that of the liquid, our study focuses on the air side. Passive techniques (use of optimised geometry) are usually adopted for this purpose. Several studies (Valencia and Mitra, 1994) (Tiggelbeck and Fiebig, 1992) characterised the thermal and dynamic flow generated by Vortex Generators (see fig. 2).

Fig. 2: Generation of longitudinal vortex

These swirl promoters are generally punched out of the fin. They generate longitudinal vortices which reduce the thickness of the dynamic and thermal boundary layer thus improving the convective heat exchange. This disturbance of the flow also generates an increase in turbulence. Valencia (Valencia and Mitra, 1994) studied exchangers with 3 rows of staggered tubes. Delta vortex generators are punched out of the fins (fig. 3(c)). This experimental work enables the local measurements of the convection on the fins with or without VG using an unsteady technique with liquid crystals thermography. The analysis of the total thermal performance showed that heat exchange could increase by 9% as a result of the addition of VG. The increase of the pressure loss due to the VG is negligible (approximately 3%).

Although its study shows the real potential of these geometries for heat transfer enhancement, it's unfortunate that the investigated Reynolds number range (Re_{2H} between 1500 and 5400) is too high for compact automotive heat exchanger. Thus, the equivalent Reynolds number Re_{2H} is generally lower than 1000. It is important to point out that, in all experimental studies, the fin temperature is constant i.e. with an infinite conductivity. However, although the efficiency of the plate is significant, no experimental work takes into account the diffusion of heat throughout the fins. Usually, calculations of the thermal performances of the exchangers take into account this conjugated effect by considering the efficiency defined by Gardner (Gardner, 1945). Nevertheless, this determination assumes that the convective heat transfer is constant on all the surface of the fin, which is absolutely not the case.

Because an experiment which takes into account the heat diffusion in the fins would be very hard, Sanchez (Sanchez and Mitra, 1991) uses a numerical code which simulates a one row exchanger with turbulence promotors. The heat diffusion in the fin is then computed. The author compares the thermal performances of the VG with the plane fin case. He finds that, the weaker the heat diffusion in the fin is, the more the increase of the thermal transfer due to the addition of the VG is significant. These results relate a new phenomenon. Indeed, if we take the theory of the fin efficiency (no conjugate heat transfer) (Gardner, 1945), the conclusions would have been different ; in fact according to this theory a reduction of the fin conductivity causes a reduction of the VG performance. In conclusion, the effect of conjugate heat transfer could have important consequences on fin performance when we use passive enhancement items like VG.

In order to study the effect of passive enhancement techniques for automotive heat exchanger, we have designed two experimental techniques. The first technique performs the measurement of local heat exchange on a fin with constant temperature field. This unsteady method is based on the work of Ireland (Ireland and Jones, 1986). In the second technique, we measure the conjugate heat flux density on the fin. Two kind of fins with different thermal conductivity are used (steel and brass). This stationary measurement technique is based on the work of Bougeard (Bougeard, 1997).

EXPERIMENTAL METHOD AND HANDLING

Experimental bench description

The two techniques use the same experimental bench shown in figure 3. The geometry of this model is similar to automotive radiators with small fin pitch (compactness of 670 fins/m). The airflow speed varies between 0.5 to 3 m/s (Re_{2H} from 450 to 2600). For a technical reason, we will study only one fin. Indeed, even if the model is not exactly a real heat exchanger, the tendencies of heat transfer and the physics of the phenomena would be representative of reality. The test bench will allow us to test various material conductivities. We will study either plane fins or punched out vortex generators. The two rows of tubes are staggered. The use of a convergent section allows a flat velocity profile to be obtained at the entry. The two investigation techniques require an accurate knowledge of the spatial distribution of the fin surface temperature. This measurement is made with an infrared camera (AGEMA 900 Long Wave). The investigation area is placed in the center of the wind tunnel to prevent wall effect. The infrared camera is placed at its minimal distance in order to give maximal space resolution. And finally, the thermal scene is viewed through an ZnSe (Zinc Selenide) IR transparent window.

The vortex generators are placed in strategic areas in order to enhance heat transfer (fig. 3(c)). The height of those corresponds to the fin pitch.

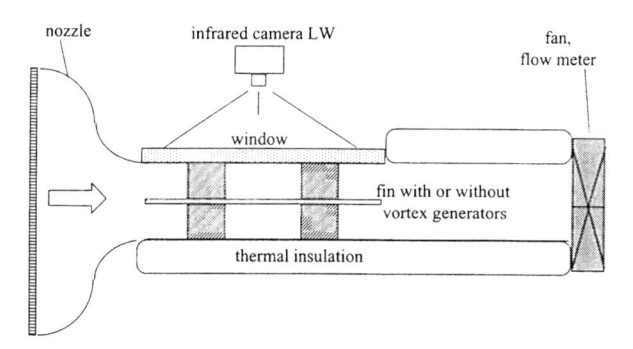

(a) Schematic diagram of wind tunnel setup

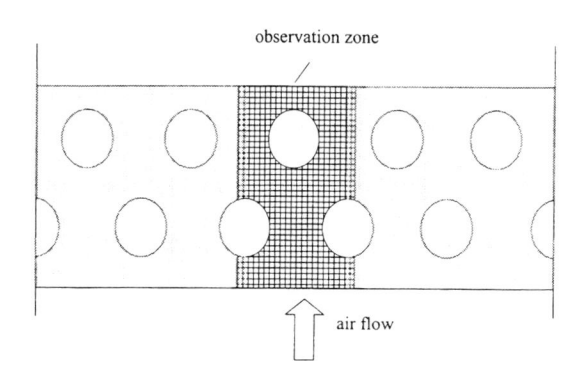

(b) Observation area

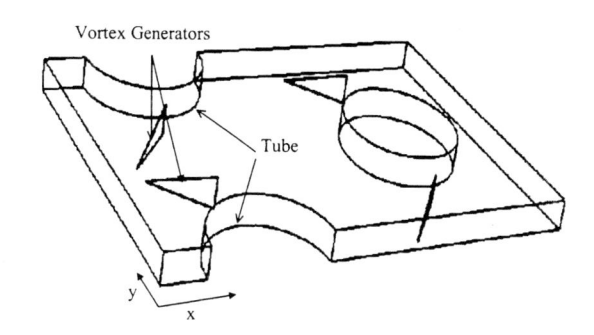

(c) View of the finned tube heat exchanger model with VG

Fig. 3: Geometry of the experimental bench

Unsteady method

This experimental method is based on the work of Ireland (Ireland and Jones, 1986). Let's consider the case of a smooth fin (fig. 4). The fin is made of polycarbonate and painted with an high emissivity coating ($\epsilon = 0.96$). Infrared emitters are used to heat this one. The heating is roughly uniform and the fin temperature reaches 45°C. After stopping the heating, an air flow is created to cool the fins. The infrared camera enables us to follow the local evolution of the surface temperatures of the plate. The heat fluxes balance in an element of the fin gives :

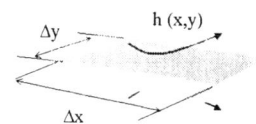

$$m_f \times \frac{\partial H_f}{\partial t} = \phi_{conv} + \phi_{cond} + \phi_{rad} \quad (1)$$

With the 3 following assumptions:

- fin thermally thin

- negligible conductive and radiative fluxes

- constant heat transfer coefficient during the experiment,

integration of the preceding equation with respect to time can be written:

$$h(x,y) = \frac{\rho_f \times Cp_f \times e_f}{2 \times \Delta t} \times ln(\frac{T_{in} - T_f(x,y,0)}{T_{in} - T_f(x,y,t)}) \quad (2)$$

An accurate measurement of the fin properties allows us to obtain a good precision of $h(x,y)$. T_{in} represents the inlet temperature. Only the knowledge of the fin surface temperature at time t is necessary to determine the heat transfer coefficient h. In our experiments, the interval Δt is constant and equal to 10s. According to the study of Ireland, this method allows the determination of the heat transfer coefficient on a constant fin surface temperature.

A good accuracy of the surface temperature is also necessary. In order to convert the thermosignal into temperature, an in-situ calibration is made using a thermal sensor. The calibration procedure uses a special high conductivity plate coated with the same black painting as the fin, and placed at the same location. This technique allows accurate measurements of the fin temperature with an overall uncertainty of about $\pm 0.3°C$.

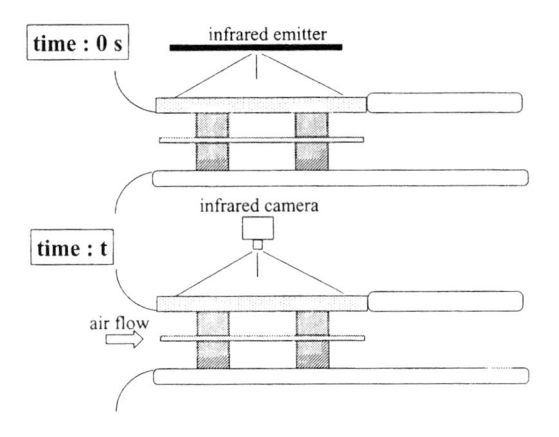

Fig. 4: Unsteady method

The various assumptions made to obtain equation (2) were checked. The temperature variation inside the thickness of the fin is negligible because the Biot number (Bi) for our tests is lower than 0.1. Because the conductivity of this one is low ($\lambda = 0.2 W/mK$), the surface averaged h differs only from 2% when we take into account the heat diffusion inside the plate. We verified that when the air velocity is equal to $0.5\ m/s$, the natural convection is not significant ($Ri << 1$).

In order to validate measurements, we compared the heat transfer on the smooth fin with the correlation given by Gray (Gray and Webb, 1986) (fig. 5). The results obtained on the surface averaged Colburn modulus j are in a good agreement. Whatever the air velocity, the heat transfer coefficient determined by this unsteady method is inside the confidence interval of $\pm 10\%$ given by Gray.

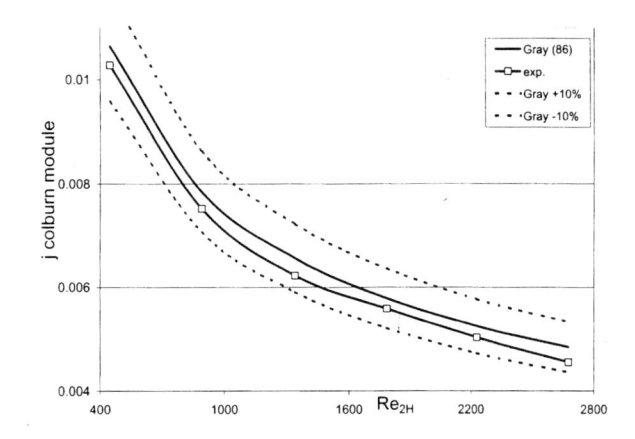

Fig. 5: Validation of the measurement on a plane fin heat exchanger

Steady method

To highlight the strong conjugate heat transfer, some modifications of the thermal bench were made. The following figure presents a cross section of the prototype.

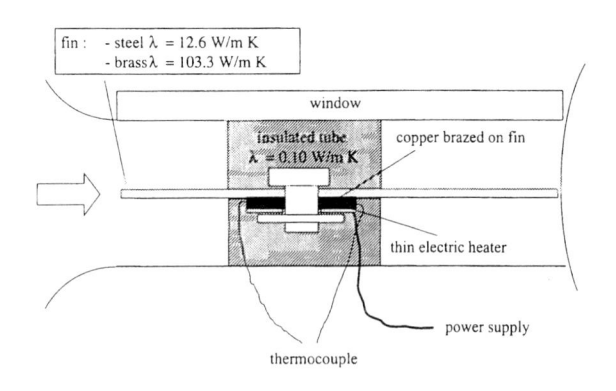

Fig. 6: Steady method

In order to discuss the influence of conduction in the fins, we kept the same boundary conditions as for the previous method. The tube is insulated and approximates an adiabatic condition. The pedestal tube temperature is kept roughly uniform using a brazed copper thin plate. A system of temperature control is adopted. As the 2 rows of tubes are completely independent, the regulation is done while keeping $(T_t - T_{in})$ constant and equal to 30°C for the 1st and 2nd rows.

Dependence of h on the material of the fin - numerical study Several previous studies (Moffat, 1998) showed that the heat transfer coefficient was dependent on the temperature field imposed on the fin. In order to understand our problem, some numerical simulations of the smooth fin prototype were carried out with the commercial computer code FLUENT. Only the nature of the fin changes: $\lambda \Rightarrow \infty$ (Tcst), $\lambda = 12.6 W/mK$ (steel), $\lambda = 103.3 W/mK$ (brass).

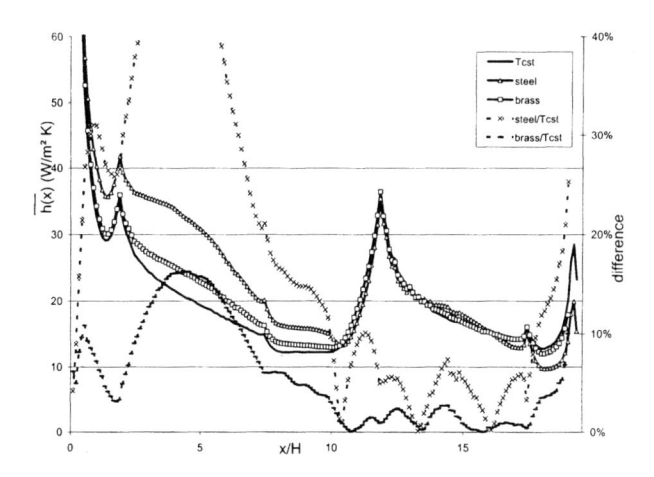

Fig. 7: Dependence of h on the material of the fin - Numerical result

Figure 7 shows the evolution of the transversal averaged heat transfer coefficient $\overline{h}(x)$ for three different material of fins. It is clear that the fin temperature field influences the convective transfer. The difference, compared to a boundary condition of constant temperature, is more significant when the thermal conductivity of the fin is low. For steel, this can reach a variation of 40% on the 1st row of tube. The difference of the average exchange is 13%. The evolution of the heat coefficient on the brass is similar to that on the fin having an efficiency equal to 1. Although there can be a variation of 15% at the beginning of the exchanger, the average transfer is only higher by 2.5%. According to these results, the determination of the surface heat flux density for our experimental work will be calculated by two different ways.

Brass We can write the assumption that $h_{brass}(x,y) = h_{Tcst}(x,y)$. Thus, the heat flux density is given in the following way:

$$\varphi_{brass}(x,y) = h_{Tcst}(x,y) \times (T_{brass}(x,y) - T_{in}) \quad (3)$$

Steel As the previous considerations are not valid for steel, another method of calculation is used. This technique is inspired by the work of Bougeard (Bougeard, 1997). By writing the heat balance in a fin element and supposing that there is no variation of temperature inside its thickness, the equation of energy becomes:

$$\varphi_{steel}(x,y) = \lambda_f \times e_f \times \Delta T_{steel}(x,y) \quad (4)$$

$\Delta T_{steel}(x,y)$ represents the laplacian of the steel surface temperature. The resolution of this equation is written in a finite difference form. The problem, in this method, is the fact that the laplacian operator acts as a high filter. The noise inherent in measurement will be enormously amplified by this calculation. A parametric study not presented in this article, shows various signal processing used to decrease the noise amplification. Indeed, in order to decrease this noise, we make several averages of temperatures in time as well as a space filtering of these temperature fields. However, this treatment reveals problems near tubes and promotors (Tourreuil, 2001).

Estimate of uncertainties

The estimate of uncertainties of the heat flux density is based on work of Moffat (Moffat, 1988). The following table presents the results of this calculation.

	Unsteady method	Steady method	
	T_{cst}	brass	steel
$\overline{\Delta \varphi / \varphi}$ (%)	7.9	10.1	14.5

The quantity $\overline{\Delta \varphi / \varphi}$ represents averaged uncertainty on the observation area. It is noted that the worst method is that used for steel. The laplacian calculation being very sensitive to the measurement noises, average uncertainty flux reaches 15%. Locally, this quantity can reach 50% behind the tubes.

RESULTS AND DISCUSSION

Local thermal transfer

Figures 8(a) and 8(b) show the measurements obtained on a constant fin temperature with and without vortex generators. In order to obtain a heat flux density, the heat transfer coefficient at constant fin temperature is multiplied by the difference temperature imposed in the case of brass and steel fin ($\varphi_{Tcst}(x,y) = h_{Tcst}(x,y) \times (T_t - T_{in})$). For the smooth fin, we find a quite well-known distribution in the exchangers. The heat transfer is significant at the entry of the field, area where the fluid is in a dynamic and thermal development (weak boundary layer). The exchange is also very significant close to the tubes. There are two explanations for this evolution. The flow rate in the exchanger being constant and the cross section between the 2 tubes of the 1st row decreasing, there is an acceleration of the flow increasing heat transfer.

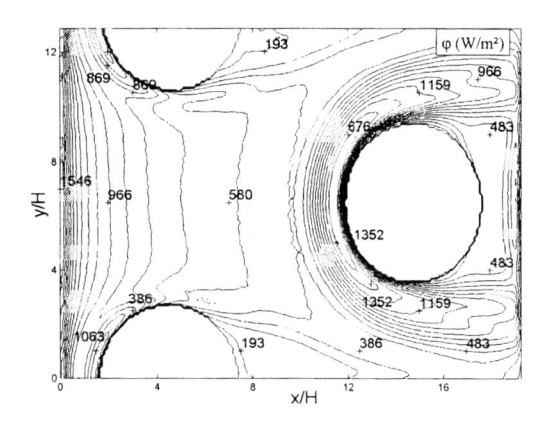

(a) Plane fin with constant temperature

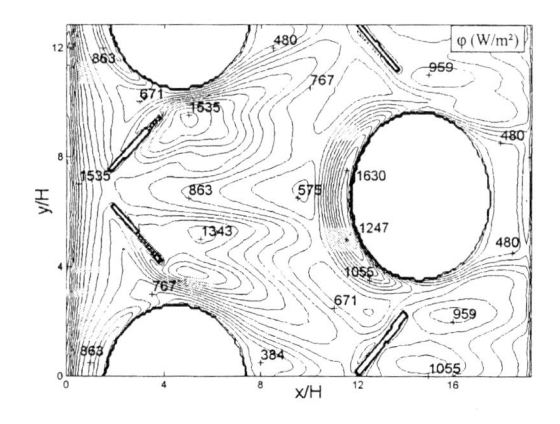

(b) Fin with VG and constant temperature

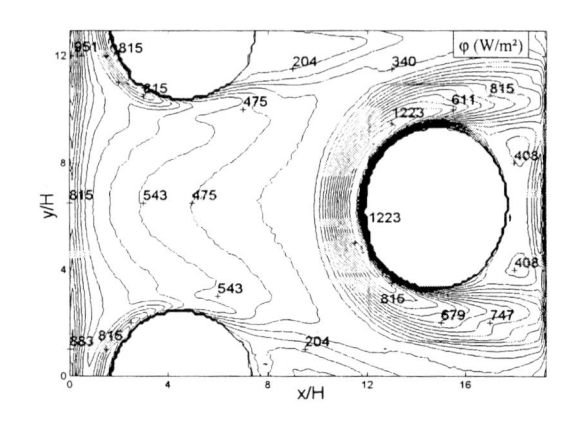

(c) Plane brass fin

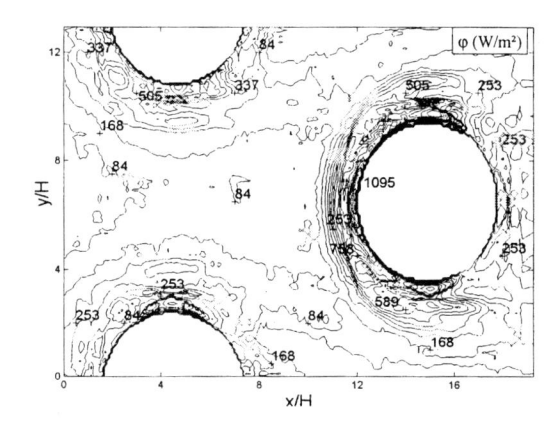

(d) Plane steel fin

Fig. 8: Local distribution of the flux density $\varphi(x,y)$ with different conductivities of fins

In addition, there is a stagnation point in front of the tube. This overpressure will involve the formation of longitudinal swirls (known as " horseshoe vortex") which also will increase the convective exchange. These explanations are completely valid for the 2nd row of tubes where the thermal transfer is significant on a very wide area. Behind the 2 rows of tubes, the fluid recirculates with very low speed producing weak heat transfer ($\varphi = 193 W/m^2$). Some vortex generators are placed close to the 1er and the 2nd rows of tubes (fig. 8(b)). The thermal flux density is strongly increased behind the promotors. The convective transfer is increased on almost the whole fin surface. Behind the 1er row of VG, the exchange is as significant as for the beginning of the fin (increase of 100% of the transfer). The wake zone is also strongly decreased. The VG of the 2nd row are effective but they also decrease slightly the strong exchange caused by the " horseshoe vortex". Figures 8(d) and 8(c) represent the local flux density for the smooth steel and brass fin. The thermal transfers are much weaker than in the case of the constant wall temperature. The more we move away from the tube i.e. of the heat source, the weaker the flux density is. This is particularly true for the steel fin where the thermal transfer on the entry of the field reaches only $84 W/m^2$. In this case, only the 2nd row of tubes is equivalent to a strong transfer. The longitudinal evolution of the span averaged heat flux density are shown in figure 9.

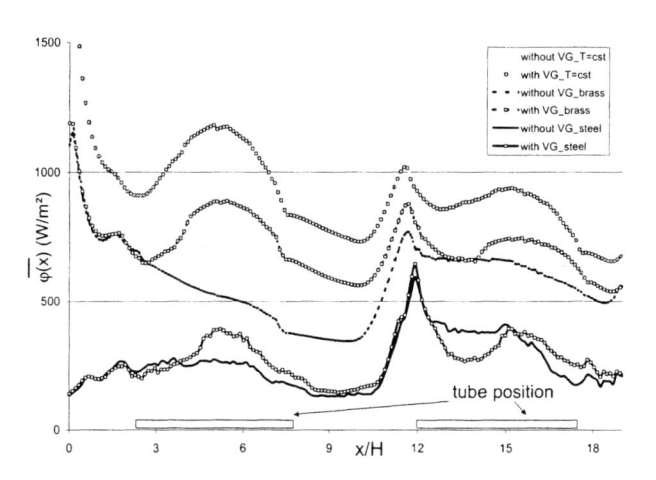

Fig. 9: Span averaged flux density $\overline{\varphi}(x)$ with and without VG for different conductivities

The augmentation of the heat transfer on the T_{cst} fin due to the addition of the VG spread out over all the investigation zone. As we saw locally, the increase transfer is much more significant for the 1st row of tubes than for the 2nd. The influence of the VG is less significant for brass. Moreover, just behind the peak of transfer of the 2nd row of tubes, the flux density with or without promotors do not change. It was shown locally that the VG broke the " horseshoe vortex". This phenomenon is more visible when the fin is in steel material. We find a reduction of the thermal transfer. When using steel fin, there is only the 2nd row of tubes which is the seat of a strong exchange. Because of the weak heat diffusion inside the fin, the VG loses almost the totality of their effectiveness. The flux density decreases dramatically at the entry. The reason is because the fin temperature is very close to the air temperature.

Total thermal transfer

The increase of the thermal dissipation of a fin with VG compared to a smooth fin $\overline{\varphi}_{VG}/\overline{\varphi}_0$ for 3 different materials tested (fig. 10) supports the preceding remarks. It is noted that the higher the conductivity of the fin is, the more the thermal performances of the disturbers decrease. These results are different than the works of Sanchez (Introduction) but no conclusion or critical remarks can be written because the geometries are too different. It is pointed out that the author studied only one row of tubes and the VG are placed differently from our experiment. In our case, the heat transfer augmentation varies between approximately 35% for T_{cst} fin, 23.5% for brass and 7% for steel. In the range of investigated Reynolds numbers, the performance of VG are not a function of air velocity.

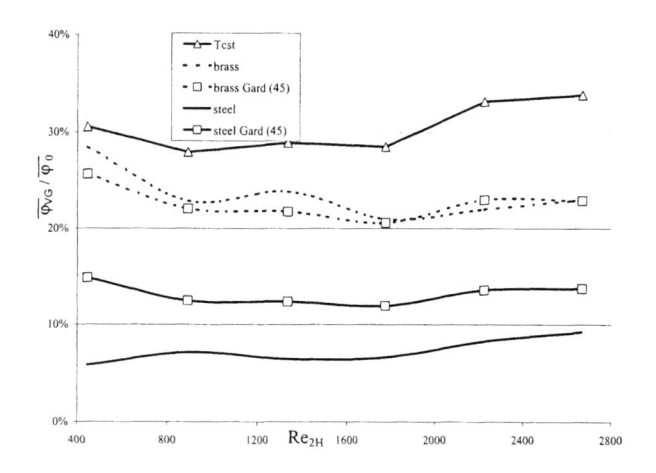

Fig. 10: Performance of Vortex Generators

The 2 last curves (brass and steel Gard (45)) represent the result that we would have if we had not taken into account the conjugate heat transfer. Their determination give the following expression:

$$\frac{\varphi_{VG}}{\varphi_0} = \frac{h_{TcstVG} \times \eta_{VG}}{h_{Tcst0} \times \eta_0} \quad (5)$$

By writing this equation, we assume that the heat transfer coefficient h_{Tcst} is spatially constant. The various efficiencies η are calculated according to the theory of Gardner (Gardner, 1945). To obtain this quantity, the author also makes the assumption that the coefficient h_{Tcst} is independent of the nature of the fin. The concept of the fin efficiency shows that the more the convective transfer is significant, the weaker the efficiency is. Thus the η_{VG}/η_0 ratio is lower than 1 ; this is checked in figure 10. The increase of the thermal transfer by the vortex generators for the brass fin is well predicted by the method of Gardner. It is not the case of the steel fin. The performance of VG are overestimated ($\varphi_{VG}/\varphi_0 = 13\%$) compared to our experimental results ($\varphi_{VG}/\varphi_0 = 7\%$). We can think that the assumption stipulating that the convective heat transfer coefficient is independent of the nature of the fin is too far from reality and our problem.

CONCLUSIONS

The aim of this work was to highlight the influence of conjugate heat transfer in compact exchangers with vortex generators. Several methods of measurement were developed. The first, unsteady, allowed to measure precisely the convective transfer on a constant temperature fin with and without vortex generators. These geometries showed their capacity to increase the exchange even in a range of weak Reynolds numbers($400 < Re_{2H} < 2600$). We can obtain up to 35% of heat transfer augmentation. A second method of calculation (stationary) was undertaken in order to highlight the conjugate heat transfer. The fin conductivity has a harmful influence on the VG performance. Finally, the results of this work showed that the theory of the fin efficiency (which enables to separate the convective and conductive transfer) is valid in the majority of the industrial applications ($\eta > 0.75$). However, when this efficiency is very low ($\eta \approx 0.4$), this theory differs largely from reality.

Acknowledgement

Financial support for this reseach work has been granted by the European community (F.E.D.E.R.).

NOMENCLATURE

Bi	Biot number (=$he_f/2\lambda_f$)	$[-]$
Cp_f	Fin specific heat	$[J/kg\,K]$
e_f	Fin thickness	$[m]$
H	Channel height	$[m]$
H_f	Fin enthalpy	$[J/kg]$
h	Heat tranfer coefficient	$[W/m^2K]$
j	Colburn module $(=(\overline{h}Pr_{air}^{2/3})/(UCp_{air}\rho_{air}))$	$[-]$
m_f	Fin weight	$[kg]$
Pr	Prandtl number	$[-]$
Re_{2H}	Reynolds number (=$U2H/\nu$)	$[-]$
Ri	Richardson number	$[-]$
t	Time	$[s]$
T_f	Fin temperature	$[K]$
T_{in}	Inlet temperature	$[K]$
T_t	Pedestal tube temperature	$[K]$
U	Inlet velocity in the main flow direction	$[m/s]$
η	Fin efficiency	$[-]$
λ_f	Fin thermal conductivity	$[W/mK]$
ν	Air kinematic viscosity	$[m^2/s]$
ρ_f	Fin density	$[kg/m^3]$
φ	Heat flux per unit area	$[W/m^2]$
ϕ	Heat flux	$[W]$

REFERENCES

Bougeard, D., 1997, Etude expérimentale du transfert thermique couplé au sein de géométries complexes, *Ph.D. thesis*, Université de Valenciennes, Ecole des Mines de Douai

Gardner, K., 1945, Efficiency of extended surfaces, *Transactions of the ASME*, 67, 621–631

Gray, D. and Webb, R., 1986, Heat transfer and friction correlations for plate finned - tube heat exchangers having plain fins, *8th Int. Heat Transfer Conference*, pp 2745–2750

Ireland, P. and Jones, T., 1986, Detailed measurements of heat transfer around a pedestal in fully developped channel flow, *Proc. 8th IHTC*, 3, 975–980

Moffat, R., 1988, Describing the uncertainties in experimental results, *Experimental Thermal and Fluid Science*, 1, 3–17

Moffat, R., 1998, What's new in convective heat transfer?, *Int. Journal of Heat and Fluid Flow*, 19, 90–101

Sanchez, M. and Mitra, N., 1991, Improvement of fin-tube heat exchangers by longitudinal vortex generators, *Proc. 18th Eurotherm Seminar*, pp 144–153

Tiggelbeck, S. and Fiebig, M., 1992, Flow structure and heat transfer in a channel with multiple longitunal vortex generators, *Experimental Thermal and Fluid Science*, 5, 425–436

Tourreuil, J., 2001, Intensification du transfert thermique couplé par une méthode passive, *Ph.D. thesis*, Université de Valenciennes, Ecole des Mines de Douai (in progress)

Valencia, A. and Mitra, N., 1994, Local heat transfer and flow losses in fin-and-tube heat exchangers with vortex generators : a comparison of round and flat tubes, *Experimental Thermal and Fluid Science*, 8, 35–45

INVESTIGATION OF A HEAT TRANSFER PROCESS IN CHANNELS WITH CYLINDRICAL INTENSIFIERS

Galitseysky B.M.

Moscow Aviation Institute (Technical University),
Volokolamskoe shosse, 4, Moscow, Russia; E-mail: heat204@mai.ru

ABSTRACT

The effectiveness of the cooling system can be raised by various methods of intensification of the heat transfer process. To one of such methods belong cylindrical intensifiers, which are located in the channel across the flow and gas flow oscillation. Until now the investigations of the heat transfer process in the channels with cylindrical intensifiers were reduced basically to determination of the middle coefficients on the channel surface of the heat transfer. The results of the investigation of the local coefficients of heat transfer in flat channels behind a grid and in a grid of the cylindrical intensifiers are presented. It is also determined how geometric parameters influence on the intensity of the local heat transfer. The results of investigation of the influence gas flow oscillation on the intensity of heat transfer process is given. The local heat transfer coefficient in the flat channels is non-uniformly distributed over the channel surface and its value depends on a longitudinal pitch of intensifiers and a distance from the channel inlet. Experimental results are generalized by criterial relations. It is established that heat transfer can be also augmented by gas flow oscillation on flat channels equipped with cylindrical intensifiers, but the level of heat transfer enhancement depends on the geometry of a grid of intensifiers.

INVESTIGATION

Improving the cooling systems of units and aggregates is one of the urgent problems in modern power plants. The efficiency of cooling systems can be improved through use of different methods for heat transfer augmentation. Among these methods are cylindrical intensifiers located across a channel flow. Up to now, the study of a heat transfer process in channels with cylindrical intensifiers has amounted to determining surface-mean heat transfer coefficient when geometrical parameters of intensifiers are constant (Kays and London, 1984; Kalinin et al, 1977). Heat transfer in flat channels may be also enhanced by producing forced oscillations of a heat carrier flow in a channel. Studies of heat transfer in cylindrical channels show that a 1.5÷5–fold increase can be seen under these conditions (Galitseyskiy et al, 1977, 1978). It is found that the phenomenon of "turbulence resonance" involving maximum heat transfer enhancement realization is seen in cylindrical channels with oscillatory gas flow. When studying heat transfer enhancement in flat channels due to forced gas flow oscillations, it is necessary to find whether "turbulence resonance" can be observed in such channels and how gas flow oscillations affect heat transfer in corner zones. The results for heat transfer enhancement by cylindrical intensifiers and forced gas flow oscillations are presented in this work.

HEAT TRANSFER ENHANCEMENT BY CYLINDRICAL INTENSIFIERS

The present work considers coefficients for heat transfer in a flat channel behind a grid and in a grid of cylindrical intensifiers and the influence of its geometrical parameters on a heat transfer process behind it as well as heat transfer enhancement under these conditions.
According to the similarity theory for a channel with cylindrical intensifiers, the heat trans-

fer process is described by the following criterial equation:

$$Nu = F(Re, Pr, m, \bar{x}, \bar{y}, \bar{s}_x, \bar{s}_y, \bar{d}) \quad (1)$$

Here Re is the Reynolds number determined from the equivalent diameter d_e; m is the number of rows of intensifiers; $\bar{x}=x/d_i$ and $\bar{y}=y/d_i$ are the relative longitudinal and transverse coordinates; $\bar{s}_x=s_x/d_i$ and $\bar{s}_y=s_y/d_i$ are the relative longitudinal and transverse pitches of intensifiers in a grid; $\bar{d}=d_i/d_e$ is the relative diameter of intensifiers, d_i is the diameter of intensifiers.

In practice, the advisability of using intensifiers to increase coefficients of heat transfer in a channel is determined from the condition of obtaining a maximum value of the heat transfer enhancement coefficient $K=\alpha/\alpha_0$ (α, α_0 are the heat transfer coefficients in a channel with intensifiers and in a smooth channel, respectively). Depending on the conditions of a particular problem, this coefficient is estimated at the same values of Reynolds numbers or pressure losses in compared channels.

For the efficiency of intensifiers to be analyzed, the criterial equation for a heat transfer process in a channel with intensifiers is represented in the form (Galitseyskiy et al, 1981):

$$Nu = A \cdot Re^n \quad (2)$$

where A and n depend on the geometrical parameters of a grid in a channel.

Let us express the Reynolds number for a channel with intensifiers in terms of the Reynolds number for a smooth channel

$$Re = Re_0 \frac{F_0}{F} \cdot \frac{G}{G_0} \cdot \frac{d_e}{d_0}. \quad (3)$$

Taking into account (2) and (3), the equation for heat transfer enhancement coefficients can be written as:

$$K = \frac{A}{A_0} Re_0^{n-n_0} \left(\frac{d_0}{d_e}\right)^{1-n} \left(\frac{F_0}{F}\right)^n \left(\frac{G}{G_0}\right)^n, \quad (4)$$

where the quantities with the subscript "0" refer to a smooth channel.

It is of interest to compare the intensities of a heat transfer process under the same values of pressure losses in a smooth channel and in a channel with intensifiers. As a heat carrier flowrate through a channel depends on a pressure drop

$$G = \frac{F}{\sqrt{\xi}} \sqrt{2\rho\Delta P}, \quad (5)$$

the equation

$$K_{\Delta P} = \frac{A}{A_0} Re_0^{n-n_0} \left(\frac{d_0}{d_e}\right)^{1-n} \left(\sqrt{\frac{\xi_0}{\xi}}\right)^n \quad (6)$$

is obtained for the enhancement coefficient under the same pressure losses in the compared channels.

Thus, the problem on heat transfer in a channel with intensifiers reduces to the experimental determination of criterial relations (1), (6). The heat transfer process behind a grid of cylindrical intensifiers in a rectangular channel was studied on the experimental section, whose schematic is shown in Fig.1. The channel was divided by means of partition 2 into 25×500 mm cavities. Cold air entered top cavity 1. Partition 2 on the side of cold air was provided with staggered threading pockets, into which cylindrical intensifiers 3 or plugs were screwed. The geometrical parameters of the grid of intensifiers ranged as follows: the number of rows of intensifiers m=1-5, the intensifier diameter d_i=25-100 mm; relative longitudinal and transverse pitches $\bar{s}_x = \bar{s}_y$=2-6. A hot gas with a temperature of 800 K that heated the partition entered bottom cavity 4.

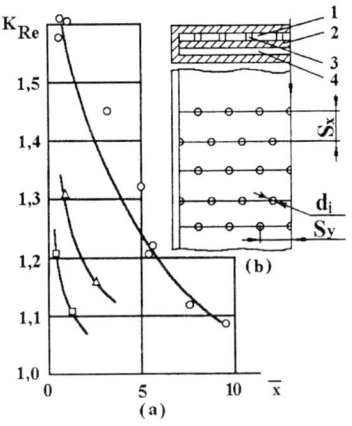

Fig. 1 Variation of the heat transfer enhancement coefficient with a relative distance behind a grid (a) and schematic of the experimental model (b): a) o - m=1; m=2; $\bar{d}$=0,525; Δ - m=2; $\bar{d}$=1,05; □ - m=2; $\bar{d}$=2,1, Re=4·10⁵, $\bar{s}_x = \bar{s}_y$=2; b) 1 - channel of cold air, 2 – partition, 3 – intensifiers, 4 - channel of hot air

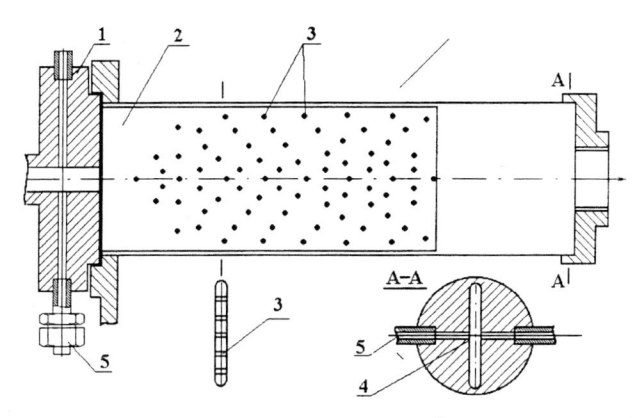

Fig. 2 Schematic of an experimental section
1 – block of aerial thermocouples and pressure pick-up, 2 – experimental channel, 3 – grid of intensifiers, 4 – aerial thermocouple, 5 – pressure pick-up

A wall surface temperature on the side of cold and hot gases was measured by chromel-alumel thermocouples in four cross-sections along the wall length and at five points across the wall width. An air temperature was measured by chromel-alumel thermocouples (the latter were made coincident with total-pressure nozzles meant for measuring a pressure drop on the grid of intensifiers).

On the second experimental section (Fig.2), the heat transfer coefficient was examined on the channel wall in a grid of intensifiers placed in rows at a distance of x/L from the channel inlet equal to 0.07; 0.266; 0.468; 0.67; 0.88. The experimental section was a slot channel with rounded off corners 200 mm long, 29.5 mm wide, and 2.5 mm high. Air entered the channel through a round 6 mm dia hole. 2 mm dia intensifiers with a relative longitudinal pitch of 5; 7.5; and 10 were press-fitted in staggered order over a thin plate. Fig. 2 shows the schematic for locating intensifiers with a pitch of 10 mm over a plate. The staggered location of intensifiers with a longitudinal pitch: 10, 15, and 20 mm ($\bar{s}_x$=5; 7.5; 10) was examined. A height of the intensifiers was equal to that of a channel. For velocity and pressure to be equalized, overflow holes were drilled on the both sides of the plate.

The experimental section was heated by alternating current supplied to it. A channel wall temperature was measured by chromel-alumel thermocouples in five cross-sections along the channel and at five points along the channel perimeter. An air temperature at the experimental section inlet and outlet was measured by aerial thermocouples.

The heat transfer process on the both experimental sections was studied over the following range of the basic parameters: Reynolds number Re=$1.2 \cdot 10^4 - 8.3 \cdot 10^5$; channel air pressure $P_0 = (1.12-12) \cdot 10^5$ N/m²; air temperature T_f=350–500 K; channel wall temperature T_w=400–800 K.

The heat transfer coefficient is calculated by the formula

$$\alpha = \frac{q_w}{T_w - T_f} \quad (7)$$

where q_w is the heat flux density; T_w is the wall temperature on the side of cold air measured by thermocouples; T_f is the mean calorimetric air temperature in a cross-section (by the equation of heat balance).

In studies of heat transfer behind a grid of intensifiers, a heat flux density was determined as the ratio of the heat flux calculated by the heat conduction equation for a flat wall to the surface area of a smooth channel. In this case, the equivalent diameter of a smooth rectangular channel was taken as a characteristic dimension.

In the slot channel the heat flux density was determined as the heat flux-to-total heat transfer surface ratio. The equivalent diameter of a cell containing two rows of intensifiers and a section of a smooth channel between them was taken as a characteristic dimension.

As a result of the experiments conducted, it is found that the coefficient for heat transfer in a channel behind a grid of cylindrical intensifiers increases 1.7 time as a maximum and its value is affected by the geometrical parameters of the grid.

It is revealed that the placement of more than two rows of intensifiers in front of a measuring cross-section practically does not enhance heat transfer. The heat transfer coefficient behind the grid decreases with increasing distance from the last row of intensifiers under all other conditions being equal (Fig. 1), and when $\bar{x}>10$ the influence of a grid on the heat transfer coefficient can be neglected.

Increasing a diameter of intensifiers decreases the heat transfer coefficient behind a grid (Fig.1). Varying a longitudinal pitch of intensifiers over the entire considered range does not exert any influence on heat transfer coefficient behind a grid. Decreasing a transverse pitch of intensifiers augments flow turbulence and heat transfer coefficient behind a grid.

The coefficient for heat transfer in a rectangular channel behind a grid of intensifiers is generalized by the criterial equation of form (2) with an error of ±20% (Fig.3) where

$$A = 0{,}0221 + 13{,}6\exp(-0{,}617\overline{x} - 1{,}15\overline{d} - 0{,}295\overline{s}_y - 0{,}068\overline{s}_x - 0{,}379m),$$

$$n = 0{,}765 - 0{,}284\exp(0{,}987 - 0{,}242\overline{x} - 0{,}357\overline{d} - 0{,}128\overline{s}_y + 0{,}018\overline{s}_x + 0{,}144m) \qquad (8)$$

Studies of the heat transfer enhancement coefficient under the same pressure losses in a smooth channel and in a channel with cylindrical intensifiers have shown that at the Reynolds numbers (Re< 10^5) the enhancement coefficient is more than unity ($k_{\Delta P} > 1$) for all geometrical parameters of a grid (Fig.4).

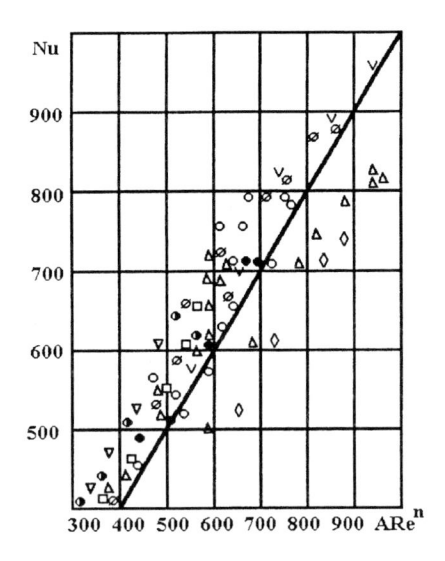

Fig. 3. Generalization of the experimental data on heat transfer in a channel behind a grid of intensifiers at $d_i = 25$ mm, $\overline{s}_x = \overline{s}_y = 2$, $\overline{x} = 1$: solid line - dependence calculated by equation (8); O - m=1, Δ - m=2, □ - m=3, ∇ - m=4, O - m=5; for m=2, $d_i = 25$ mm, $\overline{x} = 1$; 5.6; ⌀ - $\overline{s}_x = 6$, $\overline{s}_y = 2$; V - $\overline{s}_x = 2$, $\overline{s}_y = 6$; for m=2, $\overline{s}_x = \overline{s}_y = 2$, $\overline{x} = 0.5$; 2,8; ◊ - $d_i = 50$ mm, ● - $d_i = 100$ mm

The heat transfer intensity in a slot channel is nonuniform along its perimeter. That is why, the influence of intensifiers on heat transfer coefficient was considered individually for a central channel zone (flat wall) and for a corner zone (rounded off part of the channel).

Placing the cylindrical intensifiers in a slot channel causes local heat transfer coefficients to increase in both zones: in the central zone it is, as a maximum, a 3 - fold increase and in the corner zone, a 4 - fold increase, as compared to the heat transfer coefficient in a smooth channel.

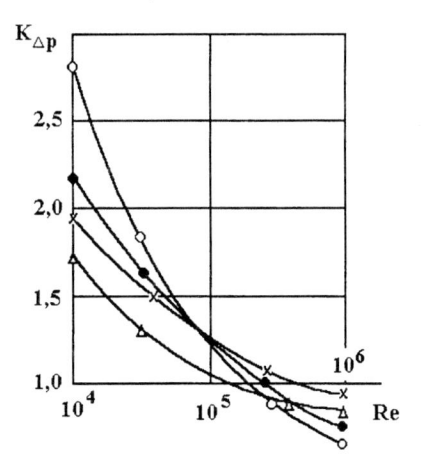

Fig. 4. Heat transfer enhancement coefficient vs. the Reynolds number m=2, $\overline{x}=1$, $\overline{d}=0.525$, $\overline{s}_x=2$:
O - $\overline{s}_y=2$; ● - $\overline{s}_y=4$; × - $\overline{s}_y=6$; Δ - $\overline{s}_y=2$; $\overline{s}_y=6$

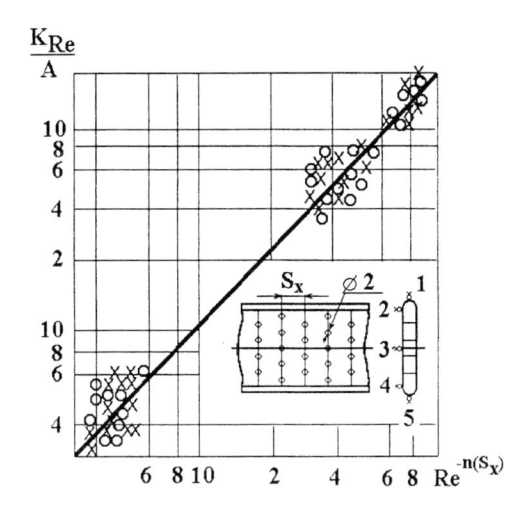

Fig. 5. Generalization of the experimental data on heat transfer in a channel in a grid of intensifiers: × - central zone, O - corner zone, 1, 2, 3, 4, 5 - thermocouples

Outward from the channel inlet, heat transfer enhancement in the central zone of the slot channel with intensifiers decreases and does not vary in the corner zone, except for the vortex zone at the channel inlet when x/l<0.26. In both channel zones, increasing the longitudinal pitch of intensifiers decreases local coefficients of heat exchange between intensifiers and behind them under all other conditions, being equal.

The results on the enhancement coefficient for a slot channel plotted in Fig.5 are generalized by the following criterial relation:

$$K_{Re} = \frac{\alpha}{\alpha_0} = A\left(\bar{s}_x, \frac{x}{\ell}\right) Re^{n(\bar{s}_x)}, \qquad (9)$$

where for the central zone

$$A = 33{,}73 \left(\frac{x}{\ell}\right)^{-0{,}25} \frac{1}{\bar{s}_x^2 - 9{,}14\bar{s}_x + 21{,}2}$$
$$n = -0{,}928 \exp(-0{,}0393\,\bar{s}_x^2), \qquad (10)$$

and for the corner zone

$$A = 102 \frac{1}{\bar{s}_x^2 - 6{,}1\bar{s}_x - 6{,}95},$$
$$n = -0{,}847 \exp(-0{,}0393\,\bar{s}_x^2), \qquad (11)$$

Thus, it is established that the intensity of a transfer process in a channel essentially depends on the geometrical parameters of a grid of intensifiers. It is shown that the intensity of the heat transfer process can be enhanced 3÷4.5 times through the use of intensifiers.

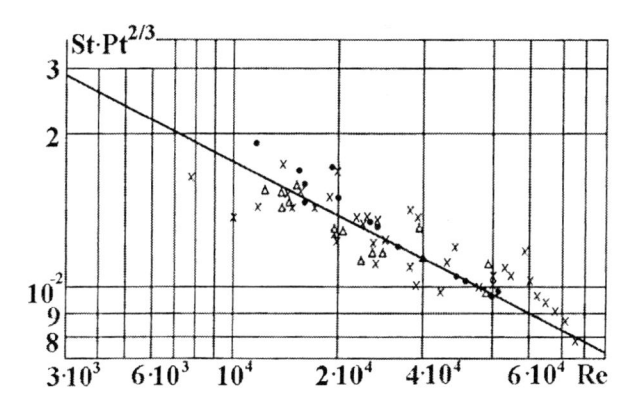

Fig. 6. The surface-mean heat transfer coefficient in a channel in a grid of intensifiers:
solid line – dependence Kays and London (1984); for $\bar{s}_x = 1{,}93$; • - $\bar{s}_x = 5$; × - $\bar{s}_x = 7{,}5$; ∆ - $\bar{s}_x = 10$ – data of the present work

The surface-mean heat transfer coefficient characterizes all processes occurring over the wall surface of an intensifier as well as heat removal from a wall due to heat conduction of intensifiers. The data for the stabilized flow region (Kays and London, 1984) well coincide with the results of the present work and are approximated by the relation (Fig.6):

$$St \cdot Pr^{2/3} = 2{,}1\, Re_0^{-0{,}42}\, \bar{s}_x^{0{,}61} \qquad (12)$$

Thus, when there develops a steady-state turbulent gas flow in a channel with intensifiers, the local heat transfer coefficient is non-uniformly distributed over the channel surface, and its value depends on a longitudinal pitch of intensifiers and a distance from the channel inlet. As for the stabilized flow region, the results on the surface-mean heat transfer coefficient satisfactorily agree with the data of other authors.

HEAT TRANSFER ENHANCEMENT DUE TO FORCED OSCILLATIONS OF A GAS FLOW

Based on the method of the similarity theory, the criterial relation for a local coefficient of heat transfer involving oscillatory gas flow in a flat channel can be given as

$$K = \frac{Nu}{Nu_0} = f\left(Re_0; Re_w; \varepsilon; \bar{x}; \bar{y}; \bar{s}_x; \bar{s}_y\right), \qquad (13)$$

where K is the relative heat transfer coefficient; Nu and Nu_0 are the Nusselt numbers for gas flow oscillation and for steady-state gas flow, respectively; $Re_0 = \frac{u_0 d_e}{\nu}$ and $Re_0 = \frac{\omega d_e^2}{\nu}$ are the time-averaged and oscillatory Reynolds numbers, respectively; $\varepsilon = \frac{\Delta(\rho u)_0}{(\rho u)_0}$ is the relative oscillation amplitude of a mass velocity; $\bar{x} = x/\ell$ is the relative longitudinal coordinate; $\bar{y} = y/z$ is the relative coordinate along the channel perimeter; z is the channel perimeter; ℓ is the channel length.

Proceeding from the results for cylindrical channel, the influence of the Reynolds number and the oscillatory Reynolds number is evaluated through the parameter representing the oscillating layer-to-viscous layer thickness ratio (Galitseyskiy et al, 1977, 1994):

$$\delta_{os}/\delta_0 = 0{,}0282 \sqrt{\frac{Re_0^{1{,}75}}{Re_\omega}} \qquad (14)$$

where $\delta_{os} = \sqrt{2\nu/\omega}$, $\delta_0 \cong 10\,\nu\sqrt{\frac{\tau_w}{\rho}}$.

In this case, criterial relation (13) is transformed to the following form:

$$K = f\left(\frac{\delta_0}{\delta_{0S}}; \varepsilon; \overline{x}; \overline{y}; \overline{s}_x; \overline{s}_y\right) \quad (15)$$

and the problem on heat transfer involving flow oscillation reduces to estimating the influence of the parameter δ_{os}/δ_o, the relative amplitude of a mass velocity ε on the heat transfer coefficient and to determining the conditions, under which the maximum intensity of the heat transfer process is being realized. In the case of oscillatory gas flow in a channel, the pressure oscillation amplitude is one of the quantities responsible for a level of the heat transfer intensity. Channel length variations of a pressure oscillation amplitude are determined using the one-dimensional linear model for high-frequency oscillations. This model assumes that a flow velocity, pressure, and temperature are equal to their values mean over the channel cross-section. The heat transfer influence on the process of wave propagation in a channel is specified by variations in a sound velocity along the channel. In studies of compression wave propagation it is important to know the attenuation factor of a pressure oscillation amplitude that is responsible for variations in a pressure (velocity) oscillation amplitude along the channel.

The heat transfer coefficient is determined by the quasi-stationary methods from the equation (7). The heat transfer process was studied using the method for direct heating of an experimental section by passing alternating current through it. The problem stated was investigated on the experimental section, whose schematic is sketched in Fig. 2.

The experimental section was represented as a flat channel with rounded off corners (b:h=11.3:2) made of 0.3 mm thick stainless steel 1X18H9T. Air was admitted into the channel via a round hole. An outer surface temperature of the experimental section was measured by chromel-alumel thermocouples. An air temperature at the channel inlet and outlet was measured by aerial chromel-alumel thermocouples. To produce forced flow oscillations, the channel inlet was provided with a mechanical pulsator. As the rotor rotated, it periodically cut off gas supply to the channel. A pressure oscillation amplitude was measured by induction pressure picks-ups mounted at the channel inlet and outlet.

The experimental set-up permitted the examination of the main parameters over the following range: Reynolds number $Re_o = 10^4 - 8 \cdot 10^4$; gas temperature at the channel outlet $T_f = 350-550$ K; wall temperature $T_w = 400-800$ K; channel gas pressure $P_o = (2-10) \cdot 10^5$ N/m^2; pressure oscillation amplitude $\Delta P_o = (0.2-2) \cdot 10^5$ N/m^2; oscillation frequency f= 30-900 Hz.

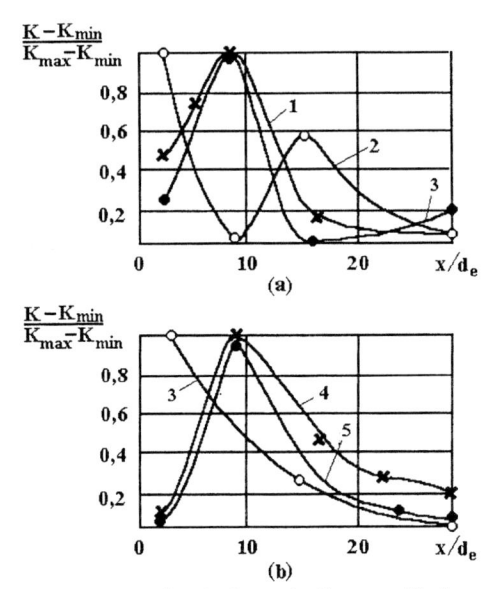

Fig.7 Influence of turbulent air flow oscillations on the heat transfer intensity distribution along a rectangular channel: a, air supply to the inlet according to Variant 3; b, air supply to the inlet according to Variants 1 and 2; 1, central zone; 2,3, corner zones; 4, central zone for air supply to the inlet according to Variant 1; 5, central zone for air supply to the inlet according to Variant 2.

Resonance heat transfer enhancement due to forced oscillations is also seen in complex-geometry channels, e.q., in flat channels with rounded off edges under different conditions at the channel inlet, in channel with intensifiers or fins, including porous channels. However, heat transfer enhancement in such channels is essentially nonuniform along perimeter and depends on inlet boundary conditions. Fig. 7 a, b plots the results on the influence of high-frequency oscillations of a turbulent air flow on heat transfer in a rectangular channel (2x11, 3x200 mm) with rounded off edges at Re=10^4-10^5, f= 800 Hz, $\Delta P_o = (0.2-2) \cdot 10^5$ Pa for three variants of inlet boundary conditions:

- air was supplied to the channel via a round hole, i.e., the shape and size of the inlet section varied sharply (Variant 1);
- boundary condition at the channel inlet was the same as in Variant 1 but air was sucked off at the inlet via a single hole (Variant 2);
- nonsymmetric velocity distribution over the inlet section was realized because of the de-

flection of the inlet device axis from the channel axis (Variant 3).

When acted upon by oscillations, heat transfer is to a greater extent enhanced in the corner zones than in the central one. The channel length distribution of the heat transfer coefficient is essentially affected by the inlet boundary conditions (Fig. 7). Maximum heat transfer enhancement K_{max} is observed at air supply to the channel inlet ($x/d_e \approx 10$). When acted upon by oscillations, the heat transfer intensity along the channel perimeter is equalized. In the case of nonsymmetric flow velocity distribution at the channel inlet, which is attained by directing an air jet from the inlet hole at angles to one of the channel corner zones under gas flow oscillations, the greatest increase of heat transfer is seen in the zones with the smallest value of a local flow velocity.

rameter $(\delta_{os}/\delta_o)_m$, at which maximum heat transfer enhancement due to oscillations is realized, is different for the central and corner zones of the channel (for the central zone of the channel $(\delta_{os}/\delta_o) \approx 1.1$) and for the corner zones $(\delta_{os}/\delta_o)_m \approx 1$).

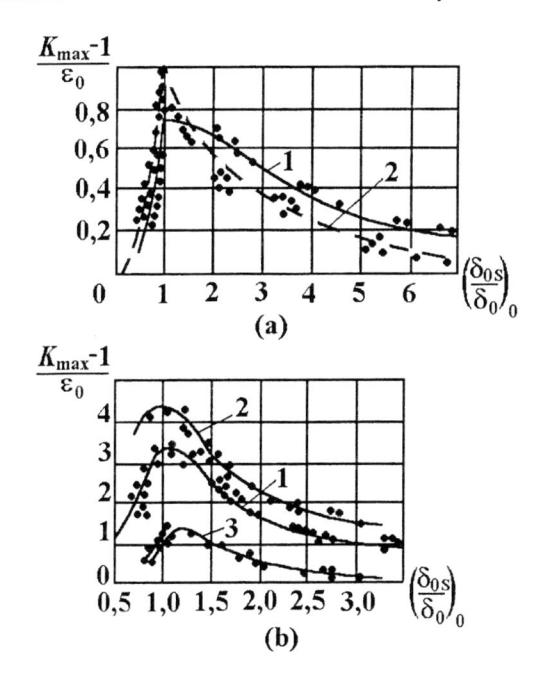

Fig.8 Heat transfer coefficient maximum along the rectangular channel vs. the parameter δ_{os}/δ_o with air supply to the inlet according to Variants 1 and 2 (a) and according to Variant 3 (b):1, central zone; 2 and 3, corner zones.

Fig.8 a, b plots the dimensionless complex $(K_{max}-1)/\varepsilon_o$ (where ε_o is the relative effective oscillation amplitude of a mass velocity at the channel inlet vs. the parameter δ_{os}/δ_o determined from the equivalent diameter of the channel. As follows from the relations plotted, the phenomenon of thermal resonance is seen both in the rectangular channel for the considered boundary conditions at the channel inlet and in the cylindrical channel. An optimum value of the pa-

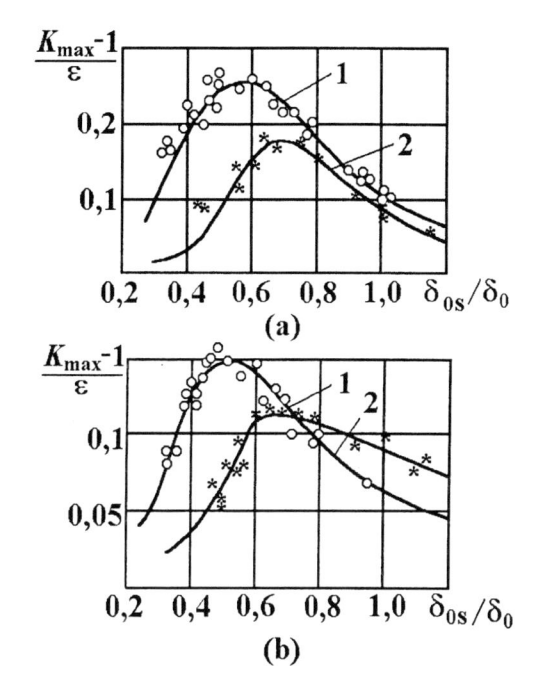

Fig.9 Relative heat transfer coefficient maximum along the channel vs. the parameter δ_{os}/δ_o under gas flow oscillations in a rectangular channel (h=9 mm) with cylindrical intensifiers: 1, $\bar{s}_x = \bar{s}_y = 5$; 2, $\bar{s}_x = 5$, $\bar{s}_y = 2.5$; central zone; b, corner zone.

Heat transfer can be also augmented by gas flow oscillation in rectangular channels equipped with cylindrical intensifiers. However, in this case, the level of heat transfer enhancement depends on the geometry of a grid of intensifiers. Maximum heat transfer enhancement due to oscillations is observed near the inlet section of the channel ($x/l \approx 0.2$). As the distance from the channel inlet increases, the level of heat transfer enhancement decreases because of oscillation damping. Fig. 9 a, b plots the maximum relative heat transfer coefficient ($K=\alpha/\alpha_o$ where α_o is the coefficient for heat transfer in a steady-state air flow) in a rectangular 9x57.5x392 mm channel with rounded off edges and with a grid of staggered cylindrical intensifiers (d_o=2-4 mm) having longitudinal ($\bar{s}_x=5$) and transverse ($\bar{s}_y=2.5$ and 5) pitches. As evident from the relations presented, thermal resonance is seen in the channel with cylindrical intensifiers (as in the smooth channel). However, the optimum value of the parameter $(\delta_{os}/\delta_o)_m$, determined from the mean equivalent diameter with regard to a surface of intensifiers, depends on a transverse pitch of a grid of intensifiers. As the transverse pitch of the grid of intensifiers increases, the parameter $(\delta_{os}/\delta_o)_m$ moves in the direction of larger values and, in the limit, tends to a corresponding value for smooth channels.

To calculate local heat transfer enhancement in channels due to oscillations, it is necessary to know a channel length distribution of a mass velocity oscillation amplitude. To do this, it will suffice to adopt the one-dimensional flow model (Galitseyskiy et al, 1977).

CONCLUSION

1. It is established that the intensity of transfer process in a channel essentially depends on the geometrical parameters of a grid of intensifiers and intensity of the heat transfer process can be enhanced 2,5-3 times through the use of intensifiers.
2. Heat transfer can be also augmented by gas flow oscillation in rectangular channels equipped with cylindrical intensifiers the level of heat transfer enhancement depends on the geometry of a grid of intensifiers.

NOMENCLATURE

$\bar{d}_i = d_i/d_e$ – relative diameter of intensifiers, dimensionless
d_i – diameter of intensifiers, m
d_e – equivalent diameter, m
F – area of cross-section channel, m^2
G – flowrate, kg/c
$K = \alpha/\alpha_0$ – relative heat transfer coefficient, dimensionless
ℓ – length of channel, m
Nu – Nusselt number, dimensionless
P – pressure, N/m^2
Pr – Prandtl number, dimensionless
q_w – density of heat flux, W/m^2
$Re = (u_f \cdot d)/\nu$ – Reynolds number, dimensionless
$Re_w = \omega d_e^2/\nu$ – oscillatory Reynolds number, dimensionless
$St = \alpha/\rho u_f c_p$ – Stanton number, dimensionless
s_x, s_y – longitudinal and transverse pitcher of intensifiers in a grid, m
$\bar{s}_x = s_x/d_i$, $\bar{s}_y = s_y/d_i$ – relative longitudinal and transverse pitches of intensifiers in a grid, dimensionless
T – temperature, K
$\Delta(\rho u)_0$ – amplitude oscillation velocity, m/c
$(\rho u)_0$ – average mass velocity, m/c
$\bar{x} = x/d_i$, $\bar{y} = y/d_i$ – relative longitudinal and transverse coordinate, dimensionless
x, y – coordinates, m
ρ – density, kg/m^3
α – heat transfer coefficient, W/(m^{2}K)
δ_{0S} – thickness of oscillating layer, m
δ_0 – thickness of viscous sublayer
$\varepsilon = \Delta(\rho u)_0/(\rho u)_0$ – mass velocity oscillation relative amplitude, dimensionless
ξ – pressure resistance coefficient, dimensionless
ω – oscillation circular frequency, 1/c
ν – kinematic viscosity, m^2/c
τ_w – wall tangential stress, N/m^2
Subscripts
w – parameter determined by a wall temperature
f – parameter determined by a flow temperature
0 – refer to a smooth channel or to a steady-state gas flow or to channel inlet.

REFERENCES

Galitseyskiy B.M., Nozdrin A.A. and Cherny M.S., 1981. Investigation of heat exchange process at channels with cylindrical intensifiers, Isvestia Vysshikh Uchebnykh Zavedenii, Aviatsionnaya Tekhnika №3, pp. 38-43.

Galitseisky B.M., Ryzhov Yu.A. and Yakush E.V., 1977. Thermal and Gasdynamic Processes in Oscillating Flows. Mashinostroyeniye, Moscow.

Galitseyskiy B.M., Sovershennyi V.D., Formalev V.F. and Chernyi M.S., 1996, Thermal protection of turbine blades, 1996, MAI publ., Moscow.

Galitseyskiy B.M., 1994, Heat Transfer in Oscillating Turbulent Flow, Proceedings of International Symposium on Turbulent, Heat and Mass Transfer, Lisbon, Portugal, vol. II, pp. 1751-1756.

Kalinin E.K., Dreitser G.A. and Yarkho S.A. 1972, Heat Transfer Enhancement in Channels. Moscow: Mashinostroyeniye.

Kays V.N. and London A.L. 1984, Compact Heat Exchangers. 3-rd ed. McGraw-Hill, New York.

ENHANCEMENT OF HEAT AND MASS TRANSFER IN HEAT EXCHANGERS WITH HELICAL TUBES

B.V. Dzyubenko and G.A. Dreitser

Moscow Aviation Institute (Technical University),
Volokolamskoe shosse, 4, Moscow, Russia; E-mail: heat204@mai.ru

ABSTRACT

At present a enhancement of heat and mass transfer in different channels due to flow swirling is among one of the more promising way to design compact heat exchangers. This method of enhancement for helical tube bundles in longitudinal flow enables to substantially decrease the size and mass of heat exchangers and to augment interchannel mixing of the heat carrier in intertube space. The heat exchangers on the base of oval twisted tubes have been developed and used in power engineering and food industry. For suppression of azimuthal nonuniformities of velocity and temperature that develop in the lateral entrance of heat carrier flow the helical-tube twisted bundle heat exchanger can be used. In this device, a helical tube bundle is twisted relative to longitudinal axis of bundle.

The goal of this paper is generalization of the comprehensive data on the heat transfer enhancement and improvement of interchannel fluid mixing as well as makes a comparison of heat and mass transfer effectiveness for the channels formed by typical helical tube and twisted bundles.

For investigating heat and mass transfer in these bundles were used the original methods of zone heating and of heat diffusion from point source. The heat transfer from a system of linear finite-size heat sources is considered for determining an effective diffusion coefficient using the homogenized flow model and statistical methods. Heat transfer in bundles was investigated by method of local similitude. The experimental data on heat transfer in a typical helical tube bundles is generalized in the form of second order multiterms in logarithm coordinates for geometrically nonsimilar bundles. The data on the interchannel flow mixing for helical tube bundles and for helical-tube twisted bundles are represented by the same relations. The cross mixing coefficients in these bundles exceed by 10 times the turbulent diffusion coefficients near the axis of circular tube. The twisting bundle of helical tubes enhances heat transfer compared with to a typical helical tube bundle. It are concluded, that a use of twisted tube allows enhancing heat transfer and transverse flow mixing and decreasing of temperature nonuniformity over the bundle cross-section.

INTRODUCTION

The method of flow swirling in helical tube bundles in longitudinal flow enables one not only to substantially decrease the size and mass of heat exchangers but also to improve interchannel mixing of the heat carrier in the intertube space, which provides a levelling of temperature nonuniformities in a cross section of bundle at nonuniform heat release (heat supply) and a lateral heat carrier flow into a heat exchanger. Because of their advantages, helical tube heat exchangers is used in different branches of industry (Dzyubenko and Yakimenko, 1997; Dzyubenko and Yakimenko, 1999). A use of helical tubes allows enhancing

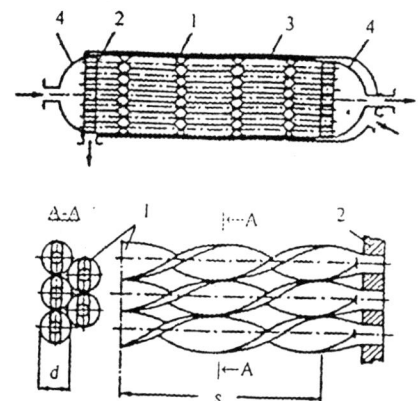

Fig. 1. Heat exchanger with a helical tube bundle: (1) helical tubes; (2) tube plates; (3) shell; (4) bottom (Dzyubenko and Vilemas, 1980).

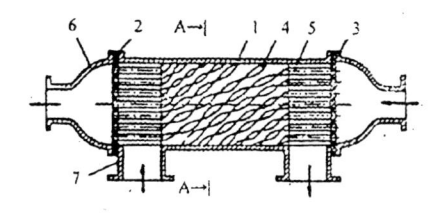

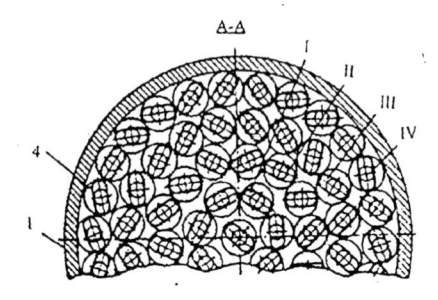

Fig. 2. Heat exchanger with twisted helical tube bundle: (1) shell; (2)-(3) tube plates; (4) helical tubes; (5) straight round ends of tubes; (6)-(7) connections; (I-IV) tube rows within a bundle (Dzyubenko et al., 1982).

heat transfer inside these tubes as well as in the intertube space of heat exchanger.

This paper describes and generalizes results of experimental study of heat transfer and interchannel mixing of the heat carrier for flows in helical tube bundles and twisted bundles of helical tubes. In the helical tube heat exchanger the tubes are located relative to each other, are in contact over a maximum size of the oval, and are fastened with straight round ends in the tube plates (Fig. 1): Dzyubenko and Vilemas (1980). Such a tube arrangement provides substantial improvement of heat and mass transfer processes in the intertube space and vibration strength of helical tube bundles.

Twisting of helical tube bundle relative to its axis (Fig. 2) provides leveling nonuniformities of heat carrier velocity and temperature fields formed by nonuniform heat supply over the bundle radius and azimuth and inlet conditions (Dzyubenko et al., 1982) as well as increasing heat transfer as compared with a straight helical tube bundle.

Enhancement of heat transfer and transverse flow mixing and decreasing of azimuth nonuniformities of temperature are a principal advantages of twisted helical tube and fuel rod bundles. These advantages of twisted bundles can be used in a nuclear reactor of space nuclear propulsion power plant intended for piloted expeditions to Mars (Dzyubenko et al., 1995). For space heterogeneous nuclear reactor the significant nonuniformities of volume heat generation intensity are characteristic. These nonuniformities may be partially uniformed by longitudinal and radial profiling of nuclear fuel concentration along and across heat generating assembly and by augmentation of interchannel mixing of heat carrier. The investigations of transport processes allow to understand the mechanisms of artificial flow turbulization and the ways of control of heat and mass transfer processes in these bundles.

The goals of this paper are to generalize and analyse the results of investigation of heat transfer and mixing of heat carrier as well as to obtain new closing empirical relations for calculation of effective turbulent heat and momentum transfer coefficients for these bundles. In this case, the dependences of the effective turbulent thermal conductivity and viscosity coefficients on the determining similarity numbers allow to be found in terms of the local flow parameters and heat release when thermal and hydraulic problems are solved.

METHOD OF ANALYSIS

Methods of experimental investigation of heat and mass transfer in the bundles of helical tubes have a number of characteristic features. Thus the mixing of working fluid in the bundles was investigated using a specially developed method of zone heating, which consisted of heating a cluster of tubes ($7 \div 37$) from 151 tube bundle so as to produce a nonuniform heat release field and to measure the field of coolant temperatures produced by it, followed by comparison with the predicted temperature fields.

The flow model was adopted to mathematically describe turbulent flow of gas heat carrier in a bundle composed of a great number of helical tubes or rods. According to this model, the real flow in the bundle was replaced by the one in a porous massif of a homogenized two-phase medium with a stationary solid phase. Sources of volume energy release q_V and hydraulic resistance $\xi \rho u^2 / 2d_{eq}$ were distributed in this massif in a certain manner (Danilov et al., 1990; Dzyubenko et al., 1990).

A system of differential equations incorporating those of energy, motion, continuity and state is solved to calculate temperature and velocity fields of heat carrier with its density. For the three-dimensional problem this system is of the form:

$$\rho u c_p \frac{\partial T}{\partial x} = q_v \frac{1-\varepsilon}{\varepsilon} + \frac{1}{r} \frac{\partial}{\partial r}\left(r \lambda_{eff} \frac{\partial T}{\partial r} \right) + \frac{1}{r^2} \frac{\partial}{\partial \varphi}\left(\lambda_{eff} \frac{\partial T}{\partial \varphi} \right) \quad (1)$$

$$\rho u \frac{\partial u}{\partial x} = -\frac{dp}{dx} - \xi \frac{\rho u^2}{2d_{eq}} + \frac{1}{r} \frac{\partial}{\partial r}\left(r \rho v_{eff} \frac{\partial u}{\partial r} \right) + \frac{1}{r^2} \frac{\partial}{\partial \varphi}\left(\rho v_{eff} \frac{\partial u}{\partial \varphi} \right) \quad (2)$$

$$G = \varepsilon \int_0^{2\pi} \int_0^{r_{shell}} \rho u r \, dr \, d\varphi \quad (3)$$

$$p = \rho RT \quad (4)$$

The system of equations (1)-(4) allows temperature fields to be calculated from experimental studies of heat and mass transfer processes on the models of the bundles of helical tubes (rods) when heated by electric current. To close the system of the equations (1)-(4), effective coefficients λ_{eff} and ν_{eff} are determined from experiment, assuming that the turbulent Lewis and Prandtl numbers are equal to unity ($Le_T = \rho c_p D_{qs}/\lambda_{eff} = 1$, $Pr_T = \rho c_p \nu_{eff}/\lambda_{eff} = 1$): $\lambda_{eff} = D_{qs}\rho c_p$, , $\nu_{eff} = D_{qs}$. The effective diffusion coefficient D_{qs} may by represented in a dimensionless form as

$$K_{qs} = D_{qs}/ud_{eq} \quad (5)$$

which is found experimentally and depends on the main similarity numbers. The governing system of equations (1)-(4) was solved numerically using the matrix factorization method (Dzyubenko et al., 2000).

A differential equations of energy and motion for twisted bundle of helical tubes are of the form:

$$\rho u c_p \frac{\partial T}{\partial x} + \rho c_p \frac{v_\tau}{r}\frac{\partial T}{\partial \varphi} = \frac{1}{r}\frac{\partial}{\partial r}\left(r\lambda_{eff}\frac{\partial T}{\partial r}\right) + \frac{1}{r^2}\frac{\partial}{\partial \varphi}\left(\lambda_{eff}\frac{\partial T}{\partial \varphi}\right) + q_v\frac{1-\varepsilon}{\varepsilon} \quad (6)$$

$$\rho u \frac{\partial u}{\partial x} + \rho \frac{v_\tau}{r}\frac{\partial u}{\partial \varphi} = -\frac{dp}{dx} - \xi\frac{\rho u}{2d_{eq}}\sqrt{u^2 + v_\tau^2} \quad (7)$$

$$\rho u \frac{\partial v_\tau}{\partial x} + \rho \frac{v_\tau}{r}\frac{\partial v_\tau}{\partial \varphi} = -\xi_\varphi \frac{\rho v_\tau}{2d_{eq}}\sqrt{u^2 + v_\tau^2} \quad (8)$$

Azimuthal heat and mass transfer into the Eqs (6)-(8) is taken into consideration by including the terms responsible for a temperature and velocity variation along the φ-coordinate. The method of heat diffusion from a point source was employed to determine the coefficient K_{twist}. This method was also adopted to find K_{qs} for straight helical tube bundles (Danilov et al., 1990). In this case, when a turbulent field is given in Lagrange's statistical representation, the mean-statistical squared displacement $\bar{y}^2$ of heated particles continuously emitted by a diffusion source is determined by the formula

$$\bar{y}^2 = 2\frac{D_{twist}}{u}(x - x_0) \quad (9)$$

which is a limiting solution to Taylor's equation for homogeneous and isotropic turbulence at large diffusion times (at large distances from a diffusion source) (Hinze, 1959)

$$\frac{d}{dt}\left(\frac{1}{2}\bar{y}^2\right) = \int_{t_0}^{t} v_1(t_0)v_1(t')dt' \quad (10)$$

Application of this method for twisted bundles is justified since the experimental temperature fields in bundle cross sections at different distances from a diffusion source are close to normal distribution law. In this case, since the mechanism for ordered heat and mass transfer with respect to a bundle azimuth is not considered because of curvilinear jet, the coefficient K_{twist} is determined by the same mechanisms as those in a straight helical tube bundle which will be responsible for mixing. This method of experimental data representation allows to compare the coefficients K_{twist} with the coefficients K_{qs} for straight helical tube bundles. In this case, the azimuthal heat and mass transfer for twisted bundles of helical tube is necessary to consider by means off mathematical grounds (Eq. (6)-(8)).

EXPERIMENTAL TECHNIQUES

Arrangement for investigating the interchannel mixing by the method of diffusion from a point source in helical tube bundles has a number of specific features. The diameter of the diffusion source was selected to be equal to the diameter of the helical tube, so as to take into account all the transfer mechanisms acting on the indicator gas which is injected into the main stream. Temperature fields were measured at different distances from the moving diffusion source.

Arrangements for study of the interchannel mixing by a method of diffusion of heat from a cluster of linear sources and for investigating the heat transfer by a method zonal heating of the helical tube bundles allow to model the conditions characterized for the helical fuel rod bundles of space nuclear reactor.

The heat transfer coefficient was determined from the measured local wall temperatures, the temperatures of the working fluid, flow velocity at the outer edge of the wall layer, and local heat flux. The main parameters were: maximum dimension of the oval of the tubes, 12.3 mm; thickness of tube walls (stainless steel tubes), 0.2÷0.3 mm; twisting pitch, 75÷330 mm; tube length, 500÷1500 mm, number of tubes in bundle, 37 and 151; coolant flow was horizontal and vertical (from the bottom up); velocity of air up to 60 m/s (Re=2·10³÷5·10⁴); inlet air temperature, 280-290 K; outlet air temperature, 450÷650 K; thermal rating, 30÷90 kW. The experimental set-up was equipped with automatic systems of control, measurement, collection and processing of experimental data which would be rapid and have a low inertia.

Study of heat and mass transfer in a twisted bundles of helical tube was made on the models of heat exchangers with different twisting principles over the bundle radius: at twisted of the helical tubes relative to the bundle axis with a constant angle γ_{twist}=const(r)=$\pi/6$ or v_τ=const(r) and at twisting with a constant pitch s_{twist}= const(r)=0.65 m or v_τ/r= const(r) (quasi-solid rotation of heat carrier). These bundles with the same dimensions of helical tubes (s=171 mm, d=12.2 mm, s/d=14) has following parameters: -ε=0.62,

d_{eq}=12.55 mm, Fr_m=190 and N=61 tubes; -ϵ=0.63, d_{eq}=14.19 mm, Fr_m=170 and N=59 tubes, respectively. The experimental technique was as follows. Heated air was continuously injected into the cocurrent longitudinal cold air flow past a tube bundle about 1.5 m long. Hot air was injected from diffusion source of 17 mm radius at the three radial orientations of source: $r_{d.s}$=0, 19.7 and 38.7 mm. Air temperatures in the bundle cross-sections at the distances of 375, 750 and 1126 mm from source were measured by chromel-alumel thermocouples welded to the walls of all the bundle tubes on the inside. A systematic error associated with a difference in temperature distributions of heat carrier in the flow core and on the tube walls is allowed for in estimating the coefficients K_{twist} for twisted bundles of helical tubes. In processing the experimental data it has been allowed for the fact that the jets injected into the bundle are curvilinear at an angle of twisting the bundle on the radii: $r_{d.s}$=19.7 and 38.7 mm. Therefore, the experimental coefficient K_{twist} is determined without regard to ordered azimuth transfer by a twisted bundle. The coefficient K_{twist} was measured within Re=$8.5 \cdot 10^3$-$3.4 \cdot 10^4$ at air temperatures at the diffusion source outlet 394-403 K and at the cocurrent flow temperature 294-308 K. Experiments were made when the main flow velocity was equal to the one of the jet flowing out of the diffusion source.

Experimental studies of heat transfer were made on a twisted bundle of helical tubes 1.5 m long with a constant twisting pitch s_{twist}=0.65 m. A relative twisting pitch of the tubes in each row with numbers: 1, 6, 12, 17 and 23 tubes, was as follows: s_{twist}/d_{twist}=∞, 12.3, 11.6, 8.1 and 6.2, respectively. The local modelling method was adopted to study the heat transfer. In this case, electric power was supplied to individual tubes of each row, and the heated length was 0.4 m. The heat transfer was investigated under hydrodynamic fully developed conditions at l/d=90. The incoming flow temperature T_f was taken as the characteristic temperature and d_{Eq} = 4F/Π = 14.19 mm as the characteristic dimension. Studies were made within the range of Re = $1.4 \cdot 10^3$-$7 \cdot 10^4$.

RESULTS

The experimental data on the interchannel flow mixing for twisted bundles of helical tube are shown in Fig. 3, where these are compared with the predicted and experimental data for straight helical tube bundles. As seen from Fig. 3, the coefficients K_{twist} for the modules of bundles with twisting of all the helical tubes with γ_{twist}=const(r) and s_{twist}=const(r) are coincided with the relations obtained for the straight helical tube bundles.

The investigation results of the coefficient K_{qs} obtained by method of heat diffusion from a system of linear sources at the Euler representation and by method of diffusion from a point source at the Lagrange representation are connected by relation for spatial turbulence scale ration L_E/L_L (Dzyubenko et al., 1990). The experimental data on the coefficient K_{qs} for straight bundles of helical tubes were well generalized for Re≥$8 \cdot 10^3$ and ϵ=0.5 by relation

$$K_{qs} = 10.35 Fr_m^{-1.4232-0.1857 \lg Fr_m} \quad (11)$$

For the range of Fr_m=43-3300 and ϵ≥0.27 in work: Dzyubenko et al. (1990), was obtained the relation

$$K_{qs} = 1.902 Fr_m^{-0.53} \cdot \epsilon^{1.086} \quad (12)$$

which described the experimental data on interchannel mixing process in finned rod bundles. In work: Danilov et al. (1990), at Re≥10^4 the relationship was obtained for straight bundles of helical tubes

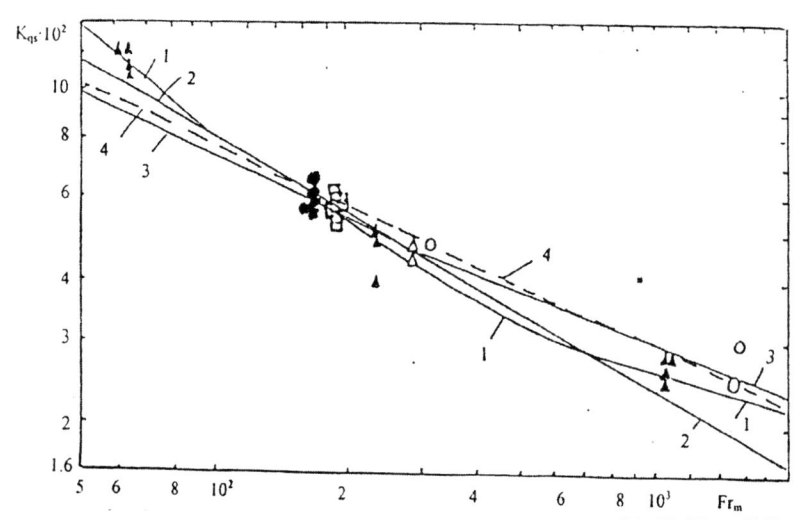

Fig. 3. Effective turbulent diffusion coefficient versus Fr_m number: (1) Eq. (11); (2)-(3) Eqs. (12) and (13), respectively; (4) Eq. (14); o, Δ, ▲, experimental data at Re ≥ $8.5 \cdot 10^3$ for helical tube bundles; □, experimental data for twisted bundle of helical tube at Re = $(8.5-34) \cdot 10^3$ and γ_{twist} = π/6=const(r); •, the same at s_{twist}=0.65, m = const(r) and Re=$(8.5-14.6) \cdot 10^3$.

$$K_{qs} = 0.136 Fr_m^{-0.256} + 10 Fr_m^{-0.66}(\varepsilon - 0.46) \quad (13)$$

The coefficients K_{qs} for straight bundle determined by the method of heat diffusion from a point source are represented by relation

$$K_{qs} = \frac{0.078484}{Fr_m^{0.254}}\left(1 + \frac{8.1}{Fr_m^{0.278}}\right) \quad (14)$$

The experimental data on the interchannel flow mixing for twisted bundles obtained the diffusion method from a point source are well agreed with Eq. (14). Within ±12% the experimental data on the K_{twist} for γ_{twist}=const(r) is equal to an arithmetic mean value K_{twist}=0.0573 in range of Re=8.5·10³-3.4·10⁴ and $r_{d.s}/r_{shall}$=0.325 and 0.638 (Fig. 3). $K_{twist}=K_{qs}$ for γ_{twist}=const(r). For a bundle with s_{twist}=const(r) the mean value of coefficient K_{twist} at Re=8.5·10³-1.46·10⁴ are agreed with Eq.(14) also. However at the same time the coefficient K_{twist} decreases with increasing the Re number at Re=8.5·10³-3.4·10⁴ (Fig. 4) in respect to relation

$$K_{twist} = 0.6 Re^{-0.243} + 0.0161 r_{d.s}/r_{shell} \quad (15)$$

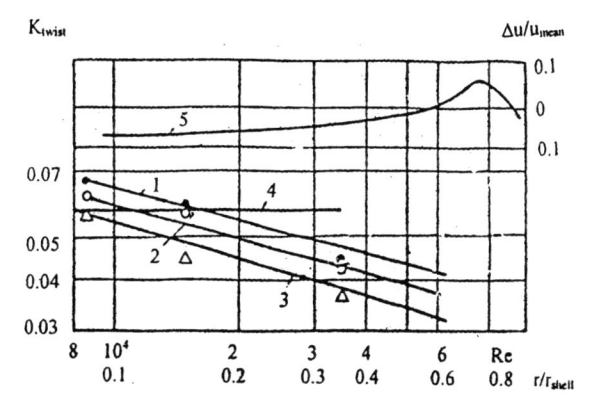

Fig. 4. Coefficient K_{twist} versus the Re number and dimensionless excess velocity $\Delta u/u_{mean}$ versus the bundle radius r/r_{shell}. •, o, Δ, experimental data for a bundle with s_{twist} = const(r) at r_{source}/r_{shell} = 0, 0.325 and 0.638 respectively; (1)-(3) relation (15) for the same positions of diffusion source; (4) K_{twist} = 0.0573 for a bundle with γ_{twist} = const(r); (5) relation $\Delta u/u_{mean}$ vs r/r_{shell} for a bundle with s_{twist} = const(r).

It is seen from Fig. 4, giving a bundle radius variation of an excess longitudinal velocity based on its mean-mass value in a bundle. Such a behaviour of velocity distributions exists behind an axial-blade swirler, near which the specific features of flow were observed. A rotational velocity component in a bundle for s_{twist}=const(r) should vary practically following the law of quasi-solid rotation.

The experimental heat transfer data in a twisted bundle of helical tubes were generalized in the form

$$Nu_{twist}/Nu_{straight} = f\left(Re_{df}, s_{twist}/d_{twist}\right) \quad (16)$$

where $Nu_{straight}=Nu_{df}$ is the Nusselt number for straight bundle of helical tube (Fig. 5). It is seen from Fig. 5 that the twisting of helical tubes in a bundle enhances the heat transfer. The smaller the relative twisting pitch of bundle of helical tubes s_{twist}/d_{twist}, the larger is the heat transfer. The smaller the Reynolds number, the larger is the heat transfer coefficient (Fig. 5)

The experimental data on heat transfer in a straight helical tube bundles is generalized in form of second order multiterms in logarithm coordinates (Dzyubenko et al., 1995), (Dzyubenko et al., 2000):

$$Nu_{df} = 6.05 \cdot 10^6 Fr_m^n Re_{df}^m Pr_f^{0.4}\left(\frac{T_w}{T_f}\right)^{-0.55} \quad (17)$$

where $\quad n=-2.494+0.2351 lg Fr_m \quad (18)$

$$m = -1.572 Fr_m^{0.01661-0.04373 lg Fr_m} + \\ + 0.269 Fr_m^{-0.0149-0.0104 lg Fr_m} lg Re_{df} \quad (19)$$

The modified number $Fr_m=s^2/dd_{eq}$ is only a function of the geometrical sizes of helical tube bundle and is reduced from the criterion for a centrifugal force field in bundle

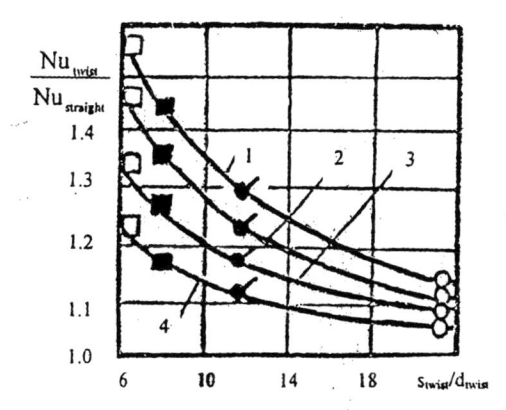

Fig. 5. Nusselt number versus relative bundle twisting pitch and Reynolds number: (1)-(4) experimental data for Re_{df}=1.4·10³, 5·10³, 10⁴, 5·10⁴, respectively.

$Fr_c=s^2/(2\pi^2 dd_{eq})$ for swirling according to the solid body law. Eq. (17) was obtained over the parameter ranges: Fr_m=60-1100, Re_{df}=2·10³÷3·10⁴ and $T_w/T_f \leq 1.75$ for bundles with a large number of helical tubes (37 and 151). Eq. (17) describes the experimental data on heat transfer over the whole range of Fr_m and Re_{df} numbers studied (Figs. 6 and 7), as compared with the relations of work: Danilov et al.

(1990), which describe the experimental data for a separate values of Fr_m only:

For $Fr_m=64$ and $Re_{df} \geq 10^4$,

$$Nu_{df} = 0.0521 Re_{df}^{0.8} Pr_f^{0.4} \left(\frac{T_w}{T_f}\right)^{-0.55} \quad (20)$$

For $Fr_m=64$ and $Re_{df} < 10^4$,

$$Nu_{df} = 83.5 Fr_m^{-1.2} Re_{df}^{0.212 Fr_m^{0.194}} Pr_f^{0.4}\left(1+3.6 Fr_m^{-0.357}\right)\left(\frac{T_w}{T_f}\right)^{-0.55} \quad (21)$$

For $Fr_m=232-2440$,

$$Nu_{df} = 0.023\left(1+3.6 Fr_m^{-0.357}\right) Re_{df}^{0.8} Pr_f^{0.4} \left(\frac{T_w}{T_f}\right)^{-0.55} \quad (22)$$

At the same time, the Eqs (20)-(22) describe well the experimental data for the concrete parameter ranges.

DISCUSSION

The flow in a straight helical tube bundle and in twisted bundle of helical tubes is spatial. Together with a longitudinal velocity vector component there exist transverse velocity components that greatly enhance interchannel flow mixing in these bundles. The turbulence in such bundles is generated by the fixed wall and due to the friction between liquid layers with different velocities. The relative contribution of turbulent diffusion to cross mixing of the heat carrier in these bundle is estimated by ~ 20% of the quantity of effective turbulent diffusion coefficient K_{qs} at $Fr_m=178$ (Dzyubenko et al., 1990). As seen from Fig. 3, with decreasing the Fr_m number the K_{qs} increases since in this case the intensity of turbulent fluctuations as well as the convective heat and mass transfer grow. Hence, the high turbulence level of the flow, convective transfer in a cells and ordered transfer in the cross section of such bundles due to spiral flow swirling by means of helical tubes are the mechanisms which effect on interchannel flow mixing and K_{qs} coefficients. The effective coefficients of turbulent diffusion responsible for interchannel mixing in such devices exceed by 10 times the turbulent diffusion coefficients on a circular tube axis (Dzyubenko et al., 1995).

For a twisted bundles of helical tubes the same mechanisms of interchannel mixing as those in a straight helical tube bundles are responsible for mixing. Therefore, the Eq. (14) may be used to close the system of equations (6)-(7) as well as the system of equations (1)-(4), where $\lambda_{eff}=\rho u c_p d_{eq} K_{qs}$. Analysis of heat and mass transfer has evidenced that in bundles of helical tubes twisted according to $\gamma_{twist}=const(r)$ and $s_{twist}=const(r)$ there exists K_{twist} as a function of Reynolds number and ratio $r_{d.s}/r_{shell}$ (Fig.4) being coefficients on a circular tube axis (Dzyubenko et al., 1995).

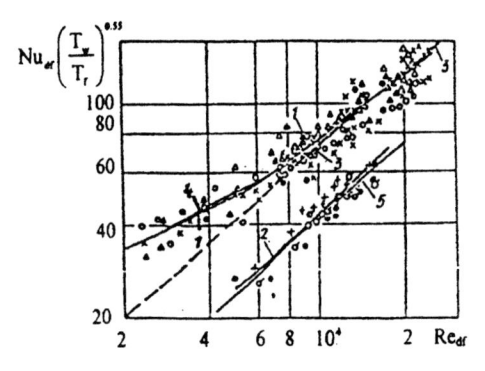

Fig. 6. Nusselt number Nu_{df} versus the Re and Fr_m: x, o, Δ, experimental data for $Fr_m=63.6$ and $x/d_{eq}=19.6$, 66.2 and 82.1, respectively; •, o, +, the same for $Fr_m=1052$ and $x/d_{eq}=20$, 83.6 and 90.5, respectively; (1)-(2) Eq. (17) for $Fr_m=63.6$ and 1052, respectively; (3) Eq. (20); (4) Eq. (21); (5) Eq. (22).

For a twisted bundles of helical tubes the same mechanisms of interchannel mixing as those in a straight helical tube bundles are responsible for mixing. Therefore, the Eq. (14) may be used to close the system of equations (6)-(7) as well as the system of equations (1)-(4), where $\lambda_{eff}=\rho u c_p d_{eq} K_{qs}$. Analysis of heat and mass transfer has evidenced that in bundles of helical tubes twisted according to $\gamma_{twist}=const(r)$ and $s_{twist}=const(r)$ there exists K_{twist} as a function of Reynolds number and ratio $r_{d.s}/r_{shell}$ (Fig.4) being attributed to a difference in the flow structure in these bundles. The discovered specific features of heat and mass transfer processes in bundles of helical tubes differently twisted relative to the bundle axis offer recommendations on practical use of bundles with a constant angle of twisting since in this case more intensive mixing of heat carrier is provided in bundle cross-section irrespective of the places, at which heat supply nonuniformities develop.

In bundles with twisting according to $\gamma_{twist}=const(r)$ the ratio $s_{twist}/d_{twist}=const(r)$. Therefore, heat transfer in such bundles has the same intensity for the all tube rows of bundle (Fig.5) at given Re and Fr_m numbers. At the same time at $s_{twist}=const(r)$ the relative twisting pitch s_{twist}/d_{twist} decreases from the bundle centre to its periphery according to the hyperbolic relation $s_{twist}/d_{twist}=5.57/(r/r_{shell})$ and the angle of twisting of helical tubes relative to bundle axis, accordingly, increases with a bundle radius. This is seen Fig.4, giving a bundle radius variation of an excess of longitudinal velocity based on its mean-mass value. A rotational velocity component in a bundle in this case should vary practically following the law of quasi-solid rotation. Thus, a use of bundles with a constant angle of twisting allow to enhance heat transfer in such bundle as well as interchannel mixing (Figs. 4 and 5).

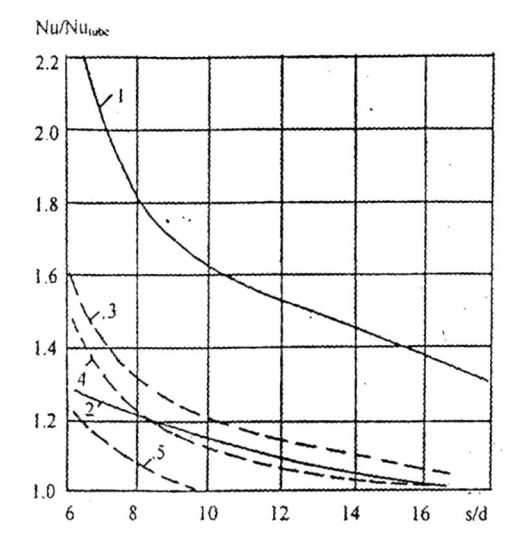

Fig. 7. The effect of s/d on the heat transfer for different channels with flow swirling: 1) Eq. (17) at Re=10^4 for helical tube bundles, 2) Eq. (23) at Re=$7 \cdot 10^3$ for helical tubes; 3-5) Eq. (24) for annular channels at Re=$5 \cdot 10^3$, 10^4 and $4 \cdot 10^4$, respectively.

For calculations of Nusselt number $Nu_{straight}=Nu_{df}$ Eq.(17) may be used. The effect of the flow swirling intensity on heat transfer in helical tube bundles, inside twisted oval tubes and in annular channels with inner helical tube is presented on Fig.7. It is seen from Fig.7 that the heat transfer inside twisted oval tube at s/d=66.2 exceed by 1.3 times the heat transfer in circular tube, where

$$Nu_{inside} = 0.0486 \left[1 + 3.0384\left(\frac{s}{d}\right)^{-0.97}\right] Re_{df}^{0.698} Pr_f^{0.4} \quad (23)$$

The heat transfer in annular channels with inner helical tube is discribed by relation (Dzyubenko et all., 1995)

$$Nu_{ann} = 1.8 \left(\frac{s}{d}\right)^{-1.48} Pr_f^{0.4} Re^{0.8[1-0.35\exp(-0.105s/d)]} \quad (24)$$

which is presented on Fig. 7. It is seen from Fig. 5 and 7, that all swirling flows (twisted helical tube bundles, straight helical tube bundles, inside twisted oval tubes and in annular channels with inner helical tube) are characterized by significant enhancement of heat transfer as compared with a flow in a circular tube.

CONCLUSIONS

1. Enhancement of heat and mass transfer in channels due to flow swirling is among one of the urgent problems and the more promising ways to design a compact heat exchangers and to augment interchannel mixing of the heat carrier in intertube space. The advantages of the twisted helical tube bundles are the more significant enhancement of heat and mass transfer and decrease of azimuth nonuniformities of temperature and velocity at nonuniform heat release (heat supply) and a lateral heat carrier flow into a heat exchanger as compared with the straight helical tube bundle.

2. The discovered specific features of heat and mass transfer processes in twisted bundles of helical tubes offer recommendations on practical use of these bundles with a constant angle of twisting of helical tubes relative to the bundle axis since in this case more intensive heat transfer and mixing of heat carrier is provided.

3. In result of enhancement of heat and mass transfer by flow swirling the use of twisted tubes allows to develop the compact heat exchangers for power engineering, chemical, food and other branches of industry and for the high temperature heat generating assemblies of space nuclear reactor.

ACKNOWLEDGMENTS

The financial support from the Russian Scientific Foundation (Grant for Supporting of leading Scientific schools of Russia No. 00-15-96654) is greatly appreciated.

NOMENCLATURE

c_p	heat capacity at constant pressure, J/(kg·K)
D	effective diffusion coefficient, m^2/s
d	maximum size of a tube oval profile, diameter, m
d_{eq}	equivalent diameter, $4F_f/\Pi_{wet}$, m
G	mass flowrate of heat carrier, kg/s
F_f	area of the flow cross section of a bundle, m^2
Fr_m	modify Frud number, s^2/(dd$_{eq}$), dimensionless
K	effective diffusion coefficient, $D/(ud_{eq})$, dimensionless
L	spatial turbulence scale, m
Le	Lewis number, $\rho c_p D/\lambda_{eff}$, dimensionless
l	length, m
lg	logarithm to the base 10
N	tube number in a bundle
Nu	Nusselt number, $\alpha d_{eq}/\lambda$, dimensionless
p	pressure, Pa
Pr	Prandtl number, $\rho v c_p/\lambda$, dimensionless
q_v	volumetric density of heat release, W/m^3
r	radial coordinate, radius, m
Re	Reynolds number, ud_{eq}/v, dimensionless
s	twisting pitch, m
T	temperature, K
t	time, s
u	longitudinal velocity, m/s
Δu	longitudinal velocity excess in the flow core, m/s
v_τ	tangential velocity, m/s

v_1	mean quadratic pulsational velocity, m/s
x	longitudinal coordinate
$\overline{y}^2$	mean statistical squared displacement, m²
α	heat transfer coefficient, W/(m²·K)
γ	angle of twisting of helical tubes relative to the bundle axis, rad
ε	bundle porosity
λ	thermal conductivity, W/(m·K)
ν	kinematic viscosity, m²/s
ξ	hydraulic resistance coefficient
Π_{wet}	wetter wall perimeter, m
ρ	density, kg/m³
φ	angle coordinate

Subscripts

ann	annular channel
d	determined in terms of d_{eq}
d.s	diffusion source
E	Euler flow
eff	effective
f	flow
inside	inside helical tube
L	Lagrange flow
mean	mean-mass value of parameter
qs	quasi-stationary
shell	heat exchanger shell
straight	straight helical tube bundle
tube	circular tube
T	turbulent
twist	twisted bundle of helical tubes
w	wall
φ	along the φ axis

REFERENCES

Danilov, Y.I., Dzyubenko, B.V., Dreitser, G.A., and Ashmantas, L.A., 1990, *Analysis and Design of Swirl-Augmented Heat Exchangers*, Hemisphere Publishing, New York.

Dzyubenko, B.V., Dreitser, G.A., and Ashmantas, L.-V.A., 1990, *Unsteady Heat and Mass Transfer in Helical Tube Bundles*, Hemisphere Publishing, New York.

Dzyubenko, B.V., Sakalauskas, A.V., Ashmantas, L.-V., and Segal, M.D., 1995, *Turbulent Flow and Heat Transfer in Channels of Power Plant*, Pradai, Vilnius.

Dzyubenko, B.V., Ashmantas, L.-V., and Segal, M.D., 2000, *Modeling and Design of Twisted Tube Heat Exchangers*, Begell House, Inc., New York.

Dzyubenko, B.V., and Yakimenko, R.I., 1997, Enhancement of Heat Transfer in Channels and Effective Choice of Heat Transfer Surface for Heat Exchangers, *Proc. Int. Conference on Compact Heat Exchangers for the Process Industries 1997*, Begell House, Inc., New York, pp. 533-540.

Dzyubenko, B.V., and Yakimenko, R.I., 1999, High Temperature Compact Heat Exchanger with Close Packed Helical Tube Bundle in Cross Flow, *Proc. Int. Conference on Compact Heat Exchangers and Enhancement Technology for the Process Industries 1999*, Begell House, Inc., New York, pp. 263-270.

Dzyubenko, B.V., and Vilemas, Yu.V., 1980, A Shell-Tube Heat Exchanger, USSR Author's Certificate □761820, Bulletin of Inventions, 1980, No. 33, p. 194.

Dzyubenko, B.V., Vilemas, Yu.V., Varshkyavicius, R.R., and Dreitser, G.A., 1982, A Shell-Tube Heat Exchanger, USSR Author's Certificate □937954, Bulletin of Inventions, 1982, No. 23, p. 189.

Hinze, J.O., 1959, *An introduction to its mechanism and theory*, McGraw-Hill, New York.

PERFORMANCE EVALUATION OF HEAT TRANSFER ENHANCEMENT BY TWISTED TAPES

Koichi Ichimiya

Department of Mechanical System Engineering, Yamanashi University, Takeda-4, Kofu, Yamanashi 400-8511 Japan
E-mail: ichimiya@ccn.yamanashi.ac.jp

ABSTRACT

This paper describes the thermal evaluation of heat transfer enhancement by three kinds of twisted tapes. These twisted tapes are the full length and full width type, the regularly spaced type and the reduced width type. Thermal evaluation was based on two methods, namely, the Nusselt number ratio Nu_a/Nu_o with and without heat transfer improvement and the entropy generation rate. In the present conditions, Nu_a/Nu_o increases in laminar region with increase of Reynolds number and the regular spaced twisted tapes give the high heat transfer enhancement and low entropy generation.

INTRODUCTION

In recent times, heat transfer control is significant in the heat transfer facilities such as heat exchangers and waste heat recovery systems to save and utilize effectively the thermal energy. There are several methods (passive technique, active technique and combined technique) to improve or control the local heat transfer. One of them is twisted-tape inserts. In practical applications, steam heating or turbulent flow liquid heating or cooling of viscous fluids can be seen in the chemical and process industry. In heat exchangers used in the process industry, substantial capital and operating cost savings can be also achieved by inserting twisted-tapes to improve their heat transfer performance.

The early research on twisted-tape inserts was conducted with turbulent flow of air and water (Smithberg and Landis, 1964 ; Date and Singham, 1972). The gas flow studies were targeted at fire-tube boiler and the water flow works were to supply the fundamental data for the flow boiling improvement. Recently, the energy conservation and abatement of environmental degradation have been focused and generated renewed interest in the use of twisted tapes as passive enhancement.

The typical twisted tape is a full-length and full-width type inside a circular tube (Thorsen and Landies, 1968 ; Lopina and Bergles, 1969 ; Watanabe et al., 1983). The phenomenological flow characteristics are arranged by Manglik and Bergles (1993). Tape insert increases the wetted perimeter, reduces the flow cross-sectional area, and increases the flow length. A centrifugal force is superimposed over the longitudinal flow due to the herically rotating fluid flow. Consequently, high pressure drop and heat transfer coefficients are obtained in comparison with those of smooth tube flows.

Several modifications in twisted-tape geometry have been investigated with the aim of restricting the increase in pressure drop and achieving material savings. In the present research, in addition to the full-length and full-width twisted tape (Manglik and Bergles, 1993 ; Manglik Bergles, 1993), the regularly spaced twisted-tape (Saha etal., 1989) and the reduced width twisted-tape elements (Patil, 2000) are applied. In these twisted tape elements, fin effect is suppressed due to the longitudinal or width space.

There are several criteria for the performance evaluation of these heat transfer enhancement devices as reported by Bergles et al. (1974). Heat transfer and pressure loss increase are compared directly by j-factor and f-factor as an area goodness factor. This method denotes the compactness of heat exchangers. In another method, heat transfer rate is expressed as pumping power using volume goodness factor.

In the present research, the performance of three kinds of geometries has been evaluated on the bases of increase of heat transfer at Reynolds number fixed, pressure drop fixed and pumping power fixed and the entropy generation rate.

TWISTED TAPES

Full Length and Full Width Twisted-Tape

The first type is the full length and full width twisted-tape in a circular tube as shown in Fig.1(a) (Manglik and Bergles, 1993 ; Manglik and Bergles, 1993). The pitch length for 180° rotation of the tape is H. and the thickness of the tape is δ. Flow regime for isothermal friction factors in a tube flows with twisted-tape elements is classified to four distinct regions where the interplay of viscous, convective inertia and centrifugal forces influences the flow resistance. These regions are (I) viscous flow, (II) swirl flow superimposed over the axial flow, (III) swirl-turbulent transition and (IV) turbulent swirl flow. Corresponding to these flow regimes, thermal behavior is also divided into four regions. In the present case, friction factor and mean Nusselt number are expressed in laminar and turbulent flows. In twisted-tape flow the centrifugal force due to the swirling flow and the inertia force is balanced by the viscous force. This force ratio is swirl parameter Sw defined by ratio of centrifugal force multiplied by inertia force to square of viscous force.

The friction factor in laminar swirl flow is expressed as follows,

$$(fRe_a) = 15.767[(\pi+2-2\delta/D)/(\pi-4\delta/D)]^2(1+10^{-6} Sw^{2.55})^{1/6} \quad (1)$$

The heat transfer data for twisted-tape inserts can be functionally correlated as

$$Nu_m = \varphi(Gz, Sw, Pr, Ra, \mu_b/\mu_w) \quad (2)$$

Where Gz, Pr and Ra are Graetz number, Prandtl number and Rayleigh number, respectively. The mean Nusselt number is expressed as following correlation,

$$Nu_m = 4.612\{[(1+0.0951Gz^{0.894})^{2.5} + 6.413 \times 10^{-9} \\ \times (Sw \cdot Pr^{0.391})^{3.835}]^{2.0} \\ + 2.132 \times 10^{-14}(Re_a \cdot Ra)^{2.32}\}^{0.1}(\mu_b/\mu_w)^{0.14} \quad (3)$$

On the other hand, the wall curvature in twisted-tape flows is expressed by twist ratio y(=H/D) as

$$Kc = [1 + (2y/\pi)^2]^{-1} \quad (4)$$

The friction factor in turbulent swirl flow including the effect of the curvature is written as

$$f = 0.0791\{1 + 2.06[1+(2y/\pi)^2]^{-0.74} Re_a^{-0.25} \\ \times [\pi/(\pi-4\delta/D)]^{1.75} [(\pi+2-2\delta/D)/(\pi-4\delta/D)]^{1.25} \quad (5)$$

Turbulent flow heat transfer with twisted-tape inserts is given by

$$Nu_m = 0.023 Re_a^{0.8} Pr^{0.4} [1+(0.769/y)] [\pi/(\pi-4\delta/D)]^{0.8} \\ \times [(\pi+2-2\delta/D)/(\pi-4\delta/D)]^{0.2} (\mu_b/\mu_w)^{0.18} \quad (6)$$

The index of Re_a is same as that for a smooth tube. The viscosity ratio exponent is for liquid heating.

Regularly Spaced Twisted-Tape

Figure 1(b) shows the regularly spaced twisted-tape (Saha etal., 1989). The twisted elements are followed by a free space of length and are connected by thin circular rods. The swirl flow decays in the space and is again augmented by the following tape element. The direction of twist imparted to each successive element is same. The fin geometry imparts a periodicity to the herical flow. The friction factor and mean Nusselt number are obtained for comparatively low Reynolds number (laminar flow).

The friction factor for regularly spaced tapes is divided by the regime of twist ratio y(=H/D) and space ratio s(= S/D). The correlation for 3.46<y<7.5 and 2.5<s<5 is presented here.

$$f = C_1 \xi_1 Re_a^{-0.7} y^{-0.3} (1 + Cs) \quad (7)$$

where

$C_1 = 8.8201y - 2.1193y^2 + 0.2108y^3 - 0.0069y^4$
$\xi_1 = (D_H^2 A)_{\delta=0}/(D_H^2 A)_{\delta \neq 0}$,
$C = -3.97 \times 10^{-3} ys + 0.01s + 0.018y - 7.15 \times 10^{-3}$.

Correlation for mean Nusselt number is given as

$$Nu_m = 5.172[1+6.7482 \times 10^{-3} Pr^{0.7} (K1 Re_a/y)^{1.25}]^{0.5} \\ \times (1 + Cs)X \quad (8)$$

where

$K1 = \pi D^2(y+s)/[(\pi D^2 - 4\delta D)y + \pi(D^2-d^2)s]$
$C = (0.057 ys + 0.3622) \exp[(-0.0296y - 0.305)s]$
$X = (1 - 4.0422 \times 10^{-2} s)$ $Re_a < 700$
$X = 1$ $Re_a > 700$

Reduced Width Twisted-Tape

Figure 1 (c) shows the reduced width twisted-tape (Patil, 2000). The gap between the tube wall and the tape is maintained constant throughout the tube length by metal pins to the edges of the tape as indicated.

The test fluid is pseudoplastic obeyed the generalized power-law model

(a) Full length and full width twisted-tape

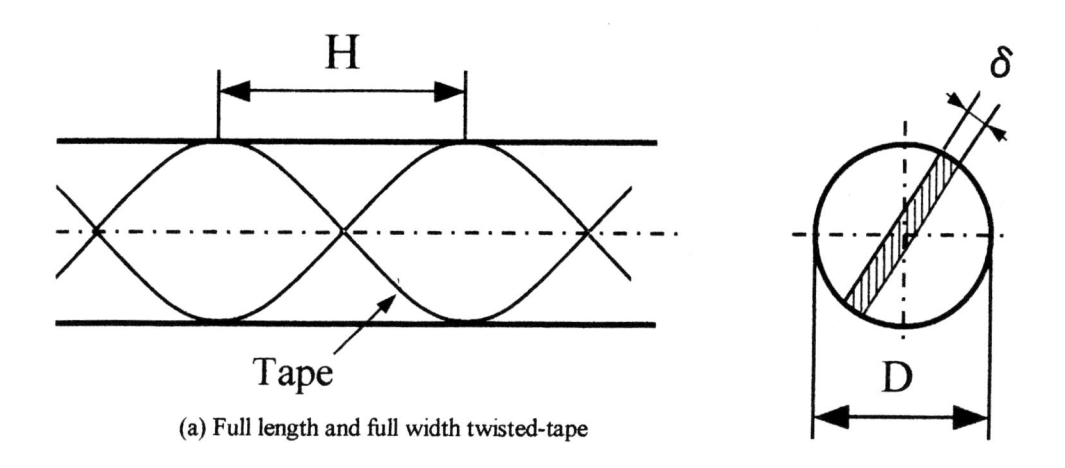

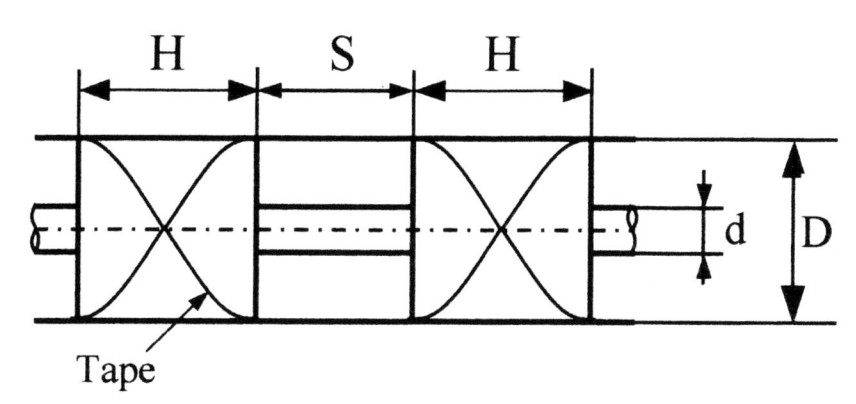

(b) Regularly spaced twisted-tape

(c) Reduced width twisted-tape

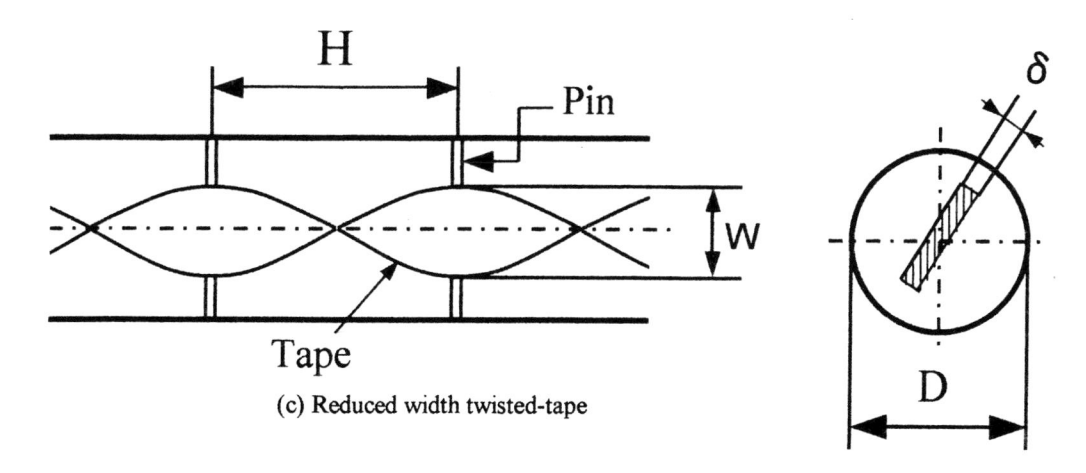

Fig.1 Illustration of the twisted-tape

$\tau_w = K'(8u/D)^{n'}$ where n'=0.75

where K' is flow consistency index.

The friction factor for reduced width tapes in laminar flow region is expressed by using the modified twist ratio y* (ratio of tape pitch to tape width).

$$(fRe_a') = C^*(Re_a'/y^*)^{0.3} \xi^* \qquad (9)$$

where

$$\xi^* = [\pi/(\pi+2)]^2[(\pi+2-2\delta/W)/(\pi-4\delta/W)]^2 \\ \times [\pi/(\pi-4\delta/W)](y/y^*) \qquad (10)$$
$$C^* = 8.8201y^* - 2.1193(y^*)^2 + 0.2108(y^*)^3 - 0.0069(y^*)^4 \qquad (11)$$

Reynolds number is defined as

$$Re_a' = GD/[K'(8u/D)^{n'-1}]$$

where G is mass flux.

Heat transfer in laminar flow region is correlated as

$$Nu_m = 4.631(K_b'/K_w')^{0.14}[0.4935\{Pr'(Re_a'/y^*)^{3.475}\}^{0.53} \\ + \{1 + 0.0954(Gz)^{0.8685}\}^{2.6316}]^{0.2} \qquad (12)$$

where Pr' is defined as

$$Pr' = (C_p K'/k)(8u/D)^{n'-1}$$

WORKING FLUID AND GEOMETRIC CONDITION

Working fluid and geometric factors were determined in advance for each twisted-tape as,
(1) Full length and full width twisted-tape
working fluid ; water (Pr=5), y=2.5, δ/D=0.05, L/D=150. Physical properties were referred at film temperature.
(2) Regularly spaced twisted-tape
working fluid ; water (Pr=5), y=3.46, δ/D=0.05, s=5, L/D=150. Physical properties were referred at film temperature.
(3) Reduced width twisted-tape
working fluid ; oil (Pr=600), y*=5.82, n'=0.75, K'=0.369-0.359, L/D=89.6, δ/H =0.1.

SMOOTH TUBE

Nusselt number and friction factor in a smooth tube for water are expressed as,

in laminar flow (JSME, 1975)

$$Nu_o=3.65+[0.0668Re_oPr(D/L)/\{1+0.004[Re_oPr(D/L)]^{2/3})]\} \qquad (13)$$
$$f=16/Re_o \qquad (14)$$

in turbulent flow (Kays and Crawford, 1993)

$$Nu_o=0.023Re_o^{0.8} Pr^{0.4} \qquad (15)$$
$$f_o=0.0791/Re_o^{0.25} \qquad (16)$$

and for oil (Metzner et al., 1957)

$$Nu_o=1.75[3n'+(1/4n')]^{1/3} Gz^{1/3} (K_b'/K_w')^{0.14} \qquad (17)$$

THERMAL PERFORMANCE

In the present study two evaluation methods were applied. The first method is to utilize (1) the first law of thermodynamics, namely, the conservation of energy and (2) the second one the second law of thermodynamics, namely, the entropy generation.
(1) The Nusselt number ratio, with and without heat transfer enhancement, Nu_a/Nu_o, was selected for three kinds of conditions:
(a) Reynolds number fixed, basic geometry fixed ;
$Re_a=Re_o$
(b) Pressure loss fixed, basic geometry fixed ;
$f_a Re_a^2 = f_o Re_o^2$
(c) Pumping power fixed, basic geometry fixed ;
$f_a Re_a^3 = f_o Re_o^3$

where f is the friction factor and subscrips a and o mean with and without heat transfer improvement.
(2) The entropy generation rate for forced convection heat transfer is generally expressed by the following equation (Bejan, 1982),

$$S_{gen} = q'\Delta T/T^2 + m/(\rho T)(-dp/dX) \qquad (18)$$

where q' is the heat flow per unit length and unit time, T the temperature, ΔT the temperature difference, m the mass flow rate, ρ the density and (-dp/dx) the pressure gradient. The first and second term in eq.(18) are the entropy generation due to heat transfer and flow, respectively.

The entropy generation rate, S_{gen}, is expressed by using the Nusselt number Nu and the friction factor f.

$$S_{gen} = \{1/(Nuk)\}(q'/T)^2(D_H^2/4A) + f(2A/T)(\rho v_f^3 Re^3/D_H^4) \qquad (19)$$

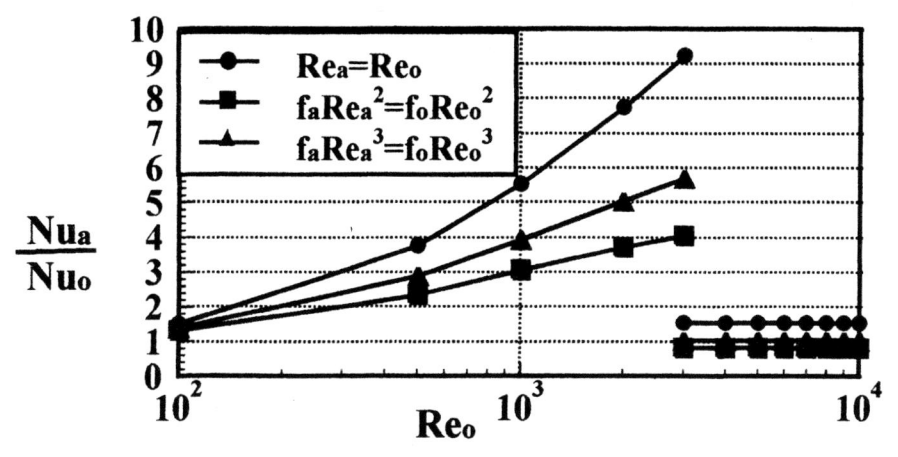

(a) Full length, full width twisted tape

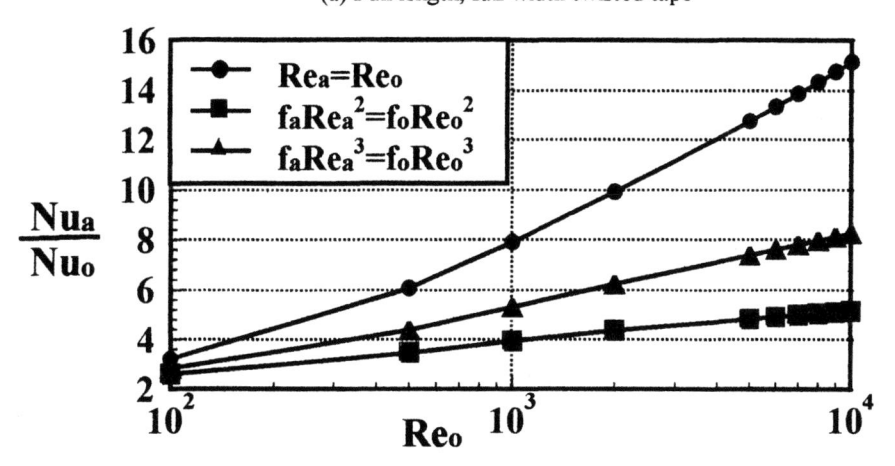

(b) Regularly spaced twisted tape

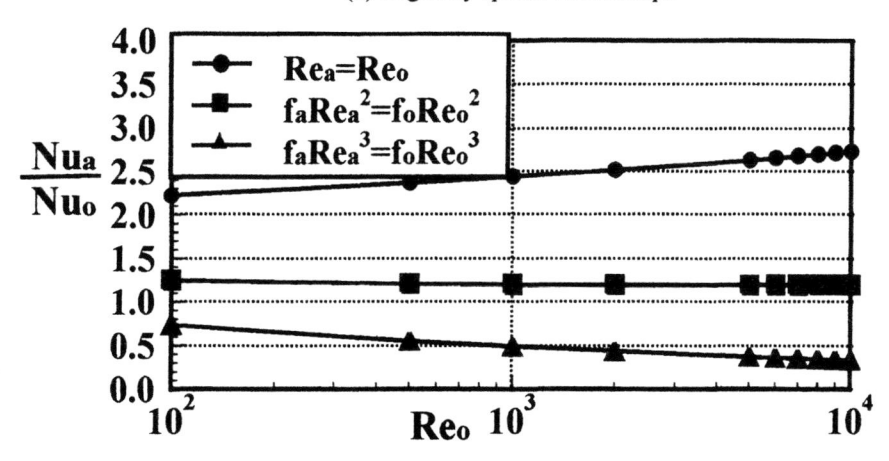

(c) Reduced width twisted tape

Fig.2 Nusselt number ratio

(a) Smooth tube

(b) Twisted-tape

(c) Entropy generation rate ratio

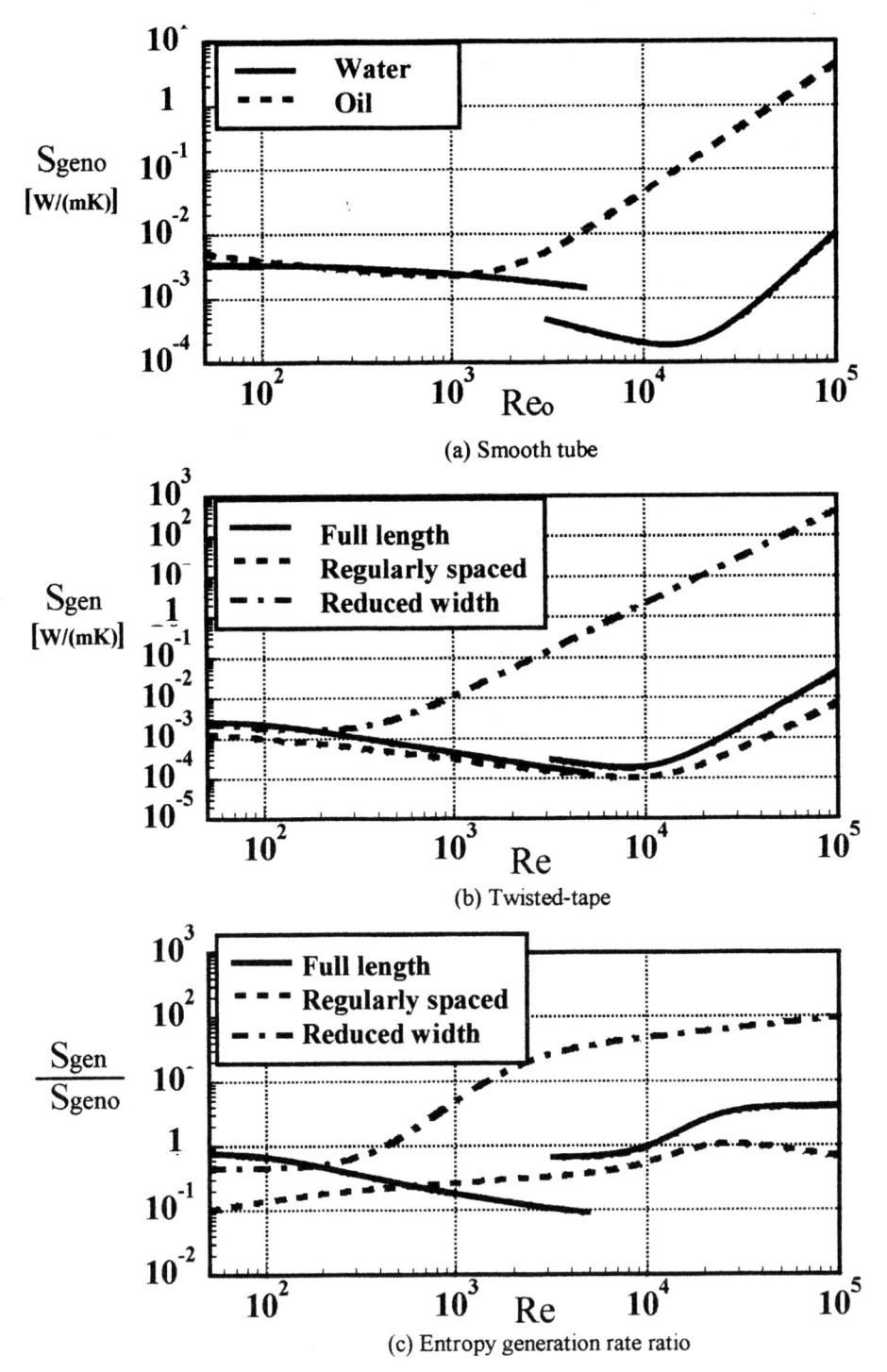

Fig.3 Entropy generation rate

where k is the thermal conductivity of fluid, A cross sectional area, D_H the hydraulic diameter, ρ the density and v_f the kinematic viscosity.

RESULTS AND DISCUSSIONS

Figure 2 (a),(b) and (c) denote the heat transfer increase Nu_a/Nu_o for three kinds of conditions [(1)$Re_a=Re_o$, (2) $f_a Re_a^2 = f_o Re_o^2$ and (c) $f_a Re_a^3 = f_o Re_o^3$] in semi-logarithmic scale. In the case of full length, full width twisted-tape and regularly spaced twisted-tape [Fig.2 (a) and (b)], Nu_a/Nu_o increases from the laminar flow region to the transition region and is higher than 1.0. However in the case of full length and full width twisted tape, both Nusselt numbers with and without twisted tapes depend on $Re^{0.8}$ in turbulent flow region. Therefore, Nu_a/Nu_o does not depend on Reynolds number and is constant. In the case of reduced width twisted tape [Fig.2 (c)], the abscissa Re_o means Re_o'. Nu_a/Nu_o does not depend Re_o in whole region. Prandtl number is higher than another two twisted tapes. However, effect of Prandtl number is eliminated by the ratio of Nusselt number. Evaluation at constant $Re_a=Re_o$ is better than those of another criteria. In the last two twisted tapes, the calculation was performed over the applied region.

Entropy generation rate is presented in Fig.3 (a),(b) and (c) in logarithmic scale. Generally, according to eq. (19), entropy generation rate due to heat transfer (the first term in right hand side) decreases and that due to flow (the second term in right hand side) increases with increase of Reynods number. Therefore, it is expected that S_{gen} takes the minimum at a certain Reynolds number. In Figs. 3, the abscissa is Reynolds number Re and the ordinate is entropy generation rate S_{gen}. In the case of a smooth tube [Fig.3(a)], S_{geno} for water (solid line) and oil (dotted line) take the minimum at $Re_o=1.5 \times 10^4$ in turbulent flow region and $Re_o(=Re_o') = 800$ in laminar flow region. However, S_{geno} for oil is remarkably higher than that for water in turbulent flow region due to the difference of flow resistance. In the next step, the entropy generation S_{gen} in tape generated swirl flow is shown in Fig. 3 (b). The behavior and absolute values of S_{gen} for full length, the full width twisted tape and the regularly spaced twisted tape are almost same, and they take the minimum at $Re_o=10000$. S_{gen} for the reduced width twisted-tape takes the minimum at Re=250. The position of the minimum moves to low Reynolds number. Figure 3 (c) presents the ratio of S_{gen} for tube with tape-inserts to S_{geno} for smooth tube. The ratio is almost less than 1.0 in low Reynolds number. The results denote that twisted tape is effective for the entropy generation in low Reynolds number. Particularly, the regularly spaced twisted tape is useful in whole region and even in turbulent flow the entropy generation is lower than that due to another twisted tapes.

CONCLUSIONS

Heat transfer increase Nu_a/Nu_o and entropy generation rate S_{gen} for assessing thermal performance were evaluated for three kinds of twisted-tape (the full length, full width type, the regularly spaced type and the reduced width type) in a smooth tube.

In the present conditions, in the case of the full length, full width twisted-tape and the regularly spaced twisted-tape, Nu_a/Nu_o for heat transfer improvement increases with Reynolds number in laminar flow region. In the case of reduced width twisted-tape, it does not depend on Reynolds number and the heat transfer enhancement is depressed for the evaluation at fixed pumping power. The entropy generation rate S_{gen} gives the operating condition which is Reynolds number at the minimum of the entropy generation. Regularly spaced twisted tape supplies the better performance. Generally the twisted-tape is effective in laminar flow region.

NOMENCLATURE

A	cross sectional area, m^2
C_p	specific heat, J/kgK
d	rod diameter, m
D	inner diameter of test tube, m
D_H	hydraulic diameter, m
f	fanning friction factor
Gz	Graetz number, mC_p/kL
h	heat transfer coefficient, W/m^2K
H	180 degree twist pitch, m
k	fluid thermal conductivity, W/mK
L	axial length, m
m	mass flow rate, kg/s
Nu	Nusselt number, hd/k
Pr	Prandtl number, $C_p \mu /k$
ΔP	pressure loss, N/m^2
q'	heat flux per unit length, W/m
Ra	Rayleigh number, $\rho^2 g d^3 C_p \beta \Delta T_w / k \mu$
Re	Reynolds number, $\rho V_o d/\mu$
Re_s	Reynolds number based on swirl velocity, $\rho V_s d/\mu$
S	space length, m
s	space ratio, S/D
S_{gen}	entropy generation, W/mK
Sw	swirl parameter, $Re_s/y^{1/2}$
T	temperature, K
ΔT_w	wall-to-bulk temperature difference
X	axial length, m
y	twist ratio, H/D
y*	modified twist ratio, H/W
V_o	mean axial velocity
V_s	swirl velocity, $V_o[1+(\pi/2y)^2]^{1/2}$

W   width of twisted tape
β coefficient of thermal expansion, 1/K
δ thickness of twisted tape, m
μ fluid viscosity, kg/ms
v_f the kinematic viscosity, m^2/s
ρ fluid density, kg/m^3

Subscripts

a   twisted-tapes inserts
b   bulk fluid temperature
o   smooth tube
w   tube wall temperature

AKNOWLEDGEMENTS

Author is grateful to Professor R.M. Manglik for sending the some references. The author thanks Mr. K. Toriyama for helping the calculation.

REFERENCES

Bejan, A., 1982, *Entropy Generation through Heat and Fluid Flow*, John Wiley and Sons, Inc., New York.

Bergles, A.E., Blumenkrantz, A.R. and Taborek, J., 1974, Performance Evaluation Criteia for Enhanced Heat Transfer Surfaces, *Proceedings of the 5th International Heat Transfer Conference*, Tokyo, Vol. 2, pp.239-243.

Date, A.W. and Singham, J.R., 1972, *Numerical Prediction of Friction and Heat Transfer Characteristics of Fully Developed Laminar Flow in Tubes Containing Twisted- Tapes*, ASME Paper No.72-HT-17.

JSME, 1975, *Data Book of Heat Transfer*, 3rd edition, pp.25, JSME, Tokyo.

Kays, W.M. and Crawford, M.E., 1993, *Convective Heat and Mass Transfer*, 3rd edition, pp.319, McGraw-Hill, Inc., New York.

Lopina, R.F. and Bergles, A.E., 1969, Heat Transfer and Pressure Drop in Tape Generated Swirl Flow of Single Phase Water, *Journal of Heat Transfer*, Vol.91, pp.434-442.

Manglik, R.M. and Bergles, A.E., 1993, Heat Transfer and Pressure Drop Correlations for Twisted-Tape Inserts in Iso-thermal Tubes : Part 1-Laminar Flows, *Journal of Heat Transfer*, Vol.115, pp.881-889.

Manglik, R.M. and Bergles, A.E., 1993, Heat Transfer and Pressure Drop Correlations for Twisted-Tape Inserts in Iso-thermal Tubes : Part 2-Transition and Turbulent Flows, *Journal of Heat Transfer*, Vol.115, pp.890-896.

Metzner, A.B., Vaughn, R.D. and Houghton, G.L., 1957, Heat Transfer to Non-Newtonian Fluids, *AIChE Journal*, Vol.3, pp.92

Patil, A.D., 2000, Laminar Flow Heat Transfer and Pressure Drop Characteristics of Power-Law Fluids inside Tubes with Varying Width Twisted Tape Inserts, *Journal of Heat Transfer*, Vol.122, pp.143-149.

Saha, S.K., Gaitonde, U.N. and Date, A.W., 1989, Heat Transfer and Pressure Drop Characteristics of Laminar Flow in a Circular Tube Fitted with Regularly Spaced Twisted-Tape Elements, *Experimental Thermal and Fluid Science*, Vol.2, pp.310-322.

Smithberg, E. and Landis, F., 1964, Friction and Forced Convection Heat Transfer Characteristics in Tubes with Twisted-tape Swirl Generator, *Journal of Heat Transfer*, Vol.86, pp.39-49.

Thorsen, R.S. and Landis, F., 1968, Friction and Heat Transfer Characteristics in Turbulent Swirl Flow Subjected to Large Transverse Temperature Gradients, *Journal of Heat Transfer*, Vol.90, pp.87-98.

Watanabe, K., Taira, T. and Mori, Y., 1983, Heat Transfer Augmentation in Tubular Flow by Twisted Tapes at High Temperatures and Optimum Performance, *Heat Transfer-Japanese Research*, Vol.12, pp.1-31.

STUDY ON CONVECTIVE HEAT TRANSFER IN FIBER-FINNED DUCTS BASED ON THE CONCEPT OF EQUIVALENT THERMAL CONDUCTIVTY

Zai-Zhong Xia[1]   Zhi-Xin Li[2] and Zeng-Yuan Guo[3]

[1] Tsinghua University, Beijing, 100084, CHINA; E-mail: xiazz@263.net
[2] Tsinghua University, Beijing, 100084, CHINA; E-mail: lizhx@tsinghua.edu.cn
[3] Tsinghua University, Beijing, 100084, CHINA; E-mail: demgzy@tsinghua.edu.cn

ABSTRACT Fins are usually used to enhance heat transfer between a primary surface and the surrounding fluid in heat exchange devices. The conventional concept of extended surface and fin efficiency are applied to analyze the heat transfer enhancement by fins. The evaluated fin performance is usually at odds with reality, especially for the fiber-finned channels where fin array is made up of very large length-to-diameter ratio fins. This paper introduces a new model for heat transfer in fiber-finned channel. Analyses on the established model reveal that the essence of heat transfer enhancement due to fiber-fins is the sharply increase of the equivalent thermal conductivity, instead of surface extension. And then, the enhancement effect due to equivalent thermal conductivity increment was studied in detail. Based on the analyses, a kind of fiber-finned duct using the dually twisted spring for increasing fluid equivalent thermal conductivity was designed and tested. The experimental results show that heat transfer in the finned duct is strongly enhanced while the additional flow resistance is relative small.

INTRODUCTION

Technology has led to a demand for high-performance, lightweight, and compact heat transfer devices. To accommodate this demand, finned surfaces are usually used to increase heat transfer rate between a primary surface and the surrounding fluid in heat exchange devices, as discussed in detail by Kern and Kraus (1972), Aziz (1992) and Kraus (1988). In general, fins are treated as a heat conduction problem, while the complex convective heat transfer between fluid and fin surface has rarely been investigated in detail. The heat flux between fin surface and the surrounding fluid is commonly considered in relation to the heat transfer coefficient and the temperature difference between fin surface and the surrounding fluid. The role of convection in conventional fin theory is confined to the introduction of an assumed uniform heat transfer coefficient and uniform fluid temperature. Thereof, the fin performance can be analyzed through a pure conduction differential equation, and then the fin effect on heat transfer is sketchily treated as extended surface.

Many investigators have analyzed the effect of finned surface using this well-known assumption attributed to Murray (1938) and Gardner (1945). However, Ghai and Jakob (1950, 1951) experimentally demonstrated that heat flux attained significantly greater value at fin tip rather than at fin base and concluded that the idealization of a uniform heat transfer coefficient and uniform fluid temperature is not necessarily realistic. Since then, a lot of papers analyzing the coordinate dependent heat conduction of fins have been published, and most of these reports were reviewed by Huang and Shah (1991). In the analysis, Razelos and Imre (1980) assumed that the heat transfer coefficient varies according to a power law of the distance from fin base.

In the analysis of heat transfer of finned surface in a channel, the disagreement with conventional fin theory, to author's knowledge, is resulted from the excessively simplified role of convection in fin system. The conventional fin model may conform much more to the conduction-convention systems in which the fluid with uniform velocity and temperature flows across fins. In this case, the heat transfer coefficients and temperature of the surrounding fluid don't vary with position from fin base to tip. However, it is noteworthy that the fluid velocity and temperature across fins in a channel are usually non-uniform especially when the flow and heat transfer have been fully developed. The conflict with the fin theory is intensified with regard to the flow and heat transfer in the fiber-finned channels where fin array is made up of very large length-to-diameter ratio fins. Many researchers, such as Brighham and Vanfossen (1984), Sparrow et al. (1984), Babus' Haq et al. (1995), and Wang et al. (1999) observed that the fin's length-to-diameter ratio is an important factor influencing the magnitude of heat transfer rate. They showed that the average fin heat transfer for short pin fins should be lower than long cylinder, and that the fin performance obtained from the conventional theory would be at odds with reality because of the forgetfulness of the fluid velocity and temperature distributions.

A new heat transfer model was introduced in this paper to analytically study the hydro-dynamically and thermally developed flow and heat transfer in fiber-finned ducts. In the analysis of fin conduction, the velocity and temperature distribution of the fluid across fins are taken into account and the convective heat transfer between fin surface and fluid is regarded as an internal heat source for analyzing fin performance. The effect of fins on convective heat transfer is treated as a special analogous internal heat source added into the energy equation, instead of the well-known extended surface. Based on the established model, the effect of fiber-fin on heat transfer enhancement is analytically studied. According to the analytical result, a kind of fiber-finned duct using the dually twisted spring for increasing fluid equivalent thermal conductivity was designed and experimented. The results show that the heat transfer in the finned duct is strongly enhanced and the additional flow resistance is relative small.

ANALYSIS

Everyone wants to design a heat transfer device with higher heat transfer and lower pressure drop. For this purpose, the heat transfer performance of an idealized fiber-finned duct, shown in Fig.1, is analytically studied with regard to the hydro-dynamically and thermally developed flow and heat transfer. Numerous cylindrical pin fins with large length-to-diameter ratio (usually more than 10), are regularly arranged normal to the wall.

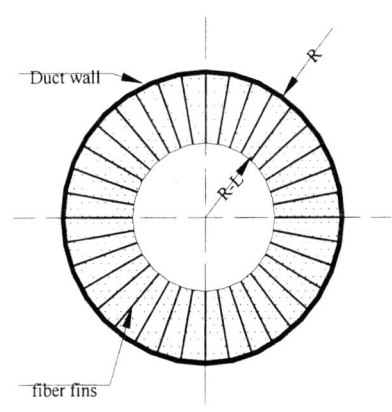

Fig. 1 Fiber-finned duct

Heat Transfer Model

It is clear that the heat flux, q, between fin surface and surrounding fluid can be expressed as $h(T - T_{fin})$, where the subscript "fin" denotes the variable of fins. Different from the conventional fin theory, the variable T, here, is the local fluid temperature not the bulk fluid temperature. It varies with the position from fin root to fin tip.

For the convenience to analyze the effect of fins on heat transfer, one-dimensional assumption is adopted. The heat conduction equation of the fiber-fin is as follow,

$$k_{fin}\frac{d^2 T_{fin}}{dr^2} + S_{fin} = 0, \quad S_{fin} = \frac{4h(T - T_{fin})}{d}, \quad (1)$$

where heat transfer between fluid and fin surface is regarded as an internal heat source.

A counteractive internal heat source, S, corresponding to S_{fin} should act on the surrounding fluid, which is expressed as,

$$S = -\varepsilon \cdot S_{fin}, \quad (2)$$

where the variable ε, varying with the radial coordinate r, is called the local volume ratio of fins to fluid, which is expressed as $\varepsilon = \varepsilon_0 r / R$. Then, the energy equation of fluid can be written as

$$\rho c_p u \frac{dT}{dx} = \frac{d\left[(k + k_t)r \frac{dT}{dr}\right]}{rdr} + S, \quad (3)$$

where k is the fluid thermal conductivity, and k_t the turbulent thermal conductivity.

Equivalent Thermal Conductivity

The effect of fiber fins on the heat transport within fluid is clearly embodied by the counteractive internal heat source S in equation (3). A novel concept of the equivalent thermal conductivity due to fiber fins is applied here to account for the effect of fiber fins on heat transfer enhancement. On the other word, we do not consider fins as an extended surface in conventional fin theory, but as slender inserts to increase the equivalent thermal conductivity of fluid. That is to say, numerous fiber-fins arranged normal to the duct surface, as shown in Fig.1, will sharply increase the thermal conductivity of the filling region.

Introducing an equivalent thermal conductivity of fluid, k_{equ}, which is resulted from the effect of fins on convective heat transfer, we can rewrite the heat source S as

$$\frac{d\left(k_{equ} r \frac{dT}{dr}\right)}{rdr} = S. \quad (4)$$

The dimensionless equations of Eq. (1) and Eq. (4) are,

$$\frac{d^2 T_{fin}^*}{dr^{*2}} + \frac{4h(T^* - T_{fin}^*)R^2}{k_{fin} d} = 0, \quad (5)$$

$$\frac{d\left(k_{equ}^* r^* \frac{dT^*}{dr^*}\right)}{r^* dr^*} + \frac{\varepsilon_0}{r^*} \cdot \frac{4h(T^* - T_{fin}^*)R^2}{k_{fin} d} = 0, \quad (6)$$

where the dimensionless parameters are defined as,

$$r^* = \frac{r}{R}, \quad T^* = \frac{T-T_w}{T_m-T_w}, \quad T_{fin}^* = \frac{T_{fin}-T_w}{T_m-T_w}, \quad (7)$$

here, T_m and T_w denote bulk fluid temperature and wall temperature respectively.

From Eq. (5) and Eq. (6), the following differential equation can be obtained,

$$\frac{d\left(k_{equ}^* r^* \frac{dT^*}{dr^*}\right)}{dr^*} + \varepsilon_0 \frac{d^2 T_{fin}^*}{dr^{*2}} = 0 \quad (8)$$

The differential equation (8) will be self-complacent when

$$k_{equ}^* = \frac{\varepsilon_0}{c_1 r} \quad \text{and} \quad T^* = c_1 T_{fin}^* \quad (9)$$

here c_1 is an undetermined constant. And then, Eq. (5) is rewritten as follow after the dimensionless temperature T^* is substituted by $c_1 T_{fin}^*$.

$$\frac{d^2 T_{fin}^*}{dr^{*2}} + m^2 T_{fin}^* = 0, \quad m = \sqrt{\frac{4h(c_1-1)R^2}{k_{fin}d}} \quad (10)$$

The differential equation (10) governs the fin temperature distribution, and can be solved at the following thermal boundary conditions,

$$T_{fin}^* = 0 \text{ at } r^* = 1 \text{ and } \frac{dT_{fin}^*}{dr^*} = 0 \text{ at } r^* = 1 - \frac{L}{R}$$

The temperature distribution of fiber-fin can be expressed as,

$$T_{fin}^* = c_2 \cos\left[mr^* - m\left(1-\frac{L}{R}\right)\right] \quad (11)$$

where c_2 is an integral constant. Meanwhile, the variable m must satisfy $mH/d = \pi/2$. As a result, the undetermined constant c_1 can be obtained according to equation (10).

$$c_1 = \left(\frac{0.6 k_{fin} d}{hL^2} + 1\right) \quad (12)$$

Substituting equation (12) into equation (9) results in the expression of the equivalent thermal conductivity,

$$k_{equ} = \frac{\varepsilon k_{fin}}{\left(\frac{0.6 k_{fin} d}{hL^2} + 1\right)} \quad (13)$$

Nusselt Number

Based on the concept of equivalent thermal conductivity corresponds to fin effect, the energy equation of fluid in fiber-finned duct can be written as

$$\rho c_p u \frac{dT}{dx} = \frac{d\left(k_{total} r \frac{dT}{dr}\right)}{rdr} \quad (14)$$

where k_{total} is composed of three parts, k, fluid thermal conductivity, k_t, the turbulent thermal conductivity and k_{equ}, the equivalent thermal conductivity corresponds to fin effect on convective heat transfer, that is, $k_{total} = k + k_t + k_{equ}$.

The dimensionless energy equation for constant wall heat flux is

$$\frac{\left(k_{total}^* r^* \frac{dT^*}{dr^*}\right)}{dr^*} + u^* \cdot Nu = 0 \quad (15)$$

where $k_{total}^* = k_{total}/k$, the thermal boundary conditions are

$$\frac{dT^*}{dr^*} = 0 \text{ at } r^* = 0, \text{ and } T^* = 0 \text{ at } r^* = 1$$

According to its definition, the dimensionless temperature of fluid satisfies the following equation,

$$\int_0^1 u^* T^* dy^* = 1 \quad (16)$$

By solving equation (15), therefore, the Nusselt number can be expressed as,

$$Nu = \frac{1}{\int_0^1 \frac{U^2}{2r^* k_{total}^*} dr^*}, \quad U = \int_0^{r^*} 2r^* u^* dr^* \quad (17)$$

It can be seen from eq. (17) that only the distribution of the dimensionless velocity (u^*) and total thermal conductivity (k^*) determine the magnitude of the Nusselt number of convective heat transfer in the fiber-finned duct.

DISCUSSION

Effects of Fluid Velocity Distribution on Heat Transfer

When numerous fiber fins are arranged normal to the wall surface in a duct, as shown in Fig.1, an increment of pressure drop is needed to drive the fluid past the fiber-fin array. As a result, the fluid velocity does vary from parabolic distribution of laminar flow in circular tube without fins.

It is difficult to obtain the velocity distribution in the fiber-finned tube. To discuss the effect of velocity profile on heat transfer, we assume that the velocity profile is as follow,

$$u^* = \frac{n+2}{n}\left(1-r^{*n}\right) \quad (19)$$

where the variable n denotes the velocity distribution feature. The velocity distribution is parabolic when $n=2$, and becomes flatter with the increment of n, as shown in Fig. 2. Then, the Nusselt number evaluated from eq. (17), when $k_{total}^* = 1$, can be expressed as follow,

$$Nu = \frac{8(n^2+6n+8)}{n^2+10n+20} \quad (20)$$

Fig.3 shows the results of Nusselt number for different velocity profiles. It is observed from Fig. 3 that the Nusselt number increases with the value of n. Nevertheless, this increment or decrement due to different velocity profile is within 40%, which is inappreciable compared with those due to the equivalent thermal conductivity. It indicates that the velocity flatness in turbulent flow plays a negligible role in heat transfer enhancement, contrasted to the turbulent conductivity k_t.

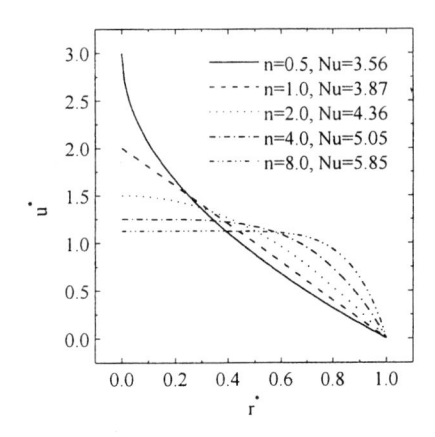

Fig.2 Velocity distribution

Effects of Total Thermal Conductivity on Heat Transfer

Compared with the velocity distribution, the dimensionless total thermal conductivity, k^*_{total}, composed of fluid thermal conductivity, turbulent thermal conductivity and equivalent thermal conductivity due to fiber-fins, have a strong effect on Nusselt number, which can be seen from eq. (17). When the distributions of total thermal conductivity and fluid velocity hold changeless, the Nusselt number will increase in proportion to the increment of total thermal conductivity. This is the reason why turbulent duct flow always takes on superior heat transfer rate to laminar flow. Besides the magnitude, the profile of total thermal conductivity has also significant effect on Nusselt number.

When $\int_0^1 2r^* k^*_{total} dr^*$ keeps a fixed value, C, there is an optimal thermal conductivity distribution obtained from Eq. (17).

$$\left(k^*_{total}\right)_{opt} = \frac{CU}{2r^* \int_0^1 U dr^*} \qquad (21)$$

With the optimal thermal conductivity distribution, the maximum Nusselt number can be realized. Eq. (21) denotes that the non-dimensional thermal conductivity times $2r^*$, $\left(k^*_{total}\right)_{opt} \cdot 2r^*$, should be a function of the variable U which is the dimensionless volumetric flow rate flowing through the cross section.

Figure 3 shows the optimal distribution relating to different velocity profiles expressed by eq. (19). It is observed that the thermal conductivity of the region near to duct wall should be greatly improved relative to close to duct center, especially for laminar flow (n > 2) for which the heat transfer should always be enhanced.

Effects of Fiber Fins on Heat Transfer

The eq. (13) gives the relation of the equivalent thermal conductivity to dimensionless parameter $k_{fin} d/(hH^2)$, it is plotted in Fig.4. The parameter $k_{fin} d/(hH^2)$ can be regarded as the ratio of heat conducted across fin cross section ($k_{fin} \pi d^2/4$) to the one transported through fin surface ($h \cdot \pi dH$) to fluid by convection.

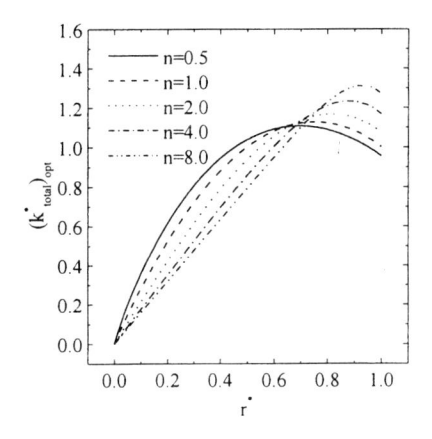

Fig.3 Optimal distribution of total thermal conductivity for different velocity profiles

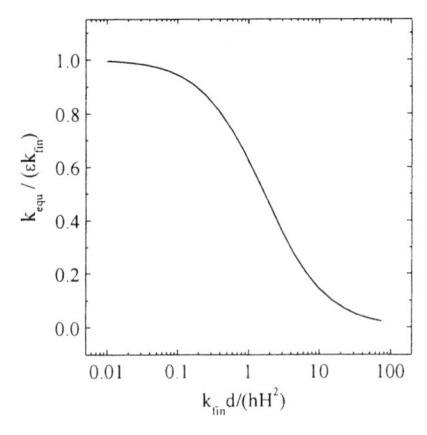

Fig. 4 Equivalent thermal conductivity versus $k_{fin} d/(hL^2)$

It can be seen from Fig. (4) that the equivalent thermal conductivity, k_{equ}, increases with the decrease of $k_{fin} d/(hL^2)$, and tends to the maximum value (εk_{fin}) which usually presents the strong effect of the high conductivity of fins. For example, copper fins filling in water can result in a thermal conductivity increment of 6 times if the fins volume is only 1% of the total space, even 15 times when the fluid is air. Increasing the heat transfer coefficient, h, and the fin length-to-diameter ratio, L/d, can reduce the value of $k_{fin}d/(hL^2)$, thereby, the equivalent thermal conductivity is increased.

For a given porosity, the fin length, L, plays an important role in heat transfer enhancement, which has strong effect on the distribution of total thermal conductivity. For example, the turbulent thermal conductivity k_t usually equals to zero at duct wall, and then increases with the decrease of radial coordinate. Approximately near to $R/2$, the value of k_t arrives at its maximum, subsequently decreases, as discussed by Gowen and Smith (1968). The bottleneck of turbulent heat transfer is the lower thermal conductivity in the region near to duct wall. For the case of turbulent convective heat transfer, therefore, the length of fins, L, usually is small, which increases the equivalent thermal conductivity and consequently improves the distribution of total thermal conductivity. But, for laminar heat transfer (k_t =0), the fiber fins should be of large length to diameter ratio to increase the total thermal conductivity in entireness flow region.

We assume that the turbulent thermal conductivity k_t can be expressed as $k_{t,max}^* \cdot \sin(r^*\pi/2)$, according to the result of Gowen and Smith. If the equivalent thermal conductivity is assumed to be equal to $10/r^*$, the effect of filling height is shown in Fig. 5, where $Nu_{L=0}$ is the Nusselt number in smooth tube.

It can be seen, from Fig. 5 that the heat transfer enhancement of laminar flow (for the case of $k_{t,max}^*$ =0) is much more than that of turbulent flow ($k_{t,max}^*$ >0). For laminar convective heat transfer, nevertheless, the increment of Nusselt number is relative smaller for the case of L/R<0.5, after L/R>0.5 the Nusselt number increase rapidly and reaches the maximum at L/R=0.8. This is in good agreement with the results offered by other researchers, such as Hu and Chang (1973), Nandakumar and Masliyan (1975).

EXPERIMENTS

Based on the above analysis, we designed a novel heat transfer ehhancement element called the dually twisted spring enhancement tube for increasing the thermal conductivity in the region near to tube wall. The metal wire of 0.3mm was twisted as a spring with a diameter of 5 mm and a pitch of 1.6mm. And then, this spring are twisted once again as the dually twisted spring with a middle diameter of 10 mm and a pitch of 6mm. Another dually twisted spring was made in light of the same steps and dimensions except the wire diameter was altered to 0.5mm. Finaly, these two enhancement elements were inserted respectively into $\phi 19 \times 2$ steel tubes after heat treatment, and were brazed on the internal surface of the tube, shown in figure 6. The dually twisted spring can be regarded as the fiber fins to increase the thermal conductivity of the filling region with fin length of about 0.66R. It is estimated that $\varepsilon \approx 0.015 r/R$ and L/d=10 for the 0.3mm wire, and $\varepsilon \approx 0.04 r/R$ and L/d=16 for the 0.5mm wire.

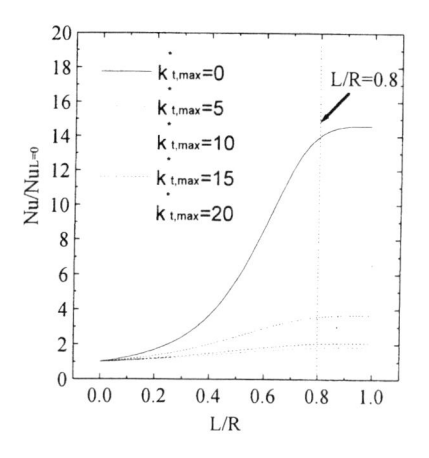

Fig. 5 Heat transfer enhancement versus filling height

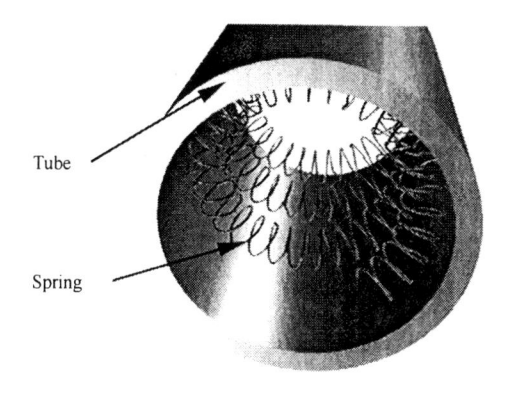

Fig. 6 The tube finned with the dually twisted spring

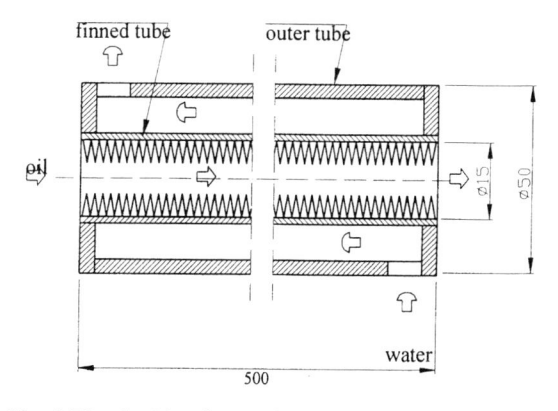

Fig. 7 The double-pipe exchanger using finned tube

The experiments for heat transfer and flow resistance were performed in a double-tube exchanger shown in Fig. 7. The transformer oil with temperature of 60℃ flows within the finned tube, and water with temperature of 30℃ flows outside. The testing Reynolds number ranges from 500 to 3000.

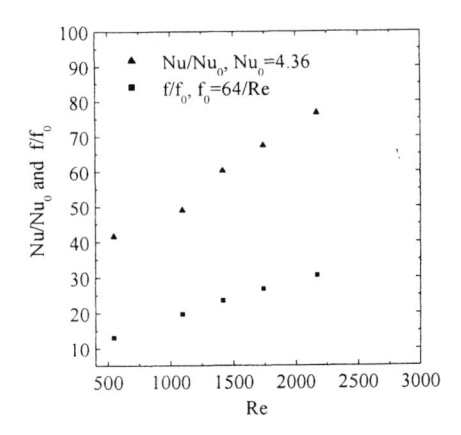

Fig. 8 Heat transfer and flow resistance of the finned tube (wire diameter of the spring is 0.3mm)

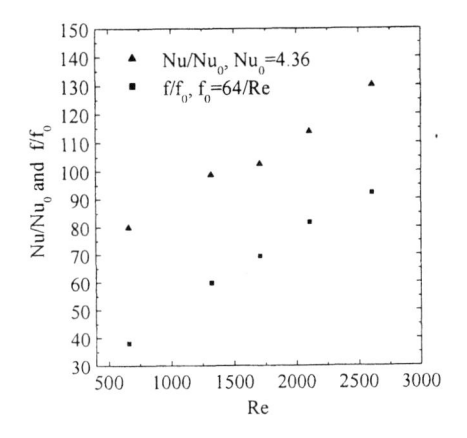

Fig. 9 Heat transfer and flow resistance of the finned tube (wire diameter of the spring is 0.5mm)

It is a common knowledge that the entrance effect on heat transfer and fluid flow is usually weak after ten rows of fiber fin. The entrance distance is about 50mm, which is much less than the tube length (500mm). As a result, the average Nusselt number and frictional resistance coefficient estimated from experimental data can be regarded as those of hydro-dynamically and thermally developed. Compared to the heat transfer and frictional factor of fully developed laminar flow in circular tube, the heat transfer enhancement and flow resistance increment were shown in Fig.8 for the finned tube using 0.3mm wire and in Fig. 9 for the finned tube using 0.5mm wire.

The considerable enhancement effect of the finned ducts can be observed from Fig. 8. The Nusselt number was enhanced approximately 40 times at Re = 600 than that of smooth tube, even 80 times when Re equals to 2500. It is interesting, however, that the increments of frictional coefficient are much lower than that of heat transfer enhancement. It means that, this kind heat transfer component is of a performance of high heat transfer enhancement and low increment of frictional resistance, especially for the case of low Reynolds number. For example, Nu/Nu_0 equals to 42 but f/f_0 is only 13. The another duct using 0.5mm wire is also of the superior heat transfer enhancement performance, as shown in Fig. 9. Nevertheless, the finned duct using 0.5mm takes on more increment of flow resistance than the one using 0.3mm wire.

CONCLUSIONS

A new heat transfer model was introduced to analytically study the hydro-dynamically and thermally developed flow and heat transfer in fiber-finned duct. In the analysis of fin conduction, the velocity and temperature distribution of the fluid across fins are taken into account and the convective heat transfer between fin surface and fluid is regarded as an internal heat source for analyzing fin performance. The analytical results show that the effect of fiber-fins on heat transfer enhancement is essentially due to the increment of the fluid equivalent thermal conductivity in the filling region, especially the great increment is in the region near to duct wall. According to the analytical result, a kind of fiber-finned duct using the dually twisted spring for increasing fluid equivalent thermal conductivity was designed and experimented. The experimental data show that the finned ducts are of good performance of heat transfer enhancement and less increment of pressure drop.

Acknowledgement This paper is funded by National Key Project of Fundamental R&D of China (Grant No. G2000026301) and the National Natural Science Foundation of China (Grant No. 59995550-2). The authors gratefully acknowledge their supports.

NOMENCLATURE

c_1	constant
c_p	specific heat, J/kgK
d	fin diameter, m
h	convective heat transfer coefficient between surrounding fluid and fin surface, W/m²K
k	fluid thermal conductivity, W/mK
k_{fin}	fin thermal conductivity, W/mK
k_t	fluid turbulent thermal conductivity, W/mK
k_{equ}	equivalent thermal conductivity, W/mK
k_{total}	total thermal conductivity, $k+k_t+k_{equ}$, W/mK
k^*_{equ}	dimensionless equivalent thermal conductivity resulted from fiber fin, k_{equ}/k_{fin}
k^*_{total}	dimensionless total thermal conductivity, k_{total}/k

L	fin length, m
Nu	Nusselt number, dimensionless
q	heat flux between surrounding fluid and fin surface, $h(T-T_{fin})$, W/m²
r	radial coordinate, m
R	tube radius, m
Re	Reynolds number, $u2R/\nu$, dimensionless
T	fluid temperature, K
T*	dimensionless fluid temperature, T/T_m
T_{fin}	fin temperature, K
T^*_{fin}	dimensionless fin temperature, T_{fin}/T_m
u	velocity in axial direction, m/s
u*	dimensionless velocity in axial direction, u/u_m
u_m	mean velocity, m/s

REFERENCES

Aziz A., 1992, Optimum dimensions of extended surfaces operating in a convective environment, *Appl. mech. rev.* Vol. 45(5), pp.155-173.

Babus'Haq R. F. et al., 1995, Thermal performance of a pin-fin assembly, *Int. J. Heat Mass Transfer*, Vol. 16, pp. 51-55.

Brighham B. A. and Vanfossen G. J., 1984, Length to diameter ratio and row number effects in short pin fin heat transfer, *J. Eng. for Gas Turbines and Power*, Vol. 106, pp. 241-245.

Gardner A., 1945, Efficiency of extended surfaces, *Trans. ASME*, Vol. 67, pp. 621-631.

Ghai M. L. and Jakob M., 1950, Local coefficients of heat transfer on fins, *ASME Paper*, 50-5-18.

Ghai M. L., 1951, Heat transfer in straight fins, *Proceedings of General Discussion on Heat Transfer*, Institution of Mechanical Engineers, London, pp. 203-204

Gowen R. A. and Smith J. W., 1968, Turbulent heat transfer from smooth and rough surfaces, *ibid*, Vol.11, pp. 1657-1667.

Hu M. H. and Chang Y. P., 1973, Optimization of finned tubes for heat transfer in laminar flow, *J. Heat Transfer*, Vol. 95, pp. 332-339.

Huang L. J. and Shah R. K., 1991, Assessment of calculation methods for efficiency of straight fins of rectangular profile, ASME HTO, *Advances in Heat Exchanger Design*, Radiation and Combustion, Vol. 182, pp. 19-30.

Kern D. Q. and Kraus A. D., 1972, *Extended Surfaces Heat Transfer*, McGraw-Hill, New York, USA.

Kraus A. D., 1988, Sixty-five years of extended surface technology (1992-1987), *Appl. mech. rev.* Vol. 41, pp. 321-364.

Murray M., 1938, Heat dissipation through an annular disk or fin of uniform thickness, *J. Appl.*, 5:A78.

Nandakumar K. and Masliyah J. H., 1975, Fully developed viscous flow in internally finned tubes, *The Chem. Eng. J.*, Vol. 10, pp. 113-118.

Razelos P. and Imre K., 1980, The optimum dimensions of circular fins with variable thermal parameters, *ASME J. Heat Transfer*, Vol.102, pp. 420-425.

Sparrow E. M., Stahl T. J. and Traub P., 1984, Heat transfer adjacent to the attached end of a cylinder in crossflow, *Int. J. Heat Mass Transfer* Vol. 27, pp. 233-242.

Wang S., Guo Z. Y. and Li Z. X., 1999, Study on heat transfer enhancement of very large length-to-diameter pin fins in duct flow, *Kyto University-Tsinghua University Joint Conference on Energy and Environment*, pp. 37-42.

COMBINED TECHNIQUES OF HEAT EXCHANGE ENHANCEMENT ALONG THE PATHS OF GAS-TO-GAS COOLERS FOR SMALL GTE

A.V. Soudarev, B.V. Soudarev, V.V. Grishaev, A.S. Leznov

[1]NPP "TARC" Ltd, JSC "CTS" Ltd, Sedova str.15, 193148 St.Petersburg, Russia,
telefax (812) 2653448, E-mail: Soudarev@Tark.spb.ru,
[2]OAO "GAZPROM", Nametkina str.16, 117884, GSP Moscow B420, Russia,
tel.: (095) 7193011, fax : (095) 7191326, E-mail : a.leznov@adm-rao.gazprom.ru

ABSTRACT

The small-size power generation gas turbine units (Ne<300 KW) are essentially superiour to the diesel units in terms of their environmental characteristics. Although, provision of their competitiveness and a steady commercialization on the power market might be achieved only by increasing their cost-effectiveness. Ensuring a notable increase in the efficiency of the small-size GTEs might be attained through complication of the thermodynamic cycle. The latter requires incorporation of the heat exchangers meant to regenerate the heat and to cool the compressed air between the compressor stages. Drive for independence and reliability of a unit paves way for application of the gas-to-gas recuperative heat exchangers having a relatively low convective heat exchange along either path. To reduce the overall dimensions and metal consumption rate in these heat exchangers, combined techniques of the heat exchange enhancement are used.

Findings of the studies over the convective heat transfer and hydraulic resistance in the internal path of the high-temperature gas ceramic cooler are presented. It was demonstrated that application of the combined technique of heat exchange enhancement using a string of split fin inserts bonded to the inner surface of tubes by the diffusive welding technique allows an increase in the compactness factor for the heat exchanger matrix up to 400-600 m^2/m^3.

INTRODUCTION

The environmental aspects of the power conversion and transportation affect everyone. On the other hand, need in power all over the world is on the persistent rise. The CO_2 content in the atmosphere rises proportionally. For the last century, it increased by as much as 14% (Shamanov N.P. and Nefedova A.V., 1999). You cannot halt this process, at the same time it is a must to alleviate its detrimental environmental effect. The target could be achieved through increasing the efficiency of the power plants and mustering waste-free and power-preserving technologies.

The gas-turbine small-size power generation units (GTUs) are being currently widely used by the power producers. The increase in the GTU efficiency and, therefore, the improvement of its environmental friendliness are implemented by elevation of the gas turbine inlet temperature (TIT), heat regeneration and reducing the power consumption on the intrinsic needs of the engine. With use of the heat-resistant and high-temperature metal materials, the TIT growth is accompanied with enlargement of the power consumption on cooling of the heat-stressed parts, the latter diminishing essentially the beneficial effect of TIT growth on the efficiency of the gas-turbine units.

The heat regeneration allows a notable increase in the efficiency of the small-size gas-turbine engines (GTE) (Fig.1). Fig.1 shows the relationships between the GTU efficiency η_e and the net power Ne both for the existing simple cycle power plants and the regenerated ones. These data have been reported in the turbomachine catalogues issued by the various companies for the last 40-50 years (Table 1).

The 10-15% increase in η_e takes place just in the small-size GTUs (Ne<0.5 MW) since the regeneration here results in the reduction in the optimum pressure ratio. This, in turn, leads to growth of the inside efficiencies of the appropriate turbomachines (Pchelkin Yu.M. et al., 1995). Along with it the TIT increase results inevitably in the elevation of the temperature of gas that enters the AH. Therefore, the matrix of the heat exchanger should be made of heat-resistant materials.

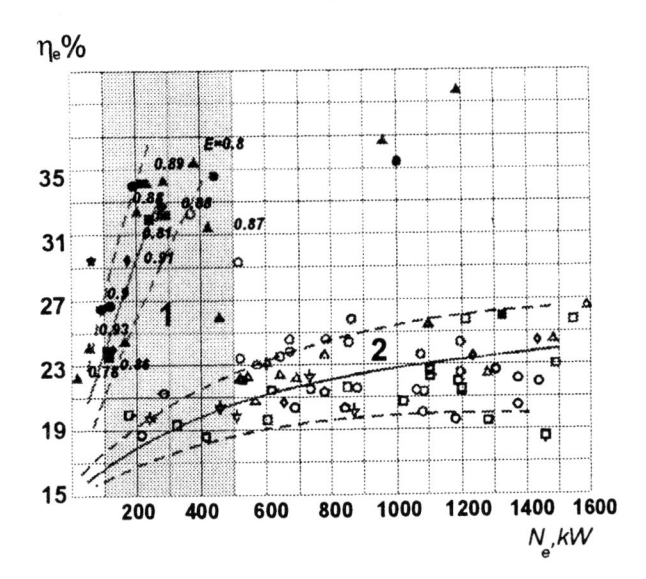

Fig.1.
Comparison of effectiveness of regenerated (1) and non-regenerated (2) GTEs of various companies.
1,2 – regenerated, simple cycle GTU (see Table 1).

Table 1
List of companies, whose existing engines supplied the characteristics presented on the graph of Fig.1.

N	As designated in Fig.	COMPANY	COUNTRY	SOURCE
\multicolumn{5}{c}{**SIMPLE CYCLE GTU**}
3	○	FIAT	Italy	Popov V.S. et al., 1987; ITH, 1996-2000.
4	☆	MAN	Germany	
5	◐	Kawasaki	Japan	
6	◐	Pratt & Whitney	Canada	
7	○	Solar	USA	
8	□	Tuma Turbomach	Switzerland	
9	□	Turbomeca	France	GTW, 1988
10	□	Ekonomaizer	USSR	Viktorov B.G. et al., 1987

N	As designated in Fig.	COMPANY	COUNTRY	SOURCE
\multicolumn{5}{c}{**SIMPLE CYCLE GTU**}
11	ο	General Electric	USA	ITH, 1996-2000; GTHC, 1999
12	△	Ebora	Japan	
13	△	Opra BV	Netherland	
14	△	Mitsui	Japan	
15	◇	US Turbine	USA	
16	△	Allied Signal	USA	
17	□	AGC Manufacturing Services	USA	
18	✿	Allison Gas Turbine	USA	GTW, 1988
19	▭	Garret	USA	
20	▭	Kongsberg Dresser	USA	GTW, 1988
21	□	Sulzer Turbosystems	USA	
22	▽	Stewart Stevenson	USA	
23	▽	Yanmar Diesel Engine	USA	
24	□	Diesel Allison General Motors	USA	GTC, 1968;69, 76.
25	△	Cantrax Limited	England	
26	✿	Mitsubischi	Japan	
27		Microturbo	France	
\multicolumn{5}{c}{**REGENERATED GTUs**}
28	●	Detroit Diesel	USA	Popov V.S. et al., 1987;
29	✱	William	USA	
30	●	Chrysler	USA	
31	✤	Solar	USA	
32	●	Avko Lykoming	USA	
33	■	Rover	England	
34	■	Leyland	England	
35	■	Daimler Benz	Germany	
36	◆	Volvo	Sweden	
37	▲	Turbomeca	France	
38	▲	Garret	USA	GTC, 1968
39	▲	Ford	USA	Popov V.S. et al., 1987;
40	◆	General Motors	USA	
41	▲	Elliott Energy Systems	USA	ITH, 1996-2000

The above statements show that the development of high-temperature small-size GTUs requires use of the structural ceramic materials (SCMs) to fabricate turbines and heat exchangers (Soudarev A.V., 1992). The alumino-boron-nitride ceramic which thermophysical and strength characteristics are presented in (Soudarev A.V. and Grishaev V.V., 1999) applies to these SCMs.

ALUMINO-BORON-NITRIDE SCM

In contrast to the heat-resistant materials and alloys, SCMs allow the simultaneous application of the beneficial effect of increase in both TIT and regeneration ratio E on the GTE efficiency. What distinguishes these materials is their excellent adaptability to machining, actually zero shrinkage at sintering, feasibility of formation of complex configuration components of the flow passage of GTE, combustors and heat exchangers.

SCMs have an excellent long-term strength at the temperature ranging 1300 to 1500K. They are resistant to the effects of the aggressive and chemically active high-temperature media and their density is less that of the high-temperature alloys by a factor of 2-3. At the same time, SCMs are more sensitive to the mechanical impact and thermal sign-variable loads, local stress concentrators.

It should be also born in mind that the negative properties of SCMs tend to increase at scaling-up of the absolute sizes of ceramic parts.

So, increasing the compactness of the ceramic heat exchangers is meant both for reducing their mass and overall dimensions and enhancement of the reliability of the heat exchanger as a whole. The recuperator segmentation leads to a reduction in sizes of its ceramic part where only a fraction of the heat output is transferred. This ensures a lowering of the gas stream temperature to the level allowing to apply the high-temperature metallic materials to manufacture its "cold" section.

The manufacture properties of the alumino-boron-nitride SCMs allow production of the tubular heat exchange elements for the recuperator matrices, these elements having both the external and internal surfaces developed which is especially important for gas-to-gas recuperated heat exchangers with their comparatively low temperature in either path.

To increase the compactness of the heat transfer matrices of the ceramic HEs, the modularity of their design and heat exchange enhancement are utilized. At the same time, given a high temperature level of the heat transfer agents, both the convective and the radiant components of the coefficient of heat transfer must be beneficial for a heat exchange enhancement.

A substantial increase in the heat pick-up could be achieved using the external washer and internal plate finning. Application of the diffusive welding of fins to the inside wall of the tube provides an ideal thermal contact at their juncture which promotes an increase in the fins effectiveness.

METAL-CERAMIC RECUPERATORS

Application of SCMs to fabricate parts of the flow passage of the GTU turbines allows a substantial elevation of the gas inlet temperature TIT and a reduction of the cooling air consumption, whereby reducing the energy consumption for the engine itself.

The TIT increase results in the increase in the gas temperature downstream of the power turbine up to 700°C and higher This makes use of high-temperature steels to manufacture HE actually unnecesary and requires application of the combined design heat exchangers (Fig.2).

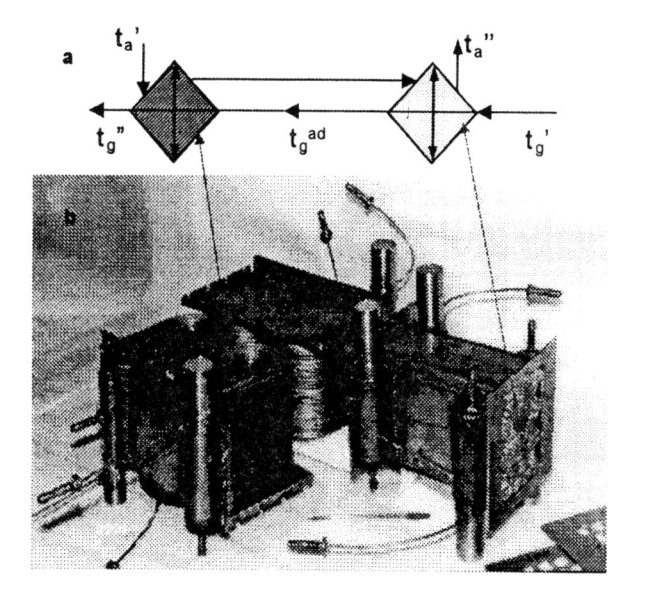

Fig.2.
Schematical diagram (a) and general view of sections (b) of combined airheater.

The high-temperature ("hot") section in the like HEs is made of SCM while the "cold" one (gas temperature at the inlet is $t_g' < 700°$) is made of a high-temperature steel or an alloy. Challenges facing the development of the ceramic section for the high-temperature combined HEs and their solutions are described below.

THERMAL-ENGINEERING STUDIES

At present, an information required to calculate the thermal-hydraulic characteristics of the ceramic tubes with internal discontinuous longitudinal finning is not available.

So, the experimental study into the hydraulic resistance and convective heat transfer at the air flow inside the ceramic tubes with the continuous and discontinuous finning of the wall was performed.

Experimental study of hydraulic resistance was carried out on the aerodynamic open type tube (Kondratiev V.V. and Lazarev M.V.,1997) at the working area where the ceramic tubes were placed (Fig.3).The number of the ceramic internal fin inserts varied. They were located in such a manner that the formation of solid or discontinuous internal channels inside the tube was ensured. The geometric characteristics of the tubes and the fin inserts are shown in Table 2.

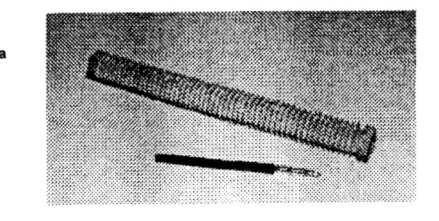

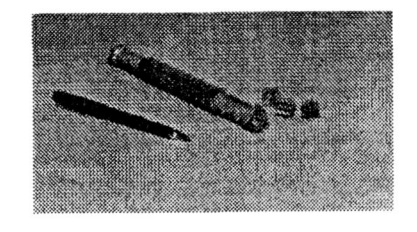

Fig.3.
Ceramic finned tubes (a,b) and fin inserts (c)

Table 2
Geometric characteristics of tubes and fin inserts

Nos	Name of value	Symbol	Unit	1	2	3
1.	Sizes of smooth tube:					
1.1	- length	l	mm	194	194	194
1.2	- ID	d	"	16	16	16
1.3	- open area	f_o	mm²	201.1	201.1	201.1
1.4	- number of fin inserts at complete fill-in	n_{max}	-	13	13	13
1.5	- area of smooth internal surface	$F\,10^3$	m²	9.75	9.75	9.75
2.	Sizes of fin insert:					
2.1	- number of fins	n_f	-	4	8	13
2.2	- diameters of sleeve	d_1	mm	0	4.2	5.4
2.3	- mean fin spacing	b	"	5.8	3.3	2.1
2.4	- height of interfin channel	h	"	7.4	5.0	4.5
2.5	- perimeter of above channel	U_{sp1}	"	26.3	16.6	13.2
2.6	- same for sleeve	U_{sl}	"	0	13.2	17.0
2.7	- open area for interfin channel	f_{sp1}	mm²	42.6	16.6	9.4
2.8	- same for internal channel of sleeve	f_{sl}	"	0.0	13.8	22.9
2.9	- same for all interfin channels	$f_{sp\Sigma 1}$	"	170.3	132.8	121.6
2.10	- same for tube as a whole	$f_{sp\Sigma 2}$	"	170.3	146.6	144.5
2.11	- porosity f/fo of tube at plugged channel of sleeve	σ_1	-	0.847	0.660	0.605
2.12	- same at open channel of sleeve	σ_2	-	0.847	0.729	0.719
	Equivalent diameter:	d_{eq}				
2.13	- at plugged channel of sleeve	d_{eq1}	mm	6.48	3.99	2.84
2.14	- at open channel of sleeve	d_{eq2}	"	6.48	4.01	3.07
3	Fin sizes					
3.1	- thickness	δ_f	mm	1.0	1.0	0.7
3.2	- length	l_f	"	7.5	5	4.5
3.3	- width	b_f	"	15	15	15
4.	Area of finned surface	$10^2 F_{f1}$	m²	2.05	2.59	3.33
5.	Same with surface of internal channel of tube accounted for	$10^2 F_{f2}$	"	2.05	2.85	3.66
6.	Finning factor	K_{f1}	-	2.1	2.66	3.42
7.	Same with internal channel of sleeve accounted for	K_{f2}	-	2.10	2.90	3.8
8.	Relative length of channels of finned tube	l/d_{eq1}	-	29.9	48.6	68.3

Nos	Name of value	Symbol	Unit	Numerical value		
				1	2	3
9.	Same with internal channel of sleeve accounted for	l/d_{eq2}	-	29.9	48.4	63.2
10.	Factor of configuration of interfin channel b/h	γ	-	0.783	0.664	0.462

During the experiment, measuring of the temperature, pressure and air flows as well as the pressure drops at the working area was carried out

$$\xi = C * \text{Re}^{-n}, \tag{1}$$

where C, n are the empirical coefficients which are a function of the geometry of the formed internal channels;

Re is the Reynolds number calculated by the mean velocity w of air, the hydraulic diameter d_{eq} of internal channels and the mean air temperature;

ξ is the coefficient of hydraulic resistance of the internal space of the finned ceramic tube equal to

$$\xi = \frac{2\Delta p}{\rho w^2} \cdot \frac{d_{eq}}{l}, \tag{2}$$

Δp is the design pressure drop at the working area

$$\Delta p = \Delta p_{spc} - \Delta p_{pipe} - (\xi_{in} - \xi_{out}) \cdot \frac{\rho w^2}{2}, \tag{3}$$

Δp_{spc} is the specified pressure drop between the bleed-off points at the working area;

ρ is the air density;

ξ_{in}, ξ_{out} are the local coefficients of resistance at inlet (outlet) (Soudarev A.V.,1992);

d_{eq}, l are the equivalent hydraulic diameter of the internal channels and the length of the experimental tube;

Δp_{pipe} is the pressure drop in the air supply (discharge) pipings

$$\Delta p_{pipe} = \xi_{pipe} \cdot \frac{l_{pipe}}{d_o} \rho \frac{w_o^2}{2}, \tag{4}$$

l_{pipe}, d_o are the length and the inner diameter of the piping;

w_o is the mean velocity of air in the piping;

ξ_{pipe} is the friction factor in pipings.

As the calculations indicated the relative error at the identification of the Reynolds number is ±7% while that for the hydraulic resistance is ±10.2%.

The experimental findings for the tubes with the internal continuous longitudinal fins are shown in Fig.4. Here, the obtained ceramic tube experimental data are compared with the similar data on the steel channels having a close form –factor b/h where b, h are the mean width and height of the channel. Furthermore, the equation of similarity by Mighay V.K. for steel tubes with the internal longitudinal finning is plotted on the graph

$$\xi = 1.624(b_o/d_{eq})^{0.16}\text{Re}^{-0.39}$$
$$5\cdot10^3 < \text{Re} \leq 75\cdot10^3$$
$$0.21 \leq b_o/d_{eq} \leq 0.49. \tag{5}$$

Here, b_o is the distance between the mating longitudinal fins.

It is obvious that for a small number of fins (n_f is 4), our test data are in good agreement with the generalization reported by Mighay V.K (1980). It allows calculation of the range of applicability of the formula (5) at the lesser values of the Reynolds number.

Fig.4 shows the test data for the ceramic tubes with discontinuous longitudinal finning. The test points are separated as a function of the number of fins in the inserts, while their location on the graph $\xi=f(\text{Re})$ is considerably higher compared to that of the ceramic tubes with the solid longitudinal tubes. The lines 13, 14, 15 plotted here and calculated using the equation of similarity by Voronin and Dubrovsky (1973) (lines 13, 14, 15)

$$\xi = 10.5(\bar{l}_f)^{-1.26}\text{Re}^{-n}$$
$$n = 0.52(\bar{l}_f)^{-0.29}$$
$$\text{Re} = 1500...10^4 \tag{6}$$

indicate the close values of the relative length of the fin $\bar{l}_f = l_f/d_{eq}$ ($\bar{l}_f = 2; 3; 5$). These tests were carried out on the steel longitudinal fin channels when the fins had a small thickness

$$d_{eq}/\delta_f = 34.8...47.1.$$

The fins on the ceramic tubes are of a relatively great thickness

$$d_{eq}/\delta_f = 6.5...4.0,$$

which tells certainly on the hydraulic resistance of the tubes. The empirical coefficients, required to calculate the coefficients ξ, incorporated into the formula (1) for the ceramic tubes with the internal discontinuous finning, are shown in Table 3.

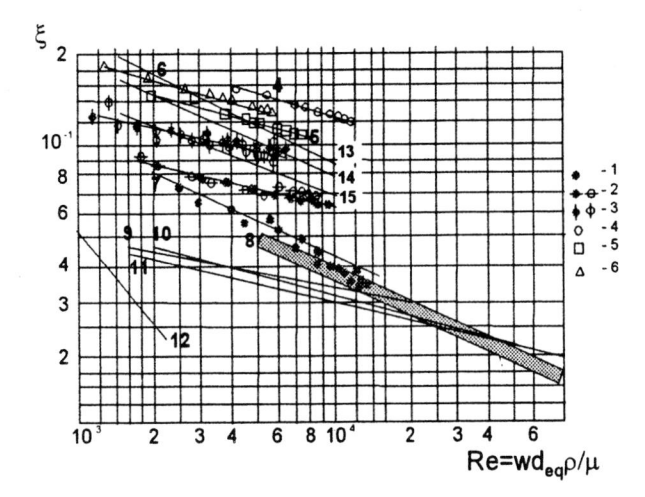

Fig. 4.
Hydraulic resistance of ceramic tubes with internal longitudinal finning.

1, 2, 3 – internal inserts with 4, 8 and 13 continuous longitudinal fins; 4, 5, 6 - same as above with discontinuous longitudinal finning (displaced fins of mating inserts across tube perimeter by half s step); 7,8 - calculation and application zone for formula (5) (Mighay V.K., 1980)]; 9, 11 - calculation for steel channels at form-parameters γ=0.664 and 0.462; 10, 12 - Blazius' and Poiseuille's relationships; 13, 14, 15 – calculation using formula (6) (Voronin G.I. and Dubrovsky E.V.,1973) for discontinuous fins with relative length $\bar{l}_f$= 2,3,5.
Light markers – hole in sleeve of fin insert is free, shaded markers – hole in sleeve of fin insert is plugged.

Table 3
Empirical coefficients C, n in formula (1)

n_p	C	N	Re	$K_\xi = \xi/\xi_{st}$
4	1.257	0.25	4200...12000	3.97
8	1.010	0.25	1900...7500	3.19
13	1.11	0.25	1250...5600	3.52

The experimental study of the convective heat transfer in the steel round section channels with the longitudinal solid and discontinuous fins was carried out in (Kryvenko A.A. and Chernyakov A.G., 1992). The experiments were carried out with the Reynolds numbers ranging $7 \cdot 10^3$ to $30 \cdot 10^3$ and with the absolute length of fins (streamwise) being l_f=8=60 mm, its length is 300 mm, the thickness of fins is δ_f=1 mm. The tubes with two and four fins were investigated. Its ideal contact with the wall was provided by welding. For calculation of the convective heat transfer along the finned channels the well-known relationship is employed (Isachenko V.P. et al., 1995):

$$\alpha_{eq} = \alpha_c \left(\frac{F_{sp}}{F_{fw}} + E\varphi \frac{F_f}{F_{fw}} \right), \quad (7)$$

where $F_{f.w.} = F_f + F_{sp}$ is the total area of the finned wall surface;

F_f, F_{sp} are the area of the fins and the area of the spacing between them;

E is the factor of fin effectiveness;

φ is the coefficient accounting for non-uniformity of distribution of heat exchange intensity upon the fin surface;

α_{eq}, α_c are the reduced and convective coefficients of heat transfer with the coefficient of convective heat transfer from the wall surface and fins being the same

$$\alpha_{c,w} = \alpha_{c,f} = \alpha_c. \quad (8)$$

For the discontinuous fins mounted in the channel with the fins displaced by half a step lengthwise the perimeter of the tube, the convective heat transfer is notably higher. Here, the adiabatic renewal of the boundary layer and variation of the direction of the heat flux are the factors of a profound importance. Nevertheless, to calculate the convective heat transfer one can employ the relationship in the form (8)

$$\alpha_{eq} = 1.1 \cdot \left(\alpha_w \frac{F_{sp}}{F_{fw}} + \alpha_f \frac{F_f}{F_{fw}} E\varphi \right), \quad (9)$$

where α_w and α_f are the coefficients of convective heat transfer from the surface of the interfin spacing and the fins themselves.

A difference between α_w and α_f manifests itself in the relative length of the formed channels

$$\varepsilon_C = 1 + \left(\frac{l}{d_{eq}} \right)^{-0.667} \quad (10)$$

The coefficient 1.1 in the formula (9) accounts for the convective heat transfer increase due to the stream perturbations on the fin edges.

For the high-temperature air or gas flows, heat transfer by radiation is important; so,

$$\varepsilon_{red} = \cfrac{1}{\cfrac{1}{\varepsilon_w} + \cfrac{F_{sp}}{F_f}\left(\cfrac{1}{\varepsilon_f} - 1\right)} \qquad (13)$$

Using the above methodology of the thermal-hydraulic calculation, the alternative calculation of the recuperator was performed (Fig.5) for CGTE-2.5 (Soudarev A.V. and Soudarev B.V., 1998). Through application of the external and internal finning, the heat exchanger compactness rose by a factor of 1.2-1.3. The compactness factor for the ceramic matrix of finned tubes is $\chi=500$-550 m^2/m^3.

SUMMARY

1. For the effective stationary and automotive gas-turbine engines, a design of the recuperator with two sections was developed:
   - high-temperature – ceramic section and
   - low-temperature – metal section.

2. The matrix of ceramic section was made of tubular heat exchange elements with the double-side finning. Here, the structural ceramics are used which allows formation inside each tube of both the solid and discontinuous fins with the ideal heat contact with the tube wall, the latter provided by the diffusive welding of ceramics.

3. The methodology of thermal-hydraulic calculation of the ceramic recuperator section of the finned tubular heat exchange elements was developed.

4. Availability of double-side finning and consideration of heat transfer by radiation along the high-temperature gas stream allows providing a high compactness of the matrix of the ceramic section ($\chi=500$-550 m^2/m^3).

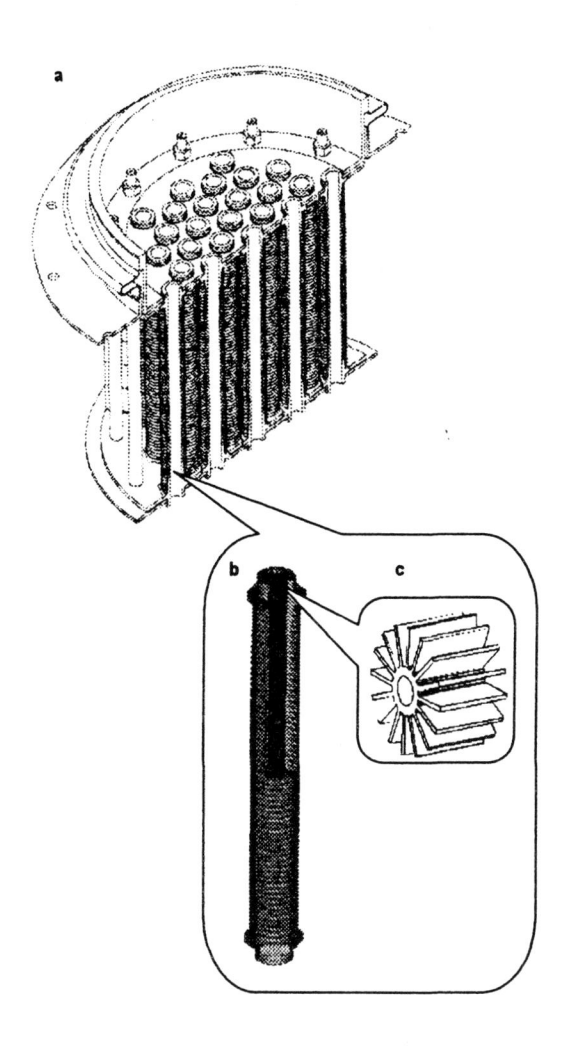

Fig. 5.
Alternative version of airheater module with ceramic matrix (a) of tubes with two-sides finning (b) for demonstration unit of CGTE-2.5 plant with gas temperature at inlet not above 1000 °C (c) internal fin insert.

$$\alpha_w = \alpha_{w,c} + \alpha_{w,r},$$
$$\alpha_f = \alpha_{f,c} + \alpha_{f,r}, \qquad (11)$$

where $\alpha_{i,c}$, $\alpha_{i,r}$ ($i=w$, $i=f$) are the convective and radiant coefficients of heat transfer

$$\alpha_r = 5.67 \cdot \varepsilon_{red} \cdot \frac{\left(\frac{T_f}{100}\right)^4 - \left(\frac{T_w}{100}\right)^4}{T_f - T_w} \qquad (12)$$

REFERENCES

Electric Power Generation (GTW) Specifications (Gas Turbine Performance Specifications), 1988.

Gas-turbine machinery catalogue (GTMC), 1999. Issued by the magazine "Gas-turbine technologies" (in Russian).

International Turbomachinery Handbook (ITH) :
v.37 N 6, 1996,
v.38, N 6, 1997
v.39, N 6, 1998
v.40, N 6, 1999/2000

Isachenko V.P. et al. Heat transfer. M., Energhia, 1995, 488 p. (in Russian).

Kondratiev V.V., Lazarev M.V. Hydraulic resistance of ceramic tubes with ball headpiece. Turbines and compressors, 1997, N 2, pp.67-69 (in Russian).

Kryvenko A.A., Chernyakov A.G. Convective heat transfer and resistance of tubes with internal discontinuous finning. Heat Transfer, MMF-92, v.1, Part 1, Minsk, 1992, pp.125-128 (in Russian).

Mighay V.K. Enhancement of effectiveness of current heat exchangers L., Energhia, L.O., 1980, 144 p. (in Russian).

Pchelkin Yu.M. et al. Ways of improvement of low power GTEs. News of MGTU, Ser. Mashinostrojenie, 1995, N 1, pp.20-24 (in Russian).

Popov V.S. et al. Automotive machines with gas-turbine engines. L., L.O., Mashinostrojenie, 1987, 259 p. (in Russian).

Sawyer's Gas Turbine Catalog (GTC). 1968; 69, 76 Edition.

Shamanov N.P., Nefedyeva A.V. Dynamics of variation of energy consumption. Mathematic model. Material of the jubilee research-engineering conference in GMTU, 18-21.05.1999, Part II, St.Petersburg, 1999, pp.281-286 (in Russian).

Soudarev A.V. Prospects of development of environmentally friendly ceramic gas-turbine engines for stationary power. News of RAN, Power, v.38, N 1, 1992, pp.49-59 (in Russian).

Soudarev A.V., Grishaev V.V. Shrinkage-free, easily adaptable to manufacture high-temperature ceramic material ANB-1. 2-4.11.1999, Hanoi, Vietnam, Proceedings of the Third International Workshop on Materials Science, Part I, p.1a-1b.

Soudarev A.V., Soudarev B.V. High-temperature compact airheaters for GTE. Turbines and compressors, 1998, N 5 (1-98), pp.21-27 (in Russian).

Viktorov B.G., Mitschenko O.A. Development of ship standardized gas-turbogenerators for 600-1500 kW power. Voprosy sudostrojenia, Ser. Ship power plants, Issue 7, 1975, pp.28-33 (in Russian).

Voronin G.I., Dubrovsky E.V. Effective heat exchangers. M., Mashinostrojenie, 1973, 96 p. (in Russian).

SUPERLEAN COMBUSTION USING COUNTERFLOW HEAT EXCHANGER FILLED WITH POROUS MEDIA

Hideo Yoshida[1], Motohiro Saito[2] and Hiroki Matsui[3]

Department of Mechanical Engineering, Kyoto University, Sakyo-ku, Kyoto 606-8501, Japan

[1] E-Mail: yoshida@mech.kyoto-u.ac.jp
[2] E-Mail: msaito@mech.kyoto-u.ac.jp
[3] E-Mail: t30y0988@ip.media.kyoto-u.ac.jp

ABSTRACT

As a crucial application of heat exchange technology to process industries, a compact counterflow combustor filled with porous media has been newly proposed to realize self-sustaining combustion of lean fuel mixtures. In the present study, combustion enhancement is based on effective heat recirculation through porous medium between burnt gas and unburnt mixture. Numerical calculations have been conducted using an analytical model consisting of energy equations for the gas phase and porous medium. Assuming a one-step Arrhenius-type reaction, the basic characteristics of the present combustor have been qualitatively investigated. In particular, by systematically varying the flow velocity, the combustor length and the channel height, the dependences of the minimum heating value required for self-sustaining combustion on these parameters were discussed.

INTRODUCTION

Heat exchange is a key technology for burning very lean fuel mixtures and low grade fuel such as waste gases produced by a wide variety of industrial processes. This is because mixtures outside normal limits of flammability can be burnt if combustion heat is effectively recirculated. Since the combustion phenomena are exclusively classified into combustible or not combustible, heat exchange technology plays a crucial role in environmental protection with respect to dealing with waste gases.

The features of the heat-recirculating scheme discussed by Weinberg (1986) are shown in Fig.1. The sensible heat of burnt gas is recovered and fed back to preheat the unburnt mixture. A flame temperature high enough to sustain combustion and higher than the adiabatic flame temperature of the mixture can be obtained by preheating, and therefore combustion can be sustained below the lean limit of flammability.

One of the most typical applications of this combustion scheme is the double-spiral "Swiss-roll" burner. In this burner, the combustion chamber consists simply of a large cavity at the center of the spiral, and heat is exchanged between extremely long double spiral channels. This is basically considered to be a counterflow heat exchanger.

In the present study, we propose a more compact counterflow combustor with a high potential for effective combustion of lean fuel mixtures, and clarify the basic characteristics on the basis of a numerical simulation.

OPERATING PRINCIPLE OF PROPOSED COUNTER-FLOW COMBUSTOR

Porous media have numerous features advantageous for use as heat-transfer materials. Their relatively large surface area per unit volume and small characteristic length lead to

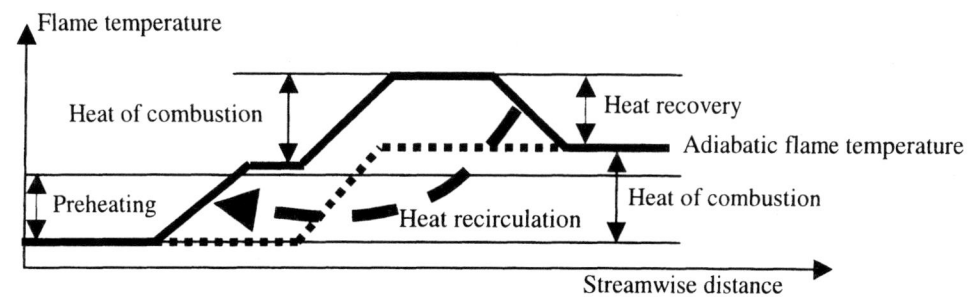

Fig. 1 Temperature profile of heat-recirculating scheme for lean mixture

high heat-transfer coefficients between the porous-medium surface and flowing gas. Furthermore, by changing the material and structure of porous medium, wide ranges of porosity and thermal conductivity are available. Also, porous media have been extensively applied for combustion enhancement (Echigo, 1991); in particular, the reciprocating-flow combustor developed by ADTEC, in which porous media are used for heat storage, has been proven to be very effective for the combustion of waste gases (ADTEC Co., Ltd., 1990).

Prompted by these features of porous media, we propose the new combustor as shown in Fig.2 for lean fuel mixtures. The difference from the conventional counterflow heat exchanger is that the channels are filled with porous media. If once the flames are stabilized in the porous media, the sensitive heat of the burnt gas is transferred to the unburnt mixture flowing through the neighboring channels. This heat exchange between the neighboring gas flows is equivalent to the heat recirculation from the downstream side to the upstream side, as explained in Fig.1.

Since the heat exchange efficiency of the present combustor is expected to be much higher than that of the Swiss-roll burner, the former would be much more compact than the latter. Furthermore, unlike the reciprocating-flow combustor which inevitably incurs exhaust loss when the flow direction changes, the present counterflow combustor has great advantages as a result of the steady operation.

PHYSICAL MODEL AND FORMULATION

Heat-Recovery System

Heat flux between neighboring channels. The heat-recovery system of the present combustor is shown in Fig.3. The channels filled with porous media are stacked. The combustor length and channel height are x_e and L, respectively. We assume that the system is steady and one-dimensional. The initial temperature of the mixtures in each channel is T_0, and flow velocity is u. Since the temperature profile for each channel is identical, the heat flux between each channel q is expressed as

$$q = K[T_p(x_e - x) - T_p(x)], \qquad (1)$$

where x is the distance from the entrance of the porous medium, K the overall heat-transfer coefficient between the neighboring channels, and $T_p(x)$ the temperature of the porous medium at position x.

Further assumptions are as follows.
- The physical properties are constant.
- Effects of thermal radiation are neglected. In principle, taking thermal radiation into account is not difficult if we introduce, for instance, the radiation thermal conductivity approximation for optically thick limit. In the present case, however, the attention is focused only on the convection and conduction heat transfer, which leads to the underestimation of combustion enhancement.
- Combustion is described by a one-step Arrhenius-type reaction, and the total numbers of moles are constant during the reaction, hence the flow velocity is constant throughout the channel.
- The porous media are noncatalytic.

Thermal contact resistance between channels. The model of thermal contact resistance shown in Fig.4 is adopted to estimate the thermal contact resistance between neighboring channels. The thickness of the contact layer L_c is assumed to be half the diameter of the cavity d. The ratio of void area to contact area is considered to be equivalent to porosity. Therefore,

$$s \text{ (void area)} : c \text{ (contact area)} = \phi : (1-\phi), \qquad (2)$$

where ϕ is defined as porosity. Under the preceding assumptions, the overall heat-transfer coefficient between the neighboring channels K is formulated as

$$\frac{1}{K} = \frac{L - 2L_c}{\lambda_p} + \frac{2(\lambda_m + \lambda_w)L_c}{2\lambda_m\lambda_w(1-\xi) + \lambda_g(\lambda_m + \lambda_w)\xi} + \frac{L_w}{\lambda_w} \qquad (3)$$

$$\xi = \frac{s}{s+c}, \qquad (4)$$

where λ_g is the thermal conductivity of gas, λ_m that of the porous-medium material, λ_p that of the porous medium and λ_w that of the partition wall. If no thermal contact resistance exists, i.e., $L_c=0$, the overall heat-transfer coefficient K_0 is as follows.

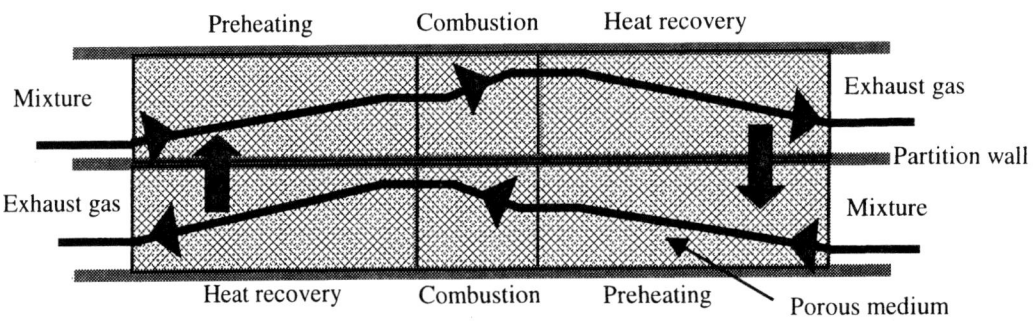

Fig. 2 A new combustor for lean mixture

Fig. 3 Physical model

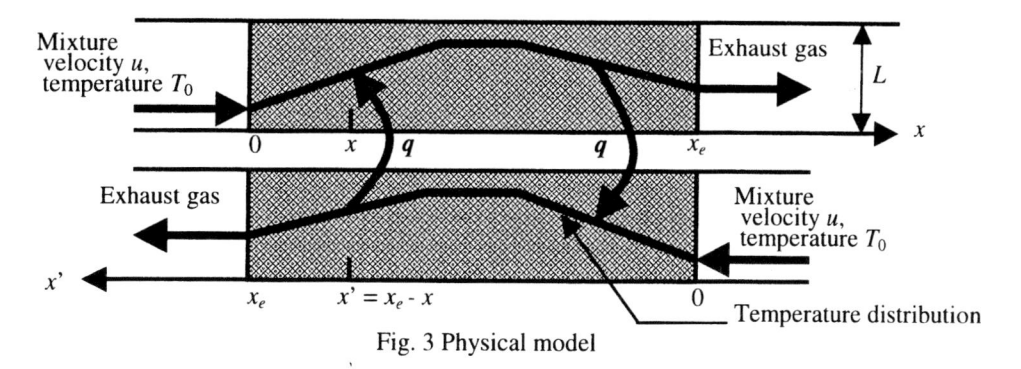

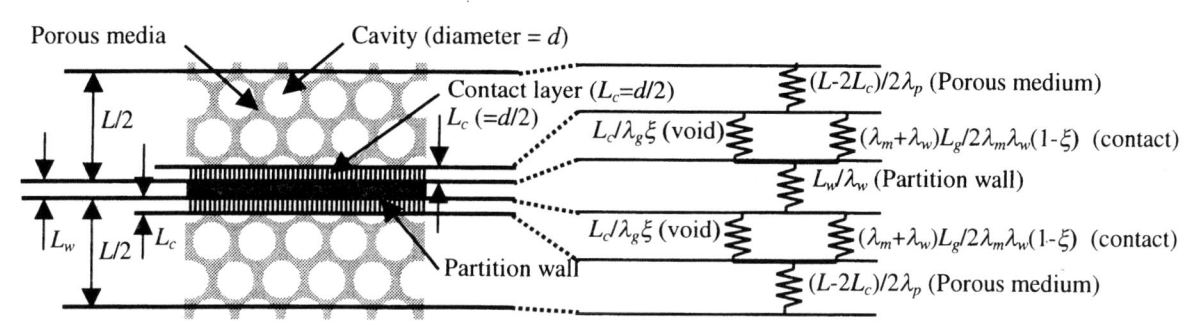

Fig. 4 Model of thermal contact resistance

$$\frac{1}{K_0} = \frac{L}{\lambda_p} + \frac{L_w}{\lambda_w} \quad (5)$$

Governing Equations

Using the above physical model, the energy equations for both the gas phase and the porous medium, as well as the conservation equation of the combustion product, are formulated.

The conservation equation for the combustion product:

$$\rho_g u \frac{\partial Y}{\partial x} - W = 0, \quad (6)$$

where Y is the mass fraction of combustion product, the reaction rate W is given by Arrhenius-type reaction kinetics.

$$W = A\rho_g (1-Y) \exp(-E/RT_g) \quad (7)$$

The energy equation for the gas phase:

$$\underbrace{\rho_g c_p u \frac{\partial T_g}{\partial x}}_{\text{(conv)}} - \underbrace{\lambda_g \frac{\partial^2 T_g}{\partial x^2}}_{\text{(cond)}} + \underbrace{\alpha A_p (T_g - T_p)}_{\text{(htra)}} - \underbrace{h_0 W}_{\text{(gene)}} = 0, \quad (8)$$

where the heat released by combustion is denoted by $h_0 W$.

The energy equation for porous medium:

$$\underbrace{\lambda_p \frac{\partial^2 T_p(x)}{\partial x^2}}_{\text{(cond)}} - \underbrace{\alpha A_p [T_p(x) - T_g(x)]}_{\text{(htra)}} - \underbrace{2\frac{K}{L}[T_p(x) - T_p(x_e - x)]}_{\text{(exch)}} = 0. \quad (9)$$

The abbreviated terms shown under Eqs. (8) and (9) corresponds to those in Eqs. (16) and (18) which appear later as the dimensionless equations. In the next chapter, the distributions of conv*, cond*, htra*, gene*, and exch* are displayed in Figs.7, 8, 12, and 17.

The computation domain covers the two free-space regions up- and downstream of the porous medium; in these two regions, only the gas-phase energy equation without the heat-transfer term between gas and porous medium,

$$\rho_g c_p u \frac{\partial T_g}{\partial x} - \lambda_g \frac{\partial^2 T_g}{\partial x^2} - h_0 W = 0, \quad (10)$$

is calculated to elucidate the effect of thermal conduction of the approaching gas at the upstream end of the porous medium.

Boundary Conditions

Boundary conditions are as follows.

Upstream far from the entrance of the porous medium:

$$x = -\infty\,; T_g = T_0,\ Y = 0 \quad (11)$$

At the entrance and outlet of the porous medium:

$$x = 0\,; \frac{\partial T_p}{\partial x} = 0, \quad (12)$$

$$x = x_e\,; \frac{\partial T_p}{\partial x} = 0, \quad (13)$$

Downstream far from the outlet of the porous medium:

$$x = +\infty\,; \frac{\partial T_g}{\partial x} = 0,\ \frac{\partial Y}{\partial x} = 0, \quad (14)$$

Nondimensionalization

The governing equations and boundary conditions derived above are nondimensionalized prior to numerical computation. Dimensionless parameters are as follows.

$X = x/x_e$, $\theta_g = T_g/T_0$, $\theta_p = T_p/T_0$, $Re = ux_e/\nu$, $Pr = \rho_g c_p \nu/\lambda_g$, $r_c = \lambda_p/\lambda_g$, $a = x_e^2 K/L\lambda_p$, $M = x_e^2 \alpha A_p/\lambda_p$, $R_s = A x_e^2/\nu$, $E_a = E/RT_0$, $H_0 = h_0/c_p T_0$

Dimensionless conservation equation for combustion product:

$$\frac{\partial Y}{\partial X} - \frac{R_s}{Re}(1-Y)\exp(-E_a/\theta_g) = 0. \quad (15)$$

Dimensionless energy equations for the gas phase. Inside porous medium ($0 \le X \le 1$):

$$\underbrace{\frac{\partial \theta_g}{\partial X}}_{(\text{conv}^*)} - \underbrace{\frac{1}{PrRe}\frac{\partial^2 \theta_g}{\partial X^2}}_{(\text{cond}^*)} + \underbrace{\frac{M}{PrRe}(\theta_g - \theta_p)}_{(\text{htra}^*)} \\ - \underbrace{\frac{H_0 R}{Re}(1-Y)\exp(-E_a/\theta_g)}_{(\text{gene}^*)} = 0. \quad (16)$$

In free-space regions upstream and downstream of the porous medium ($X < 0$, $X > 1$):

$$\underbrace{\frac{\partial \theta_g}{\partial X}}_{(\text{conv}^*)} - \underbrace{\frac{1}{PrRe}\frac{\partial^2 \theta_g}{\partial X^2}}_{(\text{cond}^*)} \\ - \underbrace{\frac{H_0 R}{Re}(1-Y)\exp(-E_a/\theta_g)}_{(\text{gene}^*)} = 0. \quad (17)$$

Dimensionless energy equation for porous medium:

$$\underbrace{\frac{\partial^2 \theta_p(x)}{\partial X^2}}_{(\text{cond}^*)} - \underbrace{\frac{M}{r_c}\left[\theta_p(X) - \theta_g(X)\right]}_{(\text{htra}^*)} \\ - \underbrace{2a\left[\theta_p(X) - \theta_p(1-X)\right]}_{(\text{exch}^*)} = 0. \quad (18)$$

Nondimensionalized boundary conditions:

$$X = -\infty\,; \theta_g = 1,\ Y = 0 \quad (19)$$

$$X = 0\,; \frac{\partial \theta_p}{\partial X} = 0, \quad (20)$$

$$X = 1\,; \frac{\partial \theta_p}{\partial X} = 0 \quad (21)$$

$$X = +\infty\,; \frac{\partial \theta_g}{\partial X} = 0,\ \frac{\partial Y}{\partial X} = 0. \quad (22)$$

Calculation Procedure

Numerical simulations are performed by solving the system of partial differential equations (15)-(18), with boundary conditions (19)-(22), using the initial values described below. The finite differences are employed to compute the derivatives for various quantities, and the upwind differences are applied to the convection terms in the basic equations. At the two ends of porous medium, the balance in the control volume is calculated. The discreticized equations are solved by the iterative method. The grids are distributed uniformly; for example, 1000 computational nodes were used for the porous medium of the length of 30 cm, and 100 nodes for free-space regions both upstream and downstream of the porous medium.

At the start of numerical computation, the initial values are given as follows.

Temperature of gas phase:
In free-space regions upstream and downstream of the porous medium:
$\theta_g = 1$.
Inside porous medium: triangular profile, i.e.,
$\theta_g = 3.5$ at the center of porous medium,
$\theta_g = 1$ at the two ends of porous medium.

Temperature of porous medium: triangular profile, i.e.,
$\theta_p = 3.5$ at the center of porous medium,
$\theta_p = 1$ at the two ends of porous medium.

Mass fraction of combustion product:
$Y = 0$ over entire computation domain.

Initially, the gas is ignited at the center of the porous medium at high temperature. During the iterations, however, the flame temperature decreases gradually, and when the gas temperature over entire region reaches unity, we judge that self-sustaining combustion is not possible under that calculation condition.

RESULTS AND DISCUSSION

Calculation Conditions and Typical Example

As the actual combustor we plan to use metallic porous media as well as ceramic ones. For metallic porous media, the channel height is designed to be less than two or three cm to ensure high heat transfer between the neighboring channels. On the other hand, for ceramic porous media with relatively low thermal conductivity, the channel height should be much lower, or an effective contribution of radiation heat transfer is desirable. Assuming these conditions are satisfied and moderate gas-flow velocity is less than 1 m/s, the numerical calculation is conducted. The calculated results, however, should be interpreted qualitatively, because the simple one-step reaction was assumed and also because assumptions made in estimating heat-transfer coefficients and other constants have some uncertainty.

Properties, conditions and constants. Unless stated explicitly in the following discussion, the properties, conditions and constants used for calculation are as follows.
porous medium:
$A_p=1700$ m^2/m^3, $L=5$ mm, $x_e=0.3$ m, $\lambda_p=1.380$ W/m K, $\phi=0.953$
gas:
$c_p=1.007*10^3$ J/kg K, $T_0=293$ K, $u=0.3$ m/s,
$\lambda_g=2.614*10^{-2}$ W/m K, $\nu=1.583*10^{-5}$ m^2/s,
$\rho_g=1.176$ kg/m^3
reaction:
$A=1.0*10^9$ s^{-1}, $E=8.3*10^4$ J/mol, $H_0=0.305$,
$h_0=90$ kJ/kg, $R=8.310$ J/mol K
overall heat-transfer coefficient between channels:
$K=2.760*10^3$ W/m^2 K

Heat-transfer coefficient at the surface of porous medium α is estimated by assuming porous medium to be a columns bank, which has porosity and specific surface area equal to those of porous medium.

The typical temperature distributions of both the gas phase and porous medium are shown in Fig.5. Owing to the high heat-transfer coefficient between the gas phase and porous medium, the temperature difference between them is negligible. A triangular temperature profile, in which the maximum temperature is high enough to sustain combustion and is much higher than the adiabatic flame temperature, is formed inside the porous medium. According to energy conservation, the temperature at the exit of the porous medium equals the adiabatic flame temperature, if the mixture could be burned completely. The distributions of the mass fraction of the combustion product and reaction rate are shown in Fig.6. Reaction occurs at the center of porous medium, where the temperature required to sustain combustion can be attained.

The energy balances in Eqs. (16) and (18) are shown in Figs.7 and 8, respectively. In Figs. 7, 8, 12, and 17, positive value means the heat to be carried into the region considered, while negative value means the heat to be carried away. As evident in these figures, the combustion heat is first transferred to the porous medium, and is effectively recirculated to the nonreacting gas flowing through the neighboring channel.

It is interesting to note that in Fig.5 the gas temperature at the entrance of the porous medium is slightly higher than unity, because the approaching gas is preheated by the thermal conduction in the gas phase. Also, at the upstream end of the porous medium, thermal conduction both in the gas phase and the porous medium is appreciable, although the region is very restricted in the vicinity of the entrance.

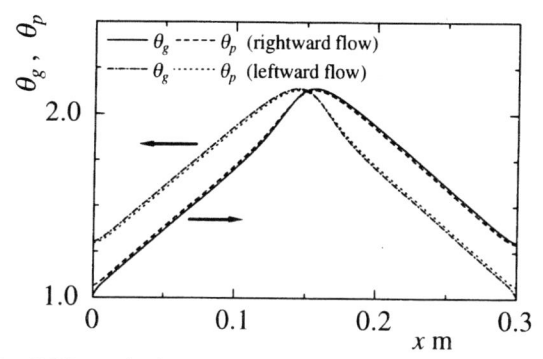

Fig. 5 Dimensionless temperature distribution in the combustor

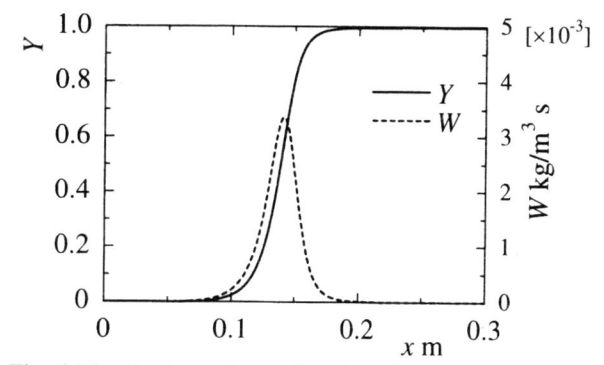

Fig. 6 Distributions of mass frantion of combustion product and reaction rate

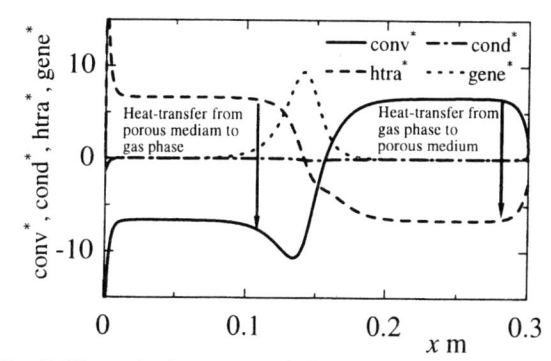

Fig. 7 Dimensionless energy balance of the energy equation for gas phase

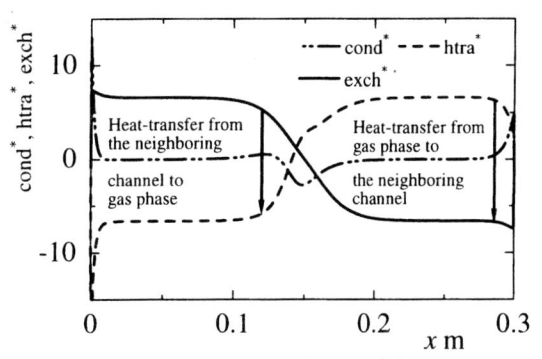

Fig. 8 Dimensionless energy balance of the energy equation for porous medium

Effect of Flow Velocity

For various flow velocities, the distributions of temperature, mass fraction of the combustion product, and reaction rate are shown in Figs.9 and 10. With decreasing flow velocity, a wider high-temperature region is obtained and reaction occurs earlier after the mixture flows into the porous medium. In case of extremely slow flow velocity, $u=0.06$ m/s, shelvy temperature distribution with low peak is obtained and reaction occurs slowly. The dependence of the dimensionless minimum required heating value $H_{0, min}$ on the flow velocity u is shown in Fig.11. It is interesting to note that there exists a minimum value of H_0. For $u > 0.14$ m/s, the minimum required heating value decreases with decreasing flow velocity. This tendency is ascribed to the enhanced heat exchange due to the relatively long passage time as the flow velocity decreases. On the other hand, for $u < 0.14$ m/s, the minimum required heating value increases as the flow velocity decreases. In Fig.12, the distribution of heat transfer between the gas phase and porous medium and the streamwise thermal conduction of porous medium are shown for flow velocities $u=0.3$ and 0.06 m/s, respectively. As illustrated in Fig.12, in this low-velocity range, the streamwise thermal conduction in the matrix of the porous medium, which is negligible in the moderate-velocity range, is not negligible and it exerts a significant influence on the flammability limit.

Effect of Combustor Length

Distributions of temperature, mass fraction and reaction rate for various combustor lengths are shown in Figs. 13 and 14. For a longer combustor, reaction occurs earlier after the mixture flows into the porous medium and proceeds faster. The dependence of the dimensionless minimum required heating value $H_{0, min}$ on combustor length x_e is shown in Fig.15. The minimum required heating value decreases as the length of the porous medium increases. This result is basically due to the enhanced heat exchange corresponding to the longer passage time. Intuitively, however, since these distributions are inactive at the plateau of the temperature profile, the authors expected the results to be independent of the combustor length, once the plateau is attained. The observed slight dependence on the combustor length is ascribed to thermal conduction within the porous medium, which is small but not completely negligible.

Effect of Channel Height

Temperature distributions for various channel heights are shown in Fig.16. In Fig.17 the distributions of heat exchange between channels, exch* term in Eq.(18), are shown for channel heights $L=5$ and 1 mm. The change of the channel height is equivalent to the change of thermal resistance between the neighboring channels. With decreasing height, i.e., decreasing thermal resistance, a higher and wider temperature region is obtained. The dependence of the dimensionless minimum required heating value $H_{0, min}$ on the channel height L is shown in Fig.18. With decreasing channel height, $H_{0, min}$ decreases.

Effect of Specific Surface Area of Porous Medium

For the parameter range of the present calculations, the specific surface area of the porous medium A_p does not exert a significant influence on the result, because it is large enough to allow heat transfer between porous medium and flowing gas.

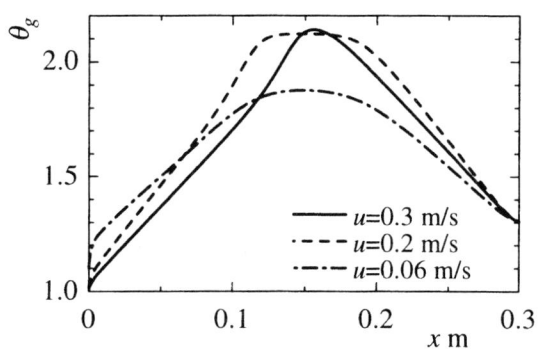

Fig. 9 Dimensionless temperature distribution for various flow velocities

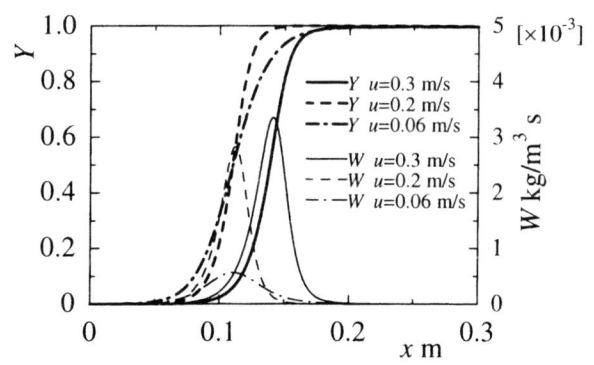

Fig. 10 Distribution of mass fraction and reaction rate for various flow velocities

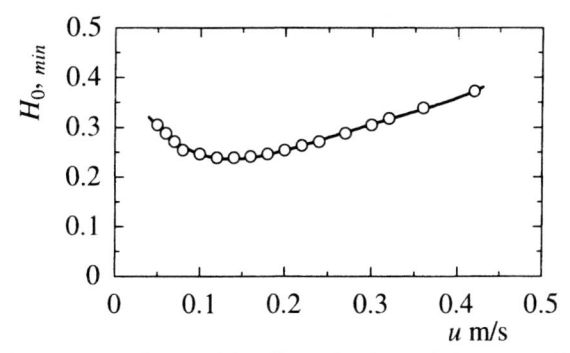

Fig. 11 Dependence of the dimensionless minimum required heating value $H_{0,\,min}$ on the flow velocity

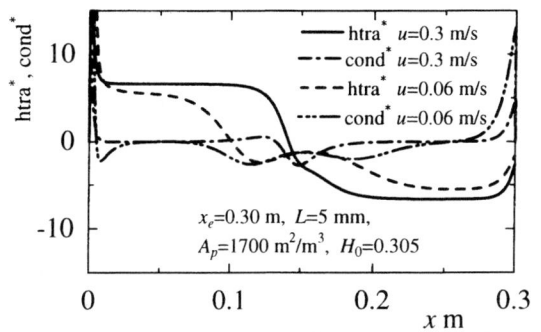

Fig. 12 Distribution of heat-transfer between gas phase and porous medium and streamwise thermal conduction

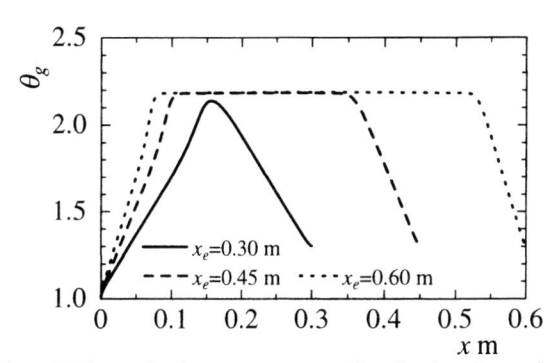

Fig. 13 Dimensionless temperature distribution for various combustor lengths

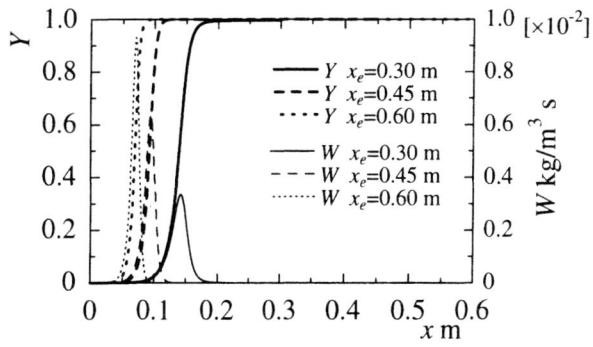

Fig. 14 Distribution of mass fraction and reaction rate for various combustor lengths

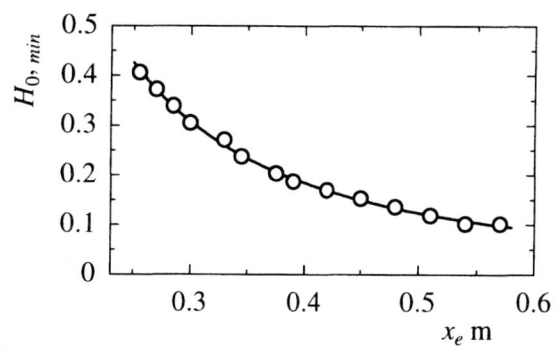

Fig. 15 Dependence of the dimensionless minimum required heating value $H_{0,\,min}$ on combustor length

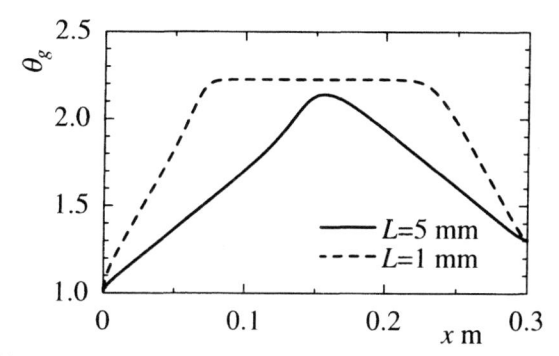

Fig. 16 Dimensionless temperature distribution for various channel heights

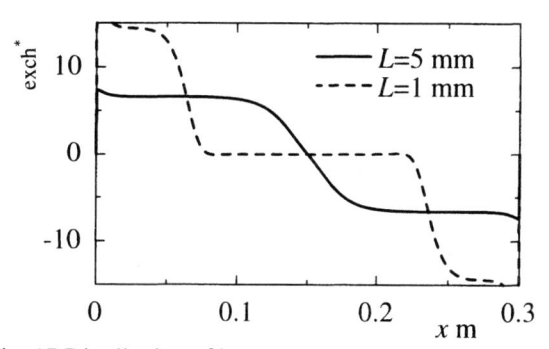

Fig. 17 Distribution of heat exchange between channels for channel heights L=5 and 1 mm

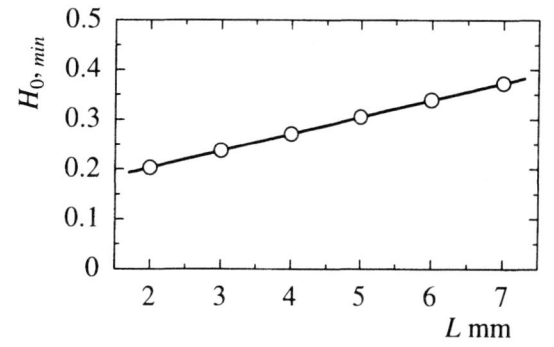

Fig. 18 Dependence of the dimensionless minimum required heating value $H_{0,\,min}$ on channel height

CONCLUSIONS

As a compact high-performance combustor for self-sustaining combustion of lean mixtures, the counterflow exchanger filled with porous media was proposed.

In order to obtain the fundamental combustion characteristics, a theoretical study was performed. The numerical calculations, assuming a one-step Arrhenius-type reaction, qualitatively clarified the basic characteristics that are summarized below.

1. A triangular or trapezoidal temperature distribution is formed within the porous medium.
2. The dependence of the minimum required heating value on flow velocity is not monotonic. There exists an optimum flow velocity at which combustion is maximally enhanced.
3. The minimum required heating value decreases with increasing combustor length and decreasing channel height.

Although these findings are promising, we are now preparing an experimental study to quantitatively confirm the limiting performance of the present combustor.

NOMENCLATURE

- A Arrhenius factor, s^{-1}
- a dimensionless number, $x_e^2 K/L\lambda_p$
- A_p specific surface area of a porous medium, m^{-1}
- c contact area on the boundary between porous medium and partition wall
- c_p constant-pressure specific heat for gas, $J\,kg^{-1}\,K^{-1}$
- d diameter of cavity, m
- E activation energy, $kJ\,mol^{-1}$
- E_a dimensionless activation energy, E/RT_0
- H_0 dimensionless heating value, $h_0/c_p T_0$
- h_0 heating value, $J\,kg^{-1}$
- K overall heat-transfer coefficient between neighboring channels, $W\,m^{-2}\,K^{-1}$
- K_0 overall heat-transfer coefficient between neighboring channels with no thermal contact resistance, $W\,m^{-2}\,K^{-1}$
- L channel height, m
- L_c thickness of contact layer, m
- L_w partition-wall thickness, m
- M dimensionless number, $x_e^2 \alpha A_p/\lambda_p$
- Pr Prandtl number, $\rho_g c_p \nu/\lambda_g$
- q heat flux between neighboring channels, $W\,m^{-2}$
- R universal gas constant, $8.314\,J\,mol^{-1}\,K^{-1}$
- r_c dimensionless number, λ_p/λ_g
- Re Reynolds number, ux_e/ν
- R_s dimensionless number, Ax_e^2/ν
- s void area on the boundary between porous medium and partition wall
- T temperature, K
- T_0 initial temperature of gas phase, 293 K
- u velocity of mixture, $m\,s^{-1}$
- W reaction rate, $kg\,m^{-3}\,s^{-1}$
- X dimensionless coordinate, x/x_e
- x coordinate, m
- x_e combustor length, m
- Y mass fraction of combustion product
- α heat-transfer coefficient at surface of porous medium, $W\,m^{-2}\,K^{-1}$
- ϕ porosity of porous medium
- λ thermal conductivity, $W\,m^{-1}\,K^{-1}$
- λ_m thermal conductivity of porous-medium material, $W\,m^{-1}\,K^{-1}$
- λ_w thermal conductivity of partition wall, $W\,m^{-1}\,K^{-1}$
- ν kinematic viscosity of mixture, $m^2\,s^{-1}$
- ρ_g density of mixture, $kg\,m^{-3}$
- θ dimensionless temperature, T/T_0
- ξ fraction of void area in the boundary area between porous medium and partition wall, $s/(s+c)$

Subscripts

- g gas phase
- min minimum
- p porous medium

REFERENCES

Commercial report, 1990, A New Method of Destroying Organic Pollutants in Exhaust Air, ADTEC Co., Ltd., Sweden.

Echigo. R, 1991, Radiation Enhanced/Controlled Phenomena of Heat and Mass Transfer in Porous Media, *ASME/JSME Thermal Engineering Proceedings*, 4, pp. xxi-xxxii.

Weinberg. F. J, 1986, *Advanced Combustion Methods*, Academic Press. pp. 183-236.

RULES OF SIMILARITY FOR THE VORTEX ELECTRO-DISCHARGED PLASMOTRON

Vyacheslav T. Volov

Samara Scientific Center of the Russian Academy of Science, Department of transport problems, Studencheski per. 3a, 443001, Samara, Russia, phone/fax (+7 846 2) 42 40 43, e-mail: sgi@samtel.ru

ABSTRACT

This report presents results theoretical and experimental investigation of a new type of the heat-mass exchanger with the strong rotated supersonic flows for gas lasers and plasmatrons.

It has been shown that heat transfer inside supersonic vortex flow can be related to convection-diffusion type and electrical power coupling in discharge can reach the level of 300 W/cm^2. It is more than two order higher than in ordinary gas discharge. The relations of similarity and scaling laws for vortex electro-discharged laser has been proposed.

The idea of using strongly rotated supersonic flows as a basis of the vortex electro-discharged lasers (VEL) and plasmotrons consist in the fact that near the axis of the vortex tube there is region with low pressure and temperature. The vortex tubes have strongly rotated flows as their basis. The investigation of the self-vacuumed vortex tube (SVVT) (Merculov, 1969) is special interest because inside the SVVT the coefficient of turbulent kinematic viscosity and diffusion are 10^4 times higher than in the corresponding laminar flows. Combining those facts, the purpose of this research was to design of new type of heat-mass transfer for the laser – the vortex electro-discharge systems, portable, water-cooled free source for biological, agricultural and forestial processing.

THE PURPOSE OF INVESTIGATION

The idea of using strongly rotated supersonic flows as a basis of the vortex electro-discharged lasers (VEL) and plasmotrons consist in the fact that near the axis of the vortex tube there is region with low pressure and temperature. The vortex tubes have strongly rotated flows as their basis. The investigation of the self-vacuumed vortex tube (SVVT) (Merkulov, 1969) is special interest because inside the SVVT the coefficient of turbulent kinematic viscosity and diffusion are 10^4 times higher than in the corresponding laminar flows. Combining those facts, the purpose of this research was to design of new type of heat-mass transfer for the laser – the vortex electro-discharge systems, portable, water-cooled free source for biological, agricultural and forestial processing.

THEORETICAL INVESTIGATION AND BACKGROUND

As shown by Volov (1992), heat transfer from the discharge area can be realized by heat conductivity to the cooling wall of the discharge tube, or by way of changing a heated mass of gas to new a new portion of gas (the convectional method).

The investigated the vortex decaying discharge inside the self-vacuumed vortex tube (SVVT) can not be referred to ether the first or the second type of cooling, because inside the SVVT there are superior values of the turbulent diffusion $D_T \sim D_L \cdot 10^4$ and the average output velocity in the potential flow area can reach $v = 50 \div 100$ m/sec.

Near the axis of the vortex tube there are only several percent of the gas mixture mass which is at the moment in the tube, because the output area is the potential flow area.

That is why heat transfer from the axis area to the periphery potential flow area in realized by turbulent diffusion, and heat transfer from the periphery of the vortex area is realized by convection.

Therefore the discussed new type of vortex decaying discharge can be referred to convectional-diffusional type of cooling.

For the determining of the vortex discharge heating we should know the share of the total gas output taking place in the periphery because of turbulent diffusion can be on the axis for the period of renewal by the new gas portion. It can be described by the following equation

$$\frac{dq}{qz} = -\beta q \qquad (1)$$

where q - is the share of the total gas output, which can not be on the vortex tube axis, β - is some dimensionless multiplayer.

This equation has shown that increasing of the vortex tube length leads to decreasing this share q.

The solution of this equation is the following formula

$$q = q_0 \exp\{-\beta(z - z_0)\} \quad (2)$$

That is why the following equation has presented the related output share of the gas mixture which had been played near the vortex tube axis:

$$\eta_{ex} = \frac{q_0 - q}{q_0} = 1 - \exp\left(-C \frac{\overline{L}_{Bk}^2 \widetilde{B}}{R_{Bk} \overline{R}_p \widetilde{v}_z}\right) \quad (3)$$

where $q = q_0$ when $z = z_0$, $C \sim 1$ - is an experimental constant, $\overline{L}_{VK} = L_{VT}/d_{VT}$ - is the related vortex tube length, d_{VT}, R_{VT} - are the vortex tube diameter and radius, $\widetilde{D}$ - is the average discharge area turbulent diffusion coefficient, $\overline{R}_d = R_d/R_{VT}$ - is the related discharge radius, $\widetilde{v}_z$ - is the average output velocity of the potential flow. For the typical regimes and geometrical sizes of the SVVT the heat exchange coefficient approximately equals $\eta_{ex} \sim 1$, at the same time the first cooling method (diffusion cooling) has the heat exchange coefficient volume $\eta_{ex}^I \ll 1$ and the second cooling way (conventional cooling) has the heat exchange coefficient volume $\eta_{ex}^{II} \gg 1$.

Thus the heat-mass transfer inside the strong rotated supersonic flows can relate to special type of the heat transfer – the convection - diffusion method of cooling. The average discharge heating can be determined in the following way:

$$\Delta \widetilde{t}_h = \frac{W^\Sigma}{C_p G_1 \eta_{ex}} \quad (4)$$

where W^Σ - is the total electrical power coupling into the vortex discharge, C_p - is the heat volume under constant pressure, G_1 - is the gas mixture output through the vortex tube. The average temperature of the discharge area equals

$$\widetilde{T}_d = \widetilde{T}_d^s + \Delta t_h \quad (5)$$

where $\widetilde{T}_d^s$ - is the average gas mixture temperature of the discharge area without discharge.

Some theoretical research has been dedicated to the VEL and plasmatron (Volov 1992, 1986, 1988; Shmelev et al., 1980). In the author's investigations (1992, 1986, 1988, 1986) the semi-empirical theory of the VEL and the vortex plasmotron has been suggested. This theory includes the mathematical model of the SVVT (Volov, 1983), the

mathematical calculations for the electronic diffusion at the vortex decaying discharge, kinetic equation of the radiation in non-uniform turbulent flows with comparative analysis between the VEL and known types of the gas electro-discharged systems. The theoretical analysis has shown that electrical power density distribution is non-uniform inside the vortex discharge (Fig. 1).

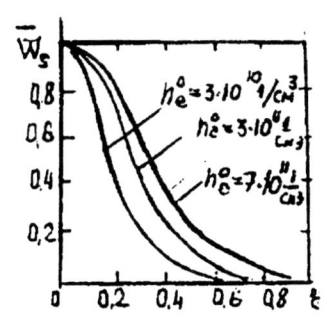

Fig. 1 Illustration of the electron power density coupling distribution as a function of the dimensionless radius.

In the case of a weak electrical power coupling into the vortex discharge adiabatic radial distribution of gas temperature (cold model) is always assumed. Under the considerable electrical power coupling into the vortex discharge we should take into account non-adiabatic gas temperature distribution (hot model) by using the following dimensionless formula

$$\theta = 1 - \zeta \int_0^\xi \frac{\int_0^\xi \overline{w}_s \rho d\rho}{\xi \overline{\lambda}_T(\xi)} d\xi, \quad (6)$$

where dimensionless constant ζ can be written as:

$$\zeta = \frac{1 - \theta_{ad}(\xi_{dis})}{\int_{\xi_{dis}}^{} \frac{\int_0^\xi \overline{w}_s \rho d\rho}{\xi \overline{\lambda}_T(\xi)} d\xi}$$

where $\theta = T/T_{ax}$ - is the related gas temperature, $\overline{\lambda}_T = \lambda_T/\lambda_{ax}$ - is the coefficient of turbulent heat conductivity, $\overline{w}_s = w_s/w_{Sax}$ - is the dimensionless radial electrical power coupling, T_{ax}, λ_{ax}, w_{ax} - are the volumes of the gas temperature, coefficient of turbulent heat conductivity, dimensionless electrical power density coupling near the axis of discharge accordingly, $\theta_{ad} = T_{ad}(\xi_{dis})/T_{ad}$, T_{ad} - is the adiabatic volume of gas temperature under condition $\xi = \xi_{dis}$.

As shown in Fig.2 the gas temperature distribution is similar to the electrical power density curve. Under the considerable electrical power coupling inside the vortex decaying discharge we can receive overheating near the discharge axis. Hence, the circular regime of radiation takes place inside the vortex discharge.

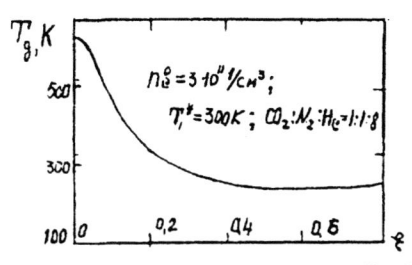

Fig. 2 The radial gas temperature distribution.

The adiabatic coefficient distribution inside vortex decaying discharge has shown in Fig. 3.

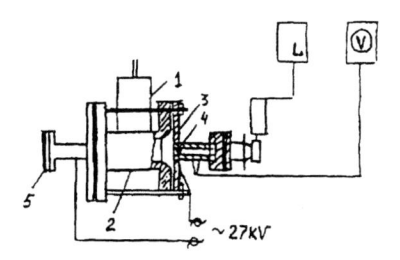

Fig. 3 The radial adiabatic the VEL.

To reduce the number of trial experiment for the making the required VEL, we have suggested the similar relation and scaling laws. The proposed relations of similarity and scaling laws of the VEL include the relations of gas discharge similarity (Konuhov, 1970), the relations of the similarity theory for heat-mass transfer, and the vortex tube properties. Those relations are shown in the Table 1. From the calculating results comparison we can see satisfactory agreement for the similar relations used (see Table 2) W_{sp} - is the electrical power coupling for the unit length of the vortex discharge and discharge and deterioration of the one under invalidity of those relation[5] (Table 3, W_{sp}).

Thus the similarity relations and scaling laws allow us to design the vortex electro-discharged laser with certain parameters without additional expensive experiments.

Table 1. Dimensionless Discharge Parameters

N Physical Volume	Law of Transformation
1. Radius of tube	$R_m = R_n \cdot b$
2. Gas density	$N_m = N_n / b$
3. Current of Discharge	$I_m = I_n \cdot b$
4. Concentration of electrons	$N_e^m = N_e^n / b$
5. Voltage of the longitudinal electric field	$E_m = E_n / b$
6. Temperature of electrons	$T_e^m = T_e^n$
7. The electrical power on unit tube length	$W_{sp}^m = W_{sp}^n$
8. The electrical power on unit tube length	$W_{sp}^m = W_s^n / b^2$
9. Average temperature of the gas in discharge	$T_m = T_n$
10. The relative settlements of power levels	$(N_n/N)_m = (N_n/N)_n$
11. Frequency of the electronic and molecular hits	$Z_m = Z_n / b$
12. Radius of the SVVT	$R_{rr}^m = R_{rr}^n \cdot b$
13. The dimensionless square of tangential nozzle entry	$F_c^m = F_c^n$
14. Diameter of the diffuser	$D_{dif}^m = D_{dif}^n \cdot b$
15. Radius of the diffuser	$R_{dif}^m = R_{dif}^n \cdot b$
16. Split of the diffuser	$\Delta m_{dif}^m = \Delta n_{dif}^n$
17. The whole gas pressure on SVVT entry	$P_m = P_n / b$
18. Available degree of the gas expansion	$\pi_m^* = \pi_n^*$
19. The gas temperature at the SVVT entry	$T_g^m = T_g^n$
20. Degree of gas expansion into swirl	$\pi_m = \pi_n$

Table 2. Numerical Modeling of the Vortex Discharge.

n_e^0	U	P	π	π^*	W_{sp}	R_{dis}	E_{error}
cm^{-3}	V/cm	Pa	-	-	W/cm³	-	%
3×10^{11}	2490	5×10^5	5	30	580	0.76	-
2.3×10^{11}	1992	4×10^5	5	30	554	0.69	4.4
1.9×10^{11}	1660	3.3×10^5	5	30	690	0.691	13.3

Table 3. Numerical Modeling of the Vortex Discharge.

n_e^0	U	P	π	π^*	W_{sp}	R_{dis}	E_{error}
cm^{-3}	V/cm	Pa	-	-	W/cm³	-	%
3×10^{11}	2490	5×10^5	4	20	200	0.5	190
2.3×10^{11}	2138	4×10^5	5	30	1468	0.73	152
1.9×10^{11}	2483	3.3×10^5	5	30	1150	0.61	106

EXPERIMENTAL METHOD OF THE VORTEX ELECTRODISCHARGED LASER INVESTIGATION

Figure 4 shows a principle scheme of the VEL. The vortex consist of the tangential inlet (1), the vortex tube (2), radial diffuser (3), circular electrons (4), Mirrors (5) and gas mixture supply systems (6). The VEL works the following way: the gas mixture (CO_2: N_2:He) is injected through the inlet (1), where velocity of flow is increasing. Then flow of gas mixture goes to the vortex tube (2), where the latter is intensively rotating. In the vortex tube the energy separation is realized: near the axis of the tube we have low pressure and temperature. On the periphery of the tube we have considerably higher pressure and temperature. At the moment of the minimum pressure voltage is applied to the circular electrodes (4). The decaying discharge was realized in the SVVT. Optical feedback is provided with mirrors (5). In our experiment resonators length was 0, 13 m, and diameter of the SVVT was d=0, 05 m. Table 4 shows a comparison between experimental data and corresponding theoretical calculations one can see that an experimental electrical power coupling into discharge can reach the level of 300 W/cm^3 without contraction. As shown (see Table 4) theory predicts the experiment with satisfactory accuracy. The optimization of the VEL is the next step of our research.

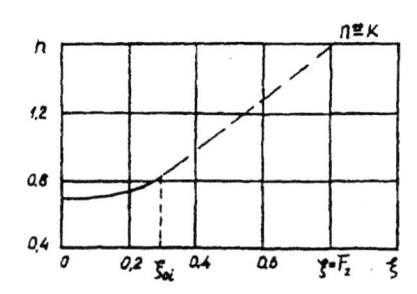

Fig. 4. The Principal scheme of coefficient distribution.

Table 4. Comparison of calculated and experimental parameters of the decaying discharge of VEL

$\overline{R}_{dis}$	p^{ex}, T_{orr}	p^{th}, T_{orr}	U, KV	I, A	W_i^{ex}, W/cm^3	W_i^{th}, W/cm^3	E_{error}, %
0.21	10	10.3	3.36	1.275	241	205	4.2
0.31	20	21.8	4.5	0,28	63	55.78	11.4
0.357	80	81.6	11.2	0.05	28	20.8	25.7

APPLICATIONS OF VORTEX LASER AND PLASMOTRON

Even without optimization of this type of laser, the VEL can find a variety of application for agricultural, biological productions plants. It is due compactness, higher efficiency and absence of cooling system. Besides, the VEL can work by using the surrounding air. In this case the VEL will be working under non-optimal mixture ratio (CO_2:N_2:He). In addition, the efficiency of the VEL will be increased due to the absence of gas mixture system and the system of cooling. The vortex plasmatron and the VEL can be used for plasma-chemical investigations, medical applications, technological processing.

SUMMARY

At has been shown that electrical power coupling into discharge can reach level of 300 W/cm^3, i.e. it is more than two orders of magnitude higher than in ordinary gas discharge. The similarity relations and scaling laws for the vortex discharge have been proposed. The vortex electro-discharge systems are related to convection-diffusion type of cooling.

ACKNOWLEDGEMENTS

I wish to thank prof. A.D. Margolin, V.M. Shmelev, Dr. H. D. Lamashapov for helpful discussions.

REFERENCES

1. A.P. Merculov, 1969, Vortex Effect and its Application in Technic, Mech. buld. puble., Moscow.
2. V.T. Volov and V.A. Shahov, 1992, Termodynamics and Heat-Mass-Transfer of Strongly Rotated Flows. Kharkov Aviation Univ., Int. Aviat. Ass. Publ., Kharkov, Ukraine.
3. V.M. Shmelev, A.D. Margolin and A. Mischenko, 1980, On a Theory of the Vortex Electrical Discharge. JETP, 4, Vol. 50.
4. V.T. Volov, 1986, Theory of the Vortex Electro-Discharged CO_2 Laser, VINITY, Moscow.
5. V.T. Volov and H.D. Lamashapov, 1988, Eelectron Diffusion into Decaying Discharge of the Strongly Rotated Compressible Turbulent Flow, JETP, 4, Vol. 58,.
6. V.T. Volov,1986, Vortex Decaying Discharge Theory, VINITY, Moscow.
7. V.T. Volov, 1983, Calculation Method the Vortex Diffuser Apparatuses, JIP, 1, Vol. 4.
8. V.K. Konuhov, Similar Gas Discharges for the CO_2 Laser, JETP, 8, v.12, 1970.

HEAT-HYDRAULIC PERFORMANCE OF ROLLED-UP THIN-WALL AND SMALL-DIAMETER TUBES AT LONGITUDINAL FLOW OF HEAT CARRIER

Svetlakov A.L., Folomeev E.A., Bezgin L.S., Galankin V.M., Ilchenko M.A., Tolmachov A.V., Filippov Y.N.

Central Institute of Aviation Motors, 111250, Moscow, Russia; Tel:+7-095-362-1381; Fax:+7-095-261-5442;
E-mail:svetlako@ciam.ru

ABSTRACT

The laboratory technique for manufacture of the rolled tubes of small-diameter with thin wall (2x0.1 mm) was developed and the set tubes with different geometrical parameters of rolling was produced. The thermo- hydraulic performance of these tubes was investigated in the model coaxial heat exchanger. The specific peculiarity of the test bed was the measurement of the tube wall temperature by its electric resistance. The method of experimental data processing was developed and verified by the tests of plain tubes. The dependencies of heat transfer enhancement and hydraulic loss growth on Reynolds and Mach numbers were experimentally obtained for investigated tubes. The possible reasons of the anomalous dependencies of heat transfer and hydraulic resistance on Mach number was proposed to be the acoustic-vortex interaction. Some influence of the mutual flow direction on heat exchange was also revealed.

INTRODUCTION

One of the most important problems for advanced aerospace propulsion with heat regeneration cycle is a problem of development of heat transfer enhancement technique in order to decrease significantly the heat exchanger mass.

The basic requirements to such heat exchangers are:
- the high degree of heat regeneration with areas of small temperature heads;
- the high pressure of coolant (up to 200-300 atm) and relatively low level of heat carrier pressure (5-50 atm);
- the severe requirements to hermiticity and reliability;
- the more severe restrictions on pressure losses in heat carrier duct than in coolant.

The tubular structure was unambiguously chosen in order to minimize the heat exchanger mass at high loading pressure. The tube diameter and walls thickness were chosen as small as possible (d=2mm and δ=0.1mm correspondingly in this research), taking into account the strength and technological restrictions; this already enhances the thermohydraulic effectiveness. Also it is expedient to use the longitudinal flow over tube bundle to assure the high level of heat regeneration in areas of low temperature head. Thus, the heat transfer enhancement must be applied to the thin-walled tubes of small diameter in longitudinal bundle at strong requirements on increase of hydraulic losses in tube space. The choice of structure definitely restricts the number of effective methods of heat transfer enhancement, which may be used here. To solve the problem, the known passive method of enhancement, which based on turbulization of wall boundary layer owing to annular ridges inside the tube and grooves outside it (Kalinin, Dreitser and others, 1998), was chosen. However, thin-walled tubes, that had been used in presented work, need other profiling (rolling) technologies which are some distinct from typical ones for the thick-walled tubes of relatively large diameter for industrial heat exchangers.

Besides, it is necessary to widen the investigated range of varying enhancement parameters, such as depth/height (H), width (b) and step (t) of ridges/grooves (Fig.1). With the aim of cost decreasing at such multi-parametric research, it was necessary to develop the cheap and simple rolling device, which would allow to re-adjust it for obtaining the sufficient number of tubes with various enhancement parameters.

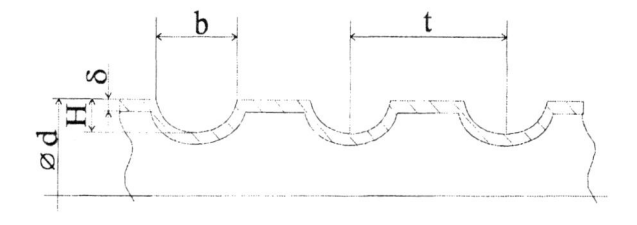

Fig.1 The geometrical parameters of rolled-up tube

THE ROLLING OF THIN-WALLED TUBES IN LABORATORY CONDITIONS

The method of tube rolling in wedge-shaped gap between plates with exchangeable forming (shaping) element on the surface of one of them was chosen for laboratory conditions, a set of hardened wire rods being used as a shaping element. The change of the diameter of wire rods d_{wr}, the step between wires in a set, the gap angle α, the number of reciprocal moves, gap height etc allows to change three enhancement parameters (H,b,t) independently and on-the-fly.

The base of this device is two massive steel plates 1 and 2 with polished working surfaces. (Fig.2) The ends of wires are fixed on the work surface of movable upper plate 2 by thin foil, that is itself spot-welded to the plate. The upper plate is propping up on two pairs of polished supports of different and controlled height, while the supports restricts the free zone ($\approx$70mm) for rolled tube displacement. The different number of foil sheets is placed under every pair of supports to control the gap angle α and the gap height. In the process of rolling, the rolled tube 3 is pressed between the set of shaping wires and work surface of immovable plate 1 and is moving translationally and rotary along wedge gap as the upper plate reciprocates. This procedure provides forming grooves on external tube surface and ridges inside tube.

It is important that the length of thin-walled tube along the generatrix doesn't change during its rolling that was defined by direct measurements of the overall tube shortening at obtained generatrix profile. It is evident that the thickness of tube and its strength characteristics in the places of rolling had been changed insignificantly. The averaged data of geometrical parameters of manufactured rolled tubes (material – stainless steel 12X18H10T, length of rolled segment - 200 mm, diameter of wire rods d_{wr}=1 mm) are presented in table 1.

Table 1. The parameters of rolling for investigated tubes

No of tube	Step t [mm]	Depth H [mm]	Width b [mm]	t/H [-]	b/H [-]	Number of rods, Z_{wr}, [-]
1	2.005	0.104	0.849	19.28	8.16	1
2	2.31	0.078	0.625	29.62	8.01	10
3	1.992	0.102	0.64	19.53	6.27	10
4	1.76	0.125	0.73	14.08	5.84	5
5	1.4	0.069	0.595	20.29	8.62	6
6	1.489	0.106	0.667	14.05	6.29	6

TEST BENCH AND OBJECT OF INVESTIGATIONS

The experimental investigation of local heat transfer and hydraulic characteristics of rolled thin-walled tubes that would be used for heat exchangers with single-phase working fluids and at longitudinal flow over tube bundle requires to provide varying independently the main governing non-dimensional parameters Reynolds numbers (Re), Mach numbers (M), temperature factor (T_w/T), enhancement parameters b/d_h, H/d_h, t/d_h in every duct.

Fig.2 The laboratory rolling device

Conditions of external longitudinal flow over tube bundle was modeled by annular channel. In this case the hydraulic diameter of the external duct was chosen equal to rolled tube diameter d_h=2 mm.

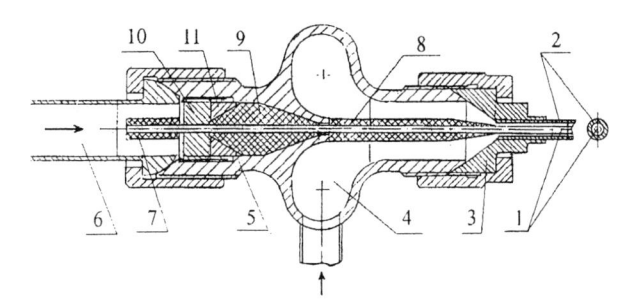

Fig.3 The structure of investigated working object

The working object is symmetric about the central cross section that simplify the reverse of the working fluid directions. Thus only one side of the working object structure is presented at Fig.3. It consists of the following elements: internal rolled tube 1, external tube 2, mounting unit 3 between external tube and manifold 4, sealing unit 5 between manifold and internal tube, inlet (outlet) pipe of an internal flow 6. The cantilever ends of an internal tube inside inlet (outlet) pipe of internal flow are heat-insulated by fluoroplastic bushes 7. The manifold with geometrically smooth outlines provides the uniformity of the heat carrier flow. The surface of external tube, manifolds and inlet/outlet ducts are heat-insulated to decrease the heat leakage (the insulation is not shown at Fig.3). To decrease the uncontrollable heat exchange between heat carriers on segments of an internal tube that are located inside manifolds, these segments are heat-insulated by fluoroplastic bushes 8. Fluoroplastic bush 9 provides an electrical insulation of an internal tube from other units. The sealing of unit of internal flow inlet is provided by griping of nuts 10 and intermediate bush 11. The investigated tube was centered by small drops of solid polymer at three points along the rolled zone (each one encumbered the annular cross section no more than by 1/4).

The independent control of 5 parameters: Re_{ext}, R_{int}, M_{ext}, M_{int}, T_{int}/T_{ext} was provided during tests and five independent control units were used for this purpose: the valves at the inlet and outlet of each channels (4 units) and the regulator of electrical heater power (W_{max} = 5.5 kW) at the preliminary heating of heat carrier. The possibility to change the co-direction of heat carrier flows from co-current to counter-current was introduced to check the influence of local heat flux distribution in micro-zones of flow separation at the rolling grooves/ridges.

The measurements of inlet and outlet temperature and their differences along working zone, and also inlet and outlet pressures for both ducts were carried out during tests. Additionally the pressures and pressure drops along annular duct were measured. The major measurements were doubled. The length-averaged temperature of an internal tube was determined by measuring its electric resistance. The necessary accuracy of this measurement was provided by use of high-precision resistance-comparator K3003. The calibration of wall temperature measurement was carried out individually for every new mounted tube. Such technique of wall temperature determining does not introduce any flow disturbances and make possible the investigation of the heat transfer in both ducts simultaneously.

To estimate the heat leakage the temperature along external surface of the external tube and manifolds were measured, as well as the temperature outside heat insulation. The flow rates of working fluids were determined by a standard technique using the measuring orifice plates at the outlets of working fluids out of test bed.

The data collection system provides the computer recording of the time-averaged parameters, their preliminary in-line processing-and-presentation (the sampling frequency f=100 Hz, the number of averaging per recorded point Z=100). The final recording of the every regime point was carried out after stationary mode settling.

TECHNIQUE OF EXPERIMENTAL DATA PROCESSING

The data of "cold" tests were used to define the hydraulic characteristics of rolled tubes. The hydraulic resistance coefficient for this case was calculated by use of the dependence that was obtained by explicit integration of the system of differential equations of mass, impulse and energy conservation for adiabatic flow:

$$\zeta = \frac{2d_h}{L}\left\{\frac{1}{2k}\frac{(M_2^2-M_1^2)}{M_2^2 \cdot M_1^2} + \frac{(k+1)}{4k}\ln\left[\left(\frac{M_1}{M_2}\right)^2 \frac{(M_2^2+2/(k-1))}{(M_1^2+2/(k-1))}\right]\right\} \quad (1)$$

here M_1, M_2 - Mach numbers at reference sections on length L.

The hydraulic resistance of external surface of rolled tube was defined by direct pressure measurements in annular duct taking into account the different friction at rolled tube surface and plain surface of external (case) tube. The friction on the last one was assumed to be of the plain smooth tube value. The hydraulic resistance inside rolled tube was determined by measurement of pressure differential on total length of a tube minus pressure losses on un-rolled (plain) segments, that were numerically defined by use of the real friction coefficient for internal surface of unrolled tube.

The definition of Nusselt numbers on the base of "hot" test data was made by numerical integration of system of unabridged one-dimensional differential equations, converted to convenient kind and written for both ducts (the integration was made by Runge-Cutta method of 5th order):

$$\frac{1-M^2}{u}\frac{du}{dx} = \left[M^2\frac{\xi}{2d_h} + \frac{\beta}{\rho u C_p}q_{\Omega K}\right]$$

$$\frac{(1-M^2)}{T}\frac{dT}{dx} = \frac{q_{\Omega K}}{\rho u C_p T}(1-M^2) - \frac{\beta C_s^2}{C_p}M^2\left[\frac{\xi}{2d_h} + \frac{\beta}{\rho u C_p}q_{\Omega K}\right] \quad (2)$$

$$\frac{1-M^2}{\rho u^2}\frac{dP}{dx} = -\left[\frac{\xi}{2d_h} + \frac{\beta}{\rho u C_p}q_{\Omega K}\right]$$

here $|q_{\Omega K}|=4\cdot|q_s|/d_h=4\cdot h\cdot|T-T_w|/d_h=4\cdot Nu\cdot|T-T_w|\cdot\lambda/d_h^2$ - convective heat per volume unit

The calculation of heat-insulated segments of internal tube outside rolled zone was made under the assumption of adiabatic flow taking into account the hydraulic losses at friction coefficient of unrolled tube. This coefficients for rolled segment were presented as power dependencies $\xi=f(Re)$ obtained from cold test data for current tube. The heat leakage into external heat insulation from external duct was taken into account on the base of measurements of the external wall temperature. This heat flux did not exceed 5-7% of a sought heat flux at rolled surface. The integration gave also the distribution of the temperature of heat exchange wall as well as its root-mean-integral value, which was experimentally determined by measurement of tube electrical resistance.

The problem of searching for the heat transfer coefficients was reduced to the minimization of sum of discrepancy squares which were between computational and experimental values of inlet and outlet temperatures of working fluids and also root-mean-integral temperature of the wall by means of the varying of sought Nusselt numbers in both ducts.

The used measurement system and processing technique was verified by tests of unrolled tube. This research had shown, that the hydraulic resistance coefficient in annular channel corresponds to its value for a hydraulically smooth surface (Fig.4). The visual inspection of internal surface of the tube has shown that it is rather rough, that explains the self-similar flow mode existence at this surface in considered range of Reynolds numbers (Fig.5), whereas the experimental data have a good correspondence to Moudy's diagram for recommended statistical values of equivalent sand roughness for internal surface of cold-drawn tubes (Kakac, Shah, Aung, 1987). The good conformity between experimental and theoretical Nusselt numbers for smooth ducts at co-current and counter-current of heat exchange in tubular and annular channel (Fig.6) was observed too.

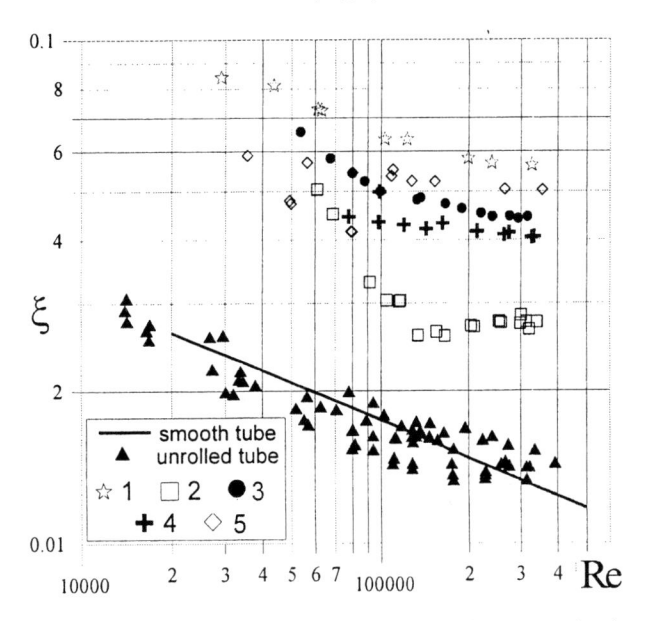

Fig.4 The hydraulic resistance coefficient for external tube surface

THE RESULTS OF EXPERIMENTAL RESEARCH

The experimental research of heat-hydraulic characteristics of rolled tubes were mainly carried out in following ranges of operating parameters:

Table2. The ranges of main flow parameters at testing

Parameter	Internal duct	External duct
Reynolds number	$4 \cdot 10^4 \div 3 \cdot 10^5$	$4 \cdot 10^4 \div 3 \cdot 10^5$
Mach number	$0.1 \div 0.3$	$0.1 \div 0.25$
Inlet temperature	$280 \div 300K$	$280 \div 450K$
Working fluid	air	air

The dependencies of hydraulic resistance coefficients on Reynolds number for every duct and different rolled tubes are shown at Fig.4,5 and the dependencies of Nusselt number on Reynolds number are shown at Fig.7,8. The dispersion of heat transfer data is illustrated by Fig.9 where two sets of experimental points for one of rolled tubes are shown as well as the approximation power functions Nu(Re).

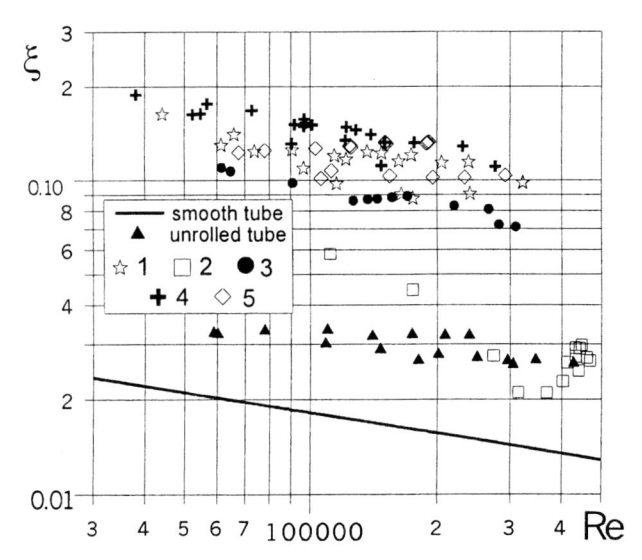

Fig.5 The hydraulic resistance coefficient for internal tube surface

The average level of growth of heat transfer and friction for obtained results basically corresponds to its level for the data reported in (Kalinin, Dreitser and others, 1998). The obvious differences may be caused by the distinction of rolling process and rolling shape for tubes of different diameter and wall thickness. It may be mentioned that influence of such parameter of groove shape as its width b, had not been considered in details in the work (Kalinin, Dreitser and others, 1998) whereas the relative groove width b/d_h for our tubes was significantly larger than at reference research.

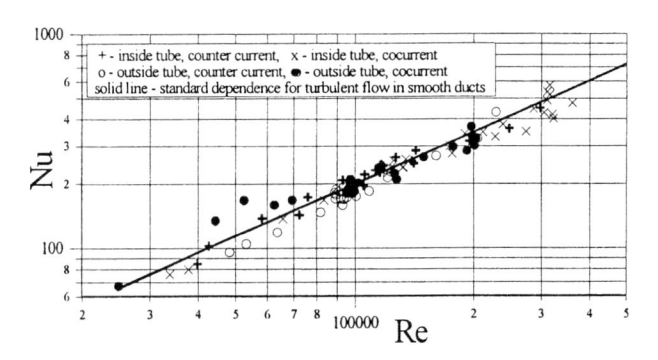

Fig.6 The verification of the Nusselt number determining technique by testing plain unrolled tubes.

The relation Nu(Re) in investigated range of numbers Re is described by strongly distinct power functions. The value of exponent "m" in approximating relations $Nu \sim Re^m$ varies ranging $m = 0.6 \div 0.9$ for different tubes. This requires more complex empirical-theoretical dependencies for further strict description of obtained results.

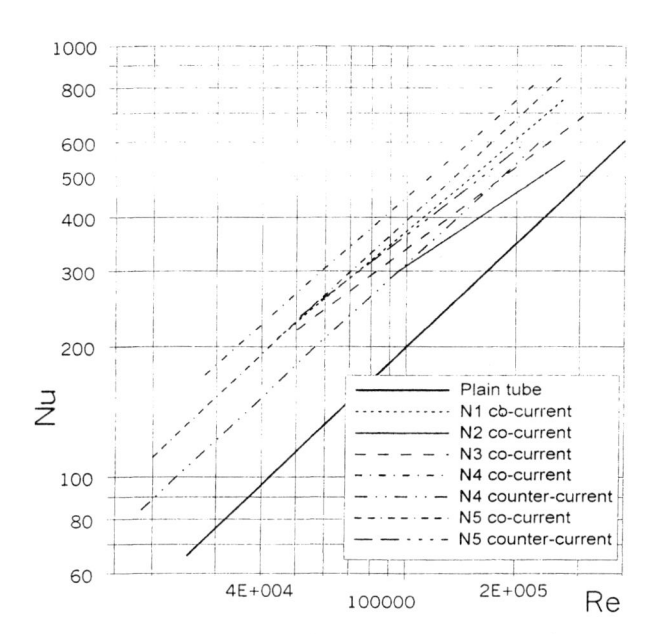

Fig.7 The Nusselt numbers at internal tube surface.

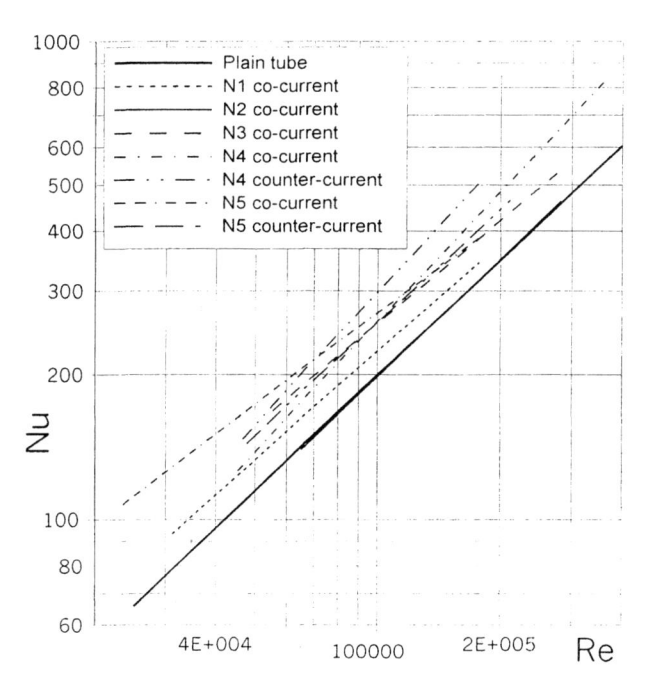

Fig.8 The Nusselt numbers at external tube surface (plot for co-current tube N2 coincide to plain tube plot)

The dependencies of hydraulic resistance coefficients on Re number demonstrate both typical behavior with transition to self-similar function, and more complex kind of the relation such as the non-monotonous character of relation $\xi(Re)$ for instance for external surface of tube N5 or internal surface of tube N2.

The essential influence of the relative direction of fluid flows in external and internal channels on heat transfer coefficients (Nusselt numbers) was noted in some cases, the most essential this influence was for tube No.4 (fig.7) with lowest values of relative step t/H and width b/H of grooves referred. This phenomenon may be caused by the influence of mutual local distribution of heat transfer coefficients at both sides of tube wall in separation zones with scale of an order of rolling element size (Fig.10).

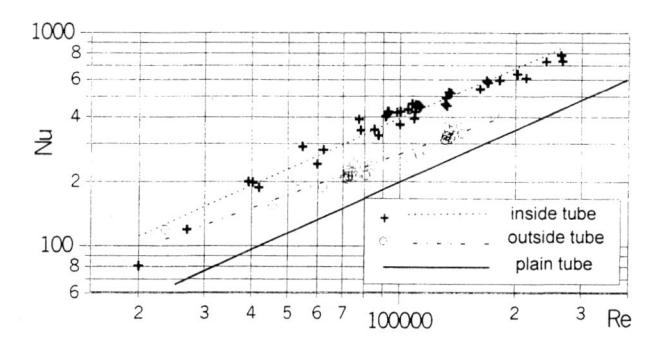

Fig.9 The typical dispersion of Nusselt numbers data (tube N5, co-current)

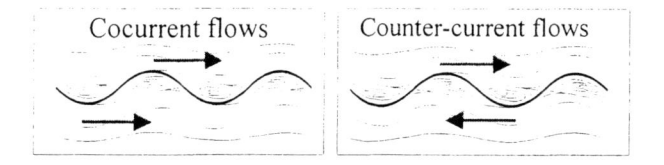

Fig.10 The scheme of different mutual positions of separation zones over a wall for co-current and counter-current flows

Then local heat exchange coefficient distribution may change with the change of co-direction of flows. The involved effect may take place at the tubes with small wall thickness (the low longitudinal smearing of heat flux by wall heat conductivity) and at relatively broad grooves, while it is unlikely for tubes with rather thick wall in relation to specific rolling element size.

It should be specially noted that some dependencies of the Nusselt number and hydraulic loss coefficient on Mach number for rolled tubes that was first revealed. When the obvious dependencies Nu (Re) and $\xi(Re)$ for the tube No.6 with maximum relative width of grooves b/t=0.45 were plotted, the essential dispersion of experimental points for the external channel appeared, while the test technique for this tube did not differ from others tested. Then tests of this tube in different combinations of regime parameters Re_{ext}, M_{ext} in essentially broadened range of a Mach number in the external channel M_{ext}=0.04-0.55 were made as well as the measurement system of test bed was re-verificated. They had shown that hydraulic loss coefficient for external surface of this tube at Re=const is sharply varying at increase of Mach number in a vicinity of M=0.1 and then remains practically constant (Fig.11). Let us mention that a plain unrolled tube at verifying tests had not demonstrated such influence. The

Nusselt number for rolled tube No.6 is also increased at Mach number increase and the growth gradient is dependent on Reynolds number (Fig.12).

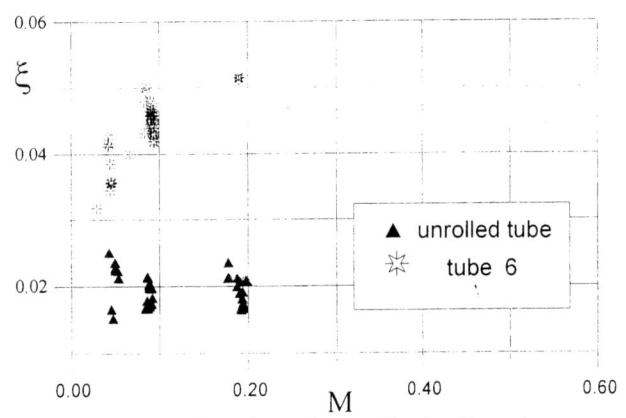

Fig.11 The anomalous dependence of hydraulic resistance on Mach number

The revealed influence of Mach number on thermo-hydraulic characteristics expands the space of parameters at heat exchanger system optimization. In this connection it is actual to understand the physical process mechanism, leading to such influence. Unfortunately, this mechanism is rather difficult to specify experimentally at the existing test bed, so it is necessary to formulate the working hypothesis to define the directions of the further research.

The authors had already seen the similar specific dependence of Nusselt number on Mach number at heat transfer enhancement by powerful acoustic-vortex auto-oscillations (Folomeev, Ilchenko, Selin, 1995), excited in a gas flow as a result of interaction between eigen acoustic oscillations in duct and vortexes shed from trailing edge of plate profile, that is mounted in the duct center (Fig.13). As the flow over grooves or ridges may generate the periodic vortex shedding too (Kalinin, Dreitser and others, 1998), it is possible to assume, that some acoustic-vortex interaction may occur on rolled surface at curtain conditions.

The influence of Mach number on excitation of acoustic-vortical auto-oscillations may be reduced to following. The eigen acoustic oscillation frequencies of gas in a duct may be written as $f_{ac} \sim C_s/L$, here L - specific duct size. The frequency of vortex shedding may be written in general case as $f_{vort} \sim Sh \cdot u/\Delta$, here Δ - specific dimension of duct element which is responsible for vortex generation. Then, the simplest resonance condition would be determined by a ratio $f_{vort}/f_{ac}=(Sh \cdot M)/(\Delta/L) \sim 1$. The excitation of auto-oscillations may occur at both different set of Strouhal numbers and different acoustic modes. However, there is always a limit Mach number, below which correlated acoustic-vortex interaction does not occur. This Mach number corresponds to the lowest eigen acoustic frequency. If it is assumed, that vortex shedding appears at rolling elements ($\Delta \approx H$), and the specific reference size L is the step of rolling t, and the same Strouhal numbers occurs, as at profiles, mounted in a duct, then the limit Mach number appears close to $M \approx 0.1$, that corresponds to results presented at Fig.11,12.

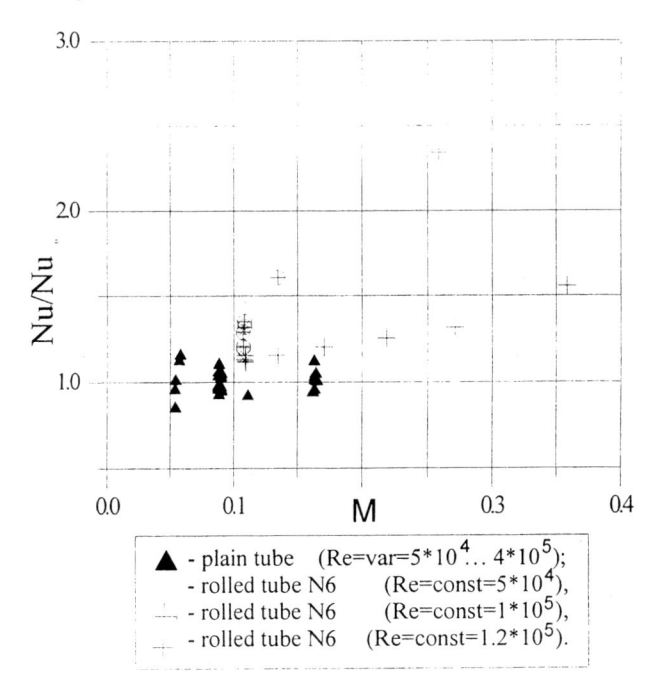

Fig.12 The anomalous dependence of Nusselt number on Mach number

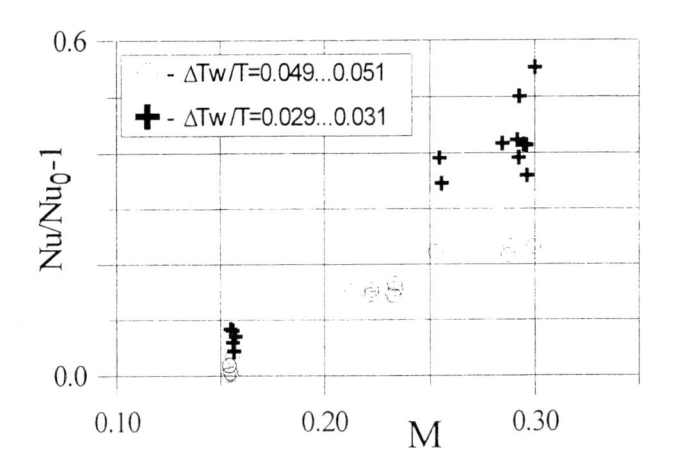

Fig.13 The dependence of relative Nusselt number on Mach number for heat transfer enhancement by acoustic-vortex ultrasonic auto-oscillations.

However there are some distinctions between heat transfer enhancement by profiles and data presented at Fig.10,11. For instance, the heat transfer enhancement Nu/Nu_0 for profile case does not depend on Re number, being strongly dependent on relative temperature heat at the wall $\Delta T_w/T$ (Fig.12) and hydraulic loss coefficient depends on Mach number non-monotonously (in that case it is mostly depend on the pressure pulsation level). The reason of these distinctions may be the level of acoustic influence for duct with profile and the rolled tubes. In the first case the pressure oscillation amplitude is up to 180 dB and it is the governing factor of the process. In the second case the level of acoustic pulsations may be rather small thus the acoustic-vortex interaction leads only to some re-ordering of the vortex structure of flow.

Thus the described anomalous phenomenon is rather complex and should be investigated separately and in detail. Authors also should note that obviously the check of Mach number influence at heat-hydraulic research of tubes is omitted and this may be a reason of the absence of similar data at previous research.

ACKNOWLEDGEMENTS

The presented work was fulfilled at the support of the International Science-and-Technology Center as a part of ISTC/CIAM Project 936-98 «Research On Fundamental Problems Of Heat Transfer Intensification In Air-Hydrogen Heat Exchanger». Authors also thank Dr. Heinz Guenter Klug (EADS Airbus) for the attention to this work.

CONCLUSIONS

1. The laboratory technology of the rolling-up of the thin-wall tubes of small diameter (2x0.1mm) with different enhancement parameters was developed. The test bed for the determining of the local thermohydraulic performances of the coaxial heat exchange element with equal hydraulic diameters of the internal and external ducts was made. The experimental technique and data processing method were verified at plain tube tests. The important peculiarities of the research were the check of Mach number influence, the reversing of mutual flow directions and the use of electric resistance of the investigated tube for wall temperature measurement.

2. The experimental research of the heat transfer and hydraulic resistance of rolled tubes with different enhancement parameters in model heat exchanger had revealed the heat transfer enhancement at external tube surface Nu_{ext}/Nu_0=1-1.5 at concomitant hydraulic resistance increase of ξ_{ext}/ξ_0=1.7-3.5, at internal tube surface Nu_{int}/Nu_0=1.5-2.3 and ξ_{int}/ξ_0=1-5.6 correspondingly (these numerical data at Re=10^5).

3. The anomalous influence of flow regime parameters on heat-hydraulic performances were revealed for some investigated tubes including the influence of mutual direction of heat carrier flows on Nusselt numbers as well as the sharp varying of the hydraulic resistance and heat transfer in vicinity of Mach number M~0.1. The hypothesis of the reasons of these phenomena were proposed. These are correspondingly the re-distribution of mutual position of small-scale separation zones at both sides of rolled tube wall and acoustic-vortex interaction at vortex shedding from rolling elements.

NOMENCLATURE

b	rolled groove/ridge width, m
Cp	specific heat capacity, J/(kg·K)
Cs	speed of sound, m/sec
d	diameter, m
f	frequency, Hz
H	rolled groove/ridge depth/height, m
h	heat transfer coefficient, W/(m²·K)
k	adiabatic exponent, dimensionless
L	specific dimension (length), m
M	Mach number, u/Cs, dimensionless
Nu	Nusselt number, h·d_h/λ, dimensionless
P	pressure, Pa
q_s	heat flux, Wt/m²
Re	Reynolds number, ρud_h/μ, dimensionless
Sh	Strouhal number, f·Δ/u, dimensionless
t	step between rolled groove/ridge, m
T	temperature, K
u	flow velocity, m/sec
W	electric power, W
Z	number (quantity), dimensionless
α	gap angle of rolling device, degree
β	coefficient of thermal expansion, 1/K
δ	tube wall thickness, m
Δ	specific dimension at vortex generation, m
λ	heat conductivity, W/(m·K)
μ	viscosity, Pa·sec
ρ	density, kg/m³
ξ	hydraulic loss coefficient, dimensionless

Subscript

int	internal
ext	internal
h	hydraulic
wr	wire rod
w	wall
max	maximum
ac	acoustic
vort	vortex
0	corresponded to plain surface
Ω	corresponded to volume
K	convective
P	at constant pressure

REFERENCES

Folomeev, E.A., Ilchenko, M.A., Selin, N.I., 1995, The Enhancement of the Heat Transfer Inside Ducts at Autooscillations of Gaseous Flow, *Izvestiya RAN, Energetika i Transport*, N4, pp.151-159.

Kakac, S., Shah, R.K., Aung, W., 1987, Handbook of Singl-Phase Convective Heat Transfer, JohnWiley&Sons, Inc.,New York.

Kalinin, E.K., Dreitzer, G.A., Kopp, I.Z., Myakochin, A.S., 1998, The Effective Heat Transfer Surfaces, Energoatomizdat, Moscow.

SINGLE-PHASE HEAT TRANSFER DESIGN DATA AND METHODS

FROM FIN SELECTION TO NETWORK DESIGN – A HOLISTIC DESIGN APPROACH FOR COMPACT HEAT EXCHANGER NETWORKS

(Frank) X.X. Zhu[1] and L. M. Pua[2]

Department of Process Integration
UMIST, P.O. Box 88, M60 1QD, England
Email[1]: F.Zhu@umist.ac.uk
Email[2]: L.Pua@stud.umist.ac.uk

ABSTRACT

Industries are starting to replace shell-and-tube heat exchangers with compact heat exchangers in several applications. The benefits have been reported by using compact heat exchanger at equipment level. However greater extent of economic benefits can be realized when applying compact heat exchangers in the context of total heat transfer system. This is not an easy task for both industries and designers who are new to compact heat exchangers. This is because the data for heat transfer conditions for each stream is not available in the early design stage to guide their overall network design. Hence method which can determine the optimal heat transfer conditions for each stream and the network configuration simultaneously is very essential in promoting the use of compact heat exchangers.

Here a novel method is developed which can achieve this objective. The new method takes the advantages of two new physical insights on heat transfer surface selection, i.e. the identical fin concept and the Z-Y graph. The concept of using same fins for all streams in the early design stage is reasoned and justified physically in this paper. The Z-Y graph is then developed to represent four important design parameters, namely fin types, volume, pressure drop and frontal area. Using these two concepts, a level-by-level methodology is developed to provide a transparent design procedure so that the designer is able to monitor and control the design process. Each design level is solved using mathematical programming to address the complex interactions and trade-offs.

In this way, the advantages of heat transfer surfaces (fins) can be fully explored at the very beginning of design stage and optimisation of fin selection is considered in the context of overall design problem. Consequently, the heat exchanger network designed using compact heat exchangers can be compared clearly and fairly with their shell-and-tube counterpart. Thus designers can have more confidence of using compact heat exchangers in the overall network. The design method starts with very little information that is already known to the system (for example, temperatures and physical properties) and systematically guide the designers towards the final design.

INTRODUCTION

It is a trend that industrial sectors are moving towards more intensified processes to achieve faster, safer, and environmental friendly manufacturing. This requires that heat transfer process, as a major part of overall system, should also be intensified and accomplished using compact heat exchangers. However, in many industrial sectors, shell-and-tube heat exchangers have been playing dominant roles due to its long history of use and the capability of handling a large range of fluids and operating conditions. In comparison, compact heat exchanger has its distinctive advantages due to its much higher heat transfer coefficient and larger extended surface area, which result in much smaller size of exchanger units. Secondly, it allows multiple streams entering a single unit, which has the great potential of reducing the number of units. Furthermore, the ability to transfer heat with smaller temperature approach requires less utility consumption and thus produces less flue gas emissions.

The question, which naturally arises, is why compact exchangers have not been applied to a larger extent for most process industries? Maybe the conservatism is the major hurdle of all. There are also technical reasons. Fouling is a major problem in many processes, which prevents the wide application of compact heat exchangers. Another major reason is the lack of design methods, in particular for design of heat exchanger networks using compact heat exchangers. For example, how to insert compact heat exchanger units into an existing heat exchanger network, which is the retrofit issue. The second one is how to design a network of compact heat exchangers featuring minimal number of units and thus minimum capital cost. This is the new design issue. This paper focuses on the second issue.

When industry sectors want to adopt compact heat exchangers into their new design, they face a very complicated problem. Firstly, they cannot apply any existing design methods for shell-and-tube exchanger network to guide their way to the optimum design. Secondly, the very essential design parameter, the heat transfer coefficient, for designing the network is unknown in the early design stage. Thirdly, simply take any arbitrary values for the heat transfer coefficients will lead to a very poor network design using compact heat exchangers. As a result of all these, the economic benefits of using compact heat exchangers in a network context cannot be fully demonstrated.

The compact heat exchanger network design problem involves issues at different levels. In the system level, we need to determine network configuration, i.e. which stream goes to which unit and enters/leaves at what position, how many units are required to fulfil all the heat transfer duties. In the exchanger level, we need make decisions on fin selection for each stream, number of plates and dimensions of each exchanger. The issues at these two levels interact with each other and cannot be separated. Strong synergy can be exploited if these issues can be taken into account in a more systematic and simultaneous way. However, the design problems for each level are very complex on their own right not to mention solving a combined problem.

In the past, the two problems of network and single exchanger design are treated separately for the design of shell-and-tube heat exchanger network. In the level of network design, heat transfer coefficients for all streams involved are assumed fixed and network configuration is determined on this basis. This assumption may be valid for plain tube type of shell-and-tube exchangers; but it is not applicable to fin type of exchangers where heat transfer conditions strongly depend on conditions of the heat transfer surfaces, e.g. fin types, fin geometry and the number of plates. Thus, network design without considering heat transfer conditions could lead to very poor overall design.

With these problems in mind, this paper will present a novel design methodology, which puts optimisation of heat transfer conditions (fin selection) into the heart of overall design procedure with the help of physical insights. Firstly, this design method optimises heat transfer conditions and stream matching simultaneously, from which optimal heat transfer conditions for each stream are determined. On this basis, the network configuration is optimised. At this stage, the optimised heat transfer conditions give minimum number of units and volume; but the streams in each exchanger may not have equal length. Thus, for each individual heat exchanger, the optimal heat transfer conditions are revised so that the lengths for each stream are the same. The next stage is to select fin types and determines the fin geometry which can best match the optimal heat transfer conditions. In this way, the issues in different levels are taken into account properly and interactions between them are fully exploited to give the best overall design.

In this paper, we focus on multi-stream plate-fin heat exchangers will be studied because there are many degrees of freedom in selecting the heat transfer surfaces (fins) and that fin thermal-hydraulic performance and costing data are readily available (Kays and London, 1984; ESDU 97006, 1997). Hence the terms multi-stream heat exchanger (MHE) and compact heat exchanger are used interchangeably.

ESSENTIAL FEATURES IN THE NEW METHODOLOGY

When developing the new method, we bear the practical applications in mind and make sure it will have the following features.
1. The method should take advantage of the insights provided from heat transfer for optimising heat transfer conditions and fin selection, and from thermodynamics for stream matching, while simultaneously utilise the power of mathematical programming for addressing complex interactions and trade-offs.
2. The method should provide a transparent design procedure. This is essential to enable a designer to monitor and control the design process. This can be achieved by adopting a level-by-level methodology in the spirit of hierarchical methods. Different levels provide a logical procedure that is easy to follow.
3. Last but not the least, the method should be able to reduce the complexity of the problem and decompose such a design problem into several relative simple sub-problems without missing interactions.

PHYSICAL INSIGHT OF DESIGNING SINGLE EXCHANGER

The design problem for single MHE exchanger involves selection of fin types for each stream, number of plates for the exchanger, dimensions of the exchanger and total pressure drops.

Fin plays an important role in a plate-fin heat exchanger. It not only provides extended surface for heat transfer but also enhances heat transfer coefficients by inducing turbulence and adds extra mechanical strength to the exchanger. When selecting fins for multi-stream heat exchanger (MHE), there are trade-offs among volumes, pressure drops, frontal areas and lengths. For example, very efficient fins are good for minimising volume but are at the expense of pressure drops. To reduce pressure drop, one can use less efficient fins or increase frontal area. This however needs to be justified by increase in capital. Similar discussions apply to the length of a MHE.

The fin selection problem is a complex combinatorial problem on its own right. Over 60 fins were reported in Kays and London (1984). In a simple three-stream heat transfer problem, the possible fin-combinations are in the order of 10^5. Hence an efficient way of screening is very necessary.

Here, it will be shown that the identical fin concept is very powerful in screening and selecting promising fins in early design stage. A newly developed Z-Y graph is used to represent four important parameters in plate-fin exchanger, which are the fin type, volume, pressure drop and frontal area. Correlations are then derived from Z-Y graph and incorporated into an optimisation model for fin selections.

Identical Fin Concept

Let us consider a simple case where fins for both sides are to be selected in order to minimise the total volume according to their pressure drop constraints (Fig.1).

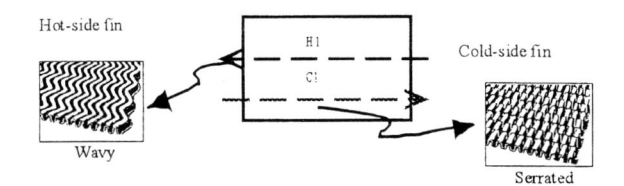

Fig.1 Two-stream plate-fin heat exchanger.

In all chemical processes, there are only two different scenarios, i.e. the properties of both streams are either similar or very different. From Eqs. (1) to (3), it is clear that the fin properties and stream physical properties are interwoven together.

$$j = St\Pr^{2/3} \quad (1)$$

where

$$St = \frac{hA_{ff}\sigma}{\dot{m}Cp} \quad (2)$$

$$\Pr = \frac{\mu Cp}{k} \quad (3)$$

The heat transfer coefficient, h, can be expressed in terms of fin property, Y'', and stream property, SP, as shown in Eqs. (4) to (6) (see Pua and Zhu (2001) for derivation):

$$h \equiv Y'' \cdot SP \quad (4)$$

where

$$Y'' = \left(j\frac{\text{Re}}{d_h}\right) \quad (5)$$

$$SP = \frac{\mu Cp}{\Pr^{2/3}} \quad (6)$$

Hence SP (Stream Property) can be used as a basis to compare physical conditions between two different streams. A large SP will resemble a liquid stream and a small SP for a gas stream. We can further define the ratios of SP between a hot stream and a cold stream as:

$$SP_R = \frac{SP_H}{SP_C} \qquad (7)$$

For two streams with similar stream properties (e.g. gas and gas), SP_R is between 0.1 and 10. For streams with differing stream properties (e.g. gas on hot side and liquid on cold), SP_R is smaller than 0.01. Otherwise for cases where liquid on the hot side, gas on the cold side, SP_R is greater than 100. The characterisation of streams using SP_R is important in the early design stage since heat transfer coefficients are unknown.

For the plate-fin heat exchanger, the volume can be expressed in terms of fin properties (Y_H' and Y_C') and stream properties (SP_H and SP_C). η and E in Eqs. (8) and (9) are fin effectiveness and the enhancement factor defined as the total heat transfer area per unit plate area.

$$V_T \approx \frac{Q}{\Delta T_{lm}} \left(\frac{1}{Y_H' SP_H} + \frac{1}{Y_C' SP_C} \right)(\delta_H + \delta_C) \qquad (8)$$

$$Y' = \eta Y'' E \qquad (9)$$

The pressure drops due to friction loss are expressed as in Eqs. (10) and (11):

$$\Delta P_H = \left(\frac{\dot{m}^2}{\rho} \right)_H \left(\frac{1}{A_{fr}} \right)^3 \left(\frac{f}{\sigma^2 d_h} \right)_H V_T \qquad (10)$$

$$\Delta P_C = \left(\frac{\dot{m}^2}{\rho} \right)_C \left(\frac{1}{A_{fr}} \right)^3 \left(\frac{f}{\sigma^2 d_h} \right)_C V_T \qquad (11)$$

Now, consider a case where SP_R is between 0.1 and 10 and frontal area, A_{fr}, is fixed. It is straightforward to see from the Eqs. (12) to (14) that if one side of the exchanger is less enhanced (e.g. decreasing in Y_C'), the match between the two streams will require higher volume and pressure drops on both sides.

$$V_T \uparrow \approx \frac{2\delta Q}{\Delta T_{lm}} \left(\frac{1}{Y_H' SP_H} + \frac{1}{Y_C' \downarrow SP_C} \right) \qquad (12)$$

$$\Delta P_H \uparrow = \left(\frac{\dot{m}^2}{\rho} \right)_H \left(\frac{1}{A_{fr}} \right)^3 \left(\frac{f}{\sigma^2 d_h} \right)_H V_T \uparrow \qquad (13)$$

$$\Delta P_C \uparrow = \left(\frac{\dot{m}^2}{\rho} \right)_C \left(\frac{1}{A_{fr}} \right)^3 \left(\frac{f}{\sigma^2 d_h} \right)_C V_T \uparrow \qquad (14)$$

For the case where SP_R is greater than 100, the followed trend is observed (Table 1). In this case, the gas side is air and the liquid side is water.

The minimum volume is found to be 5.75 m³ when plain fin 46.45 triangular (PF 46.45 T) and serrated fin SF 1/10-19.74 are used for gas and liquid sides respectively, while satisfying the gas side pressure drop constraint of 1050 Pa. It is interesting to observe that if the identical fin PF 46.45 T is used for both sides, the volume is close to the minimum. This indicates that optimal identical fin can achieve near minimum volume, which is good enough for initial fin selection.

Table 1 Different fin combinations for case $SP_R>100$

	Gas	Liquid	V_T	ΔP_{gas}	ΔP_{liquid}	h_{gas}	h_{liquid}
Optimum	PF 46.45 T	SF 1/10-19.74	**5.75**	1050	1.23	21	209
Identical fin	PF 46.45 T	PF 46.45 T	5.83	1050	4.81	22	1429
Identical fin	SF 1/10-27.03	SF 1/10-27.03	7.33	1050	0.78	32	499
Identical fin	SF 1/10-19.74	SF 1/10-19.74	11.58	1050	0.69	17	140
	SF 1/10-27.03	SF 1/10-19.74	5.80	1050	7.22	31	377
Conventional	SF 1/10-27.03	PF 19.86	11.39	1050	1.00	30	77

- The fins are named in accordance to Kays and London (1984).
- SF: Serrated Fin; PF: Plain Fin; T: Triangular.

If we apply the conventional way of thinking for fin selection, we would enhance the gas side, say, using SF 1/10-27.03 and leave the liquid side less enhanced (e.g. PF 19.86). In this case, the volume is almost doubled compared with the optimum case. This can be explained as followed: when the liquid is less enhanced, the exchanger is longer. Hence the gas side also suffers the increased length and pressure drop. To reduce the gas side pressure drop, frontal area needs to be increased which results in larger volume.

From the above discussions, we can conclude that both sides need to be enhanced and identical fins for both sides can achieve near minimum volume. This result will be used as the basis for deriving correlations for fin selection in the next section. The significance of the identical fin concept is in the great simplification of the problem for fin selection since we do not need to consider all the fin combinations in the early design stage. This is achieved by considering one fin to one stream match instead of two different fins for one match. This reduces possible fin combinations that could be beyond handling and also unnecessary. Once identical fin is determined for each stream match, the fin for cold side will be adjusted in order to fine-tune the trade-off between volume, length and pressure drop.

Z-Y Graph

Using the identical fin concept, the volume and pressure drop equations can be simplified and expressed explicitly, in Eqs. (15) to (17), as two different entities; namely stream properties (w_v, $w_{\Delta PH}$ and $w_{\Delta Pc}$) and fin related properties (Y and Z); see Pua and Zhu (2001) for derivation:

$$V_T = \frac{w_v}{Y_H} \qquad (15)$$

$$\Delta P_H = w_{\Delta P_H} Z_H \qquad (16)$$

$$\Delta P_C = w_{\Delta P_C} Z_H \qquad (17)$$

where

$$Y_H = A \cdot \text{Re}_H^B \qquad (18)$$

$$Z_H = C \cdot \text{Re}_H^D \qquad (19)$$

$$A = \frac{a \cdot a' \cdot E}{d_h \delta} \qquad (20)$$

$$B = 1 - b + b' \qquad (21)$$

$$C = \frac{1}{d_h^2} \frac{c}{a \cdot a'} \quad (22)$$

$$D = 3 - d - B \quad (23)$$

$$j = a \operatorname{Re}^{-b} \quad (24)$$

$$\eta = a' \operatorname{Re}^{b'} \quad (25)$$

$$f = c \operatorname{Re}^{-d} \quad (26)$$

Note that Y value is a better representation than j factor as Y captures both heat transfer coefficient ($a\operatorname{Re}^{-b}$) and enhanced area (E) as shown in Eqs. (18), (20) and (21).

Comparing with the original volume and pressure drops equations [Eqs. (8) to (11)], the stream process properties are lumped into known parameters, w's and fin-related properties are lumped into optimisation variables, Y and Z. In this way, the optimisation model is simplified because the inter-related fin geometry (e.g. δ, σ, f, A_{fr} in Eq. (10)) does not need to be treated as independent variables as in Eqs. (8) to (11).

However, as can be seen from Eq. (27), since each fin has different compactness, β, hence the Reynolds number is different for each fin even with same frontal area, A_{fr}.

$$\operatorname{Re} = \frac{\dot{m} d_h}{A_{fr} \sigma \mu} = \frac{4\dot{m}}{\mu (A_{fr}) \beta} \quad (27)$$

In order to put the comparison on the same basis, a reference Reynolds number, Re_{ref}, is used where β_{ref} is fixed at a constant value (value 2466 m^2/m^3 is taken in this study). Then the Reynolds numbers for each fin are modified according to Eq. (28). The Y and Z values are evaluated for the modified Reynolds numbers using Eqs. (29) and (30).

$$\operatorname{Re}_{mod} = \operatorname{Re}_{ref} \frac{\beta_{ref}}{\beta_{mod}} \quad (28)$$

$$Y = A \operatorname{Re}_{mod}^{B} \quad (29)$$

$$Z = C \operatorname{Re}_{mod}^{D} \quad (30)$$

As a result, one value of Re_{ref} means that the exchanger will have the same frontal area even for different fins. For different fins, the values of A, B, C, D and β are different. Hence putting one value of Re_{ref} (same frontal area) for each fin will generate different Y and Z values. Plotting these Y and Z pairs on the Cartesian axes, we will have a scattered diagram as in Fig. 2(a).

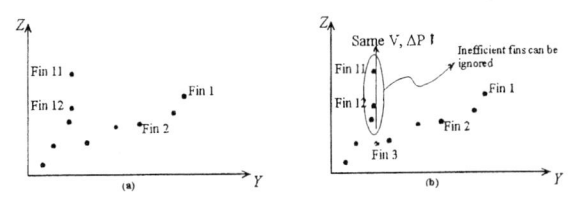

Fig. 2 Z-Y value for different fins (a) Original, (b) Highlight inefficient fins.

However, some of the fins can be ignored according to their positions on the diagram. In Fig. 2b, Fin 11 and Fin 12 are inefficient compared to Fin 3 because the former ones give the same volume (same Y) but higher pressure drops (higher Z). Thus, they should be screened out from consideration. As a result of that, a straight line can be regressed to represent all the promising fins for the reference Reynolds number in the form of $Z = RY + S$ (Fig. 3a). This procedure can then be repeated for different Re_{ref} (different frontal area). Consequently, a set of straight lines can be obtained and this graph is called a Z-Y graph (Fig. 3b).

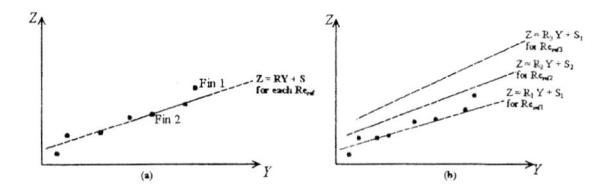

Fig. 3 Z-Y graph for (a) One Re_{ref}, (b) A set of Re_{ref}.

The Z-Y graph represents four important parameters in plate-fin exchanger design, namely fin types, volume, pressure drop and frontal area. Using $k = 160$ W/m°C and $SP = 0.06$ kg.m/s^{3}K to account for fin effectiveness, η, the following correlations Eqs. (31) to (34) are derived:

$$Z = RY + S \quad (31)$$

where

$$R = 0.1087 (\operatorname{Re}_{ref})^{1.7542} \quad \text{for } \operatorname{Re}_{ref} < 800 \quad (32)$$

$$R = 42.162 \cdot \operatorname{Re}_{ref} - 7549.7$$
$$\text{for } 800 < \operatorname{Re}_{ref} < 4000 \quad (33)$$

$$S = 2733567 \cdot (\operatorname{Re}_{ref})^{1.91}$$
$$\text{for } 10 < \operatorname{Re}_{ref} < 4000 \quad (34)$$

Fin Selection for Each Stream

Once the Z, Y and Re_{ref} for each stream have been targeted, a set of fins can be identified as suitable candidates for each stream in the MHE. This is illustrated in Fig. 4. The targeted Z, Y and Re_{ref} are 5.0577 x 10^{12}, 3.343 x 10^7 and 1504 respectively. From the Z-Y graph, the fins closest to these values are SF 1/9-24.12, SF 1/10-19.74 etc. Selecting one fin for each stream, the volume and pressure drops of the whole exchanger can be determined straightforward from the design equations (8), (10) and (11).

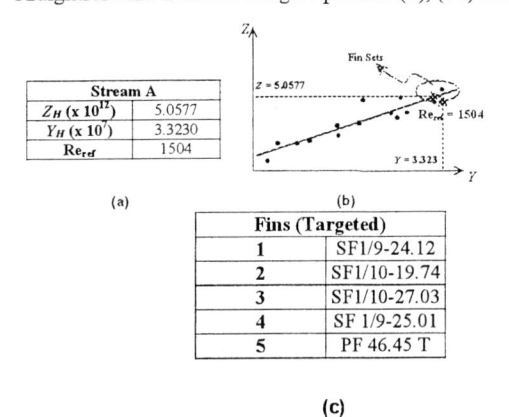

Stream A	
Z_H (x 10^{12})	5.0577
Y_H (x 10^7)	3.3230
Re_{ref}	1504

	Fins (Targeted)
1	SF1/9-24.12
2	SF1/10-19.74
3	SF1/10-27.03
4	SF 1/9-25.01
5	PF 46.45 T

Fig. 4 Fin Selection Stage (a) Targeted Z, Y and Re_{ref}; (b) Refer these values in Z-Y graph; (c) Identify fins closest to the targeted Z, Y, Re_{ref} values.

DESIGNING COMPACT HEAT EXCHANGER NETWORK

In this design level, we need to determine network issues. The first one is which stream enters/leaves which unit at what intermediate point. The second issue is what is the minimum number of units required in the network. We will explain the methods for solving these problems as follows.

Entrance Points Optimisation

Previous works (Taylor (1990), Yee *et al.* (1990) and Picon-Nunez (1995)) have proposed that streams should enter and exit a MHE according to their respective kink points on the composite curves, i.e. vertical heat transfer. This is a direct adoption of the pinch design method for shell-and-tube heat exchanger networks (Linnhoff and Ahmad (1990)) which states that close to minimum heat transfer area can be obtained if the network is arranged in this way. However, this is not true for the case of MHE network design.

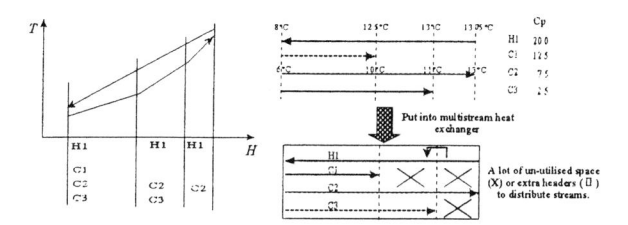

Fig. 5 Problems with vertical heat transfer.

Let us look at a simple example (Fig. 5), where the streams are arranged vertically. It can be observed that either a lot of space is wasted or extra header is required to redistribute stream H1 which in turn requires extra volume and pressure drops. Consequently, increased number of streams will increase number of intermediate headers which in turn leads to very long exchanger design. As a result, the number of intermediate headers is restricted to four in this study.

One way to overcome this problem is to extend the stream to other intervals, for example, stream C1 is extended in Fig. 6. By doing this, one can observe several effects. In one hand, when a stream is extended to other intervals, the temperature driving force for heat transfer reduces. This causes the increase in heat transfer area. On the other hand, by extending a stream to empty space, unutilised space is reduced and thus area requirement is reduced. This is a very interesting trade-off, which has been overlooked.

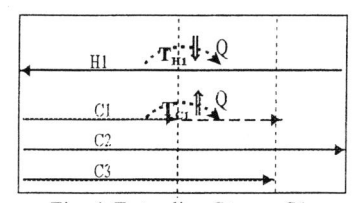

Fig. 6 Extending Stream C1.

To allow this trade-off to happen, in this work, the whole temperature range is partitioned into several blocks and partition temperatures are defined as optimisation variables (Fig. 7). Each block contains several enthalpy intervals and each interval contains a fraction of heat (K) for each stream, which is defined in Fig. 7. Thus, the summation of fraction of the heat, K, of all intervals in a block for a stream is equal to one. Any streams which are not fully flowing to another block will be indicated using binary variable, $x_{j,b}$ (for cold streams) as in Fig. 7.

When optimising entrance points, we like to minimise overall area for a network. However, we do not know heat transfer coeffi-

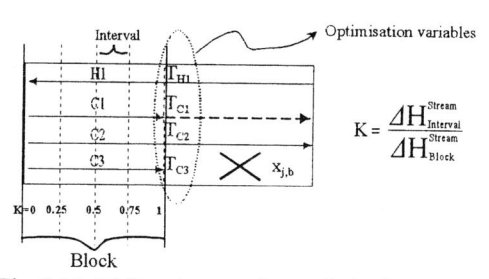

Fig. 7 Model for entrance points optimisation.

cients for each stream. Thus, it is impossible to calculate area directly. The equivalent problem for minimising area is maximising driving forces for each match and minimising the space occupied by intermediate headers. Therefore, we can obtain the objective function as Eq. (35).

$$Max\left[\sum_k DTS_k - \sum_B HD_B * 0.250\right] \quad (35)$$

where

$$DTS_k = T_{i,k} - T_{j,k} \quad (36)$$

$$HD_B = \sum_{i,j}(x_{i,B} + x_{j,B}) \quad (37)$$

Since a header takes the standard dimension of 250 mm, 0.25 is multiplied by HD_B.

Equations (38) to (41) are used to avoid transverse heat conduction across fins between same type of streams [see Pua and Zhu (2001) for detailed derivation].

$$r^H_{i,j1,j2,k} = \frac{T_{i,k} - T_{j1,k}}{T_{i,k} - T_{j2,k}}; T_{i,k} > T_{j1,k} \geq T_{j2,k} \quad (38)$$

$$r^C_{i1,i2,j,k} = \frac{T_{i1,k} - T_{j,k}}{T_{i2,k} - T_{j,k}}; T_{i2,k} > T_{i1,k} \geq T_{j,k} \quad (39)$$

$$r^H_{i,j1,j2,k} \geq r_{CRIT,j} \quad (40)$$

$$r^C_{i1,i2,j,k} \geq r_{CRIT,j} \quad (41)$$

The critical r-value, r_{CRIT}, depends on the heat transfer coefficients of a sandwiched stream (e.g. H1 in the configuration of C1-H1-C2). r^H and r^C for the sandwiched stream are used to identify if the two streams of the same type, which are allocated above and below the sandwiched stream, transfer heat cross the fin. This is called transverse heat transfer. As a first estimate, the critical r-value, r_{CRIT}, can take the value of 0.21 if the sandwiched stream is close to liquid (assuming the hot stream achieves heat transfer coefficient greater than 1000 W/m²K under standard fin geometry). On the other hand value of 0.65 will be applicable for gas stream (assuming heat transfer coefficient greater than 200 W/m²K). This r-value will be checked in later design stage to confirm the feasibility of this exchanger.

Equations (42) and (43) are used to provide feasibility conditions.

$$DTS_k \geq 1°C \quad (42)$$

$$HD_B \leq 4 \quad (43)$$

The objective function (35) together with Eqs. (36) to (43) forms an optimisation model for optimising stream entrance/exit points, which features the trade-off between the driving force and empty space.

Unit Targeting

Once the entrance and exit points for each stream are determined, they are used as the basis for targeting the minimum number of exchangers required for the network. The optimised entrance and exit points can be represented using a T-K diagram (Fig. 8).

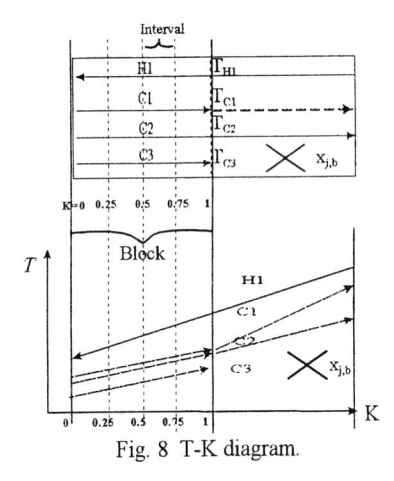

Fig. 8 T-K diagram.

From the T-K diagram (Fig. 9a), the streams are partitioned into several intervals according to their entrance and exit points determined. If some streams in the network experience phase change or any changes in their thermal-hydraulic properties (ρ, μ, k, Cp), then these variations are accounted for by taking different average values for each interval. Within each interval artificial two-stream exchangers linking hot streams and cold streams are assumed (Fig. 9b), which are modelled by Eqs. (15) to (34).

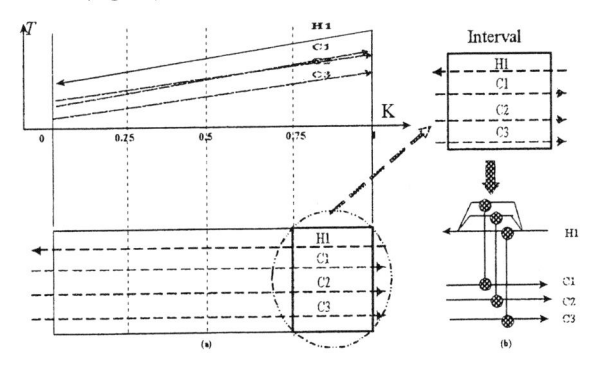

Fig. 9 Unit Targeting (a) identify intervals from T-K diagram; (b) artificial exchanger for each interval.

Since the core of an aluminium plate-fin heat exchanger is manufactured using the brazing technique, there are maximum dimensions in which the exchanger can be brazed according to the size of the furnace. In this study, the maximum allowable dimensions are set as 1m x 1m x 6m (Width x Height x Length) according to specification in Taylor (1987). Any core that exceeds any of these dimensions will require extra unit.

The objective function for unit targeting is minimisation of total cost (Eq. (44)).

$$Min\left[A_v + B_v\left(\sum V_T\right)^{C_v} + A_{\Delta P} + B_{\Delta P}\left(\sum \dot{V}\Delta P\right)^{C_n} + A_V^{Extra}\left(\sum Unit^{Extra}\right)\right] \quad (44)$$

subject to

$$\Delta P \leq \Delta P_{all} \text{ (optional)} \quad (45)$$
$$A_{fr}^{Total} \geq \Sigma(A_{fr}^{Mch}) \quad (46)$$
$$Unit^{Extra} = A_{fr}^{Total} - 1 \quad (46)$$

A, **B** and **C** in the objective function are the cost coefficients for exchangers and compressors in a network. Since the major exchanger cost comes from the number of units, any extra unit will incur a significant amount of cost. This is included in $A_V^{Extra}(\Sigma Unit^{Extra})$ where A_V^{Extra} is the extra unit cost. In Eq. (46), the total frontal area, A_{fr}^{Total}, of the MHE is taken as any value greater than the sum of frontal area of each match in one interval of the MHE. An integer variable is assigned to A_{fr}^{Total}. The extra unit, $Unit^{Extra}$, is then equals to A_{fr}^{Total} minus 1 (m^2) which is the maximum possible.

Distribution Optimisation

According to a recent cost correlation from an exchanger manufacturer (ESDU 97006), exchanger having a smaller number of streams is cheaper compared to the one having more streams if their volumes are identical. Thus, in design of MHE networks, it may be beneficial to exploit different stream distributions for MHE exchangers (Fig. 10).

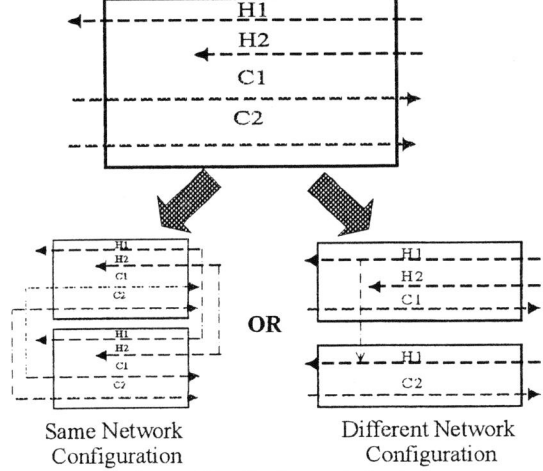

Fig. 10 Distribution optimisation.

From the previous stages, we have determined the entrance and exit points, the total volume, and the minimum number of units. At this stage, we need to determine stream distribution in each unit and connections between different units featuring minimum cost. To achieve this, an optimisation model is built, in which some of the optimisation results from the unit targeting stage are used, including the targeted number of unit, N^{Target}, the targeted matching volume between hot and cold streams, V^{Target}, and the targeted matching frontal area, A_{fr}^{Target} (Fig. 11a).

In this stream distribution model, the main optimisation variables are the volume going to be designated to each targeted exchanger, V^{MchK} and also the number of streams in each exchanger, N_{Ex} (Fig. 11b). The objective is to find the network with the least total heat exchanger cost [Eqs. (48)].

$$Min\sum_{Ex}\left[(A_v)_{Ex} + (B_v)_{Ex} V_{Ex}^{(c_v)_n}\right] \quad (47)$$

where

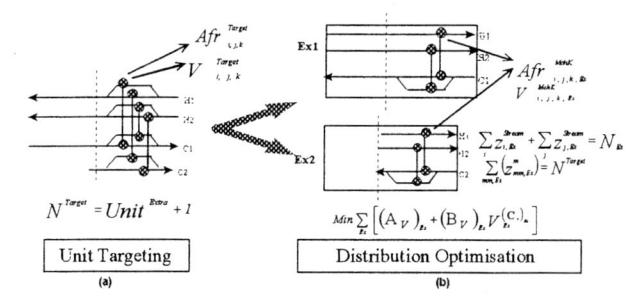

Fig. 11 (a) Results from Unit Targeting; (b) Model for distribution optimisation.

$$\sum_{i,j,k}\left[V_{i,j,k,Ex}^{MchK}\right] = V_{Ex} \quad (48)$$

$$\sum_{Ex}\left[V_{i,j,k,Ex}^{MchK}\right] = V_{i,j,k}^{Target} \quad (49)$$

A, **B** and **C** are cost coefficients according to the total number of streams in the exchanger (see below for how these coefficient are linked with V^{MchK} and N_{Ex}). Similar relationships hold for the total frontal areas.

To count the number of hot streams in one exchanger, Eqs. (50) to (55) are used:

$$\sum_{k}\left[V_{i,j,k,Ex}^{MchK}\right] = V_{i,j,Ex}^{Mch} \quad (50)$$

$$V_{i,j,Ex}^{Mch} - 5000 * z_{i,j,Ex}^{Mch} \leq 0 \quad (51)$$

$$\sum_{j} z_{i,j,Ex}^{Mch} - 5000 * z_{i,Ex}^{Stream} \leq 0 \quad (52)$$

$$z_{i,Ex}^{Stream} \leq \sum_{j} z_{i,j,Ex}^{Mch} \quad (53)$$

$$\sum_{i} z_{i,Ex}^{Stream} + \sum_{j} z_{j,Ex}^{Stream} = N_{Ex} \quad (55)$$

The indices of Eqs. (52) and (53) can be changed from i to j and vice versa to count the number of cold streams. The z's in the model are the binary variables which can only take value of 0 or 1. For example, if $z_{i,j,Ex}^{Mch}$ is designated to 1, it indicates that the match between hot stream i and cold stream j exists in exchanger Ex. Similarly, binary variable $z_{i,Ex}^{Stream}$ is used to indicate the existence or non-existence of hot stream i in exchanger Ex. The total number of streams in one exchanger, N_{Ex} can be calculated by summing the hot and cold streams as in Eq. (55). Hence the cost coefficients, which depend on N_{Ex}, can be determined.

Overall Framework

Now, we are ready to solve the overall problem, which can be stated as follows. Given a set of streams, their supplied and targeted temperatures and their process properties (including mass flow rate, density, heat capacity, conductivity and viscosity), we would like to design a network with minimum cost. The final result should tell us the following:

- The number of units
- The stream matching in each unit
- The entrance and exit points of each stream in each unit
- The number of headers
- The volumes for each unit
- The total pressure drops for each stream
- The fin used for each stream in each unit
- The total cost

To solve this overall problem, we have developed an effective design method. In this new method, first the stream entrance and exit points are optimised using mixed integer linear programming (MILP). Then the minimum number of units for the entire network is targeted considering initial cost trade-off using mixed integer non-linear programming (MINLP). The fin variations are considered in this stage to determine the best heat transfer conditions for each stream which can in turn give the network the minimum cost. If however, the number of units targeted is more than one, then the stream allocation in each unit is considered in distribution optimisation (MINLP). In the final stage, fin selection and passage arrangement in each unit are optimised using NLP on the basis of single exchanger design. In this stage, the heat transfer conditions for each stream are revised in order for lengths of streams in each exchanger to be similar. The overall procedure is shown in Fig. 12.

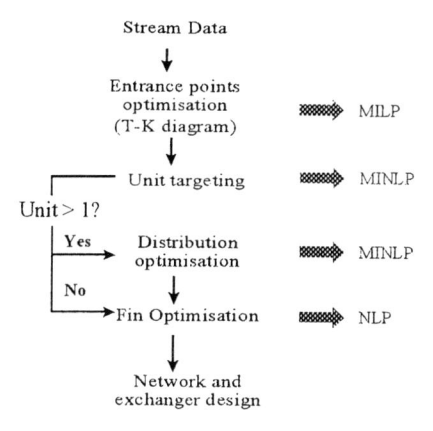

Fig. 12 Overall optimisation framework.

CONCLUSIONS

Using the new methodology, the advantages of fins can be fully explored at the very beginning of design stage and optimisation of fin selection is considered in the context of overall design problem. Consequently, the network can be designed from optimised heat transfer conditions of each stream instead of some arbitrary pre-assumed values. This design procedure provides several new features into multi-stream heat exchanger network which have been overlooked by existing methods. The two major ones are the optimisation of entrance/exit points and the optimisation of network structures. Combining these two optimisations greatly simplifies the complexity of MHE network and hence reduces the overall cost. Finally by considering heat transfer conditions at each stage, the interactions between the network level and the single exchanger level can be fully explored.

Hence with this new methodology, designers who are new to compact heat exchangers can have more confidence in designing the overall network using compact heat exchangers. This is because the design method systematically guides the designers towards the final design starting from very little information that is already known to the system (for example, temperature and physical properties). Consequently, the economic benefits of using compact heat exchangers can be fully exploited and achieved.

NOMENCLATURE

a	coefficient in heat transfer correlation
a'	coefficient in fin effectiveness correlation
A	coefficient in Y-Re relationship, m^{-2}
A, B, C	coefficients in cost functions
A_{fr}	frontal area, m^2
B	exponent in Y-Re relationship
b	exponent in heat transfer correlation
b'	exponent in fin effectiveness correlation
c	coefficient in friction factor correlation
C	coefficient in Z-Re relationship, m^{-1}
Cp	heat capacity, J/kgK
d	exponent in friction factor correlation
d_h	hydraulic diameter
D	exponent in Z-Re relationship
DTS	temperature difference in MHE, K
E	enhancement factor, ratio of the total heat transfer area to plate area
f	friction factor
h	heat transfer coefficient, W/m^{2}K
H	enthalpy, J
ΔH	enthalpy change, J
HD	header
j	Colburn factor
k	thermal conductivity, W/mK
K	fraction of interval enthalpy to block enthalpy
L	heat exchanger length, m
$\dot{m}$	mass flowrate, kg/s
N	number of units
ΔP	pressure drop, Pa
Pr	Prandtl number
r	ratio of the temperature difference between hot and cold streams in MHE
r_{CRIT}	critical ratio needed to avoid transverse heat conduction
Q	heat transfer rate, W
Re	Reynolds number
St	Stanton number
SP	stream property ($\mu Cp/Pr^{2/3}$), kg.m/s^{3}K
T	temperature, K
ΔT_{lm}	logarithmic temperature difference, K
$Unit^{Extra}$	extra unit determined in unit targeting stage
$\dot{V}$	volumetric flowrate, m^3/s
V_T	total volume, m^3
w	stream process properties in volume and pressure drop equations
x	variable to indicate the disappearance of one stream in adjacent block
Y	heat transfer related fin properties (Y'/δ), m^{-2}
Y'	heat transfer related fin properties ($\eta Y''E$), m^{-1}
Y''	heat transfer related fin properties (jRe/d_h), m^{-1}
z	binary variable
Z	pressure drop related fin properties, m^{-1}
ρ	density, kg/m^3
μ	viscosity, kg/ms
η	fin effectiveness
δ	plate thickness, m
σ	porosity of the fin surface, ratio of free flow area to plate frontal area
β	compactness, ratio of total heat transfer area to total volume of the fin surface, m^{-1}

Superscripts

m	relating to one particular number of streams
Mch	artificial two-stream heat exchangers (matches) in MHE
$MchK$	allocation in each MHE
$Stream$	relating to one particular stream
$Total$	total value by summing all the matches

Subscripts

all	allowable
b	block
C	cold stream
ex	exchanger
H	hot stream
i	hot stream
$i1$	hot stream 1
$i2$	hot stream 2
j	cold stream
$j1$	cold stream 1
$j2$	cold stream 2
k	interval
mod	modified
MHE	multi-stream heat exchanger
ΔP	pressure drop relating
ref	reference
R	ratio
$Target$	targeted
V	volume relating

REFERENCES

M. W. Kays and A. L. London, 1984, *Compact Heat Exchangers*, 3rd ed., McGraw-Hill, New York.

L. M. Pua and X. X. Zhu, 2001, Multi-stream Heat Exchanger Network Synthesis – Part I Fin Selection & Passage Arrangement, *Heat Transfer Engineering* (accepted for publication).

M. Taylor, 1990, Plate-fin Exchangers Offshore – the Background, *The Chemical Engineering,* April, pp. 19-26

T. F. Yee, I. E. Grossmann and Z. Kravanja, 1990, Simultaneous Optimisation Models for Heat Integration – I. Area and Energy Targeting and Modelling of Multi-Stream Exchangers, *Comp. & Chem. Engng.*, Vol. 14, pp 1151-1164.

M. Picon-Nunez, 1995, Development of Compact and Multi-Stream Exchangers: Design Methodology For Use in Integrated Plants, Ph.D. Thesis, UMIST, UK.

B. Linnhoff and S. Ahmad, 1990, Cost Optimum Heat Exchanger Networks – 1. Minimum Energy and Capital using Simple Models for Capital Cost, *Comp. & Chem. Engng.*, Vol. 14, No. 7, pp. 729-750.

M. A. Taylor, 1987, *Plate-Fin Heat Exchangers: Guide to Their Specification and Use*, 1st ed., HTFS, Harwell, UK.

ESDU 97006, 1997, *Selection and Costing of Heat Exchangers – Plate-fin types*, ESDU International Plc., UK.

DESIGN AND OPTIMIZATION OF COMPACT HEAT EXCHANGERS

Rongguang Jia[1], Bengt Sundén[1], Yimin Xuan[2]

[1]Division of Heat Transfer, Lund Institute of Technology, 22100 Lund, Sweden
[2]School of Power Engineering, Nanjing University of Science & Technology, 210094 Nanjing, P.R.C. China

ABSTRACT

A new and efficient software called CCHE (Computer-Aided Design of Compact Heat Exchangers), has been developed. It integrates the optimization, database, and process drawing into a software package. A strategy is developed for the optimization of compact heat exchangers (CHEs), which is a problem with changeable objective functions and constraints. The searching process combined traditional optimization methods including *Complex Method* (CM) and *Rosenbrock Method* (RM), and a database-scanning (DS) scheme. Different initial values are set at the beginning to increase the possibility of a global optimal solution. The optimization procedures are graphical and interactive. The databases of heat transfer surfaces, liquids, heat exchangers, and standards can be visited directly via *Open-Database Connectivity* (ODBC) technology. The process drawings of the designed compact heat exchanger are generated by an automatic drawing program, which is developed on the *AutoCAD-Runtime Extension* (ARX) interface of the software AutoCAD. All the computer codes are developed with *Visual C++*. A typical example of CHE design is presented to show the main function of the code. CM with penalty functions performed better than RM, and initial values are found to be important for the optimal solution.

INTRODUCTION

Compact heat exchangers (CHE) include plate-fin (PF) heat exchangers and plate heat exchangers, which are characterised by hydraulic diameters between 1 and 10 mm. A typical core of a PF CHE is shown in Fig. 1, and four kinds of PF fins are depicted in Fig. 2. Usually CHEs offer a large heat transfer area per unit volume of exchanger (compactness >1000m^2/m^3), which results in reduced space, weight, energy requirements, and cost when compared to conventional designs such as shell-and-tube heat exchangers (STHE). Their high performance has already widespread their use in automotive, aerospace, air-conditioning and refrigeration, as well as in electronic equipment industries for single phase and phase change duties. They are also suitable for industrial processes, but their use and acceptance in the process industry are not yet widespread. One reason is the absence of reliable and practical design methods and tools (Thonon 1999), unlike STHE.

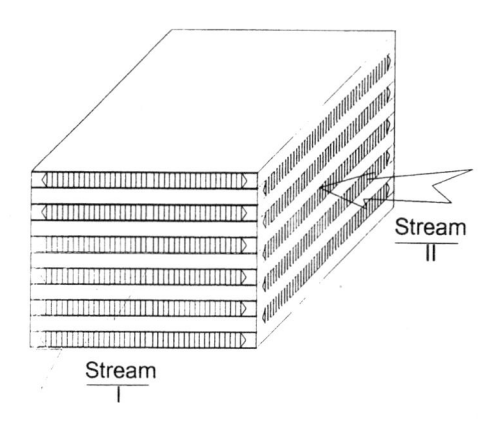

Fig. 1 A sketch of a plate-fin CHE core

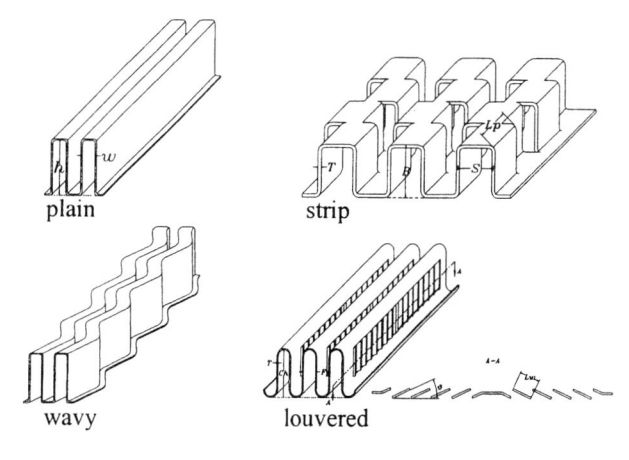

Fig. 2 Four kinds of commonly used plate-fin fins

At present, various commercial codes, such as HTRI, HTFS, B-JAC, THERM, and CC-Therm, are available for heat exchanger design. Only HTRI and HTFS provide modules for CHEs' sizing, rating and simulation: HTRI for plate heat exchangers (PHE), HTFS for PHE and plate-fin heat exchangers (PFHE). However, neither of them consists of any optimization strategies for CHEs' design, which are needed from an industrial point of view. The development of a software that combines optimization, database, and process drawing for CHEs' design seems important and urgent. This is the aim of the present paper. To assist engineers to perform more reliable and practical, optimal CHEs' design, a software package called CCHE, is developed, based on thermodynamic principles.

METHOD

Methods for the code naturally consist of four parts: thermal design, optimization, database, and process drawing.

1. Thermal design

Nearly all heat exchangers can be analysed from the following basic design equations:

$$Q = UA\Delta t \qquad (1)$$

$$Q = \{Wc_p(t_i - t_o)\}_h = \{Wc_p(t_o - t_i)\}_c \qquad (2)$$

where Δt is not independent. There are eight variables in the above equations, as shown in Table 1. In order to solve these equations, five of them have to be given.

Table 1. The variables in the heat exchanger design problem

	Sizing problem	Rating problem
UA	unknown	given
$[Wc_p]_h$	given	given
$[Wc_p]_c$	given	given
$t_{i,h}$	three of them are given	given
$t_{o,h}$		unknown
$t_{i,c}$		given
$t_{o,c}$		unknown
Q	unknown	unknown

In CCHE, the $\varepsilon - NTU$ method (Kays 1964) is employed to solve the thermal equations. In the design of a CHE, the data of heat transfer and friction factors are required for various heat transfer surfaces, which made the solving procedure complex, unlike STHE. This is done by built-in correlation curves which show the relation among Reynolds number, j- and f-factor, and fin geometry (Manglik and Bergles 1995, Sunden and Svantesson 1992, Churchill 1977, etc), or by data from experiments or manufacturers.

2. Optimization

The optimization of a technical system is the process of adjusting the different control variables to find the proper levels that achieve the best possible outcome (response or objective). Usually many conflicting responses (objectives) must be optimized simultaneously, which are characterized as multi-objective problems. The overall optimization procedure is shown in Fig. 3.

Objective functions

The optimization of CHEs is inherently multi-objective, because different objectives are needed for different situations:

- *Minimum operating cost* (C) is needed for most of the cases.

$$C = C_I + C_O \qquad (3)$$

$$C_I = C_A A a^* \qquad (3a)$$

$$C_O = \left\{k_{el}\tau\frac{\Delta p V_t}{\eta_p}\right\}_h + \left\{k_{el}\tau\frac{\Delta p V_t}{\eta_p}\right\}_c \qquad (3b)$$

The annual costs of investment C_I or capital costs are taken to be proportional to the surface area A and the amortization a^* (of say 10%/yr). The price per unit area C_A depends of course on the type of apparatus, on the material needed, and on the size (i.e., on the surface area A) itself. It is generally well known that the price of equipment is not linearly increasing with its size. Thus C_I in Eq. (3a) should be regarded as a linearization of a more appropriate empirical power law (e.g., $C = Const \cdot A^n$), with an exponent n less than unity.

The cost of operation C_O for heat exchangers with process fluids on both sides is taken as proportional to the pumping power required to overcome the flow resistances in the exchanger. It depends on the following variables: the price of electrical energy k_{el}, the hours of operation per year τ, the pressure drop Δp, the volumetric flow rate V_t, and last but not the least, the efficiency of the pump (or fan) η_p.

- *Minimum volume* (V) is needed, where the space is critical.

$$V = L_1 L_2 L_3 \qquad (4)$$

For the time being, only the volume of the core for two-stream compact heat exchangers is considered. This is dependent on the first side flow length L_1, second side flow length L_2, and no-flow height (stack height) L_3.

- *Moderate cost and volume* (CV) are needed, where the CHE size is constrained and cost also should be low, and usually these are not equally important. This might be regarded as a fuzzy optimization problem.

$$CV = \alpha_C C/C_S + \alpha_V V/V_S \qquad (5)$$

where α_C and α_V are the significance factors. Usually they are not equal, but can be selected by the user. C_S is the scale of cost, and V_S is the scale of volume, by which C and V are made compatible.

Searching methods

For the optimization of CHEs, the searching methods selected must satisfy such criteria as:
- Without derivating operation, because of the nonderivativity of the objective functions.
- Avoid unnecessary calls to the objective function, because it is time-consuming.

The searching process in the code is a combination of conventional searching methods, including CM, RM, and DS, which make the optimization more reliable and practical.

Complex method (CM) is a non-derivative searching method, which can be used for the optimization of constrained or unconstrained nonlinear programming (NLP) problems. This method, starting with a number of feasible points greater than the dimension of the problem, determines the optimum by a typical descent algorithm. Though it is capable of determining a nearly optimal feasible solution, its convergence in the general case is not guaranteed. However, this method has good convergence properties for unconstrained problems. Hence transformation of the constrained NLP problem to a series of smooth unconstrained problems by the use of a penalty function and the application of the complex method to these functions is proposed for the determination of the optimum.

The Rosenbrock Method (RM), sometimes called the Rosenbrock rotating coordinates method, is also a non-derivative searching method. It keeps probing and constructing new step directions and lengths, until all directions have been tried and the step length is less than the specified precision. It can be directly used for the optimization of unconstrained NLP problem, but for constrained problems, some transformations are needed as CM.

Database Scanning (DS) is a method where all the fins selected by the user are assembled into pairs, and each pair consists of two fins, which are used during the optimization, and then an optimal solution is selected automatically.

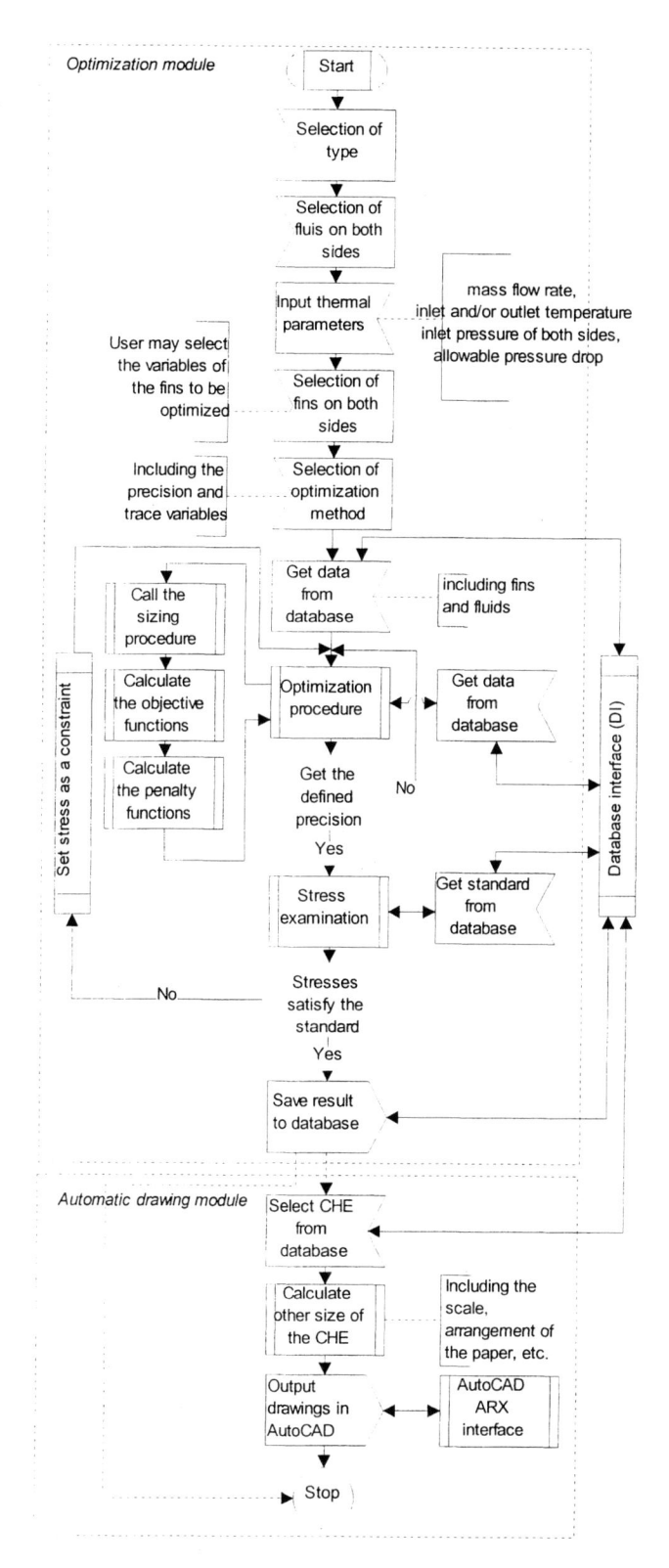

Fig. 3 The structure of optimization and automatic drawing modules

3. Database

The characteristic data of heat transfer surfaces, working fluids, standards, and heat exchangers can be obtained directly from databases via *Open-Database Connectivity* (ODBC) technology, at the beginning or during the optimization. ODBC is an Application Programming Interface (API) that allowed a programmer to extract a program from a database, which makes codes independent of the database type.

The database structure is shown in Fig. 4. The CCHE optimization module, stress examination module and auto-drawing module may visit the database through the database interface (DI). Database managing module is developed for users. Users may add new fin parameters from manufacturers and revise the existing fin parameters in the database when these are outdated. Users may also manipulate the databases of fluid, standard and heat exchangers. The DI enables DS of the selected fins during the optimization.

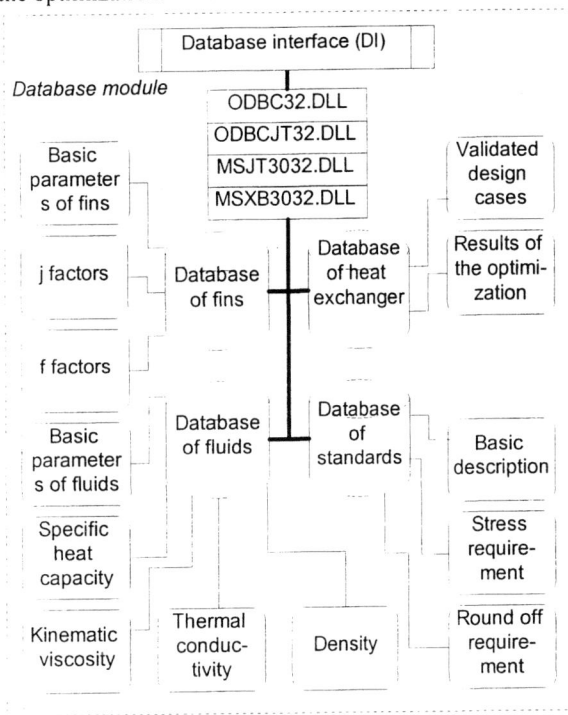

Fig. 4 The structure of the database

4. Process drawing

The process drawings of the designed compact heat exchanger are created by the automatic drawing module, which is developed using the *AutoCAD-Runtime Extension* (ARX) interface of the software AutoCAD. AutoCAD is a computer-aided design (CAD) tool, which is used across industries. It provides many secondary development interfaces, including ARX, which is used in CCHE. This is the main reason why our code output drawings are in AutoCAD.

The procedure was: First, the user selects the CHE from the database via the ODBC interface, and then the automatic drawing module would create the drafts of the selected CHE in AutoCAD automatically, including the drawings of the whole CHE and the components, such as fins, siders, and headers, as shown in Fig. 3.

RESULTS AND DISCUSSION

Problem statement

A design case of a gas-to-air single-pass cross-flow CHE is considered here.

Table 2 Process parameters

	Side A	Side B
Working Fluid	Air	Gas
Mass flow rate (kg/s)	0.8296	0.8962
Temperature, Inlet (°C)	4	240
Temperature, Outlet (°C)[*]	61.4	199.8
Pressure, Inlet (kPa)	110	110
Allowable pressure drop (kPa)[#]	1.67	0.32
Heat duty (KW)	166.886	

[*] One of the two outlet temperature is known.
[#] Allowable pressure drops can be optimization variables.

Table 3 Cost parameters

Items	Value	Description
C_A	100€/m^2	price/m^2
a^*	10%/yr	amortization
η_P	0.5	Pump efficiency
τ	6500h/yr	hours of operation
k_{el}	30€/MWh	price of electrical energy

Table 4 Optimization parameters

	Fin	Variable	Lower limit	Upper limit
A	Strip	Fin number per m	200	1200
		Fin height (m)	0.002	0.015
		Pressure drop (kPa)	0.1	3.5
B	Plain	Fin number per m	200	1200
		Fin height (m)	0.002	0.015
		Pressure drop (kPa)	0.1	0.8
Cons-traints		L1 (m)	0.3	0.7
		L2 (m)	0.3	0.7
		L3 (m)	0.3	0.7
		L1/L2	0.5	2
		L1/L3	0.5	2
		L2/L3	0.5	2

The task is to determine the dimensions of this CHE. The effect of longitudinal heat conduction is neglected and the gas is treated as air for fluid property evaluation. The calculation of j- and f-factors is from curve-fit formulas.

Fin in database

All the plain, strip, louvered, and wavy fins from *Compact Heat Exchangers* (Kays 1964), may be used in the optimization via DI, which all are stored in the database including size parameters, j- and f-factors.

Table 5 The fins in database

plain	strip	louvered	wavy
10.27T	1/2-11.94(D)	1/2(a)-6.06	11.44-3/8W
11.1	1/4-15.4(D)	1/2-11.1	11.5-3/8W
11.11(a)	1/6-12.18(D)	1/2-6.06	17.8-3/8W
11.94T	1/7-15.75(D)	1/4(b)-11.1	
12.00T	1/8-13.95	1/4-11.1	
14.77	1/8-15.2	3/16-11.1	
15.08	1/8-16.00(D)	3/4(b)-11.1	
16.96T	1/8-16.12(D)	3/4-11.1	
19.86	1/8-16.12(T)	3/8(a)-6.06	
2.0	1/8-19.82(D)	3/8(a)-8.7	
25.79T	1/8-20.06(D)	3/8(b)-11.1	
3.01	3/32-12.22	3/8-11.1	
3.97		3/8-6.06	
30.33T		3/8-8.7	
46.45T			
5.3			
6.2			
9.03			

The performance of RM and CM

Three different objective functions are used to compare the performance of RM and CM:

Firstly, annual cost (C) served as the objective function. As shown in Fig. 5, a better solution was obtained by CM, i.e., lower annual cost and less iterations.

Secondly, core volume (V) served as the objective function. The difference in performance between the two methods became more obvious, as shown in Fig. 6.

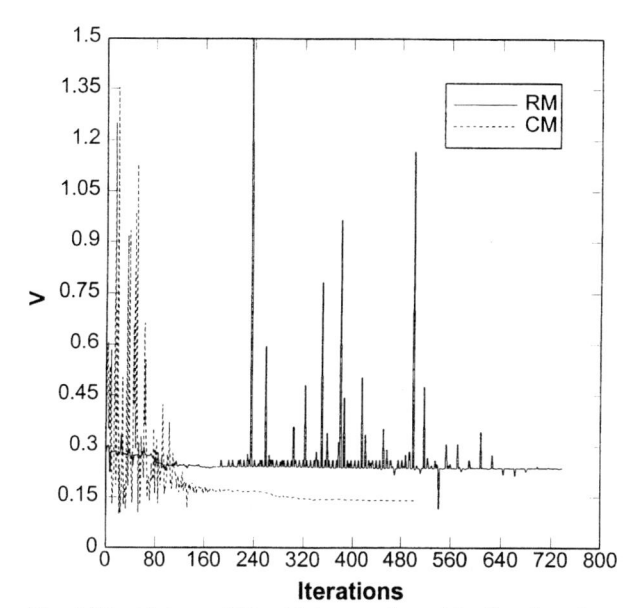

Fig. 6 The history of V, which served as objective function

Thirdly, $0.5C/C_S + 0.5V/V_S$ (CV) served as the objective function. The difference between the results from the two methods became smaller, as shown in Figs. 7 and 8. Similar annual costs were obtained from the two methods, while core volumes are a little different. A reason for this might be that the objective functions are not satisfactory.

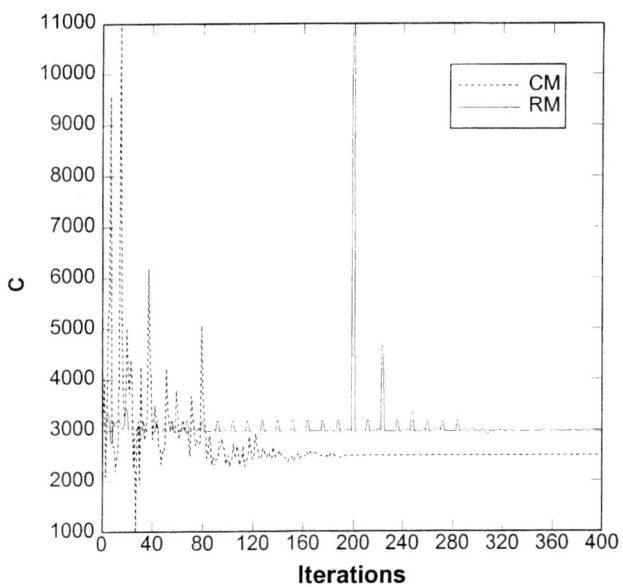

Fig. 5 The history of C, which served as objective function

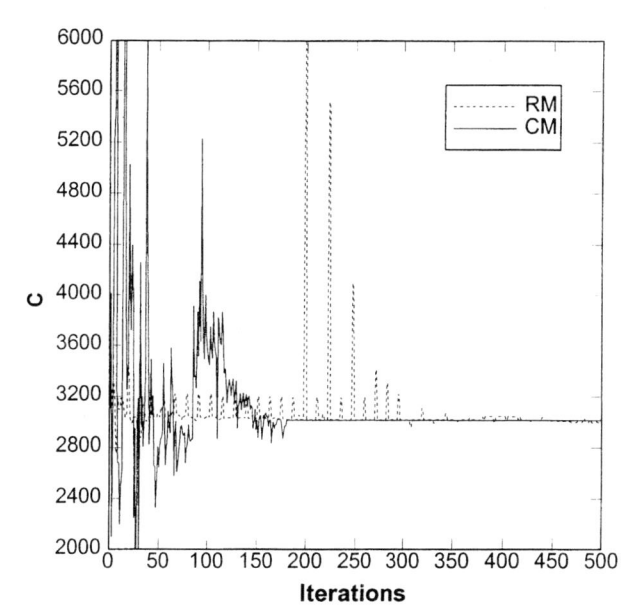

Fig. 7 The history of C, when CV served as objective function

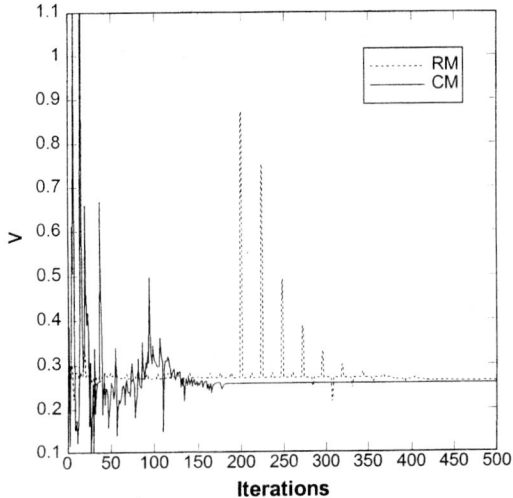

Fig. 8 The history of V, as CV served as objective function

The effect of optimization of allowable pressure drop

The design of a CHE requires a knowledge of the allowable pressure drops (ΔP_0) of the streams that can be fully used. Information about allowable pressure drops are also required as input data in all the mentioned commercial codes. However, setting the ΔP_0 from experience or engineering intuition could lead to a final solution far from the optimum design. As is shown in Fig. 9, different results are obtained whether the pressure drop is optimized or not. In the code, the user may select ΔP_0 as the optimization variable.

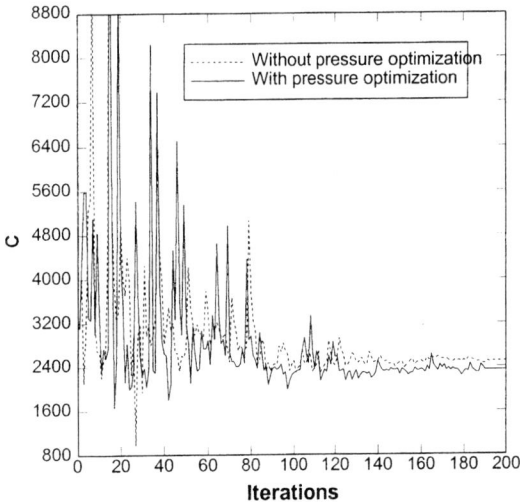

Fig. 9 The history of C, which served as objective function

The effect of initial value

All of the optimization methods need initial values, which are the starting points of the optimization process. The problem is whether the solutions are the global optimum, according to the specified initial value, objective function, and domain. It is not guaranteed, without trying different initial values. The answer is negative: Different initial values resulted in different solutions, as shown in Figs. 9, 10, and 11. Especially when RM is used, the difference is up to 150%.

The means used here to tackle this problem are straightforward. One way is to divide the domain into subdomains, and then select one point in every domain, which will serve as the initial values. Different subdomain numbers are tried here. The number of subdomains in the first case is 64, while it is 720 in the second one. As is shown, almost the same optimal solutions are obtained by CM, annual cost equals 2250 in the first case, and 2242.74 for the second case. The difference is within 0.4%, which indicates the convergence of this problem.

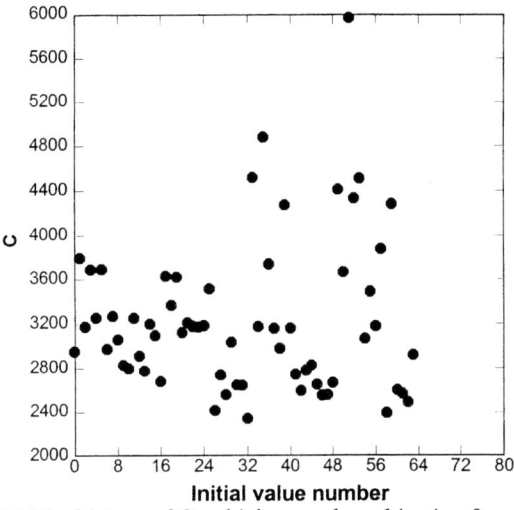

Fig. 10 The history of C, which served as objective function, using RM. Domain is divided into 64 subdomains.

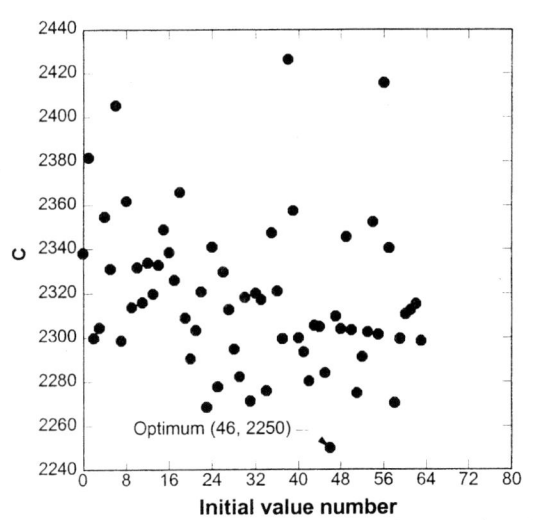

Fig. 11 The history of C, which served as objective function, using CM. Domain is divided into 64 subdomains.

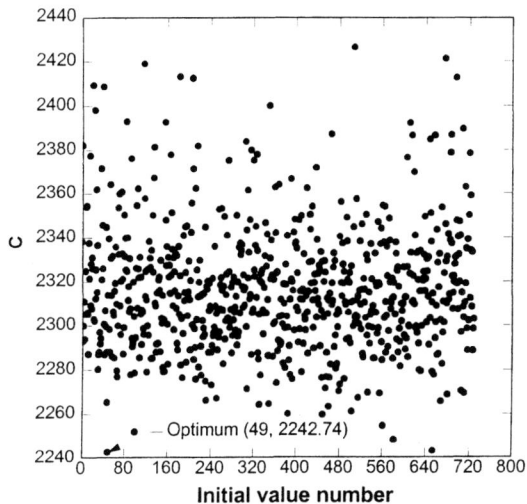

Fig. 12 The history of C, which served as objective function, using CM. Domain is divided into 720 subdomains.

Database scanning

At this stage, almost all the j- and f-factors of different fins may be calculated from empirical correlations, and this enables the continuous optimization of the fin size parameters in our code. However, are they reliable if they are used extensively? This needs to be tested. An alternative way is to store all the original data in database for different fin parameters, which are obtained from reliable experiments for the specific fins. During the optimization, scanning of those fins from the database is carried out to obtain j- and f-factors. In the current optimization problem, all the fins listed in Table 4 are scanned, and it takes more than two hours CPU time on a 800MHz PC. Optimal solution is reached.

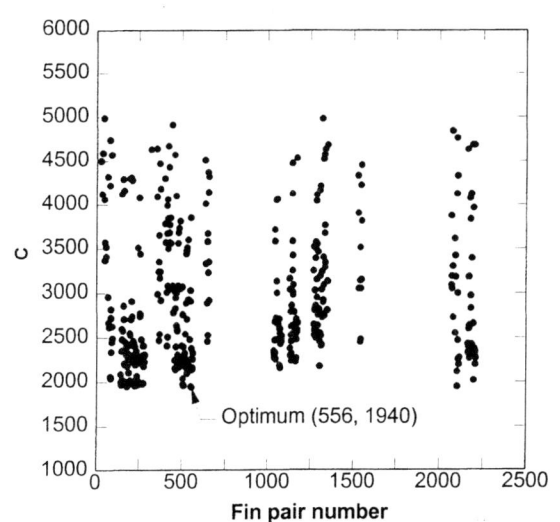

Fig. 13 The history of C, which served as objective function, using DS.

Determination of the optimum solution

Till now, a minimum annual cost 2242.74 is got by CM with different initial values, as shown in Fig. 12, and 1940 by DS in Fig. 13. The difference between these is two-fold. Firstly, four kinds of fins are used in DS, while only 2 in CM. Secondly, they gave different j- and f-factors for the same fin parameters and Reynolds number.

In the code, stress examination is carried out, and the optimal solution is selected as follows:

Table 6 The optimum solution

Items		Side A	Side B
Stream flow length (m)		0.6896	0.3992
Pressure drop (kPa)		1.2384	0.64
Fin	Type	Louver	Strip
	Name	3/8-11.1	1/8-16.00(D)
	Height (m)	0.00635	0.006477
	Thickness (m)	0.000152	0.000152
	Density (1/m)	437	630
	Length (m)	0.009525	0.003175
	Layer	49	48
Metal surface area (m^2)		131.78	
Fins stack height (m)		0.6749	
Cost €/yr	Investment cost	1317.8	
	Operation cost	624.2	

Output drawings in AutoCAD

The drawings of this CHE are created in AutoCAD by the automatic drawing program. These are simple, but at the same time helpful at the design stage. Due to the lake of space, only the drawing of the core is provided, see Fig. 14.

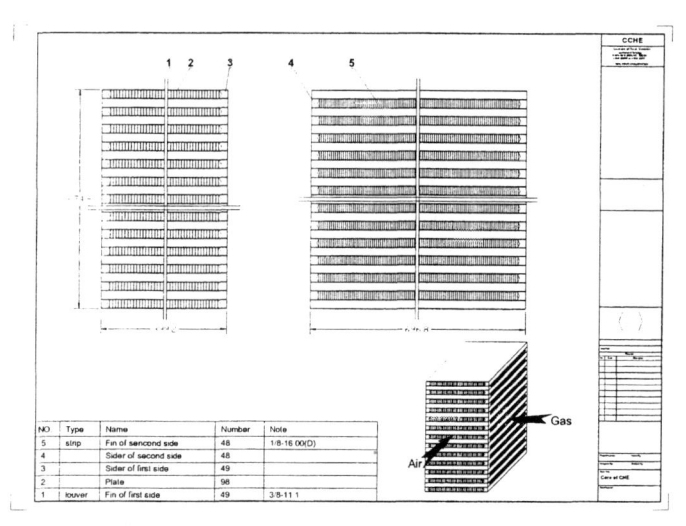

Fig. 14 The drawing of the core of the CHE, which is the solution of the optimization.

CONCLUSION

The whole process of optimization of a plate-fin CHE is presented. Different methods, objective functions, initial values, and variables were studied:

The idea to combine the conventional optimization and DS was feasible to perform more reliable and practical optimization, as shown in Fig. 13.

CM showed better performance than RM, better solutions, less iterations and greater independence of initial values, as shown in Figs. 5, 6, 10 and 11.

Allowable pressure drop should be considered carefully, because different solutions were obtained whether it was optimized or not.

Initial values were important, and the means to divide the calculation domain into sub-domains were practical, as shown in Figs. 10, 11, and 12.

Drawings could be created automatically in AutoCAD.

As a whole, this code system combined three modules of the design of CHEs, namely optimization, database, and automatic drawing. It is expected to be helpful for designers, manufacturers, users and researchers of CHEs, particularly in the process industries.

NOMENCLATURE

A	metal surface area, m^2
C	annual cost, €/yr
C_A	price per unit area, €/m^2
C_I	annual costs of investment, €/yr
C_O	cost of operation, €/yr
c_p	specific heat of fluid at constant pressure, J/kg°C
C_S	scale of C, €/yr
k_{el}	price of electrical energy, €/MWh
L_1	first side flow length, m
L_2	second side flow length, m
L_3	no-flow height (stack height), m
Δp	pressure drop, kPa
Δp_0	allowable pressure drop, kPa
Q	heat transfer rate, kW
t	temperature, °C
Δt	average temperature difference, °C
U	overall heat transfer coefficient, W/m^2°C
V_t	volumetric flowrate, m^3/s
V_S	scale of volume, m^3
V	volume of the CHE core, m^3
W	fluid mass flow rate, kg/s
a^*	amortization
α_C	significance factor of C
α_V	significance factor of V
η_p	efficiency of the pump (or fan)
τ	hours of operation per year, hour

Subscript

h	hot side
c	cold side

REFERENCE

1. W.M. Kays and A.L. London, (1964). Compact heat exchangers, 2nd Edn, McGrawHill Book Co., New York.
2. B. Thonen, (1999). Compact heat exchanger: Technology and application, Proceedings of the International Conference on Compact Heat Exchangers and Enhancement Technology for the Process Industries, Canada, June, 1-8, Begell House, Inc.
3. R.M. Manglik and A.E. Bergles, (1995). Heat Transfer and Pressure Drop Correlations for the Rectangular Offset Strip Fin Compact Heat Exchanger, Experimental Thermal and Fluid Science 10: 171-180.
4. W.M. Rohsenow, J.P. Hartnett, and E.N. Ganic, (1985). Handbook of Heat Transfer Fundamentals, 2nd Edn, McGraw-Hill, Inc.
5. B. Sunden, and J. Svantesson, (1992). Correlations of j- and f- Factors for Multilouvered Heat Transfer Surfaces, Proceedings of the 3rd UK National Conference Incorporating 1st European Conference on Thermal Sciences, I. Chem. E. 92, 805-811.
6. S.W. Churchill, (1977). Comprehensive Correlating Equations for Heat, Mass and Momentum Transfer in Fully Developed Flow in Smooth Tubes, Ind. Chem. Fundam., Vol. 16, 1:109-116.
7. H.H. Rosenbrock, (1960). An automatic method for finding the greatest or least value of a function, The Computer Journal, Vol. 3, 175-184.
8. M.A. Mazeed, T. Sunaga, E. Kondo, (1987). Optimization of nonlinear programming problems using penalty functions and complex method, Journal of the Operation Research Society of Japan, Vol. 30, 4:434-448.
9. David J. Kruglinski, (1997). Inside Visual C++, 4th Edn, Redmond, WA, Microsoft Press.

OPTIMAL DESIGN OF PLATE-FIN HEAT EXCHANGERS USING BOTH HEURISTIC BASED PROCEDURES AND MATHEMATICAL PROGRAMMING TECHNIQUES

Jean-Michel Reneaume[1] and N. Niclout[2]

[1] Laboratoire de Génie des Procédés de Pau, rue Jules Ferry, 64 000 Pau, France, E-mail : jean-michel.reneaume@univ-pau.fr
[2] NORDON CRYOGENIE SAS, 25 bis, rue du Fort, BP 87, 88 190 Golbey, France

ABSTRACT

A tool for the optimal design of Plate-Fin Heat Exchangers (PFHE) is described. This program results from the coupling of COLETH and mathematical programming techniques. COLETH is a tool for computer-aided design of PFHE. It is based on heuristic procedures. First, this program (hypothesis, modeling, limitations) is presented.

Our aim is to introduce mathematical programming techniques in order to achieve more accurate results and to increase the competitiveness of the company. The optimization problem is formulated : hypothesis, optimization variables, objective function and constraints are detailed. The Mathematical Programming techniques available for the problem solution are discussed: Successive Quadratic Programming and Simulated Annealing. The general solution strategy is described. The program abilities are illustrated with two industrial examples: capital cost reductions are greater than 15%.

INTRODUCTION

NORDON CRYOGENIE is one of the world leaders of Plate-Fin Heat Exchanger (PFHE) design and manufacturing. Brazed aluminum compact heat exchangers for cryogenic application are of particular interest.

Generally, manufacturers use proprietary heuristic-based programs for the computer-aided design of such exchangers. COLETH is the tool developed by NORDON CRYOGENIE. The use of those programs provides an approximation of the optimal design. In order to improve competitiveness, development of efficient tools for optimal design of PFHE has become a priority in the different companies. Recently, many investigations have been devoted to the development of programs based on mathematical programming algorithms in order to generate better optimized results (Abramzon et al., 1993; Abramzon, 1996; Hesselgreaves, 1993; Sunder et al., 1993). In this paper we describe how mathematical programming techniques have been introduced into COLETH. The resulting program uses both heuristic-based procedures and mathematical programming algorithms in order to achieve optimal design of PFHE.

In the first section, the PFHE model is described. Such a model is based on the well known "common wall temperature" hypothesis. Two particular steps are described in depth : the fin selection (length is also evaluated at this step) and the pressure drop calculation. In the second section, the optimization problem is formulated and the different solution methods are presented. In the third section, the general solution strategy is described. Finally the program abilities are illustrated with challenging industrial examples.

COLETH DESCRIPTION

Definitions

COLETH is an heuristic-based procedure for PFHE design. Two modes are available: the automatic sizing mode and the sizing mode. In both cases, the user must supply the following data: flow rates, inlet and outlet temperatures, enthalpies, transport properties and maximum pressure drops.

In the **automatic sizing mode**, all the geometrical parameters of the heat exchanger (parting sheet thickness, bar width, number of cores, core width, fin geometry, number of layers, distributor geometry, length of the core ...), the effective pressure drops and the cost of the heat exchanger are computed.

In the **sizing mode**, geometrical parameters of the heat exchanger (except core length) are supplied by the user. The procedure then computes the pressure drops, the cost of the heat exchanger and the necessary length of the core.

A **section** is the space between any stream inlet or stream outlet. In most cases, more than 60% of the total duty is exchanged in one section: the Main Duty Section. As discussed here after, this section is of particular interest during the optimization procedure.

Since PFHE can involve several hot and cold streams, the composite curve is built: temperature versus duty for both hot and cold streams. The total duty is divided into **intervals** of equal length. In each duty interval, the physical properties of the streams (specific heat, dynamic viscosity...) are supposed to be constant. The number of intervals (NI) is a very important parameter from the computational time point of view, especially when mathematical programming techniques are used.

Hypothesis

At a given length of the core, all parting sheets are supposed to have the same temperature. Therefore, each stream is supposed to exchange energy with the parting sheet only, regardless of the other streams. This hypothesis is known as the "Common Wall Temperature Hypothesis".

Main steps of the heuristic-based procedure

For sake of concision, the whole procedure is not presented. Only the two main steps are detailed : the Main Duty Section length and the pressure drop calculation.

Main Duty Section length calculation - fin selection.
The maximum length of the Main Duty Section, $L^{MDS,max}$, is first computed using Eq. (1).

$$L^{MDS,max} = \frac{Q^{MDS}}{Q^{Tot}} \cdot L^{max} \qquad (1)$$

where L^{max} is the maximum value of the core length (evaluated according to pressure considerations), Q^{Tot} and Q^{MDS} are respectively the duties exchanged in the whole core and the MDS.

In each duty interval and for each stream, the Colburn factor (C_j) is evaluated. For existing fins of the data bank, specific equations relating C_j to the Reynolds number are used. Those equations are based on experimental data or on estimated data. For non standard fins, proprietary correlations are used. Non standard fins are generally involved during the optimization problem solution. For confidentiality reasons, those correlations are not presented here.

Given the Colburn factor, the heat transfer coefficient is evaluated according to Eq. (2).

$$He = \Psi \cdot Cj \cdot \frac{\dot{m}}{Ac} \cdot Cp \cdot (Pr)^{2/3} \qquad (2)$$

where Ψ is an efficiency factor introduced in order to take into account the non-ideal arrangement of the different layers, axial thermal diffusion or fouling. Equation (2) is used for monophasic flows. For diphasic flows, specific correlations are available. One should note that the correlations used for diphasic flows involve discontinuities. This is of particular importance for the optimization problem solution.

The heat transfer resistance of stream i in interval j is given by Eq. 3.

$$R^{i,j} = 1 / \left(He^{i,j} \cdot A_0^{i,j} \right) \qquad (3)$$

where $He^{i,j}$ is the heat transfer coefficient of stream i in interval j and $A_0^{i,j}$ is the exchange area per unit of length. The total heat transfer resistance is calculated for both cold and hot sides. For example, for the hot side, the transfer resistance is computed according to Eq. (4).

$$R^{hot,j} = \frac{1}{\sum_{hot\ streams} 1/R^{i,j}} \qquad (4)$$

Equation (5) gives the total heat transfer resistance.

$$R^{Tot,j} = R^{hot,j} + R^{cold,j} + R^{wall,j} \qquad (5)$$

where $R^{wall,j}$ is the resistance involved by the parting sheet and the different fins. $R^{wall,j}$ is a function of the thickness of the parting sheet and the fins.

Assuming that the temperature varies linearly with respect to the resistance :

$$\Delta T_{lm}^{hot,j} = \Delta T_{ml}^{j} \cdot \frac{R^{hot,j}}{R^{tot,j}} \qquad (6)$$

In Eq. (6), ΔT^j_{lm} is the logarithmic mean temperature difference. Of course we have the same expression for cold streams.

Then the length required by stream i in interval j is computed according to Eq. (7) (hot stream for example) :

$$L^{i,j} = \frac{Q^{i,j}}{He^{i,j} \cdot A_0^{i,j} \cdot \Delta T_{lm}^{hot,j}} \qquad (7)$$

where $Q^{i,j}$ is the duty exchanged by stream i in interval j. For a given stream, the total length is:

$$L^i = \sum_{j=1}^{NI} L^{i,j} \quad (8)$$

If the sizing mode is used, the length L^i of stream i is computed using the fin supplied by the user. If the automatic sizing mode is used, for stream i, the fins of the data bank are enumerated. Heuristics allow partial enumeration only. If the calculated length (L^i) is lower than the maximum length ($L^{MDS,max}$), the current fin is accepted. With such a procedure, for stream i, several fins are selected. Finally, one fin is selected for each stream in such a way that the difference between the different lengths (L^i) is minimized. As said before this procedure is performed for the main duty section in order to select the fins. Then the selected fins are generally used for the subsequent sections. In both cases, sizing mode and automatic sizing modes, the duty requirement is satisfied.

Pressure drop calculation. For a given section, in each interval j and for each stream i, the friction factor (C_f) is evaluated. If the automatic sizing mode is used, the fins selected at the previous step (length calculation) are used. As for the Colburn factor, for existing fins of the data bank, specific equations relating C_f to the Reynolds number are used. Proprietary correlations are available when non standard fins are to be used. For a monophasic stream i in interval j, the pressure drop per unit of length is given by Eq. (9).

$$\Delta P^{i,j} = 4 \cdot C_f \cdot \frac{1}{D_h} \cdot \left(\frac{1}{2} \cdot \rho \cdot v^2\right) \quad (9)$$

Empirical based correlations are available for diphasic flow. Then, for each stream, the total pressure drop is calculated according to Eq. (10):

$$\Delta P^i = \sum_{j=1}^{NI} \Delta P^{i,j} \quad (10)$$

For stream i, the total pressure drop is evaluated adding the pressure drops of both distribution and exchange sections. An important point is that a fin is not rejected if the pressure drop constraint is not satisfied. Thus the pressure drop requirements are not necessarily satisfied.

OPTIMISATION PROBLEM FORMULATION

Hypothesis

In this work, the fin optimization is performed in the Main Duty Section only. This hypothesis results from a compromise between computational time charge and the quality of the optimal result. We will show that important capital cost reduction can be achieved in a reduce computation time.

Problem formulation

The general formulation of the optimization problem is stated on Eq. (11). An objective function (f) is minimized with respect to a set of optimization variables (x). A lower bound (x^{min}) and a upper bound (x^{max}) is defined for each variable. This minimization is submitted to a set of equality constraints (g). The variables are described in Table 1.

$$\left. \begin{array}{l} \underset{x}{Min} \; f(x) \\ s.t. \\ g(x) = 0 \\ x^{min} \leq x \leq x^{max} \end{array} \right\} (P) \quad (11)$$

ND is the number of distributors and NS* is the number of streams in the Main Duty Section. One should note that both continuous and discrete/integer variables arise in the problem formulation. For example, x_{cn} is an integer variable and typical values for x_{ft} are (in mm): 0.2, 0.25, 0.30 …

The objective function can be : the overall capital cost of the PFHE, the total volume, the number of cores and the cross section of the different cores.

The main constraints of the optimization problem are: banking limit (the ratio of hot to cold stream layers must be nearly equal to one), maximum stacking height, maximum number of layers, maximum width of the cores, maximum length of the cores, fin manufacture feasibility, header geometry and pressure drops.

Many terms which are implicit functions of the optimization variables arise in the objective function and the constraints: core length, pressure drops, capital cost … Those variables are the **interest variables** of the optimization problem.

In the proposed algorithm, pressure drop requirements are treated as constraints of the optimization problem. Pressure drop constraints will be satisfied at the solution point, but not along the solution path. At the opposite, duty requirements are implicit constraints : at each step of the optimization procedure, target temperatures are respected since, as seen before, the length of the core (an interest variable) is computed in order to satisfy the duty requirement.

Problem solution

To solve the problem, the user has to make a decision concerning the relaxation of the optimization variables. The first possibility is to relax all the discrete variables. Since all variables are assumed to be continuous, the optimization problem results in a Non Linear Programming (NLP) problem (the objective function and the constraints are non linear with respect of the optimization variables). Then the problem is solved using a Successive Quadratic Programming (SQP) algorithm (Ternet et al., 1998). Of course, at the solution, discrete/integer variables have continuous values. In order to achieve a feasible PFHE, the user has to consider the nearest acceptable value for each discrete variable.

Table 1: Variable description

Variable	Notation	Nature	Number
core width	x_{cw}	continuous	1
distributor width	x_{dw}	continuous	ND
number of layers	x_{nl}	integer	NS*
core number	x_{cn}	integer	1
fin height	x_{fh}	discrete	NS*
fin thickness	x_{ft}	discrete	NS*
fin frequency	x_{ff}	discrete	NS*
fin serration length	x_{fs}	discrete	NS*
total			5.NS*+ND+2

There is no theoretical guaranty that the solution is the optimal design. But such a method generally gives a good approximation of the optimal design. The main advantage of this method is the very little computational charge. This is a very interesting tool in order to perform sensitivity analysis and to outline major trends. Unfortunately because of non-convexities, sub-optimal solutions can be computed. As noted in the previous section, the PFHE model may involve non-continuities and non-differentiabilities. Thus convergence failures may occur.

If discrete/integer variables are not relaxed, the optimization problem results in Mixed Integer Non Linear Programming (MINLP) problem. Three algorithms have been: Branch and Bound coupled with SQP, Simulated Annealing (SA), Simulated Annealing coupled with SQP. The best results are obtained with the SA algorithm: since this is a direct method, there is no need to have a continuously differentiable problem; since this is a stochastic method, there is no need to have a convex problem (the probability to obtain the global optimum is increased). The main drawback is that the computational charge is much more important than the computational charge required for the relaxed problem solution.

GENERAL SOLUTION STRATEGY

The general solution strategy is described on figure 1. The design problem is first described in the input file. The user supplies data such as inlet and outlet temperatures and transport properties. The program is connected with a pure component data bank and a thermodynamic model library: ProPhy™. Information such as enthalpies and transport properties can be computed easily.

The second stage of the general procedure is the initialisation of the optimization variables. The automatic sizing mode of COLETH is used : the complete PHFE is designed (distributors, fins, number of cores …) in order to achieve the duty requirement.

In the subsequent stages, the optimization problem is solved : the value of the optimization variables are computed by the solver. Those values are transmitted to COLETH (sizing mode) in order to evaluate the interest variables. Then the objective function and the constraints can be calculated. Their values are transferred to the solver : convergence test is performed; if needed, new values for the optimization variables are computed.

TEST PROBLEMS

First test problem

Consider the following illustrative example (Figure 2). There are one hot stream (H_1) and two cold streams (C_1 and C_2). One should note that the stream H_1 partially leaves the core at –116°C. This creates two sections. The Maximum Duty Section (section 1) includes three streams. The total number of variables is 24 and the total number of constraints is 38.

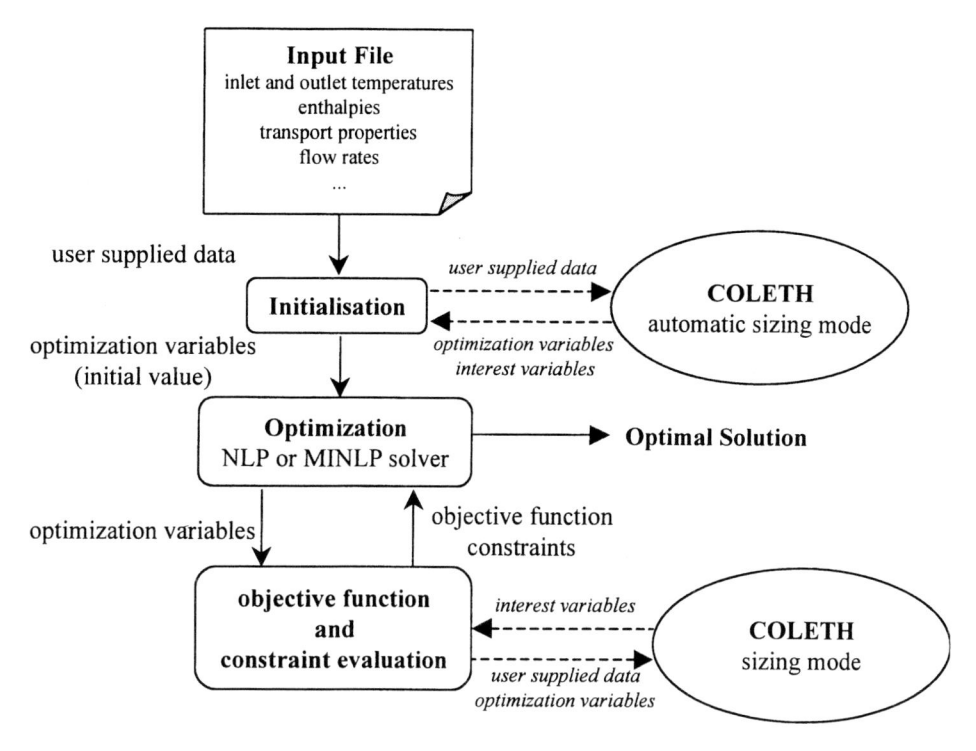

Figure 1 : General solution strategy

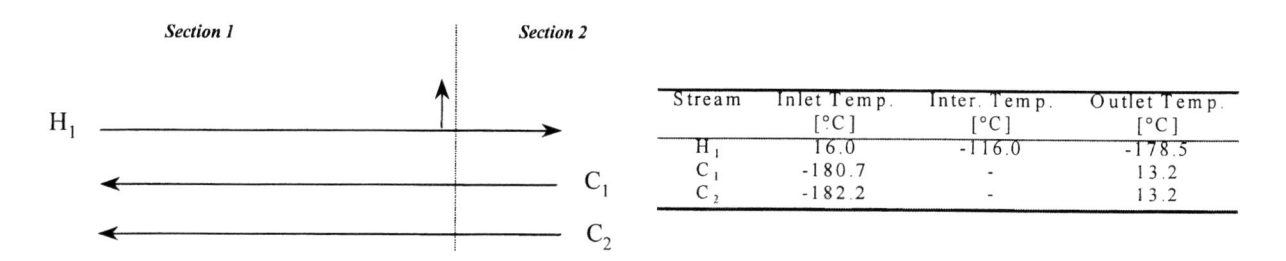

Stream	Inlet Temp. [°C]	Inter. Temp. [°C]	Outlet Temp. [°C]
H_1	16.0	-116.0	-178.5
C_1	-180.7	-	13.2
C_2	-182.2	-	13.2

Figure 2: First test problem

Capital cost is minimized. Main results are presented on table 2. The commercial proposal is presented in column "Design". The optimal result using the SQP algorithm (discrete variables are relaxed) is described in column "SQP" and the optimal result using Simulated Annealing is presented in column "SA". One should note that the optimal result using SA is better than the one using SQP. Actually, the relaxed optimum is theoretically better than the MINLP one. Solving the relaxed problem using SQP provides, on this example, a sub-optimal solution. Considering the pressure drop of stream H_1, one can note that the constraint is not saturated: the compromise between pressure drop and thermal effectiveness is not optimal (thermal effectiveness can be increased, and capital cost decreased, until H_2 pressure drop constraint is saturated). Of course the SA algorithm is much more time expansive than the SQP one.

A capital cost reduction (15%) is achieved using Simulated Annealing.

Table 2 : First test problem results

			Design	SQP	SA
Optimization variables	x_{CW} [mm]		1 300	1 300	1 300
	x_{ce} [-]		1	1.12	1
	x_{fh} [mm]	H_1	7.13	9.49	9.63
		C_1	9.63	9.63	9.63
		C_2	9.63	9.63	9.63
	x_{ft} [mm]	H_1	0.20	0.20	0.20
		C_1	0.20	0.20	0.25
		C_2	0.20	0.20	0.25
	x_{ff} [m^{-1}]	H_1	807.1	1000	897.6
		C_1	728.3	746.1	866.1
		C_2	728.3	558.8	866.1
Interest vaiables	ΔP [g/cm^2]	H_1	185	134.7	185
		C_1	140	140	140
		C_2	140	140	140
	Capital Cost*		100	92	85
	Total volume [m^3]		11.3	10.4	9.3
	CPU (Alpha Server 8200)		-	12'01"	5h46'09"

*: basis is the capital cost of the commercial proposal

Second Test Problem

The second test problem is described in figure 3. This heat exchanger is a very challenging design problem: there is an intermediary outlet on the first hot stream (H_1). This stream is also redistributed in the third section. One should note that layers occupied by stream H_2 in section 1 and 2 are occupied by the stream H_1 in section 3. This is an example of a duplex heat exchanger with redistribution. The maximum duty section is the second section. There are 11 distributors and 28 optimization variables. The total number of constraints is 49.

Here again, capital cost is minimized. Main results are presented on table 3. As in the previous example, the use of a global MINLP method leads to a better result. But the computational charge is much more important.

The main conclusion is that, with this complex example, the use of mathematical programming (SA) provides an important capital cost reduction : 21%. Comparing column "Design" to "SA", and "Design" to "SQP" one should note that fin height (x_{fh}) increases and fin thickness (x_{ft}) decreases : pressure drops are then minimized. The thermal effectiveness is augmented increasing fin frequency (x_{ff}). Those trends are quite general ones.

As in the previous example, the SQP algorithm fails to find the global optimum because of non-convexities involved in the model. To circumvent this problem, one may try to change the initialisation according to the general trends described before. This is one of the main challenging task to be done: find an automatic initialisation procedure that allows to converge to the global relaxed optimum.

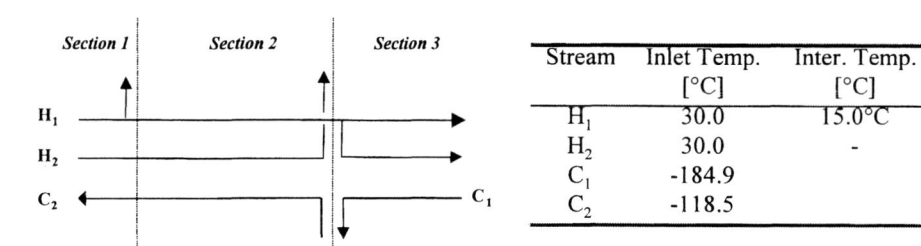

Stream	Inlet Temp. [°C]	Inter. Temp. [°C]	Outlet Temp. [°C]
H_1	30.0	15.0°C	-183.1
H_2	30.0	-	-112.6
C_1	-184.9		-127.0
C_2	-118.5		28.2

Figure 3 : Second test problem

Table 3 : Second test problem results

		Design	SQP	SA
Optimization variables	x_{CW} [mm]	915	630	914.6
	x_{cn} [-]	1	1	1
	x_{fh} [mm] H_1	7.13	8.73	8.89
	H_2	7.13	9.49	6.35
	C_1	9.63	9.63	9.63
	x_{ft} [mm] H_1	.40	0.20	0.25
	H_2	.25	0.20	0.25
	C_1	.20	0.20	0.20
	x_{ff} [m^{-1}] H_1	600.4	986.5	866.1
	H_2	930	1000	984.3
	C_1	771.7	749.2	866.1
Interest vaiables	ΔP [g/cm^2] H_1	153	153	153
	H_2	204	126.4	204
	C_1	204	204	204
	Capital Cost*	100	87	79
	Total volume [m^3]	5.4	4.0	4.1
	CPU (Alpha Server 8200)	-	3'04"	10h46'09"

*: basis is the capital cost of the commercial proposal

CONCLUSIONS

The proposed work is an example of industrial application of mathematical programming techniques. An efficient tool for computer aided design of plate fin heat exchangers is presented. Mathematical programming techniques are integrated in the COLETH program. The program allows optimization of the fins (height, thickness...), the core (width) and the distributors (widths). Numerous design or operating constraints are included : pressure drops, maximum stacking height, maximum erosion velocity.... Various objective functions can be used: capital cost, total volume.... Most of the heat exchangers configurations can be optimized: intermediate by-products, redistribution.... The user can choose to relax the discrete variables (SQP algorithm is used) or to use a global MINLP algorithm : Simulated Annealing. Two industrial examples are presented. The program abilities are illustrated : important capital cost reductions are achieved: 15% and 21%.

Future developments will include: improvement of the initialisation procedure for the relaxed NLP problem solution, improvement of the global optimization algorithm in order to reduce the CPU time, optimization of the other heat exchanger sections (in the presented version, only the maximum duty section is optimized), optimization of the layer arrangement.

NOMENCLATURE

A_0 : exchange area, m²/m
A_c : cross section area, m²
C_f : friction factor, dimensionless
C_j : Colburn factor, dimensionless
C_p : specific heat, J/kg/K
D_h : hydraulic diameter, m
f : objective function
g : optimization constraints
He : heat transfer coefficient, W/m²/K
L : length, m
$\dot{m}$: mass flow rate, kg/s
NI : number of intervals, dimensionless
Pr : Prandlt Number, dimensionless
Q : duty, W
R : heat transfer resistance, m.K/W
T : temperature, °C
v : stream velocity, m/s
x : optimization variables

λ : thermal conductivity, W/m/K
μ : dynamic viscosity, kg/m/s
ρ : mass density, kg/m³
Ψ : efficiency coefficient, dimensionless

Superscript

i : stream
j : duty interval
max : maximum value
MDS : Main Duty Section
min : minimum value
Tot : total value

Subscript

lm : logarithmic mean

ACKNOWLEDGEMENTS

This work is presented with the support and the permission of NORDON CRYOGENIE SAS.

REFERENCES

B. Abramzon and Ostersetzer S., 1993, Optimal thermal and Hydraulic Design of Compact Heat Exchangers and Cold Plates for Cooling of Electronic Components, *Aerospace Heat Exchanger Technology*, eds. R.K. Shah and A. Hashemi, Elsevier Science Publishers B.V., pp 349-368

B. Abramzon, 1996, Optimum Design of the Compact Fin-and-Plate Heat Exchangers, *Proceedings of the 26th Israel Conference on Mechanical Engineering*, Haifa, May 21-22, pp 312-338

J.E. Hesselgreaves, 1993, Optimizing Size and Weight of Plate Heat Exchangers, *Aerospace Heat Exchanger Technology*, eds. R.K. Shah and A. Hashemi, Elsevier Science Publishers B.V., pp391-399

S. Sunder and Fox V.G., 1993, Multivariable Optimization of Plate Fin Heat Exchangers, *AIChE Symposium Series*, Heat Transfer – Atlanta, Vol. 89, no. 295, pp 244-252

D.J. Ternet and Biegler L.T., 1998, Recent Improvements to a Multiplier Free Reduced Hessian Successive Quadratic Programming Algorithm, *Comp. Chem. Engng.*, Vol. 22, no. 7/8, p. 963

AN IMPROVED APPROACH TO THE DESIGN OF PLATE HEAT EXCHANGERS

Lieke Wang and Bengt Sundén

Division of Heat Transfer, Lund Institute of Technology, Box 118, 22100, Lund, Sweden
E-mails: Lieke.Wang@vok.lth.se   Bengt.Sunden@vok.lth.se

ABSTRACT

This paper presents an improved approach to the design of plate heat exchangers in single-phase applications, where full utilization of allowable pressure drops is taken as a design objective. The approach is based on a thermal-hydraulic model, which represents the relationship between heat transfer, pressure drop and exchanger area. Compared to the previous design method, the proposed approach does not require many trial iterations. Instead, all heat exchanger parameters, including plate size, number of passes, path, fluid velocity, etc., are determined in a straightforward way. In addition, optimum corrugation angle is discussed for the most common chevron-type plate heat exchangers.

INTRODUCTION

A plate-and-frame heat exchanger consists of plates, gaskets, frames and some additional devices, such as carrying and guiding bars, support column, etc. (see Fig. 1). The two streams flow into alternate channels between plates, entering and leaving via ports at the corner of the plates. Heat is exchanged between adjacent channels. This type of compact heat exchangers provides a number of advantages over shell-and-tube heat exchangers, such as compactness, effectiveness, cost competitiveness, etc. Significant developments have been achieved during last two decades. In construction, brazed plate heat exchangers, semi-welded plate heat exchangers and welded plate heat exchangers are developed in response to different operating situations (Wang, 1999). In performance, different plate patterns are developed to promote heat transfer with minimum pressure drop penalty. The chevron-type plate has proved to be the most successful design offered during last decades (Martin, 1996).

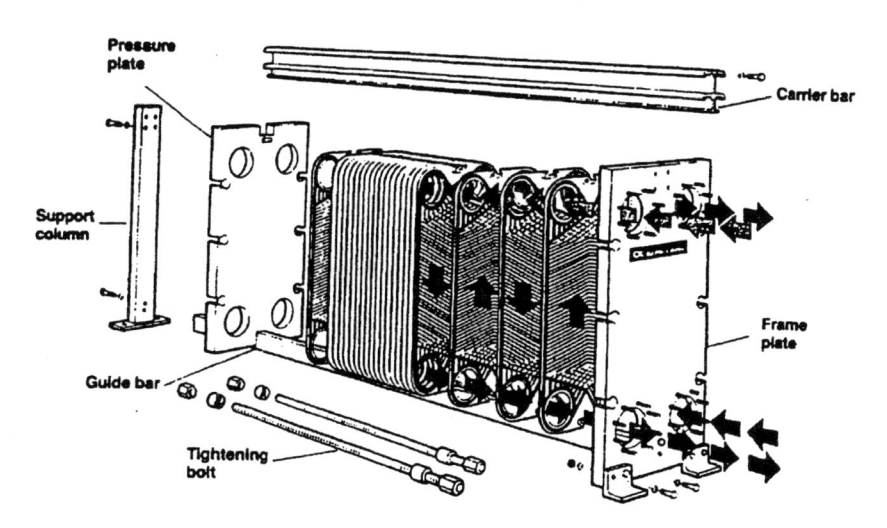

Fig. 1: An exploded view of a plate heat exchanger (courtesy of Alfa Laval)

However, the design method has not been developed in the same significant way. The design of plate heat exchangers requires a specification of the heat duty, stream pressure drops and certain aspects of exchanger geometry. In the traditional design approach, either $\varepsilon-NTU$ method or LMTD method, pressure drops are always taken as constraints. Therefore, the procedure often consists of huge testing of different geometries in order to find a suitable one, which is able to provide the specified heat duty within acceptable pressure drop bounds (Cooper and Usher, 1983; Shah and Focke, 1988). This trial method is not only very time-consuming, but does not guarantee optimum design with full utilization of allowable pressure drops. When the multi-pass plate heat exchangers are considered, the design becomes even more crucial.

In order to overcome these shortcomings, an improved approach is proposed. The proposed approach is for single-phase applications, and assumes the stream physical properties are constant. In this approach, the relationship between heat transfer coefficient, pressure drop and exchanger surface area is established. This is similar to the derivation by Jegede and Polley (1992), which is based on the objective of full utilization of allowable pressure drops. However, their approach can cause some problems if it is directly applied to the design of plate heat exchangers, which will be demonstrated later. The improved approach proposed in this paper will result in the optimum design for plate heat exchangers. It is straightforward and does not contain many trial iterations.

Because plate patterns have great influence on both thermal and hydraulic performance, the final design is certainly dependent on the initial choice of plate pattern. Although many types have been used in the past, the most successful one is the chevron-type plate heat exchanger (Martin, 1996). Therefore, the consideration of plate patterns in this paper is limited to this type. The correlations of heat transfer coefficients and pressure drops based on corrugation angles have been established by Martin (1996 & 1999). These correlations are used in this paper to investigate the influence of corrugation angles on the design. The optimum corrugation angle is obtained based on the minimum exchanger area.

THERMAL-HYDRAULIC MODEL

In order to design plate heat exchangers with full utilization of pressure drop constraints, the thermal-hydraulic model which links pressure drop and heat transfer should be developed. The considered stream in a plate heat exchanger is supposed to be the $m \times n$ arrangement, which means this stream has n passes and m paths per pass. The passage arrangement of $2 \times 3/3 \times 2$ for a plate heat exchanger is shown in Fig. 2.

The film heat transfer coefficient is often given in the following form

$$Nu = a \, Re^b \, Pr^{0.33} (\mu/\mu_w)^{0.17} \qquad (1)$$

This can be re-arranged to

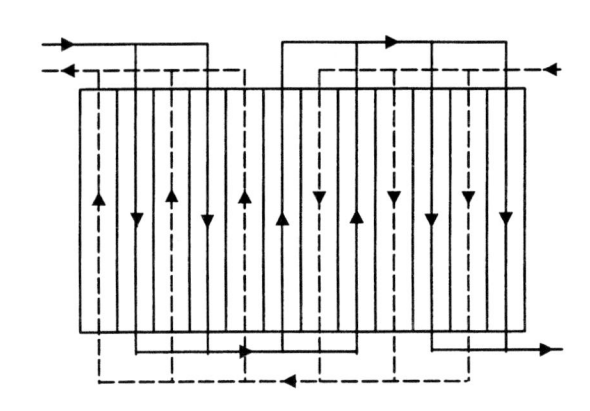

$$h = k_1 u^b \qquad (2)$$

Fig. 2: A plate heat exchanger with $2 \times 3/3 \times 2$ arrangement

where

$$k_1 = a \left(\lambda/D_e\right)\left(D_e/\nu\right)^b Pr^{0.33} (\mu/\mu_w)^{0.17} \qquad (3)$$

The Fanning friction factor is often given in the following form

$$f = c \, Re^d \qquad (4)$$

and the pressure drop is calculated as

$$\Delta P = 2f \left(nl/D_e\right) \rho u^2 \qquad (5)$$

From Eqs. (4) and (5), one finds

$$\Delta P = k_2 (nl) u^{2+d} \qquad (6)$$

where

$$k_2 = 2c \left(D_e/\nu\right)^d \left(\rho/D_e\right) \qquad (7)$$

The stream velocity is a function of the volumetric flow rate and the total channel cross section in a pass.

$$u = V/(0.5 D_e m w) \qquad (8)$$

The surface area of the heat exchanger is calculated by

$$A = mw \times nl \qquad (9)$$

Therefore, it is easy to get the following equation

$$nl = k_3 Au \qquad (10)$$

where

$$k_3 = D_e/2V \qquad (11)$$

Substituting the exchanger pass length in Eq. (6) and using Eq. (10), the pressure drop can be calculated as

$$\Delta P = k_2 k_3 A u^{3+d} \qquad (12)$$

By combining Eqs. (2) and (12), the final equation can be obtained as

$$\Delta P = \left(\frac{k_2 k_3}{k_1^{(3+d)/b}} \right) A h^{((3+d)/b)} = \sigma A h^{((3+d)/b)} \qquad (13)$$

This equation is the thermal-hydraulic model, which represents the relationship between the stream heat transfer coefficient, pressure drop, exchanger area. It will be used for the design of plate heat exchangers with the objective of full utilization of allowable pressure drops.

The values of k_1, k_2 and k_3 are dependent on the stream physical properties, the stream volumetric flow rate, the hydraulic diameter and the plate pattern. For a given design situation, the stream physical properties and the stream volumetric flow rate are fixed. In addition, the hydraulic diameter has its typical value, and the variation is usually very small. Therefore, the only unfixed parameter is the plate pattern, which affects the coefficients and exponents of heat transfer and pressure drop correlations. Minimum exchanger area can be obtained by optimizing the plate pattern.

DESIGN OF PLATE HEAT EXCHANGERS

In the design of plate heat exchangers, the allowable pressured drops are often specified prior to the design exercise. Full utilization of the allowable pressure drops can minimize the exchanger size. Therefore, the design is to find the maximum possible velocity which can achieve the maximum allowable pressure drops. This maximum velocity provides the maximum heat transfer coefficients, which in turn gives minimum exchanger size. When the above thermal-hydraulic model is applied, this design becomes easy and straightforward.

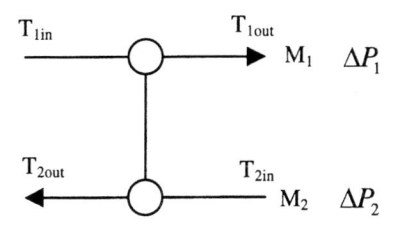

Fig. 3: A heat exchanger example

For the exchanger shown in Fig. 3, there are three governing equations given below.

$$\Delta P_1 = \sigma_1 A h_1^{((3+d)/b)} \qquad (14)$$

$$\Delta P_2 = \sigma_2 A h_2^{((3+d)/b)} \qquad (15)$$

$$Q = U A \Delta T_{LM} F_c \qquad (16)$$

where

$$\frac{1}{U} = \frac{1}{h_1} + \frac{1}{h_2} + \frac{\delta}{\lambda} + R_{f1} + R_{f2} \qquad (17)$$

Therefore, the exchanger area is given by

$$A = \frac{Q}{\Delta T_{LM} F_c} \left(\frac{1}{h_1} + \frac{1}{h_2} + \frac{\delta}{\lambda} + R_{f1} + R_{f2} \right) \qquad (18)$$

Because the heat transfer coefficients h_1 and h_2 are functions of the exchanger area, Eq. (18) is an implicit equation for the exchanger area, which can be solved by iterative methods. After the exchanger area is determined, the heat transfer coefficients on both sides can be calculated by Eqs. (14) and (15). The velocities are then determined by Eq. (2). Using Eqs. (8) and (9), the exchanger parameters, including $m_1 w_1$, $n_1 l_1$, $m_2 w_2$ and $n_2 l_2$, can be obtained.

However, the above design ignores the fact that the two stream sides are related, not separated. In fact, both sides should have the same plate size (length, width, plate pattern, etc.). Of course, the same plate pattern is already taken into account in the above calculation because the same correlations are used for both sides. But the above design can not guarantee the same length or width on both sides. When the values of $m_1 w_1$, $n_1 l_1$, $m_2 w_2$ and $n_2 l_2$ are determined, it is often impossible to obtain the same plate size for both sides because both m and n are integers. This problem is not addressed by Jegede and Polley (1992), where the calculated value of A is taken as the design value. It is sometimes totally wrong, which will be demonstrated in the examples.

In order to solve this problem, the pressure drop constrain of one side has to be relaxed. That is, only the allowable pressure drop of one side can be fully utilized. This stream can be identified as reference stream. Hence the design can be extended to the following calculation. After the optimum values of nl for both sides are obtained in the above calculation, the stream with the smallest value of nl is usually taken as reference stream in the single-pass arrangement. In addition to this, the ratio of two stream passes can also be specified. This is necessary when multi-pass plate heat exchangers are considered. These specifications should be made as close as possible to the optimum values. Let us assume that stream 1 is the reference stream, and the value of $n_1 l_1 / n_2 l_2$ is equal to c_{nl}. The heat transfer coefficient of stream 1 can be calculated by Eq. (13) because the allowable pressure drop is fully utilized. It is a function of only the exchanger area A.

$$h_1 = f_1(A) \qquad (19)$$

From Eq. (2), the heat transfer coefficient of stream 2 is calculated as

$$h_2 = h_1 \left(\frac{u_2}{u_1}\right)^b \quad (20)$$

where the value of stream velocity ratio can be calculated from the following equation.

$$\frac{u_2}{u_1} = \frac{2V_2/(m_2W_2D_e)}{2V_1/(m_1W_1D_e)} = \frac{V_2}{V_1}\frac{m_1W_1}{m_2W_2} = \frac{V_2}{V_1}\frac{n_2l_2}{n_1l_1} = \frac{V_2}{V_1}\frac{1}{c_{nl}} \quad (21)$$

Therefore, the heat transfer coefficient of stream 2 can finally be obtained by

$$h_2 = f_2(A) \quad (22)$$

Using Eq. (18), the value of exchanger area can be determined by an iterative method. Knowing the value of exchanger area A, all the exchanger details can be then obtained. Finally, the pressure drop of the non-reference stream should be checked and it should be within the pressure drop specification.

Thus far, the optimum design of plate heat exchangers has been completed. All the detailed exchanger parameters are established, such as the plate size, path number, pass arrangement, etc. The details also include the values of stream velocities or the heat transfer coefficients. In addition, the allowable pressure drop of at least one stream is fully utilized, and the pressure drop of the other stream is within its specified pressure drop bound.

PLATE PATTERN

During the above design, the plate pattern is specified before the design, and hence it is not optimized during the design. However, plate pattern has great influence on the final design. For the most common chevron-type plate heat exchangers (see Fig. 4), the corrugation angle is the most important parameter for the thermal and hydraulic performance, although many other parameters, such as wavelength and amplitude also have influences on performance. This is due to the fact that the influence from amplitude and wavelength can be summarized into one parameter, the area enlargement factor. If the hydraulic diameter, which is defined as four times amplitude divided by area enlargement factor, is used, correlations for pressure drop and heat transfer have been achieved as a function of the corrugation angle and the Reynolds number for chevron-type plate heat exchangers (Martin, 1996). Generally speaking, when the corrugation angle becomes larger, heat transfer is enhanced with the penalty of higher pressure drop. This means there is an optimum corrugation angle, which can lead to the highest heat transfer, when the total pressure drop is set. When the corrugation angle is higher than the optimum angle, the heat transfer area has to increase in order to meet the pressure drop requirement. This conclusion holds for an individual heat exchanger as well as for heat exchanger networks (Wang and Sundén, 2000). Therefore, the corrugation angle should also be optimized during the exchanger design.

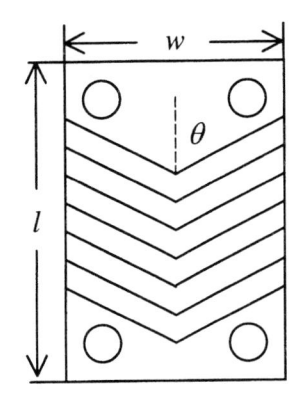

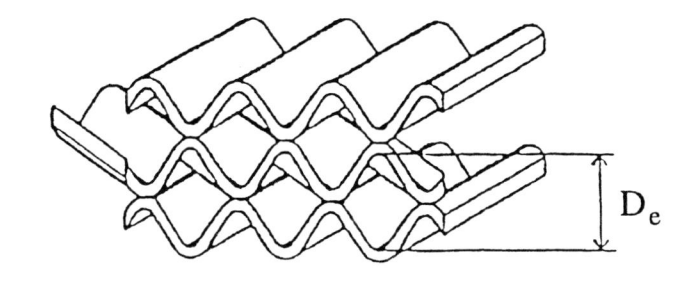

Fig. 4: A schematic drawing of a chevron-type plate heat exchanger

The correlations of heat transfer and pressure drop for plate heat exchangers often take the forms of Eqs. (1) and (4). The corresponding coefficients have specific values for specific plates. The way to obtain them is often to conduct experiments individually. However, when the corrugation angle should be optimized during the design, correlations as a function of corrugation angles should be established. Martin's correlations for chevron-type plate heat exchangers, valid for corrugation angles from 10° to 80°, are employed in this investigation (Martin, 1996 & 1999). The correlations are based on a theoretical approach, and the accuracy is decent compared to experimental data. The Fanning friction factor can be calculated as

$$\frac{1}{\sqrt{f}} = \frac{\cos\theta}{\sqrt{0.045\tan\theta + 0.09\sin\theta + f_0/\cos\theta}} + \frac{1-\cos\theta}{\sqrt{3.8f_1}} \quad (23)$$

where $f_0 = 16/\text{Re}$ for $\text{Re} < 2000$ and $f_0 = (1.56\ln\text{Re} - 3.0)^{-2}$ for $\text{Re} \geq 2000$, and

$$f_1 = \frac{149}{\text{Re}} + 0.9625 \quad \text{for} \quad \text{Re} < 2000 \quad \text{and} \quad f_1 = \frac{9.75}{\text{Re}^{0.289}} \quad \text{for}$$

$\text{Re} \geq 2000$. The correlation for the heat transfer coefficient is

$$Nu = 0.205 \, \text{Pr}^{1/3} \left(\frac{\mu_m}{\mu_w}\right)^{1/6} \left(f \, \text{Re}^2 \sin 2\theta\right)^{0.374} \quad (24)$$

However, these correlations are not in the conventional forms required by Eqs. (1) and (4). In order to make the calculation simple, Martin's correlations are converted to the forms of Eqs. (1) and (4) through curve-fit formulas for specific corrugation angles of every 5 degree. If the corrugation angle required is not the value specified, interpolation is carried out. The deviation of the revised formulas and Martin's correlations is within ±1%.

EXAMPLES

In this part, three examples are used to demonstrate the proposed method. The first example is to demonstrate the weakness of the previous method, and how the current method can obtain the optimal design. The second example is used to demonstrate how to design multi-pass plate heat exchangers. During the first two examples, the plate pattern is fixed, and the corrugation angle is not optimized. In the third example, the plate pattern is optimized, and the optimum corrugation angle is obtained.

The heat duty is the same for all three examples. (Units, see Nomenclature)

Table 1: Stream data

Fluid	T_{in}	T_{out}	M	μ	λ	c_p	ρ	R_f
1	10	50	30	8×10^{-4}	0.618	4175	980	5×10^{-5}
2	70	40	40	5×10^{-4}	0.654	4175	980	5×10^{-5}

Example 1

In the first example, the allowable pressure drops of fluids 1 and 2 are 60 kPa and 40 kPa, respectively. The values for the coefficients a, b, c and d are taken as 0.1876, 0.7179, 0.9108 and -0.0805, respectively. These values are obtained when Martin's correlations are converted to the conventional form, and it is actually equal to the corrugation angle of 60° for the chevron-type plate heat exchangers. The hydraulic diameter is 0.006 m. The thickness and thermal conductivity of the plate are 0.0005 m and 13.56 W/(m·K), respectively. In addition, the average value of inlet and outlet temperatures of both streams is taken as the wall temperature.

Using the previous design method, the optimum area is estimated to be 63.3 m². The detailed parameters for the exchanger are listed in Table 2.

Table 2: Design results of the previous method

Fluid	u (m)	nl (m)	mw (m)	ΔP (kPa)
1	0.395	2.454	26	60.0
2	0.384	1.791	35	40.0

Obviously, the plate lengths for both sides can not be equal. Therefore, the design is not finished, and should be continued using the current improved approach.

Because the optimal length of fluid 2 is smaller, fluid 2 is selected as the reference stream. The single-pass counter arrangement is selected because the optimum lengths for both sides are close. The correction factor compared to counter flow is equal to 1 if the influence by the end plates is ignored. After using the approach described above, the following design can be achieved.

Table 3: Design results of the current method

Fluid	u (m)	nl (m)	mw (m)	h	Re	ΔP (kPa)
1	0.280	1.889	36	8450	2196	23.9
2	0.374	1.889	36	11935	4183	40.0

The number of transfer units (NTU) is 1.62, and the exchanger area is about 68.8 m². Compared to the results from the previous method, the required area is bigger and this indicates that the previous result is too optimistic. In this case, the exchanger area difference is about 8.7%, but the difference may be even bigger in other situations.

This example demonstrates that the previous method can lead to unrealistic design because it assumes full utilization of allowable pressure drops for both sides. The current approach overcomes this drawback and leads to an optimum design with full utilization of the allowable pressure drops of at least one side.

Example 2

In this example, the allowable pressure drops for fluid 1 and 2 are taken as 100 kPa and 30 kPa, respectively. The other parameters are the same as in Example 1. Using the previous method, the following results can be obtained. The exchanger area is about 62.2 m².

Table 4: Design results of the previous method

Fluid	u (m)	nl (m)	mw (m)	ΔP (kPa)
1	0.474	2.89	21.5	100.0
2	0.351	1.60	38.8	30.0

Because the optimum pass length of fluid 1 is close to twice of that of fluid 2, it is preferred to have 2 passes for fluid 1. Because the optimal pass length of fluid 1 is less than twice the value of fluid 2, fluid 1 is taken as the reference stream. Using the suggested method, the following design can be achieved.

Table 5: Design results of the current method

Fluid	u (m)	nl (m)	mw (m)	h	Re	ΔP (kPa)
1	0.441	3.312	23	11697	3289	100
2	0.294	1.656	46	10044	3454	22.1

The exchanger area is about 76.6 m². If the plate height and width are taken as 1.655 m and 1 m respectively, the exchanger is 23×2/46×1 arrangement. The number of transfer units (NTU) is 1.9, and the correction factor compared to the counter flow arrangement is about 0.85.

If this exchanger is specified as single-pass and counter-flow arrangement, the heat exchanger area is 72.6 m². The pressure drops of stream 1 and 2 are 17.9 kPa and 30 kPa, respectively. It is interesting to notice that this exchange area is less than the multi-pass arrangement, although the latter makes use of more allowable pressure drops. The reason for this is that the multi-pass arrangement contains both counter and co-current flows inside, which decrease the exchanger effectiveness. The corresponding result is an increase of the exchanger area. If the correction factor is assumed to be 1 anyway, the exchanger area is 63.5 m², which is much less than that of the single-pass exchanger.

This example demonstrates that full utilization of the allowable pressure drop does not always lead to minimum exchanger area for multi-pass plate heat exchangers. It also depends on the passage arrangement.

Example 3

As stated above, the optimization of plate corrugation angle for chevron-type plate heat exchangers is necessary. There exists an optimum plate pattern when the design is carried out with fixed pressure drops. This optimal design is actually the trade-off between heat transfer and pressure drop.

In this example, the design data are taken as the same as for example 1. The optimal design is defined as the one with the minimum heat transfer area. Using the proposed method, the result is shown in Fig. 5. Obviously, the optimum corrugation angle corresponding to the minimum area is about 60° in this case. The difference between the biggest and lowest values is up to 95%. Even in the most common range between 30° to 70°, the area difference is still up to 15%. In some other cases, it can be even higher depending on the situation of the allowable pressure drops.

This example demonstrates that it is necessary to optimize the plate pattern in order to obtain minimum exchanger area. However, it should be emphasized that it is also worthwhile to optimize pressure drops if the power cost for overcoming friction forces is considered. In this situation, there are three variables, plate pattern and two pressure drops (corresponding to two heat transfer coefficients or Reynolds numbers). This is a more complicated situation, and the simplified solution is provided by Martin (1999).

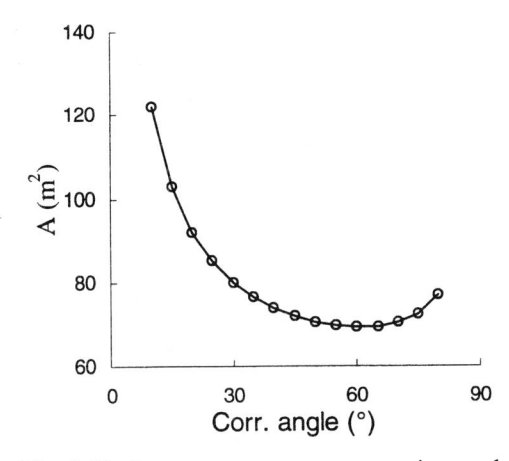

Fig. 5: Exchanger area versus corrugation angle

CONCLUSION

An improved approach to the design of plate heat exchangers is described in detail. It is intended for the design task with fixed pressure drop constraints. The design objective is to fully utilize pressured drop constraints. However, it is only possible to fully utilize the allowable pressure drop of one side in many situations. The proposed approach is able to design both single-pass and multi-pass plate heat exchangers. But the multi-pass plate heat exchanger does not always lead to less exchanger area although it can utilize more pressure drops. This is caused by the lower efficiency of multi-pass plate heat exchangers compared to single-pass counter-flow arrangement. In addition, the need for optimizing the plate pattern has been demonstrated. For the most common chevron-type plate heat exchangers, the optimal corrugation angle corresponding to the minimum heat transfer area is obtained.

ACKNOWLEDGEMENT

Financial support from STEM (the Swedish National Energy Authority) is kindly acknowledged.

NOMENCLATURE

a, b, c, d	constants
A	heat transfer area, m^2
c_{nl}	ratio of pass lengths
c_p	specific heat, J/(kg·K)
D_e	hydraulic diameter, m
f	Fanning friction factor
F_c	correction factor
h	heat transfer coefficient, W/(m^2·K)
k_1, k_2, k_3	constants in Eq. (13)
l	plate length, m
m	path number
n	pass number
LMTD	logarithmic mean temperature difference, °C
M	mass flow rate, kg/s
NTU	number of transfer units
Nu	Nusselt number
Pr	Prandtl number
Q	heat load, W
Re	Reynolds number
R_f	fouling factor, m^2·K/W
T	temperature, °C
u	velocity, m/s
U	overall heat transfer coefficient, W/(m^2·K)
V	volumetric flow rate, m^3/s
w	plate width, m
ΔP	pressure drop, Pa

Greek symbols

δ	plate thickness, m
ε	thermal effectiveness
λ	thermal conductivity, W/(m·K)
ν	kinetic viscosity, m^2/s
μ	dynamic viscosity, kg/(m·s)
θ	corrugation angle, degree
σ	coefficient

Subscripts

1	stream 1
2	stream 2
in	stream inlet
out	stream outlet
w	wall

REFERENCES

Cooper, A. and Usher, J. D., 1983, Chapter 3.7: Plate Heat Exchangers, *Heat Exchanger Design Handbook*, Hemisphere Publishing Corporation.

Jegede, F. O. and Polley, G. T., 1992, Optimum Heat Exchanger Design, *Trans. IChemE*, Vol. 70, pp. 133-141.

Martin, H., 1996, A Theoretical Approach to Predict the Performance of Chevron-Type Plate Heat Exchangers, *Chem. Eng. Process.*, Vol. 35, pp. 301-310.

Martin, H., 1999, Economic Optimization of Compact Heat Exchangers, in *Compact Heat Exchangers and Enhancement Technology for the Process Industries*, eds. R. K. Shah, K. J. Bell, H. Honda and B. Thonon, Begell House, pp. 75-80.

Shah, R. K. and Focke, W. W., 1988, Plate Heat Exchangers and Their Design Theory, in *Heat Transfer Equipment Design*, eds. R. K. Shah, E. C. Subbarao and R. A. Mashelkar, Hemisphere Publishing Corp., Washington, pp. 227-254.

Wang, L. 1999, An Experimental Investigation of Steam Condensation Performance in Plate Heat Exchangers, Publ. 99/2009, *Thesis for Licentiate of Engineering Degree*, Division of Heat Transfer, Department of Heat and Power Engineering, Lund Institute of Technology, Lund, Sweden.

Wang, L. and Sundén, B, 2000, On Plate Heat Exchangers in Heat Exchanger Networks, *34th National Heat Transfer Conference*, Pittsburgh, USA, NTHC2000-12166.

A METHOD FOR RELATIVE COMPARISON OF THERMOHYDRAULIC EFFICINCIES OF HEAT TRANSFER SURFACES AND HEAT EXCHANGERS

Dubrovsky E.V.[1] and Vasiliev V.Yu.[2]

[1] - Scientific and Research Institute for Tractors, 34 Verkhnaya Str., 125040 Moscow, Russia, E-mail: nati@ccas.ru
[2] - Astrakhan State Technical University, 16 Tatishchev Str., 414025 Astrakhan, Russia, E-mail: master@astu.astranet.ru

ABSTRACT

At present there is no reliable unified method for comparison of thermohydraulic efficiencies of different-design heat exchangers (tube and plate-fin) as well as a method for comparison of thermohydraulic efficiency of heat transfer enhancement process in channels of different designs of heat transfer surfaces. As a result, often it is very difficult and sometimes impossible to choose the most efficient heat transfer surface that would provide a minimum volume and a minimum mass of heat exchangers.

The present work was aimed at developing a generalized and reliable method for relative comparison of thermohydraulic efficiencies of different-design exchangers as well as a method for comparison and estimation of thermohydraulic efficiency of heat transfer enhancement process in different-design channels of heat transfer surfaces. These methods were developed from analytical analysis of the problem of comparing volumes of heat exchangers operating under the conditions when a value of heat transfer coefficient for a heat exchanger was equal to the smallest value of that for a heat transfer surface. The conditions for the comparison problem of heat exchanger volumes were as follows: at the same values of heat loads of heat exchangers; at the same values of heat carrier flow rates, their pressure losses, and temperature differences in heat exchangers. In addition, according the problem conditions the geometrical parameters of channels of compared heat transfer surfaces and their criterial thermohydraulic relations are considered to be known. As a result, a reliably justified generalized method for relative comparison of volumes of any-design heat exchangers was elaborated. It was based on iterative calculation of the generalized criterial condition that controls, how thermohydraulic conditions of the problem in comparing heat exchanger volumes are satisfied. Then the values of performance and design parameters of compared heat exchangers obtained for the generalized criterial condition are used to compare their volumes.

Analysis of the comparison problem of thermohydraulic efficiency of heat transfer enhancement process in channels made feasible a formulation of the analytically closed problem and necessary boundary conditions. Therefore, the derivation of the method to compare thermohydraulic efficiency of heat transfer enhancement process in channels of heat transfer surfaces was started from a number of fundamental positions of theory and practice of heat transfer enhancement in channels. This method made it possible uniquely (without iterative calculation) to compare thermohydraulic efficiency of heat transfer enhancement in different-cross-section channels with turbulators in relation to the thermohydraulic process in smooth channels (of the same geometrical parameters without turbulators) of heat transfer surfaces under the same conditions of heat carrier flow in them.

INTRODUCTION

Long-standing use of Dubrovsky's method (1977) for comparison of thermohydraulic efficiencyy of heat transfer surfaces, when volumes of heat exchangers operating at $K \approx \alpha_{min}$ are compared, ensured reliable experimental research, analysis and generalization received results for heat transfer enhancement process in different-cross-section channels with turbulators. It found use in many systematic experimental investigations of Voronin, Dubrovsky (1973), Dubrovsky (1979), Dubrovsky, Vasiliev (1988), and Dubrovsky (1995). Also, this method confirmed its versatility when comparing different-heat transfer surface constructions and heat exchangers and under different conditions of heat carrier flow in them. However, in practice, application of this method exposed necessity to specify its deduction as well as thermohydraulic and design conditions of the comparison method. As a result, the developed method allowed reliable comparison of thermohydraulic efficiency of heat transfer enhancement process in heat transfer surface constructions practically for all possible cases of comparison.

The majority of the considered existing methods of comparing heat transfer surface constructions and heat exchangers are not based on common for tubular and plate-fin heat transfer surface constructions and heat exchangers generalized criterial condition of implementation for initial thermohydraulic and design conditions of the comparison problem. Therefore these methods are an examples a particular cases of comparison problem. And so, they constitute some particular cases of the comparison method which as a rule are not enough grounded or, on the whole, are not grounded. Owing to this, use of such methods for comparing the same heat transfer surface constructions and heat exchangers yields different not only quantitative but also qualitative results.

So, in the works of Antufiev (1966), Glazer (1948), Gukhman (1938), Kays and London (1984), Kirpichev (1944), Koch (1958), Mitskevich (1967), Petrovsky and Fastovsky (1962) the comparison methods described call for complete iterative thermohydraulic calculation of compared heat exchangers.

The works of Gryaznov (1974), Mitin (1969), Novikov (1967) contain the methods of comparing heat transfer surface constructions and heat exchangers on the bases of comparison in thermohydraulic efficiency and overall dimensions of a gas-turbine engines. In fact, it is a cumbersome and compli atedmethod of solving a technical-economic problem. In addition, such methods do not permit the estimation of thermohydraulic efficiency of heat transfer enhancement process in compared heat transfer surface constructijns.. Therefore, these methods found no wide use.

The works of Cox et al. (1973), Dreitser et al. (1973), Dreitser et al. (1974), Dreitser (1999), Dubrovsky et al. (1972), Kalinin, Dreitser and Yarkho (1972), Mikhailov, Borisov and Kalinin (1962), Voronin and Dubrovsky (1973), Walger (1952), Walger (1952), Webb (1972) describe well-grounded but only particular cases of the method for comparing thermohydraulic efficiencies of tube and plate-fin heat transfer surface constructions and heat exchangers. They require no complete iterative calculation of compared heat exchanger constructions. Rapid realization and fair accuracy of such methods made them a reliable tool to compare thermohydraulic efficiencies of heat exchanger constructions and heat transfer enhancement process in heat transfer surface constructions. However, some of these methods worked out by Dreitser et al. (1973), Dreitser et al. (1974), Dreitser (1999), Kalinin, Dreitser and Yarkho (1972), Mikhailov, Borisov and Kalinin (1962), Walger (1952), Walger (1952), Webb (1972) can be applicable only for comparison of heat exchanger and heat transfer surface constructions fabricated from unfinned round and out-of-round tubes and, as a rule, only for turbulent heat carrier flow. In Dreitser's work (1999) an attempt was made to work out a generalized comparison method common to tube and plate-fin heat exchanger and heat transfer surfaces constructions. In this case, the comparison method is based on the concluded generalized criterial conditions of implementation to thermohydraulic and design initial boundary conditions. However, the design boundary conditions assumed in Dreitser's work did not fulfil the objective stated.

Dubrovsky's work (1977) presented a generalized method of comparing thermohydraulic efficiencies both of plate-fin and finned and unfinned tube heat transfer surface and of heat exchanger constructions. However, the thermohydraulic ($\Delta p_1 = \Delta p_2$) and design ($\eta = \eta_{sm}$) conditions of the problem of comparing thermohydraulic efficiency of heat transfer enhancement process in channels require a more precise. The aim of the present work was to develop a generalized and well-grounded method for comparison of thermohydraulic efficiencies of different heat exchangers constructions as well as a method of comparing and estimating thermohydraulic efficiency of heat transfer enhancement process in different channel constructions of heat transfer surfaces.

BODY

The present work is concerned with the problem of comparing heat exchanger volumes under the following initial conditions:

1. Both compared heat transfer surface constructions are considered in designs of heat exchangers operating at $K \approx \alpha_{min}$.

2. Heat transfer surfaces and heat exchangers are compared in a value of their heat transfer surface or heat exchanger constructions volumes under the following thermohydraulic conditions: at the same values of heat loads, heat carrier flow rates, total pressure losses of heat carrier flow for friction, inlet and outlet of a heat transfer surface in a heat exchanger designs, and the same values of temperature differences in compared heat exchanger constructions:

$$Q_1 = Q_2 \quad \text{or} \quad Q_1/Q_2 = 1 \qquad (1)$$

$$G_1 = G_2 \quad \text{or} \quad G_1/G_2 = 1 \qquad (2)$$

$$\Delta p_1 \geq p_2 \quad \text{or} \quad \Delta p_1 / \Delta p_2 \geq 1 \qquad (3)$$

$$\Delta t_1 = \Delta t_2 \quad \text{or} \quad \Delta t_1/\Delta t_2 = 1 \qquad (4)$$

The subscript "1" denotes reference surface and heat exchanger, whose all design and thermohydraulic parameters are known from preliminary calculations. The subscript "2" is believed to compared heat transfer surface and heat exchanger.

3. Geometrical parameters of heat transfer surfaces and their criterial thermohydraulic relations: Nu=f(Re) and ξ=f(Re) are known for compared heat transfer surface and exchanger constructions. In addition, sizes of a shape design of a core are known too.

It should be noted that unlike the usually used condition: $\Delta p_1 = \Delta p_2$, boundary condition of Eq. (3): $\Delta p_1 \geq \Delta p_2$ reflects a modern knowledge about conformity of heat transfer enhancement process in channels. In particular, at present it is grounded in theory and found from experiment that Reynolds' analogy is violated when heat transfer process in channels is enhanced due to periodic generation of wall vortices with their weak diffusion into the flow core of heat carrier.

Determine some of the relations necessary to derive a method.

According to Eqs. (2) and (4), continuity equations, and with expression $F_f = F_c f_{f1}$, the velocity ratio of heat carriers in channels of heat transfer surfaces can be given as:

$$\frac{W_1}{W_2} = \frac{F_{c2}}{F_{c1}} \frac{f_{f2}}{f_{f1}} \quad (5)$$

With comparison Eqs. (2) and (4) and Eq. (5), the Reynolds number ratio for channels of compared heat transfer surfaces can be written as:

$$\frac{Re_1}{Re_2} = \frac{F_{c2}}{F_{c1}} \frac{f_{f2}}{f_{f1}} \frac{d_1}{d_2} \quad (6)$$

or

$$\frac{Re_1}{Re_2} \frac{d_2}{d_1} = \frac{F_{c2}}{F_{c1}} \frac{f_{f2}}{f_{f1}} \quad (7)$$

From Eqs. (1), (2) and (4), with the heat transfer equation we have:

$$\frac{Q_1}{Q_2} = \frac{\alpha_1}{\alpha_2} \frac{F_{h1}}{F_{h2}} \frac{\eta_1}{\eta_2} \frac{\Delta t_1}{\Delta t_2} = \frac{\alpha_1}{\alpha_2} \frac{F_{h1}}{F_{h2}} \frac{\eta_1}{\eta_2} = 1 \quad (8)$$

Using the expression $\alpha = Nu\lambda/d$ as well as Eqs. (5) and (6), Eq. (8) may be rearranged to give:

$$\frac{Nu_1}{Nu_2} \frac{\eta_1}{\eta_2} \frac{F_{h1}}{F_{h2}} \frac{d_2}{d_1} = 1 \quad (9)$$

or

$$\frac{F_{h1}}{F_{h2}} = \frac{Nu_2}{Nu_1} \frac{\eta_2}{\eta_1} \frac{d_1}{d_2} \quad (10)$$

With the expression $F_h = V\Omega$, Eq. (9) may be rearranged to give:

$$\frac{V_1}{V_2} \frac{Nu_1}{Nu_2} \frac{\eta_1}{\eta_2} \frac{\Omega_1}{\Omega_2} \frac{d_2}{d_1} = 1 \quad (11)$$

In Eqs. (9) and (11), the design parameter, F_{h2}, of the core of a compared heat exchanger is an independent variable, and V_2, Nu_2 and η_2 are dependent variables. It is obvious that Eqs. (9) and (11) are the equivalent forms of the criterial condition that controls the fulfillment of a NECESSARY COMPARISON CONDITION of any-design heat transfer surfaces and heat exchangers. In this case, for thermohydraulic conditions of Eqs. (1), (2) and (4) of the comparison problem, the values of the main design and performance parameters of compared heat transfer surfaces and heat exchangers are uniquely defined by a possibility to fulfil conditions of Eqs. (9) or (11).

With the identity transformation, Eq. (11) may be rearranged to give:

$$\frac{V_1}{V_2} = \frac{Nu_2}{Nu_1} \frac{\eta_2}{\eta_1} \frac{\Omega_2}{\Omega_1} \frac{d_1}{d_2} \quad (12)$$

It should be noted that in Eq. (12) borrowed the values of Nu_2 and η_2 from (9) or (11) where they were determined.

It is obvious that Eq. (12) characterizes the volume ratio of compared designs of heat exchangers when *necessary comparison condition* Eq. (9) is satisfied.

According to Eqs. (2), (3) and (4) of the comparison problem and with the equation for pressure losses in Darcy's form ($\Delta p = \xi \rho w^2/2 L/D$), we have:

$$\frac{\Delta p_1}{\Delta p_2} = \frac{\xi_1}{\xi_2} \frac{W_1^2}{W_2^2} \frac{L_1}{L_2} \frac{d_2}{d_1} \geq 1 \quad (13)$$

With Eq. (5) and $L = V/F_c$, and the identity transformation, Eq. (13) may be rearranged to give:

$$\frac{V_1}{V_2} \frac{\xi_1}{\xi_2} \frac{F_{c2}}{F_{c1}} \left(\frac{F_{c2}}{F_{c1}} \frac{f_{f2}}{f_{f1}} \right)^2 \frac{d_2}{d_1} \geq 1 \quad (14)$$

In Eqs. (13) and (14), the design parameter, F_{c2}, of the core of a compared heat exchanger is an independent variable, and V_2, ξ_2, W_2 and L_2 are dependent variables. It is obvious that Eq. (14) is one of the possible forms of the criterial expression that controls the fulfillment of a *sufficient comparison condition* of heat transfer surfaces and heat exchangers. In this case, for thermohydraulic conditions of Eqs. (2), (3) and (4) of the comparison problem a range of possible values of the main design and performance parameters of compared heat transfer surfaces and heat exchangers is limited by a possibility to provide the fulfillment of Eqs. (13) or (14).

With identity transformation, Eq, (14) may be rearranged to give:

$$\frac{V_1}{V_2} \geq \frac{\xi_2}{\xi_1} \frac{F_{c1}}{F_{c2}} \left(\frac{F_{c1}}{F_{c2}} \frac{f_{f1}}{f_{f2}} \right)^2 \frac{d_1}{d_2} \quad (15)$$

It should be mentioned that in Eq. (15), the values of F_{c2} and ξ_2 are taken from Eqs. (13) or (14) where they are determined.

It is obvious that Eq. (15) characterizes the volume ratio of compared designs of exchangers when *sufficient comparison condition* of Eq. (13) or (14) is satisfied.

Substitution of Eq. (7) into Eq. (14) yields:

$$\frac{V_1}{V_2}\frac{\xi_1}{\xi_2}\frac{F_{c2}}{F_{c1}}\left(\frac{Re_1}{Re_2}\right)^2\left(\frac{d_2}{d_1}\right)^3 \geq 1 \quad (16)$$

or

$$\frac{V_1}{V_2} \geq \frac{\xi_2}{\xi_1}\frac{F_{c1}}{F_{c2}}\left(\frac{Re_2}{Re_1}\right)^2\left(\frac{d_1}{d_2}\right)^3 \quad (17)$$

Note that Eqs. Eqs. (13), (14) and (16) as well as Eqs. (15) and (17) are of equivalent expressions.

Require that initial thermohydraulic conditions of Eqs. (1)–(4) of the comparison problem be necessary and sufficient to compare thermohydraulic efficiencies of heat transfer surfaces and heat exchangers. Then a generalized criterion for satisfying initial thermohydraulic conditions of Eqs. (1)–(4) of the comparison problem of heat exchanger volumes can be obtained if the LHSs of Eqs. (11) and (14) or Eqs. (11) and (16) are equated, and simple identity transformations and cancellation of the values of V_1/V_2[†] are made.

From Eqs. (11) and (14), we have:

$$\frac{Nu_2/Nu_1}{\xi_2/\xi_1}\frac{\eta_2}{\eta_1}\frac{\Omega_2}{\Omega_1}\left(\frac{F_{c2}}{F_{c1}}\right)^3\left(\frac{f_{f2}}{f_{f1}}\right)^2 \geq 1 \quad (18)$$

From Eqs. (11) and (16), we have:

$$\frac{Nu_2/Nu_1}{\xi_2/\xi_1}\frac{\eta_2}{\eta_1}\frac{\Omega_2}{\Omega_1}\frac{f_{f1}}{f_{f2}}\left(\frac{Re_1}{Re_2}\frac{d_2}{d_1}\right)^3 \geq 1 \quad (19)$$

or with regard to (7) we arrive at:

$$\frac{Nu_2/Nu_1}{\xi_2/\xi_1}\frac{\eta_2}{\eta_1}\frac{\Omega_2}{\Omega_1}\frac{F_{c2}}{F_{c1}}\left(\frac{Re_1}{Re_2}\frac{d_2}{d_1}\right)^2 \geq 1 \quad (20)$$

Equations (18), (19) and (20) are the equivalent forms of the *generalized criterial relation for necessary and* and *sufficient conditions of comparing* thermohydraulic efficiencies of any-design heat transfer surfaces and heat exchangers for any heat carrier flow regimes and for initial thermohydraulic conditions Eqs. (1)–(4) of the comparison problem.

With the expression $F_c = V/L$ and identity transformation, (18) may be rearranged to give:

$$\frac{V_1}{V_2} = \frac{Nu_2/Nu_1}{\xi_2/\xi_1}\frac{\eta_2}{\eta_1}\frac{\Omega_2}{\Omega_1}\frac{L_1}{L_2}\left(\frac{F_{c2}}{F_{c1}}\frac{f_{f2}}{f_{f1}}\right)^2 \geq 1 \quad (21)$$

It is obvious that Eq. (21) characterizes the volume ratio of compared any types of-heat exchanger designs (both one-type and different-types) when *necessary and sufficient comparison conditions* of Eq. (18) are satisfied for initial thermohydraulic conditions of Eqs. (1)–(4) of the comparison problem. It is possible different particular cases of the comparison problem which described by different analytical forms of Eq. (21). They may be obtained, assuming that the values of ratio of any of the accepted main geometrical parameters of compared heat transfer surfaces and heat exchangers is equal to unity.

As Eq. (21) contains only one independent F_{c2} and three dependent variables - Nu_2, ξ_2 and L_2, its use for the comparison problem of heat exchanger volumes in such a form is possible when the successive approximation method is adopted. Consider its sequence.

1. A design of a compared heat transfer surface, a design type of a heat exchanger core, and its design front F_{c2}, are chosen.

2. Values of design geometrical parameters, Ω_2 and f_{f2}, of a compared heat transfer surface in a heat exchanger design are determined.

3. Re_2 is determined to according to Eq. (6)

$$Re_2 = Re_1 \frac{F_{c1}}{F_{c2}}\frac{f_{f1}}{f_{f2}}\frac{d_2}{d_1}$$

4. Values of Nu_2 and ξ_2 are determined through the value of Re_2 found in Item 3 and from the known criterial thermohydraulic relations $Nu_2 = f(Re_2)$ and $\xi_2 = f(Re_2)$.

5. A value of the coefficient α_2 in a compared heat transfer surface is determined as: $\alpha_2 = Nu_2 \lambda_2 / d_2$.

6. A total thermal efficiency coefficient of a compared heat transfer surface is determined

$$\eta_2 = 1 - F_{f2}^*/F_{h2}(1-\eta_{f2}) = $$
$$= 1 - P_{f1}/P_1(1-\eta_{f2}) = 1 - K_f(1-\eta_{f2})$$

where P_{f1}, P_1 and K_f - are the assigned design parameters, and

$$\eta_{f2} = \frac{th(b_2\sqrt{2\alpha_2/\lambda_{f2}\delta_2})}{b_2\sqrt{2\alpha_2/\lambda_{f2}\delta_2}}$$

[†] *Values of the volume ratios, V_1/V_2, of compared heat exchangers in the expressions that control the fulfillment of criterial necessary and sufficient comparison conditions of heat exchangers are equal by definition, because they are determined only by necessary criterial relation Eq.(9) or Eq. (11) for initial thermohydraulic conditions Eqs.(1), (2) and (4) of the comparison problem.*

7. A value of any of Eqs. (18)–(20) is calculated and compared with unity. In this case, a calculated value is not equal to unity, a new corresponding value of F_{c2} is predetermined unless a calculated value Eqs. (18)–(20) will be equal to unity.

8. F_{h2} is determined according to Eq. (10)

$$F_{h2} = F_{h1} \frac{Nu_1}{Nu} \frac{\eta_1}{\eta_2} \frac{d_2}{d_1}$$

9. A volume, V_2, of a compared heat exchanger is determined: $V_2 = F_{h2}\Omega_2$ and, if necessary, $L_2 = V_2/F_{c2}$.

10. With Eq. (21), a change in a volume of a compared heat exchanger is analyzed when a chosen heat transfer surface is used in it. If $V_1/V_2 \leq 1$, then it is impossible to decrease a volume of a compared design of a heat exchanger for initial thermohydraulic conditions of Eqs. (1)–(4). Because of this, it is necessary to choose a new design of a heat transfer surface and if it is necessary change geometrical parameters of core for a compared heat exchanger.

Illustrate the sequence of using the considered method by the example of comparing thermohydraulic efficiencies of air-to-water heat exchangers having louver "1" and interrupted "2" plate-fin heat transfer surfaces. The experimental criterial thermohydraulic relations (Fig. 1) and designs of heat transfer surfaces are borrowed from the work of Dubrovsky et al. (1972). The compared designs of heat exchangers had the same assembly diagrams of their cores, air design fronts ($F_{c1} = F_{c2} = 0.4434$ m^2) as well as the same height and length ($L_1 = L_2 = 156$ mm) of corrugated channels, and the same number of corrugated plates over the air cavity of a heat exchanger. But thicknesses ($\delta_1 = 0.15$ mm; $\delta_2 = 0.10$ mm) and pitches ($t_{f1} = 4.6$ mm; $t_{f2} = 4.44$ mm) of corrugated channel fins, equivalent hydraulic diameters ($d_1 = 3.25$ mm; $d_2 = 2.87$ mm), coefficients for the free area of air heat transfer surfaces in the heat exchanger design ($f_{f1} = 0.5943$; $f_{f2} = 0.5713$), and compactness coefficients for air heat transfer surfaces in the designs of compared heat exchangers ($\Omega_1 = 642$ m^2/m^3, $\Omega_2 = 724$ m^2/m^3) were different. As a result, the thermal efficiency coefficients of heat exchange surfaces were not equal ($\eta_1 \neq \eta_2$). Therefore, different values of Re_1 and Re_2 (fig.1) with their corresponding values of ξ_1, ξ_2, Nu_1, Nu_2 and $Nu_1\eta_1$, $Nu_2\eta_2$ conform to the same values of the heat load - Q of heat exchangers and air flow rates - G_{air}, in them. For illustration and convenience of comparing thermohydraulic efficiencies of air-to-water heat exchangers, at Fig. 2 presented plots their thermohydraulic relations from the Reynolds number (Re_{col}) for air supply collectors of heat exchangers. Therefore the thermohydraulic efficiency of both heat exchangers can be easy determine and compare by identical value of Reynolds number - Re_{col}.

In the analyzed example, according to Eq. (20) the criterial relation that controls the fulfillment of necessary and sufficient comparison conditions of heat exchanger volumes assumes the form:

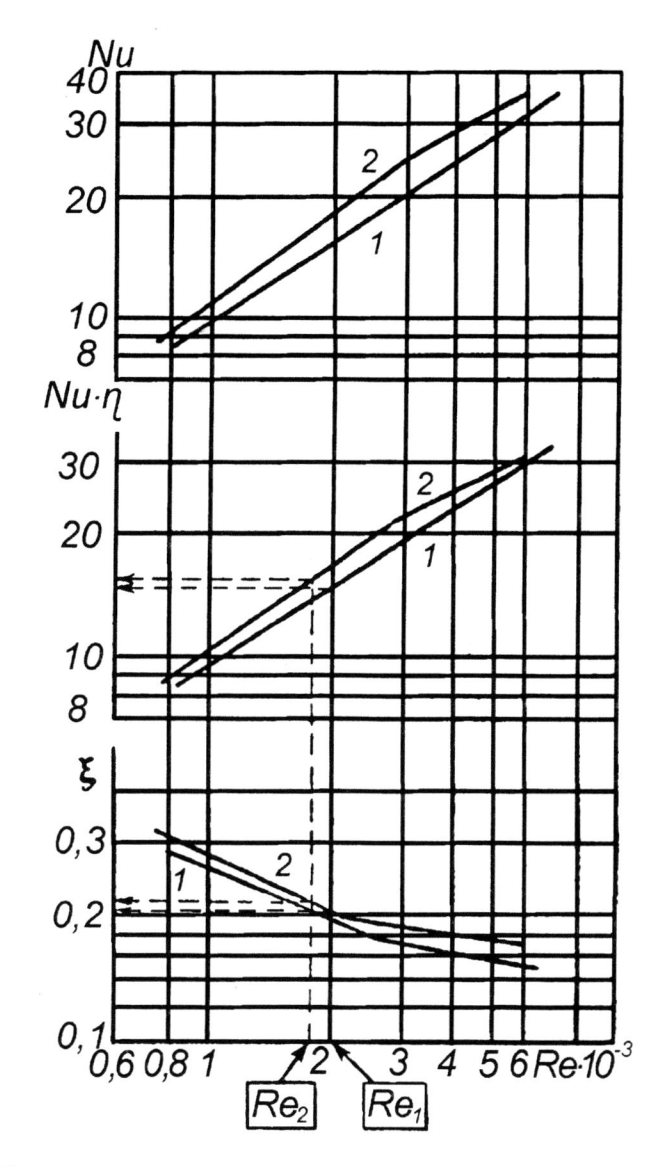

Fig.1 Relations Nu=f(Re), ξ=f(Re), and Nu*η=f(Re) for interrupted (1) and louver (2) heat transfer surfaces.

$$\frac{\left(Nu_2\eta_2 / \xi_2\right)_{Re2}}{\left(Nu_1\eta_1 / \xi_1\right)_{Re1}} \frac{\Omega_2}{\Omega_1} \left(\frac{Re_1}{Re_2} \frac{d_2}{d_1}\right)^2 \geq 1 \quad \text{(I)}$$

According to Eq. (12), the comparison of heat exchangers volumes, when the necessary comparison condition is satisfied, is written as follows:

$$\frac{V_1}{V_2} = \frac{(Nu_2\eta_2)_{Re2}}{(Nu_1\eta_1)_{Re1}} \frac{\Omega_2}{\Omega_1} \frac{d_1}{d_2} \quad \text{(II)}$$

According to Eq. (15), the comparison of heat exchanger volumes, when the sufficient comparison condition is satisfied, takes the form:

$$\frac{V_1}{V_2} \geq \frac{(\xi_2)_{Re2}}{(\xi_1)_{Re1}} \left(\frac{f_{f1}}{f_{f2}}\right)^2 \frac{d_1}{d_2} \quad \text{(III)}$$

Figure 2 plots the volumes of compared heat exchanges for necessary (II) and sufficient (III) comparison conditions and for generalized criterial condition Eq. (1) that controls the fulfillment of initial thermohydraulic conditions of Eqs. (1)–(4) of the comparison problem of heat exchangers from the Reynolds number Re_{col}. Analysis of these relations in Fig. 2 shows that over the entire range of $Re_{col}=(3-20) \times 10^4$, the favourable mechanism of heat transfer enhancement process in a heat exchanger having an interrupted heat transfer surface is realized and a heat exchanger volume is decreased $1.37 \div 1.45$ time when air pressure losses are equal or decreased up to 11% as against a heat exchanger using a louver heat transfer surface.

Further, consider a derivation of one important particular case of the problem in comparing heat exchanger volumes. This case allow as a check of modernization to substitute conventional heat exchanger designs in different engineering objects by more efficient designs that utilize heat transfer enhancement process. In this case, its derivation must employ the additional design boundary conditions that reflect certain requirements for designs of heat transfer surfaces and heat exchangers.

According to the objective stated, the expression $F_c = V/L$ is used to rearrange Eq. (18). Moreover, the parameter with the subscript "1" are replaced by those with the subscript "sm" (parameters of smooth channels of a heat transfer surface), and the parameters with the subscript "2" are written without a subscript (parameters of turbulator-equipped channels of heat transfer surface). Then

$$\frac{Nu/Nu_{sm}}{\xi/\xi_{sm1}} \frac{\eta}{\eta_{sm1}} \frac{\Omega}{\Omega_{sm}} \frac{V}{V_{sm}} \frac{L_{sm}}{L} \left(\frac{Re_{sm}}{Re} \frac{d}{d_{sm}}\right)^2 \geq 1 \quad (22)$$

Proceeding from the theory and practice of heat transfer enhancement process, the thermohydraulic efficiency of heat transfer enhancement process in turbulator-equipped channels of heat transfer surfaces (parameters without subscripts) must be compared with that in smooth channels (without turbulators and having the same geometrical parameters) of heat transfer surfaces (parameters with the subscript "sm") and under the same values of heat carrier flow. The result of these requirements is the identity of the assembly diagrams of cores of compared designs of heat exchangers and their design fronts.

The following design and operating boundary conditions obey the mentioned requirements in the comparison problem of heat exchanger volumes.

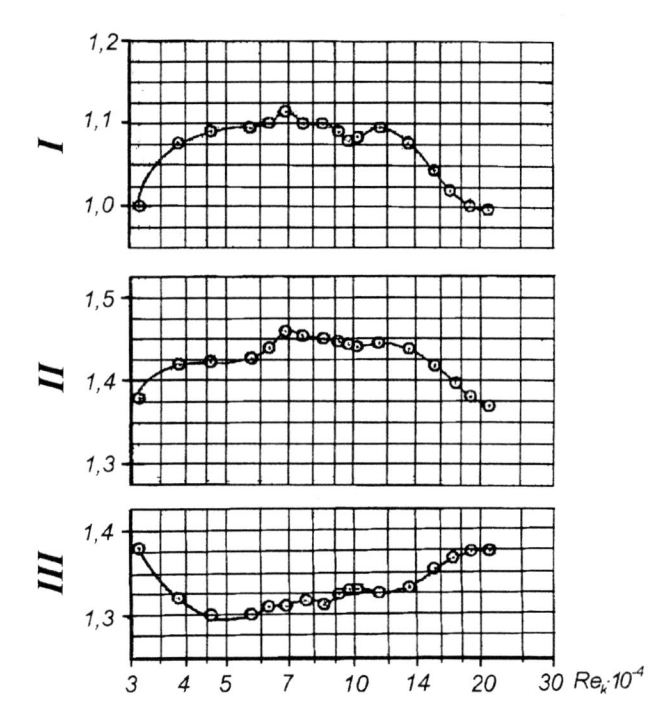

Fig.2. Volumes of compared heat exchangers for necessary (II) and sufficient (III) comparison conditions and for generalized criterial condition (I) that controls the fulfillment of thermohydraulic boundary comparison conditions for heat exchangers from the Reynolds number Re_{col} for collectors that supply air to heat exchangers.

$$Re = Re_{sm}; \quad \Omega = \Omega_{blb}; \quad d = d_{sm}; \quad F_c = F_{sm} \quad (23)$$

In this case, it should be noted that one of the necessary design boundary conditions $L = L_{sm}$, that corresponds to the theoretical requirements of comparing the thermohydraulic efficiency of heat transfer enhancement process, cannot be accepted in the considered statement of the comparison problem of heat exchanger volumes. Otherwise, there has to be a change in thermohydraulic conditions Eqs. (1), (3) and (4) in the comparison problem of heat exchangers volumes.

Then with Eq. (23), Eq. (22) assumes the form:

$$\frac{Nu/Nu_{sm}}{\xi/\xi_{sm}} \frac{\eta}{\eta_{sm}} \frac{V}{V_{sm}} \frac{L_{sm}}{L} \geq 1 \quad (24)$$

Then from Eq. (24) we have:

$$\frac{V_{sm}}{V} = \frac{Nu/Nu_{sm}}{\xi/\xi_{sm}} \frac{\eta}{\eta_{sm}} \frac{L_{sm}}{L} \geq 1 \quad (25)$$

Equation (25) allows heat exchanger volumes to be compared when the necessary and sufficient comparison conditions are implemented for thermohydraulic Eqs.(1)-(4) as well as for design and operating Eq. (23) boundary conditions. In this case, analysis of Eq. (25) shows that a heat exchanger volume under heat transfer enhancement process decreases only with decreasing length of turbulator-equipped channels of a heat transfer surface. At the same time, the design front of a heat exchanger and its overall and additional sizes remain invariable. Therefore, replacing heat exchanger constructions in a effort to modernize the latter, e.g., on various transportation facilities, requires no re-arrangement of transportation, and manufacturing the modernized heat exchangers requires no modification of the existing production process.

Now consider the problem on estimating and comparing the thermohydraulic efficiency of heat transfer enhancement process in turbulator-equipped channels of heat transfer surfaces in relation to channels (of the same geometry and having no turbulators) of heat transfer surfaces under the same conditions of heat carrier flow in them.

The above-mentioned analysis of the comparison problem of heat exchanger volumes in the considered statement does not permit the stated problem to be solved. Up to now, the formulation of the analytical closed problem and corresponding thermohydraulic initial comparison conditions for comparison of the thermohydraulic efficiency of heat transfer enhancement process in channels of heat transfer surfaces has not met with success. However, the fundamental points of the theory and practice of heat transfer enhancement process allow this problem to be solved on a level with quite reliable assumptions.

The statement of the comparison problem of the thermohydraulic efficiency of heat transfer enhancement process as well the above explanations and proofs of design boundary conditions Eq. (23) enable formulating only the following design and operating boundary conditions for the problem under consideration:

$$Re = Re_{sm}; \ \Omega = \Omega_{sm}; \ d = d_{sm}; \ F_c = F_{csm}; \ F_h = F_{hsm}; \ L = L_{sm}; \quad (26)$$

Thus, the thermohydraulic efficiencies of heat transfer enhancement process should be compared in the same (with respect to the main design parameters) designs of heat exchangers and heat transfer surfaces. In this case, the criterial thermohydraulic relations $Nu=f(Re)$, $Nu_{sm}=f(Re_{sm})$, $\xi=F(Re)$, and $\xi=F(Re_{sm})$ are known for the compared heat transfer surfaces.

Bearing this in mind, consider some fundamental points of the theory and practice of heat transfer enhancement process.

Heat transfer enhancement process is always characterized by the following values of the ratios:

$$\frac{Nu}{Nu_{sm}} = A_1 > 1 \quad (27)$$

and

$$\frac{\xi}{\xi_{sm}} = A_2 > 1 \quad (28)$$

In this situation, two cases of heat transfer enhancement process may be true:

$$\frac{Nu}{Nu_{sm}} = A_1 \geq \frac{\xi}{\xi_{sm}} = A_2 \quad (29)$$

and

$$\frac{Nu}{Nu_{sm}} = A_1 \leq \frac{\xi}{\xi_{sm}} = A_2 \quad (30)$$

The theory and practice of heat transfer enhancement process analyzes Eq. (29) as the most effective one. Therefore, analysis of Eq. (30) is not discussed in the present work.

Write case Eq. (29) in the following form:

$$\frac{\frac{Nu}{Nu_{sm}}}{\frac{\xi}{\xi_{sm}}} \geq 1 \quad (31)$$

Thus, Eq. (31) is the *criterial condition that controls the fulfillment of the necessary and sufficient comparison conditions* for the thermohydraulic efficiency of heat transfer enhancement process in channels of compared heat transfer surfaces under design boundary conditions of Eq. (26). In the case, Eq. (31) does not depend on the design parameters of compared heat transfer surfaces. Therefore, (31) provides unambiguous (without iterative calculation) comparison of the thermohydraulic efficiency of heat transfer enhancement process in channels with turbulators and in smooth channels (without turbulators) and having the same geometrical parameters) of heat transfer surfaces of any-type designs and under the same conditions of heat carrier flow in them.

Simultaneous analysis of Eqs. (27), (28) and (31) shows that Eq. (27) is the necessary condition and Eq. (28) is the sufficient condition used to compare the thermohydraulic efficiency of heat transfer enhancement process when design and operating boundary conditions of Eq. (26) are satisfied.

The process of heat transfer enhancement in channels that obeys Eq. (31) is qualitatively new and has been named by us the *process of rational heat transfer enhancement*.

RESULTS

By a process of rational heat transfer enhancement is understood such a process where in different-cross-section channels with turbulators a growth of heat transfer due to artificial turbulence of a heat carrier flow is ahead or is equal to that of hydraulic losses as against smooth channels having the same geometrical parameters under the same conditions of heat carrier flow in them. In this case, its implementation is characterized by Reynolds' analogy violation.

Reynolds analogy establishes a relationship between heat transfer and hydraulic resistance for non-separated flows in smooth channels without turbulators. In this case, hydraulic losses in smooth channels are characterized only by friction losses, while the limiting theoretical case, by an equal growth of heat transfer and energy flux expenditures only for friction.

Convective heat transfer enhancement process in channels with turbulators is always marked by separated flow of heat carriers. As pioneered by Kalinin (1962), friction pressure losses in such channels account for up to 10-15% of total pressure losses. The remaining pressure losses are stipulated by energy fluxes spent to form vortex systems and their dissipation. So, Reynolds' analogy cannot be applied to the conditions for separated flows of heat carrier that cause heat transfer intensification process in channels.

In theoretical and experimental hydrodynamics two ways are known how to generate vortices in a wall layer of flows in different-cross-section channels:

- at flow past fluff bodies at the walls of canals having geometrical parameters commensurable with a wall layer thickness;
- over the diffuser sections of channel flows given at determined values of a diffuser expansion angle and regime of heat carrier flow.

Just these two ways to generate vortices in a wall layer of channel flow were used for experimental confirmation and comprehensive systematic experimental research on the process of rational heat transfer enhancement according to criterial condition of Eq. (31). To do this, three batches of three type sizes of designs of plate-fin heat transfer surfaces were engineered and tested experimentally:

- interrupted heat transfer surfaces formed by shifted half the fin pitch rows of short corrugated channels, whose cross-section is an equilateral triangle [Dubrovsky (1995), Voronin and Dubrovsky (1973)] or a flat rectangle [Dubrovsky and Vasiliev (1988), Dubrovsky (1995)];
- heat transfer surfaces for periodic throttling of heat carrier flow in channels of isosceles triangle cross-section and with smoothly rounded off transverse protrusions and grooves located opposite to each other on channel fins [Dubrovsky (1978), Dubrovsky (1995), Voronin and Dubrovsky (1978)].

Figure 3 plots the criterial thermohydraulic relations Nu = f(Re) and ξ = f(Re) taken from the above-quoted works. Figure 4 shows their corresponding relations Nu/Nu$_{sm}$ = f(Re), ξ/ξ_{sm} = f(Re) and (Nu/Nu$_{sm}$)/(ξ/ξ_{sm})=f(Re) for the above-discussed designs of heat transfer surfaces. Analysis of these relations reliably supports the realization of a new process of rational heat transfer enhancement in the channels of the surfaces under study. This process is qualitatively and quantitatively controlled by a change in dimensionless determining geometrical parameters of turbulators in channels of heat transfer surface designs.

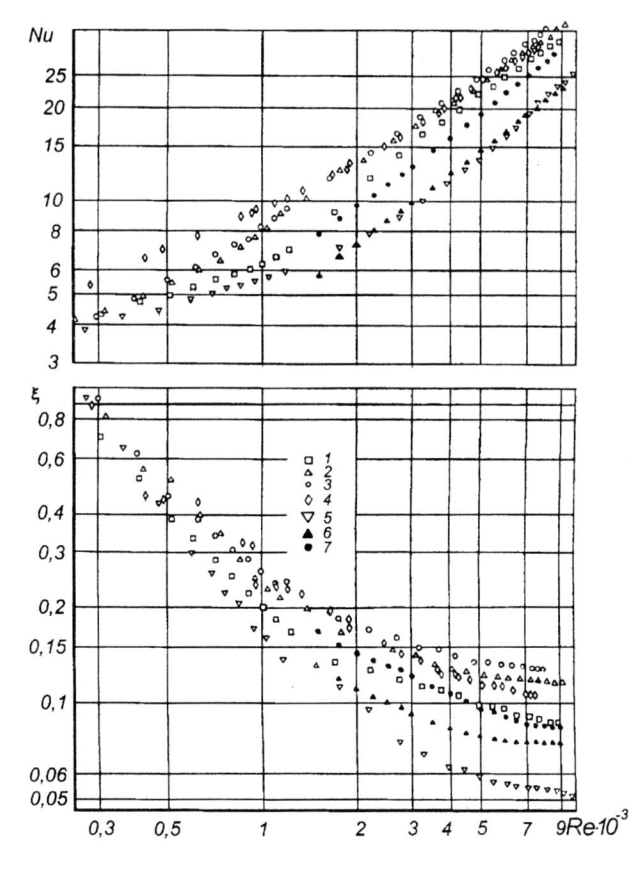

Fig. 3 Relations Nu = f(Re) and ξ = f(Re) for heat transfer surfaces:
(a) aluminium: d=3.3 mm, h=7 mm, t=4.6 mm, δ=0.15mm, L/α=23.6.
 1, throttled (d*/d=0.912, l¹/d=0.24);
 2, throttled (d*/d=0.833, l¹/d=1.52);
 3, throttled (d*/d=0.797, l¹/d=3.03);
 4, louver (l¹/d=2.91, δ/d=0.045);
 5, smooth channel surface (l¹/d=23.
(b) copper: d=3.48 mm, t=5.26 mm, h=5.56 mm, δ=0.1mm, L/d=20.1.
 6, smooth channel surface (L/d=20.1);
 7, interrupted (l'/d=1.68, δ/d=0.029)

CONCLUSIONS

1. A reliably and rapidly realizable generalized method of comparing volumes of different heat exchangers designs operating at K≈α_{min} is developed. It is founded from iterative calculations of generalized criterial condition of Eqs. (18), (19) or (20) that controls the implementation of necessary (9) or (11) and sufficient (13) comparison conditions for initial thermohydraulic conditions of Eqs. (1)-(4) of the comparison problem of heat exchanger volumes. In result of calculation of generalized criterial condition of Eq. (18),

(19) or (20) determines the corresponding values of the design and operating parameters of compared heat transfer surfaces and heat exchangers. They allow unambiguous (without iterative calculation) determination of a value of the volume ratio of compared any-design heat exchangers from Eq. (21).

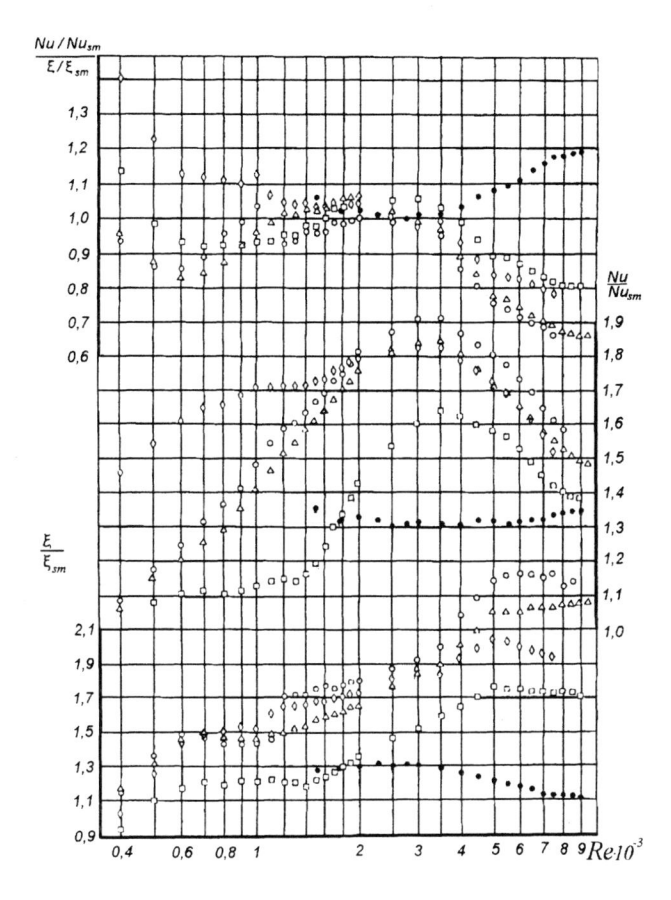

Fig.4 Relations Nu/Nu$_{sm}$ = f(Re), ξ/ξ_{sm} = f(Re) and (Nu/Nu$_{sm}$)/(ξ/ξ_{sm}) = f(Re) for heat transfer surfaces. The notations are the same as in Fig.3.

2. Analysis of the fundamental points of the theory and practice of heat transfer enhancement process has made it possible to develop a method for comparing the thermohydraulic efficiency of heat transfer enhancement process realized in turbulator-equipped channels of heat transfer surfaces of any-type of constructions. With criterial condition of Eq. (31) that control implementation of necessary [Eq. (27)] and sufficient [Eq. (28)] comparison conditions, this method provides unambiguous (without iterations) estimation and comparison of the thermohydraulic efficiency of heat transfer enhancement process implemented in turbulator-equipped channels of heat transfer surfaces of any-type designs.

NOMENCLATURE

b	fin length of a plate-fin heat transfer surface from the fin bottom to its middle; fin length of a finned tube, m
d	equivalent hydraulic diameter of a smooth section of a channel, (d=4F$_1$/P$_1$), mm
d*	equivalent hydraulic channel diameter in its most narrow section, (d*=4F*$_1$/P*$_1$), mm
d*/d	relative narrowing of a channel in its most narrow section
F$_c$	total design front of a heat exchanger over one of its cavities, m^2
F$_f$	free area of a total design front of a heat exchanger core over one cavity, F$_f$ = F$_c$f$_f$, m^2
F*$_f$	heat transfer surface falling on heat transfer surface fins, m^2
F$_h$	total heat transfer surface of a heat exchanger over one of its cavities, m^2
F$_1$	minimum cross-sectional area of one channel over its smooth section, mm^2
F*$_1$	minimum free area of one channel in its most narrow section, mm^2
f$_f$	coefficient for the free area of a heat exchanger over one of its cavities, (f$_f$ = F$_f$/F$_c$)
G	heat carrier flow rate, kg/s
h	height of one channel of a plate-fin heat transfer surface, mm
K	heat transfer coefficient of a heat exchanger, W/m^{2}K
K$_f$	finning coefficient of a heat transfer surface
L	length of heat transfer surface channel, mm
l^1	length of a short channel of an interrupted surface, mm
l*	length of a smooth channel section between the adjacent transverse protrusions and grooves
l'/d	relative length of a short channels (interruption parameter) of an interrupted surface
l*/d	relative length of a smooth channel section between the adjacent transverse protrusions and grooves
Nu	Nusselt number, Nu = αd/λ
P$_1$	wetted perimeter of a section of one smooth channel, mm
P*$_1$	wetted perimeter of the most narrow section of one channel in the section between the opposite transverse protrusions, mm
P$_{f1}$	wetted perimeter of a smooth section of a channel, mm
Q	thermal output of a heat exchanger, W
Re	Reynolds number, Re = wd/ν
t	temperature, ^{0}C
V	volume of a heat exchanger core, m^3
w	heat carrier velocity on smooth channel sections, m/s
α	heat transfer coefficient of a heat transfer surface in a heat exchanger design, (α = uλ/d), W/m^{2}K
Δp	pressure losses of heat carrier flow over one cavity of a heat exchanger (including inlet, outlet and

	friction losses in a heat transfer surface), N/m²
Δt	temperature difference in a heat exchanger, °C
δ	fin thickness, mm
δ/d	relative fin thickness
λ	thermal conductivity, W/mK
λ_f	thermal conductivity of fin material of a heat transfer surface, W/mK
ν	kinematic viscosity coefficient, m²/S
η	thermal efficiency coefficient of a heat transfer surface in a heat exchanger, $\eta = 1-F_f/F_h(1-\eta_f)$
η_f	thermal efficiency coefficient of a heat transfer surface falling on its finning
ξ	coefficient of total pressure losses of heat carrier flow for friction, inlet and outlet of a heat transfer surface in a heat exchanger design
Ω	compactness coefficient of a heat transfer surface over one cavity of a heat exchanger design, m²/m³

Subscripts

1 -	parameters of compared reference of a shape heat exchanger and heat transfer surface
2 -	parameters of compared heat exchanger and heat transfer surface
sm	parameters of compared reference of a shape smooth channel heat transfer surface

REFERENCES

Antufiev V.M., 1966, Efficiency of Different-Shape Convective Heating Surfaces, Izd. Energiya, Moscow

Cox B., and Jallouk P.A., 1973, Method of Evaluation of the Performances of Compact Heat Transfer Surface, *Trans. ASME*, Ser. C, No. 4, pp

Dzeitser G.A., Kuzminov V. A., and Neverov A. A., 1973, The simplest methods for evaluation of heat transfer enhancement in channels, Izd. VUZov, "Energetika", No. 12, pp.77-84.

Dreitser G.A., Kuzminov V.A., and Neverov A.S., 1974, Evaluation of the heat transfer enhancement due to flow swirling in tube heat exchangers, Trudy VZMI, vyp.3, Gidravlika, Vol. 10, Izd, VZMI, Moscow, pp.102-114.

Dreitser G.A., 1999., Method of evaluating the heat transfer enhancement efficiency in heat exchangers, *Izv. VUZov*, "Mashinostroyeniye", Nos 5-6, pp.67-76.

Dubrovsky E.V. and Fedotova A.I., 1972, Study of plate-fin heat transfer surfaces, Kholodilnaya Tekhnika, No. 12, pp.31-33.

Dubrovsky E.V., 1977, A method for comparison of the thermohydraulic efficiency of heat transfer surfaces, Izv. AN SSSR "Energetika i Transport", No.6, pp.118-128.

Dubrovsky E.V., 1978, Convective heat transfer enhancement in plate-fin heat transfer surfaces, Izv. AN SSSR "Energetika i Transport", No.6, pp.116-127.

Dubrovsky E.V., and Vasilev V. Ya., 1988, Enhancement of convective heat transfer in rectangular ducts of interrupted surfaces, Int. J. Heat Mass Transfer, Vol. 31, No.4, pp.807-818.

Dubrovsky E.V., 1995, Experimental investigation of highly effective plate-fin heat exchanger surfaces, J. Experimental Thermal and Fluid Science, Vol.2, No.2, pp.200-220.

Glaser H., 1948. Bewertung von Warmeaustausch System mit Hielfe einer Leistungszahle, Angew. Chem., Bd. 20, Nr.5/6

Gryaznov D.N., 1974, Methodical Instructions for Course Designing "Heat Exchangers", Izd. Bauman Moscow Technical College.

Gukhman A.A., 1938, Methods for comparing convective heating surfaces, Tekhnicheskaya Fizika, Vol. 8, vyp. 17, pp.19-26.

Kalinin E.K., Dreitser G.A. and Yarkhko S.A., 1972, Enhancement of Heat Transfer in Channels, Izd. Mashinostroyeniye, Moscow.

Kays W.M., and London A.L., 1984, Compact Heat Exchangers, 3rd ed., McGraw-Hill, New York.

Mikhailov A.A., Borisov V.V., and Kalinin E.K., 1962, Closed-Cycle Gas Turbine Plants, Izd. Mashinostroyeniye, Moscow.

Mitin B.M., 1969, Calculation of optimal heat exchangers of heat generation system of gas turbine engines, Trudy TsIAM, Heat Exchange Apparatuses of Gas Turbine Engines, No.463, Izd. TsIAM, pp1-27.

Mitskevich A.I., 1967, A method of evaluation of convective heat transfer efficiency. In "Heat Transfer and Hydraulics in Elements of Steam Generators and Heat Exchangers, Trudy TsKTI, No.78, Izd. TsKTI, Moscow, pp.9-13

Novikov M.D., 1967, Calculation of Optimal Parameters of Heat Exchangers of Gas Turbine Plants, Izd. Energiya, Moscow.

Pertovsky Yu. V., and Fastovsky V. G., 1962, Modern Efficiency Heat Exchangers, Gosenergoizdat, Moscow.

Voronin G.I., and Dubrovsky E.V., 1973, Efficient Heat Exchangers, Izd. Mashinostroyeniye, Moscow.

Voronin G.I., and Dubrovsky E.V., 1978, Highly effective heat exchanger surfaces, Proc. 14 Int. Conf. Refrigeration "Progress in Refrigeration Science and Technology", Vneshtorgizdat, Moscow, Vol. IV, pp.763-777.

Walger O., 1952, Der Wert von Wirbeleinbauten zur Steigerung des Warmeuberganges, Allg. Warmetechn, Bd.3, Nr.7/8.

Walger O., 1952, Grundlagen einer wirtschaftlichen Gestahlung vor Warmeubertragen, Chemich.-Ind.-Techn., Bd. 24, Nr.3.

Webb R.L., and Eckerd R.J., 1973, Application of rough surfaces to heat exchanger design, J. Heat Mass Transfer, Vol. 15.

NUMERICAL INVESTIGATION OF PLATE HEAT EXCHANGER DESIGN

H. Rebholz, W. Heidemann, E. Hahne, H. Müller-Steinhagen

Institute for Thermodynamics and Thermal Engineering, University of Stuttgart,
Pfaffenwaldring 6, D-70550 Stuttgart, Germany, pm@itw.uni-stuttgart.de

ABSTRACT

A 2D finite volume method is presented for the prediction of laminar flow in plate heat exchangers. The grid generation is completely automatic, and includes the settings for the boundary conditions which are dependent on the type of arrangement. The Navier-Stokes-Equations are used to compute the veloctiy and temperature fields, while Conjugate Heat Transfer is used to predict the heat transfer to or from the walls.

Validations with benchmark solutions agree well with our approach. Solutions are presented for the end effect of plate heat exchangers, for changing flow conditions in multipass arrangements and several different configurations. The results were compared to literature data.

INTRODUCTION

Plate Heat Exchangers (PHEs) are used in a wide range of applications within the fields of engineering. Although PHEs have several advantages in comparison to other types of heat exchangers, difficulty exists in determining the best flow configuration, as well as understanding the thermal behaviour of the selected PHE arrangement.

The prediction of the temperature field in PHE design is very important for classifying the arrangement. It must be decided what kind of arrangement for a given mass flow guarantees the maximum heat duty under consideration of the pressure drop. Parameters are the total number of channels for each fluid and the number of passes for each fluid.

Many PHE arrangements are poorly designed with little effort to guarantee thermal efficiency and low cost. To ensure the heat duty, PHEs are often oversized and cost more than arrangements with the same heat duty and fewer plates. Computational Fluid Dynamics (CFD) is a most effective tool for achieving a more detailed design process.

Design methods were published in the 1950s for a wide range of operating conditions, without using a uniform description. This makes it difficult to compare these methods. Bosnjakovic et. al. (1951) began to avoid these problems by defining the characteristic Φ. This characteristic sets up a connection between the inlet and outlet temperatures of heat exchangers. Difficulties arise in determining this characteristic, because it varies for different arrangements. For common flow configurations, such as pure parallel flow, counter flow or cross flow it is published in the literature.

Buonopane, Troupe and Morgan (1963) published a design method for arrangements based on the conventional log-mean temperature difference equation by using correction factors. This method has been applied to arrangements with several parallel channels and one pass or with several passes and one channel each. Domingos (1968) published a theory to compute the total effectiveness and intermediate temperatures for assemblies of heat exchangers. This leads to a method for calculating complex heat exchangers by subdividing them into smaller units of overall counterflow, overall parallel-flow or a combination of parallel and counterflow.

With the introduction of the NTU-number another two ways for calculating PHEs were developed. The first one is called the cell method by Gaddis and Schlünder (1975). The heat exchanger is divided into smaller regions, with the regions having known NTU-values. This is a common way to determine the temperatures in complex heat exchanger assemblies. The second way uses the NTU-Diagrams by Roetzel und Spang (1990). Using these diagrams, it is possible to determine the temperature when the geometry of the arrangement is known. The problem occurs, that for every type of arrangement another diagram has to be applied. Nevertheless, the NTU-Diagrams are very helpful for the reverse problem, when for given outlet temperatures the heat transfer surface demand must be determined.

The influence of multipass plate heat exchangers was researched by Kandlikar and Shah (1989/1,1989/2) as well as by Pignotti and Tamborenea (1988). Kandlikar and Shah focused on the most common arrangements by writing separate programs, while Pignotti und Tamborenea used a numerical approach to describe all possible geometries.

An advanced solution technique for the application of the cell method to any kind of heat exchanger arrangement was developed by Strelow (1997). Using coupled matrix equation networks, heat exchangers can be calculated explicitly without an iterative method.

The simplifications of these models were mentioned by (G.F. Hewitt, et al, 1994).

- The use of a constant overall heat transfer coefficient is not accurate, particularly in laminar flow. The inflow effects and the development of the boundary layers play a specific role.
- End effects are neglected, or simply implemented through a correction factor. These become important for PHE arrangements of less than 40 plates.
- The heat conduction in the fluid and in the plate in direction of the flow cannot be taken into account.
- Uniform velocities and temperatures in the channels and plates are assumed, the profiles in the channels and the distribution in the plates are neglected.

All these calculations and their results depend on the accurate determination of the overall heat transfer coefficient. If this is not well prescribed then the calculation results will be incorrect. The overall heat transfer coefficient is calculated with correlations, which are obtained experimentally.

The approximation of the overall heat transfer coefficient can be avoided, if the fluid flow is also part of the calculation. This goal can be achieved by using the tool of Computational Fluid Dynamics. Due to the need for fast results, simplifications in geometry are necessary. Otherwise the time needed for grid generation would result in a very expensive outcome. Therefore a two-dimensional approach has been used for laminar fluid flow in this paper.

NUMERICAL APPROACH

Assumptions

The flow is calculated two-dimensionally with velocity components u and v along the x and y coordinates. The flow is incompressible, and all fluid properties are constant. Heating through viscous dissipation and compression work are neglected. No internal heat generation is present in the solid or in the fluid.

Mathematical Model

The conservation equations for mass, momentum and energy, Eqs.(1)-(3), are solved in their integral form. The solutions are stationary, so the unsteady term is neglected.

$$\frac{\partial w_i}{\partial x_i} = 0 \qquad (1)$$

$$\rho w_j \frac{\partial w_i}{\partial x_j} = -\frac{\partial p}{\partial x_i} + \mu \left(\frac{\partial^2 w_i}{\partial x_j^2} \right) \qquad (2)$$

$$\rho w_i c_p \frac{\partial T}{\partial x_i} = k \frac{\partial^2 T}{\partial x_i^2} \qquad (3)$$

In the solid the convective part is zero, and only the energy equation is solved.

Computational Procedure

The governing equations were solved numerically using the control-volume-based finite-volume method. PHE geometries consist of very long and small ducts. This causes the following two numerical problems. First the multigrid approach causes problems. The calculation is based on three grids with non-uniform refinement, which are calculated one after the other from the coarse to the fine. Secondly, a discretisation of higher order in the region of the solid-fluid interfaces is not advisable, rather a second order approach is recommended [Davis 1983]. For this paper the first order discretisation is sufficent, and to avoid oscillations the Deferred Correction approach is used (Khosla and Rubin 1974).

Because of the nonlinearity and the velocity-pressure coupling an iterative solution method is necessary. Therefore the SIMPLEC approach was used (van Doormal and Raithby 1984). To avoid numerical difficulties first the mass and momentum equations and then the energy equation are solved.

For the computational domain multi-block partitions are used; one block for each fluid domain and one block for each solid domain.

The conjugate heat transfer between the solid-fluid-interface is solved applying the method described by Patankar (1980). This uses the harmonic mean of the solid and fluid thermal conductivities and other properties to derive a correct expression of the heat flux.

Configurations

PHE arrangements are designed in different ways. The highest thermal effectiveness is achieved with pure counterflow arrangements. Therefore one should focus on the variety of those configurations. A PHE arrangement is described by several parameters. The fluid is distributed on several channels, called a pass. This is shown in fig.1. The

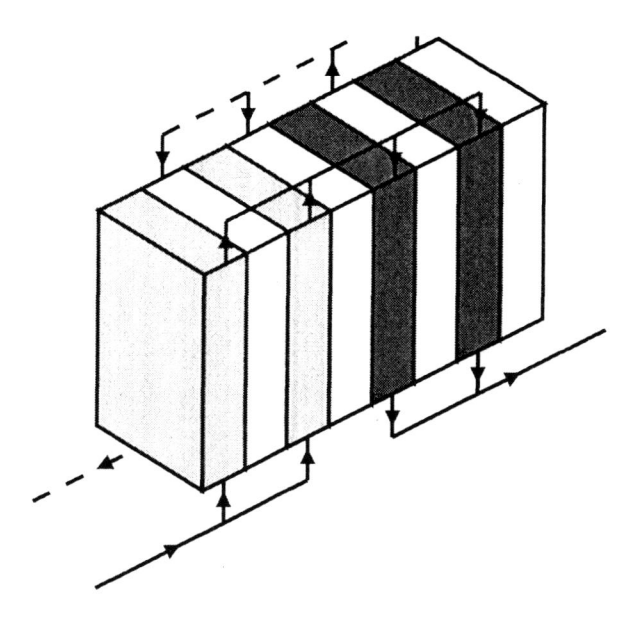

Fig.1a: Arrangement with 2 passes and 2 channels each

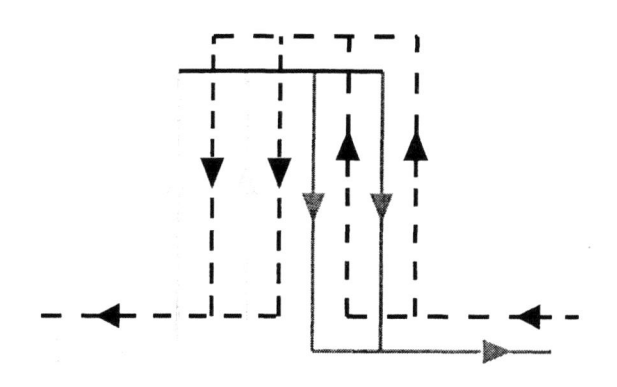

Fig.1b: 2 – 2 counterflow – counterflow arrangement

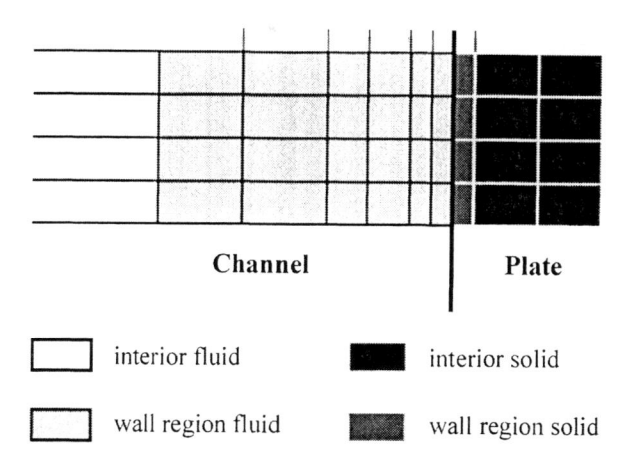

Fig.2b: Next finer grid at the Fluid-Solid Interface

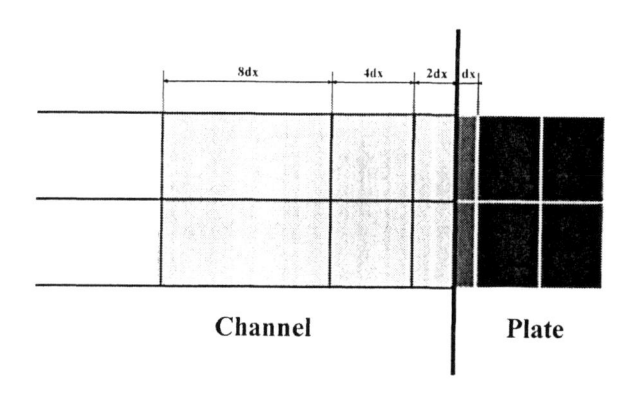

☐ interior fluid ■ interior solid
☐ wall region fluid ■ wall region solid

Fig.2a: Coarse Grid near Fluid-Solid Interface

two light grey channels represent the first pass. Between the passes adiabatic mixing occurs. The second pass is represented by the two dark grey channels. The channels for the other fluid are located alternately (white). The number of passes and the number of parallel channels for each fluid describe the size of the plate heat exchanger. For the flow configuration two more parameters must be named. First the flow configuration in the channels and secondly the flow configuration of the whole plate heat exchanger. The second one is just occuring in multipass arrangements. Fig.1b shows a 2 (pass) – 4 (parallel channels) counterflow (plate heat exchanger) – counterflow (channels) arrangement. The continuous line is comparable with the arrangement in fig. 1a. In all investigated examples, both fluids have the same number of passes and parallel channels.

Geometry and boundary conditions

The geometry and the boundary conditions are automatically linked to the input parameters. These are the number of passes and parallel channels for each fluid and the description for counterflow or parallel-flow in the channels and the whole PHE. For the geometry the input parameters are channel width, plate thickness and plate length. The algorithm determines the inflow and outflow channels as well as the connecting faces between two following passes.

Grid Generation

The grid is generated automatically. It is refined close to the fluid-solid interface to represent the correct heat flux. In order to avoid numerical difficulties and slower convergence, using too many control volumes in the solid should be avoided (fig.2a). The next finer grid is obtained by non-uniform refinement (fig.2b). In the fluid each control volume is subdivided into four finer ones. In the solid the number of control volumes perpendicular to the flow direction remains constant, therefore just two fine control volumes result.

Mesh Quality:

Resolution of the boundary layer is the main goal of mesh quality. In laminar flow the Blasius solution leads to the criterion

$$y_P \sqrt{\frac{w_\infty}{\nu x}} \leq 1. \qquad (4)$$

Here, y_P is the distance from the adjacent cell centre to the wall, while x denotes the distance along the wall from the starting point of the boundary layer.

The cell shape is, due to the geometry of the calculated flat plates, always rectangular and therefore ideal to calculate the equations with a minimum of numerical diffusion. As it follows from the geometry of the long and narrow channels the aspect ratio becomes important. The ratio is a measure for the stretching of the cells. It can exceed 5:1, but only in highly anisotropic flows.

Flow and temperature distribution for each pass

The mass flow is the same for each parallel channel. A great deal of research has been done for the manifold problem. However for plate heat exchangers the pressure drop in the channels is much higher than in the inflow- or outflow-header, respectively in-between passes. Therefore an equal distribution for the individual mass flow rates can be assumed. As a function of the pressure drop in the channels, the number of parallel channels leading to mal-distribution can exceed 100 (Kandlikar and Shah 1989/1).

The outlet temperature of every channel is calculated applying the adiabatic mixing rule of Baehr and Stephan (1994).

$$T_{out,channel} = \frac{1}{\dot{m}} \int_A \rho \cdot w \cdot T \cdot dA$$

$$= \frac{1}{\dot{m}} \sum_{n=1}^{N} \rho_n \cdot w_n \cdot T_n \cdot \Delta A_n \qquad (5)$$

The outlet temperature of a pass is determined through arithmetic averaging.

$$T_{out,pass} = \frac{1}{M} \sum_{m=1}^{M} T_m. \qquad (6)$$

Due to the end effect and parallel flow effect as discussed later in „Program Application", there occurs a non uniform temperature distribution at the outlets of each pass. For example the end channel results in a smaller temperature change, due to the one sided heating, than the channels in the interior of the same pass. Therefore adiabatic mixing in between the passes occurs. This mixing does not result in an equal temperature distribution for the next pass, especially for a few parallel channels and laminar conditions. Present investigations using a 3D-simulation show this phenomenon and the results are implemented for modelling the boundary conditions between the passes.

RESULTS

Validation

The numerical program was validated by several benchmark solutions. Kaminski and Prakash (1986) published investigations on conjugate heat transfer. The comparison of their results with the numerical values of our approach is shown here, demonstrating the good agreement for all validations. The interior domain consists of a fluid region and a solid region. The heat transfer at the interface is part of the calculation. Fig. 3 shows the temperature at the vertical interface between the fluid region and the wall. It

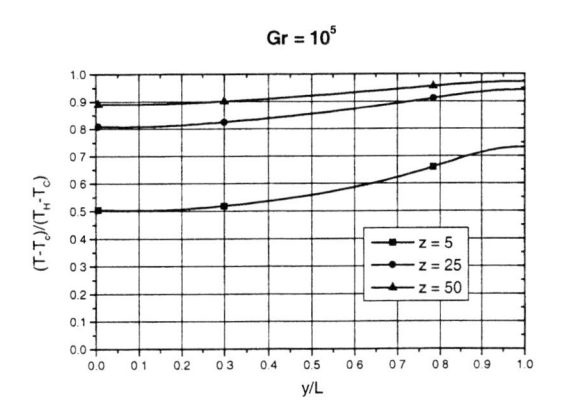

Fig.3: Temperature distribution along the solid-fluid interface

Table 1: Overall Nusselt number at the solid-fluid interface

Gr	z	Kaminski	present work
10^3	5	0.87	0.87
	25	1.02	1.02
	50	1.04	1.04
10^5	5	2.08	2.08
	25	3.42	3.41
	50	3.72	3.70
10^6	5	2.87	2.86
	25	5.89	5.87
	50	6.81	6.78
10^7	5	3.53	3.52
	25	9.08	9.09
	50	11.39	11.42

shows a good quantitative agreement with literature data (see Kaminski and Prakash (1986)). A comparison for the overall Nusselt numbers Nu at the interface as a function of the Grashof number Gr and a parameter z is shown in table 1. The parameter z is defined in eq.(7), with L as the width of the fluid region, t as the width of the solid region, k_W as the thermal conductivity of the wall and k as the thermal conductivity of the fluid.

$$z = \frac{k_W L}{k t} \qquad (7)$$

Program Application

Kandlikar and Shah (1989/2) mentioned that there are two effects in Plate Heat Exchangers leading to a decreasing heat duty between two fluids. The first one is called the end effect, which occurs in every PHE arrangement. The channels at both ends of the PHE differ from the interior channels due to adiabatic wall conditions at the ends (heat transfer losses neglected). The second one occurs within multipass PHE arrangements. The plates between two adjacent passes have different flow configurations as compared to the plates within the adjacent passes (see later fig.6). The influence of the two effects on the effectiveness is described in the following paragraphs.

Single pass arrangement: The end efffect occurs due to the fact that the channels at both ends of the plate heat exchanger are heated from one side only. The fluid experiences a smaller temperature change than in the other parallel channels. Consequently there exists a higher temperature difference to the other fluid than in the interior which leads to a higher temperature change in the adjacent channel.

Figure 4 shows these two antithetic phenomena. Fluid 1 has the minimum temperature change in channel 1, the maximum temperature change in channel 13, due to the end effect of Fluid 2. In the interior of the PHE the temperature change is approaching a constant value.

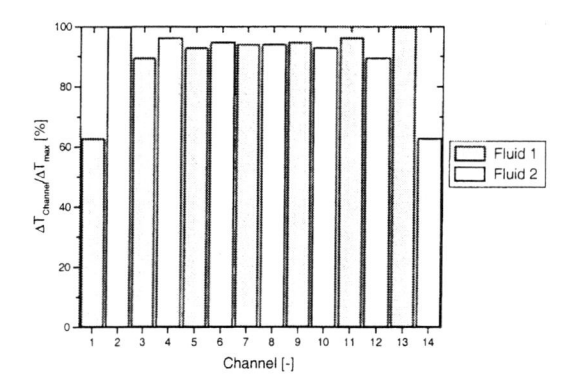

Fig. 4: Temperature change of each channel in a 1 – 7 counterflow arrangement

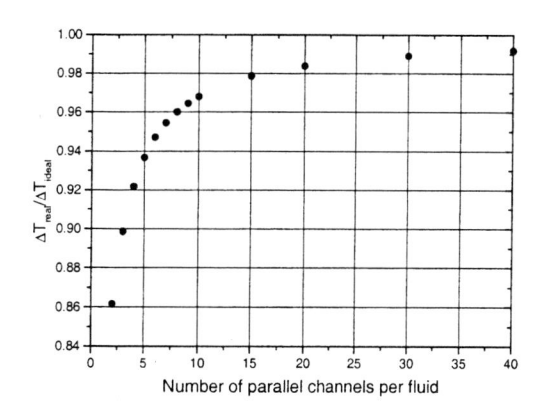

Fig. 5: End effect as a function of number of channels for a 1 pass counterflow in our calculations

With an increasing number of parallel channels, the end effect is becoming less significant. The ratio $\Delta T_{real}/\Delta T_{ideal}$ describes the reducing influence of the end effect. ΔT_{ideal} is the temperature change of the fluid which is achieved in the interior of the Plate Heat Exchanger.

Kandlikar and Shah found that the end effect becomes insignificant when the number of plates exceeds 40. This is compared to our calculations for a pure counterflow configuration and one pass. Fig. 5 shows, that with an increasing number of channels the end effect becomes less significant. For 20 channels of each fluid (equals 39 plates) the ratio $\Delta T_{real}/\Delta T_{ideal}$ reaches 0.984. A ratio of one would require a infinite number of plates. The statement of Kandlikar and Shah (1989/1) can be confirmed.

Multipass arrangements: In multipass arrangements there is another effect occuring besides the end effect. The use of more than one pass leads to channels with different flow configurations compared to the other channels. This effect is investigated for a two pass configuration and an overall counterflow arrangement of each stream in the channels and counterflow across the overall heat exchanger. The plates between these passes have parallel flow, see highlighted area in fig. 6.

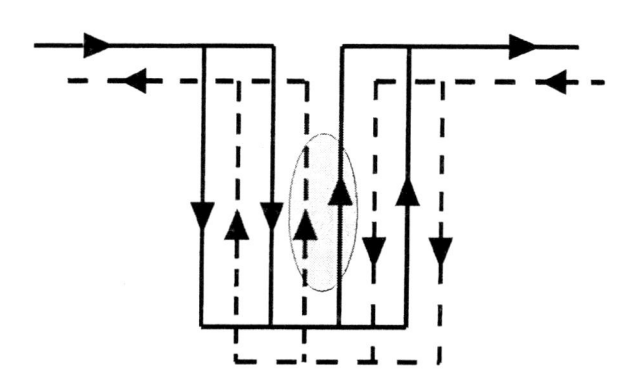

Fig.6: 2 – 2 counterflow – counterflow arrangement

The number of parallel channels is varied as for the single pass arrangement. Both effects – end effect and parallel flow effect between the adjacent passes – are occuring. In fig.7 the temperature change for each channel of Fluid 1 normalized by the maximum temperature change (channel 39), is shown (2 – 10 counterflow – counterflow).

The end effect appears in the first and last pass. A reducing effect on the heat duty is also occuring due to the parallel flow. The outline of the temperature change distribution of each pass is the same. For Fluid 1 there is one channel in each pass with a small temperature change (54.6% channel 1 – 53.4% channel 21) and on the other side of each pass a channel with a high temperature change (98.2% channel 19 – 100% channel 39).

By variation of the number of channels for each pass, the end effect and the parallel flow effect have less significance. This is shown in fig.8. One has to keep in mind that the number of channels for each fluid has to be multiplied by two (two passes), and the number of channels for the complete arrangement has to be multiplied by four (two passes, two fluids). The arrangement with 20 channels per pass shows a ratio of 0.98 and the two effects can be neglected. Such an arrangement consists of 79 plates.

Comparision of single pass and multipass arrangements

To find answers to the question of whether to design single or multipass arrangements, the result of several calculations will be shown. There are two contrary effects occuring, if the number of passes is increased. For the same mass flow and the same total number of channels, the velocity increases for multipass arrangements due to the reduced number of parallel channels. Therefore the heat transfer coefficient h increases. On the other hand due to the parallel flow in between passes, the heat transfer is reduced compared to counterflow because of the reduced temperature difference.

In pure counterflow a „P-value" (definition of the effectiveness for each fluid, see eq. (8) and (9)) of 1 is possible, whereas in parallel flow this can not be reached.

$$P_1 = \frac{T_{1,in} - T_{1,out}}{T_{1,in} - T_{2,in}} \qquad (8)$$

$$P_2 = \frac{T_{2,out} - T_{2,in}}{T_{1,in} - T_{2,in}} \qquad (9)$$

To estimate how the two effects behave under several conditions, the 2D approach is used. Four conditions are investigated, with the Reynolds number and the Prandtl number being varied. The Reynolds number is given for the arrangement with one pass, and increases with increasing number of passes.

- Re = 50   Pr = 4.5
- Re = 150   Pr = 4.5
- Re = 50   Pr = 25
- Re = 150   Pr = 250

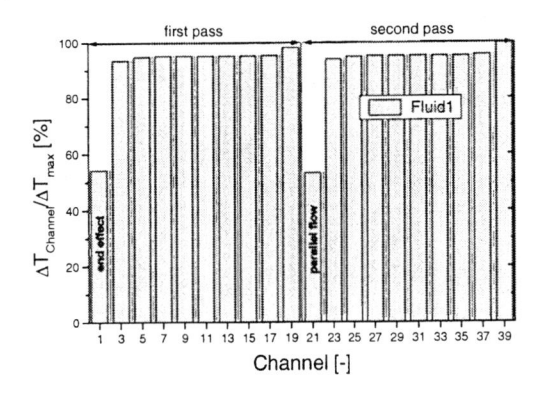

Fig.7: Temperature change in each channel for a 2-10 counterflow-counterflow arrangement

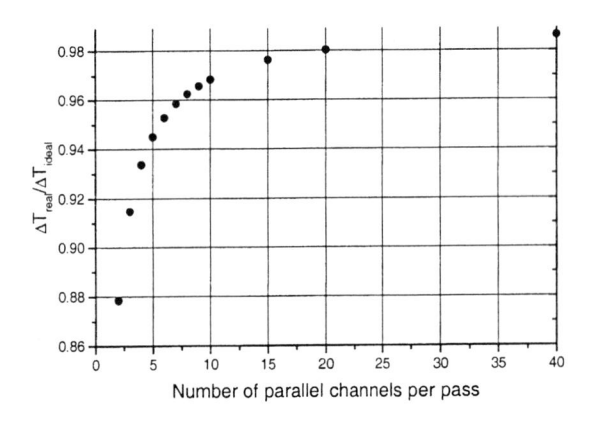

Fig. 8: End effect as a function of number of parallel channels for a 2 pass counterflow - counterflow arrangement

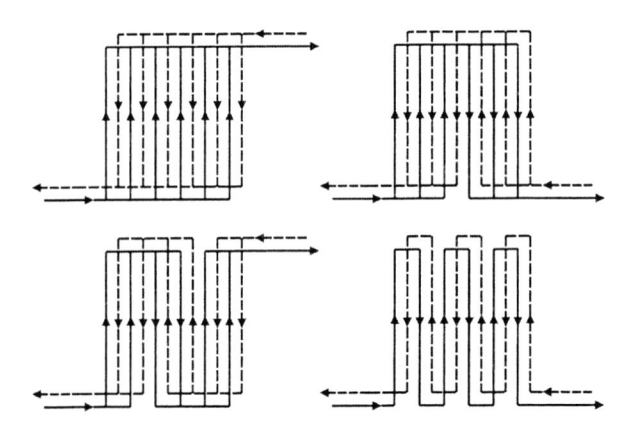

Fig.9: Configurations: one pass (upper left), two passes (upper right), three passes (lower left) and six passes (lower right)

The number of plates is fixed to 11, each fluid has 6 channels. Therefore 4 configurations exist (see fig.9):

- 1 – 6: **one pass** – 6 channels
- 2 – 3: **two passes** – 3 channels
- 3 – 2: **three passes** – 2 channels
- 6 – 1: **six passes** – 1 channel

All arrangements have counterflow in the channels as well as in the complete plate heat exchanger.

Re = 50 (one pass), Pr = 4.5

The first calculation (fig.10) shows the negative influence of the parallel flow configuration in between passes. Under laminar conditions, the reduction of the heat duty due to the parallel flow channels overcomes the increase due to higher velocities. Even a velocity that is six times higher cannot compensate the reducing effect. This occurs in cases, when the temperature change per fluid is close to unity (P=1), with the parallel flow being unable to reach 1 and therefore reducing the heat duty of the arrangement. In this case the temperature change with one pass was P = 0.824; hence there is no need for multipass arrangements.

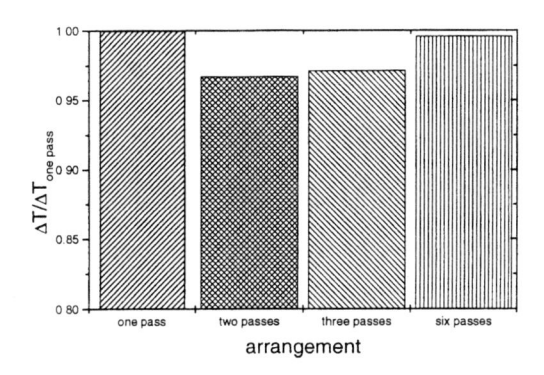

Fig.10: Comparison of the four arrangements for Re = 50 (one pass) and Pr = 4.5

The next step is to reduce the effectiveness P by either increasing the Reynolds- or Prandtl number. These two calculations are described in the following.

Re = 150 (one pass), Pr = 4.5

The increase of the Reynolds number by tripling the velocity leads to a smaller P value of 0.649 (one pass). Therefore the influence of the parallel flow in between passes is no longer as significant and does not compensate the higher heat transfer coefficient for more than one pass (fig.11). The increase of heat transfer between the two fluids

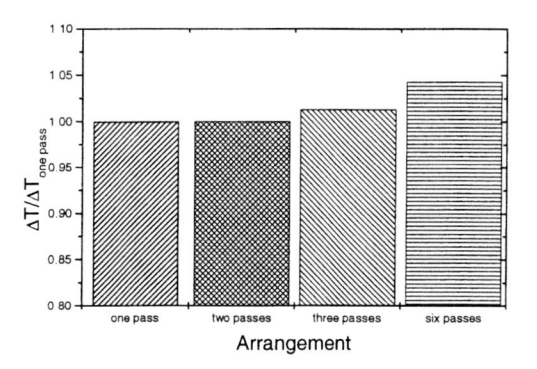

Fig.11: Comparison of the four arrangements for Re = 150 (one pass) and Pr = 4.5

is less than 5 percent, which is not justified considering the higher pressure drop.

Re = 50 (one pass), Pr = 25

The same effect is recognized for increasing the Prandtl number. This corresponds to a smaller value of the thermal conductivity of each fluid and reduces P to 0.518 for one pass. However, due to the pressure drop, arrangements with more than one pass again are not practical. The arrangement with 6 passes leads to an increase of less than 10 % (fig.12).

This reinforces the conclusions of Kandlikar and Shah (1989/1). They favoured the one pass arrangement for equal flow rates and NTU values greater than 5, since this provided the highest effectiveness. For lower NTU-values with higher flow velocities, multipass arrangements could increase the heat duty, but the pressure drop increase must be considered as well.

Higher velocities will inevitably lead into the turbulent region, and results may no longer be comparable. The only application could be fluids with a high Prandtl number, e.g. glucose solution or oil. Heat transfer by conduction is

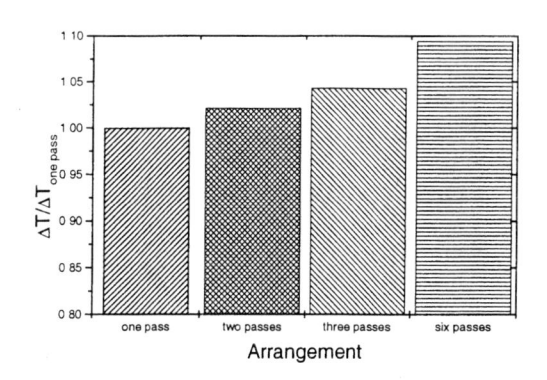

Fig.12: Comparison of the four arrangements for Re = 50 (one pass) and Pr = 25

decreased (compared to water) and therefore values for P << 0.5 can be reached. In this case arrangements with more than one pass lead to a significant improvement compared to the one pass arrangement.

Re = 150 (one pass), Pr = 250

In this case all multipass arrangements lead to a higher heat duty of at least 20 %. The 6 pass arrangement is not practicable as the 6 times higher velocitiy leads to a pressure drop which is no longer acceptable. The arrangement for three passes leads to an improvement of 43 % (fig.13).

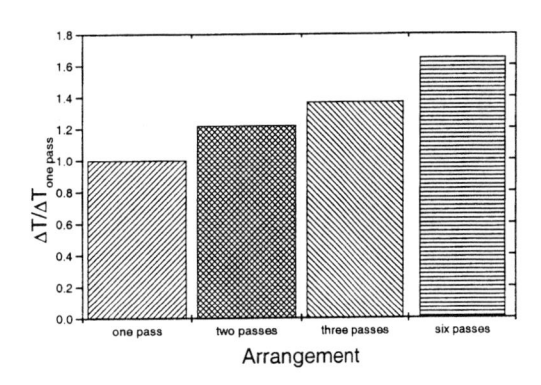

Fig.13: Comparison of four arrangements for Re = 150 (one pass) and Pr = 250

CONCLUSIONS

A 2D finite volume method is presented for the prediction of laminar flows in plate heat exchangers. The grid generation is completely automatic, this includes the settings for the boundary conditions which depend on the type of arrangement. The Navier-Stokes-Equations are used to compute the velotciy and temperature field, while Conjugate Heat Transfer is used to predict the heat transfer to or from the walls.

Validation with the benchmark solution shows a very good agreement.

The influence of the end effect as well as the influence of parallel flow between adjacent passes is investigated. Both effects are insignificant for a large number of plates. The end effect can be neglected for single pass arrangements with more than 39 plates (equals 2 percent reduction of effectiveness), while in arrangements with more than one pass the changing flow configuration leads to 79 plates (equals 2 percent reduction of effectiveness).

Similar to Kandlikar and Shah (1989/1), we found, that one pass arrangements should be preferred over multi-pass arrangements as long as the flow distribution to the parallel channels is the same. Only for arrangements with a very low P value can the heat duty be increased significantly by the use of multiple passes.

NOMENCLATURE

A	surface, m^2
c_p	specific heat, J/kg K
Gr	Grashof number, dimensionless
h	heat transfer coefficient, W/m^2 K
k	thermal conductivity, W/m K
$\dot{m}$	mass flow, kg/s
L	width of fluid region (see „Validation"), m
M	number of parallel channels, dimensionless
N	number of control volumes at outlet, dimensionless
Nu	Nusselt number, hd_h/λ, dimensionless
NTU	Number of Transfer Units, $UA/\dot{m}c_p$, dimensionless
p	pressure, N/m^2
P	effectiveness, see (Eq.(8), Eq.(9)), dimensionless
Pr	Prandtl number, dimensionless
Re	Reynolds number, wd_h/ν, dimensionless
t	width of solid region (see „Validation"), m
T	temperature, K
U	overall heat transfer coefficient, W/m^2 K
w	velocity, m/s
x,y	coordinates, m
z	Parameter see eq.(7), dimensionless

Greek symbols

ϕ	generic characteristic, dimensionless
μ	dynamic viscosity, kg/m s
ν	kinematic viscosity, m^2/s
ρ	density, kg/m^3

Subscripts

c	cold
channel	channel
h	hot
i,j,n,m	summation index
ideal	ideal temperature change without end effect or parallel flow effect
in	inlet
max	maximum
out	outlet
P	control volume center
pass	pass
real	real temperature change cause to end effect or parallel flow effect
W	wall
1	Fluid 1 (hot)
2	Fluid 2 (cold)
∞	infinity

LITERATURE

H. D. Baehr, K. Stephan, 1994, Wärme- und Stoffübertragung, 1st ed., Springer-Lehrbuch, ISBN 3-540-55086-0, p.14

F. Bosnjakovic, M. Vilicic, B. Slipcevic, 1951, Einheitliche Berechnung von Rekuperatoren, VDI-Forschungsheft 432 Ausgabe B Band 17

R.A. Buonopane, R.A. Troupe, J.C. Morgan, 1963, Heat Transfer design method for plate heat exchangers, Chemical Engineering Progress, Vol.59, No.7, pp 57-61

G. de Vahl Davis, 1983, Natural convection of air in a square cavity: a bench mark solution, Int. J. f. Numer. Methods in Fluids, Vol.3, pp. 249-264

J.D. Domingos, 1968, Analysis of complex assemlbies of heat exchangers, Int. J. Heat Mass Transfer, Vol.12, pp.537-548

E.S. Gaddis, E.U. Schlünder, 1975, Temperaturverlauf und übertragbare Wärmemenge in Röhrenkesselapparaturen und Umlenkblechen, Verfahrenstechnik 9, Nr.12

G.F. Hewitt, G.L. Shires, T.R. Bott, (1994), Process Heat Transfer, CRC Press, ISBN 0-8493-9918-1, p. 338

D.A. Kaminski, C. Prakash, 1986, Conjugate natural convection in a square enclosure: effect of conduction in one of the vertical walls, Int.J. Heat Mass Transfer, Vol.29, No. 12, pp. 1979-1988

S.G. Kandlikar, R.K. Shah, 1989/1, Multipass Plate Heat Exchangers – Effectiveness-NTU Results ans Guidelines for Selecting Pass Arrangements, Journal of Heat Transfer, Vol.111, pp. 300-313

S.G. Kandlikar, R.K. Shah, 1989/2, Asymptotic Effectiveness-NTU Formulas for Mulitpass Plate Heat Exchangers, Journal of Heat Transfer, Vol.111, pp. 314-321

P.K. Khosla, S.G. Rubin, 1974, A diagonally dominant second-order accurate implicit scheme, Computer Fluids 2, pp. 207-209

S.V. Patankar, 1980, Numerical Heat Transfer and Fluid Flow, Series in Computational Methods in Mechanics and Thermal Sciences, Hemisphere Publishing Corporation, ISBN 0-07-048740-5, pp. 44-46

A. Pignotti, P.I. Tamborenea, 1988, Thermal effectiveness of multipass plate exchangers, Int. J. Heat Mass Transfer, Vol.31, No.10, pp.1983-1991

W. Roetzel, B. Spang, 1990, Verbessertes Diagramm zur Berechnung von Wärmeübertragern, Wärme- und Stoffübertragung 25, pp. 259-264

O. Strelow, 1997, Eine allgemeine Berechnungsmethode für Wärmeübertragerschaltungen, Forsch. Ingenieurwes. 61, pp. 255-161, Springer Verlag

J.P. van Doormal, G.D. Raithby, 1984, Enhancements of the SIMPLE method for predicting incompressible fluid flows, Numer. Heat Transfer 7, pp147-164

TWO-DIMENSIONAL NUMERICAL METHOD FOR THERMAL AND HYDRAULIC DESIGN OF MULTISTREAM PLATE-FIN HEAT EXCHANGER

[1]B. Deng, [2]Q. W. Wang and [3]W. Q. Tao,

[1] Xi'an Jiaotong University, Xi'an, Shaansi 710049, China; E-mail: dengbin@xjtu.edu.cn
[2] Xi'an Jiaotong University, Xi'an, Shaansi 710049, China; E-mail: wangqw@xjtu.edu.cn
[3] Xi'an Jiaotong University, Xi'an, Shaansi 710049, China; E-mail: wqtao@xjtu.edu.cn

ABSTRACT

Presented is a numerical method for the differential thermal and hydraulic design of multistream compact heat exchanger. The entire exchanger with assumed length is first discretized into many slices, and at each slices, the fluid in every passage and the two bounding walls are taken as three nodes. The energy balance is used to set up the temperature equation for each passage, resulting in a set of algebraic equations for all the passages in one slice. This set of equations is solved from slice to slice for the entire exchanger. The arrangement of passages is improved according to the calculated temperature non-uniformity. The assumed length is modified according to the predicted averaged exit temperatures. A two-step method is proposed for passage stacking which can improve the uniformity of the fluid temperature. A numerical example shows the feasibility and effectiveness of the proposed method.

INTRODUCTION

Plate-fin heat exchangers are widely used in many engineering fields, such as cryogenics and chemical industries. When they are adopted for gas processing purpose, they are usually designed to handle up six or more different fluids(streams) in a single compact core. Thus, the design of such heat exchangers becomes quite complicated.

Generally speaking, there are two kinds of methods for the design of the plate-fin heat exchangers. One is the integral method, which has been used in the gas processing industry for years. This method has the advantage of simplicity in calculation. Its major drawback is the lack of the details in the fluid parameter variation in the heat exchanger. Recently, the so-called differential method has been developed by Prasad (1991) and Prasad and Gurukul (1992). In this method the whole plate-fin heat exchanger is first discretized into a number of sections (slices) which are perpendicular to the main flow direction. In each section there are n passages and (n-1) separating walls and they are taken as (2n-1) nodes. Based on the heat balance across a separating wall and the heat duty of a given passage a total of (2n-1) algebraic equations for the n outlet fluid temperatures and (n-1) separating wall temperatures can be formed. According to the numbering method proposed by Parsad (1991), the matrix of the algebraic equation is pentadiagonal. The solution of the algebraic equations is advanced from slice to slice until the entire exchanger is visited. Thus numerically it is a two-dimensional parabolic-type simulation. In the paper by Deng el al(2000), it is shown that by changing the numbering system of the fluids and the separating walls, the resulting algebraic equations can lead to a tridiagonal matrix, which greatly alleviate the computational work.

For the differential design method, apart from the computational work, the arrangement of fluid passages is another important issue. The high efficiency of the compact plate fin heat exchanger mainly comes from the fact that in this heat exchanger the fluids with high and low temperatures are divided into many small streams and they are appropriately arranged such that the heat exchange between the fluids of high and low temperatures may be proceeded with small temperature difference. The conventional passage arrangement method proposed by Suessmann and Mansour (1979) is based on the balance between local thermal loads. In this paper an improved passage arrangement is proposed, in which the Suessmann and Mansour method is taken to obtain a preliminary arrangement and refinement is further made to meet some requirements. An computational example is illustrated to show the feasibility of this method. In the following, a general description will first be presented briefly to show the principles of the differential method and the improved numbering system proposed by Deng et al. (2000). Then focus will be put on the passage stacking, especially the further refinement of the passage arrangement. Finally computational results of an example will be provided

GENERAL DESCRIPTION OF DIFFERENTIAL APPROACH

A pictorial view of a multistream heat exchanger is shown in Fig. 1, where x is the main stream direction and y and z are in the cross sectional direction. In the y-z plane, there are a number of passages which should be distributed to hot and cold fluids. The dashed lines shown in the figure represent the slices each of which is the computational unit and heat balance equation will be formulated.

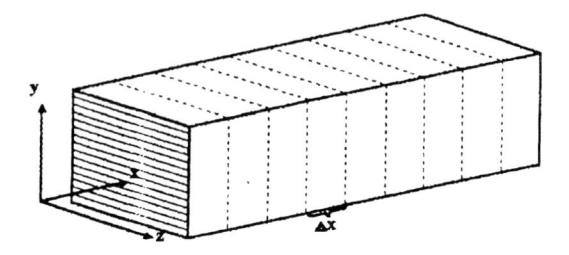

Figure 1 Multistream heat exchanger

Following assumptions are made for the mathematical model adopted in this paper. (1) The fluid flow and heat transfer are in steady state; (2). The heat loss to the environmental is neglected; (3) The hot and cold fluids are either in parallel flow or in counter flow, no extraction of fluid is considered;(4) A nominal temperature ,Ts, will be used to denote an equivalent temperature of the separating wall and the attached fins (Fig. 2); (5) For an internal passage, half of the fin length in an passage is transferring heat to the neighboring passage on the either side (Fig. 2).For the top and bottom passages, however, the entire length of the fin must transfer heat to only one side; (6) The mean temperature at the entry and exit of each slice will be used in the Newton's law of cooling to determine the convective heat transfer rate.

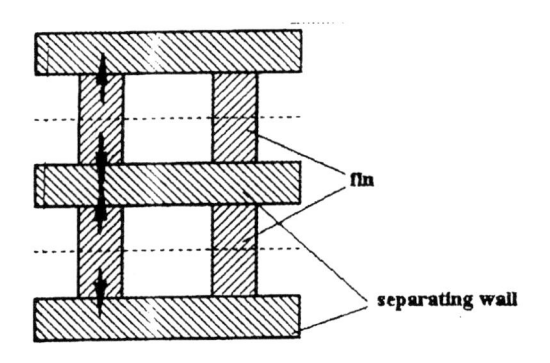

Figure 2 Heat transfer from fins

According to the above assumptions, when the heat balance equation is formulated there is only one temperature of fluid and two temperatures of the bounding walls for each passage. Thus if we take each passage as a computational node and each separating wall as a node, and number the nodes according to the sequence presented in Fig. 3, then each inner node is only related to two adjacent nodes. This implies that the resulting algebraic equation of temperatures will form a matrix of tridiagonal type, which can be expressed as

$$a_P(j)T(j) = a_N(j)T(j-1) + a_S(j+1)T(j+1) + b(j) \quad (1)$$
$$(j = 1,...2n-1)$$

where the coefficients $a_P(j), a_S(j), a_N(j)$ and the term $b(j)$ can be determined by the heat balance equation. The coefficients of the (2n-1) equations form a tridiagonal matrix, and can be solved by TDMA. For details, the paper by Deng et al (2000) may be consulted.

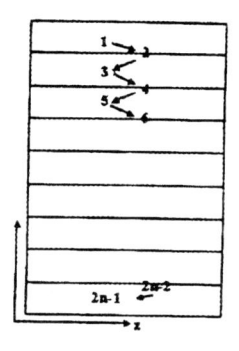

Figure 3 Node numbering system

PRELIMINARY AND FINE ARRANGEMNETS OF THE PASSAGES

As indicated above, the passage arrangement (stacking) is of essential importance for the multistream compact heat exchanger. The stacking process includes two tasks, one is to assign passage number to each fluid and the other is to segregate the passages of different streams. The first task is usually conducted by a rule of thumb, i.e., based on the specified mass flow rate of each fluid and the selected fin type and dimensions, an empirical value of the appropriate mass velocity is adopted and then the necessary passage number can be determined. If the computational results shows that this number is not adequate either from pressure drop or from heat transfer aspects, then the number should be regulated. Hence the computation is of trial-and-error nature. In the following discussion we will assume that the passage numbers for different fluids have been set and the problem left is how to segregate the passages so that the heat exchanger will have a near-optimum passage arrangement.

In the past, many methods have been proposed for stacking, among whom the Suessmann-Mansour method is widely used. In this method, the cumulative sums of the local thermal duties are plotted against the passage number with warm and cold streams having plus and minus sign, respectively. The stack pattern that could lead to the least deviation from zero is judged to be the best one. This practice is usually employed in the integral method of design, and its feasibility has to be verified by the differential computational results. In the present work, we proposed a two-step stacking method. In the first step (preliminary step), the method of Suessmann-Mansour is adopted, resulting in an initial arrangement of the passages. Differential computations are then performed with this initial arrangement for each section. Once all the sections have been visited, the temperature distribution of different fluids will be obtained. According to these temporary results we start the second step of stacking. We set the criterion for good-arrangement of the passages as follows: a good arrangement should be able to minimize the temperature non-uniformity of each fluid at the exit of the heat exchanger. If at the heat exchanger exit there is a passage of cold stream whose temperature is significantly lower than the average one of this fluid, it implies that around this passage the temperatures of hot streams were not high enough. The same conclusion can be made for the hot fluids. The second step is conducted as follows.

(1) At the end of each iteration of the entire heat exchanger, the temperature non-uniformity of each fluid at the exit of the exchanger is determined. The non-uniformity of the exit temperature for each fluid is defined as follows:

$$\Delta T^{kf} = \sqrt{\frac{\sum_{j=1}^{n}(T_{out,j}^{kf} - \overline{T^{kf}})^2}{n}} \quad (2)$$

where $\overline{T^{kf}} = \dfrac{\sum_{j=1}^{n} T_{out,j}^{kf}}{n}$

(2) If the non-uniformities of some fluids are larger than a specified value, say $4°C$, then the worst passages of the warm fluid and the worst passage of the cold fluid are found. The passage number of the warmest stream is denoted by CH, and that of the coldest passage by CL. The value of CH and CL is then compared. If CH>CL, the coldest passage is inserted to the position adjacent to the warmest passage at the side where the passage number of the cold fluids is less or the temperature is higher. If CH<CL, then the passage with CH is inserted to the position adjacent to the passage with CL at the side where the number of warm passages is less or the temperature is lower.

(3) The computational process is then re-conducted, and the above described fine rearrangement step is repeated, until the temperature non-uniformities of each fluid at the exit of the exchanger satisfy the specified values.

OVERALL COMPUTATIONAL PROCEDURE OF THE DIFFERENTIAL METHOD

In the compact heat exchanger, the warm and cold fluids are arranged, without any exception, in counterflow pattern. The counterflow arrangement leads to following two important features for the differential computations:

(1) The marching-type parabolic problem possesses some elliptic nature.

It is obvious that if the warm and cold fluids are in parallel flow arrangement, the advance of computation from section to section can be made by marching method. This is a parabolic type problem, and the downstream(along the coordinate direction) condition will not affect the upstream quantities. For the counterflow arrangement, however, if we take the inlet of the hot fluid as the start point of the axial coordinate, then the marching process along the coordinate possesses elliptic character for the cold fluid in that its downstream value have effect on its upstream quantities. The numerical computation process should take this character into account. We adopted following practice to accommodate above situation: At the first, the entire heat exchanger is taken as one slice, i.e., the section thickness equals the length of the heat exchanger, which is assumed in advance. The resulting algebraic equations of the nodes are solved. Thus all the inlet and outlet temperatures of the fluids can be obtained. Then based on these results, the inlet and outlet temperatures of different passages in each slice are determined by linear interpolation. An appropriate initial temperature distribution is thus obtained. The computation is then proceeded from the exchanger inlet of the warm fluid by marching method.

(2) The transfer of information between the inlet and outlet of two adjacent slices are different for warm and cold fluids.

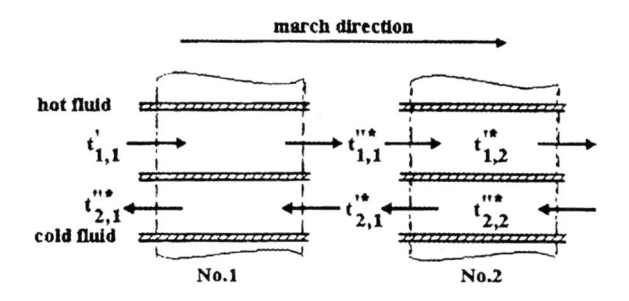

Figure 4 Information transfer between slices

As shown in Fig. 4, along the marching direction, for the warm fluid the exit temperature of one slice is the inlet temperature of the next slice. For the cold fluid, however, along the marching direction, the inlet temperature of one slice is the exit temperature of the next slice. Taking the first slice(counting the slice number from the inlet of warm fluid) as an example(Fig. 4), the known temperatures are $t_{1,1}^{'}$ (inlet temperature of the warm fluid for the first slice) and $t_{2,1}^{''*}$ (exit temperature of the cold fluid at the first section), where the symbol * stands for the interpolated or computed value. Solution of the algebraic equations of this slice will obtain the values of $t^*_{1,1}$ and $t^*_{2,1}$. The computation for the second slice can be performed in the similar way.

The general computational procedure is now summarized as follows:

(1) Input all the necessary data, including fluid number and fluid compositions; inlet temperatures of all the fluids and the required outlet temperatures, and the assumed heat exchanger length(which is preferably to be shorter than expected). It should be noted that the required outlet temperature should have at least one temperature open to the computation, which should be determined by the heat balance itself and can not be exactly fixed in advance. Because the thermophysical properties are temperature dependent, hence it is usually impossible to obtain all the outlet temperatures through simple heat balance computation in advance by using constant properties.

(2) Compute the thermal load of each fluid, and determine the passage number of each fluid according to the specified mass velocity;

(3) Arrange the passages using Suessmann-Mansour method;

(4) Take the entire exchanger as one slice, solve the resulting algebraic equations;

(5) Discretize the exchanger into slices and interpolate the temperature for warm and cold fluids to obtain the assumed temperature for each slice;

(6) Start from the first slice, solve the algebraic equations and march the process from beginning slice to the end slice;

(7) Repeat step six until the deviation of the computed temperature from the corresponding assumed temperatures is less than the specified allowance, say $0.01\,°C$;

(8) Determine the non-uniformity of each fluid at the exit. Perform the fine arrangement of the passages if the non-uniformity does not satisfy the specified value until the specified uniformity criterion is satisfied;

(9) Check the outflow temperature of each fluid. If the required outflow temperatures are not reached for all but one fluids, add one more slice to the heat exchanger and re-do the whole computational process as described above until the required outlet temperature for all but one fluids are satisfied.

ILLUSTRATING COMPUTATIONAL EXAMPLE

Numerical computations were performed for a compact heat exchanger with 116 passages and five fluids. The compositions, the specified inlet and required outlet fluid temperatures and the passage numbers are listed in Table 1. It should be noted that the outlet temperature of the second fluid is iteratively determined by the computations.

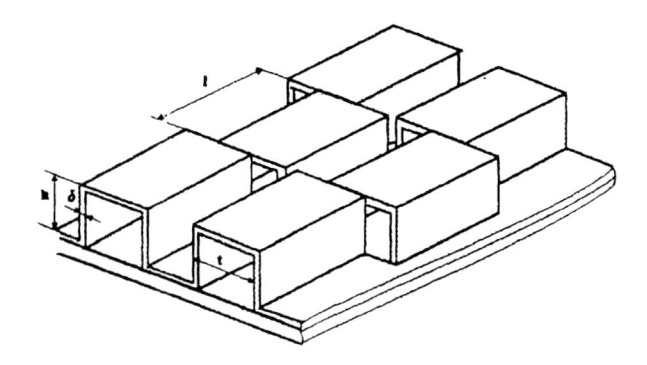

Figure 5 Off-set plate-fin

The off-set plate-fin was adopted (Fig.5). In the present paper, only the single-phase flow is considered. By comparison of several available correlations and test data, including those of Weiting (1975), Joshi (1987), Kays and London (1964), we finally adopted the correlations proposed by Weiting. These correlations are presented as follows.

(1) Laminar flow region ($\text{Re}_D \leq 1000$)

$$f = 7.661(l/D_e)^{-0.384}(\alpha)^{-0.092}\text{Re}_D^{-0.712} \quad (3a)$$

$$j = 0.483(l/D_e)^{-0.162}(\alpha)^{-0.184}\text{Re}_D^{-0.536} \quad (3b)$$

(2) Turbulent flow ($\text{Re}_D \geq 2000$)

$$f = 1.136(l/D_e)^{-0.781}(\delta/D_e)^{0.534}\text{Re}_D^{-0.198} \quad (4a)$$

$$j = 0.242(l/D_e)^{-0.322}(\delta/D_e)^{0.089}\text{Re}_D^{-0.368} \quad (4b)$$

where the characteristic length of Re_D is the equivalent diameter of the cross section:

$$D_e = \frac{2(t-\delta)\times(h-\delta)}{(t-\delta)+(h-\delta)} \quad (5)$$

and α is the aspect ratio of the cross section

$$\alpha = \frac{t-\delta}{h-\delta} \quad (6)$$

The heat transfer and friction factor for the transition region ($1000 \leq \text{Re}_D \leq 2000$) are also calculated by either Eqs.(3) or (4), depending on following reference Reynolds number:

$$\text{Re}^{ref}_{D,f} = 41(l/D_e)^{0.772}(\alpha)^{-0.179}(\delta/D_e)^{-1.04} \quad (7a)$$

$$\text{Re}^{ref}_{D,j} = 61.9(l/D_e)^{0.952}(\alpha)^{-1.1}(\delta/D_e)^{-0.53} \quad (7b)$$

If $\text{Re}_D \leq \text{Re}^{ref}_{D,f}$, Equation (3a) is used to determine the friction factor, otherwise Eq.(4a) is used; if $\text{Re}_D \leq \text{Re}^{ref}_{D,j}$, Equation (3b) is used to calculate the $j-$ factor, otherwise Eq.(4b) is used. In this paper, all the fluids are of single phase, and no phase change heat transfer is involved.

Table 1 Fluids and their compositions and passage numbers

No	Fluid composition	Inlet temperature, K	Required outlet temperature, K	Number of passages
1	Technical air, N_2: 78.8%; O_2: 10.95%; Ar: 0.97%	288.0	101.5	47
2	Expanded air, N_2: 78.8%; O_2: 10.95%; Ar: 0.97%	288.0	102.5 (determined by computations)	12
3	Oxygen, O_2: 99.6%; Ar: 0.4%	94.8	286.0	10
4	Nitrogen, N_2: 100%	98.8	286.0	10
5	Dirty nitrogen, N_2: 95%; O_2: 3.5%; Ar: 0.6%	98.8	286.0	37

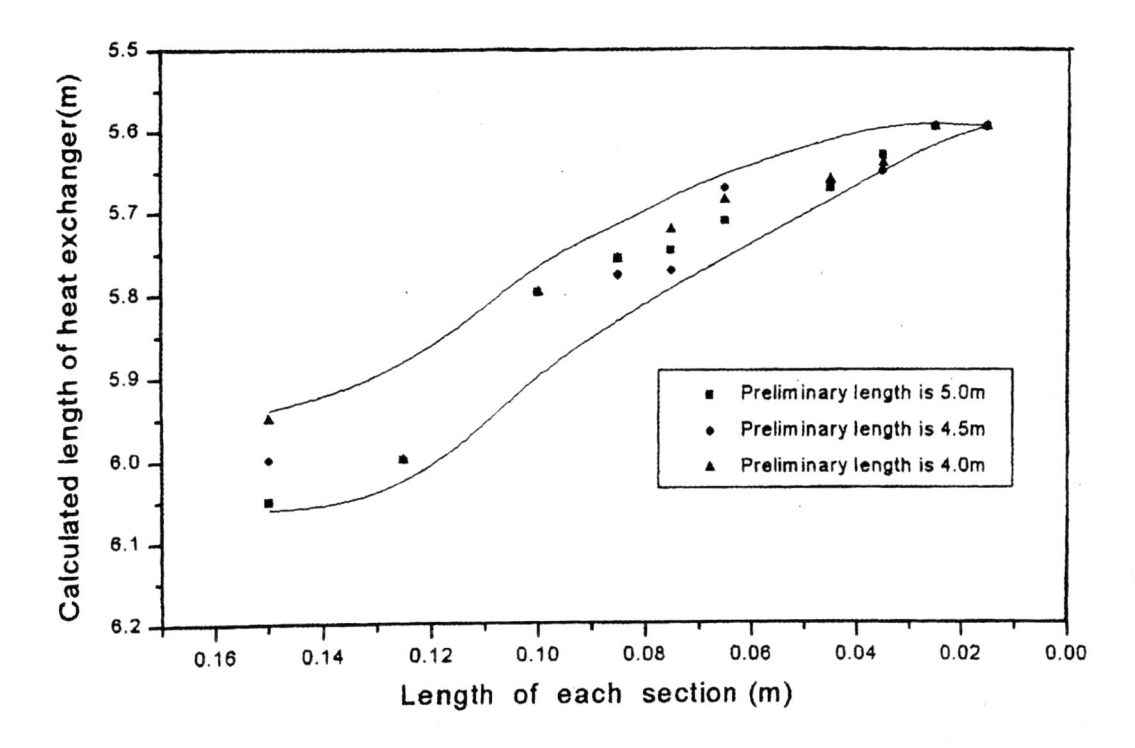

Figure 6 Convergence process of heat exchanger length

Effect of the Section Thickness and Initial Length

As indicated before, to proceed the computation an initial value of the heat exchanger length has to be assumed. In addition, the slice thickness is somewhat arbitrary. How these selections affect the final solution of the heat exchanger is of importance for the application of the differential method and the related software. Computations were conducted to reveal the effects of the two factors, and the results are presented in Fig. 6, where the converged length of the heat exchanger is displayed as the function of the slice thickness and the initial length. It can be seen that for the three initial length adopted, with the decreasing in the slice thickness, the computed length approaches a constant value (5.60 m), indicating quite good convergence character. The results presented in this paper are obtained by using 224 slices.

Results of Passage Arrangement and the Exit Fluid Temperatures

The preliminary and final arrangement of the passages, and the exit fluid temperature of each passage are listed in Table 2. The three numbers in the fluid information column are the initial fluid number, the final fluid number and the computed exit temperature. It can be seen that about 59% passages are re-arranged during the second step of stacking in order to meet the required uniformity of the exit temperatures, showing the effectiveness of the proposed two-step method for passage arrangement.

Table 2 Preliminary and final stacking and the exit fluid temperatures

Passage number	Fluid information	Passage number	Fluid information	Passage number	Fluid information	Passage number	Fluid information
1	1 – 1 – 107.58	31	5 – 5 – 287.09	61	3 – 5 – 285.92	91	1 – 1 – 103.12
2	5 – 5 – 286.10	32	1 – 1 – 101.34	62	2 – 2 – 102.87	92	5 – 5 – 285.65
3	1 – 1 – 102.91	33	5 – 5 – 286.61	63	4 – 4 – 286.12	93	1 – 1 – 102.60
4	5 – 5 – 284.08	34	1 – 1 – 101.21	64	1 – 1 – 102.02	94	5 – 4 – 285.74
5	2 – 1 – 100.42	35	4 – 5 – 286.63	65	5 – 3 – 286.16	95	1 – 1 – 101.49
6	4 – 5 – 282.64	36	1 – 1 – 101.25	66	1 – 2 – 102.55	96	5 – 3 – 285.68
7	2 – 1 – 100.44	37	5 – 5 – 286.89	67	5 – 5 – 286.07	97	2 – 1 – 101.28
8	1 – 5 – 279.97	38	1 – 1 – 101.75	68	1 – 1 – 102.01	98	1 – 4 – 285.47
9	5 – 3 – 280.27	39	5 – 5 – 287.49	69	5 – 3 – 286.17	99	5 – 1 – 101.94
10	1 – 1 – 99.82	40	2 – 2 – 103.47	70	1 – 1 – 101.92	100	1 – 4 – 285.23
11	5 – 3 – 284.35	41	3 – 5 – 287.99	71	3 – 4 – 286.25	101	5 – 1 – 102.03
12	1 – 1 – 98.80	42	1 – 1 – 110.13	72	1 – 2 – 102.55	102	1 – 5 – 284.98
13	4 – 5 – 285.26	43	4 – 2 – 110.75	73	3 – 3 – 286.32	103	5 – 1 – 102.08
14	1 – 1 – 94.80	44	2 – 5 – 286.25	74	4 – 1 – 102.46	104	1 – 5 – 284.90
15	4 – 5 – 285.79	45	3 – 1 – 104.78	75	1 – 5 – 286.30	105	5 – 1 – 101.96
16	2 – 2 – 102.21	46	1 – 3 – 285.75	76	5 – 2 – 103.66	106	2 – 5 – 284.76
17	5 – 4 – 286.52	47	5 – 1 – 103.00	77	1 – 5 – 286.54	107	5 – 1 – 101.71
18	1 – 1 – 102.07	48	2 – 5 – 285.70	78	5 – 1 – 104.16	108	1 – 5 – 284.50
19	4 – 5 – 287.77	49	3 – 2 – 103.00	79	1 – 4 – 287.21	109	5 – 1 – 101.26
20	1 – 1 – 103.76	50	1 – 5 – 286.00	80	4 – 1 – 105.31	110	1 – 5 – 284.06
21	5 – 5 – 288.36	51	5 – 2 – 104.91	81	1 – 4 – 287.78	111	5 – 1 – 100.48
22	1 – 1 – 107.45	52	2 – 1 – 104.74	82	5 – 1 – 110.71	112	1 – 5 – 283.24
23	3 – 4 – 286.96	53	5 – 5 – 285.94	83	1 – 2 – 111.27	113	5 – 1 – 98.01
24	1 – 1 – 108.79	54	1 – 1 – 101.67	84	3 – 4 – 287.81	114	1 – 3 – 279.44
25	4 – 5 – 286.85	55	5 – 5 – 285.39	85	1 – 2 – 105.84	115	5 – 5 – 280.42
26	1 – 1 – 107.18	56	1 – 2 – 101.30	86	5 – 5 – 287.16	116	1 – 1 – 99.54
27	5 – 3 – 288.47	57	5 – 5 – 284.83	87	2 – 1 – 104.38		
28	2 – 1 – 102.82	58	1 – 1 – 100.25	88	5 – 5 – 286.45		
29	3 – 5 – 288.26	59	5 – 3 – 285.37	89	1 – 1 – 103.60		
30	1 – 1 – 101.52	60	1 – 1 – 101.88	90	3 – 5 – 285.95		

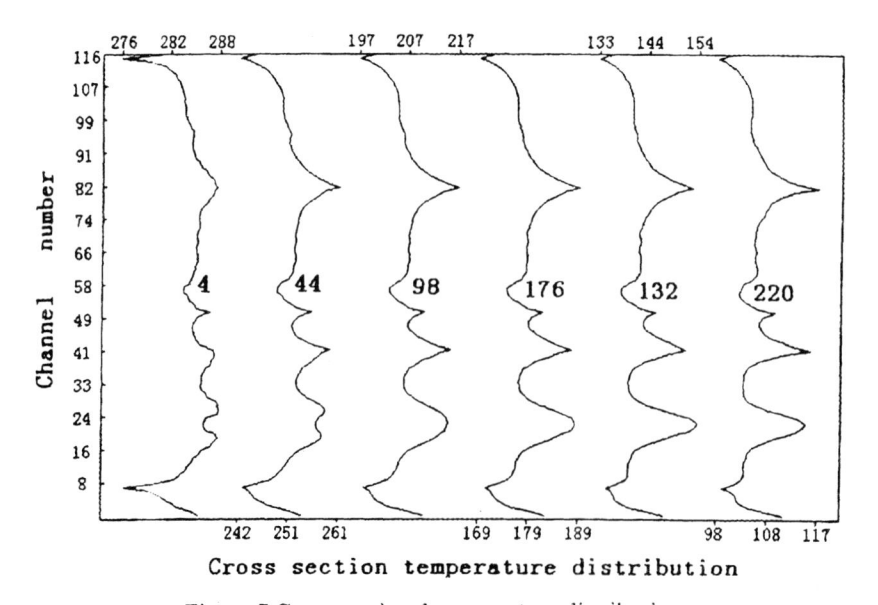

Figure 7 Cross sectional temperature distributions

Spanwise and Streamwise Temperature Distributions

The spanwise temperature distributions at six cross sections(from No.4 to 220) are shown in Fig. 7, where both the fluid and separating wall temperatures are presented by continuous lines. It is well-known that the more uniform of the spanwise temperature distribution, the more effective the multi-stream compact heat exchanger. The six cross sections presented in Fig. 7 span almost the entire length of the heat exchanger, and the spanwise wall temperature difference at different cross section is almost the same. This implies that the adopted method for the passage stacking is feasible and effective.

The mean streamwise temperatures of each fluid averaged at each cross section are shown in Fig. 8, starting from the inlet of the hot fluid. A local magnifying picture is also provided to distinguish the five fluids. It can be seen that at any cross section the temperature difference between the hot and cold fluids are nearly the same, which is a character of an ideal counterflow heat exchanger. This once again shows the advantage of the present differential method for the design of the multi-stream compact heat exchanger.

CONCLUSIONS

A two-dimensional numerical method for the differential thermal and hydraulic design is presented in detail. This method adopted the basic idea proposed by Prasad: the entire heat exchanger is split into many streamwise slices, and for each slice energy balance equation is set for every passage which is composed of the fluid in the passage and two bounding walls. The resulting algebraic equations are then solved and advanced from slice to slice for the entire exchanger. If the uniformity of the predicted exit temperatures of different fluids can not meet the specified allowances the passages are re-arranged to reduce the maximum non-uniformity. The assumed length of the exchanger is examined by the exit averaged fluid temperatures. If the predicted exit temperatures of all but one fluids can not meet required values, the length of the exchanger is added by one slice and the computations are repeated. A two-step procedure for the passage stacking is proposed, which uses the Suessmann-Mansour method to obtain a preliminary arrangement that is refined finally according to the differential computations. A numerical example is presented. The results show good convergence when slice thickness is reducing for different initial lengths. The spanwise and streamwise temperature distributions show that the proposed method is feasible and effective for the design of the multi-stream compact heat exchanger.

ACKNOWLEDGEMENT

This work is supported by the National Natural Science Foundation of China (Grant No. 50076034) and the National Key Project of Fundamental R & D of China (Grant No. 2000026303)

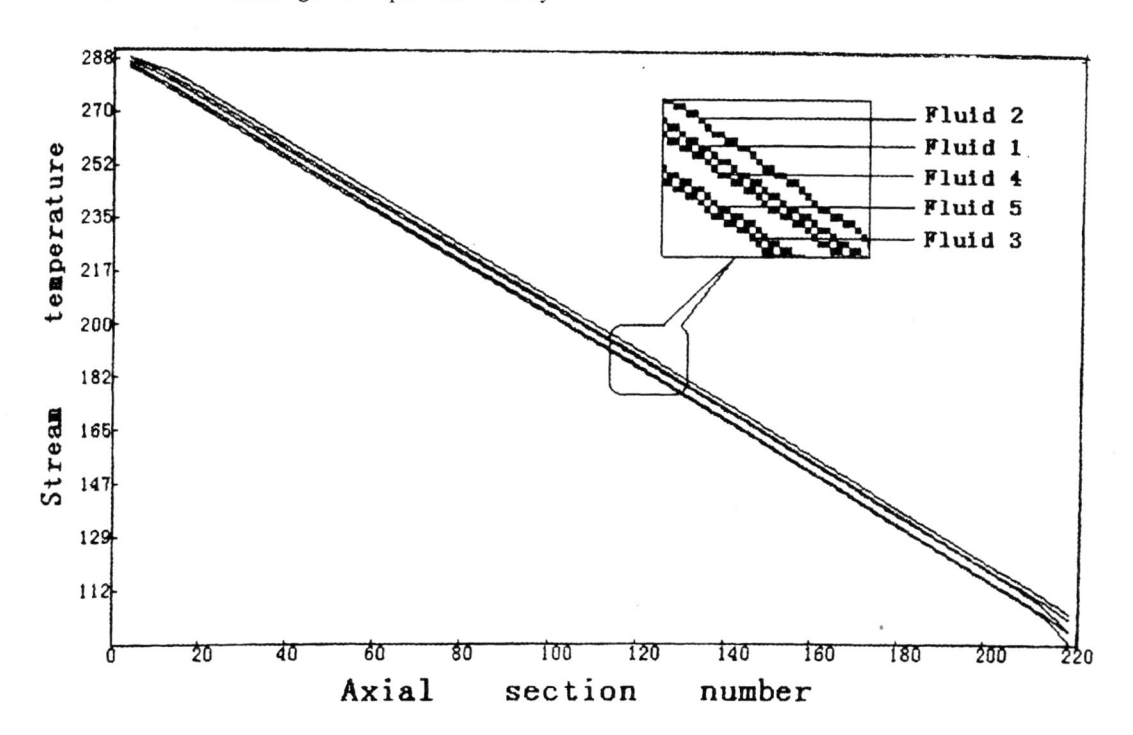

Figure 8 Streamwise fluid temperature distribution

NOMENCLATURES

a_N, a_S, a_P coefficients
b constant term in algebraic equation
D_e equivalent diameter, m
f friction factor
h fin height, m
j j-factor
l cut length of off-set fin, m
n passage number
Re Reynolds number
T temperature, K
t fin spacing, m
α aspect ratio
δ fin thickness, m

Superscripts

$'$ inlet
$''$ outlet

REFERENCES

Deng,B., Wong,T.T., Lin,M.J., Wang,Q.W., and Tao,W.Q.,2000. An Improved Differential Method for the Performance Prediction of Multistream Plate-Fin Heat Exchanger. *Heat Transfer Science and Technology 2000*, ed. B.X.Wang. Higher Education Press, Beijing, pp.708-713

Joshi, H.M.,1987. Heat Transfer and Friction in Off Set Strip-Fin Heat Exchangers. *Int J Heat Mass Transfer*, vol. 30, pp. 69-85

Kays, W.M., and London,A.L., 1964. *Compact Heat Exchangers* (2nd ed.). McGraw-Hill, New York

Prasad, B.S.V., 1991. The Performance Prediction of Multistream Plate-fin heat Exchangers Based on Stacking Patterns. *Heat Transfer Engineering*, vol. 12, pp. 58-69

Prasad, B.S.V., and Gurukul, S.M.K.A., 1992. Differential Methods for the Performance Prediction of Multistream Plate-fin Heat Exchangers. *ASME J Heat Transfer*, vol. 114, pp. 41-49

Suessmann, W., and Mansour, A., 1979. Passage Arrangement in Plate-Fin Heat Exchangers. *Proceedings of XVth International Congress of Refrigeration*, Venice. Vol. 1, pp. 421-429

Weiting, A.R., 1975. Empirical Correlations for Heat Transfer and Flow Friction. *ASME J heat Transfer*, vol. 97, pp. 488-490

Yuan,Z.X., Xia,G.P., Lin,M.J., and Tao,W.Q., 1997. Computer Implementation of the Passage Arrangement for Plate-Fin Heat Exchanger According to Local Balance Principle. *J Thermal Science*, vol. 6, pp. 190-196

SMOOTH- AND ENHANCED-TUBE HEAT TRANSFER AND PRESSURE DROP: PART I. EFFECT OF PRANDTL NUMBER WITH AIR, WATER, AND GLYCOL/WATER MIXTURES

N. T. Obot,[1] L. Das,[2] and T. J. Rabas[3]

[1] Argonne National Laboratory, Argonne, Illinois 60439, USA; E-mail: obot@anl.gov;
Professor Emeritus, Department of Chemical Engineering, Clarkson University, Potsdam, NY 13699, USA
[2] Former Graduate Student, Department of Chemical Engineering, Clarkson University, Potsdam, NY, USA
[3] Argonne National Laboratory, Argonne, Illinois 60439, USA

ABSTRACT

An extensive experimental investigation was carried out to determine the pressure drop and heat transfer characteristics in laminar, transitional, and turbulent flow through one smooth tube and twenty-three enhanced tubes. The working fluids for the experiments were air, water, ethylene glycol, and ethylene glycol/water mixtures; Prandtl numbers (Pr) ranged from 0.7 to 125.3. The smooth-tube experiments were carried out with Pr values of 0.7, 6.8, 24.8, 39.1, and 125.3; Pr values of 0.7, 6.8, and 24.8 were tested with enhanced tubes. Reynolds number (Re) range (based on the maximum internal diameter of a tube) was 200 to 55,000, depending on Prandtl number and tube geometry. The results are presented and discussed in this paper.

INTRODUCTION

Extensive experimental studies on the pressure-drop and heat-transfer characteristics of a wide range of enhanced surface geometries have been published by numerous researchers. The results of these studies indicate that varying amounts of pressure-drop and heat-transfer increases can be realized with enhanced tubes, depending on the geometric characteristics of the surfaces. Reviews of the widespread literature were presented by several researchers (Reay, 1991; Obot et al. 1990; Rabas, 1989; Webb, 1987; Ravigururajan and Bergles, 1986).

A search of the literature reveals very limited studies with commercially available enhanced tubes over the entire range of flow conditions extending from laminar through turbulent flow. With the exception of the studies by Obot *et al.* (1994), Esen *et al.* (1994), Esen (1992) and Das (1993), virtually all the investigations with tubes of the spirally fluted type were confined to either turbulent flow (Yampolsky *et al.*, 1984; Panchal and France, 1986; Ravigururajan and Bergles, 1986) or laminar and transitional flow (Shome and Jensen, 1996). Similarly, for corrugated surfaces of single and multiple helix (Withers, 1980a,b) or three-dimensional spiral ribs (Takahashi et al., 1985), pressure drop and heat transfer data were reported only for turbulent flow.

In sharp contrast to turbulent flow, limited experimental data on pressure drop and heat transfer in laminar flow and the transition region exist for enhanced passages (Shome and Jensen, 1996; Marner and Bergles, 1978; Watkinson *et al.*, 1974; Koch, 1960; Nunner, 1956). In the recent study by Shome and Jensen with internally-finned tubes, the Reynolds and Prandtl number ranges were 150 to 2,000 and 50 to 185, respectively.

Another problem that has not received much attention is the effect of fluid properties on enhanced-tube heat transfer that is usually expressed in nondimensional form by using the Prandtl number (Pr). A search of the literature revealed only a handful of studies with more than one fluid (Webb *et al.*, 1971; Marner and Bergles, 1978; Carnavos, 1980; Smith and Gowen, 1985; Gomelauri, 1964). Alternatively, in some previous studies with liquids, the Prandtl number was varied by adjusting the bulk temperature of the fluid (Shome and Jensen, 1996; Dipprey and Sabersky, 1963; Watkinson *et al.*, 1974). The drawback with this approach is that the observed heat-transfer trends with increasing Pr may not form the basis for generally valid conclusions because of the restricted range. There is a need for a comprehensive investigation of the effect of Prandtl number on pressure-drop and heat-transfer performance of enhanced tubes.

Although the problem has been studied extensively, high-quality heat-transfer and pressure-drop prediction methods for enhanced tubes are very limited and/or nonexistent. The reason is that too many variables are involved and it is difficult to develop a predictive method that accurately accounts for the effects of these variables. A recent effort focused on the *frictional law of corresponding states*, the basis of which is the transition from laminar to turbulent flow (Obot *et al.*, 1994). There is a need to validate the frictional law analysis for a range of Pr values.

In summary, pressure-drop and heat-transfer data are needed for a wide range of enhancement geometries to fill the gaps in the existing literature data base; notably, laminar and transitional flow data are needed because these regions have received very meager treatment in the past. Also, the general state of knowledge on the effect of Prandtl number on pressure drop and heat transfer is not entirely satisfactory, thus justifying a comprehensive investigation of this problem. And, further, a general approach for predicting pressure drop and heat transfer of enhanced tubes is needed.

The objectives of this work were two-fold: first, to carry out an extensive and consistent experimental investigation of pressure drop and heat transfer for laminar, transitional, and turbulent flow with smooth and internally enhanced tubes using air, water, and ethylene glycol/water mixtures as the working fluids; and second, to validate the previously developed corresponding states method for a range of Prandtl number values. This paper is devoted to the first objective; while verification of the corresponding states method by using experimental results obtained for the 0.7 to 125 Prandtl number range is considered in Part II.

EXPERIMENTAL FACILITY AND TEST PROCEDURES

General Description of Apparatus

For the air studies, the experimental facility and test procedures were described in detail elsewhere (Esen *et al.*, 1994; Esen, 1992); hence, these details are not given in this paper. The closed-loop liquid test facility is shown schematically in Fig. 1. The main components include a storage tank, a variable-speed gear pump, a surge tank, three rotameters, the test section consisting of the entrance section and the heated section, and a Basco Series 500 single-pass shell-and-tube heat exchanger having an outside surface area of 0.5 m^2. The pressure in the surge tank is limited to 310 kPa by a cut-off switch connected to the motor-controller power supply.

General Description of Enhanced Tubes

Twenty-three enhanced tubes and a smooth tube were tested in this study; the geometric characteristics of the smooth tube and the enhanced surfaces are given in Table 1. In the first column of Table 1, S, GA, HC, W, and Y denote smooth, General Atomic, Hitachi Cable, Wieland-Werke AG, and IMI Yorkshire Alloy, respectively. The last four are the suppliers of the tubes used in this study. A close-up photograph of all tubes is presented in Esen *et al.*, 1994.

In this study, the characteristic dimension in the definition of the nondimensional pressure drop, flow-rate, and heat-transfer coefficient is the maximum internal tube diameter; the mean values determined in our laboratory are given in Table 1. For helix angle, the values for the Hitachi tubes (HC-4, HC-5, HC-6) are the manufacturer's data; values for the remaining tubes were calculated from the relation $\alpha = \pi D_i/pN_s$, where p is the axial pitch and N_s is the number of starts.

It is evident from Table 1 that the twenty-three enhanced tubes have a common feature, that is, the internal surface geometries are spirally shaped. For each of the

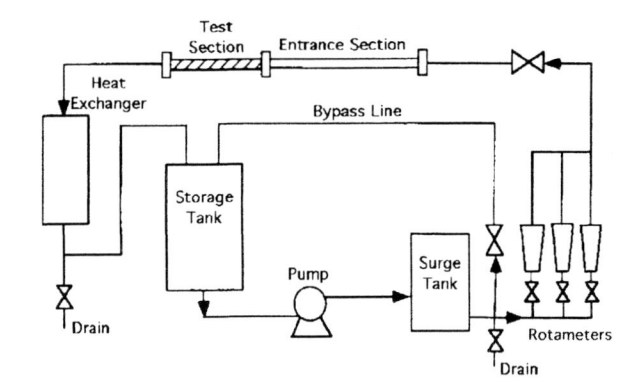

Fig. 1. Schematic diagram of closed-loop liquid test facility

spirally fluted tubes (GA-1 through GA-3, Y-22 and Y-23), the internal surface contour is similar to that on the outside surface. The HC-4 and the W-7 through W-13 tubes are basically spirally ribbed surfaces. Because the latter are also referred to by the manufacturer as spirally finned tubes, both terminologies are used interchangeably in this paper.

The two Hitachi tubes (H-5 and H-6) complement one another in that the surface protrusions are cross-cut to provide multistart three-dimensional spiral ribs. Details of the procedures used to generate these surface contours are given by Takahashi *et al.* (1985). For these tubes, the primary ribs form an angle α_1 with the tube axis, while the row of dents on the primary ribs form a different angle α_2 against the tube axis. When the ribs are oriented in the same rotational direction as the primary rib, α_2 is positive, and negative when the orientation of the dents is opposite that of the primary ribs. The height of the primary rib is e_1, while the depth of the dents is e_2. The axial pitch of the primary rib measured along the tube axis is p_1 and that of the dents is p_2.

The remaining eight tubes, (Y-14 to Y-21), were supplied by IMI Yorkshire Alloys and are referred to as spirally roped tubes by the supplier. They are characterized by indentations on the outside surfaces with ridges on the inner surfaces and are called spirally indented tubes in this paper.

Test Procedures

The basis for the design of the heated section was that of same heat transfer area. The exceptions were GA-3, Y-17, Y-19, and Y-21–Y-23, all of which were characterized by higher D_i values and larger heat transfer areas. Expressed in terms of D_i, the length of the heated section, L_h, ranged from about 9.5 D_i to 39 D_i, the lower values corresponding to the larger-diameter tubes. In other words, L_h varied slightly from tube to tube over the range of values from 475 mm for Y-19 to 313 mm for Y-16; the heated length for the smooth tube was 458 mm. Details on the experimental design are provided in the original report (Obot, 1995).

For each tube, the pressure-drop data were obtained in the presence of, as well as in the absence of, heat transfer

Table 1. Geometric Characteristics of Tubes

Tube	D_i (mm)	t (mm)	e (mm)	N_s	l (mm)	p (mm)	α (degrees)	e/D_i	p/e	Material	Roughness Description
S-0	13.39	1.2								copper	smooth
GA-1	21.45	0.7	0.95	20	82.0	4.1	39.4	0.044	4.3	stainless steel	spirally fluted
GA-2	23.96	0.98	1.33	25	141	5.6	28.1	0.056	4.2	stainless steel	spirally fluted
GA-3	28.49	1.89	1.58	31	160	5.2	29.2	0.055	3.3	aluminum	spirally fluted
HC-4	13.87	1.0	0.3	10	82.0	8.2	28.0	0.022	27.3	copper	spirally ribbed
HC-5	17.78	0.64	0.5 (0.3)	25 (25)	142.5 (93.0)	5.7 (3.7)	21.5 (-31.0)	0.028 (0.017)	11.4 (12.3)	copper	3-D spirally ribbed
HC-6	17.61	0.72	0.26 (0.14)	25 (25)	140.0 (92.1)	5.6 (3.7)	21.5 (-31.0)	0.015 (0.008)	21.5 (26.4)	copper	3-D spirally ribbed
W-7	14.10	1.07	0.42	1	2.2	2.2	87.2	0.030	5.2	copper	spirally ribbed
W-8	14.40	1.12	0.10	1	1.0	1.0	88.7	0.007	10.0	copper	spirally ribbed
W-9	15.90	1.52	0.5	41	102.5	2.5	26.0	0.032	5.0	copper	spirally ribbed
W-10	14.95	1.48	0.55	25	110.0	4.4	23.1	0.037	8.0	copper	spirally ribbed
W-11	14.45	1.49	0.45	20	76.0	3.8	30.9	0.031	8.5	copper	spirally ribbed
W-12	14.56	1.59	0.50	10	40.0	4.0	48.8	0.034	8.0	copper	spirally ribbed
W-13	14.45	1.50	0.51	25	120.0	4.8	20.7	0.035	9.4	copper	spirally ribbed
Y-14	12.68	1.08	0.38	3	15.0	5.0	69.4	0.030	13.2	copper	spirally indented
Y-15	19.16	1.24	1.27	3	30.0	10.0	63.5	0.066	7.9	copper	spirally indented
Y-16	19.53	1.23	0.51	3	30.0	10.0	63.9	0.026	19.6	copper	spirally indented
Y-17	24.22	1.67	0.31	3	15.0	5.0	78.8	0.013	16.1	copper	spirally indented
Y-18	18.81	1.66	0.36	3	7.8	2.6	82.5	0.019	7.2	copper	spirally indented
Y-19	22.88	1.04	1.5	6	198	33	20.0	0.066	22	K10	spirally indented
Y-20	16.05	1.36	1.0	3	30.0	10.0	59.2	0.062	10.0	copper	spirally indented
Y-21	23.45	0.93	0.52	1 (6)	6.0	6.0	85.3	0.022	11.5	K10	spirally indented doubly enhanced
Y-22	48.65	0.87	2.0	43	273.1	6.35	29.2	0.041	3.2	YAB	spirally fluted
Y-23	47.67	1.29	2.96	25	277.5	11.1	28.4	0.062	3.8	copper	spirally fluted

thus affording an extensive documentation of the effect of heat transfer on pressure drop. The frictional pressure coefficients (or values of the Fanning friction factor) were computed from the relation

$$f = \Delta p \rho_w A_x^2 D_i / 2m^2 L_p, \qquad (1)$$

where ρ_w is the test fluid density evaluated at the average surface temperature.

For heat transfer, each test section was heated with nichrome wire, located in small grooves machined on the outside surface of a tube, by DC power source. The temperatures close to the inner surface of a test section were measured with 24 chromel-alumel thermocouples (36 gauge) located at six axial stations, with four thermocouples equally spaced circumferentially at each station.

With regard to the test procedures, the first involved establishing steady-state conditions for a desired flow rate in the absence of heating and then recording of all temperature and pressure readings. Next, power was supplied to the test section. The power supply controls were fine-tuned periodically to obtain the desired mean surface temperature and the deviation about this mean value; both were held fixed with increasing flow rate for each tube. Steady-state conditions with heating were reached when the fluid and surface temperatures remained unchanged for about 30 min. Then, all temperature and pressure readings were recorded.

Results for the average heat transfer coefficient, expressed in nondimensional form as the values of average Nusselt number (Nu), were computed from

$$Nu = (D_i Q_c)/(k_b A_h)(T_w - T_b), \qquad (2)$$

where $T_b = (T_o + T_i)/2$ and the convective heat transfer rates were computed from

$$Q_c = Q_T - Q_L \qquad (3)$$

and

$$Q_c = m\, C_p (T_o - T_i). \qquad (4)$$

In Eq. (3), Q_T and Q_L represent the total electrical power input with and without fluid flow, respectively. The losses, Q_L, determined experimentally in the absence of the fluid flow, corresponded to the total electrical power input required to maintain the test section at the same average surface temperature as in tests with fluid flow.

For air ($Pr = 0.7$), the Q_c values were calculated from Eq. (3) for all test trials. The differences in the calculated values between Eqs. (3) and (4) were within ±10% (Esen, 1992). For water, Eq. (4) was the basis for all Q_c values used in the computation of the results in Das (1993). However, the laminar flow data of that study were reanalyzed with Eq. (3), in line with the data-reduction procedure for the glycol/water mixtures. The determination of laminar flow Q_c values from the total electrical power input is more accurate than the use of Eq. (4) due to cross-sectional variations in the fluid temperature at the exit of the test section. This method is recommended. For liquids in turbulent flow, Q_c values were computed by Eq. (4); there were almost no variations in the fluid temperature at the exit of the test section.

The Reynolds number (Re) ranges covered in the study were 200-55,000, 300-51,000, 300-12,000, 200-7,100 and 200-4,000 for $Pr = 0.7$, 6.8, 24.8, 39.1 and 125.3, respectively. For $Pr = 39.1$ and 125.3, tests were performed with only the smooth tube because laminar flow prevailed over a wide flow-rate range and it was expected that the results would afford a definite statement on the dependence of the nondimensional heat transfer coefficient on the Reynolds number. Equipment limitations precluded testing of the enhanced tubes at higher Reynolds numbers with $Pr = 24.8$.

For air ($Pr = 0.7$) and water ($Pr = 6.8$), the fluid properties used in the data reduction were based on the tabulations provided in Holman (1990). For pure glycol and glycol/water mixtures, the fluid properties were determined from the tables and graphical illustrations in the booklet provided by the Industrial Chemicals Division of Union Carbide, the supplier of the ethylene glycol. In addition, the viscosity of the glycol or glycol/water mixtures was checked periodically during the course of the experiments with a Brookfield LV Viscometer. The liquid density was also determined by weighing known volumes.

RESULTS AND DISCUSSION

Before the presentation and discussion of results, several points should be made. First, a complete tabulation of the results for friction factor, Nusselt number, and Reynolds number was given in the report that forms the basis for this paper (Obot, 1995). Second, heat-transfer tests were not carried out with the GA-1 and GA-2 tubes. Also, the two large-diameter tubes (Y-22 and Y-23) were not tested with glycol/water mixture because the attainable Reynolds numbers would be too low. Third, due to space limitations, graphical illustrations for some of the enhanced tubes are not included in this paper; these were presented in the original report (Obot, 1995).

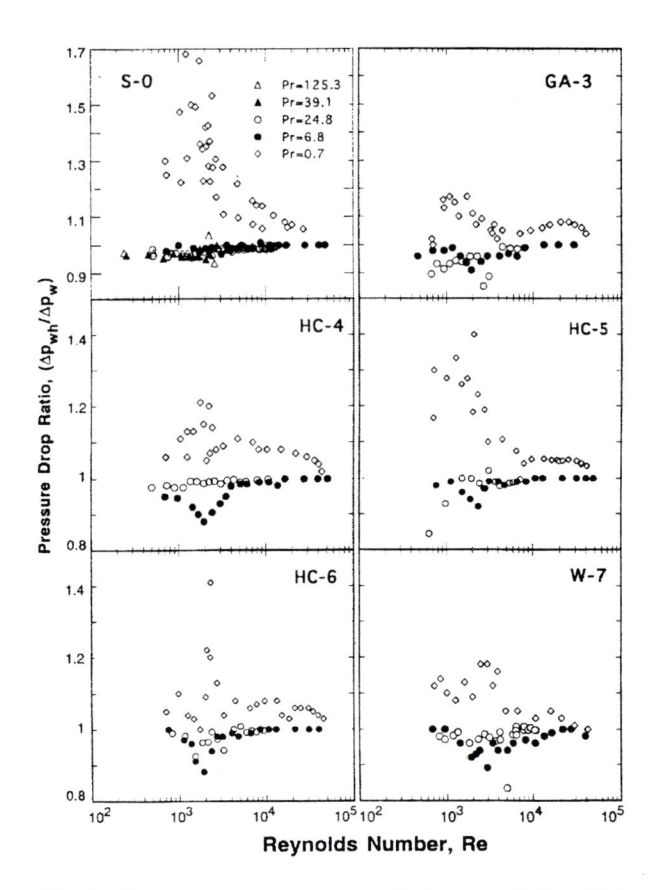

Fig. 2. Pressure drop ratio versus Re for tubes S-0 to W-7

Effect of Heat Transfer on Pressure Drop

In previous publications (Esen, 1992; Obot et al., 1994), it was established that useful information on the effect of heat transfer on pressure-drop can be obtained by using pressure-drop ratios. Accordingly, typical plots of pressure-drop ratio (• p_{wh}/• p_w) versus Reynolds number (Re) are presented compactly in Fig. 2. Similar trends were obtained with the other tubes (Obot, 1995). In Fig. 2, • p_{wh} is the steady-state pressure drop with heat transfer at the average surface temperature of the experiment, while • p_w is the corresponding value recorded in the absence of heat transfer; that is, just before the onset of heating of the test section.

For each tube, the results show that the greatest effect of fluid heating on pressure drop occurs in the transition region. For liquids ($Pr = 6.8$ and 24.8), the recorded pressure-drop with heating is generally lower than that recorded without heating in the transition regime for a given Reynolds number; however, the trend is exactly the opposite for air ($Pr = 0.7$). The viscosity increases and decreases with temperature for air and liquids, respectively, hence the reversal in the (• p_{wh}/• p_w) versus Re trends. For most tubes, the value is close to 1 in the fully turbulent regime.

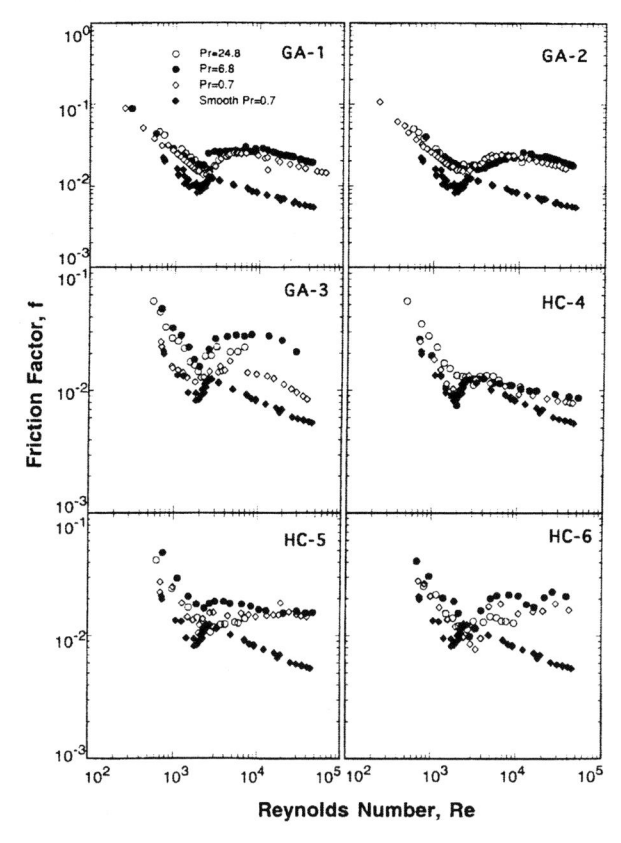

Fig. 3. Friction factor versus Re for tubes GA-1 to HC-6

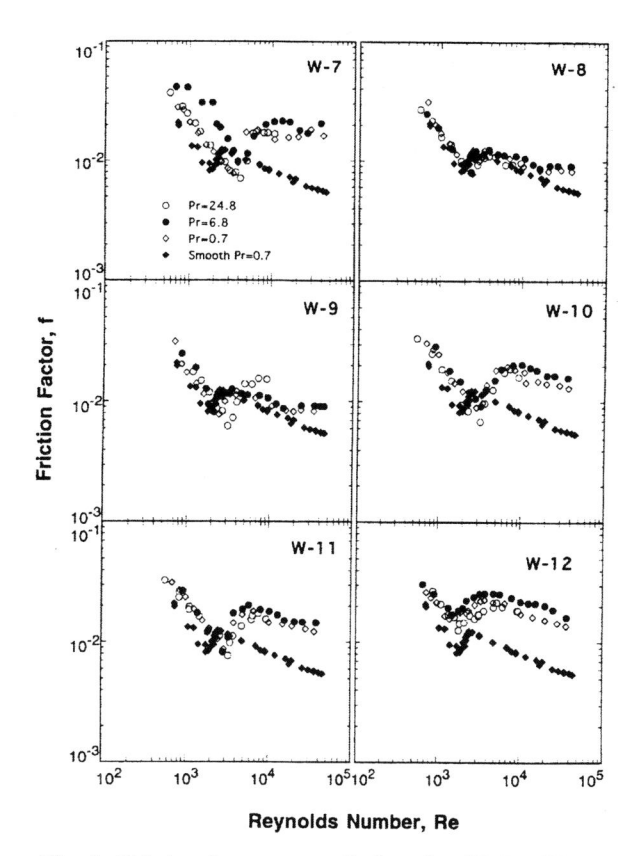

Fig. 4. Friction factor versus Re for tubes W-7 to W-12

The indication is that a plot of the pressure drop ratio against Reynolds number can be used to determine the Reynolds number at the onset of transition to turbulent flow. For liquids, the Reynolds number at transition can be determined with remarkable consistency as the location of the minimum values of ($\bullet p_{wh}/\bullet p_w$). For air, the transition Reynolds number almost corresponds to the location of the peak value of ($\bullet p_{wh}/\bullet p_w$). In passing, it is noted that the transition Reynolds numbers reported subsequently were determined from the friction factor and Nusselt number versus Re curves and not from those for pressure-drop ratio.

It should be noted that the pressure-drop ratio is dependent on the magnitude of the total electrical power input or on the average surface temperature. For liquids, the pressure-drop ratio decreases with increasing average surface temperature (Das, 1993). This contrasts sharply with the trend obtained with air ($Pr = 0.7$), the results of which indicated that the higher the average surface temperature, the higher the value of the pressure-drop ratio for a given Reynolds number (Esen, 1992).

Enhanced-Tube Friction Factor Results

The pressure-drop data reduced as values of the Fanning friction factor (f) are given compactly in Figs. 3-6. The results for each enhanced tube are compared with the $Pr = 0.71$ smooth-tube results. It is of interest to note that the smooth-tube friction factors are practically the same for all Prandtl numbers (Obot et al., 1997). The results obtained with heating are given for each tube and these are also used in all subsequent illustrations, unless stated otherwise. There are two exceptions: GA-1 and GA-2 tubes were not tested for heat transfer. Due to space limitation, the alternative representations of the friction factor/Reynolds number data as plots of the friction factor ratio, f_e/f_s, versus Re are given on Fig. 7 for W-7 through W-12.

In laminar flow, the general behavior of the results with increasing Re is the same for all tubes; f is inversely proportional to Re or the product $f Re (= C_f)$ is constant. The departure from the smooth-tube data depends on the geometric characteristics of the enhanced tube and is most pronounced for tubes of the spirally fluted type (GA-1 through GA-3, Y-22 and Y-23).

The results indicate rather complex effects of the enhanced-tube geometric details and Prandtl number on the transition process. For a number of tubes, the transition Reynolds numbers (Table 2) determined from plots of $f Re$ ($= C_f$) versus Re are significantly greater than the smooth-tube values. As a result, crossings of the enhanced and smooth-tube f versus Re curves are observed within the transition region. The friction factor at transition also varies with the tube geometry and Prandtl number (Table 2), further complicating definite statements on the effects of geometric parameters (e/D_i, p/e and α) and Prandtl number on the transition friction factor or the transition Reynolds number.

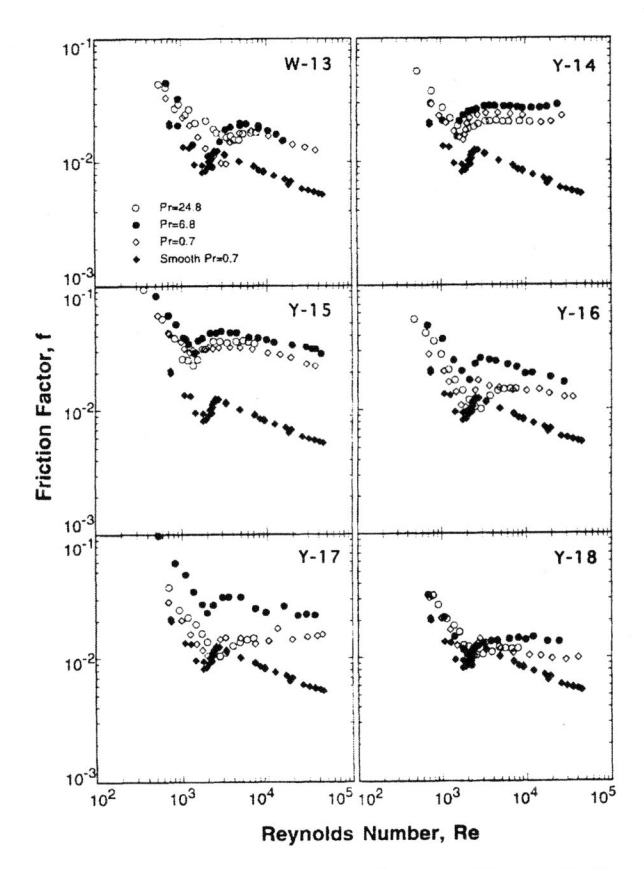

Fig. 5. Friction factor versus Re for tubes W-13 to Y-18

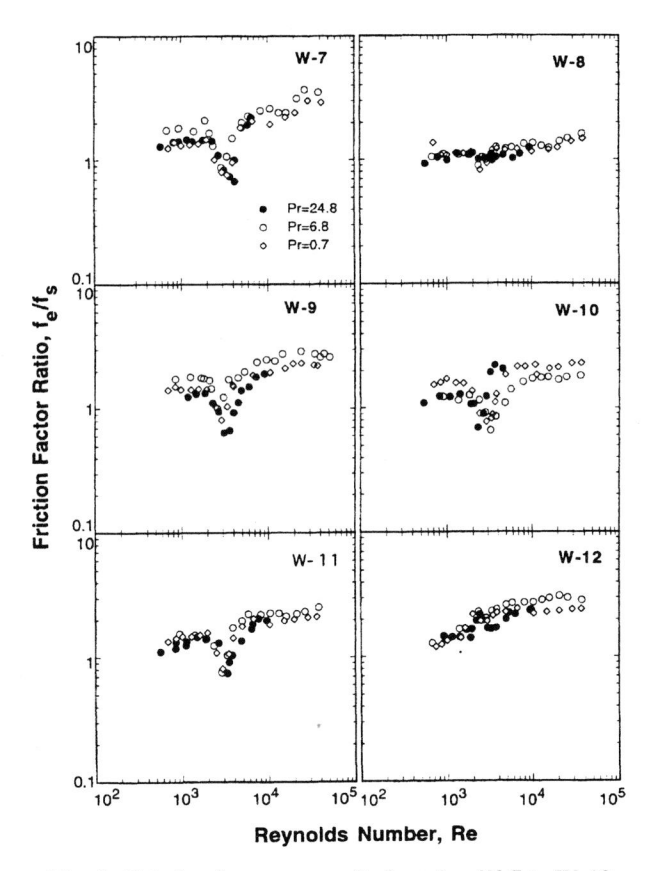

Fig. 6. Friction factor versus Re for tubes W-7 to W-12

The complications introduced by variations in the transition parameters (f_c and Re_c) are also reflected in the contrasting trends for the f_e/f_s versus Re curves (Fig. 7). For instance, the friction factor ratios for four of the seven W-tubes (W-7, 9, 10, and 11) lie well below unity; the transition Reynolds numbers for these tubes are much higher than the smooth-tube values (Table 2). By contrast, virtually all of the remaining tubes gave f_e/f_s values that were on the order of 1 or higher.

In turbulent flow, values of friction factor for any particular enhanced tube are generally greater than the smooth-tube values. Of the twenty-three enhanced tubes tested, only six (GA-1 through GA-3, Y-14, Y-15 and Y-20) are characterized by f_e/f_s values that are greater than 3 but under 5.5; f_e/f_s • 3 for the other tubes. Table 2 shows that these six tubes give transition Reynolds numbers that are generally lower than the smooth-tube values. So, the unique behavior of these six tubes is most likely the result of the inseparable effects of α, e/D_i, and p/e and the early transition to turbulence.

A study of these figures reveals that the influence of Prandtl number depends on the enhanced-tube geometric characteristics. For some of the tubes, the friction factor is essentially independent of the Prandtl number, paralleling the smooth-tube behavior (Obot et al., 1997). The variations with Pr are confined only to the laminar, transitional, or turbulent flow for some of the tubes. For four tubes (GA-3, Y-16, Y-17, and Y-21), the differences between the three sets of data are significant, and these are observed over the entire Re range covered in the study.

It is evident from Table 1 that the upper limit to the range of e/D_i tested is 0.066 and, for this case, $p/e = 2.2$ (Y-19); this value of e/D_i can be considered to be on the high side. Despite this, the friction factors for this tube are no greater than twice those for the smooth tube. This contrasts sharply with the friction factor trends for surfaces having transverse disruptions. For instance, the f_e/f_s values at comparable Re for the transverse inserts of Koch (1960) with $e/D_i = 0.045$ and 0.075 ($p/e = 9.8$ in each case) are about 17.5 and 20, respectively. These values, even that for the lower e/D_i of 0.045, are considerably greater than those for the spiral disruptions. Differences of these magnitude in the pressure-drop characteristics between the two types of surface disruptions cannot be explained in terms of the difference in p/e, because the f_e/f_s values for Y-20 ($e/D_i = 0.062$ and $p/e = 10$) are both much lower than those deduced from Koch's data for transverse ribs of $e/D_i = 0.045$.

Enhanced-Tube Heat-Transfer Results

The convective heat-transfer rates reduced as values of the average Nusselt number (Nu) are presented on Figs. 7-10. Consistent with the treatment of friction factor, the alternative representations as plots of the Nusselt number

Table 2. Critical Data at the Onset of Transition to Turbulent Flow

Tube	Re_c			$f_c \cdot 10^3$			Nu_c		
	Pr=0.7	Pr=6.8	Pr=24.8	Pr=0.7	Pr=6.8	Pr=24.8	Pr=0.7	Pr=6.8	Pr=24.8
S-0	2100	2040	1900	9.3	8.2	8.0	6.1	17.3	22.8
GA-1	1880	1520	1660	14.4	18.8	19.1	-	-	-
GA-2	1850	1870	1670	15.4	16.4	17.8			
GA-3	1790	1980	1870	14.2	15.4	13.3	9.2	21.6	34.1
HC-4	1970	1940	1640	8.9	8.4	12.5	6.4	17.1	20.9
HC-5	2000	1950	2390	10.9	15.5	10.9	7.0	20.5	29.7
HC-6	2020	1790	2160	9.1	10.1	10.7	7.1	16.9	22.6
W-7	2970	2256	3120	7.6	11.9	7.7	8.6	21.5	28.9
W-8	2440	2290	2350	7.1	9.1	7.6	7.4	18.2	26.8
W-9	2870	2600	2670	8.2	10.0	7.8	9.0	21.5	29.5
W-10	2620	2220	2270	8.8	10.7	8.6	7.7	20.7	27.2
W-11	2970	2290	2640	8.1	10.1	8.1	9.2	20.6	27.6
W-12	1500	1550	1840	13.3	15.6	13.3	5.0	16.5	21.8
W-13	2730	2300	1460	9.4	10.9	18.8	8.3	22.2	22.7
Y-14	1750	1420	1580	13.4	16.3	15.9	6.6	17.7	18.0
Y-15	1230	1090	1180	26.4	34.0	23.6	7.6	18.3	22.9
Y-16	2070	2030	2100	9.9	16.4	10.7	8.4	22.0	25.5
Y-17	1870	2040	2330	11.0	10.9	19.0	7.5	21.0	28.2
Y-18	1980	1870	1820	10.5	10.6	13.3	7.9	20.5	23.3
Y-19	2200	1980	1680	12.7	19.2	12.9	10.4	23.1	23.2
Y-20	830	1000	1070	31.7	29.4	27.5	5.0	15.1	17.4
Y-21	2170	1930	2440	9.7	13.1	21.0	10.2	29.8	31.1
Y-22	2770	3910	-	14.9	15.9	-	15.7	36.0	-
Y-23	2370	3040	-	16.6	13.5	-	12.8	33.2	-

ratio Nu_e/Nu_s versus Re are given in Fig. 10 for W-9 through Y-14.

Figures 7-9 show that Nu increases steadily in laminar flow for all tubes; each curve rises after the point of transition, in line with the friction factor behavior, and then changes direction at the onset of fully-turbulent flow. Figure 10 shows that Nu_e/Nu_s is essentially independent of Re in laminar flow, an indication that $Nu \propto Re^{1/2}$ for all enhanced tubes, as documented in Obot et al. (1997) for the smooth-tube. In laminar flow, the ratio Nu_e/Nu_s varies from slightly above unity to about 2.5, depending on the geometric details of the enhanced tube and the Prandtl number. Whereas most of tubes gave Nu_e/Nu_s values that were about the same for all Pr, the Nusselt number ratios obtained for $Pr = 24.8$ with the spirally indented tubes as a group (Y-14 through Y-21) were consistently the lowest set.

In the transition region, the trend for the most part is one of a drop in Nu_e/Nu_s value, beginning at the onset of transition to turbulent flow. As noted already in connection with the discussion of the results for friction factor ratio, this is a reflection of the crossing of the Nu versus Re curves because the transition Reynolds numbers for the enhanced tubes are much higher than the smooth-tube value (Table 2). The transition Nusselt number values determined as the limiting points with increasing Re in laminar flow on plots of $C_h = Nu/Re^{1/2}$ versus Re are presented in Table 2.

Figures 7-9 show that $Nu \propto Re^n$ in turbulent flow. For a specified Re range, the empirically determined value for n depends on the enhanced-tube geometric details and Pr. The dependence of the Reynolds number exponent on the type of surface disruptions is well documented in the literature (see, for example, Carnavos, 1980). In that study, the observation was that spirally finned tubes deviated the most from the 0.8 exponent on Re.

Given the rather contrasting trends of the friction factor versus Reynolds number curves and the effect of the transition process, some variation in the empirically determined Reynolds number exponent must be expected. Because observations based solely on the conventional Nu versus Re plots provide only a partial picture of the friction and heat transfer process, there is the need for a more general treatment in terms of Nu, f, and Re. Such a treatment is considered subsequently in Part II.

In turbulent flow, the results in Fig. 10 indicate various levels of enhancement depending on the geometric characteristics of the enhanced tube and to a lesser extent on the Prandtl number. With the exception of several tubes of the Y-series, there is no strong indication that the enhanced-tube heat-transfer results are sensitive to variations in the Prandtl number for liquids. It is of interest to note that Carnavos (1980) reached a similar conclusion for forged-finned tubes based on tests with water and glycol/water mixtures.

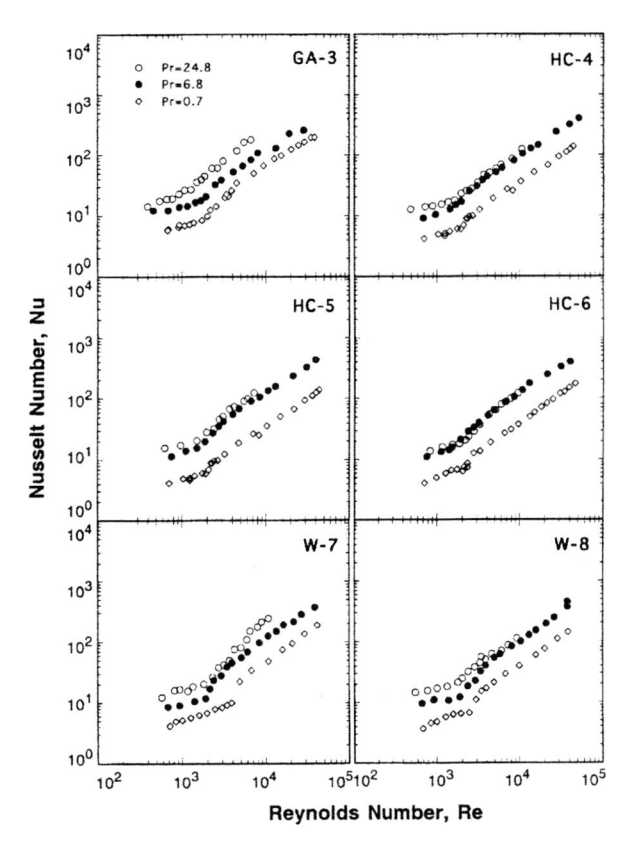

Fig. 7. Nusselt number versus Re for tubes GA-3 to W-8

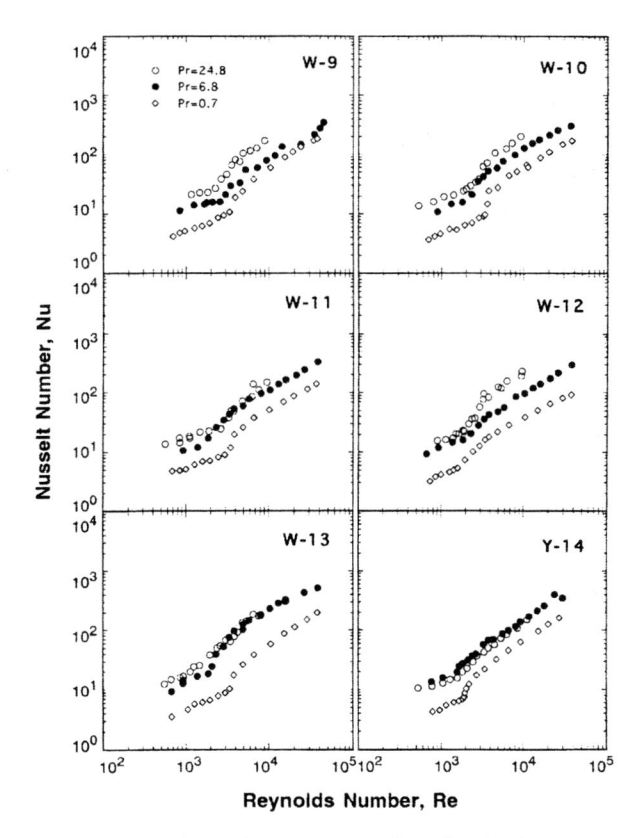

Fig. 8. Nusselt number versus Re for tubes W-9 to Y-14

A comparison of the f_e/f_s versus Re curves for W-9 through W-12 (Fig. 6) with the corresponding Nusselt number ratios of Fig. 10 reveals complete similarity in behavior between the two sets of results for any particular tube. For instance, the existence of a maximum or minimum in the f_e/f_s versus Re curve is also reflected in a maximum or minimum in the Nu_e/Nu_s versus Re plot. Also, when f_e/f_s increases with increasing Re, Fig. 10 shows that the same trend is exhibited by Nu_e/Nu_s. Although there are quantitative differences between the magnitude of f_e/f_s and Nu_e/Nu_s, the similarity in the general features of the curves begins in laminar flow and continues in the transitional and turbulent flow regimes.

Data Correlation

In principle, a generalized correlation for Nusselt number, Nu, can be presented in terms of the friction factor (f), Prandtl number (Pr), helix angle (α), relative height (e/D_i), and relative pitch (p/e). A thorough analysis of the results revealed that, expressed in terms of f and Pr, the correlation for laminar flow Nusselt number is independent of the geometric details of the enhanced tube (α, e/D_i and p/e). Accordingly, the laminar flow results for smooth and enhanced tube are closely approximated by the relation

$$\overline{C_h}/C_f = Nu/(Re^{3/2} f Pr^{0.4}) = 0.008. \quad (5)$$

The correlation, valid for $100 < Re \cdot 2100$ and $0.7 \cdot Pr \cdot 125$, predicts about 90% of the data points with errors that are mostly under 30%. There are indications that Eq. (5) is of general validity; no corrections are needed to account for the effect of transition to turbulence. A further elaboration on these observations is considered in Part II of this paper.

On a log-log plot of $\overline{C_h}/C_f$ versus Re, the data follow two distinct trends: the horizontal laminar flow segment, Eq. (5), and the decreasing trend with increasing Re for both the transition and turbulent flow regimes. For the latter, it is established that $Nu/(f Pr^{0.4}) \propto Re^n$, where $n = 1.05$. Unlike laminar flow, a single correlation of the Nu results in terms of f, Pr, and Re could not be established due to the residual effects of α, e/D_i and p/e that manifest themselves through variations in the transition parameters. The fact that $\overline{C_h}/C_f$ is constant in laminar flow for smooth and enhanced tubes, with significant spread of the data in transitional and turbulent flow, suggests a definite connection between the transition process and the attainable friction and heat transfer.

CONCLUSIONS

The results show that the effect of heat transfer on pressure drop is greatest in the transitional flow region, with moderate effect in laminar and turbulent flow. The effect of

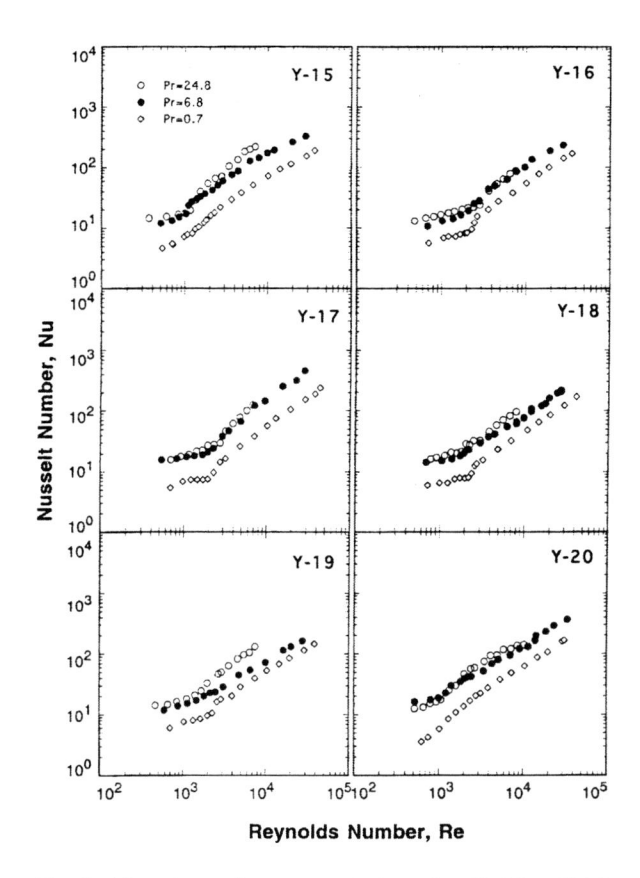

Fig. 9. Nusselt number versus *Re* for tubes Y-15 to Y-20

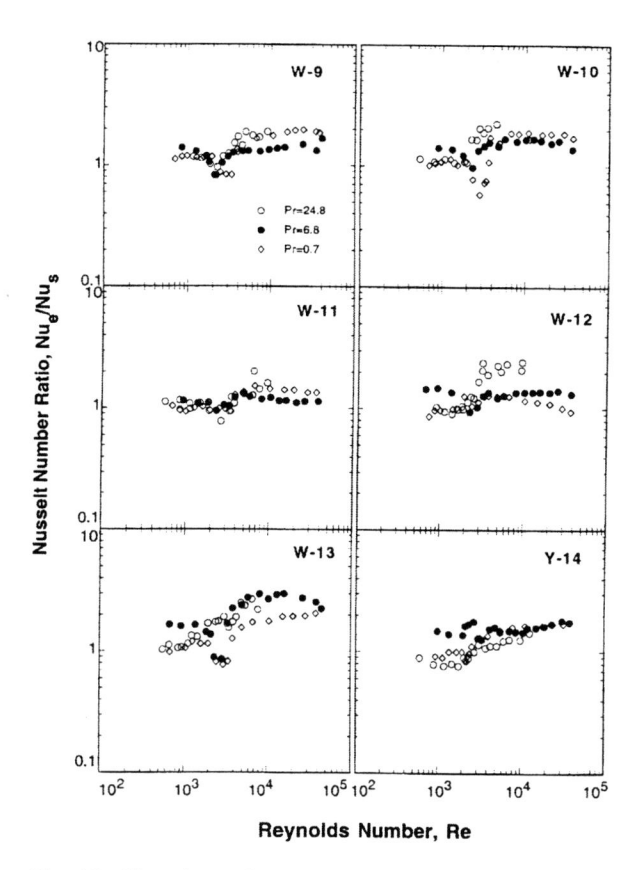

Fig. 10. Nusselt number versus *Re* for tubes W-9 to Y-14

wall cooling or fluid heating with liquids is exactly the opposite of that with air; a typical ($\bullet p_{wh}/\bullet p_w$) versus *Re* curve exhibits a well-defined minimum at the onset of transition to turbulence with liquids and a maximum with air.

The friction factor results indicate that, unlike transverse ribs or inserts, spirally shaped disruptions result in very moderate increases in pressure drop. The friction factors are usually no more than three times the smooth-tube values. The effect of Prandtl number on friction factor depends on the tube geometry and the flow type. For some tubes, the friction factors are practically the same for 0.7 $\bullet$ *Pr* $\bullet$ 25, with differences in laminar, transitional, or turbulent flow for some of the tubes.

In laminar flow, $Nu \propto Re^{1/2}$ for all the enhanced tubes with 0.7 $\bullet$ *Pr* $\bullet$ 25, paralleling the trend obtained with the smooth tube for 0.7 $\bullet$ *Pr* $\bullet$ 125.3. Each *Nu* versus *Re* curve rises after the onset of transition to turbulent flow and then changes direction at the onset of fully turbulent flow. Consistent with the low-pressure-drop characteristics of these enhanced tubes, the increases in heat transfer coefficient are no more than 2.5 times the smooth-tube values for the *Re* and *Pr* ranges covered in the study.

ACKNOWLEDGMENTS

This work, performed in the facilities at Clarkson University, was funded by the U.S. Department of Energy, Energy Efficiency and Renewable Energy, Division of Advanced Industrial Concepts, E. P. HuangFu, Program Manager, Contract No. DE-FG02-89CE90029. Technical Support was provided by Dr. T. J. Marciniak, Manager, Experimental Systems Engineering, and by Dr. T. J. Rabas, Energy Systems Division, Argonne National Laboratory.

The authors thank Roger A. Green, Daniel Vakili, and Jeremie Dalton for their important contributions to this work. John Wescott made modifications to the test facility. The ethylene glycol was donated for the research by Union Carbide at the request of John Wescott.

Part of this work has been supported by the U.S. Department of Energy, Energy Efficiency and Renewable Energy, under Contract W-31-109-Eng-38.

REFERENCES

Carnavos, T. C., 1980, Heat Transfer Performance of Internally Finned Tubes in Turbulent Flow, *Heat Transfer Eng.*, Vol. 1, pp. 32-37.

Das, L., 1993, Pressure Drop and Heat Transfer for Water Flow Through Enhanced Passages, MS Thesis, Clarkson University.

Dipprey, D. F., and Sabersky, R. H., 1963, Heat and Momentum Transfer in Smooth and Rough Tubes at Various Prandtl Numbers, *Int. J. Heat Mass Transfer*, Vol. 6, pp. 329-353.

Esen, E. B., Obot, N. T., and Rabas, T. J., 1994, Enhancement: Part I. Heat Transfer and Pressure Drop Results for Air Flow Through Enhanced Passages with Spirally-Shaped Roughness, *J. Enhanced Heat Transfer*, Vol. 1, #2, pp. 145-156.

Esen, E. B., 1992, Pressure Drop and Heat Transfer for Air Flow Through Enhanced Passages, PhD Thesis, Clarkson University.

Gomelauri, V., 1964, Influence of Two-Dimensional Artificial Roughness on Convective Heat Transfer, *Int. J. Heat Mass Transfer*, Vol. 7, pp. 653-663.

Holman, J. P., *Heat Transfer*, 7th Ed., McGraw-Hill, New York.

Koch, R., 1960, Pressure Loss and Heat Transfer for Turbulent Flow, U.S. Atomic Energy Commission, AEC-tr-3875.

Marner, W. J., and Bergles, A. E., 1978, Augmentation of Tubeside Laminar Flow Heat Transfer by Means of Twisted-Tape, Static-Mixer Inserts, and Internally Finned Tubes, *Proc. 6th Int. Heat Transfer Conf.*, Vol. 2, pp. 583-588.

Nunner, W., 1956, Heat Transfer and Pressure Drop in Rough Tubes, Atomic Energy Research Establishment (U.K.), Lib/Trans. 786.

Obot, N. T., 1995, Smooth-and-Enhanced-Tube Heat Transfer and Pressure Drop for Air, Water and Glycol/Water Mixtures, Final Report DOE/CE/90029-9.

Obot, N. T., Das, L., Vakili, D. A., and Green, R. A., 1997, Effect of Prandtl Number on Smooth-Tube Heat Transfer and Pressure Drop, *Int. Comm. Heat Mass Transfer*, Vol. 24, #6, pp. 889-896.

Obot, N. T., Esen, E. B., and Rabas, T. J., 1994, Enhancement: Part 11. The Role of Transition to Turbulent Flow, *J. Enhanced Heat Transfer*, Vol. 1, #2, pp. 157-167.

Obot, N. T., Esen, E. B., and Rabas, T. J., 1990, The Role of Transition in Determining Friction and Heat Transfer in Smooth and Rough Passages, *Int. J. Heat Mass Transfer*, Vol. 33, pp. 2133-2143.

Panchal, C. B., and France, D. M., 1986, Performance Tests of the Spirally Fluted Tube Heat Exchanger for Industrial Cogeneration Applications, Argonne National Laboratory Report ANL/CNSV-59.

Rabas, T. J., 1989, Selection of the Energy-Efficient Enhancement Geometry For Single-Phase Turbulent Flow Inside Tubes, Heat Transfer Equipment Fundamentals, Design, Applications and Operating Problems, HTD-Vol. 108, R. K. Shah, ed., American Society of Mechanical Engineers, New York, pp. 193-204.

Ravigururajan, T. S., and Bergles, A. E., 1986, Study of Water-Side Enhancement for Ocean Thermal Conversion Heat Exchangers, HTL-44/ERI Project 1718, Iowa State University.

Reay, D. A., 1991, Heat Transfer Enhancement - A Review of Techniques and TheirPossible Impact on Energy Efficiency in the U.K., *Heat Recovery Systems ~ CHP*, Vol. 11, pp. 1-90.

Shome, B., and Jensen, M. K., 1996, Experimental Investigation of Laminar Flow and Heat Transfer in Internally Finned Tubes, *J. Enhanced Heat Transfer*, Vol. 4, pp. 53-70.

Smith, J. W., and Gowen, R. A., 1965, Heat Transfer Efficiency in Rough Pipes at High Prandtl Number, *AIChE J.*, Vol. 11, pp. 941-43.

Takahashi, K., Nakayama, W., and Kuwahara, H., 1985, Enhancement of Forced Convective Heat Transfer in Tubes Having Three-Dimensional Spiral Ribs, *Trans. JSME*, Vol. 51-461, pp. 350-355.

Watkinson, A. P., Miletti, D. L., and Kubanek, G. R., 1974, Heat Transfer and Pressure Drop of Forge-Fin Tubes in Laminar Oil Flow, Noranda Research Center, Internal Report # 303.

Webb, R. L., 1987, Enhancement of Single-Phase Heat Transfer, in *Handbook of Single-Phase Convective Heat Transfer*, S. Kakac, R. K. Shah, and W. Aung, eds., John Wiley and Sons, New York.

Webb, R. L., Eckert, E. R. G., and Goldstein, R. J., 1971, Heat Transfer and Friction in Tubes with Repeated-Rib Roughness, *Int. J. Heat Mass Transfer*, Vol. 14, pp. 601-617.

Withers, J. G., 1980a, Tube-Side Heat Transfer and Pressure Drop for Tubes Having Helical Internal Ridging with Turbulent/Transitional Flow of Single-Phase Fluid. Part 1. Single-Helix Ridging, *Heat Transfer Eng.*, Vol. 2, # 1, pp. 48-58.

Withers, J. G., 1980b, Tube-Side Heat Transfer and Pressure Drop for Tubes Having Helical Internal Ridging with Turbulent/Transitional Flow of Single-Phase Fluid. Part 2. Multiple-Helix Ridging, *Heat Transfer Eng.*, Vol. 2, # 2, pp. 43-50.

Yampolsky, J. S., Libby, P. A., Launder, B. E., and LaRue, J. C., 1984, Fluid Mechanics and Heat Transfer Spiral Fluted Tubing, GA Technologies Report GA-A17833.

NOMENCLATURE

A_h	heat transfer area
A_x	cross-sectional flow area
C_f	friction parameter, $f Re$
C_h	heat transfer parameter, $Nu/Re^{1/2}$
$\overline{C_h}$	heat transfer parameter, $Nu/Re^{1/2} Pr^{0.4}$
C_p	specific heat
D_i	maximum internal diameter of tube
e	roughness height
f	Fanning friction factor
h	average heat transfer coefficient
k_b	thermal conductivity at T_b
L_e	smooth entrance length
$L_{e,r}$	rough entrance length
L_h	length of heated section
L_p	distance between pressure taps
l	lead
m	mass flow rate
N_s	number of starts
Nu	Nusselt number

p	roughness pitch
Pr	Prandtl number
Δp	pressure drop
Δp_w	pressure drop without heat transfer
$\Delta p_{w,h}$	pressure drop with heat transfer
Q_c	convective heat transfer rate
Q_L	total electrical power without flow at T_w
Q_T	total electrical power with flow at T_w
Re	Reynolds number
T	temperature
T_b	bulk mean temperature, $(T_i + T_o)/2$
T_w	average surface temperature
t	wall thickness

Greek Symbols

α	helix angle
μ	fluid viscosity
ρ_w	fluid density evaluated at T_w

Additional Subscripts

e	enhanced tube
i	condition at inlet to test section
o	condition at test section exit
s	smooth tube

SMOOTH-AND ENHANCED-TUBE HEAT TRANSFER AND PRESSURE DROP: PART II. THE ROLE OF TRANSITION TO TURBULENT FLOW

N. T. Obot,[1] L. Das,[2] and T. J. Rabas[3]

[1]Argonne National Laboratory, Argonne, Illinois 60439, USA; E-mail: obot@anl.gov; Professor Emeritus, Department of Chemical Engineering, Clarkson University, Potsdam, NY 13699, USA
[2]Former Graduate Student, Department of Chemical Engineering, Clarkson University, Potsdam, NY, USA
[3]Argonne National Laboratory, Argonne, Illinois 60439, USA

ABSTRACT

The objectives of this presentation are two-fold: first, to demonstrate the connection between the attainable coefficients and transition to turbulent flow by using the transition-based corresponding states method to generalize results obtained with smooth tubes and enhanced tubes, and second, to provide guidelines on the calculation of heat transfer coefficients from pressure-drop data and vice versa by using the transition concept or the *frictional law of corresponding states*.

INTRODUCTION

It is well established that the onset of transition to turbulent flow manifests itself through a sharp rise in pressure, whether by the pressure gradient or the static pressure in the flowing fluid. Consistent with the connection between friction and heat transfer, the results already presented in Part I show that the Nusselt number increases sharply almost at the onset of the transition to turbulence. Accordingly, there is remarkable consistency in the values of the transition Reynolds number determined from friction factor and Nusselt number results.

In the absence of transition to turbulent flow, laminar flow trends prevail for both friction factor and Nusselt number with increasing Reynolds number, *i.e.*, $f \propto Re^{-1}$ and $Nu \propto Re^{1/2}$. Thus, transition is important because it gives rise to f or Nu dependence on Re that is vastly different from that for laminar flow. Consequently, values of f or Nu differ markedly between laminar and turbulent flow for equal Reynolds number.

The origin of the corresponding states concept was the analysis of friction factor results obtained in circular and noncircular smooth passages (Obot, 1988). The conclusion was that differences in friction factor were due to the inseparably connected effects of transition to turbulent flow and the length scale used to reduce the data to nondimensional form. Allowing for the differences in values of the critical friction factor and Reynolds number, the friction factor results for circular and noncircular passages were completely generalized. Attempts to extend the analysis to heat transfer were frustrated by lack of verifiable laminar flow heat transfer data. Hence, an extensive experimental study was initiated to provide the needed data for a range of tube geometry and working fluids (Esen, 1992; Das, 1993).

In this paper, the results presented in Part I are used to demonstrate the connection between the transfer coefficients and the transition process by using the corresponding states concept. Guidelines on the use of the correlations developed herein to calculate heat transfer coefficients or pressure drop for smooth or enhanced tubes are provided and discussed.

ANALYSIS

For the convenience of the reader, the applicable equations are briefly outlined here. Additional information is provided in Obot et al. (1994). With the corresponding states method, reference critical values for the transition Reynolds number, friction factor and Nusselt number are used to scale arbitrary sets of data according to the following relationships:

$$Re_m = (Re_{c,r}/Re_{c,a})Re_a \qquad (1)$$

$$f_m = (f_{c,r}/f_{c,a})f_a \qquad (2)$$

$$Nu_m = (Nu_{c,r}/Nu_{c,a})Nu_a , \qquad (3)$$

where $Re_{c,r}$, $f_{c,r}$, and $Nu_{c,r}$ correspond to the values of the transition Reynolds number, friction factor, and Nusselt number at the onset of turbulence for a reference data set, *i.e.*, for the smooth-tube in this study, and $Re_{c,a}$, $f_{c,a}$, and $Nu_{c,a}$ are the corresponding transition parameters for any arbitrary set of results.

In the laminar regime, the well-known relationship between the friction factor and Reynolds number is

$$fRe = C_f, \quad (4)$$

where C_f is a constant for each tube. Similarly, for heat transfer in laminar flow, the expression for the Nusselt number is

$$Nu/Re^{1/2} = C_h, \quad (5)$$

where C_h is also a constant for each tube. Allowing for the Prandtl number dependence by using the 0.4 power dependence, Eq. (5) becomes

$$Nu/(Re^{1/2} Pr^{0.4}) = \overline{C_h}. \quad (6)$$

Equations (4) and (6) for C_f and $\overline{C_h}$ are combined to give

$$\overline{C_h}/C_f = Nu/(Re^{3/2} f Pr^{0.4}). \quad (7)$$

The reduced form of Eq. (7), which accounts for variations in the transition parameters, is

$$\overline{C_{h,m}}/C_{f,m} = Nu_m/(Re_m^{3/2} f_m Pr^{0.4}). \quad (8)$$

Alternatively, the results for all Prandtl numbers can be generalized without incorporating Prandtl number dependence. This is accomplished simply by selecting a common reference for all results and the applicable equation takes the form

$$C_{h,m}/C_{f,m} = Nu_m/(Re_m^{3/2} f_m). \quad (9)$$

Equations (8) and (9) are general statements of the friction-heat transfer analogy. Although the formulation is based on laminar flow observations, there is no reason that the general form of these relationships should not hold for turbulent flow, with corrections to account for the effect of transition to turbulent flow.

RESULTS AND DISCUSSION

It is emphasized that for the presentation that follows, only the final correlations are considered and shown graphically. The general trends for the individual variables (C_f, $C_{f,m}$, $\overline{C_h}$, $\overline{C_{h,m}}$, and $C_{h,m}$) were presented in graphical form and discussed in several publications (Obot et al., 1994; Esen, 1992; Das, 1993); hence, they are not shown or discussed in this paper.

To begin presentation and discussion, attention is directed to Figs. 1-3. In Fig. 1, the fit of the laminar flow results to the correlation given in Part I is illustrated graphically; the results for the smooth tube (0.7 • Pr • 125.3) and enhanced tube (0.7 • Pr • 24.8) are included in the figure. Figure 2 is an alternative representation of the same laminar flow results; Eqs. (1)-(3) are used to account for the differences in the transition parameters. Figure 3 was prepared by using the transition parameters presented in Table 2 of Part I and shows the relationship between Nu_c, Re_c, and f_c.

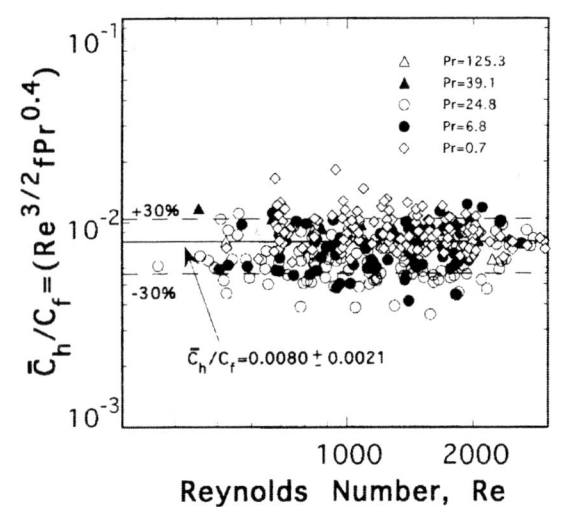

Fig. 1. Laminar flow $\overline{C_h}/C_f$ versus Re for all tubes

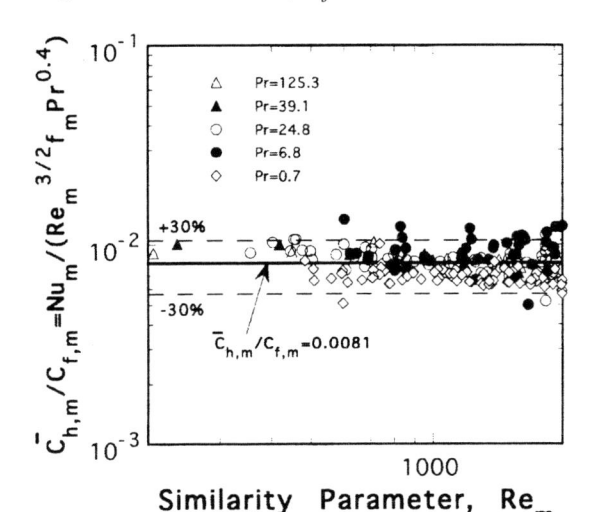

Fig. 2. Laminar flow $\overline{C_{h,m}}/C_{f,m}$ versus Re for all tubes

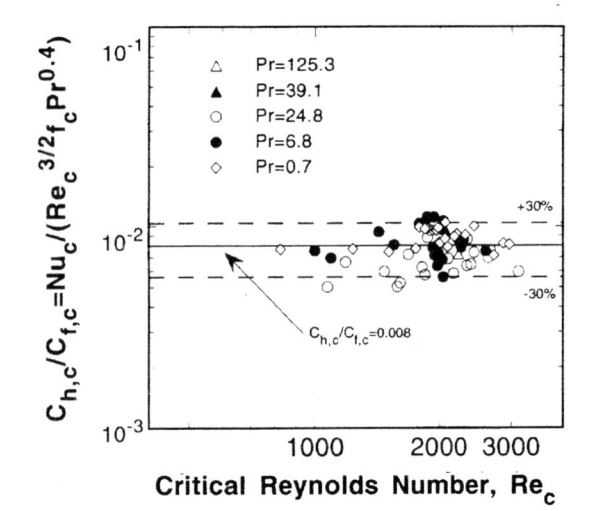

Fig. 3. Relationship between transition parameters for all tubes

It is evident that although the regression coefficient of 0.008 provides close approximation of the results in Figs. 1 and 2, the latter is characterized by less spread of the data. Accordingly, about 90% of the data points are predicted with errors that are mostly under 30% in Fig. 1; the error band is under 20% for 95% of the data in Fig. 2. The indication is that there is little influence of the transition process on laminar flow correlation. As might be expected because the Re and f values at the onset of transition must satisfy the laminar flow flow solution, the same regression constant provides satisfactory representation of the transition parameters (Nu_c, f_c, and Re_c) in Fig. 3.

If the subscripts are eliminated, the recommended correlation for calculation of laminar flow transfer coefficients, including the critical values at the onset of transition to turbulent flow, is

$$\overline{C_h}/C_f = Nu/(Re^{3/2} f Pr^{0.4}) = 0.008. \quad (10)$$

Equation (10) provides satisfactory estimates of the transfer coefficients, within the accuracy of experimental errors.

The practical significance of the trends in the laminar flow correlation relates to the calculation of transitional/turbulent-flow Nusselt numbers, inasmuch as experimental determination of all three transition parameters (f_c, Nu_c, and Re_c) is not necessary. For instance, use of laminar flow pressure-drop results, i.e., a plot of C_f (= $f Re$) versus Re, gives $C_{f,c}$ (= $f_c Re_c$), and Re_c; f_c is readily obtained by using the known Re_c. The remaining critical parameter, Nu_c, is calculated from Eq. (10). With Re_c, f_c, and Nu_c established, the friction factor/Nusselt number calculations proceed as outlined subsequently.

For the transition and turbulent regimes, it was noted in Part I that the $\overline{C_h}/C_f$ versus Re data followed the same trend for both flow types. Also, it was mentioned that correlation of the Nu results in terms of f, Re, and Pr was not possible due to the residual effects of roughness geometry. These observations can be inferred from the plot of $\overline{C_h}/C_f$ versus Re presented in Fig. 4. This figure was prepared by using results for the smooth tube and enhanced tubes, a total of 1023 data points.

Figures 5 and 6 are alternative representations of the results in Fig. 4; the data reduction formats correspond respectively to Eqs. (8) and (9). For the sake of completeness, the laminar flow results are included in Figs. 5 and 6. In Fig. 5 and for each Pr value, the values for the reference transition parameters are those for the smooth tube with the particular Prandtl number. Values of the transition parameters were presented in Table 2 of Part I. For Fig. 6, the values for the reference transition parameters were 6.0, 0.009, and 2100 for $Nu_{c,r}$, $f_{c,r}$, and $Re_{c,r}$ respectively. The results were obtained with the smooth tube and air as the working fluid. Note that in terms of $C_{h,m}$ and $C_{f,m}$, the corresponding laminar flow correlation is $C_{h,m}/C_{f,m} = 0.0068$, with no Pr dependence.

The implication of the trends in Fig. 5 or 6 is that, in terms of the expended power, laminar flow affords the best heat transfer performance relative to transitional and turbulent flow. This observation is of practical significance,

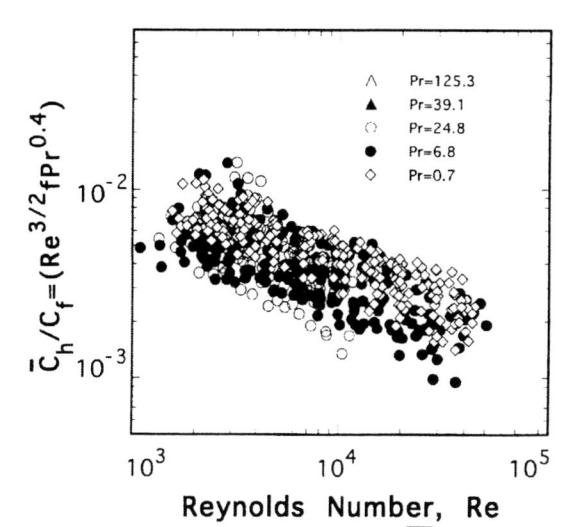

Fig. 4. Transitional and turbulent flow $\overline{C_h}/C_f$ versus Re

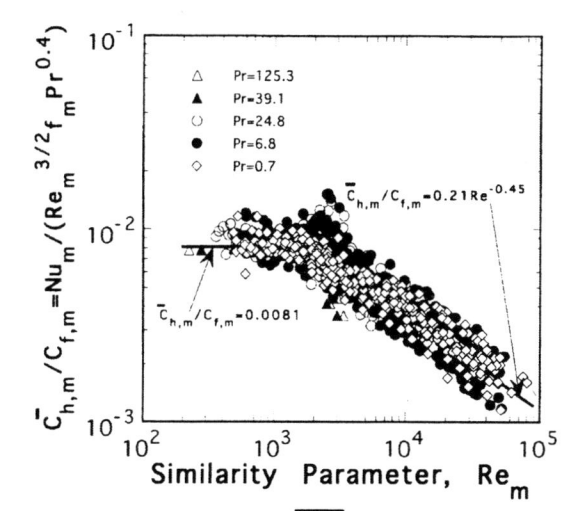

Fig. 5. Variation of $\overline{C_{h,m}}/C_{f,m}$ with Re_m

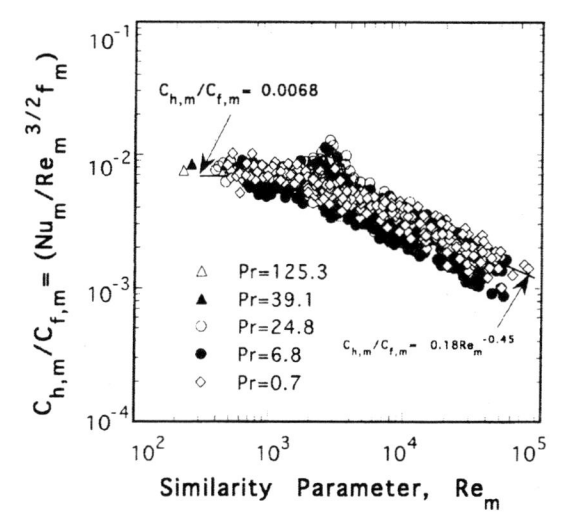

Fig. 6. $C_{h,m}/C_{f,m}$ versus Re_m with air transition parameters as reference

as it sheds light on the potential benefit of microscale heat and mass transfer. The unique feature of microheat exchangers and microreactors is the existence of mostly laminar flow; this is a primary reason for the interests in these devices.

For $Re_m \bullet 2100$, the results in Fig. 5 are closely approximated by

$$\overline{C_{h,m}}/C_{f,m} = Nu_m/(Re_m^{3/2} f_m Pr^{0.4}) = 0.21 Re_m^{-0.45}. \quad (11)$$

With a common reference for all results (Fig. 6) that does not include a Prandtl number dependence, the correlation for $Re_m \bullet 2100$ is

$$C_{h,m}/C_{f,m} = Nu_m/(Re_m^{3/2} f_m) = 0.18 Re_m^{-0.45}. \quad (12)$$

The scatter plots for Eqs. (11) and (12) are presented in Figs. 7 and 8, respectively; the appropriately calculated values for laminar flow are also included in the plots. It is evident from comparisons between Figs. 5 and 6, or between Figs. 7 and 8, that the use of a common reference gives slightly better collapse of the results. The remarkable collapse of unrelated $\overline{C_{h,m}}/C_{f,m}$ or $C_{h,m}/C_{f,m}$ versus Re_m data for $Re_m \bullet 2100$, with deviations that are much better than can be expected from this type of experiments, provides verification of the role of transition. It is apparent that the results for both transitional and turbulent flow regimes are effectively correlated with a single equation, with no additional parameters to account for the enhanced-tube geometric details.

Since Eq. (11) or (12) is valid for smooth tubes, its general form can be compared to conventional correlations, at least insofar as the dependence of Nu_m (or Nu) on Re_m (or Re) is concerned. For instance, in terms of the reduced variables, the combination of the Blasius equation ($f_m = 0.079 Re_m^{-0.25}$) and the Dittus-Boelter equation ($Nu_m = 0.023 Re_m^{0.8} Pr^{0.4}$) gives $Nu_m/f_m = 0.29 Re_m^{1.05} Pr^{0.4}$; the exponent on Re_m compares favorably with $Nu_m/f_m = 0.21 Re_m^{1.05} Pr^{0.4}$ from Eq. (11).

The comparison of Eq. (11) with the well-known Petukhov-Popov (1963) correlation was made quantitatively for the $2100 < Re_m < 10^5$ covered in this study by replacing Nu, f, and Re in the original correlation by Nu_m, f_m, and Re_m, respectively. Relative to the Petukhov-Popov correlation, Eq. (11) gives consistently lower values. Expressed as percentages, the deviations are generally within 1 to 20%, with an average value of $\bullet 15\%$. The primary reason for these deviations is that the physical properties (except for μ_b and μ_w) in the original Petukhov-Popov correlation are based on the film temperature, in contrast to the use of bulk properties (except ρ_w for friction factor) in this study. Nevertheless, this degree of concurrence demonstrates that heat-transfer coefficients for enhanced tubes can be calculated from smooth-tube correlations using the procedures outlined below in this paper.

Finally, it is worthy of note that Eq. (11) reduces to the well-known Reynolds analogy for $Pr = 0.7$. It is easily established that in terms of the reduced Stanton number

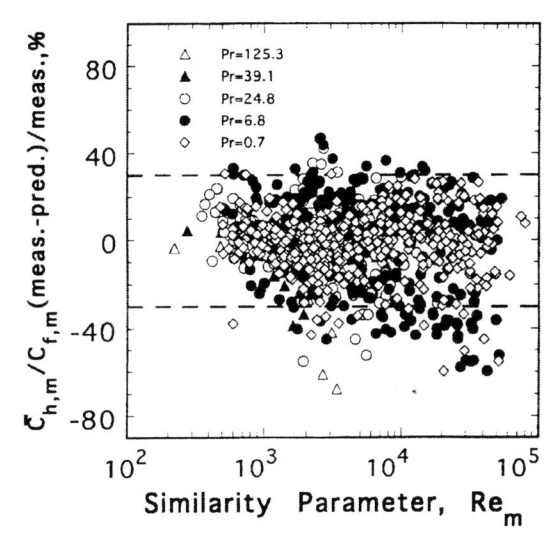

Fig. 7. Comparison of measured and predicted values of $\overline{C_{h,m}}/C_{f,m}$

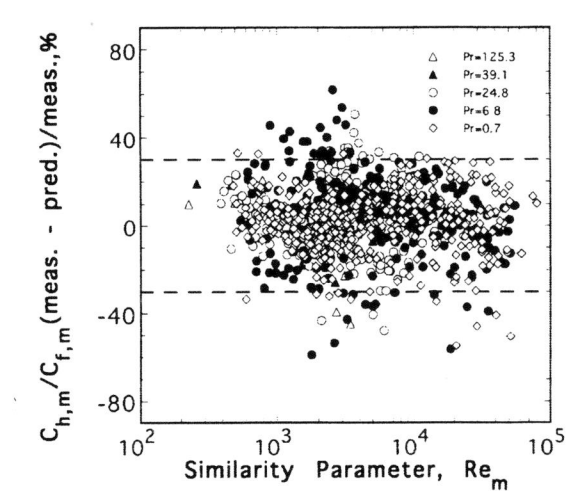

Fig. 8. Comparison of measured and predicted values of $C_{h,m}/C_{f,m}$ with air transition parameters as reference.

(St_m), reduced friction factor (f_m), and the Prandtl number (Pr), Eq. (11) is approximated by

$$St_m Pr^{0.6} = 0.4 f_m. \quad (13)$$

The validity range for this equation is $2100 \bullet Re_m \bullet 10^5$. It is evident that $St_m \bullet (1/2)f_m$ for $Pr = 0.7$, in line with the Reynolds analogy. However, unlike the familiar Reynolds analogy, Eq. (13) holds for both smooth and enhanced tubes at the same reduced conditions. The outcome of the analysis supports the 1/2 power dependence of Nu on Re in laminar flow. For $Re_m \bullet 2100$, the calculated values from Eq. (13) are generally within 30% of the experimental values. So a general form of the friction and heat-transfer analogy that is consistent with the Reynolds analogy is developed for smooth and enhanced tubes.

Calculation of Nusselt Number or Friction Factor

As noted previously, one objective of this study was to develop a general predictive method for the heat transfer and pressure drop of smooth and enhanced tubes. The correlations presented herein afford calculations of Nusselt number from friction factor data for a specified Reynolds number range, and vice versa. With the transition parameters for $Pr = 0.7$ ($f_{c,r} = 0.009$, $Nu_{c,r} = 6.0$, and $Re_{c,r} = 2100$) as the reference, the calculation procedures are outlined as follows:

1. In laminar flow and for $0.7 \leq Pr \leq 125$, the unscaled correlation (in terms of Nu, f, and Re) is the same as that at reduced conditions (in terms of Nu_m, f_m, and Re_m). For a given Pr, Nu is calculated from Re and f, or f, and hence the pressure drop is computed from Nu and Re.

2. To calculate transitional- and turbulent-flow Nusselt numbers from friction factor-Reynolds number data, the steps are (a) determine the transition Reynolds number ($Re_{c,a}$) from the C_f versus Re plot using arbitrary data set; (b) compute $f_{c,a}$ from $f_{c,a} = C_f/Re_{c,a}$ and the transition Nusselt number ($Nu_{c,a}$) from $Nu_{c,a} = 0.008 Re_{c,a}^{3/2} f_{c,a} Pr^{0.4}$; (c) Re_m and f_m are computed from Eqs. (1) and (2), respectively; and (d) Nu_m and Nu_a are calculated from Eqs. (12) and (3), respectively.

3. The calculation of transitional- and turbulent-flow pressure drop from heat-transfer data from arbitrary data is carried out in a similar manner: (a) $Re_{c,a}$ is obtained from a plot of $\overline{C_h}$ versus Re; (b) $Nu_{c,a}$ is computed from $Nu_{c,a} = \overline{C_h}/Re_{c,a}$, $f_{c,a}$ from $\overline{C_h}/C_f = 0.008$; (c) Re_m and Nu_m from Eqs. (1) and (3), respectively; and (d) f_m and the required friction factor f_a from Eqs. (12) and (2), respectively.

It is evident from the above presentation that determination of the transition parameters is the key to the use of the transitional- and turbulent-flow correlations. From a practical standpoint, the calculation of heat transfer coefficients by using pressure drop data is to be preferred for several reasons. First, it involves determination of the transition values for the Reynolds number and friction factor, and these can be obtained with errors that are smaller than the uncertainties in measured heat transfer coefficients. Second, pressure-drop measurements are less time-consuming and are inexpensive. Third, there are straightforward and easily implemented methods for determination of the transition parameters from differential pressure measurements (Obot et al., 1991). In this regard, it is worth noting that use of the pressure-drop ratio discussed in Part I gives Re_c and f_c values with a 10% error band.

For smooth circular tubes, values of Re_c and f_c can differ from those used in this paper as the reference (Obot et al., 1994). Similarly, noncircular smooth passages are usually characterized by Re_c and f_c values that are not the same as those for circular smooth tubes when the experimental data are reduced using the hydraulic diameter or an arbitrarily-defined length scale (Obot, 1988). Under these conditions, Eqs. (1)-(3) must be used to calculate the reduced parameters needed for Eq. (11) or (12).

CONCLUSIONS

The *frictional law of corresponding states* has been validated from experimental results obtained for the $0.7 \cdot Pr \cdot 125.3$ range. In laminar flow, the characteristic feature is the existence of stable flow for all Reynolds numbers. Accordingly, the results for smooth and enhanced tubes are completely generalized in terms of f, Nu, Pr, and Re, and no corrections are needed to account for the effect of enhanced-tube geometric details or transition to turbulence. Judging from the results of the analysis, it appears that the laminar flow correlation is of general validity.

In terms of Pr and the reduced variables (f_m, Nu_m, and Re_m) the transitional- and turbulent-flow results are adequately represented by a single equation, and no additional parameters are required to account for the enhanced-tube geometric details. Alternatively, with the specified set of transition parameters (f_c, Nu_c, and Re_c) as the common reference for all results, a superior correlation in terms of f_m, Nu_m, and Re_m is obtained with no Prandtl number dependence.

ACKNOWLEDGMENTS

This work, performed in facilities at Clarkson University, was funded by the U.S. Department of Energy, Energy Efficiency and Renewable Energy, Division of Advanced Industrial Concepts, E. P. HuangFu, Program Manager, Contract No. DE-FG02-89CE90029. Technical support was provided by Dr. T. J. Marciniak, Manager, Experimental Systems Engineering, and by Dr. T. J. Rabas, Energy Systems Division, Argonne National Laboratory.

Part of this work has been supported by the U.S. Department of Energy, Energy Efficiency and Renewable Energy, under Contract W-31-109-Eng-38.

REFERENCES

Das, L., 1993, Pressure Drop and Heat Transfer for Water Flow Through Enhanced Passages, MS Thesis, Clarkson University.

Esen, E. B., 1992, Pressure Drop and Heat Transfer for Air Flow Through Enhanced Passages, PhD Thesis, Clarkson University.

Obot, N. T., Esen, E. B., and Rabas, T. J., 1994, Enhancement: Part II. The Role of Transition to Turbulent Flow, *J. Enhanced Heat Transfer,* Vol. 1, #2, pp. 157-167.

Obot, N. T., 1988, Determination of Incompressible Flow Friction in Smooth Circular and Noncircular Passages: A Generalized Approach Including Validation of the Century Old Hydraulic Diameter Concept, *Trans. ASME J. Fluids Eng.,* Vol. 110, pp. 431-440.

Obot, N. T., Wambsganss, M. W., and Jendrzejczyk, J. A., 1991, Direct Determination of the Onset of Transition to Turbulence in Flow Passages, *Trans. ASME J. Fluids Eng.,* Vol. 113, pp. 602-607.

Petukhov, B. S., and Popov, V. N., 1963, Theoretical Calculation of Heat Exchange and Frictional Resistance in Turbulent Flow in Tubes of an Incompressible Fluid with Variable Physical Properties, *High Temperature,* Vol. 1, pp. 69-83.

NOMENCLATURE

C_f friction parameter, $f Re$
C_h heat transfer parameter, $Nu/Re^{1/2}$
$\overline{C_h}$ heat transfer parameter, $Nu/Re^{1/2} Pr^{0.4}$
f Fanning friction factor
Nu Nusselt number
Pr Prandtl number
Re Reynolds number
μ_b fluid viscosity based on bulk temperature
μ_w fluid viscosity at T_w
ρ_w fluid density evaluated at T_w

Additional Subscripts
a arbitrary condition
c value at onset of transition
c,a transition parameters for arbitrary condition
c,r transition parameters for reference condition
m value at reduced condition

THEORETICAL AND EXPERIMENTAL STUDY OF AXIAL DISPERSION IN PACKED-BED HEAT/MASS TRANSFER EQUIPMENT

Sarit K. Das[1], Anindya Roy[1], Wilfried Roetzel[2] and Frank Balzereit[2]

[1]Dept. of Mechanical Engg, Indian Institute of Technology, Madras 600036, India
[2]Institute of Thermodynamics, University of Federal Armed Forces Hamburg, D-22039 Hamburg

ABSTRACT

The modelling of a packed bed heat/mass exchange equipment has been carried out using the proposition of axial dispersion. The analysis indicates the sensitivity of the transient response of such packed beds to the axial dispersion which suggests an accurate method for the determination of the dispersion coefficient characterized by dimensionless dispersive Peclét number. For the determination of this parameter a mass transfer experiment has been deviced. In this experiment a tracer concentration profile has been injected at the inlet of the equipment and by fitting computed results to the recorded outlet signal, the dispersive Peclét number is estimated. The results indicate a poor performance of the conventional 'plug flow' model. The method developed in the present work is a general one since any arbitrary type of input signal can be used for it. The variation of dispersive Peclét number with Reynolds number has been presented. The strength of the proposition of axial dispersion has also been shown by analysing the response of a thermal regenerator packed bed available in literature. The results indicate that the slope of the response is the most important feature for the determination of Peclét number due to axial dispersion.

INTRODUCTION

Regenerators or regenerative heat exchangers find wide applications in power and process industry. The basic concept of thermal regeneration lies in storing energy in solids in the form of sensible heat for later use. The devices which are generally used for such storage contain solid packing in the form of metallic spheres, wool or wire mesh. From thermal regeneration point of view regenerators are attractive due to their simple construction, low cost, high compactness, self-cleaning operation and operational flexibility. This is the reason why they are used over a wide field of application ranging from cryogenic liquefaction systems dealing with very low thermodynamic temperature to gas turbine regenerators dealing with hot gases having temperature more than thousand degrees centigrade. In specific applications such as cryogenic systems, the regenerator is regarded as the heart of the system since the performance of the whole system critically depends on the performance of the regenerator. The device which is similar to regenerator in the mass transfer regime is packed bed reactor which in extensively used in chemical process industry.

The simplicity in the construction of regenerator/packed bed does not indicate simplicity in its analysis. This is because of the fact that these devices operate always in the transient mode in which the temperature is a function of both space and time. The factor that brings about even more complexity to this is the zig-zag flow path of the fluid. Hence proper modelling of the heat/mass transfer phenomena in these equipment poses a tremendous challenge. A number of studies have gone into such efforts which has been summarised by Adebiyi and Chenevert (1996).

Even though the advent of high speed computing systems has made complete numerical simulation of heat and mass transfer devices possible, still the need for simpler and more accurate simulation of response remains equally important because the numerical simulations are too complex to be directly used for the purpose of design or control. A significant breakthrough in this direction has been achieved using the concept of axial dispersion phenomenon in mass transfer during turbulent flow (Taylor, 1954). From heat and mass transfer analogy it is established (Robehorst and Law, 1970 & Backman et al, 1990) that the same concept can be used for heat transfer as well. The above studies attribute all the contribution for the deviation from conventional ‚plug flow' such as backmixing, recirculation, leakage, bypassing, flow maldistribution and stagnation to an axial heat/mass dispersion. The essence of this dispersion model is plug flow with an apparent diffusive flux propagation through the fluid due to the above deviations from normal plug flow which can be represented by a conduction like diffusive flux equation, as

$$\dot{q}_x = -\lambda \nabla \Theta \qquad (1)$$

This results in the parabolic heat conduction energy (or concentration) equation in the form,

$$\frac{\partial \Theta}{\partial \tau} = \alpha^* \frac{\partial^2 \Theta}{\partial x} \quad (2)$$

The major difference between this apparent conductivity λ and the real conduction in fluid is that the apparent conductivity λ is a function of macroscopic flow parameters because ist origin lies in macroscopic non-idealities in flow. On the other hand fluid conduction is a molecular phenomenon and is a microscopic property of fluid itself. Using this concept a number of studies have been conducted, e.g. – Spang (1991)presented analysis for multipass exchanger, Xuan and Roetzel (1992,1993)initiated work on transient analysis of heat exchangers, Das and Roetzel (1995) presented a model for plate heat exchanger which was subsequently confirmed by experimental evidence from Das et al (1995).

It is evident from the above studies that for equipment related to heat transfer, the concept of axial dispersion has already been applied successfully. However, a judicious way of determining the axial dispersion coefficient is to carry out transient experiments with injected concentration profile (Das et al, 1998) which suppresses the lateral transfer (like heat transfer) and gives the accurate values of parameters related to axial dispersion, since dispersion of heat and mass concentration are analogous.

The present study is aimed at the modelling of thermal regenerator/packed bed effectively with the help of the axial dispersion model. The major issue of experimental determination of dispersion parameters has been addressed here and a much easier concentration tracer experiment has been devised. The experiment uses a generalized concept without the requirement of an ideal (impulse or oscillation) input function. For simplicity only single blow regenerator has been considered and 'impulse like' functions are used at the observed inlet. The effect of flow rate on dispersion has been observed. The present model has also been used to evaluate the dispersion in a thermal regenerator taken from literature which suggests that the slope of the response is the most important criterion for the determination of dispersion parameters. The present paper on one hand takes the axial dispersion theory to analyse complex heat transfer equipment (viz. regenerator). On the other hand it use a mass transfer experimental technique as used in (Balzereit and Roetzel, 1997) for describing a heat transfer equipment which can act as a powerful tool in future.

ANALYTICAL MODEL

The regenerator/packed bed can be mathematically modelled by considering a general approach where the scalar parameter Θ can be temperature or concentration depending on whether heat transfer or mass transfer is considered. If this scalar quantity has an axial dispersive flux of $\dot{q}_x$, then under the assumption of constant fluid properties and absence of axial conduction in solid the following governing coupled equations can be derived. The fluxes are shown in the solid and liquid control volumes in Fig. 1(here the scalar is shown as temperature T and as a general case solid conduction is shown).

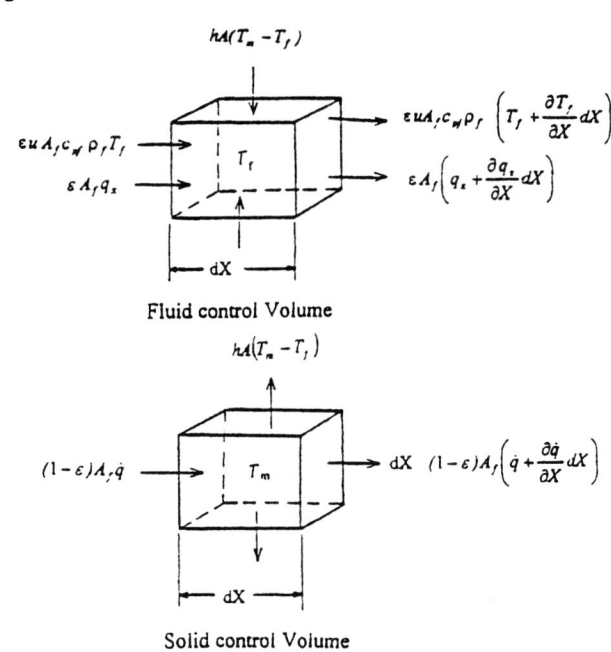

Fig. 1 The control volume for the liquid and solid and the scalar fluxes associated with them.

It should be mentioned here that the molecular conductivity of the fluid is neglected with respect to axial dispersion, the initial values of the scalar quantity is taken as uniform over the entire apparatus, the convective transfer coefficient has been taken to be constant over the entire apparatus and over the scalar range. The flow has been taken to be one dimensional since all the deviations from the plug flow are incorporated in the axial dispersion term. Thus the flux equations read as follows.

For fluid,

$$\psi \varepsilon \frac{D\Theta_f}{D\tau} = -\varepsilon \frac{\partial q_x}{\partial x^*} + h\sigma (\Theta_m - \Theta_f) \quad (3)$$

For solid,

$$(1-\varepsilon)\rho_m c_{p,m} \frac{\partial \Theta_m}{\partial \tau} = h\sigma (\Theta_f - \Theta_m) \quad (4)$$

where the substantative operator is given by

$$\frac{D}{D\tau} = \frac{\partial}{\partial \tau} + u\frac{\partial}{\partial x^*} \quad (5)$$

Application of diffusion flux equation of type (1) for scalar quantity Θ transforms equation (3) to

$$\frac{1}{\alpha^*}\left[\frac{\partial \Theta_f}{\partial \tau} + u\frac{\partial \Theta_f}{\partial x^*}\right] - \frac{\partial^2 \Theta_f}{\partial x^{*2}} = \frac{h\sigma}{\psi\alpha^*\varepsilon}(\Theta_w - \Theta_f) \quad (6)$$

The boundary conditions can be taken after Danckwert (1953) for dispersive system as

$$\Theta_f^+ - \Theta_f^- = -\frac{\lambda}{u\psi}\frac{\partial \Theta_f^+}{\partial x^*} \quad at\ x^* = 0 \quad (7)$$

$$\frac{\partial \Theta_f}{\partial x^*} = 0 .. at.. x^* = L \quad (8)$$

Equations (4) and (6) can be non-dimensionalized using the parameters.

$$\theta = \frac{\Theta_f - \Theta_o}{\Theta_{in} - \Theta_o}$$

$$\theta_m = \frac{\Theta_m - \Theta_o}{\Theta_{in} - \Theta_o}$$

$$t = \frac{\tau}{L/u} \quad (9)$$

$$x = x^*/L$$

$$B = \frac{\rho c_p A_c \varepsilon L}{(1-\varepsilon)\rho_m c_{p,m} L A_c}$$

$$Pe = \frac{uL\psi}{\lambda}$$

$$N_{tu} = \frac{h\sigma A_c}{\sigma c_p u A_c \varepsilon}$$

With these non-dimensional quantities Eqns. (6) and (4) are transformed to

$$\frac{\partial \theta}{\partial t} + \frac{\partial \theta}{\partial x} = \frac{1}{Pe}\frac{\partial^2 \theta}{\partial x^2} + N_{tu}(\theta_m - \theta) \quad (10)$$

$$\frac{\partial \theta_m}{\partial t} = BN_{tu}(\theta - \theta_m) \quad (11)$$

The boundary conditions given by Eqns. (7) and (8) are transformed into

$$\theta^- - \theta^+ = \frac{1}{Pe}\frac{\partial \theta^+}{\partial x} \quad at\ x = 0... \quad (12)$$

$$\frac{\partial \theta}{\partial x} = 0 \quad at\ x = 1... \quad (13)$$

The equations (10) and (11) along with boundary conditions (12) and (13) can taken to frequency domain by Laplace transformation $(as\ \theta = \theta_m = 0\ at\ t = 0)$

$$s\bar{\theta} + \frac{d\bar{\theta}}{dx} = \frac{1}{Pe}\frac{d^2\bar{\theta}}{dx^2} + N_{tu}(\bar{\theta}_m - \bar{\theta}) \quad (14)$$

$$s\bar{\theta}_m = BN_{tu}(\bar{\theta} - \bar{\theta}_m) \quad (15)$$

$$\bar{\theta} - \frac{1}{Pe}\frac{d\bar{\theta}}{dx} = F(s) \quad at\ x = 0 \quad (16)$$

$$\frac{\partial \bar{\theta}}{\partial x} = 0 \quad at\ x = 1 \quad (17)$$

The differential equations (14) and (15) can be solved using boundary conditions (16) and (17) to give

$$\bar{\theta} = \frac{\left[(G-D)e^{-D(1-x)} - (G+D)^{D(1-x)}\right]e^{Gx}}{\left[(G-D)e^{-D}\left(1-\frac{G+D}{Pe}\right)\right] - \left[(G+D)e^{D}\left(1-\frac{G-D}{Pe}\right)\right]}F(s)$$

(18)

where

$$G = \frac{Pe}{2}$$

$$D = \sqrt{\left(\frac{Pe}{2}\right)^2 + sPe\left(1 + \frac{N_{tu}B}{s + N_{tu}B}\right)}$$

The solution obtained using the above method which is in frequency domain should be inverted back to the time domain. Obviously, it can be done only by using a numerical method. There are several methods available for numerical inversion of Laplace transform. In the present case, the numerical inversion is done using Fourier series approximation of Crump (1976) because it can be used successfully to both monotonous and periodic functions. The algorithm reads,

$$f(t)=\frac{\exp(at)}{t}\left[1/2\bar{f}(a)+\mathrm{Re}\sum_{k=1}^{x}\bar{f}\left(a+\frac{ik\pi}{t}\right)(-1)^k\right] \quad (19)$$

where the constant 'a' is generally chosen in the range of 4 < at < 5, so that the truncation error can be considered to be small enough.

With the help of this mathematical model, Fig. 2 and Fig. 3 are two typical computed responses for impulse and step input. It can be seen in both the cases that in absence of axial dispersion (Plug flow Pe → ∞) the inlet signal remains unaltered in form but gets shifted by one residence time. Thus all other situations where the input signal gets deformed, an axial dispersion can be said to be present which is at the root of the present proposition. In the present case the figures have been drawn with $N_{tu} = 0$ which leaves out the effect of heat transfer so as to facilitate the study of the nature of axial dispersion alone.

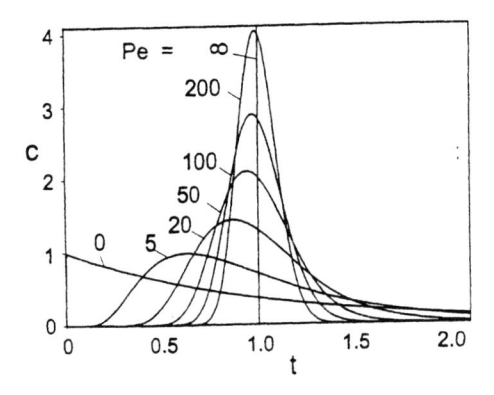

Fig. 2 Computed result for an impulse input for $N_{tu} = 0$.

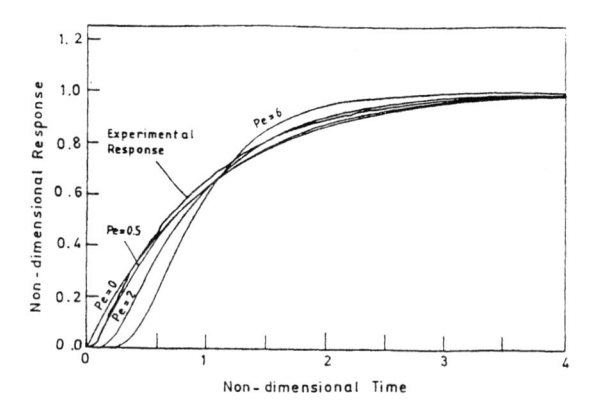

Fig. 3 Computed result for a step input for $N_{tu} = 0$.

EXPERIMENTAL INVESTIGATION

The experimental investigation was aimed primarily to evaluate the dispersion parameter. Hence mass transfer experiment was chosen so that uncertainty in heat transfer calculations does not interfere with the axial dispersion measurement. The experiments were carried out in the laboratory of the Institut für Thermodynamik der Universität der Bundeswehr Hamburg. Following sections give the details of the measurement technique.

Principle of RTD measurement

The dispersion coefficient characterized by the Peclét number Pe is evaluated from the Residence Time Distribution (RTD). The procedure followed is same as that in (Balzereit and Roetzel, 1997) used for plate heat exchanger. Since the flow is considered to be incompressible and property variation is neglected the governing equations are linear. The systems is thus said to be linear which will give an output given by the convolution integral

$$\theta_{out}(t) = \int_{o}^{t} E(t^*)\,\theta_{in}(t - t^*)\,dt^* \quad (20)$$

The analytical residence time distribution function E(t) is calculated by the integration of the flow equation containing dispersive Peclét number. This function combined with the experimental input signal θ_{in} in the convolution integral will yield computed response of the apparatus. It is interesting to note that it is not necessary to have a particular type of inlet function of the tracer solution since it is difficult to generate a particular type of mathematical function by physical injection processes. The opening and closing of the injection valve and the dispersion between point of injection and entry to the equipment contribute to non-ideal form of concentration function at the inlet.

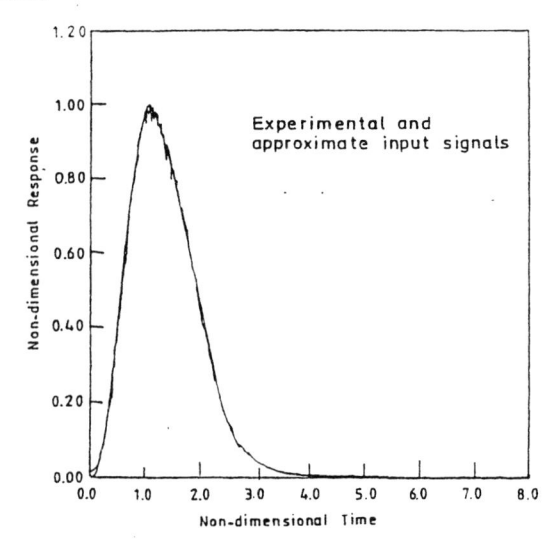

Fig. 4 A typical experimental input curve and the Mathematical approximation for it.

Fig. 4 shows such an inlet concentration profile with the corresponding approximated mathematical curve. The general form of the fitted function in this case is

$$\theta_{in}(t) = \left[a_o + a_1 t + a_2 t^2 + a_3 t^3\right]\left[1 - u(t-\alpha)\right] \quad (21)$$
$$+ \left[b_o + b_1 t + b_2 t^2 + b_3 t^3\right]\left[u(t-\alpha) - u(t-\beta)\right]$$

Which essentially means fitting two different curves for ascending ($t \leq \alpha$) and descending ($\alpha < t < \beta$) portions.

Experimental Setup and Measurement Technique

The set up used for the experiment is shown schematically in Fig. 5.

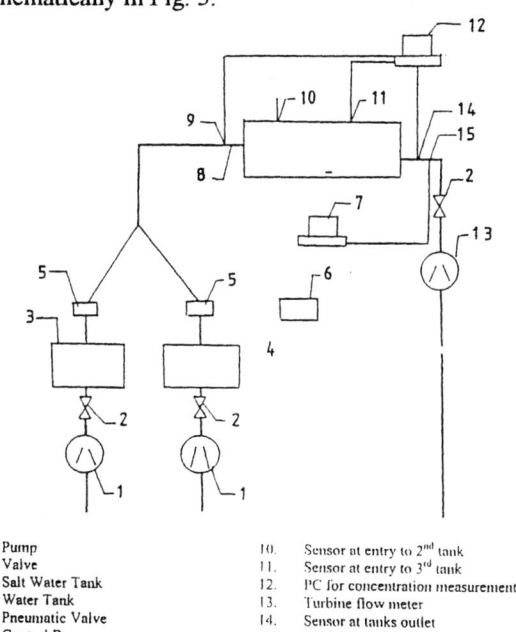

1. Pump
2. Valve
3. Salt Water Tank
4. Water Tank
5. Pneumatic Valve
6. Control Box
7. PC for Temperature Measurement
8. Thermocouple at entry
9. Sensor at entry to 1st tank
10. Sensor at entry to 2nd tank
11. Sensor at entry to 3rd tank
12. PC for concentration measurement
13. Turbine flow meter
14. Sensor at tanks outlet
15. Thermocouple at exit
16. Transducer for RPM measurement

Fig. 5 Schematic of the experimental setup.

The solenoid valve used for injecting saline water at the inlet is activated by electrical signal triggered by a computer (PC) and A/D converter. Turbine flow meter was used for flow measurement and K type thermocoax thermocouple was used to ensure an isothermal measurement which is essential for electrical conductivity of the fluid to be a function of concentration alone.

For the measurement of concentration, the special probe measuring concentration from the electrical conductivity of the fluid developed by Köllmann and Balzereit (1996) was used. This probe has been used in (Balzereit and Roetzel, 1997) and has the distinction of being able to capture fast changes in the electrical conductivity of the fluid. It consists of electrodes made from 0.5 mm platinum wire extending parallel with an approximate centre distance of 3 mm, intruded 5 mm within the flow cross section. The probe scans the AC electrical signal in discrete time and amplitude steps and process them digitally. It has been found that over the conductivity range of 0.02 to 200 mS the probe behaves linearly at a given temperature as shown in Fig. 6 a. The small time constant of the probe can be seen in Fig. 6 b.

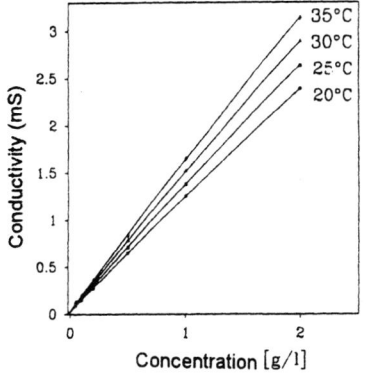

Fig. 6a The relation between conductivity, concentration and temperature for the probe.

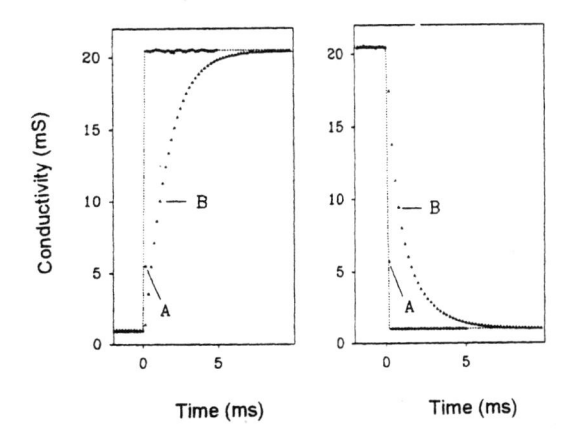

Fig. 6b Transient response of the concentration probe.

The test regenerator section was made of 16 x 16 wire mesh tightly packed inside 40 mm tube to an axial length of 100 mm such that a porosity of 0.766 is achieved. Fig. 8 gives a picture of the test section.

Data Reduction

Classically the matching of the data can be done by taking moments of residence time after reducing the data in convenient non-dimensional form. Since the variation of concentration is linear with electrical conductivity in this regime the nondimensional concentration can be obtained by directly non-dimensionalizing the electrical conductivity as

$$\theta = \frac{C - C_o}{C_{in} - C_o} = \frac{\Lambda - \Lambda_o}{\Lambda_{in} - \Lambda_o} \quad (22)$$

The individual residence times of all tracer particles give a distribution of the residence times. If θ (t) is the output response of a linear system then the concept of temporal moment gives

$$\bar{t}_R = \int_0^\infty t\,\theta(t)\,dt \Big/ \int_0^\infty \theta(t)\,dt \quad (23)$$

In this case residence time is evaluated from Eqn. (24) since the form of the signal is arbitrary here.

$$\bar{t}_R = \bar{t}_{out} - \bar{t}_{in} \quad (24)$$

Experimental Procedure and Error Estimates

Desalinated water (with electrical conductivity less than 0.02 mS.cm^{-1}) flows at constant rate through the wire mesh packed test section which is measured by a turbine flow meter. The thermocouples at inlet and outlet checks the validity of isothermal condition. A salt solution (NaCl) of approximate concentration 200 g/l is stored in pressurised vessel (600 kPa) as shown in Fig. 5. The salt solution is injected to the main stream by momentary opening of a solenoid valve controlled by a monostable timer to synchronically open for a short pulse duration of 50 - 70 ms. Using common hardware (PC and AD card) and appropriate sampling rate successive averaged measurement of eletrical conductivity was obtained for time intervals of 1 - 10 ms.

The pulse length for injection was so chosen that it injects enough salt for an 'impulse like' input on one hand, the pulse duration is restricted to 2 to 3 % of the system residence time on the other to avoid longer pulse.

The flow rate was kept between 5 and 14 l/s with an uncertainty of 3 %. Over the range of conductivity of 0.02 - 200 mS the uncertainty in conductivity measurement was found to be less than 3 % for all concentrations larger than 2 % of the average response peaks. The accuracy of the calibrated thermocouples was 0.1 K for water temperature between 20 and 35 °C within which experiments have been performed. The temperature increasing slowly through dissipation stayed constant and equal (within 0.1 K) on both sides during each experiment. Fig. 7 shows typical inlet-outlet curve for two different flow rates.

RESULTS AND DISCUSSION

Results for the theoretical simulation has already been presented in Fig. 2 and Fig. 3 which brings out the general effects of axial dispersion under varying input condition. However the main task for the simulation of a packed bed remains in determining the dispersion coefficient experimentally. The first step in this direction is to estimate the residence time.

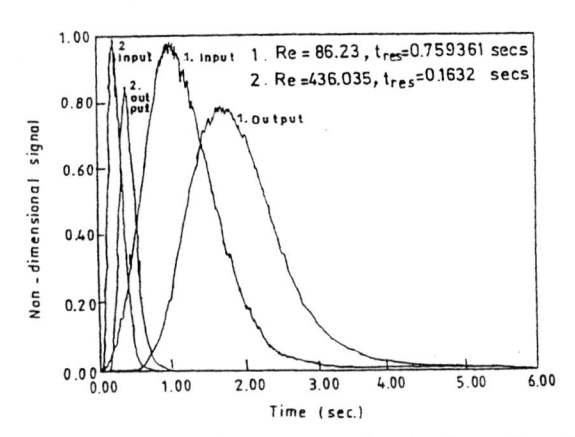

Fig. 7 Inlet and outlet concentration pulses for the different fluid flow rates.

After evaluating the residence time the dispersion parameters are determined by matching the computed curve with the experimental response. For example Fig. 8 shows the three simulated curves for Re = 96 for three different Peclét numbers (Pe = 5, 8 and 12).

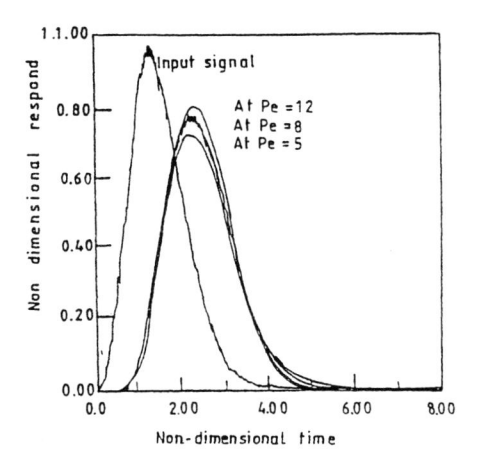

Fig. 8 The experimental output (Re = 96) and the three computed outputs at different Peclét number.

In this simulation the actual inlet concentration profile shown in the same figure has been approximated by a function of the form Eqn. (21) and this has been used for computation. Obviously the experimental response matches with the computation at Pe = 8 and hence in this case Peclét number can be taken as 8.0. It must be mentioned here that both amplitude and the phase of the curve has been chosen as matching criterion. Fig. 9, 10 and 11 show similar matching which indicate how by the present experimental technique accurate determination of dispersion can be effected.

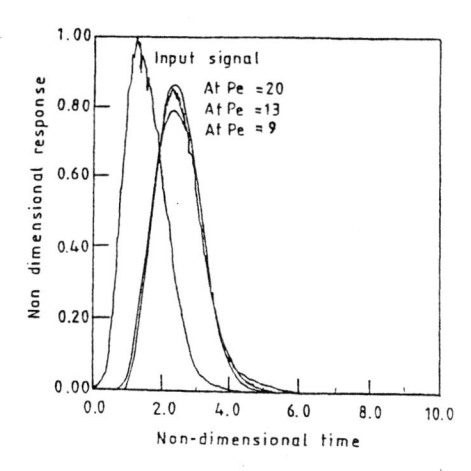

Fig. 9 Determination of Peclét number (Re=32).

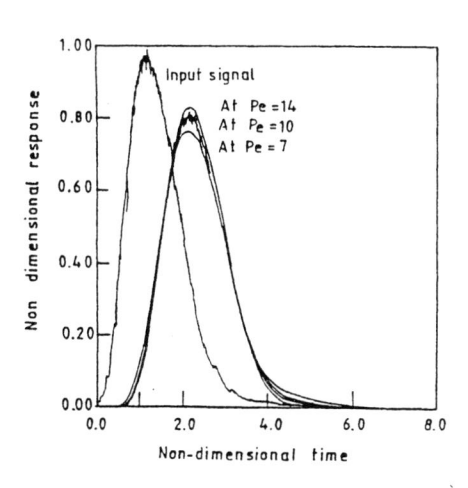

Fig. 10 Determination of Peclét number (Re=220).

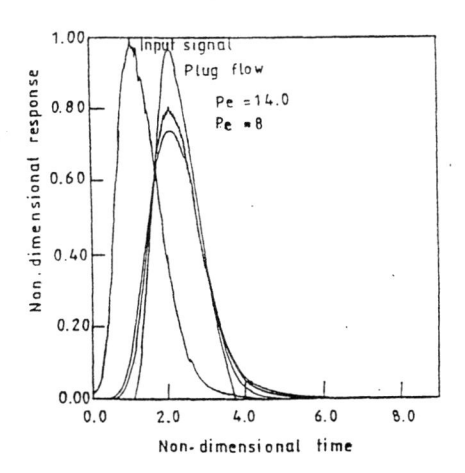

Fig. 11 Comparison of predicted dispersed flow and plug flow (Re = 490).

Fig. 11 shows specifically how, in spite of all the efforts over few decades to simulate regenerators/packed columns by plug flow model, they fail to simulate the transient response of a scalar quantity (concentration in this case). Thus the results comprehensively prove the proposition of axial dispersion and indicates the method for determination of axial dispersion parameters in a packed bed.

It has already been proclaimed that axial dispersion is a flow property and hence it is likely to be altered with flow parameters. Fig. 12 gives the variation of dispersive Peclét number with Reynolds number. It can be readily observed that the variation of dispersive Peclét number is physically consistent because in presence of the packing the flow path is zig zag the increase of Reynolds number flatten the concentration profile making it nearer to plug flow.

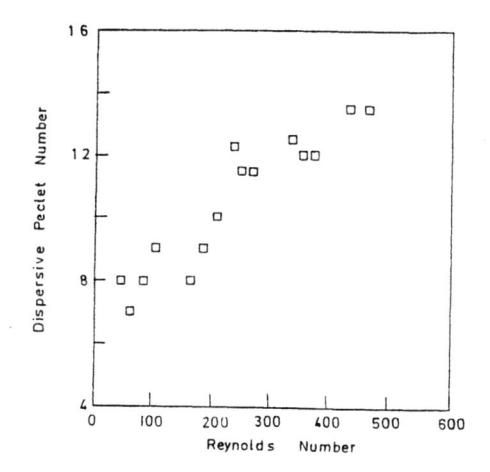

FIg. 12 Variation of dispersive Peclét number with Reynolds number.

Finally the response of a thermal regenerator as given by Cheng and Chang (1998) has been simulated by the present model as shown in a Fig. 13.

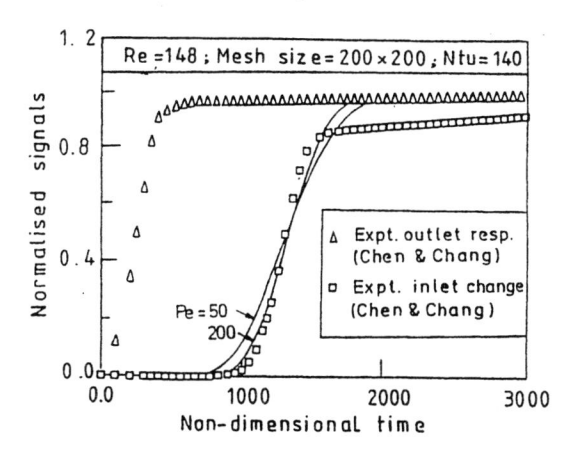

Fig. 13 Comparison with the experiments of Cheng and Chang (1998).

It can be seen that even in presence of heat transfer the dispersion model describes the temperature transient quite well. The mismatch at the steady part of the response is primarily due to the heat loss which was conceded by the authors themselves while comparing with their own computation. The figure indicates that the slope of the response is the most important criterion for matching response and determining dispersive Peclét number which in this case was found to be 200. For this case the mash size was 200 x 200 and a Reynolds number of 148 giving N_{tu} (Kays and London, 1984) of 140.

CONCLUSION

A transient measurement technique has been presented for the evaluation of axial dispersion parameter in a packed bed heat/mass exchanger. The theoretical background clarifies the philosophy of introducing axial dispersion as a measure for the deviations from the so-called 'plug flow'. However the theoretical results indicate the need for evaluation of dispersive Peclét number accurately. To cater this need, a reliable mass transfer experiment has been devised in the present work which determines the dispersion parameter by measuring the tracer concentration profile through a specially designed electrical conductivity probe which is capable of capturing fast responses. The present technique can be used for any general form of input function. The dispersion in fluid is found to decrease with Reynolds number. The present formulation is found to be equally important in the case where heat transfer takes place within the bed. However since it has already been (Balzereit and Roetzel, 1997) shown that heat and mass dispersions are analogous, hence the present mass transfer technique can also be used for a regenerative heat exchanger for the determination of its axial dispersion alone. Thus the present proposition along with the experimental technique presents a comprehensive method for evaluation of the response of packed beds which always operates in the transient regime.

NOMENCLATURE

A   heat transfer Area, m^2
A_c cross sectional area, m^2
B   heat capacity ratio = $mc_p/m_m c_{pm}$, dimensionless
c_p specific heat of fluid, J/Kg K
$c_{p,m}$ specific heat of solid wall, J/kg K
C   concentration, kg/m^3
D   variable given by Eqn. (18)
F(s)   Laplace transform of the input temperature/ concentration function
G   variable given by Eqn. (18)
h   heat transfer coefficient for thermal case, $W/m^2 K$
L   length of the equipment, m
m   mass of hold up fluid over the bed at a particular instant, kg
m_f Mass flow rate of the fluid, kg/s
m_m mass of the wire mesh packing, kg
N_{tu} Number of Transfer Units, $hA/m_f c_p$, dimesionless
Pe   dispersive Peclét number, uL/α^*, dimesionless
q_x heat transfer by dispersion in the axial direction, W/m^2
s   transformed time variable in Laplace domain
t   time, $u\tau/L$, dimesionless
u   Velocity of the fluid, m/s
u( )   unit step function
x   space coordinate, x^*/L, dimensionless
x^* axial coordinate, m
Λ electrical conductivity, mS
α thermal diffusivity, m^2/s
α^* thermal diffusivity of axial dispersion, $\lambda/\rho c_p$, m^2/s
λ axial dispersion coefficient of fluid, W/mK
ρ density of the fluid, kg/m^3
θ fluid concentration/temperature, dimensionless
$\bar{\theta}$ fluid concentration/temperature in Laplace domain, dimensionless
Θ scalar (temperature or concentration), K or kg/m^3
τ time, s
ε matrix porosity
ψ ρc_p for heat transfer
ψ 1 for mass transfer
σ heat transfer area per unit volume of the matrix, m^2/m^3

Subscripts

0   initial
f   fluid
in   inlet
m   solid wall
-     Just before entry
+   Just after entry

REFERENCES

Adebiyi, G.A. and Chenevert, D.J., 1996, An appraisal of one-dimensional analytical one-dimensional analytical models for packed bed thermal storage system utilizing sensible heat storage material, J. Energy Res. Tech., Trans ASME, 118, 44-49.

Backman, L.V., Law, V.J., Bailey R.V. and von Rosenberg, D.U., 1990, Axial dispersion for turbulent flow with a large radial heat flux, AICh.E.J., 36, 598-604.

Balzereit, F. und Roetzel, W., 1997, Determination of axial dispersion coefficients in plate heat exchangers using residence time measurements. Snowbird, 26. June 1997, Engineering Foundation Conference on Compact Heat Exchangers for the Process Industries, Snowbird, Utah, USA, June 22-27, 1997

Cheng, P.H. and Chang, Z., Measurement of thermal performance of cryocooler regenerators, using an improved single blow method, 1998, Int. J. Heat Mass Transfer, 41, 2857-2864.

Crump, K.S., Numerical inversion of Laplace transforms using a Fourier series approximation, 1976, J. Assoc. Comput. Mech, 23, 89-96.

Danckwert, P.V., 1953, Continuos flow systems - distribution of residence times, Chem. Engng. Sci., 2, 1-13.

Das, S.K., Roetzel, W., 1995, Dynamic analysis of plate heat exchangers with dispersion in both fluids, Int. J. Heat Mass. Transfer; 38, 1127-40.

Das, S.K., Spang, B., Balzereit, F. and Roetzel, W., 1998, Experimental determination of axial dispersion parameters during enhanced heat transfer in compact heat exchanger surfaces, proceedings of at the conference on „Heat exchangers for sustainable development", June 15-18, Lisbon, 1998.

Das, S.K., Spang, B., Roetzel, W., 1995, Dynamic behaviour of plate heat exchangers - Experiments and modelling, J. Heat Transfer, Trans ASME; 117, 859-64.

Kays, W.M. and London, A.L., , 1984, Compact Heat Exchangers, 1984, McGraw-Hill Book Company.

Köllmann, K. and Balzereit, F., 1996, Meßverfahren zurzeitaufgelösten Konzentrationsbestimmung in strömenden electrolytischen Flüssigkeiten, Technisches Messen, 63, 458-464.

Robehorst, C.W., and Law, V.J., 1970, One dimensional dispersion model for transient heat transfer in turbulent pipe flow, Proceedings of the 5th International Heat Transfer Conference, Paris.

Roetzel, W. and Xuan, Y., 1992, Analysis of transient behaviour of multipass shell-and-tube heat exchangers with dispersion model, Int.J.Heat Mass Transfer,35(11),2953-2963.

Spang, B., 1991, Über das thermische Verhalten von Rohrbundelwärmeübertragern mit Segmentumle-nkblechen, Fortschritt-Berichte VDI, Vol.19, No 48, VDI Verlag, Düsseldorf.

Taylor, G., 1954, The dispersion of matter in turbulent flow through a pipe, Proceedings of the Royal Society of London, 223, 446-468.

Xuan, Y. and Roetzel, W., 1993, Stationary and dynamic simulation of multipass shell and tube heat exchangers with dispersion model for both the fluids, Int.J.Heat Mass Transfer, 36, 4221-4231.

HEAT TRANSFER ENHANCEMENT AND FLOW VISUALIZATION OF WAVY-PERFORATED PLATE-AND-FIN SURFACE

Yimin Xuan, Houlei Zhang and Rong Ding

Nanjing University of Science & Technology, Nanjing 210094, China
E-mail: ymxuan@mail.njust.edu.cn

Abstract

Compact enhanced surfaces are frequently applied to transportation, refrigeration and other industries. In this paper a new wavy-perforated plate-and-fin surface is developed and investigated by using the single blow method and visualization technique. In order to take the effect of non-uniform thermal capacity distribution of the test core into account, an approximate model considering transient heat conduction in the side bars is built. In the whole test range the experimental results show better heat transfer performance than the conventional wavy fin. Only when the Reynolds number is high, the pressure drop for wavy-perforated fin is higher than that of the wavy fin. Flow visualization experiments are also performed for both wavy fin and wavy-perforated fin channels in order to observe the flow patterns. The images for different geometrical parameters are acquired and the effects of holes as well as the wavy angle, wave length and wave spacing on the flow pattern are briefly discussed.

INTRODUCTION

Compact heat exchanger surfaces such as wavy and offset surfaces are frequently applied in transportation, refrigeration and other industries. The heat transfer coefficient and pressure drop data are essential for design and operation. A lot of work has been done in this field and many theoretical and experimental results have been reported (Kays and London, 1984). This paper aims at developing a new wavy-perforated plate-and-fin surface and testing its performance. The wavy-perforated surface is a hybrid of integrating the wavy fin and perforated plate-fin. It is manufactured by punching a number of spatially distributed holes in the fin material before it is formed into wavy channels.

The single blow method with direct numerical curve matching technique similar to that described by Mullisen and Loehrke (1986) is used to determine the heat transfer coefficient in this paper. In a previous work, Cai et al.(1984) have taken the effect of the longitudinal heat conduction in the wall into account and a quantitative criterion whether this effect should be considered is given. The effect of the lateral heat conduction resistance is also discussed by Zhou and Cai (1988). In order to take the effect of non-uniform thermal capacity distribution of the test core into consideration, an approximate mass correction model considering transient heat conduction in the side bars is built which is similar to the work done by Luo et al. (1990).

Flow visualization is an experimental tool for revealing flow distribution, flow pattern and flow structure. Webb and Trauger (1991) applied a dye injection technique to flow visualization study of the louvered fin geometry in a closed-circuit open water channel. Nishimura and Kawamura (1995) conducted a flow visualization of oscillatory flow in a two-dimensional symmetric sinusoidal wavy-walled channel and investigated transition mechanism from a two-dimensional to three-diemnsional flow. Boilider and Sunden (1995) carried out flow visualization on a laminar flow in the helical square duct with finite pitch and found the dependence of vortex appearance. In a recent work done by Rush et al.(1999), flow observations in a wavy-walled channel indicated that the flow behaviour-the onset of unsteadiness in particular-depends on streamwise location in the channel. In this paper, visualization experiments are performed for the wavy-perforated surface and conventional wavy surface by using the dye injection and aluminum powder together with CCD technique. The effects of holes as well as the wavy angle, wave length, wave spacing and Reynolds number on the flow are briefly discussed.

HEAT TRANSFER AND PRESSURE DROP TEST

This part has been presented in an early paper by the authors (Zhang et al., 1999). The geometry of the test plate-and-fin surface is illustrated in Table 1 and Fig.1. Apparently, the thermal capacity per unit heat transfer surface area is different for bars and thin plates or fins, so the temperature history is also different in the single blow process. Thus, a modification is introduced into the

traditional energy equation for the wall. Only one passaage, represented in Fig.1, is analyzed with the following idealizations applied:
- steady flow
- finite wall heat conduction in the flow direction for plates and bars
- infinite wall conduction perpendicular to the flow direction for fins
- zero fluid heat conduction or dispersion in the flow direction
- zero thermal capacity of the residing fluid
- adiabatic side walls

Table 1. Dimensions of the test plate-and-fin surface.

Fin thickness (mm)	0.1
Fin pitch (mm)	2.8
Fin height (mm)	9.8
Bar width (mm)	10.0
Perforated area/total area (%)	10.0
Wavy angle (deg)	15~16
Mass of bars/total mass (%)	37.0

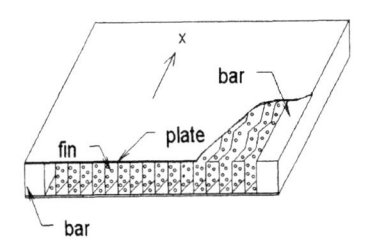

Fig.1. A flow passage (27 passages for the test core)

Under these assumptions the energy equation for the wall is

$$(\Delta m_{plate} + a\Delta m_{bar} + \Delta m_{fin})c_{pw}\frac{\partial T_w}{\partial t} = k_w(A_{plate} + aA_{bar})\Delta x\frac{\partial^2 T_w}{\partial x^2} + h\Delta F(T_f - T_w) \quad (1)$$

where a is the non-uniform thermal compactness factors of bars, which is defined as follows.

$$a = \frac{m_{bareff}}{m_{bar}} = \frac{\frac{F_{bar}}{L}\int_0^{t_m}\int_0^L q_{bar}dxdt}{m_{bar}c_{pw}[\frac{1}{t_mL}\int_0^{t_m}\int_0^L T_w dxdt - T_0]} \quad (2)$$

where t_m is the time length of curve matching. Firstly, in the side bars, transient heat conduction equations are solved numerically to determine their temperature histories. Then, numerical integration in equation (2) is done to get the values of a. In fact, due to the much smaller heat transfer area of bars than the total heat transfer area of the test core, the effect of the bars is very limited. The mass correction in this paper is an approximate method.

The energy equation for the fluid is

$$\frac{\partial T_f}{\partial x} + \frac{hF}{m_f c_{pf} L}(T_f - T_w) = 0 \quad (3)$$

The governing equations (1) and (3) are subjected to the following initial and boundary condition

$$T_w(x, t=0) = T_f(x, t=0) = T_0 \quad (4)$$

$$T_f(x=0, t) = T_1(t) \quad (5)$$

$$T_f(x=L, t) = T_2(t) \quad (6)$$

$$(\frac{\partial T_w}{\partial x})_{x=0} = 0 \quad (7)$$

$$(\frac{\partial T_w}{\partial x})_{x=L} = 0 \quad (8)$$

The initial temperature is known from the experiment. The conditions at the test core inlet and outlet are also known from the experiment.

Equations (1) and (3) are converted into finite-difference equations and solved numerically. The solution process consists of guessing a value of h, solving the finite-difference equations and obtaining a theoretical exit fluid temperature history. The theoretical result is then compared with the experimental exit fluid temperature history. If these two temperature histories match each other within the specific limits, then the originally assumed value of h is the correct average heat transfer coefficient of the test core. If not, h is iteratively changed until the agreement is achieved. The average square root error between the two histories is chosen as the objective function and the trial-error method is applied to find the optimal value of h.

The experimental setup consists of the wind tunnel with instrumentation, data acquisition and processing systems shown in Fig.2.

In fact, for the matching time length under consideration (no more than 30s) and Re from 350 to 4000, the factor a

varies from 0.025 ~0.08 which means that only a little effective mass of bars influences the fluid temperature history. So if only the mass of the fins and plates are deemed as the effective mass, the error is small. But if the whole mass of bars are included in the effective mass, i.e. $a=1.0$, unreal results will be obtained.

The heat transfer coefficient data from the present experiment and the literature (Kays and London, 1984) vs. Re are plotted in Fig.3. According to Fig.3, only for low values of Re, the measured Nu of considering longitudinal heat conduction is larger than that of zero longitudinal heat conduction. When Re>1000, they are nearly the same, so the effect of longitudinal heat conduction can be neglected. Compared with the wavy heat transfer surface 11.5-3/8w reported by Kays and London(1984), the wavy-perforated heat transfer surface gives 20-40% higher heat transfer coefficients in the whole range of Re from 350-4000.

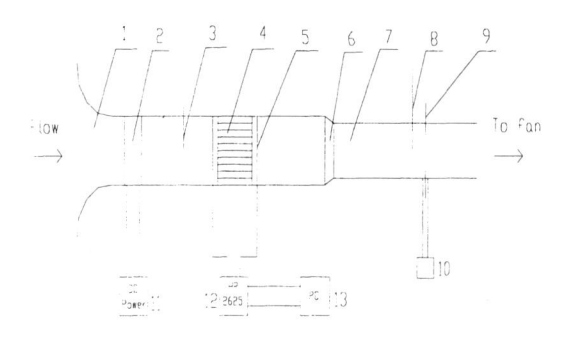

1,inlet section; 2,heater; 3,pitot tube; 4,test section; 5,thermocouples; 6,square-to-round section; 7,velocity measuring section; 8,pitot tube; 9,ISO orifice; 10,flow rate meter; 11,DC power; 12,HP2625 data logger; 13,computer.

Fig.2. Scheme of the experimental setup for heat transfer and pressure drop test

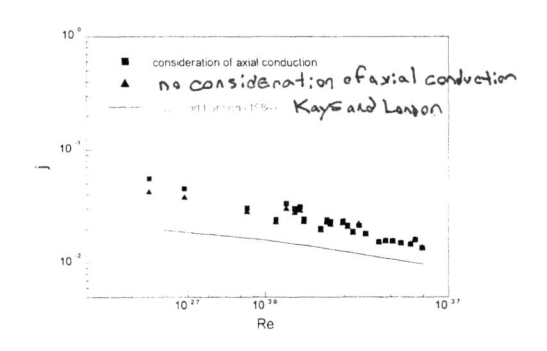

Fig.3 Variation of convective heat transfer with Re

The friction factors computed from the measured pressure drops are plotted in Fig.4. For Re<2000, the friction factors of the wavy-perforated surface are lower than or equal to those of the wavy surface 11.5-3/8w in literacy (Kays and London,1984). But if Re>2000, a 15-40% higher pressure drop is necessary for the test unit.

FLOW VISUALIZATION

In this paper, the experimental setup of flow visualization is schematically shown in Fig.5 which consists of a closed circuit open water tunnel and the image acquisition system. The water channel is comprised of test section, stabilized section, contraction section, tank, pump, injection unit and valves. Either dye or aluminum powder can be used as the tracer. In case of dye injection, the selected dye, i.e. Chinese ink, is injected into the flow with a hypodermic needle. The image system consists of a personal computer, CCD camera, focusing control unit and a white/black image prodecessor. Video signals are transformed into digital signals after wave-filtering and A/D modulation. The digitized images are stored in the computer for further ananysis. The experimental Reynolds number ranges from 100 to 1000.

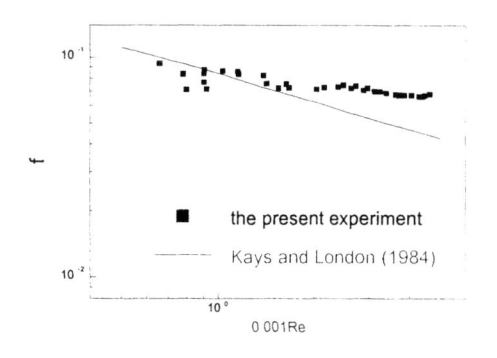

Fig.4 Variation of friction factor with Re

In the light of geometrical similarity principle, nine models of the wavy fin and wavy-perforated fin are constructed using plexiglass according to the scale 5:1. The geometrical parameters of the fin models are listed in Table 2.

Figure 6 shows the flow patterns in the conventional wavy fin channels. The images indicate that the flow fields in the channels are similar to each other. In the case of a small Reynolds number (Re=180), the streamlines in the channels are clearly discernable and the flow distribution is even, stable and laminar. Small vortices are formed on the inner side of the wavy fin and a small reverse flow area appears near the turning point on the other side of the fin. As the Reynolds number increases, the vortex is intensified. For Re=880, the vortex intensity remarkably increases and the vortex area is extended. Appearance of vortices strengthens fluid mixing and induces flow turbulence. The flow near the fin surface can be divided into the vortex flow and the sublayer flow. From the turning point on the facing-flow side to the crossing-point between the facing point and the lee side, the thickness of the boundary layer increases along the flow direction. At the cross-point, the boundary layer is separated subject to the reverse pressure gradient and viscosity.

Figures 7,8 and 9 give the flow pattern in the perforated wavy-wall channels whose angles are $\alpha =18°$, $\alpha =22.5°$ and $\alpha =33°$, respectively. It is due to the perforated holes on the fin surface that the flow in the wavy-perforated fin channels shows some different characteristics besides the common features existing in the conventional wavy fin channels. Because of the effect of the holes, the vortex does not appear at the corner region and scours holes into the neighboring channels as a result of the inertial effect. The vortex begins to form near the hole region rather than at the corner of the fin. The holes interrupt the boundary layer and hinder development of the boundary layer, which is advantageous for heat transfer enhancement. Since there is a small reverse pressure gradient nearby the holes on the lee side of the wavy-perforated fin, the small vortices appear and boundary layer separation occurs later. It results from the effect of the pressure difference and centrifugation that part of the fluid enters into the adjacent channel through the holes, which promotes fluid mixing and flow turbulence.

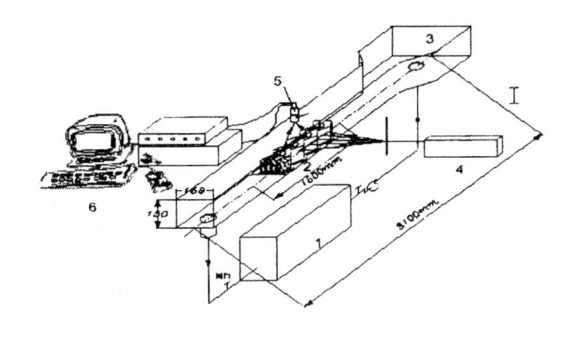

1,water tank; 2,test section; 3,water channel; 4,laser generator; 5,CCD camera; 6,computer system.
Fig.5 Scheme of the experimental setup for visualization

Although both the conventional wavy fin and wavy-perforated fin force the fluid to alter its flow direction in the zigzag channels, the holes on the wavy-perforated fin surface make it accessible that the fluid in the adjacent channels is mixed and promotes early transition from the laminar to turbulent flow which is quite different from what happens in the conventional wavy fin channels. Whereas the fluid regularly flows over the wavy-perforated fin surface in the zigzag channels, the fluid mixing among the adjacent channels is chaotic. Recent research work (Sen and Chang, 1994) revealed that chaotic mixing (or advection) of the fluid remarkably enhances heat and mass transfer. The streamlines in Fig.8b and Fig.9b distinctly illustrate such a mixing process and stronger disturbance which results in complicated flow in the channels. The holes thin the boundary layer, suppress development of the boundary layer and interrupt the viscous sublayer. Therefore, the enhancement effect of the wavy-perforated fin becomes more remarkable for high Reynolds numbers. It means that the enhancement behavior of the wavy-perforated fin is superior over that of the conventional wavy fin, if the porosity of the wavy-perforated fin is reasonably controlled. Such transverse mixing of the fluid through the perforated holes depends upon the Reynolds number. Comparison between Fig.10a and Fig.10b shows that as the Reynolds number increases, the effect of centrifugation becomes evident, so that transverse mixing of the fluid through the holes is strengthened and more fluid eddies enter into the adjacent channels. The higher the Reynolds number is, the stronger the transverse mixing of the fluid between the neighboring channels appears.

Table 2 Geometrical parameters of the fin models

No.			l (mm)	m (mm)	a (°)	H (mm)
Real geometry			10.0	2.5	33.0	10.0
Model geometry	I	1	77.3	12.5	22.5	50.0
		2	98.5	12.5	18.0	50.0
	II	3	77.3	12.5	22.5	50.0
		4	69.9	12.5	24.6	50.0
		5	50.0	12.5	33.0	50.0
		6	139.8	12.5	24.6	50.0
		7	150.0	14.3	33.0	50.0
		8	150.0	14.3	24.6	50.0
		9	150.0	14.3	22.5	50.0

Note: I -wavy fin. II -wavy-perforated fin.

Figure 11 gives three images of the flow structure in the channels which correspond to the angles $\alpha =18°$ and $\alpha =24.6°$ with the same amplitude. It is discernable that the vortices near the inner side are intensified with increasing the angle of the wavy fin. It means that a larger wave angle results in intenser vortices. In the mean time, the friction loss increases when the wavy fin angle increases.

The wave length l and the channel spacing m are also two important parameters of affecting performance of the wavy fin. Figure 12 shows the effect of the aspect ratio (l/m) on the fluid flow. For the wavy-perforated fin with a longer wave length and larger amplitude, the effect of the holes on the flow field becomes evident. The reason is that in this case a spacious space relaxes confinement of vortex development and makes vortices free move and revolve. For the channel with a small aspect ratio, which means a wide channel, the fluid flows gently and the flow turbulization is small, so that enhancement performance is rather poor and the compactness is limited. However, excessively narrow channel spacing may lead to high friction loss, so a compromise should be made in practical application.

CONCLUSIONS

A wavy-perforated plate-and-fin surface has been proposed and tested quantitatively by using the single blow technique and the flow pattern is visualized and observed qualitatively. The heat transfer data have shown better performance of the fin. The images of the flow visualiztion have indicated the possible enhancement mechanism of the wavy-perforated fin which is compared with the conventional wavy fin. Geometrical parameters of the wavy-perforated fin should be optimised when it is used in practical heat exchanger design.

Nomenclature

A	cross-sectional area
a	equation (2)
	specific heat
F	heat transfer area
H	height of test units for visualization
h	convective heat transfer coefficient
k	thermal conductivity
L	total length of test core along flow direction
l	wave length
m	mass, mass flow rate, channel spacing
q	local heat transfer rate
T	temperature
t	time
t_m	time length of curve matching
x	axial flow coordinate
α	wavy channel angle
Δ	finite increment

Subscripts

bar	side bar
eff	effective
f	fluid
fin	fin of test core
p	isobaric
$plate$	seperating plate
w	wall
0	initial
1	inlet
2	outlet

REFERENCES

Bolinder, C.J., and Sunden, B., 1995, Flow Visualization and LDV Measurements of Laminar Flow in a Helical Square Duct with Finite Pitch, *Experimental Thermal and Fluid Science*, Vol.11, pp.348-363

Cai, Z.H., Li, M.L., Wu,Y.W. and Ren, H.S., 1984, A Modified Selected Point Matching Technique for Testing Compact Heat Exchanger Surfaces, *International Journal of Heat and Mass Transfer*, Vol.27, No.7, pp.971-978

Kays W.M., London A.L., 1984, *Compact Heat Exchangers*, 3rd ed., McGraw-Hill Book Company, New York

Luo X. et al. , 1990, The Influence of the Side-bars of Plate-fin Heat Exchangers on Transient Test Matching and its Correction , *Journal of Shanghai Institute of Mechanical Engineering* (in Chinese), Vol.12, pp.40-44

Mullisen R.S., Loehrke R.I., 1986, A Transient Heat Exchanger Evaluation Test for Arbitrary Fluid Inlet Temperature Variation and Longitudinal Core Conduction , *Trans. ASME J. Heat Transfer*, Vol. 108, pp.370-376

Nishimura, T., and Kawamura, Y., 1995, Three-Dimensionality of Oscillatory Flow in a Two-Dimensional Symmetric Sinusoidal Wavy-Walled Channel, *Experimental Thermal and Fluid Science*, Vol.10, pp.62-73

Nishimura, T., Kunitsugu, K, and Morega, A.L., 1998, Fluid Mixing and MassTransfer Enhancement in Grooved Channels for Pulsatic Flow, *J. Enhanced Heat Transfer*, Vol.5, pp.23-27

Rush, T.A., Newell, T.A., and Jacobi, A.M., 1999, An Experimental Study of Flow and Heat Transfer in Sinusoidal Wavy Passages, *Int. J. Heat Mass Transfer*, Vol.42, pp.1541-1553

Webb, R.L., and Trauger, P., 1991, Flow Structure in the Louver Fin Heat Exchanger Geometry, *Experimental Thermal and Fluid Science*, Vol.4, pp.205-217

Zhang H., Xuan, Y. and Roetzel, W., 1999, A Single Blow Test for a Wavy-Perforated Plate-and-Fin Surface Considering Non-Uniform Thermal Compactness, *Progress in Engineering Heat Transfer*, B.Grochal, J.Mikielewicz and B.Sunden (Editors), Institute of Fluid-Flow Machinery Publishers

Zhou, C.W. and Cai, Z.H., 1988, The Single-Blow Transient Testing Technique Considering the Lateral Heat Conduction Resistance of the Wall , *Proceedings of the National Cqnference on Heat and mass Transfer*, Shanghai, China

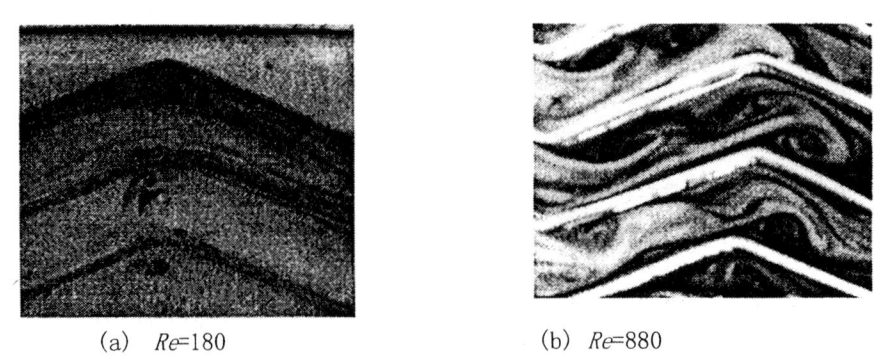

(a) *Re*=180 (b) *Re*=880
Fig.6 Flow patterns in the conventional wavy fin channels(l/m=6, a = 22.5°)

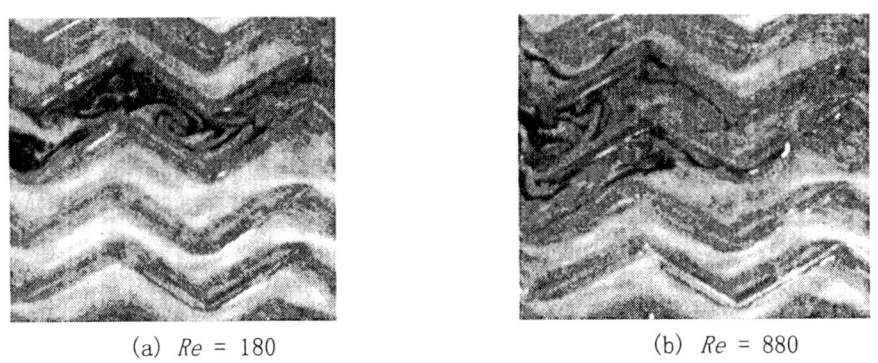

(a) *Re* = 180 (b) *Re* = 880
Fig.7 Flow patterns in the wavy-perforated fin channels (l/m=8, a =18°)

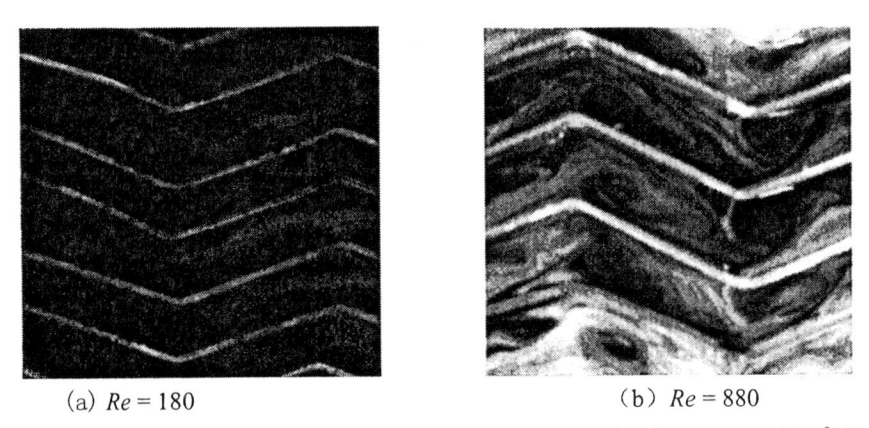

(a) *Re* = 180 (b) *Re* = 880
Fig.8 Flow patterns in the wavy-perforated fin channels (l/m=6, α = 22.5°)

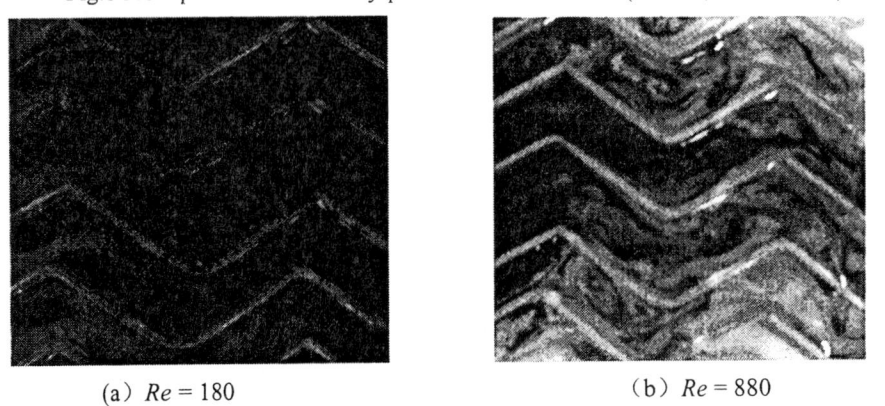

(a) *Re* = 180 (b) *Re* = 880
Fig.9 Flow patterns in the wavy-perforated fin channels (l/m=4, α = 33°)

(a) *Re* = 100 (b) *Re*=300

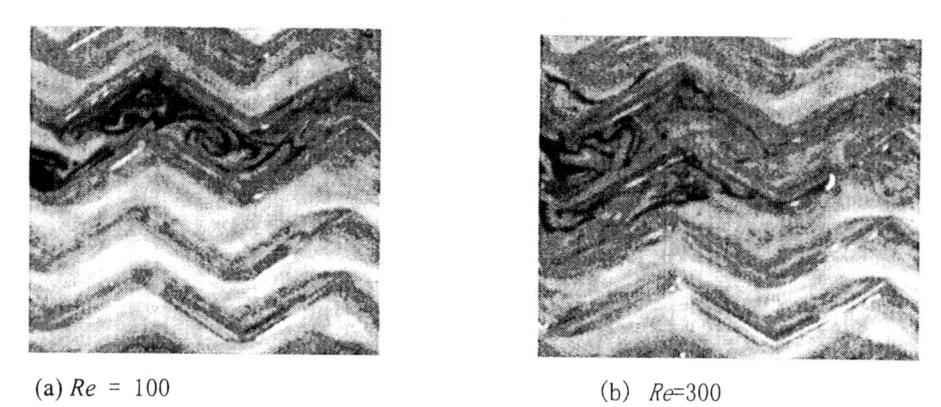

Fig.10 Transverse mixing effect of the fluid through the perforated holes (*l/m*=4, *a* = 33°)

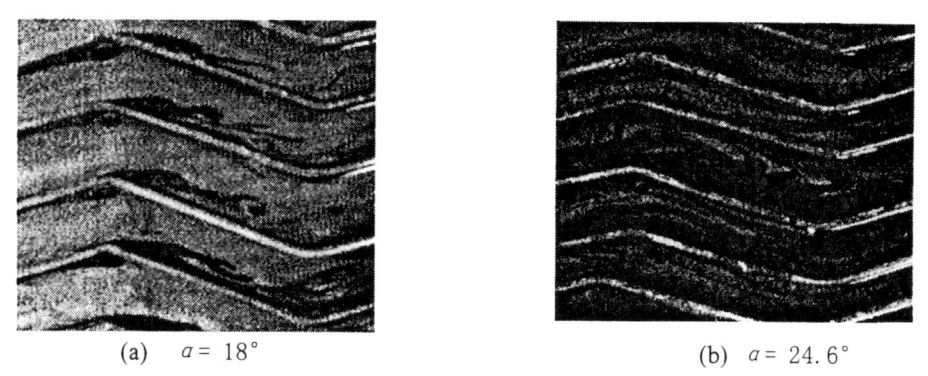

(a) *a* = 18° (b) *a* = 24.6°

Fig.11 The effect of the perforated wavy fin angle (*l/m*=6, *Re*=180)

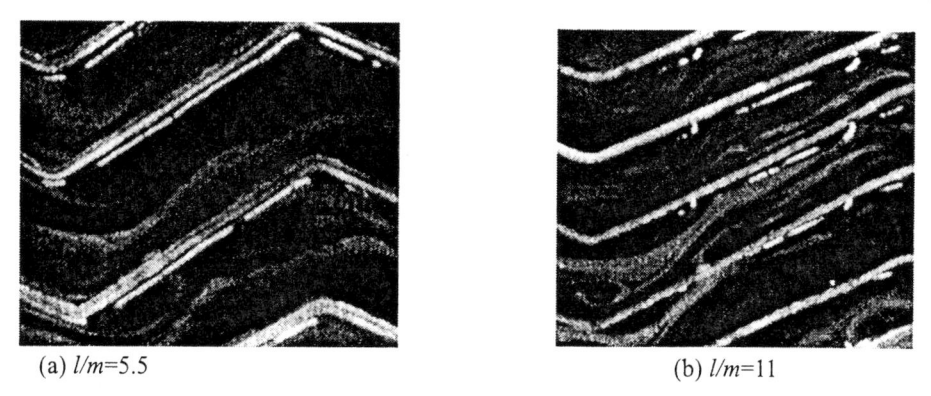

(a) *l/m*=5.5 (b) *l/m*=11

Fig.12 The effect of the aspect-ratio *l/m* (*Re*=180, *a* = 22.5°)

POSITRON EMISSION PARTICLE TRACKING (PEPT) – A NEW TECHNIQUE TO INVESTIGATE THE FLOW PATTERN IN A PLATE AND FRAME HEAT EXCHANGER WITH CORRUGATED PLATES

H.U. Zettler[1], H. Müller-Steinhagen[2], R.N.G. Foster and P. Fowles[3]

[1] Department of Chemical & Process Engineering, University of Surrey, Guildford, Surrey, UK
[2] Institute for Thermodynamics and Thermal Engineering, University of Stuttgart
Institute for Technical Thermodynamics, German Aerospace Center, Stuttgart, Germany
[3] Positron Imaging Center, School of Physics and Astronomy, The University of Birmingham, Edgebaston, Birmingham, UK

ABSTRACT

The flow pattern of water in a corrugated heat exchanger channel has been investigated using the non-invasive technique of Positron Emission Particle Tracking (PEPT) in a plate heat exchanger (PHE). PEPT allows an individual neutrally buoyant particle moving at speeds of up to 10m/s to be tracked in three dimensions to within 1mm or less. The technique produces particle trajectories and velocity profiles, and has also been used to detect low flow velocity and re-circulation zones in the flow channel. In addition, it allows for the determination of residence time distributions over a variety of flow rates, plate corrugation angles, and plate set-ups (side- and diagonal flow). PEPT data obtained at the University of Birmingham is compared to data obtained by visual observation and fouling experiments obtained at the University of Surrey, to identify how different flow paths influence the fouling behaviour in a PHE.

Keywords: Plate heat exchanger, PHE, flow pattern, PEPT

INTRODUCTION

Plate and frame heat exchangers, more commonly known as plate heat exchangers (PHEs), owe their high heat transfer efficiency to the specific plate design, and in particular to the presence of corrugations. These corrugations force the flow in the plate channels to experience a continuous change in direction and cross-sectional flow area. As a result, turbulence is induced at lower flow velocities than one would normally expect. A high level of turbulence leads to higher heat transfer coefficients and, as a consequence, to lower heat transfer area requirements for the same heat duty. This means that generally less material and space are needed for the installation of PHEs, resulting in substantial savings for the industrial user [Raj80].

Compared to some other heat exchanger types, PHEs have a lower tendency to foul due to the presence of higher shear forces capable of removing fouling deposits attached to the sides of the plate channels. However, recent investigations have shown that similar commercial heat exchanger plates with different corrugation patterns and different flow path are observed to foul at very different rates under identical process conditions. These differences are attributed to the effects of flow distribution on fouling rates in plate channels [Kho97],[Mul97].

Obviously, heat transfer, pressure drop and fouling performance of PHEs is highly dependent on the local velocities and temperatures. Unfortunately, these local data are difficult to obtain. In this investigation, heat transfer fouling experiments and flow visualisation experiments using polycarbonate plates and a high-speed camera were performed. With the high-speed camera however, only two-dimensional results could be obtained and quantitative information was difficult to obtain [Kho97]. Therefore, Positron Emission Particle Tracking (PEPT) has been used in order to obtain more detailed quantitative information. This technique relies upon the detection in coincidence of the two gamma rays emitted following annihilation of an electron with a positron emitted from a single radioactive tracer particle. This tracer particle could be accurately located up to 500 times per second to within 0.5 mm, allowing the tracking of particles inside the PHE for flow velocities of several meters per second. Examining the change in tracer position with time provides an estimate of its instantaneous velocity with an accuracy of approximately 10%. Using PEPT, velocity maps of the fluid over a range of flow rates, corrugation angles and flow paths can be determined. In addition, occupancy plots may be obtained, which indicate the fraction of total run time during which the tracer particle was found at each point The new data may be compared with data obtained with a high-speed camera and with experimental fouling data. An important application of these results is to provide validation of CFD (Computational Fluid Dynamics) model predictions.

FLOW VISUALISATION IN THE PAST

The local flow patterns in industrial plate heat exchangers (PHE) are not only affected by the corrugation of the plates but also by the overall geometry of the flow channel induced by the gaskets. An investigation on the improvement of flow distribution requires a method for visualisation in order to investigate possible re-circulation and mal-distribution zones.

Conventional visualisation methods can be broadly divided into two groups, direct injection methods and chemical methods as can be seen in Table 1.

Table 1. Compilation of Flow Visualisation Methods [Tri98]

Method	Visualisation	Application in PHE
Direct Injection	- Dye Injection	At low Reynolds numbers
	- Tracers (solid)	Tracer Particle must be very small
	- Air Injection	At low Reynolds numbers
Chemical Methods R-12	- Indicator Method	Effective even at higher Reynolds numbers
	-Electrochemical Method	At low Reynolds numbers
	-Hydrogen Bubble Method	At low Reynolds number, because buoyancy effects can interfere

By using a direct injection method, the flow is visualised by injection of solid particles or fluids. The use of air or dye is limited to low Reynolds numbers as the more turbulent state of the flow at higher Reynolds numbers renders it impossible to follow the path lines under such conditions.

The use of solid particles is possible at high Reynolds numbers but the particles must be very small to prevent them lodging in small channel gaps. Different kinds of particles can be used for the flow visualisation in plate heat exchangers e.g. Aluminium powder, Acronytril Butadiene Styrene particles or Polystyrene particles. Particle traces can even reveal vortices and re-circulation zones.

When using a chemical method, the flow is visualised by a colour change of an indicator caused by a change in pH. A solution of a pH-indicator is used as the working fluid and acid or alkali is injected into the flow. Alternatively, hydrogen ions can be produced by an electrical field within the heat exchanger geometry, producing the same effect.

The injection of acid or alkali is applicable even with mixing effects experienced at higher Reynolds numbers, because the colour does not vanish.

This is not the case with the electrochemical method where hydrogen ions are produced by an electrical field. The concentration of hydrogen ions is less than the concentration of injected acid, resulting in a less intensive colour change, which is further reduced by mixing effects.

For the hydrogen bubble method, water is used as the working fluid and the flow is visualised by generating hydrogen bubbles in an electrical field. This method gives only a rough idea of the flow structure and furthermore buoyancy effects must be taken into account.

All the previously described methods are problematic in that they do not allow the investigation of process system using opaque apparatus. In addition, they only give a two dimensional representation of the flow and are mostly limited to lower Reynolds number flows in their use due to mixing effects at higher Reynolds numbers.

In this study, the flow in a plate heat exchanger was experimentally observed using the technique of Positron Emission Particle Tracking (PEPT). The advantage of PEPT in comparison with other visualisation methods is that full 3D flow field information may be obtained within optically opaque apparatus.

PRINCIPAL FLOW PATTERN IN PHEs

The flow patterns in the channel for diagonal flow and side flow are illustrated in Figure 1, where the length of a given arrow is an approximation of the flow velocity at that point. The solid arrows represent the first channel (which is the solution channel) and the broken arrow describes the water channel on the backside of the plate.

In the diagonal flow arrangement, the solution enters at the bottom left side and leaves at the diagonal top right side and the water (adjacent channels) enters at the top left side and leaves the bottom right side. For side flow plates however, the inlet and outlet ports for the solution are on the left side and for water (adjacent channels) are located on the right side.

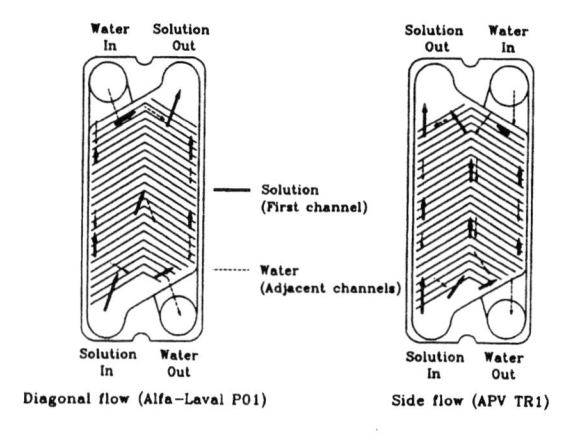

Fig. 1 Different flow arrangements in a plate heat exchanger [Ban94]

VISUALISATION OF FLOW PATTERNS IN PHEs

The most important investigation of flow visualisation is reported by Focke and Knibbe [Foc86]. They investigated the flow in a plate heat exchanger with an electrochemical indicator method in a range of Reynolds between 10 and 1000. Hydrogen ions produced by an electrical field dissolve into the water and change the colour of an indicator solution, which flows through the investigated geometry. With this technique they found two flow patterns to occur, **crossing flow** along a furrow and **longitudinal wavy flow** which exhibits a strong dependence on the corrugation angle. At a corrugation angle of 45°/45° the flow was found to follow mainly the furrows on each walls as can been seen in Figure 2 (**B**).

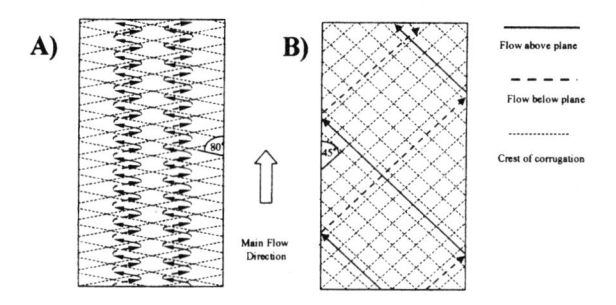

Fig. 2 Flow path existing in corrugated channels. (A) longitudinal wavy flow; (B) Crossing flow [Foc86].

After reaching the side walls the flow returns along the furrow on the opposite wall. Tribbe suggested that the crossing flow path is independent of Reynolds number.

This is confirmed by Martin [Mart95] who based his theoretical approach for calculating the pressure drop in plate heat exchangers on these investigations.

Tribbe [Tri98] observed a crossing flow distribution for a corrugation angle of 30°/30° but mentioned that the flow paths are not independent of the Reynolds number. He used the same experimental method as Focke and observed similar results for a 30°/60° combination.

For a corrugation angle of 80°/80° Focke found that the flow does mainly not follow the furrow, but is reflected close to the contact points, so that the flow motion has a more longitudinal wavy character as illustrated in Figure 2 (**A**).

Tribbe observed in a 60°/60° plate arrangement that a superposition of crossing and longitudinal flow exists. As other authors reported before [Foc86], [Gai89], [Cio96] he verified the existence of a secondary flow path, the corkscrew pattern. Furthermore he reported the existence of the stagnation zone directly after the contact point. This vortex is illustrated in Figure 3.

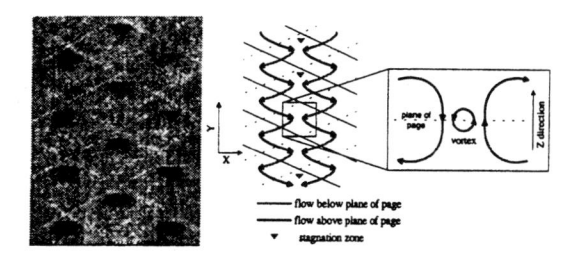

Fig. 3 Vortex generation in wavy flow [Tri98]

The secondary flow induces a rotation on both sides after the contact point causing parts of the fluid to rotate. A similar observation for the 30°/30° plate arrangement could not be made.

POSITRON EMMISION PARTICLE TRACKING (PEPT)

The technique of Positron Emission Particle Tracking (PEPT), enables a single radioactive tracer particle moving inside a piece of equipment to be tracked accurately at speeds of up to 5m/s. The technique is most often used in granular flow studies, where the tracer is typically composed of the same material as the bulk or comparable with it in size or density [Par93], [For01].

Following irradiation within the University of Birmingham's Cyclotron, the active component of the tracer particle decays by positron emission. Once emitted from the nucleus, the positron annihilates with an electron, leading to the production of two near collinear 'back to back' gamma rays. The detection of these and subsequent pairs of gamma rays enable the tracer to be located in three dimensions by triangulation, as shown in Figure 4 below.

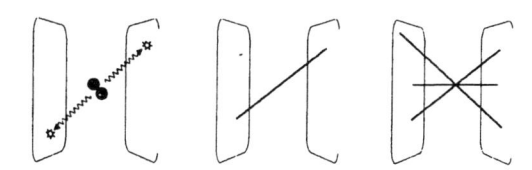

Fig. 4 Principles pf PEPT: Detection of gamma rays via two large position sensitive detectors (left). Definition of a line, were the tracer is located (middle). After detection of several pairs, tracer may be located via triangulation (right).

The emitted gamma rays are quite penetrating, with 50% transmitted through 11mm steel, and thus tracking is possible inside real laboratory scale process equipment. Typical applications of PEPT to Industrial Processes are the investigation of the motion of dry granular materials in gas fluidised beds, [Ste00], vibrating beds [Wildman] and mixers and blenders [Mar98], [Kuo01]. In addition, the use of neutrally buoyant tracers allows flow field visualisation within closed pipes and stirred tanks.

The investigation of the isothermal flow in a plate heat exchanger duct for this study was carried out at the Positron Imaging Centre of the University of Birmingham where the technique of Positron Emission Particle Tracking was developed. This is the first time this technique has been applied to flow in a plate heat exchanger as described in the following sections.

TEST RIG AND EXPERIMENTAL PROCEDURE

An experimental set-up was constructed at the Department of Chemical and Process Engineering in the University of Surrey, and then installed at the Positron Imaging Centre in the University of Birmingham.

Figures 5 and 6 show the test-rig with one closed flow loop for water, which was stored in a 15-litre tank. In this study the plate heat exchanger consists of one diagonal flow channel of two Alfa Laval M3 plates (60°/60°). However, the PHE-test section can be used with different Alfa Laval M3 plates of various corrugation angles in future investigations.

The experimental investigation of flow distribution of water in a duct of a plate heat exchanger was carried out under isothermal conditions. The use of a conventional centrifugal pump resulted in some problems with the tracer particle shattering or becoming trapped due to the action of the impeller, and therefore a peristaltic pump was employed in these preliminary investigations.

Figure 5 Photograph of the experimental set-up

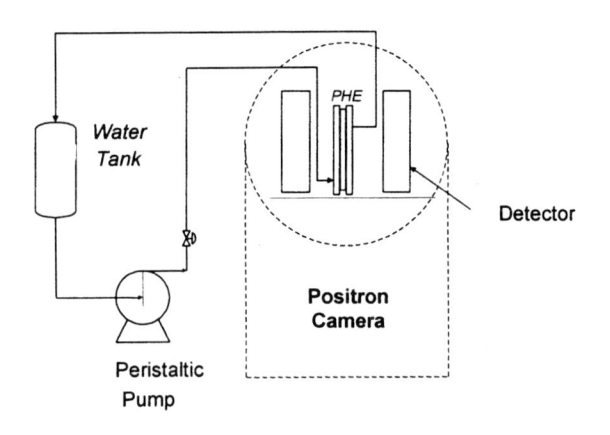

Figure 6 Flow chart of the experimental test-rig

EXPERIMENTAL PROCEDURE

The tracer used in this work was a resin bead with a diameter of approximately 240 μm. Tracers were made active by mixing them with radioactive water for approximately 30 minutes, resulting in up to 90% of the water activity being transferred to the beads. The beads were subsequently coated with rapid curing paint in order to restrain activity transfer to the impure water in the test-section during the experiments. [Fowles, to be published].

The tracer particle was placed in the flow loop and the pump was started whilst the data acquisition of the positron camera was begun. The flow velocity for the experiments in this study was 0.35 m/s.

RESULTS AND DISCUSSION OF THE PEPT ANALYSIS

The method produces a series of x, y and z coordinates, along with a corresponding time for each position. Typically, the position could be obtained several hundred times per second with an accuracy of approximately 0.5mm. The flow enters at the bottom right side and leaves at the diagonal top left side. The particle was also traced outside the heat exchanger in the tubing section from and to the pump, which can be seen as detected positions outside the plate channel in Figure 7 where all detected locations during one test run are illustrated. It can be seen that in some regions the tracer was rarely detected. These regions are mainly in the corner on the bottom left side and the upper right side. This shows that the flow is not uniformly distributed across the corrugated section.

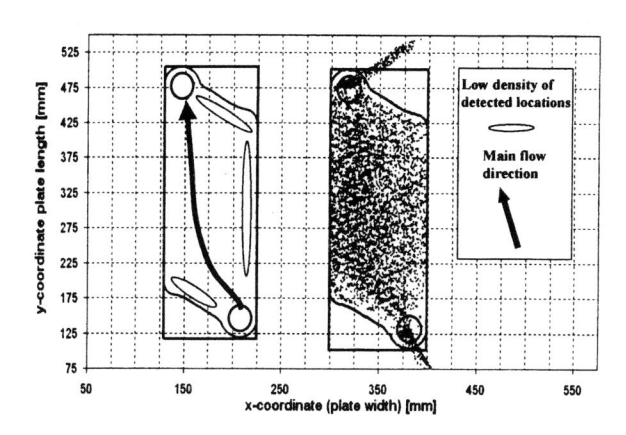

Figure 7: Distribution of all detected positions of the tracer in the PHE

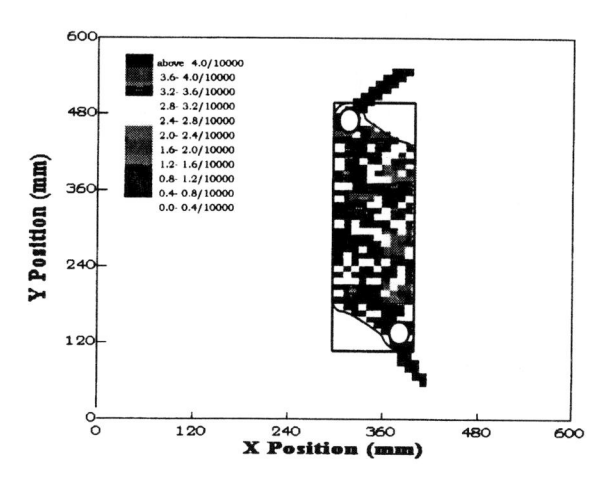

Fig. 8 Occupancy distribution

Furthermore it can be assumed that the flow follows a slightly curved direction after the inlet to the outlet since in Figure 7 it can be seen that the location density is higher on the left than on the right side.

In Figure 8 an occupancy plot is illustrated. This quantity represents the fraction of the selected time range that the tracer was located in each location of the investigated geometry.

Occupancy O is defined as

$$O = \frac{t_{pixel}}{t_{total}} \times 10,000 \qquad (1)$$

where t_{pixel} is the time spent in one defined volume in the geometry and t_{total} is the total time range.

It can be seen that the assumption of a main flow direction in a diagonal, slightly curved line between the inlet (bottom right side) and the outlet (upper left side) is correct. This leads to a non-uniform flow distribution as can be seen by comparing the occupancy values on the left and right sides of the plate.

The velocity distribution and vectors are illustrated in Figure 9. Re-circulation zones could not be detected. The reason might be that the tracer density was too different to the water density or that the tracer was not close enough to the vortices behind the contact points to be involved in the re-circulation. This would lead to the assumption that the flow characteristics obtained by this investigation are mainly representative for the bulk flow. However the tracer was affected by the wavy structure of the duct where a change in cross sectional area induces a zig zag flow pattern.

At the right and left side at the wall-adjacent zones of the inflow, two low flow velocity zones can be seen. One is on the right side behind the inlet (A) and the other one is in the lower corner on the left side of the plate (B). Further low flow velocity regions are detected on the upper right side in the outlet distributor section of the heat exchanger (C).

It must be noted that this velocity distribution gives less accurate results in the wall adjacent zones, because here the tracer was not located at every position of the duct.

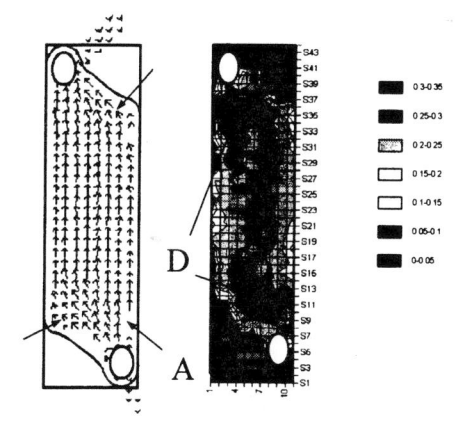

Figure 9 Velocity vector distribution and velocity contour distribution

Higher flow velocities are detected in the middle of the plate in the main direction from the inlet to the outlet. Two zones

227

with higher velocities can be seen at the transitions areas of the inlet- and outlet distributor sections to the corrugated section (D).

This might imply that the transition from the distributor zone to the corrugated zone leads to a non-constant flow distribution. Low flow velocities occur on the left and right side, whereas the flow velocity reaches high values in the centre.

These results are in good agreement with former investigations. The highest velocities are in a diagonal line from the inlet to the outlet [Ban94]. The low flow velocity zones in the lower left corner and the upper right corner, which were observed in several studies could be identified. The lower flow velocity zone at (A) is probably caused by air bubbles, which influenced the flow velocity in this small area at the inlet and could also be observed in experiments [Zet97].

The average velocity of 0.35 m/s, which was adjusted during the experiments, could not be detected as can be seen in Figure 9, because flow velocities in the range of 0.30-0.35 are scarce. The reason might be that the tracer was only located in the bulk flow, where the average velocity is slower.

The distribution of the tracer particle locations at different times makes it possible to visualise path lines. A combination of crossing flow along the furrows and wavy flow along the rows of contact points can be assumed from the illustration of some path lines in Figure 10. This result contradicts what has been published in the literature, where longitudinal wavy flow is reported in corrugated ducts with a corrugation angle of 60°/60° [Foc86][Tri98].

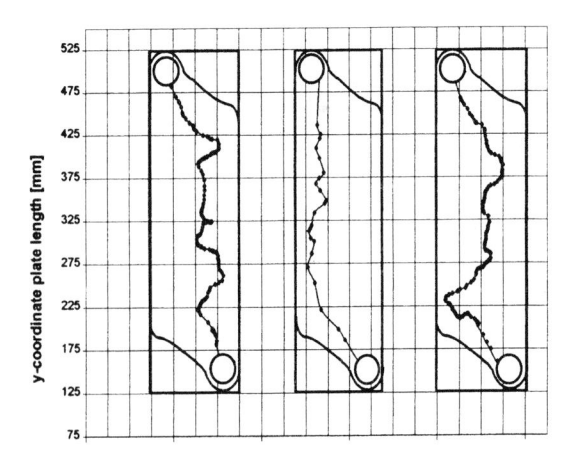

Fig. 10 Different flow path in the heat exchanger duct

The difference is probably caused by the use of a peristaltic pump in this study, whereas in conventional flow visualisation studies centrifugal pumps produce a more uniform flow in the plate heat exchanger.

CONCLUSIONS

To investigate the flow pattern in a real heat exchanger a test rig was built and the flow in a diagonal flow heat exchanger duct (Alfa Laval M3, 60°/60°) was visualised with the PEPT- method (Positron Emission Particle Tracking). Since the method is based on the detection of a radioactive particle the heat exchanger had not to be made of transparent material. Therefore, for the first time the flow paths in a real plate heat exchanger duct could be investigated. This method allowed the localisation of the tracer and produced a set of 3 cartesian coordinates and time. Conventional flow visualisation techniques usually only produce 2-dimensional representations of flow patterns. Whereas the PEPT technique allowed the 3-dimensional representation of pathlines, occupancy and velocity distributions. These results gave a good insight in the prevailing flow conditions. Low flow velocity zones could be discovered, which are mainly located in the corners of the inlet and outlet distributors. This is caused by the uneven flow distribution of the duct in a plate heat exchanger.

These results can be compared with numerical simulations carried out with the laminar flow model in a corrugated geometry with flat in and outlet distributors. It could be shown that the experimental flow velocity distribution represents mainly the bulk flow velocity. The velocity of the tracer particle never reached high flow velocities up to 1 m/s.

NOMENCLATURE

U	Total surface energy,	J/m^2
T	Temperature,	K
S	Entropy,	$J/K\,m^2$
γ	Interfacial tension,	J/m^2
θ	Contact angle,	°
W	Work of adhesion,	J/m^2
I	Interaction term,	J/m^2

Subscripts

sl Solid/liquid
sv Solid/vapour
lv Liquid/vapour
drop Drop phase
bulk Bulk phase

Superscripts

d   Disperse
p   Polar
S   Surface
nd   Non-disperse

REFERENCES

[Bec96]   Becker-Balfanz, C.D., Hopp, W.-W., Königsdorf, W., Maier, K.H., and Pletka, H.D., Erfahrungen mit Platten- und Spiral-Wärmeübertragern, *Gaswärme International*, vol. 45, no. 6, pp. 276-284, 1996.

[Bon81]   Bond, M.P., Plate Heat Exchangers for Effective Heat Transfer, *The Chemical Engineer*, April 1981, pp. 162-167, 1981.

[Bot95]   Bott, T.R., *Fouling of Heat Exchangers*, Elsevier Science B.V., Amsterdam, The Netherlands, 1995.

[Cio96]   Ciofalo, M., Stasiek, J,. Collins, W., Investigation of Flow and Heat Transfer in Corrugated Passages, Part I and II, Int. J. Heat and Mass Transfer, Vol 39, pp 149-192

[Foc86]   Focke, W.W., Knibbe, P.G., Flow Visualisation in Parallel-plate ducts with corrugated walls, J. Fluid Mechanics., Vol. 165, pp. 73-77

[For01]   Forster R.N., Seville J.P.K., Parker D.J., Ding Y.
Tracking Single Particles in Process Equipment Kona, to be published 2001

[Gai89]   Gaiser, G., Kottke, V., Flow Phenomena and Local Heat and Mass Transfer in Corrugated Passages, Chemical Engineering Technology, Vol 12, pp. 400-405

[Ker87]   Kerner, J., Sjogren, S., Svensson, L., Where Plate Exchangers Offer Advantages Over Shell-and-Tube, *Power*, May 1987, pp. 53-58, 1987.

[Kho97]   Kho, T., Zettler, H.U., Müller-Steinhagen, H., and Hughes, D., Effect of Flow Distribution on Scale Formation in Plate and Frame Heat Exchangers, IChemE, 1997.

[Kuo01]   Kuo H.P., Burbidge A.S., Parker D.J., Tsuji Y., Adams N.J., Seville J.P.K., Studies of Particle Motion in a V-Blender using DEM simulation and the PEPT technique, Partec 2001, Nuremburg 27-29th March, 2001-02-21

[Mar95]   Martin, H., A theoretical approach to predict the performance of chevron-type plate heat exchangers, Vol. 35, pp 301-310

[Mar98]   Martin T W, Seville J P K.
Dispersion of Particulate Materials in Mixers. Proc. World Cong. on Particle Technol. IchemE, No.331 1998

[Mul93]   Müller-Steinhagen, H., Fouling: The Ultimate Challenge for Heat Exchanger Design, *The Sixth International Symposium on Transport Phenomena in Thermal Engineering, Seoul, Korea*, pp. 811-823, 1993.

[Mul97]   Müller-Steinhagen, H., Plate Heat Exchangers: Past – Present – Future, *7. International Congress on Engineering and Food, Brighton*, part 1, edited by Jowit, R., Sheffield Academic Press, April 1997.

[Par93]   Parker D J, Broadbent C J, Fowles P, Hawkesworth M R, McNeil P A. Positron Emission Particle Tracking – Atechnique for studying flow within engineering equipment. Nucl. Instrum. & Meth. A326 1993 pp592-607

[Raj80]   Raju, K.S.N., and Chand, J., Consider the Plate Heat Exchanger, *Chemical Engineering*, August 11, pp. 133-144, 1980.

[Ste00]   Stein M., Ding Y., Seville J.P.K., Parker D.J., Solids Motion in Gas Bubbling fluidised Beds, Chemical Engineering Science, vol. 55, pp5291-5300, 2000

[Tri98]   Gas/Liquid Flow in Cylindrical and Corrugated Channels, PhD Thesis, department of Chemical and Process Engineering, Uniersiy of Surrey, Guildford, England

[Zet97]   Influence of Operation Conditions and Flow Distribution on Scale Formation in Plate and Frame Heat Exchangers, Diploma Thesis, Department of Chemical and Process Engineering, University of Surrey, Guildford

A SIMPLE AND INDUSTRIAL TEST FACILITY FOR CORRUGATED FINS EVALUATION

F. Châtel and E. Werlen

Air Liquide, Claude-Delorme Research Center, Les Loges en Josas, 78354 Jouy-en-Josas, France
E-Mail : fabienne.chatel@airliquide.com, etienne.werlen@airliquide.com

ABSTRACT

This paper describes Air Liquide experimental facility, called Cj-f test bench, to determine the heat transfer (Colburn factor Cj), and flow friction (Fanning factor f), characteristic of corrugated fins built in plate fins heat exchangers. As a matter of fact, management of heat exchanger design, as well as new fins development, require accurate values of Colburn and Fanning factors; and though CFD calculations are very useful for fins studies, experimental measurements are absolutely necessary to validate numerical approaches, and to account for manufacturing defauts. In the following study, Cj-f test core and test bench are described, and their design or operating principle are compared with some existing test facilities. The main advantages of this new method consists in a simple set-up which allows for quick operation, and where pressure drop and heat balance control guaranties large-scale representation. High accuracy results are obtained thanks to suitable instrumentation and calculation procedure taking all specific working features of the test bench into account. Hydraulic and thermal performances of many fins implemented in plate fins heat exchangers for cryogenic plants have been measured and, most of the time, a good agreement in the heat transfer data for Cj-f test sample has been found, when compared with manufacturer correlations; on the contrary, prediction of Fanning is very difficult, because of significant manufacturing effects on pressure drops, and thus, experimental friction measurements are essential. Finally, such an accurate, quick and low cost test bench is very useful for evaluation of new designed fins, for validation of manufacturers data, and for characterization of fins built in a specific industrial heat exchanger.

INTRODUCTION

Corrugated fins used in compact plate fins heat exchangers are characterized by their thermal and hydraulic performances, generally represented by two graphs : dimensionless heat transfer coefficient (Colburn factor Cj) and pressure drop coefficient (Fanning factor f) versus Reynolds Number.

Accurate values of Cj and f in single phase flow are key values for an optimized heat exchanger design. Moreover, developing new fins steadily comes up against performance evaluation difficulties : numerical studies are an halfway solution, which can bring out satisfactory performance trends versus fin geometry, but those kind of results are not accurate enough to replace experimental data and they are not taking into account manufacturing defauts (tool wear, burrs, tolerances...).

Regarding the two above statements, optimized heat exchanger design as well as efficient fin developments require an experimental way which has to be able to characterize accurately, quickly and with cost respect, a great number of fins.

This paper describes Air Liquide original experimental facility, dedicated to fins evaluation and located in Air Liquide's Claude-Delorme Research Center (CRCD), near Paris.

MEASUREMENT METHOD

Test facility novelty

Because of the constant requirement of experimental data for design or development, test facilities already exist and are likely to give accurate values of fin thermal and hydraulic performances.

However, most of existing experimental facilities meet some of the following limits, which the test bench described in this paper should improve :

- Complex test unit to be measured (many layers, welded pieces...) results in significant cost and long manufacturing time.

- When there are more than one stream (often gas and water) to manage, the testing process is more complex and the test duration is increased.
- With one water side, thermal balance can not be checked and test unit length is limited because of temperature measurement accuracy

Description of the original experimental bench

Test unit. The main advantage of this bench consists in a test sample easy to manufacture, since corrugated fins to be characterized are brazed into the simplest plate fin heat exchanger : a simple double layers heat exchanger (Fig.1). Like in most of air separation unit main exchangers, layers are in counter-current flow. Using this test unit design, any existing or newly designed fins are likely to be built quickly and at low cost. Moreover, the test unit is long enough (length about 0.4m) in order that inlet effects remain negligible. Also, distribution zones are designed without special fins, and an empty chamber insure a proper distribution of flows in the test zones. Doing so, extra cost in manufacturing prototypes are avoided, as well as correction in Cj and f for the contribution of distribution zones to overall measurements.

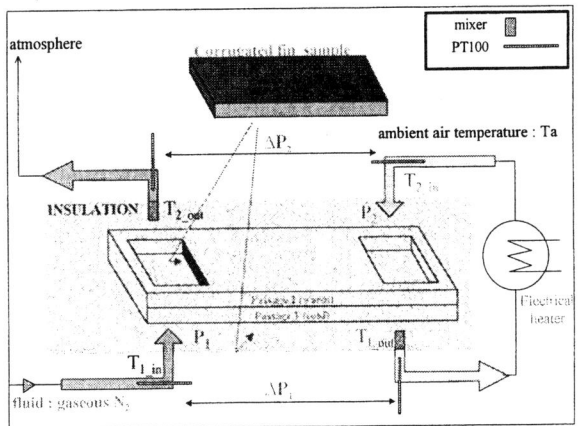

Fig. 1 Corrugated fins brazed in test sample in order to be characterized on Cj-f test bench.

Bench process. With an original process, one single gaseous fluid (pressured gaseous nitrogen) is flowing into the unit, so that flow rates in both layers are strictly identical : outlet flow from lower layer 1 of the test unit is heated through an electrical heater, and is re-injected at upper layer 2 inlet before being rejected to the atmosphere (Fig. 1 and Fig. 2).

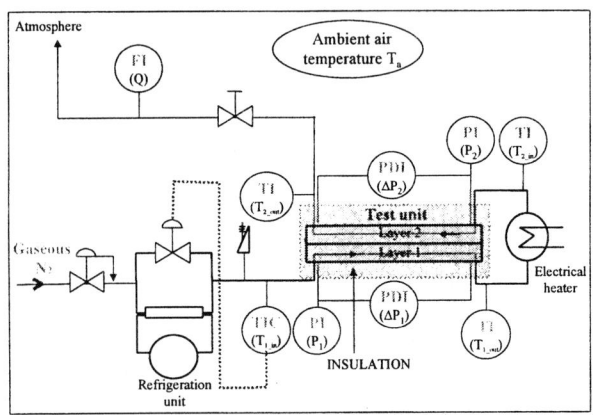

Fig. 2 Cj-f test bench principle

Experimental accuracy. Experimental accuracy is increased by the test unit design and a strictly parallel heat transfer diagram (Fig. 3) resulting from bench process : drop or rise in the fluid temperature, as well as pressure drops are measured twice (once per layer), and discrepancies between passages indicate manufacture defaults, abnormal heat losses, or, generally speaking, experimental problems. Such a diagram is easily adjusted by varying exchange length, inlet temperature in cold layer and power input at warm end, in order to improve results accuracy.

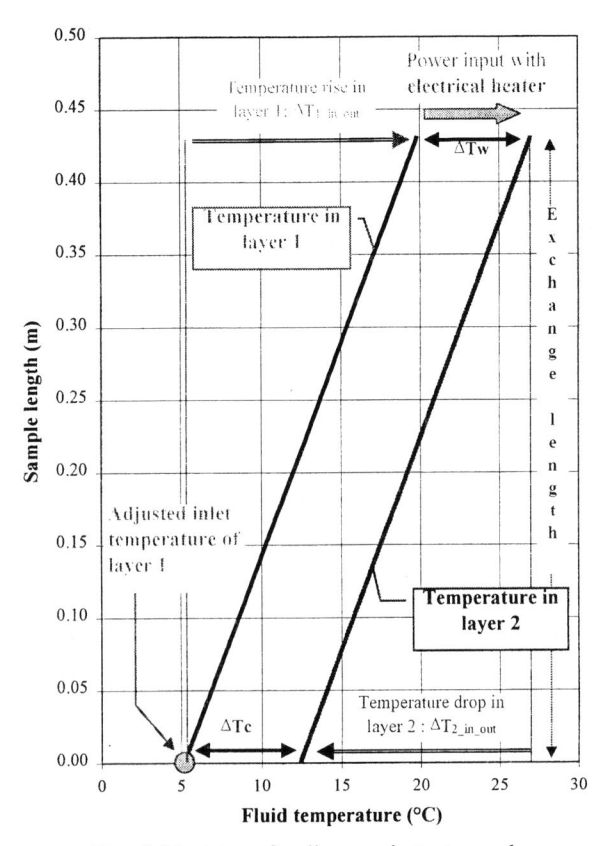

Fig. 3 Heat transfer diagram in test sample

In addition to the above checks, high accuracy instrumentation (temperature platinum devices, two ranges of pressure sensors and flow rate measurement devices) is installed on the bench (Table 1), so that fin performances precision is about 4% for friction factor and 10% for thermal efficiency, depending on flow rate and fin geometry.

Table 1 Instrumentation and result accuracy on Cj-f experimental bench

G	relative accuracy ΔG/G (max)		absolute accuracy ΔG		Range of measurements	
Q	1% - 5%	1%	0.2 m3/h	1.15 - 15m3/h	3 - 15 m3/h	15 - 150 m3/h
ΔP	0.1% - 3%	0.1% - 2%	0.1 mb	2 mb	3 - 100mb	100 - 3000mb
ΔT_{in_out}	0.4% - 2%		0.2 °C		10 - 50°C	
ΔT_c or ΔT_w	2% - 4%		0.2 °C		5 - 10°C	
f	2% - 6%					
Cj	7% - 10%					

Therefore, and exactly because of the simplified unit and process, a test unit does not work like a piece of "ideal" exchanger : some operating specificities are to be considered to obtain the right performances of the tested fins. A suitable calculation method has been developed which refers to the exact test unit operation, which involves :

- asymmetrical behavior of the layers : insulation on external plates prevent from heat losses, and the separating sheet is working under the complete heat flux.
- longitudinal conduction which has to be considered especially at low Reynolds Number

CALCULATION METHOD

Hydraulic and thermal phenomena are very specific to the test unit design and to the operating conditions set up on Cj-f test bench. The following method describes those specificities, as well as relationships to be used to derive Friction factor (Fanning factor) f and Colburn factor Cj from the measurements. In all the equations below, fluid physical properties are calculated at the mean temperature and pressure in each layer of the test sample.

Friction factor coefficient f (Fanning factor) calculation

Relationship between Fanning factor and linear pressure drop is :

$$\Delta P_{lin} = \frac{4.f}{D_h} \cdot \frac{1}{2} \cdot \rho . v^2 \qquad (1)$$

Where fluid velocity into fin sample, v, is determined from the mass flow rate measurement Q and the cross sectional area for fluid flow A_c.

As a result, Fanning factor is a function of the Reynolds Number, defined in Eq. 2 :

$$R_e = \frac{Q.D_h}{\mu.A_c} \qquad (2)$$

Colburn coefficient C_j calculation method

Calculation of Colburn coefficient derived from measurements on Cj-f test sample require an accurate knowledge of thermal phenomena in test unit :

1. Test sample design involves an "asymmetrical" working of both layers because of external plate insulation. Therefore, effective heat transfer area can not be expressed from usual addition of primary and secondary surface areas corrected by normal fin efficiency (Eq.7). An accurate calculation of effective heat transfer area (Eq.8) has to be performed to derive Colburn factor.

2. Counter current flow involves longitudinal conduction through sample material cross section. Such thermal effect are carefully calculated in measurement post running. As a matter of fact longitudinal conduction overlook would result in underestimated Colburn factor at low Reynolds Number by 30% (Fig. 4)

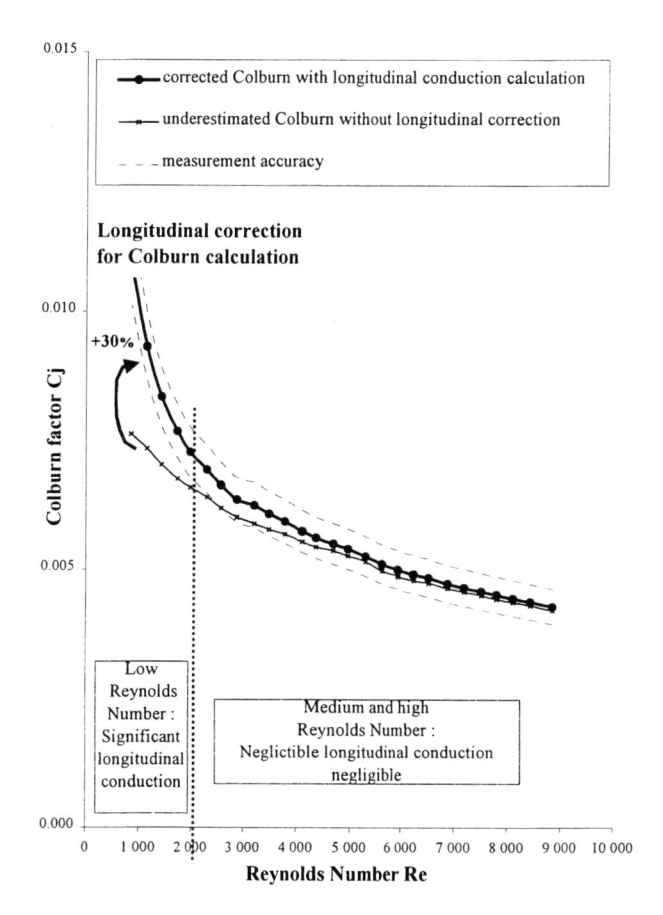

Fig. 4 Experimental measurements on CRCD Cj-f test bench : Colburn factor Cj versus Reynolds Number Re. Correction of longitudinal correction in tested sample.

To make the post running of thermal measurements easier, those two phenomena are considered separately; equations used to calculate Colburn coefficient versus Reynolds Number are detailed as follow.

Colburn factor Cj. Colburn Factor Cj is derived from heat transfer coefficient k_{cv} :

$$k_{cv} = C_j \cdot \frac{Q}{A_c} \cdot c_p \cdot P_r^{-\frac{2}{3}} \qquad (3)$$

Heat transfer coefficient k_{cv}. To calculate k_{cv}, we assume that heat flux $d\varphi$ through a small element of surface ds located on either primary or secondary heat transfer areas, is related to the difference between the temperature of ds, called θ, and the average temperature of the fluid in the corresponding cross section, called T(x), by the heat transfer coefficient called k_{cv}, constant on the whole surface of the test unit :

$$d\varphi = k_{cv} \cdot ds \cdot (\theta - T(x)) \qquad (4)$$

Heat transfer efficiency ka. In order to simplify the calculations, the heat transfer efficiency ka is introduced : in a small slice dx of layer i (Fig. 5), ka relates the heat flux $d\Phi_i$, exchanged between the fluid i and the fins, to the difference between the mean temperature of the parting sheet $T_p(x)$ and the average temperature of the fluid $T_i(x)$ [1]:

$$d\Phi_i = ka \cdot (T_i(x) - T_p(x)) W \cdot dx \qquad (5)$$

Relationship between global coefficient ka and heat transfer coefficient k_{cv} has the general following expression derived from heat balance around a parting sheet element :

$$\frac{1}{ka} = \frac{L \cdot W}{k_{cv} \cdot A_e} + \frac{e_p}{2 \cdot \lambda} \qquad (6)$$

Where A_e is called "effective heat transfer area" and is expressed as a relationship between primary and secondary surface areas (Fig. 5) in addition to "normal" and "by-pass" fin efficiencies as defined in HTFS DR50 (1980). This relationship is due to specific heat flux pattern :

In industrial heat exchangers, the more commonly implemented arrangement of layers is known as "single banking" (HTFS, 1980), and refers to patterns where each cold layer has a hot layer on either side of it, and vice versa. Such a layer is working "symmetrically" since the same heat flux is entering through both parting sheets which are delimiting the layer; in single banking arrangement, thanks to symmetrical heat flux dispatching, all plates have the same temperature T_p and the effective heat transfer area A_e^{sym} is only dependent on normal fin efficiency η, thanks to well known formula :

$$A_e^{sym} = A_p + \eta \cdot A_s \qquad (7)$$

Contrary to symmetrical layers, the two passages of an unit build to be tested on Cj-f test bench are working "asymmetrically" : heat flux between both passages is completely transferred through the parting sheet whereas bottom and top plates are insulated. Therefore, test sample bottom plate, top plate and parting sheet are at three different temperatures and thus, by-pass fin efficiency value η^* is needed in addition to usual normal fin efficiency η for effective surface area calculation. Effective heat transfer area per layer A_e on Cj-f test sample is called A_e^{sample} and depends on both normal and by-pass fin efficiency coefficient η and η^*:

$$A_e^{sample} = \frac{A_p + \eta \cdot A_s}{2} \cdot \left(1 + \frac{\eta^* \cdot A_s}{A_p + (\eta + \eta^*) A_s}\right) \qquad (8)$$

It follows from this formula that effective area on Cj-f test sample is lower than the effective area which would exists in symmetrical layers, but higher than effective area calculated with the classical formula Eq. 7 applied to an half channel.

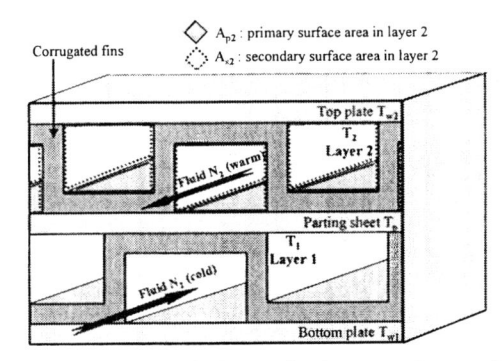

$$\frac{A_e^{sym}}{2} < A_e^{sample} < A_e^{sym} \qquad (9)$$

Fig. 5 Geometric definition of primary and secondary surface areas A_p and A_s.

Longitudinal conduction. In order to estimate the effect of longitudinal conduction in the test unit, we assume that the physical model to be considered, described in Fig. 6, can be split into :

---

[1] Note : If longitudinal conduction remains negligible in test sample, ka can be easily determined from global measurement of total heat exchanged Φ and inlet and outlet temperatures used in ΔT_{log} :

$$\Phi = \frac{ka}{2} \cdot L \cdot W \cdot \Delta T_{log}$$

- On the first side, heat transfer equations in a elementary dx long cross section where longitudinal conduction is neglected :

$$d\Phi_1 = ka.W.(T_1(x) - T_p(x)).dx = Q.c_{p1}.dT_1 \quad (10)$$

$$d\Phi_2 = ka.W.(T_2(x) - T_p(x)).dx = -Q.c_{p2}.dT_2 \quad (11)$$

- On the other side, equations of longitudinal conduction where the conduction effects in the cross section are neglected, so that the temperature in the cross section is taken as a constant value $T_p(x)$:

$$d\Phi_1 - d\Phi_2 = \lambda_m.s.\frac{d^2T_p}{dx^2}dx \quad (12)$$

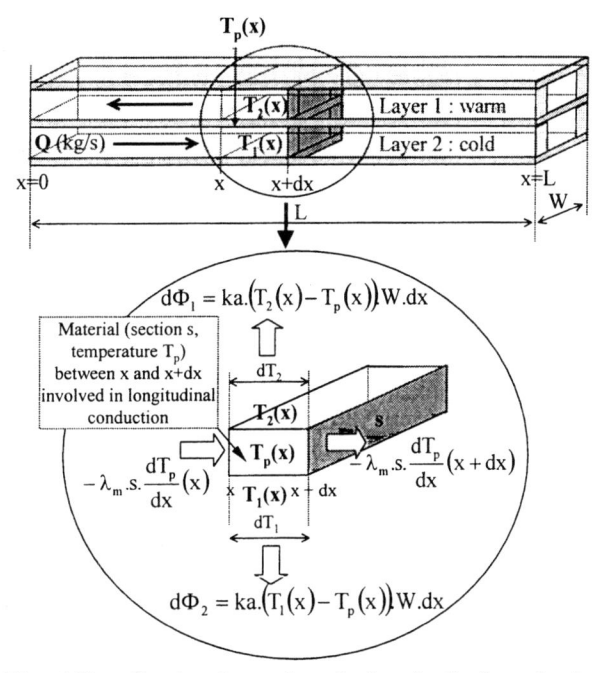

Fig. 6 Heat flux local equations for longitudinal conduction assessment.

Numerical resolution of this system (Eq.10, Eq.11, Eq.12) gives outlet temperatures T_{1_out} and T_{2_out} as functions of :

- s : the material section to be considered as taking part in longitudinal conduction, which includes the sections of parting sheet, external plate and side bars.
- T_{1_in} and T_{2_in} : layers inlet temperatures
- ka : heat transfer efficiency

$$T_{1\_out} = \text{function}(T_{1\_in}, T_{2\_in}, s, ka) \quad (13)$$

$$T_{2\_out} = \text{function}(T_{1\_in}, T_{2\_in}, s, ka) \quad (14)$$

Algorithm for Colburn calculation. For a specific operating point, Equations 1 to 14 are computed following the algorithm summarized in Fig. 7.

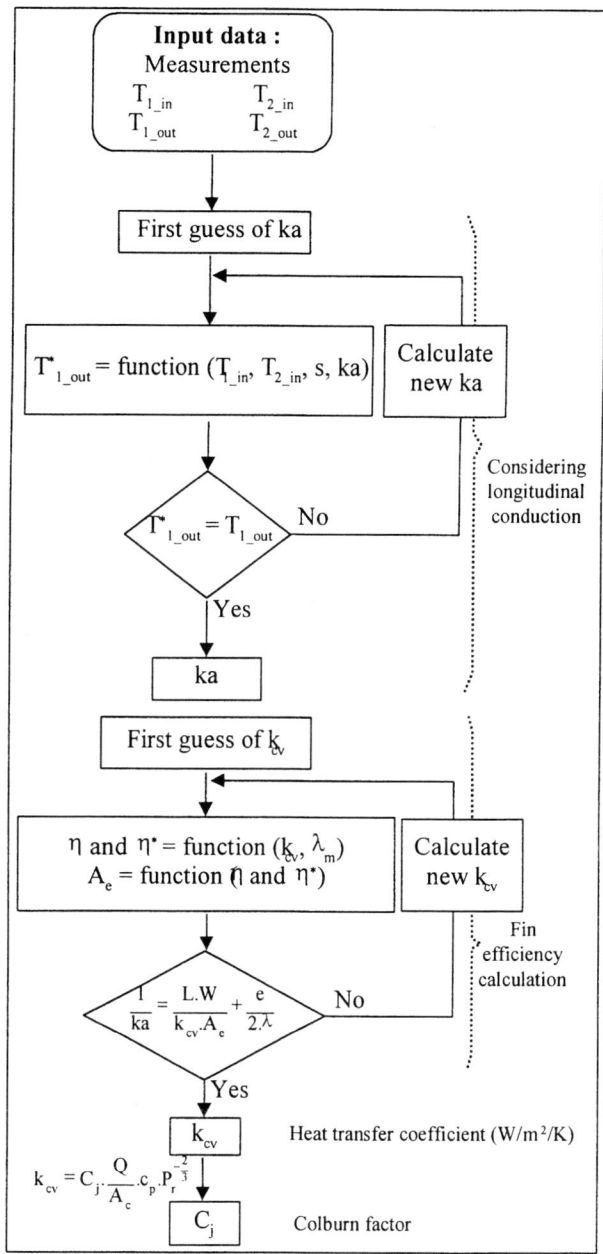

Fig. 7 Numerical algorithm for Colburn factor calculation with longitudinal conduction assessment.

RESULTS

As an example of experimental results obtained on Cj-f test bench, measurements of a "classical" serrated fin, one of those widely installed in cryogenic heat exchangers, are given in Fig.8. Experimental error in hydraulic and thermal results are shown, and measurements are compared with manufacturers' data.

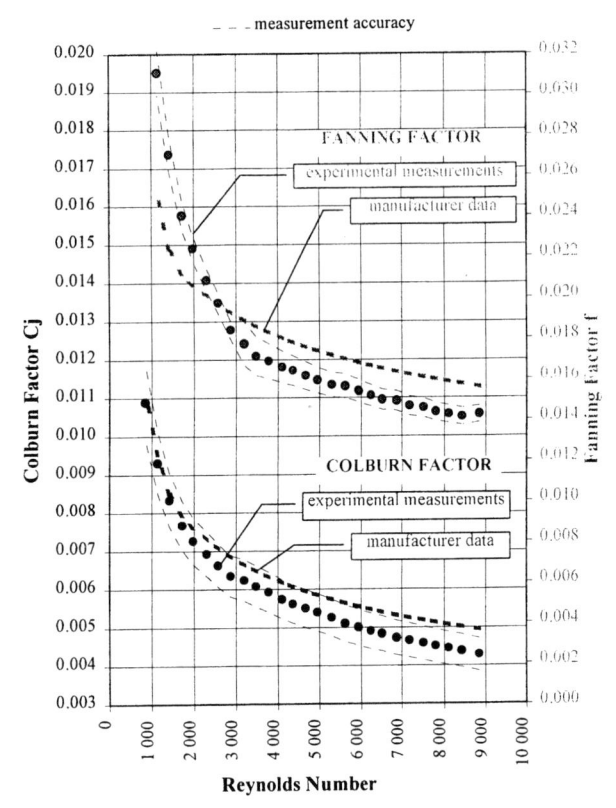

Fig. 8 Example of experimental results compared with manufacturer data

General trend is that heat transfer is in good accordance with manufacturers' values, while f factor is more scattered; This is not surprising, as variations in manufacturing process (tool wear, burrs...) may affect significantly pressure drops, without significant change in overall heat exchange surface. In this case, experimental measurements give the opportunity to validate or correct manufacturers' data and to improve thermal and hydraulic predicting models.

CONCLUSIONS

Air Liquide Cj-f test bench described in this paper is commonly working for some years in CRCD Research Center.

Cj-f testing procedure was validated through characterizations of corrugated fins widely implemented in industrial cryogenic heat exchangers, and cross checked on other water/gas facilities.

In addition to accurate results, a main advantage of this test bench consists in the short time needed to complete a whole test : measuring hydraulic and thermal performances of a unit within the range of Reynolds Number from about 1000 to about 10 000 (depending on the fin hydraulic diameter and cross-sectional area for flow), takes only a few hours (30 operating points).

Such a test bench is used as an easy way to address three main concerns :
- validation manufacturers' data regarding fins thermal and hydraulic performances (accurate process data for design)
- performance measurements of fins installed in a particular plate fin heat exchanger (manufacturing control and troubleshooting)
- Evaluation of new fin designs and accelerated fins development by efficient combination of modeling and experimental approach.

NOMENCLATURE

A_c cross-sectional free flow area, $w.n.h_s.W$, m^2

A_e effective heat transfer area per layer, $L.W.(1/ka-e/(2\lambda))/k_{cv}$, m^2

A_e^{sym} effective heat transfer area per layer in single banking pattern, $A_p+\eta.A_s$, m^2

A_e^{sample} effective heat transfer area per layer in Cj-f test sample, $(A_p+\eta.A_s).[1+\eta^*.A_s/(A_p+(\eta+\eta^*).A_s)]/2$, m^2

A_p primary surface area of an half layer, $2.w.n.W.L$, m^2

A_s secondary surface area of an half layer, $2.n.h_s.(1-p).W.L$, m^2

Cj Colburn factor, $k_{cv}.A_c/(\varphi.cp.P_r^{-2/3})$, dimensionless

c_p specific fluid capacity, $j.kg^{-1}.K^{-1}$

D_h fin hydraulic diameter, $2.h_s.w/(h_s+w)$, m

ds element of heat transfer surface, m^2

dT temperature difference in the fluid between x and x+dx, °C

$d\varphi$ heat flux through ds, $k_{cv}.ds.(\theta-T)$, W

e_p parting sheet thickness, m

f Fanning factor (friction factor), $\Delta P_{lin}.D_h/(2.\rho.v^2)$, dimensionless

h fin height, m

h_s layer height, h-t, m

k_{cv} heat transfer coefficient, $W.m^{-2}.K^{-1}$

ka heat transfer efficiency, $W.m^{-2}.K^{-1}$

L layer length, m

n fin frequency, m^{-1}

p fin porosity, dimensionless

P absolute pressure, Pa

P_r fluid Prandtl Number, $\mu.c_p/\lambda$, dimensionless

Q fluid mass flow rate, $kg.s^{-1}$

R Reynolds Number, $Q.D_h/(\mu.A_c)$, dimensionless

t fin thickness, m

T fluid temperature, °C

T_p parting sheet mean temperature, °C

T_{w1} bottom plate temperature, °C

T_{w2} top plate temperature, °C

v fluid velocity within corrugated fins, $Q/(\rho.A_c)$, $m.s^{-1}$

W layer width, m

w	channel with, $1/n-t$, m
β	$h \cdot [2 \cdot kcv/(\lambda_m \cdot t)]^{0.5}/2$, dimensionless
ΔP	pressure drop in the layer, Pa
ΔP_{lin}	linear pressure drop, $\Delta P/L$, Pa.m^{-1}
ΔT_c	temperature difference at cold end, $T_{2\_out}-T_{1\_in}$, °C
ΔT_{in-out}	temperature rise or drop in passages, $(T_{2\_in}-T_{2\_out})$ or $(T_{1\_out}-T_{1\_in})$, °C
ΔT_{log}	logarithmic temperature difference, $(\Delta T_c - \Delta T_w)/Ln(\Delta T_c/\Delta T_w)$, °C
ΔT_w	temperature difference at warm end, $T_{2\_in}-T_{1\_out}$, °C
η	normal fin efficiency, $th(\beta)/\beta$, dimensionless
η^*	by-pass fin efficiency, $[coth(\beta)-th(\beta)]/(2\beta)$, dimensionless
θ	local temperature of primary or secondary surfaces, °C
λ	fluid thermal conductivity, W.m^{-1}.K^{-1}
λ_m	thermal conductivity of fin and parting sheet material, W.m^{-1}.K^{-1}
μ	dynamic fluid viscosity, kg.m^{-1}.s^{-1}
ρ	fluid density, kg.m^{-3}
Φ	exchanged heat flux between layers, W

Superscripts

sym   related to symmetrical layer (single banking)
sample related to test unit measured on Air Liquide Cj-f test bench

Subscripts

i   in layer number i = 1 or 2
_in   entry value
_out exit value

REFERENCES

L. E. Haseler, 1980, Layer pattern effects in plate-fin heat exchangers, Heat Transfer and Fluid Flow Service, pp. 119-125.

HTFS, 1984, single-phase heat transfer and pressure drop in plate-fin heat-exchangers, *HTFS Handbook*, sheet YP1.

HTFS, 1984, fin efficiency in a plate-fin heat exchanger, *HTFS Handbook*, sheet YM1.

L. E. Haseler, 1980, Multi-stream heat exchangers, Heat Transfer and Fluid Flow Service, DR50, Part 1.

SINGLE-PHASE HEAT EXCHANGER
DEVELOPMENT AND APPLICATIONS

RESEARCH AND DEVEROPMENT ON THE HIGH PERFORMANCE CERAMIC HEAT EXCHANGERS

M. Kumada[1] and K. Hanamura[2]

[1] Gifu University, 1-1 Yanagido, Gifu 501-1193, Japan; E-mail: kumada@cc.gifu-u.ac.jp
[2] Gifu University, 1-1 Yanagido, Gifu 501-1193, Japan; E-mail: hanamura@cc.gifu-u.ac.jp

ABSTRACT

Research and development of ceramic heat exchangers, especially for gas turbines, were summarized. As the turbine inlet temperature rises, the thermal efficiency of the system becomes higher. Therefore, ceramics had been utilized for elements of the turbine and heat exchangers for regeneration; it has a high heat resistance and a high corrosion resistance. However, it is not easy to make any complicated surfaces due to the manufacturing difficulty of the ceramics; then, the ceramic heat exchanger becomes huge. Using a high temperature fluidized bed the outside heat transfer coefficient increased even in the case of smooth surface, and in the case of a ceramic finned tube it was twenty times higher than that for a single-phase flow. On the other hand, inside heat transfer coefficient for the finned tube was less than that for the smooth tube because of the relaminarization in the fan-shaped flowing-duct formed by inner radial ribs. As a result, a twist tape, that can be made of ceramics, was proposed for the inside heat transfer enhancement. Furthermore, a ceramic finned tube, that had a fin thickness and a fin pitch similar to those of general metal finned tubes, was developed for a more compact and higher performance ceramic heat exchanger; it was a first try in all over the world.

IINTRODUCTION

Recently, higher temperature working-gases have been required to obtain a high performance of many kinds of heat transfer facilities. In Japan we have a project to develop a high performance gas turbine of a power of 300kW; it finished in 2000. In the project, many elements of the turbine and heat exchangers were made of ceramics to increase the thermal efficiency up to 42%. In the ceramic gas turbine (CGT), the target turbine-inlet-temperature (TIT) was set up to 1350 deg. C. Some useful results were obtained in the project. On the other hand, the target temperature of 1750 deg. C is proposed for the TIT of a hydrogen-combustion gas turbine in the project of World Energy Network (WE-NET) that is continued as a part of the New Sun Shine Project in Japan. In the abovementioned ceramic gas turbine (CGT301), a hybrid ceramic heat exchanger, which can be used under the condition of an inlet gas temperature of 825 deg. C, had been developed (Yoshimura, 1995). On the other hand, a rotary heat exchanger that is made by a ceramic honeycomb core had been developed to apply to a 100kW-class ceramic gas turbine for vehicles; it was also a national project in Japan. In this case, the inlet gas temperature was set up at 950 deg. C (Nakazawa, 1997) as the final target.

As shown in the abovementioned projects, in order to obtain a high performance, the ceramics is necessarily used in the high temperature region: because it has a high heat resistance and a high corrosion resistance. However, from a viewpoint of the reliability it is not easy to realize the ceramic gas turbine in near future. Furthermore, it is difficult to make any complicated surface configurations due to the manufacturing difficulty, compared with metals. Therefore, it is a key technology for the ceramic heat exchangers to make it compact.

A more compact and higher heat-resistance heat exchanger is required for the gas turbine. However, this is the fact that the compactification of the heat exchanger is restricted by the requirement of the utilization in a high temperature region. On the other hand, the heat transfer enhancement for air-conditioning heat exchangers is advanced to a considerable extent in the recent decade. Now, the rate of development is almost saturated since the manufacturing technology for the heat transfer tube achieves to the limitation. As a result, the utilization of the fluidized-bed and the electric and/or magnetic fields are proposed as the active heat transfer enhancement methods; but much effort is required to realize.

Provided that the heat transfer is enhanced, the heat exchangers applied to the gas turbines as well as the air-conditioners becomes more compact theoretically. As a result, the ceramics should be necessarily used in the high temperature region. However, until the reliability of the ceramics is confirmed, some metallic alloys with a high heat resistance will be used. Therefore, we have to select the materials, which one is better for the heat exchangers, from a viewpoint of the cost and/or the required working-gas temperature. In a micro gas turbine that is useful to a dispersed power generation, in order to increase the thermal efficiency and to make a cogeneration system integrated with it, many elements of the system should be made of ceramics. Especially, an application of the ceramics to a 70kW-class of micro gas turbine is a

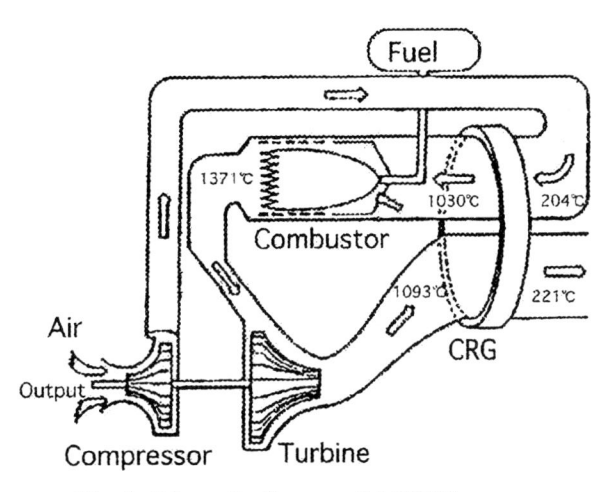

Fig. 1 Schematic diagram of AGT101

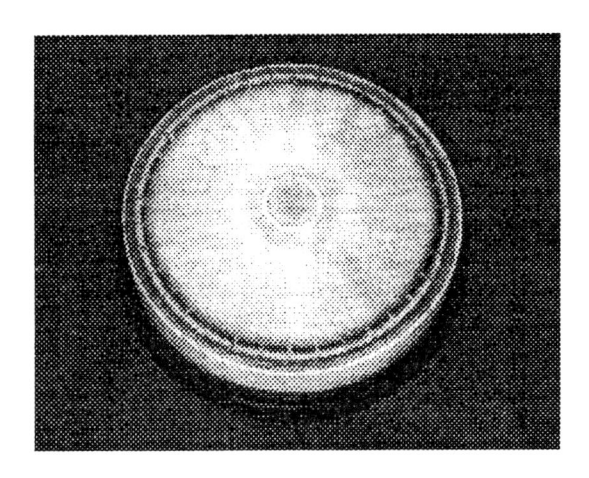

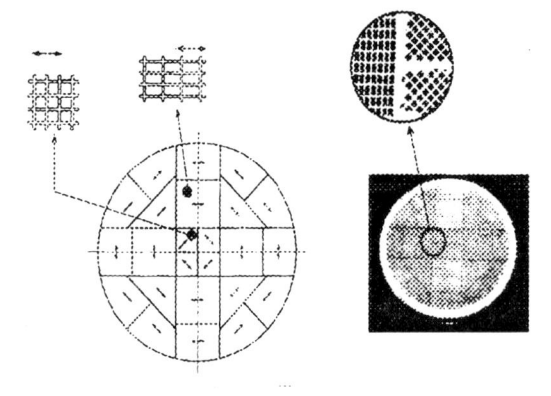

Fig. 2 LAS ceramic honeycomb core (Upper) and MAS ceramic honeycomb core

Significant target for the development of the ceramic heat transfer facilities since the reliability is not sufficiently high.

In the current review and study, the research and the development of the ceramic heat exchangers for the gas turbines are summarized and the experimental results of the high-temperature fluidized-bed ceramic heat exchanger are presented. Furthermore, a ceramic finned tube that has a configuration similar to that of general metal fin is also shown.

HEAT EXCHANGERS FOR GAS TURBINES FOR VEHICLES

On the basis of a high heat resistance of ceramics, a rotary heat exchanger consisted of ceramic honeycomb cores had been mainly developed for gas turbines for vehicles. It was a storage type heat exchanger and was extremely compact. A schematic diagram of the system, which has been developed in projects entitled by AGT100 and 101, is presented in Fig.1 (Garrett Co. Ltd., 1986). In this system the cordierite ceramic honeycomb was utilized and its target temperature-efficiency was 93%. On the other hand, a rotary heat exchanger for a 100kW-class gas turbine for vehicles had been developed in a national project in Japan (Nakazawa, 1997 and Sakai, 1991). For the heat exchanger, the LAS material (Fig.2, Upper) produced by Corning Co. Ltd. and MAS material (Fig.2, Lower) produced by NGK Co. Ltd. were adopted as the core material and its target temperature-efficiency was also 93%. The main subject for the development was a sealing method, especially, an inner-sealing method in the high temperature side. This system achieved almost the aim of the performance and the endurance. On the contrary, there is a report, in which the heat transfer type heat exchanger is recommended (Avran, 1991). Figure 3 shows the counter-flow cylindrical heat exchanger developed by ONERA. The heat exchanger was made of a metal or ceramics. Basically, in the partial load the temperature efficiencies were much

Fig. 3 Counter-flow heat exchanger

the same between the heat storage type and the heat transfer type heat exchangers; then, the performance of the total gas turbine system with heat transfer type heat exchanger was better since the leakage was free. On the other hand, if the leakage of the working gas is large, the loss becomes larger even if the system temperature is increased using ceramics; in practice, a leak of 7% is estimated. In a 100kW heat engine, the improvement of the system is proposed, introducing the concept of "OVEREXPANSION", that means the recompression after the working gas expanded in the turbine is cooled in the heat exchanger. When a finned tube ceramic heat exchanger is used, the performance of the system under the condition of the compression ratio of 5.5 is much the same as that with the rotary heat exchanger.

CERAMIC HEAT EXCHANGERS FOR A 300KW-CLASS GAS TURBINE

Two kinds of ceramic heat exchangers were developed in a national project CGT. The one was a hybrid type heat exchanger developed in the project CGT301, as shown in Fig.4. The other was a plate-fin type heat exchanger made of a super heat-resistance alloy (HA230) and developed in the project CGT302, as shown in Fig.5. Furthermore, there was one more project, i.e., CGT303, to develop a rotary heat exchanger similar to that for vehicles; but it was stopped. In the plate-fin type heat exchanger, there was no originality on the structure. Furthermore, whether the structural quality and the brazing quality are maintained or not is the subject remained in the development because of the manufacturing difficulties of Heins alloys. The target of the development and the operating conditions for CGT301 are shown in Table 1. The targets for both CGT301 and CGT302 are much the same. In the ceramic heat exchangers, the ceramic heat transfer tube has six ribs on the inside surface. The ceramic tube made of a silicon nitride was manufactured by NGK Co. Ltd. The specifications of the heat transfer element for one module are shown in Table 2. In practice, in the heat exchanger six modules were installed in parallel. In this case, the total number of tubes is more than 1000. Consequently, the compactness is the final subject in the development of the ceramic heat exchanger.

FLUIDIZED-BED CERAMIC HEAT EXCHANGER

As mentioned above, it is useful to utilize the ceramics in a high temperature since it has a high heat resistance and a high corrosion resistance. Therefore, the ceramic heat exchanger will be realized in the near future; though there are some problems on the fragility and on the

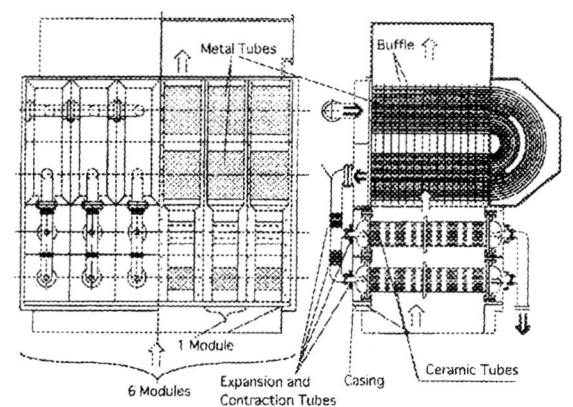

Fig. 4 Hybrid ceramic heat exchanger

Fig. 5 Plate-fin heat exchanger

Table 1 Target and operating conditions of heat exchanger

Step of Development (Turbine Inlet Temperature, deg C)			Intermediate Target (1200)	Final Target (1350)
Target		Temperature Efficiency ε_c(%)	>84	>84.5
		Total Pressure Loss $\Delta P/P$(%)	<4.5	<4.5
Operating Condition	Air	Inlet Temp. Ta,in (deg C)	280	281
		Inlet Press. Pa,in (kPa)	732	740
		Flow Rate Ga (kg/s)	0.945	0.889
	Gas	Inlet Temp. Tg,in (deg C)	696	825
		Inlet Press. Pg,in (kPa)	107	107
		Flow Rate Gg (kg/s)	1.03	0.929

Table 2 Specification of heat transfer element

	Ceramic Tube	Metal Tube
Outside Diameter of Tube (mm)	8	6
Inside Diameter of Tube (mm)	6	5.2
Number of Tube (1 module)	163*2	797
Effective Length of Tube (mm)	550	590
Material	SN-84	SUS316
Inside Configuration	Straight Fin (Radial)	Twist Plate (Insertion)
Thickness of Fin (mm)	0.3	0.2
Tube Arrangement (Pitch/Outside dia.)	Triangle(1.33) with Buffles in part of Metal Tube	

difficulties of connecting the ceramics with metals. On the other hand, it is not now expected that the ceramic heat exchanger becomes compact since the difficulty of manufacturing complicated surfaces has prevented heat transfer from improving. By applying a fluidized-bed to the ceramic heat exchanger for high temperature, improvement in heat transfer is anticipated due to, not only the renewal effect by particles, but also the radiation from particles. The results are presented, as follows.

Figure 6 shows a schematic diagram of the experimental apparatus for the investigation of heat transfer characteristics of the ceramic heat transfer tube bundle immersed in the fluidized-bed of the maximum temperature with 1200 deg. C (Himeji, 1997). The fluidization tower has a rectangular cross section (320x120mm). The tubes that are the same configuration as those used in CGT301 were arranged in a staggered arrangement with 16mm and 21mm spacing between centers, respectively. The tube bundle consists of 35 tubes (seven rows). Particles are made of alumina with a 1.0mm average diameter. Air-propane mixture is supplied through the distributor plate and a flame is stabilized on the plate. The bed temperature is controlled by the equivalence ratio. An electric heater controls the cooling air temperature.

Figure 7 shows the temperature distributions for both the fluidized bed and the tube wall along the axis of the tube. The tube wall temperature in the vicinity of the inlet of the cooling air is low. However, the bed temperature is kept at almost constant in entire region of the bed. The tube wall temperature is not dependent on the position of the tube. As a result, it is possible to use the fluidized bed to such a high temperature region.

Figure 8 shows the heat transfer coefficient with the fluidizing velocity. The heat transfer coefficient slightly decreases with increasing fluidizing velocity, as well as the case of the low temperature fluidized bed. On the other hand, the heat transfer coefficient increases with a bed temperature because of the increase in the effect of the radiation. The result shows that the heat transfer coefficient increases by 7% when the bed temperature increases by 100K. The heat transfer coefficient calculated by Zukauskas

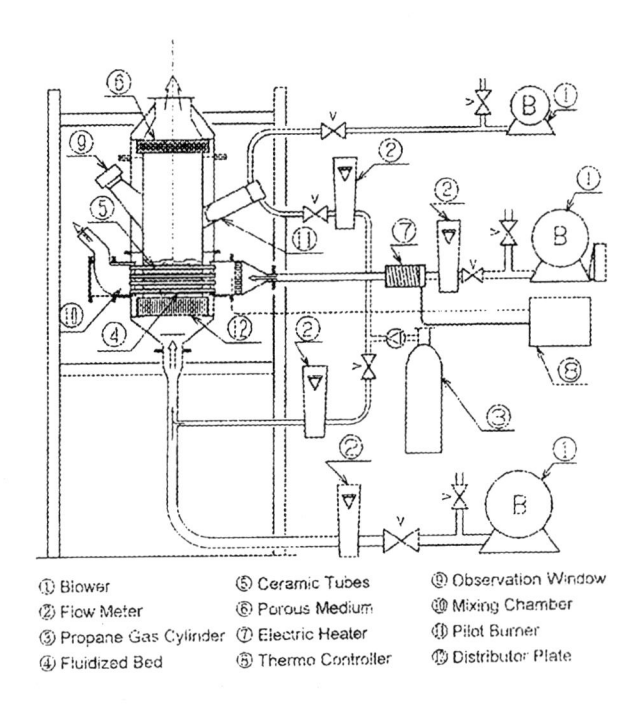

Fig. 6 Experimental Apparatus of fluidized bed ceramic heat exchanger

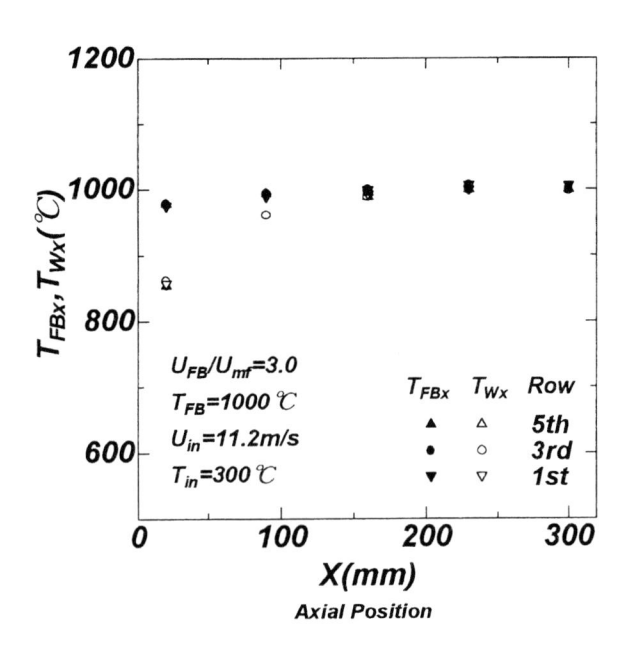

Fig. 7 Bed and tube wall temperature distribution

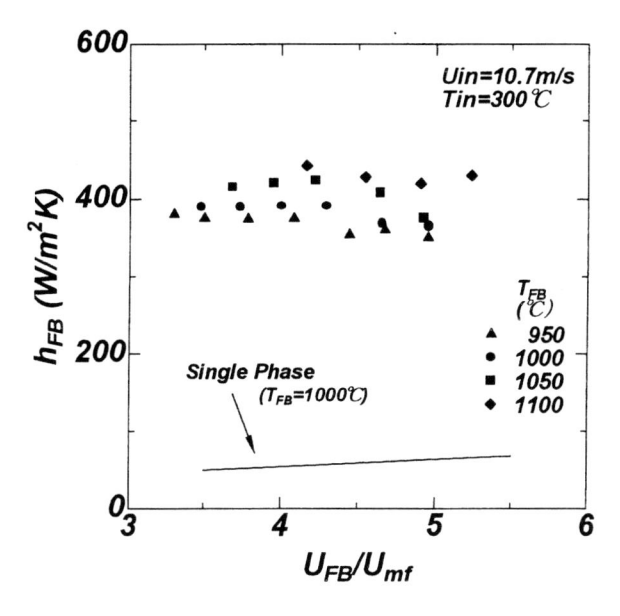

Fig. 8 Outside average heat transfer coefficient

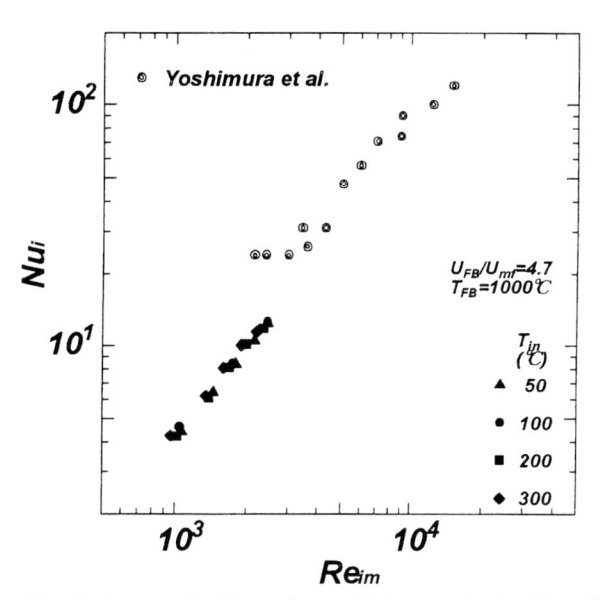

Fig. 9 Average heat transfer coefficient on the inside wall of a tube (smooth tube)

correlation (Zukauskas, 1997), which is in the case of a single-phase flow, is also shown in Fig.8. The heat transfer coefficient of the tube in the fluidized bed is 8 times higher than that in free space.

Figure 9 shows the variation of the Nusselt number on the inside wall of the tube with the Reynolds number in the tube. The result of Yoshimura, et al. using the same shaped smooth ceramic tube is also shown, where the wall temperature is less than 650 deg. C (Yoshimura, 1995). The Nusselt number increases with the Reynolds number and is in proportion to about 1.1 power of the Reynolds number. This tendency is not similar to the conventional one of the circular tube at low temperature. The results of the experiment and Yoshimura, et al. show a similar tendency.

Figure 10 shows the variation of the temperature efficiency with the fluidizing velocity. The temperature efficiency slightly increases with the fluidizing velocity and is kept at almost constant with the bed temperature.

Because the overall heat transmission coefficient is controlled by the inside heat transfer coefficient. The higher temperature efficiency than about 85% will be expected, depending on the operating condition.

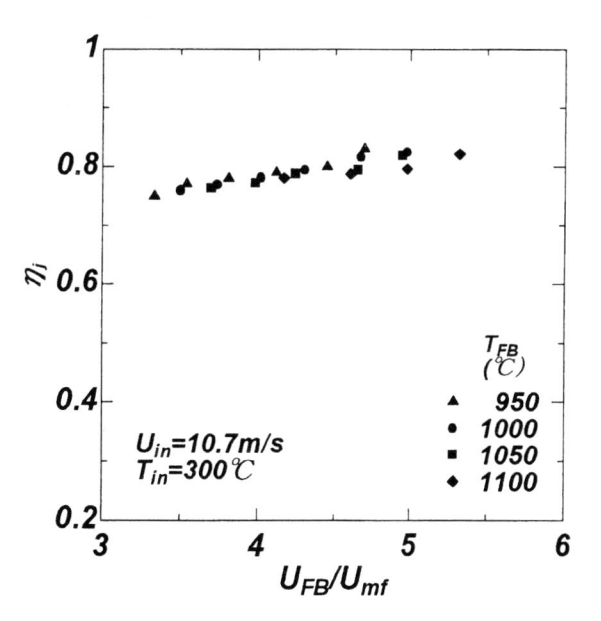

Fig. 10 Temperature efficiency

DEVELOPMENT OF CERAMIC FINNED-TUBE

As mentioned above, the outside average heat transfer coefficient increases to a considerable extent, using the fluidized bed heat exchanger. However, the fluidized bed has many problems, such as the handling difficulty and its maintenance. As a result, it is desired to develop a ceramic finned-tube shaped like general metal fins.

Figure 11 shows a ceramic finned tube developed by NGK Co. Ltd. in 1995. The fin shape was determined from optimizing the fin efficiency after many trial manufactures for the fin configuration, the surface

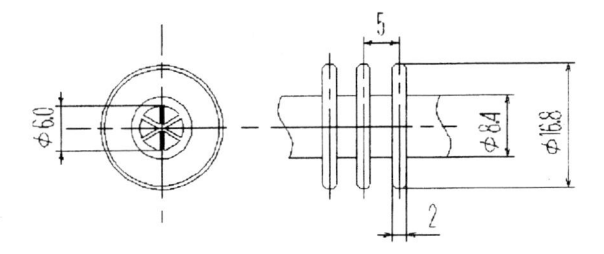

Fig. 11 Ceramic finned tube

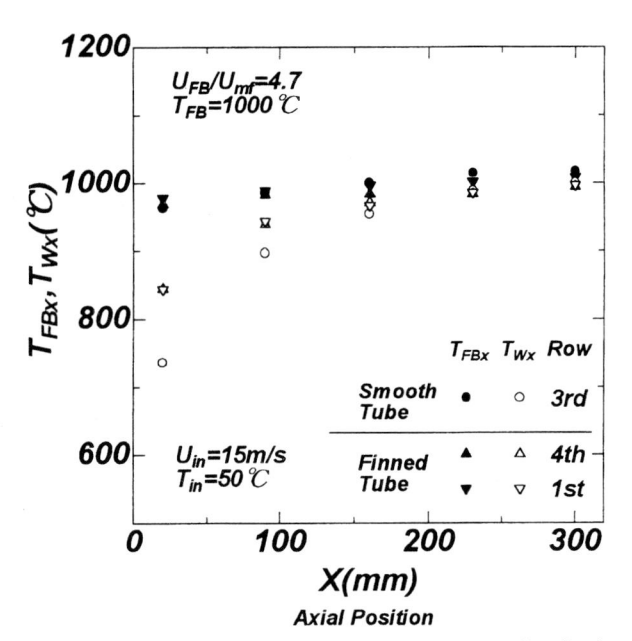

Fig. 12 Bed and tube wall temperature distribution

manufacturing properties and the interference thickness between fin and tube. It is a first time to manufacture such a kind of ceramic finned tube in all over the world. It was made of silicon nitride. The finned tube consists of a smooth tube, used for the above fluidized bed heat exchanger, and many beads. The beads and the smooth tube were fitted together by heating the beads and allowing them to shrink to the smooth tube. The diameter and the thickness of the circular fins were 16.8mm and 2.0mm, respectively. The fin pitch was 5.0mm. The interference-thickness of 0.18mm was finally selected. The heat transfer surface of the finned tube is four times higher than that of the smooth tube. The fin efficiency is 0.85.

Figure 12 shows temperature distributions for both the fluidized bed and tube wall along the axis of the tube (Himeji, 1998). The tube wall temperature of the finned tube is higher than that of the smooth tube, because of heat transfer enhancement on the outside wall of the tube. On the other hand, the bed temperature is kept at almost constant.

Figure 13 shows the variation of the Nusselt number on the outside wall of the tube with fluidizing velocity. The Nusselt number of the finned tube is about three times as large as that of the smooth tube. The Nusselt

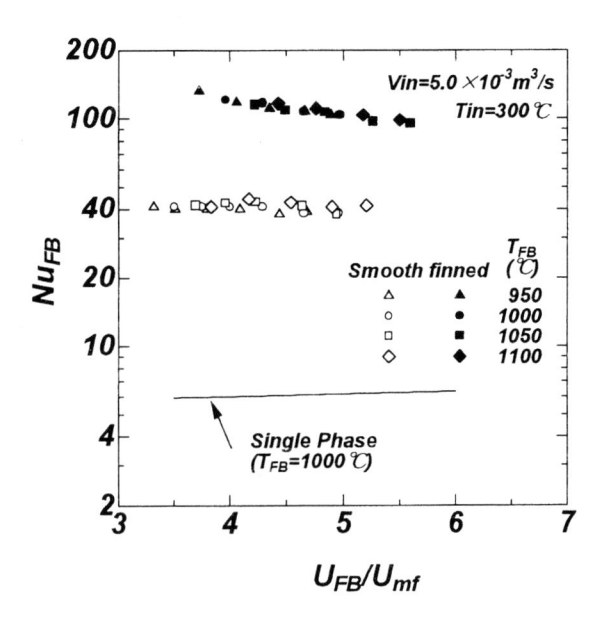

Fig. 13 Variation of outside heat transfer coefficient with fluidizing velocity

number for the finned tube slightly decreases with the fluidizing velocity, while the Nusselt number for the particle movement by fins. Furthermore, the decreasing tendency of the Nusselt number with fluidizing velocity of smooth tube is kept at almost constant. The reason of the decrease in the fin effect is due to the obstruction to the the finned tube is larger than that of the smooth tube. The reason is that the density of fluidizing particles decreases with fluidizing velocity. The effect of the radiation emitted from particles also decreases with increasing tube wall temperature. The Nusselt number of the finned tube immersed in a fluidized bed is three times as large as that of the smooth tube and is twenty times as large as that of a single-phase flow.

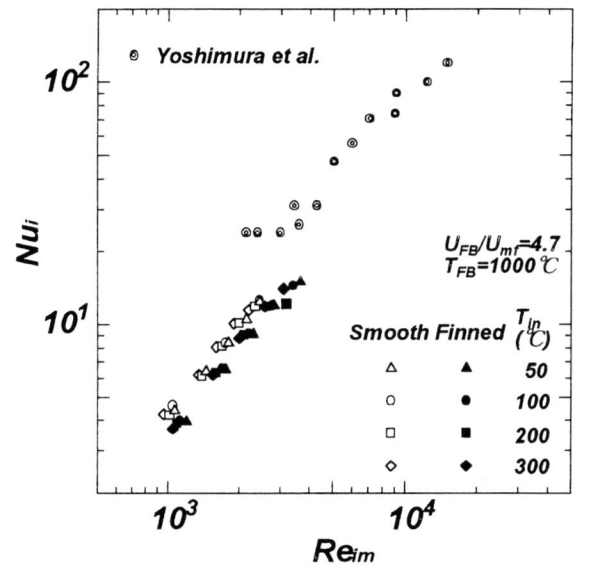

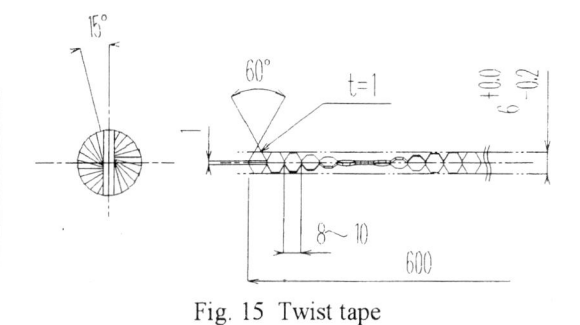

Fig. 14 Average heat transfer coefficient on the inside wall of a tube (smooth and finned tubes)

Figure 14 shows the variation of the Nusselt number on the inside wall of the tube with the Reynolds number in the tube. The inside Nusselt number for the finned tube is smaller than that for the smooth tube. This is because of occurring relaminarization of the cooling-air flow. By the outside heat transfer enhancement, the wall temperature becomes higher; then, the outlet Reynolds number is drastically decreased by a large change of the thermophysical properties, as shown in Table 3. As a result, the relaminarization occurs. Especially, in the fan-shaped flowing duct formed by inner ribs installed for inside heat transfer enhancement, both the laminar and the turbulent flows exist in the vicinity of the apex. In this region the laminarization is enhanced as the temperature rises; then the inside heat transfer coefficient is decreases. A 5% decrease in the temperature efficiency was also observed. As a result, a new method for heat transfer enhancement, such as a turbulence promoter, is desired in spite of the ribs since the relaminarization can't be prevented as the temperature rises.

POSSIBILITY AND REALIZATION OF COMPACT AND HIGH PERFORMANCE CERAMIC HEAT EXCHANGER

The relaminarization yielded by the inside radial rib configuration results in the decrease in the performance. However, it is not easy to change the inside rib configuration in manufacturing the heat transfer tube by means of an extrusion molding. Figure 15 shows a twist tape to prevent the laminarization in the tube. Now, it was made of a stainless steel; but it can be made of ceramics. Figure 16 shows the variation of the Nusselt number on the inside wall of the tube with the Reynolds number in the tube. Although the configuration of the twist tape is not sufficiently optimized, the inside heat transfer is slightly improved by the effect of turbulence and radiation.

Table 3 Thermophysical properties and Reynolds number

	Heat Transfer Tube Inlet (50 deg C)	Heat Transfer Tube Outlet (1000 deg C)
Density (kg/m^3)	1.06	0.268
Kinetic Viscosity (m^2/s)	1.85x10^{-5}	1.84x10^{-4}
Thermal Conductivity (W/mK)	2.78x10^{-2}	7.62x10^{-2}
Reynolds Number d=6.0mm (Inside dia.)	5100	2070
Reynolds Number d=2.17mm (Fan-shaped Duct)	1870	760

Fig. 15 Twist tape

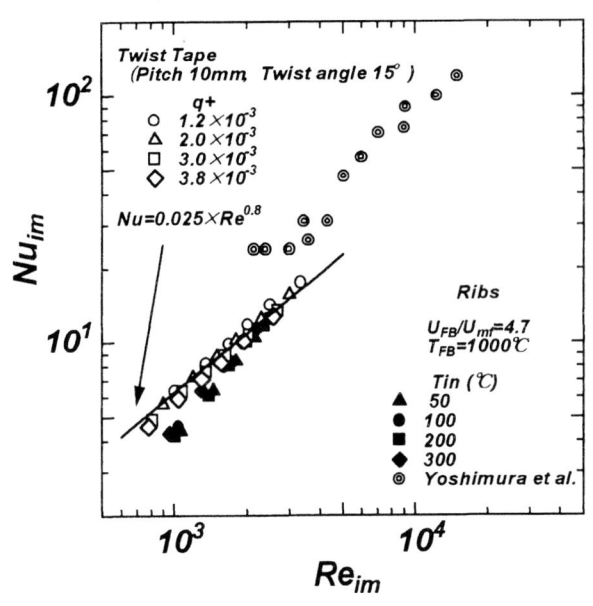

Fig. 16 Average heat transfer coefficient on the inside wall of a tube with a twist tape

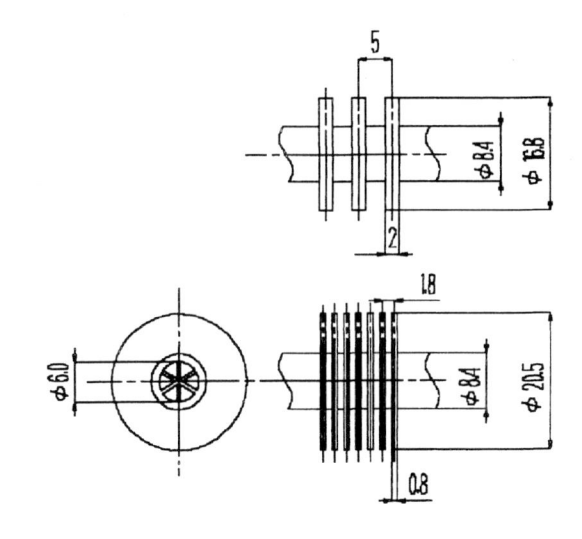

Fig. 17 Finned tubes manufactured in 1995 (Upper) and in 2000

Furthermore, the relaminarization is suppressed. As a result, it is recognized that the ceramic finned tube with the inside twist tape has a high potential to increase the performance of ultra high temperature heat exchangers; though the twist tape with the optimized configuration must be sintered together with the ceramic heat transfer tube.

A comparison is made between ceramic finned tubes manufactured in 1995 and in 2000, as shown in Figs.17 and 18. Under existing conditions, the ceramic finned tube manufactured in 2000 has the highest performance; it is manufactured by NGK Co. Ltd. The thickness of the circular fins is 0.8mm that is almost the same as that of conventional metal fins. The fin pitch is 1.8mm. The entire length of the tube is 600mm. The finned tube has a larger surface area than the smooth tube by a factor of sixteen.

Figure 19 shows the variation of the temperature efficiency with the superficial velocity. This is for the ceramic finned tube manufactured in 2000. In this case, the fluidized bed is not used. The results obtained in the case of the ceramic finned tube manufactured in 1995 are also shown; where it is immersed in the fluidized bed. Using the ceramic finned tube manufactured in 2000, a high temperature efficiency was obtained even if it is not

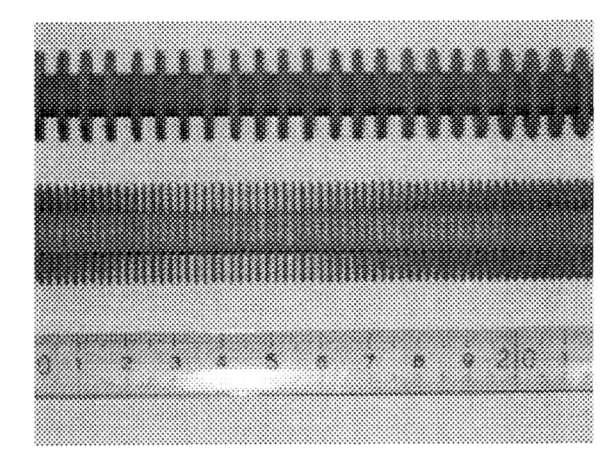

Fig. 18 Photograph of finned tubes manufactured in 1995 (Upper) and in 2000

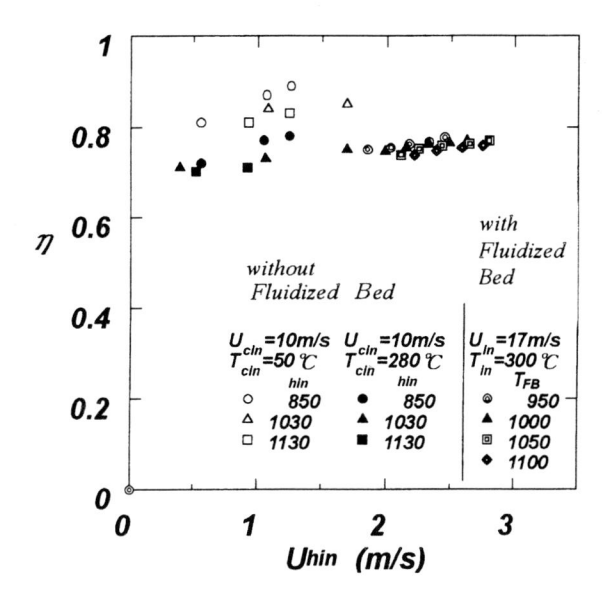

Fig. 19 Temperature efficiency of finned tubes

immersed in the fluidized bed. As far as the authors know, in all over the world there is no information that such a kind of ceramic finned tube and/or heat exchanger was manufactured. In the near future, both the inside and the outside heat transfer characteristics of the finned tube will be clarified in detail. Furthermore, the reliability and the manufacturing difficulties will be also investigated.

CONCLUDING REMARKS

More than decade passed since ceramics was applied to heat exchangers. One can state that the development of the ceramics in the recent decade is fast amazingly; while someone may state that it is too late. Anyway, this is the fact that the development is advanced step by step. Especially, the advances in the sintering technology are out of expectation. The authors are planning to manufacture a ceramic heat exchanger for a micro gas turbine and to make an operating test of the total system. On the other hand, a new gas turbine system, in which the synthesis gas produced by coal gasification is supplied for combustion, is proposed for an effective utilization of the coal (Robertson, 1994). Provided that the above heat exchanger with the ceramic finned tube is applied to the gas turbine cogeneration system with the coal gasification combustion, the thermal efficiency of the total system is expected to become extremely high. Therefore, The ceramic heat exchanger for an ultra high temperature has a high potential to improve the total system performance.

NOMENCLATURE

h_{FB}	:	Outside heat transfer coefficient, W/m²K
Nu_{FB}	:	Outside Nusselt number
Nu_i	:	Inside Nusselt number
q^+	:	Heat flux parameter
Re_{im}	:	Inside Reynolds number
T_{FB}	:	Fluidized bed temperature, deg.C
T_{in}	:	Inlet temperature of cooling air, deg.C
T_w	:	Heat transfer tube wall temperature, deg.C
U_{FB}	:	Fluidizing velocity, m/s
U_{mf}	:	Minimum fluidization velocity, m/s
U_{in}	:	Inlet velocity of cooling air, m/s
X	:	Coordinate, mm

REFERENCES

Himeji, Y., Kumada, M. and Hanamura, K., 1997, Studies on High-Performance Ceramic Heat Exchanger for Ultra High Temperatures (1st Report, Heat Transfer of Bare Tube Bundle Immersed in Fluidized Bed), Trans. Japan Soc. of Mech. Engineers, Vol.63, 3700-3705.

Himeji, Y. and Kumada, M, 1998, Studies on High-Performance Ceramic Heat Exchanger for Ultra High Temperatures (2nd Report, Heat Transfer of Finned Tube Bundle Immersed in Fluidized Bed), Trans. Japan Soc. of Mech. Engineers, Vol.64, 871-876.

Nakazawa, N., 1997, Development of Engine Elements: Heat Exchanger, J. Gas Turbine Society of Japan, Vol.25, 29-32.

Robertson, A and Bonk, D., 1994, Effect of Pressure on Second-Generation Pressurized Fluidized Bed Combustion Plants, Trans. ASME, J. Gas Turbine and Power, Vol.116, 345.

Yoshimura, Y., Itoh, K., Ohori, K. and Hori, M., 1995, The Research and Development of a 30kW-class Ceramic Gas Turbine, J. Gas Turbine Society of Japan, Vol.23, 55-60.

Zukauskas, A., 1972, Heat Transfer from Tubes in Cross Flow, Advances in Heat Transfer 8, Academic Press, 93-160.

Garrett Co. Ltd., 1986, Report 31-6435.

Avran, P. and Boudigues, S., 1991, Comparison of Regenerators and Fixed Heat Exchangers for Automotive Applications, YOKOHAMA-IGTC, Vol. II, 283-290.

Sakai, I., Kawasaki, K. and Ozawa, T, 1991, Heat Transfer Characteristics of Rotating Ceramic Regenerator -Measurement of the Heat Exchanger Performance-, YOKOHAMA-IGTC, Vol. II, 291-297.

INFLUENCE OF HEADER DESIGN ON PRESSURE DROP AND THERMAL PERFORMANCE OF A COMPACT HEAT EXCHANGER

Ch. Ranganayakulu[1] and A. Panigrahi[2]

[1] Aeronautical Development Agency, P.B.No. 1718, Bangalore – 560017, India, E-mail: r_chennu@hotmail.com
[2] R&D Division, Bharat Heavy Plate and Vessels Limited, Vishakapatnam - 530012, India

ABSTRACT

An analysis of a cross flow two pass plate-fin compact heat exchanger, accounting for the effects of non-uniform inlet fluid flow due to headers is carried out using a finite element method. The heat exchanger effectiveness and pressure drops have been measured in the test rig. Using inlet flow non-uniformity models, the exchanger effectiveness and pressure drops and its deterioration due to the effects of flow non-uniformity are calculated for design case of the exchanger. Based on the flow non-uniformity effects, the exchanger hot side inlet and outlet headers have been modified to improve the exchanger effectiveness and pressure drop.

INTRODUCTION

The demand for high performance heat exchange devices having small spatial dimensions is increasing due to their need in applications such as Aerospace industry (Darius A. Nikanpour and Bai Zhou, 1997). The accurate prediction of the pressure drop and thermal performance of an exchanger in the design stage is highly desirable for most applications. In most heat transfer and pressure drop analysis of heat exchangers, it is presumed that the inlet fluid flow distributions across the exchanger core are uniform. This assumption is generally not realistic under actual operating conditions. A significant reduction in heat exchanger performance may result when the flow distribution through the core is non-uniform. In gross mal-distribution, the fluid flow distribution at the core inlet is non-uniform. This occurs because of either poor header design or gross blockage in the core during manufacturing. The passage-to-passage mal-distribution occurs with in a core in a highly compact heat exchanger when the small flow passages are not identical as a result of manufacturing tolerances. The design of headers and inlet ducts significantly affects the velocity distribution approaching the face of the exchanger core. In this type of flow mal-distribution, the variations in flow at the inlet of exchanger core mainly depends on the location of the inlet duct, the ratio of core frontal area to the inlet duct cross-sectional area, the distance of transition duct/header between the core face and the inlet duct and the shape of headers i.e oblique flow headers or normal flow headers and with/with out manifolds.

The objective of this paper is to measure the influence of header design on pressure drop and thermal performance of a compact heat exchanger. The ratio of heat transferred to the pumping power required is an important factor in economizing the heat exchanger. Pumping power depends upon the pressure drop required to pump the fluid at the required velocity. Hence, it is necessary to estimate the pressure drop along with thermal performance. The influence of pressure drop and the thermal performance deterioration in the cross flow plate-fin heat exchanger have been analyzed and published (Ranganayakulu et all., 1997; Chiou, 1978; Shah, 1985; Paffenbarger, 1990) with effects of flow non-uniformity. The objective of the thermal design of a heat exchanger is to determine the most favorable size and configuration of the exchanger core, which meets the demand of the required heat transfer rate within the specified fluid pressure drops, space and cost limitations. A serious deterioration in heat exchanger performance may result when the flow distribution through the core is not uniform, particularly when the flow area increase from the pipe to the core face is 5 to 50 times. In these cases, generally, the pressure gradient is much higher at the center than that of the edge points of exchanger core if the inlet duct/pipe is straight. Uniformity of flow distribution over the core is the primary function of the headers.

Using the flow non-uniformity equation (Ranganayakulu et all., 1997), different types of flow non-uniformity (FN) models were generated by considering the different magnitudes of pressure gradients. Two types of flow models, which are required for present subject, are tabulated in Table 1. The velocity at the wall of inlet duct is zero. The non-zero velocity values in the proposed models are at the points away from the wall of transition duct. The local inlet flow non-uniformity parameter (α) is defined as,

$$\alpha = \frac{actual\ inlet\ flow}{average\ inlet\ flow} \quad (1)$$

In each model, there are 10 x 10 local flow non-uniformity dimensionless parameters (α's), which correspond to the 10 x 10 subdivisions on the x-z plane perpendicular to the direction of non-uniform fluid flow. In view of the symmetry of the values with respect to o-x and o-y, only one-fourth of flow non-uniformity parameters (α's) are presented in Table 1

Table 1. Flow Non-uniformity Parameters (α's)

Model – A1

I =	1 ; 10	2 ; 9	3 ; 8	4 ; 7	5 ; 6
J=1;10	0.500	0.500	0.500	0.500	0.500
2 ; 9	0.500	0.639	0.776	0.899	0.998
3 ; 8	0.500	0.776	1.045	1.291	1.489
4 ; 7	0.500	0.899	1.291	1.655	1.956
5 ; 6	0.500	0.998	1.489	1.956	2.356

Model – A2					
J=1;10	0.900	0.900	0.900	0.900	0.900
2 ; 9	0.900	0.923	0.952	0.978	0.999
3 ; 8	0.900	0.952	1.009	1.062	1.104
4 ; 7	0.900	0.978	1.062	1.139	1.203
5 ; 6	0.900	0.999	1.104	1.203	1.289

FINITE ELEMENT METHOD

A discredited model of a cross flow plate-fin heat exchanger is shown in Fig. 1(a). It is divided into a number of elemental strips. Each element consists of a number of pairs of 7-noded stacks which carry hot and cold fluids as shown in Fig. 1(b). A 2-noded element has been considered for both hot and cold fluids. These are the basic elemental exchangers for which the finite element equations are formulated as similar to Paffenbarger, (1990). Under the basic assumptions of constant material properties, and of no heat exchange between the exchanger and its environment, the basic equations governing the temperature distribution are formulated as shown below:

$$-(\vartheta h a')_h (T_h - T_{w,t}) = q_t \quad (2)$$

$$-(\alpha M C_p)_h \frac{\partial T_h}{\partial x} + \frac{1}{a}(\vartheta h a')_h (T_h - T_{w,t}) + \frac{1}{a}(\vartheta h a')_h (T_h - T_{w,m}) = 0 \quad \ldots(3)$$

$$-(\vartheta h a')_h (T_h - T_{w,m}) + (\vartheta h a')_c (T_{w,m} - T_c) = 0 \quad (4)$$

$$(\alpha M C_p)_c \frac{\partial T_c}{\partial y} + \frac{1}{b}(\vartheta h a')_c (T_{w,m} - T_c) + \frac{1}{b}(\vartheta h a')_c (T_{w,b} - T_c) = 0 \quad \ldots(5)$$

$$(\vartheta h a')_c (T_{w,b} - T_c) = q_b \quad (6)$$

The boundary conditions are,

$$T_h(0, y) = T_{h,in} ;\ T_c(x, 0) = T_{c,in} \quad (7)$$

The temperature variation of the hot fluid (T_h), and cold fluid (T_c) in the element are approximated by a linear variation as,

$$T_h = N_i T_i + N_j T_j \quad (8)$$

$$T_c = N_k T_k + N_l T_l \quad (9)$$

where N_i, N_j, N_k and N_l are shape functions, $N_i = 1-x/a$, $N_j = x/a$ for cold fluid and $N_k = 1-y/b$, $N_l = y/b$ for hot fluid. The boundary conditions are to be satisfied are

$$(\alpha M C_p)_h T_h = Q_h \quad (10)$$

$$(\alpha M C_p)_c T_c = Q_c \quad (11)$$

Substituting the approximations (10) and (11) into equations (2-6) and applying the sub-domain collocation procedure to minimize the error over the entire domain, the final set of element matrices can be obtained as (Ranganayakulu et all., 1997):

$$\begin{bmatrix} j_h & -j_h/2 & -j_h/2 & 0 & 0 & 0 & 0 \\ 0 & i_h & 0 & 0 & 0 & 0 & 0 \\ j_h & -i_h-j_h & -j_h+i_h & j_h & 0 & 0 & 0 \\ 0 & -j_h & j_h & 2j_h+2i_c & j_c & -j_c & 0 \\ 0 & 0 & 0 & 0 & i_c & 0 & 0 \\ 0 & 0 & 0 & j_h & i_c-j_c & -i_c+j_c & j_c \\ 0 & 0 & 0 & 0 & -j_c/2 & -j_c/2 & j_c \end{bmatrix} \begin{bmatrix} T_1 \\ T_2 \\ T_3 \\ T_4 \\ T_5 \\ T_6 \\ T_7 \end{bmatrix} = \begin{bmatrix} -q_t \\ q_2 \\ 0 \\ 0 \\ q_5 \\ 0 \\ q_b \end{bmatrix}$$

where $j_h = (\vartheta h a)_h$, $i_h = (\alpha M c_p)_h$, $j_c = (\vartheta h a)_c$ and $i_c = (\alpha M c_p)_c$.

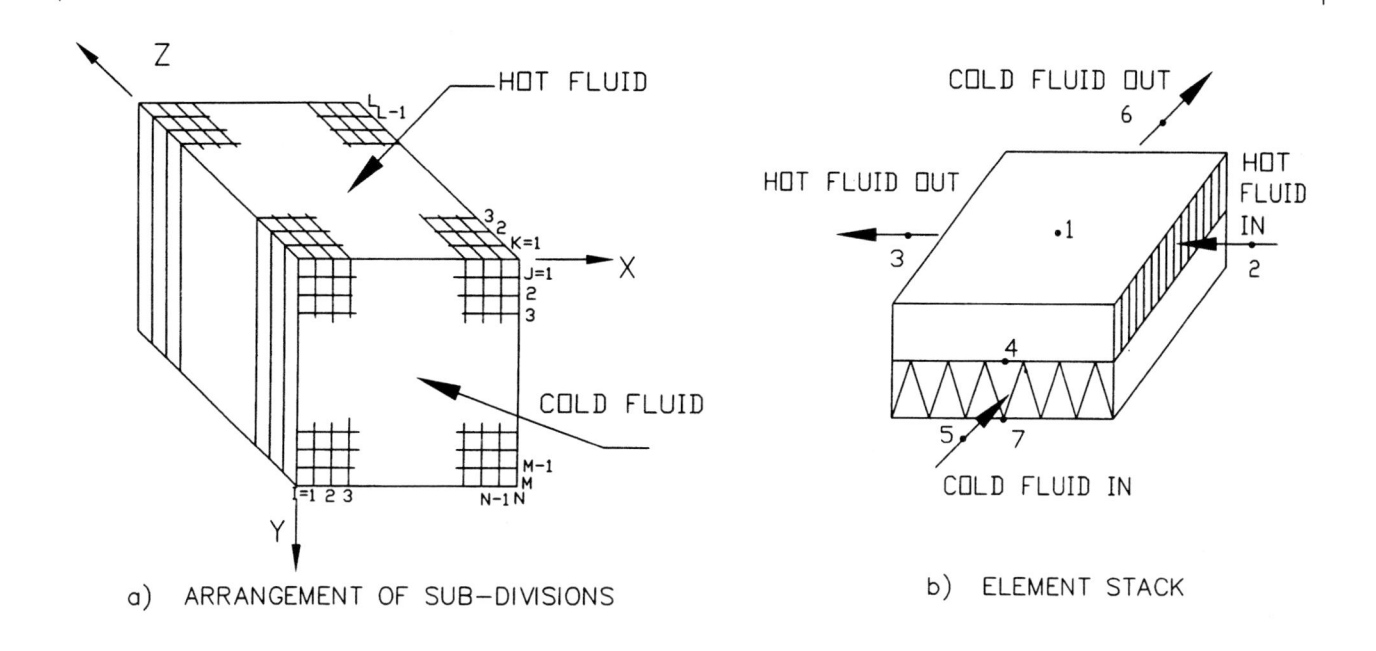

Fig. 1 – CROSSFLOW PLATE-FIN HEAT EXCHANGER

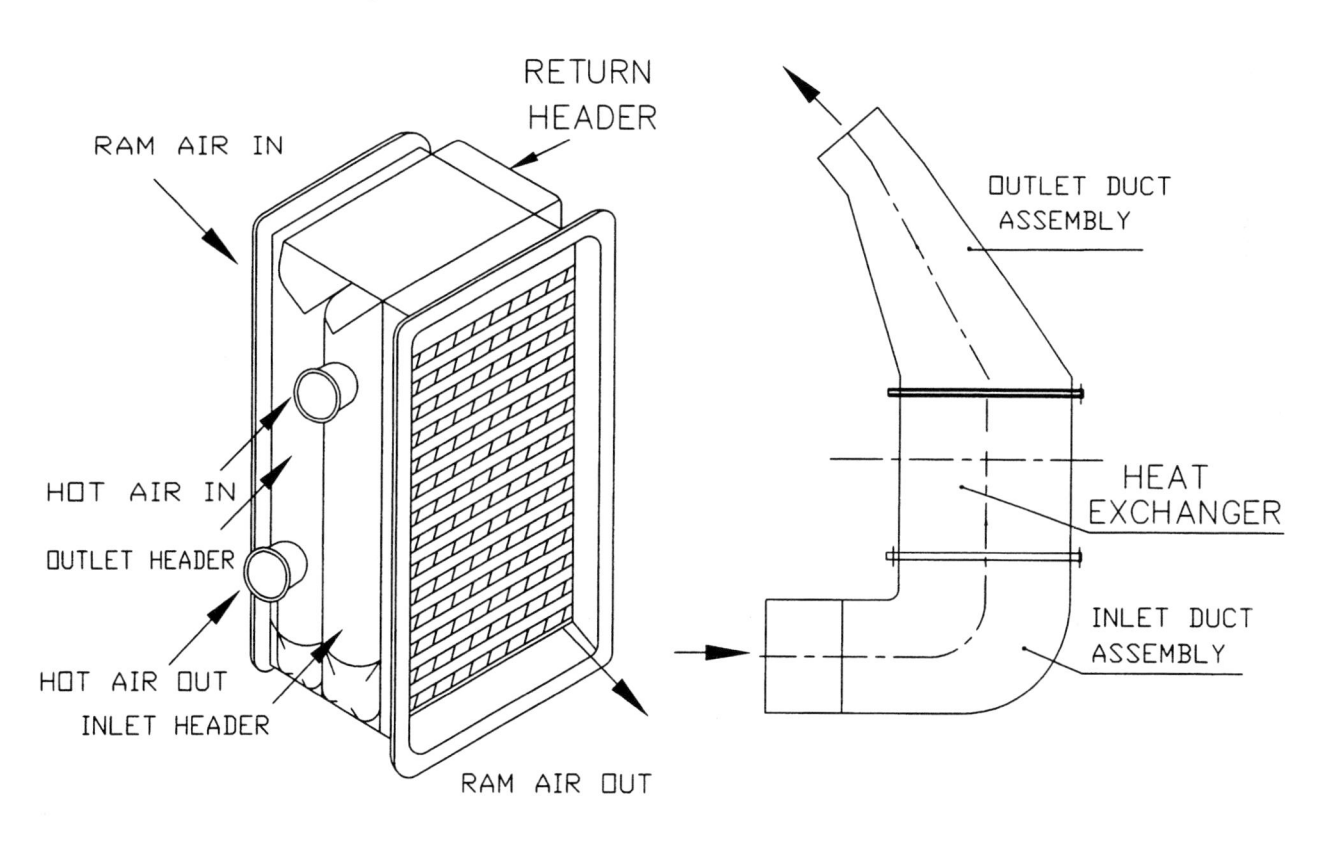

a) ARRANGEMENT OF SUB-DIVISIONS

b) ELEMENT STACK

Fig. 2 TWO PASS CROSSFLOW HEAT EXCHANGER

Fig. 3 RAM AIR DUCT ASSEMBLY

The element matrices for other pairs of stacks are evaluated and assembled into a global matrix. The term q_b on the right-hand side of matrix gets cancelled when the adjacent element is assembled. It remains on the right-hand side of the global matrix only for the bottom pair of stacks. Similarly, q_t remains only for the top pair of stacks. If top and bottom surfaces are insulated, then q_t for top pair of stacks and q_b for the bottom pair become zero. The final sets of simultaneous equations are solved after incorporating the known boundary conditions (inlet temperatures) and fluid capacity rates. The outlet temperatures of first strip will be the inlet temperatures for adjacent strips. Thus by marching in a proper sequence, the temperature distribution in the exchanger is obtained.

Analytical solutions without considering the effects of flow non-uniformity are obtained using the solution procedure given by Kays and London (1984). If the temperatures are not known a priori, the iteration is started with assumed outlet temperatures. The new outlet temperatures are calculated and compared with assumed outlet temperatures. The iterations are continued until the convergence is achieved to the fourth digit for all cases. In this analysis, the fluid properties and heat transfer coefficients have been varied from element to element, depending on their local bulk mean temperatures. The following fin surface characteristics have been selected from Kays and London (1984). A wavy fin surface, 17.8-3/8W for ram air side and a strip fin surface, 1/8-16.12(D) for hot air side have been selected from Kays and London (1984). The exchanger is assumed to be ideal with no heat loss to the ambient and the longitudinal wall heat conduction effects are neglected in this analysis. The temperature distribution on the outlet surfaces of cold and hot fluids are averaged to get the representative outlet mean temperatures for uniform flow cases. In case of non-uniform flow cases, the temperature distribution in the elemental heat exchangers on the outlet surfaces are obtained using the mass balance.

A one-dimensional non-uniformity concept in heat exchanger is simple and straight forward. The one-dimensional inlet velocity distribution can be divided into N step functions, and correspondingly the exchanger is divided into N parallel small exchangers each representing uniform velocity distribution at the entrance. The analysis of N exchangers will then provide the overall exchanger effectiveness, which can be compared with the effectiveness if the flows are totally uniform at the exchanger inlet. However, in the present analysis, a two-dimensional non-uniformity concept is considered. When a more complete analysis is attempted, a numerical procedure is feasible.

DESIGN PROBLEM

Using this finite element analysis program, a two pass cross flow heat exchanger has been designed and validated with experimental results. The schematic diagram of subject heat exchanger is shown in Fig. 2. This is a air to air two pass plate-fin heat exchanger. In this heat exchanger, hot air is being cooled in two passes by cold air in single pass, as shown in the figure. This plate-fin heat exchanger, which was designed to use in the Environmental Control System (ECS) of an Aircraft, was tested for pressure drop and thermal performance using ground test rig. The test data is given below in Table 2 as a sample case for reference.

Table 2. Exchanger performance – measured data

	F kg/min	T_{in} °C	T_{out} °C	P_{in} bar	P_{out} bar	ΔP_a bar	ε_{rc} %	ε_{rq} %
Hot air	33	201	104	9.4	8.6	0.3	88	92
Cold air	252	91	104	0.6	0.1	0.5		

The pressure drop across the hot air side measured (0.8 bars) very much beyond the allowable limits (0.3 bars). It was observed that, at some design points, the measured pressure drops were 3 times higher than the corresponding allowable values and thermal performance was 4 to 5% less than the design value. To tackle this problem, all possible causes leading to abnormally high pressure drops across the plate-fin heat exchanger was studied. A small test rig with suitable flow and pressure drop measuring facility was setup for hot air side. It was understood that all the excessive pressure drop is contributed by headers and nozzles welded to the core on hot side. In subsequent trails and analysis one fact that came out was that the non-uniform flow distribution of hot fluid (hot air) across the exchanger. For the hot stream, the inlet and outlet cross sections are about 300 mm x 60 mm each, and the inlet and outlet nozzles are of 50 mm diameter each due to space limitations of Aircraft installation. Because there are two passes on the hot side, a return header is also provided between the passes. The nozzles are welded to semi-cylindrical headers which are in turn welded to the exchanger core.

TESTING

The pressure distribution in the ram air duct was also measured earlier using a dedicated test set up. The schematic details of the ram air duct is shown in Fig. 3. The ram air enters the inlet duct via a rectangular duct of cross-section 200 mm x 110 mm and passes through two 90 degrees bends (located at 90 degrees with respect to each other) before leading to the heat exchanger duct of 285 mm x 265 mm cross section. The air is then led to the outlet duct assembly which is inclined in two directions. The entire ducting was made of aluminum sheet of thickness 1.5 mm. The test set up is shown in Fig. 4. A constant area section (285 mm x 265 mm), having a length of 800 mm, was introduced in

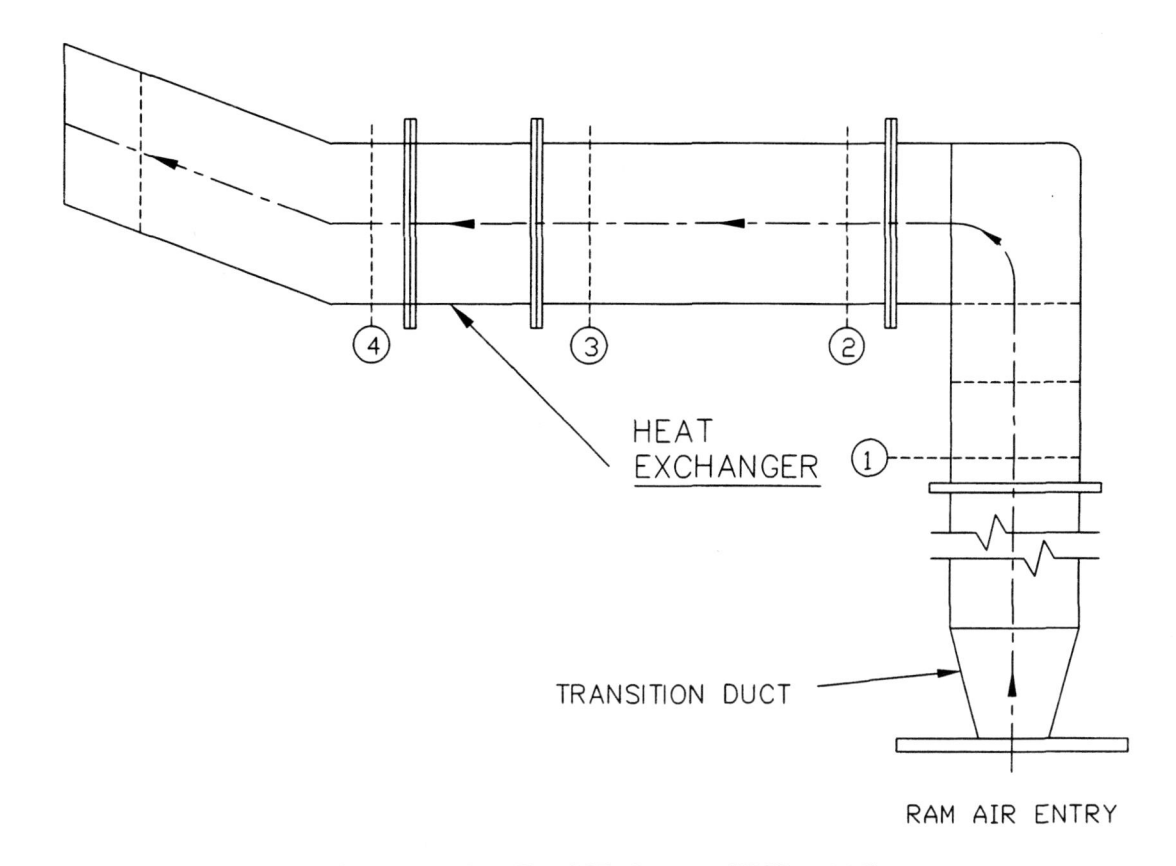

Fig. 4 RAM AIR DUCT ASSEMBLY – TEST MODEL

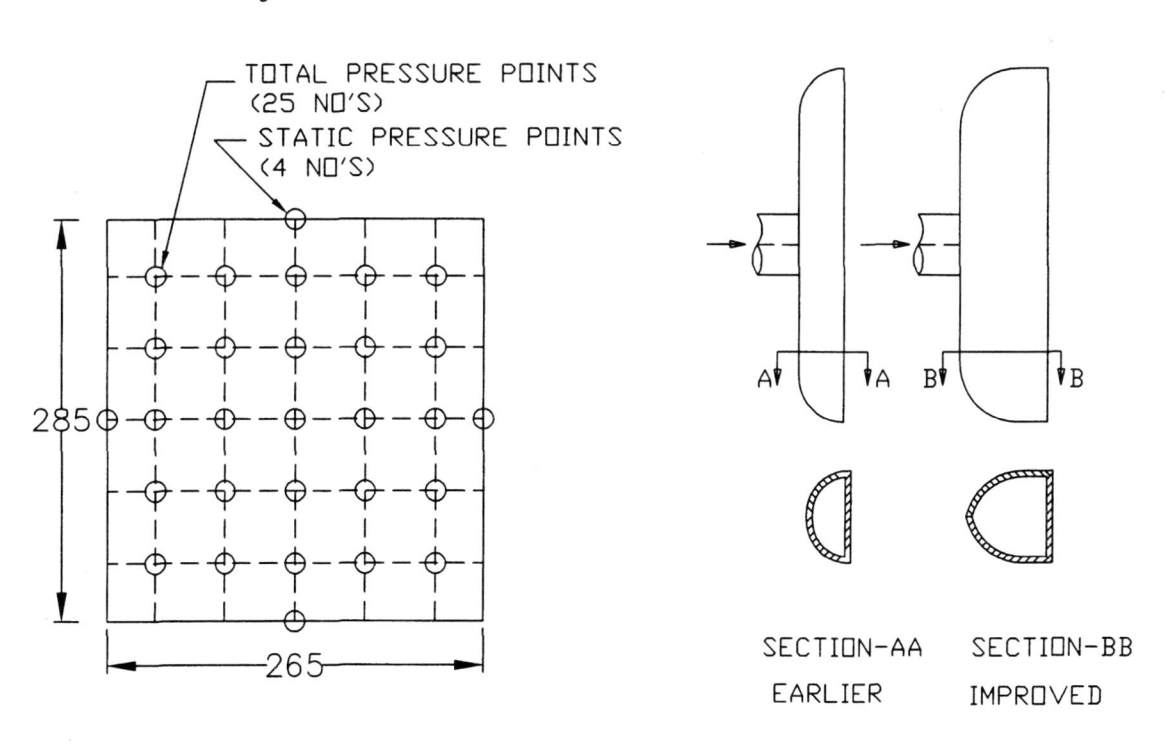

Fig. 5 – TOTAL AND STATIC PRESSURE MEASUREMENT POINTS – RAM AIR DUCT

Fig. 6 – TWO VIEWS OF HEADERS

between the ducts 2 and 3 as shown in Fig. 4, to facilitate measurement of total pressure drop across the heat exchanger. This helped to make the flow that came out of the double bend more uniform. In order to measure the ram air flow rate entering the duct 3, a constant area section of 200 mm x 110 mm (1450 mm long) was introduced ahead of this duct as shown in Fig. 4. The other end of this constant area duct had a transition piece (from 154 mm diameter to 200 mm x 110 mm) to connect to the high pressure air line of combustion laboratory. This constant area rectangular duct helped to achieve uniformity in the flow that entered the double bend. A 25 channel total pressure rake was located at section-1 just downstream of this constant area duct to estimate the air flow rate passing through. The details of total and static pressure measurement locations at heat exchanger exit are shown in Fig. 5. The pressure drop across the heat exchanger was based on the total/static pressures measured at sections 3 and 4. The total pressure at section 4 was taken to be the average of 25 pressures measured by the rake at the section. All the total and static pressures at all the sections were recorded by using two type 'J' scanni-valve differential pressure transducers. All the pressure data were recorded to an accuracy of +/- 0.7 mbar. The scanni-valve data was recorded with a frequency of 1 port/sec and the collected data were stored in the PC. The total pressure profile for a typical flow case is shown in Table 3 for reference.

Table 3. Total Pressure Measurement of Heat Heat Exchanger Ram Air Duct

	Total pressure values in mbar				
Location in mm	# 44	# 88	# 133	# 177	# 221
# 238	134	165	189	210	202
# 190	142	170	200	205	203
# 143	159	180	209	215	206
# 95	165	188	211	216	217
# 48	154	188	212	217	218

ANALYSIS

The exchanger thermal performance and pressure drops were studied using the above velocity profile. The air had been taking a least resistance path and more of the flow had been passing through a limited portion of the heat exchanger area at high velocities, causing high pressure drops. This, however, did not bring down the thermal performance significantly, because the higher velocities resulted in higher heat transfer coefficients to some extent. Then it was decided to minimize frictional losses due to non-uniform flow distribution in the headers wherever possible. The shallow headers were replaced with perfectly rounded semi-cylindrical ones as shown in Fig. 6 and additional space was provided by adding some straight flange to the headers.

With each of the above improvements introduced, the pressure drop performance got better and better, until finally, with about 45 mm total depths of headers, the measured pressure drops fell down to within the allowable

Table 4. Heat Exchanger Performance – Finite Element Results

Case No.	Flow (Kg/min)		Temperatures (°C)			ε	NTU
	Hot	Cold	Hot In	Hot Out	Cold In		
1	33	252	201	98.8	91	0.929	3.25
2	33	252	201	99.5	91	0.923	3.35
3	33	252	201	100	91	0.918	3.36
4	33	252	201	104	91	0.881	3.37

Case 1: Both hot and cold fluids are uniform
Case 2: Hot fluid uniform and cold fluid as per measured flow non-uniformity.
Case 3: Hot side with model A2 and Cold side with measured flow non-uniform.
Case 4: Hot side with model A1 and Cold side with measured flow non-uniformity.

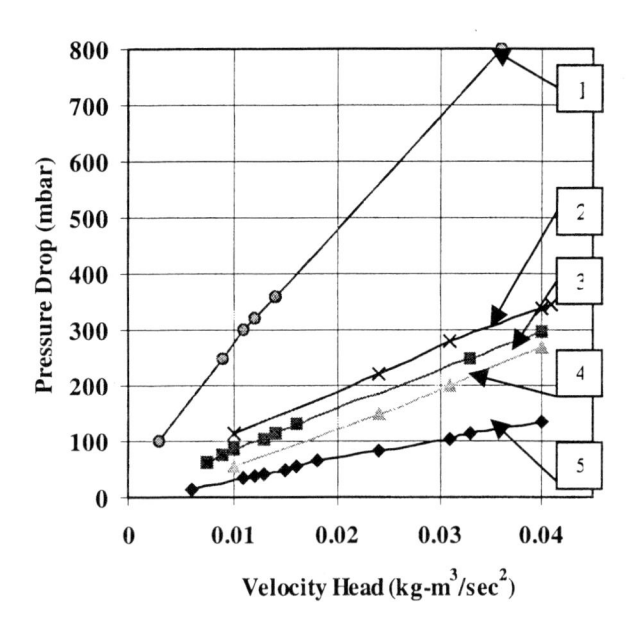

Legend : 1 - Pressure loss with old headers
2 – Allowable pressure loss
3 – Pressure loss with improved headers
4 – Predicted total pressure loss including headers
5 – Predicted core pressure loss

Fig. 7 Heat Exchanger Pressure Drop values

values. The results of pressure drop experiments are plotted in Fig. 7. The pressure drop versus flow rate values recorded turned out to be well within allowable limits with the header modifications. The observations were repeated a number of times to confirm adequacy of the improved design. The thermal performance was improved by about 5% as indicated in Table 4.

The present finite element analysis results of flow non-uniformity effects on pressure drops and deterioration in thermal performance have been validated with experimental results. It is understood from Table 4 that the thermal performance is better than required (Case.No.2) when both hot and cold fluid flows are uniform as shown in Case No. 1 of Table 4 above. The exchanger effectiveness is 0.923 when the hot fluid flow is uniform and cold fluid distribution as measured. It was observed that the performance variation is negligible with cold side flow non-uniformity. However, the performance deteriorates with the intensity of flow non-uniformity on hot side as analyzed above. The effectiveness of the exchanger with flow model A1 is about 0.877, which is very close to the measured value of 0.88. Similarly the effectiveness of the exchanger with flow model A2 is about 0.918, which is very close to the measured value of 0.92. The information obtained in this study clearly indicates that significant variations in pressure drop values and moderate deteriorations in thermal performance due to flow non-uniformity caused by headers of heat exchanger.

CONCLUSIONS

A compact two pass cross flow plate-fin heat exchanger has been designed using finite element analysis. The exchanger performance has been validated with experimental results. The thermal performance deteriorations and pressure drop variations in cross flow plate-fin heat exchanger have been reviewed with effects of flow non-uniformity due to header design. Using the fluid flow mal-distribution models, the exchanger effectiveness and its variations due to the effects of flow non-uniformity are calculated on both the cold and hot fluid sides of an exchanger. Based on the analysis, the exchanger headers have been modified to improve pressure drops. Also, the thermal performance has been improved by about 5% with uniform flow distribution in the exchanger. The information obtained in this study clearly indicates that the deteriorations in thermal performance and pressure drop due to flow non-uniformity should not be ignored in the analysis and design of heat exchangers. The information presented in this study can provide the designer a means of estimation of effective performance deterioration when fluid flow is not uniformly distributed on exchanger core.

Acknowledgements – The authors are grateful to the managements of M/s Aeronautical Development Agency, Bangalore and M/s Bharat Heavy Plate and Vessels Ltd., Visakhapatnam for allowing this paper for publication.

NOMENCLATURE

a'   elemental heat transfer area per unit core area, dimensionless
a   distance in x – direction, meters
b   distance in x – direction, meters
c_p specific heat of the fluid at constant pressure, J/kg K
C   (=Mc_p) - fluid heat capacity rate, J/s K
ΔP Pressure drop, bars
F   Flow rate, kg/min
FN   Flow Non-uniformity case
h   convection heat transfer coefficient, W/m^2 K
I,J divisions in the x or y directions (1,2,3 ....n)
L,l the length of the elemental exchanger in x or y directions respectively, meters
M   mass flow rate, kg/s
NTU(= AU/C_{min}) - number of transfer units, dimensionless
N   number
P   pressure, bars
q   enthalpy/heat entering/leaving the plate per unit area, W/m^2
Q   heat load, KW
T   temperature of hot/cold fluids and metal wall, °C

Greek Symbols

α flow non-uniformity parameter as defined in eq. (1)
ε exchanger effectiveness, dimensionless
ϑ overall surface efficiency, dimensionless

Subscript

a   allowable value
c   cold side
b   bottom plate
h   hot side
in   inlet
m   middle plate
out outlet
rc   recorded data
rq   required value
t   top plate
w   metal wall.
1-7 node numbers

REFERENCES

Darius A. Nikanpour and Bai Zhou, 1997, Trends in Aerospace Heat Exchangers, International conference on "Compact Heat Exchangers for the process industries", Snowbird, Utah, USA, June 22-27.

Ch. Ranganayakulu, K.N. Seetharamu and K.V. Sreevatsan, 1997, The effects of inlet flow non-uniformity on thermal performance and pressure drops in cross flow plate-fin compact heat exchangers, Int. J. Heat & Mass Transfer, Vol. 40, No. 1, pp. 27-38.

Ch. Ranganayakulu and K.N. Seetharamu, 1997, The combined effects of inlet fluid flow and temperature non-uniformity in cross flow plate-fin compact heat exchanger using finite element method, Heat and Mass Transfer, No. 32, pp. 375-383, Springer-Verlag.

J. P. Chiou, 1978, Thermal performance deterioration in a cross flow heat exchanger due to the flow non-uniformity, J. Heat Transfer Trans., ASME 100, pp. 580-587.

J. P. Chiou, 1978, Thermal performance deterioration in a cross flow heat exchanger due to the flow non-uniformity on both hot and cold sides, Sixth International Heat Transfer Conference, Vol. 4, pp. 279-284.

R. K. Shah, 1985, Flow mal-distribution – Compact Heat Exchangers, in Hand book of Heat Transfer Applications, Edited by W.M.Rohsenow, P.J.Hartnett and E.N.Ganic, McGraw-Hill, New York, pp. 266-280.

J. Paffenbarger, 1990, General computer analysis of multi-stream, plate-fin heat exchangers, Edited by R.K.Shah, A.D.Kraus and D.Metyger, Compact Heat Exchangers, Hemisphere, pp. 727-746.

W. M. Kays and A. L. London, 1984, Compact Heat Exchangers, 3rd Edn.,McGraw-Hill, New York.

PORT FLOW DISTRIBUTION IN PLATE HEAT EXCHANGERS

L. Huang

Heat Transfer Research, Inc.
150 Venture Dr., College Station, Texas 77845, USA
E-mail: LH@HTRI.net

ABSTRACT

This paper discusses port pressure drop and flow distribution in plate heat exchangers with different arrangements. A new grouping of parameters affecting port flow distribution permits development of guidelines for minimizing port flow maldistribution. The effects of flow maldistribution on heat transfer and fouling are also discussed. The HTRI PHE program, using a data-based model, provides illustrative example calculations.

INTRODUCTION

Plate heat exchangers (PHEs) experience three main types of flow maldistribution:

- Port flow maldistribution: Port flow maldistribution is caused by port pressure variation in a given pass that produces flow rate variations from channel to channel along the ports.
- Flow maldistribution resulting from different plate groups arranged in a given pass: The plate group is defined as a group of plates consists of channels having the same geometric and operating characteristics. This type of flow maldistribution occurs primarily when hydraulic flow resistance between two plate groups differs. Fluid passes more easily through the plate group with the lower hydraulic flow resistance.
- Channel flow maldistribution: Channel flow maldistribution, which occurs within any channel, results from asymmetrical parallel and diagonal channel flows. Channel flow maldistribution is more likely to occur in parallel flow. Proper design of a plate's inlet and outlet flow distribution regions minimizes this problem.

Because the port flow distribution model described here cannot predict channel flow maldistribution, this paper examines only the first and second maldistribution types.

Flow distribution among channels in a given pass is determined by pressure profiles in the inlet and outlet ports and by hydraulic resistance in the channels. Two factors, namely, fluid friction and momentum change, affect pressure profiles in PHE manifolds. Different models [1-4] have been developed to predict port flow distribution in a plate heat exchanger.

HTRI has also developed models for predicting port pressure distributions, channel flow rate distributions in a plate heat exchanger based on its port maldistribution data. In this paper, HTRI PHE computer program is used to illustrate port pressure and flow distributions in plate heat exchangers with different arrangements. The program prediction of HTRI port flow distribution data is discussed. Parameters which affect port flow distribution are also presented, as well as, guidelines to minimize port flow maldistribution.

EXPERIMENTAL DATA

Over the past several years, HTRI has compiled a port flow distribution data bank for PHEs with different plate packs and working fluids. The data cover a wide range of operating conditions and provided an excellent basis for developing and testing port flow distribution models for PHEs.

Figure 1 shows an example of this data set, together with the prediction by the HTRI PHE computer program. It shows a U-arrangement with inlet and outlet ports on the same side of the exchanger. The inlet port pressure increases along the port because the momentum gain from the decrease in flow rate is higher than the sum of friction and turnaround losses. While pressure in the outlet port decreases due to both friction and momentum losses. As a result of this, the channel pressure drop, that is, the pressure difference from inlet port to outlet port, decreases along ports for the U-arrangement, which in turn causes the channel flow rate to decrease along the ports.

Figure 2 shows three sets of channel pressure drop profiles for three different total flow rates with water as a working fluid for U-arrangement exchanger. Flow in the channels of all these data is in the fully developed turbulent flow range with channel Reynolds numbers ranging from 800 to 3000. As the figure demonstrates, the difference of channel pressure drops between two ends of the exchanger increases rapidly as flow rate (or velocity head, as defined by Equation (1)) increases. However, the ratio of this channel pressure drop difference and total pressure drop does not increase significantly as flow rate increases, as the experimental data indicated. For example, the variation in channel pressure drop is about 53% for velocity head of 1.24 kPa, the bottom line in Figure 2. As velocity head increases significantly to 11.1 kPa, the top line in Figure 2, the variation in channel pressure drop only increases to about 60%.

$$Velocity\ head = N_p \frac{G_p^2}{2\rho} \quad (1)$$

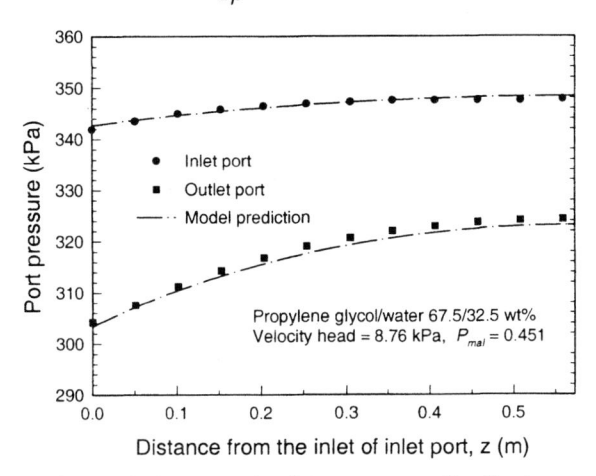

Figure 1. An example of port pressure distributions from experimental data and model prediction.

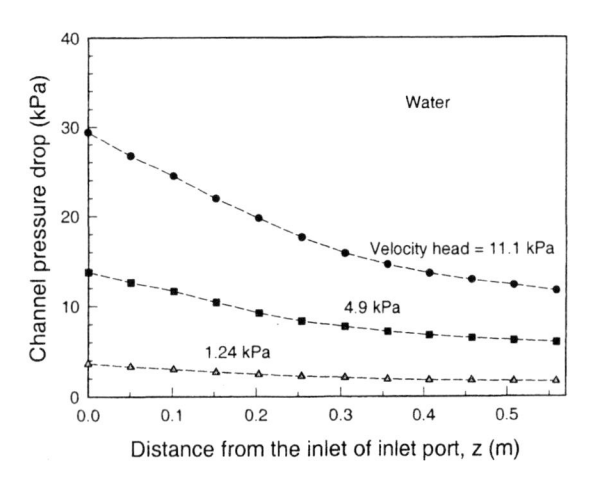

Figure 2. Experimental data for channel pressure drop distribution

FLOW DISTRIBUTIONS FOR DIFFERENT ARRANGEMNTS

In this section, the HTRI PHE computer program is used to demonstrate port pressure profile and flow distribution for different arrangements in PHEs. Figures 3 – 6 provide the results of PHE's prediction on port pressure and channel velocity distributions for U and Z arrangements. The same process conditions and plate pack were used for both arrangements. The inlet port pressure increases along the port for both arrangements, as shown in Figures 3 and 4, although with different curve shapes. This increase suggests that momentum gain from the decrease in flow rate is higher than the sum of friction and turnaround losses. As a result of friction and momentum losses, pressure in the outlet port decreases in the direction of fluid flow for both U and Z arrangements. However, because fluid flow direction in the outlet port differs for these two arrangements, the channel pressure drop distributions also differ. Channel pressure drop decreases in the direction of inlet port fluid flow for a U- arrangement and increases for a Z arrangement, as shown in Figures 3 and 4.

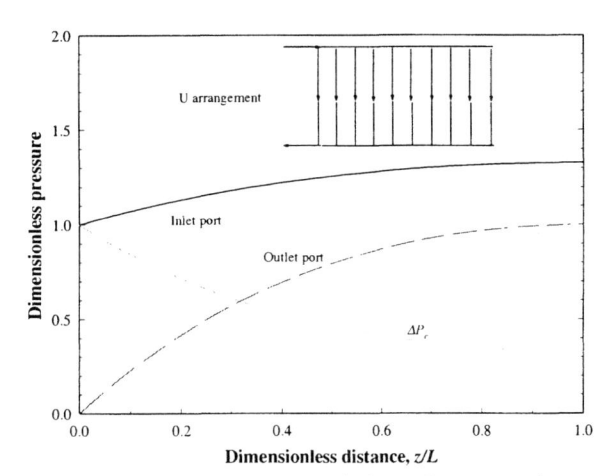

Figure 3. Port pressure and channel pressure drop distributions for U arrangement

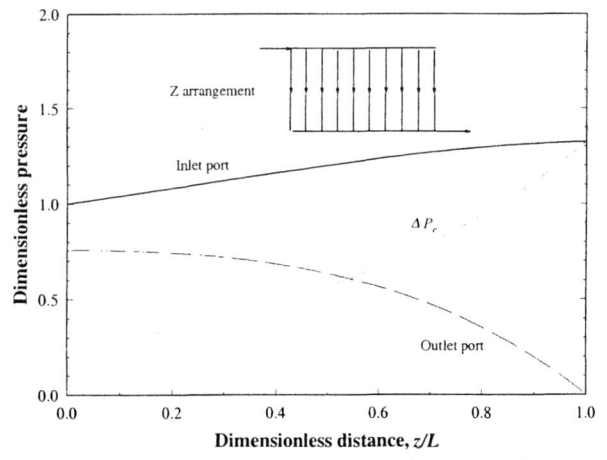

Figure 4. Port pressure and channel pressure drop distributions for Z arrangement

The channel velocity distributions for U- and Z-arrangements (Figures 5 and 6, respectively) also reflect the channel pressure drop distribution. Z-arrangements have higher flow maldistribution than do U-arrangements. A Z-arrangement's inlet port momentum gain and outlet port momentum loss are in the same direction, causing channel pressure drop and channel velocity to vary more along the ports.

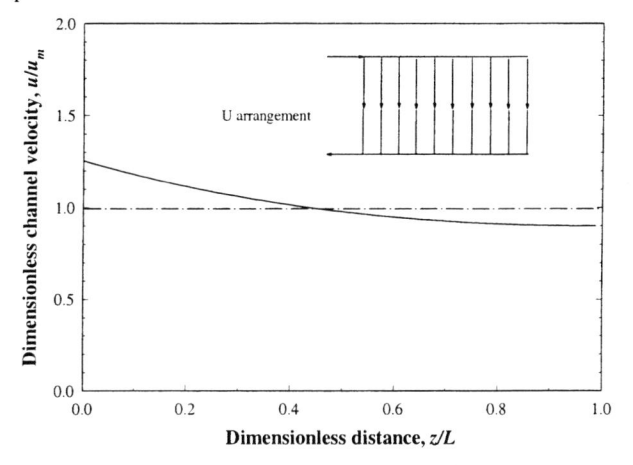

Figure 5. Channel velocity distribution for U arrangement

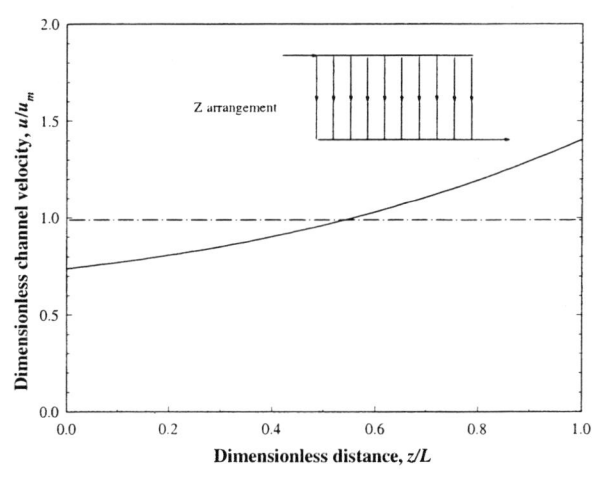

Figure 6. Channel velocity distribution for Z arrangement

For two plate groups in one pass (see Figure 7), plate heat exchangers experience another type of flow maldistribution.

Figures 8 – 11 show the prediction by the HTRI PHE computer program for two plate groups in one pass for a U-arrangement. In Figures 8 and 9, the plate group with the lower chevron angle and higher hydraulic flow resistance was placed in the first group. The decrease in channel velocity within these two groups results from the decrease in channel pressure drop caused by port pressure variation. However, the channel velocity jumps at the connecting point from group one to group two, as shown in Figure 9, due to the fluid's tendency to flow through the plate group having lower hydraulic resistance.

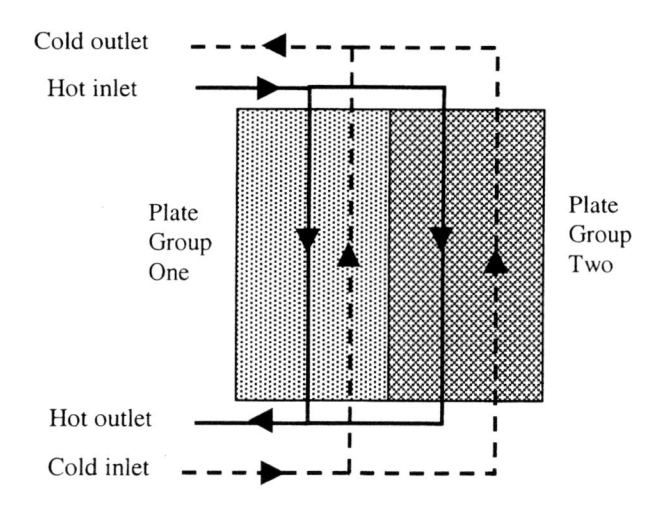

Figure 7. Two plate groups in one pass

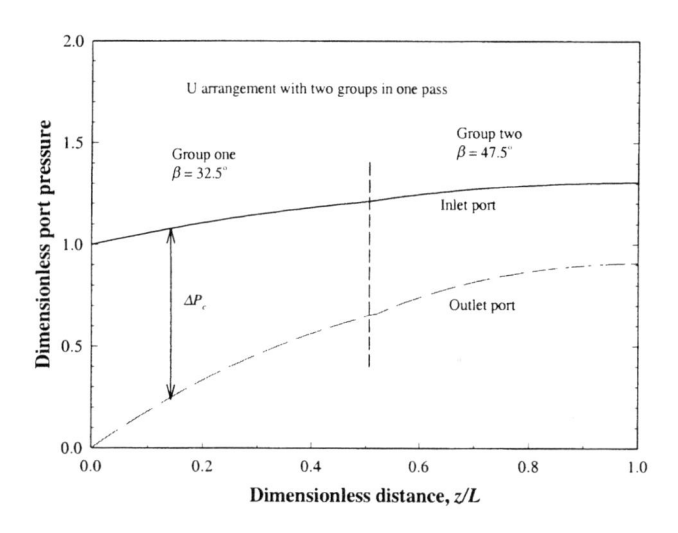

Figure 8. Port pressure distribution in two plate groups for the first group arrangement

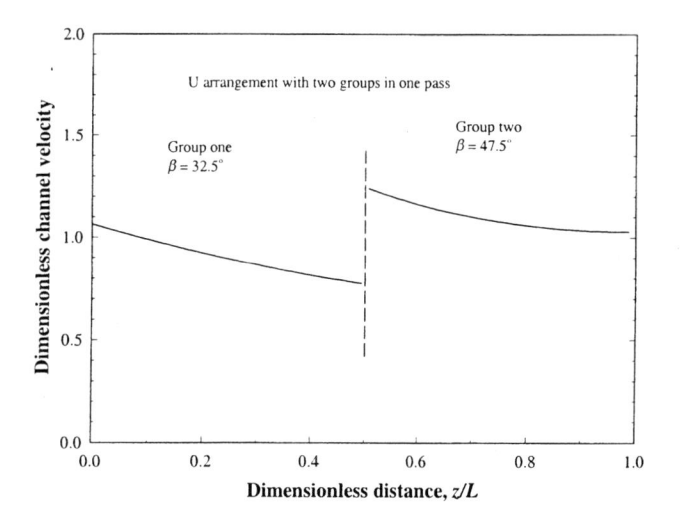

Figure 9. Channel velocity distribution in two plate groups for the first group arrangement

Placing the plate group with lower hydraulic flow resistance and higher chevron angle in the first group results in a different flow distribution between the two groups. Most flow maldistribution occurs in the first group, shown in Figures 10 and 11, due to the following:

- Lower hydraulic flow resistance in the channels, which increases port flow maldistribution
- Rapid decrease in channel pressure drop in the front channels for the U arrangement, due to higher port flow rate change in those channels

The combined port flow maldistribution within the two groups for this group arrangement (Figures 10 and 11) is higher than that for the first group arrangement (Figures 8 and 9). Also, the channel velocity drop at the transition from group one to group two corresponds with the decrease in channel velocity, making the second arrangement's port flow maldistribution much worse than that of the first (see Figures 11 and Figure 9).

Therefore, a U arrangement should include the plate group with higher hydraulic flow resistance in the first group. The reverse is true for a Z arrangement, in which channel pressure drop distribution is opposite that of a U arrangement, as Figures 3 and 4 illustrate.

PARAMETERS AFFECTING PORT FLOW MALDISTRIBUTION

This section discusses parameters affecting port flow maldistribution and presents guidelines for minimizing it.

Maldistribution Parameter P_{mal}

We can use a single parameter to estimate port flow maldistribution in a plate group. This parameter, derived from model analysis, is defined by

$$P_{mal} = \frac{\theta_2 - \theta_1}{4} \frac{A_R^2 D_e}{f_c L_e} \quad (2)$$

where θ_1 and θ_2 are the overall momentum correction factors of inlet and outlet ports, respectively. Values of θ_1 and θ_2 have been experimentally determined by HTRI. Alternatively, Bajura and Jones [1] obtained the overall momentum correction constants $\theta_1 = 1.05$ and $\theta_2 = 2.6$, using manifolds composed of small pipes connected to larger inlet and outlet pipes.

The severity of port flow maldistribution in a plate group can be measured by

$$S_{mal}, \% = \frac{\Delta P_{c,max} - \Delta P_{c,min}}{\Delta P_{c,max}} \times 100 \quad (3)$$

Port flow maldistribution can be estimated if the relationship between S_{mal} and P_{mal} is known. This relationship for both U- and Z-arrangements was generated using the HTRI PHE computer program, as shown in Figure 12. The parameter S_{mal} increases as P_{mal} increases. At the same value of P_{mal}, S_{mal} is higher for a Z-arrangement than for a U-arrangement, which is consistent with the findings in the previous section.

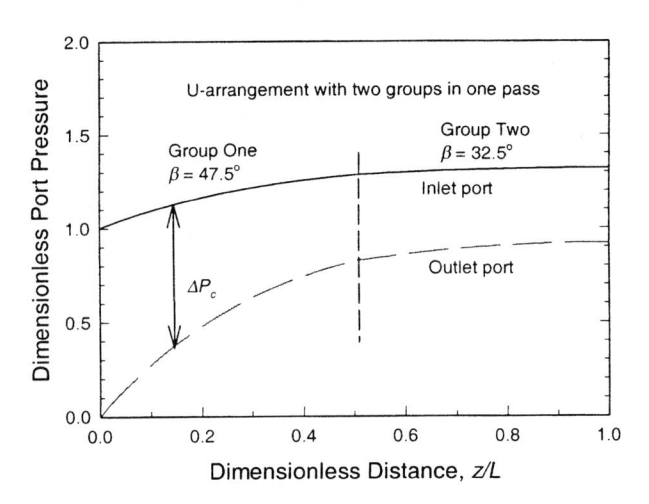

Figure 10. Port pressure distribution in two plate groups for the second group arrangement

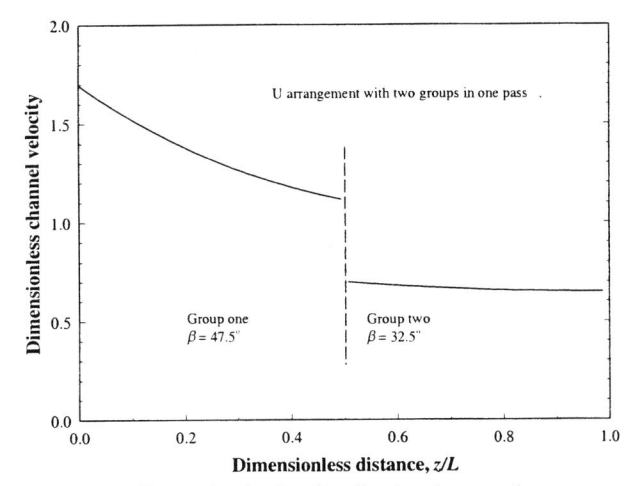

Figure 11. Channel velocity distribution in two plate groups for the second group arrangement

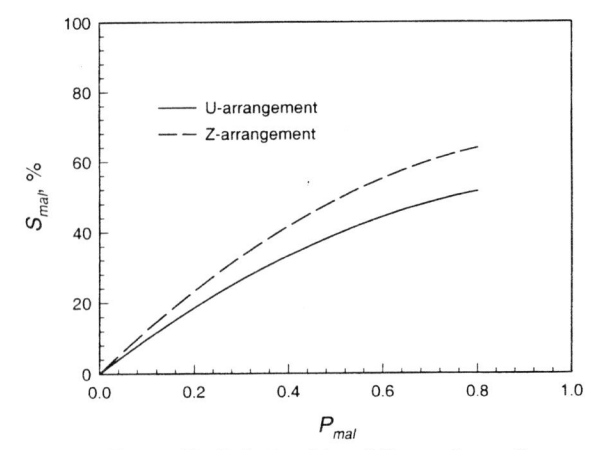

Figure 12. Relationship of P_{mal} and port flow maldistribution

The maldistribution parameter P_{mal} can be used as a guideline for proper PHE design. As Equation (2) shows, it is a function of a PHE's geometric dimensions, number of plates, and channel friction factor. Thus, for a specified plate heat exchanger, the severity of port flow maldistribution depends on the channel friction factor, which is related to channel geometry, flow rates, and fluid physical properties.

To minimize port flow maldistribution, P_{mal} should be kept low. As shown in Figure 12, a U-arrangement PHE designed with P_{mal} less than 0.2 keeps the variation in channel pressure drop within 20 percent. When plate types and process conditions have been selected, Equation (4)—an alternative form of Equation (2)—permits calculation of the number of channels allowed in a plate heat exchanger with only one plate group.

$$N_c = \frac{A_p}{A_c} \sqrt{\frac{4 P_{mal} f_c L_e}{(\theta_2 - \theta_1) D_e}} \quad (4)$$

P_{mal} can be determined from Figure 12 based on the permitted percentage of variation in channel pressure drop.

Each term in Equation (1) exhibits a different effect on port flow maldistribution. The effect of each term is discussed in detail in the following sections.

Area Ratio

The term A_R, ratio of the total channel cross-sectional area to the cross-sectional area of the port, may be interpreted as the porosity of the port. P_{mal} increases rapidly as the area ratio, A_R^2 in Equation (1), increases. Because A_R is proportional to the number of channels and the reciprocal of the port cross-sectional area in a PHE, increasing the number of plates or decreasing the port diameter increases port flow maldistribution rapidly.

Channel Flow Resistance

P_{mal} decreases as the channel flow resistance, represented by the channel friction factor, f_c, and the hydraulic length to diameter ratio of a channel, L_e/D_e, increases. Large channel flow resistance is desirable for even flow distribution. An infinite channel flow resistance would cause even a small diameter port with large porosity to act as an infinite reservoir. In such a situation, port pressure drop is negligible comparing to the very high channel pressure drop, which almost equals to total pressure drop. However, large channel resistance results in a high total pressure drop for the PHE, which may be unacceptable if pumping costs are an important design consideration.

Overall Momentum Correction Factors

Overall momentum correction factors account for all effects that relate to momentum changes in the ports. Different types of plates may exhibit different values of •₁ and •₂ because of varying momentum changes. However, •₁ and •₂ are relatively fixed for fully developed flow [1].

Port Flow Resistance

Port flow resistance, $f_p L/D_p$, is the only term that is not included in P_{mal}. Port flow resistance, which is usually negligible, increases as the port flow rate, port wall roughness, and the length/diameter ratio increase. As a result, port pressure drop increases.

FLOW MALDISTIRBUTION EFFECT ON HEAT TRANSFER

When the flow distribution through the various channels of a PHE is not uniform the loss in the heat transfer efficiency will increase with the flow maldistribution parameter, P_{mal}. HTRI's data showed that the resulting decrease in overall heat transfer coefficient is less than 5% for a single-group single-pass plate heat exchanger with U-arrangement if the value of the maldistribution parameter is less than 0.8.

The fouling data of Bossan et al. (1995) showed dramatically lower asymptotic fouling rates for plate heat exchangers than for shell-and-tube exchangers. However, plate heat exchangers designed with severe flow maldistribution will have a high tendency to foul. Plates in the channels with lower flow rates tend to foul more than those with higher flow rates. HTRI's PHE computer program allows the calculation of the flow distribution in plate heat exchangers based on the model, and backed by extensive experimental data. It provides a reliable tool for designing well-distributed plate heat exchangers in order to increase heat transfer efficiency and reduce the tendency for fouling.

CONCLUSION

The HTRI PHE computer program, which predicted HTRI port maldistribution data well, was used to demonstrate port pressure and flow distributions in a plate heat exchanger. The program can predict accurately the port pressure distributions and channel flow rate profiles and provides a reliable tool for designing well distributed plate heat exchangers. It also provides more accurate heat transfer calculations because the local channel flow rate is used to calculate local heat transfer coefficients.

From HTRI's experimental data, we found that port pressure drop results mainly from momentum loss in the ports; port friction loss is negligible. Higher port flow maldistribution occurs in plate heat exchangers with Z arrangements than for those with U arrangements.

For the arrangement of two plate groups in one pass, it is recommended to place the plate group with the higher hydraulic flow resistance in the first group for U arrangements and in the second group for Z arrangements.

The parameters affecting port flow maldistribution were defined and discussed. Guidelines for minimizing port flow maldistribution were presented. It was found that a single parameter, P_{mal}, could be used to accurately estimate port flow maldistribution.

NOMEMCLATURE

A_c	Crossflow area of one channel, m^2
A_p	Port flow area, m^2
A_R	Area ratio ($= N_c A_c/A_p$)
D_e	Equivalent hydraulic diameter of a plate channel, m
D_p	Port diameter, m
f_c	Channel Fanning friction factor
G_p	Port mass velocity, kg/s m^2
L	Length of compressed plate pack, m
L_e	Effective flow length between inlet and outlet ports, m
N_c	Number of channels of a plate group
N_p	Number of passes
P_{mal}	Maldistribution parameter, defined by Equation (1)
S_{mal}	Variation of channel pressure drop, defined by Equation (2)
u	Channel fluid velocity, m/s
u_m	Averaged channel fluid velocity, m/s
z	Coordinate pointed at flow direction in the inlet port, m
β	Chevron angle (angle of corrugations measured from horizontal direction), degree
ΔP_c	Channel pressure drop, Pa
$\Delta P_{c, max}$	Maximum channel pressure drop in a plate group, Pa
$\Delta P_{c, min}$	Minimum channel pressure drop in a plate group, Pa
θ_1	Overall momentum correction factor on inlet port
θ_2	Overall momentum correction factor on outlet port

REFERENCES

1. R. A. Bajura and E. H. Jones, Jr., Flow distribution manifolds, *Journal of Fluids Engineering*, 654 – 666 (December 1976).
2. M. K. Bassiouny and H. Martin, Flow distribution and pressure drop in plate heat exchangers – I: U-type arrangement, *Chemical Engineering Science*, 39(4), 693 – 700 (1984).
3. M. K. Bassiouny and H. Martin, Flow distribution and pressure drop in plate heat exchangers – II: Z-type arrangement, *Chemical Engineering Science*, 39(4), 701 – 704 (1984).
4. W. L. Wilkinson, Flow distribution in plate heat exchangers, *The Chemical Engineer*, 280 – 293 (May 1974).
5. D. Bossan, J. M Grillot, B. Thonon, and S. Grandgeorge, Experimental Study of Particulate Fouling in and Industrial Plate Heat Exchanger, *Journal of Enhanced Heat Transfer*, 2(1), 167-175 (1995).

THE SPIRAL HEAT EXCHANGER CONCEPT AND MANUFACTURING TECHNIQUE

Nathalie Bacquet

Acte S.A., B-4430 Ans, Belgium; E-mail: acte@euronet.be

ABSTRACT

The Spiral Heat Exchanger (SHE) has been developed for gas turbines because of a lack of satisfactory proposal in term of reliability, compacity and cost. The annular shape allows a gain of place in relation to plate heat exchangers and reduces the number of connecting pipes which lower the final cost of the turbine.

INTRODUCTION

Though numerous applications utilize coiled heat exchangers, the gaz turbine recuperator is among the most demanding. In any application, and especially when used as a gaz turbine recuperator, the heat exchanger should be compact, efficient, reliable, and relatively inexpensive to manufacture.

THE EXCHANGER IN THE TURBINE

The role of the exchanger in a gas turbine is to increase the temperature of the combustive air between the compressor and the combustion chamber by recovering the heat of combustion gases at the output of the turbine. This recovery of heat allows to increase the output of the cycle of the gas turbine when the compression ratios are not too high. For the more significant compression ratios, the difference in temperature between the outgoing air of the compressor and the outgoing gases of the turbine is too small to recover heat and the output of the cycle is not improved.
The recuperator has annular shape which is compact and easy to fit on any turbomachinery that has cylindrical axisymmetry. Axisymmetric temperature field reduces the thermal stress and coiled construction reduces production costs. The exchanger is formed and coiled in a continuous process.

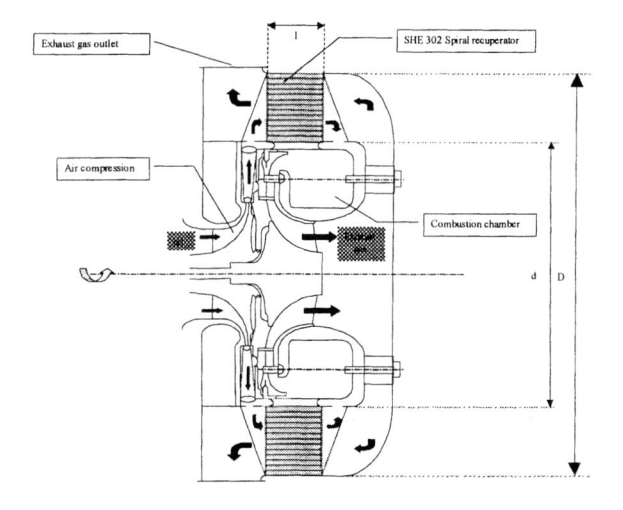

Fig. 1 : gaz turbine

DESCRIPTION OF THE CONCEPT

The heat exchanger has a cylindrical shape. It is a primary surface exchanger, it is compact and stationary. The exchange area is as high as 1600 m2/m3 and the hydraulic diameter is as small as 1.03 mm. The heat transfer fluids enter, circulate, and exit the heat exchanger in a counterflow manner in a direction substantially parallel to the coil longitudinal axis. The flow is distributed inside the coil which includes two distributing areas and a heat transfer area like a plate heat exchanger. The high degree of turbulence created by the corrugated sheets promotes the heat transfer while considerably reducing the fouling.

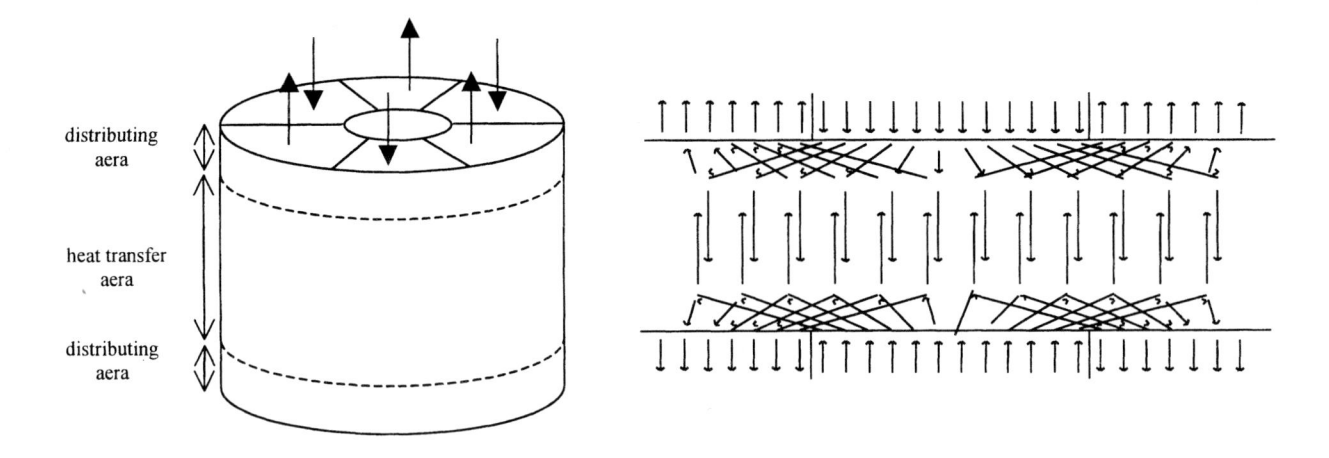

Fig. 2 : Spiral Heat Exchanger and flow distribution

MANUFACTURING TECHNIQUE

The material is an austenitic stainless steel. The core of the S.H.E. is formed by coiling a pair of corrugated sheets. Each sheet a and b have the following sizes : 300 mm wide and 0.1 mm thickness. Sheet a is corrugated in roller a. The sheet a has corrugations extending parallel to the longitudinal axis of the exchanger core and two edges. The corrugated sheet b has corrugations extending perpendicularly to the longitudinal axis. The sheet b also has two edges. From corrugated rollers, sheet a and b are passed through a roller that connects the sheets for seam welding along their edges. The double sheet is then dot welding on the contacts points. The double sheet has side walls 1.2 mm high. A first set of openings is formed by flattening the wall along periodic intervals of the lenght of the double sheet. The double sheet is coiled and welded to form a cylindrical core, with the core having the inner face formed by a wall. The second set of openings is created by cutting the edges of the wall. The S.H.E is 100 % austenitic stainless steel and 100% laser welded.

The cold heat transfer fluid (air) passes inside the double sheet (because of the second set of openings) and hot fluid (gas) passes between two double sheets (because of the first set of openings). The use of sheets as thin as 0.1 mm lowers the metal mass of the recuperator, allowing a reduced weight as well as a faster respond time to operating condition changes.

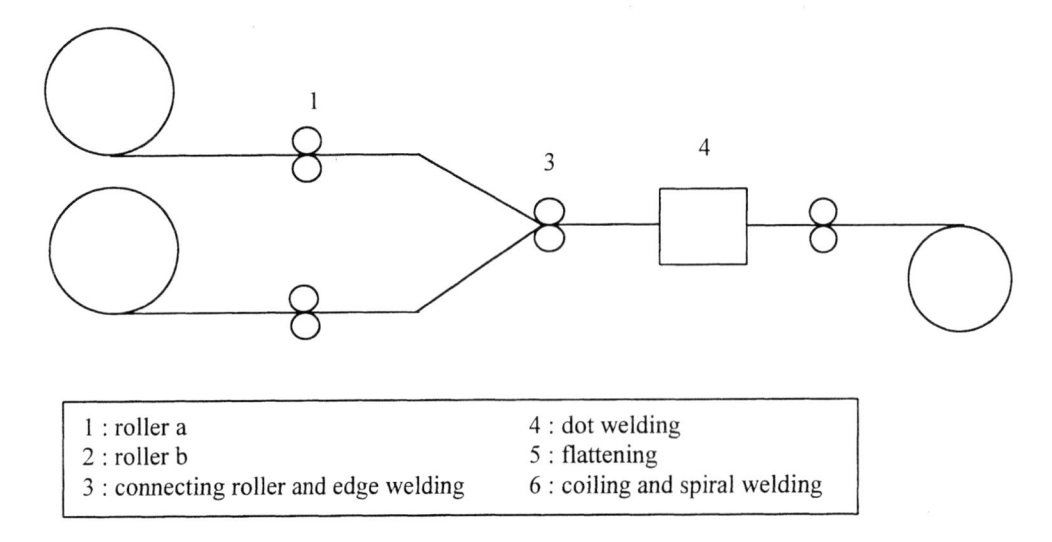

1 : roller a
2 : roller b
3 : connecting roller and edge welding
4 : dot welding
5 : flattening
6 : coiling and spiral welding

Fig. 3 : Manufacturing procedure

WELDING TECHNIQUE

Welding is performed with a laser that allows a small spot weld. The width of the bead at the top of the weld (0.45 mm) is greater than that deeper into the weld (0.12 mm). The material melts because of the absorption and the thermal conduction of the laser beam. This technique doesn't need additional material. Consequently, the maximum operating temperature for the SHE only depends on the stainless steel behaviour and not on the brazing material. The spot is generated by a single laser pulse. The advantage of the laser welding is the quality of the weld that doesn't change, the beam fires always the same way. There is no contamination from an electrode material. A continuous weld is formed by a succession of overlapped spots. The laser fires pulses continuously and the sheet is moved relative to the focused laser spot. The spot welds formed by each pulse overlap to produce a seam weld.

CONCLUSION

The concept of the SHE is perfectly adapted to the gas turbines but it can also serve other applications mechanically less demanding. A significant point of this concept is that the exchanger is manufactured continuously

REFERENCES

Antoine H., 1997, Echangeur de chaleur spirale, European Patent EP 0798527

Mc Donald C.F., 2000, Low cost recuperator concept for microturbine applications, ASME paper 2000-GT-0167

REDUCED TOTAL LIFE CYCLE COSTS USING HELIXCHANGER™ HEAT EXCHANGERS

Bashir I. Master[1], Krishnan S. Chunangad[2], Bert Boxma[3], Graham T. Polley[4] and Mohamed B. Tolba[5]

[1] ABB Lummus Heat Transfer, 1515 Broad Street, Bloomfield, NJ 07003, USA; bashir.i.master@us.abb.com
[2] ABB Lummus Heat Transfer, 1515 Broad Street, Bloomfield, NJ 07003, USA; krishnan.s.chunangad@us.abb.com
[3] ABB Lummus Heat Transfer B.V., Oostduinlaan 75, 2596 JJ The Hague, The Netherlands, bert.boxma@nl.abb.com
[4] Pinchtechnology.com, 96 Park Road, Swarthmoor, Ulverston LA12 0HJ, UK, gtpolley@pinchtechnology.com
[5] ABB Lummus Heat Transfer, 1515 Broad Street, Bloomfield, NJ 07003, USA; mohamed.b.tolba@us.abb.com

ABSTRACT

Initial Capital Cost is usually the primary concern to project managers of new plants or major plant upgrades. On Lump Sum TurnKey (LSTK) projects, major effort is made to control the capital costs of new equipment. The purpose of this article is to present two arguments. The first is that the full economic benefit of heat exchanger technology can only be achieved if consideration of that technology is made at the very onset of process design. The second is that full economic benefit involves consideration of the Life Cycle Costs of a heat exchanger and not just the cost at the shipping dock of the fabrication shop. We will substantiate this using the experience of the application of one exchanger type: the Helixchanger heat exchanger. However, the same arguments may apply to a range of heat exchanger technologies.

INTRODUCTION

A very large percentage of global heat transfer processes are served by the industry work-horse called the 'shell and tube' heat exchanger. A shell and tube heat exchanger offers reliable mechanical structure while retaining the flexibility in diverse process conditions. System pressures of up to 250 bar and process temperatures of up to 800°C are not uncommon in shell and tube heat exchanger applications. The mechanical design of a shell and tube heat exchanger offers the flexibility in operation and the ease of maintenance during downtime. This type of heat exchanger also offers one of the most economic housing of heat transfer surface.

While rugged in design, the conventional shell and tube heat exchanger uses segmental plate baffle geometry on the shellside to support the vulnerable tube bundle from flow-induced forces while achieving high convective flow velocity across the tubes. Conflicting goals of achieving high performance and reliability compromises the overall economy achieved in using the conventional heat exchanger for many process applications.

In a HELIXCHANGER™ heat exchanger, quadrant-shaped baffle plates are positioned on the shellside at an angle to the tube axis creating a helical flow pattern. Higher conversion of pressure drop to heat transfer is achieved at an optimum helix angle. The goals of achieving higher performance to achieve economy and reliability do not work against each other, as in conventional heat exchangers using segmental baffles. Near plug flow conditions are achieved with helical baffle geometry reducing significantly the fouling rates on the shellside. Longer run lengths are achieved when using a Helixchanger heat exchanger for sustained high performance and greater economy over a life cycle.

IDENTIFICATION OF APPROPRIATE HEAT EXCHANGER TECHNOLOGY

Here we will briefly address two features of proper identification of exchanger technology for a given task: the need to consider the alternatives at an early stage of a project and the need to understand the limitations of the technologies under consideration.

The process engineer is becoming ever more dependent upon computer software. Unfortunately, the programs available (process simulators, pinch analysis programs and exchanger design programs) nearly all center on the use of conventional shell and tube heat exchangers. The result can be that process design is conducted on the basis that this is the type of exchanger that will be used for the plant. The possibility of using other technologies is then only considered when the process specifications are sent out to suppliers of these alternative exchanger types.

What is the problem here? Well, the heat recovery level for a plant is often set on the trade-off between equipment capital cost and operating cost. The identified heat recovery level then has an influence on the heat recovery network structure. If the initial cost assumptions are wrong, the final design is wrong.

The real capital cost of a heat exchanger involves a consideration of the cost of installation. This cost is a function of equipment size and complexity (for example, how many individual heat exchangers in series and/or parallel are required for the heat recovery unit). Complexity is a function of the chosen technology. For instance, when a conventional shell-and-tube unit is used, the size limitations can result in the need for additional shells and velocity considerations can result in a need to use multiple tube-passes with temperature cross considerations then requiring the use of multiple shells-in-series. Both of these problems can be overcome through the use of heat transfer enhancement. The Helixchanger heat exchanger provides improved shell-side heat transfer for a given pressure drop, so higher duties can be accommodated in a single shell. This advantage can be consolidated by using Helifin heat exchanger (a design which combines the benefits of helical baffles and low finned tubes). On some applications we have used tube inserts to improve tube-side performance as well as helical baffles to improve shell-side performance. The following two cases illustrate such applications.

Case-1 Platform Gas Cooling with Sea Water using HELIFIN™ Heat Exchanger

This case represents a classic well-stream hydrocarbon two-phase flow cooling application using sea water as coolant. See Table 1.

Unit is proposed for installation on an offshore platform with very expensive plot space. Sea water cooling also requires using titanium tube bundle.

Preliminary design prepared for this unit was with bare tube bundle and conventionally baffled unit. Further optimization lead to the use of integrally-finned tube bundle using still the segment baffles on shellside.

A Helifin option was developed in this case which reduced the equivalent bare tube surface by one half with significant reduction in the required plot space. Helical baffles on the shellside offered an additional advantage by reducing the gas pressure drop by one half of that required in the original segment baffled bare tube design. See Fig. 1.

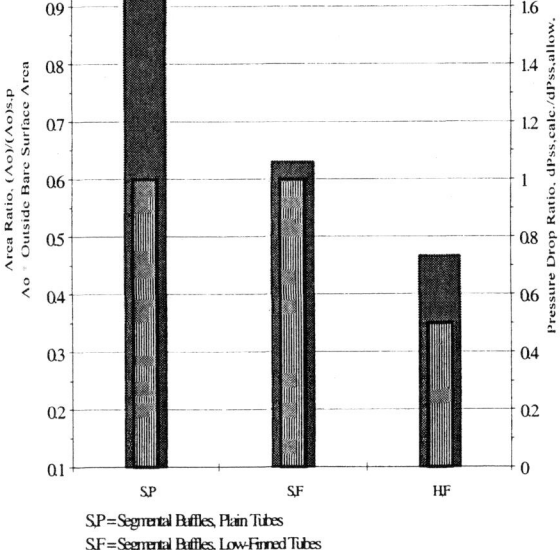

Fig. 1 Platform Gas Cooling with Sea Water

S,P = Segmental Baffles, Plain Tubes
S,F = Segmental Baffles, Low-Finned Tubes
H,F = Helical Baffles, Low-Finned Tubes

Table 1. Platform Gas Cooling with Sea Water

PERFORMANCE																
Unit Type[1]	Shellside Fluid (WellStream HCs)						Tubeside Fluid (Sea Water)						Heat Duty Q (MMBtu/hr)	Q / Q$_{S,P}$	dP$_{SS,calc}$ / dP$_{SS,allow}$
	Flowrate (1000 lb/hr)	Wt. Fr. Vapor In	Wt. Fr. Vapor Out	Temp. (°F) In	Temp. (°F) Out	dP$_{SS}$ (psi) calc. / allow.	Flowrate (1000 lb/hr)	Wt. Fr. Vapor In	Wt. Fr. Vapor Out	Temp. (°F) In	Temp. (°F) Out	dP$_{TS}$ (psi) calc. / allow.			
S,P	240	0.43	0.35	240	110	10 / 10	1200	0	0	75	95	7 / 10	23.0	1.0	1.0
S,F	240	0.43	0.35	240	110	10 / 10	1200	0	0	75	95	10 / 10	23.0	1.0	1.0
H,F	240	0.43	0.35	240	110	5 / 10	1200	0	0	75	95	10 / 10	23.0	1.0	0.5

GEOMETRY												
Unit Type[1]	Orientation	No. of shells Series / Parallel	Shell Type	Shell ID (in.)	Shell Matl.	Tube Type	Tube OD/Avg. wall/Pitch/Layout (in)	Tube Length (ft)	Tube Matl.	Baffle Type	Outside Bare Surface, A$_o$ (ft²)	A$_o$ / (A$_o$)$_{S,P}$
S,P	Horizontal	1 / 1	AEU	36	CS	Plain	0.75 / 0.049 / 1.0 / 45°	24	Titanium	Single Seg.	3685	1.00
S,F	Horizontal	1 / 1	AEU	32	CS	Finned[2]	0.75 / 0.049 / 1.0 / 45°	20	Titanium	Double Seg.	2325	0.63
H,F	Horizontal	1 / 1	AEU	31	CS	Finned[2]	0.75 / 0.049 / 1.0 / 45°	16	Titanium	Single Helix	1722	0.47

(1) S,P = Segmental Baffles, Plain tubes
    S,F = Segmental Baffles, Low-finned tubes
    H,F = Helical Baffles, Low-finned tubes
(2) 36 fins per inch

The characteristics of helical baffles with minimum back-mixing and no stagnant flow areas on shellside is expected to increase the run-length for this unit between scheduled downtime for cleaning the tube-bundle.

Case-2 Crude Preheating with Combined Enhancement Technologies – HiTran inserts and Helical Baffles

The use of advanced heat transfer technologies offers the prospect of both substantially improving heat transfer and eliminating fouling due to the deposition and thermal degradation. For new heat exchanger design, not only the exchanger capital cost can be reduced, but also the decreased size, weight and complexity can reduce associated expenditure, contributing to reduced total life cycle costs. Furthermore, if technological improvements can be applied to the design of both sides of a shell and tube heat exchanger, maximum operational and installed cost benefits are achieved.

In a first collaboration of its kind, two enhancement technologies, each proven in its own right, were combined to deliver an innovative design for a refinery in Germany, in exchanging heat from vacuum residue to crude oil.

Tube inserts called HiTran (supplied by Cal Gavin Process Intensification Engineering) were used in this case with helical baffles on the shellside to deliver the heat duty in a single shell where otherwise two shells were required using conventional options. See Table 2.

Table 2. Crude Preheating with Combined Enhancements

	Conventional S&T Exchanger	HiTran-Helixchanger
Type	AES	AES
Shellside fluid	Vac. resid-gas oil	Crude oil
Number of shells	2	1
Shell ID, mm	1050	991
Baffle type	Single segmental	Single helix
Baffle arrangement	27 x 150 mm	10°
Shellside ΔP, kPa	50	40
Tubeside fluid	Crude oil	Vac. resid-gas oil
Tube OD, mm	25	25
Tube length, mm	4880	4880
Tube passes	4	2
Tube elements	None	HiTran
Tubeside ΔP, kPa	50	50
Total surface, m^2	502	236
Exchanger cost, $	140,000	120,000

The opportunity to combine tubeside enhancement with helical baffle technology should be carefully considered on a case-by-case basis. Particularly, where a high degree of fouling due to thermal degradation or solids deposition may be anticipated both inside and outside the tubes, the combination is very cost effective.

Case-3 Crude Preheat Application: Reduced Fouling

Figures 2 and 3 represent a case in which a refinery in the Netherlands employed one Helixchanger tube-bundle in a heat exchanger, in series with two conventionally baffled heat exchangers, in 1996. Over a three-year operation, this refinery saved approximately US$ 18,000 by not having to clean the Helixchanger tube bundle each year as required for the conventional tube-bundles.

Fig. 2 Conventional Segmental Tube-Bundle with Fouling Accumulation in the Stagnant Areas

Fig. 3 Helixchanger Tube-Bundle with little or no Fouling

Now let's consider the importance of appreciating the limitations of exchanger technology. We will consider the case of an exchanger used to cool lube oil. Conventional wisdom would suggest that the lube oil should be flowing through the shellside and water through the exchanger tubes. This results in better heat transfer on the lube oil side.

We know of one shipboard unit that performs very well in tropical waters. However, in the North Sea the cooling

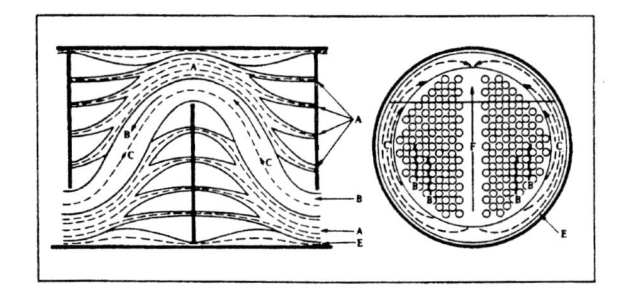

Fig. 4 Conventional S&T Exch.: Shellside Stream Analysis

system does not perform well. What is the problem?

The mechanical clearances required for the fabrication of segmentally baffled exchangers result in the production of a number of alternative flow paths through the shell-side of the exchanger. See Fig. 4. In terms of this problem, the most significant of these is the clearance between the bundle and the shell. This clearance can give rise to a significant quantity of the oil flow bypassing the tube bundle. The viscosity of lube oil increases sharply as the temperature of the oil falls and the oil flowing through the bundle can be much more viscous than that flowing through the bypass. This encourages more bypassing. The effect becomes more pronounced as the temperature within the bundle falls. So, an exchanger that works well with warm water can perform very badly with cold water.

The flow through the Helixchanger heat exchanger is very different to that through a segmentally baffled unit and is such that it is ideal for use in lube oil cooling duties. Fluid from the periphery of the bundle is systematically drawn into the bundle.

Two lessons can be drawn from the above examples:
- the process engineer needs to consider which exchanger technology should be used, at an early stage
- the process engineer should seek expert advice in making these considerations

ECONOMIC ANALYSIS

Petroleum and chemical process companies make decisions regarding large capital investments only after extensive economic analysis has been made. Such analysis can typically involve determination of investment required, determination of revenues and expenditures over the life time of the plant and evaluation of the cost of decommissioning the plant at the end of its working life. Such analysis often involves a consideration of the 'time value' of money. Decisions on overall investment are made on the basis of Total Life Cycle Cost.

If heat exchanger users make overall investment decisions on this basis, surely there is a strong argument for the smaller decisions, required for the execution of a project, to be made on a similar basis. As we will show, this approach does not require repetition of complex economic analysis. Neither does it require the use of process simulation and the analysis of complex operating data. The objective is to gain an appreciation of the total cost picture by deriving cost numbers that are used for comparative purposes, for example, comparing the cost of one exchanger technology with another. The engineer uses his own judgement to assess the accuracy of the figures.

In our view, the Total Life Cycle Cost Analysis (TLCCA) involves evaluation of the following:
- purchase cost
- installation cost
- operating costs
- maintenance/downtime costs
- disposal cost

The above cost figures often use separate bases. For example, purchase price is an absolute cost figure and operating benefit is a cost per unit time. We can overcome this problem by annualizing the capital cost. The disposal cost is a 'future' demand for capital. We can deal with this by determining its present worth and then annualizing the resultant value.

Purchase Cost. This can be obtained from the supplier as a budget cost per unit for a given duty.

Installation Cost. This is the cost associated with installing the exchanger as part of a plant. It covers piping costs and the costs of foundations and supporting structures. It would be inaccurate to estimate this cost by simply multiplying the exchanger purchase price by a installation factor. The cost is dependent upon unit size and complexity.

We suggest that factors be developed which relate exchanger size with installation cost and these factors be applied to each exchanger making up the overall unit.

Operating Costs. The cheaper the heat exchanger, the higher is the economically justifiable heat recovery. This cost trade-off can be quickly determined using process integration software (such as the INTEGRITY computer program issued by ESDU International Ltd.)

Operating benefit should be established over the operating life of the unit. In crude oil applications, the Helixchanger heat exchanger exhibits a much lower fouling propensity than that of conventional units. This means that heat recovery is maintained at a higher level. See Fig. 5.

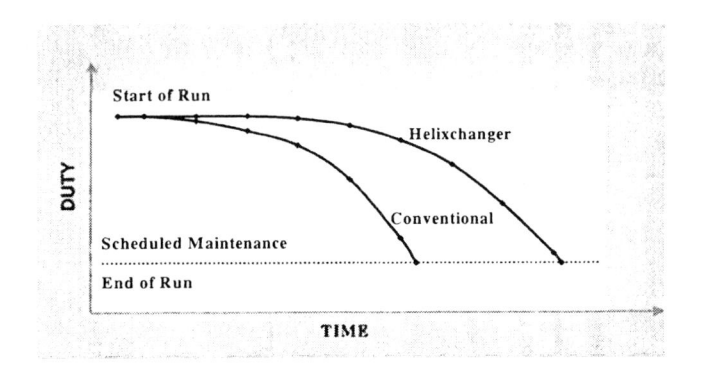

Fig. 5 Run-Length Performance: Helixchanger v/s Conventional S&T Heat Exchanger

Maintenance Costs. Reduced fouling results in longer run times between maintenance outage. The costs associated with maintenance are not just the costs of cleaning and repair. They should take into account the costs of reduced plant throughput. This can be done by determining the throughput that is achieved for both the conventional unit and the alternative technology. The total capital cost of the plant is then divided by each value to determine 'capitalization per unit of throughput'. This figure can be converted to an annual capital cost figure.

Disposal Cost. The cost of disposing of a unit, like installation cost, is dependent upon unit size and complexity. The approach suggested above for installation costs can be adapted for the estimation of disposal costs.

RESULTS

The development of the above described technique is in its infancy. We have explored its potential through case study work. The results of those studies have demonstrated its importance. In Fig. 6, we show the results of an exercise involving a heat exchanger for a new plant. We found the following:

- The purchase price of the Helixchanger was marginally lower than that of the conventional unit.
- The subsequent capital expenditure was very much lower for the Helixchanger.
- The operating costs (here we considered operating cost in terms of energy consumed rather than heat saved) of the Helixchanger were much lower.
- Given the additional run time, the maintenance costs were much lower for the Helixchanger. For this study, we were not able to take into consideration the credit for the gain in production advantage.
- Not unexpectedly, given that the installation cost was cheaper, there would also be benefit in disposal cost.

Figure 7 shows the results of a similar exercise in which a case involving a replacement of a conventional heat exchanger tube bundle with a helically baffled tube bundle are presented. In this case, we found the following:

- The purchase price of the Helixchanger replacement tube-bundle was marginally higher since the surface optimization was restricted by the use of the existing exchanger shell.
- Frequency of replacement of tube-bundle was reduced by using the Helixchanger tube bundle, which reduced subsequent capital cost expenditure.
- The operating costs of the unit with the new Helixchanger tube bundle were much lower due to smaller pressure drops on the shellside and less downtime.
- Longer run time achieved with the Helixchanger tube bundle reflected in much lower maintenance costs. Production gain due to longer run lengths could not be established in this case.

- Again, considering the fact that in the total life cycle, less replacement Helixchanger tube bundles will be required, the associated disposal costs would be significantly lower.

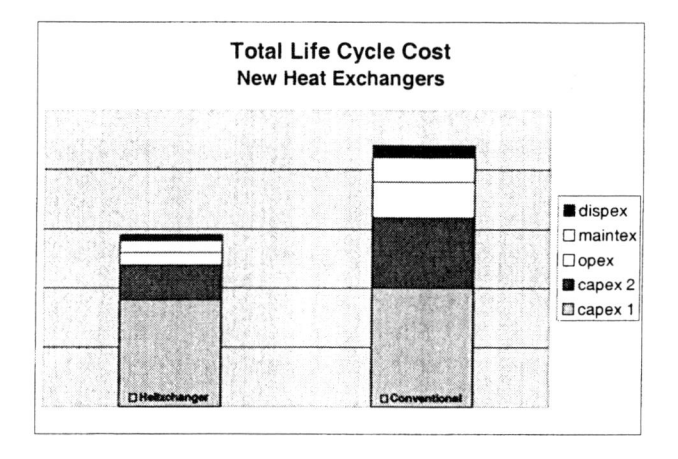

Fig. 6 Total Life Cycle Cost: New Heat Exchangers

The glossary of abbreviated cost terms in the figures is:

CAPEX 1   initial purchase price
CAPEX 2   installation cost and cost of replacement bundles plus associated spares over life cycle
OPEX   operating costs over life cycle, which includes pumping or compressor power costs and costs of chemical dispersants and anti-fouling agents
MAINTEX   maintenance costs over life cycle, which includes costs of cleaning bundles, costs of tube plugging and replacement required due to corrosion or flow-induced vibration damage and loss of production costs
DISPEX   disposal costs over life cycle

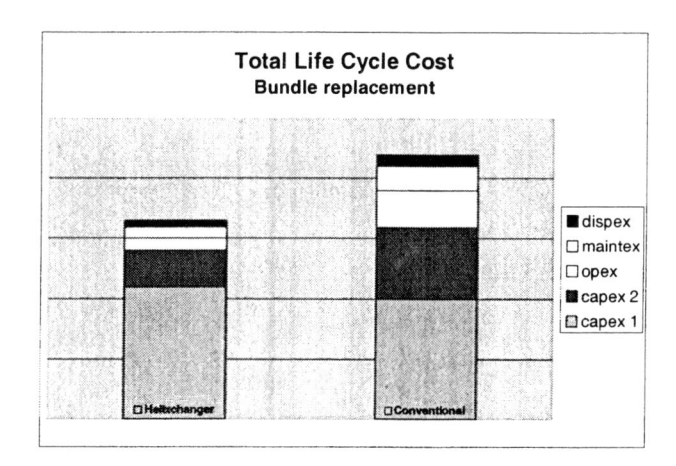

Fig. 7 Total Life Cycle Cost: Bundle Replacement

CONCLUSIONS

A number of important steps are necessary to make the Total Life Cycle Cost analysis become a useful tool in actual decision making while preparing the Basic Engineering Package (BEP). The following actions are identified:
- develop simple means of estimating installed costs from exchanger size and complexity
- develop a similar approach for estimating disposal costs
- consideration of available heat exchanger technology into an integration analysis (such as the program developed by ESDU)
- collect information on the exchanger run-time advantage
- develop a simple software that can bring the elements together in such a way that 'what if' studies can be conveniently conducted

The cost of crude oil fouling was the prime mover in the development of the HELIXCHANGER™ heat exchanger. At the start of the new Millenium, it was estimated that this fouling costs US refineries alone approximately $2.0 billion per year.

Heat transfer conversion advantage for a given pressure drop is further enhanced when using the Helixchangers for turbulent flow applications. More than 500 Helixhangers are currently in service globally in applications ranging from crude preheat trains and feed-effluent exchangers to quench water coolers and secondary transfer-line exchangers. The Helixchanger advantage has been clearly substantiated by the industry feedback obtained on operating units.

For new installations, the Helixchanger heat exchanger can save capital costs as well as operating and maintenance costs. In revamps and upgrades, it can offer a cost-effective solution by replacing the tube bundles in existing shells. See Fig. 8 for a Helixchanger tube-bundle in fabrication. The Total Life Cycle Cost Analysis confirms that using Helixchangers improves the reliability and availability of process plants in a cost-effective way.

REFERENCES

Kral, D., Stehlik, P., van der Ploeg, H. J., and Master, B. I., 1996, Helical Baffles in Shell-and-Tube Exchangers, Part I: Experimental Verification, *Heat Transfer Engineering*, Vol. 17, No. 1, pp. 93-101.

Lutcha, J., and Nemcansky, J., 1990, Performance Improvement of Tubular Heat Exchangers by Helical Baffles, *Trans. IChemE*, Vol. 68, Part A, pp. 263-270.

Palen, J., Taborek, J., and Yarden, A., 1977, Stream Analysis Method for Prediction of Shellside Heat Transfer and Pressure Drop in Segmentally Baffled Exchangers, *Report S-SS-3-1, Heat Transfer Research, Inc.*, College Station, Texas.

van der Ploeg, H. J., and Master, B. I., 1996, Helically Baffled Heat Exchangers – Design Option for Reactor Feed/Effluent Heat Exchange, *Advances in Industrial Heat Transfer*, IChemE, UK.

Storey, D., and van der Ploeg, H. J., 1997, Compact exchanger to reduce refinery fouling, *Petroleum Technology Quarterly*, Autumn 1997, pp. 88-89.

van der Ploeg, H. J., and Master, B. I., 1997, A new shell-and-tube option for refineries, *Petroleum Technology Quarterly*, Autumn 1997, pp. 91-95.

Chunangad, K., O'Donnell, Jr., J., and Master, B., 1997, Helifin Heat Exchanger, *Proc. of the International Conference on Compact Heat Exchangers for the Process Industries, 1997*, Begell House, New York, pp. 281-289

Chunangad, K., and Master, B., 1998, Helixchanger: Performance Enhancement with Greater Reliability, *First Int. Symp. on Innovative Approaches For Improving Heat Exchanger Reliability, 1998*, Materials Tech. Inst. of the Chemical Process Industries, Houston, Texas, pp. 12-21.

Chunangad, K. S., Master, B. I., Thome, J. R., and Tolba, M. B., 1999, Helixchanger Heat Exchanger: Single-Phase and Two-Phase Enhancement, *Proceedings of the International Conference on Compact Heat Exchangers and Enhancement Technology for the Process Industries, 1999*, Begell House, New York, pp. 471-477.

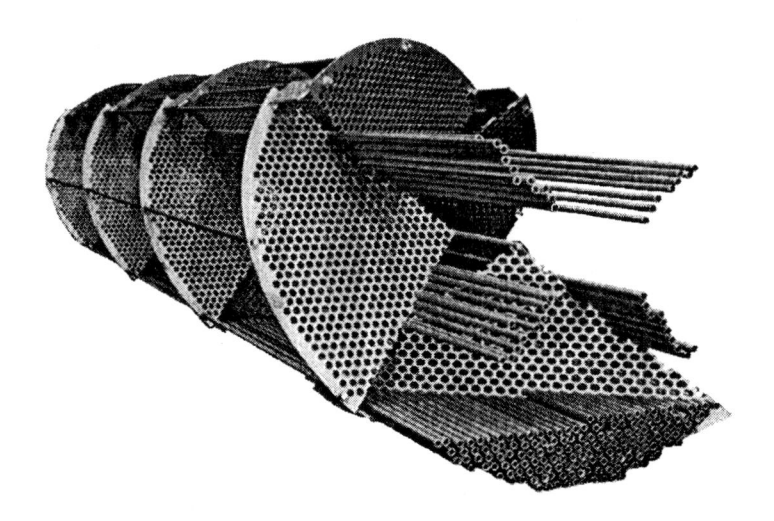

Fig. 8 HELIXCHANGER™ Tube-Bundle in Fabrication

Micro Channel Heat Exchanger for Cooling Electrical Equipment

Koichiro Kawano[1], Masayuki Sekimura[1], Ko Minakami[1], Hideo Iwasaki[1] and Masaru Ishizuka[2]

[1] Toshiba R&D Center, Kawasaki, Japan; E-mail: koichiro.kawano@toshiba.co.jp
[2] Toyama Prefectural University, Toyama, Japan; E-Mail: ishizuka@pu-toyama.ac.jp

ABSTRACT

The micro channel heat exchanger, which has many small channels (10-100 μm order), is a promising device to cool an integrated circuit, which generates a large amount of heat. In order to investigate the performance of the micro channel heat exchanger, three-dimensional numerical simulations and experiments on heat transfer behavior and pressure loss were carried out. So far as the heat transfer phenomena is concerned, results obtained using a silicon chip micro channel model a very small thermal resistance, about 0.1(Kcm^2/W) at Re=300. And, in the designing of heat exchanger, the pressure loss phenomenon is of practical importance. In the present study, measured pressure loss was good agreement with that of analytical result obtained on the basis of fully developed laminar pipe flow assumption. And, statistics analysis was also performed in order to confirm the manufacture accuracy of the micro channel size by KOH anisotropy etching. And, based on the statistics result, the error range of pressure loss data was determined.

Furthermore, a practical setup was made with a micro channel heat exchanger to clarify the possibility of using the micro channel heat exchanger in electrical equipment. As a result, it was confirmed that the performance of the micro channel heat exchanger system is sufficient to cool a silicon chip which generates a large amount of heat, and the scale of the system is compact compared to that of the whole setup of electrical equipment. .

INTRODUCTION

In recent years, the amount of heat generated from electronic circuits in LSI chips has been increasing. Therefore, it is important to cool LSI chips in order to ensure high reliability of electrical equipment. The micro channel heat exchanger, which was introduced by Tuckerman and Pease (1981, 1982), is a promising high performance cooling method. Because coolant directly contacts the reverse side of a chip, the reduction in thermal resistance is much greater than in the case of other techniques applied on an LSI Package.

In the design of a heat exchanger, the heat transfer and the pressure loss phenomenon are of practical importance. After the introduction of the micro channel by Tuckerman and Pease (1981), various researchers reported on the thermal and flow phenomena of the micro channel heat exchanger, and proposed the methods to design it. For example, Weisberg et al. (1992) provided the analytical results of heat transfer phenomena by solving the energy equation, to design and optimize the micro channel heat exchanger cooling system.

Some researchers reported the eccentric phenomena with the micro channel which has an influence on the design of the practical heat exchanger. Pfahler et al. (1990) reported that the pressure loss of a micro channel could not be predicted by the Navier-Stokes equations, because of non-Newtonian flow behavior or other micrometer order effects. Not only the pressure loss but also the heat transfer phenomena were different from the behavior of the large-scale channels (Choi et al. (1991), Peng et al. (1996)). Peng et al. (1995) provided that flow and heat transfer phenomena are strongly affected by the coolant temperature and velocity, and turbulent transition was shown in the low Re region in the micro channel whereas it was not shown in the large scale channel. Mala et al. (1997) explained these eccentric phenomena in the micro channel by the interfacial effects such as interfacial electric double layer (EDL).

The micro channel heat exchanger has the weak points such as the high pressure loss and the large temperature gradient, in applying it to a practical cooling system. To overcome these problems, the manifold micro channel heat exchanger was introduced by Harpole and Eninger(1991). And, Copeland et al. (1995) proposed the theory and reported on experiments for this type.

In the present work, experiments and numerical simulations were performed to investigate the above-mentioned problems (the eccentric heat transfer and flow phenomena, the high pressure loss, the large temperature gradient, etc.) which were concerned with the design of the micro channel heat exchanger.

In the experiments, the micro channel heat exchangers (15mm*15mm area) were made which have 110 channels (width and Height are about 60μm* 150~370μm), and water was used as the coolant. To clarify the precise performance of heat exchanging, thin Fe-Ni thermocouples on the surface of a silicon chip were made by the sputtering technique. Based on the assumption of fully developed laminar flow, three-dimensional numerical simulations were carried out by using the energy equation, and compared the results with those of experiments. The experimental data of pressure loss was also compared with the theoretical prediction. And, the manufacture accuracy of micro channels is also necessary, when developing practical micro channel heat exchanger. Since micro scale machining is needed in the manufacturing process of micro channels, it is important to know this manufacture accuracy when evaluating the pressure loss and heat transfer behaviors of micro channel heat exchangers. Therefore, the manufacture accuracy of the micro channels at the time of using KOH anisotropy etching of silicon chips was estimated by statistics analysis.

Furthermore, to clarify the practical application of the micro channel heat exchanger to the heated chip, the experimental setup of the cooling system by using the micro channel heat exchanger was made for cooling electrical equipment such as VLSI or MCM (Multi Chip Module). The setup consisted of the micro channel heat exchanger, a pump which supplies coolant to micro channels, and other electrical components. The performance of heat exchanging and the reliability of the cooling system were tested using this setup.

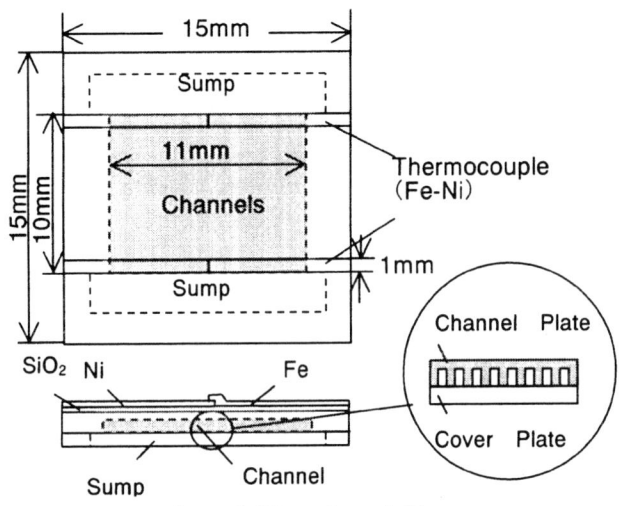

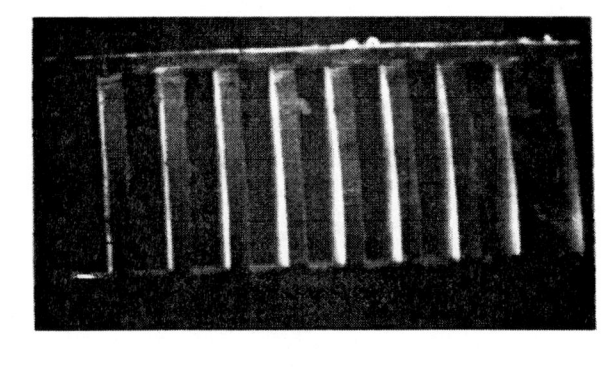

Figure 1 Micro channel chip

Figure 2 Cross-sectional view of micro channel chip

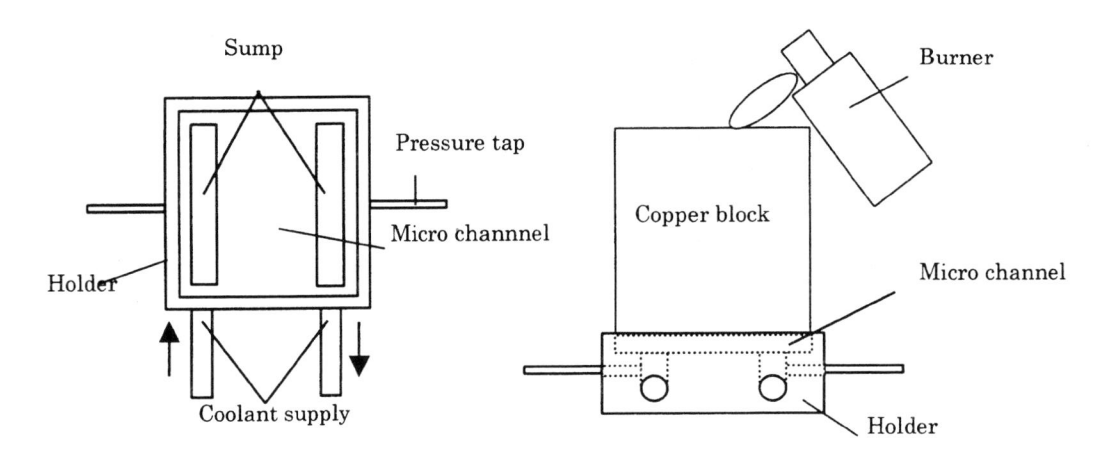

Figure 3 Experimental apparatus

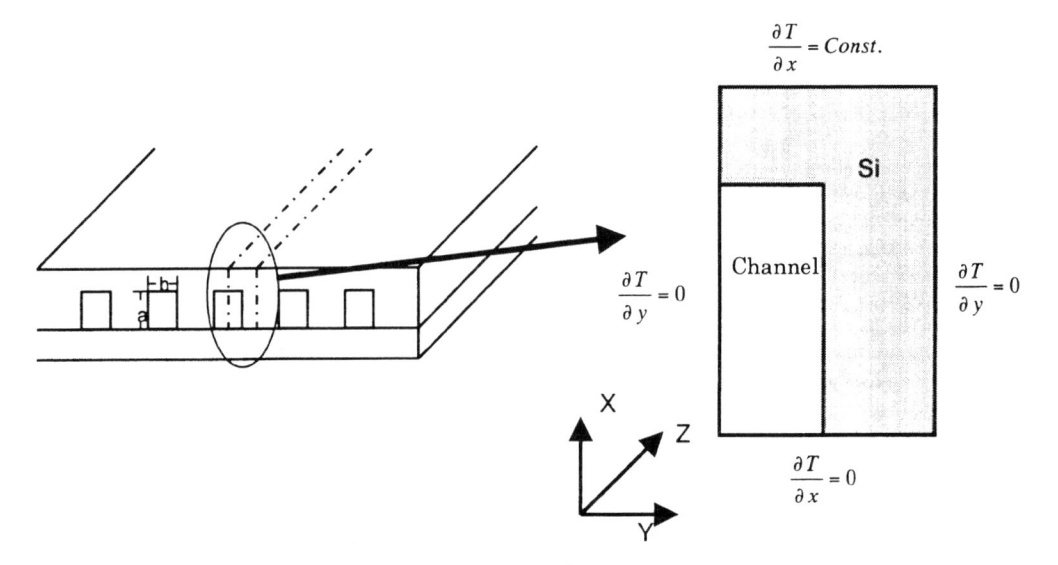

Figure 4 Numerical model

BASIC BEHAVIOR OF HEAT TRANSFER AND PRESSURE LOSS

Experimental setup

Experiments and three-dimensional numerical simulations were performed to obtain heat transfer behavior and pressure loss of the micro channel heat exchanger. Figure 1 shows a silicon chip with the micro channel heat exchanger. This consists of two chips, a silicon chip with micro channels and a cover plate with sumps. The micro channel chip (15mm*15mm area) has 110 channels with 100 μm pitch, and the size of the channels are 57μm*180μm and 57μm*370μm. The cover plate chip, which is silicon, has two sumps to supply coolant. The cover plate and the micro channel chip were made using KOH anisotropy etching. These chips were directly attached together by the direct bonding (molecular diffusion) technique. Figure 2 shows a photograph of a cross-sectional view of the micro channel chip. To obtain the precise temperature distribution on the surface of the micro channel chip, two thin film thermocouples were made at the inlet and the outlet by using the sputtering technique. The size of a thermocouple was 1mm in width and 0.2μm in thickness. For electrical insulation, a thin layer of SiO_2 was made between the silicon chip and the thermocouples. Temperature distribution can be measured along a channel by these thermocouples, and the thermal stress in the chip can be estimated.

Figure 3 shows the experimental apparatus for basic behavior of heat transfer and pressure loss. The micro channel heat exchanger was set on the holder. And, the coolant (water) was supplied from the waterworks. Flow rates of the coolant was controlled by the valve and the flow meter. To input a large amount of heat to the chip, a copper block, which was contacted to the chip, was heated by a gas burner. Large heat flux (about 100 W/cm^2) could be input by this method. The precise value of input heat was calculated by measuring flow rates and temperature rise of the coolant. The pressure loss between the inlet and the outlet of the channel was measured using a differential pressure transducer through the pressure taps.

Table 1 Properties used in numerical simulation

Property	Silicon	Water(coolant)
ρ (Kg/m^3)	2330	996.62
Cp (kJ/(kg·K))	0.713	4.179
λ (W/(m·K))	148	0.610

at T=300(K)

Numerical Simulation and theoretical prediction

In order to test these experimental data, three-dimensional steady state numerical simulations were performed by solving the energy equation.

Energy equation (Steady state)

$$\rho Cp\, w \frac{\partial T}{\partial z} = k \left(\frac{\partial^2 T}{\partial x^2} + \frac{\partial^2 T}{\partial y^2} + \frac{\partial^2 T}{\partial z^2} \right) + q \quad (1)$$

The temperature field of silicon and coolant was calculated together. Figure 4 shows the geometry and boundary conditions that were used in the numerical model of the micro channel. The velocity distribution of coolant flow in the channel was assumed to be fully developed laminar flow in a rectangular duct (Weisberg et al. 1992).

Velocity distribution

$$w = -\frac{\rho 16 b^2 Pz}{\pi^3 \nu} \sum_{n=1}^{\infty} \frac{(-1)^{n-1}}{(2n-1)^3} \frac{\cosh(mx)}{\cosh(ma)} \cos(my) + \frac{\rho Pz}{2\nu}(b^2 - y^2) \quad (2)$$

$$m = (2n-1)\pi / 2b \quad (3)$$

Properties of silicon and coolant (water) used in the calculations are shown in Table 1. As for kinematic viscosity of coolant, the average of the inlet and the outlet value of channel was used in the calculations, since kinematic viscosity of coolant (water) is strongly dependent on temperature. Using these properties, equation (1) was calculated by the SIMPLE method (Patankar, 1980).

The measured pressure loss (friction factor) was compared with the theoretical predictions which assume fully developed laminar flow in rectangular duct. Friction factor is defined as (Ward-Smith(1980)),

$$f = \frac{2 \cdot \Delta P \cdot de}{\rho L w_m^2} = \frac{C}{Re} \quad (4)$$

C is the Poiseuille number that depends on aspect ratio of cross section of the channel. When the flow field is fully developed laminar flow, C is a constant value.

$$C = \frac{96}{(1+\varepsilon)^2 \left[1 - \frac{192\varepsilon}{\pi^5} \sum_{n=1}^{\infty} \frac{\tanh\{(2n-1)\pi / 2\varepsilon\}}{(2n-1)^5} \right]} \quad (5)$$

Table 2 Average value and standard deviation of the size

Width×Height	57 μm×180 μm	57 μm×360 μm
Average (μm)	56.6, 180.9	57.4, 360.4
Standard dev. σ	2.86, 4.49	3.19, 5.95
95% ±2σ	±5.72, ±8.98	±6.38, ±11.90

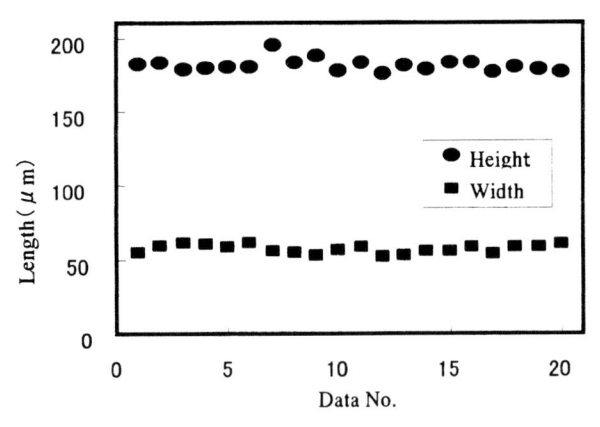

Figure 5 Height and width of channels (57×180 μm)

Figure 6 Regular distribution of channel size (57×180 μm)

Manufacture accuracy

In order to investigate the manufacture accuracy of the channel size, the height and width of channels at several points were measured under a microscope. Figure 5 shows the dispersion of the channel size data about the height and width. This dispersion was caused on the roughness of channel wall which originated in the slight directional deviation between mask pattern of channels and crystal direction of Si in KOH anisotropy etching process. This dispersion of channel size will be included in the data of pressure loss as errors gathered using the average value of the data. In order to determine the range of the channel size accuracy from the dispersion, the average and the standard-deviation σ were calculated. Table 2 shows the channel size used in the experiments with error estimations. Figure 6 shows the normal-distribution from the standard-deviation of Table 2 in setting the average to 0. If assumed that the measurement data follows the normal distribution, 95% of the channel size data will be contained in the range of the average value ± 2σ. From figure 6, the dispersion in measured channel size will become large if the aspect ratio of channel becomes large.

However, since the average and standard-deviation σ which were shown in the present study may change with the manufacture procedure of the channels, they do not become this value necessarily by KOH anisotropy etching process.

Results and discussion

Figure 7 shows the Poiseuille number C (Eq. (4)) which indicates pressure loss of channel flow. The micro channel chip was not heated when pressure loss was measured. In Fig. 7, for 0<Re<200, the measured pressure loss was in good agreement with that of the analytical result obtained based on the assumption of fully developed laminar pipe flow. Slash parts in the Figure 7 are the error ranges (95%) derived from the standard deviation shown in Table 2. The range of the pressure loss data including error values was decided from where aspect ratio becomes large (average value+2σ) and small (average value -2σ) in Table 2. This

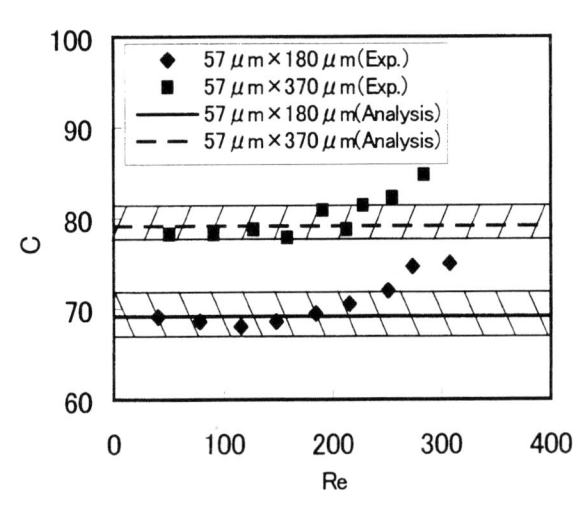

Figure 7 Pressure loss

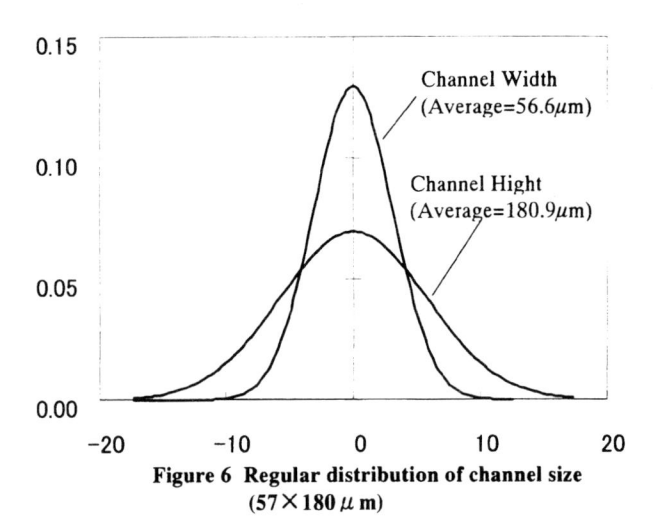

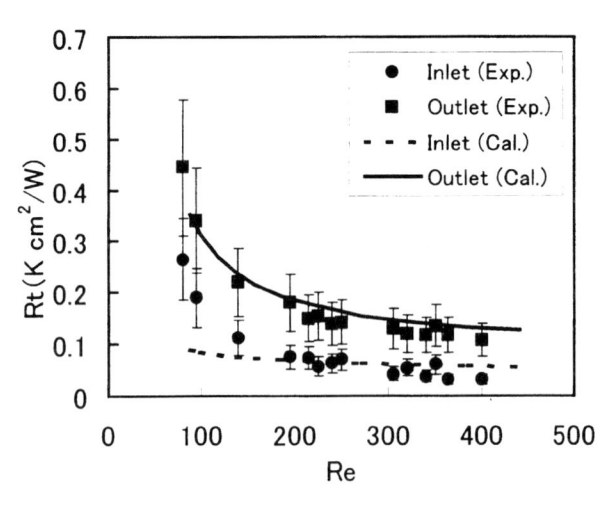

Figure 8 Thermal resistance (57 μm × 180 μm)

result suggested that the flow pattern in the microchannel could be assumed to be fully developed until Re=200. For Re>200, C value of the experiments is slightly higher than that of theoretical prediction.

Figure 8 shows the thermal resistance of the experiment and the numerical simulations, which is derived from the temperature distribution on the surface of the chip. The thermal resistance is defined as,

$$R_t = \frac{\Delta T}{q} = \frac{T_s - T_i}{q} \quad [Kcm^2/W] \quad (6)$$

As shown in Fig.8, small thermal resistance, about 0.1(K cm^2/W) at Re=300, can be observed. For 0<Re<200, thermal resistance of experiment at inlet portion was quite different from that of numerical result. Because the viscosity of the coolant greatly depends on the temperature, it is difficult to estimate it's precise value at the inlet portion where the temperature gradient is large. The difference of the thermal resistance between the inlet and outlet portion was about 0.1 Kcm^2/W, therefore the temperature difference was about 10K when the heat flux was 100 W/cm^2. If 790W/cm^2 (Tuckerman and Pease, 1981) can be input to the micro channels, the mean temperature gradient in the chip will be 79 K/cm. From these results, the thermal stress in the chip is large, and has a possibility to beak an electric pattern of the chip. This must be taken care to design the micro channel heat exchanger.

PRACTICAL APPLICATION OF MICRO CHANNEL HEAT EXCHANGER

Experimental setup

On the basis of the discussion in section 2, practical LSI cooling equipment using the micro channel heat exchanger was set up. In this section, the heat exchanger with the channels (57μm*300μm) was used. Figure 9 shows a coolant-supplying holder of the micro channel heat exchanger. Four micro channel chips were attached to this holder by epoxy adhesive, and this simulated an MCM which had several LSI chips. Coolant flow in the holder is also shown in Fig.9.

In choosing the parts of the cooling system, thermal resistance and pressure loss must be considered. In the present study, the cooling system was designed so that the mean value of thermal resistance of the chip was about 0.1 (K cm^2/W). To estimate the thermal resistance of the heat exchanger, the numerical simulation was performed. From the numerical results shown in Fig.8, the mean value of the thermal resistance is achieved to 0.1 (K cm^2/W) at Re =150. At this condition, the pressure loss and the flow rate of the channels were about 0.1 (MPa) and 2.5 (cm^3/s), respectively. Therefore, those of the holder were about 0.2 (MPa) and 5.0(cm^3/s) (2*2 chips, Fig.9). The pressure loss value of the whole cooling system was thought to be almost the same as that of the holder, and the pump was chosen. Table 3 shows the specifications of the parts. The tube was chosen as compact as possible.

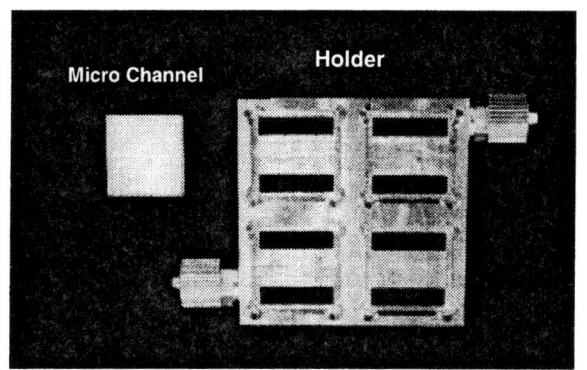

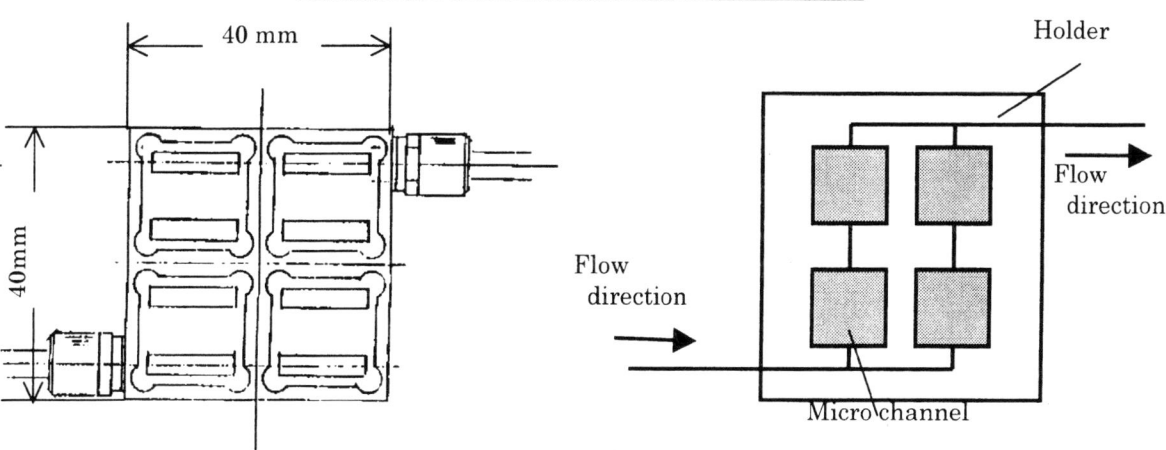

Figure 9   Micro channel holder

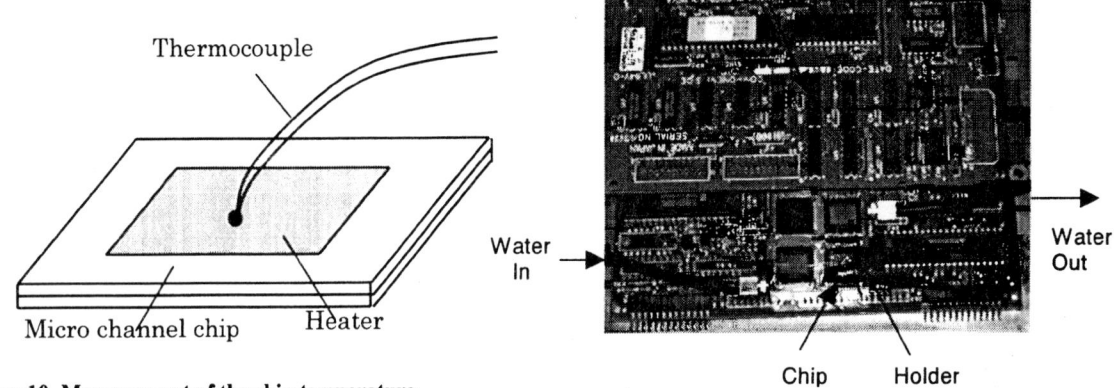

Figure 10 Measurement of the chip temperature

Figure 11 The micro channel heat exchanger

To verify the mean value of the thermal resistance of this heat exchanger, a thin film heater (Ni, 10mm*10mm in area, 0.2μm in thickness) was sputtered on the surface of the chip. And, a thermocouple (K type, 0.2mm in diameter) was contacted at the center of the channel as shown in Fig. 10. The micro channel heat exchanger set up on the motherboard is shown in Fig.11. The heat exchanger is almost the same size as the MCM, and is sufficiently small compared with that of the motherboard. Figure 12 shows the whole setup of electrical equipment using micro channel heat exchanger with the motherboard, the pump, a flow meter and the tube. The heat exchanger could be set up in a main board, and the pump or other parts also were compact compared to the size of the whole setup of electrical equipment.

In the experiments, coolant flow rate and input heat flux were about 3 cm^3/s and 30 W/cm^2, respectively. At this flow rates, Re ranged from 150~200. The cooling system was working for 7days, and the thermal resistance of the heat exchanger was maintained at 0.1 Kcm^3/W without any troubles such as water leak or short circuit.

SUMMARY AND CONCLUSION

To investigate the possibility of applying the micro channel heat exchanger to the cooling of electrical equipment, experiments and numerical simulations were performed. In the present study, the basic behaviors of heat transfer and pressure loss were clarified.

As for the pressure loss, experimental results of the micro channel (57μm*180μm, 57μm*370μm) showed good agreement with the theoretical predictions. The measured thermal resistance

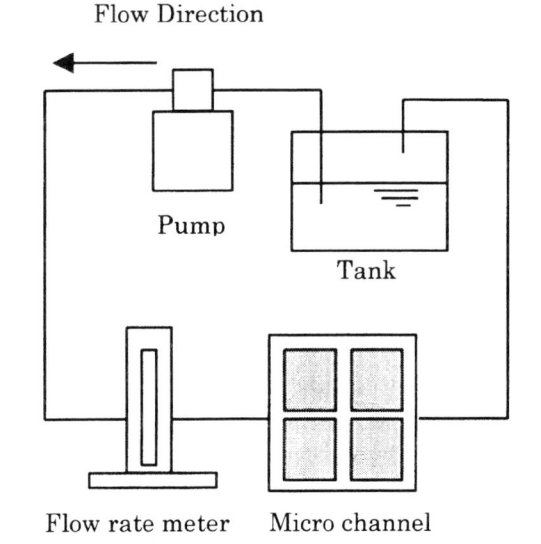

Figure 12 Cooling system using the micro channel

Table 3 Specifications of cooling equipment

	Type	Max. Pressure	Max. Flow Rate
Pump	Magnetic gear pump	0.3 MPa	33.3 cm^3/s

	Type	Diameter
Tube	Flexible	3 mm

of the micro channel chip (57μm*180μm) was about 0.1 Kcm^2/W above Re=200. In the present study, no eccentric phenomena in the heat transfer and the pressure loss could be seen. And, the difference of the thermal resistance between the inlet and the outlet portion in the chip was also about 0.1 Kcm^2/W at Re=200. These results indicate that when large heat flux is input to this chip, the thermal stress in it also will be large enough to break an electric pattern.

On the basis of the basic behavior of micro channel mentioned above, the practical cooling system using the micro channel heat exchanger was set up. The size of the cooling system is compact compared with that of the whole electrical equipment. Although the cooling system was working for 7 days, no troubles such as water leak or short circuit occurred.

NOMENCLATURE

a	Height of channel
b	width of channel
C	Poiseuille number
Cp	specific heat
de	hydraulic diameter of channel (=2ab/(a+b))
f	friction factor
k	thermal conductivity
L	length of channel
Pz	pressure gradient
q	mean heat flux
Re	Reynolds number (=$u_m \times$ de / ν)
Rt	thermal resistance
S	heat source
t	time
T	temperature
Ts	temperature of a chip
Ti	temperature of coolant at inlet portion
w	velocity of coolant
w_m	mean velocity of coolant
x,y	coordinates (cross-sectional directions)
z	coordinate (streamwise direction)
ΔT	temperature rise
ΔP	pressure loss
ε	aspect ratio (=a/b)
μ	viscosity
ρ	density
ν	kinematic viscosity

REFERENCES

Patankar S.V., 1980, *"Numerical heat transfer and fluid flow"*, McGraw-Hill, New York, NY.

Ward-Smith, A.J., 1980, *" Internal Fluid Flow"*, Clarendon Press, Oxford.

Tuckerman, D.B. and Pease, R.F., 1981, "High-performance heat sinking for VLSI", IEEE Elec. Dev. Let., Vol.EDL-2, pp126-129.

Tuckerman, D.B. and Pease, R.F., 1982, "Optimized convective cooling using micromachined structure", J. Electrochemical Society, 129(3), 98C.

Pfahler, J., Harley, J., Bau, H.H. and Zemel, J., 1990, "Liquid transport in micron and submicron channels", J. Sensors Actuators, Vol.21, pp431-434.

Harpole, G.M. and Eringer, J.E., 1991, "Micro channel heat exchanger optimization", Proceeding of the seventh IEEE semi-therm symposium, pp.59-63.

Choi S.B., Barron R.F. and Warrington R.F., 1991, "Fluid flow and heat transfer in microtubes", DSC-Vol.32, Micromechanical Sensors, Actuators, and Systems, ASME.

Weisberg A., Bau H.H. and Zemel J., 1992, "Analysis of microchannels for integrated cooling", Int. J. Heat Mass Transfer, vol.35, pp2465-2474.

Peng X.F. and Peterson G.P., 1995, "The effect of thermofluid and geometrical parameters on convection of liquids through rectangular microchannels", Int. J. Heat Mass Transfer, vol.38, pp755-758.

Copeland, D., Takahira, H., Nakayama, W. and Pak, B.C., 1995, "Manifold microchannel heat sinks: Theory and Experiment", Thermal science and engineering, Vol.3, No.2, pp.9-15.

Peng X.F. and Peterson G.P., 1996, "Convective heat transfer and flow friction for water flow in micro-channel structures", Int. J. Heat Mass Transfer, vol.39, pp2599-2608.

Mala G.M., Li D. and Dale J.D., 1997, "Heat transfer and fluid flow in microchannels", Int. J. Heat Mass Transfer, vol.40, pp3079-3088

SUSTAINABILITY ASSESSMENT OF ALUMINIUM HEAT SINK DESIGN

Suzana Prstic, Avram Bar-Cohen, *Naim Afgan, *Maria Graca Carvalho
University of Minnesota
*Instituto Superior Tecnico, Lisbon, Portugal

ABSTRACT

The sustainability assessment is based on the criteria which are designed to reflect: resources use, environment capacity use, social welfare and economic justification. In order to apply adapted methodology for the sustainability assessment the respective sustainability indicators are defined, which include: resource indicator, energy indicator and efficiency indicator. of the computer heat sink design.

The development of forced convection heat sink for microelectronic application, which are compatible with sustainable development involve the achievement of a subtle thermal design, minimum material consumption and minimum pumping power. The demonstration of the adapted procedure is performed on the different design of heat sink in personal computers.

Three options are taken into a consideration reflecting different optimal design of heat sink Thee heat sink designs, each dissipating 100W were chosen, reflecting a following options: smallest pumping power, least mass of material and use the lowest total energy. Using Sustainability Index defined in the paper the rating among options is obtained reflecting different constrains among indicators. It has been shown that option with lowest energy consumption is under any constrain between indicators will have highest rating between option under consideration

1. INTRODUCTION

Continued rapid industrialization and unfettered technology advancement is threatening to deplete valuable resources and exact an unacceptable toll on the Earth's environment. As a result of the explosion in information technology, an incredible number of computers is already in use all around the world, estimated approaching 600 million units by year-end 2000 [1]. Approximately, half of these are personal computers, using a heat sink and fan for cooling a microprocessor dissipating some 30W. In addition to the 300 million microprocessor-driven personal computers currently in use, sales of such units are expected to exceed 100 million in 2000. The requisite thermal management of such high-performance desk-top computers is most often achieved via an aluminum heat sink (fin structure) and a small fan.

The substantial material stream and energy consumption rate associated with the cooling of these desktop computers, as well as other categories of computers and electronic equipment, lends urgency and importance to the "perfection" of these air-cooled heat sinks. The effort discussed herein deals with the development of a design and optimization methodology for "sustainable" forced convection cooled, plate fin, electronic heat sinks. This methodology seeks to maximize the thermal energy that can be extracted from a specified space, while minimizing the material and energy consumed in the fabrication and operation of the specified heat sink.

Two, forced convection air cooling configurations are in common use. In the "fan-sink" configuration, a fan is used to directly impinge air on the fins of a heat sink attached to the microprocessor packages as in Fig. 1[2]. In the second configuration, a fan is used to develop a pressure head, which pushes or pulls airflow through a heat sink, as in Fig. 2. For a microprocessor with a typical heat generation of 30W, both of the configurations are required between 1.0W and 1.6 W of electricity to operate the fan and reduce the chip junction-to-ambient thermal resistance to the commonly encountered value of 1.1 ~ 1.7 K/W [3]. While the power used for computing creates unique opportunities for the enhancement of human activity, use of power for these cooling system has no inherent value.

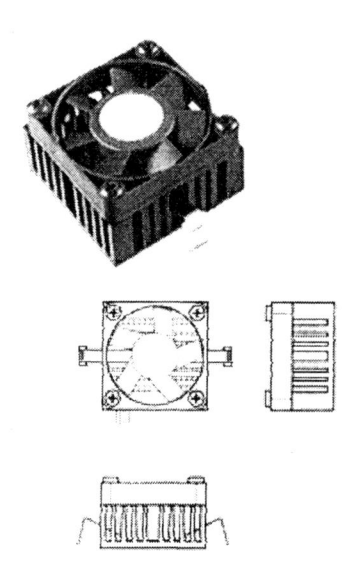

Figure 1, Example Fan/Heat sink

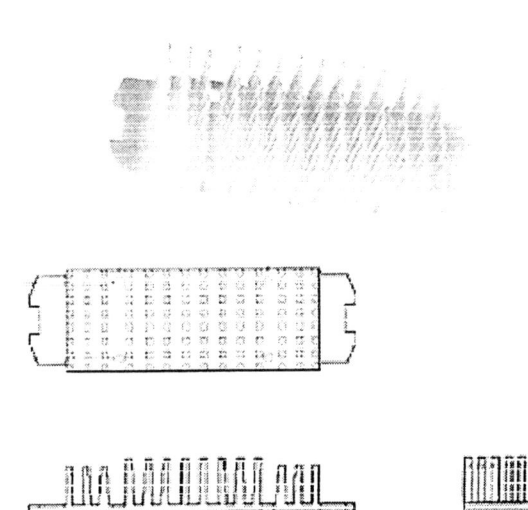

Figure 2, Example of CPU Heat Sink

Based on information available from the aluminum industry [4], approximately 85 kWh/kg are required to form, assemble, and transport aluminum heat sinks, and also a waste in the formation process is included, see Table 1. Combining this data with the energy consumed in the operation of the fans, the energy required for the thermal management of a typical microprocessor can be expressed as:

$$E_T = 85\,M + W_{PP}\,t_l \qquad (1)$$

where, E_T [kWh] is energy used for cooling, M [kg] is the heat sink mass, W_{PP} [W] is pumping power of fan, and t_l [hrs] is the lifetime operating hours of the fan/heat sink combination in life time.

To determine the total energy consumption for cooling of desk-top microprocessors, it is convenient to examine a "best-practices" personal computer heat sink design presented at the most recent ITHERM Conference [5]. In this paper the authors determine that an 80g heat sink operating with a 0.18W of pumping power by fan can adequately cool a 30W microprocessor. Using the values of Table 1, the formation and fabrication of such a heat sink thus requires some 6.8 kW-hr. Assuming 2000 hours per year of fan operation some 0.36kW-hr in pumping power will be needed annually for this fan. Thus, assuming an average of 3 years of life for such personal computer, the total amount of energy consumed by the heat sink will reach some 7.9 kW-hrs, or - on average - approximately 2.6 kW-hr/yr.

Thus, the amount of energy used for the cooling of the approximately 400 million units anticipated to be in use in 2001 can be expected to exceed 1 Terra (1×10^{12}) Watt-hours per year.

Table 1. Energy used for 1kg heat sink

Aluminum formation*	74.77 kWh/kg
Heat Sink fabrication	7.91 kWh/kg
Transport **	2.3 kWh/kg

* Based on 3% re-cycled material
** Inland case example

2. ANALYTICAL MODEL FOR THERMAL CHARACTERIZATION OF PLATE FIN HEAT SINKS

The analytical model developed by Holahan et al in [6] for calculating the thermal performance and pressure drop in fully- shrouded, laminar, parallel plate heat sinks has been utilized to characterize the thermofluid performance of the present heat sinks. Results obtained with this approach, which evaluates fin conduction by successive superposition of a kernel function determined from the method of images, was shown to give good agreement with experimental results, e.g. Iwasaki et al [7]. The simple side-inlet-side-exit (SISE) configuration considered in the current study is depicted in Fig. 3, showing the nomenclature of the array geometry, including the fin height, H, fin thickness, □, inter-fin spacing, S, width of base, W, and the length of the heat sink base, L.

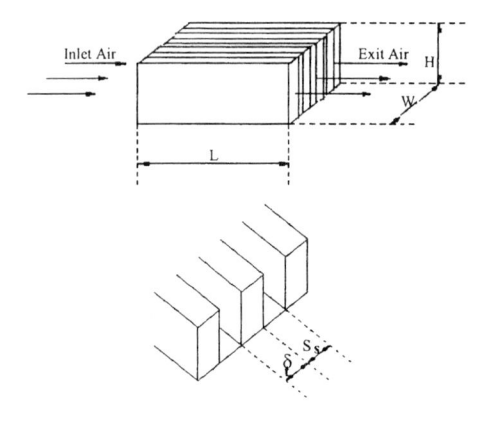

SISE Configuration

Figure 3, Side-Inlet-Side-Exit (SISE) Rectangular Plate Fin Heat Sink Configuration

In this approach, the local heat transfer coefficient needed to evaluate heat transfer rate from individual segments of the fin surface area, is obtained from the correlation for developing thermal and hydrodynamic laminar flow in parallel plate channels, with uniform wall temperature, provided in Kakac et al [8] and given by,

$$h_{fin,local} = \frac{k_{air}}{2s}[7.55 + (0.024X^{-1.14})\frac{0.0179Pr^{0.17}X^{-0.64} - 0.14}{(1 + 0.0358Pr^{0.17}X^{-0.64})^2}] \quad (2)$$

where, k_{air} is the air thermal conductivity, and $Pr = \nu/\alpha$ is the Prandtl number with ν as the mean kinematic viscosity of air and α the thermal diffusivity. In eq. (2), X is the dimensionless axial distance, and is given by,

$$X = \frac{[x\nu]}{4s^2 U_m Pr} \quad (3)$$

where x is the distance along the stream tube from the fin entrance to the patch, and U_m is the mean air velocity. The heat transfer rate from a single segment is then found using the superposition of a kernel function determined from the method of images. This heat flow divided by the local temperature difference yields the fin temperature for the segment of interest. A simple heat balance on the fluid flowing in each inter-fin channel can then be used, in an iterative fashion, to determine the local air temperature. The heat dissipation from the heat sink array, q, is then found by the summation of patch heat transfer from all the segments. A careful analysis of the resulting heat dissipation rates, throughout the parametric space of interest, can then be used to guide the designer to the most thermally advantageous combinations of air flow characteristics and fin geometries.

3. ASSESSMENT OF HEAT SINK DESIGNS

In succeeding sections of this paper an advanced heat sink application, occupying a volume of 500cm³ [0.1m×0.1m×0.05m, Aluminum], with an excess base temperature of 25 K, is used to illustrate the design methodology. Using the analytical model [6] described in the earlier section, volumetric flow rates and pressure drops across the heat sinks were varied from 0.01 to 0.04 m³/s and 20 to 80 Pa, respectively, yielding pumping powers ranging from 0.2 W to 3.2 W. This methodology has been described in detail by Bar-Cohen at al. [9]. It should be noted that the range of ΔP and V_{air} have been deliberately extended beyond common practice to allow for future fan capabilities, as well as to identify the opportunity for enhancing the heat transfer performance. Three heat sink designs, each dissipating 100 W (within ± 5 W) were chosen to which require the smallest pumping power (D1), consume the least mass (D2), and utilise the lowest total energy (D3).

Options under consideration

Options	Option definition
D –1	Smallest pumping power
D –2	Least mass use
D -3	Lowest total energy

Data for each of these designs has been represented in Table 2.

Tables 2, Summary of Heat Sink Designs

Option	Resource Indicators [Kg/comp]	Energy Indicators [kWh/comp]			Economy Indicators [Cents/comp]		
		Operating Energy	Formation Energy	Total Energy	Formation Cost	Operating Cost	Total Cost
D 1	0.664	1.2	54.9	56.1	284	12	296
D 2	0.124	9.6	10.2	19.8	48	96	144
D 3	0.163	3	13.5	16.5	70	30	100

* Approximate cost estimates used based on per pound heat sink vendor data and 1kWh cost
** Estimation based on 6000 hours of operation

As may be seen from Table 2, the best design with respect to the total energy consumption and economy indicator is the lowest energy (D3) design. The least mass design (D2) is the second best design because it requires the highest operating energy though it has the lowest formation energy. With very high formation energy of 54.9 kWh/comp and the smallest pumping power of 1.2 kWh, the worst design in all aspects is the smallest pumping power (D1) design.

4. MULTI-CRITERIA SUSTAINABILITY ASSESSMENT

Indicators to be used in this analysis are based on the data from Table 2. For the multi-criteria analysis a following Indicators are defined: RI – Resource Indicator, OEI – Operation Energy Indicator, Formation Energy Indicator, Formation Cost Indicator and Operation Cost Indicators.

Table 3 – *Sustainability Indicators fro Heat Sink Design*

Options	Resource Indicators [kg/comp]	Operation Energy Indicators [kWh/comp]	Formation Energy Indicators [kWh/comp]	Formation Cost Indicators [cents/comp]	Operation Cost Indicators [cents/comp]
D 1	0.664	1.2	54.9	284	12
D 2	0.124	9.6	10.2	48	96
D 3	0.163	3	13.5	70	30

Multi-criteria sustainability assessment is based on the Sustainability Index defined as the additive aggregative function which meets the condition of monotonicity [10].

The weighted arithmetic mean is the most popular type of synthesis function. This function is most simple and easy interpreted synthesis function. The Sustainability Index is an additive aggregative function defined as follows

$$Q(q,w) = \sum_{i=1}^{m} w_i q_i \quad (4)$$

where
w_i – weight-coefficients
q_i – normalised specific indicators

The normalisation of indicators is achieved by the use of respective membership function in presenting the actual values of indicators. Assuming maximum and minimum values of indicator will correspond to 0 and 1, respectively and using linear membership function the set of indicators for all option under consideration will be converted to the fuzzy set of the respective indicators.

The membership function are defined as follows

$$q_i(x_i) = \begin{cases} 1, & \text{if } x_i \leq MIN(i), \\ \dfrac{MAX(i) - x_i}{MAX(i) - MIN(i)}, & \text{if } MIN(i) < x_i \leq MAX(i), \\ 0, & \text{if } x_i > MAX(i) \end{cases} \quad (5)$$

Membership function value corresponding to each actual value will lead us to the normalised specific indicators as given in the Table 4.

Table 4 – *Normalised Sustainability Indicators for Heat Sink Design*

Options	RI	OEI	FEI	FCI	OCI
D – 1	0.000	0.971	0.000	0.000	0.971
D – 2	0.892	0.000	0.893	0.904	0.000
D – 3	0.812	0.722	0.813	0.803	0.722

The idea of randomisation of uncertainty developed by Thomas Bayes [11] where he models uncertain choice of a values w of a probability from the set by random objects uniformly distributed in the interval [0,1] was used to define the weight coefficients. The weight coefficients w_i is a measure of relative significance of the corresponding specific indicator q_i for aggregative estimation of Q (q,w). The outlined approach to uncertainty (deficiency of information) for the aggregative indices construction under uncertainty is named ASPID(Analysis and Synthesis of Parameters under Information Deficiency) [12,13,14,15].

In order to use this method we have to suppose that the measurement of weight coefficients is accurate to within a steps h = 1/n, with n a positive integer. In this case the infinite set of all possible weight-vectors may be approximated by finite set. Of all possible weight-vectors with discrete components. Under assumption that the number of indicators m = 5 and n = 40, the number of elements of the set W(m,n) may be determined by

$$N(m,n) = \frac{(n+m-1)!}{n!(m-1)!} = 135751 \quad (6)$$

Now non-numerical information may be used for the reduction of the set W(m,n) of all possible vectors $w^{(t)} = (w^{(t)}_1, w^{(t)}_2, w^{(t)}_3, w^{(t)}_4, w^{(t)}_5)$ with discrete components set to a W(I;m,n) of all admissible weight-vectors which meet the requirements implied by the additional information in form of mutual relation between criteria with respective indicators.

The measure of reliability of the revealed preference relation between options is defined by

$$P(j,l,I) = \frac{\left|\left\{s : Q^s\left(q^{(j)}\right) > Q^s\left(q^{(l)}\right)\right\}\right|}{N(I,m,n)} \quad (7)$$

where $\left|\left\{s : Q^s\left(q^{(j)}\right) > Q^s\left(q^{(l)}\right)\right\}\right|$ is number of elements in the finite set $\{s:...\} \leq \{1...N(I,m,n)\}$.

The non-numerical information in our analysis is defined as constrain between the indicators. In our analysis we will take cases as defined in Table 5.

Table 5 – *Cases for Non-numerical Constrains*

Cases	Non-numerical constrains
CASE 1	RI = OEI = FEI = FCI = OCI
CASE 2	RI > OEI = FEI = FCI = OCI
CASE 3	OEI > RI = FEI = FCI = OCI
CASE 4	FEI > RI = OEI = FCI = OCI
CASE 5	FCI > RI = OEI = FEI = OCI
CASE 6	OCI > RI = OEI = FEI = FCI
CASE 7	FCI = OCI > RI = OEI = FEI

Legende: = - equal ; > - larger ; < - smaller

Case 1

The Case 1 represent situation when we assume that all weighting coefficients are having the same value. This is not realistic situation since there is only one combination among total number of the weight-coefficient vectors generated in this analysis. In all diagrams colour lines are: black - value of Sustainable Index; red – Standard Deviation of Sustainable Index ; blue – Probability of Dominancy.

Fig.3

It can be noticed that priority for this case is obtained for Option 3 – the lowest total energy, followed by Option 2 and Option 1. The high value of the probability as the measure of reliability proves that this combination is realistic case within the total number of combination under consideration

Case 2

Fig.4

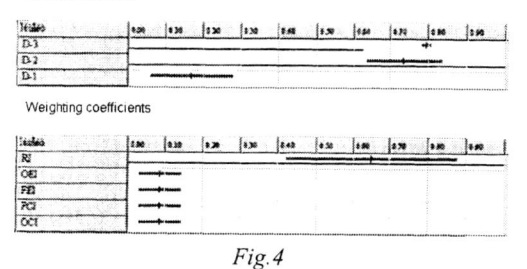

In order to investigate effect of different cases with priority given to one of the indicators , it was constructed number of situations which reflecting priority of individual indicators with other indicator keeping the same. Among the first in this group is the Case 2 with priority given to Resource Indicator

Sustainability Index rating for this case gives priority to the Option 3 with small differences between Option 2. It will lead us to the conclusion that in this case there is small difference between two options under consideration. It should emphasised that this case has rather high value of probability for mutual relation among the option 1 and option 2 , what is result of compatibility of the two options in this case.

Case 3

The Case 3 is designed to investigate situation if priority is given to the Operation Energy Indicator. In this case first on the priority list of Sustainability Index is Option 3 – option with lowest total energy.

Fig.5

With the low value for the reliability of the preferences between options 1 and 3, it can be concluded that this case belongs to the unrealistic case since the value of reliability factor is close to 0.50. Also ,the high value of standard deviation for Option 1 and Option 2 is imposing high uncertainty in prediction of this case.

Case 4

Since the Formation Energy Indicator - FEI is obtained as the formation energy needed for component production it is of interest to obtain assessment of its affect to the rating of the option under consideration.

CASE 4
FEI > RI = OEI = FCI = OCI
Sustainability Index

Weighting coefficients

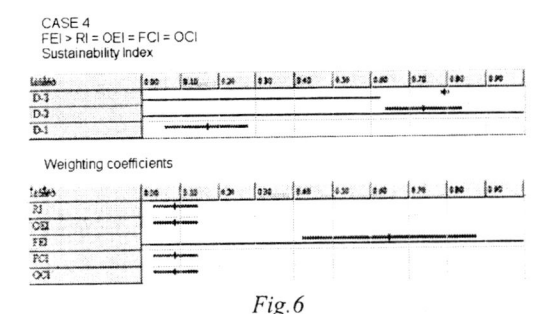

Fig.6

Result obtained as seen lead to the conclusion that Option 3 has high priority in comparison with other two options.

Case 5

If the Formation Cost Indicator -FCI has priority as it is designed for Case 5 there will be marginal difference in the priority among Option 2 and Option 3.

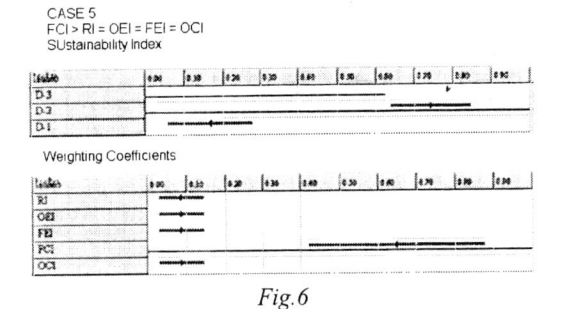

CASE 5
FCI > RI = OEI = FEI = OCI
SUstainability Index

Weighting Coefficients

Fig.6

The high value of probability among the Options as shown in Fig.6 is leading to the conclusion that this Case is a realistic case. From the other side it may be derived that Formation Cost Indicator priority constrain is imposing small differences between Option3 and Option 2.

Case 6

This Case is designed to investigate situation when priority is given to the Operation Cost Indicator.

CASE 6
OCI > RI = OEI = FEI = OCI
Sustainability Index

Weighting coefficients

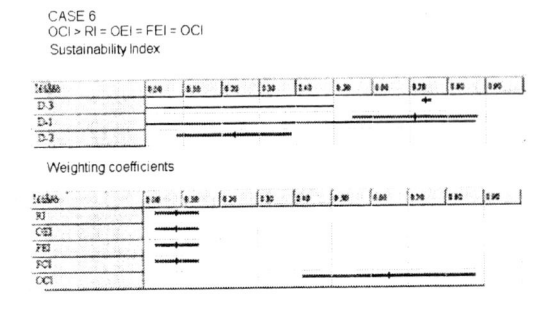

Fig.7

As shown the Sustainability Index Rating gives priority Option 3 and Option 1 have marginal difference. The low value of probability of dominancy among the options 1 and 3 prove that this case is not realistic since the value is close to 0.50.

Case 7

In order to investigate the effect of total cost on the priority list these, the Case 7 is designed with priority given to the Formation Cost Indicator and Operation Cost Indicator in comparison with other three indicators

It can be noticed that this Case is realistic with Sustainability Index for Option 3 having the highest value.

CASE 7
FCI = OCI > RI = OEI = FEI
Sustainability Index

Weight Coefficients

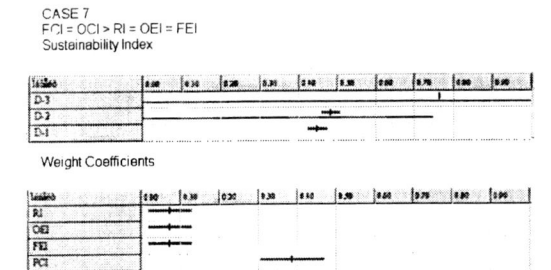

Fig.8

CONCLUSIONS

Multi-criteria assessment is proved to be powerful tool for the assessment of options under consideration. It was shown that even optimised design of heat sink are different in its characteristic represented in the form of respective indicators, the selection among the option will be byes if not taken by multi-criteria assessment . From the selected options and respective constrain imposed on the mutual relation between indicators it can be obtained different Sustainability Index Rating as measure of multi-criteria parameter for the assessment of the options.

For majority of the Cases under consideration the first place on the rating list was taken by the Option 3 – Least total energy. This is conform that under different constrain the Option 3 the best choice. It should be noticed that there are differences in reliability of the cases as well a different standard deviation among the selected option within individual cases.

It is of interest to notice that there are cases which are leading to the selection of Option 2 or Option 1 but with different reliability and standard deviation.

There are number of other Cases which could be analysed if all constrain have to be studied. It may lead to the better understanding of the effect of different indicators to the priority list obtained by the Sustainability Index Rating.

REFERENCES

[1] Computer Industry Almanac Inc., 1999, "Nearly 600 Million Computers-in-Use in Year 2000", Press release 11/03 1999

[2] AAVID Thermal Product Inc., 2000, Products data sheet

[3] Intel Corporation, 1999, AP-905 Pentium® III Processor Thermal design Guidelines Application Note

[4] LCA committee report, 1999, Japan Aluminum Association

[5] J.Y.Chang, C.W.Yu and R. L. Webb, 2000, Identification of Minimum Air Flow Design for a Desktop Computer Using CFD Modeling, Proceedings of ITHERM 2000.

[6] Holahan, M. F. , Kang S. S., Bar-Cohen, A., "A Flowstream Based Analytical Model for Design of Parallel Plate Heat Sinks", 1996 ASME HTD-Vol. 329 Volume 7, 1996 National Heat Transfer Conference-Vol 7. pp 63-71.

[7] Iwasaki, H., Sasaki, T., Ishizuka, M., 1994, Cooling Performance for Plate Fins for Multichip Modules, Proceedings of the Intersociety Conference on Thermal Phenomena, pp. 144-147.

[8] Kakac, S., Shah, R. K., Aung, W., 1987, Handbook of Single-Phase Convective Heat Transfer, Second Edition, Wiley & Sons, New York, pp. 3-35, 3-42.

[9] Bar-Cohen, A., Prstic, S., Yazawa, K., and Iyengar, M., 2000, Design and Optimization of Forced Convection Heat Sinks for Sustainable Development, Proceedings of EURO Conference,

[10] Naim H. Afgan, Maria G.Carvalho, Sustaianbility Assessment Method for Energy Systems, Kluwer Academic Publisher , Boston, 2000

[11] Bayes T. An ssey toward solving a problem in doctrine of chances, Biometrika,45,pp.296-315,1958

[13] Hovanov N., Fedotov Yu., Zakharov V., The making of index numbers under uncertainty //,Environmental Indices: Systems Analysis Approach, EOLSS Publishers Co., Oxford, 1999, P.83-99

[14] Hovanov N. Analysis and Synthesis of Parameters under Information Deficiency. St. Petersburg, St. Petersburg State University Press, 1996 (Monograph, in Russian).

[15] Hovanov N., Kornikov V., Seregin I. Qualitative information processing in DSS ASPID-3W for complex objects estimation under uncertainty, Proceedings of the International Conference "Informatics and Control". St. Petersburg (Russia), 1997, pp.808-816.

[16] Hovanov A.N., Hovanov N.V., DSSS "ASPID – 3W" Decision Support System Shell, Registered by Federal Agency for Computer Programms Copyright Protection Russia Federation, 22 September 1996, Num. 960087

A NEW APPROACH TO THE THERMAL ANALYSIS OF ROTARY AIR PREHEATER

H. Shokouhmand*, A. Rezai**

*Professor of Mechanical Engineering, Tehran University, Shokouhmand@yahoo.com
**Research Engineer, Tehran University, Alirezain@yahoo.com

ABSTRACT

In this paper a new approach is presented for analyzing the thermal behavior of rotary regenerator heat exchangers used in utility boilers. In this type of heat exchangers, due to the rotation of the solid matrix, the temperature of the fluids as well as heat exchange elements are functions of position and time. Energy equations for hot stream (gas) and cold stream (air) and the solid matrix have been solved using the Heat Balance Integral Method. The variation of physical properties of the cold and the hot fluids as well as the solid matrix has been considered. The results have been compared with existing data from lambertson (1958) and the agreement appears to be reasonably good.

INTRODUCTION

The advantages of rotary periodic-flow heat exchangers for industrial waste heat recovery systems and vehicular gas turbine-regenerator applications have been known for many years. Storage and release of thermal energy is the unique feature of the regenerative heat exchanger. The basic rotary type air preheater regenerative principle is very simple. The cylindrical rotor containing the heat transfer surface rotates at 1 to 3 rpm inside housing and is divided into two sectors. Waste heat is absorbed from the flue gas and transferred to the incoming cold air by means of the continuously rotating heat transfer surfaces formed from metal plates (matrix). As the heat transfer surfaces pass through the counter-flowing of parallel-flowing exhaust gas and incoming air streams, heat is absorbed uniformly and released to the combustion air. Unlike recuperative type air preheaters, heat need not pass through the tube or plate walls. Instead, the heat is simply absorbed by and released from the same heat transfer surfaces of the entire mass as it rotates. This storage type heat exchanger is generally refereed to as a rotary type heat regenerator. Its tree major advantages relative to the recuperative air preheaters are: (*i*) a much more compact heat transfer surface can be employed, (*i*) the heat transfer surface is substantially less expensive per unit of transfer area, and (*iii*) because of the periodic flow reversals, the surface tends to be self-cleaning. Researchers have obtained a number of approximate solutions by employing different mathematical techniques. One of the more complete solutions has been obtained by lambertson (1958). Razelos and Benjamin (1978) developed a new computer model based on the Gauss-Seidel method to calculate the thermal performance of a blast furnace stove regenerator without the effect of conduction in the wall along the flow direction. The important differences in their solution as compared to previous works are variable flow-rate and variable thermal properties of the walls and the fluids. Their excellent analysis is one of the current industry standards for performance predictions of blast furnace stove regenerators. A major disadvantage of the otherwise attractive Gauss-Seidel method is that its convergence is slows especially when a large number of grid points are involved. The reason for slowness is easy to understand: the method transmits the boundary condition information at a rate of one grid interval per iteration. Alternatively, Nahavandi and Weinstein (1961) and Iliffe (1984) solved the steady periodic problem in terms of the solution to an integral equation. The purpose of this investigation is to study an effective calculation method for the rotary periodic-flow regenerator. In this paper a simpler method is developed for the solution by approximating the time variation of temperature at any point as a cosine function. The method allows for the use of variable thermal properties such as the specific heat of the metal, and the specific heats, densities, and convection heat transfer coefficients of the fluids.

GOVERNING DIFFERENTIAL EQUATIONS

The one-dimensional form of the energy equation for the moving fluid and storage material governs the transient response of the heat storage unit. To consider the fluid and metal physical properties variations, the regenerator is splitted into a number of zones and the appropriate fluid and metal physical properties is taken to the mean value of the inlet and outlet temperature in each zone. Considering an incremental volume can derive governing differential equations. The idealization assumed in the derivation of governing differential equations for each zone of the rotary regenerator are as follow (Schmidt and Willmott, 1981)

1- Constant fluid and material properties in each zone.
2- Infinite thermal conductivity for the solid in direction normal to the fluid flow.
3- Zero thermal conductivity for the solid in the direction of the fluid flow.
4- Uniform heat transfer coefficient in each zone.
5- Uniform initial temperature distribution in the storage material.
6- No heat transfer through the side of the storage unit.
7- Constant fluid velocity.
8- Step change in inlet fluid temperature.

In the gas side, The differential equations describing the thermal behavior of the regenerator are:

$$\frac{\partial T_g}{\partial x} = -\frac{h_g A}{\dot{m}_g c_g l}(T_g - T_m) \qquad (1)$$

$$\frac{\partial T_m}{\partial t} = \frac{h_g A}{M_m c_m}(T_g - T_m) \qquad (2)$$

Similar to gas side equations, the differential equations for the air side are:

$$\frac{\partial T_a}{\partial x} = \frac{h_a A}{\dot{m}_a c_a l}(T_m - T_a) \qquad (3)$$

$$\frac{\partial T_m}{\partial t} = \frac{h_a A}{M_m c_m}(T_m - T_a) \qquad (4)$$

Setting can make Eq. (1) to Eq. (4) dimensionless

$$\pi = \frac{hAt}{M_m c_m} \quad \Lambda = \frac{hA}{\dot{m}c} \quad y = \frac{x}{l}$$

$$u = \frac{T - T_{ai}}{T_{gi} - T_{ai}} \quad \tau = \frac{t}{t_g}$$

Hence, gas dimensionless equations can be obtained as: side

$$\frac{\partial u_g}{\partial y} = -\Lambda_g(u_g - u_m) \qquad (5)$$

$$\frac{\partial u_m}{\partial \tau} = \pi_g(u_g - u_m) \qquad (6)$$

for $0 < y < 1$ and $-1 < \tau < 0$

In a similar way, air side dimensionless equations can be obtained as:

$$\frac{\partial u_a}{\partial y} = \Lambda_a(u_a - u_m) \qquad (7)$$

$$\frac{\partial u_m}{\partial \tau} = \pi_a(u_a - u_m) \qquad (8)$$

for $0 < y < 1$ and $0 < \tau < 1$

BOUNDARY CONDITIONS

There are two boundary conditions to be considered. Firstly, the entrance gas temperatures in both heating and cooling periods are constant. Secondly, the solid temperatures at the end of a heating/cooling period are the same as those at the beginning of the succeeding cooling/heating period. Hence, appropriate boundary conditions are

$$u_g = 1 \quad \text{at} \quad y = 0 \quad \text{and} \quad -1 < \tau < 0 \qquad (9)$$

$$u_a = 0 \quad \text{at} \quad y = 1 \quad \text{and} \quad 0 < \tau < 1 \qquad (10)$$

Periodicity conditions which describe the quasi-steady periodic operation of the regenerator are:

$$u_m \quad \text{continuos across} \quad \tau = 0$$

and

$$u_m(\tau = -1) = u_m(\tau = 1) \qquad (11)$$

for $0 < y < 1$

METHOD OF SOLUTION

Approximate solutions to transient heat transfer problems may be obtained relatively easy by the use of what is commonly called Heat Balance integral Method (THEBIM) (Langford, 1973). THEBIM is applicable to one-dimensional linear or nonlinear problems involving temperature dependent thermal properties. The accuracy of this method is in general unknown and the results must be compared with an exact or numerical solution to the problem. Consider the governing differential equations by the form of Eq. (12)

$$L(u) = 0 \qquad (12)$$

A function V is accepted as an approximate solution of the problem if it satisfies the boundary conditions together with a weak form of Eq. (12), namely

$$\int_0^1 L(V) X_j(\tau) d\tau = 0 \qquad (13)$$

for some sequence of weight functions

$$X_j, \; j = 1, 2, ..., J \qquad (14)$$

Hence, instead of solving Eq. (5) to Eq. (8) an approximate solution will be accepted (Atthey, 1988). The weaker or integrated forms of Eq. (5) to Eq. (8) are:

$$\int_{-1}^{0} \left(\frac{\partial u_g}{\partial y} + \Lambda_g (u_g - u_m) \right) X_j(\tau) d\tau = 0 \qquad (15)$$

$$\int_{-1}^{0} \left(\frac{\partial u_m}{\partial \tau} - \pi_g (u_g - u_m) \right) X_j(\tau) d\tau = 0 \qquad (16)$$

$$\int_{0}^{1} \left(\frac{\partial u_a}{\partial y} - \Lambda_a (u_a - u_m) \right) X_j(\tau) d\tau = 0 \qquad (17)$$

$$\int_{0}^{1} \left(\frac{\partial u_m}{\partial \tau} - \pi_a (u_a - u_m) \right) X_j(\tau) d\tau = 0 \qquad (18)$$

The approximate solution will be chosen so that the temperature distributions are cosine functions in τ at any point along the length of the regenerator and the coefficients are functions of y. Because of the time derivative in Eq. (6) and Eq. (8), the functions for u_m are required to be one degree higher in order than those for u_g and u_a. Therefore, the temperature distributions in gas and air sides have been considered as follow:

$$u_g = G(y) + g(y) \cos\left(\frac{\pi}{2} \tau - \frac{\pi}{4} \right) \qquad (19)$$

$$u_m = M(y) + m(y) \cos\left(\frac{\pi}{2} \tau - \frac{\pi}{4} \right) + \frac{3\sqrt{2}}{4} \mu(y) \left\{ \cos^2\left(\frac{\pi}{2} \tau - \frac{\pi}{4} \right) + \frac{2-\pi}{2\pi} \right\} \qquad (20)$$

for $0 < y < 1$, $-1 < \tau < 0$

and

$$u_a = A(y) + a(y) \cos\left(\frac{\pi}{2} \tau + \frac{\pi}{4} \right) \qquad (21)$$

$$u_m = N(y) + n(y) \cos\left(\frac{\pi}{2} \tau + \frac{\pi}{4} \right) + \frac{3\sqrt{2}}{4} v(y) \left\{ \cos^2\left(\frac{\pi}{2} \tau + \frac{\pi}{4} \right) + \frac{2-\pi}{2\pi} \right\} \qquad (22)$$

for $0 < y < 1$, $0 < \tau < 1$

The space functions G, g, etc. are unknown and can be obtained from weak form equations. Weight functions in weak form equations are considered as a form Eq. (23)

$$X_j = \cos^{(j-1)} \beta \qquad j = 1, 2 \qquad (23)$$

with the choice $j = 1$ and $X_1(\tau) = 1$, four equations are obtained

$$\frac{\partial G}{\partial y} = -\Lambda_g (G - M) \qquad (24)$$

$$\frac{\partial A}{\partial y} = \Lambda_a (A - N) \qquad (25)$$

$$m = \pi_g (G - M) \qquad (26)$$

$$n = -\pi_a (A - N) \qquad (27)$$

Although, with the choice $j = 2, X_2(\tau) = \cos\left(\frac{\pi}{2} \tau - \frac{\pi}{4} \right)$ for gas side and $X_2(\tau) = \cos\left(\frac{\pi}{2} \tau + \frac{\pi}{4} \right)$ for air side, Further four equations can be obtained

$$\frac{\partial g}{\partial y} = -\Lambda_g (g - m) \qquad (28)$$

$$\frac{\partial a}{\partial y} = \Lambda_a (a - n) \qquad (29)$$

$$\mu = \pi_g (g - m) \qquad (30)$$

$$\nu = -\pi_a (a - n) \qquad (31)$$

Boundary conditions become simply

$$\begin{aligned} G = 1, g = 0 & \quad \text{at } y = 0 \\ A = a = 0 & \quad \text{at } y = 1 \end{aligned} \qquad (32)$$

The periodicity Eq. (11) give

$$m = n \qquad (33)$$

$$M + \frac{\mu}{\pi} = N + \frac{\nu}{\pi} \qquad (34)$$

DETERMINATION OF SPACE FUNCTIONS (G, A, etc.)

Governing differential equations (Eq. (6) to Eq. (8)) are regarded as a fourth order system of differential equations (Eq. (24) to Eq. (31)). As the system is linear, it is appropriate to obtain the solution in terms of superposition of some fundamental solutions. For convenience a solution of the system is denoted by the sequence

$$F = \{G, A, M, N, g, a, m\} \qquad (35)$$

Because of the length derivative in Eq. (24), Eq. (25), Eq. (28) and Eq. (29), it is appropriate to assume that F has the exponential form $Ce^{\alpha y}$. With the choice $\alpha = 0$ Eq. (24) to Eq. (31) and the Eq. (33) are satisfied by

$$F = \{1,1,1,1,0,0,0\} \qquad (36)$$

For any value of α

$$F = \left\{ \frac{\Lambda_g}{\pi_g}, \frac{\Lambda_a}{\pi_a}, \frac{\Lambda_g + \alpha}{\pi_g}, \frac{\Lambda_a - \alpha}{\pi_a}, \frac{-\Lambda_g \alpha}{\Lambda_g + \alpha}, \frac{-\Lambda_a \alpha}{\Lambda_a - \alpha}, -\alpha \right\} e^{\alpha y} \qquad (37)$$

While Eq. (34) gives

$$\frac{\pi(\Lambda_g + \alpha)^2 + \pi_g^2 \alpha^2}{\pi_g (\Lambda_g + \alpha)} = \frac{\pi(\Lambda_a - \alpha)^2 + \pi_a^2 \alpha^2}{\pi_a (\Lambda_a - \alpha)} \qquad (38)$$

Equation (38) gives three values for α and by considering Eq. (36) and Eq. (37), four fundamental solutions are obtained. Hence, general solution of Eq. (23) to Eq. (30) is given by

$$F = \sum_{j=1}^{4} b_i F_i \qquad (39)$$

To obtain general solution, coefficients b_i must be determined. As mentioned previously, the regenerator is splitted into N zones. Since for each zone, four coefficients b_i are unknown, thus 4N coefficients b_i must be determined. At the interface between the two zones it is clear that the gas and air temperatures must be equal at all times. Hence, at the interface G, A, g and a are taken to be continuous. Using the four boundary conditions Eq.(32) and the continuity conditions of functions G, A, g and a gives a system of 4N linear equations from which the 4N constants b_i in Eq. (39) may be determined numerically.

RESULTS AND ANALYSIS

In order to demonstrate the heat transfer phenomena in a regenerator with multiple layers as well as variable thermal properties of fluid and matrix, a numerical example for rotary air preheater is computed using the following data.

Table. 1 Operating conditions

	Gas side	Air side
Inlet temperature (°C)	365	39
Mass flow rate (kg/s)	256.87	243.02
Gas composition	$N_2 = 71.39\%$	$O_2 = 0.84\%$
	$H_2O = 7.57\%$	$CO_2 = 20.20\%$

Table. 2 Information on the layered regenerator

	Layer 1	Layer 2	Layer 3
Layer depth (mm)	975	950	300
Surface thickness (mm)	0.6	0.6	1.2
Heat transfer surface (m²)	11150	10830	3090

Rotor speed is 3 RPM

Rotor free area per each sector is 22.26 (m²)

To consider the fluid and metal physical properties variations, the air preheater is splitted in to 23 zones and the

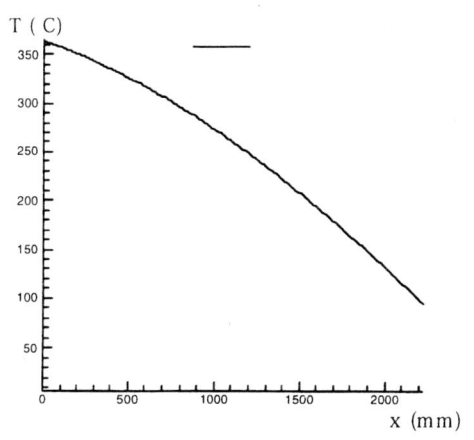

Fig 1. time averaged gas temperature

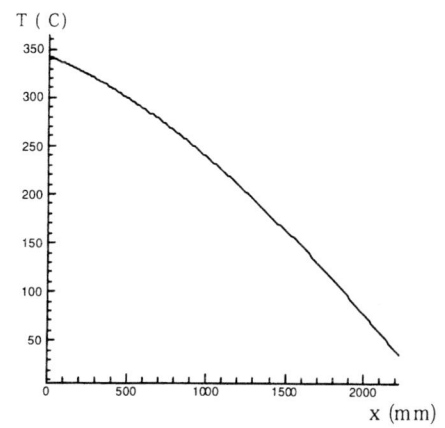

Fig 2. time averaged air temperature

physical properties is taken to the mean value of the inlet and outlet temperature in each zone. The abscissa (flow length) of Fig. 1 to Fig. 4 is based on the gas flow direction. The entry point of gas is at flow length = 0 Fig. 1 and Fig. 2 show the time averaged fluids temperature in gas and air sides. Fig. 3 and Fig. 4 show the gas and air temperature distribution in hot and cold period. To verify the results obtained by the present method, we have compared our results with those given by lambertson (1958). It is seen in table (3) the agreement is excellent.

Table 3. Qualification of the Method

	Lambertson Results	This Study
Exit Gas Temp. ($°C$)	125.4132	125.7559
Exit Air Temp. ($°C$)	316.8172	316.3541

CONCLUSIONS

The steady state temperature distribution in a rotary regenerator has been determined by an approximate analytical method based on the Heat Balance Integral method. The solution has been shown to be a good approximation to the temperature distribution in rotary regenerator in power station boilers. A novel feature of the method is that it can readily cope with a number of regenerators in series. This has applications to power station air preheaters where different heat exchange elements are usually fitted in the hot and cold zones of the air heaters. Another application is in high-temperature regenerators where it is convenient to approximate the temperature variation of the fluid properties by splitting the regenerator into a number of zones, in each of which the fluid properties are assumed to be constant.

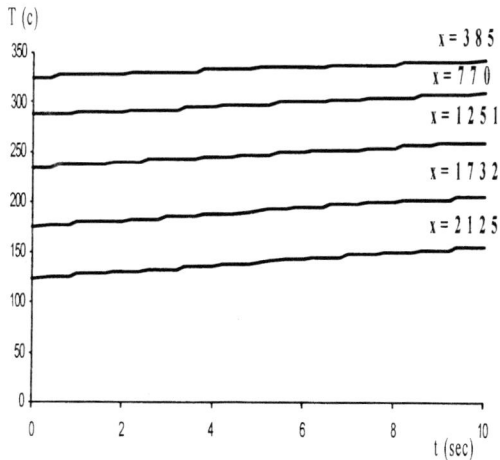

Fig 3. gas temperature in hot period

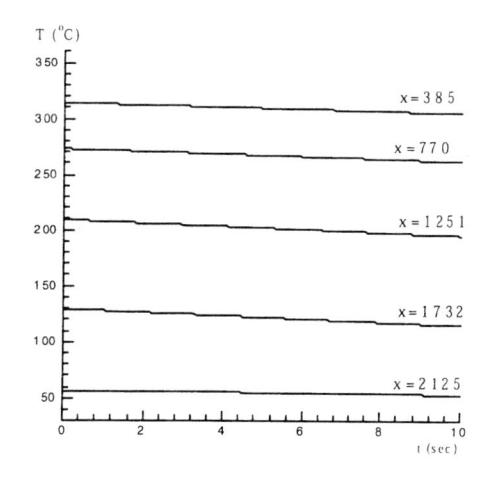

Fig 4. air temperature in cold period

NOMENCLATURE

A	time averaged air temperature function
a	amplitude of air temperature oscillation function
b	constant
c	specific heat, J/kg °C
F	vector of fundamental solution
G	time averaged gas temperature function
g	amplitude of gas temperature oscillation function
h	heat transfer coefficient, W/m² K
l	length of channel, m
M_m	mass of heat transfer surface, kg
M	time averaged metal temperature function
m	amplitude of metal temperature oscillation function during gas cycle
$\dot{m}$	mass flow of fluid, kg/s
N	time averaged metal temperature function
n	amplitude of metal temperature oscillation function during air cycle
T	temperature, °C
t	time, s
t_g	period of gas cycle, s
u	dimensionless temperature
x	length, m
X	weight function for heat balance integral
y	dimensionless length

Greek Symbols

α	exponent in general solution, dimensionless
Λ	reduced length,
μ	Coefficient of quadratic term in metal Temperature function during gas cycle
ν	Coefficient of quadratic term in metal Temperature function during air cycle
π	Reduced period, dimensionless
τ	Dimensionless time

Subscripts

a	air
g	gas
i	inlet
j	index
m	metal

REFERENCES

Atthey, D. R., 1988, An Approximate Thermal Analysis for Regenerative Heat Exchanger, *Int. J. Heat Mass Transfer*, Vol. 31, No. 7, pp. 1431-1441.

Iliffe, C. E., 1984, Thermal Analysis of Counterflow Regenerative Heat Exchanger, *J. Instn Mech. Engrs*, 159, pp. 363-372.

Lambertson, T. J., 1958, Performance Factors of a Periodic Flow Heat Exchanger, *ASME Transaction*, vol. 80, pp. 586-592.

Langford, D., 1973, The Heat Balance Integral Method, *Int. J. Heat Mass Transfer*, Vol. 16, pp. 2424-2428.

Nahavandi, A. N., and Weinsten, A. S., 1961, A Solution to the Periodic Flow Regenerative Heat Exchanger Problem, *Appl. Scient. Res*, Section A, Vol. 10, pp. 335-348.

Razelos, P., and Benjamin, M. K., 1987, Computer Model of Thermal Regenerators with Variable Mass Flow Rates, *Int. J. Heat Mass Transfer*, Vol. 21, pp. 735-743.

Schmidt, F. W., and Willmott, A. J., 1981, *Thermal Energy Storage and Regeneration*, McGraw-Hill, New York.

SIMULATION OF REGENERATORS UTILIZING ENCAPSULATED PHASE CHANGE MATERIALS (PCM)

M. Sadrameli*, N. S. Tabrizi

Tarbiat Modarres University, Chemical Engineering Department, P. O. Box 14115-143, Tehran, Iran,
Email: Sadramel@modares.ac.ir

ABSTRACT

The operational behaviors of thermal storage systems utilizing the latent heat of fusion of phase change materials (PCM) are simulated for symmetric-balanced/unbalanced conditions. A unified two-phase model, which accommodates the performance prediction of cyclic thermal regenerators with encapsulated PCM or conventional packing, is solved numerically by finite difference method. Extensive numerical results revealed the effect of various parameters on the performance of such systems.

INTRODUCTION

Much waste heat is emitted continuously or periodically from metallurgical and chemical industries. Fixed-bed heat regenerators are used to recover, store and reuse waste energy. Large regenerators are used to recover waste heat from power and steam plants and incinerators. These regenerators are cycled between heat storing and heat releasing modes. First, a stream containing waste heat (e.g., exhaust gas from a furnace) is passed through the bed where the heat is transferred to the inert packing. Later a cold stream (e.g., feed air to the furnace) is passed through the hot bed thereby heating the stream. Conventional regenerators utilize the sensible heat of packing to absorb and store energy. It has been recognized that, compared to the sensible heat storage, larger amount of energy per unit volume can be stored by utilizing the latent heat of phase change. Energy storage schemes utilizing latent heat of fusion are the most practical because the change in specific volume for most solid-liquid phase transitions is minimal.

Selection of the storage material depends largely on the mean storage temperature. Some desirable characteristics of a PCM for use as a heat storage medium are a high latent heat of transformation (usually fusion), a narrow freezing range or a congruent melting point, long term chemical stability, inertness with respect to a suitable container material, low cost, and ready availability. Thermal performance of cyclic thermal regenerators using sensible heat of the storage medium has been well established and documented (Sadrameli, 1988). Early work on PCM systems mainly focused on the thermal properties and selection strategies of suitable PCM shot and encapsulating shell materials (Farkas and Birchenall, 1985). Ananthanayanan et al., 1987 performed numerical simulation of two particular Al-Si PCM beds with a simple convective model. Ma, 1994 reported some preliminary characterization results and revealed that the theoretical analysis of the chronological performance characteristics of cyclic PCM thermal regenerator systems had been inadequately represented. Recently Wang (1999) studied on phase change temperature distributions of composite PCMs in thermal energy storage systems. Dincer (1996) has also studied an application of PCM for solar energy storage systems. In this investigation the operational behavior of a fixed bed regenerator packed with PCM are studied. A unified non-thermal equilibrium model is formulated with solid phase enthalpy as a dependent variable in the energy conservation equation of the packing matrix. This accommodates both sensible and latent heat utilization systems. The finite difference techniques have been show to be more convenient for the model solution. The effects of model parameters have also been investigated and the results have been compared with the experimental results obtained from the literature.

MATHEMATICAL MODEL

For a bed undergoing a possible phase change process, a temperature-based two-phase model results in a discontinuity in the numerical solution to the solid phase conservation equation at the phase transformation temperature.

Ma (1994) derived a more reliable model, based on the solid phase enthalpy to provide unified formulation for mathematical modeling of thermal regenerators utilizing sensible and/or latent heat of the storage medium. In dimensionless form, the energy equation for fluid phase and the solid packing can be written as:

For fluid phase,

$$\frac{\eta_{fs}}{1-\bar{\varepsilon}}\frac{\partial T_f}{\partial \tau} + \frac{\partial T_f}{\partial \hat{x}} = \Lambda(T_s - T_f) \quad (1)$$

For solid phase,

$$\frac{\partial H_s}{\partial \tau} = \Lambda \cdot Ste \cdot (T_f - T_s) \quad (2)$$

where

$$Ste = \frac{c_s^s(t_{fi}' - t_{fi}'')}{h_{sl}} \qquad T = \frac{t - t_{fi}''}{t_{fi}' - t_{fi}''} \qquad H_s = \frac{h - h_s^*}{h_{sl}}$$

$$\tau = \frac{\rho_f c_f V_0 \theta}{\rho_s c_s^s(1-\bar{\varepsilon})L} \qquad \eta_{fs} = \frac{\rho_f c_f \varepsilon}{\rho_s c_s^s} \qquad \Lambda = \frac{\overline{h_c} a_v L(1-\bar{\varepsilon})}{\rho_f c_f V_0} \quad (3)$$

h_s^* and h_l^* Are the critical enthalpy at the beginning and end of a phase change process and $h_l^* = h_s^* + h_{sl}$. The dimensionless solid temperature can be related to dimensionless solid enthalpy by the following equation (3):

$$T_s = \begin{cases} T_m + H_s/Ste & H_s < 0 \\ T_m & 0 \leq H_s \leq 1 \\ T_m + (H_s - 1)/(Ste\, c_{fs}) & H_s > 1 \end{cases} \quad (4)$$

Clearly values of local H_s indicates the physical status of the PCM shot in the bed and during the phase transformation, the values of H_s lies between 0 and 1 representing the percentage of phase change.

The above model is valid subject to the following assumptions:
- Axial thermal conduction in the fluid and solid phase is negligible.
- The physical properties of the fluid and solid are independent of temperature, except that the specific heat capacity of the encapsulated PCM at solid and liquid forms may be different.
- The thermal resistance of encapsulating shell is negligible.
- Velocity and voidage are uniform.
- Radial thermal conductivity is infinite.
- For the above mentioned differential equations, inlet boundary conditions, for the fluid are:

$$T_{fi}' = 1 \text{ and } T_{fi}'' = 0 \quad (5)$$

For the solid phase, the inlet conditions can be explicitly derived for hot and cold periods as:

For hot period,

$$H_s = \begin{cases} Ste(1-T_m) - [Ste(1-T_m) - H_s(0)]Exp(-\Lambda\tau_h) & H_s^\circ < 0 \\ H_s^\circ + \Lambda Ste(1-T_m)\Delta\tau_h & 0 \leq H_s^\circ \leq 1 \\ 1 + SteC_{fs}(1-T_m) - [SteC_{fs}(1-T_m)]Exp(-\Lambda(\tau_h - \tau_h^*)/C_{fs}) & H_s^\circ > 1 \, \& \, H_s(0) \leq 1 \\ 1 + SteC_{fs}(1-T_m) - [SteC_{fs}(1-T_m) - (H_s(0)-1)]Exp(-\Lambda\tau_h/C_{fs}) & H_s^\circ > 1 \, \& \, H_s(0) > 1 \end{cases} \quad (6)$$

For cold period,

$$H_s = \begin{cases} 1 - SteC_{fs}T_m + [SteT_m + (H_s(0)-1)]Exp(-\Lambda\tau_c/C_{fs}) & H_s^\circ > 1 \\ H_s^\circ - \Lambda SteT_m \Delta\tau_c & 0 \leq H_s^\circ \leq 1 \\ -SteT_m[1 - Exp(-\Lambda(\tau_c - \tau_c^*))] & H_s^\circ < 0, H_s(0) \geq 0 \\ -SteT_m + [SteT_m + H_s(0)]Exp(-\Lambda\tau_c) & H_s^\circ < 0, H_s(0) < 0 \end{cases} \quad (7)$$

where $H_s(0)$ is the value of H_s at the beginning of a hot or cold period, τ_h^* ç τ_{cO}^* represent the dimensionless time corresponding, respectively to $H_s = 1$ and $H_s =$. These conditions are valid for both co-current and counter-current. The reversal conditions associated with the PCM thermal regenerators are identical to those applied for conventional regenerators. For co-current and counter-current, respectively, they can be written as:

$$T_f''(\hat{x}, \tau=0) = T_f'(\hat{x}, \tau=U_h), \quad H_s''(\hat{x}, \tau=0) = H_s'(\hat{x}, \tau=U_h)$$
$$T_f''(\hat{x}, \tau=0) = T_f'(1-\hat{x}, \tau=U_h) \quad H_s''(\hat{x}, \tau=0) = H_s'(1-\hat{x}, \tau=U_h) \quad (8)$$

The system is assumed symmetrical and unbalance factor is defined as the utilization ratio of hot to cold period ($\beta = U_h/U_c$). The utilization factor (Π/Λ) is consistent with the definition of dimensionless time.

PERFORMANCE CHARACTERISTICS

While the traditional thermal effectiveness may still be used as one of the objective performance characteristics, more attention has been focused on the extent to which the bed is being operated around the phase change temperature. Relating to this, α factor is defined (3) as the chronologically averaged volumetric fraction of the PCM shot, of which the solid temperature is equal to the transformation temperature of the employed PCM. The following mathematical equation is used to calculate the α factor for either heating or cooling periods:

$$\alpha_{h(c)} = \frac{1}{U_{h(c)}} \int_0^{U_{h(c)}} \frac{\sum_{K}^{N_c} \Delta V_k}{V} d\tau \qquad 0 \leq H_{sk} \leq 1 \quad (9)$$

In addition, to quantitatively describe the efficiency of non-sensible heat utilization, the latent heat utilization factor, ξ, which is the ratio of the actual amount of latent heat utilized to the maximum possible value at cyclic thermal equilibrium (Ma, 1994) is computed by the following equations:

$$\xi_{local} = \begin{cases} H_s(U) & 0 \leq H_s(U) \leq 1 \text{ and } H_s(0) \leq 0 \\ |H_s(U) - H_s(0)| & 0 \leq H_s(U) \leq 1 \text{ and } 0 \leq H_s(0) \leq 1 \\ 1 - H_s(U) & 0 \leq H_s(U) \leq 1 \text{ and } H_s(0) > 1 \\ H_s(0) & H_s(U) < 0 \text{ and } 0 \leq H_s(0) \leq 1 \\ 1 - H_s(0) & H_s(U) > 1 \text{ and } 0 \leq H_s(0) \leq 1 \\ 1 & H_s(U) > 1 \text{ and } H_s(0) \leq 0 \\ & \text{or } H_s(U) \leq 0 \text{ and } H_s(0) \geq 1 \\ 0 & \text{otherwise} \end{cases} \quad (10)$$

Here, $H_s(0)$ and $H_s(U)$, are the local H_s value at the beginning and the end of a hot or cold period.

The bed utilization factor (BUF) for a phase change bed, can be computed by:

$$BUF = \frac{\int_0^{U_{h(c)}} (T_{fi} - T_{fo}) d\tau}{T_m + 1/Ste + C_{fs}(1 - T_m)} \quad (11)$$

If the BUF of the successive hot and cold period are similar, the system is said to have reached cyclic thermal equilibrium.

NUMERICAL SOLUTION

Finite difference method is used to solve the model equations. The dimensionless time step less than 0.1 and dimensionless length step less than $1/3\Lambda$ are found to produce sufficiently accurate results. The numerical results are validated by:
- Checking the BUF at cyclic thermal equilibrium to ensure the satisfaction of the overall energy balance.
- Comparing the computed effectiveness of symmetrical-balanced regenerators over a wide range of parameters for $T_m > 1$, with published data (Table 1).
- Comparing the computed performance characteristic factors, with the results published by Ma (1994). See Table 2.

DISCUSSION

Co-current cyclic PCM thermal regenerators operating under a variety of conditions with the specified PCM encapsulation are simulated. Typical enthalpy profiles of solid packing at various dimensionless time of the forth cycle is shown in Fig. 1. It can be seen that at cyclic equilibrium, the solid phase enthalpy of the majority part of the bed remains between 0 and 1, which indicates that the system is being operated largely around T_m throughout the heating and subsequent cooling periods.

The H_s profiles of a counter-current bed, operating under the same conditions are demonstrated in Fig. 2. It is

Table 1. Comparison of regenerator effectiveness, $\beta = 1$, $T_m > 1$.

Present work	Ma	Schmidt	U	Λ	Mode
0.564	0.563	0.562	1	5	Parallel
0.613	0.616	0.613	1.5	10	-
0.474	0.473	0.474	0.5	15	-
0.658	0.658	0.658	1	10	-
0.250	0.252	0.250	4	5	Counter-Current
0.333	0.336	0.333	3	10	-
0.845	0.846	0.845	1	30	-
0.774	0.778	0.773	1.25	40	-

Table 2 Comparison of performance characteristic factors: $\beta = 3$, $Ste = 1.25$, $T_m = 0.75$, $C_{fs} = 1$.

Λ	U_H	α_H		ξ_H		η_H	
		Ma	this work	Ma	this work	Ma	this work
5	0.5	0.978	0.938	0.271	0.282	0.373	0.372
5	2	0.617	0.631	0.673	0.685	0.372	0.371
20	0.5	0.902	0.911	0.162	0.172	0.375	0.375
20	2	0.265	0.262	0.735	0.737	0.412	0.415

clear that, in comparison with its parallel mode counter part, only a small part of the bed is involved in the phase change process. It signifies the fact that it would be more difficult to achieve a near isothermal heat recuperator condition in the counter-current mode than the co-current mode.

Quantitative comparisons between co-current and counter-current arrangements in terms of various performance characteristics at cyclic thermal equilibrium showed that:
- Counter-current beds usually possess higher thermal effectiveness.
- Co-current mode usually offers higher α values, indicating that a relatively large portion of the packing matrix is working around the phase change temperature.
- With co-current arrangement extremely low temperature swing can be achieved if the reduced length, utilization/ unbalanced factors are appropriately selected.
- With counter-current mode, more cycles may be needed to reach cyclic thermal equilibrium.

The performance characteristics of a PCM thermal regenerator depend on operational and configuration parameters, i.e. Λ, U_H, β, as well as the parameters that are closely associated with the properties of the employed PCM encapsulation, i.e. T_m, Ste, C_{fs}.

Characterization is conducted in Λ, U_H, β domain for a typical phase change material with fixed value of $T_m = 0.75, Ste = 1.25, C_{fs} = 1$. The range of parameters considered is: $U_H = (0.5-3)$, $\beta = (0.5-4)$ and $\Lambda = (5-20)$.. In T_m, Ste, C_{fs} domain, characterization is carried out for a number of combinations of Λ, U_H, β.

The results show that, for a given PCM, there is a maximum α_H and ξ_H region with regard to U_H and for various Λ. Fig. 3 shows that, for the given reduced length, the maximum α_H is largely located at $\beta = 3$ for low value of U_H. The latent heat utilization factor, ξ_H, increases with an increase in U_H and the maximum ξ_H are around $\beta = 3$ for any fixed U_H, Fig. 4. The maximum harmonic mean effectiveness is shown in Fig. 5 to be basically located at $\beta = 1$, this means that the effectiveness of balanced systems are always higher than the corresponding unbalanced ones. Numerical results also show a weak dependency of η_H on Ste. Stefan number is commonly around 2. Reduced length, was found to have insignificant effect on the pattern of the concerning characters with respect to U_H and β. The reduced length will, however, change the magnitude of the characters. The α_H values increase with the decrease of Λ. From the value of α_H, the level of temperature swing can be inferred. The higher the α_H, the smaller is the temperature swing.

Furthermore, the location of maximum α_H with respect to the physical properties of PCM shot, depends on the operational parameters, with the unbalance factor, β, being the most influential one. The unbalanced factor should be reduced as the phase change temperature decreases. Fig. 6 show that for the specified conditions the desirable thermal-physical property of PCM should have $T_m = 0.78$.

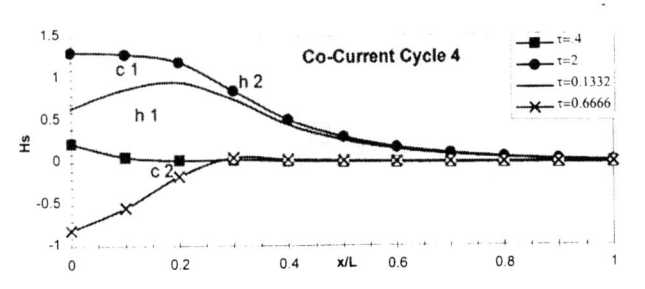

Fig. 1 Axial profiles of solid phase enthalpy in the 4th cycle with heating period (h) followed by cooling period (c).

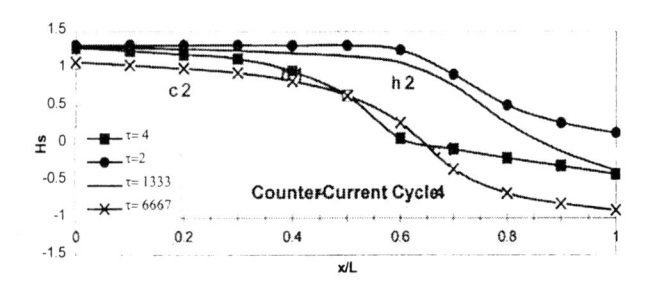

Fig. 2 Axial profiles of solid phase enthalpy in the 4th cycle of a counter-current with heating period (h) followed by cooling period (c) ($\Lambda = 5, U_H = 1, \beta = 3$)

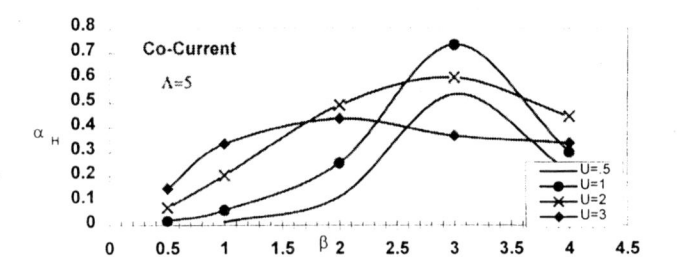

Fig. 3 Effect of U_H and β on α_H.

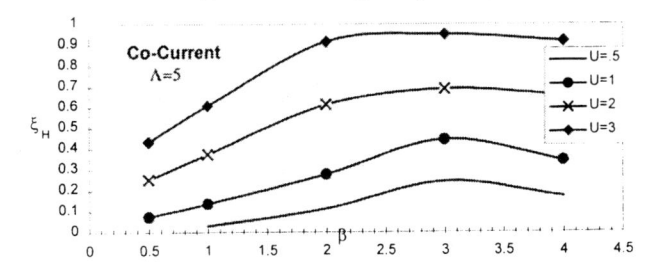

Fig. 4 Effect of U_H and β on ξ_H.

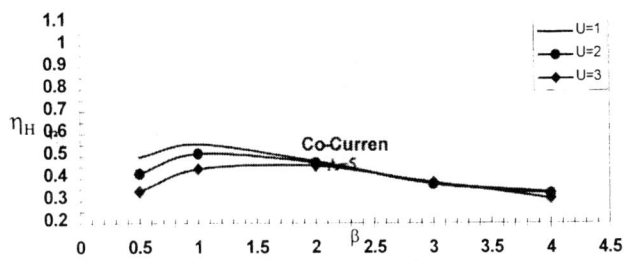

Fig. 5 Effect of U_H and β on η_H.

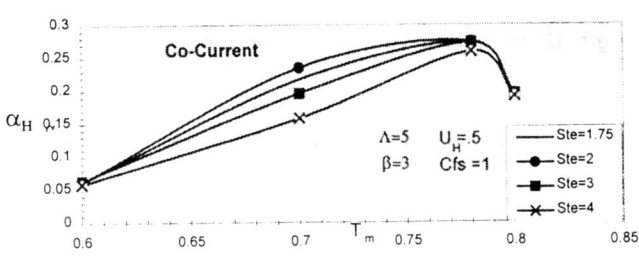

Fig. 6 Effect of T_m and Ste on α_H.

CONCLUSIONS

A unified model for performance prediction of conventional and encapsulated PCM cyclic thermal regenerator systems is solved numerically. The performance characterization carried out in this study provides useful information on optimum design of PCM regenerator systems.

NOMENCLATURE

a_v	Volumetric surface area of packing, 1/m
BUF	bed utilization factor
c, c_{fs}	Specific heat capacity and heat capacity ratio, J/kgC
h_c	convective heat transfer coefficient, W/m² K
h_{sl}	latent heat of fusion of encapsulated PCM, J/kg
L	bed length, m
Ste	Stefan number
t,T	dimensional and dimensionless temperature
T_m	dimensionless fusion temperature
U	utilization factor
V, ΔV	total and control volume occupied by solid packing
V_o	superficial velocity, m/s
α	the α factor defined in equation 9
β	unbalance factor
η_{fs}	non dimensional parameter defined in equation 3
$\bar{\varepsilon}$	mean voidage
η	thermal effectiveness, $\eta_h = 1-(T_{fo}^r), \eta_c = (T_{fo}^r)$
Λ	reduced length
Π	reduced period
ρ	Density, kg/m³
θ	Time, s
ξ	latent heat utilization factor

Superscript

0 , *	previous time step, critical status
h/c	hot/cold period
f/s , fi/fo	fluid/solid phase, fluid inlet/outlet
H	harmonic mean value

REFERENCES

Ananthanarayanan, Y., 1987, Modeling of fixed bed heat storage units utilizing phase change materials, *Metal Trans.* B18, 339.

Dincer, I., Dost, S., 1996, A perspective on thermal energy storage systems for solar energy applications, *Int. J. Energy Res.*, Vol. 20, Issue 6, pp. 547-557.

Farkas, D., Birchenall, C. E., 1985, *Metal Trans.* A. Vol. 16A, pp. 323-328,

Ma, J., 1994, Numerical Simulation of Transport Processes in Packed Bed Systems, PhD Thesis, University of Leeds, Leeds, UK.

Sadrameli, M., 1989, Investigation of Asymmetric and Unbalanced Regenerators, PhD thesis, University of Leeds, U.K.

Schmidt, F. W., Willmott, A. J., 1981, *Thermal Energy and Regeneration*, McGraw Hill, New York.

Wang, J., et al., 1999, Study on phase change temperature distributions of composite PCMs in thermal energy storage systems, *Int. J. Energy Res.*, Vol. 23, Issue 4, pp. 277-285.

PERFORMANCE OF MATERIAL SAVED FIN-TUBE HEAT EXCHANGERS IN DEHUMIDIFYING CONDITIONS

Hie Chan Kang[1] and Moo Hwan Kim[2]

[1]Kunsan National University, Kunsan 573-701, South Korea; E-mail: hckang@ks.kunsan.ac.kr
[2]POSTECH, Pohang 701-784, South Korea; E-Mail: mhkim@postech.ac.kr

ABSTRACT

This work discusses the pressure drop, heat and mass transfer of the finned-tube heat exchangers having 7 mm tubes and offset strips in dehumidifying applications. It focuses on the fin material saving and the reduction of pressure drop. The experiment was conducted using three times scaled-up models to simulate the performance of the prototype. Eight kinds of fins having different strips and S shape edges were tested. The area density of the strip was a major factor and its shape and the location were secondary factors on the pressure drop, the heat and mass transfer. The reduced-area fin can almost equal the non-reduced fin in the aspect of heat and mass transfer. The strip fins proposed in the present work can considerably reduce both the pressure drop and the fin material for a similar thermal load.

INTRODUCTION

The role of the fin is very important in heat exchangers, especially in the finned-tube heat exchangers of air conditioners since the major thermal resistance is located in the air-side. A lot of work has been focused on the development of fin shape in air conditioning applications (Kang and Webb, 1998a). The basic plain fin was modified to a wavy fin to promote turbulence and extend the effective flow length between the fins (Beecher and Fagan, 1987, Ali and Ramadhyani, 1992, Wang et. al., 1997a, Kim, et. al., 1997, Kang and Webb, 1998b). The other concept to enhance the heat transfer was to use strips, which can obtain multiple edge effects at each strip along the flow passage. The strip patterns varied their side view patterns to produce the louver fin, the offset strip fin and the convex strip (Nakayama and Xu, 1983, Tanaka et al., 1984, Hiroaki, et al., 1989, Hatada et al., 1989, Wang et al., 1996, Wang et al., 1997b). Recently, the alignments of strips were studied to reduce the pressure drop and maximize the heat transfer. (Kang and Kim, 1999a, 1999b)

Experiments on the evaluation of heat exchangers are classified into two major kinds: full-scale test or calorimeter test (Hiroaki, et al., 1989, Wang et al., 1996, Wang et al., 1997a, 1997b), and scaled-up test (Kang and Kim, 1999a, 1999b). The calorimeter test gives reliable results. However it is very expensive and needs much time in preparing the prototype for the test. Kang and Kim (1999a, 1999b) proposed the scaled-up model test and showed that it could be a very powerful technique in the development of fin patterns. They showed that their technique could replace the full-scale test in performance evaluation. Moreover, it saves great deal of time and money compared to the full-scale test. Recently Kang et al. (2000a) extended the technique to the dehumidifying condition. The test results of the enlarged model reasonably agreed with those of the prototype in the cases of hydrophilic and hydrophobic fin surfaces.

This work proposes eight kinds of fin patterns to reduce the pressure drop and fin material. The performance of the fin shapes was tested using the enlarged model. In addition, the heat and mass transfer analogy was checked.

EXPERIMENTAL METHOD

Similarity Law

Table 1 Comparison of parameters between the prototype and the model heat exchangers in the present work

Parameter	Prototype	Model
Re, Pr, Sc	1	1
Bo (dropwise)	1	1
Re$_f$ (filmwise)	1	1
η, f, j	1	1
Scale factor	1	3 (n)
Fin material	Aluminum	
Thermal conductivity, W·m^{-1}·°C^{-1}	222	222
Fin thickness, mm	0.1	0.3
Declination angle, °	90.0	2.1
Frontal velocity, m/s	0.7-1.5	0.23-0.5
Inlet temperature, °C	27.0	27.0
Inlet relative humidity, %	50	50
Pressure drop, Pa	1	1/9(1/n^2)
Heat transfer coefficient, W·m^{-2}·°C^{-1}	1	1/3(1/n)
Heat transfer rate, W	1	3 (n)
Condensation rate, kg·s^{-1}	1	3 (n)
Condensate thickness, m	1	3 (n)

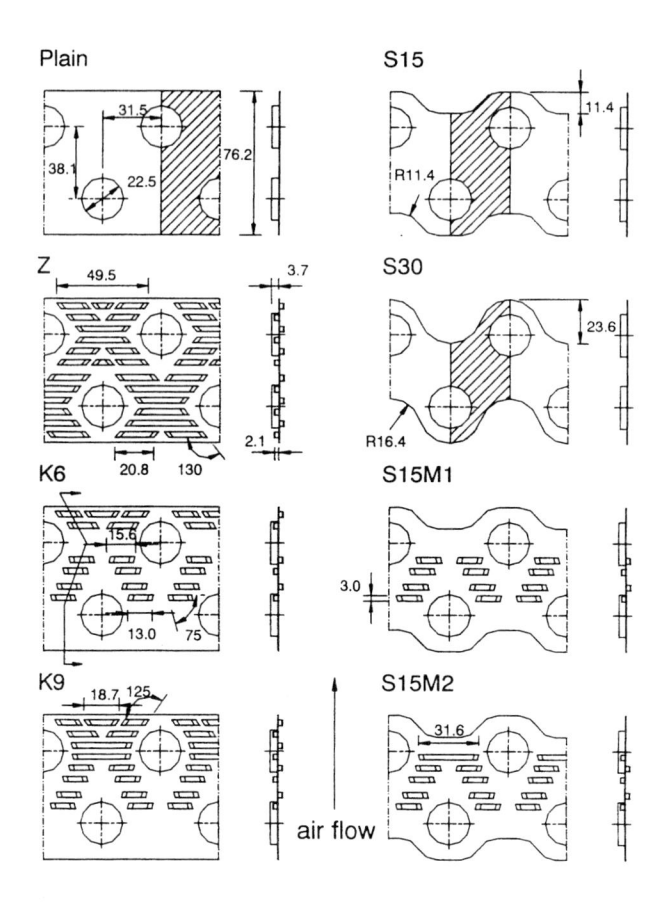

Fig. 1 Fin configuration tested in the present work (unit in mm)

A detailed consideration is needed to get similitude between the prototype and the scaled-up heat exchanger model. First, mass, momentum, heat and species conservations must be satisfied to get the air-side similarity. The previous results of Kang and Kim (1999a) showed that the major flow regime was laminar when the Reynolds number based on the tube diameter was near 1,000 in the finned-tube heat exchanger. Therefore, the velocity, temperature and vapor concentration fields in the scaled-up model can be similar to those in the full scale if Reynolds, Prandtl and Schmidt numbers are the same. If we use the same working fluid, air, in the similar range of temperature then we can get the same Prandtl and Schmidt numbers. The Reynolds numbers of both cases can be made the same by changing the air velocity of the model. In addition, the heat conduction in the fin of the model must be similar with that of the prototype. We can get similarity of the normalized fin temperature if we use the same fin material. These are necessary for similarity of sensible heat transfer.

The shape and size of the condensate should be scaled as much as the ratio of the prototype and model sizes in dehumidifying conditions. In the case of filmwise condensation, such as the case of the hydrophilic-coated fin, the film thickness in the scaled-up model must be enlarged as much as the geometric scale factor of the prototype. The thickness of condensate film depends on the film Reynolds number (Re$_f$) as below.

$$\mathrm{Re}_f = \frac{4g \sin\theta\, \rho_w \delta^3}{3\nu_w} \quad (2)$$

The δ and ν are the thickness and the kinematic viscosity of the condensate, respectively. For the three times enlarged model, the similitude angle is 2.1 degrees to get a condensate film three times as thick, when the angle of the fin in the prototype is 90 degrees. The relations between the three times model and the prototype are listed in Table 1. The detail information and its validity can be found in Kang et al. (2000a).

Test Heat Exchangers

The present test used a three times enlarged model to evaluate the heat transfer performance of the air side of the heat exchanger. Aluminum rings were used to simulate the tubes and fin colors. The aluminum fins shown in Fig. 1 were sandwiched between the rings. The unit heat exchanger was stacked with nine fins and ninety aluminum rings having 22.5 mm outer diameter. The unit model was tightly fixed at the end by bolts to reduce contact resistance. A 0.075 mm diameter of T type thermocouple was welded to the center aluminum ring to measure the wall temperature of the model. The overall dimensions are shown in Fig. 1. The test fin was made of 0.3 mm thick, 76.2 mm wide and 315 mm

Table 2 Dimensions of heat exchanger in the present test model

Parameter	Plain, Z, K9, K6	S15, S15M1, S15M2	S30
Tube diameter (D)		22.5	
Traversing tube pitch (P_t)		63.0	
Longitudinal tube pitch (P_l)		38.1	
Fin pitch (P_f)		4.01	
Fin thickness (t)		0.3	
Bare tube area (A_{tube})		231	
Fin surface area (A_{fin})	4005	3285	2522
Total surface area (A)	4236	3516	2753
Fin area reduction ratio	0%	18.0%	37.0%
Material reduction ratio	0%	15.0%	30.9%

Dimensions are in mm and mm^2. The fin surface area is the unit fin area as shown in Fig. 1. The reduction ratios are based on the plain fin.

long aluminum plate and was three times enlarged in all dimensions. The total number of test tubes was ten, and they consisted of staggered alignment with two rows. Fin pitch, distance between fins, was 4.01 mm. A summary of the heat exchanger dimensions is listed in Table 2.

Fig. 1 shows the configurations of eight fins tested in the present work. The plain fin is a basic frame of the other fins. The heat exchanger consisted of two-row tube banks having 22.5 mm diameter to simulate a 7 mm tube. The diameter was decided after considering the fin color and tube expansion. In other fins, the pattern layout was the same, but the location and density of the strip were modified. Fin Z is a commercial fin tested by Hiroaki et al. (1989). Twelve steps of strips were shifted alternatively along the flow direction. The patterns in the plane view is an X shape. Each strip of the fins in Fig. 1 was punched and shifted with a 3 mm width and a 2.4 mm depth from the base plate. Fin K9 had no strip in the front one-fourth part of fin. The fin had eight steps of strips, the front four steps of strips were the same. The four strips of the front row were designed so that the air flow enveloped the rear tube to reduce the rear wake zone due to the tube. Fin K6 was similar to fin K9 except that the fifth and sixth strips of fin K9 were eliminated.

Fins S15, S30, S15M1 and S15M2 show the fin configuration developed to save fin material. The basic concept to save fin area is shown in Fig. 2. In the production process, each fin is cut from aluminum sheet. The concave edge in the second fin became the convex edge of the first fin as shown in the right figure of Fig 2. As this process is repeated, we can reduce the amount of material used in the fin. The zone A of the original fin in Fig. 2 is not good for heat transfer, because the temperature difference between fin and air is relatively small compared to the other region. Kang and Kim (2000b) reported the zone is not very useful

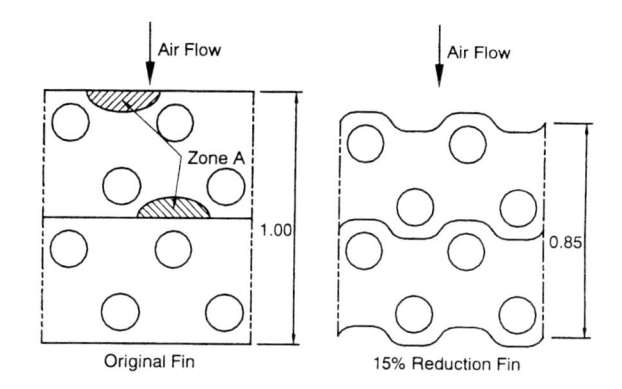

Fig. 2 Conceptual drawing for the fin area reduction

from the measurement of fin temperature. This concepts were applied to four fins in the present work. S15 series fins and fin S30 reduced 15% and 30% of the fin material and 18% and 37% of the fin area respectively. Fin S15M1 was modified from fin S15, which had four steps of strips at the middle of tube banks such as fin K9 and K6. Fin S15M2 had one more step of strips at the rear of the strips. All of the fins were treated by the same Pre-Coated Material (PCM) so that the fins showed hydrophilic characteristics, fully wetting.

Experimental Apparatus

Fig. 3 shows a low speed wind tunnel to test the model heat exchanger. The wind tunnel was a closed loop type. It consisted of a fan for airflow, an electric heater and a humidifier to acquire constant inlet temperature and humidity, a settling chamber and a contraction for uniform flow of air, a main test section and an exit chamber for measuring the air flow rate. The air velocity was controlled by adjusting the fan speed. Two screens and a honeycomb were installed to enhance the flow quality of the air. The dimensions of the main test section were 315.0 mm wide, 29.2 mm high and 623 mm long. The test section had 100 mm long upstream chamber. All outer walls of the main test section were insulated with at least 40 mm of styrofoam to minimize heat loss. In the exit chamber, a 32.2 mm diameter flow nozzle was installed to measure the air flow rate. It was made according to the British Standard (1964). Cooling tubes were used to cool the model heat exchanger. Cold water from the constant water bath fed inside of the cooling tube. Ten cooling tubes were inserted into each heat exchanger model.

Experimental Method

The assembled heat exchanger was installed at the main test section and the fin surface was declined 2.1 degrees from horizontal. The error of angle was less than 0.05 degree. The air temperature of the inlet was controlled as 27±0.3°C and its relative humidity was 50±0.5%. The

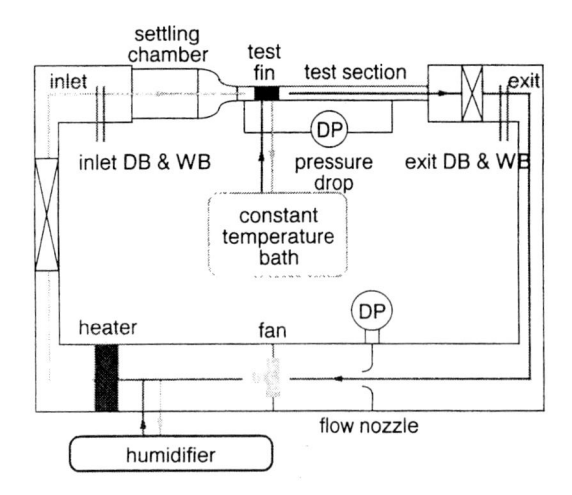

Fig. 3 Experimental apparatus used in the present work

temperatures of the tubes were 7.5±1.0°C and the individual valve installed at each cooling tube controlled them. The difference of tube temperatures was less than 0.2°C. Inlet and exit enthalpies were calculated by the dry and wet bulb temperatures. Four platinum resistance thermometers were used for the air temperature measurement and they were calibrated within ±0.05°C. Air flowrate and pressure drop across the test heat exchanger were measured by two pressure transducers at the flow nozzle and the main test section. The transducers were calibrated by a U-tube micro manometer within 0.06 Pa. We made two sets of the static pressure taps flush on the wall of the test section. One was located at two times of tube diameter upstream and the other at twenty times of tube diameter downstream from the edge of the test sample. We measured the average pressure drop on the heat exchanger from each set of pressure taps. The airflow rate was calculated from the measured pressure differences at the flow nozzle. Eight static holes were made to obtain the average pressure difference. The mass flow rate at the flow nozzle was calibrated by means of a Pitot tube downstream from the flow nozzle. The calibration agreed well with the equation given by the British Standard (1964) within 0.5%. About two hours were needed to obtain the steady state at each test run. All data were acquired and calculated by a data acquisition system (AOIP SA32). The error of energy balance was less than 5% at the frontal velocity, $V > 0.4$ m s^{-1}, and that was less than 8% at $V < 0.4$ m s^{-1}.

Data Reduction

The net heat transfer rate was estimated as follows.

$$q = \dot{m}_a (h_{a,i} - h_{a,e}) - \dot{m}_d c_{p,w} T_{a,e} - q_{a,\text{loss}} \quad (3)$$

Each term of the right hand side is the heat transferred to the air, the heat of the condensate and the heat loss from the air. The heat loss was acquired by a separated experiment and it was less than 2% of total heat transfer. The rate of condensation was obtained as below.

$$\dot{m}_d = \dot{m}_a (W_{a,i} - W_{a,e}) \quad (4)$$

The W is the absolute humidity. The total heat transfer is consisted of the sensible and latent heat transfer.

$$q = q_s + q_l = \dot{m}_a c_{p,a} (T_{a,i} - T_{a,e}) + \dot{m}_d h_{fg} \quad (5)$$

The heat transfer coefficient is written as below;

$$h_s = \frac{q_s}{A_s \Delta T_{LM}}, \quad (6)$$

where the logarithmic mean temperature difference is

$$\Delta T_{LM} = \frac{T_{a,i} - T_{a,e}}{\ln[(T_{a,i} - T_w)/(T_{a,e} - T_w)]}. \quad (7)$$

The T_w, $T_{a,i}$ and $T_{a,e}$ are the wall, air inlet and air exit temperatures. The mass transfer coefficient was defined as the same way as blow:

$$h_m = \frac{\dot{m}_d}{A_s \Delta W_{LM}} \quad (8)$$

$$\Delta W_{LM} = \frac{(W_{a,i} - W_w) - (W_{a,e} - W_w)}{\ln[(W_{a,i} - W_w)/(W_{a,e} - W_w)]}. \quad (9)$$

The measured data was calculated to those at the standard conditions, $T_{a,i} = 27°C$, $RH_{a,i} = 50\%$, $T_w = 8.0°C$. All properties were obtained at the film temperature.

RESULTS AND DISCUSSION

Pressure Drop

Fig. 4 compares the pressure drops for the eight fins of Fig. 1. The horizontal and left vertical axes denote the frontal velocity and pressure drop in the prototype respectively. The strip fins (fin Z, K9, K6) showed high pressure drop while the plain fin series (plain fin, fin S15, S30) showed low values. Among the plain fins, the fin area affected on the pressure drop. Fin S30 showed the lowest pressure loss because its fin area is the smallest. Results of fin S15 is located between those of fin S30 and the plain fin. The pressure drops of the strip fins were affected by the ratio of strip area to the fin area. This result generally agreed with Kang and Webb's model (1998a) that the pressure drop of the strip fins was nearly proportional to the ratio, i.e. the strip area to the fin area. Fin Z showed the highest pressure

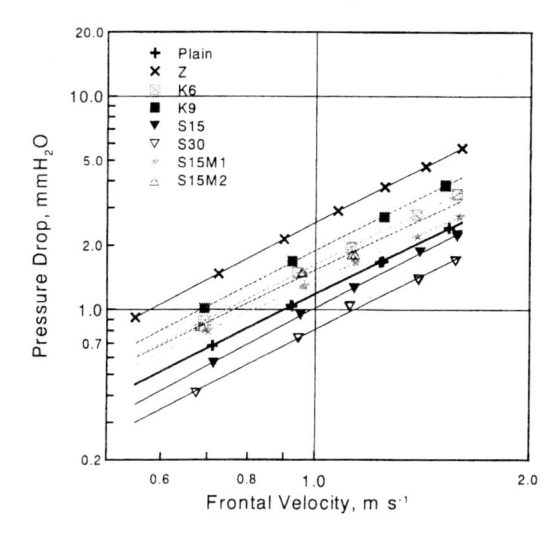

Fig. 4 Pressure drops versus frontal velocity

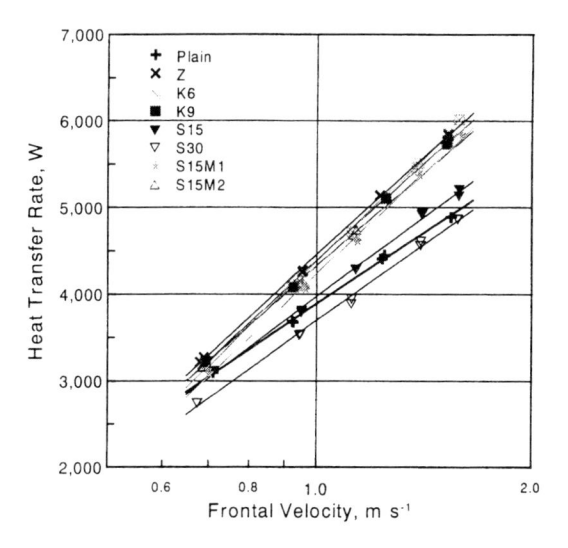

Fig. 5 Heat transfer rate versus frontal velocity

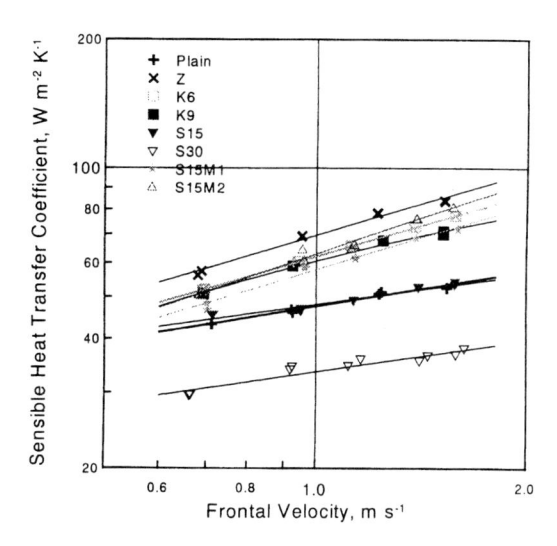

Fig. 6 Sensible heat transfer coefficient versus frontal velocity

drop. It is very interesting that fin S15M2 showed relatively higher pressure drop than fin S15M1 even though only the fifth strip was added in fin S15M2. Also, fin S15M2 was similar to fin K6 in the pressure drop. Fin K6 has more fin and strip areas than fin S15M2. From these results, we can say that fin and strip areas are the primary factors, and the shape and the location of the strips in their plane view yields a secondary effect, on the pressure drop.

Heat Transfer

Fig. 5 shows the comparison of heat transfer rates of the test fins. The measured data in the model test were transformed to the value in the prototype heat exchanger having 32 tubes and 322 fins, i.e. an evaporator of one ton of refrigeration. The performance of the strip fin group (fin Z, K9, K6, S15M1 and S15M2) is better than the plain fin group (plain fin and fin S15 and S30). Fin Z showed the best heat transfer rate, however the other fins are also within ten percents in heat transfer compared with fin Z. We have to note that fin S15M1 and S15M2 showed the comparable performance in heat transfer even though the fin area is 18% less than fin Z, K9 and K6. Among the plain fin series, fin S30 was less than the plain fin, however fin S15 was similar with the plain fin in heat transfer. The data of fin S30 are normally acceptable because its fin area is 37% less than that of the plain fin. For fin S15, the heat transfer rate was similar to that of the plain fin. From these results, we can say that the strip strongly affect the heat transfer, and the area-reduced fin can be useful in heat transfer.

Fig. 6 shows the comparison of sensible heat transfer coefficients of test fins. We defined the heat transfer coefficient as the pure heat transfer coefficient multiplied by surface efficiency, since we do not have an adequate theory for the fin efficiency of the strip part. Fin Z showed the highest heat transfer coefficient. The other strip fins (fin K9, K6, S15M1 and S15M2) showed similar values. The plain fin and fin S15 showed similar low heat transfer coefficients while fin S30 showed the lowest heat transfer coefficient. The slope of heat transfer coefficient to the frontal velocity depends on the fin depth along the flow direction. The exponents of velocity for the plain fin series (plain fin, fin S15, S30) were 0.20-0.25 and those of the strip fins were near 0.5. Therefore, the flow regime in the strip fins is similar to a laminar flow at the leading edge, i.e. each strip, and flow in the plain fins is similar to a laminar flow in the parallel plate.

Mass Transfer

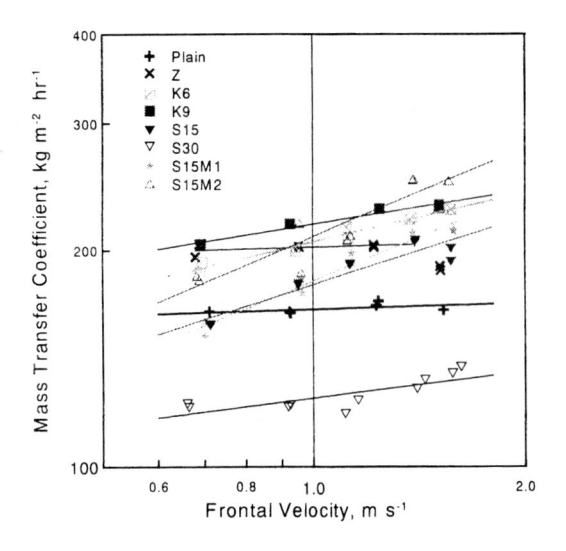

Fig. 7 Mass transfer coefficient versus frontal velocity

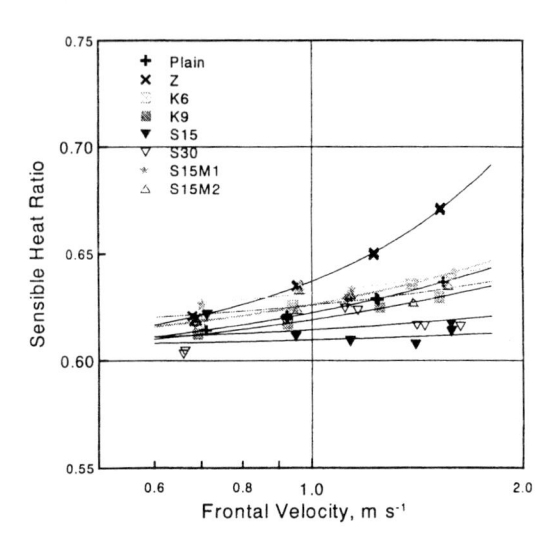

Fig. 8 Sensible heat ratio versus frontal velocity

Fig. 7 shows the comparison of mass transfer coefficients. The strip fin series (fin Z, K9, K6, S15M1, S15M2) showed higher mass transfer coefficients than the plain fin series (plain fin, fin S15 and S30) Fin S30 was the smallest in mass transfer. From the point of view of heat and mass transfer, too much of the area of fin S30 had been removed. However, fin S15 was favorable in mass transfer. The heat transfer rate of fin S15 was similar to that of the plain fin as shown in Fig. 5 even though the fin area of fin S15 was less than that of the plain fin. The main reason for the result is believed the augmentation of the latent heat transfer rather than the sensible heat transfer. Fin Z showed the highest performance in the sensible heat transfer as shown in Fig. 6, however the mass transfer was not the highest. The reason for this is supposed that fin Z was holding much more condensate at the strip because the fin had so many strips. The strip fins proposed in the present work (fin K9, K6, S15M1, S15M2) showed good performance in latent heat transfer. This is mainly due to the increase of heat and mass transfer behind the second row tube because of the guard strips located at the first row. Krückels and Kottke (1970) and Saboya and Sparrow (1974) measured the local heat transfer coefficient on the finned-tube bank. Their data showed that the local heat transfer coefficient behind the tube was one fourth or less than that of the front or inter tube region. Therefore, the design of the strip is very important in the finned tube heat exchanger.

Fig. 8 shows the sensible heat ratio of test fins. The sensible heat ratios were around 0.6-0.7 in the present experimental range. The sensible heat portion of the strip fin series increased with increase of air velocity. However, the increase of the plain fin series was less. The area reduced fin showed low sensible heat ratio.

Heat and Mass Transfer Analogy

It is very useful if we know the information for the analogy of heat and mass transfer. The simplified Chilton-Colburn analogy for the heat and mass transfer can be expressed as below.

$$\frac{h_s}{h_m c_p} = \text{Le}^{2/3} \qquad (10)$$

In the equation, the Le is Lewis number. The ratio of heat and mass transfer, $h_s/h_m c_p$, for the test fins is shown in Fig. 9. The present data ranged at 0.8-1.2 and increased for increase of Re_{D_h}. The Re_{D_h} is Reynolds number based on the D_h and the frontal velocity, and the hydraulic diameter; D_h is defined as four times of the contained air volume in the fin to the total fin area. The parameter, $h_s/h_m c_p$, of the plain fin series (plain fin and fin S15, S30) was lower than the strip fin series (fin Z, K9, K6, S15M1, S15M2). The Le number in the present work was around 0.84, and the present data were larger than the theory. Recently Wang (1997c) proposed a model for the parameter as shown in Fig. 9, here we recalculated according to their definition. The value of the plain fin series was slightly larger than the model, however the strip fin series showed higher values than the model. The parameter only for the strip fin series was correlated as below.

$$\frac{h_s}{h_m c_p} = 0.188 \, \text{Re}_{D_h}^{0.26} \qquad (11)$$

Performance Comparison

Fig. 10 shows the comparison of pressure drop and heat transfer rates of test fins at $V = 1.2$ m s^{-1}. In the figure, we chose the plain fin as the reference fin. Fin S15M1 and S15M2 was about 50-55% of fin Z in pressure drop.

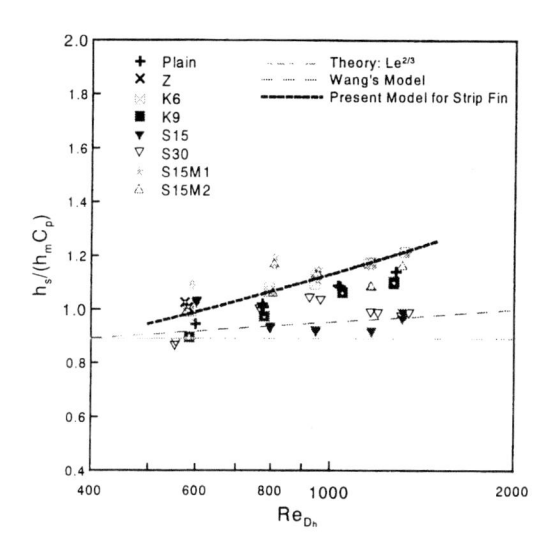

Fig. 9 The ratio of sensible heat transfer to mass transfer coefficients.

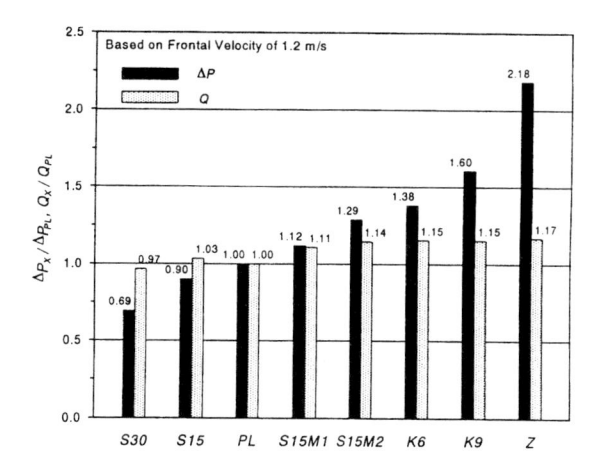

Fig. 10 Performance comparison of pressure drop and heat transfer rate at the frontal velocity of 1.2 m s^{-1}

Therefore, we can say that the pressure drop can be greatly reduced by the adequate design of the strips. The strip fins were similar in heat transfer rate. Fins S15M1 and S15M2 especially have strong merit based on pressure drop and material saving with negligible loss of heat transfer.

CONCLUDING REMARKS

An experimental work was performed to investigate the thermal-hydraulic characteristics of two-row finned-tube heat exchanger by using enlarged models. This work discusses the pressure drop and heat transfer characteristics of the offset strip fins in dehumidifying applications. The eight fin patterns having different strip locations were tested for the same pre-coated condition and tube layout. We obtained the following conclusions.
1. The strip density was the major factor and its shape and location were secondary factors on the pressure drop, the heat and mass transfer.
2. The strip pattern and fin shape affect the heat and mass transfer analogy. Excessive strip density could reduce the latent transfer.
3. The reduced-area fin could almost equal the non-reduced fin in the aspect of heat and mass transfer with the adequate design of the shape and strips.
4. The previous theories for the heat and mass transfer analogy normally agreed with the present experiment. However, the theories under-predicted the ratio of the sensible to mass transfer coefficients.
5. The strip fins proposed in the present work can considerably reduce both the pressure drop and fin material for a similar thermal load.

ACKNOWLEDGMENT

This work was supported by grant No. 2000-1-30400-009-2 from the Basic Research Program of the Korea Science & Engineering Foundation.

NOMENCLATURE

A_s surface area or total surface area, m^2
c_p heat capacity, J kg^{-1} K^{-1}
D tube diameter including fin color, m
D_h hydraulic diameter, $4V/A_s$, m
f friction factor for the fin, dimensionless
g gravity of acceleration, m s^{-2}
h enthalpy, J kg^{-1}
h_{fg} latent heat, J kg^{-1}
h_m mass transfer coefficient, kg m^{-2} s^{-1}
h_s sensible heat transfer coefficient, W m^{-2} K^{-1}
j Colburn factor, dimensionless
k thermal conductivity of fin, W m^{-1} K^{-1}
Le Lewis number, dimensionless
$\dot{m}$ mass flow rate, kg s^{-1}
n geometric scale factor, dimensionless
P_f fin pitch, m
P_l longitudinal tube pitch, m
P_t traversing tube pitch, m
Pr Prandtl number, dimensionless
ΔP pressure drop, Pa
q heat transfer rate, W
Re$_{D_h}$ Reynolds number based on hydraulic diameter, $\rho D_h V/\mu$, dimensionless
Re$_f$ film Reynolds number, dimensionless
RH relative humidity
Sc Schmidt number, dimensionless
t fin thickness, m
T temperature, K
V volume in side heat exchanger, m^3 or frontal air velocity, m s^{-1}

W absolute humidity, kg kg^{-1}

Subscripts

a air
d condensate
e exit
i inlet
l latent
LM logarithmic mean
s sensible or surface
w tube wall or water

Greek Letters

θ declination angle from horizon, rad
δ film thickness, m
η fin efficiency, dimensionless
ν kinematic viscosity, m^2 s^{-1}
ρ density, kg m^{-3}

REFERENCES

Ali, M. M. and Ramadhyani, S., 1992, Experiment on Connective Heat Transfer in Corrugate Channels, *Experimental Heat Transfer*, Vol. 5, 175-193.

Beecher, D. T. and Fagan, T. J., 1987, Effect of Fin Pattern on the Air-Side Heat Transfer Coefficient in Plate Finned-Tube Heat Exchangers, *ASHRAE Trans.*, Vol. 93, Part 2., pp. 1961-1984.

British Standard Institution, 1964, Method for the Measurement of Fluid Flow in Pipes: Part 1: Orifice Plates, Nozzles and Venturi Tubes, *British Standard 1042*.

Hatada, T. E., Ueda, H., Oouchi, T. and Shimizu, T., 1989, Improved Heat Transfer Performance of Air Coolers by Strip Fins Controlling Air Flow Distribution, *ASHRAE Trans.*, Vol. 95, Part 1, pp. 166-170.

Hiroaki, K., Shinichi, I., Osamu, A. and Osao, K., 1989, High-Efficiency Heat Exchanger, *National Technical Report*, Vol. 35, No. 6, pp. 653-661.

Kang, H. C. and Webb, Ralph L., 1998a, Evaluation of the Wavy Fin Geometry Used in Air-cooled Finned-Tube Heat Exchangers, *Proc. 11th International Heat Transfer Conference*, Vol. 6, pp. 95-100.

Kang, H. C. and Webb, Ralph L., 1998b, Evaluation of the Wavy Fin Geometry Used in Air-cooled Finned-Tube Heat Exchangers, *Proc. 11th Int. Heat Transfer Conference*, Vol. 6, pp. 95-100.

Kang, H. C. and Kim, M. H., 1999a, Effect of Strip Location on the Air-Side Pressure Drop and Heat Transfer in Strip Fin-and-Tube Heat Exchanger, *Int. J. of Refrig.*, Vol. 22, pp. 302-312.

Kang, H. C. and Kim, M. H., 1999b, Alignment of Strips to Reduce the Pressure Drop in the Finned-tube Heat Exchanger Having Offset Strip Fins, *Proc. International Conference on Compact Heat Exchangers and Enhancement Technology for the Process Industries*, pp. 217-223.

Kang, H. C., Kim, M. H. and Ha, S. C., 2000a, Experimental Method to Evaluate the Performance of Finned-Tube Heat Exchanger under Dehumidifying Conditions by Using the Scaled Model, *Proc. 4th JSME-KSME Thermal Engineering Conference*, Vol. 3, pp. 769-774.

Kang, H. C., Kim, M. H. and Kim, M. S., 2000b, A Study on the Thermal Characteristics of Finned-tube Heat Exchanger by Using the Liquid Crystal Technique, *J. of the Society of Air Conditioning and Refrigerating Engineers of Korea*, Vol. 4, pp. 414-421.

Kim, N. H, Yun, J. H. and Webb, R. L., 1997. Heat Transfer and Friction Correlations for Wavy Plate Fin-and - Tube Heat Exchangers, *Trans of the ASME*, Vol. 119, pp. 560-567.

Krückels, S. W. and Kottke, V., 1970, Investigation of the Distribution of Heat Transfer on Fins and Finned Tube Models, *Chemical Engineering Technology*, Vol. 42, pp. 355-362.

Nakayama, W. and Xu, L. P., 1983, Enhanced Fins for Air-Cooled Heat Exchangers-Heat Transfer and Friction Factor Correlations, *Proc. 1983 ASME-JSME Thermal Engineering Conference*, Vol. 1, pp. 495-502.

Saboya, F. E. M. and Sparrow, E. M., 1974, Local and Average Heat Transfer Coefficients for On-Row Plate Fin and Tube Heat Exchanger Configurations, *J. of Heat Transfer*, Vol. 96, pp. 265-272.

Tanaka, T., Itoh, M., Kudoh, M. and Tomita, A., 1984, Improvement of Compact Heat Exchangers with Inclined Louvered Fins, *Bulletin of JSME*, Vol. 27, No. 224, pp. 219-226.

Wang, C. C., Chen, P. Y. and Jang, J. Y., 1996, Heat Transfer and Friction Characteristic of Convex-Louver Fin-and-Tube Heat Exchanger, *Experimental Heat Transfer*, Vol. 9, pp. 61-78.

Wang, C. C., Fu, W. L. and Chang, C. T., 1997a, Heat Transfer and Friction Characteristics of Typical Wavy Fin-and-tube Heat Exchanger, *Experimental Thermal and Fluid Science*, Vol. 14, No.2, pp. 174-186.

Wang, C. C., Lee, C. J. and Chang, C. T., 1997b, Some Aspects of Plate Fin-and-Tube Heat Exchangers: with and without Louvers, *J. of Enhanced Heat Transfer*, Vol. 14, pp. 174-186.

Wang, C. C., Lee, W. S. and Chang, C. T., 1997c, Heat and Mass Transfer for Plate Fin-and-tube Heat Exchangers with and without Hydrophilic Coating, Working paper.

FRICTIONAL PERFORMANCE OF RECTANGULAR FIN IN PARTIALLY WET CONDITIONS

Yur-Tsai Lin[1], Kuei-Chang Hsu[1], Ruey-Jong Shyu[2], and Chi-Chuan Wang[2]

[1] Department of Mechanical Engineering, Yuan-Ze University, Taoyuan, Taiwan
[2] Energy & Resources Laboratories, ITRI, Hsinchu, Industrial Technology Research Institute, Taiwan 310,
ccwang@itri.org.tw

ABSTRACT

An experimental study concerning the performance of a rectangular fin in partially wet condition was carried out. The visual observation of the dehumidifying phenomenon on a rectangular fin can be classified into four regions, including the fully dry, very fine droplet, larger droplet, and film-like region. If the frictional data of "completely wet" and "completely dry" are known, the frictional performance of "partially wet" surface can be well described by the associated dry-wet area ratio. However, if the frictional data of "completely wet" is not available, a correlation taking into account the influence of inlet relative humidity is proposed. The proposed correlation is applicable to "completely wet", "completely dry", and "partially wet" conditions with a mean deviation of 9.8%

INTRODUCTION

Air-cooled heat exchangers used for air-conditioning and refrigeration applications are generally consisted of plate fin-and-tube heat exchangers. For heat exchangers operated as the evaporators or cooling coils of air-conditioning equipments, the surface temperature of the fins may be below the dew point temperature. As a result, moisture is condensed on the fins. For many applications that may operate under higher air frontal velocities or lower air inlet relative humidity conditions, part of the fin temperature may be higher than the dew point temperature, causing this portion "dry". As a result, the surface conditions in air-conditioning applications can be classified as "completely dry", "completely wet", and "partially wet" conditions.

Figure 1 depicts the difference between "completely wet" and "partially wet" surface conditions. As seen, for completely wet conditions, the fin surface temperature T_{fin} is lower than its corresponding dew point temperature on the heat transfer surface. For "partially wet" surface condition, part of the heat transfer surface temperature is higher than their corresponding dew point temperature (namely $T_{fin} > T_{dew}$). For partially wet surface, the overall heat transfer performance is a combination of "completely-dry" and "completely-wet" conditions. Many experimental and theoretical studies were performed on the "completely-dry" and "completely-wet" conditions. However, only limited studies were mad to the subject of "partially wet" condition of extended surfaces (e.g. McQuiston, 1975; Wu and Bong 1994; Kazeminejad, 1995; Salah, 1998; Rosario and Rahman, 1999).

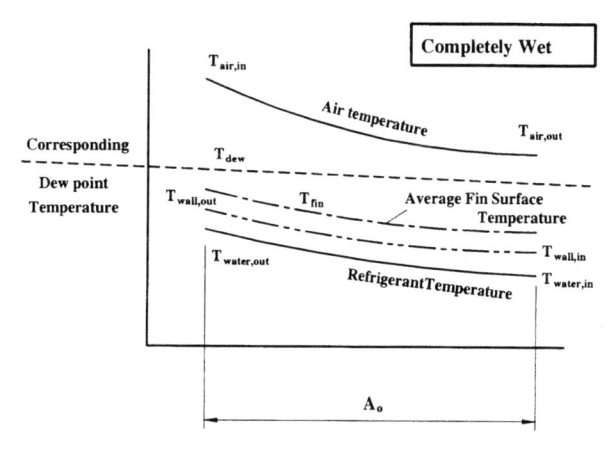

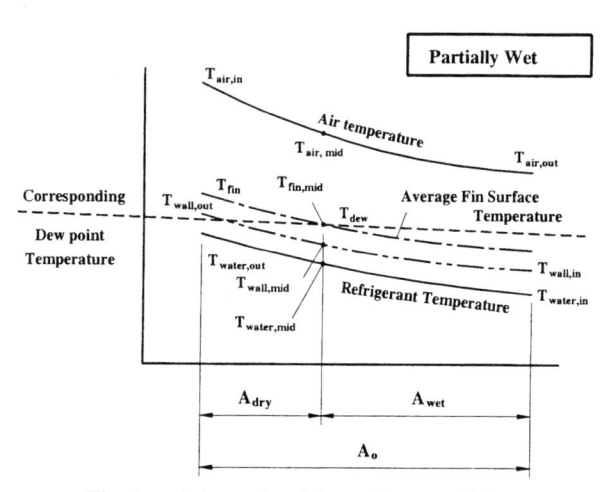

Fig. 1 Schematic of the surface condition

Furthermore, these above-mentioned studies were all based on analytical approach. Experimental information about this basic phenomenon is not available. As a consequence, in view of lack of experimental evidence to describe this phenomenon, it is the objective of the present study to explore this basic condensation flow pattern and the related frictional behavior of a "partially wet" surface via experimental approach.

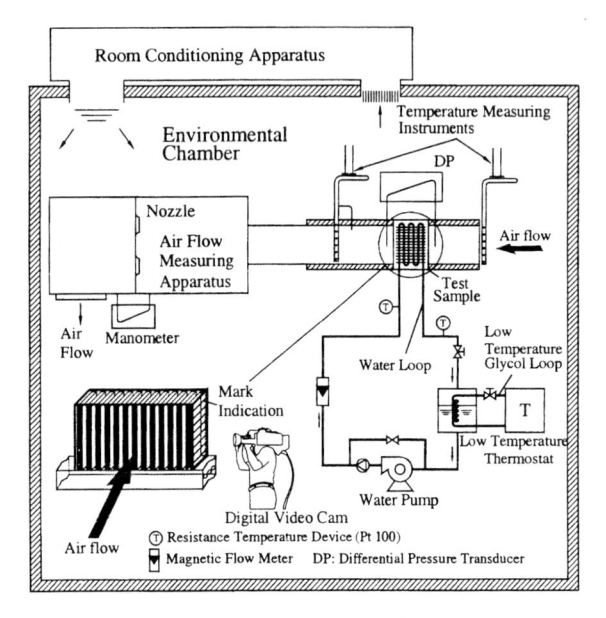

Fig. 2 Schematic of the experimental setup

EXPERIMENTAL APPARATUS

Experiments were performed in an environmental chamber as shown in Figure 2. The test apparatus is based on the air-enthalpy method proposed by ANSI/ASHRAE Standard 37 (1978). Cooling capacity was measured from the enthalpy difference of the air flowrate across the test sample. The airflow measuring apparatus is constructed from ASHRAE Standard 41.2 (1986). The dry and wet bulb temperature measurement devices of the airflow are constructed based on ASHRAE 41.1 Standard (1987). Schematic diagram of the experimental air circuit assembly is shown in Fig. 2. Experiments were performed in an environmental chamber. The environmental chamber can control the ambient conditions in the range of 10 °C $\leq T_{DB} \leq$ 50 °C and 40% $\leq$ RH $\leq$ 95%. Control resolution for the related dry bulb and wet bulb temperatures is 0.1 °C.

For the experimental investigations, a model extended surface was designed and manufactured as depicted in Figure 3. The Aluminum alloy 601 is chosen as the base fin material because of its relatively high thermal conductivity and rigidity. The fin is 100 mm deep, 100 mm wide, and 2 mm thick. The fins were placed on a stainless steel block as shown in Figure 3. The fin spacing can be 3 or 8 mm. To measure the fin temperature distribution in both directions of transverse and longitudinal to the airflow, a total of eighteen thermocouples were mounted on one of the fins that are located near the center position. Detailed locations of the thermocouples were shown in Figure 3. These 'T'-type thermocouples were pre-calibrated with a resolution of 0.1 °C. To heat and cool the fin, water was circulated via four holes drilled beneath the fin base in the stainless steel block. The water inlet temperature of the test section was controlled by low temperature thermostat. During the experiments, water was circulated with a sufficiently high velocity (> 3 m/s) to maintain the fin base temperature at a constant level. Variation of the fin base temperature between the inlet and the outlet can be maintained to be less than 0.2 °C. The test conditions are given as follows:

Inlet dry bulb temperature: 20 ~ 27 °C
Inlet relative humidity: 40 ~ 90 %
Water temperature at the inlet: 2 - 9 °C
Base fin temperature: 3 ~13 °C
Fin spacing (F_s): 3 and 8 mm.

The fins were mounted vertically to the support. Care was taken to align the fin so that it was parallel to the air flow. Droplet formation on the fin was recorded by a JVC digital video cam DVM50u with a speed of 30 frames/s. Experimental uncertainties for temperatures on the fin and airflow are 0.1 °C whereas the uncertainties of the pressure drops across the fin are 0.1~1 Pa.

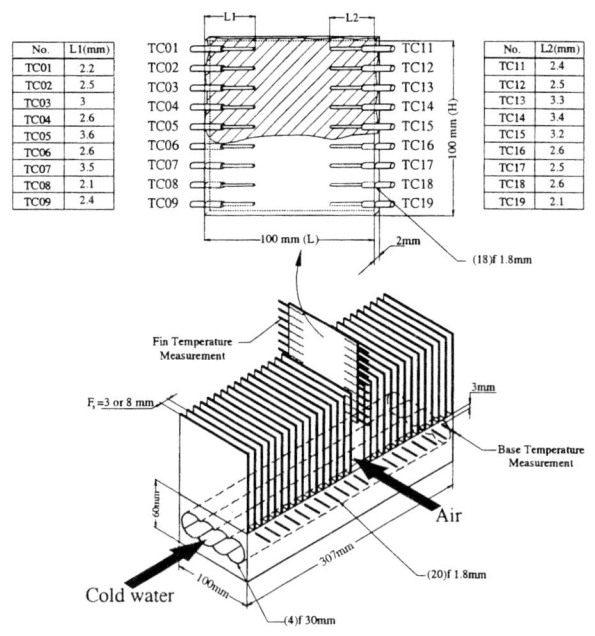

Fig. 3 Enlarged view of the test section

RESULTS AND DISCUSSION

In the visualization study, photographs were taken for frontal velocities ranging from 0.3 m/s to 6 m/s. The corresponding fin base temperature and inlet dry bulb temperature were maintained at 9 °C and 27 °C, while the inlet relative humidity ranged from 50% to 90%. Selected photographs from the observations for F_s = 8 mm at RH$_{in}$ =

50, 70, and 90% are shown in Figs, 4a, 4b, and 4c, respectively.

In Fig. 4a, at a high frontal velocity like 6 m/s or 4 m/s and an inlet relative humidity of 50%, formation of the droplet can be seen only at the vicinity of the fin base. One can clearly see the boundary separating the dry and wet portions. Note that the droplet size at the boundary is very fine and the droplet size increases as it approaches to the fin base. The height of the boundary increases as the frontal velocity is further decreased to 2.5 m/s or 1.5 m/s. As clearly seen in Fig. 4a, the wet portion on the fin is increased. Notice that the boundary line is not horizontal, the dry portion near the leading edge is larger than that of trailing region. The observation is quite different from those common assumptions made by the analytical approaches. The analytical model of the "partially wet" fin always assumed a horizontal boundary line and one-dimensional fin temperature distribution. Apparently, the dry-wet boundary is not horizontal as shown in Fig. 4a. The inclination of the dry-wet boundary is related to the fin temperature distribution.

Airflow direction ←

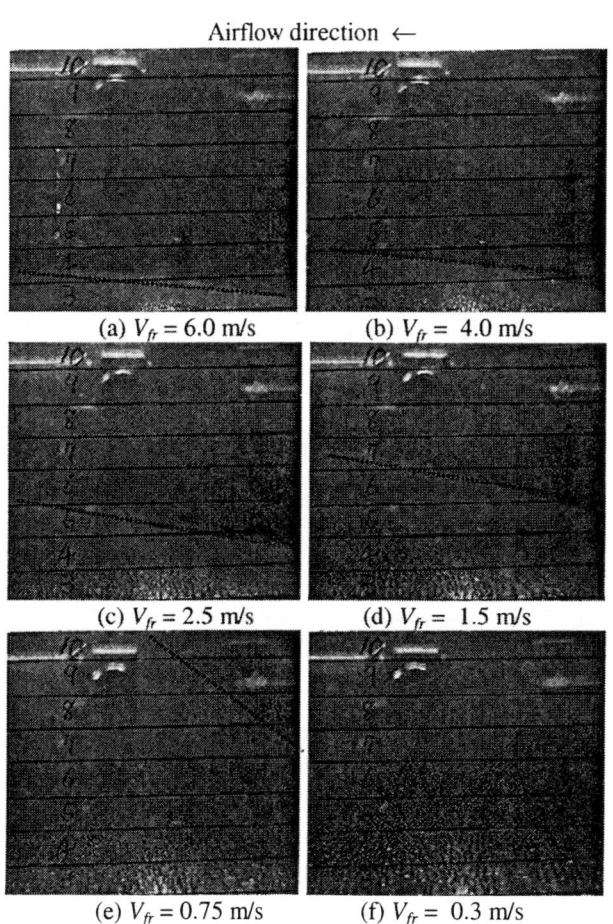

(a) V_{fr} = 6.0 m/s (b) V_{fr} = 4.0 m/s

(c) V_{fr} = 2.5 m/s (d) V_{fr} = 1.5 m/s

(e) V_{fr} = 0.75 m/s (f) V_{fr} = 0.3 m/s

Fig. 4a Droplet formation on a fin with RH_{in} = 50%, F_s = 7 mm, T_{fb} = 9 °C.

The slope of the inclination boundary line increases with the decrease of velocity. Although the mean dew point temperature decreases along the airflow direction, the boundary line is located at the fin temperature that is close to the inlet dew point temperature. This is because the cross-flow arrangement of the present fin geometry. A further decrease of frontal velocity to 0.75 m/s and 0.3 m/s causes the boundary line to move up even further. For V_{fr} = 0.3 m/s, the entire fin is fully wet because all the fin temperature is below the inlet dew point temperature.

Airflow direction ←

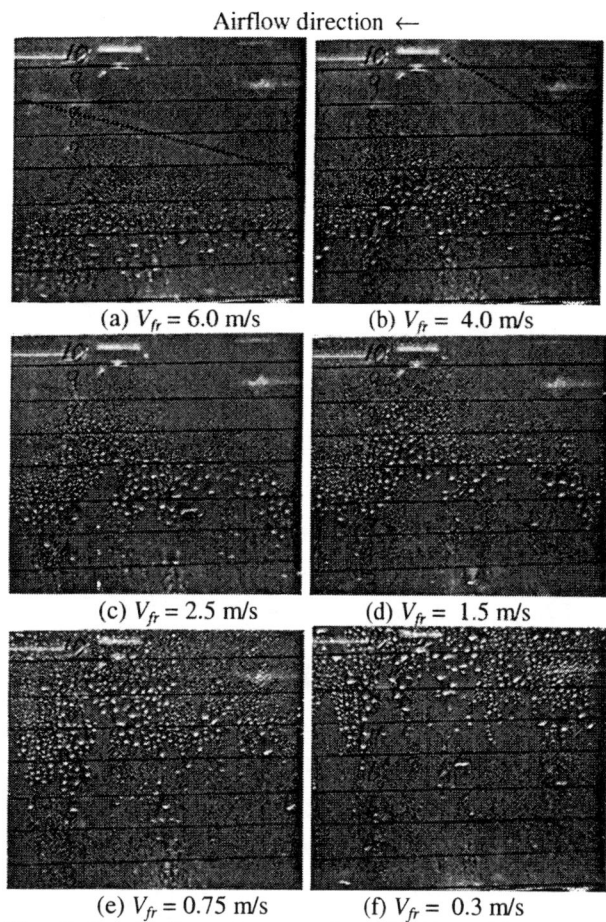

(a) V_{fr} = 6.0 m/s (b) V_{fr} = 4.0 m/s

(c) V_{fr} = 2.5 m/s (d) V_{fr} = 1.5 m/s

(e) V_{fr} = 0.75 m/s (f) V_{fr} = 0.3 m/s

Fig. 4b Droplet formation on a fin with RH_{in} = 70%, F_s = 7 mm, T_{fb} = 9 °C.

The results of RH_{in} = 70% are analogous to those of RH_{in} = 50% with the dry-wet boundary being moved further upward. Basically, the dehumidification phenomenon of a rectangular fin can be roughly classified into four regions. For region (I) the fin temperature is above the inlet dew point temperature, the surface is fully dry. For region (II) that is near the dry-wet boundary, the droplet size is very small (of the order of 0.1~ 0.3 mm). In region (II), the heat transfer is similar to dropwise condensation. Eventually, the very fine size of droplet may grow up in the down stream of airflow and join with the neighboring droplet to become larger droplets (region III). In this region, the larger droplet grew more quickly to a critical size, overcoming the force due to surface tension. They then rolled randomly down the fin due to gravity, dragging other droplets and drained the condensate close to the base of the fin. Notice that region (III) is quite unsteady when comparing to region (I) and (II). In region (IV) that is close to the base portion, a thin film was formed that is formed by the "washed away" droplet drained from region (III). Notice that a wave-like droplet may deposit on the film. It should be pointed out that region (I) will disappear when the surface temperature is below inlet dew point temperature and region (II) may also disappear when the air velocity is low and the inlet humidity is sufficiently high. As can be seen from Fig. 4c, the region (II) is not seen for $V_{fr} < 1.5$ m/s. The condensate drainage phenomenon is especially pronounced in the leading portion of the fin. One can see that the frequency of droplet falling off from region (II) is more rapid when comparing to the trailing edge of the fin. This phenomenon can be interpreted as follows. As is known, when the condensation take place, the non-condensable air will form at the interface between water condensate and air flow, the mass transfer resistance will inhibit the heat transfer. In addition, the fall of local dew point temperature will also contribute to reduce the heat transfer process.

Airflow direction ←

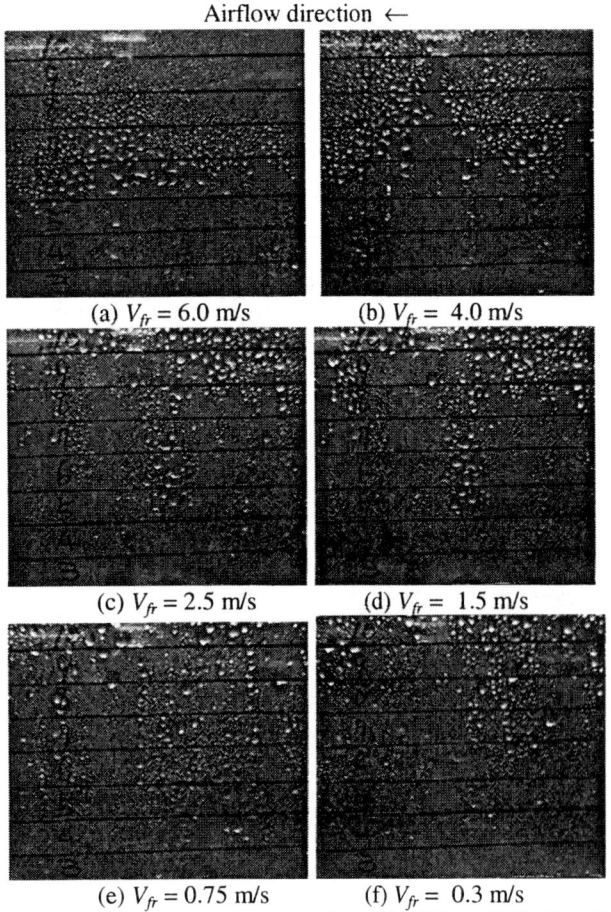

(a) $V_{fr} = 6.0$ m/s   (b) $V_{fr} = 4.0$ m/s

(c) $V_{fr} = 2.5$ m/s   (d) $V_{fr} = 1.5$ m/s

(e) $V_{fr} = 0.75$ m/s   (f) $V_{fr} = 0.3$ m/s

Fig. 4c Droplet formation on a fin with $RH_{in} = 90\%$, $F_s = 7$ mm, $T_{fb} = 9$ °C.

In addition to the condensate flow pattern, pressure drops for $F_s = 3$ mm and $F_s = 8$ mm for $RH_{in} = 50\%\sim 90\%$ are shown in Fig. 5. As expected, the pressure drops for a close fin spacing ($F_s = 3$ mm) are higher than those of $F_s = 8$ mm. In Figure 5, one can see that the effect of relative humidity on the pressure drops is comparatively pronounced for $F_s = 3$ mm. For $F_s = 8$ mm, the influence of inlet relative humidity is quite minor. However, for $F_s = 3$ mm, one can see the pressure drop for $RH_{in} = 90\%$ is about 50% higher than that of $RH_{in} = 50\%$ at $V_{fr} = 3$ m/s. This is due to the associated roughness effect caused by the condensate is amplified for a smaller fin spacing.

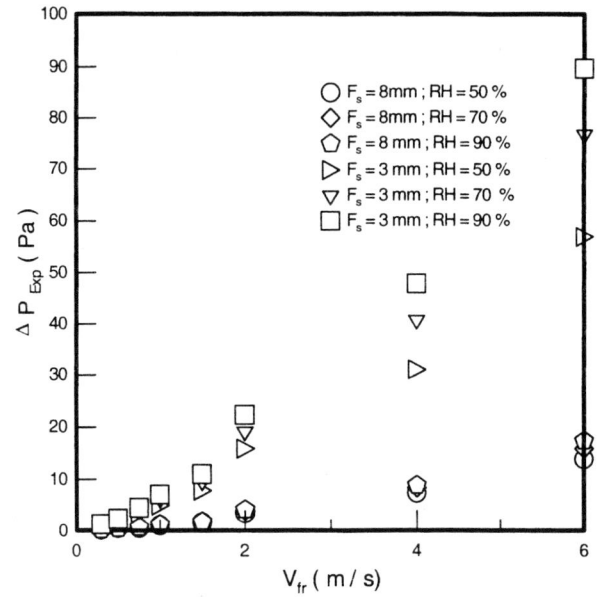

Fig. 5 Pressure drops for $F_s = 3$ mm and $F_s = 8$ mm in wet condition.

In order to quantify the frictional performance in this study, two approaches are adopted in this study. Firstly, the frictional performance of "partially wet" rectangular fin is predicted by the following approach:

$$\Delta P = \Delta P_{dry} \frac{A_{dry}}{A} + \Delta P_{wet} \frac{A_{wet}}{A} \qquad (1)$$

Notice that the subscript dry and wet denote the condition in "completely dry" and "completely wet" condition. The conditions had been justified from the measurements of the fin temperature along the rectangular fin. If all fin temperatures are below the corresponding inlet dew point temperature, the surface is regarded as "completely wet". Conversely, if all the fin temperatures are above the corresponding inlet dew point temperature, the surface is regarded as "completely dry". If the information are available, we can use these data to predict the associated "partially wet" characteristics. Results of comparisons for both $F_s = 3$ mm and $F_s = 8$ mm is shown in Figure 6. As seen, "partially wet" data can be well predicted by Eq (1). The results implies that the "partially wet" frictional performance can be well described by use of "completely wet" and "completely dry" friction data. The experimental results may substantiate the complex modeling developed by Liaw et al. (1998) who use the existing "completely wet" and "completely dry" data to predict the performance of "partially wet" characteristics.

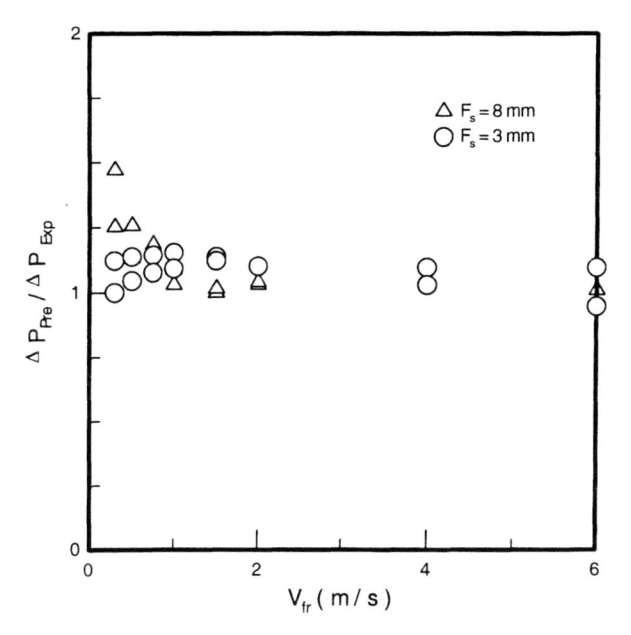

Fig. 6 Comparison of the pressure drops between experimental data and Eq. (1)

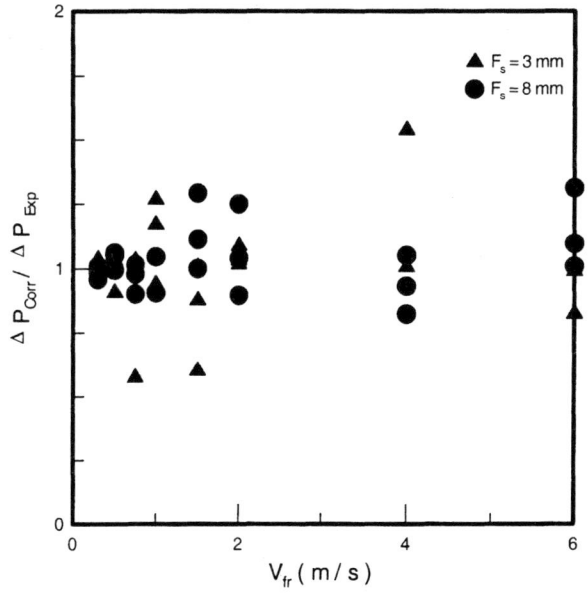

Fig. 7 Comparison of the pressure drops between experimental data and proposed correlation

However, in practice it is very hard to have both frictional data of "completely wet" and "completely dry" in hand. Further, it is not likely to know the exact fin surface temperature distribution along the fins without detailed measurements or complex calculation. As a consequence, this will make the simple model of Eq. (1) physically impractical. Thus, an alternative approach was also given to correlate the wet frictional performance in connection to the "completely dry" conditions. Based on the present test results, the proposed frictional correlation is given as follows:

For $F_s = 8$ mm,

$$\frac{\Delta P_{wet}}{\Delta P_{dry}} = \begin{cases} \dfrac{63 RH^{1.5}}{Re^{0.75}} + 1 & \text{if } Re < 400 \\ \dfrac{1.6 - 0.3 RH^{-2}}{7.9 - 0.0009\,Re - 2830\,Re^{-1}} & \text{if } Re \geq 400 \end{cases} \quad (2)$$

For $F_s = 3$ mm,

$$\frac{\Delta P_{wet}}{\Delta P_{dry}} = (1.46 - 1.8 RH^2)(0.036 + 30 Re^2) + (0.24 - 0.12 RH^{-1}) Re^{0.217} + 1 \quad (3)$$

Results of comparison of Eqs. (2-3) with the experimental data is show in Figure 7. The mean deviation of the above-mentioned equation is 9.8%. Note that the proposed equations are applicable to "completely wet", "completely dry", and "partially wet" conditions.

CONCLUSIONS

An experimental study concerning the condensate flow pattern on a rectangular fin in "completely wet" and "partially wet" condition was examined. Focus is made especially on the frictional characteristics in "partially wet" condition. Based on the test results from the observations the following conclusions are made.

Visual observation of the dehumidification along the enlarged aluminum fin indicated the droplet formation can be roughly classified into four regions. In region (I), the fin temperature is above the inlet dew point temperature, the surface is fully dry. In region (II) that is nearby the dry-wet boundary, the drop size is very small (of the order of 0.1~0.3 mm) and the heat transfer mode is similar to dropwise condensation. In region (III), the droplet of very fine size may grow and join with the neighboring droplet to become a larger droplet. In region (IV) that is close to the base portion, a thin film was formed.

If the frictional data of "completely wet" and "completely dry" are known, the frictional performance of "partially wet" surface can be well described by the associated dry-wet area ratio. However, if the frictional data of "completely wet" are not available, a correlation taking into account the influence of the inlet relative humidity is proposed. The proposed correlation is applicable to "completely wet", "completely dry", and "partially wet" conditions with a mean deviation of 9.8%.

NOMENCLATURE

A	surface area, m²
A_{dry}	dry fin surface, m²
A_{wet}	wet fin surface, m²
F_s	fin spacing, mm
k_w	thermal conductivity of water, W/m·K
L	fin depth, m
Re	Reynolds number based on hydraulic diameter
RH_{in}	inlet relative humidity
T	temperature, °C
T_{DB}	dry bulb temperature, °C
T_{dew}	dew point temperature, °C
T_{fin}	fin temperature, °C
T_{fb}	fin base temperature, °C
T_{∞}	ambient temperature, °C
V_{fr}	frontal velocity, m/s
x	distance from the fin base, m
y	distance from the leading edge of fin, m
ΔP	total pressure drop, Pa
ΔP_{Corr}	pressure drop predicted by the proposed correlation of Eqs. (2)~(3), Pa
ΔP_{dry}	pressure drop of "completely dry" condition, Pa
ΔP_{EXP}	experimental pressure drop, Pa
ΔP_{Pre}	pressure drop predicted by Eq. (1), Pa
ΔP_{wet}	pressure drop of "completely wet" condition, Pa

ACKNOWLEDGMENTS

The authors are indebted to the National Science Committee (NSC 89-2212-E-155-007) and the Energy R&D foundation funding from the Energy Commission of the Ministry of Economic Affairs of Taiwan for supporting this study.

REFERENCES

McQuiston, F.C., 1975, Fin Efficiency With Combined Heat and Mass Transfer, *ASHRAE Transactions*, Vol. 81, part 1, pp. 350-355.

Wu, G., and Bong, T.Y., 1994, Overall Efficiency of a Straight Fin With Combined Heat and Mass Transfer, *ASHRAE Transactions*, Vol. 100, part 1, pp. 367-374.

Kazeminejad, H., 1995, Analysis of One-dimensional fin assembly Heat Transfer With Dehumidification, *Int. J. of Heat and Mass Transfer*, Vol. 38, pp. 455-462.

Salah El-Din, M.M., 1998, Performance Analysis of Partially-wet fin Assembly, *Applied Thermal Engng.*, Vol. 18, pp. 337-349.

Rosario, L., and Rahman, M.M., 1999, Analysis of Heat Transfer in a Partially wet Radial fin Assembly During Dehumidification, *Int. J. Heat and Fluid Flow*, Vol. 20, pp. 642-648.

ASHRAE Standard 33-78, 1978, Method of Testing Forced Circulation Air Cooling and Air Heating Coils, Atlanta: American Society of Heating, Refrigerating and Air-Conditioning Engineers, Inc.

ASHRAE Standard 41.1-1986, 1986, Standard Method for Temperature Measurement. Atlanta: American Society of Heating, Refrigerating and Air-Conditioning Engineers, Inc.

ASHRAE Standard 41.2-1987, 1987, Standard Methods for Laboratory Air-flow Measurement. Atlanta American Society of Heating, Refrigerating and Air-Conditioning Engineers, Inc.

Liaw, J.S., Jang, J.Y., and Wang, C.C., 1998, A Rationally Based Model for Air-cooled Evaporator/Cooling Coils; Accounting the Effect of Partially Wet Condition, *Proc. of the Int. Conf. and Exhibit on Heat Exchangers for Sustainable Development*, Portugal, pp. 303-312.

PHASE-CHANGE HEAT TRANSFER
FUNDAMENTAL STUDIES

TWO-PHASE FLOW PATTERNS, PRESSURE DROP AND HEAT TRANSFER DURING BOILING IN MINICHANNEL AND MICROCHANNEL FLOW PASSAGES OF COMPACT EVAPORATORS

Satish G. Kandlikar

Mechanical Engineering Department, Rochester Institute of Technology, Rochester, NY 14623; E-mail: sgkeme@rit.edu

ABSTRACT

The small hydraulic diameters employed during flow boiling in compact evaporator passages are becoming more important as they are employed in diverse applications including electronics cooling and fuel cell evaporators. The high pressure drop characteristics of these passages are particularly important as they alter the flow and heat transfer, especially in parallel multi-channel configuration. The pressure drop oscillations often introduce dryout in some passages while their neighboring passages operate under single-phase mode. This paper presents a comprehensive review of literature on evaporation in small diameter passages along with some results obtained by the author for water evaporating in 1-mm hydraulic diameter multi-channel passages. Critical heat flux is not covered in this paper due to space constraint.

1. INTRODUCTION

Major progress in compact evaporator development has been made by the automotive, aerospace, and cryogenic industries over the last fifty years. The thermal duty and the energy efficiency increased during this period, while the space constraints became more vital. The trend was toward greater heat transfer rates per unit volume. The hot side of the evaporators in these applications was generally air, gas or a condensing vapor. The airside fin geometry also saw significant improvements resulting from increased heat transfer coefficients as well as greater surface area densities.

As the airside heat transfer resistance decreased, more aggressive fin designs were employed on the evaporating side, resulting in narrower flow passages. The narrow refrigerant channels with large aspect ratio were brazed in small cross-ribbed sections to provide a better refrigerant distribution along the width of the channel. The major changes in the recent evaporator designs involve individual, small hydraulic diameter flow passages, arranged in multi-channel configuration for the evaporating fluid.

Figure 1 shows a plate-fin evaporator geometry widely used in compact refrigerant evaporators. The refrigerant side passages are made from two plates brazed together, and airside fins are placed between two refrigerant flow passages. The plates have cross ribs that are oriented in opposite directions so that the top and bottom plates forming a refrigerant passage have only contact points at the intersection of these ridges. The two-phase flow of refrigerant gets distributed across the width of the flow passage. This feature is important in that the localized flow oscillations, caused by nucleate boiling and expanding vapor bubbles, is dissipated across the width and does not affect the upstream flow.

Figure 2 shows two geometries being used more widely in compact evaporators. The parallel minichannel geometry shown in Figure 2a is used extensively in condensation applications, whereas the geometry shown in Figure 2b has received quite a bit of attention for boiling applications. The channels are fabricated by a variety of processes depending on the dimensions and plate material. Conventional machining and electrical discharge machining are two typical options, while semiconductor fabrication processes are appropriate for microchannel fabrication in chip cooling application.

2. FLOW PATTERNS IN SMALL DIAMETER TUBES

2.1 Previous Studies

Visualization of boiling phenomena inside flow channels provides insight into the heat transfer mechanisms. Bubble formation, coalescence, formation of slugs or plugs, and local dryout conditions, are all important in understanding the heat transfer phenomena. Although some of the heat transfer and pressure drop equations employed in the design of commercial equipment are derived from flow pattern based models, the major benefit of flow pattern information lies in understanding the causes for premature dryout or CHF condition in an evaporator. Another major benefit is in the design of the inlet and the outlet manifolds in multi-channel evaporators.

The flow pattern maps available in literature were first developed for the petrochemical industry (Baker, 1954) for flow of oil and gas in large diameter pipes. Subsequently, the adiabatic flow pattern maps were developed as general flow pattern maps (for example, Hewitt and Roberts, 1969, and Taitel et al., 1980). In recent years, a number of flow pattern maps have been developed for specific conditions such as small diameter tubes, evaporation or condensation, and compact heat exchanger geometries.

Earlier investigators extensively studied flow patterns for gas-liquid flows in channels with small hydraulic diameters. A representative survey of the flow patterns was presented by Fukano et al. (1989). They identified bubbly, plug, slug, and annular flow patterns and compared the flow pattern transitions with the available flow pattern maps. Subsequently, a detailed study by Wambsganss et al. (1991) provides a more comprehensive summary and representation of gas-liquid flow patterns. The role of surface tension becomes more important in smaller diameter channels. Triplett et al. (1999) explain that due to the dominance of surface tension, stratified flow is essentially absent, slug (plug) and churn flow patterns occur over extensive ranges of parameters, and the slip velocity under these patterns is small. Stratified flow can exist at very low flow rates, as observed by Kasza et al. (1997) for a mass flux of 21 kg/m^{2}s.

Hewitt (2000) gives a comprehensive summary of flow pattern studies available in literature. For large diameter tubes, the generalized flow pattern map for an air-water system, developed by Mandhane et al. (1974), was quite representative of other flow conditions as well. However, the theoretically based transition

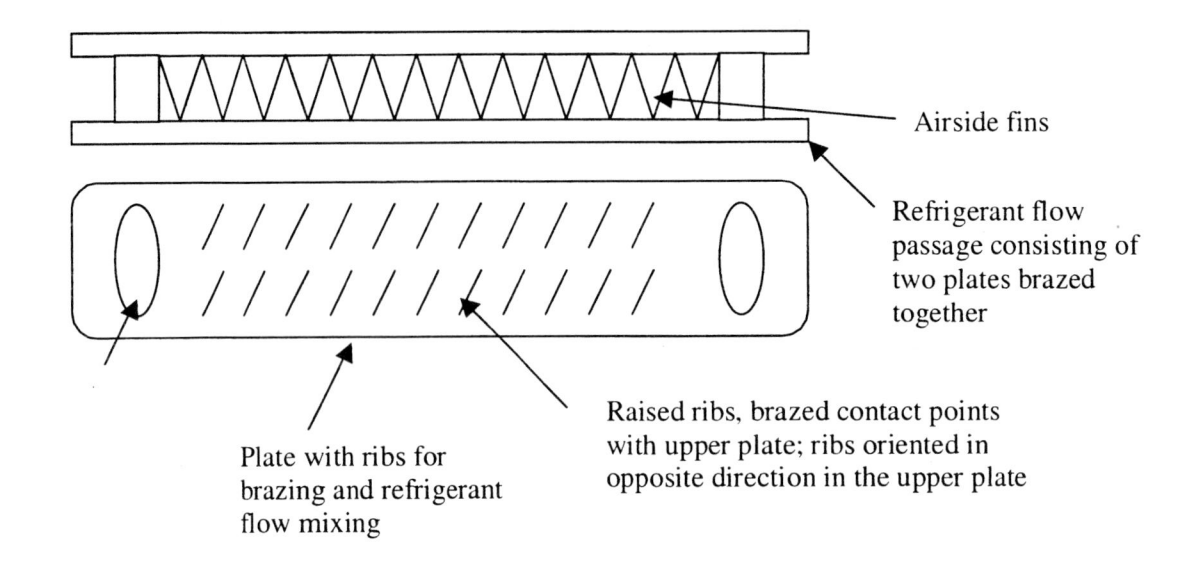

Figure 1 Schematic of refrigerant and airside flow passages in a compact plate-fin type evaporator

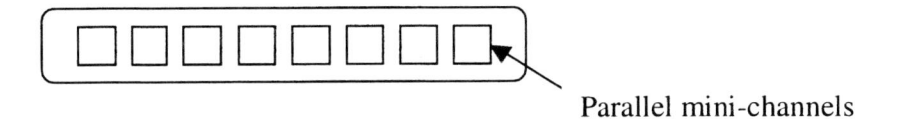

(a) Schematic cross sectional view of a multichannel evaporator with parallel mini-channels

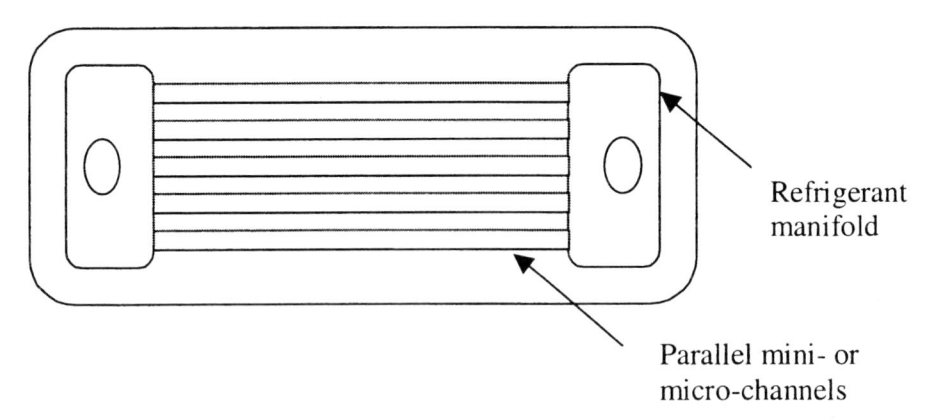

(b) Schematic of a multichannel evaporator with parallel mini- or micro-channels

Figure 2 Recent developments in multichannel evaporators

TABLE 1 SUMMARY OF INVESTIGATIONS ON EVAPORATION IN MINI- AND MICROCHANNELS

Author/ Year	Fluid and Ranges of parameters, G-kg/m^{2}s, q-kW/m^2	Channel Size/ D$_h$, mm, Horiz. (unless otherwise stated)	Heat Transfer	Pressure Drop	Flow Patterns	Remarks
Lazarek and Black, 1982	R-113, G=125-750, q=14-380	Circular, D=3.1, L=123 and 246	Heat transfer coeff. and CHF	pressure drop measured and correlated	Not observed	Subcooled and saturated data obtained, h almost constant in the two-phase region, dependent on heat flux. Behavior similar to large dia. tubes-combination of nucleate and convective boiling. Pressure drop correlations obtained.
Cornwell and Kew, 1992	R-113, G=124-627, q=3-33	Parallel rectangular, 75 channels-1.2x0.9 deep 36 channels-3.25x1.1 deep	Heat transfer coeff	Not reported	Isolated bubble, Confined bubble, annular-slug	The heat transfer coefficient was found to be dependent on the existing flow pattern. In the isolated bubble region, h~q$^{0.7}$, lower q effect in confined bubble region, convection dominant in annular-slug region
Moriyama and Inoue, 1992	R-113, G=200-1000, q=4-30	Narrow rectangular channel, 0.035-0.11 gap, width=30, L=265	Heat transfer coeff.	Pressure drop measured and components calculated	Flattened bubbles, w/coalescence, liquid strips, liquid film	Two-phase flow boiling data in narrow gaps obtained and correlated with an annular film flow model. Nucleate boiling ignored, although a dependence of h on q was observed.
Wambsganss et al. (1993)	R-113 G=50-100, q=8.8-90.7	Circular, D=2.92 mm	h as a function of x, G and q	Not reported	Not reported	Except at the lowest heat and mass fluxes, both nucleate boiling and convective boiling components were present.
Bowers and Mudawar, 1993, 1994	R-113, q=1000-2000, 0.28-1.1 ml/s	Mini and micro channels, D=2.54 and 0.51	Heat transfer rate	Pressure drop components	Not studied	Analytical (1993) and experimental (1994)studies comparing the performance of mini and micro channels. Mini-channels are preferable unless liquid inventory or weight constraints are severe.
Mertz, et al., 1996	Water and R-141b, G=50, 100, 200, 300 q=3-227	Rectangular channels, 1, 2 and 3 mm wide, aspect ratio up to 3	Heat transfer coeff and heat flux	Not measured	Nucleate boiling, confined bubble and annular	Single and multi-channel test sections. Flow boiling pulsation was observed in multi-channels, with reverse flow in some cases. Nucleate boiling dominant.
Ravigururajan et al., 1996	R-124, 0.6-5 ml/s, 20-400W	270 μm wide, 1 mm deep. 20.52 mm long,	Heat transfer coeff.	Not studied	Not studied	Experiments were conducted over 0-0.9 quality and 5 °C inlet subcooling. Wall superheat from 0-80 °C
Tran, et al., 1996	R-12, G=44-832, q=3.6-129	Circular, D=2.46, Rectangular, D$_h$=2.4	Heat transfer coeff.	Not studied	Not studied	Local heat transfer coeff. obtained up to x=0.94. Heat transfer in nucleate boiling dominant and convective dominant regions obtained. New correlation for nucleate boiling dominant region
Kasza et al., 1997	Water, G=21, q=110	Rectangular 2.5x6.0x500	Not reported	Not reported	Bubbly, Slug	Increased bubble activity on wall at nucleation sites in the thin liquid film responsible for high heat transfer in small channels
Tong, Bergles and Jensen, 1997	Water, G=25,000-45,000, CHF of 50-80 MW/m^2	Circular, D=1.05-2.44	Subcooled flow boiling	Focus on pressure drop	Not studied	Pressure drop measured in highly subcooled flow boiling, correlations presented for both single phase and two-phase
Bonjour and Lallemand, 1998	R-113, q=0-20	Rectangular, 0.5 to 2 mm gap, 60 wide, 120 long, Vertical	Not studied	Not studied	3 flow patterns with nucleate boiling	Nucleate boiling with isolated bubbles, nucleate boiling with coalesced bubbles and partial dryout, criteria proposed for transitions

Table 1 continued

Author/Year	Fluid and Ranges of parameters, G-kg/m²s, q-kW/m²	Channel Size/ D_h, mm, Horiz. (unless otherwise stated)	Heat Transfer	Pressure Drop	Flow Patterns	Remarks
Peng and Wang, 1998	Water, ethanol and mixtures,	Rectangular, width=0.2-0.4, height=0.1-0.3, L=50; Triangular, D_h=0.2-0.6, L=120	Heat transfer coeff., heat flux	Not reported	Not observed	The heat transfer results indicate that both, nucleate and forced convection boiling are present. No bubbles were observed, and the authors propose a fictitious boiling. The authors did not use proper microscope and high-speed video techniques resulting in contradictory conclusions.
Peng et al., 1998	Theoretical	Theoretical	Bubble nucleation model	Not studied	Not studied	Bubble nucleation model uses a vapor bubble growing on a heater surface with heat diffusion in the vapor phase. This corresponds to post- CHF heat transfer and is an inaccurate model of heat transfer during nucleation and bubble growth.
Kamidis and Ravigururajan, 1999	R-113, power 25-700 W, Re=190-1250	Circular, D=1.59, 2.78, 3.97, 4.62	Single and two-phase, max h~11 kW/m²C	Not reported	Not studied	Extremely high heat transfer coefficients, up to 11 kW/m²C were observed. Fully developed subcooled boiling and CHF were obtained.
Kuznetsov and Shamirzaev, 1999	R-318C, G=200-900, q=2-110	Annulus, 0.9 gap x 500	h~1-20 kW/m²C	Not studied	Confined bubble, Cell, Annular	Capillary forces important in flow patterns, thin film suppresses nucleation, leads to convective boiling
Lin et al., 1999	R-141b, G=300-2000 q=10-150	Circular, D=1	Heat transfer coeff.	Not studied	Not studied	Heat transfer coeff. obtained as a function of quality and heat flux. Trends are similar to large tube data
Downing et al., 2000	R-113, Ranges not clearly stated	Circular multi-channels in helical coils, D_h= 0.23-1.86, helix dia=2.8-7.9	Not studies	Single and Two-phase	Not studied	As the helical coil radius became smaller, pressure drop reduced – possibly due to rearrangement in flow pattern
Hestroni et al., 2000	Water, Re=20-70, q=80-360	Triangular parallel chann, θ=55° n=21,26 D_h=0.129-0.103, L=15	Measured, but data not reported	Not reported	Periodic annular	Periodic annular flow observed in microchannels. There is a significant enhancement of heat transfer during flow boiling in microchannels.
Kennedy, et al., 2000	Water, G=800-4500, q=0-4 MW/m²	Circular, D=1.17 and 1.45, L=160	Onset of nucleate boiling starts instability	Not reported	Not studied	Heat flux at the Onset of Flow Instability was 0.9 of the heat flux required for saturated vapor at exit, similarly, G at OFI was 1.1 times G for saturated exit vapor condition
Lakshmi-narasimhan, et al., 2000	R-11, G=60-4586,	Rectangular, 1x20x357,	Subcooled and saturated flow boiling	Not measured	Boiling incipience observed through LCD	Boiling front observed in laminar flow, not visible in turbulent flow due to comparable h before and after, saturated flow boiling data correlated by Kandlikar (1990) correlation
Kandlikar, et al., 2001	Water, G=80-560 kg/m²s	Rectangular, 1x1x60 mm	Subcooled and saturated flow boiling	Measured instanta-neous values of pressure drop	High-speed photography on single and multiple channels	Flow oscillations and flow reversal linked to the severe pressure drop fluctuations, often leading to flow reversal during boiling.

criteria presented by Taitel and Dukler (1976) is one of the most widely used flow pattern maps. Hewitt (2000) notes that both evaporation and condensation processes have a significant effect on the flow patterns.

The effect of evaporation on the flow pattern transitions was considered to be quite small in large diameter tubes. This is one of the reasons why the flow pattern studies from gas-liquid systems, such as air and water, were extended to the case of evaporation. In smaller diameter tubes, the effect of evaporation could be quite dramatic.

The evaporation of the liquid phase affects the flow in two ways. Firstly, it alters the pressure drop characteristics by introducing an acceleration pressure drop component that can be quite large at higher heat fluxes. Secondly, the tube dimension is quite small, and the effect of surface tension forces become more important in defining the two-phase structure.

Table 1 also includes the flow pattern studies for circular tubes and narrow rectangular channels with hydraulic diameters on the order of 3 mm or smaller. Cornwell and Kew (1992) conducted experiments with R-113 flowing in 1.2 mm x 0.9 mm rectangular channels. The parallel channels were machined in aluminum and a 6-mm thick glass plate was used to observe the flow pattern. In the flow ranges investigated, three flow patterns were observed as shown in Fig. 3. They also observed a strong link between the flow pattern and the heat transfer coefficient. In the isolated bubble region, $h \propto q''^{0.7}$, indicating the dominance of nucleate boiling. When the bubbles occupy the entire channel cross-section in the confined bubble region, h dependence on q decreased. The convection effects were dominant in the annular-slug region. In a subsequent study, Lin et al. (1998) compared the flow region transitions with those predicted by Barnea et al. (1983), developed for evaporating steam-water system in 4 mm diameter tubes, and Mishima and Hibiki (1996), developed for adiabatic air-water system in 2.05 mm diameter tube. The results indicate that the flow pattern maps developed for air-water flow are in general applicable, but the transition boundaries need to be refined for flow boiling in small diameter tubes and channels.

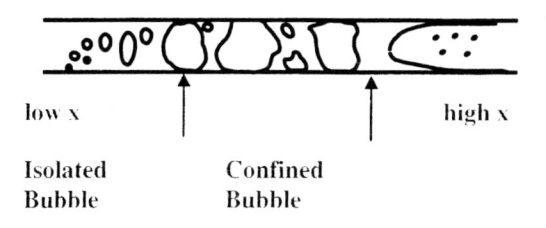

Figure 3 Flow Patterns Observed by Cornwell and Kew (1992) in 1.2 mm x 0.9 mm parallel rectangular channels

Moriyama and Inoue (1992) conducted experiments in single narrow rectangular channels with R-113. They observed (i) flattened bubbles, confined by the flow channel (some of them coalescing), (ii) liquid strips flowing along the wall, and (iii) liquid film flow.

Mertz et al. (1996) investigated single and multi-channel test sections with water and R-141b flowing in rectangular multi-channels, 1, 2 and 3 mm wide. They observed the presence of nucleate boiling, confined bubble flow and annular flow and discovered that the bubble generation in the channels was not a continuous process. In addition, the vapor seemed to stay in the channels blocking the flow, and in some cases, causing a reverse flow to occur in the channels. In both single-channel and multi-channel cases, large pressure pulsations were noticed.

Kasza et al. (1997) present a detailed study on flow visualization of nucleation activity in a rectangular flow channel of 2.5 mm x 6 mm cross section. They viewed the flow using a high speed video camera with a maximum frame rate of 12,000 frames per second. The mass flux was 21 kg/m²s. They observed nucleate boiling on the wall, as the individual bubbles nucleated and grew on the wall. The bubble growth rates were similar to those in pool boiling case when the bubbles grew without interacting with the wall or any liquid-vapor interface.

Kasza et al. made interesting observations on individual bubbles and their interaction with other bubbles and vapor slugs. The vapor slugs were separated from the wall by a thin liquid film with a 0.67-mm average thickness. Nucleation was observed in this liquid film. The bubbles growing in this film did not easily coalesce with the vapor in the slug. Bubbles growing under the liquid slug were flattened and covered a larger wall area compared to those growing in the liquid flow. The heat transfer in the thin liquid microlayer underneath the bubbles improved the heat transfer; the bubble frequency and the vapor volume both increased for such bubbles. Their findings clearly indicate the occurence of nucleate boiling in thin liquid films that exist in both slug flow and annular flow conditions.

Bonjour and Lallemand (1998) report flow patterns of R-113 boiling in a narrow space between two vertical surfaces. The flow patterns observed are similar to those observed by other investigators: isolated bubbles, coalesced bubbles, and partial dryout. Comparing the bubble dimensions with the channel size is vital in determining whether the small channel size influences the bubble growth and leads to the confined bubble flow pattern. Following the work of Yao and Chang (1983), the Bond number,

$$= e\left(\frac{\sigma}{g(\rho_L - \rho_G)}\right)^{-1/2}$$

, along with the Boiling number, $= q/(Gh_{fg})$, provided the basis for the transition from individual to confined bubble flow. For smaller Bond number, the channel dimension is smaller than the departure bubble diameter resulting in confined bubble flow pattern. For large Bond numbers, the channel size does not interfere with the bubble flow. A more detailed treatment of the forces acting on the bubble is needed to clearly identify this boundary.

Kuznetsov and Shamirzaev (1999) studied the flow patterns during flow boiling of R318C in a 0.9-mm annular gap between two circular tubes. This work was an extension of previous research on air-water systems by Kuznetsov and Vitovsky (1995). The isolated bubble region was called as the small bubble region. It was followed by long Taylor's bubbles, similar to confined bubbles (defined by earlier investigators), which were elongated in the flow direction. These bubbles spread along the periphery of the annular gap and formed a cell structure referred to as the cell flow regime. As the vapor quality increased, the annular flow pattern was established with a ripple of waves.

Hestroni et al. (2000) studied the evaporation of water in multi-channel evaporators. The evaporators consisted of 21-26 parallel flow passages with channel hydraulic diameters of 0.103-0.129 mm. They observed periodic behavior of the flow patterns in these channels. The flow changed from single-phase flow to annular flow with some cases of dryout. The dryout, however, did not result in a sharp increase in the wall temperature. This clearly indicates that there is still some evaporating liquid film on the channel walls that could not be observed in the video images. They also reported the presence of vapor phase in the inlet plenum.

Lakshminarasimhan et al. (2000) studied the flow boiling in a narrow rectangular channel, 1 mm high × 20 mm wide × 357 mm long. They used LCD crystal display on the heated wall to observe the nucleation front and locate the onset of nucleate boiling. The ONB was clearly identified in the laminar flow region with subcooled R-11 entering the flow channel. As the flow rate increased to the turbulent region, the ONB could not be identified through this technique due to the high heat transfer coefficient in the subcooled single phase region. However, it is possibile that the ONB may have occurred at discrete locations rather than appearing as a clear identifiable front.

2.2 Discussion on Flow Patterns and Flow Pattern Transitions in Small Diameter Channels

The three flow patterns shown earlier in Fig. 1 represent the characteristic flow patterns associated with two-phase flow in minichannels. It is clear that discrete bubbles, resulting from nucleation activity on the wall, are present in the subcooled boiling and low quality regions. The presence of nucleation in the small diameter tubes is also evident through the available studies. The observations of nucleating bubbles in the thin liquid films by Kasza et al. (1997), shown in Fig. 4, are also very revealing.

The experimental conditions employed by Kasza et al. represent low Reynolds number conditions. With their hydraulic diameter of 3.53 mm and a mass flux of 21 kg/m^{2}s, the single-phase Reynolds number is 262 at one atmosphere pressure. The shear stress effects under these conditions are expected to be quite low, and it should not come as a surprise that the growth rate exponent for nucleating bubbles reflects that under pool boiling conditions. In general, most of the visual studies are for low flow rates as the bubble activity cannot be easily traced at higher mass fluxes.

Kandlikar and Stumm (1995), and Kandlikar and Spiesman (1997) demonstrated presence of nucleating bubbles under highly sheared flow conditions in a rectangular channel. It was noted that bubble nucleation occurred when the nucleation criterion for available cavity sizes was satisfied. The effect of flow and wall temperature on bubble characteristics is illustrated by Kandlikar (2000) and is shown in Figs. 5 and 6.

Figure 5 shows the effect of flow Reynolds number on the bubble growth rate. As the flow velocity increased, flow changed from laminar to transition region with an associated increase in the single phase heat transfer coefficient. This caused the bubble to

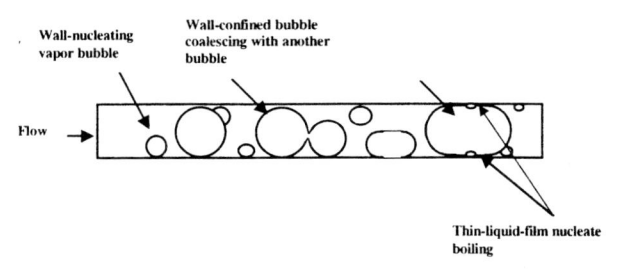

(a) Flow patterns

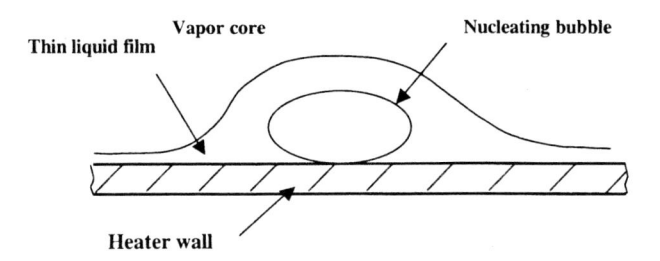

(b) Nucleation in the liquid film under a vapor slug

Figure 4 – Flow patterns and bubble nucleation in the liquid film observed by Kasza et al. (1997)

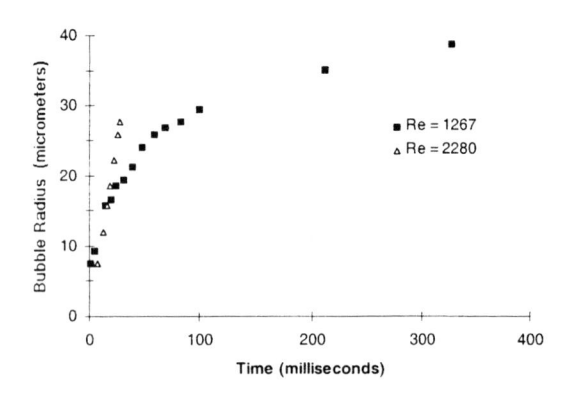

Figure 5 – Effect of flow on bubble growth, subcooled flow of water at 1 atmospheric pressure in 3 mm × 40 mm rectangular channel, T$_{wall}$=108 °C, T$_{bulk}$=80 °C, cavity radius 3.2 μm, Kandlikar (2000).

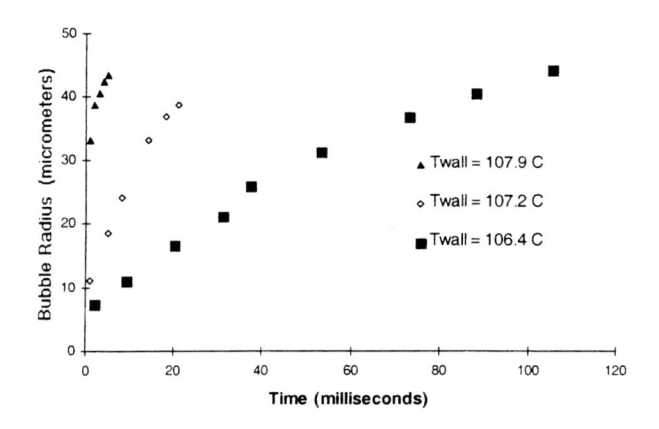

Figure 6 – Effect of wall temperature on bubble growth, subcooled flow of water at 1 atmospheric pressure in 3 mm x 40 mm rectangular channel, T$_{wall}$=108 °C, T$_{bulk}$=80 °C, cavity radius 3.2 µm, Kandlikar (2000).

grow much faster, reaching a departure condition in about 25 millisecond, at Re=2280, as opposed to 300+ millisecond at Re=1267. The departure bubble diameter is also influenced by the flow. As the flow velocity increases, the drag forces cause the bubbles to depart at smaller diameters.

Figure 6 further illustrates the sensitivity of the wall temperature conditions on the bubble growth rates. A higher wall temperature, with a greater associated heat flux, causes the bubbles to grow rapidly and reach the departure conditions much sooner. The departure sizes appear to be more dependent on the flow velocity for the conditions depicted in these figures.

From the observations above, it may be concluded that the nucleating bubbles are present in flow boiling under high shear conditions. Kasza et al. (1997) observed such bubbles in thin films, shown in Fig. 4b, confirming that nucleation can occur in annular flow as well. Under these conditions, the nucleating bubble size decreases, and bubble departure frequency increases. This further confirms the conclusions made by Kandlikar and Stumm (1995) that a high speed camera with suitable magnification is needed to observe the nucleating bubbles (a) under high shear stress conditions, and (b) at high wall temperatures. In fact, the use of high-speed photography is essential in clearly observing the flow patterns in small diameter tubes.

The bubbles departing in the flow can exist as individual bubbles unless their size is smaller than the channel dimension normal to the nucleating surface. Further growth of these bubbles results in their confinement by the channel walls under the confined flow pattern. In reality, the confined flow pattern is similar to the early stages of the plug flow pattern seen in the conventional two-phase flow patterns for larger diameter tubes (>3 mm). Annular flow pattern then follows at higher qualities.

Flow Instabilities:

The flow instability is another concern in the design of evaporators that employ small channels. The flow instabilities can be reduced considerably by increasing the upstream pressure drop prior to the flow entering a channel. A large diameter flow section however generally precedes the test section in a number of experimental studies conducted on gas liquid flows in small diameter tubes. In the study conducted by Lin et al. (1998), air was introduced in a large mixing chamber upstream of the test section to reduce the disturbances resulting from gas injection. The presence of such low pressure drop regions immediately before the test section leads to significant flow instabilities that cause large pressure drop excursions, and occasionally result in a negative pressure drop with a corresponding flow reversal in the channel.

The instability occurs in the negative pressure drop region of the demand curve plotted as the pressure drop versus inlet flow velocity of the subcooled liquid. Kennedy et al. (2000) studied the onset of flow instability and noted that instability sets in at a slightly lower mass flux than the onset of significant void condition.

Cornwell and Kew (1992) conducted experiments on flow boiling of R-113 in parallel microchannels. They observed that the flow was unstable at lower flow rates. The pressure drop fluctuations were not reported.

Kandlikar et al. (2001) viewed the flow boiling of water in electrically heated multi-channel evaporators consisting of six parallel channels. The flow structure was visualized using high speed video camera up to a maximum speed of 1000 frames per second. The typical bubbly flow, slug flow and annular flow patterns were observed. Nucleation was also observed in the bulk liquid as well as in the liquid film. However, the most interesting discovery made, in an attempt to understand the severe pressure fluctuations (described later under Pressure Drop section), was a visual confirmation of complete flow reversal in some of the channels.

Figure 7 shows the schematic of the multi-channel evaporator investigated by Kandlikar et al. (2001). The evaporator was heated electrically from the back wall of the test section. The front part was covered with a high temperature resistant glass for flow visualization.

Figure 8 shows the sequence of the events that lead to flow reversal in the flow channels. Two adjacent central channels are shown at 2 ms intervals in frames (a) through (e). Both channels (1) and (2) exhibit slug flow in the visible section.

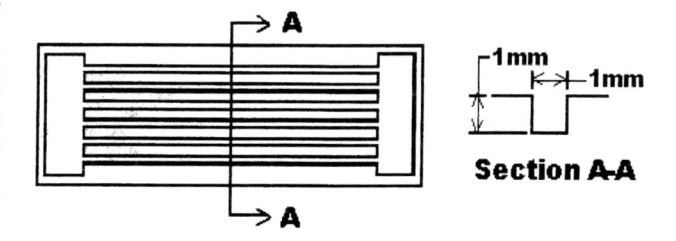

Figure 7 Multi-channel evaporator investigated by Kandlikar et al. (2001)

Vertical lines (y) and (z) indicate the initial boundaries of a vapor slug in channel (2) of Fig. 8(a). Vertical lines (x) and (w) are reference lines to aid visualization of slug motion throughout the frames.

In Fig. 8(b), channel (1) has flow in the direction of bulk flow (left to right). The vapor slug in channel (2) has expanded in the direction of bulk flow yet the inlet-side fluid/vapor interface has not moved. The fluid/vapor interface on the inlet side of the slug in channel (2) is still stationary in Fig. 8(c) although the outlet-side interface moves in the bulk flow direction. In Fig. 8(d) the flow in channel (1) moves along the direction of bulk flow, but the inlet-side fluid/vapor interface of the slug in channel (2) progresses in the direction counter to the bulk flow. In Fig. 8(e), the inlet fluid/vapor interfaces in both channels move in the direction counter to bulk flow. For this particular case, it appears that both of the channels experience a vapor-clogging condition where the differential pressure across the channels increases due to vapor generation, the mass flow rate is consequently reduced through these two channels. The reaction to this condition in the other four channels would be an increased flow rate.

Concluding Remarks on Flow Patterns:

The flow patterns observed in the small channels indicate that the nucleating bubbles play an important role in small diameter channels. The three predominant flow patterns are- (a) Isolated bubbles, (b) Confined bubbles, and (c) Annular-Slug. Note that the flow patterns under high mass flux conditions (G>500 kg/m²s) have not been studied in literature due to difficulty in capturing the high-speed movement under these conditions. Further work in this area is recommended.

3. PRESSURE DROP IN SMALL DIAMETER FLOW BOILING CHANNELS

3.1 Single Channel

Pressure drop in small diameter tubes has been studied by a number of investigators. Lazarek and Black (1982) conducted systematic experiments to evaluate the three components of pressure drop. The desired quality was generated in the heated inlet section and the frictional pressure drop was measured under adiabatic conditions in the discharge section. The frictional pressure was correlated using a correlation recommended by Collier (1981) with the Martinelli parameter χ_{tt}:

$$\frac{\Delta p_{TP}}{\Delta p_{LO}} = 1 + \frac{C}{\chi_{tt}} + \frac{1}{\chi_{tt}^2} \qquad (1)$$

The subscript TP corresponds to the two-phase value, while LO corresponds to the value with the total flow in the liquid phase. Lazarek and Black found that using a value of C=30 produces results that are in good agreement with their experimental data. The value of C recommended by Collier is 20 for large tubes.

The acceleration pressure drop was accurately predicted using the Martinelli and Nelson's (1948) separated flow model with a multiplier K_{sa}.

$$\frac{\Delta p_{sa}}{G^2/(2\rho_L)} = K_{sa}\left[\frac{\rho_L}{\rho_G}\frac{x_{ex}^2}{\alpha_{ex}} + \frac{(1-x)^2}{1-\alpha_{ex}} - \frac{\rho_L}{\rho_G}\frac{x_{in}^2}{\alpha_{in}} - \frac{(1-x_{in})^2}{1-\alpha_{in}}\right] \qquad (2)$$

Here x is the vapor quality and α is the void fraction. K_{sa} is an empirical constant. Lazarek and Black found that a value of $K=2.5$ correlated most of their data within ±20 percent.

Moriyama and Inoue (1992) measured pressure drop of R-113 boiling in narrow annular gaps of 35-110 μm. Their experimental values for frictional pressure drop were correlated by slightly modifying eq. (1). From their study, it is evident that the separated flow model is applicable for the narrow gaps typically encountered in microchannel applications.

On the other hand, Bowers and Mudawar (1994) employed a homogeneous flow model with f_{TP}=0.003 as recommended by Collier (1981). Their results for both minichannels and microchannels were correlated to within ±30% with this model.

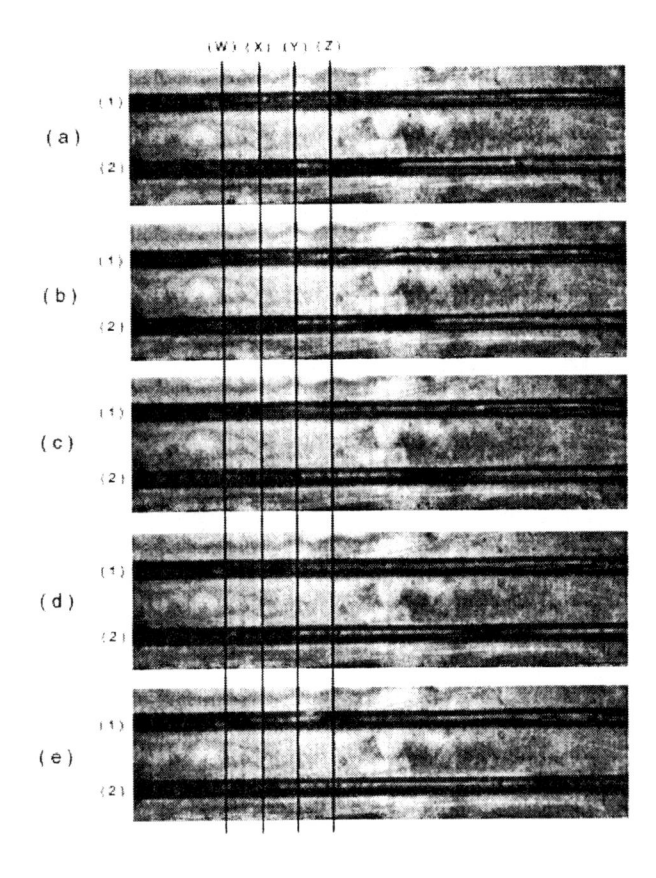

Figure 8 – Successive Frames (a) through (e) at 2ms Intervals of Two Channels Interacting- G=40kg/m²s, Surface Temperature=110.6°C, Entering Water Temperature=24.7°C, Outlet Water Temperature=99.3°C, x>0

Tong et al. (1997) present an exhaustive treatment of pressure drop during subcooled flow boiling in minichannels. In addition, they presented a correlation to predict the two-phase pressure drop. Since the void fraction was very small, a two-phase friction factor was applied. They observed a roughness effect on the single-phase laminar to turbulent transition in these tubes. The effect of tube diameter to length ratio was also noted to be quite significant.

From the studies available in literature, the effect of channel dimension on two-phase pressure drop is not clearly established. Although several investigators provide different correlation schemes to correlate their data, they do not provide a clear indication of the effect of small passage sizes on pressure drop. The added effect of channel wall roughness on pressure drop, seen in the single phase data, is also not incorporated while analyzing the two-phase pressure drop parameters. These effects will become more important as the channel size becomes decreases from minichannel to microchannel geometries.

3.2 Multi-channels

As the tube diameter decreases, the vapor slugs fill the tubes. Under two-phase flow conditions, flow instabilities occur when the pressure drop in the upstream section is relatively small. Introducing a large pressure drop through a throttle valve in the liquid line immediately prior to the test section considerably reduces the instabilities. These instabilities have a significant effect on pressure drop and heat transfer under flow boiling conditions.

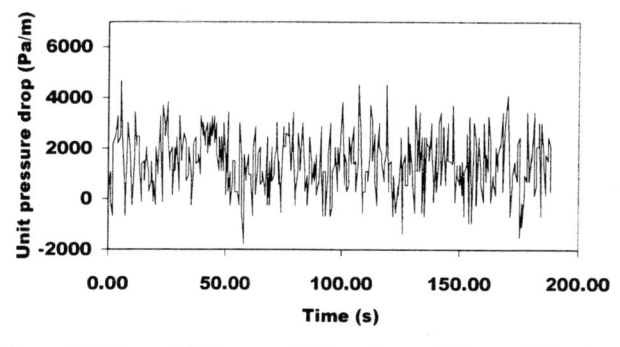

Figure 9 Differential Pressure History for a 6 Channel (1mm x 1mm) Parallel Configuration- G=48kg/m^{2}s, $\Delta P_{max}/L$ = 4688Pa/m, $\Delta P_{min}/L$ = -1793Pa/m, Average Surface Temperature = 125°C, q"=74.3kW/m^2, Water Inlet and Outlet Temperatures 25.0 and 90.2°C, Kandlikar et al. (2001).

Figure 9 shows the pressure drop fluctuations measured in a multi-channel evaporator with six parallel 1mm × 1mm square microchannels. Similar observations were made by Kew and Cornwell (1996) during flow boiling of R-141b in 2-mm square channels and in 2.87-mm diameter circular tubes. The pressure drop fluctuations observed by Kandlikar et al. (2001) are quite large and result in flow reversals as discussed earlier under flow patterns in Section 2 of this paper. The compressibility of the two-phase mixture, in adjacent channels, acts in a manner similar to the negative slope in the upstream section of a single evaporator tube.

The large pressure drop fluctuations lead to instantaneous localized flow reversal in some of the parallel channels. There are no models currently available that predict the pressure drop fluctuations and the flow reversals under flow boiling conditions. Knowledge of these conditions is essential for safe operation of evaporators employing minichannels and microchannels.

4. HEAT TRANSFER IN SMALL DIAMETER FLOW BOILING CHANNELS

Flow boiling in small diameter tubes and compact heat exchanger passages has been a subject of interest in automotive, aerospace, air liquefaction, chemical, and petroleum industries, and in electronics cooling applications. Nakayama and Yabe (2000), and Kew and Cornwell (2000) present a good overview of the recent advances in this area. Table 1 includes some of the recent works on flow boiling heat transfer in minichannels.

Flow boiling heat transfer in 1-3 mm diameter channels has been a subject of investigation for a long time. In one of the earlier papers, Bergles (1964) studied the critical heat flux in 0.584, 1.194, and 2.388 mm diameter electrically heated tubes. They indicated that when the bubble diameter approaches the tube diameter, considerable non-equilibrium vapor volume exists in the evaporator section, and flow oscillations cause a premature burnout in the small diameter tubes.

Bowers and Mudawar (1993) studied flow boiling pressure drop and CHF in a minichannel of 2.54-mm diameter, and a microchannel of 510 µm diameter. Boiling curves for the two channels were obtained at nearly equal values of liquid Reynolds number. Their results are reproduced in Fig. 10. Despite the large variation in the tube diameter, the two curves overlap in the boiling region. It is believed that these experiments fall under the fully developed nucleate boiling regime. The differences between the two boiling curves are only evident at low heat flux (near single phase region) and high heat flux values (approaching CHF condition). This indicates that in spite of the differences in the flow characteristics of the channels, the flow boiling behavior is quite similar in the two geometries.

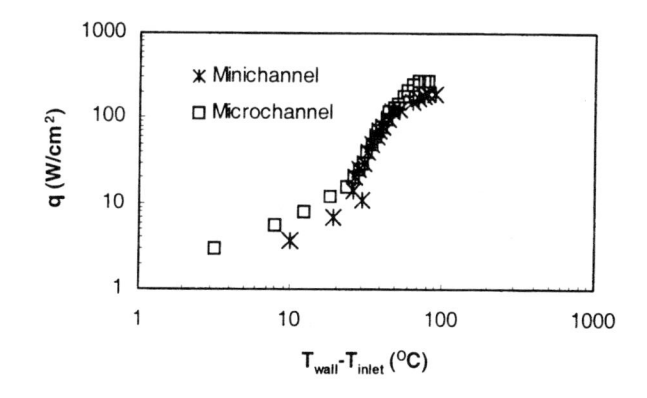

Figure 10 Flow Boiling Characteristics in Minichannel and Microchannel Evaporators, Bowers and Mudawar (1993)

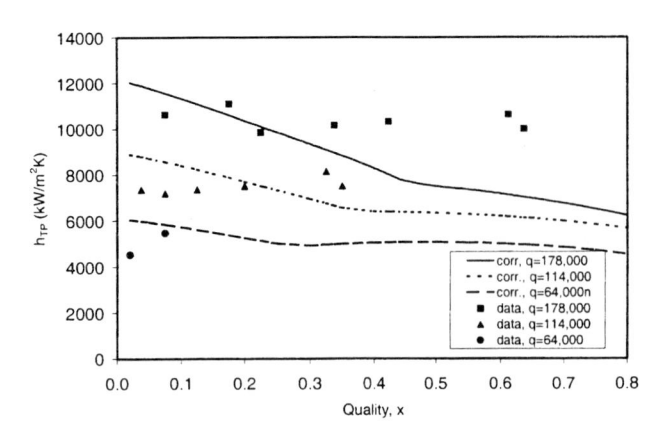

Figure 11 Comparison of Flow Boiling Heat Transfer Coefficient Data by Lazarek and Black (1982) with Kandlikar (1990) correlation

The detailed flow boiling data by Lazarek and Black (1982) provides a clear comparison between the flow boiling characteristics of minichannels and regular tubes (>3mm diameter). Figure 11 shows a comparison of Lazarek and Black's flow boiling data in a 3.1 mm diameter tube and Kandlikar's (1990) flow boiling correlation. The correlation represents the data quite well, although some differences exist in the high quality region. Although a detailed study is warranted, as a first order approximation, one may use the correlations developed for the large diameter tubes for predicting the heat transfer coefficients in minichannels.

Cornwell and Kew (1992) conducted experiments in two sets of parallel channel geometries, 1.2 mm x 0.9 mm deep, and 3.25 mm x 1.1 mm deep. Their results indicate that the flow boiling in such small channels exhibits fully developed nucleate boiling characteristics in the isolated bubble region at lower qualities. At higher qualities (when the bubbles fill the entire cross section), and in the annular flow region, convective effects dominate heat transfer. These characteristics are similar to those observed for the large diameter tubes. The isolated and confined bubble regions exhibit the characteristics similar to the nucleate boiling dominant region, while the annular-slug region exhibits the convective dominant trend seen in large diameter tubes (Kandlikar, 1991).

Another aspect of flow boiling heat transfer in small channels is the oscillatory nature of the flow. The time averaged value between two regions (i.e. between the confined bubble and the annular regions) would yield a combination of nucleate boiling dominant and convective boiling dominant regions. Purely flow pattern based correlations need to include this averaging effect. Since the large diameter correlations combine these effects, rather than using distinct boundaries, they are expected to provide the basis for accurate correlation schemes for small diameter tubes and channels.

Continuing the work in this area, Lin et al. (1999) obtained flow boiling data with R-141b in 1-mm diameter tubes. Their results indicate that the heat transfer coefficient exhibits behavior similar to that reported by Cornwell and Kew (1992). The role played by bubbles is clearly an important one. Specifically they presented detailed results at G=500 kg/m^{2}s with q from 18 to 72 kW/m^2. At high heat fluxes, they observed considerable fluctuations in the wall temperature readings, indicative of flow oscillations that cause changes in the instantaneous values of local saturation temperature and heat transfer rate.

Wambsganss et al. (1993) conducted extensive experiments on flow boiling of R-113 in a 2.92-mm diameter tube. Their results indicate that the heat transfer coefficient was sensitive to both heat flux and mass flux changes, except for the lowest mass flux result. At the lowest value, G=50 kg/m^{2}s, changing the heat flux from 8 to 16 kW/m^2 did not have any influence on the heat transfer coefficient. One possible explanation is that for this case, the nucleate boiling is in the partial boiling region at low heat fluxes, and the effect of heat flux is therefore quite small. For their other test conditions, the mass flux was varied from 100 to 300 kg/m^{2}s, and the heat flux was varied from 16 to 63 kW/m^2. For these tests, the heat transfer coefficient exhibited a dependence on both heat flux and mass flux. This indicates the contributions from both nucleate boiling and convective boiling mechanisms. They also compared their data with the available correlations and found that the correlations by Liu and Winterton (1988), Shah (1976), and Kandlikar (1990) predicted their results with a mean deviation of less than 20 percent. In particular, Wambsganss et al. found that the specific correlation developed by Lazarek and Black (1982), who used their own small diameter tube data with R-113 in the correlation development, predicted the data with a mean error of 12.7 percent. The Chen (1966) correlation predicted the results with a mean deviation of 36 percent.

Mertz et al. (1996) conducted extensive experiments with water and R-141b boiling in six different minichannel configurations. The flow boiling was observed as oscillatory phenomena in multi-channels. It is interesting to note that although

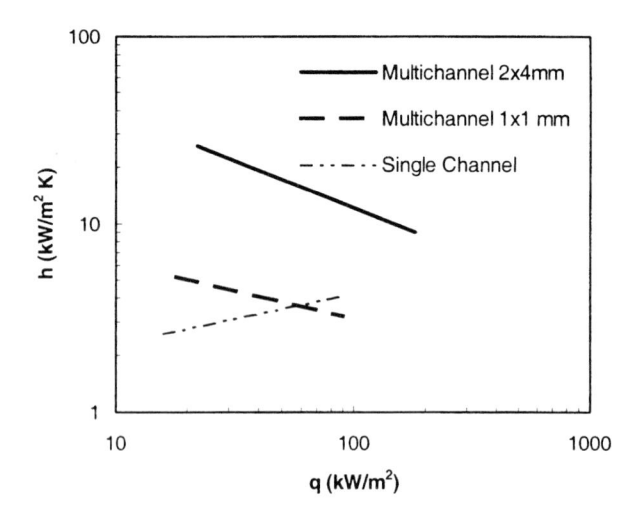

Figure 12 Comparisons of the average heat transfer coefficient trendlines for water in 2mm x 4mm single and multi-channel configurations, and 1mm x 1mm multi-channel configuration, G=200 kg/m^{2}s, T$_{SAT}$=120°C, Mertz et al. (1996)

the heat transfer coefficient increased with heat flux in almost all cases for single channels, the trend for water flowing in the multi-channel configuration at G=200 kg/m²°C was different. For all channels in multi-channel configuration, the heat transfer coefficient decreased with increasing heat fluxes. It is suspected that the flow oscillations and reversals observed in multi-channel are responsible for the degradation in heat transfer. For R-141b boiling in multi-channel configuration, the heat flux effects were weak and somewhat mixed.

Mertz et al. observed that the heat transfer coefficient for both fluids in the multi-channel configuration was considerably higher than the corresponding single channel values under the same operating conditions. Figure 12 shows the comparison for water (at 200 kg/m²s in 2 mm x 4 mm channels) in single and multi-channel configurations under the same operating conditions. The oscillatory behavior found in the multi-channel evaporator is therefore regarded to improve the heat transfer, although the improvement decreases at higher heat fluxes. Another fact to note is that the heat transfer coefficient is lower for the 1mm x 1mm multi-channel configuration than for its 2 mm x 4mm counterpart as shown in Fig. 12.

Tran et al. (1996) conducted experiments with R-12 in small circular and rectangular channels both of 2.46 mm hydraulic diameter. Their results indicated two distinct regions, convective boiling dominant region at lower wall superheat values, and nucleate boiling dominant region at higher wall superheat values. They compared their data with the Kandlikar (1990) correlation, and found that it exhibited similar trends, but underpredicted the results. One reason for this difference is the fact that the single phase heat transfer coefficients in small diameter tubes is generally higher than those predicted by the Dittus-Boelter type correlations. It is recommended that the measured single phase heat transfer coefficients be used in the correlation as recommended by Kandlikar (1991).

It is interesting to note that the heat transfer coefficients obtained by Tran et al. at higher qualities exhibited a dependence on heat flux alone. The mass flux had virtually no influence on the heat transfer coefficient. These observations are supported by the visual observations made by Kasza et al. (1997) who studied flow boiling of water in rectangular channels of 2.5 mm x 6 mm cross section. They concluded that the increased bubble activity at nucleation sites in the thin liquid film is responsible for high heat transfer coefficient in small hydraulic diameter tubes and channels. Kuznetsov and Shamirzaev (1999) conducted experiments with R-318C in an annular gap of 0.9 mm. They observed that at higher values of quality, nucleation was seen to be suppressed. However, their experimental results were in agreement with the correlation by Tran et al. (1999) which was developed for nucleate boiling dominant region.

Kamidis and Ravigururajan (1999) conducted flow boiling experiments with R-113 in circular tubes of 1.59, 2.78, 3.97 and 4.62 mm diameter tubes. Their results indicate that very high heat transfer coefficients of the order of 10 kW/m²C are obtained in the flow boiling region. Figure 13 shows a comparison of their data for 1.59 mm diameter tube with the Kandlikar (1990) correlation. For the Reynolds number of 5720, the agreement is excellent. For Re=2370, the correlation underpredicts the data. It is suspected that the Reynolds number is in the transition region where the turbulent single phase heat transfer correlation is not applicable.

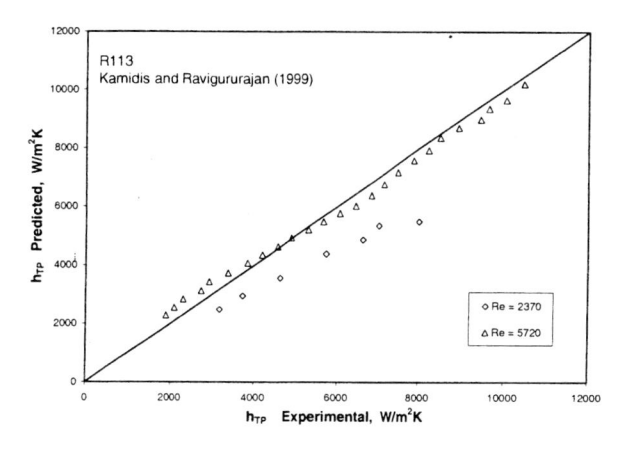

Figure 13 Comparison of Kamidis and Ravigururajan (1999) data with Kandlikar correlation (1990) for 1.59 mm diameter tube, R-113

Kennedy et al. (2000) studied the instability during flow boiling of water in circular tubes of 1.17 and 1.45 mm diameter. Based on their results, they proposed that the instability is initiated in the tube when the impressed heat flux is 90 percent of the value required to obtain saturated vapor condition at the exit section.

Lakshminarasimhan et al. (2000) conducted experiments with R-11 in 1 mm x 20 mm rectangular channels. They noted that the their saturated flow boiling data was accurately predicted using the large tube correlation by Kandlikar (1990). They also observed a clear boiling front with liquid flow in the laminar region. In the turbulent region, the boiling front could not be clearly viewed due to the high heat transfer coefficient in the single-phase region prior to nucleation.

The experimental studies available in literature provide some preliminary data on flow boiling heat transfer in small diameter tubes and channels. In the case of Lazarek and Black's (1982) and Lakshminarasimhan et al. (2000) data, the Kandlikar (1990) correlation for large diameter tubes predicted the results satisfactorily. However, Tran et al. (1996) indicated a significant enhancement over the large diameter correlations. Pressure fluctuations and multi-channel effects are not clearly understood for small diameter tubes. As noted by many investigators, including Lin et al (1999) and Mertz et al. (1996), pressure fluctuations have a significant effect on the flow characteristics and associated heat transfer performance. The periodic filling of the flow channel with large vapor plugs, followed by all liquid flow in the channel, make it very difficult to apply flow pattern based models to predict the heat transfer rates. Further research in this area is highly recommended.

Another class of flow channels that have received some attention in literature are those with hydraulic diameters below 500 μm. These channels are referred to as microchannels, although the precise boundary is not well defined. There are very few quantitative studies available for the microchannel geometry under flow boiling conditions. Further efforts are needed in this area to generate high-quality experimental data in microchannels under flow boiling conditions. This geometry has been investigated with its potential application in microelectronics cooling. Table 1 clearly indicates that there is very little quantitative data available for microchannels.

Ravigururajan et al. (1996) studied flow boiling in a microchannel 270 μm wide and 11 mm deep. The working fluid was R-124, and was tested over the entire quality range. They found that the heat transfer coefficient decreased from a value of 11 kW/m^{2}C at x=0.01 to about 8 kW/m^{2}C at x=0.65. Although no conclusions were drawn, this behavior may be the results of the two trends: a) the nucleate boiling heat transfer is dominant, leading to its suppression at higher qualities, or b) the higher vapor fraction leads to flow oscillations in multi-channels with a consequent change (increase?) in the heat transfer coefficient.

A number of investigators (for example, Peng and Wang, 1998, Peng et al., 1998) have indicated that the flow boiling heat transfer in microchannels may be quite different than that in larger diameter tubes. They also indicated that the regular nucleate boiling phenomenon does not exist in microchannels.

Peng and Wang (1998) conducted experiments with water, ethanol, and their mixtures in different shaped microchannel geometries (listed in Table 1). They noted the presence of both nucleate boiling and convective boiling in various regimes. They did not observe any bubble activity in the rectangular and triangular passages with hydraulic diameters between 0.1 to 0.6 mm. In turn, they called it a fictitious boiling phenomenon.

It is difficult to accept the notion of the fictitious boiling presented by Peng and Wang (1998). Similar studies reported by Kandlikar and Stumm (1995) and Kandlikar and Spiesman (1997) with a channel height of 3 mm indicates that bubbles as small as 10 μm are seen to depart from the nucleating sites. The key to observing bubble activity in small channels is to employ high speed photography along with a high resolution microscope.

The theoretical analysis presented by Peng and Wang (1998) considers a bubble nucleus that completely fills the tube. The microchannel dimension is on the order of 100 μm, while the cavity sizes for active nucleation are on the order of a few μm or smaller. It is expected that the nucleation criterion for flow boiling, established for large diameter tubes, will hold true unless the tube diameter approaches the cavity dimensions. Such a condition may exist only in submicron-sized tubes.

In conclusion, flow boiling in microchannels is an area where further research is needed. The difficulty in observing the bubbles and in the accurate measurement of heat fluxes at the wall make it very difficult to understand the mechanism of flow boiling heat transfer in this geometry. With the availability of more accurate data, we may be able to find some of the answers in the near future.

5. DESIGN CONSIDERATIONS IN MINI-CHANNEL EVAPORATORS

Mehendale et al. (1999,2000) present a good overview of the design consideration for heat exchangers employing mini- and microchannels.

Flow Instability in Multi-channel Evaporators

The small diameter multi-channel evaporators differ from the small hydraulic diameter compact heat exchangers in one vital aspect: there are no cross flow connections available for the fluid to flow across the width of the flow channel as it passes through the evaporator. This cross connection helps the nucleating bubbles grow in the cross-wise direction without blocking the entire flow passage, as in the case of small diameter channels. This flow structure is clearly illustrated by the flow pattern investigation conducted by Kuznetsov and Shamirzaev (1999) in an annular gap between two concentric tubes. They observed a cell pattern that effectively allows the bubbles and vapor to grow in the crossflow direction without blocking the flow. In the case of tubes, as shown by Kandlikar (2001), the vapor bubble growth leads to large pressure fluctuations that are not desirable for stable operation of the evaporator. The presence of fins or ribs in the gap is expected to further provide stability to the flow by increasing flow resistance.

In the case of multi-channel evaporators employing individual small-diameter tubes or channels, the channels running parallel to any given channel (which is experiencing vapor expansion in a direction opposite to the flow) act in a manner similar to reducing the upstream pressure drop characteristics in a two-phase system. The severe pressure drop fluctuations, coupled with the backflow of vapor into the inlet manifold, are not desirable. It could lead to premature CHF in some of the channels where vapor may flow, preferentially, without being accompanied by the liquid flow. Some of the systems that employ such evaporators may not tolerate such severe fluctuations in the flow rate.

With these considerations, it is necessary to design multi-channel evaporators that avoid severe pressure fluctuations found in the parallel channels. Further research in this area is warranted.

5.1 Design Guidelines for Sizing Small Diameter Multi-channel Evaporators

The small channels present a number of advantages making them attractive for specific systems. Their compact size, low weight, low liquid/vapor inventory, and fast response are just some of the desirable features. In this section, preliminary guidelines are presented for designing the multi-channel evaporators with small diameter channels. The design conditions considered in this analysis are as follows.

Design Conditions

It is assumed that an existing single or multi-channel evaporator employing large diameter (D) tubes is to be replaced with a multi-channel evaporator employing small diameter (d) tubes. Capital letter subscripts refer to the large diameter tube while lower case subscripts refer to smaller diameter tubes. For the qualitative analysis presented here, it is assumed that both geometries employ circular tubes, and that one larger diameter tube is being replaced by n number of smaller diameter tubes. The total mass flow rate and the total heat transfer rate are identical in both cases. The objective is to arrive at the number of small diameter tubes needed to replace each large diameter tube, and the new length the evaporator. Another consideration is the need for equal pressure drop in the two cases. The analysis is presented for a single-pass evaporator. It can be extended to a multipass evaporator configuration, but the treatment will introduce many additional parameters. The purpose of the following exercise is to provide some simple guidelines to help in designing the new evaporator. Extensive design efforts will be needed to arrive at the final design and all of the associated details.

5.2 Design Comparison

If n parallel channels are employed in the new evaporator (for each tube in the original design), the mass flux and heat flux for the new heat exchanger with small diameter (d) tubes would be different than those of the original heat exchanger with large diameter (D) tubes. Since the total mass flow rate remains constant between the two designs, we get:

$$n = \frac{G_D}{G_d} \frac{D^2}{d^2} \qquad (3)$$

The total heat transfer rates in the two cases are also identical since the evaporators are being designed for the same heat duty. If we apply the heat transfer rate equations with the respective average heat transfer coefficients, and assume that the operating temperature difference in the two cases is identical, we can write:

$$\frac{L_d}{L_D} = \frac{\overline{h}_D}{\overline{h}_d} \frac{1}{n} \frac{D}{d} \qquad (4)$$

Substituting n from eq. (3) into eq. (4), the ratio of the lengths in the flow direction is given by:

$$\frac{L_d}{L_D} = \frac{\overline{h}_D}{\overline{h}_d} \frac{G_d}{G_D} \frac{d}{D} \qquad (5)$$

For the same exit quality conditions, the total heat transferred in the heat exchanger remains the same. The heat fluxes in the two cases are related by the following equation:

$$\frac{q_d}{q_D} = \frac{1}{n} \frac{D}{d} \frac{L_D}{L_d} \qquad (6)$$

The flow boiling in the small diameter tubes exhibits a nucleate boiling dominant region at low qualities. Here the heat transfer coefficient varies as $q^{0.7}$. In the higher quality region, the flow becomes convective boiling dominant, with the heat transfer coefficient independent of q, and varying as $G^{0.7}$. In general, the dependence of h may be expressed as $h \propto q^m G^p$. The ratio of the two heat transfer coefficients is given by:

$$\frac{\overline{h}_d}{\overline{h}_D} = \left(\frac{q_d}{q_D}\right)^m \left(\frac{G_d}{G_D}\right)^p \qquad (7)$$

The exponents m and p depend on the flow boiling heat transfer characteristics under the prescribed operating conditions.

Their values range from $m=0$ to 0.7, and $p=0$ to 0.8. Substituting the heat flux ratio from eq. (6) into eq. (7), combining it with eq. (3), and then substituting the heat transfer coefficient ratio into eq. (5), the ratio of the two lengths is obtained as:

$$\frac{L_d}{L_D} = \left(\frac{G_d}{G_D}\right)^{\frac{1-m-p}{1-m}} \frac{d}{D} \qquad (8)$$

Equation (8) provides a preliminary estimate of the length of the heat exchanger needed to obtain the same vapor generation rate using a small diameter evaporator. The number of small diameter tubes replacing each large diameter tube is given by eq. (3).

A few observations can be made regarding the effect of boiling characteristics on the length ratio presented in eq. (8). If the nucleate boiling was the dominant mode in both cases, then $p=0$. This results in a mass flux ratio dependence given by $(L_d/L_D) = (G_d/G_D)(d/D)$. On the other hand, if the heat transfer was convective dominant, then $m=0$ and $p=0.8$. In the latter case, the mass flux ratio has a weak effect on the length ratio, $(L_d/L_D) = (G_d/G_D)^{0.2}(d/D)$. However, the actual variation lies between the two extreme cases discussed here.

Another important factor to be considered in the design of heat exchangers is the pressure drop. The design guideline provided by eq. (8) does not address this issue. A similar analysis is now presented for pressure drop comparison in the two cases.

The pressure drop analysis is a bit more complex. To simplify the analysis, the gravitational pressure drop is considered negligible when compared to the friction and acceleration pressure drop terms. Assuming the respective exit quality and exit void fractions to be equal in both cases, the pressure drop may be expressed in terms of the mass flux, tube length, tube diameter, exit quality, and fluid properties. The frictional pressure drop varies as $\Delta p_f \propto f_{TP} L G^2 / D$, while the acceleration pressure drop varies as $\Delta p_f \propto G^2$. In addition, the void fraction plays a role in the pressure drop terms. The two-phase friction factor may be considered to vary as $f_{TP} \propto G^{-0.25}$. Assuming the same equations to apply for the small and large diameter tubes, the ratio of pressure drops may be expressed as

$$\frac{\Delta p_d}{\Delta p_D} = \frac{(L_d G_d^{1.75}/d) C_1 F_1(\alpha_d) + G_d^2 C_2 F_2(\alpha_d)}{(L_D G_D^{1.75}/D) C_1 F_1(\alpha_D) + G_D^2 C_2 F_2(\alpha_D)} \qquad (9)$$

The constants C_1 and C_2 include additional variables in the respective pressure drop terms. F_1 and F_2 are functions of the void fractions in the two cases. Although the effects of mass flux, diameter and length on pressure drop are not immediately obvious, assuming the frictional pressure drop to be dominant provides some degree of guidance. In addition, assuming the void fractions and their effects to be similar, eq. (9) may be simplified as follows.

$$\frac{\Delta p_d}{\Delta p_D} = \frac{\left(L_d G_d^{1.75}/d\right)}{\left(L_D G_D^{1.75}/D\right)} \qquad (10)$$

For the case of equal pressure drop between the two configurations, the length ratio is obtained as,

$$\frac{L_d}{L_D} = \left(\frac{G_d}{G_D}\right)^{-1.75} \frac{d}{D} \qquad (11)$$

Comparing eqs. (8) and (11), it is clear that the effect of mass flux is more severe on the pressure drop than on the heat transfer. The diameter effect is same in both cases. If the mass flux is held constant for the two configurations, then the length ratio is identical to the diameter ratio, $L_d/L_D = d/D$. However, in practical system designs, a higher pressure drop is generally accepted with evaporators employing small channels, and tube lengths larger than that given by eq. (11) are employed.

The negative exponent in the mass flux ratio of eq. (11) indicates that increasing the mass flux results in shorter tube lengths for the same pressure drop. In other words, increasing G_d causes the pressure drop to increase, and shorter tube lengths are needed to meet the pressure drop requirements. From the heat transfer perspective, a larger tube length may be needed to accommodate higher mass fluxes. Consequently, the design mass flux is a compromise between these considerations and other system requirements.

The preceding discussion provides a preliminary basis for the selection of a mass flux value for the smaller tube diameter heat exchanger being designed to replace an existing larger tube diameter evaporator. Needless to say, a number of additional parameters, including fluid properties, the local heat transfer coefficient and pressure drop relationships, differences in manifolding and number of passes, and the differences in allowable pressure drop will affect the design of the new evaporator with smaller diameter tubes. Another major consideration in the design of the evaporator is the performance on the hot fluid side. In the analysis presented here, the wall temperatures were considered to be identical in both evaporators. The comparisons should, therefore, be treated as qualitative in determining first-order effects.

5. CONCLUSIONS

On the basis of a critical literature review and the work consucted by the author, following conclusions can be drawn:

1. Three flow patterns are commonly encountered during flow boiling in minichannels: isolated bubble, confined bubble or plug/slug, and annular. The visual studies available in literature have been generally conducted for low mass flux values in tubes of 1-mm or larger hydraulic diameters.
2. The literature on flow patterns in microchannels is insufficient to draw any conclusions at this stage.
3. Large pressure drop fluctuations are noted in multi-channel evaporators. Flow pattern observations revealed a flow reversal in some channels with expanding bubbles pushing the liquid-vapor interface in both upstream and downstream directions.
4. Heat transfer studies in the microchannels indicate that, as a first order estimate, heat transfer may be predicted using the flow boiling correlations developed for large diameter tubes.
5. The heat transfer rate in multi-channel evaporators is different from that in single-channel evaporators under same set of operating conditions. The role flow rate fluctuations due to flow instabilities is not clearly understood at this stage.
6. The severe pressure drop fluctuations are not included in any pressure drop prediction schemes for minichannel and microchannel evaporators. Both separated and homogeneous flow models have been used with some degree of success by previous investigators.
7. In designing the evaporators with small diameter channels, the length to diameter ratio depends on the heat transfer and pressure drop characteristics. Larger pressure drops are generally accepted in evaporators with small diameter channels.

6. FUTURE RESEARCH NEEDS

The future research needs are summarized below:
   a) Conduct high speed video studies to obtain flow pattern information under high mass flux conditions in small diameter tubes and channels.
   b) Compare the performance of single tube evaporators and multi-channel evaporators under the same operating conditions and identify the reasons for the differences in their performance.
   c) Study the effects of inlet flow conditions and manifold design on the performance of the multi-channel evaporator.
   d) Conduct more experiments with microchannel evaporators to obtain accurate flow boiling heat transfer and pressure drop data as a function of quality, heat flux, mass flux, and tube/channel hydraulic diameter.
   e) Critical heat flux is an important factor in the design of evaporators. Although not covered in this paper due to space constraint, there is a need for obtaining more experimental data for CHF in single and parallel minichannel and microchannels.

7. NOMENCLATURE

C constant in eq. (1)
C_1 and C_2 constants in eq. (9)
D diameter of
e gap size, m
F_1 and F_2 constants in eq. (9)
G mass flux, kg/m^{2}s
g acceleration due to gravity, m/s^2
K_{sa} pressure drop multiplier in acceleration pressure drop eq. (2)
h_{ffg} latent heat of vaporization, J/kg

$\bar{h}$ average heat transfer coefficient in the evaporator, W/m² °C
L length of the evaporator tube
n number of parallel small diameter tubes for each large diameter tube
Δp pressure drop, N/m²
q heat flux, W/m²
x quality

Greek Letters

α void fraction
ρ density, kg/m³
σ surface tension, N/m
χ_{tt} Martinelli parameter, $= \left(\dfrac{\rho_V}{\rho_L}\right)^{0.5}\left(\dfrac{\mu_L}{\mu_V}\right)^{0.1}\left(\dfrac{1-x}{x}\right)^{0.9}$

Subscripts

D large diameter tube
d small diameter tube
ex exit
f friction
in inlet
L liquid
LO all flow as liquid
TP two-phase
tt turbulent-turbulent
V vapor

8. REFERENCES

Baker, O., 1954, "Simultaneous Flow of Oil and Gas," *Oil and Gas Journal*, Vol. 53, pp. 185.

Barnea, D., Luninsky, Y., and Taitel, Y., 1983, "Flow Pattern in Horizontal and Vertical Two-Phase Flow in Small Diameter Pipes," *Canadian Journal of Chemical Engineering*, Vol. 61, pp. 617-620.

Bergles, A.E., 1964, "Subcooled Burnout in Tubes of Small Diameter," *ASME Paper No. 63-WA-182*, ASME.

Bonjour, J., and Lallemand, M., 1998, "Flow Patterns during Boiling in a Narrow Space between Two Vertical Surfaces," *International Journal of Multiphase Flow*, Vol. 24, pp. 947-960.

Bowers, M.B., and Mudawar, I., 1994, "High Flux Boiling in Low Flow Rate, Low Pressure Drop Mini-channel and Microchannel Heat Sinks," *International Journal of Heat and Mass Transfer*, Vol. 37, No. 2, pp. 321-334.

Chen, J.C., 1966, "A Correlation for Boiling Heat Transfer to Saturated Fluids in Convective Flow," *I & EC Process Design and Development*, Vol. 5, No. 3, pp. 322-329.

Coleman, J.W., and Garimella, S., 2000, "Two-phase Flow Regime Transitions in Microchannel Tubes: The Effect of Hydraulic Diameter," HTD-Vol. 366-4, Proceedings of the ASME Heat Transfer Division-2000, Vol. 4, ASME IMECE 2000, pp. 71-83.

Collier, J.G., 1981, *Convective Boiling and Condensation*, McGraw-Hill, London.

Cornwell, K., and Kew, P.A., 1992, "Boiling in Small Parallel Channels," *Proceedings of CEC Conference on Energy Efficiency in Process Technology*," Athens, October 1992, Paper 22, Elsevier Applied Sciences, pp. 624-638.

Downing, R.S., Meinecke, J., and Kojasoy, G., 2000, "The Effects of Curvature on Pressure Drop for Single and Two-phase flow in Miniature Channels," Proceedings of NHTC2000: 34th National Heat Transfer Conference, Pittsburgh, PA, August 20-22, 2000, Paper No. NHTC2000-12100.

Fukano, T., Kariyasaki, A., and Kagawa, M., 1989, "Flow Patterns and Pressure Drop in Isothermal Gas-Liquid Flow in a Horizontal Capillary Tube," ANS Proceedings, 1989 National Heat Transfer Conference, ISBN 0-89448-149-5, ANS, Vol. 4, pp. 153-161.

Hetsroni, G., Segal, Z., Mosyak, A., 2000, "Nonunifrom temperature distribution in electronic devices cooled by flow in parallel microchannels," Packaging of Electronic and Photonic Devices, EEP-Vol. 28, pp.1-9.

Hewitt, G.F., 2000, "Fluid Mechanics Aspects of Two-Phase Flow," Chapter 9, *Handbook of Boiling and Condensation*, Eds. Kandlikar, S.G., Shoji, M., Dhir, V.K., Taylor and Francis, NY.

Hewitt, G.F., and Roberts, D.N., 1969, "Studies of Two-Phase Flow Patterns by Simultaneous X-Ray and Flash Photography," UK AEA Report ASRE-M2159.

Kamidis, D.E., and Ravigururajan, T.S., 1999, "Single and Two-phase Refrigerant Flow in Mini-channels," Proceedings of NHTC2000: 33rd National Heat Transfer Conference, Albuquerque, NM, August 20-22, 2000, Paper No. NHTC2000-12100, pp. 1-8.

Kandlikar, S.G., 1990, "A General Correlation for Saturated Two-Phase Flow Boiling Heat Transfer Inside Horizontal and Vertical Tubes," *ASME Journal of Heat Transfer*, Vol. 112, pp. 219-228.

Kandlikar, S.G., 1991, "Development of a Flow Boiling Map for Saturated and Subcooled Flow Boiling of Different Fluids in Circular Tubes," *Journal of Heat Transfer*, Vol. 113, pp. 190-200.

Kandlikar, S.G., and Stumm, B.S., 1995, "A Control Volume Approach to Predict Departure Bubble Diameter in Flow Boiling," *ASME Journal of Heat Transfer*, Vol. 117, pp. 990-997.

Kandlikar, S.G., and Spiesman, 1997, "Effect of Surface Characteristics on Flow Boiling Heat Transfer," Paper presented at the Engineering Foundation conference on Convective and Pool Boiling, Irsee, Germany, May 18-25.

Kandlikar, S.G., 2000, "Flow Boiling in Circular Tubes," Chapter 15, Handbook of Boiling and Condensation, Editors, S.G. Kandlikar, M. Shoji, and V.K. Dhir, Taylor and Francis, 2000.

Kandlikar, S.G., Steinke, M.E., Tian, S., and Campbell, L.A., 2001, "High-Speed Photographic Observation Of Flow Boiling Of Water In Parallel Mini-Channels," Paper presented at the ASME National Heat Transfer Conference, June, 2001, ASME.

Kasza, K. E., Didascalou, T., and Wambsganss, M. W., 1997, "Microscale flow visualization of nucleate boiling in small channels: mechanisms influencing heat transfer," *Proceeding of International Conference on Compact Heat Exchanges for the Process Industries*, Ed. R.K. Shah, New York, Begell, House, Inc., pp. 343-352.

Kennedy, J.E., Roach, Jr., G.M., Dowling, M.F., Abdel-Khalik, S.I., Ghiaasiaan, S.M., Jeter, S.M., and Quereshi, Z.H., 2000, "The Onset of Flow Instability in Uniformly Heated horizontal Microchannels," *ASME Journal of Heat Transfer*, Vol. 122, pp. 118-125.

Kew, P.A., and Cornwell, K., 1996, "On Pressure Drop Fluctuations During Boiling in Narrow Channels," *2nd European Thermal Sciences and 14th UIT National Heat Transfer Conference*, Eds. Celata, G.P., Di Marco, P., and Mariani, A., Edizioni ETS.

Kew, P.A., and Cornwell, K., 2000, "Flow Boiling in Compact Heat Exchangers," *Handbook of Phase Change: Boiling and Condensation*, Chapter 16.2, Taylor and Francis, pp. 412-427.

Kuznetsov V. V., and Shamirzaev A. S., 1999, "Two-phase flow pattern and flow boiling heat transfer in noncircular channel with a small gap," *Two-phase Flow Modeling and Experimentation*, pp. 249-253.

Kuznetsov, V.V., and Vitovsky, O.V., 1995, "Flow Pattern of Two-Phase Flow in Vertical Annuli and Rectangular Channel with Narrow Gap," *Two-Phase Flow Modelling and Experimentation 1995*, Editors G.P. Celata and R.K. Shah, Edizioni ETS.

Lakshminarasimhan, M.S., Hollingsworth, D.K., and Witte, L.C., 2000, "Boiling Incipience in Narrow Channels," HTD-Vol. 366-4, Proceedings of the ASME Heat Transfer Division 2000, Volume 4, ASME IMECE 2000, pp. 55-63.

Lazarek, G.M., and Black, S.H., 1982, "Evaporative Heat Transfer, Pressure Drop and Critical Heat Flux in a Small Diameter Vertical Tube with R-113.," *International Journal of Heat and Mass Transfer*, Vol. 25, No. 7, pp. 945-960.

Lin, S., Kew, P. A., and Cornwell, K., 1998, "Two-Phase Flow Regimes and Heat Transfer in Small Tubes and Channels," *Heat Transfer 1998, Proceedings of 11th International Heat Transfer Conference*, Kyongju, Korea, Vol. 2, pp. 45-50.

Lin, S., Kew, P. A., and Cornwell, K., 1999, "Two-phase Evaporation in a 1mm Diameter Tube," *6th UK Heat Transfer Conference in Edinburgh*, September 1999.

Liu, Z., and Winterton, R.H.S., 1988, "Wet Wall Flow Boiling Correlation with Explicit Nuclear Term," presented at the 5th Miami Symposium on Multiphase Transport and Particulate Phenomena.

Mandhane, J.M., Gregory, G.A., and Aziz, K., 1974, "A Flow Pattern Map for Gas-Liquid Flow in Horizontal Pipes," *Chemical Engineering Progress*, Vol. 45, pp. 39-48.

Martinelli, R.C., and Nelson, D.B., 1948, "Prediction of Pressure Drop during Forced Convection Boiling of Water," *ASME Transactions*, Vol. 70, 695-702.

Mehendale, S.S., Jacobi, A.M., and Shah, R.K., 1999, "Heat Exchangers at Micro- and Meso- Scales," *Compact Heat Exchangers and Enhancement Technology for the Process Industries*, Ed. R.K. Shah, Begell House, New York, pp. 55-74.

Mehendale, S.S., Jacobi, A.M., and Shah, R.K., 2000, "Fluid Flow and Heat Transfer at Micro- and Meso-Scales with Applications to Heat Exchanger Design," *Applied Mechanics Review*, Vol. 53, pp. 175-193.

Mertz, R., Wein, A., and Groll, 1996, "Experimental Investigation of Flow Boiling Heat Transfer in Narrow Channels," Calore e Technologia, Vol. 14, No. 2, pp. 47-54.

Mishima, K., and Hibiki, T., 1996, "Some Characteristics of Air-Water Two-Phase Flows in Small Diameter Tubes," *International Journal of Multiphase Flow*, Vol. 22, No. 4, pp. 703-712.

Moriyama, K., and Inoue, A., 1992, "The Thermohydraulic Characteristics of Two-Phase Flow in Extremely Narrow Channels (The Frictional Pressure Drop and Heat Transfer of Boiling Two-Phase Flow, Analytical Model)," 1992, *Heat Transfer-Japanese research*, Vol. 21, No. 8, pp. 838-856.

Nakayama, W., and Yabe, A., 2000, "Flow Boiling in Narrow Channels for Thermal Management of Microelectronic Equipment," *Handbook of Phase Change: Boiling and Condensation*, Chapter 16.1, Taylor and Francis, pp. 403-411.

Peng, X.F., and Wang, B.X., 1998, "Forced Convection and Boiling Characteristics in Microchannels," *Heat Transfer 1998, Proceedings of 11th IHTC*, August 23-28, Kyongju, Korea, Vol. 1, pp. 371-390.

Peng, X.F., Hu, H.Y., and Wang, B.X., 1998, "Boiling Nucleation During Liquid Flow in Microchannels," *International Journal of Heat and Mass Transfer*, Vol. 41, No. 1, pp. 101-106.

Ravigururajan, T.S., Cuta, J., McDonald, C.E., Drost, M.K., 1996, "Effects of Heat Flux on Two-Phase Flow Characteristics of refrigerant Flows in a Micro-Channel Heat Exchanger," 1996, HTD-Vol. 329, *National Heat Transfer Conference*, Vol. 7, ASME, pp. 167-178.

Shah, M.M., 1982, "Chart Correlation for Saturated Boiling Heat Transfer: Equations and Further Study," *ASHRAE Transactions*, Vol. 88, pp. 185-196.

Taitel, Y., Barnea, D., and Dukler, A.E., 1980, "Modeling Flow Pattern Transition for Gas-Liquid Transitions for Steady Upward Gas-Liquid Flow in Vertical Tubes," *AIChE Journal*, Vol. 26, No. 3, pp. 345-354.

Taitel, Y., and Dukler, A.E., 1976, "A Model for Predicting Flow Regime Transitions in Horizontal and Near Horizontal Gas-Liquid Flow," *AIChE Journal*, Vol. 22, pp. 47-55.

Tong, W., Bergles, A.E., Jensen, M.K., 1997, "Pressure Drop with Highly Subcooled Flow Boiling in Small-Diameter Tubes," *Experimental Thermal and Fluid Science 1997*, Elsevier Science Inc., Vol. 15, pp. 202-212.

Tran, T.N., Wambsganss, M.W., France, D.M., 1996, "Small Circular- and Rectangular-channel Boiling with Two Refrigerants," *International Journal of Multiphase Flow*, Vol. 22, No. 3, pp. 485-498.

Triplett, K.A., Ghiaasiaan, S.M., Abdel-Khalik, S.I., LeMouel, A., and McCord, B.N., 1999, "Gas-Liquid Two-Phase Flow in Microchannels Part II: Void Fraction and Pressure Drop," *International Journal of Multiphase Flow*, Vol. 25, pp. 395-410.

Wambsganss, M.W., Jendrzejczyk, J.A., France, D.M., 1991, "Two-Phase Flow Patterns and Transitions in Small, Horizontal, Rectangular Channels," *International Journal of Multiphase Flow*, Vol. 17, No. 3, pp. 327-342.

Wambsganss, M.W., Jendrzejczyk, J.A., France, D.M., 1992, "Two-Phase Flow and Pressure Drop in Flow Passages of Compact Heat Exchangers, *SAE Technical Paper 920550*.

Wambsganss, M.W., France, D.M., Jendrzejczyk, J.A., and Tran, T.N., 1993, "Boiling Heat Transfer in a Small Diameter Tube," *ASME Journal of Heat Transfer*, Vol. 115, No. 4, pp. 963-972.

Yao, S-C, and Chang, Y., 1983, "Pool Boiling Heat Transfer in a Confined Space," *International Journal of Heat and Mass Transfer*, Vol. 26, No. 6, pp. 841-848.

KEYWORDS: Multi-channel evaporation, compact evaporators, flow patterns, heat transfer, pressure drop, microchannels, minichannels

REVIEW OF ASPECTS OF TWO-PHASE FLOW AND BOILING HEAT TRANSFER IN SMALL DIAMETER TUBES

X. Huo, Y.S. Tian, V.V. Wadekar[*] and T.G. Karayiannis

South Bank University, 103 Borough Road, London SE1 0AA, E-mail: karayitg@sbu.ac.uk
* HTFS, AEA Technology-Hyprotech, B392.7 Harwell, Oxfordshire OX11 0RA;
E-mail: Vishwas.Wadekar@hyprotech.com

ABSTRACT

Previous studies on flow boiling in small tubes/channels are reviewed in this paper mostly with reference to flow patterns, heat transfer rates and critical heat flux (CHF). It was found that research in flow boiling in small diameter channels is at the early stages and much effort is needed to clarify the fundamental physical phenomena and obtain working correlations.

INTRODUCTION

Boiling heat transfer, as one of the most efficient techniques for removing high heat fluxes, has been studied and applied in thermal systems for a very long time. Most of the early applications were in the petrochemical and other large-scale plants. These studies focused on boiling heat transfer in large diameter tubes. Now, the rapid development of practical engineering applications for micro-devices, micro-systems, advanced material designs and manufacturing, i.e. compact heat exchangers, high capacity micro heat pipes for spacecraft thermal control, electronic chip and devices (Mishima and Hibiki 1996, Xu et al. 1996, Wadekar 1997) is increasing the demand for better understanding of small and micro-scale transport phenomena. For example, the very large-scale integration chips in computers can generate very high heat fluxes, so the temperature control is very important in protecting chips and allowing proper operation. Single phase convective heat transfer, which has been used in all computers so far, was shown to be inadequate in removing such high heat fluxes because of its low heat transfer coefficient. This encouraged engineers to consider boiling heat transfer for such applications. However, because of the size of chips, only very small heat exchangers can be used in this case. Increased effort in research in small to micro and nano-scale transport processes is expected in the next few years with progress in the physical understanding of these phenomena. In technologies where fluid flow and heat transfer play an important part, this understanding will allow the development of new, advanced systems, see also Yang and Zhang (1992).

Research in small-scale systems relates more to applications in compact heat exchangers. One of the most important characteristics of compact heat exchangers is the high surface area-to-volume ratio. To quantify this characteristic, Shah (1986) defined a compact heat exchanger as an exchanger with a surface area volume ratio >700 m²/m³, which translates into a hydraulic diameter of <6 mm. The higher heat transfer surface area density of compact heat exchangers allows attainment of significantly higher heat transfer rates in two-phase flows relative to conventional circular tube exchangers. At the same time, the small hydraulic diameter produces a thin thermal boundary layer and results in large heat transfer coefficients. Further, small heat exchangers will result in smaller overall plant size and require less working fluid charges. This is important in refrigeration and other applications where the release of the fluids to the atmosphere may have undesirable effects. In addition, the reduction of temperature differentials that can result from the use of highly efficient compact heat exchangers will give higher plant efficiencies thus not only responding to regulatory pressure but also offering long-term benefits of resource conservation and environmental protection.

Therefore, it is imperative that research effort is directed towards examining the two-phase flow patterns of commonly used fluids including the new refrigerants in small diameter channels and their effect on heat transfer. The effects of geometry and size on two-phase flow and heat transfer were examined by Kew and Cornwell (1997). They reported that when the confinement number, Co, defined by Eq. (1) below, was in excess of 0.5, two-

$$\mathrm{Co} = \frac{\left[\sigma/(g(\rho_l - \rho_g))\right]^{1/2}}{d_h} \quad (1)$$

phase flow in such small hydraulic diameters exhibited different flow characteristics and heat transfer results compared with corresponding flow in traditional size passages. For example, for typical working fluids like water and R134a, this equates to hydraulic diameters less than 5.4 and 2.1 mm, respectively (calculated at 30°C for water and −20°C for R134a). Normally, changes in the working pressure can result in changes in the flow patterns and heat transfer mechanisms of flow boiling; additionally the confinement number itself can be directly influenced by pressure mainly through vapour density and to a lesser extent through surface tension.

As stated above, experimental data available in the literature demonstrate that the flow patterns and boiling characteristics in small diameter tubes/channels differ from those in conventional situations. The objective of the paper is to review the findings of past research works on the flow patterns and heat transfer, reveal the importance of relevant parameters and recommend further study.

FLOW PATTERNS

Numerous experiments (e.g. Alves 1954, Derbyshire et al. 1964, Chaudry et al. 1965, Hosler 1968, Wallis 1969) have been conducted to study the flow patterns in normal size tubes, and four main flow patterns for the particular case of upward flow in vertical tubes were summarized from these experiments by Hewitt and Hall-Taylor (1970), namely: bubbly flow, plug/slug flow, churn flow and annular flow. However, the experimental data relating to two-phase flow in narrow channels are very limited and the results are very different compared to gas-liquid flow in large channels. Oya (1971) investigated the upward two-phase flow patterns in tubes of 2, 3 and 6 mm in diameter using mixtures of air and tap or distilled water, gasoline, solvent and surfactant solution. The flow patterns observed in their experiments are shown in Figure 1. Amongst their flow patterns, granular-lumpy bubbly flow and fish-scale type slug flow have never been discussed in any other papers. They concluded that these different flow patterns compared to large tubes were attributed to the effect of the small diameter and the liquid surface tension.

However, Barnea et al. (1983) reached very different conclusions with air-water system. They studied flow pattern transition in 5 glass tubes of 4, 6, 8.15, 9.85 and 12.3 mm diameter. The flow regime maps are depicted in Figure 2. After comparing their experimental results with the theoretical models published by Taitel et al. (1980), they concluded that the effect of pipe diameter was small for vertical upward flow. But they emphasized the effects of surface tension on the stratified non-stratified transition boundary of horizontal flow.

Xu et al. (1999) investigated adiabatic air/water two-phase flow in vertical rectangular channels (12 mm high x 260 mm long) with narrow gaps of 0.3, 0.6 and 1.0 mm. Flow regimes were observed using a CCD camera and were identified by examining the video images. The flow regimes for gaps of 1.0 and 0.6 mm were found to be similar with those in conventional channels, but the transitions from churn flow to annular flow and from slug flow to annular occur earlier with a decrease of the channel gap. These observations agree with those of Barnea et al. (1983); they were considered to be due to the increased interfacial shear stress and the wall shear stress. For the 0.3 mm gap, the flow regimes were quite different from the previous ones, in which bubbly flow was never observed even at very low gas flow rates. In channels with a 0.3 mm gap, it was also found that, due to the increased influence of the surface tension force and the shear stress, the liquid droplets adhered to the wall surface and were pushed by the gas phase.

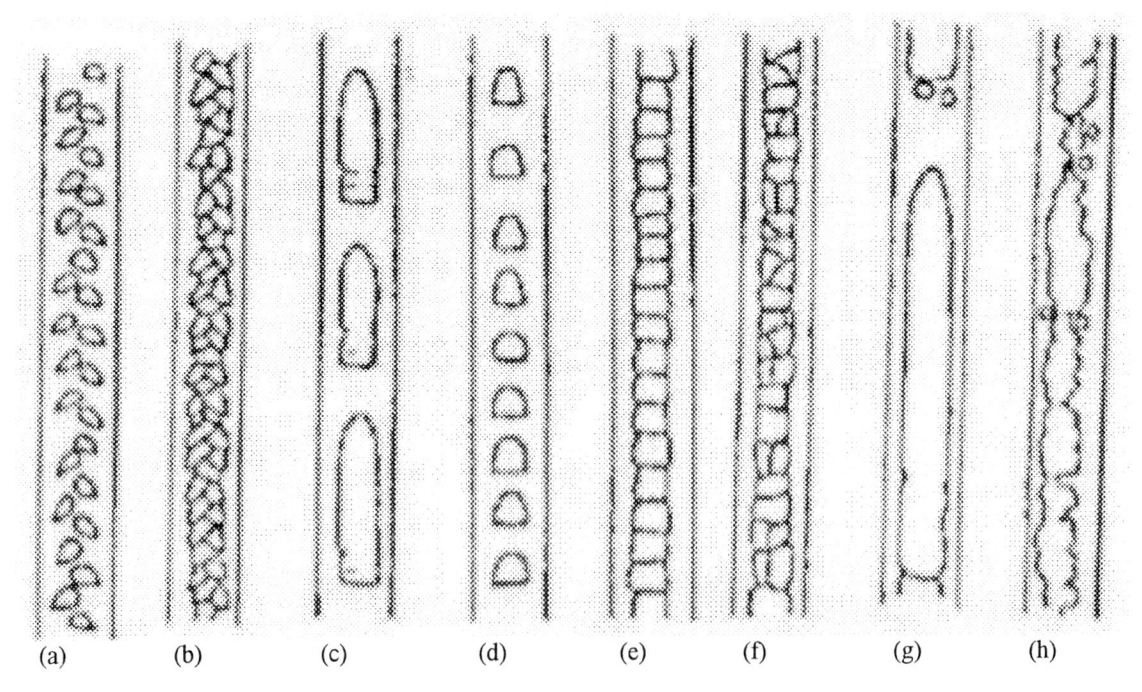

(a) Simple bubbly flow (b) Granular-lumpy bubbly flow (c) Piston type slug flow (Piston flow) (d) Simple slug flow
(e) Single fish-scale type slug flow (f) Double fish-scale type slug flow (g) long piston type slug flow (h) Froth flow

Fig. 1 Flow patterns reported by Oya (1971).

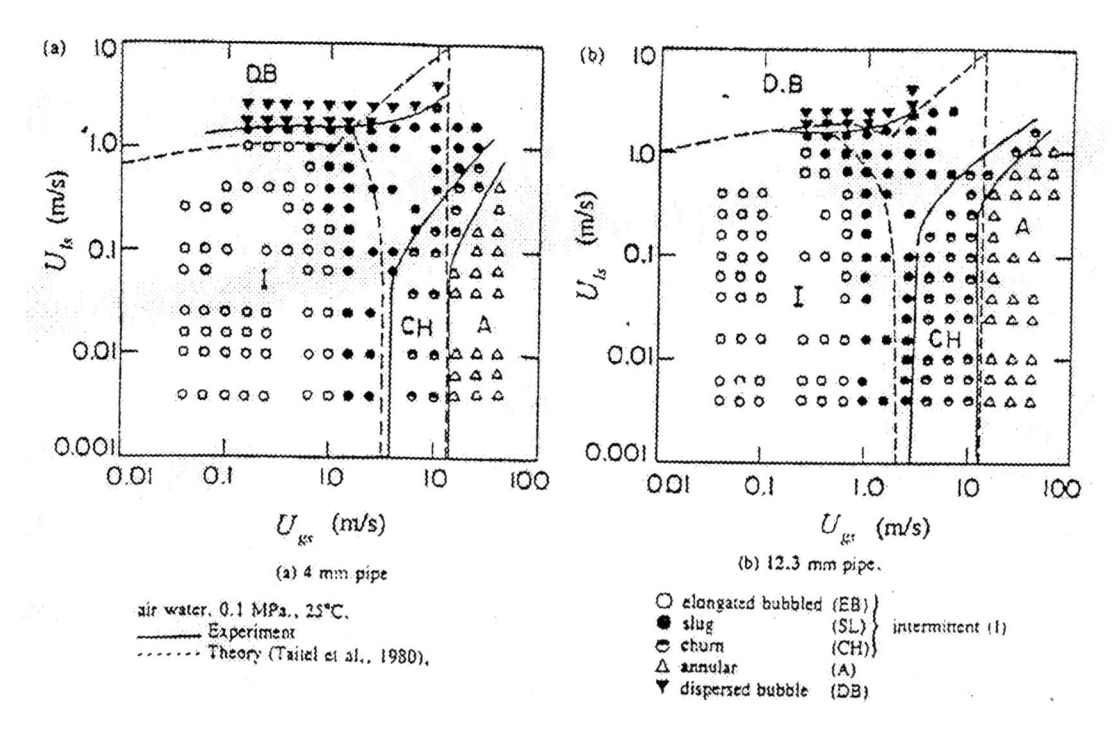

Fig. 2 Flow pattern maps reported by Barnea et al. (1983) comparing data with the predictions of Taitel et al. (1980).

Mishima and Hibiki (1996) presented flow pattern maps for air/water flow in 1-4 mm diameter vertical tubes. Based on their experiments, they reported that the capillary force largely affected the bubble shape. They also observed some peculiar flow regimes, which had not been observed in normal diameter tubes. A sketch of typical flow regimes observed by Mishima and Hibiki (1996) is shown in Figure 3. Some flow regimes, which are denoted by the asterisks in the figure, are particular to capillary tubes. However, these are different from the flow patterns observed by Oya (1971), shown in Figure 1, i.e. the bubble shapes and their agglomeration are very different in the bubbly flow and slug flow.

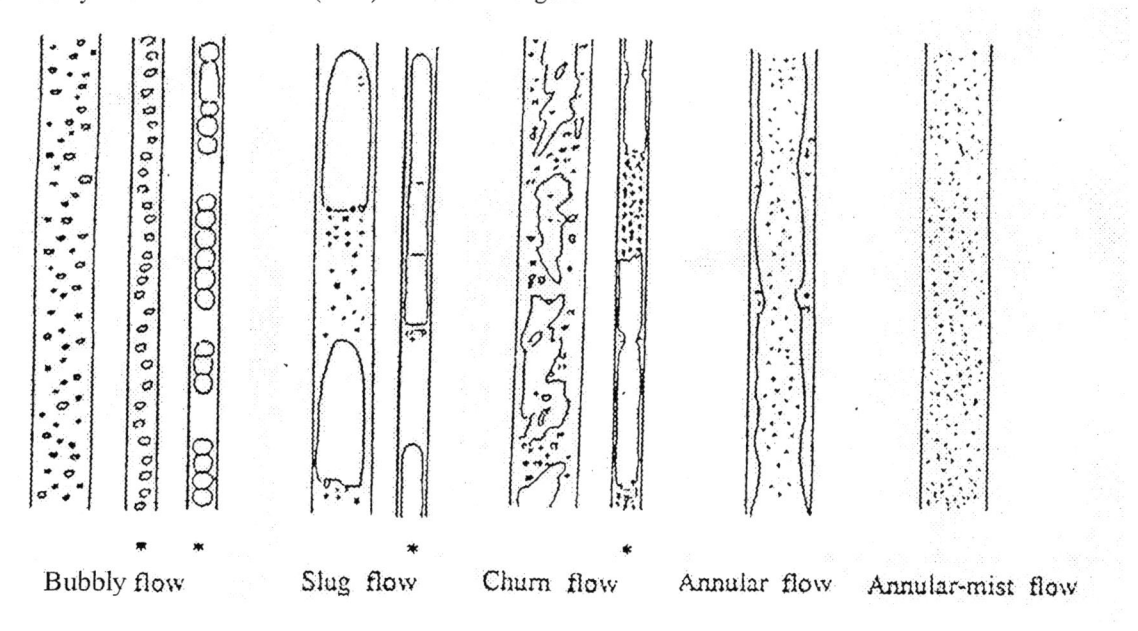

Fig. 3 Flow regimes in small diameter tubes reported by Mishima and Hibiki (1996).

Lin et al. (1998) carried out their experimental study with air/water mixtures in a small vertical tube 2.1 mm in diameter. It was concluded that in a relatively small flow rate range, four distinct flow patterns are presented: bubbly, plug, churn and annular; the annular flow occurred at a relatively low air flow rate compared to air-water flow in larger channels/tubes. In comparison with the experiments of Mishima and Hibiki (1996), the annular flow was reported to occur at a higher air flow rate and the bubbly flow at a lower water flow rate. The effect of the separator used in their experiment was offered as a possible explanation.

In conventional tubes/channels the flow patterns in horizontal flow are different from those in vertical flow due to the effect of gravity. In horizontal flow, the top and bottom may be vapour and liquid rich respectively and stratified flow is observed. However, this is not the case for small size passages. Gas-liquid two-phase flow patterns in long horizontal micro-channels with circular and semi-triangular (triangular with one corner smoothed) cross-sections, were experimentally investigated by Triplett et al. (1999). The test section hydraulic diameters of the circular channels and the semi-triangular cross-section channels were 1.1 and 1.45 mm, and 1.09 and 1.49 mm respectively. The gas and liquid superficial velocity ranges were 0.02-80 m/s, and 0.02-8 m/s, respectively. Five major flow patterns were reported: bubbly, slug, churn, slug-churn and annular. These flow patterns occurred in all test sections and the flow pattern map using gas and liquid superficial velocities as co-ordinates was similar overall, which meant that for the micro channels, there was no geometry effect on flow patterns. Compared with flow patterns published for conventional channels, no stratified flow occurred in micro channels. The precise conditions at which stratified flow became impossible were not pointed out in these experiments, though the researchers emphasized the effect of surface tension. Available relevant flow regime transition models were compared with the experimental data with poor agreement.

Coleman and Garimella (1999) investigated the effects of tube diameter and shape on flow regime transitions for two-phase flow in tubes with small hydraulic diameters. Flow patterns of air/water mixtures in horizontal round and rectangular tubes were determined by high-speed video camera. The hydraulic diameter ranges were from 1.3 to 5.5 mm, gas and liquid superficial velocities from 0.1 to 100 m/s and 0.1-10 m/s, respectively. The results differ from those reported by Triplett et al. (1999), i.e. bubbly, dispersed/elongated bubble, slug, stratified, wavy, annular-wavy and annular were observed in this investigation, and geometry affected flow regime transitions. However, similar to Triplett et al. (1999), they also emphasized the effect of surface tension on flow regime transitions. They found the combination of surface tension and decreasing tube diameter shifted the dispersed flow regime to a higher liquid superficial velocity, suppressed the stratified regime and increased the size of the intermittent regime.

Mertz and Groll (2000) carried out flow visualization and heat transfer tests on vertical single rectangular and semi-circular channels (2 mm wide and 4 mm deep) and multi-channel arrangements (1, 2 and 3 mm wide with aspect ratios up to 3) using water and R141b. The heat flux range was from 1 to 17% of the critical heat flux depending on geometry. They reported that, in the multi-channel heat exchanger, the boiling process follows the sequence of generation and flow of isolated bubbles to the generation of confined bubbles and slug flow and further to annular flow. The heat flux range for different boiling modes depended on the channel geometry. They also observed that nucleation in the channels was not a continuous process (bubbles continuously generated and released) but was characterized by a pulsating release of bubbles through the exit of the channels. At low mass flux and low heat flux some channels were completely filled with vapour bubbles. In single channels, carrying water, bubbles, confined bubbles and annular flow were observed at low, medium and high heat flux respectively. For R141b, the change to confined bubbles and annular flow seems to occur at lower heat flux values than for water.

Sekoguchi et al. (1992) reported that pressure can influence the fluid density and viscosity, affecting the boundary between different flow regimes, e.g. the effect of increased vapour density (higher pressure) was to move the bubbly-plug, churn-plug and churn-annular transitions to higher gas mass flow rates. This is in line with the observations of Weisman and Kang (1981).

Some general observations can be made from the above brief review of flow pattern work. Firstly, there is no general flow pattern map which could be applicable to a wide range of hydraulic diameters. Secondly, most of the work was carried out using air-water mixtures in adiabatic flow. Thirdly, although pressure is a key parameter influencing the flow pattern transitions, most of the experiments were carried out at a fixed pressure near to one atmosphere.

HEAT TRANSFER

The two-phase flow boiling region in tubes/channels can be characterized by either the nucleate or convective component or both. All three possibilities have been reported based on experiments under different system parameters. In large diameter tubes/channels, the flow patterns are usually annular for the largest range of quality and the convective heat transfer mechanism dominates (Reid et al. 1987, Jung and Radermacher 1991 and Carey et al. 1992). In contrast, conclusions differ among the various researchers as to the boiling heat transfer mechanisms in small-diameter tubes/channels over the entire quality range.

The local heat transfer coefficient, pressure drop, and critical heat flux were measured by Lazarek and Black (1982) for flow boiling of R113 in a round vertical tube with an internal diameter of 3.1 mm, and heated lengths of 12.3 and 24.6 cm. The heat flux was varied from 1.4×10^4 to 3.8×10^5 W/m^2, the mass flux from 125 to 725 kg/m^{2}s and

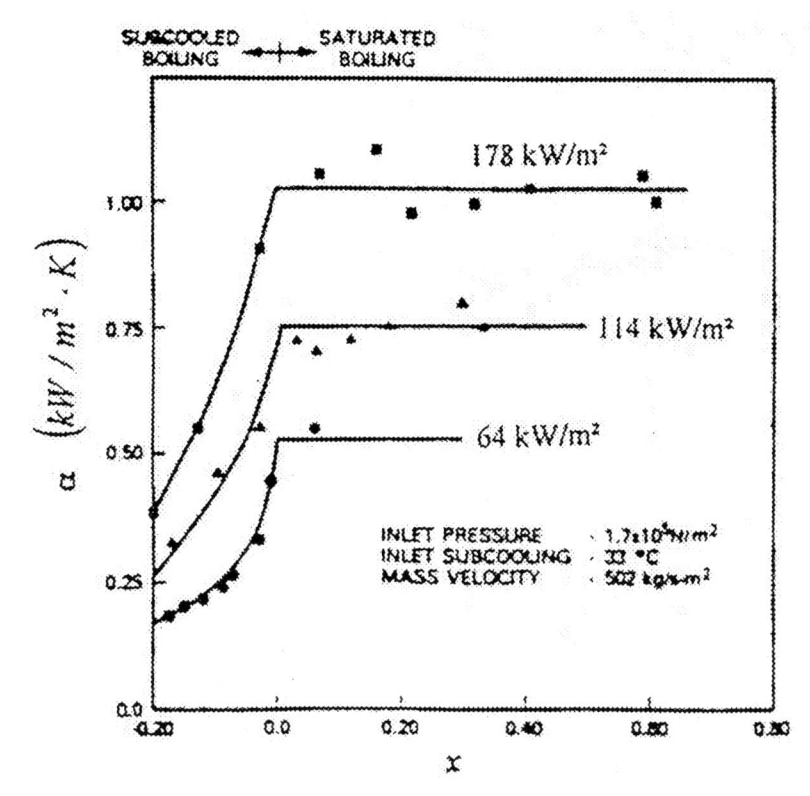

Fig. 4 Local heat transfer coefficient as a function of local vapour quality for $d_i = 3.1$ mm and different heat flux values, Lazarek and Black (1982).

the pressure from $1.3 \times 10^5 - 4.1 \times 10^5$ Pa. It can be seen very clearly in Figure 4 obtained from their experiments that: (1) the subcooled and saturated boiling heat transfer coefficients are a strong function of heat flux, (2) the local heat transfer coefficient increases rapidly during subcooled boiling, (3) beyond saturated boiling, i.e. $x > 0.0$, the boiling heat transfer coefficient is independent of quality. From the strong dependence of the saturated boiling heat transfer coefficient upon heat flux and negligible influence of quality, they concluded that the mechanism of nucleate boiling controlled the wall heat transfer process. The saturated local boiling heat transfer coefficient data were correlated in the following simple form:

$$Nu = 30 \cdot Re^{0.875} \cdot Bo^{0.714} \qquad (2)$$

The results of a study on boiling heat transfer of refrigerant R113 in a 2.92 mm diameter horizontal tube were reported by Wambsganss et al. (1993). All the data from these experiments are shown in Figure 5. It is clear, from the figure that the heat transfer coefficient is a function of heat flux. As reported by the researchers, it is also independent of x over most of the quality range. Therefore, they suggested that nucleate boiling is the dominant heat transfer mechanism. The flow patterns indicated in this figure show that, for the small diameter tube of this study, the predominant flow regime is slug flow until qualities of 0.6 to 0.8. As mentioned above, the predominant flow regime in larger diameter tubes is annular. Therefore, they supposed the thick-liquid regions of slug flow are more likely to support nucleation than the thin liquid films of annular flow.

Boiling heat transfer experiments were performed by Tran et al. (1996) in a small circular channel ($d_i = 2.46$ mm) and a small rectangular channel ($d_h = 2.40$ mm) with R12. It is obvious from Figure 6 and 7, which summarize their experimental data, that the local heat transfer coefficient is a function of heat flux and effectively independent of quality and mass flux for $x \geq 0.2$. Their results were applicable for wall superheats above 2.75 K. At lower wall superheats, the mass flux effect became important. Therefore, they concluded that over a broad range of heat flux, nucleation was the dominant heat transfer mechanism for flow boiling in the small passages of their study, and at sufficiently low values of heat flux (very low wall superheat), forced convection dominates. There was very little difference between the heat transfer coefficients in rectangular and circular channels. Their mean heat transfer data for R12 and R113 (studied previously, Wambsganss et al. 1993) were correlated in the nucleation dominant region as:

$$\alpha = 8.4 \times 10^{-5} \cdot Bo^2 \cdot (We_l)^{0.3} \cdot \left(\frac{\rho_l}{\rho_v}\right)^{0.4} \qquad (3)$$

Recently, Bao et al. (2000) studied flow boiling heat transfer coefficients for R11 and R123 in a copper tube

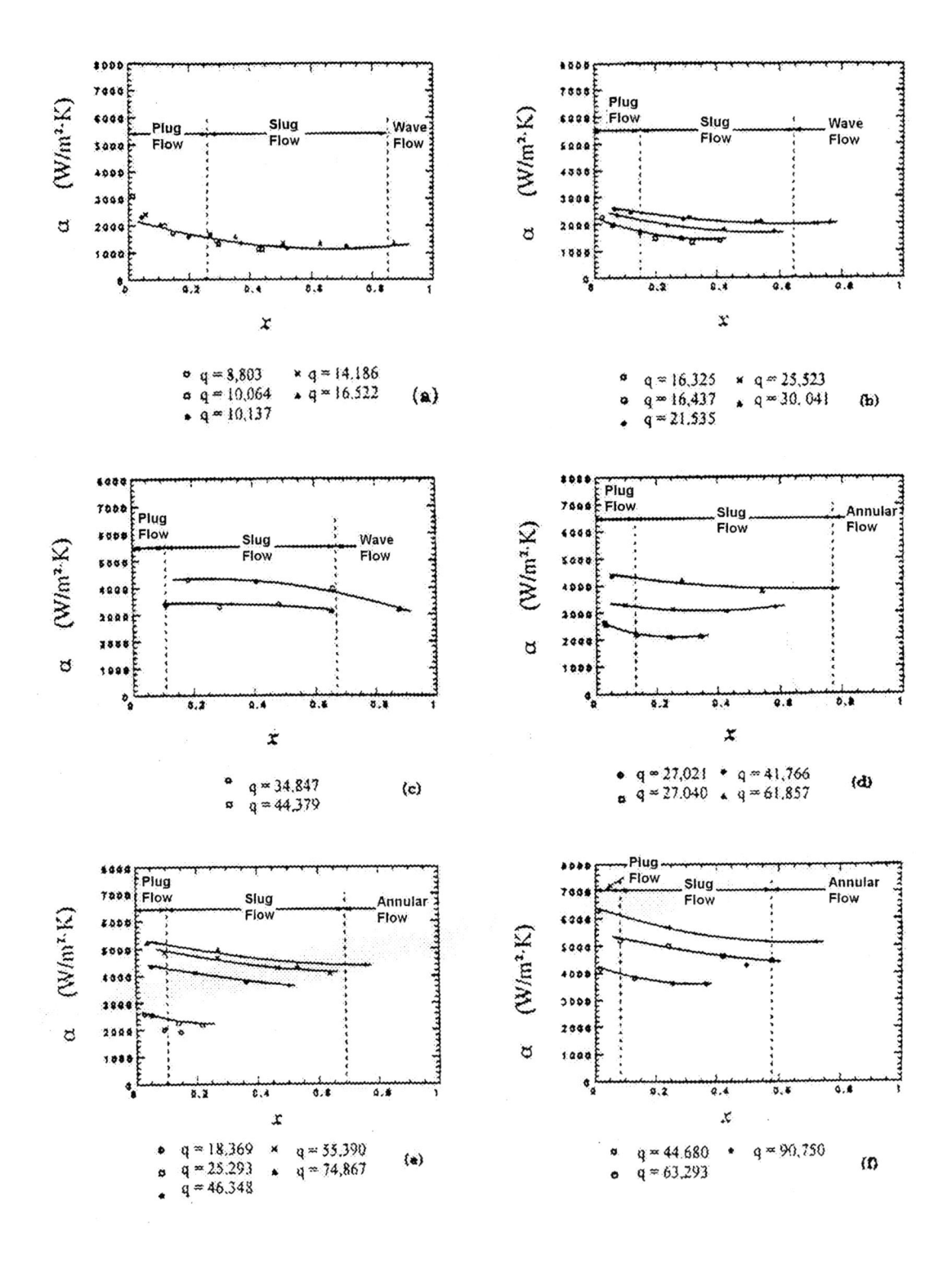

Fig. 5 Heat transfer coefficients at constant mass fluxes of: (a) G = 50 kg/m²s (b) G = 100 kg/m²s
(c) G = 150 kg/m²s (d) G = 200 kg/m²s (e) G = 242 kg/m²s (f) G = 300 kg/m²s (Note q is in W/m²).

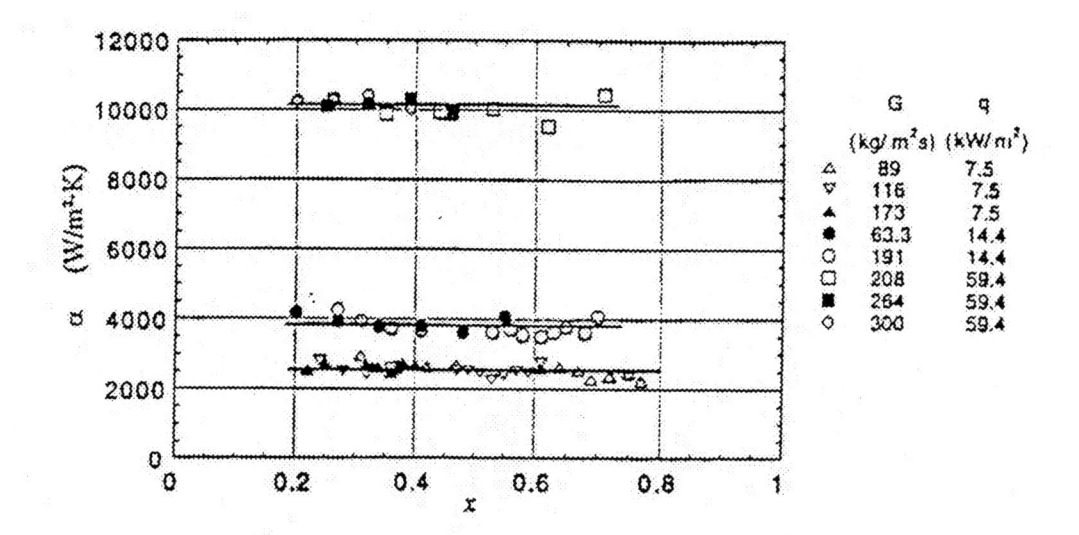

Fig. 6 Local heat transfer coefficients as a function of quality for various combinations of mass flux at three constant heat fluxes and $\Delta T_{sat} > 2.75\,K$; $p_{sat} \approx 825\,kPa$, Tran et al. (1996).

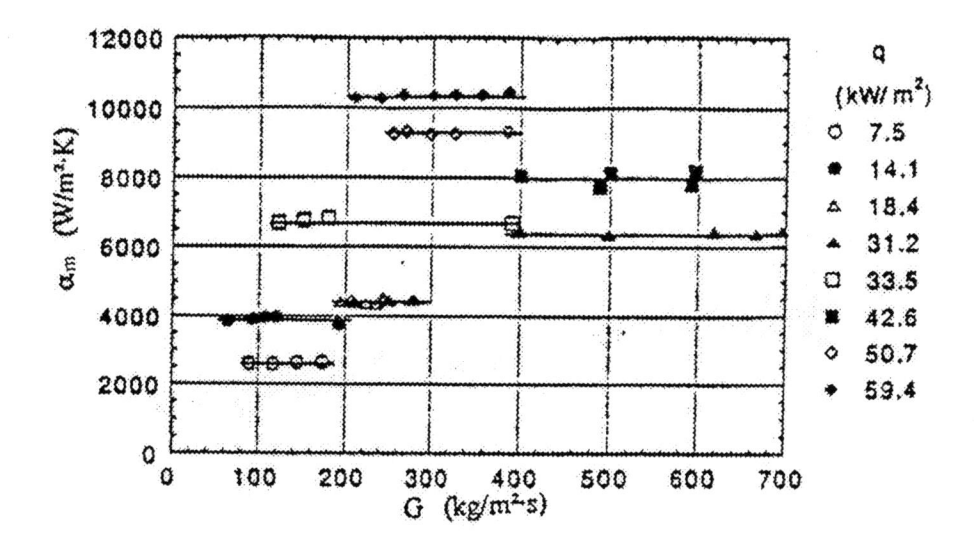

Fig. 7 Average heat transfer coefficients as a function of mass flux for selected values of approximately constant heat flux and $\Delta T_{sat} > 2.75\,K$; $p_{sat} \approx 825\,kPa$, Tran et al. (1996).

with an inner diameter of 1.95 mm. The range of parameters examined was: heat flux from 5 to 200 kW/m^2; mass flux from 50 to 1800 kg/m^{2}s; vapour quality from 0 to 0.9; system pressure from 200 to 500 kPa; and experimental heat transfer coefficients from 1 to 18 kW/m^{2}K. They too reported that the heat transfer coefficient is a strong function of the heat flux and the system pressure, while the effects of mass flux and vapour quality are very small in the range examined.

During nucleate boiling, heat transfer from the surface occurs due to: (i) latent heat of bubble, (ii) continuous evaporation at the root of the bubble and condensation at the top of bubble, while the bubble is still attached to the wall, (iii) liquid-vapour exchange caused by bubble agitation of the boundary layer and (iv) single-phase convection between patches of bubbles. Therefore, nucleate boiling has an extremely high heat transfer coefficient, which tends to maintain the heating surface temperature within a narrow range while dissipating a wide range of heat flux. Thus, the dominance of nucleate boiling, as proposed by the above experiments, is useful in explaining the heat transfer enhancement in small diameter tubes. However, some other researchers reported that the high heat transfer coefficients were not attributed to nucleate boiling. Robertson (1979, 1983), Robertson and Wadekar (1988) and Wadekar (1992) concluded from their tests with multi-channel arrangements that convection rather than nucleation was the important mechanism. Robertson and Wadekar (1988) carried out upward flow boiling tests on a vertical plate-fin test section, with perforated fins, of $d_h = 2.4 \text{ mm}$ hydraulic diameter. Figure 8 summarizes their experiments and it is clear that the results indicate significantly different trends at the two mass fluxes of 105 and 290 kg/m²s and negligible influence of heat flux leading to the conclusion that the convective mechanism of heat transfer was dominant. However, the low values of the wall superheat in their experiments may explain the dominance of convection, Tran et al. (1996). Moreover, the apparently conflicting observations made for hydrocarbons and refrigerants could also be explained on the basis of differences in reduced pressures (i.e. pressure/ critical pressure) for the fluids and pressures employed in the respective studies.

Experiments were carried out by Yan and Lin (1998) to investigate the characteristics of boiling heat transfer and pressure drop for refrigerant R134a flowing in a horizontal small circular pipe with 2.0 mm inside diameter. The heat transfer coefficients for the small pipes in their experiments were about 30-80% higher than that in larger pipes ($d_i \geq 8.0 \text{ mm}$) reported in the literature. Moreover, they noted that in the small pipe the boiling heat transfer coefficient was higher at a higher imposed wall heat flux except in the high vapour quality region, and also, the boiling heat transfer coefficient was higher at a higher mass flux and saturation temperature when the imposed heat flux was low. However, from their results, it was very difficult to conclude which regime was dominant, nucleation or forced convection.

Geometry effects on flow boiling heat transfer in narrow channels were reported by Mertz and Groll (2000), e.g. their multi-channel arrangement of aspect ratio 2 gave the best results. They also observed a different behaviour between single and multi-channel arrangements, i.e. the heat transfer coefficient increased with increasing heat flux in the single channel while it decreased or remained about constant with increasing heat flux in the multi-channel heat exchanger.

Critical Heat Flux

Subcooled flow boiling can accommodate significantly high heat transfer rates due to the nucleate boiling dominance. The upper limit of heat flux is determined by the critical heat flux (CHF), the occurrence of which can

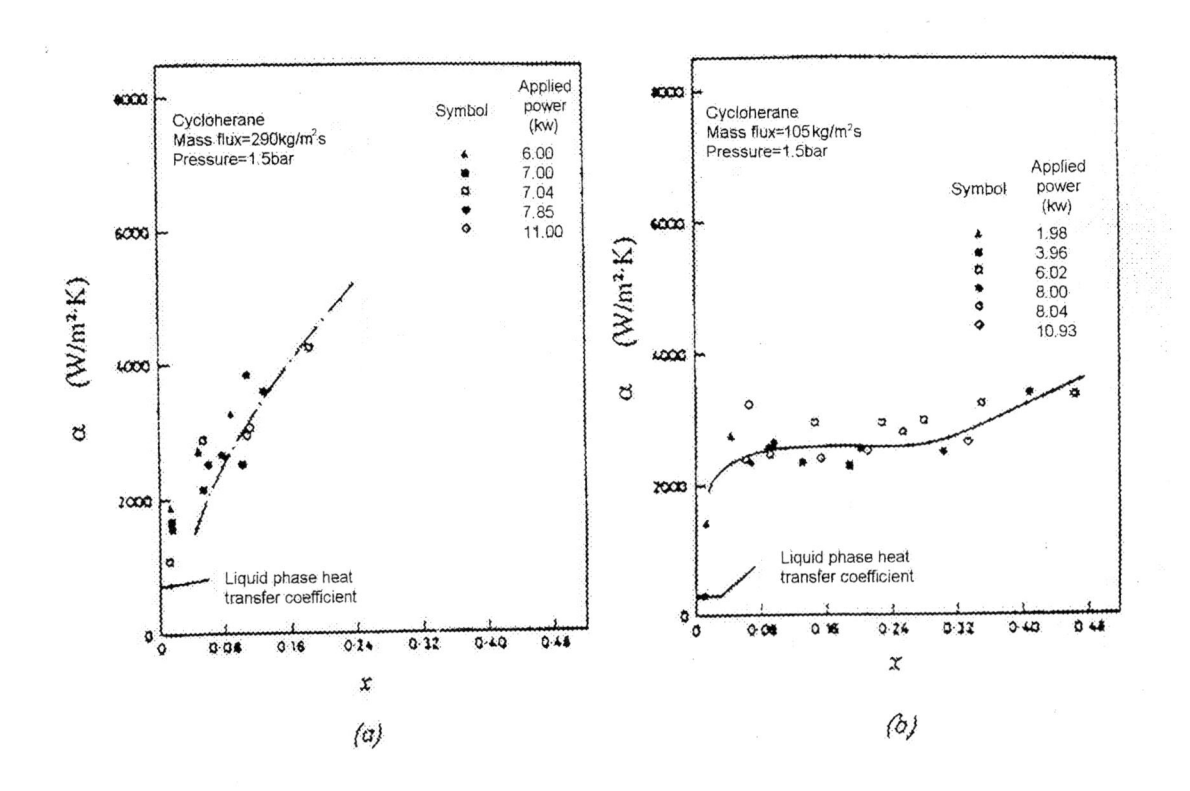

Fig. 8 Plot of boiling heat transfer coefficient against quality for two different mass flux values, Robertson and Wadekar (1988).

result in a sharp reduction in heat transfer rates. Therefore, the determination of the CHF and its dependence, among others, on diameter is important when flow boiling is used in compact heat exchangers. The CHF during subcooled flow boiling in narrow one-side heated rectangular channels was investigated experimentally by Nakajima et al. (1996) using channels of 1.3 mm, 2.0 mm, and 3.0 mm. Critical heat flux was found to increase by 38-79 % with decreases in the channel width. The influence of the channel width was stronger for higher mass velocity. An experimental investigation dealing with the CHF in subcooled flow boiling in a short tube was performed by Celata et al. (1993). It was carried out with a 2.5 mm inside diameter stainless steel tube. Very high values of the CHF were achieved, ranging from 12.1 to 60.6 MW/m². The CHF exhibited the usual increasing dependence on the degree of inlet subcooling and mass flux within the above range of parameters. In a technical note, Celata et al. (1993) concluded that the CHF may become very high and was supposed to be inversely related to the tube inside diameter. They stated that this effect could be ascribed to the reduction of the void fraction with the decrease of the channel diameter, because of smaller diameter of bubbles (or thickness of the two-phase boundary layer) and stronger action of bubble entrainment effect due to the increase of the velocity gradient in the two-phase boundary layer. Celata et al. (1996) described the CHF dependence on the diameter using Eq. (4), where (CHF)$_d$ is the CHF for

$$\frac{(CHF)_d}{(CHF)_{d=8mm}} = \left(\frac{d}{8}\right)^n \qquad (4)$$

a diameter of interest (d in mm) and (CHF)$_{d=8mm}$ is the CHF for an 8 mm tube (i.e. typical reference diameter from the CHF table, Groeneveld et al. (1993)). The exponent was given the value of –0.3 for a range of diameters from 0.5 to 32 mm. Equation (4) was proposed earlier by Doroshchuk et al. (1975a, 1975b) with n=-1/2 for a range of diameters between 4 and 16 mm. Groeneveld et al. (1986) later proposed a value of –1/3 for d = 4 – 20 mm and Smith (1986) extended the range to 32 mm with the same value of n.

More recently, Celata et al. (1997) tested stainless steel tubes with inner diameters ranging from 0.25 to 1.5 mm and water as the working fluid. They concluded that the CHF increased as the channel diameter decreased until a value of 0.7 mm, and became independent of diameter thereafter. They explained that this may be due to a change of CHF mechanism, i.e. from the liquid sub-layer dryout model to flooding of the channel as the size of the bubbles were of the same order of magnitude as the channel diameter.

Cheng et al. (1997) carried out an experimental study on CHF using R12 in tubes of different diameters ranging from 2 to 16 mm, pressure ranging from 1.0-3.0 MPa, mass flux 1.0-6.0 Mg/m²s and exit steam quality –0.75 to +0.59. They stated that their data revealed that the effect of the tube diameter on CHF cannot be described by the equations or models available in literature (e.g. simple exponential function of Eq. (4) above) with sufficient accuracy. They also explained that the CHF depended on thermal-hydraulic (exit vapour quality) and geometric conditions. They recommended that further research work was needed to clarify the effect of diameter on CHF.

Cheng et al. (1997) observed the normal relationship between CHF and exit vapour quality, i.e. a linear decrease in the CHF with exit quality except for in their tests with the 2 mm tube. In this case, at low pressure and mass flux, the CHF increased slightly with exit quality. They attributed this strange observation to the effect of the heated length (L/d = 100 in this case) and recommended further work to study this especially in small diameter tubes. CHF decreases with increasing channel heated length and investigators use the ratio L/d as the characteristic non-dimensional length, but this, claims Celata (1997) still needs to be established. The dependence of CHF on L/d is shown in Figure 9 (a) from the work of Nariai et al. (1987) on subcooled flow boiling in small tubes 1-3 mm. The relationship between CHF, tube inside diameter and heated length is shown as Figure 9 (b) from Inasaka and Nariai (1987). In the low heat flux region (LOW) their experimental data could be predicted well by the existing correlations for ambient pressure, while in the high heat flux region (HIGH) the experimental data were higher than the predictions by the existing correlations. The high heat flux region appeared when the tube inside diameter and the tube length decreased. Celata (1996) stated the fact that the length effect is more significant for L/d < 20 indicates that the CHF is related to the state of development of the bubble boundary layer.

Bowers and Mudawar (1994) performed an experiment to study the pressure drop and critical heat flux (CHF) in a small (d_h = 2.54 mm) and micro-channel (d_h = 510 μm) heat sink of 1 cm heated length using R113. In agreement with earlier reports, a larger heat transfer coefficient for the micro-channel was found in this experiment, which also gave an increase in CHF as compared to the small channel. Bowers and Mudawar (1994) explained that the reason for this result was that a small hydraulic diameter assured a thin thermal boundary layer, resulting in a larger heat transfer coefficient. However, as illustrated in Figure 10, the increase in CHF in the micro-channel required a substantial increase in pressure drop. Thus, they suggested a need for predictive tools for both the two-phase hydrodynamic and thermal behaviour of mini- and micro-channels.

CONCLUSIONS - RECOMMENDATIONS

The following conclusions can be reached based on the above discussion:

(i) Flow patterns in small diameter channels are significantly different than those observed in traditional size tubes due to the increased influence of surface tension. The confined geometry may even result in flow patterns that were not reported in larger channels such as the granular-

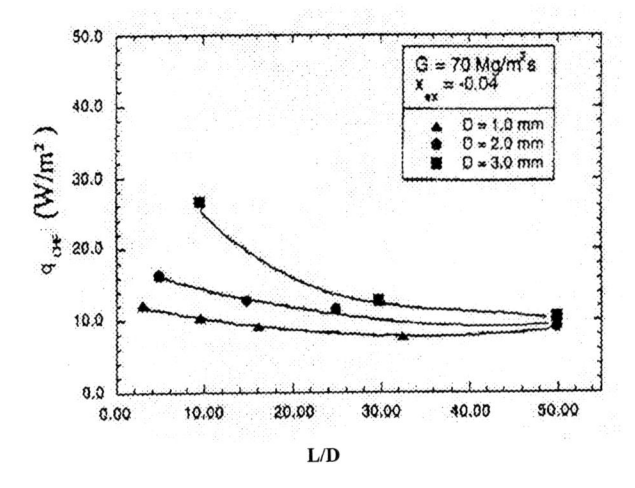

(a) CHF versus length for different diameters, Inasaka et al. (1987)

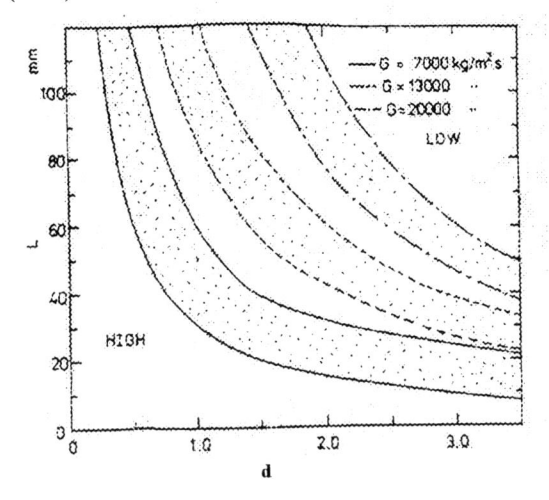

(b) High and low heat flux regions, Nariai et al. (1987)

Fig. 9 The dependence of CHF on tube diameter and heated length.

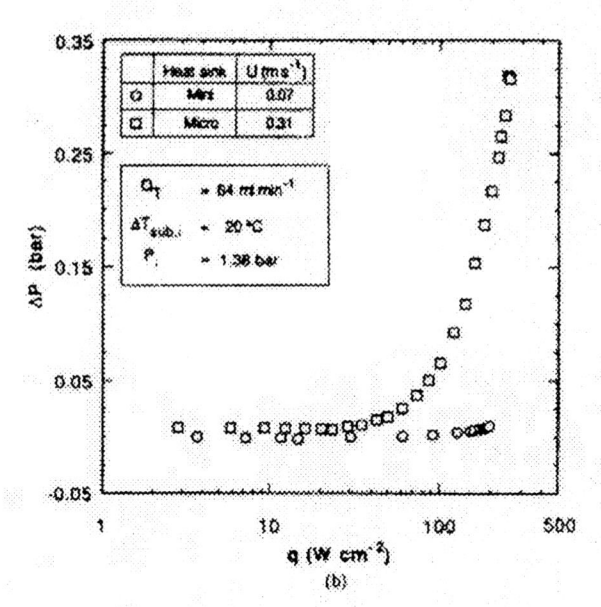

Fig. 10 Pressure drop of mini- and micro-channel for a flow rate of 64 ml/min and 20°C inlet subcooling, Bowers and Mudawar (1994).

lumpy bubbly flow and fish scale type slug flow. In very small passages bubbles may not be observed. On the other hand, there are some claims of flow regimes similar to those in traditional tubes for a certain range of hydraulic diameters (> 0.6 mm). However, even in these cases, the transition boundaries are different and in some cases poor agreement was reported when comparing with large size passages. In addition, the effect of the channel geometry was not conclusively clarified with some researchers reporting similar results for channels of different cross section and others observing no differences. Surface tension effects are more pronounced in rectangular cross section channels and thus differences may be expected in flow regimes in small passages where surface tension effects may dominate. The flow complexity and differences with larger diameter channels are more pronounced in small multi-channel passages.

(ii) The literature review indicated that nucleate boiling is the dominant heat transfer process in these small passages. A limited number of correlations have been proposed. Again, the effect of channel geometry was not conclusively clarified with some researchers claiming no effect while others reported differences. Also, the use of single channel data to predict the behavior of multi-channel arrangements still needs to be clarified.

(iii) The CHF was found to increase as the channel diameter decreases in an exponential form. This occurs down to a certain diameter (0.7 mm) after which there is no dependence on the diameter. However the need for further research was identified to include the effect of thermohydraulic as well as geometric conditions on CHF. The effect of heated length was also identified as a parameter to be included in such studies. Further work is also necessary to determine the effect of channel diameter on pressure drop and obtain generalized correlations.

From the above literature review, it can be concluded that the results obtained so far for small diameter tube/channels are far from sufficient to allow a deep and comprehensive analysis. This is exacerbated because of: (i) very limited research has been carried out so far on flow boiling patterns and heat transfer in small diameter tubes; (ii) different experimental conditions were used (such as different temperature, mass flux, heat flux and pressure), which makes it difficult to compare results directly.

A comprehensive study is needed which may include the following aspects:

(1) A detailed and systematic study of the effect of diameter (range from conventional size to nano size) on flow patterns, heat transfer rates and pressure drop;
(2) A large range of controlling variables such as pressure, degree of subcooling, mass flux, heat flux, quality and tube length;
(3) The effect of passage geometry, such as circular, rectangular and triangular, on flow patterns and heat transfer rates should be examined;
(4) Study and compare behavior and results for different fluids;
(5) Single and multiple parallel passages should be tested for practical applications.

NOMENCLATURE

Bo	Boiling number, q/Gh_{fg}
Co	Confinement number, defined in Eq. (1)
CHF	Critical heat flux, W/m^2
D	Diameter, m
G	Mass flux, kg/(m^2 s)
g	Gravitational acceleration, m/s^2
h_{fg}	Enthalpy of vaporization, W/kg
L	Length (heated), m
n	Factor, Eq. (4)
Nu	Nusselt number, $\alpha D/\lambda_l$
P	Pressure, Pa
Q_T	Volume flow rate ml/min, in Figure 10
q	Heat flux, W/m^2
Re	Reynolds number, GD/μ_l
t	Temperature, °C
T	Temperature, K
x	Quality
U	Velocity, m/s
We	Weber number, $G^2 d/(\rho_l \sigma)$
α	Heat transfer coefficient, W(m^2 K)
Δ	Finite increment
λ	Thermal conductivity, W/mK
μ	Dynamic viscosity, Pa·s
ρ	Density, kg/m^3;
σ	Surface tension, N/m;

Subscripts

CHF	Critical heat flux, W/m^2
E	Equivalent
g	Gas
h	Hydraulic
i	Inside, inlet
l	Liquid
m	Mean
s	Superficial
sat	Saturated
sub	Subcooled
v	Vapour
w	Wall

REFERENCES

Alves, G. E., 1954, Co-current liquid-gas flow in a pipeline contactor, Chem Process. Engng, Vol. 50, No. 9, pp. 449-456.

Bao, Z. Y., Fletcher, D. F. and Haynes, B. S., 2000, Flow boiling heat transfer of Freon R11 and HCFC123 in narrow passages, Int. Journal of Heat and Mass Transfer, Vol. 43, pp. 3347-3358.

Barnea, D., Luninski, Y. and Taitel, Y., 1983, Flow pattern in horizontal and vertical two phase flow in small diameter pipes, The Canadian Journal of Chemical Engineering, Vol. 61, No. 10, pp. 617-620.

Bowers, M. B. and Mudawar, I., 1994, High flux boiling in low flow rate, low pressure drop mini-channel and micro-channel heat sinks, Int. J. Heat Mass Transfer, Vol. 37, No. 2, pp. 321-332.

Groeneveld, D.C. Leung, L.K.H., Erbacher, F.J., Kirillow, P.L., Bobkov, V.P. and Zeggel, W., 1993, An improved table look-up method for predicting critical heat flux, Proc. NURETH-6, 6th Int. Topical meeting on Nuclear Reactor Thermal Hydraulics, pp. 223-230.

Carey, V. P., Tervo, P. and Shullenberger, K., 1992, Partial dryout in enhanced evaporator tubes and its impact on heat transfer performance, SAE Technical Paper 920551.

Celata, G. P., Cumo, M. and Mariani, A., 1993, Burnout in highly subcooled water flow boiling in small diameter tubes, Int. J. Heat Mass Transfer, Vol. 36, No. 5, pp. 1269-1285.

Celata, G. P., Cumo, M., Mariani, A., Nariai, H. and Inasak, F., 1993, Influence of channel diameter on subcooled flow boiling burnout at high heat fluxes, Int. J. Heat Mass Transfer, Vol. 36, No.13, pp. 3407-3410.

Celata, G. P., 1996, Critical heat flux in water subcooled flow boiling: Experimentation and modeling, 2nd Eur. Thermal Sciences and 14th UIT Nat. Heat Transfer Conf., Italy.

Celata, G. P., Cumo, M. and Mariani, A., 1996, The effect of the tube diameter on the critical heat flux in subcooled flow boiling, Int. J. Heat Mass Transfer, Vol. 39, No. 8, pp. 1755-1757.

Celata, G. P., Cumo, M. and Mariani, A., 1997, Geometrical effects on the subcooled flow boiling critical heat flux, 4th World Conf. On Experimental Heat Transfer, Fluid Dynamics and Thermodynamics, Belgium, pp. 867-872.

Chaudry, A. B., Emerton, A. C. and Jackson, R., 1965, Flow regimes in the co-current upwards flow of water and air, Paper B2 presented at Symposium on Two-phase Flow, Exeter, June.

Cheng, X., Erbacher, F.J. and Muller, U., 1997, Critical heat flux in uniformly heated vertical tubes, Int. J. Heat Mass Transfer, Vol. 40, No.12, pp. 2929-2939.

Coleman, J. W. and Garimella, S., 1999, Characterization of two-phase flow patterns in small diameter round and rectangular tubes, Int. J. of Heat and Mass Transfer 42, pp. 2869-2881.

Derbyshire, R. T., Hewitt, G. T. and Nicholls, B., 1964, X-radiography of two-phase gas-liquid flow, AERE-M 1321.

Doroshckuk, V. E., Levitan, L. L. and Lantzman, F. P., 1975a, Investigation into burnout in uniformly heated tubes, ASME paper, 75-WA/HT-22.

Doroshckuk, V.E., Levitan, L.L. and Lantzman, F.P., 1975b, Recommendations for calculating burnout in a round tube with uniform heat release, Teploenegetica 22, pp.66-70

Hewitt, G. F. and Hall-Taylor, N. S., 1970, Annular two-phase flow, Pergamon Press.

Holser, E. R., 1968, Flow pattern in high pressure two-phase (steam-water) flow with heat addition, AIChE Symposium Series, Vol. 64, pp. 54-66.

Inasaka, F. and Nariai, H., 1987, Critical heat flux and flow characteristics of subcooled flow boiling in narrow tubes, JSME International Journal, Vol. 30, No. 268, pp. 1595-1600.

Jung, D. S. and Radermacher, R., 1991, Prediction of heat transfer coefficient of various refrigerants during evaporation, ASHRAE Transactions, Vol. 97, Pt. 2.

Kew, P. and Cornwell, K., 1997, Correlations for the prediction of boiling heat transfer in small diameter channels, Applied Thermal Engineering Vol. 17, No. 8-10, pp. 705-715.

Lazarek, G. M. and Black, S. H., 1982, Evaporative heat transfer, pressure drop and critical heat flux in a small vertical tube with R-113, Int. J. Heat Mass Transfer, Vol. 25, No. 7, pp. 945-960.

Lin, S., Kew, P. A. and Cornwell, K., 1998, Two-phase flow regimes and heat transfer in small tubes and channels, Heat Transfer 1998, Proceedings of 11th IHTC, Vol. 2, pp. 45-50.

Mertz, R. and Groll, M., 2000, Flow boiling heat transfer in narrow channels, Heat and Technology, Vol. 18, Supplement n. 1, pp. 75-79.

Mishima, K. M. and Hibiki, T., 1996, Some characteristics of air-water two-phase flow in small diameter vertical tubes, Int. J. Multiphase Flow, Vol. 22, No. 4, pp. 703-712.

Nakajima, R., Ramanujapu, N. K. and Dhir, V. K., 1996, Critical heat flux during subcooled flow boiling in narrow rectangular channels, Proceedings of the ASME Heat Transfer Division, Vol. 3, pp. 93-100.

Nariai, H. Inasaka, F. and Shimura, T., 1987, Critical heat flux of subcooled flow boiling in narrow tube, ASME-JSME Thermal Eng. Joint Conf. Honolulu.

Oya, T., 1971, Upward liquid flow in small tube into which air streams (1st report, experimental apparatus and flow pattern), Bulletin of JSME, 14, No. 78, pp. 1320-1329.

Reid, R. S., Pate, M. B. and Bergles, A. E., 1987, Evaporation of refrigerant 113 flowing inside smooth tube, ASME Paper No. 87, HT. 51.

Roberston, J. M. 1979, Boiling heat transfer with liquid nitrogen in brazed-aluminum plate-fin heat exchanger. In Heat Transfer-San Diego 1979, AIChE Symposium Series, Vol. 85, pp. 90-97.

Robertson, J. M., 1983, The boiling characteristics of perforated plate-fin channels with liquid nitrogen in upflow, Heat Exchanger for Two-Phase Application, HTD, Vol. 27, pp. 35-40.

Robertson, J. M. and Wadekar, V. V., 1988, Boiling characteristics of Cyclohexane in vertical up flow in perforated passages, Heat Transfer - Houston 1988, AIChE Symposium Series, Vol. 84, pp. 120-125.

Sekoguchi, K., Nakazatomi, M., M. Takeishi, H. Shimizu, K. Mori and G. Miyake, 1992, Pressure effect on velocity of liquid lumps in vertical upward gas-liquid two-phase flow, JSME International Journal, Series II, Vol. 35, No. 3.

Shah, R. K., 1986, Classification of heat exchangers. In Heat Exchangers: Thermal-Hydraulic Fundamentals and Design (Edited by S. Kakac, A. E. Bergles and F. Mayinger), pp. 9-46, Hemisphere Publishing Corp., Washington, DC.

Smith, R. A., 1986, Boiling inside tubes: critical heat flux for upward flow in uniformly heated tubes, ESDU Data Item No. 86032, Engineering Science Data Unit International Ltd, London.

Taitel, Y., Bornea, D. and Dukler A. E., 1980, Modelling flow pattern Transitions for Steady upward gas-liquid flow in vertical tubes, AIChE Journal, Vol. 26, No. 3, pp. 345-354.

Tran, T. N., Wambsganss M. W. and France D. M., 1996, Small circular- and rectangular-channel boiling with two refrigerants, Int. J. Multiphase Flow, Vol. 22, No. 3, pp. 485-498.

Triplett, K. A., Ghiaasiaan, S. M., Abdel-Khalik, S. I. and Sadowski, D. L., 1999, Gas-liquid two-phase flow in microchannels Part I: two-phase flow patterns, Int. J. of Multiphase Flow 25, pp. 377-394.

Wadekar, V. V., 1992, Flow boiling of heptane in a plate-fin heat exchanger passage, ASME HTD Vol. 201, pp. 1-6, New York.

Wadekar, V. V., 1997, Boiling hot issues - some resloved and some not-yet resolved, Trans IChemE, Vol 76, Part A, pp. 133-142.

Wallis, G. B., 1969, One-dimensional two-phase flow, McGraw-Hill.

Wambsganss, M. W., France, D. M., Jendraejczyk, J. A. and Tran, T. N., 1993, Boiling heat transfer in a horizontal small-diameter tube, Journal of Heat Transfer, Vol. 115, pp. 963-972.

Weisman, J. and Kang, S. Y., 1981, Flow pattern transitions in vertical and upwardly inclined lines, Int. J. Multiphase Flow Vol. 7, pp. 271-291.

Xu, G. P., Tso, C. P. and Tou, K. W., 1996, A review on cooling by channel flow boiling for electronic systems, Journal of Electronics Manufacturing, Vol. 6, No. 3, pp. 193-207.

Xu, J. L., Cheng, P. and Zhao, T. S., 1999, Gas-liquid two-phase flow regimes in rectangular channels with mini/micro gaps, Int. J. of Multiphase Flow 25, pp. 411-432.

Yang, W. J. and Zhang, N. L., 1992, Micro and nano-scale heat transfer phenomena science and technology, Ed. B. X. Wang, pp. 1-15, Higher Education Press, Beijing.

Yan, Y. Y. and Lin, T. F., 1998, Evaporation heat transfer and pressure drop of refrigerant R-134a in a small pipe, Int. J. Heat Mass Transfer, Vol. 41, pp. 4183-4194.

TWO-PHASE FLOW DISTRIBUTION IN A COMPACT HEAT EXCHANGER

Pierre BERNOUX[1], Pierre MERCIER[1] and Michel LEBOUCHÉ[2]

[1] GRETh-CEA Grenoble – 17, rue des Martyrs – 38 054 GRENOBLE Cedex 9 – FRANCE
tel 33 (0) 4 76 88 37 66 – fax 33 (0) 4 76 88 54 35 – pierre.mercier@cea.fr

[2] LEMTA – 2, avenue de la Forêt de Haye – 54 504 VANDOEUVRE Cedex – FRANCE

Abstract

This paper presents experimental results about two-phase flow distribution in the inlet port of a compact heat exchanger. A test section consisting in an horizontal manifold and 8 downward rectangular channels was instrumented. The distribution of vapour and liquid phases is obtained by measurement of the total mass flow rate and the calculation of the mass quality. The presence of transparent windows allows the observation of two-phase flow pattern for different operating conditions. Comparison between quantitative and qualitative results gives a general understanding of the process of phase distribution and is the starting point of other studies with more complex geometries.

INTRODUCTION

Flow maldistribution in heat exchangers is a cause of deterioration of both thermal and hydraulic performances. According to Mueller & Chiou (1988), there are many causes of maldistribution in exchangers: mechanical causes (design, tolerances), self-induced maldistribution due to the heat transfer process itself, fouling and/or corrosion, or the use of predisposed flows such as two-phase flows.

This paper deals with liquid-vapour two-phase flow distribution in compact heat exchangers. Improvement of phase distribution requires a good understanding of physical phenomena involved. But two-phase flow structure in compact heat exchangers is very complex and only a few authors had been interested in the subject of two-phase flow distribution in such geometries (Asoh *et al.* 1991, Rong *et al.* 1995, Watanabe *et al.* 1995).

The main objectives of this study were to understand how mass flow rate and vapour quality at the inlet of the manifold affect the flow distribution in a plate heat exchanger, and to develop a data bank which could be used in further studies (such as the development of a suitable predictive model).

Two-phase distribution profiles in the channels are obtained by the use of mass flow rate and mass quality measurements in each channel. Observation of the two-phase flow pattern in the inlet nozzle and in the manifold is presented. Comparison of quantitative results of the distribution with qualitative results of the observations allows to understand how maldistribution occurs in the geometry studied.

EXPERIMENTAL SET-UP

A specific experimental set-up was designed, related to compact heat exchangers. The test section is made of stainless steel and consists in an horizontal manifold and 8 parallel downward branches (**Figure 1**). The manifold is 96 mm long and its diameter is 50 mm. It is horizontally supplied by a 17.3 mm in diameter and 100 mm long glass pipe with a 1,500 mm upward tube made of stainless steel of the same diameter. The end of the manifold is closed by a transparent polycarbonate plug. Each branch is 2 × 50 mm rectangular. The channels are regularly 10 mm spaced along the manifold.

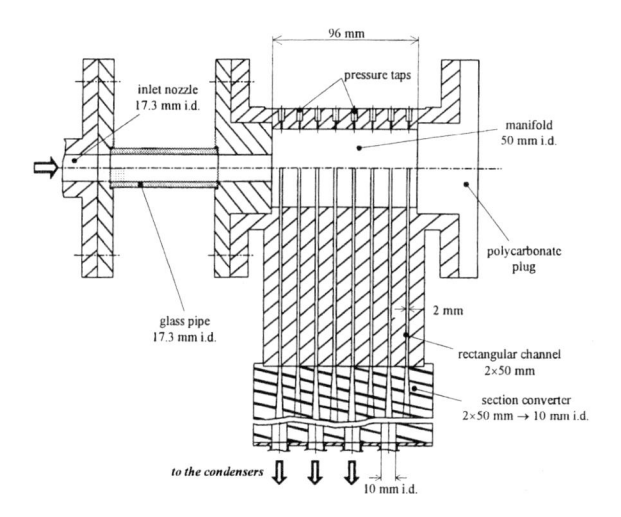

Figure 1. *Test section.*

Figure 2 shows the scheme of the experimental apparatus and especially measurement devices used for this study. Nine heat exchangers are placed in the main loop of the experimental set-up: a pre-evaporator (upstream the test section) and eight condensers (one downstream each channel). Four RTD sensors are located at the inlet and the outlet of each heat exchanger (*ie* two on the refrigerant side and two on the heating/cooling water side). The mass flow rate in the test section is controlled through a regulation valve upstream the pre-evaporator and is measured by a mass flow-meter. The mass flow rate in each channel is deduced from a volumetric flow-meter situated downstream each condenser.

Refrigerant 113 is used in two-phase flow conditions. The loop operates at saturation temperature conditions and the mass quality at the inlet of the test section is regulated by the pre-evaporator and by a set of valves downstream the condensers. The mass quality at the inlet of the manifold is deduced from an enthalpy balance using measured temperatures and mass flow rates. In a parallel way, the quality of the two-phase flow in each branch is deduced from an enthalpy balance on the condensers downstream each branch.

In addition, an absolute pressure transducer is used to follow the pressure evolution in the manifold. Differential pressure transducers are used to study pressure drop/gain in the channels. The results related to pressure measurements are not shown in this paper.

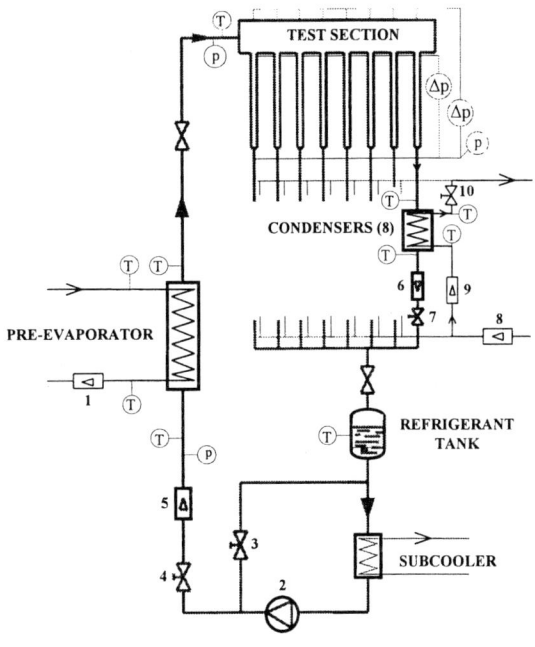

1	Flowmeter/heating water	8	Main flowmeter/cooling water
2	Refrigerant pump	9	Flowmeters/cooling water (8)
3	Bypass valve	10	Regulation valves/cooling water (8)
4	Main regulation valve		
5	Main flowmeter/refrigerant	T	Temperature probe
6	Flowmeters/refrigerant (8)	p	Pressure transducer
7	Regulation valves/refrigerant (8)	Δp	Differential pressure transducer

Figure 2. *Schematic diagram of the experimental set-up.*

RESULTS

The manifold inlet mass quality was varied from 0.1 to 0.8. The mass flow rate at the inlet nozzle of the test section was 100, 200 and 290 kg/h, so the mean mass velocity in the channels was 35, 70 and 100 kg/m² s. The operating saturation pressure at the inlet of the test section was fixed between 0.07 and 0.15 MPa according to the general equilibrium of the experimental set-up.

Two-phase flow distribution in the channels

All quantitative results presented hereunder are non-dimensional values.

The non-dimensional flow rate $\dot{M}_i^*$ in channel No i is the ratio of the total (liquid + vapour) flow rate $\dot{M}_i$ measured in channel No i over the mean flow rate measured if the distribution would have been uniform:

$$\dot{M}_i^* = \frac{\dot{M}_i}{\sum_{j=1}^{8} \dot{M}_j / 8} \quad [1]$$

In the same way, the non-dimensional liquid (resp. vapour) flow rate in channel No i is the ratio of the liquid (resp. vapour) flow rate measured in channel No i over the mean liquid (resp. vapour) flow rate if liquid distribution would have been uniform:

$$\dot{M}_{ki}^* = \frac{\dot{M}_{ki}}{\sum_{j=1}^{8} \dot{M}_{kj} / 8} \quad [2]$$

where k = ℓ (liquid) or k = υ (vapour).

The mass quality, or vapour mass quality, x_i in channel No i is the ratio of the vapour flow rate $\dot{M}_{\upsilon i}$ over the total flow rate $\dot{M}_i$:

$$x_i = \frac{\dot{M}_{\upsilon i}}{\dot{M}_i} \quad [3]$$

where

$$\dot{M}_i = \dot{M}_{\ell i} + \dot{M}_{\upsilon i} \quad [4]$$

Moreover, the total flow rate $\dot{M}$ (kg.s⁻¹) in the inlet nozzle and the mean mass velocity $\overline{\dot{m}}$ (kg.m⁻².s⁻¹) in the channels are related as follows:

$$\dot{M} = n S \overline{\dot{m}} \quad [5]$$

where n is the number of channels and S is the cross section of each channel.

Mass quality effect.

Figure 3 shows the evolution of the phase distribution along the 8 channels according to the mass quality at the inlet of the test section, when the mean mass velocity in the channels is equal to 70 kg m^{-2} s^{-1}.

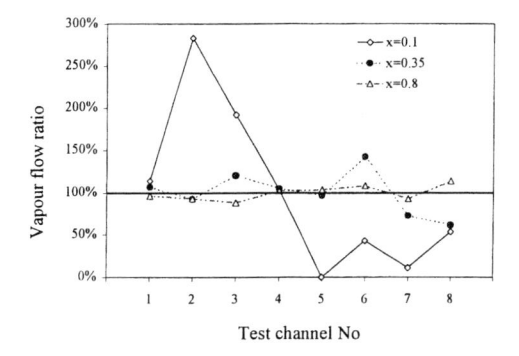

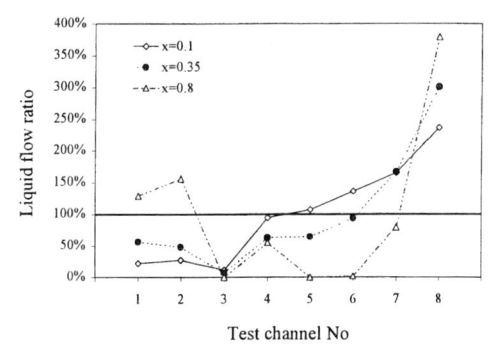

Figure 3. Influence of inlet mass quality on phase distribution ($\overline{\dot{m}} = 70$ kg/m^2 s).

At low quality ($x = 0.1$), the first channels (No 1 to 3) are largely underfed with liquid phase whereas the last ones (No 5 to 8) are largely underfed with vapour phase. The channels No 1 to 4 take between 100 % and 300 % of the MVFR (Mean Vapour Flow Rate in the channels). The channel No 5 take no vapour at all and the channels No 6 to 8 take between 15 % and 50 % of the MVFR. Simultaneously, the channels No 1 to 3 take between 10 % and 30 % of the MLFR (Mean Liquid Flow Fate in the channels) while the channels No 6 to 8 take between 140 % and 240 % of the MLFR. The channels No 4 and 5 are quite well fed with liquid phase.

At medium quality ($x = 0.35$), the vapour distribution is better than at low quality. The first channels (No 1 to 5) receive between 95 % and 120 % of the MVFR. Only the three last channels are badly fed and take between 60 % and 140 % of the MVFR. The liquid distribution is very uneven. The five first channels receive between 5 % and 60 % of the MLFR whereas the three lasts receive between 160 % and 300 % of the MLFR.

At high quality ($x = 0.8$), the vapour distribution is good whatever the channel considered may be. The vapour flow ratio noticed is always between 90 % and 110 %. But liquid distribution reaches now its worse level of non-uniformity. The six first channels take less than 60 % of the MLFR and the channels No 3, 5 and 6 receive no liquid at all while the last channel take 380 % of the MLFR.

→ The vapour distribution in the channels is considerably improved by the increase of the mass quality at the inlet of the manifold: the vapour flow ratio goes from 0 % to 280 % for $x = 0.1$, to 90 % to 110 % for $x = 0.8$. But the liquid distribution, which is always uneven, is significantly deteriorated by this increase: for $x = 0.1$, the most overfed channel (No 8) receives 240 % of the MLFR and no channel receives only vapour whereas for $x = 0.8$, the most overfed channel (No 8) takes 380 % of the MLFR (*ie* almost half of the total liquid flow rate) and three channels receive no liquid. So the increase of the mass quality improves the vapour distribution and deteriorates the liquid distribution.

Mass velocity effect.

Figures 4 and 5 present the influence of an increase of the mean mass velocity in the channels for various mass qualities at the inlet of the test section. **Figure 4** deals with the vapour distribution and **Figure 5** deals with the liquid distribution.

At low mass quality ($x = 0.1$), the increase of the mean mass velocity in the channels weakens the vapour and liquid maldistributions. The vapour flow ratio is between 0 % and 280 % for 70 kg/m^2 s and between 0 % and 230 % for 100 kg/m^2 s with only one channel without vapour (No 7) instead of two channels without vapour (No 5) or nearly without vapour (No 7). The liquid flow ratio is between 15 % and 240 % for 70 kg/m^2 s and between 50 % and 210 % for 100 kg/m^2 s. Both phase distribution profiles keep the same aspect when increasing mass velocity, but they are flattened.

At medium mass quality ($x = 0.35$), the increase of the mass velocity (from 35 kg/m^2 s to 70 kg/m^2 s) shows an significant improvement of the vapour distribution from 0 %-200 % to 60 %-140 %, but no change for the liquid distribution is noticed.

At high quality ($x = 0.8$), the level of the mean mass velocity in the channels has no influence anymore. The vapour flow ratio is kept between 90 % and 110 % while the liquid flow ratio is between 0 % and 320 % for 35 kg/m^2 s and between 0 % and 370 % for 70 kg/m^2 s.

→ In the range of inlet mass flow rate studied (directly related to the mean mass velocity in the channels, see [5]), the influence of this parameter is not so significant than that of the mass quality. Its effects are much more noticeable on the vapour distribution than on the liquid distribution. At low quality, both vapour and liquid distribution are improved by the increase of the mass flow rate. At medium quality, the vapour distribution is improved by this increase but without important change in the liquid distribution. At high quality, neither the vapour distribution nor the liquid distribution is affected.

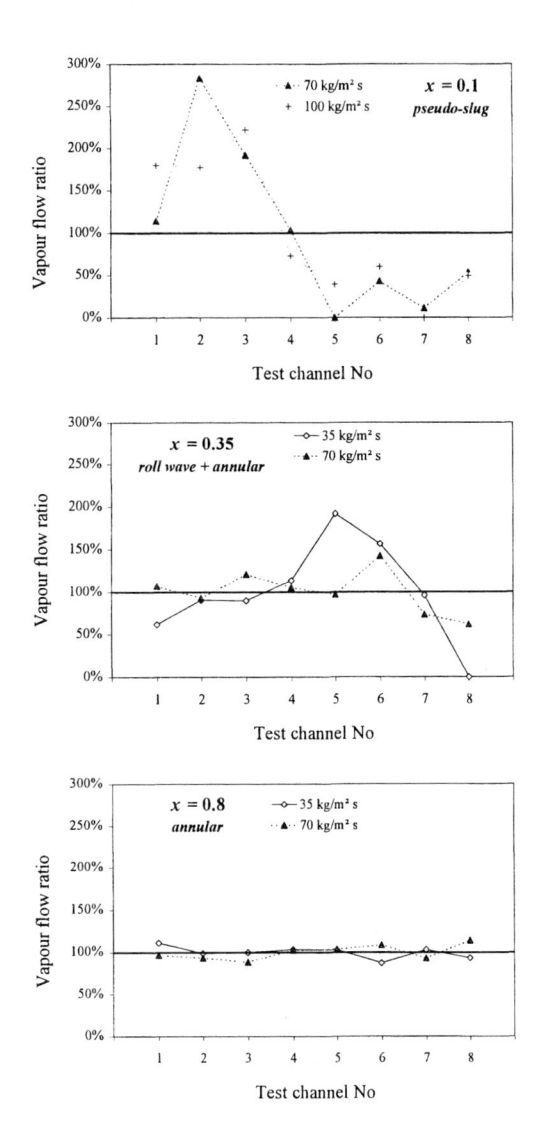

Figure 4. *Influence of mean mass velocity in the channels on vapour distribution.*

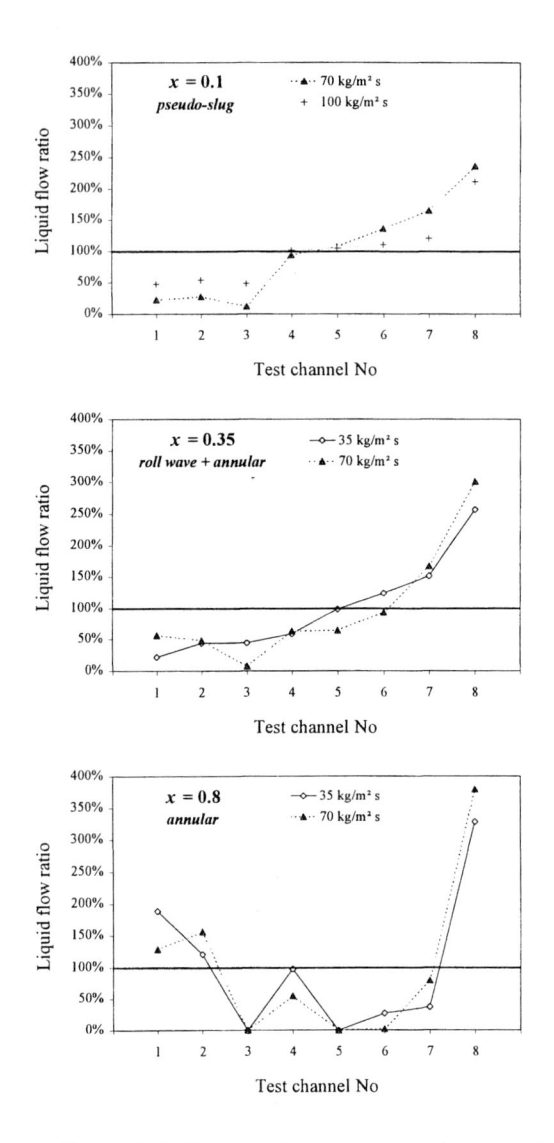

Figure 5. *Influence of mean mass velocity in the channels on liquid distribution.*

Flow pattern at the inlet

Different flow patterns have been observed through the glass pipe placed upstream the test section. These flow patterns are related to the operating conditions: mass flow rate, quality and pressure. The designation used in this paper to qualify flow pattern is given in a recent study of Wong & Yau (1997) and presented in **Figure 6**.

Mass quality effect.
For low qualities ($x = 0.1$~0.2), pseudo-slug pattern was noted. For the lowest quality (0.1), pseudo-slug corresponds to a strong stratification of the liquid phase with the passage of slugs of liquid carried by the vapour phase; the wall of the pipe is wetted continuously by a thin film of liquid (= pseudo-slug + thin annular). At $x = 0.2$, the strong stratification is kept while the slugs are propelled very fast and the wall film becomes significant (= pseudo-slug + annular).

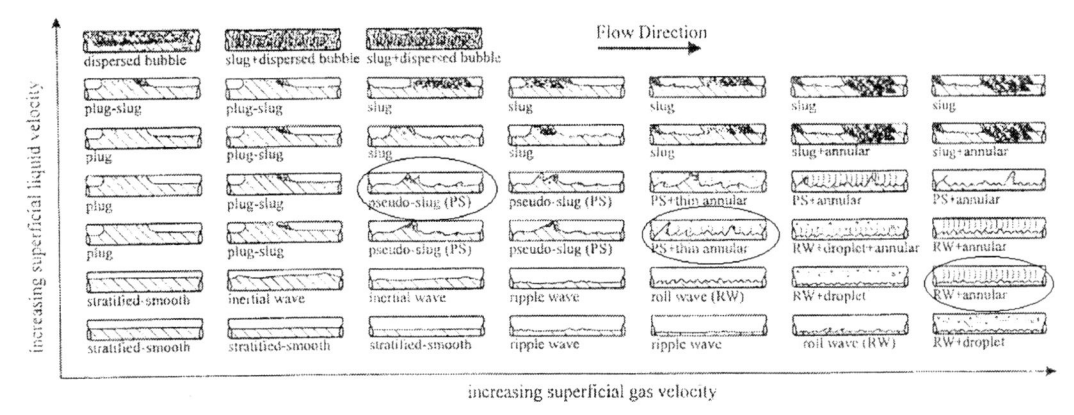

Figure 6. *Two-phase flow pattern in horizontal tube (Wong & Yau, 1997).*

For medium qualities ($x = 0.35~0.5$), the flow pattern evolves and reaches the roll wave + annular flow. No more liquid slugs are perceptible. Liquid phase stratification remains more or less, but the annular pattern is hereafter the major characteristic of the flow.

For the highest quality studied ($x \sim 0.8$), the annular flow pattern is fully developed. No more stratification of the liquid phase is identifiable.

Inlet mass flow rate effect.

The inlet mass flow rate appeared to have no significant effect on the flow pattern in the range it was studied.

Flow structure in the manifold

The liquid phase flows slower than the vapour phase in the nozzle. When entering the manifold, the sudden enlargement provides a jet effect of the phases and especially of the liquid. Gradation has to be put in this very general description: the flow structure in the manifold is closely related to the two-phase flow pattern in the inlet nozzle.

Figure 7 presents the two-phase flow structure in the manifold according to 3 different mass qualities in the inlet nozzle.

For pseudo-slug pattern (pseudo-slug + thin annular or pseudo-slug + annular) in the inlet nozzle (**Figure 7a**), the velocities of the phases are quite low so that the liquid falls on the ground of the manifold before reaching the end of the manifold (because of the competition between inertia and gravity effects). The liquid jet collides with the ground of the manifold on the level of the 5th when the quality is equal to 0.1 (and 7th if $x = 0.2$). Then, the liquid runs on the ground and crashes to the end of the manifold. So a great part of the bottom of the manifold is continuously immersed under several millimetres of liquid. More or less large splashes are generated. At $x = 0.1$, the splashes reach the mid-height of the manifold. At $x = 0.2$, they go up until the top of the manifold.

For roll wave + annular pattern in the inlet nozzle (**Figure 7b**), a fixed impacting point of the liquid phase is noticed on the transparent plug of the manifold. At $x = 0.35$, the flow remains very frothing. The liquid jet reaches straight the end of the manifold and is splitted by the impact. When increasing the mass quality until 0.5, the impacting point is still noticeable but a smooth film of liquid flows on the wall surface of the plug.

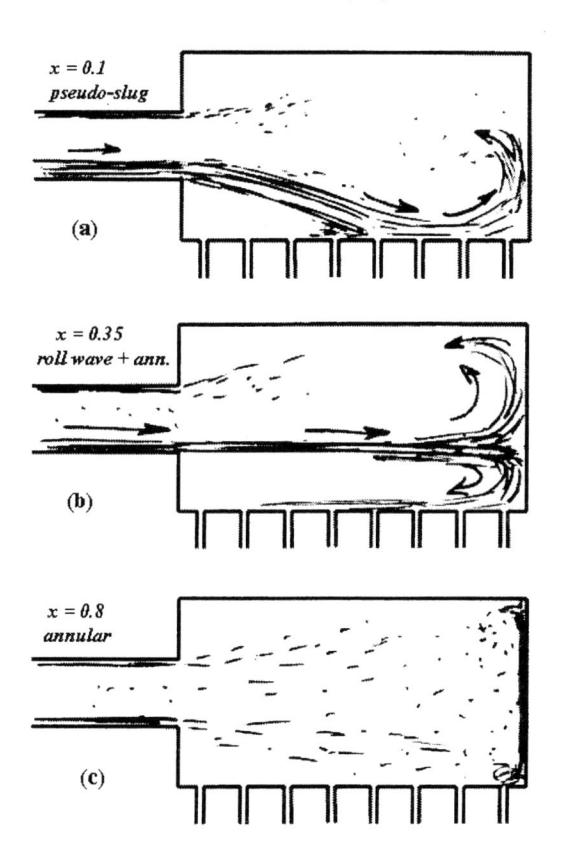

Figure 7. *Influence of inlet mass quality on structure in the manifold.*

For annular pattern in the inlet nozzle ($x = 0.8$, see **Figure 7c**), no more impacting point is noticeable on the end of he manifold. The liquid film of is sprayed when going through the sudden enlargement at the inlet of the manifold. A very thin film of liquid permanently covers the end of the manifold.

SYNTHESIS

The simultaneous use of flow observations in the manifold and of quantitative distribution results allows to explain how distribution of phases takes place in the channels. We use only the results obtained for 70 kg/m^2 s because experiments for other mass velocities do not show very different results in the general behaviour of the two-phase flow.

At low mass quality, the collision of the liquid jet on the inlet of a channel (see channel No 5 and 6) fills this channel with liquid and forbids the vapour entering it. The permanent level of liquid on the bottom of the manifold obstructs the passage of the vapour, explaining the lack of vapour in the last channels and their overfeeding with liquid.

At medium mass quality, the liquid phase impacts on the end of the manifold and flows on the bottom of the manifold: it fills first the channels placed nearest the end of the manifold. The vapour phase has difficulties to enter the two last channels, but however it is no more as important as at low quality.

At high quality, the whole liquid phase is spread and overlays the end of the manifold. So its major part flows in the last channel. The vapour phase meets no significant obstacle to enter the channels even the last one because of the lowness of the volume of liquid.

CONCLUSION

A representative test section operating with refrigerant 113 in saturation conditions has been used.

Experimental results have been obtained through measurements of flow rates of each phase in channels by varying the quality at the inlet and the total mass flow rate in the loop (directly related to the mean mass velocity in the channels). Observation of the flow structure in the manifold has allowed the understanding of the phase segregation process and the taking of the two-phase flow by the different channels.

In the range of parameters studied, the mass quality in the inlet nozzle has the most significant influence and governs the supplying flow pattern.

Whatever the flow pattern in the inlet nozzle and the flow structure in the manifold may be, the phase distribution is never satisfactory. The flow maldistribution is different for each main type of inlet flow pattern. The vapour distribution is excellent at high mass quality but it deteriorates strongly when quality decreases. Inversely, the liquid distribution is very uneven at high quality while it is much better at low quality, although not very good.

From these first results obtained on a quite simple geometry, other experiments are now possible. In particular, it will be possible to study the effect of the use of specific devices disturbing the inlet flow pattern. Comparison with results obtained with the simple geometry, quantitative as well as qualitative, should allow the improvement of phase distribution by adjusting such devices according to the type of inlet flow pattern.

NOMENCLATURE

$\bar{\dot{m}}$ mean mass velocity (kg m^{-2} s^{-1})
$\dot{M}$ mass flow rate (kg s^{-1})
$\dot{M}^*$ mass flow ratio
n number of channels
S channel cross section (m^2)
x vapour mass quality

Subscripts

i channel No i
k   phase (liquid or vapour)
ℓ liquid
υ vapour

REFERENCES

ASOH M., HIRAO Y., AOKI Y., WATANABE Y. & FUKANO T., 1991, Phase separation of refrigerant two-phase mixture flowing downward into three thin branches from a horizontal header pipe, *ASME/JSME Thermal Engineering Proceedings*, **2**, 159-164.

MUELLER A.C. & CHIOU J.P., 1988, Review of Various Types of Flow Maldistribution in Heat Exchangers, *Heat Transfer Engineering*, **9**, n°2, 36-50.

RONG X., KAWAJI M. & BURGERS J.G., 1995, Two-phase header flow distribution in a stacked plate heat exchanger, *ASME Conf. on Gas-Liquid Flows*, FED-**225**, 115-122.

WATANABE M., KATSUTA M., NAGATA K. & SAKUMA K., 1995b, General characteristics of two-phase flow distribution in a multipass tube, *Heat Transfer-Japanese Research*, **24**, n°1, 32-44.

WONG T.N. & YAU Y.K., 1997, Flow patterns in two-phase air-water flow, *Int. Comm. Heat Mass Transfer*, **24**, n°1, 111-118.

TECHNIQUE FOR MEASUREMENT OF VOID FRACTION IN STRATIFIED FLOWS IN HORIZONTAL TUBES

L.Wojtan, T. Ursenbacher and J.R. Thome

Laboratory of Heat and Mass Transfer (LTCM)
Department of Mechanical Engineering
Swiss Federal Institute of Technology Lausanne (EPFL)
Lausanne, Switzerland CH-1015
e-mail: john.thome@epfl.ch

ABSTRACT

Significant advances in modeling of intube evaporation and intube condensation have been made by the introduction of flow pattern regime based methods into prediction models. These methods provide much higher accuracy than older methods by using a heat transfer model specific to the particular flow pattern. Thus, a good representation of the two-phase interfacial flow structure, an accurate prediction of the void fraction, and a reliable flow pattern map are basic requirements for their continued development.

In response, a new optical technique is being developed to observe and measure dynamic void fractions and two-phase flow structures in stratified types of two-phase flows, which tend to be the most difficult to model. The technique utilizes a laser sheet to illuminate the cross-section of the flow channel (a plain horizontal glass tube) and a CCD video camera to obtain the images through the glass tube wall. A specially developed image analysis subroutine (now nearly fully automatic) transforms and corrects the refraction distorted image to its true shape, thus obtaining the interfacial profile between the liquid and vapor. The technique has been successfully applied to round horizontal tubes of 10.0, 13.6 and 46.0 mm internal diameter. The most recent setup measures void fractions to within about ±0.015 or less.

INTRODUCTION

A new optical image analysis technique utilizing a laser sheet and a video camera to measure void fractions and the interfacial flow structure for two-phase flows inside horizontal glass tubes is under development. The technique is presently applicable to stratified types of two-phase flows only. The measurement system will be used to measure void fractions in tubular sight glasses located at the end of heat transfer test sections, allowing concurrent observation of flow patterns and measurement of void fractions, heat transfer coefficients and two-phase pressure drops in a single test facility for intube boiling and condensation tests. The dry perimeter around the top of the tube in stratified flows is also now measured with this technique.

For stratified types of flows, both the local void fraction and the local wetted perimeter of the horizontal tube are important parameters in two-phase flow models. For example, Kattan et al. (1998) and Zürcher et al. (1999) proposed a flow regime based flow boiling model for evaporation inside horizontal tubes, assuming simplified geometries representative of the two-phase flow structures for describing annular flow, stratified-wavy flow and fully stratified flow as shown in Fig. 1.

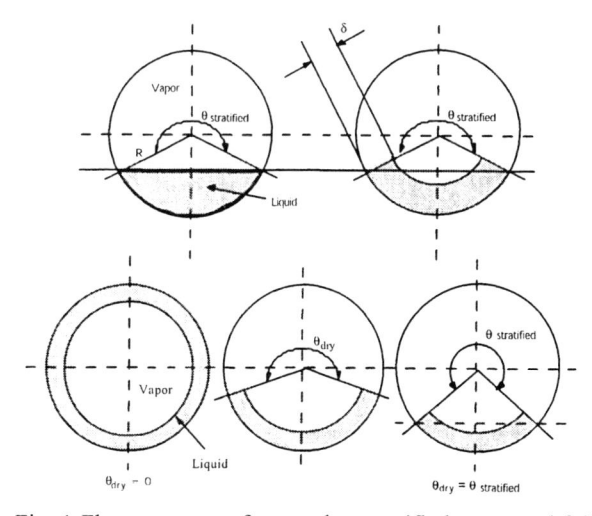

Fig. 1 Flow structures for annular, stratified-wavy and fully stratified flows (left to right in bottom three diagrams) and for fully stratified flow and its film flow equivalent (top two diagrams).

Utilizing a void fraction equation, the area occupied by the vapor phase is determined and then that of the liquid phase, such that the film thickness is calculable from the cross-sectional area of the tube and the dry angle. Knowing the total flow of liquid plus vapor and the local vapor quality, the mass flow rate of the liquid is easily calculated and together with its density, the mean velocity of the liquid can be determined in the film. Hence, turbulent flow heat transfer in the film can be correlated based on the mean velocity of the liquid film and film thickness. Subsequently, wet wall and dry wall heat transfer coefficients can be applied to the respective perimeters of the tube using the dry angle of the flow.

For intube condensation at low mass flow rates, the flow takes on a stratified flow structure. The condensate drains to the bottom of the tube via gravity and collects as a stratified liquid layer while the upper perimeter of the tube functions in the falling film condensation mode, i.e. similar to Nusselt condensation on a vertical plate. Using a configuration like that for fully stratified flow in Fig. 1, Dobson and Chato (1998) proposed a flow pattern based intube condensation heat transfer model that utilizes the local void fraction to determine θ_{strat}. Hence, here again void fraction plays an important role in predicting local heat transfer coefficients.

As a further example, a change in void fraction from 0.96 to 0.98 changes the liquid film thickness by a factor of two and hence significantly affects the prediction of the local two-phase heat transfer coefficient. Likewise, at small vapor qualities the void fraction rises very rapidly, increasing the liquid-phase velocity several fold and thus the convective contribution in heat transfer. Thus, accurate measurement of void fraction is of paramount importance to the development of improved void fraction models and in turn for the reliable prediction of flow pattern transitions, two-phase heat transfer coefficients and two-phase pressure drops. For a recent review of modeling of two-phase heat transfer based on flow pattern, refer to Thome (2000).

Numerous methods have been developed for measuring void fractions in tubular two-phase flows over the years: quick-closing valves, resistance/conductance sensors, refractive index fiber optic probes, radiation attenuation and multi-beam gamma or x-ray methods, multi-beam optical intensity techniques, capacitance meters, etc.

For instance, Costigan and Whalley (1997) recently developed an improved type of conductance probe for measuring cross-sectional void fractions for air-water flows in vertical tubes. The probe was installed in a 32 mm internal diameter transparent acrylic tube and used copper electrodes to measure the conductance through the water phase across the channel. The method was particularly accurate at low void fractions (< 0.3), with *maximum* errors in the range from about 0.005 to 0.03. Similarly, the maximum error was about 0.05 for void fractions between 0.3 and 0.8. For larger void fractions, the errors were even smaller.

Also, Krepper E. et al. (1999) recently developed a new method for measuring local and cross-sectional void fractions using a matrix of fine wires (0.120 mm diameter) extended as a grid across a 51.2 mm internal diameter tube, obtaining 242 local measurement points. Using this method, they achieved *maximum* errors of ±0.05 over the entire range of void fractions. The method was applied to two-phase flows in vertical pipes to obtain gas-phase distributions for various flow patterns.

DEFINITION OF VOID FRACTION

The most useful form of void fraction for modeling purposes is the *cross-sectional void fraction*, which is defined as the ratio of the cross-sectional area occupied by the gas or vapor phase to that of the channel. For the special situation of fully stratified flow with a constant liquid level along a horizontal channel, the *volumetric void fraction* is equal to the cross-sectional void fraction. It is the cross-sectional void fraction that is measured using the optical image analysis technique while the volumetric void fraction is measured for static conditions in horizontal test sections to determine the accuracy of the new method.

OPTICAL MEASUREMENT TECHNIQUE

This technique is being prepared for making dynamic measurements of void fractions in two-phase flows. As a first step to achieving this goal, static tests with stratified liquid and gas inside horizontal cylinders of several diameters are being made to perfect the method, i.e. the void fraction obtained from the optical measurement technique can be compared to those that can be very accurately measured using a gravimetric balance (in the present case the mass of the air is negligible with respect to that of the liquid) for the fixed volume of the closed cylinder.

In the present static tests, a highly uniform 46.0 mm internal diameter glass tube with a 2 mm wall of 400 mm length was used as the first test section, a similar 10.0 mm internal diameter glass tube with a 1.5 mm wall of 300 mm length was used as the second and most recently a 13.6 mm internal diameter tube with a 1.2 mm wall has been used. Appropriate flat-faced teflon plugs were machined for closing the ends of the tubes. A precision grid was glued to a third plug of each size that could be viewed through the glass cylinder and be used for transformation of the distorted image back to its true image.

Ethanol and air are used presently as the test fluids. A trace amount of fluorescent powder (Rhodamin B) is mixed with the ethanol to improve the contrast in the image at the liquid-air interface.

The gravimetric balance used to obtain the volumetric void fractions has an operating range from 0-4000 g with an accuracy of ±0.01 g. The void fraction is obtained by measuring the mass of liquid added to the cylinder divided by the mass of liquid required to completely fill the cylinder.

Hence, the gravimetric void fraction is potentially accurate to about 0.004% for the 46 mm tube and 0.11% for the 10 mm tube when full of liquid. However, some evaporation of the alcohol during the transfer process is inevitable. Including this effect, preliminary tests showed that the gravimetric void fraction was precise to within about ±0.009% for the 46 mm tube but was only about ±3.0% for the 10 mm tube because of its much smaller mass of liquid when full.

The test facility is depicted in Fig. 2 with its principle components. It is composed of a base plate for the setup plus a vertical support plate for the tube. A monochromatic laser beam from a fiber optic cable is transformed into a laser sheet that cuts across the tube normal to the tube's axis from above. A CCD video camera is used to record images of the cross-section of the tube.

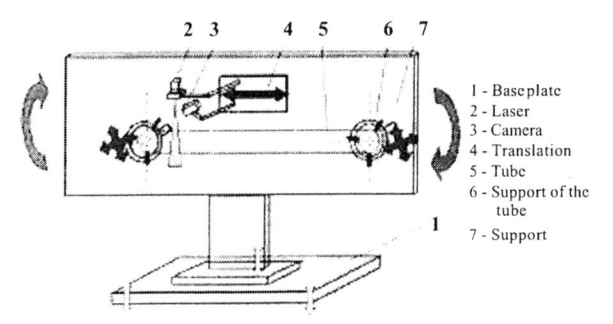

Fig. 2 Diagram of the test facility.

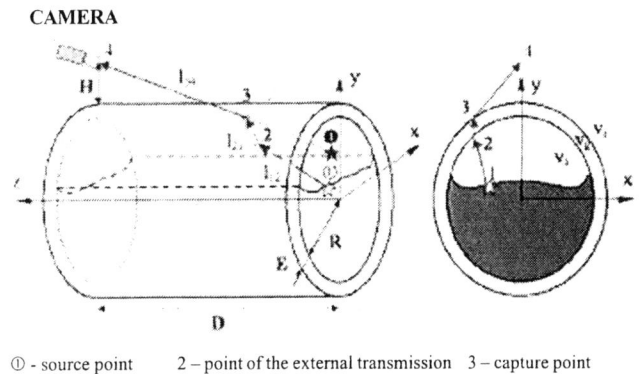

Fig. 3 Schematic of the optical setup.

① - source point   2 – point of the external transmission   3 – capture point
❶ - apparent point   3 – point of the internal transmission

Figure 3 shows a schematic of the optical setup with the refraction of the light from the external air into the glass and then from the glass into the air stratified above the liquid. Since an additional refraction takes place at the interface of the liquid, only the image of the cross-section occupied by the air is analyzed. To overcome the distortion of the image by refraction through the curved glass wall, a grid of known dimensions is used to transform the image observed by the video camera to its true shape. A nearly automatic computer subroutine was written for use within the LabView5.1 image analysis software for transforming, viewing and analyzing the images.

Figure 4 depicts a grid of splines that are fit to the grid of the distorted image, whose grid points are then mathematically transformed back into their analogous Cartesian co-ordinates. The first stage of development used a coarser grid as illustrated in Fig. 4. The image analysis system is used interactively to designate each grid line and then the spline is calculated for the particular grid line. Once this is done for all the grid lines, this grid of splines of known location are used to transform the image back to its real rectalinear shape. The most recent version of the method uses a much higher density grid in order to better capture the distortion of the image near the internal perimeter of the glass tube, as shown in the top of Fig. 5.

Fig. 4 Computer image showing splines fit to the grid lines.

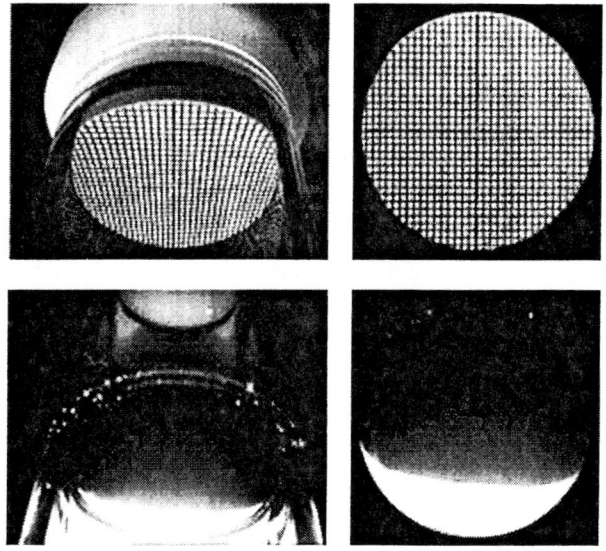

Fig. 5 Distorted images (left) compared to transformed images (right).

Applying the same transformation to an equivalent image of the liquid-vapor interface at a location about 10 cm from the end of the cylinder, the locus of the interface is obtained. Figure 5 shows a comparison of the original distorted image to the transformed image for both the grid itself and for the liquid-air interface. Surface tension and a low contact angle create a capillary effect at the tube wall that affects the entire interface in the smaller diameter tube.

The image can now be processed to determine the void fraction. After converting the image to black and white and then selecting a contrast threshold at the interface between the black and white areas, i.e. setting a grayness factor for the transition in the image, the cross-sectional area occupied by the air is determined by counting pixels in that zone. Dividing this value by the total number of pixels occupied by the cross-section of the tube, which is determined by fitting a circle to the image of the tube, the cross-sectional void fraction is determined. Also, the image in the bottom right of Fig.5 can be used to determine θ_{dry} in Fig.1.

This approach has been applied to a large number of air-ethanol mixtures in the the largest and smallest glass tubes over the entire range of void fraction in the first series of tests and more recently to the 13.6 mm tube, which has smoother inner and outer surfaces that are better for the optics involved and can also withstand high pressures that will be involved in the tests with two-phase flows of refrigerants. All the present tests were done at room temperature. A very accurate Stiefelmayer level (0.04 mm per meter accuracy) was utilized to adjust the tubular test sections to a very precise horizontal position.

STATIC VOID FRACTION MEASUREMENTS

Figure 6 depicts the absolute errors in the void fractions measured with the optical image analysis technique, which are those measured with the optical technique minus those obtained with the gravimetric balance for the 46 mm diameter glass tube. Void fractions were measurable over the range from 0.02 to 0.96. Because there is a very thin "grey" boundary between the black and white regions of the digitized image, the location of the boundary is dependent on the contrast threshold chosen to define the boundary. The absolute errors have thus been plotted to illustrate the effect of contrast threshold values across the interface ranging from 66 to 116 (in the range from 0 to 255 of the 8-bit grey scale). The worst case gives a spread of about -0.012 in void fraction at a void fraction of 0.70 while typically the spread is smaller than -0.008. Hence, the gravimetric balance measurements can be used to determine the best contrast threshold and thereby obtain absolute errors within a bandwidth as small as -0.0038 to -0.0162 over the entire range of void fraction. However, even *without* this the gravimetric tests, the method gives absolute errors within a bandwidth of only -0.0085 to -0.028 for the worst constrast threshold. The "spike" in the error curves is caused by the need to tilt the tube between successive measurements to add more liquid at void fractions lower than 0.66, something which will not be required in dynamic void fraction measurements of two-phase flows.

Figure 7 shows the corresponding relative errors for the measurements in Fig. 6. The relative errors become large at low void fractions since an absolute error of only -0.007 at a void fraction of 0.026 becomes a -25.3% relative error.

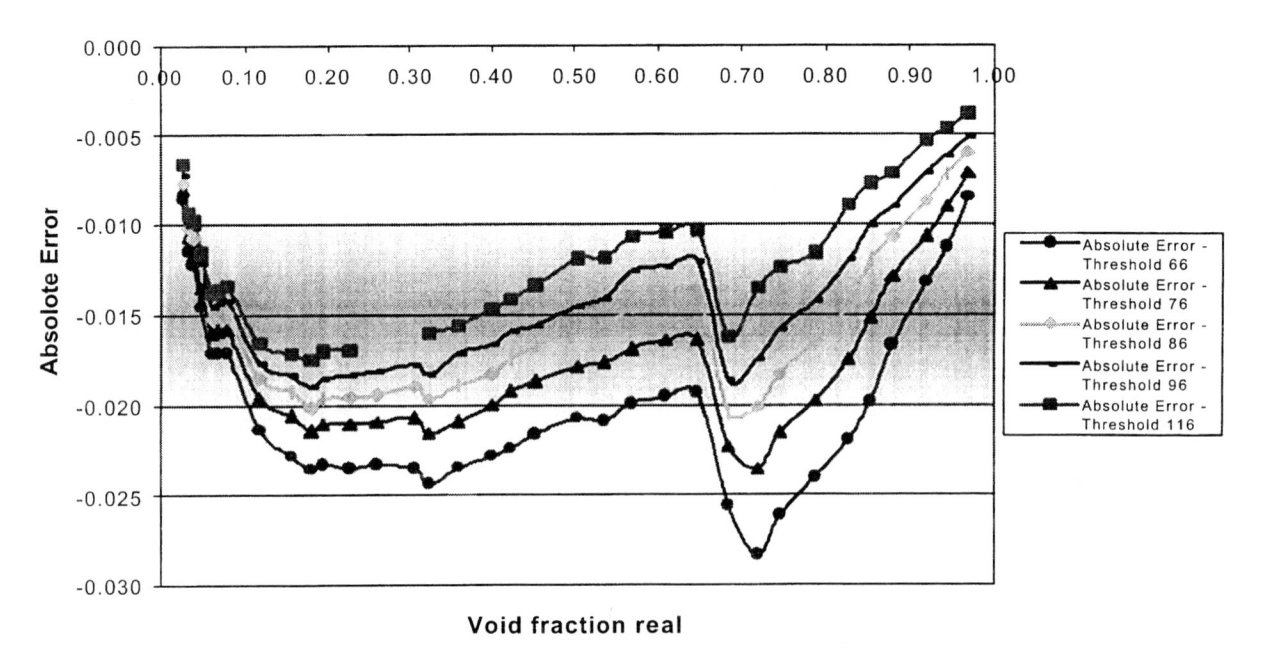

Fig. 6 Absolute errors in void fractions measured in a 46 mm tube showing the effect of choosing contrast thresholds from 66 to 116.

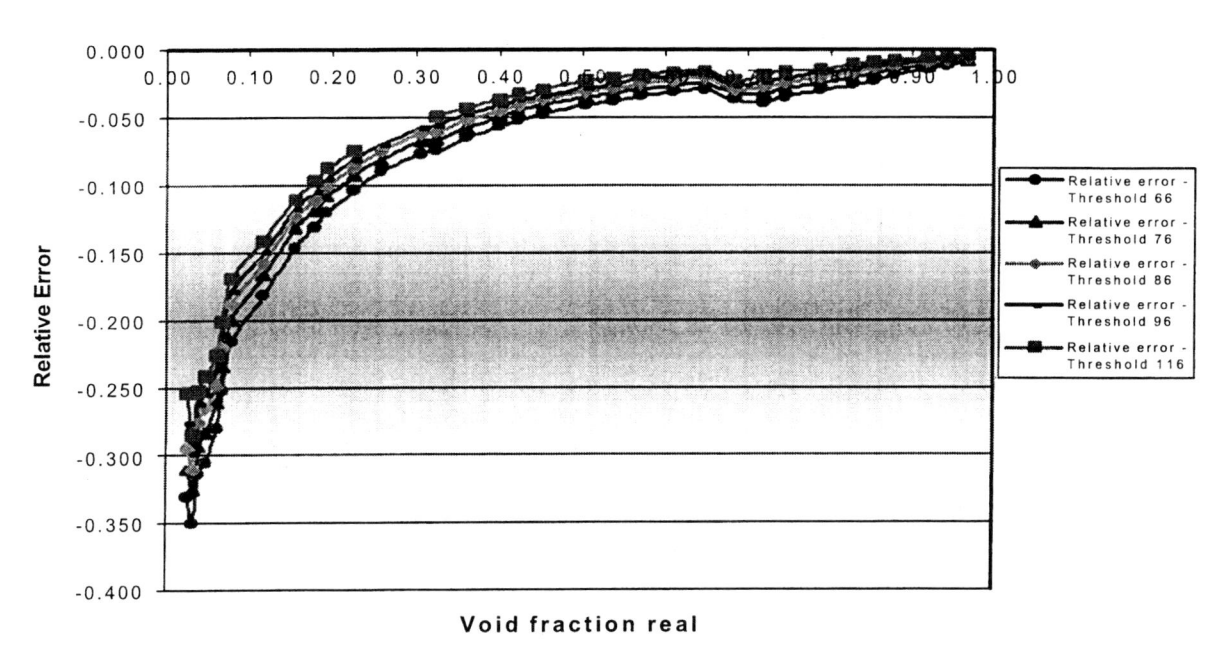

Fig. 7 Relative errors in void fractions measured in a 46 mm tube showing the effect of choosing contrast thresholds from 66 to 116.

Figures 8 and 9 present analogous measurements for the 10 mm tube, whose smaller diameter magnifies the effect of refraction on distortion of the registered image and hence increases the error in the transformation of the image. Here, absolute errors within a bandwidth as small as -0.0083 to -0.0493 were still obtainable for void fractions greater than 0.26 (a smaller error range would be obtainable by applying a calibration factor of 0.97 to the data which was *not* done here nor is planned to be done).

For void fractions below 0.30, void fractions were not measureable because gas slugs formed along the top of the 10 mm tube rather than a continuous vapor zone. This would require step-wise measurements to be made along the length of the tube to determine the void fraction in these static tests. Since the application of the optical technique is to dynamic flow measurements, this will not be a problem since the slugs will flow past the laser sheet and be recorded on video.

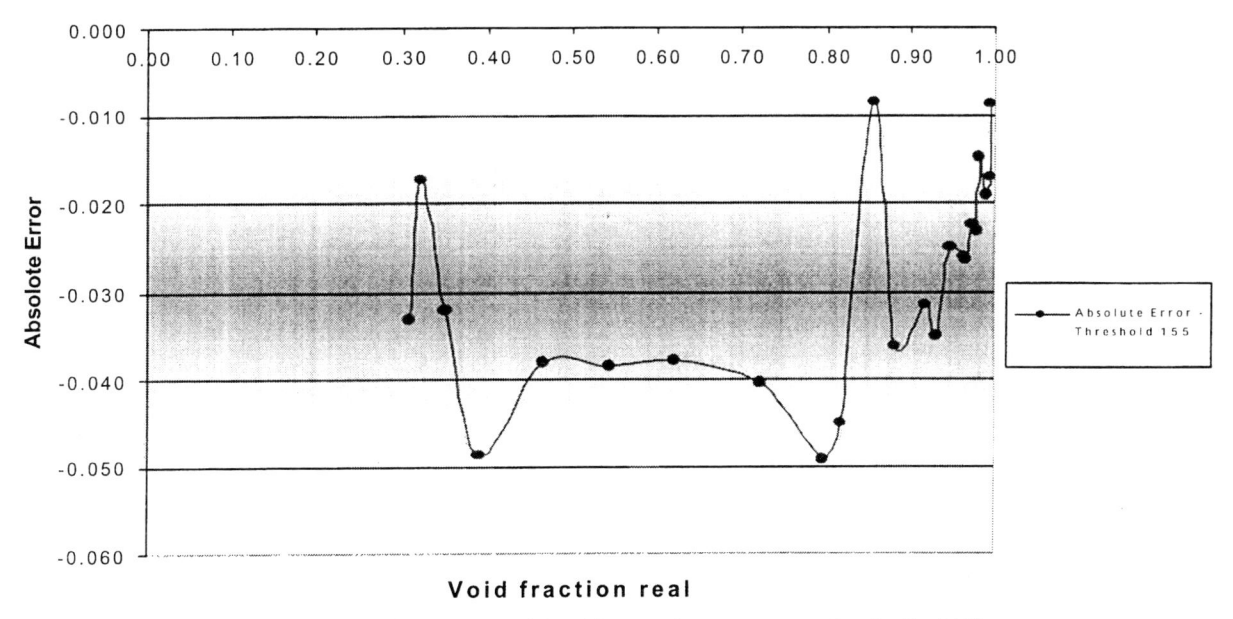

Fig. 8 Absolute error in void fractions measured in a 10 mm tube at a contrast threshold of 155.

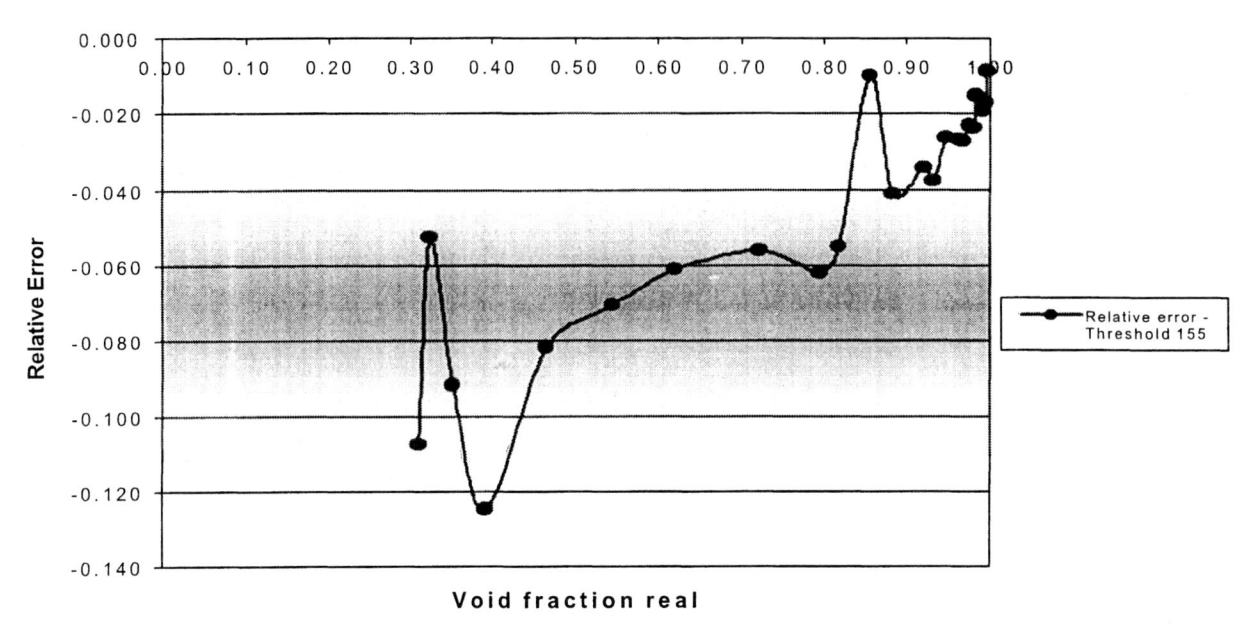

Fig. 9 Relative error in void fractions measured in a 10 mm tube at a contrast threshold of 155.

Figure 10 and 11 depict the absolute and relative errors in void fraction measured for the better optical quality 13.6 mm tube, respectively. The images were taken using a much higher laser intesity than for the previous tubes, which improved the interfacial contrast and thus the definition of the interface of the liquid. Line spacing of the reference grid was two times denser than the previously used grid. The quality of the images improved with the better quality surface finish of the glass. The graph shows a comparison between the previous and the new image transformation routines using the same high density grid. For void fractions below 0.7, it was necessary to turn the tube to an upright position after each test to add the liquid so some changes in the position of the tube were unavoidable with this turning back and forth. This has a minor effect on the accuracy in the present static tests but will not occur under flow conditions since the tube will not have to be moved. Even so, the absolute error in void fraction varied within a maximum bandwidth as small as ±0.014 and with a mean absolute error of only -0.007 using the newest transformation software routine.

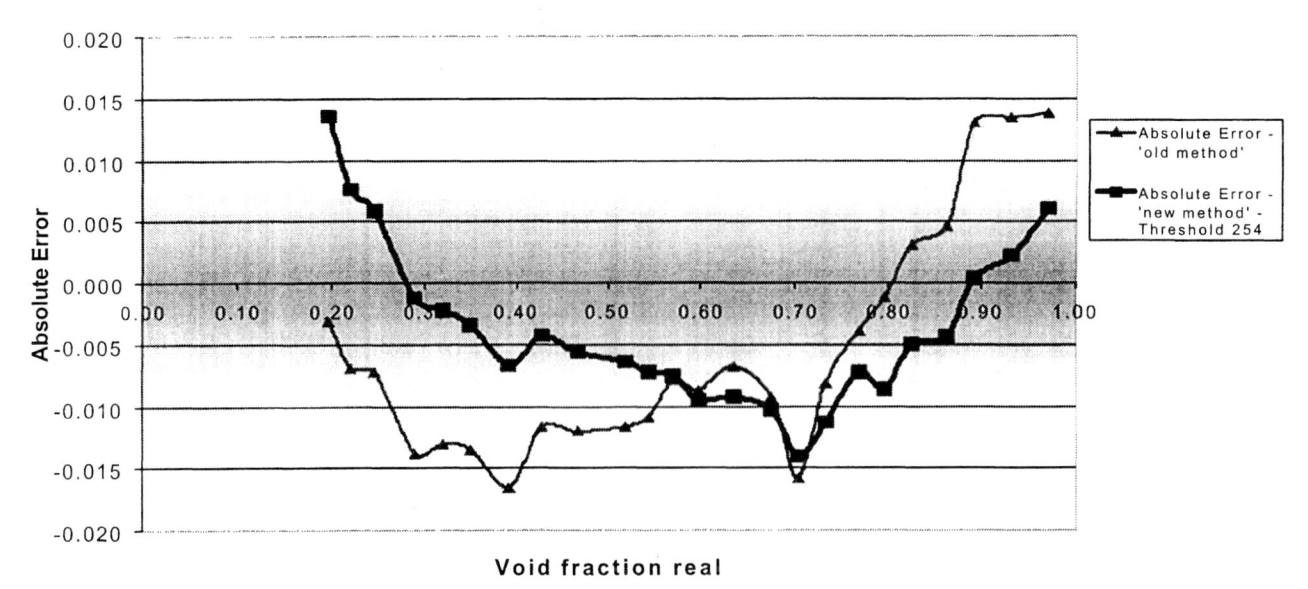

Fig. 10 Absolute error in void fractions measured in a 13.6 mm tube for the new and old method of image transformation – line spacing of the grid 0.5mm.

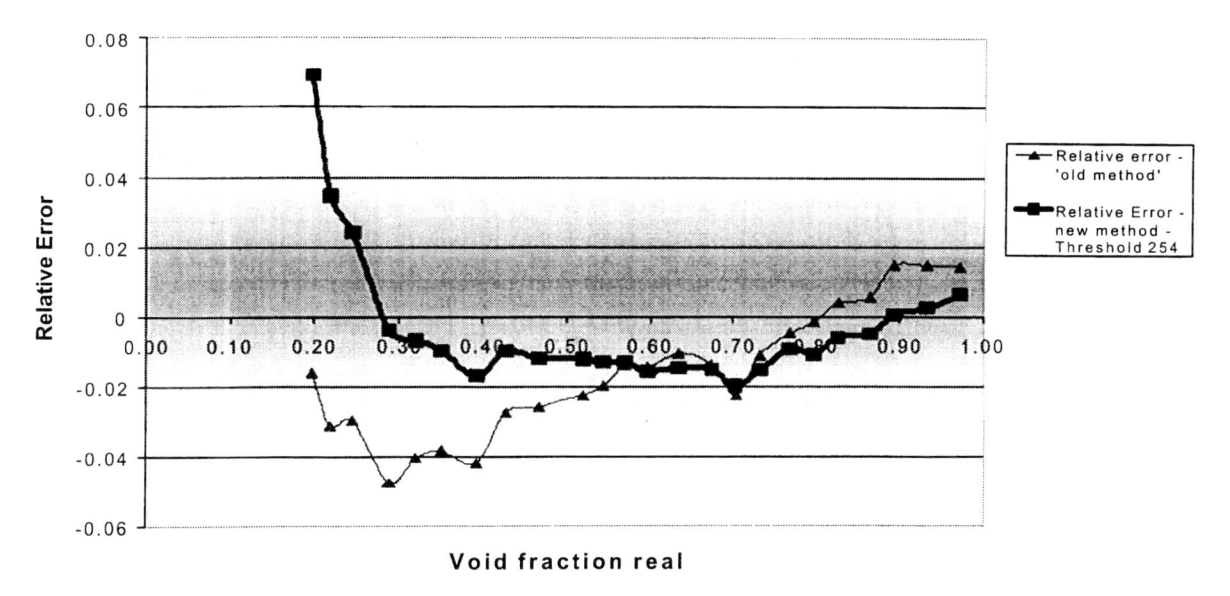

Fig. 11 Relative errors in void fractions measured in a 13.6 mm tube for the new and old method of image transformation - line spacing of the grid 0.5mm.

This new version of the image processing software allows use of a much higher black-to-white threshold level, even if the side parts of the interface are removed due to the lower light intensity. Direct analytical transforming of the refraction distorted images has also been developed. Our software for image transformation is now nearly completely automatic and needs approximately 2 seconds for complete transformation and calculation of void fraction of one image. Hence, CCD videos taken at 25 images per second will be able to be rapidly transformed and processed.

FUTURE DEVELOPMENTS

Besides instantaneous and time-averaged void fractions, the optical image analysis technique has the potential to provide valuable information about the two-phase flow structure of stratified types of flows. For instance, the contour of the two-phase interface, the mean liquid height and the dry angle around the top perimeter of the tube are obtainable directly from such a transformed digital video image. Secondly, analyzing a sequence of images, the height and frequency of interfacial waves will be measurable and thus provide valuable information on interfacial roughness, for instance. Work is currently proceeding on these aspects using a new horizontal test loop with air-water flows, and then the technique will be applied to two-phase flows of refrigerants exiting from heat transfer test sections. Development are also underway to avoid use of the fluorosent dye, whose use is not always convenient.

The angle of the dry perimeter around the top of the tube in stratified flows is an important physical parameter, not only important to the heat transfer model but also to accurate heat exchanger design. For instance, evaporation in the stratified-wavy flow regime leaves the top of the tube completely dry and results in a very low refrigerant-side heat transfer coefficient on this part of the tube wall. Consequently, in a *compact heat exchanger* the air-side fin geometry on the outside of the tube is less effective around the upper perimeter of the tube where the inside wall is dry, which may represent more than half the circumference of the tube.

The optical technique will then be further developed to handle void fractions in non-circular tubes, such as flattened tubes and rectangular channels, geometries which are of particular importance to compact heat exchangers. It is also planned to extend this method to microchannels.

CONCLUSIONS

A new optical image analysis technique using a CCD video camera and a monochromatic laser to illuminate a cross-section of the flow containing a trace amount of fluorescent powder has been developed for measuring cross-sectional void fractions in horizontal tubes. A computerized image analysis routine has been developed to reconstruct the true cross-sectional image of the two-phase flow from the original image distorted by refraction through the glass tube walls. Applying the technique so far to a static stratified liquid with air inside a horizontal tube and comparing to gravimetric void fraction measurements, it has been found that the method yields very good accuracy over most of the void fraction range. For instance, for a 13.6 mm bore glass tube the absolute error in void fraction varied within a maximum bandwidth as small as ±0.014 with a mean absolute error of only -0.007. Thus, very accurate void fraction measurements are achievable using this non-intrusive optical measurement technique.

ACKNOWLEDGEMENTS

This work was supported by the Swiss National Science Foundation under contract 21/57210.99.

REFERENCES

Costigan G. and Whalley P.B., 1997, Slug Flow Regime Identification from Dynamic Void Fraction Measurements in Vertical Air-Water Flows, *Int. J. Multiphase Flow*, vol. 23, no. 2, 263-282.

Dobson M.K. and Chato J.C., 1998, Condensation in Smooth Tubes, *J. Heat Transfer*, vol. 120(1), 193-213.

Kattan N., Thome J.R. and Favrat D., 1998, Flow Boiling in Horizontal Tubes. Part 3: Development of a New Heat Transfer Model Based on Flow Patterns, *J. Heat Transfer*, vol. 120(1), 156-165,.

Krepper E., Krüssenberg A.-K., Prasser H.-M. and Schaffrath A., 1999, High Resolution Void Fraction Measurements for the Validation of Flow Maps and CFD Codes, *Two-Phase Flow Modelling and Experimentation*, Edizioni ETS, Pisa, 1371-1378.

Thome J.R., *2000,* Flow Regime Based Modeling of Two-Phase Heat Transfer, *Proc. of Boiling: Phenomena and Emerging Applications*, Anchorage, Alaska, April 30-May 5, 2000.

Zürcher O., Thome J.R. and Favrat D., 1999, Evaporation of Ammonia in a smooth Horizontal Tube: Heat Transfer Measurements and Predictions, *J. Heat Transfer*, vol. 121(1), 89-101.

DIVIDING TWO-PHASE ANNULAR FLOW WITHIN A SMALL VERTICAL RECTANGULAR CHANNEL WITH A HORIZONTAL BRANCH

Jun Kyoung Lee and Sang Yong Lee

Department of Mechanical Engineering
Korea Advanced Institute of Science and Technology
373-1, Kusong-Dong, Yusong-Gu, Taejon 305-701, Korea
E-mail: e_hyunny@cais.kaist.ac.kr

ABSTRACT

The objective of the present experimental study is to investigate the characteristics of the dividing two-phase annular flow in a vertical, small T-junction (less than 10 mm in hydraulic diameter). The T-junction consists of a vertical rectangular main channel and a horizontal rectangular branch. Three different aspect ratios (1, 0.5, 0.125) for the branch were tested. Air and water were used as the test fluids. The superficial velocity ranges of air and water were 17.2 – 30 m/s and 0.02 – 0.32 m/s, respectively. Dividing flow characteristics in the small T-junctions are different from those in large T-junctions. Smaller values of the branch aspect ratio and the inlet quality reduce the fraction of the liquid separated. Shoam et al.'s model was modified to predict the fraction of the liquid separation for the present size range. The proposed model represents the experimental data within the deviation of ± 20%.

INTRODUCTION

Recently, the flow distribution from the header to the parallel tubes is becoming of interest in predicting the heat transfer performance of compact heat exchangers. In many cases, the flow rates through the tubes are not uniform and occasionally, there is almost no flow through some of the tubes. Especially with a two-phase flow, the phase separation occurs in the dividing junctions and the situation becomes even more complicated.

Figure 1 shows a tube array of a compact heat exchanger. The header/channels configuration can be simulated as an accumulation of T-junctions with rectangular cross sections. Therefore, as a basic study on the flow distribution, the two-phase flow behavior in a vertical rectangular T-junction with a horizontal branch was examined in detail experimentally. There have been substantial amount of works published on the dividing two phase flow at T-junctions. Those cover the effects of flow pattern (Azzopardi and Whalley, 1982; Shoam et al., 1987), inlet (main tube) diameter (Azzopardi, 1994; Stacey et al, 2000), main to branch size ratio (Azzopardi, 1984), and fluid viscosity (Hong, 1978), etc.

The experiments performed by Hong (1978) and Stacey et al. (2000) are of particular relevance to the current study since their hydraulic diameters are close to those considered in the present work. Stacey et al. (2000) used a 5 mm diameter T-junction while Hong (1978) used a 9.5 mm diameter T-junction. However their data are for horizontal T-junctions. Stacey et al. (2000) compared their data with

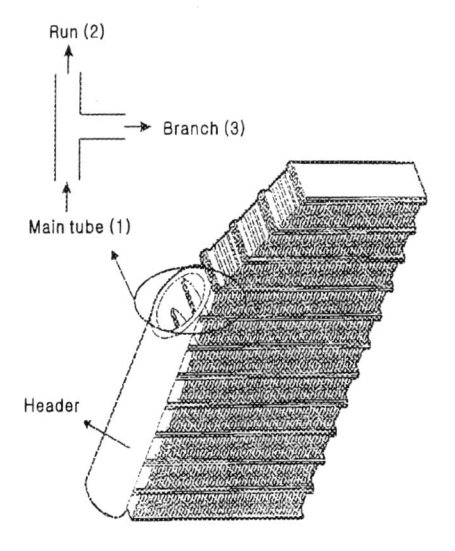

Fig. 1 Tube array of a compact heat exchanger.

those for larger diameter junctions but with similar superficial inlet velocities for annular flow. They showed that the decrease of the pipe diameter increases the fraction of the liquid separation. Hong (1978) reported that the fraction of liquid separation increased with the liquid viscosity.

However, those previous results cannot be applied to designing of conventional compact heat exchangers for the following reasons. Firstly, the hydraulic diameters of the headers in compact heat exchangers are smaller (D_h = 5 - 10 mm) compared to those in the published data (D_1 = 30 - 127 mm). Secondly, characteristics of a dividing two-phase flow in a T-junction with rectangular cross sections are rarely reported. Thirdly, the existing data on the dividing two-phase flow at small T-junctions are only for the horizontal flows. Furthermore, rectangular channels of the compact heat exchangers may have various gap sizes (or aspect ratios). But there is little data available on the effect of the branch aspect ratio for small T-junctions.

Therefore, in the present study, the characteristics of the dividing two-phase flow at a small, vertical T-junction with rectangular cross sections were investigated. The experiments were limited to the annular flow regime in the main channel. This is because the flow pattern in a compact heat exchanger is mostly an annular flow. The experimental data were compared to that of Stacey et al. (2000) and Azzopardi (1994) to check the effect of the main-channel size on the dividing two-phase flow. The effect of the branch aspect ratio was presented as well. The experimental data were also compared to the prediction model of Azzopardi and Whalley (1982), and Shoam et al. (1987). Then a modified model was proposed to characterize the dividing two-phase annular flow within the small T-junctions.

EXPERIMENTAL SETUP

Figure 2 illustrates the experimental setup. Air and water were used as the test fluids. Water was pressurized by the compressed air passing through the pressure regulator and the flow rate was metered by the calibrated rotameters. Also, the air flow rate was measured by the calibrated rotameters. As shown in Fig. 2 (b), water and air entered the mixer (④) which consists of concentric tubes ; air passed through the inner tube while water entered the two sides of mixer and flowed through the outer tube. Therefore, an annular flow was formed at the entrance of the test section. The test section is made of transparent acrylic plates for flow visualization. It consists of a 1 m-length main (①), 320 mm- length run (②), and a 300 mm-length branch (③). The cross sections of the main and the run were fixed to 8 mm × 8 mm while three different branch sizes (8 mm × 8 mm, 8 mm × 4 mm and 8 mm × 1 mm) were tested. The flow rates through the branch and the run were controlled with the valves (⑤, ⑥) at the downstream locations. Air/water mixture through the branch flowed into the air/liquid separator. The air flow rate through the branch was obtained

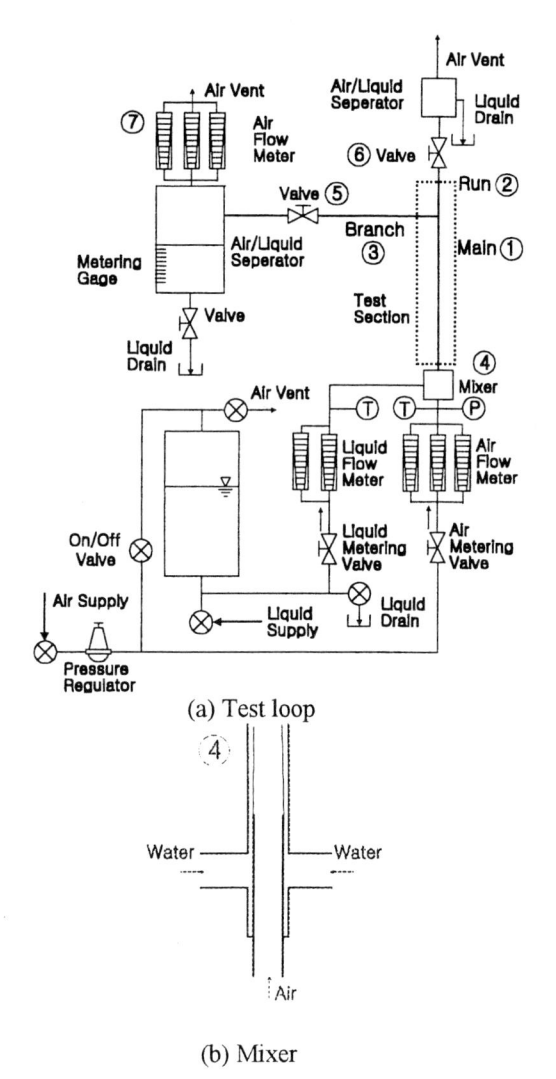

(a) Test loop

(b) Mixer

Fig. 2 Experimental Setup.

by measuring the flow rate through the air vent of the separator by using the calibrated rotameters (⑦). The water flow rate was estimated from the rate of the volume change of the water accumulated in the separator.

Superficial velocity ranges of air and water were 17.2 – 30 m/s and 0.02 – 0.32 m/s, respectively. The annular flow pattern was maintained in those ranges. The flow pattern was identified by visualization and also confirmed with the flow pattern map of Leon and Roman (1984) that had been constructed to identify flow pattern in a vertical rectangular channel with the hydraulic diameter range of 7.4 – 13.3 mm. The pressure within the T-junction was always maintained at 200 ± 5 kPa.

EXPERIMENTAL RESULTS

Before looking into the effect of the gap size of the branch, it would be instructive to see the flow split behavior for the T-junction with the same square cross sections in the main tube, run and branch. This is because it gives a basic idea to understand the dividing flow characteristics.

Effects of the Flow Direction and Cross-Sectional Shape

Figure 3 shows the variation of W_{f3}/W_{f1} with W_{g3}/W_{g1} which stand for the fractions of liquid and gas separated to the branch, respectively. The straight line in the figure represents the cases with the same fractions of separation; in other words, the quality at the branch (and also at the run) is the same with that of the main flow, i.e., $x_1 = x_3$. The dividing annular flow generally shows the S-shape curve. This trend depends strongly on the film momentum and liquid entrainment within the main channel. When the film momentum is low (i.e., low liquid flow rate) in the main channel, the fraction of the liquid separation increases. On the other hand, with the high liquid entrainment, the fraction of liquid separation decreases since the entrained liquid (drops) tends to pass through the run. These will be discussed in detail in the next section.

Figure 3 compares the present data with those of Stacey et al. (2000) with small (5mm in diameter) horizontal T-junction for similar inlet superficial velocities. Little difference is found between two cases. This implies that the flow split phenomenon is almost independent of the flow direction of the main flow when the channel size is small. This is due to the uniform distribution of the liquid film in the main. For the dividing annular flow behavior, there are two important factors to be considered. The local momentum fluxes of the fluids (Azzopardi, 1989), and the liquid film distribution (Roberts et al., 1997). For the cases shown in Fig. 3, the momentum fluxes of the fluids between the present work and Stacey et al. (2000) are similar. That is, the inlet gas and liquid superficial velocities, and hydraulic diameter are similar in both cases. Hence, the circumferential distribution of the liquid film along the periphery of the main pipe is the only important factor remains to be considered. For vertical annular flows, the liquid film is uniformly distributed along the periphery of the main pipe. For horizontal annular flow, asymmetry of the liquid film decreases with the decrease of the pipe diameter. That is, for the small tubes, the effect of the tube orientation becomes minor due to the predominance of the surface tension effect over the gravitational effect for similar inlet superficial velocities.

Another thing to consider is on the shape of the cross sections. The dividing annular flow within the small T-junction with circular cross sections (Stacey et al., 2000) is not much different from that with rectangular (square) cross sections (the present work). For the single phase flow, Lemonnier and Hervieu (1991) also reported that the flow structure was unchanged whether the cross sections were square or circular.

Effect of the T-Junction Size (Diameter)

Azzopardi (1994) compared the cases between the junction (pipe) diameters of 32 mm and 125 mm for a vertical annular flow. He reported that the dividing flow results are almost the same when the inlet superficial velocities are fixed. On the other hand, Stacey et al. (2000) compared their data for 5 mm diameter T-junction with those for a horizontal annular flow in 38 mm (Buell et al., 1994) and 127 mm (Roberts et al., 1995) pipes. They reported that the fraction of the liquid separated to the branch for their case (5mm in diameter) is higher than that with the larger T-junction for the fixed fraction of the gas separated. The results by Azzopardi (1994) and Stacey et al. (2000) seem to be contradictory; however, it should be noted that their size ranges of interest are different (i.e., 32 – 125 mm with Azzopardi (1994) and 5 mm with Stacey et al. (2000)).

Figure 4 compares the present work with those of Azzopardi (1988, 1994) for the vertical flows, respectively. This figure shows that more liquid is separated into the branch with the smaller T-junction. According to Stacey et al. (2000), the increase of the fraction of liquid separation to the branch for a small diameter T-junction is due to the less entrainment. In Fig. 4 (a), for the inlet conditions in the present work, the liquid film Reynolds number ($Re_f = 187$) remains below the critical value ($Re_{f,CRIT} = 420$) for the onset of entrainment suggested by Govan et al. (1988). Thus, the entrainment is almost negligible. But, for large diameter ($D_1 = 125$ mm), the entrainment rate is very high ($E = 0.5$) ; thus the fraction of liquid flowing as a film is small, and the fraction of liquid separated is lower than the equal quality line ($x_1 = x_3$) at the low range of W_{g3}/W_{g1}. In other words, the fraction of the liquid film increases with the decrease of the pipe diameter for the same superficial velocities at the inlet. According to Azzopardi (1994), at low inlet liquid flow rate, 'film stop' phenomenon occurred at the run. That is, the liquid film velocity at the run decreases with increasing of W_{g3}/W_{g1} and eventually falls to zero beyond a critical value as follows:

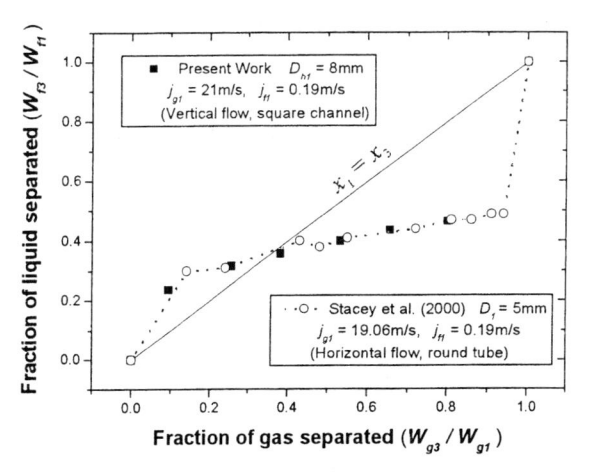

Fig. 3 Effects of the flow direction and the cross-sectional shape of the channel.

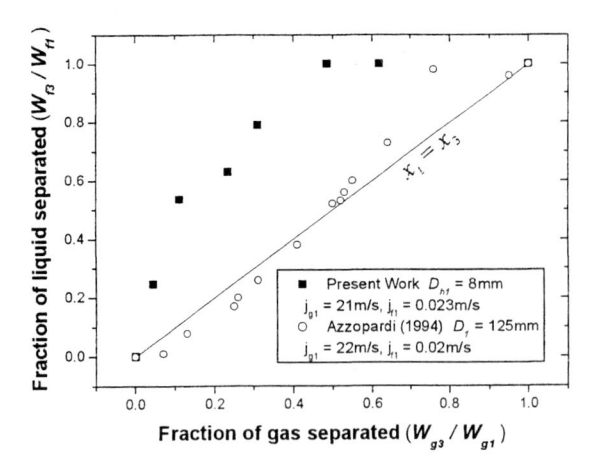

(a) Comparison with the results ($D_1 = 125$ mm) of Azzopardi (1994).

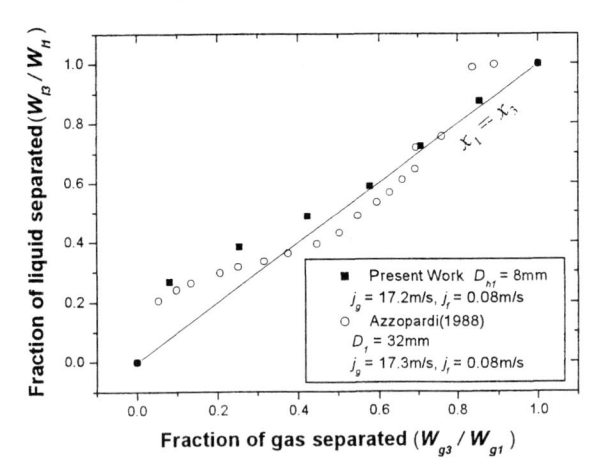

(b) Comparison with the results ($D_1 = 32$ mm) of Azzopardi(1988).

Fig. 4 Effect of the T-junction size.

$$(W_{g3}/W_{g1})_{CRIT} = 0.715 - \sqrt{0.493 - 0.633\frac{\rho_{f1}v_{f1}^2}{\rho_{g1}v_{g1}^2} + \frac{1.266L}{\rho_{g1}v_{g1}^2}}$$

(1)

Here, L is the dissipation of energy and generally is equal to zero. According to Eq. (1), $(W_{g3}/W_{g1})_{CRIT}$ is 0.2 for the present case, while for the large diameter case (D_1 = 125mm, Azzopardi (1994)), $(W_{g3}/W_{g1})_{CRIT}$ is 0.6. In Fig. 4 (a), the value of W_{f3}/W_{f1} is not the unity at $(W_{g3}/W_{g1})_{CRIT}$, because the liquid flowing as drops is not considered in Eq. (1). Therefore, more liquid is separated with the smaller T-junction.

Consequently, it is not appropriate to apply the large diameter T-junction data to predict the fraction of liquid separation within a small size T-junction.

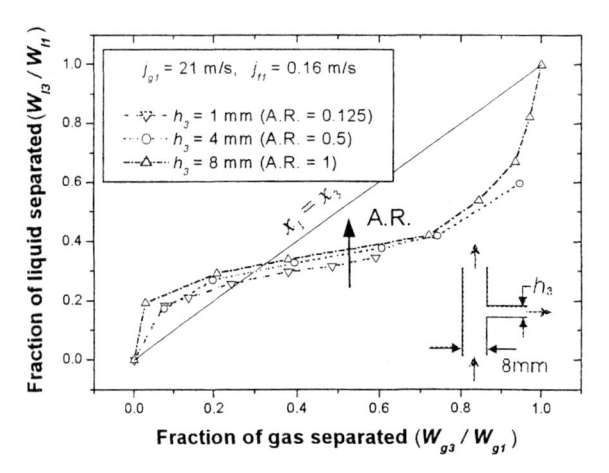

Fig. 5 Effect of the aspect ratio of the branch.

Effect of the Aspect Ratio of the Branch

Figure 5 shows the effect of the branch aspect ratio on the liquid separation. As the aspect ratio of the branch decreases, the fraction of the liquid separation decreases, though slightly. In addition, the range of W_{g3}/W_{g1}, shown in the figure depends on the aspect ratio. That is, the values of W_{g3}/W_{g1} can be raised up to 0.9 for an aspect ratio of 0.5, while up to 0.6 for 0.125. This is because, when the aspect ratio becomes small, it is impossible to maintain the constant pressure (200kPa in the present case) within the T-junction with the higher fraction of gas separation.

Azzopardi (1984) reported the similar trend for vertical large junctions (D_1 = 32 mm) with various aspect ratios of the branch (0.2 – 1.0). It was explained that the smaller region of the liquid film in the main was separated through the branch when the branch aspect ratio becomes small. The reason for the minor effect of the aspect ratio is considered to be the competition of two opposing effects ; the increase of the suction force to the branch on the liquid phase and the decrease of the flow passage. For a fixed value of W_{g3}/W_{g1}, the gas velocity through the branch becomes higher with the smaller branch size. Hence, the pressure in the branch becomes lower and the suction force to the branch appears to be stronger (Bernoulli effect). This feature has been explained by Walters et al. (1998) for the 38.1 mm horizontal junction. They reported that decreasing of the diameter ratio between the main and the branch from 1.0 to 0.5 decreased the fraction of the liquid separation for a fixed value of W_{g3}/W_{g1}. However, the effect gradually diminishes and further reduction in the diameter ratio (down to 0.2) increases the fraction of the liquid separated.

In summary, the smaller aspect ratio of the branch results in the smaller axial distance of the liquid film in the main channel separated through the branch. At the same time, the suction force to the branch increases. Consequently, due to those competing effects, the smaller aspect ratio of the branch induces smaller liquid separation through the branch, but only slightly.

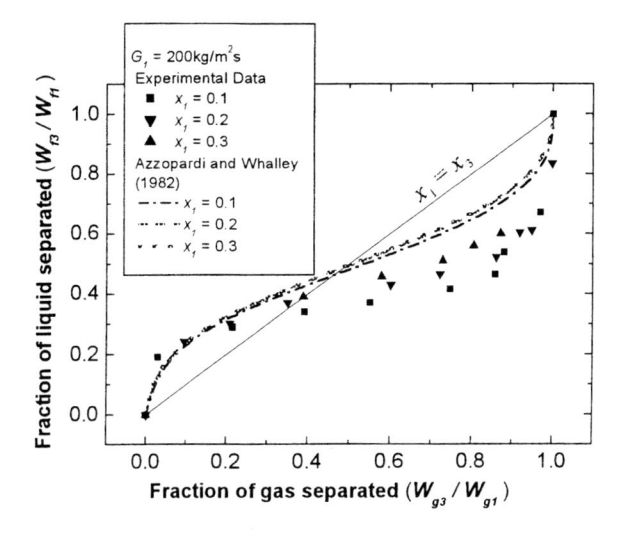

Fig. 6 Effect of the inlet quality(x_1) and comparison with the model of Azzopardi and Whalley (1982)

Effect of the Inlet Quality

Figure 6 shows the effect of inlet quality at the main (x_1) on the fraction of liquid separation. For a fixed value W_{g3}/W_{g1}, the fraction of
liquid separation increases with the increase of the inlet quality. This result is the same with that of Collier (1976) (x_1 = 0.17 – 0.5). It was explained that gas momentum increased while liquid momentum decreased with the increase of the inlet quality, and the liquid was more likely to be separated to the branch. In the same figure, the analytical results of Azzopardi and Whalley (1982) were plotted, which will be discussed in the later section.

ANALYSIS

Comparison with the Previous Models

There are several models available to predict the flow separation at dividing T-junctions. Among them, the models by Azzopardi and Whalley (1982), Azzopardi (1984) and Shoam et al. (1987) are the most relevant to be compared. For a vertical annular flow, Azzopardi and Whalley (1982) developed a phenomenological model as illustrated in Fig. 7. They suggested that most of the liquid flow through the branch is from the film portion in the main pipe rather than from the entrained drops. The proportion of the liquid film entering the branch was considered linearly dependent on the gas flow rate flowing into the branch. This result is represented by the following equations provided that the liquid film thickness is small compared to the diameter of the main.

$$\frac{W_{f3}}{W_{f1}} = \frac{\theta}{2\pi}(1-E) \quad (2)$$

$$\frac{W_{g3}}{W_{g1}} = \frac{1}{2\pi}(\theta - \sin\theta) \quad (3)$$

Here, E is the fraction of the liquid entrainment and θ is the angle of inclusion which covers the portion of the liquid film flowing into the branch. Later, Azzopardi (1984) modified Eq. (2) by introducing an empirical correction factor K as follows to take account of the effect of the diameter ratio of the branch to the main :

$$\frac{W_{f3}}{W_{f1}} = K\frac{\theta}{2\pi}(1-E) \quad (4)$$

$$K = 1.2 \cdot \left(\frac{D_3}{D_1}\right)^{0.4} \quad (5)$$

Stacey et al. (2000) reported that Eq. (2) well represented the dividing two phase flow within a small T-junction. The present experimental data were compared to Eqs. (2) and (3) as in Fig. 6. The inlet quality (x_1) has been changed from 0.1 to 0.3 at constant inlet mass flux (G_1) of 200 kg/m²s. In Eq. (2), E can be obtained from Martin et al.'s entrainment correlation (1997). This correlation is a modified form of Kataoka and Ishii (1982)'s entrainment correlation which has been developed for vertical pipes with their diameters ranging from 9.5 mm to 40 mm. In applying the Azzopardi and Whalley's model to the present case, the hydraulic diameter is used since the cross sectional shape is rectangular. Figure 6 shows the model generally overpredicts the fraction of the liquid separation. However the trend of the inlet quality effect is predicted to be correct though the effect appears to be minor.

Based on the similar physical concept, Shoam et al. (1987) developed a flow pattern dependent model illustrated as in Fig. 8. They assumed a constant film thickness with no entrainment, and the dividing gas streamline has an arc-shape with the radius of curvature R as:

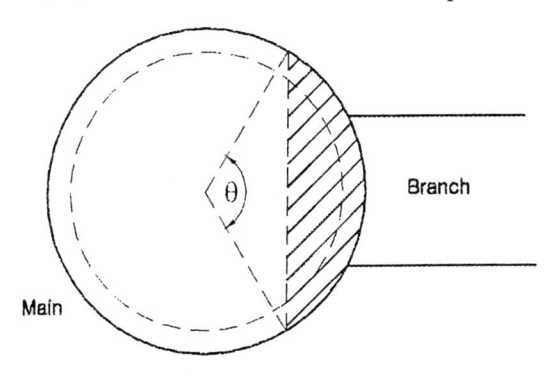

Fig. 7 Flow geometry used in Azzopardi and Whalley model (1982).

(a)

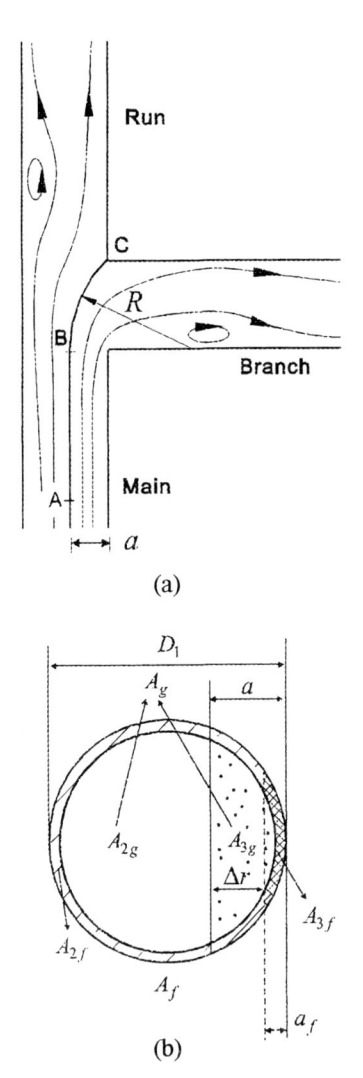

(b)

Fig. 8 Schematic flow description illustrating the model of Shoam et al. (1987).

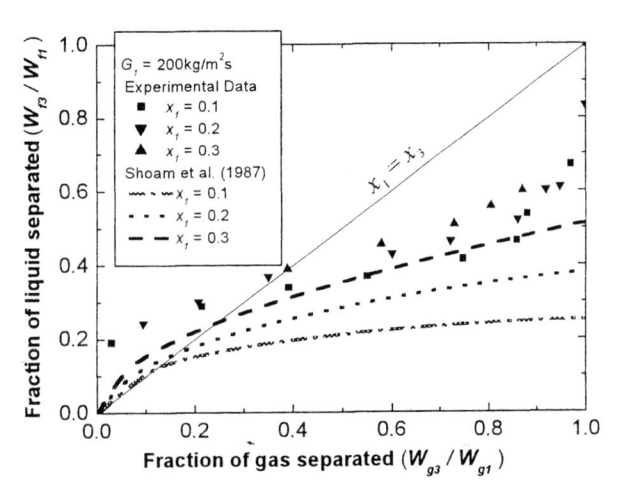

Fig. 9 Effect of the inlet quality(x_1) and comparison with Shoam et al. model (1987).

$$R = \frac{D_1^2 + a^2}{2a} \quad (6)$$

In Eq. (6) and Fig. 8 (b), "a" denotes the distance of streamline A-B from the branch-side wall of the main pipe. The fraction of the gas separation was obtained by considering the gas flow configuration as follows:

$$\frac{W_{g3}}{W_{g1}} = \frac{A_{3g}}{A_{3g} + A_{2g}} = \frac{A_{3g}}{A_g} \quad (7)$$

Here, in Fig. 8 (b), A_{2g} and A_{3g} are expressed with the white and the dotted areas of the gas core, respectively.

The location of the liquid boundary line (a_f) is determined by the balance between the inertial, centrifugal and damping forces as in the following equation,

$$a_f = a - \Delta r \quad (8)$$

with the displacement of the liquid boundary from the gas stream line "a" as

$$\Delta r = \frac{f}{k}\left[t - \frac{1}{k}(1 - e^{-kt})\right] . \quad (9)$$

This means the dividing lines are different between the gas and liquid phase. Here $f \; [= (\rho_{f1} v_{f1}^2 - \rho_{g1} v_{Lg1}^2)/(\rho_{f1} R)]$ is the centrifugal force, $k \; [= 3\mu_{f1}/(\rho_{f1}\delta^2)]$ is the damping coefficient and $t \; [= (R\sin^{-1}(D_1/R))/v_{f1}]$ is the residence time of the liquid between points B and C. Then the fraction of the liquid separation can be determined as follows.

$$\frac{W_{f3}}{W_{f1}} = \frac{A_{3f}}{A_{3f} + A_{2f}} = \frac{A_{3f}}{A_f} \quad (10)$$

A_{2f} and A_{3f} are shown as the single-hatched and the cross-hatched areas in Fig. 8 (b), respectively.

Figure 9 shows that Shoam et al.'s model generally underpredicts the fraction of the liquid separation. However it predicts the dependence on the inlet quality better than the model by Azzopardi and Whalley (1982).

The Modified Model

In the present work, Shoam et al.'s model has been improved to predict the flow separation at small, dividing T-junctions for the two-phase annular flow. The liquid boundary line shown in Eq. (8) was modified by introducing a correction factor similar to K (in Eq. (5)) suggested by Azzopardi (1984).

Instead of using Eq. (8), a modified correlation for a_f with empirical correction factor is proposed as follows:

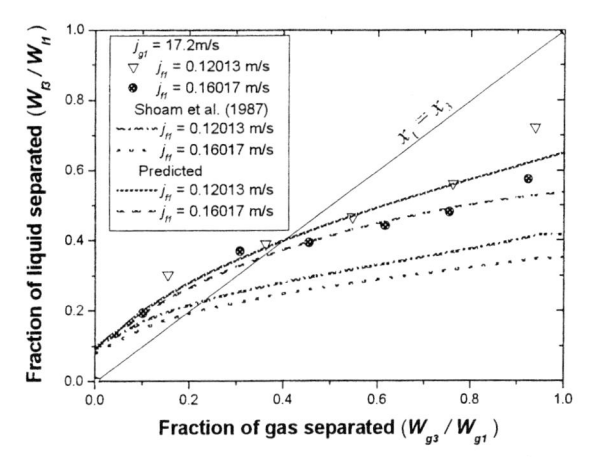

Fig. 10 Comparison between the present experimental data and the models

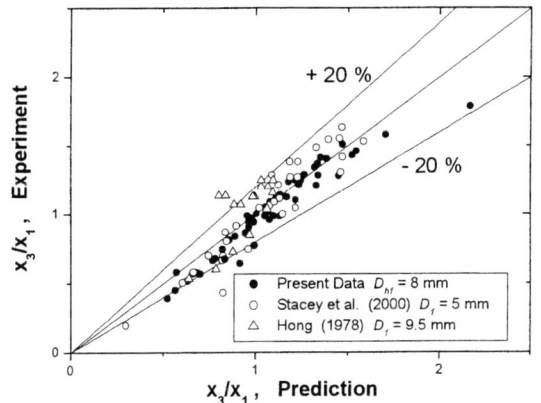

Fig. 11 Overall evaluation of modified model

$$a_f = K' \cdot (a - \Delta r) \quad (11)$$

Variation of the film distribution near the branch was already counted in Δr (through f and k). However, the momentum flux effect also should be considered. In addition, decreasing of the aspect ratio of the branch decreases the fraction of the liquid separation, and the aspect ratio effect also should be taken into account. That is,

$$K' = fn\left(\frac{\rho_{f1} j_{f1}^2}{\rho_{g1} j_{g1}^2}, \frac{h_3}{D_{h1}}\right) = C \cdot \left(\frac{\rho_{f1} j_{f1}^2}{\rho_{g1} j_{g1}^2}\right)^p \cdot \left(\frac{h_3}{D_{h1}}\right)^q \quad (12)$$

Constant C and exponents p and q could be determined from the data regression process as follows :

$$C = 5.3, \quad p = 0.25, \quad q = 0.05 \quad (13)$$

The value of K' in Eq. (12) is larger than the unity. This implies more amount of the liquid separates through the branch than that predicted by Shoam et al.(1987).

Figure 10 compares experimental data and the modified model. In the same figure, the Shoam et al.'s model (1987) was also plotted for comparison. The Shoam et al.'s model underpredicts the fraction of the separation. On the other hand, the modified model predicts rather accurately except for W_{g3}/W_{g1} in the range of 0.9 – 1.0. At the high value of W_{g3}/W_{g1}, even the modified model underpredicts the fraction of the liquid separation. This is because the present model did not consider the flooding phenomena. In practice, at the high value of W_{g3}/W_{g1}, the amount of the gas flowing into the run decreases below the flooding condition and the liquid film falls down to the junction point.

An overall evaluation of modified model is exhibited in Fig. 11 using the present data along with those by Stacey et al. (2000) and Hong (1978). The data cover the small hydraulic diameter range of 5 – 10 mm T-junctions in the main tube, run and branch with the same cross sections, and also the wide range of the aspect ratio of the branch (0.125 – 1.0) for small rectangular junctions. It can be confirmed that the model is capable of predicting more than 90% of available data within an accuracy of ± 20%.

CONCLUDING REMARKS

- For the annular flow in the main channel of a small T-junction with the same cross sections in the main, run and the branch, the flow split appears to be independent of the channel orientation and the cross-sectional shape.
- The fraction of the liquid separation through the branch is larger with the smaller T-junctions.
- As the branch aspect ratio or the inlet quality decreases, the fraction of the liquid separation decreases.
- A modified model was proposed to represent the liquid separation of the dividing annular flow at a small T-junction for various aspect ratios of the branch. The model encompasses more than 90% of available data within the deviation of ± 20%.

ACKNOWLEDGEMENTS

This work was supported by a grant from the Critical Technology 21 Project of the Ministry of Science and Technology, Korea and in part by the Brain Korea 21 Project.

NOMENCLATURE

A	area, m^2
a	distance of streamline A-B (in Figure 8) from the pipe wall on the side of the side arm, m
D	diameter, m
D_h	hydraulic diameter, m
E	the fraction of the liquid entrained
f	centripetal force per unit mass, N/kg

G	mass flux, kg/m^{2}s
h	gap size, m
j	superficial velocity, m/s
k	damping coefficient
K, K'	empirical correction factor
L	dissipation of energy
P	pressure, kPa
R	radius of curvature, m
Re	Reynolds Number
v	velocity, m/s
W	mass flow rate, kg/s
x	quality, $W_g/(W_g+W_f)$
θ	angle, rad
δ	film thickness, m

Subscripts

1	main tube
2	run
3	branch
CRIT	critical
g	gas
f	liquid

REFERENCES

Azzopardi, B. J. and Whalley, P. B., 1982, The effect of flow patterns on two phase flow in a T junction, *Int. J Multiphase Flow*, Vol.8, pp. 491-507.

Azzopardi, B.J., 1984, The effect of side arm diameter on two phase flow split at a T junction, *Int. J Multiphase Flow*, Vol.10, pp. 509-512.

Azzopardi, B. J., 1988, Measurements and observations of the split of annular flow at a vertical T-junction, *Int. J Multiphase Flow*, Vol. 14, pp. 701-710.

Azzopardi, B. J., 1994, The split of vertical annular flow at a large diameter T junction, *Int. J Multiphase Flow*, Vol. 20, pp. 1071-1083.

Azzopardi, B.J., 1999, Phase separation at T junctions, *Multiphase Science and Technology*, Vol. 11, pp.223-329.

Buell, J. R., Soliman, H. M., Sims, G. E., 1994, Two-phase pressure drop and phase distribution of film thickness at a horizontal tee junction, *Int. J Multiphase Flow*, Vol.20, pp. 819-836.

Charron, Y. and Whalley, P. B., 1995, Gas-Liquid annular flow at a vetical tee junction – Part 1.Fow separation, *Int. J Multiphase Flow*, Vol.21, pp.569-589.

Collier, J.G. 1976, Single-phase and two-phase behavior in primary circuit components, Proc. N.A.T.O. Advanced Study Institute on Two-phase Flow and Heat transfer, Istanbul, Turkey, Hemisphere, Washington, D.C., pp.322-333

Govan, A. H., Hewitt, G. F. Owen, D. G., and Bott, T. R., 1988, An improved CHF modelling code, 2nd UK National Conference on Heat Transfer, Inst. of Mechanical Engineers, London, ,Vol.1, pp.33-48.

Hong, K. C., 1978, Two phase flow splitting at a pipe tee, *J. Pet. Technol.*, pp.290-296.

Kataoka, I., Ishii, M., 1982, Mechanism and correlation of droplet entrainment and deposition in annular two phase flow, NUREG/CR-2885, ANL-82-44

Lemonnier, H. and Hervieu, E., 1991, Theoretical modelling and experimental investigation of single-phase and two-phase division at a tee junction, *Nucl. Eng. and Des.*, Vol.125, pp.201-203.

Leon, T. and Roman, U., 1984, Two phase gas liquid flow in rectangular channels, *Chem. Eng. Sci.*, Vol.39, pp. 751-765.

Martin, A. L. B., Cheng, S, J., Stephen, G.B., 1997, Annular flow entrainment in a small vertical pipe, *Nucl. Eng. and Des.*, Vol.178, pp.61-70.

Roberts, P. A., Azzopardi, B. J. and Hibberd, S., 1995, The Split of horizontal semi annular flow at a large diameter T junction, *Int. J Multiphase Flow*, Vol.21, pp. 455-466.

Riemann, J., Brinnkmann, H. J. and Domanski, R., 1988, Gas-liquid flow in dividing tee-junction with horizontal inlet and different branch orientations and diameters, Kernforschungszentrum Karlsruhe, Institue fur Reaktorbauelemente Report KfK 4399, pp. 105-119.

Stacey, T., Azzopardi, B. J. and Conte, G., 2000, The split of annular two phase flow at a small diameter T-junction, *Int. J Multiphase Flow*, Vol. 26, pp. 845-856.

Shoam, O., Brill, J. P. and Taitel, Y., 1987, Two-phase flow splitting in a tee junction - experimental and modelling, *Chem. Eng. Sci.*, Vol .42, pp. 2667-2676.

Shoam, O., Arirachakaran, S. and Brill, J. P., 1989, Two-phase flow splitting in a tee junction - experimental and modelling, *Chem. Eng. Sci.*, Vol .44, pp. 2388-2391.

Walters, L. C., Soliman, H. M. and Sims, G. E., 1998, Two-phase pressure drop and phase distribution at reduced tee junctions, *Int. J Multiphase Flow*, Vol.24, pp. 775-792.

HEAT TRANSFER DURING REFLUX CONDENSATION OF R134A IN A SMALL DIAMETER INCLINED TUBE

S. Fiedler, H. Auracher

Institut für Energietechnik, TU Berlin, Marchstr. 18, 10587 Berlin, Germany

ABSTRACT

Heat transfer coefficients for reflux condensation of R134a in a small diameter inclined tube are determined. The inner diameter of the tube is 7 mm and the length is 500 mm. Experiments are carried out at condensate film Reynolds numbers between 50 and 165 and at pressures between 0.67 MPa and 0.74 MPa. It has been found that the inclination angle has a strong effect on the heat transfer coefficient. At an inclination angle of about 45° the heat transfer coefficient is increased by a factor of nearly 2 when compared with the heat transfer coefficient for reflux condensation in the vertical position. Existent correlations from the open literature do not agree well with experimental data for the small diameter inclined tube. A modified form of a semi-empirical correlation proposed by Wang and Ma is presented that predicts well the experimental data.

INTRODUCTION

In reflux condensation a vapour enters a vertical or inclined mounted condenser at the bottom and flows upward. The condensate flows downward countercurrently to the vapour due to gravity. Typical applications of reflux condensation are in overhead condensers of distillation columns and vent condensers of reactors or stirred vessels. Other applications are in the vent cooling section of air-cooled steam condensers and in two-phase closed thermosyphons. Compact plate heat exchangers are increasingly used for reflux condensation applications in process industries. In such condensers the hydraulic diameter of the flow channels formed between two plates is 5-10 mm, and the flow channels are inclined to the vertical. The fundamental mechanisms of heat and mass transfer as well as of two-phase flow in these small channels are not well understood. In this experimental investigation a single sub-channel of a compact condenser is simulated by a small diameter inclined stainless steel tube. The inner diameter of the tube is 7 mm and the length is 500 mm. To the authors' best knowledge investigations on heat transfer during reflux condensation in small diameter inclined tubes have not been carried out before.

Heat transfer with cocurrent, single-component condensation inside vertical or horizontal tubes has been intensively investigated for the last 25 years. However, very little effort was devoted to the reflux condensing arrangement [1, 2]. Reflux condensation is gravity controlled, the countercurrently flowing vapour retards the condensate flow and thickens the condensate film. However, the velocity of the entering vapour is relatively low - especially in small diameter tubes - as reflux condensation is limited by the phenomenon of flooding: There is a maximum vapour velocity above which condensate will be carried upward rather than draining to the bottom. Chen et al. [3] developed correlations for local and mean Nusselt numbers for reflux condensation in vertical tubes on the basis of analytical and empirical results from the literature. These correlations incorporate the effect of interfacial shear stress, interfacial waviness and turbulent transport in the condensate film. However, the minimal tube diameter for which the correlations are still valid is not given. They obtained the following correlation for the average Nusselt number:

$$\overline{Nu} = \left(0.54\, Re^{-0.44} + \frac{Re^{0.8} Pr^{1.3}}{5.67 \cdot 10^4} - \frac{C\, Pr^{1.3} Re^{1.8}}{6.38 \cdot 10^3} \right)^{\frac{1}{2}} \quad (1)$$

where C is defined by

$$C = \frac{0.023}{d^2 g^{2/3}} \frac{\eta_l^{1.133} \eta_g^{0.2}}{\varrho_g \varrho_l^{0.333}}. \quad (2)$$

The average Nusselt number, the Reynolds number and the Prandtl number are defined as follows:

$$\overline{Nu} = \frac{\overline{\alpha}(\nu_l^2/g)^{1/3}}{\lambda_l}, \quad (3)$$

$$Re = \frac{\dot{M}/b}{\eta_l}, \quad (4)$$

$$Pr = \frac{\nu}{a} = \frac{\eta\, c_p}{\lambda}. \quad (5)$$

In the ESDU data item on reflux condensation [4], data of the heat transfer coefficient of reflux condensation experiments in vertical tubes have been compared with results of correlations for gravity controlled cocurrent flow of vapour and condensate. The correlations for different ranges of the condensate film Reynolds number are, respectively, those of Nusselt [5] (Re ≤ 7.5), Kutateladze [6] (7.5 < Re < 400) and Labuntsov [7] (Re ≥ 400). Due to the proximity of the experimental reflux condensation data to the respective correlations for cocurrent flow, it is recommended in the ESDU data item to use these correlations also in the reflux situation. The recommended correlations for the condensate film Reynolds number ranges Re ≤ 7.5 and 7.5 < Re < 400 are as follows:

$$Re \leq 7.5 : \overline{Nu} = 0.925 \, Re^{-\frac{1}{3}} \quad (6)$$

$$7.5 < Re < 400 : \overline{Nu} = \frac{Re}{1.47 \, Re^{1.22} - 1.3} \quad (7)$$

Groß and Hahne [8] investigated the influence of the inclination angle on heat transfer with reflux condensation in a two-phase thermosyphon with an inner diameter of 40 mm. Refrigerant R13B1 was the working fluid. They found that by tilting the tube from the vertical the mean heat transfer coefficient increases significantly and then decreases slightly for inclination angles of less than 30° from the horizontal. If the tube is inclined, the film thickness is no longer axisymmetric. Therefore the mean heat transfer coefficient increases.

Groß [9] evaluated data for the heat transfer with reflux condensation in two-phase thermosyphons from 18 research works and proposed a correlation for the heat transfer in the condensate film with stagnant vapour phase. This correlation incorporates the influence of the inclination angle. It is based on a theoretical study by Hassan and Jakob [10] who analysed heat transfer with condensation on the outside of an inclined tube. However, the validity of this correlation has not been checked for tube diameters less than 14 mm.

Wang and Ma [11] carried out theoretical and experimental studies on vertical and inclined thermosyphons. They pointed out that no final conclusion can be drawn on the optimum inclination angle at which the maximum heat transfer coefficient occurs because it is not independent from operating conditions. Wang and Ma presented the following semi-empirical correlation for reflux condensation heat transfer:

$$\frac{\overline{Nu}}{\overline{Nu_n}} = \left(\frac{L}{R}\right)^{\cos\beta/4} (0.54 + 5.86 \cdot 10^{-3} \beta) \quad (8)$$

where L is the length of the condenser section, R is the inner radius of the tube and β is the inclination angle from the horizontal. $\overline{Nu_n}$ is the average Nusselt number according to the classical Nusselt film theory [5] for condensation in a vertical tube (Eq. (6)). For condensation in a vertical tube ($\beta=90°$) this semi-empirical correlation predicts:

$$\overline{Nu} = 1.07 \, \overline{Nu_n} \quad (9)$$

EXPERIMENTAL SETUP

Refrigerant R134a was used as the test fluid because it represents fluid properties that are similar to those of process fluids, e.g. the density ratio of vapour and liquid. Due to the relatively low surface tension of R134a a continuous condensate film is formed, i.e. film condensation occurs.

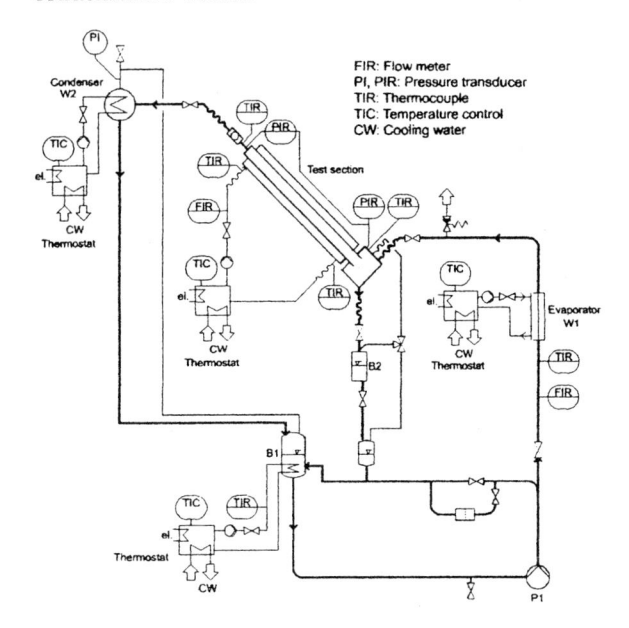

Figure 1: Schematic of experimental setup

A schematic of the experimental setup for the reflux condensation heat transfer experiments is depicted in Fig. 1. The forced flow in-tube evaporator W1 provides saturated or superheated vapour. It is a coaxial coiled evaporator. The entering liquid is completely evaporated when passing through the evaporator. The mass flow rate of vapour that exits the evaporator is therefore the same as of the liquid that enters the evaporator. The mass flow rate of the liquid is determined by a coriolis mass flow meter before entering the evaporator. The vapour enters the test section at the bottom. The part of the vapour that is not condensed in the test section is liquefied in the secondary condenser W2. The condensate is returned to the flow return circuit which consists of the header tank B1, the pump P1 and a filter/dryer for the refrigerant.

The condensate that exits the test section at the bottom passes through the volumetric cylinder B2. The flow rate of the condensate was determined by collectiong condensate in this cylinder for a known period of time. From the volumetric cylinder the condensate is as well returned to the flow return circuit. From there the liquid is pumped again to the evaporator.

The test section is a 500 mm long double pipe heat exchanger made of stainless steel. Condensation takes place in the inner tube. The inner diameter of the inner tube is 7 mm and the outer diameter is 8 mm. Temperature controlled cooling water flows in the annulus of the test section counter-currently to the vapour in the inner tube. The cooling water flow rate is determined with a calibrated impeller flow meter. To be able to vary the inclination angle the test section is connected by flexible pressure hoses. Experiments were carried out at inclination angles of 30°, 45°, 60° and 90° from the horizontal. 20 K-type thermocouples (∅ 0.25 mm) were embedded in the inner tube wall around the circumference of the test section to be able to measure local wall temperatures. The thermocouple leads were silver-soldered into axial grooves (10 mm long, 0.3 mm wide, 0.3 mm deep) and taken out through the coolant and fittings in the outer tube wall located 15 mm downstream in flow direction of the coolant. As can be seen in Fig. 2 the thermocouples are located at five different axial positions of the test section, at each axial position four thermocouples are embedded around the circumference of the tube: one at the top, one at the bottom and one at each side.

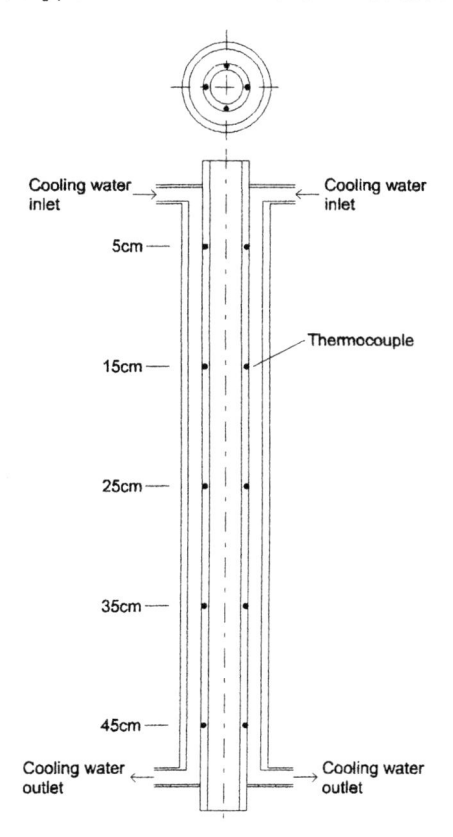

Figure 2: Location of the thermocouples along the test section

RESULTS AND DISCUSSION

All reflux condensation heat transfer experiments were carried out at system pressures between 0.67 MPa and 0.74 MPa at the inlet of the test section and at inclination angles of 30°, 45°, 60° and 90° from the horizontal. The vapour entered the test section slightly superheated. The mean condensation heat transfer coefficient was calculated from

$$\overline{\alpha} = \frac{\dot{Q}}{\pi d L (T_S - T_W)}. \tag{10}$$

T_S is the saturation temperature of the vapour. The wall temperature T_W is the arithmetic mean value of the local inner wall temperatures. The thermocouples are embedded only 0.2 mm beneath the inner tube surface. Therefore the arithmetic mean value of the measured local inner wall temperatures can be taken directly as the inner wall surface temperature. The temperature difference between the location of the thermocouples and the inner wall surface is only 0.1 K or less for the heat fluxes at which the experiments were performed. This uncertainty was assumed to be tolerable.

The total heat transfer $\dot{Q}$ was determined from the flow rate of the cooling water and the temperature rise as well as computed from the measured condensate flow rate:

$$\dot{Q} = \dot{M}_{CW} c_{p,CW} \Delta T, \tag{11}$$

$$\dot{Q} = \dot{M}_{cond} \Delta h_v. \tag{12}$$

The difference between the two heat flow rates was usually smaller than ±10%. A test run was repeated if the deviation was larger than ±12%.

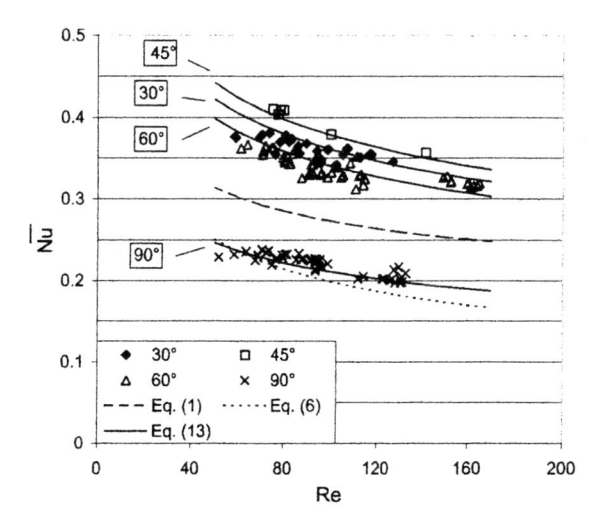

Figure 3: Average Nusselt number vs. film Reynolds number for different inclination angles

Once the mean condensation heat transfer coefficient was established the mean Nusselt number could be obtained according to Eq. (3). The condensate film Reynolds number was calculated from Eq. (4). The experimental reflux condensation heat transfer data for the 0.5 m long tube with an inner diameter of 7 mm are shown in Fig. 3. They are presented - as usual in gravity controlled condensation - in terms of the mean condensate film Nusselt number and the condensate film Reynolds number at the bottom of the tube.

It can be seen that the inclination angle has a strong effect on the condensation heat transfer. The average Nusselt number is smallest for reflux condensation in the vertical position ($\beta=90°$). The Nusselt number increases when the tube is tilted from the vertical because then the film thickness is no longer axisymmetric. The mean film thickness decreases and therefore the mean heat transfer coefficient increases. Fig. 4 shows the effect of the inclination angle on the condensation heat transfer for constant film Reynolds numbers of Re$\approx$75 and Re$\approx$100. In this diagram also trendlines are depicted. It can be seen that according to the experimental data the optimum inclination angle for the heat transfer lies between 35° and 50° from the horizontal.

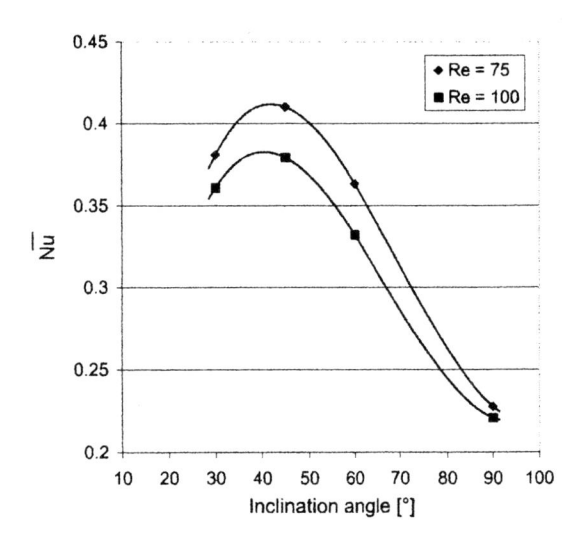

Figure 4: Effect of inclination angle on reflux condensation heat transfer

Fig. 5 shows a comparison between the own reflux condensation heat transfer data in the vertical position and data obtained by Onda et al. [12] for partial reflux condensation of methanol in a vertical tube (d = 21 mm) at near atmospheric conditions (length of condensing section: 0.31 m, 0.81 m, 1.32 m, 11.81 m). In this study the condensate film thickness was not zero at the top of the reflux tube, condensate was introduced at the top. Therefore the mean condensate film Nusselt number is lower than if the condensate film thickness had been zero at the top. This can be seen in Fig. 5, the own experimental data lie above the data of Onda et al.. According to this comparison a strong influence of the tube diameter cannot be noticed for a vertical tube arrangement.

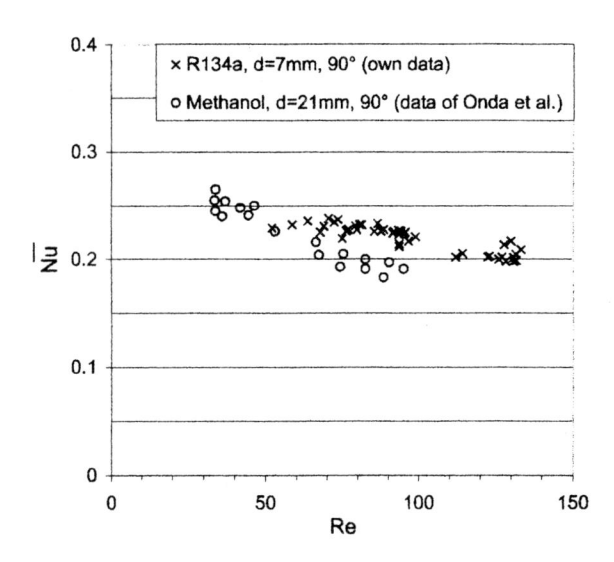

Figure 5: Comparison between own data and data of Onda et al. [12], data taken from [4]

A comparison of data for the inclined tube where the heat transfer coefficient was found to be strongly augmented when compared with the vertical position (see Fig. 3) was not possible. All investigations on reflux condensation in inclined tubes that can be found in the open literature were carried out with closed two-phase thermosyphons. These investigations focus mainly on the influence of the inclination angle on the maximum heat transport capability, i. e. on the maximum performance which is limited by flooding (e. g. Huanzhuo et al. [13], Groll and Rösler [14], Chen [15]). Comparable systematic studies on the heat transfer coefficient in dependence of film Reynolds number and inclination angle have not been carried out. The condensation heat transfer coefficient of closed two-phase thermosyphons is also strongly influenced by effects that are not present in systems like ours as for example by the amount of liquid filling of the thermosyphon.

The experimental data for reflux condensation heat transfer were compared with values predicted by correlations for gravity controlled condensation. As can be seen in Fig. 3 the correlation proposed by Chen et al. (Eq. (1)) for reflux condensation in a vertical tube overpredicts the average Nusselt numbers determined experimentally for the vertical tube. The classical Nusselt correlation (Eq. (6)) only very slightly underpredicts the experimental data. The slope of this correlation is marginally too big (see Fig. 3). Correlations for reflux condensation heat transfer in inclined tubes are sparse. They were developed based on data collected for the condensation zone of inclined two-phase thermosyphons, and diameters of thermosyphons are usually significantly above 10 mm. The own experimental data for the inclined tube were compared with correlations proposed by Semena and Kiselev [16], Stoyanov [17] as well as correlations proposed by Groß [9] following Hassan and Jakob [10] and Uehara et al. [18] for reflux condensation inside thermosyphons. It was found that some correlations overpredicted the average Nusselt number by a factor of about 1.6 or underpredicted it by a factor of 0.7. The semi-empirical correlation by Wang and Ma (Eq. (8)) showed the best agreement with the own experimental data. However, this correlation predicts the optimum inclination angle in a range less than 30° from the horizontal whereas the optimum inclination angle was found to be close to 45° in the own experiments. Therefore the Wang/Ma correlation for the average Nusselt number was modified based on the own experimental data. The following modified semi-empirical Wang/Ma correlation is proposed:

$$\frac{\overline{Nu}}{\overline{Nu_k}} = \left(\frac{L}{R}\right)^{\cos\beta/4} (0.125 + 1.46 \cdot 10^{-2}\beta - 7.27 \cdot 10^{-5}\beta^2) \quad (13)$$

Unlike in the original Wang/Ma correlation (Eq. (8)) in this correlation the average Nusselt number was not related to the one predicted by Nusselt [5] for condensation in a vertical tube (Eq. (6)) but to the one predicted by Kutateladze [6], $\overline{Nu_k}$, according to Eq. (7). As can be seen in Fig. 3 the modified Wang/Ma correlation agrees well with the experimental data for all inclination angles. Fig. 6 shows a comparison of the experimental data for the average Nusselt number and the values predicted by the proposed modified Wang/Ma correlation (Eq. (13)). The correlation predicts the experimental data in a range of ±10%.

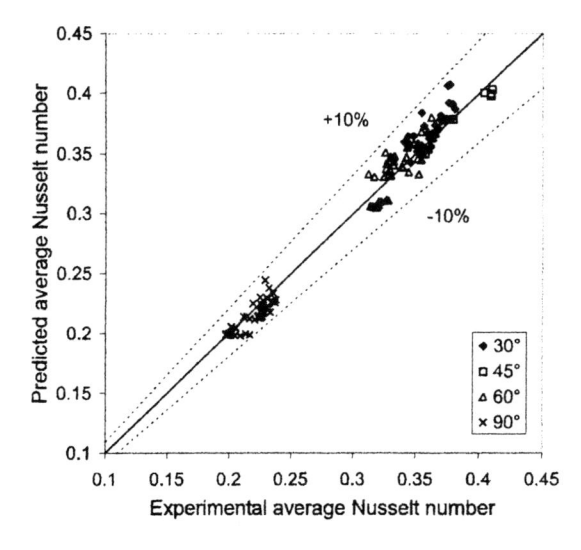

Figure 6: Comparison of experimental and predicted (Eq. 13)) average Nusselt numbers

In Fig. 7 the average Nusselt number is depicted in dependence of inclination angle and film Reynolds number according to the modified Wang/Ma correlation (Eq. (13)). The optimum inclination angle is predicted between 40° and 45° from the horizontal.

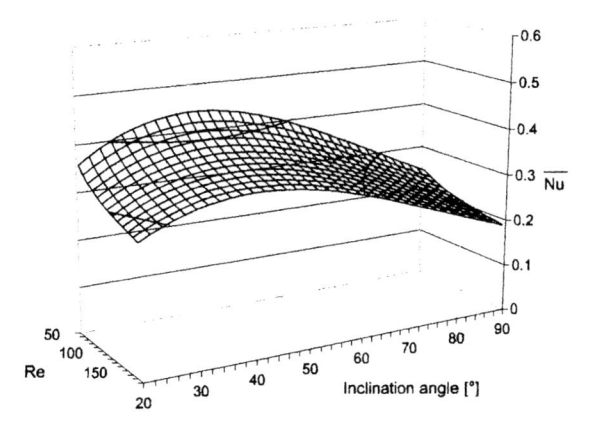

Figure 7: Average Nusselt number vs. inclination angle and film Reynolds number according to Eq. (13)

PRACTICAL IMPLICATIONS

The inner diameter of 7 mm of the test section is in the range of the hydraulic diameter of flow channels of compact plate heat exchangers that are used for reflux condensation applications. Although it is certainly not possible to directly transfer the results for the single tube to a single flow channel formed between two plates, the results indicate that there is an optimum inclination angle of the plate corrugation of compact plate heat exchangers for the heat transfer in gravity controlled reflux condensation. Furthermore the results imply that an inclined arrangement of an in-tube reflux condenser (e. g. shell-and-tube condenser, thermosyphon) can be beneficial to the condensation heat transfer coefficient.

However, the results cannot directly be extended to condensation with cocurrent flow of vapour and condensate in small/mini diameter tubes as for example in innovative air conditioning and heat pump applications. But also in such applications an influence of the inclination angle on the heat transfer coefficient is observed at certain operating conditions. For these applications refer for example to the recent study of Wang and Du [19].

CONCLUSIONS

Dimensionless heat transfer coefficients have been determined in terms of the average Nusselt number for reflux condensation of R134a in a small diameter inclined tube (d = 7 mm). It has been found that the inclination angle has a significant effect on the reflux condensation heat transfer coefficient. When the tube is tilted from the vertical the condensate film thickness becomes non-uniform around the circumference of the tube, the mean film thickness decreases and consequently the heat transfer coefficient increases. The optimum inclination angle for the heat transfer was found to be close to 45°. At this inclination angle the heat transfer was increased by a factor of nearly 2 when compared with reflux condensation in a vertical tube. A semi-empirical modified Wang/Ma correlation was proposed to predict the heat transfer in dependence of the film Reynolds number and the inclination angle. This correlation agrees well with the experimental data.

ACKNOWLEDGEMENTS

It is gratefully acknowledged that financial support for this work has been provided in part by the EU (JOULE III-project, contract JOE-CT97-0062). Solvay Fluor und Derivate GmbH is thanked for supplying the test fluid.

NOMENCLATURE

a	Thermal diffusivity, m^2/s
b	Width, m
c_p	Specific heat at constant pressure, J/kgK
C	Dimensionless group
d	Tube diameter, m
Δh_v	Specific enthalpy of vaporization, J/kg
L	Tube length, m
$\dot{M}$	Mass flow rate, kg/s
$\overline{Nu}$	Average Nusselt number
Pr	Prandtl number
$\dot{Q}$	Total heat transfer, W
R	Tube radius, m
Re	Reynolds number
T	Temperature, K
ΔT	Temperature difference, K

Greek letters

$\overline{\alpha}$	Average heat transfer coefficient, W/m^{2}K
β	Inclination angle from the horizontal, °
ϱ	Density, kg/m^3
λ	Thermal conductivity, W/mK
ν	Kinematic viscosity, m^2/s
η	Dynamic viscosity, kg/ms

Subscripts and superscripts

$cond$	Condensate
CW	Cooling water
g	Gas, vapour
k	According to Eq. (7)
l	Liquid, condensate
n	According to the Nusselt film condensation theory, Eq. (6)
S	Saturation
W	Wall

REFERENCES

1. A. N. Abdelmessih, T. J. Rabas, C. B. Panchal: Reflux Condensation of Pure Vapors With and Without a Noncondensible Gas Inside Plain and Enhanced Tubes, *AIChE Symposium Series 314*, vol. 93, pp. 227-232, 1997.
2. S. Thumm, Ch. Philipp, U. Groß: Experimental Investigation of the Influence of Countercurrent Flow of the Phases on Condensation Heat Transfer in a Vertical Tube, *Proc. 3rd Europ. Thermal Sciences Conf.*, vol. 1, pp. 933-938, 2000.
3. S. L. Chen, F. M. Gerner, C. L. Tien: General Film Condensation Correlations, *Exp. Heat Transfer*, vol. 1, pp. 93-107, 1987.
4. ESDU Data Item No. 89038: Reflux Condensation in Vertical Tubes, Heat Transfer Sub-series, vol. 6, 1989.
5. W. Nusselt: Die Oberflächenkondensation des Wasserdampfes, *VDI-Zeitschr.*, vol. 60, no. 27, pp. 541-546 and no. 28, pp. 568-578, 1916.
6. S. S. Kutateladze: Fundamentals of Heat Transfer, Academic Press, New York 1963.
7. D. A. Labuntsov: Heat Transfer in Film Condensation of Pure Steam on Vertical Surfaces and Horizontal Tubes, *Teploenergetica*, vol. 4, no. 7, pp. 72-80, 1957.
8. U. Groß, E. Hahne: Experimentelle Untersuchung des Wärmeübergangs bei der Rückstrom-Kondensation in einem geneigten Rohr, *Chem.-Ing.-Tech.*, vol. 59, no. 2, pp. 168-169, 1987.
9. U. Groß: Reflux Condensation Heat Transfer Inside a Closed Thermosyphon, *Int. J. Heat Mass Transfer*, vol. 35, no. 2, pp. 279-294, 1992.
10. K.-E. Hassan, M. Jakob: Laminar Film Condensation of Pure Saturated Vapours on Inclined Circular Cylinders, ASME Paper No. 57-A-35, 1957.
11. J. C. Y. Wang, Y. Ma: Condensation Heat Transfer Inside Vertical and Inclined Thermosyphons, *J. Heat Transfer*, vol. 113, no. 3, pp. 777-780, 1991.
12. K. Onda, E. Sada, K. Takahashi, K. Ito: Film Condensation in a Vertical Tube (in Jap.), *Kagaku Kogaku*, vol. 32, no. 12, pp. 1215-1221, 1968.
13. C. Huanzhuo, M. Tongze, M. Groll: Performance Limitation of Micro Closed Two-Phase Thermosyphons, *Proc. 10th Int. Heat Pipe Conf.*, 1997.
14. M. Groll, S. Rösler: Operation Principles and Performance of Heat Pipes and Closed Two-Phase Thermosyphons, *J. Non-Equilib. Thermodyn.*, vol. 17, pp. 91-151, 1992.
15. M.-M. Chen: Heat Transfer Performance of Two-Phase Closed Thermosyphons With Different Lengths, *Proc. 6th Int. Heat Pipe Conf.*, vol. 2, pp. 647-653, 1987.

16. M. G. Semena, Y. F. Kiselev: Study on Heat Transfer in the Condensation Part of Two-Phase Thermosiphons, *Teploobmen Energ. Ustanovkakh*, pp. 68-74, 1978.
17. N. M. Stoyanov: Effect of the Angle of Inclination of a Closed Evaporative Thermosyphon on Heat Transfer, *Teploenergetika*, vol. 15, no. 3, pp. 74-76, 1968.
18. H. Uehara, H. Kusuda, T. Nakaoka, M. Yamada: Filmwise Condensation for Turbulent Flow on a Vertical Plate, *Heat Transfer - Jap. Res.*, vol. 12, no. 2, pp. 85-96, 1983.
19. B.-X. Wang, X.-Z. Du: Study on Laminar Film-Wise Condensation for Vapour Flow in an Inclined Small/Mini-Diameter Tube, *Int. J. Heat Mass Transfer*, vol. 43, pp. 1859-1868, 2000.

EXPERIMENTAL STUDY OF CONDENSATION ON TWO FLAT VERTICAL PARALLEL PLATES

S. Russeil, C. Tribes, B. Baudoin

Département Energétique Industrielle, École des Mines de Douai
941, rue Charles Bourseul - B.P. 838 - 59508 DOUAI Cedex
Serge.Russeil@ensm-douai.fr

ABSTRACT: In this paper experimental results for the convective heat transfer in a narrow two-dimensional vertical duct, for both dry and humid conditions, are presented. The walls of the channel are covered by aluminium foils such those used to manufacture industrial heat exchangers fins. Operating conditions are similar to usual air-conditioning conditions. The results show that the relative increase of the sensible heat transfer in wet conditions compared to dry mode depends on the inter-fin spacing and on the air flow rate.

INTRODUCTION

In the literature, a great number of studies deals with heat transfer performances of finned-tube heat exchangers encountered in air-conditioning systems or dehumidification process. The aim of most of those studies, (Jacobi and Goldschmidt, 1990), (Khalfi, 1998), (McQuiston, 1978), (Wang and Chang, 1998), is to determine the heat transfer coefficient on the airside for dry (without condensation) and wet (with condensation) modes.

Authors report either an increase or a decrease of the sensible Colburn j-factor j_{sens} obtained in wet mode compared to dry surfaces. The relative increase is often explained by the action of the condensate on the walls which acts as a surface roughness and increases the heat transfer coefficient. In the other case, the decrease is supposed to be due to the condensate retention in the inter-fin spacing. According to the shape of the fin, the value of the fin pitch, the arrangement of the tubes, or even the Reynolds number, one of the two trends will be preponderant.

The purpose of this study is to quantify the influence of the inter-fin spacing and of the air velocity on the heat transfer between humid air and surfaces, without being influenced by the phenomena related to the presence of the tubes. An experimental test bench has been built in order to study thermal performances of continuous aluminium fins in a parallel plate heat exchanger for various inter-fin spacings.

EXPERIMENTAL SET-UP AND PROCEDURE

Description of the test-bench

The convective heat and mass transfer of a developing airflow is studied in a flat rectangular vertical duct simulating a plate fins and tube heat exchanger. The test section is made of two support plates on which the two heat-transferring surfaces are pasted, see Fig. 1. The support plates are cooled by means of an internal flow (in a groove) of a water-glycol mixture at low temperature.

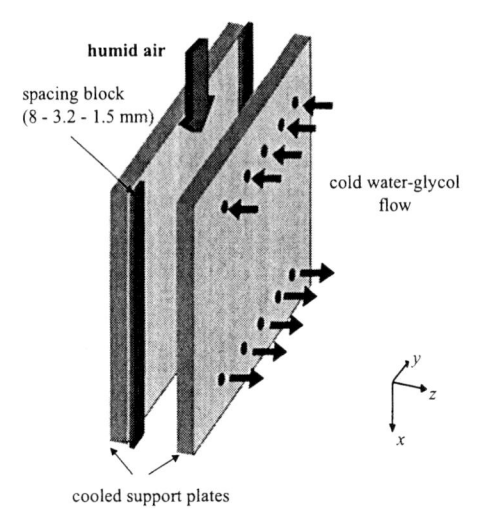

Fig. 1: Condensing test section

The test section has the following dimensions:

- length, in the x-direction: $L = 370\ mm$
- width, in the y-direction: $l = 367\ mm$
- clearance of the channel (fin pitch), in the z-direction: $e = 8,\ 3.2$ and $1.5\ mm$

The values of the channel clearance, 8, 3.2 and 1.5 mm respectively correspond to fin densities of 1.2, 3 and 6.3 $fins/cm$. Due to these values of inter-fin spacing, the test section can be considered of infinite width ($e/l < 2.2\%$) and then the hydraulic diameter can be written: $Dh \cong 2e$.

In order to measure the condensate mass flow rate, the mass of water flowing downward along the fins is collected at the bottom of the test section in a small container and weighed periodically.

The aluminium foils pasted on the support plates have a thickness of 0.2 mm. These foils are strictly similar to those usually used to manufacture the fins of industrial heat exchangers. So the humid air condensation occurs on real industrial surfaces; and the same type of condensation as in real finned-tube heat exchangers is expected.

Instrumentation and operating conditions

In order to measure the surface temperature, fifteen thermocouples of small diameter ($\phi = 0.13\ mm$) are positioned underneath each aluminium foil. For the range of values of heat flux obtained during our various tests, the temperature gradient in the thickness of the aluminium foil is negligible and thus we can consider that the thermocouples measure indeed the fin surface temperature. For each test, the mean fin temperature is kept at 5 °C $\pm$ 0.25 °C.

For $e = 8\ mm$ and $e = 3.2\ mm$, the variation of the fin surface temperature in the flow direction, i.e. x-axis, has been shown to be limited to the entry of the test section. Therefore, the boundary condition is of "constant temperature" type.

For the configuration $e = 1.5\ mm$, the temperature gradient is more significant and the "constant temperature" boundary condition is not observed.

Owing to its uniformity over the cross section at the entry of the test section, the inlet air temperature is measured by a single thermocouple.

To determine the air temperature at the outlet of the test section, an adiabatic and airtight convergent prolongs the test section and homogenizes the air flow. Thus, the temperature measurement made at the exit of the convergent, gives us the air temperature at the exit of the test section.

Air humidities are measured at the entry and the exit of the condensing test section by means of a dew-point hygrometer. The sampling lines are in stainless steel in order to limit the water vapour adsorption and desorption phenomena. Condensation in the sampling lines is avoided by an appropriate heating device. By means of this system, investigations with water contents up to 25 g_w/kg_a have been performed with a good accuracy.

The air mass flow rate is deduced from a differential pressure measurement made on both sides of the orifice plate of a flow meter located upstream the test section.

The operating conditions of our investigations, similar to those usually observed in air-conditioning coils and in dehumidification process, are as follows:

- Humidity ratios, w: 5, 10, 15, 20, 25 g_w/kg_a
- Reynolds numbers, Re_{Dh}: from 700 to 4000, corresponding to mean inlet velocities from 1 m/s to 10 m/s (depending on the fin pitch)
- Air temperature at the inlet, T^{in}: 35 °C
- Mean fin temperature, $\overline{T_{wall}}$: 5 °C

Heat flux calculation

An enthalpy balance on the humid air between the entrance and the exit of the test section allows the calculation of the total heat flux between the humid air and the cold surfaces.
The sensible heat power can be written:

$$\dot{Q}_{sens} = \dot{M}_a\, Cp^{out}\, (T^{in} - T^{out}) \quad (1)$$

where $Cp = Cp_a + w\, Cp_v$.
The sensible heat evacuated by the condensate has been shown to be negligible for our conditions. The latent heat rate can be obtained by:

$$\dot{Q}_{lat} = \dot{M}_a\, L_v(T^{in})\, (w^{in} - w^{out}) \quad (2)$$
$$\text{or} \quad \dot{Q}_{lat} = \dot{M}_{cond}\, L_v(T^{in}) \quad (3)$$

The sensible heat transfer coefficient can be calculated from the following expression with the Logarithmic Mean Temperature Difference (ΔT_{LM}):

$$h_{sens} = \frac{\dot{Q}_{sens}}{A\, \Delta T_{LM}} \quad \text{where} \quad (4)$$

$$\Delta T_{LM} = \frac{\left(T^{in} - T^{out}\right)}{\ln\left(\frac{T^{in} - \overline{T_{wall}}}{T^{out} - \overline{T_{wall}}}\right)}$$

In the same way, the latent heat transfer coefficient comes from:

$$h_{lat} = \frac{\dot{Q}_{lat}}{A\, L_v\, \Delta w_{LM}} \quad \text{where} \quad (5)$$

$$\Delta w_{LM} = \frac{\left(w^{in} - w^{out}\right)}{\ln\left(\frac{w^{in} - w^{sat}(\overline{T_{wall}})}{w^{out} - w^{sat}(\overline{T_{wall}})}\right)}$$

EXPERIMENTAL RESULTS

Validation of heat flux measurements

For all the investigations carried out, uncertainty calculations have been performed in order to quantify the errors resulting from the sensors accuracy. For the three spacings considered in this study, the relative uncertainty of the total heat power is at most 7%.

Another validity criterion of the overall experimental facility and procedure, comes from the comparison between the condensate mass flow rate measured by weighing (direct method) and the condensate mass flow rate calculated from the enthalpy balance (indirect method). The comparisons, not presented in this paper (see (Tribes, 1998) for more details), show that for the three fin pitches, the two values of condensate mass flow rate are in very good agreement; the largest deviation observed remaining smaller than uncertainties of measurement.

Dry mode -without condensation-

First of all, we present our experimental results for a developing laminar air flow in dry mode (without condensation) in a straight channel with flat parallel walls of length L.

Comparisons are made between our results and a correlation established numerically by *Stephan* for this kind of problem (Stephan, 1959). For this configuration, the author proposes the following correlation for the average Nusselt number:

$$Nu_{Dh} = 7.55 + \frac{0.024(x^*)^{-1.14}}{1 + 0.0358(x^*)^{-0.64} Pr^{0.17}} \quad (6)$$

where $x^* = \dfrac{L}{Dh\, Re_{Dh}\, Pr}$.

Figure 2, which presents a comparison in term of the sensible heat transfer coefficient, shows that there is a good agreement between this correlation and our experimental data for inter-fin spacings $e = 8\ mm$ and $e = 3.2\ mm$ in dry mode.

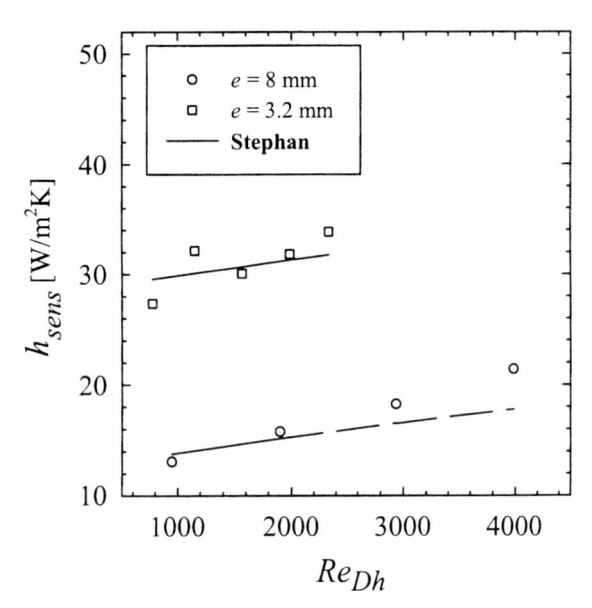

Fig. 2: Sensible heat transfer coefficient in dry mode

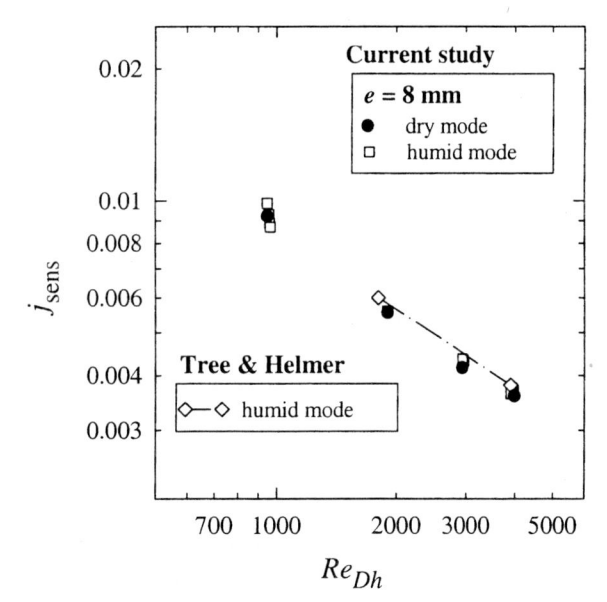

Fig. 3: Sensible Colburn j-factor, $e = 8\ mm$

Humid mode -with condensation-

When studying a developing air flow in a duct, for both dry and humid modes, the dynamic and thermal inlet lengths (which are defined as the location of the cross-section downstream the channel inlet where the two boundary layers join each other and form a fully developped flow) are of great importance.

Among all the works found in the literature, only one configuration (Tree and Helmer, 1976) has inlet lengths similar to this study. Hence, we have confronted our data to the experimental results of *Tree and Helmer* in term of the sensible Colburn j-factor in humid mode: $j_{sens}^{cond} = (h_{sens}^{cond}\ Pr^{2/3})/(\rho\ u_m\ Cp)$, where h_{sens}^{cond} is calculated from Eq. 4.

For $e = 8\ mm$ the comparison leads to a perfect agreement between our results and those of *Tree and Helmer*, see Fig. 3.

For $e = 3.2\ mm$, see Fig. 4, we can notice that our values of j_{sens}^{cond} are slightly lower than those of *Tree and Helmer*.

These results show, as mentioned above, the importance of the inlet lengths when comparing experimental data. And hence the two previous observations can be explained by the fact that considering the ratio *inlet length/channel length* our configuration is closer to that of *Tree and Helmer* for $e = 8\ mm$ than for $e = 3.2\ mm$.

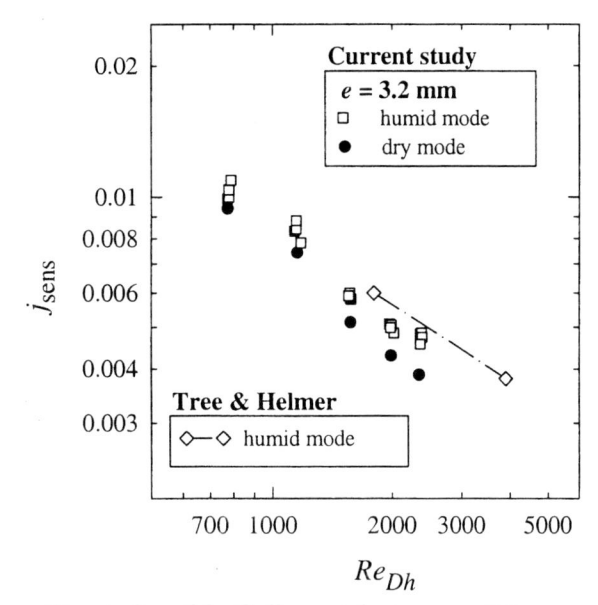

Fig. 4: Sensible Colburn j-factor, $e = 3.2\ mm$

Furthermore, depending on the inter-fin spacing, our experimental results indicate a slight influence of the condensation phenomena on the values of the sensible Colburn j-factor j_{sens}.

For $e = 3.2\ mm$, the values of sensible Colburn j-factor in humid mode, j_{sens}^{cond}, are indeed greater than those corresponding to the dry mode, j_{sens}^{dry}.

For the inter-fin spacing $e = 8\ mm$, no significant influence is noticed, the two values of the sensible Colburn j-factor are nearly the same for both dry and humid mode.

Examining more precisely our experimental results, the rise of j_{sens} in the condensation mode compared to the dry mode, appears to be proportional to Re_{Dh}.

For example, for a given inlet humidity ratio of 10 g_w/kg_a, while the increase of j_{sens} is 5% for $Re_{Dh} = 770$, it rises up to 24% for a Reynolds number of 2300.

On the contrary, no significant influence of the air humidity on the increase of the sensible heat transfer coefficient has been observed for our operating conditions.

Figure 5 presents the latent Colburn j-factor, $j_{lat} = (h_{lat} Sc^{2/3})/(\rho u_m)$, with h_{lat} being calculated from Eq. 5, versus the Reynolds number for various humidity ratios for $e = 3.2$ mm. The latent Colburn j-factor, j_{lat}, has been shown to be independent of the water content of the air at the entry of the test section for the three inter-fin spacings.

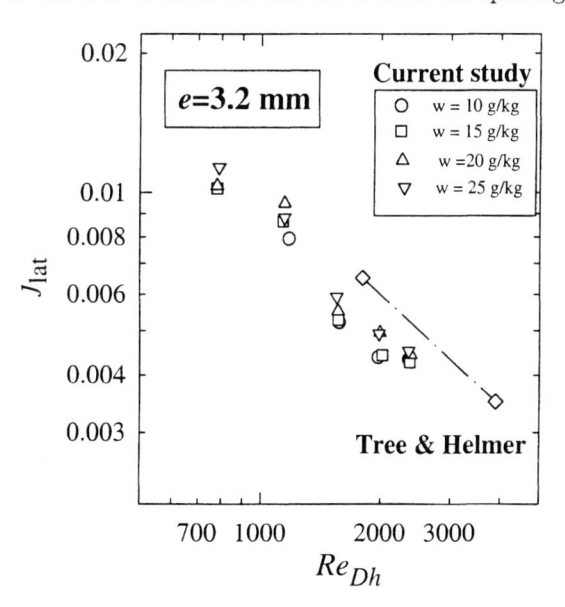

Fig. 5: Latent Colburn j-factor

As written above, for the inter-fin spacing $e = 1.5$ mm, the wall temperature is not constant and the boundary condition is not of "constant temperature" type and thus Eq. 1 which gives the sensible heat transfer coefficient cannot always be used. Indeed, for this fin pitch, the air temperature at the exit can be lower than the mean temperature of the wall ($\overline{T_{wall}}$) and the Logarithmic Mean Temperature Difference is then undefined, and j_{sens} can not be calculated.

Nevertheless, in order to highlight the influence of the condensation phenomena on the sensible heat transfer for the three fin pitchs, we decide to present our experimental data in the form of heat powers transfered between the humid air and the wet fins.

Thus, for the three inter-fin spacings, ratios of "heat rate exchanged in humid mode" on "heat rate exchanged in dry mode", that is: $\dot{Q}_{sens}^{humid} / \dot{Q}_{sens}^{dry}$, are presented versus the Reynolds number. And, in order to show only the influence of the fin pitch, see Figure 6, averaging of the values obtained for different humidity ratios have been made.

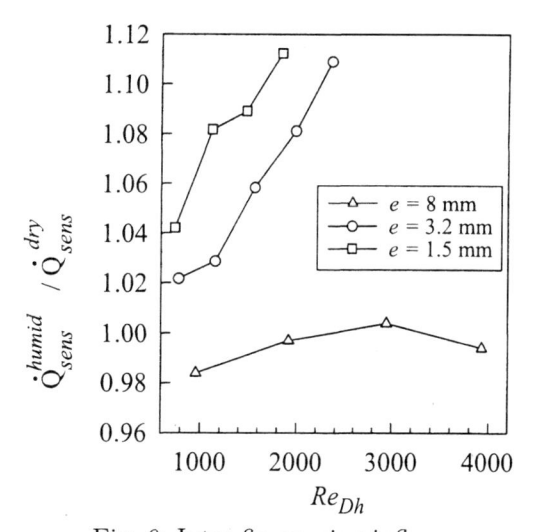

Fig. 6: Inter-fin spacing influence

The results show clearly that the relative increase of the sensible heat transfer in humid mode compared to the dry condition, is obtained only for the two lowest inter-fin spacings. We can also notice that higher the Reynolds number is, higher this increase is, and that the most significant rise is observed for the lowest fin pitch.

CONCLUSION

In this paper, experimental data for a developing humid air flow in a narrow vertical channel (model of a parallel plate heat exchanger) are presented. The walls are covered by aluminium fins such those used to manufacture industrial heat exchangers. Investigations have been carried out for three plate spacings $e = 8$, 3.2 and 1.5 mm. Thus the type of condensation and the heat and mass transfer studied in this work are similar to those occuring in real industrial plate fins and tube heat exchangers.

These results highlight how the fin pitch and the Reynolds number influence the sensible heat transfer coefficient in humid mode.

However, contrary to some results (Khalfi, 1998), (McQuiston, 1978) concerning finned-tube heat exchangers with inter-fin spacing of 1.5 or 2 mm, we never observed a relative decrease of the sensible heat transfer in humid mode compared to the dry mode. Generally, this decrease is explained by the flooding of the inter-fin spacing that obstructs the air flow. Thus, we can suppose that in the configuration here presented there is no flooding inside the channel but that, on the contrary, the condensate disturbs the flow and leads then to ratios $\dot{Q}_{sens}^{cond}/\dot{Q}_{sens}^{dry} \geq 1$ due to a relative increase of the sensible heat transfer.

In order to confirm this assumption, an additional study on the drainage and the retention of the condensate has been carried out by the mean of visualisations (Tribes et al., 2000). This work has confirmed the absence of flooding and has shown that large droplets on the fins interact with the flow for low fin pitches ($e = 1.5$ mm and $e = 3.2$ mm).

The presence of the tubes in real finned-tube heat exchangers and the cross-flow configuration apparently strongly contribute to the flooding of the inter-fin spacing in regions with low wall shear stress. Further investigations are presently carried out to study the role of the tubes during the drainage and retention of the condensate process.

NOMENCLATURE

A	surface	$[m^2]$
Cp	specific heat	$[J/(kg.K)]$
Dh	hydraulic diameter	$[m]$
e	inter fin spacing	$[m]$
h	heat tranfer coefficient	$[W/(m^2.K)]$
j	Colburn j-factor	$[-]$
k	thermal conductivity	$[W/(m.K)]$
L_v	enthalpy of evaporation	$[J/kg_{water}]$
$\dot{M}$	mass flow rate	$[kg/s]$
Nu	Nusselt number	$[-]$
Pr	Prandtl number	$[-]$
$\dot{Q}$	heat rate	$[W]$
Re	Reynolds number	$[-]$
ρ	density	$[kg/m^3]$
Sc	Schmidt number	$[-]$
T	temperature	$[^\circ C]$
$\overline{T_{wall}}$	mean wall temperature	$[^\circ C]$
u_m	mean velocity	$[m/s]$
w	humidity ratio	$[kg_w/kg_a]$

a	dry air
$cond$	condensation, condensate
in	inlet
lat	latent
out	outlet
sat	saturation
$sens$	sensible
v	water vapour
w	water

REFERENCES

Jacobi, A. M. and Goldschmidt, V. W., 1990, Low Reynolds number heat and mass transfer measurements of an overall counterflow, baffled, finned-tube, condensing heat exchanger, *Int. J. Heat and Mass Transfer*, 33(4), 755–765

Khalfi, M. S., 1998, Étude de l'influence de l'humidité de l'air sur le coefficient de transfert de chaleur d'une batterie froide en présence ou non de condensation, *Ph.D. thesis*, Université Henri Poincaré Nancy-1

McQuiston, F., 1978, Correlation of heat, mass and momentum transport coefficient for plate-fin-tube heat transfer surfaces with staggered tubes, *ASHRAE Transactions*, 84(1), 294–309

Stephan, K., 1959, Wärmeübergang und Druckabfall bei nicht ausgebildeter Laminarströmung in Röhren und in ebenen Spalten, *Chem. Eng. Tech.*, 31, 773–778

Tree, D. and Helmer, W., 1976, Experimental heat and mass transfer data for condensing flow in a parallel plate heat exchanger, *ASHRAE Transactions*, 82(2402), 289–299

Tribes, C., 1998, Étude du transfert de chaleur et de masse lors de la condensation de vapeur d'eau en présence d'incondensable, *Ph.D. thesis*, Université de Valenciennes

Tribes, C., Russeil, S., and Baudoin, B., 2000, Retention and draining of condensate on heat-exchangers surfaces, *3rd European Thermal Sciences Conference, Heidelberg*, 2, 893–898

Wang, C.-C. and Chang, C.-T., 1998, Heat and mass transfer for plate fin-and-tube heat exchangers, with and without hydrophilic coating, *Int. J. Heat Mass Transfer*, 41, 3109–3120

CIRCUMFERENTIAL TEMPERATURE DISTRIBUTIONS ON PLAIN AND FINNED TUBES IN POOL BOILING

Peter Hübner[a], Dieter Gorenflo[b] and Andrea Luke[b]

[a] ISE Fraunhofer Institute for Solar Energy Systems, D-79100 Freiburg, Germany
[b] Wärme- und Kältetechnik, Universität Paderborn, D-33098 Paderborn, Germany
e-mail: digo@wkt.uni-paderborn.de

ABSTRACT

The local wall superheats ΔT along the circumference of a finned tube with large diameter (90 mm) have been measured with a great number of miniaturised thermocouples and have been compared with the superheats of a plain tube of the same wall material (mild steel) and approximately the same diameter and surface roughness as the finned, in order to separate convective from evaporative contributions to heat transfer in nucleate boiling. The comparison demonstrates the significant influence of convective effects caused by the sliding bubbles streaming upwards along the tube surfaces.

At intermediate heat fluxes – between beginning and fully developed nucleate boiling – and at intermediate reduced saturation pressures p*, i.e. within approx. 5 to 20% of the critical pressure p_c, heat transfer is enhanced most near the bottom parts of the finned tube and less towards the top. The effect is smaller at high heat fluxes or reduced pressures, respectively, thus demonstrating that the *relative effect* of convection on the enhancement is highest when a comparatively small number of big bubbles is generated.

Neglection of the temperature drop within the fins – due to the comparatively poor thermal conductivity of mild steel – is of little importance at low heat fluxes, but it influences the superheat of the tube significantly at higher, and may lead to higher superheats at the bases of the fins than on the surface of the plain tube at very high reduced pressures, and if the two kinds of tubes are compared for the same heat fluxes on their *total outer* surfaces.

INTRODUCTION

For the prediction of pool boiling heat transfer on high performance evaporator tubes to be based on the processes connected with production and release of vapour in the immediate vicinity of the wetted tube surface, it is essential to separate the contributions of evaporation and convection, and to understand the local interferences of both effects near the bubbles growing, detaching and sliding along the tube surface.

As the features of the macro geometry (dimensions ≥ 0.1 mm) of commercially available enhanced evaporator tubes are highly complex and differ very much between each other, the development of *generalized* prediction methods for pool boiling heat transfer on tubes of this kind has not proceeded very far up to date, except with severe simplifications (cf eg Thome, 1990; Pais and Webb, 1991; Chien and Webb, 1996; Bergles, 1997; Smirnov, 2000; Kuzma-Kichta et al., 2000).

In order to identify the local evaporative or convective contributions to heat transfer enhancement on high performance tubes, it is important to investigate local heat transfer, surface conditions, and bubble formation and motion *on plain and enhanced* tubes equally thoroughly by the *same* equipment and procedure.

The combination of high-speed liquid crystal thermography with high-speed video recording of local heat transfer and bubble formation and motion on surfaces without modifications for enhancement by DBR Kenning and his group at Oxford in recent years has contributed very much to our knowledge of the fundamental local events (cf eg Kenning et al., 2000; McSharry et al., 2000).

Our own work on local effects started with horizontal *plain* tubes with big diameters (≈ 88 mm, for getting pronounced convection around the circumference), which contain a great number of miniaturized thermocouples in the wall to monitor temperature variations from the bottom to the top of the tube (cf eg Gorenflo, 1993; Buschmeier et al., 1993; Sokol, 1994; Gorenflo et al, 1998).

In a second stage, experiments were extended to a finned tube with simple shape of the fins, made of the same material as the plain, with the same surface treatment, measuring equipment and almost the same outer diameter (≈ 90 mm; Hübner, 2000), some results of which are presented in this paper. The investigations presently continue with measurements on plain tubes with defined roughness structure and additional grooves for heat transfer enhancement (Gorenflo et al., 2000).

The following discussion focusses on comparatively low reduced pressures $p^* = p_s/p_c$ (p_c = Critical Pressure), because this domain with low heat transfer coefficients for organics boiling on plain tubes is particularly attractive for the use of enhanced tubes. Where necessary for better explanation, also results at high reduced pressures are compared, although enhanced tubes are not as favourable as at low, because pool boiling heat transfer is already very good on the plain tubes under these circumstances.

APPARATUS AND TEST TUBES

The measurements were performed in a so-called standard apparatus for pool boiling heat transfer with the following main features:

- natural circulation loop of the test fluid between evaporator and condenser, both mounted in a chamber the air temperature of which is carefully adjusted to the saturation temperature T_s of the test liquid in the evaporator,
- dc-heated, horizontal test tube,
- direct measurement of the wall superheat by miniaturized thermocouples with their individual reference junctions in the pool of liquid, approx. 40 mm below the test tube.

Further informations about the apparatus and test procedure, with detailed discussion of the experimental limits of error are given in Sokol (1994), Hübner (2000) and are essentially the same as for the earlier versions (Gorenflo et al., 1982).

The main features of the test tube design are (a) three coaxial tubes (the two outer: mild steel St 37.8; the inner: copper), (b) miniaturized thermocouples in longitudinal grooves of the middle tube, (c) longitudinal resistance heaters in the inner tube, and (d) the whole assembly being soft soldered together over the entire length in a glove box with reducing atmosphere.

The shape of the fins is given in *Figure 1*: h = 2.04 mm, t = 2.00 mm, d_a = 90.04 mm, d_K = 85.96 mm, d_i = 81.6 mm. 20 thermocouples with 0.5 mm O.D. are placed in 0.6 mm deep grooves in the middle tube with $d_{a,m}$ = 81.4 mm. The augmentation of the outer finned surface area A over a plain tube with diameter d_K is A/A_K = 2.91. (The corresponding data for the plain tube are : d_a = 88.4 mm, d_i = 84.2 mm, 17 thermocouples with 0.5 mm O.D. in 0.55 mm deep grooves in the middle tube with $d_{a,m}$ = 84.0 mm.)

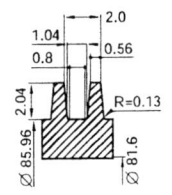

Figure 1: Finned tube, shape of the fins

Table 1: Roughness parameters acc. to DIN EN ISO 4287(10.98)

Tubes		Standardized roughness parameters (µm)					
		P_a	P_q	P_p	P_{pm}	P_t	P_z
Plain tube emery ground [a] D = 88.4 mm	average	0.18	0.23	0.64	0.46	1.61	1.13
	max	0.25	0.35	1.92	0.77	3.29	1.55
	min	0.14	0.18	0.42	0.33	1.13	0.91
	σ	0.03	0.04	0.23	0.08	0.43	0.16
Finned tube top of the fins [b] fine sandblasted D = 90.0 mm	average	0.19	0.25	0.67	0.45	1.49	1.01
	max	0.40	0.51	1.30	0.78	2.46	1.61
	min	0.12	0.16	0.46	0.28	0.86	0.59
	σ	0.08	0.10	0.24	0.13	0.50	0.27

number of runs: [a] 80, gauge length l_m = 0.5 mm ; [b] 11, l_m = 0.2 mm.

The roughness of the test tube surfaces is characterized by the standardized roughness parameters given in *Table 1* according to the new international standard DIN EN ISO 4287 (10.98) and by typical roughness profiles in *Figure 2*. The plain tube used in the first stage of our investigations had been emeried applying a mechanical device and exactly defined procedure (see eg Luke, 2000). The result was a very uniform surface roughness with P_a = 0.18 µm and small standard deviations σ, particularly for the integral parameters P_a and P_q, see *Table 1*. For scanning the surface, a new stylus instrument without mechanical contact of stylus and probe was used, with the advantage that tip radius and cone angle of the stylus can be smaller than before (r ≈ 1 µm, ß = 60°).

For the determination of the enhancement effect, it is obvious that the optimum conditions for comparison should be identical surface structures of the plain and finned tubes. Unfortunately it would have been very difficult to produce homogeneous roughness on the entire outer surface of the finned tube by grinding. So the tube was sandblasted with

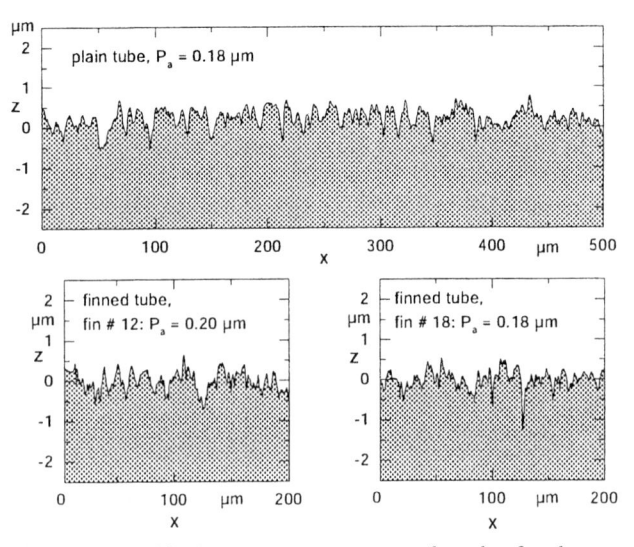

Figure 2: Profile departure z over gauge length x for the plain tube (top) and for the tops of two fins (bottom). Enlargement z/x=36.

fine corundum grain (with particle sizes between 20 and 30 µm) using a mechanical facility with motorized shifting of the grain injector along the tube surface, and – as in the case of grinding – with exactly defined working procedure (see Luke, 2000).

As intended, almost the same surface roughness as on the plain tube has been achieved, with an average value of $P_a = 0.19$ µm on top of the fins, the standard deviation σ, however, being somewhat greater (*cf Table 1*). To a certain extent, this may be due to the fact that the gauge length for scanning the tops of the fins with the stylus in axial direction could not be as long as for the plain tube (see *Figures 1 & 2*). The flanks and bases of the fins could not be scanned, but it can be assumed from a comparison of visual observations of the flanks and bases with the tops of the fins, that the values of the integral roughness parameters will be similar, taking also into account the high injection pressure (1.5 bar) and the small particle sizes of the grains with narrow size distributions (cf Luke, 2000).

CIRCUMFERENTIAL TEMPERATURE DISTRIBUTION AND DISCUSSION

The variation of the local wall superheats ΔT at the bases of the fins with the azimuthal angle φ and heat flux q is represented in *Figure 3* for n-Hexane boiling on the finned tube at 10 % or 3.7% of the critical pressure (ΔT in logarithmic scales). As can be seen, the superheat of the tube wall may vary significantly with φ. At $p/p_c = 0.10$, the maximum *absolute* increase of ΔT from bottom to top of approx. 2 K occurs near $q = 10$ kW/m², and the maximum *relative* increase of slightly more than 50% between approx. 1 and 10 kW/m² (left hand side of *Figure 3*).

Obviously, heat transfer on the lower parts of the tube surface is enhanced against the upper parts at intermediate heat fluxes – which correspond to intermediate densities of active nucleation sites – by the bubbles growing and sliding upwards along the tube surface, and improving heat transfer by convection and additional evaporation. At high heat fluxes, the sliding bubbles loose importance and the differences in ΔT between bottom and top of the tube tend to vanish, because the tube surface is more or less entirely covered by bubbles growing at active sites.

At low heat fluxes, the number of bubbles growing on the lowest parts of the tube is not sufficient to maintain the heat transfer improvement, and only near the flanks (90° or 270°) where all the bubbles produced on the lower half of the tube are sliding along the tube surface, heat transfer is improved by their combined effect, see the three lower curves at $p/p_c = 0.10$ in *Figure 3*.

For the smaller reduced pressure, the same tendency occurs already at approx. five times higher heat flux (1.7 instead of 0.34 kW/m²), due to the smaller number of large bubbles existing at lower reduced pressures. As can be concluded from a comparison with *Figure 4*, the change in the $\Delta T, \varphi$-curves starts at roughly the same distance from the onset of nucleation, i.e. the line for free convection without bubble formation in *Figure 4*. Without bubbles, the $\Delta T, \varphi$-variation from the mean superheat is smaller than ±0.03 K and has disappeared within the experimental scatter, cf the lowest line or smallest experimental heat flux at $p/p_c = 0.037$ in *Figures 3 & 4*.

The interpolating curves in *Figures 3* (*and 7, 8*) correspond to functions of the general structure

$$\Delta T(\varphi) = A + B \cdot \cos \varphi + C \cdot \cos(2\varphi)$$

which was developed by Kaupmann (2000) for plain tubes. This kind of function was used for all of the pressures and heat fluxes investigated, after fitting the parameters A-C to the experimental data.

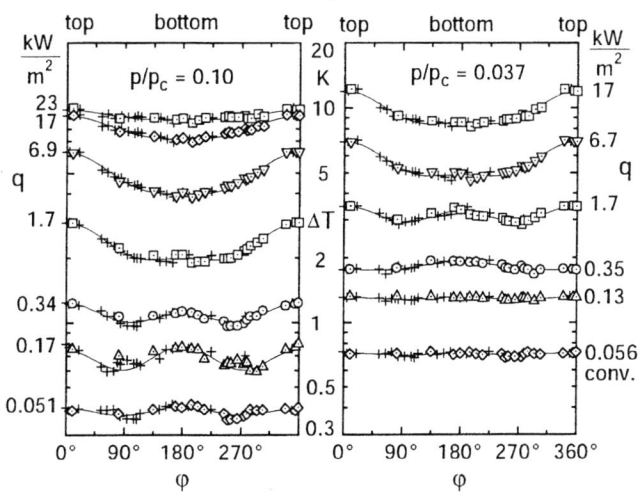

Figure 3: Variation of the local wall superheat ΔT with the azimutal angle φ and heat flux q for n-Hexane boiling on a finned tube at $p/p_c = 0.10$ or 0.037.

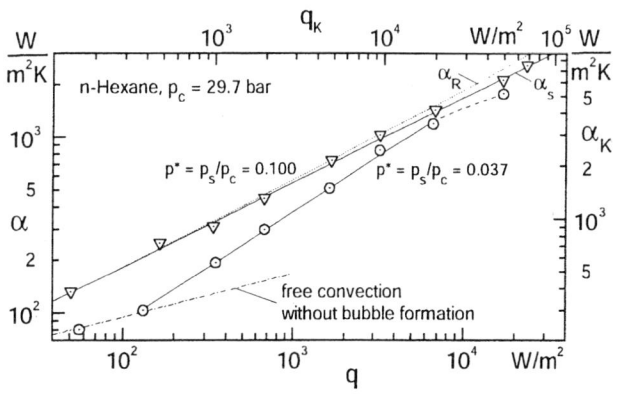

Figure 4: Heat transfer coefficient as function of heat flux for the measurements of Figure 3.
α, q related to the total outer surface of the finned tube; α_K, q_K related to the surface of the tube without fins.

Various trends in the average heat transfer results of *Figure 4* and in the local data of *Figure 3* can be interpreted by connecting the heat transfer data with visual observations of bubble formation, but the detailed interactions between bubble formation, convection and heat transfer are highly complex, as is demonstrated by selected photos. In *Figure 5*, sections of slightly more than the lower half of the 90 mm tube (each with 40 fins) are shown for two heat fluxes at the two pressures of *Figure 3*.

On the one hand, the differences in the $\Delta T, \varphi$-relationships of *Figure 3* for $q = 1.7$ kW/m² at the two pressures are

Figure 6: Enlarged detail of bubble formation near the height of the tube axis ($\varphi = 90°$ or $270°$) at $p/p_c = 0.037$, $q = 6.7$ kW/m² (as in Figure 5, #c).

plausible, because the many small bubbles produced near the bottom of the tube and sliding upwards between or on the fins at the higher pressure (b) will enhance local heat transfer on the lower parts of the tube against the situation at the lower pressure (a) with a much smaller number of bubbles growing at the bottom (cf the 'empty' gaps between the fins). As most of the bubbles are considerably larger, however, the vigorous motion induced by their high velocities near the flanks and upper parts of the tube reduces the differences in local heat transfer that exist at the bottom of the tube between the two pressures.

On the other hand, the two upper photos show that the almost identical $\Delta T, \varphi$-relationships for $q = 6.7$ or 6.9 kW/m² in *Figure 3* are obviously caused by distinctly different heat transfer mechanisms: At the higher pressure it will mostly be evaporation at many active nucleation sites and into slowly rising tiny bubbles (d), while at the lower pressure almost the same average (*Figure 4*) and local (*Figure 3*) amount of heat is transferred by convection and evaporation into a small number of big bubbles starting from the lowest parts of the tube (c). And at low reduced pressures – particularly interesting for enhanced surfaces – the situation is further complicated by tiny and big bubbles existing simultaneously (at least on tubes with large diameter), and by events as the burst of big bubbles on the upper parts of the tube, as has been caught on the detail of *Figure 6* showing the moment immediately after such a burst (upper corner on the right).

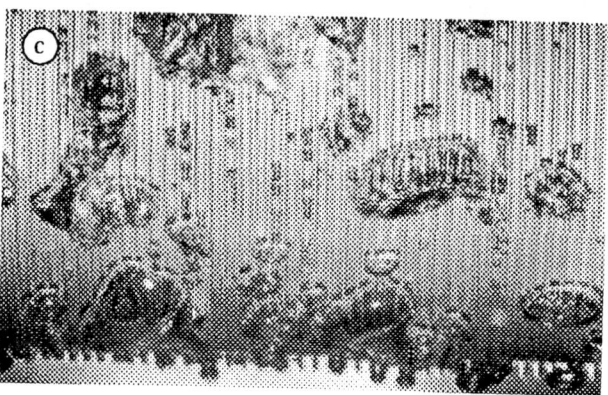

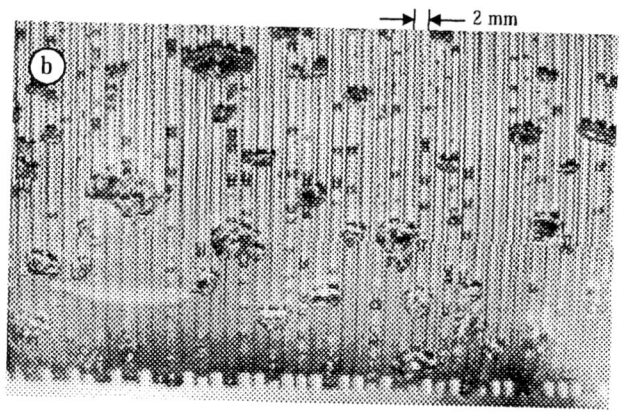

Figure 5: Bubble formation on 40 fins from bottom to slightly above tube axis for 1.7 (a,b) or 6.7 (c,d) kW/m² and $p/p_c = 0.037$ (a,c) or 0.10 (b,d).

Thus the problem arises, in which way a "characteristic" parameter for the bubble size should be defined to enter correlations predicting *average* heat transfer, but being based on *local events*. A conclusion to be drawn beforehand is that using 'average' values for the bubble size would not meet the real conditions and could only be misleading.

Enhancement of local heat transfer of the finned tube against a plain tube with (almost) the same diameter and surface roughness as the finned will be discussed in the following, using different conditions for the comparison:

In *Figure 7*, the results of the local wall superheats ΔT are compared under the condition of the same heat flux q on the total outer surface areas of the two kinds of tubes, and in *Figure 8*, the condition of the same heat flux q_K related to the surface area A_K of a plain tube with diameter $d_K = 86$ mm at the bottom of the fins (cf *Figure 1*) is used for the finned tube, i.e. the results of the *plain* tube for 20 kW/m² are identical in both figures, while the ΔT-values of the *finned* tube in *Figure 8* belong to heat fluxes q (on the outer surface) which are reduced by the factor of the surface ratio $A/A_K = 2.91$ of the finned tube. In other words, the curves for the finned tube in *Figure 8* belong to heat fluxes q only slightly higher than according to the curves for 5 kW/m² in *Figure 7*.

Results of Propane for higher reduced pressures have been added to the results of n-Hexane from *Figure 3* for the lowest reduced pressure, in order to have a similar broad *relative* variation for p/p_c as for q in the comparison. The data points represent experimental superheats ΔT of the plain tube from the earlier investigation (Sokol, 1994), and the solid lines have been interpolated for the finned tube from representations as *Figure 3* at the same heat fluxes that had existed on the plain in the experiments. (The data of q = 20 kW/m² for n-Hexane at $p/p_c = 0.037$ had to be slightly extrapolated for this purpose, cf *Figure 4*).

Starting the discussion at the intermediate reduced pressure $p/p_c = 0.10$ (*Figure 7*, lower diagram on the left), the superheat ΔT of the finned tube is diminished below the plain particularly at the bottom of the tube and with the lowest heat flux. This heat transfer enhancement is caused by nucleation sites getting activated mainly by the superheat near the bases of the fins being somewhat higher than on the rest of the fin surface, and by the bubbles sliding upwards *between* the fins. With increasing heat flux and pressure, nucleation increases and the gaps between the fins are filling

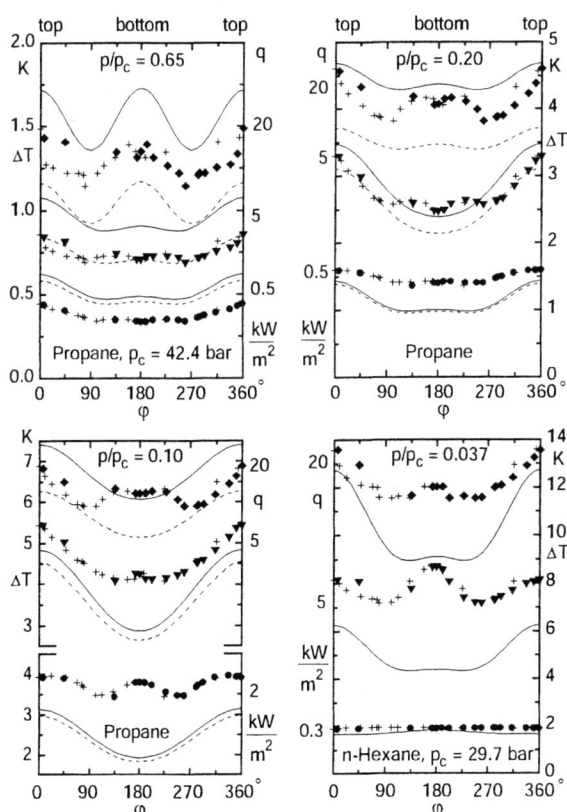

Figure 7: Local wall superheat ΔT for plain and finned tube at different heat fluxes and pressures.
Comparison for $q_{plain} = q_{finned}$.

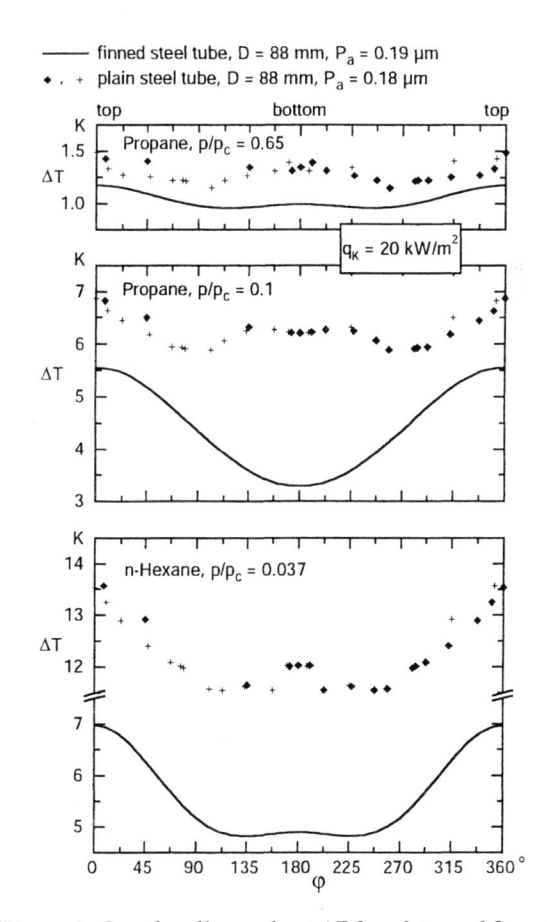

Figure 8: Local wall superheat ΔT for plain and finned tube at different pressures.
Comparison for $q_{plain} = q_{K,finned} = 20$ kW/m².

up with vapour, so this effect looses importance – and the diminishing buoyancy forces at higher pressures act in the same direction. At the lowest reduced pressure, the same *relative* heat transfer enhancement occurs at higher heat fluxes (cf the symbols and solid curves for 5 or 20 kW/m^2 with the corresponding for 2 and 5 kW/m^2 at p/p$_c$ = 0.10), because less and bigger bubbles are generated, and therefore a smaller part of the vapour-liquid interface is exposed to the superheated liquid layer near the finned wall.

Different to the lower pressures where the fins enhance heat transfer most at small heat fluxes, they *diminish* heat transfer at the highest (upper diagram of *Figure 7*, on the left, cf symbols and solid lines). At the lowest heat flux of 0.5 kW/m^2, this is mainly due to the superheated liquid boundary layer between the fins being thicker than with the plain tube. The additional temperature drop by heat conduction within the fins – which has been accounted for by 2D finite element calculation – is of little importance at this small heat flux, as can be seen from the small distance of the dashed curve for the superheat of the *outer tube surface* from the solid curve for the superheat of the wall *at the fin bases*.

With increasing heat flux, two contrary effects are competing with each other: The temperature drop within the fins is augmented (distances between dashed and solid lines increase), but simultaneously convection between the fins is enhanced against the plain tube, thus the superheat of the finned tube *surface* is smaller than of the plain for the highest heat flux investigated (cf symbols and highest dashed line).

Looking at the advantages of the finned tube investigated here, over a plain tube *in practical application*, only the comparison of the *solid* lines for the finned tube with the symbols for the plain is important, because the *overall* superheat from the bases of the fins to the boiling liquid should be smaller than the superheat of the plain tube. Furthermore, the superheats necessary for transferring the same amount of heat *per tube length* by the two kinds of tubes should be compared, i.e. the comparison should be conducted under the condition of the heat flux q$_K$ at the bases of the fins being the same as on the surface of the plain tube, as has been done in *Figure 8*. Now we see that the driving superheat at low and intermediate reduced pressures p/p$_c$ is much smaller for the finned tube than for the plain (lower two diagrams), while at the highest pressure the differences tend to vanish – because heat transfer is excellent already on the *plain* tube, cf upper diagram of *Figure 8*, with $\alpha_{m,plain}$ = q/$\Delta T_{m,plain}$ = 20/1.3 ≈ 15.4 kW/m^{2}K.

CONSIDERATION OF THE TEMPERATURE DROP WITHIN THE FINS

The superheats ΔT in *Figure 3* and the ΔT-values according to the solid lines in *Figures 7 & 8* represent the temperature differences between the bases of the fins and the saturation temperature T$_s$ of the boiling liquid in the pool at the pertaining saturation pressure p$_s$. Thus, they contain the temperature drop within the fins by conduction through the wall material from the bases of the fins to the outer surface together with the superheat of the tube surface over T$_s$; the same holds for the heat transfer coefficients α_s in *Figure 4*.

In order to eliminate the temperature drop by conduction from ΔT and to get the heat transfer coefficients α_R representing only the effect of evaporation and convection on heat transfer from the outer surface of the finned tube to the boiling liquid, heat conduction within the fins has been calculated by various approximate calculation methods: The method of Th.E. Schmidt (1950; 1958) and Kern/Kraus (1972) for straight fins of rectangular shape – referred to by '1D-approximation' in *Figure 9* – the method of Gutwald (1987) for the real fin shape, *Figure 1*, with numerical integration and different boundary conditions – referred to by '1D-calculation' in *Figure 9* – and a 2D-calculation method based on finite element approximation with control volumes proposed by Baliga/Patankar (1983) and applied by Solodov (1996) and Figge (1999) for axial and radial conduction within the fins.

The results of *Figure 9* for the lower pressure of *Figure 3* show that the relative influence of heat conduction within the fins is comparatively small – due to the low heat

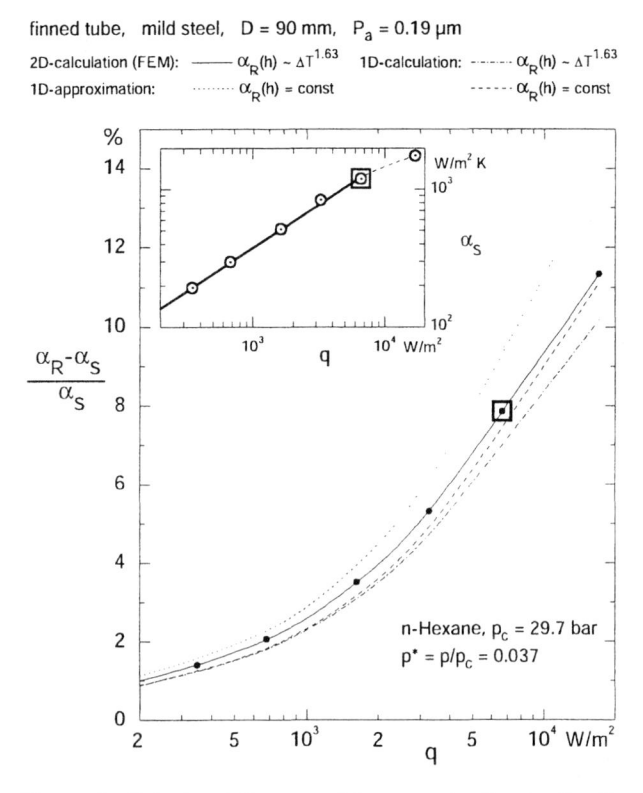

Figure 9: Relative influence of heat conduction in the fins on the overall heat transfer coefficient at low reduced pressure.
Comparison of various calculation methods.

transfer coefficients between 0.2 and 2 kW/m²K at this low reduced pressure – and amounts up to approximately 11% at the highest heat flux investigated (solid curve). By the 1D-approximation, the effect is somewhat overestimated (dotted) and by the 1D-calculation slightly underestimated (dot-dashed). The good agreement of the dashed curve with the 2D-calculation is obviously caused by different simplifications compensating each other.

The relative influence of conduction within the fins increases with increasing reduced pressure because of the improving outer heat transfer and amounts up to 60% for the highest heat flux investigated at $p/p_c = 0.65$ (Hübner, 2000). This is also reflected in the increasing *relative* differences between the ΔT-values according to the solid and dashed curves in *Figure 7* with rising pressure.

The isotherms within two fins plotted in *Figure 10* for the boiling conditions marked by squares in *Figure 9* demonstrate the axial part of heat conduction by the distances between the solid and dot-dashed lines increasing from bottom to top for the same superheats and by the curvatures of the solid lines. From the curvature of the lines *beneath* the fins, it can be concluded that the heat is flowing mainly into the fins (because of the comparatively poor heat transfer on the surface of the base tube between the fins). At the highest pressure and heat flux investigated, this has significantly changed (Hübner, 2000). The same holds for the variation of the local superheats ΔT_R, the heat transfer coefficients α_R and axial heat fluxes q_{ax} shown in *Figure 11* for the example of *Figure 10* (squares in *Figure 9*).

At the low reduced pressure $p/p_c = 0.037$, the relative decrease of ΔT along the fin height h of 13% (according to the 2D-calculation, solid line, diagram in the middle) leads to a stronger relative decrease of α_R (22%) and to more than the double (30%) in q_{ax}. (The corresponding values for the highest pressure and heat flux are 66.5% / 80% / 92.5%, Hübner 2000).

Having in mind that the superheat varies not only in radial and axial direction but also very distinctly in the circumferential – at least for higher heat fluxes – (*Figures 3,7,8*), it is obvious that the results of *Figures 9-11* would be modified by taking this into account via extension to a 3D heat conduction model. Calculations of this kind are currently running.

CONCLUSIONS

The development of calculation methods for pool boiling heat transfer on tubes with enhanced surfaces to be based on the local events of heat transfer, more detailed experimental information about local heat transfer and bubble formation is necessary because of the complex interactions of convective and evaporative effects in the immediate vicinity of the vapour–liquid boundaries near the heated wall.

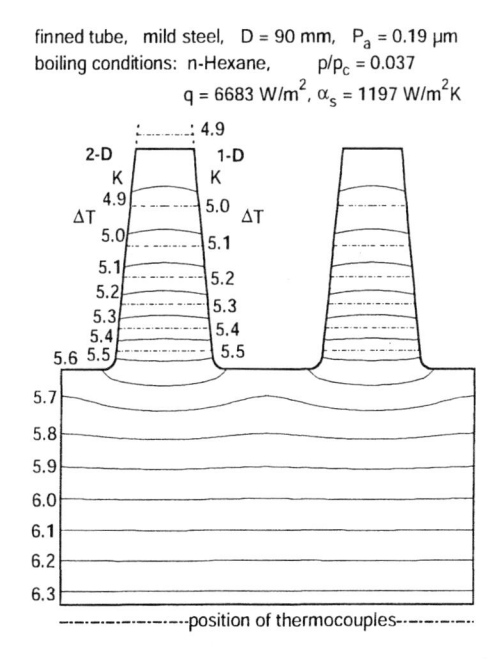

Figure 10: Typical example for isotherms within the finned tube: Comparison of 2D-FEM calculation (solid lines) with 1D-calculation (dot-dashed lines) for the heat transfer conditions indicated by squares in Figure 9.

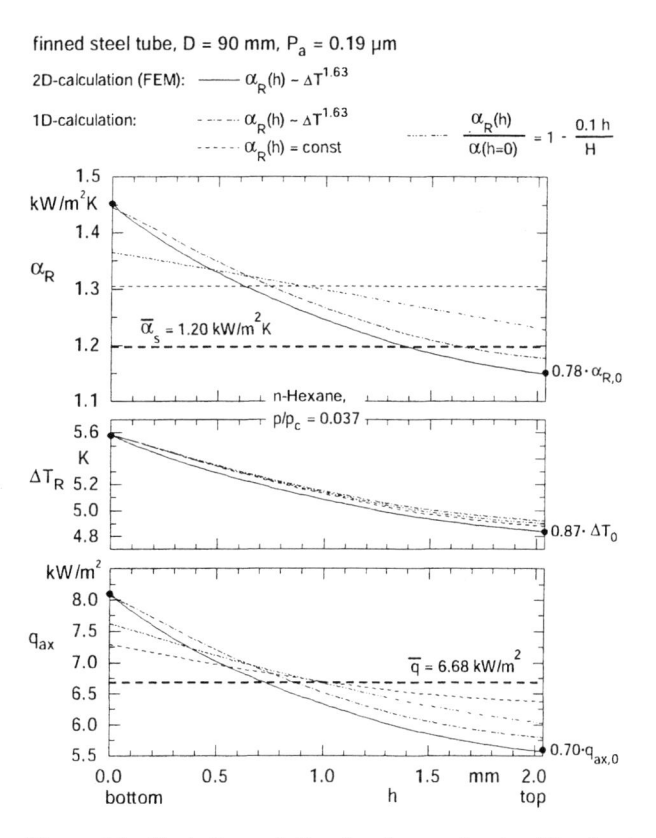

Figure 11: Variation of the local superheats ΔT_R, heat transfer coefficients α_R and axial heat fluxes q_{ax} with fin height h on the surface of the finned tube for the example of Figure 10.

ACKNOWLEDGEMENTS

The authors appreciate financial support of Deutsche Forschungsgemeinschaft and thank Dipl.-Phys.-Ing. Elmar Baumhögger and Dipl.-Ing. Untung Chandra for their assistance in the preparation of the Manuscript.

REFERENCES

Baliga, B. R. and Patankar, S. V., 1983, A Control Volume Finite-Element Method for Two-dimensional Fluid Flow and Heat Transfer, *Numerical Heat Transfer 6*, pp. 245-261.

Bergles, A.E., 1997, Enhancement of Pool Boiling, *Int. J. Refrig.*, Vol. 20 (8), pp.545-551.

Buschmeier, M., Luke, A., Sokol, P. and Gorenflo, D., 1993, Wärmeübergang beim Blasensieden von Propan/ n-Butan-Gemischen mit Queranströmung, *DKV-Tagungsberichte*, Bd. II/1, pp.341-353.

Chien L.-H. and Webb, R.L., 1996, A Parametric Study of Nucleate Boiling on Structured Surfaces, Part I: Effect of Tunnel Dimensions, Part II: Effect of Pore Diameter and Pore Pitch, *ASME HTD*, Vol. 326 (4), pp.129-143.

Figge, V., 1999, *Berechnung des zweidimensionalen Temperaturfeldes in Rippenrohren beim Blasensieden*, Studienarbeit, Energie-Institut Moskau und Universität Paderborn.

Gorenflo, D., Goetz, J. and Bier, K., 1982, Vorschlag für eine Standardapparatur zur Messung des Wärmeübergangs beim Blasensieden, *Wärme- und Stoffübertragung*, Vol. 16, pp.69-78.

Gorenflo, D., 1993, Pool Boiling Heat Transfer from Horizontal Tubes to Alternative Refrigerants; *Heat Transfer, Thermophysical Properties and Cycle Performance of Alternative Refrigerants*, T. Fujii ed., Jap. Assoc. Refrig., Tokio, pp.91-106.

Gorenflo D., Luke, A. and Danger, E., 1998, Interactions between Heat Transfer and Bubble Formation in Nucleate Boiling, *Proc. 11th Int. Heat Transfer Conf.*, Kyongju, Vol.1, pp. 149-174.

Gorenflo, D., Fust, W., Luke, A., Danger, E. and Chandra, U., 2000, Pool Boiling Heat Transfer from Tubes with Basic Surface Modifications for Enhancement, *Proc. 3rd European Thermal Sciences Conference 2000*, Heidelberg, Vol. 2, pp. 743-748.

Gutwald, Th., 1987, Untersuchungen zum Wärmeübergang beim Blasensieden an Glatt- und Rippenrohren, Diplomarbeit, Universität Paderborn.

Hübner, P., 2000, Zum Wärmeübergang beim Blasensieden an Rippenrohren, Diss., Universität Paderborn, Germany.

Kaupmann, P., 2000, Durchmessereinfluß und örtlicher Wärmeübergang beim Blasensieden an horizontalen Stahlrohren, Diss. Universität Paderborn, Germany.

Kenning, D. B. R., Wienecke, M., 2000, Investigation of Boiling Heat Transfer by Liquid Crystal Thermography, *Thermal Sciences 2000*, Proc. ASME-ZSITS Intern. Seminar, Bled, pp. 35-46.

Kern, D. Q. and Kraus, A. D., 1972, *Extended Surface Heat Transfer*, New York, McGraw-Hill.

Kuzma-Kichta Yu. A., Komendatov, A.S., Bakunin, V.G., Bartsch, G., Goldschmidt, R. and Stein, M., 2000, Enhancement of Heat Transfer at Boiling with Porous Coated Surfaces, *Proc. 3rd European Thermal Sciences Conference 2000*, Heidelberg, Vol. 2, pp. 809-814.

Luke, A., 2000, New Methods of Characterization for the Microstructure of Heated Surfaces in Boiling, *Proc. 3rd European Thermal Sciences Conference 2000*, Heidelberg, Vol. 2, pp. 737-742.

McSharry, P. E., Smith, L.S., Kono, T. and Kenning, D.B.R., 2000, Nonlinear Analysis of Site Interactions in Pool Nucleate Boiling, *Proc. 3rd European Thermal Sciences Conference 2000*, Heidelberg, Vol. 2, pp. 725-730.

Pais, C. and Webb, R.L., 1991, Literature Survey of Pool Boiling on Enhanced Surfaces, *ASHRAE Transactions*, Vol. 97 (1), pp.79-89.

Schmidt, Th. E., 1950, Die Wärmeleistung von berippten Oberflächen, *Abhandlungen des Deutschen Kältetechnischen Vereins Nr. 4*, Verlag C.F. Müller, Karlsruhe.

Schmidt, Th. E., 1966, Verbesserte Methoden zur Bestimmung des Wärmeaustausches an berippten Flächen, *Kältetechnik 18*, pp. 135-168.

Smirnov, H. F., 2000, Heat and Mass Transfer when Boiling on Coated Surfaces and in Porous Structures, *Proc. 3rd European Thermal Sciences Conference 2000*, Heidelberg, Vol. 2, pp. 803-808.

Sokol, P., 1994, Untersuchungen zum Wärmeübergang beim Blasensieden an Glatt- und Rippenrohren mit großem Außendurchmesser, *Forschungsberichte des Deutschen Kälte- und Klimatechnischen Vereins, Nr. 46*, Diss. Universität Paderborn.

Solodov, A.. and Eroshenko, E., 1996, Calculation of the Two-dimensional Temperature Distribution in Finned Tubes at Pool Boiling, *Proc. Eurotherm Seminar No. 48*, Gorenflo, D., Kenning, D. B. R., Marvillet, Ch., eds., ETS, Pisa, pp. 141-147.

Thome, J.R., 1990, *Enhanced Boiling Heat Transfer*, Hemisphere, Washington D.C.

FLOW PATTERNS AND PHENOMENA FOR FALLING FILMS ON PLAIN AND ENHANCED TUBE ARRAYS

J.F. Roques and J.R. Thome

Laboratory of Heat and Mass Transfer (LTCM)
Department of Mechanical Engineering
Swiss Federal Institute of Technology Lausanne (EPFL)
Lausanne, Switzerland CH-1015
e-mail: john.thome@epfl.ch

ABSTRACT

Falling film heat transfer on enhanced tube bundles is particularly important for condensation in refrigeration systems and recently falling film evaporators using enhanced boiling tubes have been successfully introduced into large refrigeration systems (in replacement of flooded evaporators) to achieve more compact and efficient designs. In the present study, adiabatic falling films on horizontal tubes arranged in a vertical array have been studied for several types of tubes: plain, low fin, enhanced boiling and enhanced condensation. Types of intertube flow patterns from the bottom of the upper tube onto the top of the lower tube observed are: droplet flow, column flow and sheet flow and have been recorded utilizing a high speed digital camera (videos up to 1000 images/sec).

A unique photographic (and video) presentation of these flow patterns and interesting flow phenomena is presented here. For instance, the transition from a large number of successive droplets into a continuous column of liquid is captured using the high speed digital camera, which provides a strong visual representation of the elongation of a droplet near the transition into a column. Transition modes where both droplets and columns or column and sheets occur side by side are also observable. The effect of intertube flow patterns is thought to play an important role on the tuberow effect during condensation. Analogously, they are presumed to have an equally important effect on falling film evaporation, i.e. fluid distribution, dryout, etc.

INTRODUCTION

The tube row effect in falling film condensation on a vertical array of horizontal plain tubes was first modeled by Jakob (1936) by applying the Nusselt (1916) vertical plate theory. Jakob assumed that the condensate flowed from one tube to the next as a continuous sheet of liquid from the bottom of the upper tube onto the top of the lower tube, such that

$$\frac{\alpha(N)}{\alpha(N=1)} = N^{3/4} - (N-1)^{3/4} \quad (1)$$

where $\alpha(N)$ is the local coefficient on tube N from the top and $\alpha(N=1)$ is the Nusselt solution for the top tube. Kern (1958) later modified the Jakob tube row expression to reduce the tube row effect based on his practical experience with actual condensers, changing the exponent to arrive at the following expression:

$$\frac{\alpha(N)}{\alpha(N=1)} = N^{5/6} - (N-1)^{5/6} \quad (2)$$

Referring to Figure 1, Marto (1986) compared these two expressions to the tube row effect measured experimentally for plain tubes in various independent studies, noting a significant scatter relative to that predicted. Here, the Jakob equation for sheet flow seems to be the lower limit to the tube row effect measured experimentally while the Kern equation is more representative of the mean trend in the data. Thus, the question arises as to why there is such a large scatter in the tube row data. The most likely answer is that this scatter is caused by the different intertube film flow modes from tube to tube. For instance, sheet mode should have the most severe tube row effect. The tube row effect for the column flow mode should be less since lower tubes are only partially inundated by condensate from above. Similarly, the droplet mode should have an even weaker tuberow effect since the flow is also intermittent. (Photographs of these modes are shown later Figure 4).

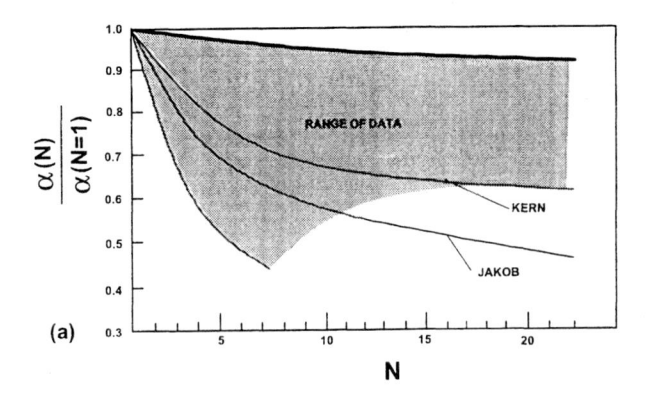

Fig. 1 Comparison by Marto (1986) of Jakob and Kern tube row expressions to experimental data.

Rather than applying a tube row expression, falling film condensation can also be modeled in terms of the film Reynolds number of the condensate. The film will be laminar up to a film Reynolds number of about 1600-2000, and then become turbulent. The film may also have interfacial ripples. Therefore, some of the scatter in Figure 1 may also be due to these effects. For example, at large row numbers the film may be turbulent where the local condensation heat transfer coefficient increases with Reynolds number, and hence these data would fall above the Jakob curve. The data falling below the Jakob curve at small tube row numbers are probably the adverse effect of non-condensables in the vapor phase in those tests.

For low fin tubes, experimental data typically show much less tube row degradation than plain tubes. For example, many tests show less than 10% fall in heat transfer on the first 4 tube rows. For condensation on enhanced condensing tubes, the degradation is typically reported to be more severe than for low finned tubes; this may or may not be true depending on how the comparison is made. Since their heat transfer coefficients are much larger, they produce much more condensate and hence more inundation occurs if the condensing temperature difference is held constant. Apparently no generalized flow mode map is currently available for predicting flow mode transitions for either plain or enhanced types of tubes under condensing or evaporating conditions. Honda et al. (1987), however, have presented several transition expressions for individual fluids condensing on low finned tubes.

Falling film transitions for liquid under adiabatic conditions has been studied extensively by Hu and Jacobi (1996a) for a variety of fluids, tube diameters, tube pitches and flow rates and with/without cocurrent gas flow. Based on their observations, they proposed a flow mode transition map with coordinates of film Reynolds number (Re) versus the Galileo number (Ga). The map delineates the transitions between the three dominant modes (sheet, column and droplet) with two mixed mode zones (column-sheet and droplet-column) in which both modes are present. The modified Gallileo number of the liquid, which is the ratio between the gravity and the viscous force based on the capillary length scale, is defined as:

$$Ga = \frac{\rho \sigma^3}{\mu^4 g} \quad (3)$$

and the liquid film Reynolds number is defined as:

$$Re = \frac{2\Gamma}{\mu} \quad (4)$$

where Γ is the total flow rate on both sides of the tube [kg/m s], ρ is the density of the liquid, μ is liquid dynamic viscosity, σ is the surface tension and g is the acceleration due to gravity. Their transition map is applicable to plain tubes for cocurrent air velocities less than 15 m/s.

An investigation on falling film flow mode transitions of adiabatic and non-phase change films was presented for plain tubes by Mitrovic (1986). Various other studies have investigated the dynamics of transitions, droplet formation and column departure wavelengths, such as those of Taghavi and Dhir (1980), Tang and Lu (1991) and Mitrovic and Ricoeur (1995). Hu and Jacobi (1996b) obtained subcooled heat transfer data, and fit them with a distinct correlation for each flow mode. Hu and Jacobi (1998) have recently reported additional data on tube spacing effects on column and droplet departure wavelengths.

Roques et al. (2000) have reported measurements for flow mode transitions for plain, 26 fpi low fin, Turbo-Bii and Thermoexcel-C tube arrays for water, glycol and a water-glycol mixture for intertube spacings from 3.2 to 24.9 mm, typical of inline and staggered tube layouts. They noted that the Hu and Jacob transition map was quite successful for describing their plain tube observations. The present paper presents a selection of the flow mode observations and phenomena made during that study.

DESCRIPTION OF THE TEST FACILITY

The test facility is shown in Fig. 2 and is comprised of three main parts:
- *Fluid circuit*: The fluid in the tank is pump by a small centrifugal pump through a filter and then one of two rotameters to measure the flow rate. The fluid then goes on to the top of the tranquilization chamber.
- *Flow tranquilization chamber*. This is a rectangular box of 300x260x10mm (inside dimensions: *Height* x *Width* x *Depth*). Its function is to insure a uniform flow onto the top tube and hence onto the second tube. The fluid leaves at the bottom through 1-mm diameter holes spaced 2mm apart. The length of this distribution system (and the resulting film visualized) is 200mm.
- *Test section*. It is comprised of three tubes on which the fluid flows. They are held at one end and the distance between them is fixed with tube spacers. Care is taken

to obtain a very precise alignment of the horizontal tubes in a vertical array.

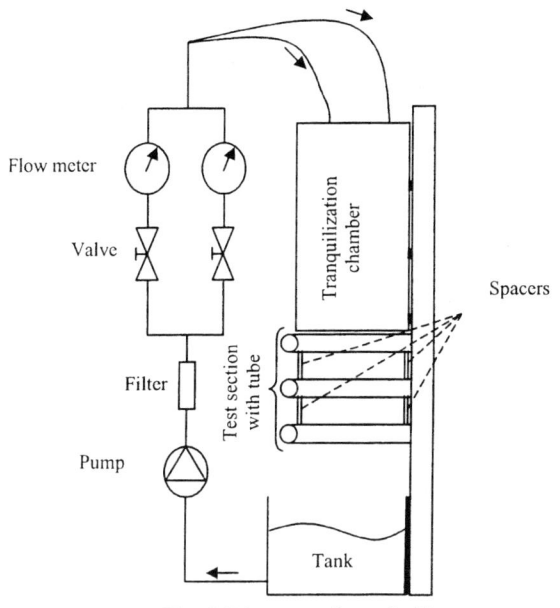

Fig. 2 Diagram of test facility.

The temperature in the liquid bath is measured with a thermocouple for calculation of the liquid properties.

TYPES OF TUBES TESTED

Figure 3 depicts closeup photographs of the Wolverine Tube Turbo-BII enhanced boiling tube, Wolverine Tube Turbo-Chil low fin tube (26 fpi) and the Hitachi Thermoexcel-C enhanced condensing tube that were tested. The integral low fin tube is often used as a condensing tube but not as a falling film evaporation tube. The first two tubes and the plain tube are 19.05 mm in diameter while the Thermoexcel-C is 12.7 mm in diameter. Looking closely at the Thermoexcel-C tube, some spines are seen to be compressed, which results from a helical deformation during corrugation of the tube for tube-side water enhancement. As will be noted latter, this period deformation has an effect on the flow mode.

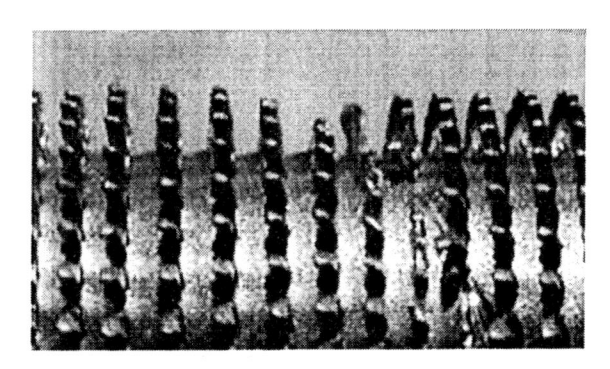

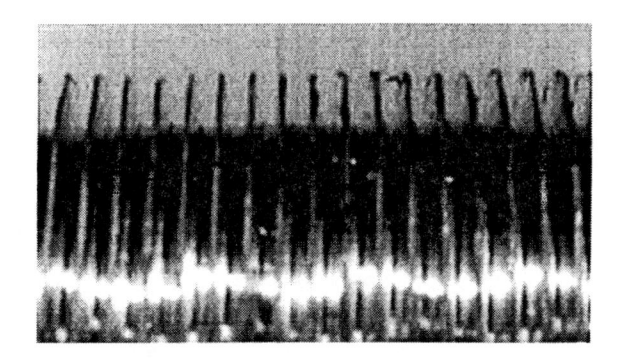

Fig. 3 Photographs of the enhanced tubes: Thermoexcel-C (top), low fin (middle), Turbo-BII (bottom).

DEFINITION OF INTERTUBE FLOW MODES

Selected images from digital videos are presented in Figure 4 to illustrate the five different modes, which are defined as follows:

Droplet mode. The flow is in droplet mode when only droplets are falling between the tubes. Note that the videos are essential to distinguishing a rapid sequence of droplets under some test conditions from a continuous column of liquid in others.

Droplet-Column mode. This is an intermediate mode with both droplets and colums occurring side by side from the upper tube to the lower tube.

Column mode. A column is a continuous fluid link between tubes. This mode is when there are only liquid columns between the tubes. Columns can occur either inline from tube to tube or staggered, as illustrated in Fig. 4.

Column-Sheet mode. This is an intermediate mode with both columns and short liquid sheets occurring simultaneously side by side along the tubes. Small sheets are formed by the merging of two neighboring columns and have a triangular profile and illustrate in Fig. 4.

Sheet mode. This mode occurs when the fluid flows uniformly between the tubes as a continuous liquid film. In Fig. 4 the edge of a sheet is observable at the end of the test section. Surface tension pulls the edge and narrows the film from top to bottom.

These transitions are encountered in the above order as the mass flow rate, and hence film Reynolds number, is increased.

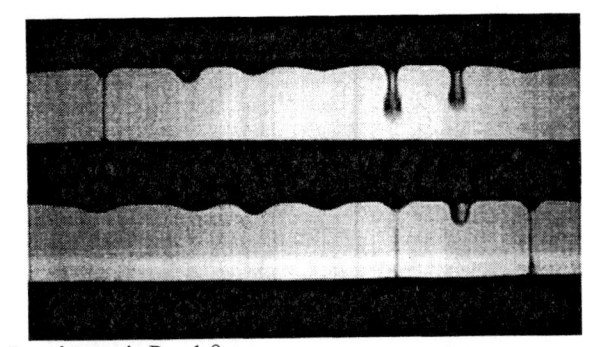

Droplet mode Re<1.8

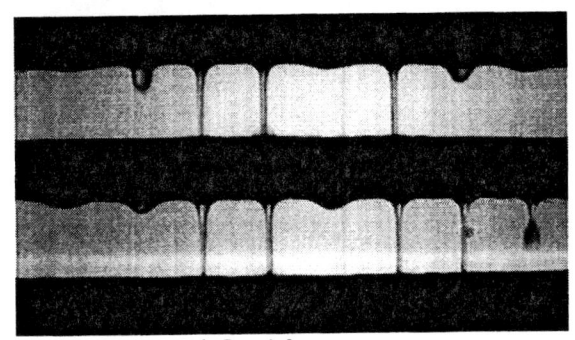

Droplet-Column mode Re=1.9

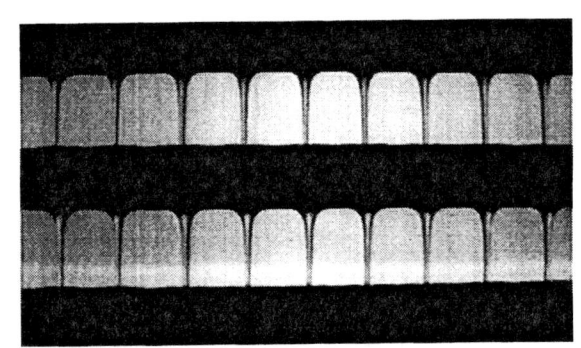

Inline Column mode Re=2.9

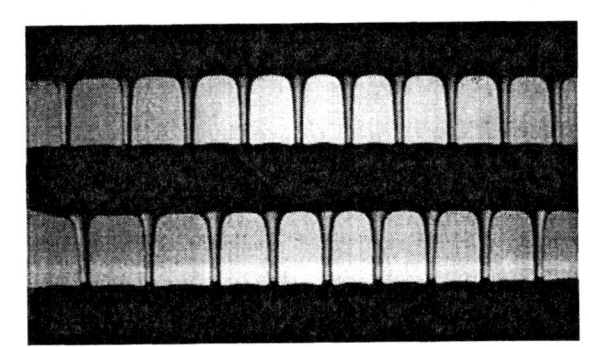

Staggered-Column mode Re=12

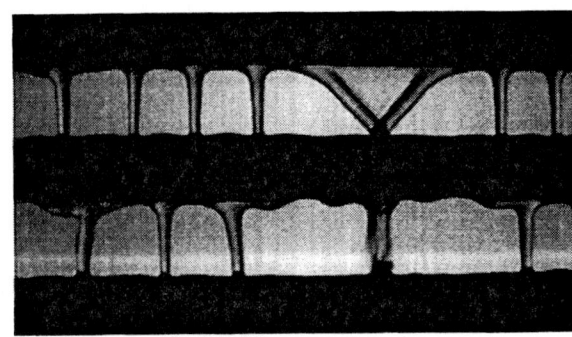

Column-Sheet mode Re=17

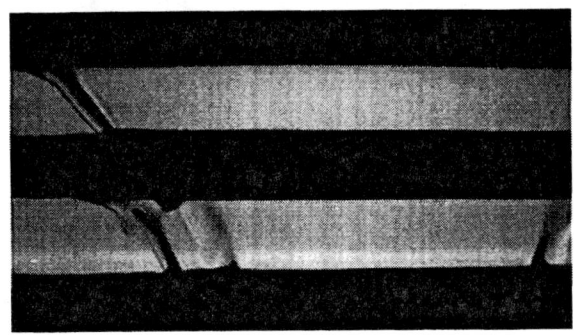

Sheet mode Re=24

Fig. 4 Illustration of mode transitions for glycol on plain tube with tube spacing of 24.9 mm.

Flow Mode Observations

Figure 5 depicts a sequence of images for the droplet mode near the transition to column mode. The images were taken at a speed of 1000 images/sec and a shutter speed of 1/2000 sec. The fluid is water. The bottom of the upper tube and the top of the lower tube are just visible in this close up view of the phenomena where the spacing between the tubes is 19.4 mm. Rather than breaking off on the top tube, the droplets are seen to elongate into unstable columns which pinch into segments as they become very thin. Increasing the liquid flow rate, sufficient liquid becomes available to avoid surface tension forces from pinching the film into segments and the transition to column flow occurs (not shown). Similar to Fig. 4, the characteristic wavelength along the tube between droplets remains nearly constant from droplet to droplet.

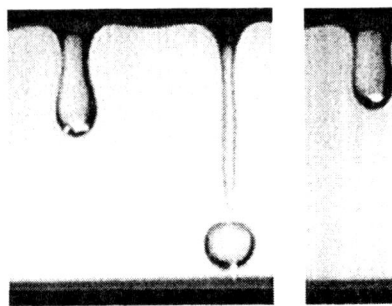

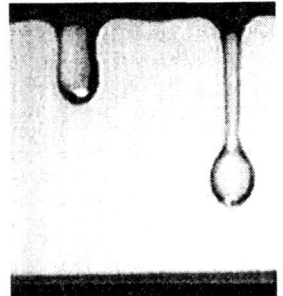

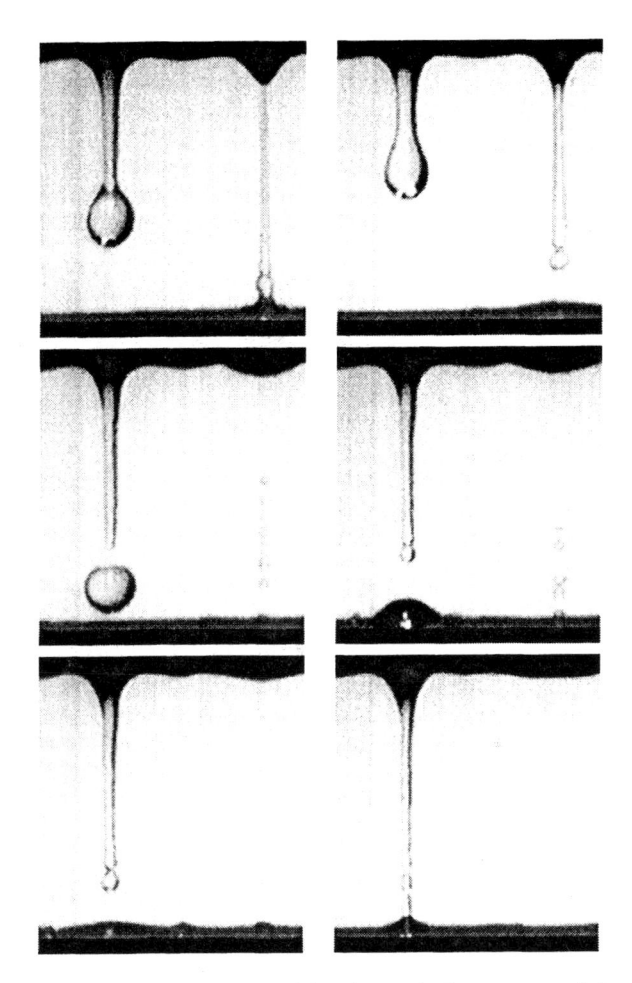

Fig. 5 High speed images of droplet mode for water on plain tube at tube spacing of 19.4 mm.

Figure 6 depicts the effect of spacing between the tubes on the shape of columns for the 50%water-50%glycol mixture. At large spacings of 19.4 and 24.9 mm, the traditional shape of a long column is observed. At smaller spacings, the columns to not have the necessary free height to elongate and hence form these "light bulb" and inverted triagular shapes. At very small spacings, the columns look like a series of small sheets.

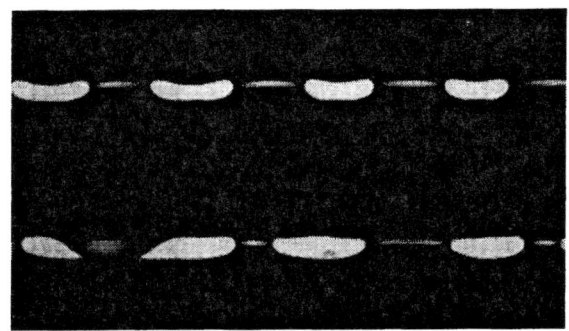

Spacing=3.2mm, Re=15

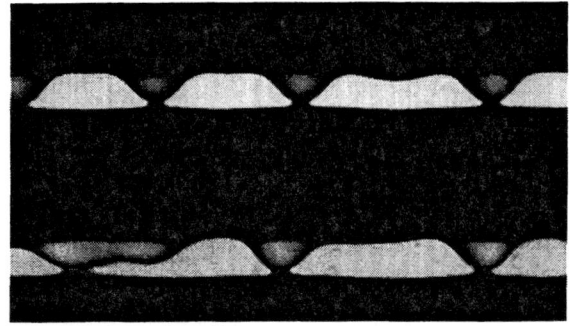

Spacing=6.4mm, Re=43

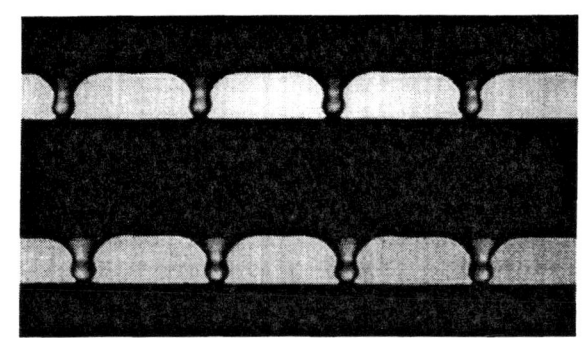

Spacing=9.5mm, Re=42

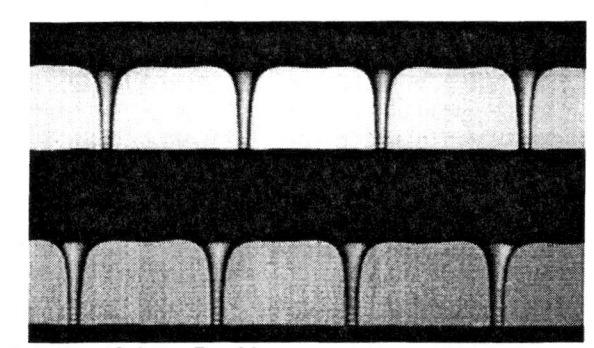

Spacing=19.4mm, Re=30

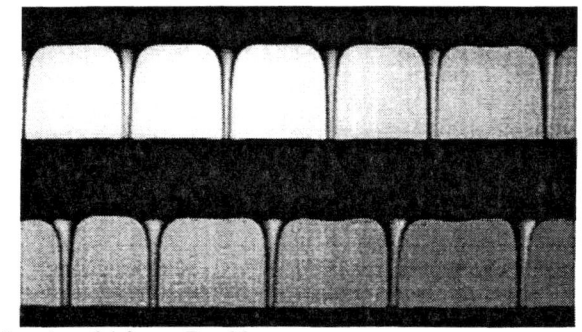

Spacing=24.9mm, Re=29

Fig. 6 Comparison of column shape for plain tube with water-glycol mixture for spacings from 3.2 to 24.5 mm for column mode.

Figure 7 shows an example of inline columns and staggered columns for a tube spacing of 19.4 mm for glycol. The flow mode changes from inline to staggered as the flow rate of the film is increased. Note the nearly identical characteristic wavelength between the columns along the tube. Just barely visible in the lower image, the liquid tends to form a crown on the top of the tube between the impinging columns in the staggered mode. This comes from lateral spreading of the liquid from the columns along the tube and their headon confrontation with one another. The inline mode does not have this crown as the liquid that spreads is dispersed before reaching the midpoint to its neighbor. When a crown is formed, the mode changes from inline to staggered as now the liquid flow rate below the crown towards the bottom of the tube is locally higher than elsewhere along the tube.

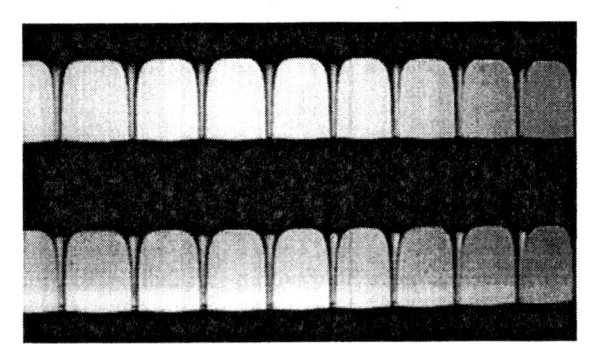

Spacing=19.4mm, Re=3.5

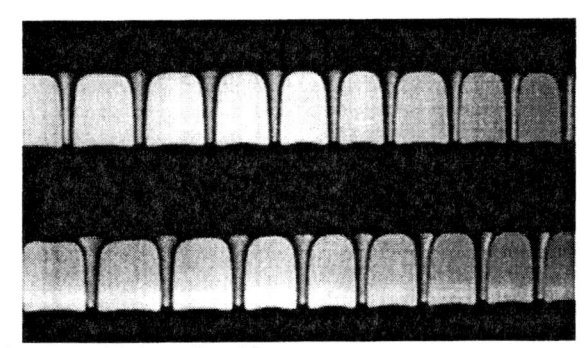

Spacing= 19.4mm, Re=15

Fig. 7 Inline columns (top) vs. staggered columns (bottom) for glycol on plain tube.

Figure 8 depicts the column mode of the Thermoexcel-C tube for a 50%water-50%glycol mixture. The helical corrugations effect on the fin profile is observable here. In fact, the columns have a definite tendency to form at the location of the corrugation. Hence, the pitch of the corrugation, when small enough, may be able to be used to control the pitch of column formation.

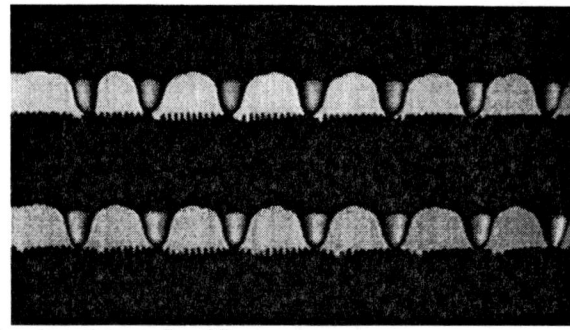

Spacing=6.4mm, Re=27

Fig. 8 Inline column mode for water-glycol mixture on Thermoexcel-C tubes.

Figure 9 illustrates a comparison of the shapes of columns on the low fin tube at two different tube spacings for the 50%water-50%glycol mixture. The flow is nearly inline in both images. The dark zones at the attachment point of the columns on the top of the tubes is a build up of liquid, which cannot laterally dissipate because of the low fins. Hence, the inline column mode is stable here and the staggered column mode is never observed for the low fin tube in this study. Clearly, the portion of the lower tube affected by inundation is quite limited for this low fin tube.

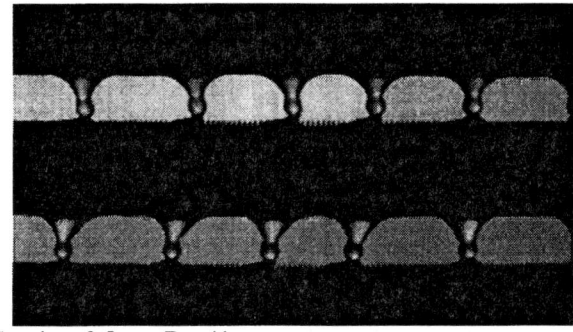

Spacing=9.5mm, Re=41

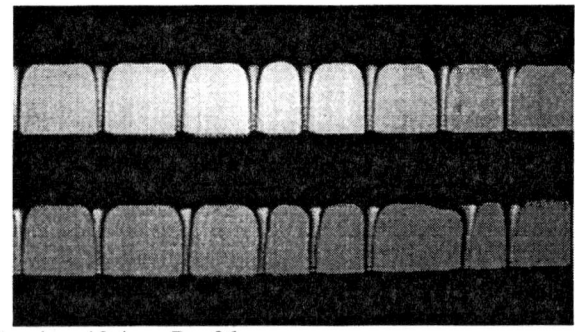

Spacing=19.4mm, Re=36

Fig. 9 Inline column mode for water-glycol mixture on low fin tube at two different spacings.

Flow Mode Transitions

The flow mode transition data were already presented in Roques et al. (2000) for these tubes and here only a composite of those results is presented, using two dimensionless groups. The fluid type is characterized by the modified Gallileo number (Ga). The mass flow rate is included in the film Reynolds number (Re). Neglecting the effect of tube spacing, the Reynolds number at transition can be correlated as a function of the Galileo number as:

$$\text{Re} = a\,\text{Ga}^b \quad (5)$$

where a transition is the change from one flow mode to another one. As there are five modes, there are four transitions (or eight if one takes into account the hysterisis effect for increasing or decreasing flow conditions). Since we observed only a small degree of hysteresis, the observations of these two cases are put together to find the transitions.

Figure 10 shows a composite diagram of the transitions observed for the Turbo-Chil low fin tube compared to those for the plain tube. The effect of the low fins is to increase the Re threshold from Column-Sheet mode to Sheet mode and from Column mode into Column-Sheet mode. Hence, this type of tube remains in the column flow mode to much higher flow rates than the plain tube, which should increase falling film condensation heat transfer coefficients at these conditions. The effect of the fins on the other two transitions is not very significant.

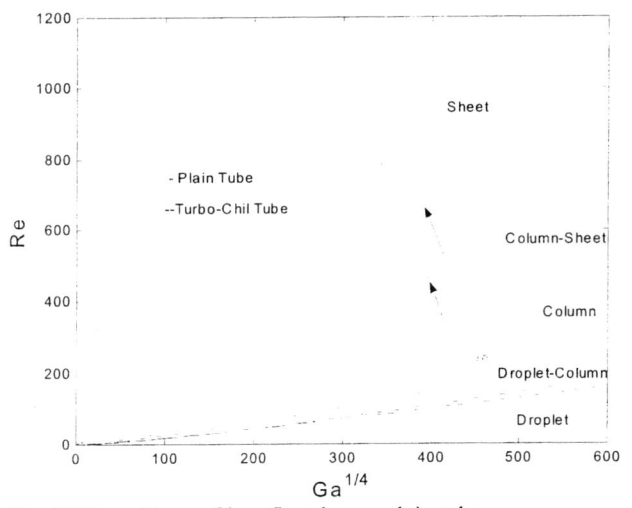

Fig. 10 Transitions of low fin tube vs. plain tube.

Figure 11 shows a composite diagram of the transitions observed for the Thermoexcel-C tube compared to those for the plain tube. The effect of the notched fins is to increase the Re threshold from Column-Sheet mode to Sheet mode and from Column mode into Column-Sheet mode, similar to that for the low fin tube. The effect of the fins on the other two transitions is to lower their thresholds. Hence, this enhanced condensing tube remains in the column flow mode over a much broader range of flow rates than the plain tube, which should increase falling film condensation heat transfer coefficients at these conditions.

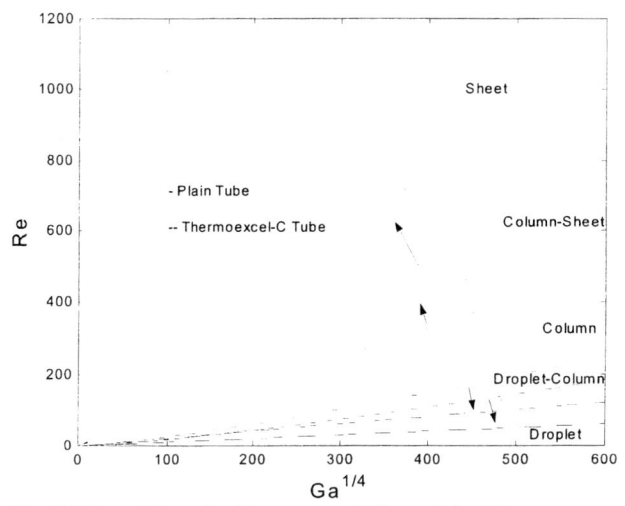

Fig. 11 Transitions for Thermoexcel-C vs. plain tube.

Figure 12 shows a composite diagram of the transitions observed for the Turbo-BII tube compared to those for the plain tube. The effect of the deeply notched and compressed fins is negligible on the Re threshold from Column-Sheet mode to Sheet mode and less pronounced from Column mode into Column-Sheet mode compared to Figures 10 and 11. The effect of the fins on the Droplet-Column to Column transitions is similar to that of the Thermoexcel-C while the other threshold remains unaffected. Hence, this enhanced boiling tube remains in the column flow mode over a broader range of flow rates than the plain tube, but less so when compared to the two other enhanced tubes.

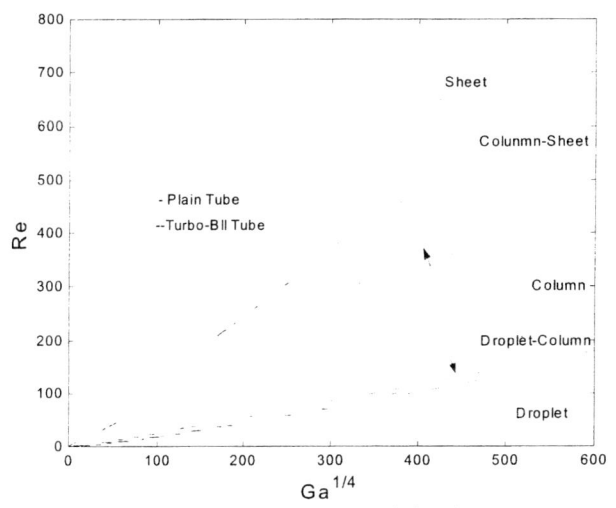

Fig. 12 Transitions from Turbo-BII vs. plain tube.

Revisiting Fig. 1, the graph can tentatively be explained as follows. The Jakob expression represents condensation when the film mode is present, the Kern expression apparently captures condensation in the column mode and the data at the top are apparently for condensation in the droplet flow mode. For actual condensation on a vertical array of horizontal tubes, the intertube flow mode will be dependent on the film Reynolds leaving the bottom of each tube. Starting from the top tube and going down through the bundle, one would expect the flow to begin in the droplet flow mode on the first few tube rows, then pass into the droplet-column intermediate flow mode, then pass into the column flow mode, then go into the column-sheet flow mode and finally into the sheet flow mode.

Hence, the tube row effect is not given by one expression, such as Eq. (1) for sheet flow, but requires similar expressions to be derived for each flow mode. Similarly, converting the Nusselt condensation expresion into its analogous Nu vs. Re expression does not capture the flow mode effects on heat transfer either. Research is currently underway to investigate the effects of flow mode on falling film evaporation and falling film condensation on plain, low fin and enhanced tubes as part of this project.

CONCLUSIONS

Images of intertube flow modes between horizontal tubes oriented in a vertical array have been shown for plain, low fin, enhanced boiling and enhanced condensation tubes. The effect of intertube spacing, fluid, type of tube surface and film Reynolds number on the aspect of the flow modes have been illustrated. The effect of tube surface on the transitions from one flow mode to another have been shown in composite graphs and related to the expected effect on condensation in tube bundles.

ACKNOWLEDGEMENTS

This research was sponsored by the EPFL/LTCM Falling Film Research Club together with a financial contribution by the EPFL. Club members are: Dunham-Bush Inc., Sulzer Friotherm SA, UOP Inc., Wieland-Werke and Wolverine Tube Inc.

NOMENCLATURE

a   multiplicative factor in relations
b   power of Ga number
g   acceleration due to gravity (9.81 m/s^2)
Ga   modified Gallileo number
N   tube number
Re   film Reynolds number
α(N) heat transfer coefficient on tube N (W/m^2 K)
Γ liquid flow rate per unit length on both sides (kg/m s)
μ dynamic viscosity (N s/m^2)
ρ density (kg/m^3)
σ surface tension (N/m)

REFERENCES

Armbruster, R. and Mitrovic, J. (1995). Heat transfer in falling film on a horizontal tube, *Proc. of National Heat Transfer Conference,* ASME HTD-Vol. 314 (12), 13-21.

Honda, H., Nozu, S. and Takeda, Y. (1987). Flow characteristics of condensation on a vertical column of horizontal tubes, *Proc. 1987 ASME-JSME Thermal Engineering Joint Conference*, Honolulu, Vol. 1, 517-524.

Hu, X. and Jacobi, A.M. (1996a). The intertube falling film Part 1 – Flow chararcteristics, mode transitions and hysteresis, *J. Heat Transfer,* Vol. 118, 616-625.

Hu, X. a nd Jacobi, A.M. (1996b). The intertube falling film Part 2 – Mode effects on sensible heat transfer to a falling liquid film, *J. Heat Transfer,* Vol. 118, 626-633.

Hu, X. and Jacobi, A.M. (1998). Departure-site spacing for liquid droplets and jets falling between horizontal circular tubes, *Experimental Thermal and Fluid Science*, Vol. 16, 322-331.

Jakob, M. (1936). *Mech. Engng.,* Vol. 58, 163.

Kern, D.Q. (1958). Mathematical development of loading in horizontal condensers, *AIChE J.*, Vol. 4(2), 157-160.

Marto, P.J. (1986). Recent progress in enhancing film condensation heat transfer on horizontal tubes, *Heat Transfer Engng.*, Vol. 7, 53-63.

Mitrovic, J. (1986). Influence of tube spacing and flow rate on heat transfer from a horizontal tube to a falling liquid film, *Proc. 8th International Heat Transfer Conf.*, San Francisco, Vol. 4, 1949-1956.

Mitrovic, J. and Ricoeur, A. (1995). Fluid dynamics and condensation heating of capillary liquid jets, *Int. J. Heat Mass Transfer*, Vol. 38, 1483-1494.

Nusselt, W. (1916). Die oberflächenkondensation des wasserdampfes, *Zeitschr. Ver. Deutch. Ing.*, Vol. 60, 541-569.

Roques, J.F., Dupont, V. and Thome, J.R. (2000). Falling Film Transition on Plain and Enhanced Tubes, *J. Heat Transfer*, submitted for review.

Taghavi, K. and Dhir, V.K. (1980). Taylor instability in boiling, melting, condensation or evaporation, *Int. J. Heat Mass Transfer*, Vol. 23, 1433-1445.

Tang, Z. and Lu, B.Y.C. (1991). Droplet spacing of falling film flow on horizontal tube bundles, *Proc. 18th International Congress of Refrigeration*, Montreal, Vol. 2, 474-478.

FLOW DYNAMICS AND INTENSIFICATION OF THE HEAT TRANSFER IN PRECRITICAL REGIMES OF INTENSIVELY EVAPORATING WAVY LIQUID FILM

A.N. Pavlenko, V.V. Lel, A.F. Serov, and A.D. Nazarov

Institute of Thermophysics SB RAS Lavrentiev ave. 1, Novosibirsk, 630090, Russia; pavl@itp.nsc.ru

ABSTRACT

Here we present experimental results on the behavior of a laminar-wavy liquid film at its intensive evaporation under conditions of a gravity flow over the vertical locally heated surface. It was determined that at some certain heat fluxes, the shape of a residual layer changes significantly. It is shown that due to intensive heat release, the relative wave amplitude increases drastically, and "dry" spots appear when the limit heat fluxes are reached. For the first time, data on probability density alteration were obtained for the local film thickness depending on the heat flux density in the range of Reynolds numbers from $Re_{in.}=32$ to $Re_{in.}=103$. The effect of wave amplitude, heat flux density and thickness of the residual layer on the phase velocity and shape of large waves was shown. The measured heat transfer coefficient for regime before formation of large "dry" spots with evaporating liquid jets between them ($0.3 < q/q_{f.b.} < 1$) are higher calculated heat transfer coefficient for the model of laminar-wavy flow of evaporating liquid film at the low heat flux densities.

INTRODUCTION

To intensify heat transfer processes in the industrial apparatuses, the film flow of liquid over a surface of the heat exchanger elements is widely used. By the present, a lot of papers on theoretical and experimental investigations of the wavy flow of evaporating liquid film were published. These works are thoroughly analyzed in monographs (Vorontsov and Tananaiko, 1972; Gimbutis, 1988; Alekseenko, Nakoryakov and Pokusaev, 1992). The main part of these works deals with the study of a film flow at experimental setups of a relatively high length, where water, water-glycerin mixture and oils are used as the working liquid. Due to this fact, the flow regimes of evaporating film of saturated liquid in these experimental studies were limited by low densities of heat fluxes.

The influence of the wave hydrodynamics on the liquid film was investigated numerically and experimentally in the work (Adomeit, Leefken and Renz, 2000.). The time-dependent Navier-Stokes equations and the energy equation of a falling laminar liquid film with two-dimensional waves was solved numerically with a spectral element technique. It is shown that the heat transfer in wavy films is enhanced by effective film thinning and convection effects depending on the Prandtl number. The convection enhancement has been separated to explain the phenomena of hydrodynamic mixing. Experiments were carried out to support the numerical findings. In an experimental set-up two-dimensional waves on the surface of a falling film were produced by a loudspeaker located within the liquid distributor. The thickness distribution is measured by a fluorescence technique. The velocity profiles within the wavy film with a typical film thickness were measured by a microscopic particle image velocimetry (MPIV).

Experimental study of evaporating turbulent-wavy water films at $Re_{in.} > 3000$ was carried out in (Lyu and Mudawar, 1991), using the statistical methods. We do not know any works on experimental study of flow dynamics of intensively evaporating laminar-wavy films of saturated liquid.

Stability of the waveless flow of evaporating film over a vertical surface is studied in (Bankoff, 1971; Spindler, 1982). It is shown that evaporation destabilizes the flow and expands the range of single-wave perturbations, growing in time. The integral approach for the description of wave behavior on the evaporating film of saturated liquid is developed in (Trifonov, 1993) without assumption about low amplitude. The authors of this work describe generation of "dry" spots on the heat-releasing surface via consideration of the non-linear regimes for various wave types in the range of low Reynolds numbers, when boiling in evaporating liquid film is suppressed.

Taking into account that the plug height in finned apparatuses, used by cryogenic techniques, is not high, and viscosity, surface tension and limiting wetting angle of cryogenic liquids are small, application of published approximate dependencies for calculation of characteristics of evaporating liquid film should be amplified with experimental results. However, the latter is hardly available because it is very difficult to perform experiments in the range of cryogenic temperatures.

Experimental study of flow dynamics of intensively evaporating laminar-wavy liquid nitrogen film, falling down the vertical heated surface was the goal of our work.

EXPERIMENTAL SETUP

Experiments were carried out at the experimental setup, shown schematically in Fig.1. The detailed description of this experimental setup is presented in (Pavlenko and Lel, 1997). The film of liquid from a slot distributor fell down the vertical duralumin plate1 with the size of 280 mm*75 mm and run against the heat-releasing part of the surface. A distance from the slot to the area of local heating, equal to 160 mm, provided liquid film flow over the heat-releasing surface in the regime of hydrodynamic stabilization. Using the three-channel capacity probe, the local

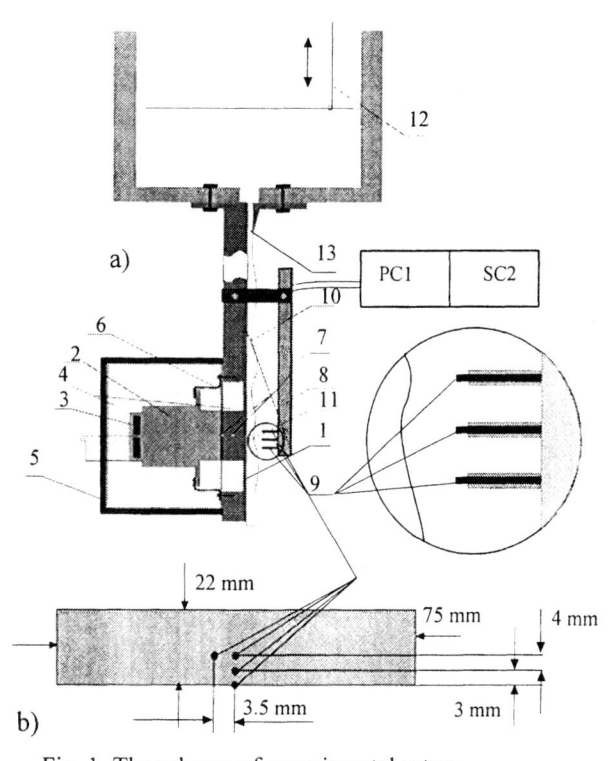

Fig. 1. The scheme of experimental setup
1 – duralumin working plate, 2 – copper heat conductor, 3 – resistive heaters, 4 – thermal insulator, 5 – protective glass textolite cover, 6 – pressure cramps, 7 – thermal resistors, 8 – microprobe plate, 9 – four capacity microprobes, 10– liquid nitrogen film, 11 – heat-releasing surface, 12 – movable cleat with level meters, 13 – distribution slot; b – frontal view of the heater.

film thickness was measured in these experiments; the surface temperature and heat flux density were measured by the thermal resistors. Stability of irrigation degree at experimental setup was obtained due to a constant liquid level above the distributor slot. The local heating was performed with the help of resistive heaters from the stabilized current source. Power of heaters provided heat flux densities from zero to $4 \cdot 10^4$ W/m^2 with an error of ~5% (considering heat losses).

The heat flux to the working plate was send through the copper heat conductor 2, where 24 heaters 3 were mounted. The required contact between the copper heat conductor 2 and the plate was obtained with the help of pressure cramps 6. To minimize heat release from the zone of heat liberation, some grooves with residual thickness of the working plate of 0.6 mm were made by special mechanical treatment. The copper heat conductor was thermally insulated by the polyurethane insertion 4 with a glass textolite cover 5. Three local thermal resistive temperature probes of PPT mark were installed at the heated section of the plate. Their temperature calibration was carried out in the temperature range of (70-150 °K) at the specialized meteorological institution VNIIFTRI. To provide maximal contact between the thermal probes and the heat-releasing wall and minimize the cooling effect of the output probe wires, the holes, where these probes were installed, were completely filled with copper powder. The temperature of heat-releasing surface was determined by the method of heated wall due from the readings of thermal probes.

The block from three capacity probes 9, which allowed us to study the change in local thickness of the liquid film and wavy characteristics along the flow, was mounted above the heated area (see Fig. 1). The main advantage of the capacity method is that the used probes do not disturb the flow of liquid and have high sensitivity. The method is based on registration of changes in electric capacity of the probe condenser depending on thickness of the liquid film, flowing between probe coats.

A high-frequency electric signal, proportional to the local film thickness, is formed by a primary converter (PC1), which includes two similar LC-generators, operating at the frequency of ~ 30 MHz. The oscillatory circuit of one generator includes the capacity probe, situated above the heat exchanger surface. An outlet signal of the primary converter is a signal, whose amplitude has the low-frequency component, equal to the difference in generators' frequency (beat frequency ~100 kHz) and proportional to the film thickness beneath the probe. The following signal treatment takes place in a secondary converter (SC2), where it is detected, filtered and amplified. A change in instantaneous frequency $f(\tau)$ is performed by a special module in the set of IBM PC bus. Together with the service routine, this module allows us to measure beat frequency $f(\tau)$ during minimal time $\tau=0.5$ ms, and this corresponds to the clock rate of the film thickness measurer F=2 kHz. The volume of on-line storage (4000 values) for three channels, registered simultaneously at available beat frequency allows us to study the wavy process at the film surface with maximal frequencies up to 700 Hz. Depending on investigation goal, we can optimally set the parameters of registration equipment. In the work presented, the studied regimes of the film flow are characterized by the frequencies up to 200 Hz. Therefore, the discretization range of measured frequency $f(\tau)$ was assumed equal to 1 ms. For this case, the beat frequency of discretization was $F=1$ kHz, and this was quite sufficient for the study of given flows because we could investigate the wavy processes on the film surface in the range up to 300 Hz. The determination accuracy for the frequency was $f(\tau) < 10^{-3}$.

Calculation of the local film thickness was performed with the help of alteration dependencies of the probe capacity vs. nitrogen film thickness $C(\delta)$ and generator frequency vs. the probe capacity $f(C)$, introduced into the general program. Reliability of transition from function $f(\tau)$ to $\delta(\tau)$ was checked experimentally, and it was shown that for the chosen size of the probe, accuracy of film thickness measurement made up ± 2.5 µm. The detailed description of modified method is presented in (Alekseenko, Nazarov, Pavlenko, Serov and Chekhovich, 1997).

EXPERIMENTAL RESULTS

The film profile was simultaneously measured in three points along the flow at distances $x_1 = 15$ mm, $x_2 = 19$ mm, $x_3 = 22$ mm from the upper edge of the heated area; for instance, this makes up $x_n/\lambda_{aver.} \sim$ 1.6, 2.0, 2.3 for $Re_{in}=38$. This allowed us to get information about alteration in the flow character and wave shape at intensive evaporation of liquid. Typical profiles of the liquid film thickness are shown in Fig. 2 depending on time in point x_3 for various heat flux densities. The profile of the wavy liquid film without heat release is shown in the lower part of the figure. It is obvious that the wavy structure of film makes up periodical succession of solitary waves, separated by a lengthy residual layer of almost constant thickness. Small capillary waves are clearly seen in the residual layer. It follows from data analysis that at low heat flux densities ($q/q_{f.b.} \leq 0.6$, q is the heat flux density, $q_{f.b.}$ is the heat flux density, when time-stable "dry" spots appear), a decrease in the wavy film thickness occurs due to evaporation of the residual layer without significant change in the wave shape.

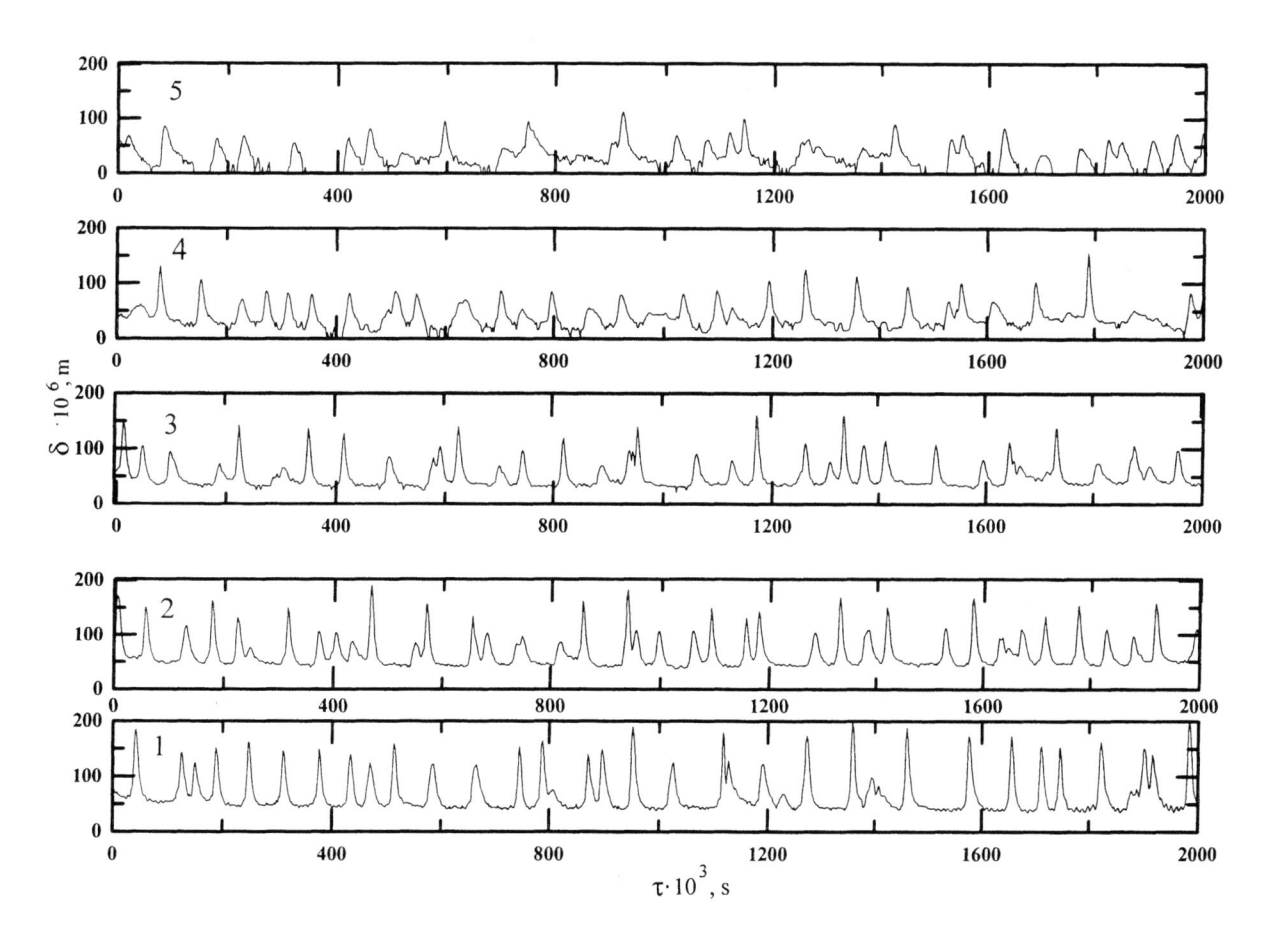

Fig. 2. Instantaneous thickness of liquid nitrogen film at the film flow over the vertical heat-releasing surface. The distance from the beginning of heat-releasing surface $x_3 = 22$ mm $Re_{in.} = 103$. 1 - q=0, $\delta_{aver.} = 65.8 \cdot 10^{-6}$ m, $\delta_{res.} = 51.4 \cdot 10^{-6}$ m; 2 - q=$6.2 \cdot 10^3$ W/m², $\delta_{aver.} = 65.0 \cdot 10^{-6}$ m, $Re_f=85$; 3 - q=$7.8 \cdot 10^3$ W/m², $\delta_{aver.} = 50.9 \cdot 10^{-6}$ m, $Re_f=80$; 4 - q=$9.9 \cdot 10^3$ W/m², $\delta_{aver.} = 39.1 \cdot 10^{-6}$ m, $Re_f=74$; 5 - q=$1.2 \cdot 10^4$ W/m², $\delta_{aver.} = 28.6 \cdot 10^{-6}$ m, $Re_f=68$.

Insignificant alteration of the average thickness of film and residual liquid layer is observed due to evaporation. With a rise in the heat flux density $q/q_{f.b.}$ up to the values close to that, which corresponds to the moment of formation of unstable local "dry" spots, a considerable change in wave amplitude and shape is observed. According to Fig. 2, at high heat flux densities $q=9.9\cdot10^3$ W/m^2 and $q=1.2\cdot10^4$ W/m^2, significantly non-uniform thinning of the residual layer occurs at a period between two large waves. The minimal thickness of the residual layer is reached just before running of the steep front of large waves. In the studied alteration range of the heat load, local Reynolds number Re_l changes essentially at a scale of several typical wavelengths

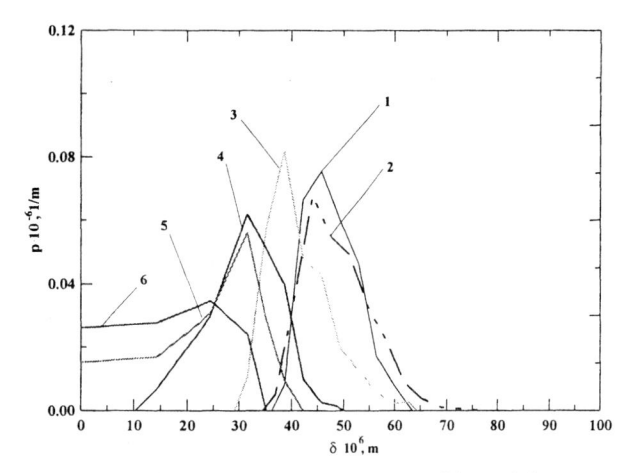

Fig. 3. Probability density of liquid film thickness at various heat flux densities. The distance from the beginning of heat-releasing surface $x_1 = 15$ mm. $Re_{in.} = 38$. 1 - q = 0, 2- q=2.4·10^3 W/m^2, 3- q=3.9·10^3 W/m^2, 4- q=5.4·10^3 W/m^2, 5- q=6.7·10^3 W/m^2, 6- q=8.2·10^3 W/m^2.

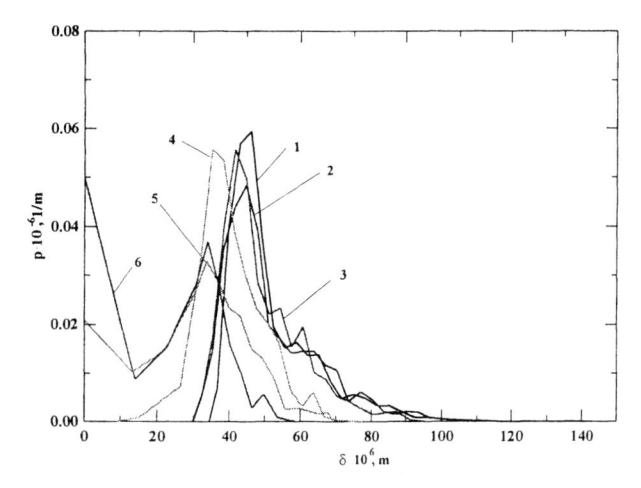

Fig. 4. Probability density of liquid film thickness at various heat flux densities. The distance from the beginning of heat-releasing surface $x_3 = 22$ mm. $Re_{in.} = 51$.
1 - q = 0, 2- q=1.9·10^3 W/m^2 3 - q=3.3·10^3 W/m^2, 4- q=4.8·10^3 W/m^2, 5- q=6.8·10^3 W/m^2, 6- q=9.5·10^3 W/m^2.

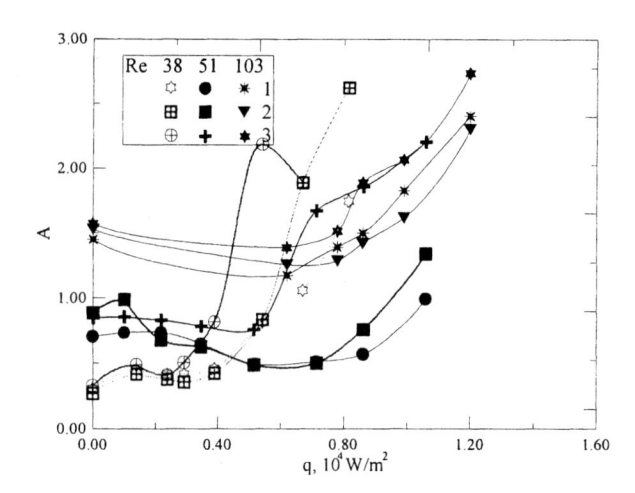

Fig.5. Dependence of the relative amplitude of large waves at various heat flux densities. $1 - x_1$, $2 - x_2$, $3 - x_3$,

due to intensive evaporation. It follows from calculation of the simplest heat balance that at maximal heat flux density, local Reynolds number changes from 103 to 68 during propagation of the liquid film above the area of heating to the measurement point.

Curves in Figs. 3, 4, correspond to calculation of probability density of the liquid film p(δ) in points (x_1=15 mm, $Re_{in.}$=38) and (x_3=22 mm, $Re_{in.}$=51) at various heat flux densities. With an increase in the heat flux density within the area of low values, probability density function moves to the left without considerable alteration in its shape. While approaching to the critical values of the heat flux density, the probability density function becomes more gentle. Location of the point, which corresponds to the maximum of function p(δ), decreases and moves towards the area of thinner liquid films. An increase in the heat flux density increases the possibility of liquid film with very small film thickness. For high heat flux densities $q \sim q_{f.b}$, the final value of function p(δ) at δ=0 can be explained by the beginning of formation of time-unstable "dry" spots. The following increase in the heat flux density leads to formation of time-stable "dry" spots on the heat-releasing surface ($q=q_{f.b}$) and presence of delta-function in zero point in the diagram of probability density for the liquid film thickness. Formation conditions of these crisis regimes in this range of Reynolds number alteration were studied in detail by (Pavlenko and Lel, 1997).

Treated experimental data on the relative amplitude of large waves $A = \dfrac{\delta_{max} - \delta_{min}}{\delta_{aver}}$ are shown in Fig. 5 for various heat fluxes at different distances from the point of heat release starting. It is clear that for low heat fluxes, the relative wave amplitude decreases unconsiderably. Such a behavior is typical for the dependency of relative wave amplitude on the current Reynolds number

($Re_l = Re_{in} - \frac{4qx_n}{v'\rho'r}$) at weak evaporation (low heat fluxes at lengthy surfaces along the film flow). With the following increase in the heat flux, the relative wave amplitude grows fast. Finally this leads to formation of "dry" spots on the heat releasing surface, unstable at first, and then stable.

According to the analysis of above figures, with an increase in the heat flux density from $q/q_{f.b.}$ =0.6 to 1, relative wave amplitude A increases. Here the growth of relative wave amplitude during instability development at the flow of intensively evaporating film of saturated liquid over the vertical heated surface was determined experimentally. This effect was predicted theoretically in (Trifonov, 1993). Calculations were performed for T=const at the flow of a laminar-wavy liquid film ($Re_{in.}$=20) under conditions of slowly changing solution, i.e., approximation of a weak change in the current Reynolds number was used at a scale of large wave length. This work demonstrates calculation result for the shape of wavy surface of evaporation liquid film for corresponding temperature differences ΔT=0.23 K, ΔT=2.3 K in liquid nitrogen. Analysis of experimental results, obtained in this paper, and calculations from (Trifonov, 1993) demonstrates similar points of most possible film thinning and "dry" spot formation just in front of the steep incident front of large waves.

Experimental results on the average thickness of the residual liquid layer between large waves, obtained under adiabatic conditions are shown in Fig. 6 together with dependencies, recommended in literature for its calculation. Measured values of undisturbed residual layer $\delta_{res.}$ turn out to be higher than those, calculated by the following relationships in corresponding with (White and Tallmadge, 1965; Levich, 1959; Brauner and Maron, 1983). Empirical dependency from (Vorontsov, 1999), based on result treatment for water at $Re_{in.}$<800 :

$$\delta_{res.} = 1.351(v'^2/g)^{1/3} Re^{0.31}(x^+/S)^{-0.06} \quad (1)$$

is in a good agreement with the above relationship. Curve 6 demonstrates the interpolation dependency, describing experimental points:

$$\delta_{res.} = 18.9 \cdot 10^{-6} Re^{0.22} \quad (2)$$

Experimental results on the phase velocity of large waves are shown in the following figures. Experimental data under adiabatic conditions is shown in Fig. 7. In this figure, the phase velocity of large waves c, assigned to characteristic velocity $V = \frac{g\delta_{res.}^2}{3v'}$, is shown in the form of dependency on large wave amplitude. Due to increased measurement error for the low residual layer thickness at such a method of normalization, the high data scattering is being obtained during processing. Curves 1, 2, in Fig. 7. correspond to calculations by theoretical dependencies, shown below:

$$c = \frac{g\delta_{res}^2}{v'}\left(1 + \widetilde{A}\right) \quad (3)$$

was obtained in (Alekseenko, Nakoryakov and Pokusaev, 1992) for slightly non-linear positive steps;

$$c = \frac{g\delta^2_{res}}{3v'}\left(3 + 2,324\widetilde{A}\right) \quad (4)$$

was obtained by the dependency, suggested in (Tsvelodub,

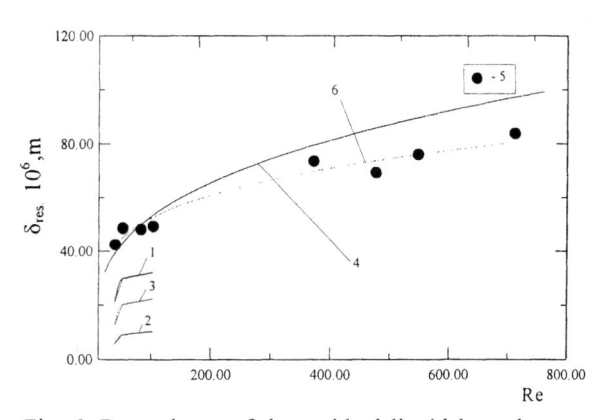

Fig. 6. Dependence of the residual liquid layer between large waves $\delta_{res.}$ on Reynolds number.
1, 2, 3, - calculated dependencies by (White and Tallmadge, 1965; Levich 1959; Brauner and Maron, 1983). 4 - empirical dependency from (Vorontsov, 1999), (1); 5 – experimental data, 6 –calculation by (2).

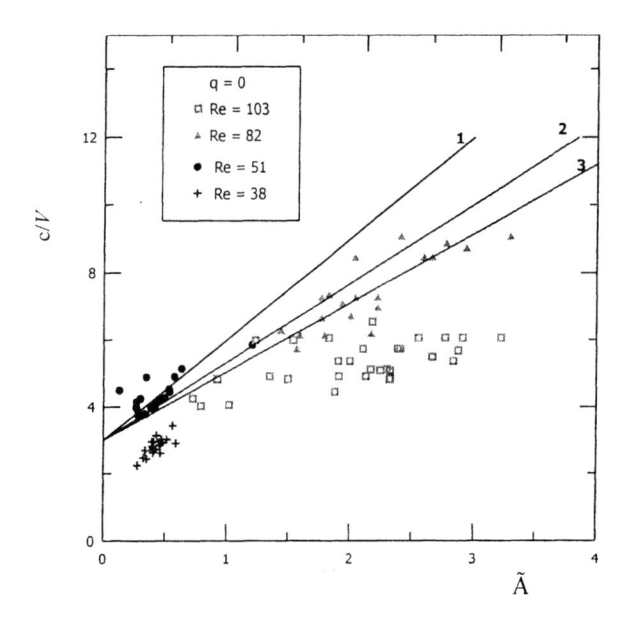

Fig. 7. The dimensionless phase velocity of large waves c/V vs. relative amplitude Ã.
1,2,3 – calculated dependencies by (3) (Alekseenko, Nakoryakov and Pokusaev, 1992), (4) (Tsvelodub, 1980), (Chang, 1989) respectively.

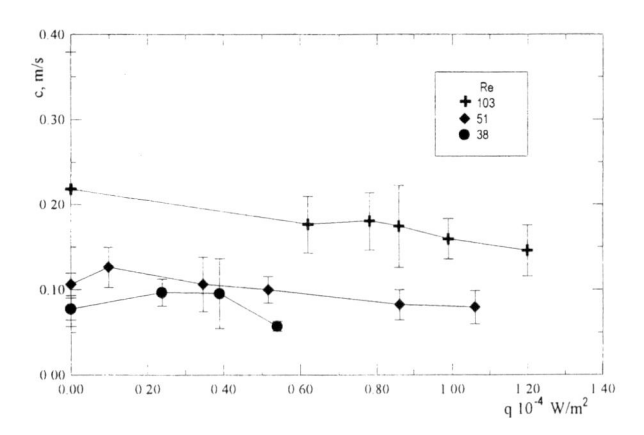

Fig. 8. The large wave velocity vs. the heat flux density at various Reynolds numbers.

1980) for succession of solitary waves with a smooth residual layer between them, under conditions of Re~1;
$$\tilde{A} = \frac{\delta_{max} - \delta_{res}}{\delta_{res}} \ll 1.$$

Curve 3 corresponds to analogous dependency from (Chang, 1989) for succession of solitary waves.

It is clear from the figure that these experimental data on the phase velocity of real waves under adiabatic conditions satisfactorily coincide with theoretical curves and in area $\tilde{A} > 1$.

Intensive evaporation effects not only the shape of wave surface, but also the phase velocity of large waves. Experimental results on the average phase velocity vs. the heat flux density are shown in Fig. 8 for various Reynolds numbers. The phase velocity was determined via registration of wave propagation time from one probe to another in

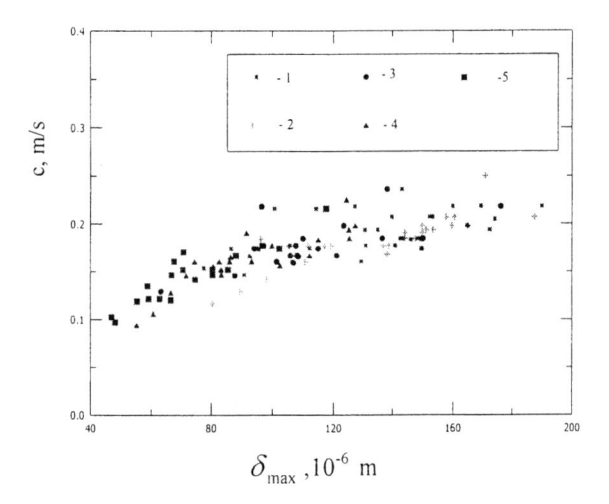

Fig. 9. The phase velocity of large waves vs. absolute amplitude at various heat flux densities. $Re_{in} = 103$.
1 - q=0, 2 - q=0.62·10⁴ W/m², 3- q=0.78·10⁴ W/m², 4- q=0.86·10⁴ W/m² 5 - q=1.2·10⁴ W/m² (formation of unstable "dry" spots).

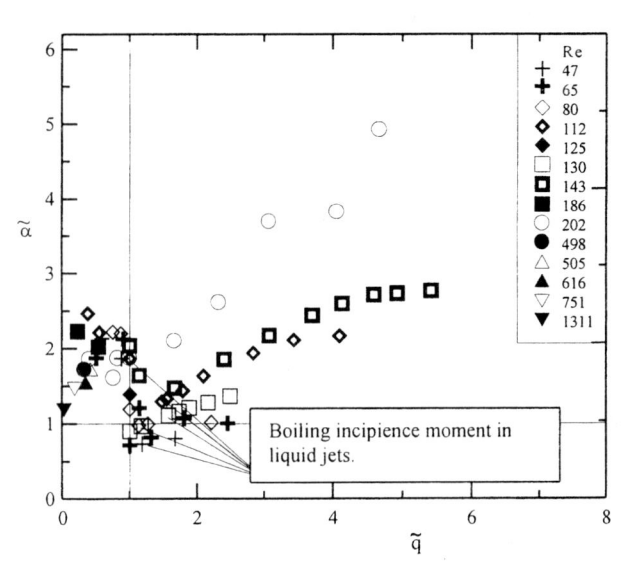

Fig.10. Dependence of dimensionless heat transfer coefficient $\tilde{\alpha} = \alpha_{exp}^{aver} / \alpha_{evap}^{aver}$ on heat flux density $\tilde{q} = q/q_{f.b.}$ at various Reynolds numbers.

measurement points x_2 and x_3. According to this figure, with a rise in the heat flux density, the average phase velocity of large waves decreases significantly. The most considerable alteration of the phase velocity is observed in the area of high heat flux densities, where residual layer thickness decreases and local time-unstable "dry" spots appear.

Results of experimental data generalization on the phase velocity of large waves are shown in Fig. 9. depending on their amplitude at various heat flux densities. It is clear that with a decrease in absolute amplitude of large waves at intensive evaporation, reduction of their phase velocity occurs. More drastic change in the phase velocity of large waves is observed under the regimes, when local unstable "dry" spots are formed at the highest heat flux density (black squares in Fig. 9.). Under these conditions, propagation of large waves takes place over the non-wetted surface. Perhaps, drastic deceleration of wave crest propagation can be explained by the additional effect of reaction forces of vapor release at intensive liquid film evaporation within the meniscus zone.

When reaching the threshold values of the heat flux density $q_{f.b.}$, formation of stable (non-vanishing) "dry" spots occurs. Stationary cylinders are formed in the incident liquid in front of "dry" spots. In the area of low Reynolds numbers ($Re_{in} < 60 \div 80$), monotonous spreading of "dry" spots at $q = q_{f.b.}$, can lead to complete drying of the heat-releasing surface.

Experimental data on average heat transfer intensity is shown in Fig. 10 in a dimensionless view. These results are presented like dependency $\alpha_{exp}^{aver} / \alpha_{evap}^{aver} = f_i(q/q_{f.b.})$, where α_{evap}^{aver} is calculation of the heat transfer coefficient, average for the whole heat-releasing surface, at the laminar-wavy flow in the regime of evaporation according to the dependency from (Gimbutis, 1988) for q=const:

$$\frac{\alpha_{evap.}^{aver.}(v'/g)^{1/3}}{\lambda'} = 2.27(\text{Re}_{in.})^{-1/3} \times$$

$$\times \left[1 + 0.022\left(\frac{\text{Re}_{in} \cdot \text{Pr} \cdot \delta_{aver.}}{\ell_{in.r.}}\right)^{4/3}\right]^{1/4} \frac{\ell_{in.r.}}{\ell} +$$

$$+ 1.1\text{Re}_{aver.}^{-1/3}\left(1 + 0.02\text{Re}_{aver.}^{0.2} + 0.0009\text{Re}_{aver.}^{0.85}\text{Pr}^{0.65}\right) \times$$

$$\times \frac{\ell - \ell_{in.r.}}{\ell}, \qquad (5)$$

where the first term from the left corresponds to calculation of the heat transfer coefficient at the initial region of heat stabilization $\ell < \ell_{in.r.}$, the second term corresponds to calculation of the heat transfer coefficient at the stabilized region $\ell \geq \ell_{in.r.}$; the length of initial region was calculated by dependency from (Nakoryakov V.E., Grigorieva, 1980); averaged Reynolds number $\text{Re}_{aver} = \frac{\text{Re}_{in} + \text{Re}_l}{2}$, where local Reynolds number Re_l was calculated in the lower point of the heated zone. It is obvious from the figure that in the regime before formation of large "dry" spots with evaporating liquid jets between them ($0.3 < q/q_{f.b.} < 1$), experimental data on heat transfer coefficient is situated considerably higher than calculation data on heat transfer for the laminar-wavy flow of evaporating liquid film, obtained for the low heat flux densities. Perhaps, intensification of heat transfer under pre-crisis regimes is caused by significant local thinning of a lengthy residual layer between large 3D waves of intensively evaporating liquid film.

According to the figure, formation of "dry" spots ($\tilde{q} = q/q_{f.b.} \geq 1$) leads to a drastic decrease in the average heat transfer coefficient. Drying of the main part of the heat-releasing surface causes 2-3-fold reduction of the average heat transfer coefficient. With the following rise in the heat flux density, corresponding increase of the heat transfer coefficient is caused by boiling development in regular liquid jets. With a rise in temperature of the heat-releasing surface, expansion of zones of developed nucleate boiling both over the height and width within liquid jets provides multiple increase of the heat transfer coefficient. It is clear that at high heat flux densities up to crisis development, even in the presence of large unwetted zones, the average heat transfer coefficient can significantly exceed the average heat transfer coefficient at the laminar-wavy film evaporation without "dry" spots at the heat-releasing surface. The most considerable increase is observed for the highest Reynolds numbers.

We should note that quantitative contribution of the swirl flow within wave crests under conditions of intensive evaporation is still unknown. As it is shown in (Miyara, 1998), at Prandtl number, corresponding to liquid nitrogen at the saturation line under the atmospheric pressure, the current lines in a wave crests at the laminar flow can have the swirl-like character. It is shown theoretically that for the given Prandtl numbers at development of the swirl flow, an increase in heat transfer intensification can reach 50% in comparison with the waveless flow. To determine the relative contribution into average heat transfer from the components related to swirl formation in the wave crests and intensive evaporation of the residual layer with significantly non-uniform thickness, experimental studies on local heat transfer in the falling films of saturated liquid are required.

CONCLUSION

For the first time, experimental results on dynamics of the laminar-wavy flow of intensively evaporating film of saturated liquid over a vertical heated surface were obtained by measurements of the local liquid film thickness. For the studied heat transfer regimes, a considerable change in the average film thickness takes place along the stream at the scale, comparable with typical wavelengths of large waves. It was shown that:

- In the area of low heat flux densities, wavy film thinning occurs uniformly without considerably changes in the wave shape. It is shown that at low heat fluxes, an insignificant decrease is observed in the relative wave amplitude. In the area of high heat fluxes, an increase in the relative wave amplitude and non-uniform thinning of the residual layer at the scale of characteristic wavelength take place. This characterizes by a significant change in the form of probability distribution of the film thickness. Formation of the local zones with a very thin film in the residual layer will lead to intensification of heat transfer in the pre-crisis regimes. This effect should be considered, when developing different simulation models of heat transfer in intensively evaporating wavy films of liquid.

- When reaching the threshold values of the heat flux density, arising instability in intensively evaporating residual layer leads to formation of "dry" spots.

- Experimental data on the phase velocity of natural large waves under adiabatic conditions can be satisfactory described by the known dependencies for succession of stationary solitary waves.

- A most considerable decrease in the phase velocity of large waves with a rise in the heat flux density is observed in the regimes with formation of unstable "dry" spots.

The results obtained are important for the study of interconnection between wavy characteristics and local heat transfer, conditions of "dry" spot and crisis development in the falling laminar-wavy films of intensively evaporating liquid.

This work was financially supported by the Russian Fund for Basic Research (Grant No. 00-02-17923).

NOMENCLATURE

A – relative wave amplitude
c – phase velocity of waves, m/s
C – probe capacity, μF
g – acceleration of gravity, m/s^2
f – beat frequency, 1/s
f_i – function
F – clock rate, 1/s
$l_{in.r.}$ – the length of initial region, m
q – heat flux density, W/m^2
$q_{f.b.}$ – heat flux density, when time-stable "dry" spots appear
p – probability density, m^{-1}
Re – Reynolds number (Re = $4\Gamma/\nu'$)
$Re_{in.}$ – Reynolds number at the inlet of the heated area
Re_l – Reynolds number calculated by the heat balance
S – the slot width, m
T – temperature, K
ΔT – difference between the temperature of heat-releasing surface and the temperature of liquid saturation at a given pressure, K
x_n – distance from the upper heater edge to the measurement point, where n = 1, 2, 3; m
x^+ – the distance from the outlet slot to measurement point, m

Greek symbols

Γ – liquid flow rate per a length unit, m^2/s
δ – liquid film thickness, m
$\delta_{res.}$ – average thickness of a residual layer of liquid film, m
$\delta_{aver.}$ – average thickness of the liquid film, m
$\delta_{max.}$ – liquid film thickness at the top of the crest, m
λ' – thermal conductivity coefficient, W/mK
$\lambda_{aver.}$ – average wave length, m
ν' – kinematics viscosity coefficient, m^2/s
τ – time, s

Subscript

aver.   average value
in.r.   initial region
evap.   evaporation
exp.   experimental
in.   inlet
f.b.   film breakdown
max.   maximal value
min.   minimal value
res.   residual layer

REFERENCES

Alekseenko, S.V., Nazarov, A.D., Pavlenko, A.N., Serov, A.F., Chekhovich, V.Yu., 1997, The Flow of Cryogenic Liquid Film over a Vertical Surface, *Thermophysics and Aeromechanics*. Vol. 4, № 3. pp. 307-318.

Bankoff, S.G., 1971, Stability Study of Liquid Flow down a Heated Inclined Plane, *Int.J. Heat Mass Transfer*, V. 14. P. 377

Brauner, N., Maron, D.M., 1983, Modeling of Wavy Flow in Inclined Thin Films, *Chem. Eng. Sci.*, Vol. 38. pp. 775-788.

Chang, H.-C., 1989, Onset of Nonlinear Waves on Falling Folms, *Physics Fluids A.*, Vol.1 №8. pp. 1314-1327.

Lyu, T.N., Mudawar I., 1991, Statistical Investigation of the Relationship Between Interfacial Waviness and Sensible Heat Transfer to a Falling Liquid Film, *Int. J. Heat Mass Transfer*, Vol. 34. № 6. pp. 1451-1464.

Nakoryakov, V.E., Grigorieva, N.I., 1980, Calculation of Heat Transfer at Isothermal Absorption at the Initial Region of the Falling Liquid Film, *Teoreticheck. osnovy khim. Tekhnol*, Vol. 14, №4, pp. 483-488.

Pavlenko, A. N., Lel', V.V., 1997, Heat Transfer and Crisis Phenomena in Falling Films of Cryogenic Liquid, *Russian Journal of Engineering Thermophysics*, № 3-4, Vol. 7. pp. 177-210.

Spindler, B., 1982, Linear Stability of Liquid Films with Interfacial Phase Change, *Int.J. Heat Mass Transfer*, Vol. 25. № 2

Trifonov, Yu.A., 1993, The Effect of Waves with Finite Amplitude on Evaporation of a Liquid Film, Falling down a Vertical Wall, *J. Appl. Mech. Tech. Phys*, Vol. 34. № 6. pp. 64-71.

Tsvelodub, O., 1980, Solitary Waves on a Falling Film at Moderate Flow Rates of Liquid, *J. Appl. Mech. Tech. Phys*, №3. pp. 64-66.

Vorontsov, E.G., 1999, Thermal Conductivity of Falling Films, *Teoreticheck. osnovy khim. tekhnol.*, Vol. 33, №2, pp. 117-127.

White, D.A., Tallmadge J.A., 1965, Theory of Drainage of Liquids on Flat Plates, *Chem. Eng. Sci.*, Vol. 20. – pp. 33.

Alekseenko, S.V., Nakoryakov V.E., Pokusaev B.G., 1992, *Wavy Film Flow*. Nauka, Sib. otd, Novosibirsk.

Gimbutis, G., 1988, *Heat Transfer at Gravitation Flow of a Liquid Film*, Mokslas, Vilnius.

Levich, V.G., 1959, *Physical-Chemical Hydrodynamics*. Fizmatgiz, Moscow.

Vorontsov, E.G., Tananaiko, Yu.M., 1972, *Heat Transfer in Liquid Films*, Tekhnika, Kiev.

Adomeit, P., Leefken, A., Renz, U., 2000, Experimental and Numerical Investigation on Wavy Films, *Proc. of 3rd European Thermal-Science Conference*, 2000, Heidelberg, Germany, 10-13 September 2000. Vol.2. pp. 1003-1009.

Miyara, A., 1998, Numerical Analysis on Heat Transfer of Falling Liquid Films with Interfacial Waves, *Proc. of the 11th IHTC (Heat Transfer Conference)*, 1998, Korea. Kyondju. August 23-28. Vol. 2. pp. 57-62.

VAPORIZATION, CONDENSATION AND ABSORPTION AUGMENTATION TECHNIQUES

POOL BOILING EXPERIMENTS WITH LIQUID NITROGEN ON ENHANCED BOILING SURFACES

Vijayaraghavan Srinivasan, J. D. Augustyniak & M. J. Lockett

Praxair Technology Center, Praxair Inc., Tonawanda, NY 14150, USA; e-mail:vijay_srinivasan@praxair.com

ABSTRACT

The primary purpose of this work was to compare the boiling performance of the UOP High Flux™ surface with the Wolverine Turbo BII LP surface at liquid N_2 temperatures (~80K). Pool boiling experimental data were obtained for the two surfaces in the heat flux range, 3 to 60 kW/m², applicable to air separation plants. Kedzierski at NIST, using R-123 as the working fluid, carried out a similar comparison of the boiling performance of the two surfaces. Analysis of the results shows that (1) the boiling heat transfer coefficient in liquid N_2 was much larger than for R-123, (2) Turbo BII LP was found to have equal or superior performance over the High Flux surface, at low ΔTs and (3) a main condenser of an air separation plant could be designed, in theory, with Turbo BII LP tubes that would result in roughly 20% savings in capital cost.

INTRODUCTION

Enhanced boiling surfaces are extensively used in cryogenic applications. For example, the main condenser/reboilers of Praxair's cryogenic air separation plants employ UOP's High Flux™ tubes for enhancing the boiling and condensation heat transfer. High Flux tubes have a porous coating on the inside surface for enhancing boiling heat transfer and have longitudinal flutes on the outside surface to enhance condensation (see Figs. 1a and 1b). Due to this double enhancement, it is possible to obtain close temperature approaches (ΔTs) in the main condenser. It is very important in air separation plants to operate heat exchangers at a small temperature difference (below 2K) to minimize power requirements for compression of air.

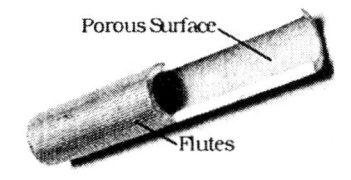

Fig. 1a. Doubly Enhanced High Flux Tube

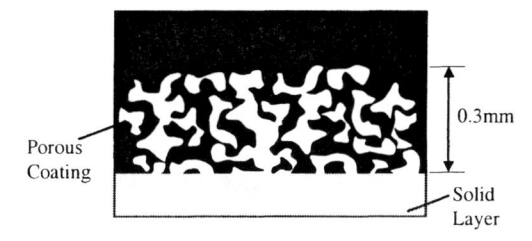

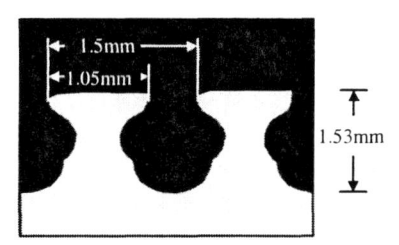

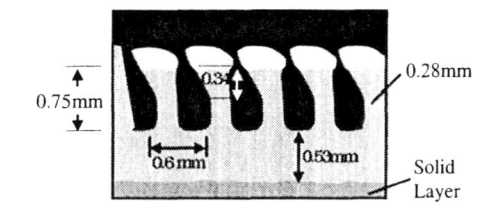

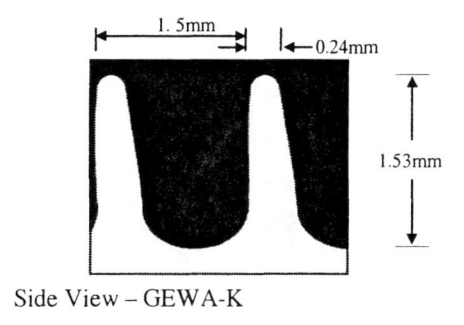

Fig. 1b. Typical Enhanced Surfaces for Boiling

© Copyright 2001 Praxair Technology, Inc.

Similar to porous surfaces for boiling enhancement, a number of other structured surfaces have been successfully used in the refrigeration industry. Some of these structured surfaces are commercially known as GEWA-K, GEWA-T, Turbo BII LP, etc. See Fig. 1b for details. Unlike the porous surfaces, these structured surfaces have well defined cavity sizes and shapes to enhance boiling heat transfer. However, almost all of these surfaces are manufactured on the external surface of the tubes.

A study by Kedzierski (1995) at NIST compared the performance of structured surfaces such as Turbo BII LP, GEWA-K and GEWA-T with the High Flux surface, where pool-boiling experiments were conducted with R-123. His work showed that the performance of Turbo BII LP at low ΔT's was superior to all other enhanced surfaces, including High Flux tubes.

Since performance at low ΔTs is of great interest to the air separation industry, the objective of the present work was to compare the performance of the Turbo BII LP structured surface with the High Flux™ surface using a fluid that boils at cryogenic temperatures (80 to 100K).

BOILING ON ENHANCED SURFACES

Enhanced boiling surfaces can be classified as structured surfaces or porous surfaces. The heat transfer associated with the boiling process is strongly influenced by two parameters — "boiling nucleation" and "boiling site density." Boiling nucleation, or the formation of vapor bubbles at the heated surface, is fundamental to the boiling process. Similarly, the parameter "boiling site density" represents the number of active boiling nucleation sites per unit area on the heated surfaces. Enhanced boiling surfaces improve the boiling heat transfer by enhancing boiling nucleation and boiling site density.

Structured Surfaces

Bankoff (1957) studied the boiling performance on plain surfaces and surmised that boiling initiated from pre-existing vapor embryos that are trapped beneath the scratches and cavities in the heated surface. Understanding the mechanism by which the vapor embryos are trapped beneath the flowing liquid film forms the basis for designing structured surfaces for enhancing boiling.

Among the various parameters, Griffith & Wallis (1960) demonstrated that the geometry of a cavity is important for containing stable vapor nuclei. This led them to study the boiling process in re-entrant cavities similar to those shown in Fig. 2. They found that the superheat required to initiate generation of vapor bubbles was a function of the mouth diameter. The stability of the trapped vapor nucleus was dependent on the shape of the cavity as well as the contact angle of the liquid at the vapor interface inside the cavity. Benjamin and Westwater (1961) reported similar results. Thus, the design of enhanced boiling surfaces focused on developing efficient re-entrant cavities. Commercially available structured surfaces employ metal working processes to form re-entrant nucleation cavities. These cavities are interconnected below the metal surface to permit inter-communication between the cavities.

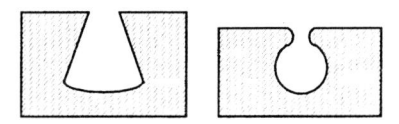

Fig. 2. Typical Reentrant Cavity Designs

A very useful chronological summary of various enhanced boiling surfaces has been provided by Thome(1990).

In a recent study by Kim et al. (2001), structured surfaces were classified into three groups: (a) those having pores, (b) those having narrow gaps or, (c) those having pores with connecting gaps. Kim, et al., as did Chien and Webb (1998), systematically investigated the effect of geometry on the boiling performance of the structured surfaces. It was found that the performance of tubes under category (a) or (b) was dependent on heat flux, whereas (c) was independent of heat flux. The study by Kim also found that the performance of the tube was dependent on fluid properties, in particular the reduced pressure.

Porous Surfaces

Porous surfaces for enhancing boiling heat transfer were first developed by Milton (1971) at Union Carbide. Tubes with porous surfaces are commercially known as High Flux™ tubes. Different techniques, such as sintering, brazing, etc., are available to obtain a porous surface and result in a wide variation in the cavity size and cavity shapes. The initial particle size, the thickness of the porous layer, and the resulting porosity are some of the parameters that govern the performance of the enhanced surface.

O'Neill et al. (1972) first studied the mechanism of boiling on porous surfaces. It was explained that entrapped vapor in the pores initiated the boiling process. The pores within the matrix are interconnected, establishing communication, so that any vapor that is generated can find its way to the free liquid interface. Similarly, the cavities are replenished with liquid flow from the surface for further evaporation.

EXPERIMENTAL APPARATUS AND PROCEDURES

Two sets of experimental apparatus were used for the pool boiling experiments. This was necessitated by the fact that High Flux tubes have boiling enhancement on the inside surface of the tube and Turbo BII LP tubes have it on the external surface.

Fig. 3 shows a schematic view of the experimental apparatus used for testing High Flux tubes. The apparatus used for testing Turbo BII LP tubes is shown in Fig. 4. The location of the RTDs and thermocouples are also shown in the Figures. The tubes were tested in the vertical orientation.

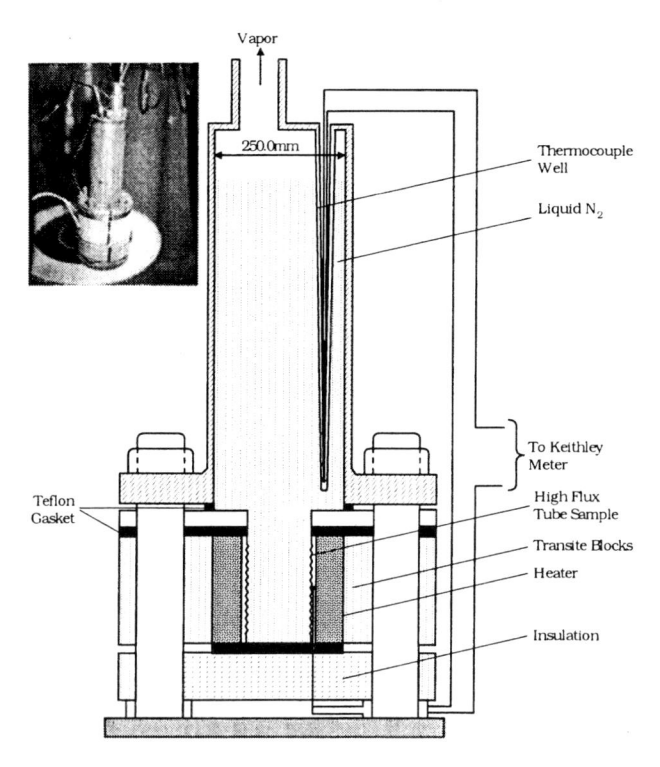

Fig. 3. Pool Boiling Test Apparatus for High Flux Tubes

The apparatus shown in Fig. 3 consisted of a 250mm cylindrical vessel with a vapor outlet at the top and a flanged bottom. The cylindrical vessel was filled with liquid nitrogen. The bottom of the vessel was open and connected to the test tube so that liquid could continuously flow from the vessel to the test tube. The test tube (75.0mm long) was heated externally with electrical heaters and the entire assembly was insulated to minimize heat leaks.

The experimental arrangement for Turbo BII LP tubes is schematically shown in Fig. 4. It consisted of a large cylindrical vessel about 500mm in diameter and 450mm tall and it contained about 25 liters of liquid N$_2$ during the experiments. The large volume of liquid ensured that the liquid level in the storage vessel remained almost constant during a steady-state test run. The test tube was electrically heated by inserting an appropriate electric cartridge heater. The tube specimens were 150mm long. The tube wall temperature in both cases was measured by inserting a Type E thermocouple inside the tube wall. The hole through the tube wall was drilled parallel to the tube axis (half way down the length of the tube) using the EDM technique, so that the thermocouple junction was in contact with the tube surface at the mid-point of the tube. In order to provide good thermal contact, the gap between the thermocouple wires and the tube hole was filled and glued with thermal epoxy that had similar thermal conductivity to that of the base tube material. This technique ensured accurate measurement of the wall temperature. The thermocouple junction did not come into contact with either the liquid pool or the electrical heater. As shown in Fig. 4, the temperature of the liquid nitrogen pool was measured, using a platinum RTD, at the same vertical location in the vessel where the wall temperatures in the test tube were measured. The measured liquid pool temperature was typically 78K, which corresponds to the boiling temperature of liquid nitrogen at the pressure of operation (104.0kPa). Liquid pool temperature was not measured for the High Flux tests.

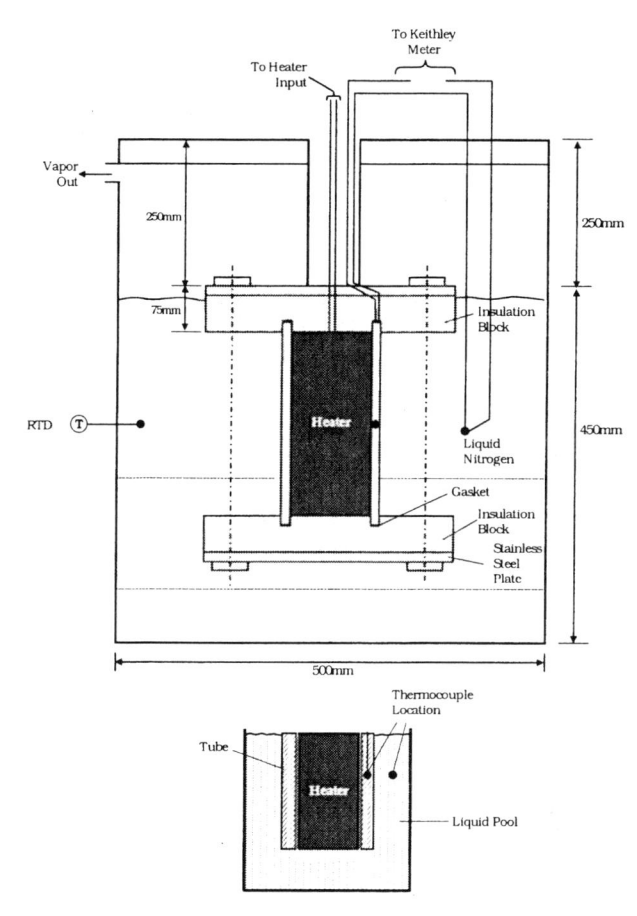

Fig. 4. Pool Boiling Test Apparatus for Turbo BII LP Tubes

RESULTS AND DISCUSSION

The heat flux from the electrically heated tube surface was calculated from the measured values of the power input as follows:

$$\dot{q} = \frac{Q}{A}$$

where Q is the heat input and A is the tube surface area (based on smooth surface diameter, inside or outside).

The wall superheat (ΔT) was obtained by differentially connecting the thermocouple outputs from the wall temperature, T_W and the saturation temperature T_{sat} of the liquid.

Thus,

$$\Delta T = T_W - T_{sat}$$

and the boiling heat transfer coefficient $h = \dfrac{\dot{q}}{\Delta T}$

The heat-input (Q) to the heater was measured using a Nu-Watt digital Wattmeter.

Accuracy of Measurements

Thermocouples and RTDs were calibrated in liquid nitrogen and oxygen baths. They were calibrated to an accuracy of ±0.025K. The thermocouples that were differentially connected to measure $T_W - T_{sat}$ were first initialized (zeroed) in the instrument. The uncertainty in the heat transfer surface area was 2%. Similarly, the uncertainty in the electrical output measurement was ±5%. The procedure outlined in Kline & McClintok (1953) was used to perform an error analysis. It was found that the maximum uncertainty in h was ±7% at a heat flux of 120 kW/m^2 and ± 25% at 10 kW/m^2.

High Flux Results

The ID of the tube was 16.97 mm and the length was 71.33 mm so that the surface area was 0.003824 m^2. The heat input was varied from 0.0 to 150.0 Watts, giving a heat flux range from 0 to 41 kW/m^2. The reproducibility of the experimental data was verified, first by increasing and then by decreasing the heat flux. The results are shown in Fig. 5. The data shows no significant difference in the values of ΔT at a given heat flux, irrespective of how the heat flux value was approached.

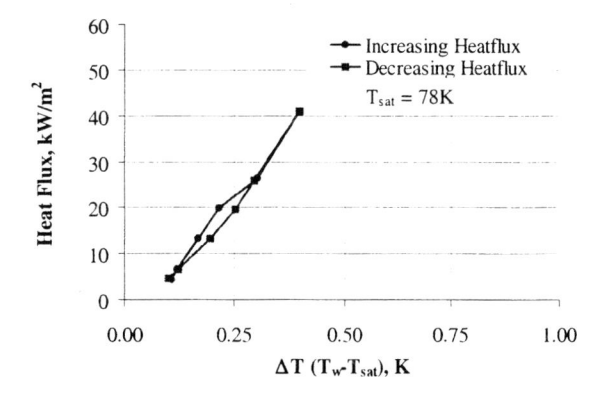

Fig. 5. Pooling Boiling Results in Liquid N$_2$ – High Flux Tube 1

Results of tests with liquid N$_2$ for three sample tubes are shown in Fig. 6. Even though there is a slight difference in the performance of the individual tubes, all the tubes exhibit a similar trend or slope of the pool boiling curve. The differences in performance between the sample tubes are attributed to the difference in porous surface characteristics arising from variations in the manufacturing process. Kedzierski's results with refrigerant R123 are also shown in Fig. 6. The pool boiling curve for liquid N$_2$ is shifted to the left compared to the curve for R123, clearly implying a strong interaction between the properties of the fluid and the geometry of the cavity.

The curves (Figs. 5 & 6) imply that in order to initiate nucleate boiling, liquid nitrogen requires a wall superheat of the order of 0.2K, compared to R-123 that requires about 1.5K (Fig. 6). The bubble nucleation superheat is strongly influenced by the surface tension of the fluid. A smaller value of surface tension provides less resistance to bubble formation/growth and therefore requires less wall superheat. Carey (1992) has shown that the minimum wall superheat required to initiate nucleate boiling can be determined from the following expression:

$$(T_l - T_{sat})_{min} = \frac{2\sigma\, T_{sat}\, \vartheta_{fg}}{h_{fg}\, r_{min}}$$

The above equation is based on a number of idealizations. However, the equation serves the purpose of explaining the dependence of wall superheat on surface tension. The surface tension of nitrogen is 8mN m^{-1} whereas for R-123 it is 18mN m^{-1} at the experimental condition.

The reduced pressure is also said to strongly influence the pool boiling characteristics of the fluid (Kim et al., 2001). For the experimental data shown in Figs. 5 thru 9, the reduced pressure for R-123 is 0.0107 and for liquid N$_2$ it is 0.0304. Therefore, in addition to the influence of surface tension, the superior performance of both the High Flux and the Turbo BII LP surfaces in liquid N$_2$ compared to R-123 can be attributed to the higher value of reduced pressure for liquid N$_2$ at the experimental conditions.

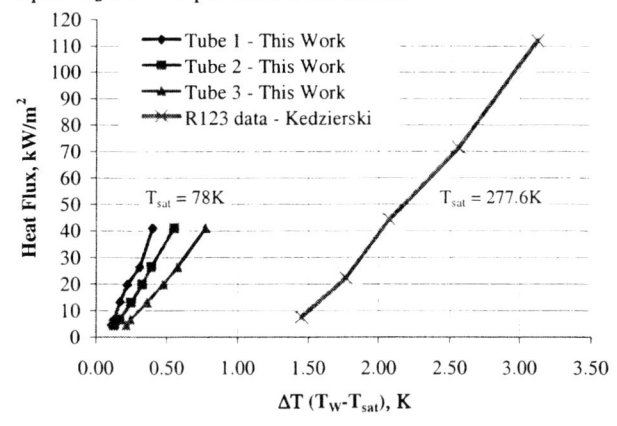

Fig. 6. Pool Boiling Results in Liquid N$_2$ – High Flux Tubes

Turbo BII LP Results

The OD of the tube was 19.05 mm and the length of tube 1 was 184.1 mm giving a surface area of 0.011 m^2. The length of tube 2 was 152.4 mm giving a surface area of 0.091 m^2. The heat input was varied form 0 to 1250 W for tube 1 and 0 to 1500 W for tube 2, so that the heat flux in the experiments ranged between 0 to 162 kW/m^2. Similar to the High Flux tubes, the reproducibility of the experimental data was verified by measuring the wall superheat, first by increasing and then by decreasing the heat flux and the results are shown in Fig. 7.

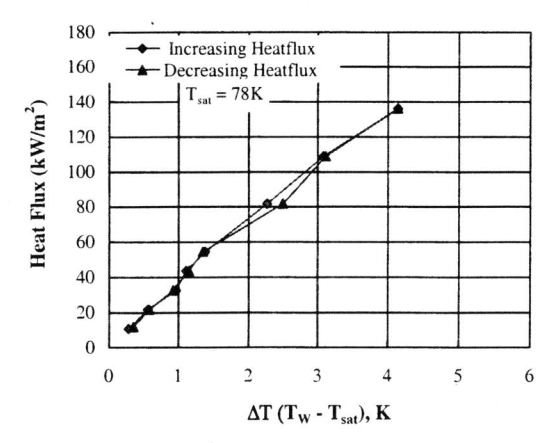

Fig. 7. Pool Boiling Results in Liquid N_2 – Turbo BII LP Tube 1

The pool boiling curves obtained for the two sample tubes tested are shown in Fig. 8. The variation in the performance of the Turbo BII LP tubes is significantly less than it was for the High Flux tubes. In the case of Turbo BII LP, the enhanced surface is obtained by machining and forming. The cavity size and geometry is expected to be more uniform and, hence, the performance is subjected to only a small variation. Similar to the High Flux tube results, the pool-boiling curve for liquid N_2 is shifted to the left compared to the curve for R123.

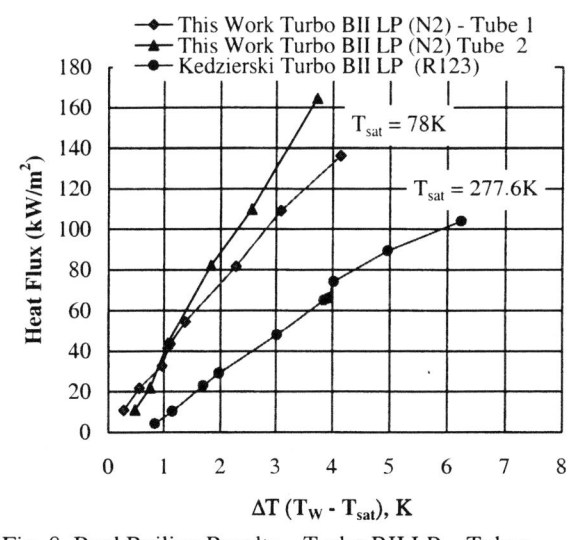

Fig. 8. Pool Boiling Results – Turbo BII LP – Tubes

It is interesting to note that Thome (1990) reached similar conclusions, where the boiling curves for enhanced tubes varied from one fluid to another.

In Fig. 9, the pool boiling curves for High Flux (tube 2) and Turbo BII LP (tube 1) are plotted for liquid N_2 and R123. The performance of Turbo BII LP is superior to High Flux tubes for R123 in the low range of ΔT (below 1.5K). The performance with liquid N_2 shows that both types of enhanced surfaces perform equally well at very low ΔT (below 0.3K), but otherwise High Flux is superior. However, if the performance of High Flux (tube 3) is compared to Turbo BII LP tubes (not shown in Figs.), the performance of the latter is superior at low ΔTs.

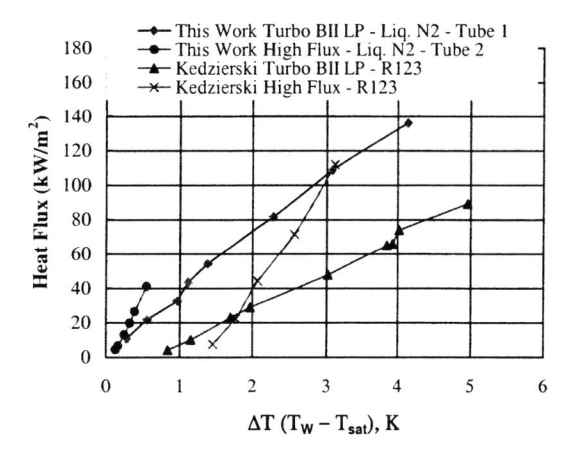

Fig. 9. Pool Boiling Results – Turbo BII LP and High Flux Tubes

APPLICATION TO MAIN CONDENSER/ REBOILERS

The results of these pool-boiling tests show the possible suitability of Turbo BII LP tubes for obtaining low ΔTs in several applications involving boiling and two-phase flow in air separation plants. One of the important applications is the main condenser/reboiler which usually is placed at the bottom of the distillation column in a pool of liquid oxygen. Low pressure liquid oxygen is vaporized to provide the vapor boil-up and high pressure vapor nitrogen is condensed to provide reflux for the distillation columns. Main condensers designed with High Flux tubes are manufactured as standard modules for air separation applications. A typical air separation plant with a production capacity of 100,000 cfh (NTP) of oxygen, depending on the ΔT (overall), requires at least one standard High Flux module. A cost break-down of the different components shows that the tubes account for more than 50% of the total cost and substantial savings can be achieved if alternative lower cost tubes can be used.

Table 1. Design Comparison of Main Condenser with High Flux and Turbo BII LP Tubes

	Standard High Flux Module	Main Condenser With Turbo BII LP Tubes
Module diameter, mm	711.0	813
Module height, m	2.3	2.3
Tube surface area, m^2	67.7	84.5
Number of tubes	384	481
Typical design heat flux, kW/m^2	6.3	5.0

In Table 1, the conceptual design details of a main condenser with both High Flux and Turbo BII LP tubes are shown. The performance of the modules is equivalent. However, boiling takes place on the shell-side and condensation inside the Turbo BII LP tubes. Boiling oxygen outside the tubes (shell-side) could give rise to the possibility of liquid deficient regions in the tube bundle, which could result in oxygen boiling to dryness. In such situations hydrocarbon accumulation could occur on the tube surface affecting the operational safety of the main condenser. One solution could be the provision of flow distribution baffles on the shell side. Another alternative could be to develop techniques for forming structured surfaces on the inside surface of the tubes while remaining cost competitive. In the absence of such developments, High Flux tubes are likely to still be preferred for thermosyphon main condenser/reboilers. Finally, if down-flow boiling of oxygen is adopted on the outside of the tubes, the use of machined surfaces, such as Turbo BII LP, could be attractive and a Praxair patent is pending on this concept.

CONCLUSIONS

1. Turbo BII LP tubes were found to have equivalent, or in some cases superior, performance at low ΔTs to the High Flux surface when boiling liquid nitrogen, in agreement with the previously reported results of Kedzierski for R-123.

2. The heat transfer coefficient in liquid nitrogen was much larger than for R-123, although the trend was similar.

3. The plot of heat flux vs. ΔT showed a horizontal shift in the pool boiling curve for liquid nitrogen compared with R-123, implying that boiling nucleation occurs at a lower ΔT for liquid nitrogen.

4. Turbo BII LP can offer significant cost advantage over High Flux tubes. However, safety considerations with regard to boiling liquid oxygen restricts its use in air separation plants, as currently boiling of oxygen on the outside tube surface is not practiced. If the Turbo BII LP could be economically manufactured on the inside surface, there could be an opportunity for significant cost savings in air separation.

REFERENCES

1. Bankoff, S. G., 1957, Ebullition from Solid Surfaces in the Presence of a Pre-existing Gaseous Phase, Trans. ASME, Vol. 79, pp. 735-741

2. Benjamin, J.E., and Westwater J. W., 1961, Bubble Growth in Nucleate Boiling in a Binary Mixture, Int. Dev. Heat Transfer, pp. 212-218.

3. Chien, L. H., and Webb, R. L., 1998, A Parametric Study of Nucleate Boiling on Structured Surfaces, Part I: Effect of Tunnel Dimensions, J Heat Transfer, Vol. 120, pp. 1042-1048.

4. Griffith, P. and Wallis, J. D., 1960, The Role of Surface Conditions in Nucleate Boiling, Chem. Engr. Progress Symp. Series, Vol. 56, pp. 49-63

5. Kline, S. J., and McClintock, F. A., 1953, The Description of Uncertainties in Single Sample Experiments, Mech. Engg., Vol. 75, pp. 3-9.

6. Nae-Hyun Kim, and Kuk-Kwang Choi, 2001, Nucleate Pool Boiling on Structured Enhanced Tubes having Pores with Connecting Gaps, Int. J of Heat Mass Transfer, Vol. 44, pp. 17-28.

7. O'Neill, P.S., Gottzmann, C. F., and Terbot, J. W., 1972, Novel Heat Exchanger Increases Cascade Cycle Efficiency for Natural Gas Liquefaction, Adv. Cryo. Engg., Vol. 17, pp. 420-437

8. Kedzierski, M. A., 1995, Calorimetric and Visual Measurements of R-123 Pool Boiling on Four Enhanced Surfaces, National Institute of Standards and Tech. Rept. NISTIR 5732, Gaithersburg, Maryland.

9. Carey, V. P., 1992, Liquid-Vapor Phase-Change Phenomena, Hemisphere, Washington D.C.

10. Thome, J. R., 1990, *Enhanced Boiling Heat Transfer*, Hemisphere, New York.

11. Milton, R. M., 1971, Heat Exchange System with Porous Boiling Layer, US Patent 3,587,730 assigned to Union Carbide.

NOMENCLATURE:

A = Tube surface area, m^2

h = Heat transfer coefficient, W/m^{2}K

h_{fg} = Latent heat of vaporization, J/kg

$\dot{q}$ = Heat flux, W/m^2

Q = Heat transferred, W

r_{min} = Minimum radius of curvature vapor-liquid interface, m

T_l = Liquid pool temperature, K

T_{sat} = Liquid saturation temperature, K

T_W = Tube wall temperature, K

σ = Surface tension, N/m

ΔT = $T_W - T_{sat}$, K

ϑ_{fg} = Specific volume at T_{sat} (difference between vapor and liquid), m^3/kg

BOILING OF HYDROCARBONS ON TUBES WITH SUBSURFACE STRUCTURES

R. Mertz, R. Kulenovic, P. Schäfer and M. Groll

Institute for Nuclear Technology and Energy Systems (IKE), University of Stuttgart,
Pfaffenwaldring 31, 70550 Stuttgart, Germany
E-mail: mertz@ike.uni-stuttgart.de

ABSTRACT

New enhanced tubular evaporation surfaces with subsurface structures, made of carbon steel St35.8, have been investigated in the frame of a r&d project funded by the European Union. The experiments are carried out in the pool boiling mode using the hydrocarbon propane as working fluid at saturation temperatures of T = 253 K to 293 K and heat fluxes in the range from 2 kW/m^2 to 100 kW/m^2. Single tubes and mini bundles (two tubes inline) were tested. A high speed video system was used to visualize the boiling phenomena and two-phase flow from the surface. For the evaluation of the visualization records a software for digital image processing, developed at IKE, was modified to extract important boiling parameters, e.g. bubble departure diameter and bubble frequency.

The single tube and mini bundle experiments carried out at IKE showed distinct improvements of the heat transfer up to a factor of 3, compared with the results of a smooth reference tube. For the given operating conditions the best enhanced tube was identified and recommended for tube bundle experiments by another project partner.

Finally, after successful tests with a tube bundle with 45 tubes in a staggered arrangement a compact heat exchanger prototype with a thermal duty of 5 MW was manufactured and installed in a petrochemical plant. The industrial prototype is integrated in the production line and the first results show an increase of the overall heat transfer coefficient of 75 %.

INTRODUCTION

Today the improvement of heat transfer equipment means to deal with heat transfer surfaces comprising very small structures with dimensions typically smaller than 1 mm. Such microstructures have originally been designed to increase the available heat transfer area, thereby providing an enhanced heat transfer, but the greater contribution to an enhanced heat transfer results from the influence of these very small structures on the boiling phenomena. The evaporation processes in such small structures are much more complicated than those occurring on smooth or low-finned surfaces and they cannot be described with models or correlations generated for those simple surfaces. This lack of knowledge concerning the heat transfer efficiency in general, the influence of working fluid, mixtures and contaminated fluids, etc., is an obstacle for the acceptance of enhanced surfaces by industrial users. On the other hand, the advantages of compact heat exchangers, e.g. reduced volume, less material and energy consumption, less fluid inventory leads to remarkable cost reductions and an increase in safety for the industrial users. To increase the acceptance of compact heat exchangers in the process industries, it is neccessary to provide experimental data from basic investigations with enhanced surfaces, as well as to prove their usability for industrial applications.

In the frame of several EU projects enhanced tubular heat exchanger surfaces with subsurface structures (in the submillimeter range) were developed, investigated and optimized over the last years by a group of international partners from industry, research centers and universities.

The task of IKE in these projects consists of the experimental investigation of the boiling performance of single tubes and two-tube mini-bundles in the pool boiling mode using hydrocarbons, especially propane, as working fluids. By a qualitative comparison of various tube structures the best surface for the given test parameters, e.g. working fluid and heat flux range, was identified and used for further investigations at the project partners. A major achievement of the work was the design and testing of an optimized structured evaporator tube and the design and construction of an industrial compact heat exchanger prototype. This prototype has been installed in a petrochemical plant and is successfully operated in the production line as a reboiler of a propane/propylene splitter.

The results of the experimental investigations with the working fluids propane and n-pentane (Mertz and Groll, 1998) showed, that there is a strong influence of the working fluid on the heat transfer performance. To study this influence, experimental investigations with other hydrocarbons, viz. iso-butane and mixtures of propane/iso-butane, are carried out in the frame of a current project.

Visualization techniques are used for the documentation of the boiling phenomena and the two-phase flow from the enhanced surfaces. Video sequences recorded with a high speed video camera provide important information about boiling parameters like bubble departure diameter and bubble frequency. Also the generation and growing of the bubbles at the outlet of the subsurface structures to the surrounding fluid and the interaction of bubbles at the same or different nucleation sites can be visualized. The evaluation of the videos is carried out with the digital image processing software IKEDBV.

EXPERIMENTAL SETUP

The test apparatus (Fig. 1) at IKE is designed for high pressure experiments using flammable and explosive liquids, e.g. hydrocarbons.

Fig.1. View of test rig

A standard apparatus for nuleation boiling investigations was described by Götz (1981) and Gorenflo et al. (1982). For the basic design of the IKE test rig some of their suggestions have been adopted. However, due to the more industrial application aimed experiments at IKE, the setup of the apparatus is simpler and the precision of the employed measurement devices is lower. A detailed description of the experimental apparatus is given in Groll et al. (1994-1995).

The visualization of the boiling phenomena and the two-phase flow from the evporation surfaces was carried out with a Kodak Motion Analyser 4540 (Fig. 2); for the illumination, the most difficult and important part of high speed recording, cold light sources with fibre optic cables are used.

Fig.2 High speed video system

The main heat transfer area of the enhanced surfaces is located in the subsurface structure. These channels and cavities are in the sub-millimeter range with pores and slits as openings to the surrounding working fluid (Fig. 3 and 4).

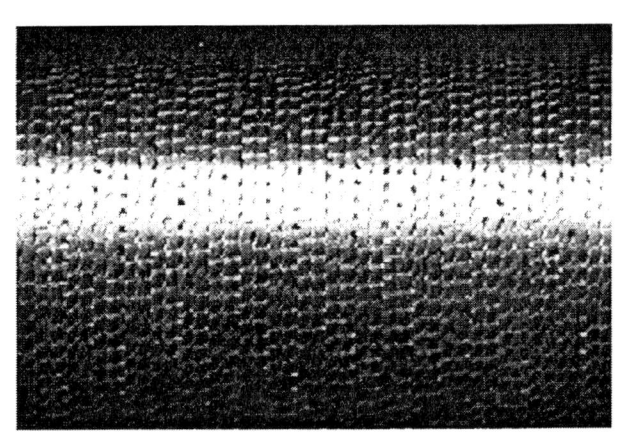

Fig.3 Enhanced surface, tube diameter 19.05 mm

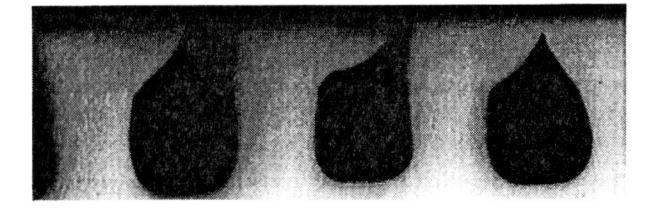

Fig.4 Subsurface structure

The experimental investigations are carried out with one smooth reference tube ($R_a = 1.09$ μm) and four structured tubes, all made of carbon steel St35.8, at saturation temperatures from T = 253 K to 293 K. The employed heat fluxes are in the range from q = 2 kW/m² to 100 kW/m². The tubes are 115 mm long and have an outer diameter of 19.05 mm. They are heated by a cartridge heater.

RESULTS

Single Tube Results

Figures 5 and 6 describe the results of the single tube experiments with the smooth reference tube (Fig.5) and the enhanced tube variant 4 (Fig.6).

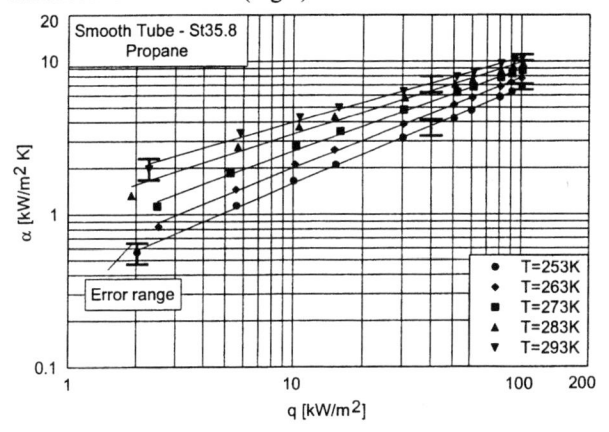

Fig.5 Smooth reference tube, working fluid propane

The results of the smooth tube show increasing heat transfer coefficients with increasing heat fluxes. This behaviour is well know from literature. The results can be compared with other investigations on boiling of propane from smooth tubes (Sokol et al., 1990, Pinto et al., 1996, Luke and Gorenflo, 1996, Gorenflo et al., 1998), taking into account the different surface roughness and wall material of the smooth tube used at IKE.

The heat transfer coefficients are in the range from (0.6 to 2 kW/m²K) to (7 to 11 kW/m²K) for saturation temperatures of 253 K and 293 K, respectively.

The enhanced tubes show a different behaviour. For low heat fluxes up to 30 kW/m² the heat transfer coefficients are increasing linearly, but for higher heat fluxes q > 30 kW/m² the slope of the heat transfer coefficients starts to decrease or even become negative. The degree of the slope decrease depends on the saturation temperature, for higher temperatures a more distinct decrease can be found.

The reason for the decrease of the heat transfer coefficient is the deactivation of the subsurface heat transfer area with increasing vapour production. Larger amounts of vapour cannot leave the subsurface structures fast enough through the small openings to the fluid and block parts of the subsurface cavities and channels.

The heat transfer coefficients of the enhanced surface variant 4 are between 1.5 kW/m²K and 5 kW/m²K for low heat fluxes (2 kW/m²), 10 kW/m²K and 21 kW/m²K for medium heat fluxes (up to 40 kW/m²) and between 16 kW/m²K and 18 kW/m²K for high heat fluxes (up to 100 kW/m²).

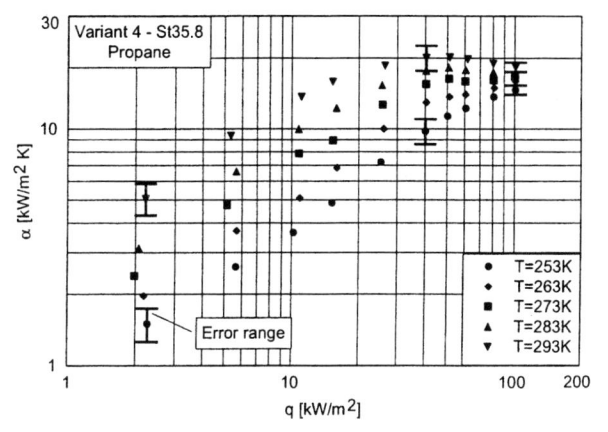

Fig.6 Enhanced tube variant 4, working fluid propane

The improvement factors $F = \alpha_{enh.}/\alpha_{ref.}$ of the variant 4 heat transfer surface are depicted in Fig. 7 for all tested saturation temperatures. Improvements up to a factor of 3 are found, especially for heat fluxes in the range from 10 kW/m² to 30 kW/m², which is of major interest for various industrial applications.

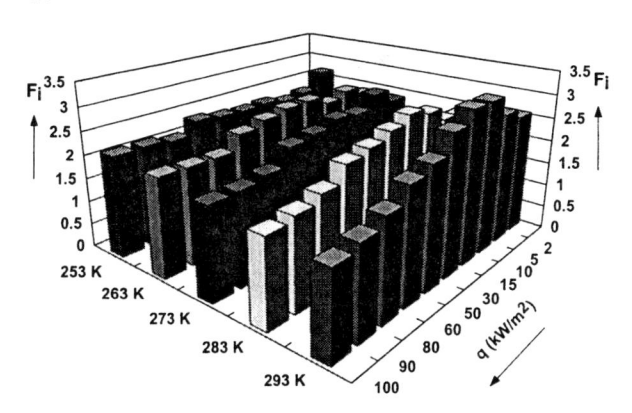

Fig.7 Improvement factors F_i, enhanced surface variant 4

Mini Bundle Results

To investigate the influence of the two-phase flow occurring in a tube bundle on the heat transfer of the individual tubes, experiments with two tubes in inline were carried out. Mini bundles with enhanced tubes variant 1 and 4 and for reference a smooth tube bundle were tested under the same conditions like the single tubes. Only the maximum employed heat flux was reduced to 70 kW/m², caused by the limited cooling systems.

Figures 8 and 9 show the results of the smooth tube bundle and the variant 4 tube bundle.

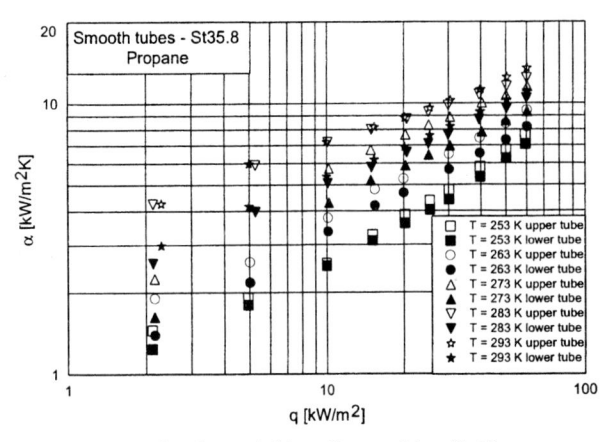

Fig.8 Smooth tubes mini bundle, working fluid propane

The individual tubes of the smooth tube bundle have higher heat transfer coefficients than the single tube (table 1). The upper tube always has higher heat transfer coefficients than the lower tube.

Table 1 Improvement factors of the smooth tube mini bundle. Comparison with the single smooth tube results for heat fluxes of 2 kW/m² and 70 kW/m²

T_{Sat}	253 K	263 K	273 K	283 K	293 K
lower tube	2 - 1.3	1.8 - 1.3	1.5 - 1.3	2 - 1.3	1.5 - 1.2
upper tube	2.5 - 1.4	2.4 - 1.5	2 - 1.6	3.3 - 1.5	2.1 - 1.5

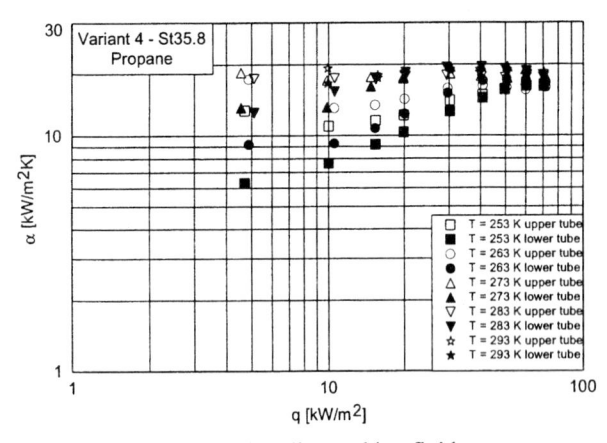

Fig.9 Variant 4 mini bundle, working fluid propane

The results of the variant 4 mini bundle are depicted in Fig. 9. Again for low heat fluxes the upper tube has higher heat transfer coefficients than the lower tube, especially for the lower saturation temperatures, but with increasing heat flux the differences become smaller and for high heat fluxes (q > 40 kW/m²) almost no differences can be found (table 2).

Table 2 Improvement factors of the variant 4 mini tube bundle. Comparison with the variant 4 single tube results for heat fluxes of 5 kW/m² and 70 kW/m².

T_{Sat}	253 K	263 K	273 K	283 K	293 K
lower tube	2.4 - 1.3	2.5 - 1.1	2.8 - 1.1	1.95 - 1	1.7 - 0.9
upper tube	4.8 - 1.3	4.6 - 1.1	3.9 - 1.1	2.7 - 1	2.1 - 0.9

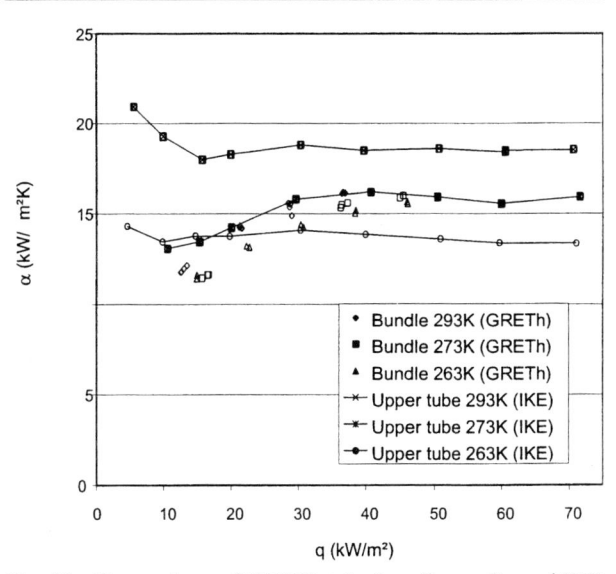

Fig. 10 Comparison of GRETh tube bundle results and IKE mini bundle results (upper tube), enhanced surface variant 4, working fluid propane

The results of the experiments carried out at GRETh (Groupement pour la Recherche sur les Echangeurs Thermiques), CEA Grenoble with a variant 4 tube bundle (Mertz et al., 2000) with 45 tubes in staggered arrangement show a good agreement with the results of the upper tube of the IKE variant 4 mini bundle (Fig. 10), especially for heat fluxes in the range of industrial interest (20 kW/m² - 50 kW/m²). These results indicate that, for the given tube and working fluid, experiments with small numbers of tubes, even single tubes, are fairly representative for the tube performance of a big bundle.

VISUALIZATION

To investigate the boiling phenomena of the tested heat transfer surfaces, a high speed video camera was used. Video sequences were recorded with 1125 frames/s. For the evaluation of the video data the software for digital image processing IKEDBV was modified, i.e. modules to determine boiling parameters like the bubble departure diameter and the bubble frequency were added to the software (Kulenovic et al., 2000a - 2000d).

For the evaluation of the bubble parameters (Fig. 11) from the high speed video at first a region of interest (ROI) was selected, close to the tube surface and of course including an active bubble generation site. For the detection of he bubble contours a 2D-gradient-filtering (Laplace-operator, 7×7 filter-masque) was carried out. Additionally to the bubble contour detection a classification of the bubble quality was done, corresponding to the bubble position compared with the focus plane. Bubbles inside the focus plane are sharply imaged, bubbles outside the focus plane have unsharp contours. If the visualization plane, i.e. the focus plane is positioned exactly above a bubble generation site, only bubbles of this site will be taken into account for the evaluation, which improves the quality of the results. For the separation of the bubbles inside and outside of the focus plane the filtered image is tresholded (bilevel image).

As next step the bubbles were labelled and coloured and their cross-sectional areas, gravity centers, maximum/ minimum diameters, etc. were calculated. With these data relevant bubble parameters, e.g. the bubble departure diameter, the bubble frequency, the bubble velocity can be determined.

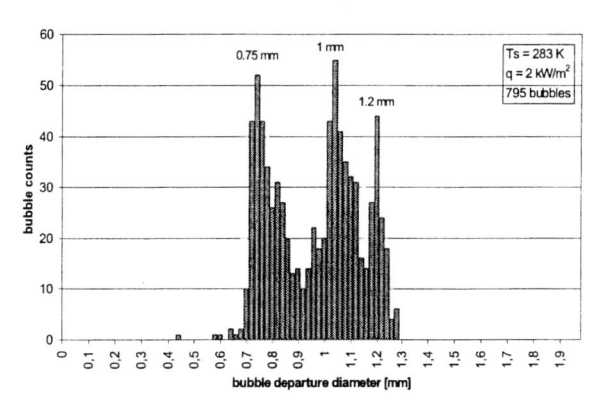

Fig.12 Bubble departure diameters of enhanced surface variant 4, working fluid propane, saturation temperature T = 283 K, heat flux q = 2 kW/m^2

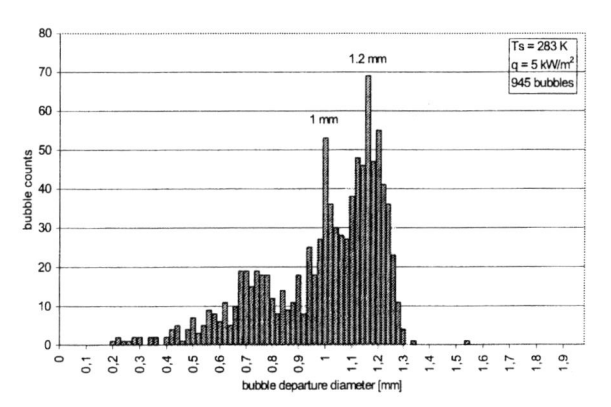

Fig.13 Bubble departure diameters of enhanced surface variant 4, working fluid propane, saturation temperature T = 283 K, heat flux q = 5 kW/m^2

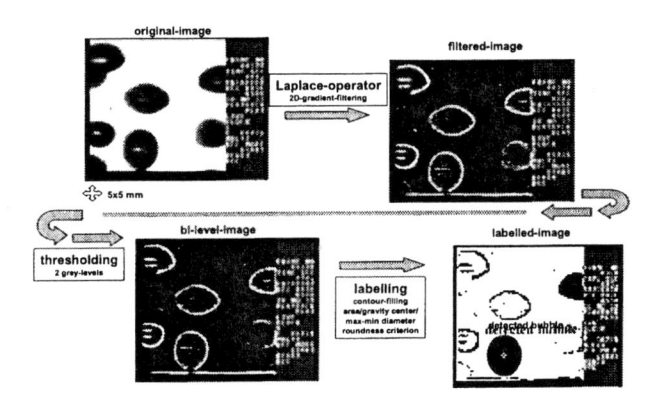

Fig.11 Description of the used evauation method for high speed videos

Figures 12 and 13 show the bubble departure diameters obtained with the enhanced surfaces variant 4 at a saturation temperature of 283 K and heat fluxes of 2 kW/m^2 and 5 kW/m^2.

For a heat flux of 2 kW/m^2 three dominant bubble departure diameters of 0.75 mm, 1 mmm and 1.2 mm were found. With increasing heat flux the number of bubbles increases and the smaller bubble departure diameters become less important.

The bubble frequencies obtained by testing the enhanced tubes varaint 4 are depicted in Figs. 14 and 15; Fig. 14 shows the bubble frequencies at the saturation temperature of 293 K for an employed heat flux of 2 kW/m^2, Fig. 15 for 5 kW/m^2.

Both graphs show two main frequency ranges, low frequencies in the range from 5 Hz to 10 Hz and higher frequencies of 100 Hz to 120 Hz (q = 2 kW/m^2) and 195 Hz to 205 Hz (q = 5 kW/m^2). The lower frequencies describe the interval between active periods of the nucleation sites (= non active period or waiting time plus active period, where bubbles are generated). The higher frequencies are the real bubble generation frequencies during the active periods of the nucleation sites.

With increasing heat flux the bubble frequencies are shifted to higher values.

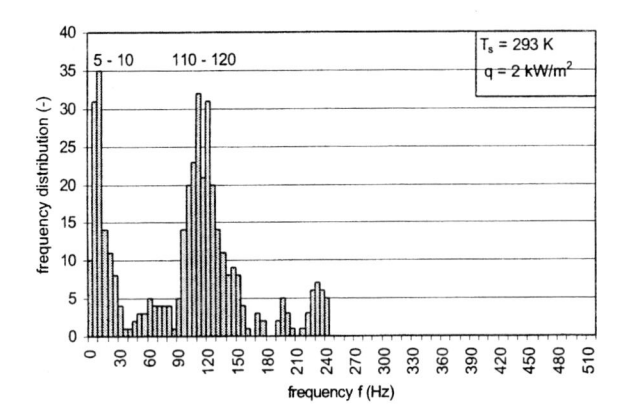

Fig.14 Bubble frequencies of enhanced surface variant 4, working fluid propane, saturation temperature T = 293 K, heat flux q = 2 kW/m^2

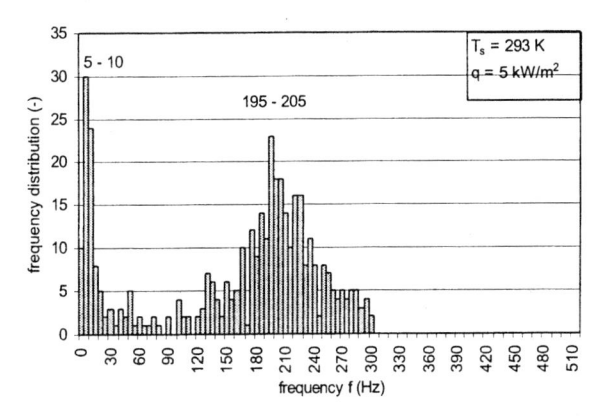

Fig.15 Bubble frequencies of enhanced surface variant 4, working fluid propane, saturation temperature T = 293 K, heat flux q = 5 kW/m^2

Both boiling parameters, the bubble departure diameter and the bubble frequency are very important for generation of an empirical correlation and/or a numerical model and should always be determined experimentally if the necessary equipment is available. Additional parameters, e.g. the bubble rising velocity and the bubble volume (in our case, assuming axis-symmetry of the bubbles) can be used for a thermofluiddynamic description of the boiling phenomena and the two-phase flow from the evaporation surface.

SUMMARY

In collaboration with project partners from universities, research centers and industries, a new enhanced heat transfer tube for compact heat exchangers was developed and tested. The new evaporation tubes will be used in high efficient reboilers for the oil and gas processing industries.

At IKE pool boiling experiments with single tubes (smooth tube, four enhanced tubes) and mini bundles (2 tubes inline, smooth tubes, enhanced tubes) were carried out. The operating conditions were: working fluid propane, saturation temperatures from T = 252 K to T = 293 K and heat fluxes ranging from q = 2 kW/m^2 to 100 kW/m^2.

The visualization of the boiling phenomena and the two-phase flow from the evaporation surface provided additional information about the boiling behaviour on an enhanced surface. The visualization was carried out by a high speed video system and the video records were evaluated by a software generated at IKE.

Compared with the smooth reference tube the best enhanced tube variant 4 showed improvements of the heat transfer coefficients up to a factor of 3 and absolute heat transfer coefficients up to 20 kW/m^{2}K. Especially in the range which is of interest for industrial applications in the process industry (10 kW/m^2 - 40 kW/m^2) very good results are found.

The mini bundle experiments are carried out with the two best enhanced tubes, variant 1 and 4, to investigate the influence of the two-phase flow from the lower tube on the heat transfer of the upper tube. The upper tube always had higher heat transfer coefficients than the lower one. The upper tube heat transfer coefficients are comparable with those from a bigger tube bundle tested at GRETh. That means, to obtain first information about the quality of an evaporation surface, small tube bundle experiments can be used.

The experimental results at IKE were used to identify the best enhanced surface for further tube bundle experiments (45 tubes, staggered arrangement) and for the manufacturing of an industrial compact heat exchanger. This prototype with a thermal duty of 5 MW is installed in the production line of a petrochemical plant and is successfully operated as substitute for the original heat exchanger (Martin et al., 2000), with an increase of the overall heat transfer coefficient of 75% (Thonon, 2001).

The work carried out in the frame of this project covered the development of a compact shell-and-tube heat exchanger from surface design to the manufacturing of an industrial prototype. All necessary steps of the work were carried out by specialized project partners and this cooperation was the main reason of the success of the project.

In the frame of a current project, experiments with pure fluid i-butan and with propane/i-butan mixtures will be carried out. The results will help to describe the influence of the thermophysical properties of the working fluid on the boiling phenomena. First the structured tubes used in the former project will be tested and if necessary new enhanced sturctures will be investigated.

REFERENCES

Goetz, J., Entwicklung und Erprobung einer Normapparatur zur Messung des Wärmeübergangs beim Blasensieden. Diss. Univ. Karlsruhe (TH), 1981

Gorenflo, D., Goetz, J., Bier, K., Vorschlag für eine Standard-Apparatur zur Messung des Wärmeübergangs beim Blasensieden. Wärme- und Stoffübertragung 16, 1982

Gorenflo, D., Luke, A., Danger, E., Interactions between heat transfer and bubble formation in nucleate boiling. 11th Int. Heat Transfer Conf., Kyongju, Korea, 1998

Groll, M., Mertz, R., Improved evaporation heat transfer surfaces for cost-effective heat exchangers for the process industries, Contract No. JOU2-CT94-0362, Progress Reports 1-3, 1994-1995

Kulenovic, R., Mertz, R., Groll, M., High speed video visualization of pool boiling from structured tubular heat transfer surfaces. ASME-ZSITS Int. Thermal Science Seminar, Bled, Slowenien, 2000a

Kulenovic, R., Mertz, R., Groll, M., High speed video flow visualization and digital image processing of pool boiling from enhanced tubular heat transfer surfaces., 9th Int. Symposium on Flow Visualization, Edinburgh, UK, 2000b

Kulenovic, R., Mertz, R., Groll, M. High speed flow visualization of pool boiling from enhanced evaporator tubes. 3rd European Thermal Sciences Conference, Heidelberg, Germany, 2000c

Kulenovic, R., Mertz, R., Schäfer, P., Groll, M., Quantitative microscale high speed visualization of pool boiling phenomena from enhanced evaporator tubes. Engineering Foundation Conference: Heat Transfer and Transport Phenomena in Microsystems. Banff, Canada, 2000d

Luke, A., Gorenflo, D., Pool boiling heat transfer from horizontal tubes with different surface roughness. Eurotherm Seminar No 48, Pool Boiling 2, Paderborn, Germany, 1996

Martin, F., Delorme, J.J., Portz, A., Zoetemeijer, L., Technical and economic studies of intensified reboilers. European Seminar on Heat Equipment for the Process and Refrigeration Industries, Grenoble, France, 2000

Mertz, R., Groll, M., Evaporation heat transfer from enhanced industrial heat exchanger tubes, Eurotherm Seminar No. 62, Grenoble, 1998

Mertz, R., Kulenovic, R., Groll, M., Lang, T., Thonon, B., High performance boiling tubes. European Seminar on Heat Equipment for the Process and Refrigeration Industries. Grenoble, France, 2000

Pinto, A.D., Gorenflo, D., Künstler, W., Heat transfer and bubble formation with pool boiling of propane at a horizontal copper tube. 2nd European Thermal Sciences and 14th UIT National Heat Transfer Conference, Rome, Italy, 1996

Sokol, P., Blein, P., Gorenflo, D., Rott, W., Schömann, H., Pool boiling heat transfer from plain and finned tubes to propane and propylene. 9th Int. Heat Transfer Conf., Jerusalem, Israel, 1990

Thonon, B., Enhanced reboilers for the process industry. 3rd International Conference on Compact Heat Exchangers and Enhancement Technology for the Process Industries. Davos, Switzerland, 2001

ENHANCEMENT OF BOILING HEAT TRANSFER OF ETHANOL BY PTFE COATING ON HEATING TUBE

Vijaya Vittala, C. B[1], Bhaumik, S[2], Gupta, S.C[2] and Agarwal, V.K[2]

[1] Mechanical Engineering Department, NERIST, Arunachal Pradesh-791 109, India ; E-mail : cvittala@nerist.ernet.in
[2] Chemical Engineering Department, University of Roorkee, Roorkee-247 667, India ; E-mail : satisfch@rurkiu.ernet.in

ABSTRACT

This paper presents an investigation on nucleate pool boiling of ethanol on a polytetrafluoroethylene (PTFE) coated brass heating tube at atmospheric and sub-atmospheric pressures. Three thicknesses of PTFE coating viz. 21, 39 and 51 µm are used here. Boiling heat transfer characteristics on such surfaces are compared with an uncoated one and enhancement of heat transfer due to coating is observed. Effectiveness of a coated tube is evaluated in terms of heat flux, pressure and coating thickness and thereby criteria for the enhancement of heat transfer coefficient for the boiling of ethanol on PTFE coated tubes are established.

INTRODUCTION

Enhancement of boiling heat transfer is an active area of research because of its academic and industrial importance. Webb (1981), Bergles (1983) and Thome (1990) have reviewed all the enhancement techniques developed over the last few decades. All the techniques fall into the categories of active and passive or combinations of both. Literature indicates the dominance of passive techniques over active ones as they do not require any power from an external source. Basically, it involves the alteration of surface characteristics of the heating surface. A few researchers namely Young et. al (1964), Marto et. al (1968), Vachon et. al (1969), Coeling et. al (1969), Hinrichs et. al (1981) and Kenning (1992) have studied boiling of liquids on PTFE coated surfaces and observed significant improvement of heat transfer rate. However, the results of these studies are not inclusive, as none of them has included the effect of heat flux and pressure on the boiling characteristics of such systems. Hence, the present paper is an attempt in this direction.

EXPERIMENTAL SETUP

Fig. 1 depicts an experimental setup for this investigation. Its main components are a vessel (1), a heating tube (2), a heater (3), a condenser (8), a liquid level indicator (4), a vacuum pump (12), wall and liquid thermocouples and an instrumentation panel board containing a wattmeter, a digital multimeter, volt- and ampere- meters, etc. The vessel is a 304 ASIS stainless steel cylinder of 240 mm inside diameter and 470 mm height. The vessel top has mountings for a knock-out condenser (8), a vacuum/ pressure gauge (9) and a pipeline leading to a bubbler (10). Two view ports are provided at diametrically opposite positions to observe bubble dynamics over the heating tube. It is adequately insulated by wrapping it with asbestos rope and thick layers of paste of plaster of Paris and magnesia powder.

The heating tube is a brass cylinder of 18 mm inside diameter, 32 mm outside diameter and 228 mm length. One end of it is open whereas the other end is closed with a 25 mm thick plug. The plug reduces the possibility of heat flow longitudinally. The outer surface of the tube is made smooth by turning, polishing and rubbing against a standard o/o grade emery paper. A home-made cartridge heater, prepared by winding a 24 gauge nichrome wire over a porcelain tube of 16 mm diameter, is placed inside the tube. Thin mica sheet with glass tape is used to provide electrical insulation between heater and the tube. The surface of the tube is coated by Dupont green PTFE enamel (850-314, 850-214) according to the prescribed guidelines. Coating thickness is measured by Minitest 2000 having 1µm accuracy at different circumferential positions randomly over the entire length of coated surface. However, no significant variation is observed.

Calibrated PTFE coated copper-constantan thermocouples of 30 gauge are used to measure surface and liquid temperatures. The thermocouples are placed inside the four holes provided in the heating tube for measuring

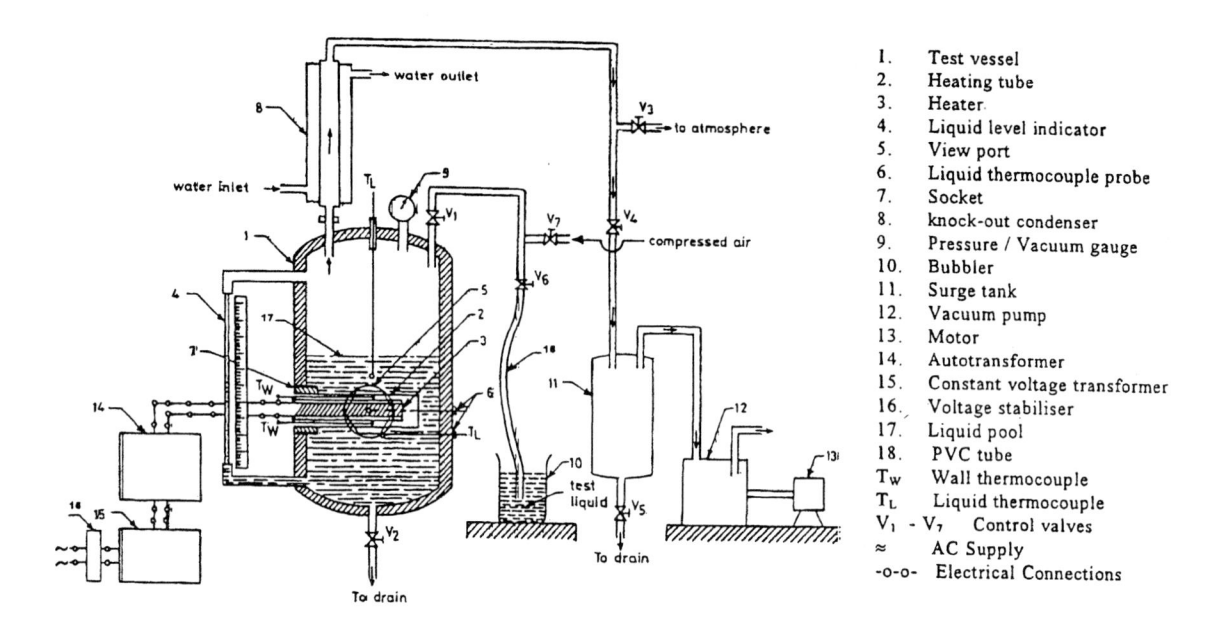

Fig. 1. Schematic diagram of experimental setup.

1. Test vessel
2. Heating tube
3. Heater
4. Liquid level indicator
5. View port
6. Liquid thermocouple probe
7. Socket
8. knock-out condenser
9. Pressure / Vacuum gauge
10. Bubbler
11. Surge tank
12. Vacuum pump
13. Motor
14. Autotransformer
15. Constant voltage transformer
16. Voltage stabiliser
17. Liquid pool
18. PVC tube
T_w Wall thermocouple
T_L Liquid thermocouple
$V_1 - V_7$ Control valves
$\approx$ AC Supply
-o-o- Electrical Connections

surface temperature. Thermocouples are placed in the liquid pool corresponding to top, bottom and the two side-positions of the heating tube probes. The thermocouple probes in the liquid are located at a sufficient distance from the heating tube to ensure that they are outside the superheated boundary layer surrounding the tube. All thermocouples are connected to a 24 point selector switch and their e.m.f.s are measured by a digital multimeter of Thurlby model 1905a having a least count of 0.1 mV in 20 mV range. Melting ice bath at 0^0 C is used as reference junction.

EXPERIMENTAL PROCEDURE

The heating tube is placed in the vessel in such a manner that its closed end remains floating in the liquid pool while the open end is fixed to the vessel wall by a PTFE gland and nut (7). The vessel is filled with the liquid through inlet valves V1 and V7. The heater is energized with single phase, 200 V, 50 Hz AC stabilized supply. An autotransformer modulates input voltage to heater. The heating tube is thermally stabilized by submerging it in the pool of liquid for 24 hours and then vigorously boiling for about 72 hours. Readings of thermocouples are monitored. Experiments are repeated under the same operating conditions till the data are found to be reproducible. The system is degassed by vigorous boiling of liquid for several hours and passing the vapour through the bubbler, which acts as liquid seal. The liquid is boiled at a fixed power input. At steady state condition, readings of the wattmeter and all thermocouples are noted. At a constant pressure, power input to the heater is increased progressively from 260 W to 700 W in six steps and experiments are repeated. The pressure in the system is increased from 26.33 kN/m² to 97.60 kN/m² in six steps and the above procedure is repeated. Experiments are conducted in the same manner using heating tubes of different coating thicknesses.

DATA REDUCTION

Since the thermocouples are located inside the tube wall, they do not represent the outer surface temperature of the tube. It is obtained by incorporating a correction for the temperature drop from the thermocouple tip to the outer surface, δT_w. It is calculated by the following equation for a thin cylinder :

$$\delta T_w = q\, (D_2/2k)\, \ln(D_2/D_h) \qquad (1)$$

The heat flux, q is obtained from the following equation:

$$q = Q/(\pi D_2 L) \qquad (2)$$

The average value of the wall temperature, T_{wa} is calculated by taking arithmetic mean of the wall temperatures at the top, the two sides and the bottom positions on the heating surface.

$$T_{wa} = (T_{w,t} + T_{w,s} + T_{w,b} + T_{w,s})/4 \qquad (3)$$

Similarly, the liquid temperatures as monitored by liquid thermocouple probes are processed to provide the average liquid temperature, T_{la}

$$T_{la} = (T_{l,t} + T_{l,s} + T_{l,b} + T_{l,s})/4 \qquad (4)$$

Heat transfer coefficient, h is obtained from the following equation:

$$h = q / (T_{wa} - T_{la}) \qquad (5)$$

Data are processed by using a computer program developed in Turbo C. Analysis of the data is carried out to determine the error due to the inherent limitations of the measuring instruments and the method of measurement. The results indicate errors of 0.00001 m, 10 W and 0.021 ^{0}C in the measurement of length/ diameter, power input and wall/ liquid temperature respectively. The cumulative maximum error in the value of the heat transfer coefficient is found to be ± 3.85 % only.

The temperature drop across PTFE coating is not included in the calculation of wall temperature. As such, the calculated values of heat transfer coefficients are based on temperature difference through the tube material only.

RESULTS AND DISCUSSION

Fig. 2 represents a plot of heat transfer coefficient and heat flux for the boiling of ethanol on an uncoated (plain) tube surface. In this plot, pressure is a parameter. From this plot, the following important features can be inferred:

a. At a given pressure, heat transfer coefficient increases with increase in heat flux. The variation between the two can be described by the relationship, $h \propto q^{0.70}$.
b. At a given heat flux, increasing the pressure increases the heat transfer coefficient.

Both the above features are consistent and can be explained as follows:

At a given pressure, an increase in heat flux raises the wall superheat, ΔT_w, which-, in turn-, reduces the value of the minimum radius of curvature of the nucleation sites on the tube, as can be seen from the following equation:

$$r_m = \frac{2\sigma}{\left(\frac{dp}{dT}\right)_{st} \Delta T_w} \qquad (6)$$

As the population of small size nucleation sites is greater than the large ones a greater number become active as the heat flux increases. Consequently, larger numbers of vapour bubbles take birth, grow, collapse/ detach and travel in the pool. All this leads to increased turbulence and thereby higher heat transfer coefficient.

An increase in pressure alters the physico-thermal properties of the liquid, the most significant effect being on the surface tension. In fact, raising pressure lowers the value of surface tension. This, according to Eq. (6), causes the value of the minimum radius of curvature of nucleation sites to decrease and-, for the reasons explained above the-, heat transfer coefficient to increase.

Figs. 3-5 shows the plots of heat transfer coefficient versus heat flux for the boiling of ethanol on a heating tube coated with 21, 39 and 51 μm thick PTFE, respectively. In these plots, pressure is a parameter. As can be seen, the features of these plots are similar to those observed on the uncoated tube. However, the functional relationship between heat transfer coefficient and heat flux at a constant pressure for coated tube changes to $h \propto q^r$, where the value of exponent, r depends upon the thickness of coating. It has the values of 0.45, 0.50 and 0.38 for the coating thicknesses of 21, 39 and 51 μm, respectively.

Fig. 6 compares the boiling characteristics of a PTFE coated heating tube with those of an uncoated one. Basically, it is a plot of heat transfer coefficient versus heat flux for the boiling of ethanol on PTFE coated tube at atmospheric pressure. Coating thickness is taken as a parameter in this plot. The data for boiling on the uncoated tube are also included in it for comparison. From this plot-, following salient features are observed:

a. At a given value of heat flux, the heat transfer coefficient on a 21 μm thick coated tube is more than that on an uncoated one for heat fluxes less than 36.85 kW/m^2 and is smaller for other values of heat flux. Further, an

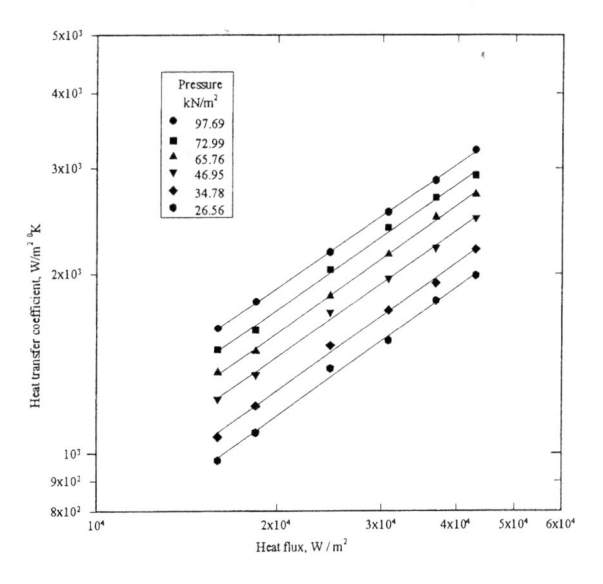

Fig. 2. Heat transfer coefficient vs heat flux for the boiling of ethanol on plain tube.

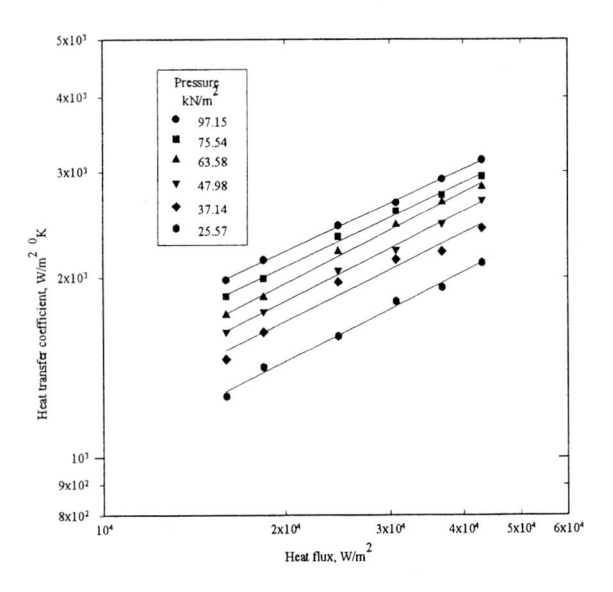

Fig. 3. Heat transfer coefficient vs heat flux for the boiling of ethanol on 21 μm thick PTFE coated tube.

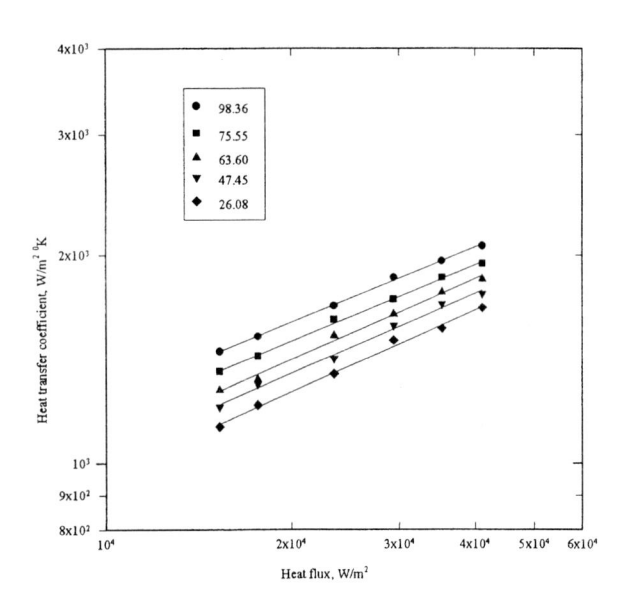

Fig. 5. Heat transfer coefficient vs heat flux for the boiling of ethanol on 51 μm thick PTFE coated tube.

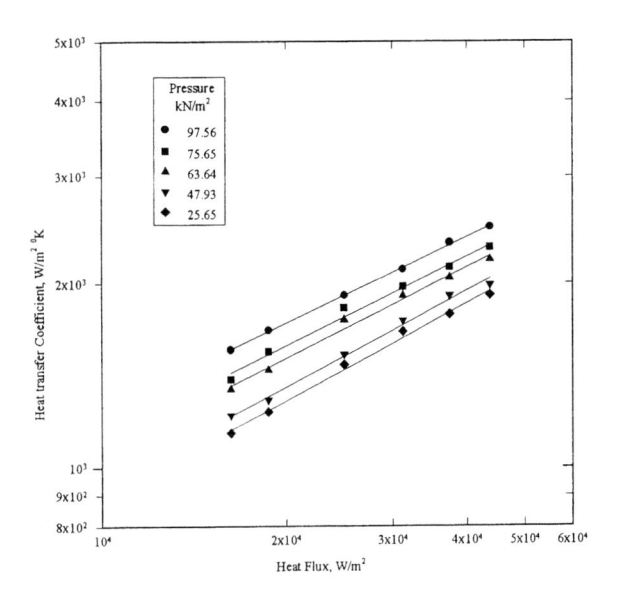

Fig. 4. Heat transfer coefficient vs heat flux for the boiling of ethanol on 39μm thick PTFE coated tube.

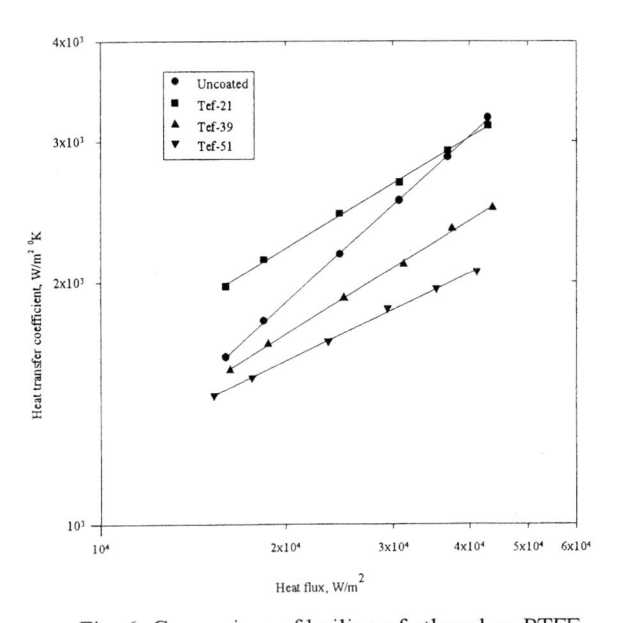

Fig. 6. Comparison of boiling of ethanol on PTFE coated tubes with that of plain tube atmospheric pressure.

increase in thickness of coating from 21 to 39 to 51 μm decreases the heat transfer coefficient.

b. For both the cases, the heat transfer coefficient increases with increased heat flux. However, the increase in heat transfer coefficient on the coated tube is less when compared to that on an uncoated tube for a given increase in heat flux.

The above features have consistently been obtained for the boiling of ethanol at sub-atmospheric pressure too (Fig. 7). Possible reasons for the above behaviour are as follows:

PTFE is a microporous material whose application on the tube leads to the formation of an interwoven matrix,

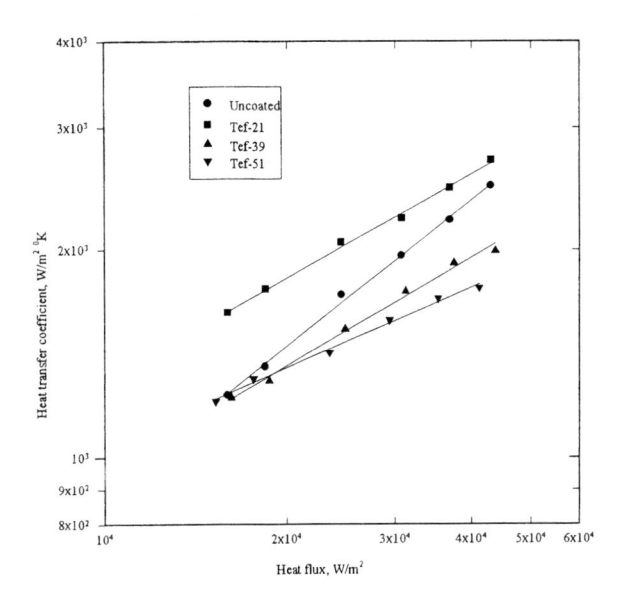

Fig. 7. Comparison of boiling of ethanol on PTFE coated tubes with that of plain tube at sub-atmospheric pressure.

consisting of microporous layers. Depending upon the method of coating, some pores of the inner layers are partially or fully exposed to liquid and thus, contribute to multiply the number of small size nucleation sites on the heating tube surface. The coating also alters the contact angle significantly. Measurement of contact angle by the Sessile drop technique has indicated it to be 27.5^0 on a 21 μm thick PTFE coated brass tube as against 24^0 on an uncoated tube surface for ethanol at atmospheric pressure. Therefore, for a given heat flux, the bubble population increases formings a cloud of vapour which hampers bubble initiation and development. This is also accompanied by an increase in the intensity of liquid re-circulation. Hence, a reduced number of vapour bubbles are emitted from a PTFE coated surface with higher emission frequency than that from uncoated one. So heat transfer coefficient on a 21 μm thick coated surface is found to be greater than that on uncoated one for heat fluxes up to 36.85 kW/m^2. Raising the heat flux increases the bubble population but does not increase it as much on a coated surface as on uncoated one because of the bubble development hampering effect noted above. Therefore, the heat transfer coefficient of a 21 μm thick coated surface is less than that on the uncoated one for heat fluxes greater than 36.85 kW/m^2.

Since the PTFE coated surface has a large number of small sized sites, an increase in heat flux leads to a higher population of vapour bubbles originating in the inner surface layers. However, they do not get a free path for their development and movement due to restriction imposed by the interwoven type of structure. Hence, vapour bubbles formed on inner surface collapse and the resultant vapour escapes. The vapour, in fact, offers a resistance to the formation of vapour bubbles on the outermost and adjacent layers of coating. In this way, formation and growth of vapour bubbles are hampered, resulting in poor heat removal from the surface. Consequently, the increase in heat transfer coefficient with heat flux on coated tube is not as much as it is on an uncoated tube. Hence, the value of exponent, n for the coated tube is found to be less than that for an uncoated one.

An increase in coating thickness causes the surface structure to be more interwoven. This improves nucleation site density which affects the heat removal rate adversely and thereby the heat transfer coefficient.

As the heat transfer coefficient is significantly influenced by the coating, the effectiveness of the heating tube is determined by the ratio of the heat transfer coefficient on a coated tube to that on an uncoated one at the same values of heat flux and pressure. This is required to select the values of heat flux and pressure at which coated tube can provide enhancement. Since the variation of the heat transfer coefficient with heat flux and pressure on a coated surface is different to that on uncoated one, effectiveness also depends upon these parameters. In fact, effectiveness of the coated tube decreases with heat flux and pressure. Using regression analysis, effectiveness is related to heat flux and pressure by the following equation within a maximum error of ± 10 %:

$$\tau = K\, q^r\, p^s \qquad (7)$$

where the value of the constant, K and indices r and s depend upon the thickness of the PTFE coating. These values are given in Table 1.

Table 1. Values of K, r and s for various coating thicknesses

Sl. No.	Coating thickness, μm	K	r	s
1	21	22.20	-0.25	-0.10
2	39	13.03	-0.20	-0.15
3	51	50.29	-0.32	-0.20

Equation (7) is empirical in nature and is applicable to the boiling of ethanol on a PTFE coated brass tube at atmospheric and sub-atmospheric pressures only. Computations yield some values of τ to be smaller than unity. Naturally, a PTFE coated tube can not provide enhanced heat transfer for the situations of τ<1. Hence, the condition of τ >1 is employed to determine the range of heat flux and pressure at which the PTFE coated tube is preferred over an uncoated tube. With this condition the following criteria is obtained:

$$q^r\, p^s < K \qquad (8)$$

Equation (8) is a simple and convenient equation to calculate the range of heat flux at which a PTFE coated tube can be advantageously used for the boiling of ethanol at a given pressure. Alternatively, it can also be used to determine the range of pressures for boiling of ethanol at a specified value of heat flux. Further, it also brings out that a coating of a 21 μm thick layer of PTFE on a tube is not always beneficial. It depends upon the range of heat flux and pressure employed for a specified fluid.

CONCLUSIONS

1. The boiling heat transfer coefficient for ethanol on a plain uncoated brass tube at atmospheric and sub-atmospheric pressures varies with heat flux according to the relationship, $h \propto q^{0.7}$. However, it changes to $h \propto q^r$ when boiling occurs on a PTFE coated tube.
2. Coating of a tube with PTFE enhances heat transfer coefficient. However, the enhancement depends upon the thickness of the coating, heat flux and pressure. Furthermore, at a given value of heat flux, the heat transfer coefficient decreases with an increase in thickness of coating on tube.
3. The effectiveness of PTFE coated heating tube is a function of heat flux and pressure. Criteria have been established for the determination of operating conditions for boiling of ethanol on a PTFE coated heating tube.

NOMENCLATURE

D diameter, m
D_h thermocouple pitch circle diameter, $(D_1 + D_2)/2$, m
k thermal conductivity of heating tube material, W/ m K
L effective length of heating tube, m
q heat flux, W/ m^2
Q heat input, W
T temperature, K
h heat transfer coefficient, W/ m^2 K
p pressure, N/ m^2
σ surface tension, N/m
τ effectiveness, dimensionless

Subscripts

1 inner
2 outer
a average
b bottom
l liquid
m minimum
s side
t top
w wall

REFERENCES

Bergles,A.E, 1983, Techniques to Augment Heat Transfer, in *Hand Book of Heat Transfer Applications*, eds.W.M.Rohsenow,J.P.Hartnett and E.N.Ganic, 2nd ed., Mc-Graw Hill, New York,pp. 134-153.

Coeling, K.J and Merte, H.Jr., 969, Incipient and Nucleate Boiling of Liquid Hydrogen, *J Engg. for Industry*,Trans. ASME Ser-B, Vol. 91, pp. 513-518.

Hinrichs,T , Hennecke,E and Yasuda,H, 1981, The Effect of Plasma Deposited Polymers on the Nucleate Boiling Behaviour of Copper Heat Transfer Surfaces, *Int. J. Heat Mass Transfer*, Vol.24,pp. 1359-1368.

Kenning, D. B. R, 1992, Wall Temperature Patterns in Nucleate Pool Boiling, *Int. J. Heat Mass Transfer*, Vol. 35, pp. 73-86.

Marto, P. J, Moulson, J, A and Maynard, M, D, 1968, Nucleate Pool Boiling of Nitrogen with Different Surface Conditions, *J. Heat Transfer*, Vol. 90, pp. 437-444.

Thome, J. R, 1990, *Enhanced Boiling Heat Transfer*, Hemisphere, Washington D.C.

Vachon, R. I, Nix, G. H, Tanger, G. E and Cobb, R. O, 1969, Pool Boiling Heat Transfer from Teflon Coated Stainless Steel Surface, J. Heat Transfer, Vol. 91, pp. 364-370.

Young, R. K and Hummel, R. L, 1964, Improved Nucleate Boiling Heat transfer, *Chem. Engg. Prog.*, Vol. 60, no. 7, pp. 53-58.

Webb, R. L, 1981, Enhancement Techniques, *Heat Transfer Engg.*,Vol. 2, pp. 40-69.

HEAT TRANSFER ENHANCEMENT WITH A SURFACTANT IN HORIZONTAL BUNDLE TUBES ON AN ABSORBER

J.I. Yoon[1], E. Kim[2], W.S. Seol[3], C.G. Moon[3] and D.H. Kim[3]

[1] Pukyong National University, Namgu, Pusan 608-737, Korea; E-mail: yoonji@dolphin.pknu.ac.kr
[2] Pukyong National University, Namgu, Pusan 608-737, Korea; E-mail: ekim@mail.pknu.ac.kr
[3] Graduate Student, Pukyong National University, Namgu, Pusan, 608-737, Korea; E-mail: mchg@mail1.pknu.ac.kr

ABSTRACT

This paper is concerned with the enhancement of heat transfer by surfactants added to the aqueous solution of LiBr. Three different kinds of tubes in horizontally stagged arrangement are tested with and without an additive of normal octyl alcohol. The test tubes are a bare tube, a floral tube and a hydrophilic tube. The additive mass concentration is about 0.05~5.5%. The heat transfer coefficient is measured as a function of the solution flow rate in the range of 0.01~0.034 kg/m-s. Among three kinds of tubes, the hydrophilic tube shows the highest wettability. It shows 4 to 73% higher wetted area ratio than that of the bare tube, and 10 to 22% higher than that of the floral tube. Without surfactants, the hydrophilic tube is in the range of 10 to 35% higher heat transfer coefficient than that of the bare tube, and 5 to 25% higher than that of the floral tube. With surfactants, the increase of the heat transfer coefficient is about 35~90% for the bare tube, 40~70% for the floral tube, and 30~50% for the hydrophilic tube.

INTRODUCTION

In recent years, the usage of an absorption chiller/heater is positively encouraged for the unused energy application, the preservation of earth environment, and the settlement of unbalanced demand between electric power and natural gas. An absorption chiller/heater can reduce the demand of electric power by using unused natural gas during the summer time. However, an absorption chiller/heater has a disadvantage. The size of an absorption chiller/heater is larger than that of a vapor compression type chiller/heater based on the same capacity. An absorption chiller/heater is composed of an absorber, an evaporator, a condenser, a generator, and a solution heat exchanger. Among these components the absorber has the largest volume. The absorber has about 33% of the total heat transfer area, and about 27% of the total volume (Inoue, 1988). Therefore, to have a high efficiency of an absorption chiller/heater, the absorber must be designed carefully.

The absorber, which is the heat exchanger of the falling film type, is used to minimize the performance decrease caused by the pressure loss. To improve the efficiency of the absorber, the development of tubes having the high efficiency and the supply of proper surfactants are considered. It is reported that the two cases enhance the heat and mass transfer and improve the performance of the absorber (Hoffmann, 1996 and Yoon et al., 1999). The development of the high efficiency tubes needs the new investment and increases the production rate. Thus, many use the second case simply supplying surfactants for improving the system performance. Many researchers reported that the performance enhancement by supplying surfactants decreases the surface tension of the absorption solution and the surface disturbance by the Marangoni convection (Yoon et al., 1994; Seol et al., 1998; Kim and Lee., 1998; Kashiwagi, 1985; Kim et al., 1993). However, the mechanism for improving the performance is not clear.

In this paper, the enhancement of the heat transfer by using three kinds of tubes, a bare, a floral, and a hydrophilic, are tested with the experiment. In the absorber horizontally stagged tubes are arranged to investigate the absorption enhancement. The experimental set-up has a commercial size. Thus, the error caused by the reduction of the test section can be minimized. Also, the effect of additives can be seen clearly. The fundamental experimental data for falling film absorption with surfactants are limited. In many commercial equipments, additives are supplied by experience. This research is to investigate the effects of three different tubes and surfactant additives on the absorption process.

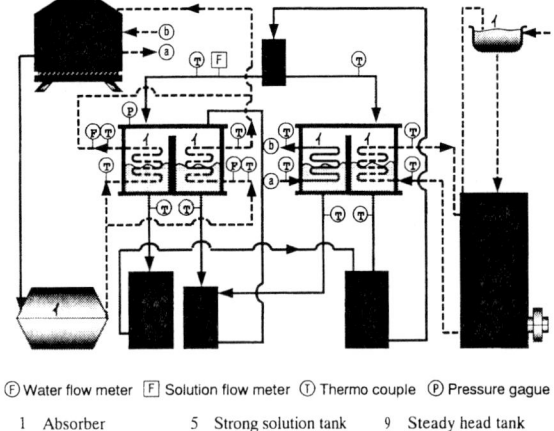

Table 1 Experimental conditions

Items	Parameters	Conditions
Refrigerant	Evaporating temperature (°C)	6
LiBr solution	Inlet concentration (wt%)	60
	Inlet temperature (°C)	45
	Mass flow rate (kg/ms)	0.010~0.034
Cooling water	Inlet temperature (°C)	32
	Velocity (m/s)	1.0
Surfactant	n-octanol (ppm)	500~5500

Fig. 1 Experimental apparatus

1 Absorber
2 Evaporator
3 Condenser
4 Generator
5 Strong solution tank
6 Weak solution tank
7 Refrigerant tank
8 Expansion tank
9 Steady head tank
10 Cooling water tank
11 Cooling tower
12 Hot water tank

(F) Water flow meter (F) Solution flow meter (T) Thermo couple (P) Pressure gague

EXPERIMENTAL APPARATUS AND METHOD

Experimental apparatus. Fig. 1 shows a schematic diagram of the experimental apparatus. As shown in the figure, the apparatus consists of an absorber, an evaporator, a condenser, a generator, strong/weak solution tanks, and a refrigerant tank, etc. These components are connected with pipes. Pipes and tanks are made of glass fibers. The glass areas are installed to see the inner phenomena of these components. In the absorber 48 horizontal tubes of 400 mm length are installed. The tubes are arranged with 6 columns and 8 rows. At the inlet of the absorber, the steady head tank is installed to prevent the variations of the solution flow rate caused by pulsate phenomena of the solution pump. A vacuum pump (0.5 PS) is then installed to keep the vacuum pressure constant in the apparatus. A circulation pump is used to circulate cooling water in the absorber and chilled water in the evaporator. At the inlet/outlet of the tubes of the absorber, the evaporator, the generator, the condenser, and the tanks, thermocouples are installed to measure temperature variations. Flow meters are installed to measure the flow rate of cooling water, chilled water, and solutions.

Experimental method. By the solution pump, the strong solution in the strong solution tank flows down the outer surface of horizontal tubes of the absorber through the tray installed at the upper part of the absorber. The refrigerant stored at the refrigerant tank flows over the outer surface of the tubes of the evaporator heat exchangers. And at the inner surface of the tubes of the evaporator, the cooling load is given by flowing cooling water. The refrigerant vapor evaporated from the evaporator is injected to the absorber through the eliminator located between the evaporator and the absorber. The rest of the refrigerant vapor returns to the refrigerant tank. The absorber is kept about the vacuum pressure of 7 mmHg. The solution temperature is adjusted depending on the saturated

Type	Appearance	Dimensions
Bare tube		d_o=15.88 d_i=14.05 L=400
Hydrophilic tube		d_o=15.88 d_i=14.05 L=400
Floral tube		d_o=15.88 d_i=13.88 N=11 L=400

Fig. 2 Specifications of the test tubes

temperature with the change of the solution concentration. The strong solution absorbs the refrigerant vapor evaporated from the evaporator falling down the horizontal tubes. The steady head tank is installed at the inlet of the absorber to have a constant flow rate. The rest of the strong solution returns to the strong solution tank through a by-pass tube. The weak solution weakened by absorbing the refrigerant is stored at the weak solution tank. The cooling water in the cooling water tank is supplied to the evaporator and the absorber. The cooling water supplied at the absorber exchanges heat flowing though the heat transfer tubes. The cooling water obtained heat at the absorber and the chilled water deprived of heat at the evaporator combine at the exit. The working fluid cools to the appropriate temperature at the cooling tower. The n-octanol is used as surfactant. The experiments are conducted for the six cases of additives of 500 ppm, 1500 ppm, 2500 ppm, 3500 ppm, 4500 ppm, and 5500 ppm.

Three different types used in this study are as follows. The bare tube has an outer diameter of 15.88 mm and an inner diameter of 14.05 mm. The hydrophilic tube has the same diameters of the bare tube. The floral tube has an outer diameter of 15.88 mm and an inner diameter of 13.88 mm.

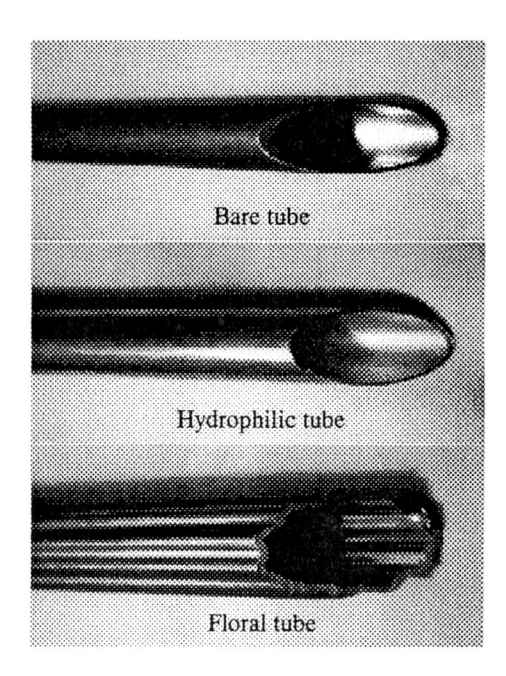

Fig. 3 Photographs of the test tubes

The hydrophilic tube has enhanced surface to improve the wettability, and has the same geometry with the bare tube. Experimental conditions are shown in Table 1. Fig. 2 and Fig. 3 show the specifications and the photographs of test tubes, respectively.

Calculation of the heat transfer coefficient. The log mean temperature difference for a bank of tubes is defined as follows:

$$\Delta T_{lm} = \frac{\{(T_{Asi} - T_{Acoo}) - (T_{Aso} - T_{Acoi})\}}{ln\{(T_{Asi} - T_{Acoo})/(T_{Aso} - T_{Acoi})\}} \quad (1)$$

where T_{Acoi} and T_{Acoo} are the average cooling water temperatures for the inlet and outlet tubes in the bank, respectively, and T_{Asi} and T_{Aso} are the average equilibrium temperatures at the absorber for the inlet and outlet solutions, respectively.

The heat quantity transferred to the cooling water is

$$Q = G_{co} \cdot c_{co} \cdot (T_{coo} - T_{coi}) \\ = U \cdot A_T \cdot \Delta T_{lm} \quad (2)$$

where $A_T = \pi \cdot d_o \cdot L_T$.

An experimental correlation for the convection heat transfer coefficient of the cooling water in the inner tube surface, h_i, given by Dittus-Boelter is

$$Nu = 0.023 Re^{0.8} \cdot Pr^{0.4} = \frac{h_i L}{\lambda} \quad (3)$$

The heat transfer coefficient of the outer tube surface of

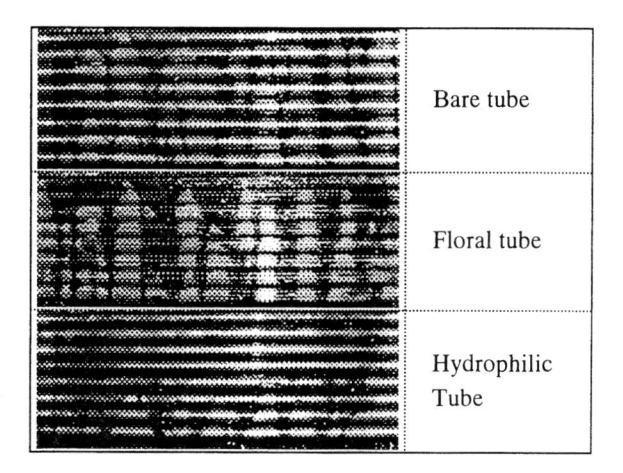

Fig. 4 Photograph of flow pattern

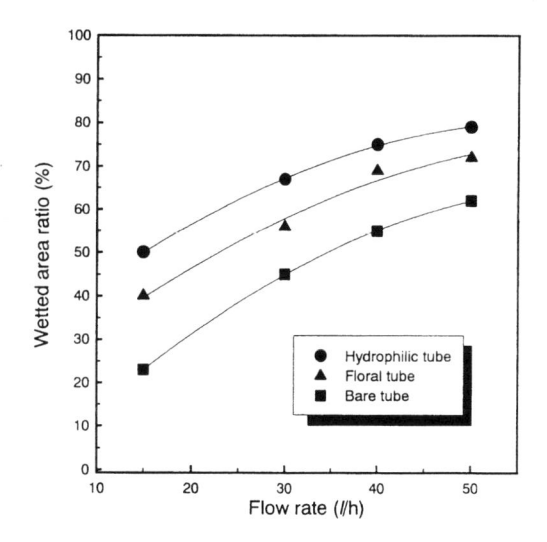

Fig. 5 The variation of wetted area ratio on flow rate

the absorption solution, h_o, can be obtained from the equation (4) and the thermal resistance of the tube wall is assumed to be negligible.

$$h_o = 1/\{1/U - d_o/(d_i \cdot h_i)\} \quad (4)$$

Finally, the mass flow of the solution per tube length, Γ_s, given by equation (5), is

$$\Gamma_s = G_s/(2 \cdot L \cdot P) \quad (5)$$

EXPERIMENTAL RESULTS AND DISCUSSION

Fig. 4 shows the photographs visualized by the mixture of water and ink at the atmosphere pressure for the case of flow rate 30 *l/h*. Experimental results for the variation of the wetted area ratios with the solution flow rate for the previous figure are plotted in Fig. 5. Comparing the bare tube and the hydrophilic tube, the wetted area ratio of the hydrophilic tube

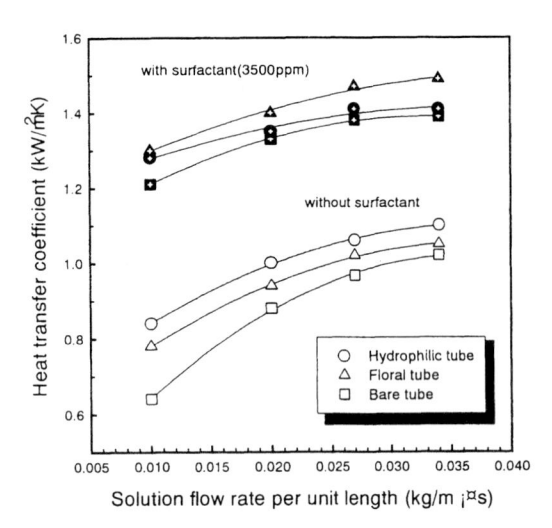

Fig. 6 Comparison of heat transfer coefficients for three types of tubes

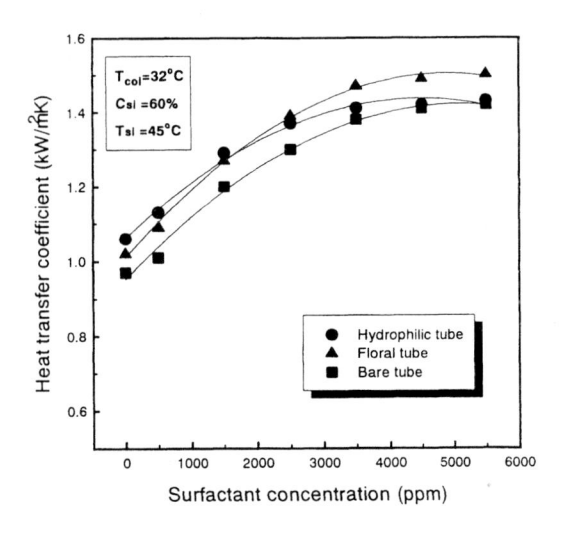

Fig. 7 Effect of surfactant concentration on heat transfer coefficient

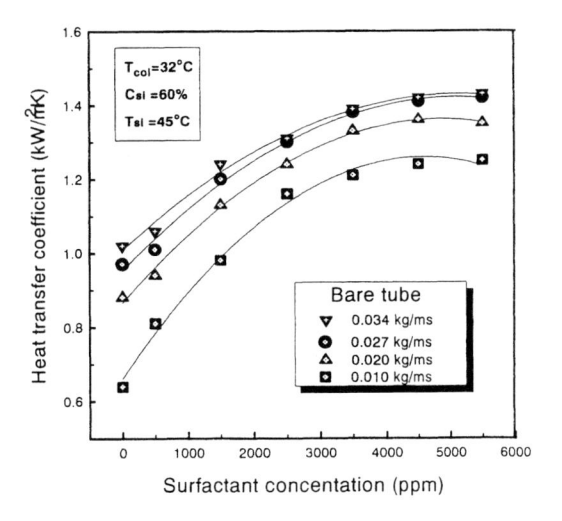

Fig. 8 Effect of surfactant concentration for the bare tube

is about 110% higher than that of the bare tube when the flow rate is 15 l/h. As the flow rate increases, the rate of the wetted area ratio decreases. At the flow rate 50 l/h, the wetted area ratio of the hydrophilic tube is 30% higher than that of the bare tube. The floral tube shows 20 to 70% higher wetted area ratio than the bare tube does as the flow rate decreases. The results are concluded that the hydrophilic tube shows better wettability than the other tubes. The improvement in wet ability is obtained that the surface of the hydrophilic tube has the surface treatment to increase the wettability.

Fig. 6 compares the heat transfer coefficients for three different types of tubes in terms of solution flow rates with/without surfactants. For the experiment the solution temperature supplied at the absorber is set to have equilibrium with the pressure. The additive of a surfactant is 3500 ppm. Within the experimental ranges the heat transfer coefficients of all three types of tubes increase as the solution flow rate does.

Without the additive, the heat transfer coefficient of the hydrophilic tube shows about 35% better performance than that of the bare tube at the solution flow rate of 0.01 kg/m-s while only shows 10% difference in 0.034 kg/m-s. In the case of the floral tube, the heat transfer coefficients increase from 5 to 25% although the increase rate decreases when the flow solution rate increases. With the insertion of a surfactant the heat transfer coefficients increase about 35-90% in the case of the bare tube, about 40-70% in the case of the floral tube and about 30-50% in the case of the hydrophilic tube. The increase rate is higher with the small range of the solution flow rate than with the large range of the solution flow rate. Without a surfactant, the hydrophilic tube shows higher heat transfer coefficients than the floral tube does. With a surfactant, it is opposite showing higher heat transfer coefficients in the case of the floral tube than in the case of the hydrophilic tube.

Fig. 7 shows the heat transfer coefficients for three different tubes varying surfactant concentrations with the solution flow rate of 0.027 kg/m-s. The experiments are conducted with the additive concentrations of 500 ppm, 1500 ppm, 2500 ppm, 3500 ppm, 4500 ppm and 5500 ppm. The absorption heater/chiller usually has the concentration of 2000-3000 ppm. The heat transfer coefficients increase until the surfactant concentration reaches 3500 ppm while it shows no noticeable difference above 3500 ppm for all three types of tubes. With the surfactant concentration below 1500 ppm, the hydrophilic tube shows higher heat transfer coefficient than the floral tube does. However, with the surfactant concentration above 1500 ppm, the floral tube shows higher heat transfer coefficient than the hydrophilic tube does. With the increase of the surfactant concentration, the hydrophilic tube shows almost the same or slightly higher heat transfer coefficient than the bare tube when the additive concentration is above 3500 ppm.

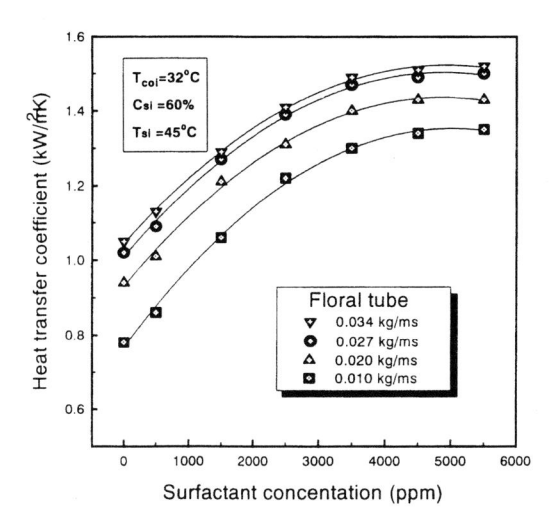

Fig. 9 Effect of surfactant concentration for the floral tube

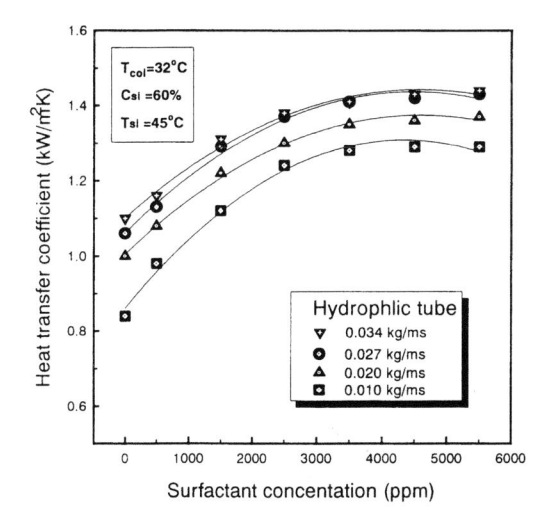

Fig. 10 Effect of surfactant concentration for the hydrophilic tube

Figs. 8, 9, and 10 show the heat transfer coefficients for three types of tubes as a function of the surfactant concentrations for different solution flow rates. With the solution flow rates of 0.01 kg/m-s, 0.02 kg/m-s, 0.027 kg/m-s, and 0.034kg/m-s, the heat transfer coefficients decrease 7%, 5%, 1% in the case of the bare tube; 8%, 6%, 2% in the case of the floral tube; 5%, 4%, 1% in the case of the hydrophilic tube. This result leads to the conclusion that when the solution flow rate is low the additive effect of surfactant is high. Among three kinds of tubes the floral tube shows the best improvement of the heat transfer coefficient. The enhancement of the floral tube is obvious. For low flow rate, the surface geometry of the floral tube increases the heat transfer rates. However, when the solution flow rate increases, the effect of the outer surface diminishes.

CONCLUSIONS

The wetted area ratio for three kinds of tubes such as the bare tube, the floral tube, and the hydrophilic tube is measured by the use of the visualization operation and the heat transfer properties are investigated according to the different surfactant concentrations. The conclusions are shown as follows.

(1) Among three kinds of tubes, the hydrophilic tube shows the highest wettability. It shows 4 to 73% higher wetted area ratio than the bare tube does, and 10 to 22% higher than the floral tube does.
(2) Without surfactants, the hydrophilic tube is in the range of 10 to 35% higher heat transfer coefficients than the bare tube is, and 5 to 25% higher than the floral tube is.
(3) Irrespective of the tubes, the addition of a surfactant more than 3500 ppm does not show the improvement of the heat transfer coefficients. It is an important factor for the decision of a surfactant addition.
(4) With surfactants, the floral tube shows the highest heat transfer coefficient, and the heat transfer coefficient increases about 35 to 90% for the bare tube, about 40 to 70% for the floral tube and about 30 to 50% for the hydrophilic tube.

NOMENCLATURE

A Heat transfer area, m^2
c   Specific heat at constant pressure, J/(kg·K)
d Diameter of the heat transfer tube, m
G Mass flow rate, kg/s
h Convection heat transfer coefficient, kW/(m^2·K)
L Length of the heat transfer tube, m
N Number of hill
Nu Nusselt number
P Number of heat transfer tube
Pr Prandtl number
Q Quantity of heat, kW
Re Reynolds number
U Overall heat transfer coefficient, kW/(m^2·K)
T Temperature,
λ   Conduction, kW/(m·K)
Γ   Solution flow rate, kg/(m·s)

Subscript

A Average
co Cooling water
i Inlet (inside)
lm Logarithmic mean temperature difference
o   Outlet (outside)
s   Solution
T   Total

REFERENCES

Hoffmann, L., 1996, "Experimental Investigation of Heat Transfer in Horizontal Tube Falling Film Absorber with Aqueous Solutions of LiBr Solution Without Surfactants," Int. J. Refrig., Vol. 19, N0. 5, pp. 331–341.

Inoue, N.U., 1988, "Practical Studies on Absorbers in Japan," Refrigeration Engineering Division EBARA Corporation, pp. 18–9.

Kashiwagi, T., 1985, "The Activity of Surfactant in High-Performance Absorber and Absorption Enhancement," Refrigeration, Vol. 60, No. 687, pp. 72–9.

Kim, B. J., and Lee, C. W., 1998, "Effects of Non-Absorbable Gases on the Absorption Process of Aqueous LiBr Solution Film in a Vertical Tube(I) –Experimental Studies", Transactions of KSME(B), Vol. 22, No. 4, pp. 489–498.

Kim, K. J., Berman, N. S., and Wood, B. D., 1993, "Experimental Investigation of Enhanced Heat and Mass Transfer Mechanisms using Additives for Vertical Falling Film Absorber," International Absorption Heat Pump Conference, ASME, pp. 41–7.

Seol, W.S., Kwon, O. K. and Yoon, J. I., 1998, "Experimental Investigation of Enhanced Heat and Mass Transfer for LiBr/H_2O Absorber", Transactions of SAREK, Vol. 10, No. 5, pp. 581–588.

Yoon, J. I., et al., 1999, "The Effect of Surfactant in the Absorptive and Generative Processes", KSME International Journal, Vol. 13, No. 3, pp. 264-272

Yoon, J. I., Oh, H. K., and Takao Kashiwagi, 1994, "Characteristics of Heat and Mass Transfer for a Falling Film Type Absorber with Insert Spring Tubes", Transactions of KSME, Vol. 19, No. 6, pp. 1501–509.

VAPORIZATION AND CONDENSATION
DATA AND METHODS

A NEW PROCEDURE FOR TWO-PHASE THERMAL ANALYSIS OF MULTI-PASS INDUSTRIAL PLATE-FIN HEAT EXCHANGERS

Foluso Ladeinde [1] and Kehinde Alabi [2]

[1] Thaerocomp Technical Corporation, Stony Brook, NY 11790, USA; E-mail: ladeinde@thaerocomp.com
[2] Thaerocomp Technical Corporation, Stony Brook, NY 11790, USA; E-mail: alabi@thaerocomp.com

ABSTRACT

A simple alternative to the NTU and/or F-factor (MTD) method of analyzing plate-fin heat exchangers is presented in this paper. The new procedure discretizes the flow path in each stream into finite difference-type grid points. Analysis is carried out in the cells defined by adjacent grid points. Defining the geometric and thermal relationships between cells in the two fluid streams is a key to the success of the method. Unlike the procedures that the new one is intended to replace, there are no limitations regarding the flow arrangement, number of passes, banking type, or phase change. Furthermore, judicious choice of the grid points along a passage allows the treatment of strong dependence of property on temperature. Finally, the new procedure is validated by comparing its predictions with the NTU method and experimental measurements of a multi-pass plate-fin heat exchanger. Heat transfer between R134a and a proprietary fluid is calculated with the Kandlikar correlations (Kandlikar, 1990), to demonstrate the capability for multi-pass, two-phase, plate-fin systems.

INTRODUCTION

Plate-fin heat exchanger rating analysis usually involves the determination of the total heat transfer rate, the pressure drop and the outlet temperatures (Anonymous, 1994). In its most basic form, plate-fin heat exchangers involve the transfer of heat between two fluids – termed the hot and cold fluid. The fluids are separated from each other in passages with a large amount of surface area between them. The exchange of heat takes place between fluids moving in parallel or cross-flow directions. Parallel flow arrangements may also be either co-current or counter-current. In the co-current arrangement, both the hot and cold fluids are flowing in the "same" direction. In this case, the hottest section of the hot fluid is exchanging heat with the coldest section of the cold fluid and temperature difference progressively becomes smaller along the flow direction. Counter-current parallel flow is the opposite of this arrangement and the flow of the fluids is in opposite directions. The use of cross counter-current plate-fin heat exchangers is also quite common. Finally, realistic systems often contain multiple passes or they may involve a phase change.

Because of the foregoing, the analysis of realistic heat exchanger systems represents a challenging task. The modern technique of computational fluid dynamics (CFD) (Ladeinde and Nearon, 1997; Shah et al., 2000), although very promising, is not yet at the stage where it can routinely be applied to the analysis of two-phase heat exchanger systems with realistic geometries. Therefore, empirical methods currently provide the means of analysis. The text by Hewitt et al. (1994) provides the basic analytical procedures for thermo-hydraulic performance and design analysis of various types of heat exchangers. By far the two most common empirical approaches are the F-factor and NTU methods (Bowman et al., 1940; Shah and Sekulic, 1997). The F-factor of a given heat exchanger is the ratio of the effective temperature difference in the exchanger to that in a pure counter-current flow arrangement with identical terminal temperatures. Closed-form formulas for F-factors were presented by Bowman et al. (1940) for a few plate-fin arrrangements. For other selected group of heat exchangers, one must calculate the F-factor from NTU and other parameters. The functional relationship is known only for a relatively small selection of configurations and is not by any means explicit because it depends on the unknown outlet temperatures. The NTU method provides curves or functional relationships for the effectiveness, which is the ratio of heat transfer rate obtained from a system to that obtainable from an ideal system operating between the maximum temperature difference.

The F-factor and NTU methods do not offer much help in the analysis of many heat exchanger systems, particularly those involving a phase change and/or configurations for which no correlations or graphs are available. Moreover, these overall methods are restricted to constant properties and overall heat transfer coefficient, which is obviously not the case for phase change duties.

The present paper presents a versatile finite difference-type procedure for single phase and two-phase heat transfer analysis in a multi-pass plate-fin heat exchanger. The procedure is easy to implement and is readily applicable to various plate-fin flow arrangements, including arbitrary cross counter-current flow and cross co-current flow arrangements. The results from the proposed technique and from the NTU method are compared with experimental measurements of a single-phase, multi-pass, plate-fin system. Two-phase calculations with the method are also presented. The experimental measurements were carried out by Lytron, Inc., MA (USA).

NUMERICAL PROCEDURE

Our method is illustrated in this section with a two-stream, plate-fin problem. The flow passages are divided into small sections. The calculation tracks the flow from the inlet to the outlet of a fluid stream as the fluid goes through the passage. The fluid passage being tracked is termed the primary passage while the complementary passage of the other fluid is the secondary passage. In Figure 1, for instance, for every section along the cold fluid passage, there is an exchange of heat with the hot fluid through the thin separating plate. The heat transfer rate is computed in each section of the primary passage. The total heat transfer rate is an aggregate of the values in the various sections that make up the passage. The temperatures and other states of the fluid in the secondary passage are calculated at the end of every iteration.

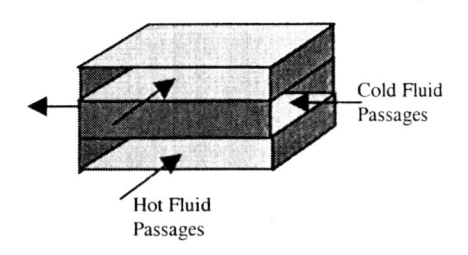

Figure 1. Schematic of the cross flow arrangement in a plate-fin heat exchanger

For plate-fin heat exchangers, heat transfer occurs in its basic form between a hot and cold fluid separated by plates. This is illustrated in Figure 1 for the cross flow arrangement.

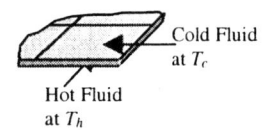

Figure 2. Infinitesimal area of hot and cold fluid exchanging heat

The heat flux between a sub-section of hot and cold fluid shown in Figure 2 can be expressed as

$$q'' = U \Delta T,$$

where q'' is the heat flow rate per unit area, ΔT or $(T_h - T_c)$ is the local temperature difference between the fluids, while U is the overall heat transfer coefficient for the small strip, which can be calculated from

$$\frac{1}{UA} = \frac{1}{(A\alpha)_h} + R_w + \frac{1}{(A\alpha)_c}, \qquad (1)$$

where A is the effective heat transfer surface area which includes the effect of fin area, α is the effective heat transfer coefficient which depends on the Reynolds number, the Prandtl number and fluid fouling factor. Note that R_w is the wall resistance which, for a plate-fin heat exchanger, is given by

$$R_w = \frac{a}{A_w k_p},$$

where a is the plate thickness, A_w is the total wall area for conduction, and k_w is the conductivity of the plate material. Note that $A_w = L_1 L_2 N_p$ where L_1, L_2 and N_p are the length, width, and the total number of separating plates.

From the basic nature of the cell in Figure 2, it is obvious that this technique is independent of whether the flow is cross or parallel. An important aspect of our procedure is the tracking of the flow segments for a variety of flow arrangements. Segments are groups of sections, which are exchanging heat with the fluid on the opposite side of the wall at the same station. In Figure 3, for example, the sections that make up segment 1 of the primary passage, for instance, will exchange heat with the same sections of fluid in the secondary passage at segment 16 of the secondary passage.

Tracking the Flow Segments

For flow through a passage of four passes, the primary passage segments can be easily tracked by a loop. However, the segments of the fluid on the opposite side of the plate are not apparent. This is illustrated in Figure 3.

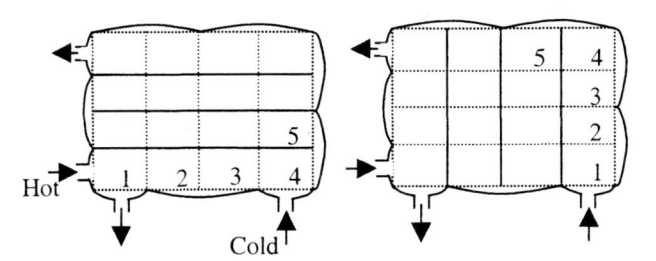

(a) Primary Passage Labels   (b) Secondary Passage Labels

Figure 3. Labeling of the fluid segments for a 4 pass by 4 pass cross counter-current flow heat exchanger

In the example in Figure 3, there are 16 flow segments. Assuming that the hot fluid, with inlet as shown, is the primary fluid, then this fluid will initially exchange heat with the 16th segment of the cold fluid. The temperature difference across a segment is of interest. To calculate the temperature difference across a segment, the temperature of the fluid in the secondary passage must be determined. This will be obtained only after the complementary segment of the secondary fluid corresponding to the primary fluid segment under consideration has been determined. Denoting the segment numbers of the primary fluid by p and that of the secondary fluid by q, the mapping between p and q can be expressed as

$$q = 2N_q \times \text{int}\left(\frac{i}{2}\right) + j \times \text{mod}(i,2) - (j-1) \times \text{mod}(i+1,2)$$

where

N_p = number of passes of the hot fluid
N_q = number of passes of the cold fluid

and (i, j) is a matrix of indices, with i running from right to left in Figure 3 and j from the bottom to the top.

The proposed procedure is described below.

Iterative Procedure

1) The passage for the primary stream is discretized into a total of, say, N_1 sections. The N_1 sections are further grouped into segments, depending on the intersection with the secondary stream. Thus, a segment contains more than one section. Note that segments are labeled 1, 2, ... in Figure 3.
2) The passage for the secondary stream is also discretized into a total of, say, N_2 sections. The N_2 sections are further grouped into segments, depending on the intersection with the primary stream.
3) An initial guess for the heat flux is made, and a compatible initial temperature distribution is generated.
4) For each section j of the primary passage, the following steps are carried out:
   a) obtaining fluid properties at the current value of the temperature.
   b) updating the flow quality, x, at the downstream end of the section
   c) calculating the temperature for the section using, for illustration only, procedures such as

   $T_{j+1} = T_{sat}$ (phase change)
   $T_{j+1} = T_j + \dfrac{q_j dxW}{\dot{M}C_p}$ (single-phase),

   where W is the flow width of the section and dx is the section or strip length
   d) calculating single-phase or two-phase pressure drop, as the case may be
   e) calculating the heat transfer coefficient, α (single-phase or two-phase) for the current section j
   f) computing fin efficiency, $(\eta_f)_j$ and the effective heat transfer area A_j'
   g) for each section k of the secondary fluid which exchanges heat with the current jth section:
      i) determine the segment to which k belongs and hence the temperature of the secondary fluid
      ii) calculate ΔT
      iii) get the fluid property for the secondary fluid at the secondary fluid's temperature
      iv) calculate α, A' and η_f for the secondary fluid
      v) calculate overall U
      vi) calculate $q'' = U\Delta T$.
5) For every section k of the secondary fluid
   a) retrieve the q'' values from Step 4 above
   b) compute the temperature for the section using similar equations to 4(c).
6) Go back to (4) and iterate until convergence.
7) Compute the overall heat transfer rate by summing the section values in the primary or secondary passage. Also compute the overall ΔP by integrating dp/dx across the entire passage.

In item (6) above, convergence is assumed when ε_s, defined as

$$\varepsilon_s = \sum \frac{\delta T_i}{\Delta T},$$

where δT_i is the residual and ΔT is the temperature difference that corresponds to the driving force.

The procedure involves the integration of certain quantities along the flow passage. For example, the pressure loss is given by

$$\int \frac{dp}{dx} dx$$

where dp/dx depends on the passage coordinate x. This integration is carried out using Simpson's 1/3 rule. Other quantities that need to be integrated along x include the U and the total fin efficiency.

RESULTS

The procedure reported in this paper has been incorporated into the general-purpose thermal analysis software called INSTED®. As is apparent from the description above, the procedure is equally applicable to both single-phase and two-phase problems. It is not limited by the number of passes of the hot or cold stream, as the results below illustrate.

Single-phase, Multi-pass Applications

The method described has been used for performance calculations of a plate-fin exchanger for streams of hot oil and air. The results are compared with those obtained from the NTU method and experimental measurements.

The oil has properties $\rho = 922.633$ kg/m^3 (57.6 lb/ft^3), $\mu = 4.274 \times 10^{-3}$ Ns/m^2 (10.342 lb/hrft), $k = 0.1267$ W/mK (0.0732 BTU/hrft°F), $C_p = 2110.86$ J/kgK (0.5042 BTU/lb°F). The fin description on either side is outlined in the table below:

Table 1. Fin description for single-phase calculation

	Oil	Air
Fin type	Triangular, with lanced offset	Triangular, louvered
Fin thickness	0.000152m 0.006"	0.000152m 0.006"
Offset pitch	0.003175m 1/8"	0.003175m 1/8"
Fin density	1102 fins/m 28 FPI	452.7 fins/m 11-12 FPI
Fin height	0.0032m 0.126"	0.0106m 0.420"

The number of passes on the oil side is two. Total number of oil passages is 12 while the number of air passages is 13. The core height is 0.04064m (1.6") with core width of 0.19685m (7.75"). This gives a flow length of 0.3937m (15.50") for the oil and 0.04064m (1.6") for air. The separator bar thickness is 0.0094m (0.37"), giving a flow width of 0.01613m (0.635") for the oil and 0.1872m (7.37") for air. The plates are made of aluminum with thickness 0.0004064m (0.016").

Seven combinations of inlet temperatures and flow rates were computed. The results are shown in the Table 3.

Table 2. Process Conditions for the single-phase heat transfer calculations

Case No.	Air Flow	Oil Flow	Air Inlet Temp.	Oil Inlet Temp.
A	0.49 kg/s 3892.1 lb/hr	0.32 kg/s 2539.4 lb/hr	300.61 K 81.41 °F	380.85 K 225.84 °F
B	0.2146 kg/s 1703.3 lb/hr	0.437 kg/s 3472.1 lb/hr	299.054 K 78.61 °F	390.76 K 243.69 °F
C	0.44 kg/s 3487.0 lb/hr	0.4363 kg/s 3462.8 lb/hr	300.61 K 81.41 °F	384.83 K 233.01 °F
D	0.489 kg/s 3881.3 lb/hr	0.4375 kg/s 3472.1 lb/hr	300.8 K 81.74 °F	383.22 K 230.12 °F
E	0.443 kg/s 3515.0 lb/hr	0.32 kg/s 2539.4 lb/hr	302.17 K 84.22 °F	383.89 K 231.31 °F
F	0.212 kg/s 1683.0 lb/hr	0.32 kg/s 2539.4 lb/hr	298.81 K 78.17 °F	385.08 K 233.45 °F
G	0.44 kg/s 3498.1 lb/hr	0.32 kg/s 2539.4 lb/hr	300.4 K 81.02 °F	381.81 K 227.57 °F

Table 3. Results of single-phase heat transfer calculations compared with experiments

Case No.	Q	ε	Q	ε	Q	ε
	(NTU)		(New Method)		Experiment	
A	8725 W 29773.5 Btu/hr	0.22	9486 W 32368.4 Btu/hr	0.24	10607W 36197.3 Btu/hr	0.27
B	8509.6W 29037.3 Btu/hr	0.43	9057.9W 30908.3 Btu/hr	0.457	8132W 27726.6 Btu/hr	0.41
C	10959W 37396.1 Btu/hr	0.29	11939W 40742.6 Btu/hr	0.321	11048W 37701.0 Btu/hr	0.30
D	11207W 38241.7 Btu/hr	0.276	12772W 43581.7 Btu/hr	0.315	11431W 39006.5 Btu/hr	0.28

E	8605W 29365.9 Btu/hr	0.236	9375.4W 31991.9 Btu/hr	0.257	10289W 35109.8 Btu/hr	0.28
F	6684.9W 22810.9 Btu/hr	0.363	7271.8W 24708.8 Btu/hr	0.373	7350W 25080.7 Btu/hr	0.40
G	9214.7W 31443.4 Btu/hr	0.255	10015W 34175.3 Btu/hr	0.277	10214W 34855.7 Btu/hr	0.28

The temperature variation along the length of both passages for Case A, non-dimensionalized by each stream's flow length is shown in Figure 4.

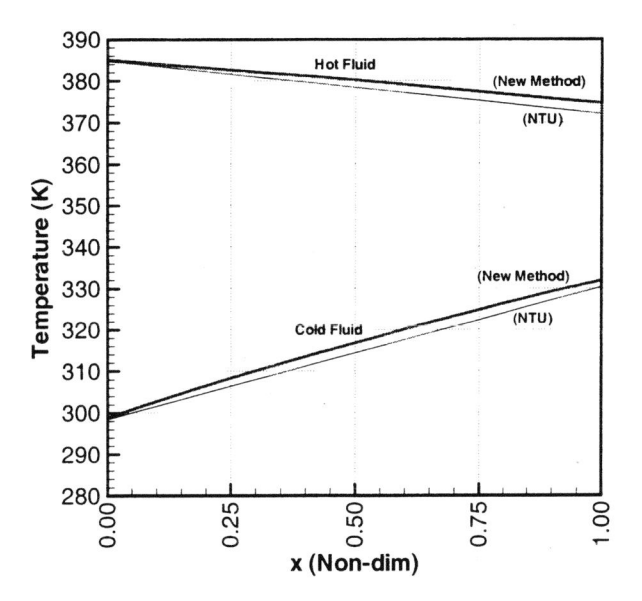

Figure 4. Temperature profile for single-phase heat transfer calculation

Deviation from a linear temperature profile is evident in Figure 4. The NTU method gives a straight-line temperature profile in its normal usage, although it is possible to employ the method in a temperature-dependent framework. However, the beauty of the new method lies in the natural way in which property variation along the heat exchanger is handled and the generality with respect to phase change, number of passes, flow arrangement and banking. The NTU and F-value methods are proven methods and Table 3 shows that they provide fairly accurate results. The results also show that the new method is quite competitive. Note that the motivation for the new method is in the ability to analyze new designs for which no F-value or empirical effectiveness relations are available.

Application to Two-Phase Heat Transfer Between R134a and a Proprietary Fluid

The method described in this paper has been used to compute the heat transferred between a hot fluid and R134a. The hot fluid has the properties: ρ = 1818 kg/m^3 (113.5 lb/ft^3), μ = 0.00455 Ns/m^2 (11 lb/hrft), k = 0.0066 W/mK (0.0381 BTU/hrft°F), C_p = 900.11 J/kgK (0.215 BTU/lb°F). The description of the fins on either side is outlined in the table below:

Table 4. Fin description for two-phase calculation

	Hot Fluid	R 134a
Fin type	Triangular, with lanced offset	Triangular, with lanced offset
Fin thickness	0.000152m 0.006″	0.000152m 0.006″
Offset pitch	0.003175m 1/8″	0.003175m 1/8″
Fin density	709 fins/m 18 FPI	709 fins/m 18 FPI
Fin Height	0.002184m 0.086″	0.002184m 0.086″

The number of passes of the hot fluid is four and there are 44 passages. The number of refrigerant passes is two; the number of passages is 20. The core length is 0.2286m (9″) with core width of 0.05715m (2.25″). This gives a flow length of 0.05715m (2.25″) for the hot fluid and 0.2286m (9″) for the refrigerant. The separator bar thickness for both streams is 0.003175m (0.125″), giving a flow width of 0.0532m (2.094″) for the hot fluid passage and 0.0238m (0.9375″) for the refrigerant passage. The plates are made of aluminum with thickness 0.0004064m (0.016″). The flow rate of the refrigerant is 0.0019656 kg/s (156 lb/hr) and that of the hot fluid is 1.053 kg/s (8358 lb/hr). The inlet quality of the refrigerant is 10%. Several methods were used to calculate α for sections undergoing two-phase heat transfer but the results shown are for the boiling procedures in Kandlikar (1990). The two-phase pressure drop was calculated using the Martinelli procedure.

Calculations using the method outlined in this paper give a total heat flow rate of 2802 W (9561 Btu/hr) with an exit quality of 78% for the refrigerant. The temperature variation along the length of both passages, non-dimensionalized by each stream's flow length, is shown in Figure 5.

Comparison of the two-phase calculations with limited experimental measurements shows good agreement.

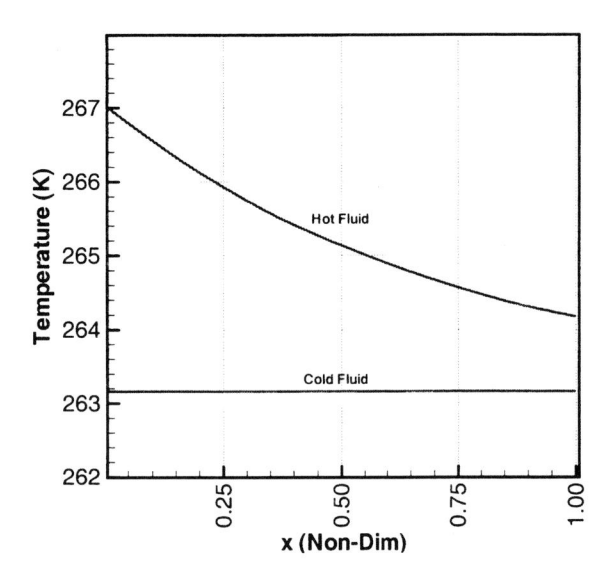

Figure 5. Temperature profile for two-phase heat transfer calculation

Figure 5 shows a nonlinear temperature variation for the hot fluid while the temperature of the refrigerant is fairly constant, indicating that the passage is not long enough to cause a sensible heat transfer in the refrigerant.

Figure 6. Profile of the pressure drop and the heat transfer coefficient, α, along the cold stream passages

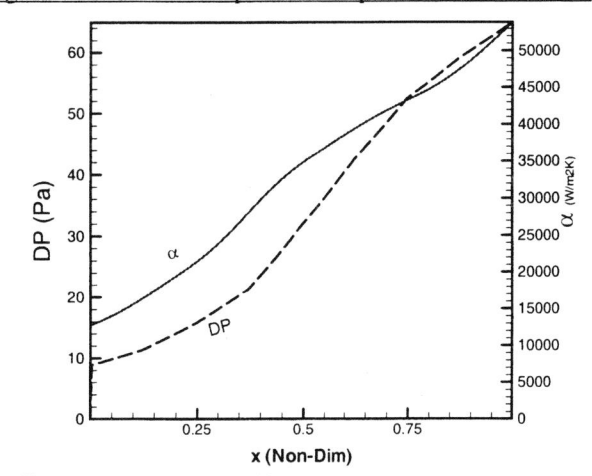

Figure 6 represents a plot of the pressure drop and heat transfer coefficient along the cold fluid passages. The pressure drop increases along the passages. This is expected because of the dependence of Δp on quality, which increases from the inlet value of 10% to 78% at the outlet.

Figure 6 also shows a significant increase in the two-phase heat transfer coefficient, α, as the passage is traversed. This can be explained in terms of the increasing quality, which appears explicitly in the convection number, Co:

$$Co = \left(\frac{1-x}{x}\right)^{0.8}\left(\frac{\rho_g}{\rho_l}\right)^{0.5}.$$

Note that in the Kandlikar procedure, Co is raised to the power of a negative number.

CONCLUSIONS

In this paper a method is presented for thermo-hydraulic analysis of plate-fin heat exchangers. This method is presented as an alternative to the F-factor and NTU methods. The new method is not limited by phase change, strong temperature dependence of properties or the number of passes. The method is also readily applicable to various flow arrangements and single and multiple banking.

To validate the procedure, computed results are compared with experimental measurements for single-phase and boiling in plate-fin heat exchangers.

Further development of the procedure reported in this paper is being undertaken to address the potential problems that might compromise the robustness of the method. For example, when both streams involve two-phase flow (such as evaporation on the cold side and humid air on the hot side), the numerical stability of the method becomes an issue. In this case, the heat transfer coefficient depends on the heat flux in both streams. There are ways to resolve this stability issue, however.

The paper assumes uniform distribution of the inlet fluid into the passages. A mal-distribution situation can also be analyzed although the specifics of the distribution has to be provided to the program.

The present version of the program assumes the saturation temperature is a function of some average pressure. The case of dependence with the local segment pressure complicates the procedure somewhat, although calculations are still possible.

NOMENCLATURE

a	plate thickness, m
A'	effective heat transfer area, m^2
A	area of plate, m^2
C_p	specific heat, J/kgK
h_{lg}	latent heat, J/kg
k_p	plate thermal conductivity, W/mK
$\dot{M}$	mass flow rate, kg/s

$N_{passages}$ number of passages
q'' heat flux, W/m²
R   wall resistance, K/W
T   temperature, K
U   overall heat transfer coefficient, W/m²K
ρ density, kg/m³
α heat transfer coefficient, W/m²K

Subscript

i   stream
h   hot fluid
c   cold fluid
g   gas
l   liquid
1   primary
2   secondary

ACKNOWLEDGEMENT

This study was funded by Thaerocomp Technical Corp. The authors appreciate the permission from Thaerocomp to publish the work.

The authors would like to express their appreciation to Lytron, Inc. (Lonnie Fultz and Richard Goldman) for providing the experimental data for validating the procedure in this paper.

REFERENCES

Anonymous, 1994, Standards of the Brazed Aluminium Plate-Fin Heat Exchanger. Manufacturer's Association *(ALPEMA)*, First Edition.

R. A. Bowman, A. C. Mueller, W. M. Nagle, 1940, Mean Temperature Difference in Design, Transactions of the ASME, Vol. 41, pp. 283-294.

G. F Hewitt and G. L. Shires, 1994, *Process Heat Transfer*, CRC Press, Boca Raton, New York, USA.

S. G. Kandlikar, 1990, A General Correlation for Saturated Two-phase Boiling Heat Transfer Inside Horizontal and Vertical Tubes. *Journal of Heat Transfer*, Vol. 112, pp. 41-44.

F. Ladeinde and M. Nearon, 1997, CFD Applications in the HVAC & R Industry. *ASHRAE Journal*, pp. 41-44.

R. K. Shah, M. R. Heikel, B. Thonon and P. Thonon, 2000, Progress in the Numerical Analysis of Compact Heat Exchanger Surfaces. Advances in Heat Transfer. Vol. 34, pp. 363-441.

R. K. Shah and D. P. Sekulic, 1997, Heat Exchangers., Handbook of Heat Transfer. Edited by W. M. Rohsenow, J. P. Hartnett, and Y. I. Cho, Chapter 17.

A SIMPLE METHOD FOR EVALUATION OF HEAT TRANSFER ENHANCEMENT IN TUBULAR HEAT EXCHANGERS UNDER SINGLE-PHASE FLOW, BOILING, CONDENSATION AND FOULING CONDITIONS

G.A. Dreitser, A.S. Myakotchin, and I.E. Lobanov

Department of Aviation and Space Thermal Technics,
Moscow Aviation Institute,
4, Volokolamskoe shosse, Moscow, 125993, Russia
E-mail: heat204@mai.ru

ABSTRACT

For heat transfer enhancement to be used effectively, it is important to possess a simple and reliable method of evaluating the efficiency of heat transfer enhancement. Now there is no adequate approach to evaluating heat transfer enhancement efficiency. Therefore, there raises the problem how to choose optimal parameters of heat exchangers when heat transfer enhancement is used.

By using any method for heat transfer enhancement, not only increase in heat transfer coefficients but also in hydraulic resistance coefficients and pumping pressure losses is obtained.

Therefore, there is a need to provide a means for thermal and hydraulic evaluation of any method without detailed design of heat exchangers. It is important to take into account that compared channels may have different perimeters and different equivalent diameters.

A method of evaluating the efficiency of heat transfer enhancement in channels of heat exchangers is introduced. The method is the simplest and allows estimating one or another procedure of heat transfer enhancement without detailed design of exchangers. This method has been developed both for tubular heat exchangers with longitudinal flow in tube bundles and for plate-and-fin heat exchangers. The method for calculation of heat transfer enhancement efficiency can be used at any heat transfer coefficient ratio on hot and cold sides of heat exchangers. Compared channels may have different geometry, different perimeters, and different cross-section areas.

Various methods of heat transfer enhancement are analyzed in the paper describing tubular heat exchangers with single-phase liquid and gas flow, at boiling and condensation of heat carriers as well as heat exchangers operating under fouling conditions. As a result, practical recommendations on an optimal choice of heat transfer enhancement parameters are obtained.

INTRODUCTION

Heat exchangers are widely used in power engineering, chemical industry, petroleum refineries, and food industry; in heat engines, cars, tractors, boats, and ships; in aviation and space vehicles; in refrigerating and cryogenic engineering; in space conditioning, heating, and hot water supply systems; and in many other fields of technology.

The problem of decreasing a mass and size of heat exchangers is urgent. Enhancement of convective heat transfer processes is a promising means for solving this problem. It should be noted that use of heat transfer enhancement enables solving some other, no less important problem, such as lowering maximum temperatures of heat exchanger working surfaces, improving the operational reliability, and reducing the fouling and other contamination.

At present, different methods of convective heat transfer enhancement in heat transfer channels have been proposed and studied [Dreitser et al, 1990]. These include flow turbulators on a surface, a rough surface, a developed surface due to finning, flow swirling by spiral fins, worm devices, swirlers mounted at the channel entrance, gas bubble mixing in the liquid flow, mixing of particles or liquid drops in the gas flow, rotation of a heat transfer surface, surface vibration, heat carrier pulsation, action of an electrostatic field on the flow, flow suction from the boundary layer. Use of jet cooling systems is a means for heat transfer enhancement in single-phase heat carrier flow.

Along with turbulators, twisting and finning, low heat conducting and porous coatings called nonisothermal fins are also utilized for boiling heat transfer enhancement.

Turbulators or fins that break up a condensate film, nonwetted coatings, liquid stimulators to produce drop condensation, flow swirling, and rotation of a heat transfer surface are proposed as the ways to enhance condensation heat transfer from heat carriers.

The number of publications on heat transfer enhancement progressively increases. These publications are mentioned in [Dreitser et al., 1990; Kalinin et al., 1998,1; Kalinin et al., 1998,2]. However, the results of these investigations are often contradictory, and the enhancement methods proposed are not always adaptable to production and/or efficient. In a number of cases, the choice is not supported and is accidental in nature. This situation hampers a substantiated choice of efficient methods of heat transfer enhancement as well as the evaluation of different methods. This is attributed to weak use of heat transfer enhancement methods, which undoubtedly causes large economic losses and retards the reduction of metal consumption for heat exchangers.

There are many different methods for calculation of heat transfer enhancement efficiency [Antufuev, 1966; Bajan et al., 1989; Dedusenko, 1960; Dubrovsky, 1977; Dzyubenko et al., 1993; Gukhman, 1977; Kalafati et al., 1986; Kovalenko et al., 1986; Mitin et al., 1975; Mitskevich, 1967]. These methods are enough complex and, as rule, require detailed design of heat exchangers. By this reason, these methods are no usable.

For effective use of heat transfer enhancement, it is important to possess a simple and reliable method of evaluating the efficiency of heat transfer enhancement. Now we have no adequate approach to evaluating heat transfer enhancement efficiency. Therefore, there raises the problem how to choose optimal parameters of heat exchangers when heat transfer enhancement is used.

By using any heat transfer enhancement method, not only increase in heat transfer coefficients but also in hydraulic resistance coefficients and pressure losses for pumping of heat carriers is obtained. Therefore, there is a need to provide thermal and hydraulic evaluation of any method without detailed design of heat exchangers. It is important to take into account that compared channels may have different parameters and different equivalent diameters.

STATEMENT OF THE PROBLEM

Use of the analyzed heat transfer enhancement method allows overall sizes and mass of tubular and other-type heat exchangers to be decreased, and their temperatures and flow rates to be reduced for a given heat transfer surface size. Some other problems can be solved. Optimum parameters of turbulators will be chosen for specific conditions and, accordingly, the chosen method for heat transfer enhancement is evaluated. For example, in the system of an engine (or a vehicle) a heat exchanger that provides the extreme of engine or vehicle parameters can appear to be optimal. The specific fuel flow rate as a function of both the heat power of a heat exchanger and the level of hydraulic losses can be a criterion for heat exchangers used in the heat recovery system of a gas turbine plant. A minimum flow rate of cooling air at a given temperature of a blade or a minimum temperature for a specified flow rate can be a criterion for turbine blades to be cooled.

However, the efficiency of heat transfer enhancement can be evaluated in a general sense. Here, three criteria for evaluation of the efficiency of heat transfer enhancement can be mentioned:
1. Comparison of heat transfer surfaces or volumes of two heat exchangers: one heat exchanger has smooth surfaces and another is equipped with facilities to enhance heat transfer. In this case, both compared heat exchangers must have the same heat power, heat carrier flow rate, and pumping pressure loss.
2. Comparison of the heat power of heat exchangers with and without heat transfer enhancement for the same volumes, flow rates of heat carriers, and pressure losses for heat carrier pumping, i.e., the same pumping power.
3. Comparison of heat powers or pressure losses for pumping of fluids of heat exchangers with and without heat transfer enhancement when the volumes, heat power and flow rates are the same.

Comparison of the efficiency of heat transfer enhancement is made for heat exchangers with longitudinal flow near heat transfer surfaces. This category of heat exchangers includes tubular heat exchangers with longitudinal flow in the tube bundles and plate-and-fin heat exchangers. Compared channels may have different geometry, different perimeters, and different cross-sections.

We shall compare different channels: a smooth channel has perimeter U_{sm}, cross-sectional area F_{sm}, equivalent diameter $d_{eq\,sm}=4F_{sm}/U_{sm}$; length l_{sm}; a channel with facilities to enhance heat transfer has U, F, d_{eq}, l, respectively.

COMPARISON OF HEAT EXCHANGERS WIHT HEAT TRANSFER COEFFICIENT FOR ONE SIDE MUCH GREATER THAN FOR ANOTHER

Comparison of the efficiency of heat exchangers is made individually both for flow inside the tubes and in the intertube space of tubular heat exchangers and for flow on one side of plate-and fin exchangers.

Such an attack at the problem allows mathematical manipulations to be made illustrative. For this, it will suffice to assume that the heat transfer coefficient will be essentially lower for a compared surface than for any other heat transfer surface. Then, a knowledge of the efficiency of heat transfer enhancement inside the tubes and in the intertube space enables evaluating the efficiency of the heat exchanger as a whole depending on the heat transfer coefficient ratio on both sides of the heat exchanger.

Criteria for evaluation of the heat transfer enhancement efficiency are derived for tubular in-line heat exchangers in the intertube space, considering that the heat transfer coefficient in the intertube space is much higher than inside the tubes. As shown below, the obtained conclusions are also true for the cases when heat transfer in the intertube space specifies the sizes of heat exchangers and when different-geometry heat exchangers are compared. The diameter of the tubes and their pitch in a bundle are chosen minimum possible, i.e., heat transfer enhancement is the only way to decrease heat exchanger sizes. We shall consider that heat exchangers are single-pass.

First, let us derive a criterion for heat transfer surfaces or volumes of heat exchangers with the same heat power, flow rates in the tubes and hydraulic resistance in the tube flow:

$$Q = Q_{sm} \quad (1)$$

$$G = G_{sm} \quad (2)$$

$$\Delta p = \Delta p_{sm} \quad (3)$$

Since it is assumed that thermal resistance in the intertube space and normal resistance of the wall are small, the heat power for compared heat exchangers $Q = \alpha \cdot \Delta T \cdot U \cdot l \cdot N$ and $Q_{sm} = \alpha_{sm} \cdot \Delta T_{sm} \cdot U_{sm} \cdot l_{sm} \cdot N_{sm}$ where α is the heat transfer coefficient inside the tubes; ΔT is the driving temperature difference; N is the number of tubes or channels in a heat exchanger. As for both heat exchangers the temperature differences are the same ($\Delta T = \Delta T_{sm}$). According to Eq.(1) we have

$$\alpha \cdot U \cdot l \cdot N / (\alpha_{sm} \cdot U_{sm} \cdot l_{sm} \cdot N_{sm}) = 1 \quad (4)$$

Friction pressure losses are

$$\Delta p = \xi \frac{l}{d_{eq}} \frac{\rho w^2}{2} \quad \text{and}$$

$$\Delta p_{sm} = \xi_{sm} \frac{l_{sm}}{d_{eq\,sm}} \frac{\rho_{sm} w_{sm}^2}{2}$$

Here w is the mean fluid velocity in the tubes; ρ is the density; ξ is the hydraulic resistance coefficient.

By Eq. (3) ($\rho = \rho_{sm}$) we arrive at

$$\frac{\xi \cdot l \cdot w^2 \cdot d_{eq\,sm}}{\xi_{sm} \cdot l_{sm} \cdot w_{sm}^2 \cdot d_{eq}} = 1 \quad (5)$$

From Eq.(2) the Reynolds number ratio is

$$\frac{Re}{Re_{sm}} = \frac{w \cdot d_{eq}}{w_{sm} \cdot d_{eq\,sm}}$$

$$= \frac{F_{sm} \cdot N_{sm} \cdot d_{eq}}{F \cdot N \cdot d_{eq\,sm}} = \frac{N_{sm} \cdot U_{sm}}{N \cdot U} \quad (6)$$

as the fluid velocity in the tubes for a given total flow rate is inversely proportional to their number.

It is should be noted that when the heat transfer coefficient α is determined in the turbulator-equipped tube or channel a possible increase of the heat transfer surface is not allowed for individually, i.e., it refers to a smooth tube or channel surface. A mean velocity is determined by the tube (or channel) cross-section, and a smooth channel equivalent diameter is taken as a determining size, i.e., comparison is made under the some mass flow rates of heat carriers in the compared channels. This very mass velocity and very diameter are adopted to determine ξ in the turbulator-equipped channel. When the coefficients α and ξ are determined in the above manner, evaluating the efficiency by either this or that heat transfer enhancement method is extremely simplified.

Unfortunately, in many heat transfer enhancement investigations these simple conditions are not being carried out, a determining size, a determining velocity, a heat transfer surface are defined by another ways, and as a result it is very difficult to compare the efficiencies of heat transfer surfaces and to make practical calculations.

For smooth channels $Nu_{sm} = C_1 Re_{sm}^n$ and $\xi_{sm} = C_2 Re_{sm}^m$. For turbulator-provided channels, the increase in heat transfer and hydraulic resistance is allowed for by the ratios Nu/Nu_{sm} and ξ/ξ_{sm}, that depend on the Reynolds number for a given turbulator geometry.

Therefore,

$$Nu = (Nu/Nu_{sm})_{Re} \cdot C_1 \cdot Re_{sm}^n$$

$$\xi = (\xi/\xi_{sm})_{Re} \cdot C_2 \cdot Re_{sm}^m$$

$$\alpha/\alpha_{sm} = (Nu/Nu_{sm})_{Re} \cdot (Re/Re_{sm})^n \cdot (d_{eq\,sm}/d_{eq}) \quad (7)$$

$$\xi/\xi_{sm} = (\xi/\xi_{sm})_{Re} \cdot (Re/Re_{sm})^m \quad (8)$$

Here the subscript "Re" means that the ratios Nu/Nu_{sm} and ξ/ξ_{sm} are taken so that the Reynolds number for a smooth channel and a channel with turbulators are the same and in the considered case are equal to the Reynolds number for the channel with turbulators.

From Eqs. (4)-(8), it is possible to obtain the number of channels and the length ratio of the compared heat exchangers

$$\frac{N}{N_{sm}} = \frac{\left(\frac{\xi}{\xi_{sm}}\right)_{Re}^{\frac{1}{3+m-n}}}{\left(\frac{Nu}{Nu_{sm}}\right)_{Re}^{\frac{1}{3+m-n}} \cdot \frac{U}{U_{sm}} \cdot \left(\frac{d_{eq}}{d_{eq\,sm}}\right)^{\frac{2}{3+m-n}}} \quad (9)$$

$$\frac{l}{l_{sm}} = \frac{\left(\dfrac{d_{eq}}{d_{eq\,sm}}\right)^{\frac{5+m-2n}{3+m-n}}}{\left(\dfrac{Nu}{Nu_{sm}}\right)_{Re}^{\frac{2+m}{3+m-n}} \cdot \left(\dfrac{\xi}{\xi_{sm}}\right)_{Re}^{\frac{1-n}{3+m-n}}} \quad (10)$$

The heat transfer surface ratio for compared heat exchangers is

$$\frac{S}{S_{sm}} = \frac{U \cdot l \cdot N}{U_{sm} \cdot l_{sm} \cdot N_{sm}}$$

$$= \frac{\left(\dfrac{d_{eq}}{d_{eq\,sm}}\right)^{\frac{3+m-3n}{3+m-n}} \cdot \left(\dfrac{\xi}{\xi_{sm}}\right)_{Re}^{\frac{n}{3+m-n}}}{\left(\dfrac{Nu}{Nu_{sm}}\right)_{Re}^{\frac{3+m}{3+m-n}}} \quad (11)$$

If the compared channels (for example, the tubes) are located in a bundle having the same pitch, then the heat exchanger cross-sectional area ratio is

$$\frac{F_\Sigma}{F_{\Sigma\,sm}} = \frac{F \cdot N}{F_{sm} \cdot N_{sm}}$$

$$= \frac{\left(\dfrac{d_{eq}}{d_{eq\,sm}}\right)^{\frac{1+m-n}{3+m-n}} \cdot \left(\dfrac{\xi}{\xi_{sm}}\right)_{Re}^{\frac{1}{3+m-n}}}{\left(\dfrac{Nu}{Nu_{sm}}\right)_{Re}^{\frac{1}{3+m-n}}} \quad (12)$$

and the heat exchanger volume ratio is

$$K_v = \frac{V}{V_{sm}} = \frac{F \cdot N \cdot l}{F_{sm} \cdot N_{sm} \cdot l_{sm}}$$

$$= \frac{\left(\dfrac{d_{eq}}{d_{eq\,sm}}\right)^{\frac{6+2m-4n}{3+m-n}} \cdot \left(\dfrac{\xi}{\xi_{sm}}\right)_{Re}^{\frac{n}{3+m-n}}}{\left(\dfrac{Nu}{Nu_{sm}}\right)_{Re}^{\frac{3+m}{3+m-n}}} \quad (13)$$

If the compared channels have the same perimeters ($U=U_{sm}$) but different equivalent diameters ($d_{eq} \neq d_{eq\,sm}$), then Eq. (9) for N/N_{sm} becomes more simple

$$\frac{N}{N_{sm}} = \frac{\left(\dfrac{\xi}{\xi_{sm}}\right)_{Re}^{\frac{1}{3+m-n}}}{\left(\dfrac{Nu}{Nu_{sm}}\right)_{Re}^{\frac{1}{3+m-n}} \cdot \left(\dfrac{d_{eq}}{d_{eq\,sm}}\right)^{\frac{2}{3+m-n}}} \quad (14)$$

Eq.(10) for l/l_{sm}, Eq. (11) for S/S_{sm}, Eq.(12) for $F_\Sigma/F_{\Sigma\,sm}$ and Eq. (13) for V/V_{sm} become invariable.

In turbulent flow, when $n=0.8$; $m=-0.2$, we have

$$K_v = \frac{V}{V_{sm}} = \frac{\left(\dfrac{d_{eq}}{d_{eq\,sm}}\right)^{0.2} \cdot \left(\dfrac{\xi}{\xi_{sm}}\right)_{Re}^{0.4}}{\left(\dfrac{Nu}{Nu_{sm}}\right)_{Re}^{1.4}} \quad (15)$$

As indicated by the above ratios, use of heat transfer enhancement always makes a heat exchanger length (l/l_{sm}) smaller. The number of tubes in a bundle decreases if $\xi/\xi_{sm} < (Nu/Nu_{sm})$ and increases for $\xi/\xi_{sm} > (Nu/Nu_{sm})$. The size of a heat exchanger reduces if $\xi/\xi_{sm} < (Nu/Nu_{sm})^{3.5}$.

Heat powers of heat exchangers are compared when their flow rates, pressure losses, and volumes are equal. Since compared heat exchangers are composed of tubes with the same perimeters in a bundle with the same pitch, they have the same heat transfer surfaces, i.e., $U \cdot l \cdot N = U_{sm} \cdot l_{sm} \cdot N_{sm}$, from which it follows that

$$\frac{l}{l_{sm}} = \frac{N_{sm}}{N} \quad (16)$$

and also

$$\frac{Q}{Q_{sm}} = \frac{\alpha}{\alpha_{sm}} \quad (17)$$

(when $\Delta T = \Delta T_{sm}$ as in the first case). Taking into account Eqs. (5)-(8) and (17), we obtain

$$K_Q = \frac{Q}{Q_{sm}} = \frac{\left(\dfrac{Nu}{Nu_{sm}}\right)_{Re} \cdot \left(\dfrac{d_{eq}}{d_{eq\,sm}}\right)^{\frac{3n-m-3}{3+m}}}{\left(\dfrac{\xi}{\xi_{sm}}\right)_{Re}^{\frac{n}{3+m}}} \quad (18)$$

In turbulent flow ($n=0.8$, $m=-0.2$), we obtain

$$K_Q = \frac{Q}{Q_{sm}} = \frac{\left(\frac{Nu}{Nu_{sm}}\right)_{Re} \cdot \left(\frac{d_{eq}}{d_{eq\,sm}}\right)^{-0.143}}{\left(\frac{\xi}{\xi_{sm}}\right)_{Re}^{0.286}} \quad (19)$$

Use of heat transfer enhancement increases the number of tubes and, accordingly, decreases their length. When flow rates, pressure losses, and volumes of heat exchangers are equal, use of heat transfer enhancement increases the heat power if $\xi/\xi_{sm} < (Nu/Nu_{sm})^{3.5}$.

Let us compare heat exchangers with respect to pressure losses for fluid pumping. In this case, the fluid flow rates, the volumes of the heat exchangers, and the heat transfer surfaces as well as the heat powers of the heat exchangers are assumed to be the same. Hence, it follows that for the same temperature differences we obtain

$$\alpha = \alpha_{sm} \quad (20)$$

From Eqs. (6), (8) and (20) we have

$$K_{\Delta p} = \frac{\Delta p}{\Delta p_{sm}} = \frac{\left(\frac{\xi}{\xi_{sm}}\right)_{Re} \cdot \left(\frac{d_{eq}}{d_{eq\,sm}}\right)^{\frac{3+m-3n}{n}}}{\left(\frac{Nu}{Nu_{sm}}\right)_{Re}^{\frac{3+m}{n}}} \quad (21)$$

In turbulent flow (n=0.8, m= -0.2), we arrive at

$$K_{\Delta p} = \frac{\Delta p}{\Delta p_{sm}} = \frac{\left(\frac{\xi}{\xi_{sm}}\right)_{Re} \cdot \left(\frac{d_{eq}}{d_{eq\,sm}}\right)^{0.5}}{\left(\frac{Nu}{Nu_{sm}}\right)_{Re}^{3.5}} \quad (22)$$

In this case, use of heat transfer enhancement also increases the number of tubes and decreases their length. The heat power for heat carrier pumping decreases if $\xi/\xi_{sm} < (Nu/Nu_{sm})^{3.5}$ and increases if $\xi/\xi_{sm} > (Nu/Nu_{sm})^{3.5}$.

It should be noted that heat transfer enhancement is efficient according to all three criteria if $(Nu/Nu_{sm})^{3.5} > \xi/\xi_{sm}$.

COMPARISON OF HEAT EXCHANGERS FOR ANY RELATIONSHIPS BETWEEN HEAT TRANSFER COEFFICIENTS FOR HEAT AND COLD SIDES

In the calculations performed, heat transfer was considered on one side of a heat transfer surface and thermal resistance was not allowed for on another. An expression was obtained for V/V_{sm} when heat transfer enhancement was considered on both sides of the heat transfer surface.

With the thermal resistance of tube walls neglected, the volume ratio of heat exchangers is

$$K_V = \frac{V}{V_{sm}} = \frac{K_{sm}}{K} = \frac{1/\alpha_1 + 1/\alpha_2}{1/\alpha_{1sm} + 1/\alpha_{2sm}} =$$

$$= \frac{\alpha_{1sm}}{\alpha_1} \cdot \frac{1 + \alpha_1/\alpha_2}{1 + \alpha_{1sm}/\alpha_{2sm}} \quad (23)$$

where the subscript "1" stands for the inner surface and the subscript "2" is the outer surface.

Eq. (23) can be transformed by using $(V/V_{sm})_1 = \alpha_{1sm}/\alpha_1$ and $(V/V_{sm})_2 = \alpha_{2sm}/\alpha_2$ where the subscripts "1" and "2" show that the volume ratios V/V_{sm} are taken when heat transfer is considered to occur either only from the inner surface or only from the outer surface by ignoring the thermal resistance of another heat transfer surface. Then

$$K_V = \frac{V}{V_{sm}} = \left(\frac{V}{V_{sm}}\right)_1 \left\{\frac{1 + \alpha_1/\alpha_2}{1 + [(V/V_{sm})_1/(V/V_{sm})_2](\alpha_1/\alpha_2)]}\right\} \quad (24)$$

α_1/α_2 can be respected as

$$\frac{\alpha_1}{\alpha_2} = \frac{\alpha_1}{\alpha_{1sm}} \cdot \frac{\alpha_{1sm}}{\alpha_{2sm}} = \left(\frac{Nu}{NU_{sm}}\right)_1 \cdot \frac{1}{(Nu/Nu_{sm})} \cdot \frac{\alpha_{1sm}}{\alpha_{2sm}} \quad (25)$$

Because the ratios $(Nu/Nu_{sm})_1$ and $(Nu/Nu_{sm})_2$ for the compared surfaces are known, it is possible to obtain the relation for K_v if the heat transfer coefficient ratio for heat exchangers with smooth surfaces $\alpha_{1sm}/\alpha_{2sm}$ is known, too.

As seen from Eq. (24), if $\alpha_1 \ll \alpha_2$, then $K_v = K_{v1}$, and if $\alpha_1 \gg \alpha_2$, then we obtain $K_v = K_{v2}$.

The relations permit a comparatively simple comparison of the efficiency of heat transfer enhancement when the compared channels have different cross-sectional areas and perimeters and also when $d_{eq} = d_{eq\,sm}$ and $U = U_{sm}$ for these channels.

As mentioned before, the proposed method is applied for calculation of heat transfer enhancement in tubular heat exchangers with longitudinal flow in tube bundles and for plate-and-fin heat exchangers.

EXPERIMENTAL AND COMPUTED RESULTS FOR HEAT TRANSFER ENHANCMENT EFFICIENCY

Turbular Heat Exchangers with One-Phase Heat Carriers

At the Moscow Aviation Institute the high-effective method was developed to enhance heat transfer in tubular heat exchangers, and comprehensive studies were made of the efficiency of this method when applied to gas and liquid flow in tubes, circular channels, and in tube bundles over a wide range of performance parameters.

The essence of the method proposed is as follows. Equidistant annular grooves are rolled on an outer surface of a heat transfer tube. In this case, the annular smooth-wall diaphragms are formed over on an inner tube surface. Annular grooves and diaphragms swirl the flow in a wall layer and provide heat transfer enhancement outside and inside the tubes (Fig.1).

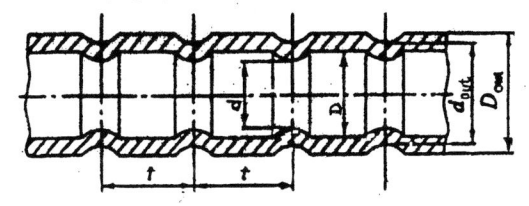

Fig. 1 A tube with annular turbulators

The outer diameter of the tubes does not increase, which permits their use in compact bundles and does not require change in the existing way of assembling heat exchangers.

The technology worked out at the Institute VNIMetMash for manufacturing knurled tubes is not complicated and makes use of standard equipment. The knurling cost is not more than several percent of that of the tubes. Knurling can be done at a rate of one to two meters per minute using a facility installed on a lathe. A special facility provides knurling at a rate of 9 m/min.

The tubes with annular turbulators can be installed in heat exchangers using gases and liquids with boiling and condensation of heat carriers; that is, they possess the versatility required for practical applications. Moreover, as shown below, these tubes are less susceptible to fouling. Thus, tubes with annular turbulators satisfy virtually all of the requirements necessary for their widespread practical application.

Use of the developed method for heat transfer enhancement permitted both the 1.5÷2.5 –fold reduction of a heating surface, at present heat power and pumping power of heat carriers as well as the improvement of the performance parameters of heat exchangers due to reducing fouling.

For laminar flow in a tube with diaphragms, the space near the wall between the diaphragms is filled with a stagnant liquid that reduces heat transfer; the higher the height of a diaphragm, the more the heat transfer reduction. However, the diaphragm itself reduces the critical Reynolds numbers because the separated flow zones formed by the diaphragms disturb the stability of viscous flow, which extends the range (Re=2000÷5000) over which the most effective relationships are obtained between the growth of heat transfer and pressure drop coefficients (Nu/Nu$_{sm}$=2.83, with ξ/ξ_{sm}=2.85). The interaction mechanism was evolved between specially designed turbulators and the flow at a transition to weakly developed turbulent flow. On the basis of this mechanism, it is established that reasonable augmentation can be achieved under these conditions with sufficiently high diaphragms (d/D=0.92) and an optimum pitch of t/D=1 (Fig.1).

In the developed turbulent flow zone, the best results are obtained with low diaphragms (d/D=0.94) and small pitches (t/D=0.25…0.5). At higher heights of diaphragms (when d/D becomes smaller), the ratio Nu/Nu$_{sm}$, at first, sharply rises and then levels off, reaching a value of 3 to 3.1 with gas flow and 2.3 with liquid flow. As the height of a diaphragm increases, the pressure drop rises first gradually. In the range of low diaphragms (d/D=0.96÷0.993), values of d/D and t/D exist, over which the growth of heat transfer is equal to or leads to an increase in pressure drop, that is, Nu/Nu$_{sm}$≥ξ/ξ_{sm}. The ratio d/D, which Nu/Nu$_{sm}$=ξ/ξ_{sm}, becomes smaller as Re increases and t/D becomes greater. With smaller d/D, we have Nu/Nu$_{sm}$<ξ/ξ_{sm}. When t/D=0.25, the ratio Nu/Nu$_{sm}$=ξ/ξ_{sm} increases with Re and attains a value of 2 when Re=4×10^5. When t/D=1, the ratio Nu/Nu$_{sm}$=ξ/ξ_{sm} falls as Re increases and attains 1.78 when Re=2×10^4.

It is interesting to mention that received values of Nu/Nu$_{sm}$ for tubes with annular diaphragms are close to limiting heat transfer enhancement in a tube. This means that the possibilities of further heat transfer enhancement increase are not so high.

We have calculated limiting heat transfer enhancement in a tube at different values of Re and Pr numbers (Fig. 2).

V.K. Migai's model (1990) has been used, according to which the turbulent flow is divided into four layers: turbulent core, vortex core in a groove, intermediate layer, viscous sublayer. Limiting flow swirling assumes that each component of flow thermal resistance can be ultimately decreased, namely: a magnitude of a viscous sublayer under any external flow swirling is conserved; an intermediate region practically cannot be larger than the half a protrusion (physically, this means that vortices, breaking away from their tops at heights of elements larger than the half a turbulator height, do not reach the surface); the flow core practically cannot undergo flow swirling to a greater extent than in jet flow.

Unlike V.K. Migai's model, we managed to improve calculation because of the modern level of computing machinery. Also, we eliminated some drawbacks peculiar to this model. The above-mentioned enabled us to obtain data for a more wide range of Re and Pr numbers.

A: Re = 10 000
B: Re = 50 000
C: Re = 100 000

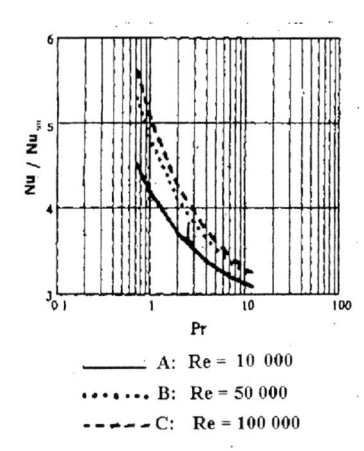

Fig.2 Limiting Nusselt number ratio Nu/Nu$_{sm}$ vs. Prandtl number.

In previous section, it is shown that if in a tubular heat exchanger the heat transfer coefficient inside tubes is significantly smaller than that in the intertube space, then the volume ratio of compared heat exchangers (in one of them, facilities to enhance heat transfer are located inside tubes and in another, tubes are smooth) in the turbulent flow is

$$\frac{V}{V_{sm}} = \frac{(\xi/\xi_{sm})^{0.4}_{Re_0}}{(\alpha/\alpha_{sm})^{1.4}_{Re}} \quad (26)$$

In this case, to calculate the Reynolds number for a turbulator-equipped tube, the mean velocity is determined from the smooth tube cross-section, and the smooth tube diameter is taken as a determining size, i.e., comparison is made under the same mass flow rates of heat carriers in the compared tubes.

Continuous flow swirling can be arranged by twisted strips or screw inserts located along the entire tube (Fig.3). Unlike local flow swirling, these are simple in manufacturing and offer a larger increase of mean heat transfer since the flow swirling degree along the channel is not diminished. However, the hydraulic resistance in this case also increases because of additional friction pressure

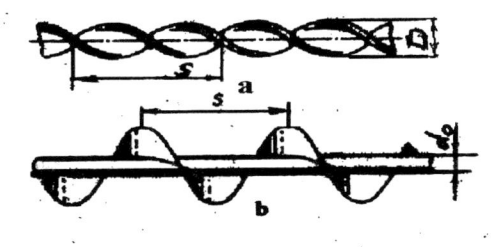

Fig. 3 Spiral inserts: a, twisted tape; b, spiral screw.

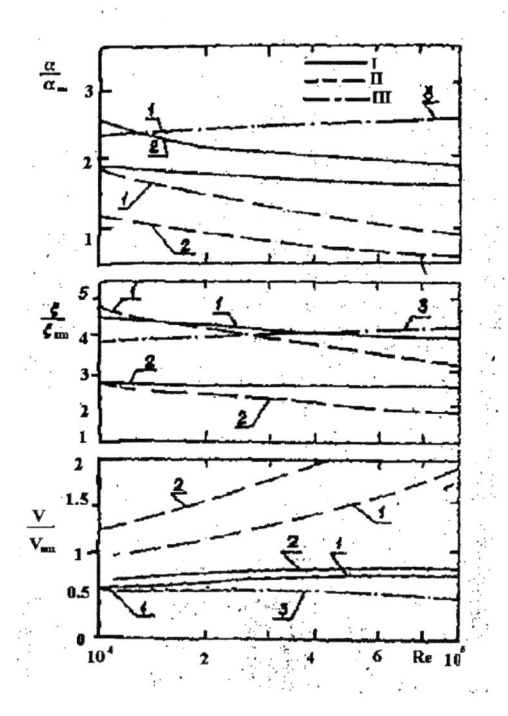

Fig 4. Influence of the Reynolds number on heat transfer enhancement α/α_{sm}, on the hydraulic resistance coefficient ratio ξ/ξ_{sm} and on the heat exchanger volume ratio V/V$_{sm}$: I, twisted tape; II, spiral screw; III, circular diaphragms; 1, S/D=4; 2, S/D=10; 3, circular diaphragms with d/D=0.94 and t/D=0.25

losses on the surface of a strip or a screw.

Shchukin (1970,1980) obtained the appropriate relations that allow numerous experimental data of various researchers to be generalized. But their application for analyzing the flow swirling efficiency is difficult, as according to these data the flow velocity in insert-provided tubes taken with regard to the insert blocking diameter of the insert-containing tube serves as a determining size spiral screw.

Having used these results, Dreitser (1997) obtained the formulas for $(\xi/\xi_{sm})_{Re}$ and $(\alpha/\alpha_{sm})_{Re}$ convenient for practical calculations.

The hydraulic resistance coefficient, ξ, of the strip insert-equipped tube is determined in terms of the flow velocity w in the insert-free tube. The tube diameter D serves as a determining size and the local hydraulic resistances in the flow at the entrance and exit of the turbulator-provided tube are allowed for.

Then for screw-shaped inserts an increase of heat transfer and hydraulic resistance coefficients can be easily determined as against the tube having no inserts at given geometry of inserts and constant flow rates of heat carrier and, hence, the efficiency of heat transfer enhancement can be found. Use of these relations allows one to immediately answer the question of interest to practical workers: how will the location of helical inserts inside tubes enhance hydraulic resistance and heat flux under the same flow rate and other conditions being equal.

At $Re_D=10^4$ when the strip twisting pitch S/D varies between 4 and 12, the heat transfer increase $\alpha/\alpha_{sm}=2.34\div1.8$ is accompanied by the hydraulic resistance growth $\xi/\xi_{sm}=4.05\div2.5$. This enables decreasing the heat exchanger volume $V/V_{sm}=0.53\div0.64$, i.e., the heat exchanger volume can be decreased by a factor of 1.8 to 1.51.

As the Reynolds number grows, the efficiency of heat transfer enhancement by means of strip inserts substantially degrades: at $Re = 10^5$ we have $\alpha/\alpha_{sm} = 1.88\div1.49$, $\xi/\xi_{sm} = 5.55\div1.65$, which yields $V/V_{sm} = 0.822\div0.70$, i.e., the heat exchanger volume can be decreased by a factor of 1.21 to 1.43.

It should be noted that at no values of the Reynolds number and strip twisting pitches we could not obtain $\alpha/\alpha_{sm}>\xi/\xi_{sm}$, i.e., we could not achieve an advancing increase in the heat transfer coefficient as against the hydraulic resistance growth.

Fig. 4 plots α/α_{sm}, ξ/ξ_{sm} and V/V_{sm} as a function of Reynolds number for different values of a twisting pitch S/D.

Use of strip inserts is seen to be most effective at $Re_D=10^4\div2\cdot10^4$ and at small twisting pitches $S/D\cong4\div6$.

Compare the obtained data on the efficiency of heat transfer enhancement by a twisted strip with the results for annular smooth-wall diaphragms (Section 4). Having utilized such turbulators, we could strive for a larger increase of heat transfer coefficient in tubes as against that of hydraulic resistance.

Fig. 4 plots the appropriate data for an annular-diaphragmed tube with a relative diameter d/D=0.94 and a diaphragm pitch t/D=0.25. As seen from this figure, use of an annular turbulator allows one to gain a stable $2.3\div2.43$-fold increase of heat transfer over the range $Re=10^4\div10^5$ typical of heat exchangers when hydraulic resistance grows $3.8\div4.15$ times, yielding $V/V_{sm}=0.52\div0.5$, or to decrease the heat exchanger volume $1.95\div2$ times (in such tubes an advancing growth of heat transfer is larger $d/D=0.97\div0.98$ with $V/V_{sm}=0.5\div0.6$). Thus, the efficiency of heat transfer enhancement by strip inserts is somewhat lower at $Re=10^4$ and is substantially lower at $Re=10^5$ as against the one initiated by annular turbulators.

As seen from Fig.4, the efficiency of screw inserts is much worse than that of strip ones. Even at minimum values of $d_0/D=0.33$ and $\delta/D=0.05$ for $S/D=4\div12$, we have $\alpha/\alpha_{sm}=1.75\div1.16$ and $\xi/\xi_{sm}=4.74\div2.64$ at $Re_D=10^4$ and $\alpha/\alpha_{sm}=0.88\div0.58$ and $\xi/\xi_{sm}=3.4\div1.38$ at $Re_D=10^5$. In this case, at $Re_D=10^4$ we obtain $V/V_{sm}=0.84\div1.19$, and at $Re_D=10^5$ we have $V/V_{sm}=1.9\div2.67$. Thus, a moderate improvement of efficiency ($V/V_{sm}<1$) can be attained only at $S/D\cong4$ and $Re_D\cong10^4$.

As the Reynolds numbers and the twisting pitches grow, we have $V/V_{sm}>1$, i.e., the application of screw inserts yields an adverse result since it does not improve but degrades heat exchanger parameters. Screw inserts with large relative screw core diameters d_0/D are still less effective.

All the above data for helical swirlers describe the case where twisted strips and screw swirlers are tight against the inner walls of tubes. If an annular gap is formed between the inserts and the tube, then the heat transfer enhancement efficiency markedly decreases (Shchukih, 1970, 1980). Other ways of flow swirling (spiral channels, flow swirling at the channel entrance) are less effective than continuous flow swirling.

High efficiency of annular smooth-wall turbulators used to enhance heat transfer inside tubes with liquid and gas flow in them as well as ease of their manufacture and good operating characteristics allows this heat transfer enhancement method to be recommended for heat transfer improvement inside tubes.

Turbular Evaporative Heat Exchangers

As seen from foregoing section, tubes with annular turbulators are very effective for single-phase heat carriers, especially for heat transfer enhancement inside a tube. There, when a choice is being made of a method for heat transfer enhancement in heat exchangers with boiling of heat carriers, obviously preference should be given to these tubes. Tubes having annular turbulators are particularly effective for this-type heat exchangers with single-phase heat carrier flow inside tubes.

Analysis was made by the example of a tubular evaporative heat exchanger of air cooling system meant for a conditioning system of aircrafts. Cooled air flowed inside the tubes, and boiling water filled the intertube space of the heat exchanger. Nucleate boiling occurred, a distance between the tubes was assigned and chosen such as not to hinder a vapor bubble rise to a liquid surface. A tubular smooth-wall heat exchanger and several variants of tubular heat exchangers with annular turbulators (Fig. 1) were calculated at the same value of heat power, heat carrier flow rate, air hydraulic resistance, inlet and outlet temperatures of heat carriers, and heat carrier pressures. Calculations adopted the method (Dreitser, 1986) of calculating a tubular heat exchanger for a given hydraulic resistance.

A preliminarily assigned value of α_h/α_c was made more precise when calculated. To calculate the heat transfer coefficient on the cold side, we used the relation for water nucleate boiling

$$\alpha_c=38.7\cdot\Delta T_c^{2.33}\cdot p^{0.5} \ (W/m^2/s) \quad (27)$$

where $\Delta T_c=T_w-T_s$, p is the pressure, bar. A wall temperature T_w can be determined from the relation for heat fluxes on the cold and hot sides

$$\alpha_h\cdot D\cdot(T_h-T_w) = A\cdot D_{out}\cdot(T_w-T_s)^{3.33} \quad (28)$$

where $A=38.7\ p^{0.5}$. Eq. (28) is being solved graphically. To

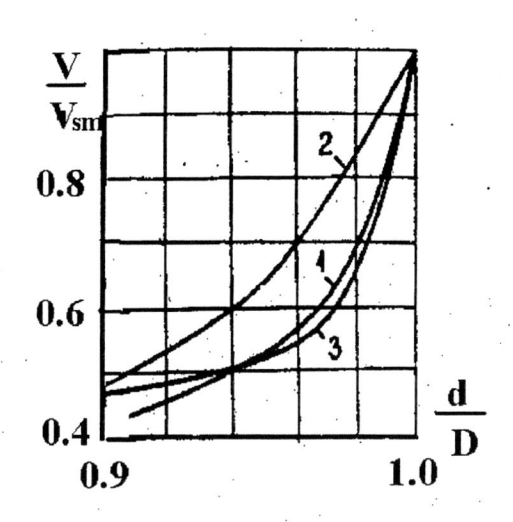

Fig.5 V/V$_{sm}$ for evaporates with different d/D and t/D.
1-3 t/D = 0.25; 1; 05 respectively

do this, the LHS and RHS of Eq. (28) are being built as a function of T$_w$ Increasing the heat transfer coefficient on the hot side results in a T$_w$ growth. According to Eq. (27), this leads to a substantial growth of the heat transfer coefficient on the cold side. It should be noted that calculations did not allow for a possible increase in the constant A at boiling over a turbulator-equipped surface.

Fig. 5 plots the volume ratio of the compared heat exchangers V/V$_{sm}$ vs. the diaphragm-to-inner tube diameter ratio d/D at t/D=0.25÷1.

Thus, use of heat transfer enhancement permits an evaporative heat exchanger volume to be decreased more than 2 times at the same values of heat power, cooled air flow rate, and hydraulic resistance of a heat exchanger. The largest efficiency is seen for tubes with d/D=0.9÷0.92 and t/D=0.25÷0.5.

Turbular Condensers

Kalinin and Dreitser (1998,1,2) investigated heat transfer enhancement involving vapour condensation on the outer surface of tubes with annular grooves. For horizontal tubes heat transfer enhancement was attained equal to α/α_{sm}=2÷2.65. If account is taken of the fact that heat transfer inside these tubes with cooling water flow in them also substantially increases up to 2.3 times, then the high efficiency of tubes with annular turbulators is true for condensers.

Analysis was made by the example of a tubular condenser of an industrial heat power plant. Calculations were performed for tubes with D$_{out}$/D=20/18 mm under saturated water vapour condensation at pressures p$_s$=0.1; 1; 2 bars (T$_s$=45.83; 94.63; 120.23°C), at vapour flow rate G=5 kg/s and inlet temperature of cooling water T$_c$=15°C. Cooling water flowed inside the tubes horizontally located in a bundle, and condensation of saturated water vapour occurred outside the tubes. A tubular smooth-wall heat exchanger and several variants of tubular heat exchangers with annular turbulators (Fig.1) were calculated at the same values of heat power, heat carrier flow rate, cooling water hydraulic resistance, and temperatures and pressures of heat carriers. Design was based on the methods (Dreitser, 1986) of calculating a tubular heat exchanger for a specified hydraulic resistance. Calculations were preformed at d$_{out}$/D$_{out}$ = 0.85÷0.95; t/D$_{out}$ = 0.37; in this case, d/D = 0.86÷0.97; t/D = 0.41. In this range of the parameters of turbulators outside the tubes, α_h/α_{sm}=2.11÷3.16; inside the tubes, (Nu$_c$/Nu$_{csm}$)$_{Re}$ = 1.8÷2.3. However, for the same hydraulic resistance inside the tubes with turbulators, it is necessary somewhat to decrease a water velocity and to increase the number of tubes; in this case, a real value of $\alpha_c/\alpha_{c\;sm}$ decreases. The calculation results for V/V$_{sm}$ are plotted in Fig.6. The quantity V/V$_{sm}$=0.50÷0.65; in this case, V/V$_{sm}$ decreases with increasing T$_s$ and depends weakly on d/D.

Thus, use of tubes with annular turbulators to enhance heat transfer outside and inside the tubes allows the condenser volume to be decreased up to 2 times at the same values of heat power, cooling water flow rate, and hydraulic resistance of a heat exchanger. The largest efficiency is seen for tubes with d/D = 0.86÷0.9 and t/D = 0.4 (d$_{out}$/D$_{out}$ = 0.85÷0.89; t/D$_{out}$ = 0.35÷0.4).

It should be noted that the efficiency of the investigated tubes with annular turbulators used in condensers is much higher than that of helical tubes that enhance the heat transfer coefficient by 25%. Also, use of different fins welded to the outer surface of a tube is less effective because in this case heat transfer inside the tubes is not enhanced.

Heat Exchangers Operating under Fouling Conditions

In many fields of engineering, the fouling control on a heat transfer surface is a serious problem. Use of water containing temporary-hardness salts as a cooling medium yields their deposition on heat transfer surfaces, as a cold water temperature grows. Therefore, in designing high-efficiency

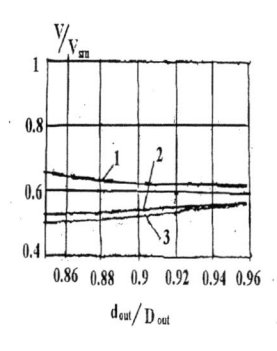

Fig.6 V/V$_{sm}$ for condensators with different T$_s$ and d/D
1-3 T$_s$ =120.23; 99.63; 45.85°C respectively.

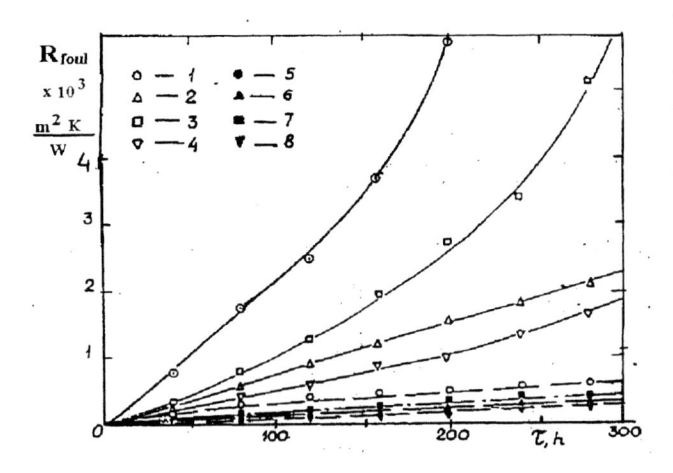

Fig. 7 Time variations of the fouling thermal resistance inside the smooth tubes (1-4), annular diaphragm-equipped tubes with d/D=0.91 and t/D=0.5 (5-6)
1, 5 – Re=3200;C=20 mg eq./l;
2, 6 – Re=16000;C=20 mg eq./l;
3, 7 – Re=3200;C=10 mg eq./l;
4, 8 – Re=16000;C=10 mg eq./l;

refrigerators and water-cooled condensers, it is necessary, along with heat transfer enhancement, to prevent or decrease a rate of the fouling growth on a heat transfer surface.

The existing methods of controlling salt deposition mainly amount to preliminary water treatment mostly by chemical reagents. These methods are inapplicable for those productions that require a great amount of water for cooling. Open-cycle cooling is made. Of late, great interest has been paid to the problem how to reduce salt deposits on the heat transfer surface from flow swirling.

Tubes with annular turbulators turned out to be very effective in service when water had elevated bicarbonate hardness (up to C=20 mg-equiv./l). This was proved in special experiments, in which water with elevated hardness flowed over the outside and inside surfaces of tubes with turbulators of different parameters. A water velocity was varied from 0.1 to 1.5 m/s, a temperature was in the range 50÷90°C, and experiments lasted for about 360 h. As a result, the thermal resistance of a scale layer R_{foul} outside and inside the tubes was obtained as a function of the parameters of the turbulators, water velocity, and time.

Fig.7 shows the increase of fouling thermal resistance R_{foul} inside smooth tubes versus time.

Turbulators reduce salt deposition on both surfaces of the tubes by a factor of 3÷5. The time dependence of R_{foul} is asymptotic; after 100÷150 h, the value of R_{foul} becomes constant.

Less scale is formed on the tubes with turbulators when the height of a diaphragm increases, the depth of grooves grows or their pitch decreases. Typically, after water of elevated hardness flowed in a tube with diaphragms for 100 h, the heat transfer coefficient fell by a factor not more than 10%, and the pressure drop did not change at all. During this time, if the tube had smooth walls, the heat transfer coefficient fell by 30% and the pressure drop rose by 25%. The effectiveness of tubes with turbulators was increased in the presence of scale. When there was no scale (at τ=0), the ratio of heat transfer coefficients k_l/k_{ls}=2.5÷3; then at τ/τ_∞=1 we have k_l/k_{lsm}=3.5÷5.

A generalization of the experimental data has led us to the following dependences for the thermal resistance of a scale layer [R_{foul},(m^{2}K)/W] that were reported by Kalinin and Dreitser (1998, 1, 2)

Under the fouling conditions the turbulator parameters are as follows: d/D=0.9÷0.92; t/D=0.25÷0.5.

CONCLUSIONS

1. A simple method of evaluating the efficiency of heat transfer enhancement in the channels of tubular heat exchangers with longitudinal flow in tube bundles and in the channels of plate-and-fin heat exchangers is introduced. This procedure permits the estimation of one or another method for heat transfer enhancement without detailed design of heat exchangers.

2. Tubes with annular turbulators can substantially augment heat transfer involving gas and liquid flow in the tubes under the conditions of surface and film boiling inside the tubes as well as condensation on the outside surface of horizontal and vertical tubes.

3. Tests of commercial heat exchangers with the tubes have supported that equipment is highly effective and can reduce a heat transfer surface 1.5-2.5 times, as compared to a similar apparatus with smooth-wall tubes.

4. Much less scale is formed inside and outside the tubes with turbulators as against that on smooth-wall tubes; moreover, the scale grows asymptotically in time. Therefore, heat exchangers having no special facilities to clean their surface can be used.

NOMENCLATURE

A   coordinate, dimensionless;
C   water hardness, mg.eq/l;
C_1, C_2 constants, dimensionless;
D, D_{out} inner and outer diameter respectively, m;
d   annular diaphragm diameter, m;
d_{eq} equivalent channel diameter;
d_o screw core diameter, m;
d_{out} annular groove diameter, m;
F   channel cross-sectional area, m^2;
F_Σ heat exchanger cross-section area, m^2;
G   mass flow rate, kg/s;
K   heat transfer coefficient, W/m^2 K;
K_V ratio of the volumes of compared heat exchanger;
K_Q ratio of the heat powers of compared heat exchangers;
$K_{\Delta p}$ ratio of the pressure drops of compared heat exchangers;
l   heat exchanger length, m;

m, n   constants, dimensionless;
N   channel or tube quantity;
Nu   Nusselt number, dimensionless;
$\left(\dfrac{Nu}{Nu_{sm}}\right)_{Re}$ ratio of heat transfer coefficients for the same Re number;
p   pressure, Pa;
Δp   pressure drop, Pa;
Pr   Prandtl number, dimensionless;
Q   heat flux of heat exchanger, W;
R_{foul} thermal resistance of a fouling layer, m²K/W;
Re   Reynolds number, dimensionless;
S   heat transfer surface, m²;
S   spiral inserts pitch, m;
t   turbulizer pitch, distance between the grooves, m;
T   temperature, K;
ΔT   temperature drop, K;
U   perimeter of the channel, m;
V   heat exchanger volume, m³;
w   mean velocity, m/s;
α   heat transfer coefficient, W/ m² K;
ρ   density, kg/m³;
ξ   hydraulic resistance coefficient, dimensionless;
$\left(\dfrac{\xi}{\xi_{sm}}\right)_{Re}$ ratio of hydraulic resistance coefficients for the same Re number;
τ   time, hr;

Subscripts:

c   cold side of heat exchanger;
foul   fouling;
eq   equivalent;
h   hot side of heat exchanger;
Re   at the same Re number;
sm   smooth;
w   wall;
1,2   inner and outer surfaces of tubular heat exchanger;
0, ∞   initial and final moment of process;

ACKNOWLEDGEMENTS

The support of the Russian Research Foundation under its Project No.00-15-96654 (The Program of Support of Leading Scientific Schools of Russia) is gratefully acknowledged.

REFERENCES

Antufiev V.M., 1966, *Efficiency of Different Shapes of Convective Heat Transfer Surfaces*, Energiya, Leningrad.

Bajan P.I., Kanevets C.I., and Seliverstov V.M., 1989, *Handbook on Heat Exchangers*, Mashinostroyeniye, Moscow.

Dedysenko Yu.M., 1960, *Regenerative Diagrams and Regenerators of Gas Turbine Setups (Theory and Calculation)*, Ukrainian Academy of Sciences Publishers, Kiev.

Dreitser G.A., 1986, *Compact Heat Exchangers*, MAI Publishers, Moscow.

Dreitser G.A., Dubrovsky E.V., Dzybenko B.V., Iyevlev V.M., Kalinin E.K., Shimonis V.M., Shlenchiauskas A.A., Vilemas J.V., Voronin G.I., Zukauskas A.A., and Yarkho S.A., 1990, *Enhancement of Heat Transfer*, Heat Transfer Soviet Reviews, Vol.2, Hemisphere Publishing, New York.

Dreitser G.A., 1997, *Modern Problems in Creating Compact Turbular Heat Exchangers. Compact Heat Exchangers for the Process Industries*, Snowbird USA, Ed. by N.K. Shan, Begell House Inc., New York, Wellington (UK), pp.121-140.

Dubrovsky E.V., 1977, Method of Relative Comparison of Thermohydraulic Efficiencies of Heat Transfer Surfaces, *Izv. AN USSR, Energetika i Transport*. No.6, pp.118-128.

Dzybenko B.V., Dreitser G.A., and Yakimenko R.I., 1993, The Methods of Choosing Effective Heat Transfer Surfaces for Heat Exchangers. *Proceedings of International Symposium on New Developments in Heat Exchangers*, Portugal, Lisbon, Paper 5.2, September 6-10, 1993.

Gukhman A.A., 1977, Heat Transfer Enhancement and Problem of Heat Transfer Surface Comparative Evaluation, *Teploenergetika*, No.4, pp. 5-8.

Kalafati D.D., and Popalov V.V., 1986, *Heat Exchangers Optimization on Heat Transfer Efficiency*, Energoatomizdat, Moscow.

Kalinin E.K., and Dreitser G.A., 1998, 1, Heat Transfer Enhancement in Heat Exchangers. *Advances in Heat Transfer*, Vol. 31, pp.159-332, Academic Press, New York.

Kalinin E.K., Dreitser G.A., Kopp I.Z., and Myakotchin A.S., 1998,2, *Effective Heat Transfer Surfaces*, Energoatoizdat, Moscow.

Kovalenko L.M., and Glushkov A.F., 1986, *Heat Exchangers with Heat Transfer Enhancement*, Energoatonizdat, Moscow.

Migai V.K., 1990, About Maximum Heat Transfer Enhancement in the Tubes with Flow Turbulization, *Izv.AN USSR,Energetika i Transport (Proceedings of the USSR Academy of Sciences, Energy and Transport)*, No.2.

Mitin B.M., Baranov Yu.F., and Mezdjiel E.K., 1975; Comparative Evaluation of Heat Transfer Surfaces, In Heat Exchangers of Gas Turbine Engines, *Trudy TsIAM (Proc. Central Intitute of Aviation Engines)*,Moscow, No. 646, pp. 97-113.

Miskevish A.I., 1967, Method of Convective Heat Transfer Evalution, *Trudy TsKTI(Proc. Central Institute of Boilers and Turbines)*, Leningrad, No.78, pp.75-80.

Shchukin V.K., 1970, 1980, *Heat Transfer and Hydrodynamics of Internal Flows in Mass Force Fields*, Mashinostroyeniye, Moscow, 1st ed., 2nd ed.

THEORETICAL STUDY ON THE EFFECTS OF TUBE DIAMETER AND TUBESIDE FIN GEOMETRY ON THE HEAT TRANSFER PERFORMANCE OF AIR-COOLED CONDENSERS

H. S. Wang and H. Honda

Institute of Advanced Material Study, Kyushu University, Kasuga, Fukuoka 816-8580, Japan
E-mail: hhonda@cm.kyushu-u.ac.jp

ABSTRACT

A previously proposed stratified flow model of film condensation inside helically grooved, horizontal microfin tubes has been modified to take account of vapor-liquid interface curvature due to surface tension effect. A better agreement with available experimental data for refrigerants was obtained by the new model. By applying this model, an extensive numerical calculation of vapor to coolant heat transfer has been made to study the effects of tube diameter and tubeside fin geometry on the heat transfer performance of air-cooled condensers in which R410A is condensed in the horizontal microfin tubes. The numerical results show that the effects of tube diameter, fin height, fin number and helix angle of groove are significance, whereas the effects of other parameters are small.

INDRODUCTION

Helically grooved, horizontal microfin tubes have been commonly used in air conditioners and refrigerators due to their high heat transfer performance. Many experimental studies on the effects of tube diameter, fin geometry, refrigerant, oil etc. on the condensation heat transfer and pressure drop of the microfin tubes have been reported in the recent literature. Webb (1994) and Newell and Shah (1999) have given comprehensive reviews of relevant literature.

Cavallini et al. (1995) and Shikazono et al. (1998) have developed empirical equations of the circumferential average heat transfer coefficient. Yang and Webb (1997) and Nozu and Honda (2000), respectively, proposed a semi-empirical model and an annular flow model of condensation in microfin tubes that considered the combined effects of vapor shear and surface tension forces. Honda et al. (2000) proposed a stratified flow model that considered the combined effects of surface tension and gravity. In this stratified flow model, the height of stratified condensate was determined by a modified Taitel and Dukler model (Honda et al., 2000) in which a flat vapor-liquid interface was assumed. Available experimental data for four refrigerants and three tubes with the fin height h of 0.16~0.24 mm and outside diameter d_o of 9.5~10.0 mm were predicted within $\pm 20\%$ by the higher of the two theoretical predictions of the annular flow model and stratified flow model.

In order to meet the need for more compact heat exchangers, microfin tubes with smaller diameters have been developed for the air-conditioners and refrigerators. Currently, the microfin tubes with d_o = 6~8 mm are commonly used. The microfin tubes with d_o = 4 mm have also been tested. For these tubes, the assumption of flat vapor-liquid interface does not hold any more, because the surface tension acts to form a round interface.

The objectives of the present work are twofold. The first is to develop a modified stratified flow model in which the effect of surface tension on the profile of stratified vapor-liquid interface is taken into account. The second is to apply this model to the thermal design of air-cooled condensers and study the effects of tube diameter and fin geometry on the heat transfer performance during condensation of R410A. By assuming typical operating conditions of the air-cooled condenser, numerical calculations are conducted with systematically changed values of tube diameter, fin height, fin number, helix angle of groove, fin half-tip angle, etc.

MODIFIED STRATIFIED FLOW MODEL

Physical Model

Figure 1 shows the physical model of stratified condensate flow in a helically grooved, horizontal microfin tube. In Fig. 1(a), the shape of vapor-liquid interface is assumed to be a circular arc centered at O_1. The angle φ is measured from the top of the tube. The φ_s denotes the

(a) Tube Cross Section   (b) A-A Cross Section

(c) Fin Cross Section   (d) Fin Cross Section
$(0 \leq \varphi \leq \varphi_f)$ $(\varphi_f \leq \varphi \leq \varphi_s)$

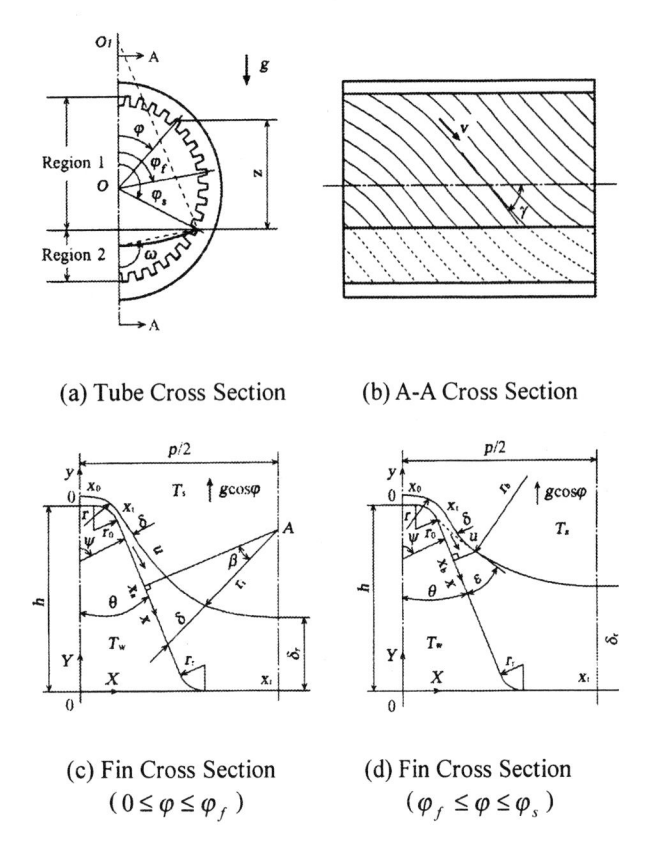

Fig. 1 Physical model and coordinates

angle below which the tube is filled with stratified condensate. The coordinate z is measured vertically upward from the surface of stratified condensate at $\varphi = \varphi_s$. The tube surfaces at the angular portions $0 \leq \varphi \leq \varphi_s$ and $\varphi_s \leq \varphi \leq \pi$ are denoted as regions 1 and 2, respectively. In the region just above the angle φ_s, condensate is retained in the groove between adjacent fins due to the capillary effect. As a result, a relatively thick condensate film is formed in the groove. The angle below which the condensate is retained in the groove is denoted as the flooding angle φ_f. Figures 1(c) and 1(d) show the condensate profiles in the fin cross-section in the regions of $0 \leq \varphi \leq \varphi_f$ and $\varphi_f \leq \varphi \leq \varphi_s$, respectively. The fin profile is assumed to be a trapezoid with round corners at the fin tip and fin root. The fin height and fin pitch are h and p, respectively, and the fin half tip angle is θ. The coordinate x is measured along the tube surface from the center of fin tip and y is measured vertically outward from the tube surface. The condensate on the fin surface is drained by the combined gravity and surface tension forces toward the fin root and then it flows down the groove by gravity. Thus the condensate film thickness δ is very small near the fin tip and it is relatively thick near the fin root. The effect of vapor shear force on the condensate flow on the fin surface is assumed to be negligible.

Profile of Vapor-Liquid Interface

The profile of vapor-liquid interface is estimated by the combination of the modified Taitel and Dukler model (Honda et al.,2000) and the model of interface configuration proposed by Brauner et al. (1996). The basic equation for the stratified flow with a curved interface is written as

$$f_v \frac{\rho_v U_v^2}{2} \frac{S_v}{A_v} - f_l \frac{\rho_l U_l^2}{2} \frac{S_l}{A_l} + f_i \frac{\rho_v U_v^2}{2}\left(\frac{S_i}{A_v} + \frac{S_i}{A_l}\right) = 0 \quad (1)$$

where f_v and f_l are the friction factors in regions 1 and 2, respectively, f_i is the interfacial friction factor, ρ_v and ρ_l are the densities of vapor and condensate, respectively, $U_v = GAx/\rho_v A_v$, $U_l = GA(1-\chi)/\rho_l A_l$, $S_v = Fd\varphi_s$, $S_l = Fd(\pi - \varphi_s)$, $S_i = d\sin\varphi_s(\pi - 2\omega)/\sin(2\omega)$,

$$A_l = \frac{d^2}{4}\left[\frac{A}{A_n}(\pi - \varphi_s) + \frac{\sin(2\varphi_s)}{2} + \sin^2\varphi_s \frac{\pi - 2\omega + \sin(4\omega)/2}{\sin^2(2\omega)}\right],$$

$A_v = \pi d^2/4 - A_l$, A is the actual cross sectional area of the tube, A_n is the nominal cross sectional area based on the fin root diameter d, F is the surface area enhancement as compared to a smooth tube. The expressions for f_v, f_l and f_i are given in Honda et al. (2000).

Following Brauner et al. (1996), the interface curvature is determined by assuming the sum of gravitational potential and surface energy Δe minimum. The Δe is given by

$$\Delta e = \frac{1}{8}(\rho_l - \rho_v)gd^3[\sin^3\varphi_s(\cot(2\omega) + \cot\varphi_s)$$
$$\times \frac{\pi - 2\omega + \sin(4\omega)/2}{\sin^2(2\omega)} + \frac{2}{3}\sin^3\varphi_s^p + \frac{8}{\text{Bo}}\{\sin\varphi_s$$
$$\times \frac{\pi - 2\omega}{\sin(2\omega)} - \sin\varphi_s^p + \cos\varsigma(\varphi_s - \varphi_s^p)\}] \quad (2)$$

where $\text{Bo} = (\rho_l - \rho_v)gd^2/\sigma$ is the Bond number, φ_s^p is the value of φ_s for a plane interface ($\omega = \pi/2$), and ς is the wettability angle. It is relevant to note here that $\varsigma = 0$ for condensation.

Profile of Thick Condensate Film

In the angular portion $\varphi_f \leq \varphi \leq \varphi_s$, the condensate velocity in the thick film is considered to be very small. Thus its profile is approximated by a static meniscus that touches the fin flank (shown by a dotted line in Fig. 1(d)). Then the radius of curvature of the thick film r_b is given by

$$\frac{\sigma}{r_b} = (\rho_l - \rho_v)gz = \frac{(\rho_l - \rho_v)gd}{2}(\cos\varphi - \cos\varphi_s) \quad (3)$$

Profile of Thin Condensate Film

In the thin film region $0 \leq \varphi \leq \varphi_f$, δ is assumed to be sufficiently smaller than h and p. The condensate on the fin surface is drained in the x direction by the combined surface tension and gravity forces. At the same time, it is drained along the groove by the gravity force. The condensate flow is assumed to be laminar. Substituting the solutions of condensate velocities in the x and φ directions into the continuity equation yields

$$-\frac{(\rho_l - \rho_v)g\cos\varphi}{3v_l}\frac{\partial}{\partial x}(\sin\psi\delta^3) - \frac{\sigma}{3v_l}\frac{\partial}{\partial x}\left\{\frac{\partial}{\partial x}\left(\frac{1}{r}\right)\delta^3\right\}$$
$$+\frac{2(\rho_l - \rho_v)g\sin^2\gamma}{3v_l d}\frac{\partial}{\partial \varphi}(\sin\varphi\delta^3) = \frac{\lambda_l(T_s - T_{w1})}{h_{fg}\delta} \quad (4)$$

where $r = r(\delta, d\delta/dx, d^2\delta/dx^2)$ is the radius of curvature of the condensate surface in the fin cross-section. The expression for r is given in Honda et al. (2000). The boundary conditions are

$$\partial\delta/\partial\varphi = 0 \quad \text{at} \quad \varphi = 0 \quad (5)$$
$$\partial\delta/\partial x = \partial^3\delta/\partial x^3 = 0 \quad \text{at} \quad x = 0 \text{ and } x_r \quad (6)$$

For $\varphi_f \leq \varphi \leq \varphi_s$, where the condensate film is consisted of a thin film region near the fin tip and a thick film region near the fin root, the boundary conditions at the connecting point between the thin film and thick film are given by

$$\partial\delta/\partial x = \tan\varepsilon, \quad r = -r_b \quad \text{at} \quad x = x_b \quad (7)$$

where ε is the angle shown in Fig. 1(d).

The solution of Eq. (4) subject to the boundary conditions (5) and (6) for $0 \leq \varphi \leq \varphi_f$, and (5) and (7) for $\varphi_f \leq \varphi \leq \varphi_s$ was obtained numerically by a finite difference scheme. The description of the numerical scheme is given in Honda et al. (2000).

Wall Temperature and Heat Transfer Coefficients

For region 1, the average heat transfer coefficient for the fin cross section α_φ is defined on the projected area basis as

$$\alpha_\varphi = \frac{2}{p}\int_0^{x_r}\alpha_x dx = \frac{2\lambda_l}{p}\int_0^{x_r}\frac{1}{\delta}dx \quad (8)$$

where $\alpha_x = \lambda_l/\delta$ is the local heat transfer coefficient. The average heat transfer coefficient for region 1, α_1, is defined on the projected area basis as

$$\alpha_1 = \frac{1}{\varphi_s}\int_0^{\varphi_s}\alpha_\varphi d\varphi = \frac{2\lambda_l}{p\varphi_s}\int_0^{\varphi_s}\int_0^{x_r}\frac{1}{\delta}dxd\varphi \quad (9)$$

The heat transfer coefficient in region 2, α_2, is assumed to be uniform. The α_2 is estimated using the following empirical equation for forced convection in internally finned tubes developed by Carnavos (1980)

$$\alpha_2 = 0.023\frac{\lambda_l}{d_l}\left(\frac{\rho_l d_l U_l}{\mu_l}\right)^{0.8}\Pr_l^{0.4}\left(\frac{A}{A_c}\right)^{0.1}F^{0.5}(\sec\gamma)^3 \quad (10)$$

where $A_c = \pi(d - 2h)^2/4$ is the core flow area and γ is the helix angle of the groove.

The condensation temperature difference $(T_s - T_{wk})$ and the heat flux q_k for region $k(=1,2)$ are obtained from

$$q_k = \left\{\frac{1}{\alpha_k} + \frac{d}{2\lambda_w}\ln\left(\frac{d_o}{d}\right) + \frac{d}{\alpha_o d_o}\right\}^{-1}(T_s - T_c)$$
$$= \alpha_k(T_s - T_{wk}) \quad (11)$$

where α_o is the heat transfer coefficient at the outer surface and T_{mk} is the inside tube wall temperature for region k. Then the circumferential average heat transfer coefficient $\alpha_{m\chi}$ is obtained from

$$\alpha_{m\chi} = q_{m\chi}/(T_s - T_{wm\chi}) \quad (12)$$

where

$$q_{m\chi} = \{\varphi_s q_1 + (\pi - \varphi_s)q_2\}/\pi \quad (13)$$
$$T_s - T_{wm\chi} = \{\varphi_s(T_s - T_{w1}) + (\pi - \varphi_s)(T_s - T_{w2})\}/\pi \quad (14)$$

$q_{m\chi}$ is the circumferential average heat flux, $T_{wm\chi}$ is the circumferential average wall temperature. It is relevant to note here that T_{w1} in Eq. (4) is not known a priori. Thus Eqs. (4), (9) and (11) were solved iteratively starting from an appropriate assumption of T_{w1}.

Comparison with Experiments

Figure 2 shows a comparison of the theoretical predictions with available experimental data for R22 condensing in a 7.0 mm dia. tube (Uchida et al., 1997). The tube and fin dimensions are listed in Table 1. In Figs. 2(a) and 2(b), the values of $\alpha_{m\chi}$ obtained from the modified stratified flow model and the previously proposed annular flow and stratified flow models are compared with the experimental data for $G = 217$ and 387 kg/m²s, respectively. The predictions of the annular flow model are shown only for the region where the grooves are not flooded with condensate. This model gives a much higher $\alpha_{m\chi}$ than the measured values. In the flooded region, on the other hand, the annular flow model predicts a much smaller $\alpha_{m\chi}$ (about one half) than the measured values. The modified stratified flow model predicts 5-16% lower $\alpha_{m\chi}$ than the previously proposed stratified flow model that assumed a flat vapor-liquid interface and gives a better agreement with the measured value irrespective of χ and G.

Figure 3 shows a similar comparison for R134a condensing in a 9.5mm dia. tube (Hayashi, 1997). The tube and fin dimensions are listed in Table 1. Figures 3(a) and 3(b) show the cases of $G=102$ and 299 kg/m²s, respectively. Generally, the annular flow model predicts a

Table 1 Fin and tube dimensions

		Uchida	Hayashi		Simulation	
d_o	mm	7.0	9.52	9.5	7.0	4.0
d	mm	6.50	8.88	8.74	6.49	3.52
d_e	mm	6.34	8.75	8.63	6.34	3.24
n	-	50	60	67	50	27
γ	deg	12	18.7	13	13**	13
p^*	mm	0.40	0.44	0.40	0.40**	0.40
h	mm	0.22	0.19	0.24	0.24**	0.24
θ^*	deg	12.7	22.3	15.6	15.6**	15.6
r_0^*	mm	0.02	0.025	0.021	0.021**	0.021
x_0^*	mm	0.032	0.015	0.01	0.01**	0.01
F	-	1.83	1.51	1.87	1.87	1.87

*dimension in a cross-section normal to groove
**standard value

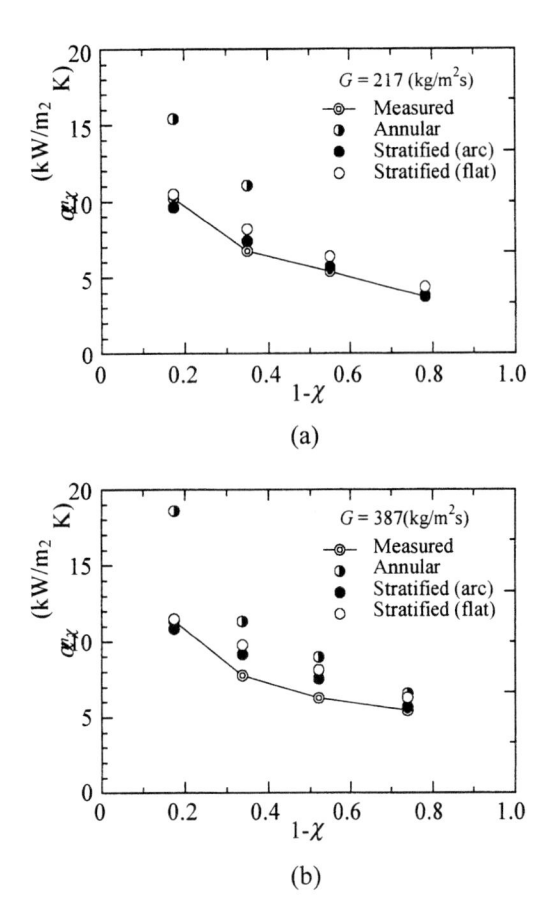

Fig. 2 Comparison of measured and predicted circumferential average heat transfer coefficients; R22, d_o=7.0mm.

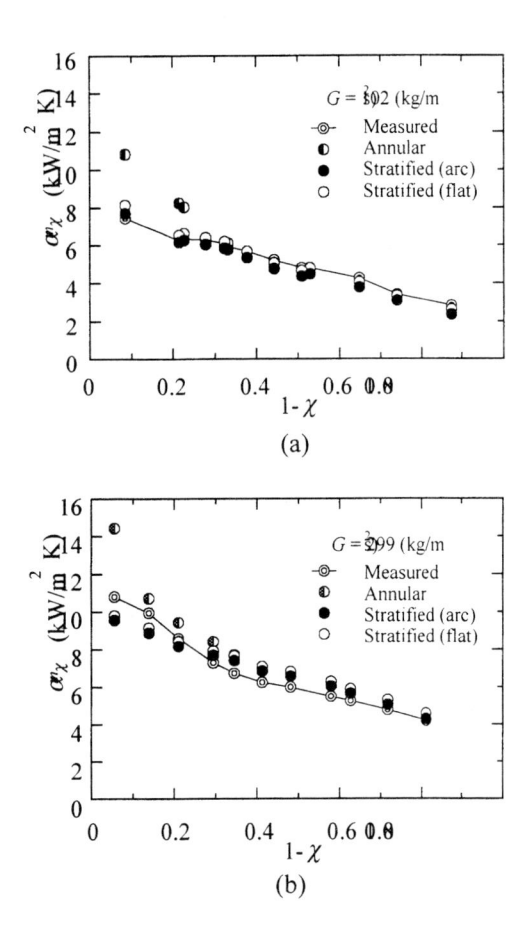

Fig. 3 Comparison of measured and predicted circumferential average heat transfer coefficients; R134a, d_o=9.5mm.

higher $\alpha_{m\chi}$ than the experimental data. The deviation is larger for smaller G. The modified stratified flow model predicts 2-8% lower $\alpha_{m\chi}$ than the previously proposed stratified flow model and gives a good agreement with the experimental data except for the case of G=299 kg/m²s and $1-\chi<0.15$. For the latter case, the theoretical predictions is lower than the experimental data by a maximum of 12%.

The foregoing results indicate that the effect of surface tension on the profile of stratified vapor-liquid interface is not negligible for condensation of refrigerants in horizontal microfin tubes with $d_o \leq 9.5$ mm.

EFFECTS OF TUBE DIAMETER AND TUBESIDE FIN GEOMETRY

Thermal Design of Air-Cooled Condenser

We consider a horizontal air-cooled condenser as shown in Fig. 4 in which R410A is condensed. The refrigerant is assumed to be dry saturated ($\chi=1$) at the condenser inlet and $\chi=0$ at the condenser outlet. The refrigerant mass velocity G is assumed to be 50, 100 and 300 kg/m²s, and the air and refrigerant temperatures are assumed to be 20 and 50°C, respectively. Following the current design practice of manufacturers, the air-side heat transfer coefficient is assumed to be 135 W/m²K, the outside-to-inside surface area (based on the core diameter) ratio F_{ac}=17.5, the fin efficiency η_f=0.8 and the thermal contact resistance of fins R_f=6×10⁻⁵m²K/W. Thus the heat transfer coefficient α_o in Eq. (11) is given by

$$\alpha_o = \{R_f + 1/(\alpha_a \eta_f F_{ac})\}^{-1} \quad (15)$$

Three tubes with d_o = 9.5, 7.0 and 4.0mm are considerd. For these tubes, d is assumed to be 8.74, 6.49 and 3.52mm, respectively. The standard values of the fin geometry parameters h, p, x_0, r_0, γ and θ are assumed to be 0.24mm, 0.4mm, 0.01mm, 0.02mm, 13° and 16°, respectively (see Table 1).

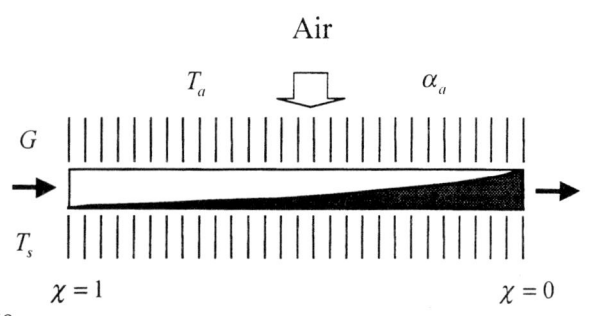

Numerical calculation was conducted for the range of $1-\chi = 0.025 \sim 0.975$ with an increment of 0.05. On the basis of Figs. 2 and 3, in which the theoretical predictions of the modified stratified flow model gave a good agreement with available experimental data for R22 and R134a, the modified stratified flow model was applied to the whole range of χ. The average heat transfer coefficient of the condenser was defined as

$$\alpha_m = \frac{1}{4} dG h_{fg} / \int_0^L (T_s - T_{wm}) dl \qquad (16)$$

where L is the tube length required for complete condensation. The L is given by

$$L = \frac{1}{4} dG h_{fg} \int_0^L (1/q_{m\chi}) dl \qquad (17)$$

Numerical Results

Figure 5 shows examples of the variations of T_{w1}, T_{w2} and $T_{wm\chi}$ with the wetness fraction $(1-\chi)$ for a microfin tube with $d_o = 7.0$mm and the standard values of fin geometry given in Table 1. As a result of neglecting the effect of vapor shear on the thin condensate film, T_{w1} takes almost the same value for 100 and 300 kg/m²s. The condensation temperature difference in region 1, $(T_s - T_{w1})$, is almost constant in the region of $(1-\chi) = 0 \sim 0.9$ and it increases slightly with further increasing $(1-\chi)$. The condensation temperature difference in region 2, $(T_s - T_{w2})$, increases gradually as $(1-\chi)$ increases. The value of $(T_s - T_{w2})$ is much smaller for $G = 300$ kg/m²s than for 100 kg/m²s, which is due to the higher value of α_2. Thus $T_{wm\chi}$ decreases more quickly for $G = 100$ kg/m²s than for 300 kg/m²s. The temperature drop at the air-side $(T_{wm\chi} - T_a)$ ranges from 43 to 79% of the overall temperature drop $(T_s - T_a)$ depending on G and χ.

Figure 6 compares the variation of $\alpha_{m\chi}$ with $(1-\chi)$ for three tubes with $d_o = 9.5$, 7.0 and 4.0mm and the identical fin geometry (see Table 1). The numerical results are presented for the cases of $G = 100$ and 300 kg/m²s. The $\alpha_{m\chi}$ value decreases quickly near the inlet. Then it decreases gradually with further increasing $(1-\chi)$ and then decreases quickly near the outlet of the condenser. It is also seen from Fig. 6 that $\alpha_{m\chi}$ is higher for smaller d_o. This is due to the fact that the flow path of condensate along the groove is shorter for smaller d_o. This causes a decrease in the condenste film thickness in the groove, resulting in an increase in the thin film region near the fin tip.

Figure 7 shows the values of $q_{m\chi}$ corresponding to Fig. 6. It is seen that the effect of d_o on $q_{m\chi}$ is much smaller than the case of $\alpha_{m\chi}$, and $q_{m\chi}$ is smaller for smaller d_o. This is due to the fact that the air-side heat transfer resistance is 42 to 81 % of the total resistance and the thermal contact resistance term $R_f d/d_o$ is larger for smaller d_o.

Figure 8 shows α_m plotted as a function of the length of flat portion at the fin tip x_0, with G and the radius of curvature at the corner of fin tip, x_0, as parameters. The α_m decreases slightly as x_0 increases for $x_0 < 0.01$mm and it is

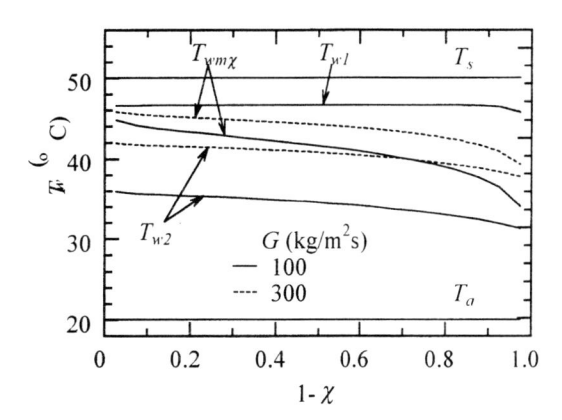

Fig. 5 Variation of T_{w1}, T_{w2} and $T_{wm\chi}$ with $1-\chi$

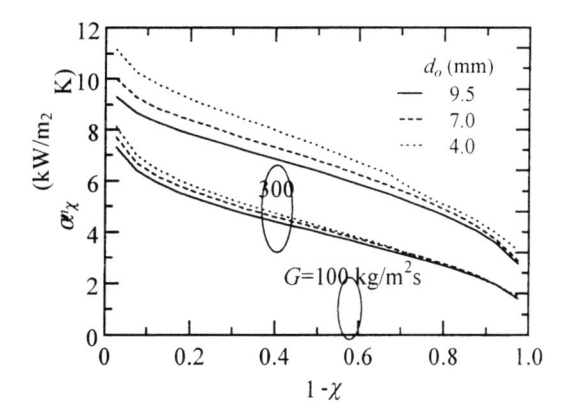

Fig. 6 Variation of $\alpha_{m\chi}$ with $1-\chi$

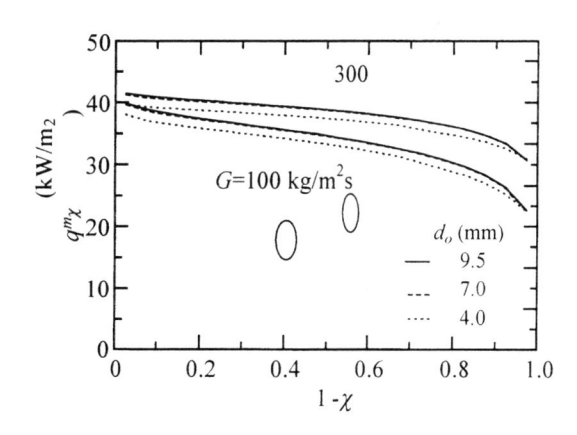

Fig. 7 Variation of $q_{m\chi}$ with $1-\chi$

almost constant for $x_0 > 0.01$mm. The α_m is slightly higher for larger x_0.

Figure 9 shows α_m plotted as a function of x_0 with G and x_0 as parameters. For $x_0 = 0$, α_m takes a nearly constant value. For $x_0 = 0.01$ and 0.03mm, on the other hand, α_m increases slightly as x_0 increases. The α_m is slightly higher for smaller x_0. It is seen from Figs. 8 and 9 that the effects of x_0 and x_0 are negligible.

Figure 10 shows α_m plotted as a function of the fin half tip angle θ with G as a parameter. The α_m decreases

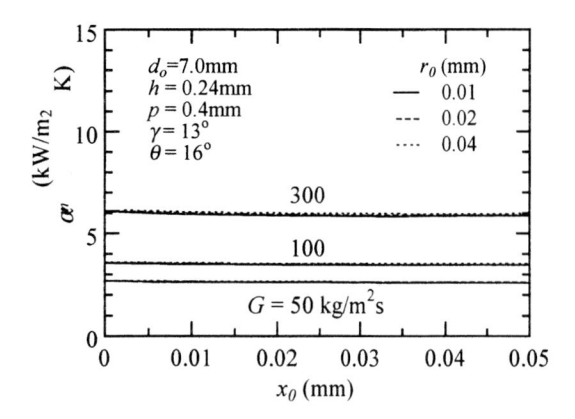

Fig. 8 Variation of α_m with x_0

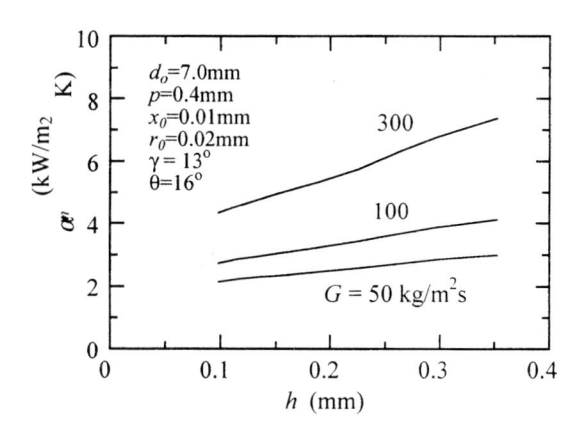

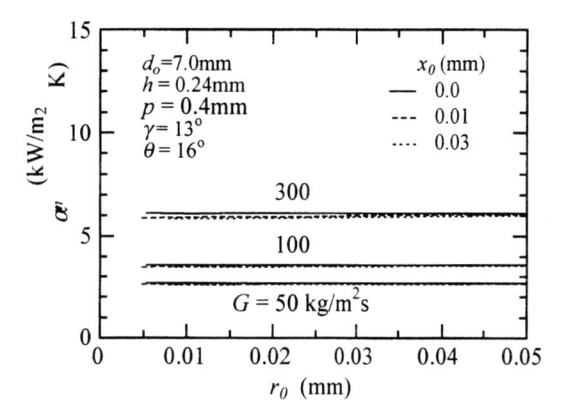

Fig. 9 Variation of α_m with r_0

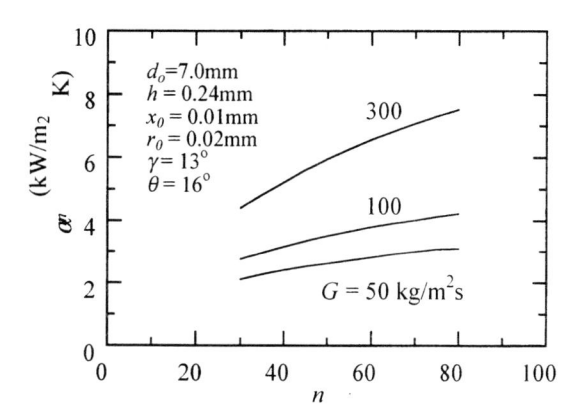

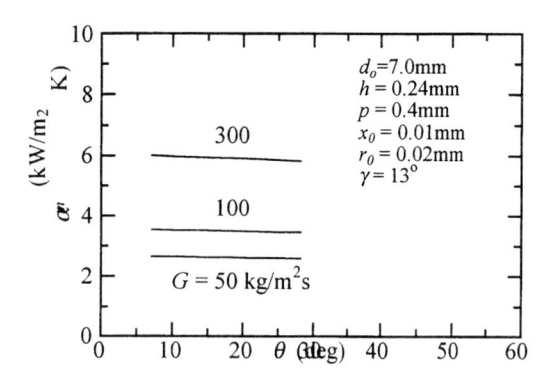

Fig. 10 Variation of α_m with θ

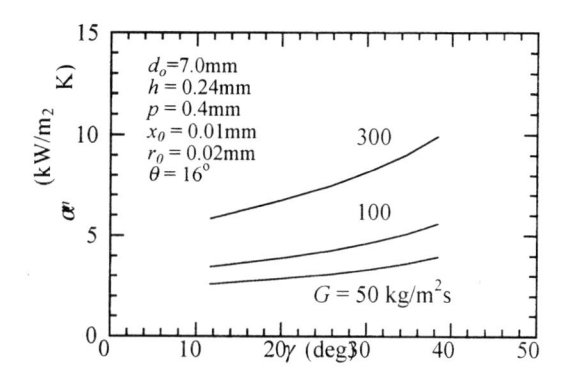

Fig. 11 Variation of α_m with h

Fig. 12 Variation of α_m with n

Fig. 13 Variation of α_m with γ

slightly as θ increases.

Figure 11 shows α_m plotted as a function of the fin height h with G as a parameter. As expected, α_m increases as h increases. The increase is more significant for larger G.

Figure 12 shows α_m plotted as a function of the fin number n with G as a parameter. The α_m increases first quickly, and then gradually as n increases. This is due to the combined effects of surface area increase and thickening of condensate film near the fin root which is caused by the decrease in the groove width.

Figure 13 shows α_m plotted as a function of the helix angle γ with G as a parameter. The α_m increases as γ increases, with the increase being more pronounced for large γ. This is due to the combined effects of the increase in the gravity component and the decrease in the groove length along the groove (see Figs. 1(a) and 1(b)), which act to augment the drainage of condensate in the groove, thereby increasing the effective heat transfer area.

As seen from Figs. 6, 11-13, the effects of d_o, h, n and γ on $\alpha_{m\chi}$ and α_m are more significant for larger G.

According to the present theoretical model, α_1 is not affected by G. On the other hand, α_2 increases as G increases. This results in an increase in q_2 and a decrease in (T_s-T_{w2}) given by Eq. (11). Consequently, $\alpha_{m\chi}$ and α_m (defined by Eqs. (12) and (16), respectively) are more sensitive to the variation of α_1 with d_o, h, n and γ at larger G.

It is of interest to compare the present results with previous experimental studies reporting on the effects of tube diameter and fin geometry. Table 2 summarizes the tube and fin dimensions tested in the previous studies. In these studies α_m was obtained from the overall heat transfer data by subtracting the coolant side resistance using an appropriate correlation for the coolant side heat transfer coefficient. As for the effect of d_o, Hori and Shinohara (1990) and Ohtani et al. (1994) have reported that α_m decreased as d_o decreased. On the other hand, Morita et al. (1993), Chiang (1993) and Mori et al. (2000) have reported that α_m increased as d_o decreased. In the former studies, the value of h was generally smaller for smaller d_o. Thus there is a possibility that the decrease in α_m was due to the decrease in h. In the latter studies, the values of α_m for different tubes are plotted on the coordinates of α_m versus G. Thus the effect of d_o on α_m can be confirmed by comparing the data for tubes with almost the same fin dimensions. Generally, the experimental results show that α_m first increases with increasing h and n. Satoh and Nosetani (1990), Ohtani et al. (1994) and Ishikawa et al. (2000) have reported the optimum value of n that depended on d_o and γ. According to Ohtani et al. (1994) and Ishikawa et al. (2000), the optimum value of n for d_o =7.0mm is 60 and 80, respectively. Ohtani et al. (1994) have reported the optimum value of h that depended on d_o. For d_o =7.0mm, the optimum value of h is 0.18 mm. On the other hand, Morita et al. (1993) and Mori et al. (2000) have reported that α_m increased monotonically as h increased in the ranges of 0.10-0.20 mm and 0.15-0.26 mm, respectively. Satoh and Nosetani (1990) have reported a slight decrease in α_m with increasing γ in the range of 0-23 deg for d_o =4.0mm. On the other hand, Ohtani et al. (1994) and Koseki et al. (2000) have reported an increase in α_m with increasing γ in the ranges of 8-25 deg and 15-30 deg for d_o =7.0mm, respectively. Since the uncertainty of the measured α_m value is not described in these papers, the foregoing results are not conclusive. However, it may be concluded that the present numerical results are in accord with the majority of the previous experimental results.

CONCLUSION

A modified stratified flow model taking account of vapor-liquid interface curvature due to surface tension effect has been developed to predict condensation heat transfer inside helically grooved, horizontal microfin tubes. A better agreement with available experimental data was obtained by the new model as compared to a previously proposed model that assumed a flat vapor-liquid interface. An extensive numerical calculation of vapor to coolant heat transfer has been made to study the effects of tube diameter and tubeside fin geometry on the heat transfer performance of air-cooled condensers in which R410A is condensed in the horizontal microfin tubes. The numerical results show that the effects of tube diameter d_o, fin height h, fin number n and helix angle of groove γ are of significance, whereas the effects of the length of flat portion at the fin tip x_0, radius of round corner at the fin tip r_0 and the fin half tip angle θ are small as far as d_o, h, n and γ are held constant.

ACKNOWLEGMENT

This study was partly supported by the Japan Society of Refrigerating and Air Conditioning Engineers through the R&D Project of fiscal 1999.

NOMENCLATURE

A = cross-sectional area of tube, m^2
Bo = Bond number
d = diameter at fin root, m
d_l = equivalent diameter of liquid space, m
d_o = outside diameter, m

Table 2 Tube and fin dimensions tested in the previous works

Reference	d_o mm	h mm	n	γ deg	Refrigerant	G kg/m^{2}s
Hori & Shinohara	4.0-12.8	0.14-0.24	30-75	8-18	R22	250
Satoh & Nosetani	4.0	0.09, 0.15	40-60	0-23	R22	100-400
Ohtani et al.	4.0-9.5	0.12-0.26	30-60	1-10	R22	200-500
Morita et al.	4.0-9.52	0.10-0.20	34-60	0, 8	R22	120-400
Chiang	7.5, 10	0.17-0.19	43-72	0, 18	R22	270-1100
Tang et al.	9.52	0.2	60, 72	0, 18	R22, R134a, R410A	260-800
Koseki et al.	7.0	0.10-0.25	55, 65	15, 30	R410A	170
Mori et al.	6.35-9.53	0.15-0.27	45-60	13-18	R410A, R407C	80-380
Ishikawa et al.	7.5	0.21-0.24	55-85	16	R22	160-320

e	=	energy, J/m
F	=	surface area enhancement as compared to a smooth tube
F_{ac}	=	outside-to-inside surface area ratio
f	=	friction factor
g	=	gravitational acceleration, m/s^2
G	=	refrigerant mass velocity, kg/m^{2}s
h	=	fin height, m
h_{fg}	=	specific enthalpy of evaporation, J/kg
l	=	tube length, m
n	=	number of fins
p	=	fin pitch, m
Pr	=	Prandtl number
q	=	heat flux, W/m^2
R_f	=	thermal contact resistance, m^{2}K/W
r	=	radius of curvature of condensate surface in fin cross-section, m
r_0	=	radius of curvature at corner of fin tip, m
r_b	=	radius of curvature of condensate surface in thick film region, m
r_r	=	radius of curvature at corner of fin root, m
S	=	perimeter length, m
T	=	temperature, °C
U	=	velocity in axial direction, m/s
x, y	=	coordinates, Fig.1
x_b	=	coordinate at connecting point between thin and thick film regions, Fig.1, m
x_0, x_t	=	coordinates at connecting points between straight and round portions of fin, Fig.1, m
x_r	=	mid point between adjacent fins, m
z	=	vertical height measured from condensate surface, Fig.1, m

Greek Symbols

α	=	heat transfer coefficient, W/m^{2}K
β	=	angle, Fig.1, deg
γ	=	helix angle of groove, deg
δ	=	condensate film thickness, m
ε	=	angle, Fig.1, deg
ς	=	wettability angle, deg
η_f	=	fin efficiency
θ	=	fin half tip angle, deg
λ	=	thermal conductivity, W/mK
ν	=	kinematic viscosity, m^2/s
ρ	=	density, kg/m^3
σ	=	surface tension, N/m
φ	=	angle measured from tube top, deg
χ	=	mass quality
ψ	=	angle, Fig.1, deg
ω	=	angle, Fig.1, deg

Subscripts

a	=	air side
b	=	boundary of thin and thick film regions
c	=	coolant side
f	=	flooding point
i	=	liquid-vapor interface
l	=	liquid
m	=	average value of tube
$m\chi$	=	circumferential average value
r	=	fin root, fin root mid point
s	=	saturation
v	=	vapor
w	=	wall
x	=	local value
φ	=	average value for fin cross-section
1	=	region 1
2	=	region 2

REFERENCES

Brauner, N., Rovinsky, J. and Maron, D. M., 1996, Determination of the Interface Curvature in Stratified Two-Phase Systems by Energy Considerations, *Int. J. Multiphase Flow*, Vol.22, pp.1167-1185.

Carnavos, T. C., 1980, Heat Transfer Performance of Internally Finned Tubes in Turbulent Flow, *Heat Transfer Engineering*, Vol. 4, pp. 32-37.

Cavallini, A., Doretti, L., Klammsteiner, N., Longo, G. A. and Rosetto, L., 1995, Condensation of New Refrigerants Inside Smooth and Enhanced Tubes, *Proc. 19th Int. Cong. Refrigeration*, Vol. IV, pp. 105-114.

Chiang, R., 1993, Heat Transfer and Pressure Drop During Evaporation and Condensation of Refrigeration-22 in 7.5 mm and 10 mm Diameter Axial and Helical Grooved Tubes, *AIChE Symposium Series*, Vol. 89, pp.205-210.

Hayashi, T., 1998, Ehancement of Condensation of HFC-134a in Horizontal Tubes, M.Eng. Thesis, Kyushu University.

Honda, H., Wang, H. S. and Nozu, S., 2000, A Theoretical Model of Film Condensation in Horizontal Microfin Tubes, *Proc. 34th Nat. Heat Transfer Conf.*, Pittsburgh, Pennsylvania, NHTC-12213.

Hori, H., and Shinohara, Y., 1990, *J. Japan Copper and Brass Research Assoc.*, Vol. 29, pp.65-70.

Ishikawa, S., Nagahara, K., and Sukumada, S., 2000, Heat Transfer Characteristics of Inner Grooved Tubes with Large Groove Numbers, *Proc. of 40th Annual Meeting of Japan Brass Maker's Association*, pp.32-33.

Koseki, K., Kobayashi, T., Saeki, C., and Higo, T., 2000, Development of Light Weight, High Performance Inner Grooved Tube for New Refrigerants, *Proc. 40th Japan Copper and Brass Research Assoc. Conf.*, 27-28.

Mori, Y., Yamamoto, K., Sumitomo, T., and Hashizume, T., 2000, Development of High Performance Inner Grooved Tube, *Proc. 40th Japan Copper and Brass Research Assoc. Conf.*, 29-31.

Morita, H., Kito, Y. and Satoh, Y., 1993, Recent Improvement in Small Bore Inner Grooved Copper Tube, *Tube and Pipe Technology*, Nov./Dec., pp.53-57.

Newell, T. A., and Shah, R. K., 1999, Refrigerant Heat Transfer, Pressure Drop, and Void Fraction Effects in Microfin Tubes, *Proc. 2nd Int. Symp. on Two-Phase Flow and Experimentation*, Pisa, Italy, Vol. 3, pp. 1623-1639.

Nozu, S. and Honda, H., 2000, Condensation of Refrigerants in Horizontal, Spirally Grooved Microfin Tubes: Numerical Analysis of Heat Transfer in Annular Flow Regime, *ASME J. Heat Transfer*, Vol. 122, pp. 80-91.

Ohtani, T., Tsuchita, T., Yasuda, K. and Hori, N., 1994, *Hitachi Densen*, Vol. 13, pp.115-120.

Satoh, Y. and Nosetani, K., 1990, *J. Japan Copper and Brass Research Assoc.*, Vol. 29, pp.233-239.

Ozeki, K., Kobayashi, T., Saeki, C., and Higo, T., 2000,

Development of Light Weight, High Performance Inner Grooved Tube, *Proc. of 40th Annual Meeting of Japan Brass Maker's Association*, pp.27-28.

Shikazono, N., Itoh, M., Uchida, M., Fukushima, T. and Hatada, T., 1998, Predictive Equation Proposal for Condensation Heat Transfer Coefficient of Pure Refrigerants in Horizontal Microfin Tubes, *Trans. JSME*, Vol. 64, pp. 196-203.

Taitel, Y. and Dukler, A. E., 1976, A Model for Predicting Flow Regime Transitions in Horizontal and Near Horizontal Gas-Liquid Flow, *AIChE J.*, Vol. 22, pp. 47-55.

Uchida, M., Itoh, M., Shikazono, N., Hatada T. and Ohtani, T., 1997, *Proc. 31th Air Cond. Refrig. Eng. Conf.*, pp. 81-84.

Webb, R. L., 1994, *Principles of Enhanced Heat Transfer,* Chapter 14, John Wiley and Sons.

Yang, C. Y. and Webb, R. L., 1997, A Predictive Model for Condensation in Small Hydraulic Diameter Tubes Having Axial Microfins, *ASME J. Heat Transfer*, Vol. 119, pp. 776-782.

EFFECTS OF SEVERAL FACTORS IN TUBE ON THE CAPACITY OF AIR-COOLED PLATE-FIN AND ROUND-TUBE HEAT EXCHANGERS

N. Sasaki[1], S. Kakiyama[1], and H. Morita[2]

[1] R&D Center, Sumitomo Light Metal Industries, Ltd., 1-12, 3-chome, Chitose, Minato-ku, Nagoya 455-8670, Japan;
E-mail: Naoe_Sasaki@mail.sumitomo-lm.co.jp
[2] Copper Works, Sumitomo Light Metal Industries, Ltd., 100 Ougishinmichi, Ichinomiya-chou, Hoi-gun 441-1295 Japan;
E-mail: Hiroyuki_Morita@mail.sumitomo-lm.co.jp

ABSTRACT

The performance of air-cooled heat exchangers consisting of several kinds of tubes were experimentally investigated and compared in this study. The experiments were performed with the heat exchanger installed in the inner unit of an actual air-conditioner for residential use. It was confirmed that the heat capacity of the heat exchanger consisting of inner grooved tubes during condensation was evidently higher than that of smooth tubes and significantly affected by the deformation of the groove profile due to the tube expanding with bullets. For evaporation, the capacity of the heat exchanger consisting of inner grooved tubes was approximately equal to that of the smooth tubes and the effect of the deformation on the capacity was smaller than that during condensation.

INTRODUCTION

Recently, the global environmental issues such as stratospheric ozone depletion and global warning are of worldwide concern. The revised law for energy-saving was enforced in 1999 to control the global warning in Japan. Air-conditioners were one of the specified machinery and tools regulated by this law in which the "Top-runner" method was applied and an objective coefficient of performance (COP) was established for every range of cooling capacity. In this method, air-conditioners with a cooling capacity of 2.2 to 2.8 kW have to realize a mean COP weighted by shipped sets of 5.27 by 2004.

Thus the manufacturers of air-conditioners have realized that they have to tackle the difficult problem of improving the COP (Ebisu et al., 1996 and 1997, Torikoshi et al., 1998). Therefore, the manufacturers of heat transfer tube need to develop high performance tubes for air-conditioners because this can improve the performances of the air-conditioner without a big change in the system design. The degree of improvement in the performance of the air-conditioner only by applying a high performance tube becomes lower year by year due to the unbalance of the heat transfer coefficient between the air side and refrigerant side. For the manufacturers of heat transfer tubes, improvement in the performance of the heat exchanger is more important than that of the heat transfer tube (Kakiyama et al., 1999, Sasaki et al., 2000).

In this paper, we present the experimentally investigated data for the unit heat exchanger performance and consider the effect of the tube shape, i.e., smooth tube and several kinds of inner grooved tubes, the change in tube shape by the tube expanding with bullets, and the heat transfer properties inside the tube sampled from the tested heat exchanger on the unit heat exchanger capacity.

EXPERIMENTS

Experimental Apparatus

A schematic diagram of the experimental apparatus used for the capacity measurement of the tested heat exchangers is shown in Fig. 1. The air under fixed conditions was drawn over the heat exchanger by a wind tunnel installed in the hermetic testing room and the refrigerant was supplied to the heat exchanger by the refrigerant supply device.

The wind tunnel was composed of the test section, the diffusion section, and the blower section. The blower sucked the air from the wind tunnel, and the air flow rate used to calculate the frontal air velocity was determined from the measurement of the static pressure differences across the calibrated nozzles.

The refrigerant supply device consisted of a compressor, a condenser, two expansion valves, an evaporator, and a mass flow meter.

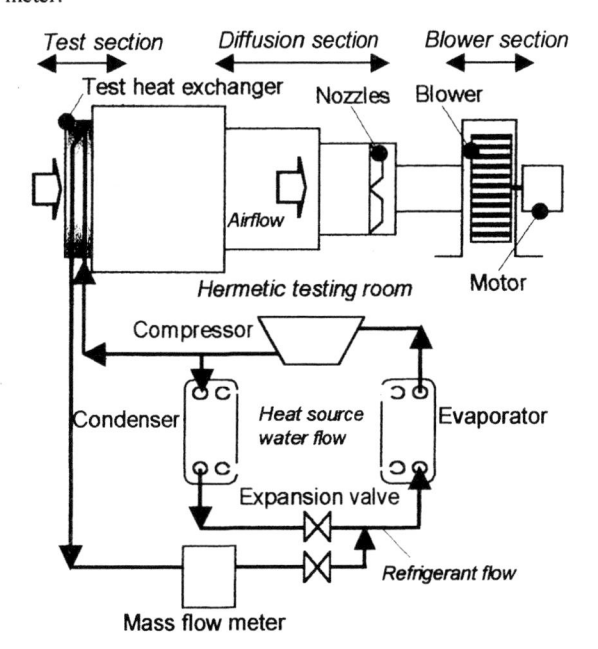

Fig. 1 Schematic diagram of the experimental apparatus

Test Heat Exchanger

Specifications of the heat exchanger tested in these experiments are shown in Table 1. The table lists information on the fin and tube that composed the test heat exchanger. The heat exchanger was placed on the test part of the wind tunnel and connected to the refrigerant supply device.

Dimensions of the heat transfer tubes that are component parts of test heat exchanger are shown in Table 2. The heat transfer tube No.1, No.2, and No.3 are inner spirally grooved tubes, and the No.4 heat transfer tube is a smooth tube. The tube arrangement of the test heat exchanger is schematically shown in Fig. 2. The heat transfer tubes were positioned in a staggered array with two rows. Parallel and counter refrigerant flows were applied to the test heat exchanger during evaporation and condensation, respectively, to evaluate the heat exchanger performances for HCFC-22.

Experimental Conditions

The experimental conditions during evaporation and condensation are shown in Table 3. The frontal air velocity ranged from 0.6 to 1.2 m/s. The evaporation and condensation temperatures of the refrigerant were determined from the saturation temperature corresponding to the refrigerant pressures measured at the outlet and inlet of the test heat exchanger during evaporation and condensation, respectively.

Data Reduction

The capacity of the heat exchanger, Q, was defined by the following equation:

$$Q = G \Delta h \quad (1)$$

where G and Δh are the refrigerant mass flow rate and the enthalpy difference between the inlet and outlet of the refrigerant, respectively. Δh was found by the following equations (2) for evaporation and (3) for condensation.

$$\Delta h = h_o - h_i \quad (2)$$
$$\Delta h = h_i - h_o \quad (3)$$

On the other hand, Q was given by the following equation:

$$Q = \rho c_p u A \Delta T \quad (4)$$

where ΔT is the temperature difference of the air between the inlet and outlet of the test section.

The heat balance between the refrigerant side and the air side was found to be in good agreement within 3%. Therefore, the capacity of the heat exchanger was defined by the capacity on the refrigerant side.

RESULTS AND DISCUSSION

Evaporation

The relationship between the capacities during evaporation of the test heat exchangers and the frontal air velocity is shown in Fig. 3. Comparisons between the capacities of the test heat exchangers at 1.0m/s for the frontal air velocity that is almost

Table 1 Specifications of the test heat exchangers

Constructions	Width/Height/Length: 400/168/18mm 16 tubes in the 1-path staggered array with 2rows
Fins	Hydrophilic pre-coated slit fin Fin pitch: 1.0mm
Tubes	Outside diameter: 6mm 3 kinds of inner spirally grooved tubes & smooth tubes Mechanical tube expanding

Table 2 Dimensions of installed tubes

No.	Wall thickness [mm]	Groove depth [mm]	Apex angle [deg]	Helix angle [deg]	Number of grooves [-]
1	0.25	0.19	23	14	45
2	0.25	0.19	30	13	50
3	0.25	0.17	23	14	40
4	0.30	---	---	---	---

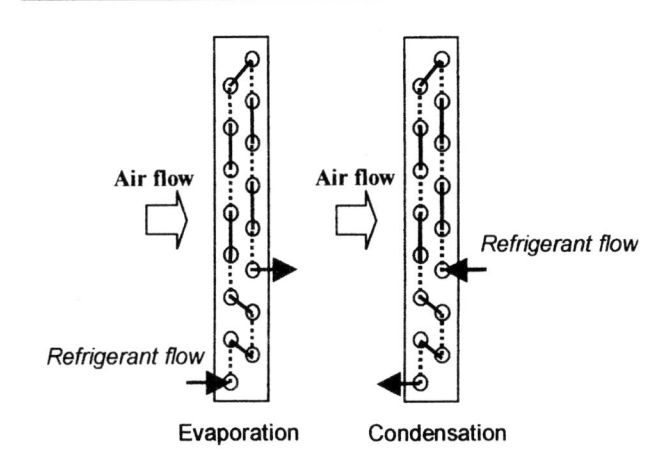

Fig. 2 Tube arrangement of the test heat exchangers

Table 3 Experimental conditions

Test		Evaporation	Condensation
Air side	Dry bulb temp.	300[K]	293[K]
	Wet bulb temp.	292[K]	288[K]
	Frontal air velocity	0.6 to 1.2[m/s]	
Ref. side	Saturation temp.	281[K]	319[K]
	Inlet	Quality=0.22	Degree of superheat=34[K]
	Outlet	Degree of superheat=2[K]	Degree of subcooling=13[K]

equal to that in an actual heat exchanger installed in the inner unit of residential air conditioners reveal the following results:
(a) The capacity of the highest performing heat exchanger consisting of the inner grooved tubes is about 4% higher than that consisting of the smooth tubes.
(b) The maximum capacity difference between the heat exchangers consisting of the inner grooved tubes with similar groove shapes is about 6%.

As one of the factors for the capacity of the heat exchanger consisting of the smooth tubes was higher than our expectation, we regard the effect as a decrease in the pressure drop in the smooth tubes. It is considered that a good influence was exerted on the capacity of the heat exchanger because the temperature difference between the air and refrigerant around the inlet of the refrigerant increased as much as the refrigerant side pressure drop decreased.

Condensation

The relationship between the capacities during condensation of the test heat exchangers and the frontal air velocity is shown in Fig. 4. Comparisons between the capacities of the test heat exchangers at a 1.0m/s frontal air velocity reveal the following results:
(a) The capacity of the highest performing heat exchanger consisting of the inner grooved tubes is about 16% higher than that consisting of the smooth tubes.
(b) The maximum capacity difference between the heat exchangers consisting of the inner grooved tubes with similar groove shapes is about 8%.

As one of the factors for the capacity difference between the heat exchangers consisting of the inner grooved tubes with similar groove shapes was greater than our expectation, we regard the effect of the deformation groove shape due to the tube expanding using bullets.

CONCLUSIONS

Evaluation of the unit performance of the air-cooled heat exchangers consisting of three kinds of inner spirally-grooved tubes and the smooth tubes was experimentally carried out in this study. We obtained the following conclusions.
(a) The capacity of the cross-fin-tube-type heat exchanger consisting of the inner grooved tubes during condensation was evidently higher than that of the smooth tubes and significantly affected by the deformation of the groove shape due to the tube expanding using bullets.
(b) The capacity of the cross-fin-tube-type heat exchanger consisting of the inner grooved tubes during evaporation was approximately equal to that of the smooth tubes, and the effect of the deformation of the groove shape on the capacity was smaller than that during condensation.

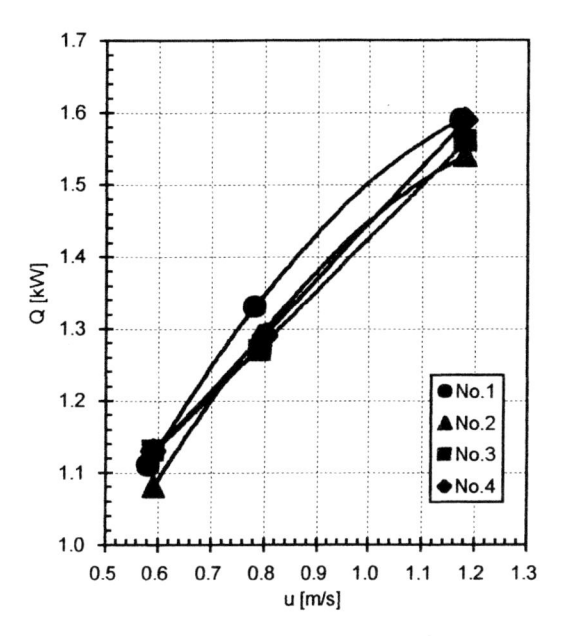

Fig. 3 Capacity of the heat exchangers during evaporation

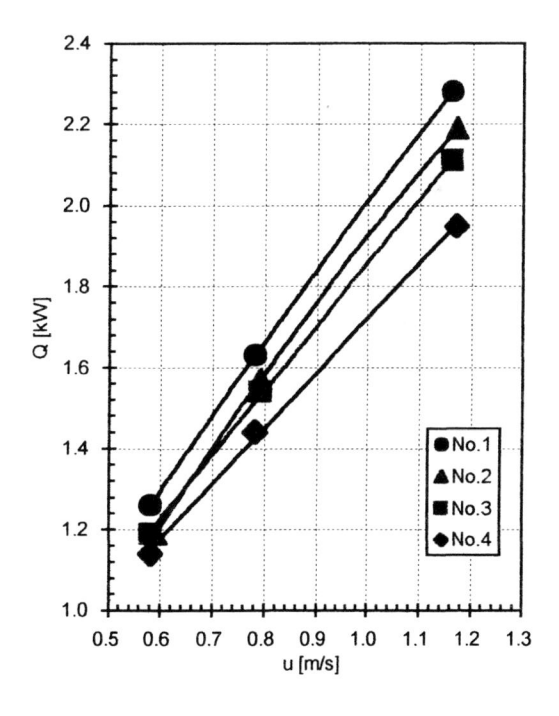

Fig. 4 Capacity of the heat exchangers during condensation

NOMENCLATURE

- A : Area of the heat exchanger
- G : Refrigerant mass flow rate
- Q : Capacity of the heat exchanger
- ΔT : Temperature difference
- c_p : Specific heat of air
- h : Enthalpy
- Δh : Enthalpy difference
- u : Frontal air velocity
- ρ : Density of air

Subscript

- i : inlet
- o : outlet

REFERENCES

Sasaki, N., Kakiyama, S., and Morita, H., 2000, Evaluation of Capacity of cross-fin-tube-type heat exchanger, Proceedings of 37th National Heat Transfer Symposium of Japan, Vol. 3, pp.673-674 (in Japanese).

Kakiyama, S., Sasaki, N., and Morita, H., 1999, Unit performance of fin-tube heat exchanger using the heat transfer tubes of 6mm in outside diameter, Proceedings of 39th Copper Technological Research Meeting Lecture Meeting, pp.25-27 (in Japanese).

Ebisu, T., Kasai, K., and Torikoshi, K., 1996, Air-cooled heat exchanger performances using alternative refrigerants, Proceedings of 3rd KSME/JSME Thermal Engineering Conference, 3, pp. 179-185.

Ebisu, T., Yoshida, K., Torikoshi, K., and Ohkubo, Y., 1997, Development of compact heat exchanger using $\phi 6$ heat transfer tube installed into residential air-conditioner, Proceeding of the 30th Japanese Joint Conference on Air-conditioning and Refrigeration, pp. 65-68 (in Japanese).

Torikoshi, K. and Ebisu, T., 1998, Japanese advanced technologies of heat exchanger in air-conditioning and refrigeration applications, Proceedings of the International Conference on Compact Heat Exchangers and Enhancement Technology for the Process Industries, pp. 17-24.

ALTERNATIVE REFRIGERANTS PERFORMANCE IN EVAPORATORS OF PLATE HEAT EXCHANGER TYPE

Gino Boccardi[1], Gian PieroCelata[2]

[1] ENEA - Institute of Thermal-Fluid Dynamics - Rome, Italy; e-mail: boccardi@casaccia.enea.it
[2] ENEA - Institute of Thermal-Fluid Dynamics - Rome, Italy; e-mail: celata@casaccia.enea.it

ABSTRACT

The present paper reports on the results of experimental research carried out on brazed plate heat exchangers, used as evaporators in refrigeration circuits. The research was aimed at getting information on the thermal efficiency of the evaporators, under commercial conditions, when using new ozone-friendly refrigerants to replace CFCs and HCFCs.

At first, the influence of some thermal-hydraulic parameters on the heat flux and the overall heat transfer coefficient were investigated, using R290 (propane) in a heat pump prototype. Hydrocarbons, initially set aside for safety problems in non-industrial applications, have been thoroughly re-considered thank to availability, cheapness and ODP and GWP values very interesting. Benchmark tests with R22 allow the comparison with the heat exchanger design fluid.

To complete the comparison with other fluids proposed as R22 alternative, the data have been compared with experimental points obtained in an earlier campaign with an other evaporator, similar but with a different number of plates, using R407C, R410A, R134a and R22 as test fluids.

The results show that R290 has a good performance in evaporator designed for R22, better than R407C and R134a. R410A has similar performance but it works at greater pressure.

KEY WORDS: Propane, R134a, R407C, R410A, R22, and evaporation plate heat exchangers

INTRODUCTION

The research on fluids alternative to CFC and HCFC, not dangerous for Ozone Layer, has led over the last years to testing some new fluids, typically mixtures of HFC, characterized by ODP (Ozone Depletion Potential) equal to zero; in air conditioning, R407C and R410A are proved the most interesting alternative to R22 (Boccardi et al., 2000, Cavallini, 1998), but these fluids have a not negligible GWP (Global Warming Potential).

Together with other considerations pertaining to fluid production and availability, all that has made the research on alternative refrigerants going on. In particular, hydrocarbons, initially set aside for safety problems in non-industrial applications, have been thoroughly re-considered (Granryd, 2001). Among these, propane (R290) has characteristics such as to be considered a very interesting refrigerant for air conditioning applications. In table 1, ODP and GWP values of above-mentioned fluids are shown.

Table 1. ODP and GWP values for refrigerants tested

	R22	R407C	R410A	R290
ODP (CFC11=1)	0.055	0	0	0
GWP(CFC11=1)	0.44	0.34	0.37	0.03

Of course, the use of this substance has to be combined with adequate precautionary measures; as a first step, in this respect, the goal is the reduction in the refrigerant volume stored per power unit, coupled with an increase in the efficiency and the reduction of volume components.

The present paper deals with the results of an experimental campaign carried out on a prototype of propane heat pump. The thermal performance of heat pump prototype is described elsewhere (Corberan et al., 2000) while this work refers, specifically, to the performance of the compact heat exchanger when it has been used as evaporator. The influence of thermalhydraulic parameters on the heat flux and on the overall heat transfer coefficient has been determined. Benchmark tests with R22 allow the comparison with the heat exchanger design fluid.

In order to get the best performance under winter conditions, unlike from operating conditions forecast in the design of this component, heat transfer with water (secondary side) in the evaporator takes place in co-current flow. Counter current flow tests have been also performed in order to ascertain the influence of such a choice.

Present data have been compared with data points obtained in an earlier campaign with an other evaporator, similar but with a different number of plates, using R407C, R410A, R134a and R22 as test fluids. Some geometric characteristics of two evaporators are shown in table 2; a secrecy agreement with the manufacturer does not allow giving the values of others data.

Table 2. Evaporators geometric data

Evaporator	Number of plates	Surface m²
Heat Pump tests	44	2.1
Old tests	80	3.9

EXPERIMENTAL SETUP

The test facility consists mainly of three cycle loops: a refrigerant cycle, a water loop for the evaporator and an air loop for the condenser (Fig. 1). The refrigerant flow is regulated by an inverter that allows a variation in the frequency between 30 and 60 Hz of a reciprocating compressor power supply. There is only a thermostatic valve for the two fluids; its behaviour with propane was studied in preceding tests (Boccardi et al., 2000). The control of the ΔT_{sup} at the inlet of the compressor is obtained by regulating by hand its set point.

The water circuit supplies the thermal load and allows the control of the inlet and outlet water temperatures of the exchanger; the water flow rate through the evaporator and the discharge flow rate are controlled by two-regulation valve.

The airflow rate in the air loop is obtained regulating by an inverter the fan rotation speed; the air temperature is controlled by warming and cooling auxiliary loops. Three valves, two at inlet and outlet of air loop and the other one in a by-pass, allow another control on air temperature and flow rate.

Two mass flow meters are installed in the refrigerant loop and in the water loop; a Wilson grid measures the flow rate in the air loop. The temperature and pressure values are taken at the inlet and outlet of all the main components of the facility.

TEST MATRIX AND CORRELATIONS

As reported in Corberan et al. (2000), an aim of experimental tests, carried out on a prototype of propane heat pump, was to verify a heat exchange model; for this objective, it was chosen to keep constant the water mass flow rate and the evaporation temperature. These values were about 2820 kg/h for the water mass flow rate and 3.9 °C for R22 and 2.7 °C for propane for the evaporation

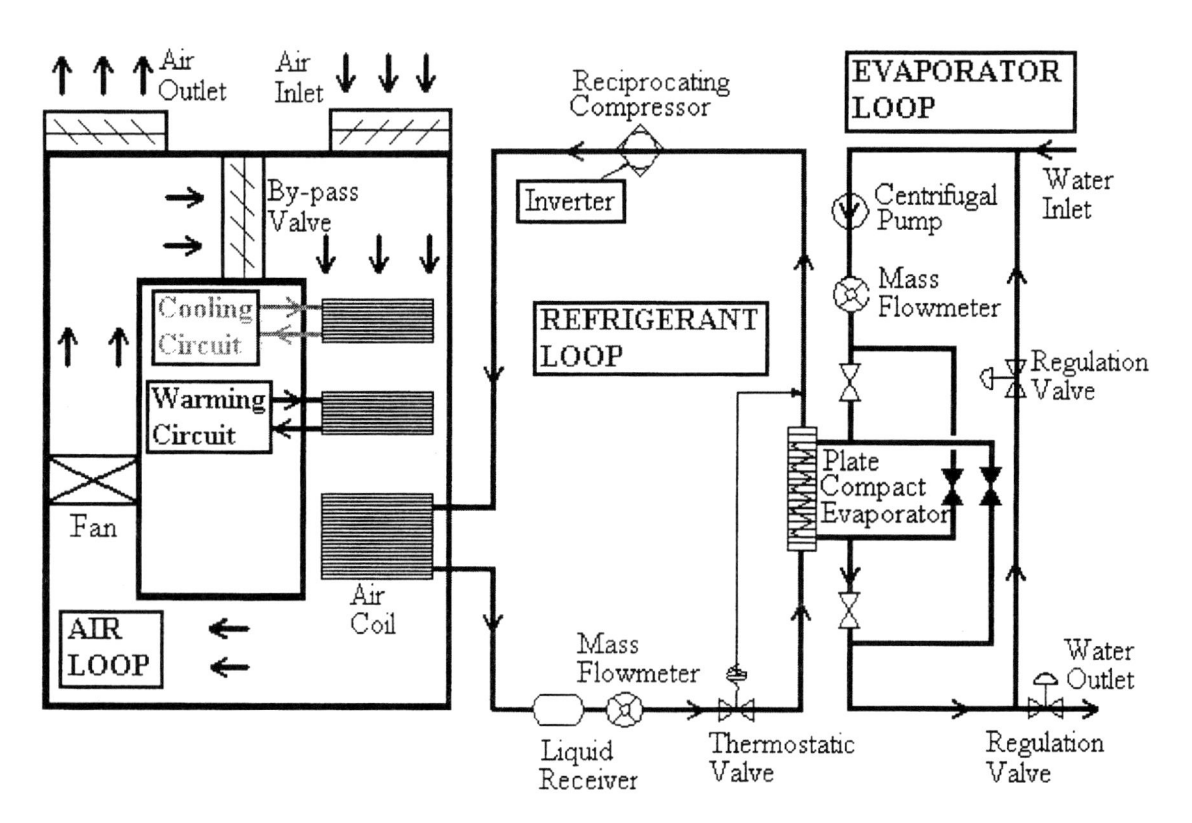

Fig.1: Test Facility

temperature. Table 3 shows the working ranges of the other main parameters observed during the tests.

We made three groups of tests on propane: the first two are different only for values of ΔT_{sup}, in the third we changed the water flow from co-current to counter current. The counter current flow, with the same T_{eva} and G_w, involves ΔT_{sup} quite greater than the cases 1 and 2.

The saturation temperature at the evaporator depends on the pressure and therefore changes in accordance with the pressure drops inside the component. In tested plate heat exchangers the distribution of the mass flow through the plates is regulated by small nozzles (called distributors) at the entrance to the channels. For this reason the inlet pressure in the evaporator is different to the inlet pressure in the channels, where the heat exchange occurs.

Table 3. Evaporator Test Conditions for Heat Pump

Test	G_r kg/s		W_r kW		X_{in} chan. %		ΔT_{sup} K	
	min	max	min	max	min	max	min	max
R22	0.102	0.125	15.7	19.4	23.7	24	4.4	5.1
Propane_1	0.043	0.066	11.8	17.9	28	28.9	5.1	5.2
Propane_2	0.05	0.072	14.3	20.5	26.3	27	6.1	6.8
Propane_3	0.049	0.073	14.6	21.3	26.1	26.8	9.6	11.1

According to data provided by the manufacturer, we can consider that the pressure losses are concentrated in the distributors and can thus assume the pressure inside the channels to be equal to the outlet pressure of the evaporator. So we assume for evaporation:

$$T_{sat} = T_{sat}(p_{roe}) \qquad (1)$$

The quality value "x" is calculated considering an isoenthalpic expansion at the thermostatic valve, a typical situation of the refrigeration cycle. Therefore

$$H_{rie} = H_{roc} \qquad (2)$$

If we also consider T_{eva} as constant during the evaporation, ΔT_{lm} becomes:

$$\Delta T_{lm} = \frac{(T_{wie} - T_{woe})}{\ln((T_{wie} - T_{eva})/(T_{woe} - T_{eva}))} \qquad (3)$$

The thermal power is

$$W_r = G_r (H_{roe} - H_{rie}) \qquad (4)$$

but also

$$W_r = U S \Delta T_{lm} \qquad (5)$$

Knowing the heat transfer surface, S, and calculating W from Eq. (4), and ΔT_{lm} from Eq. (3), it is possible to calculate U from Eq. (5) which, for heat transfer between the refrigerant and water separated by a flat surface with a thickness "s" and having a thermal conductivity "k", is given by

$$U = \frac{1}{1/\alpha_r + s/k + 1/\alpha_w} \qquad (6)$$

where α_r and α_w are the refrigerant and water heat transfer coefficient, respectively.

In the experimental tests, we have s/k and α_w practically constant (thickness temperature, water temperature and water mass flow rate almost constant) and, therefore, variations in U will be mainly due to variations in α_r.

Reynolds number at inlet of channels is defined as

$$Re_{ch} = \frac{\rho u_{ch} d_h}{\mu} \qquad (7)$$

where

$$u_{ch} = \frac{Q_{ch}}{bA} \qquad (8)$$

and the thermodynamic properties are calculated at outlet pressure of the evaporator. The geometric definitions for the plate exchanger are shown in Fig. 2.

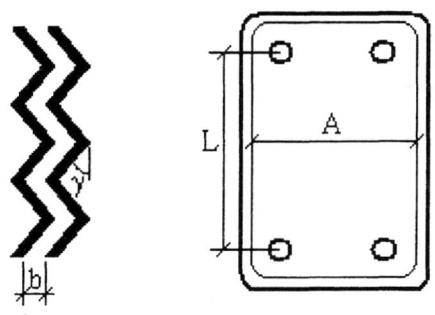

Fig. 2. Plate geometric dimensions.

The hydraulic diameter is (Edwards and Wilkinson, 1984)

$$d_h = 4 \frac{bA}{\left(\frac{2A}{\cos\gamma} + 2b\right)} \cong 2b\cos\gamma \qquad (9)$$

Substituting U and d_h in Eq. (7), we have

$$Re_{ch} = \frac{\varrho \frac{Q_{ch}}{bA} 2b\cos\gamma}{\mu} = \frac{2G_{ch}\cos\gamma}{A\mu} \quad (10)$$

Fluid viscosity is calculated using the homogeneous model

$$\mu = \varrho_{tp}(x \cdot \mu_v/\varrho_v + (1-x)\mu_l/\varrho_l) \quad (11)$$

with

$$\varrho_{tp} = \varrho_l(1-F) + \varrho_v F \quad (12)$$

The thermodynamic properties of the fluids are calculated by using Refprop 6.0. (McLinden et al., 1998).

RESULTS AND DISCUSSION

The analysis has been divided in three steps:
1. Comparison of evaporator performance with propane and R22, the fluid of the heat exchanger design.
2. To ascertain the influence of the water flow (co-current or counter current) on the evaporator performance.
3. Comparison of propane performance respect on other refrigerants, R407C, R410A, R134a and R22, using data points obtained in an earlier campaign with an other evaporator, only different in number of plates

In every case, the test were carried out in real working conditions, with the water temperature at outlet of evaporator practically constant and, therefore, the data allow a comparison of evaporator performance with the five refrigerants in an operative situation.

Comparison of Evaporator Performance with Propane and R22

Figure 3 depicts the global heat transfer coefficient (U) as a function of the Reynolds number calculated at the inlet of the evaporator, after distributors, for R22, propane_1 and propane_2 tests.

The effect of different ΔT_{sup} on U is evident from the comparison of propane_1 and propane_2 data with the U difference becoming less for Reynolds number increasing. R22 shows the best behaviour with ΔT_{sup} a bit lower than propane_1 data. To reduce the influence of differences of ΔT_{sup} on data analysis, afterwards propane_1 data will be used.

In Fig. 4, average heat flux (q") versus Reynolds number is shown. Propane and R22 experimental points are very close and look the same trend.

Figure 5 depicts q" as a function of pressure loss in the evaporator (Δp). Propane has a better behaviour than R22, but the differences are very low, almost 50 mmbar for the same q".

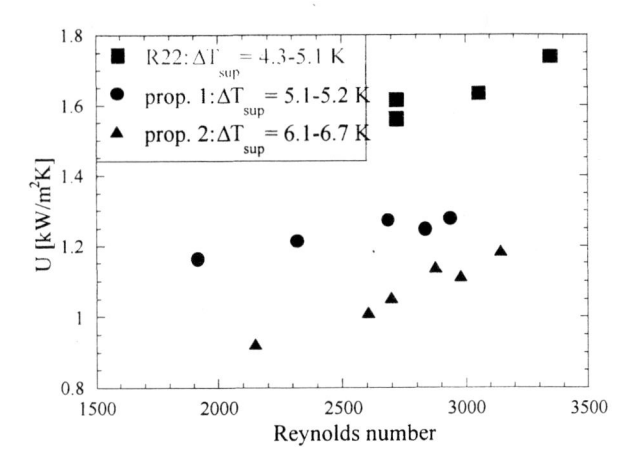

Fig. 3 Global overall heat transfer coefficient versus Reynolds number.

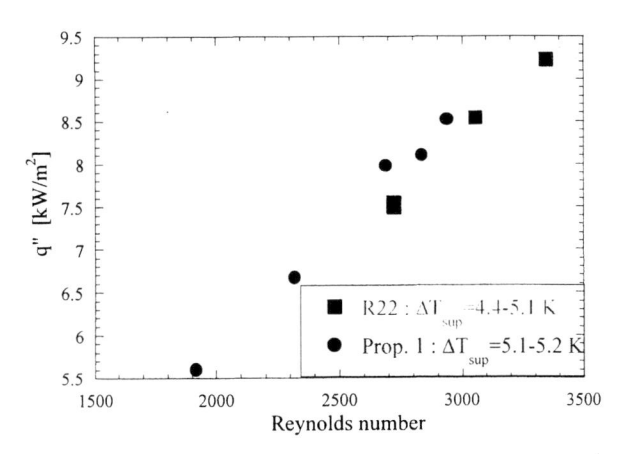

Fig. 4 Heat flux versus Reynolds number.

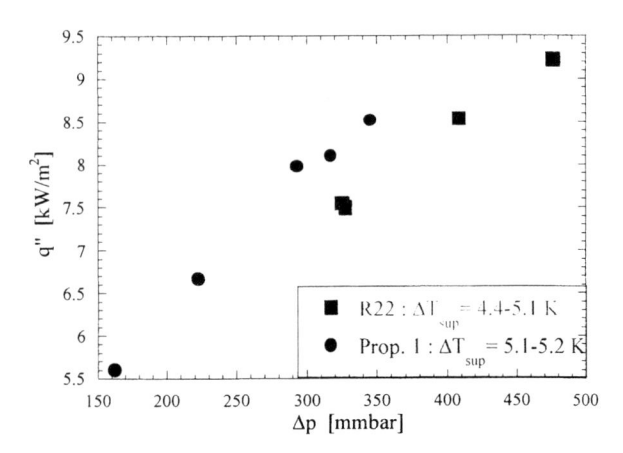

Fig. 5 Heat flux versus pressure loss.

These data show that, in real working conditions, propane, in spite of its lower U values, has the same performance of R22 with an evaporator designed for this refrigerant.

Influence Of the Water Flow (Co-Current or Counter Current) On the Evaporator Performance

In the heat pump, in cooling conditions, the plate heat exchanger is used as evaporator while, in heating conditions, it works as condenser; the change of employment involves the inversion of refrigerant flow across the exchangers. Because the water flow is usually not inverted, in the first case the plate heat exchanger works in co-current flow, in the other one in counter current flow. This choice depends on the optimization of global heat pump performance. Prohpheta circuit is designed to test the evaporator in both flow conditions; in the following experimental data, obtained in these two case (propane 2 for co-current flow and propane 3 for counter current flow), are shown.

All the tests have the same evaporation temperature and almost the same water temperature at outlet of evaporator (~6.8 °C); consequently ΔT_{sup} is higher in counter current case.

These test conditions allow the comparison of the two fluids when greater refrigeration power is required at the same conditions of utilization.

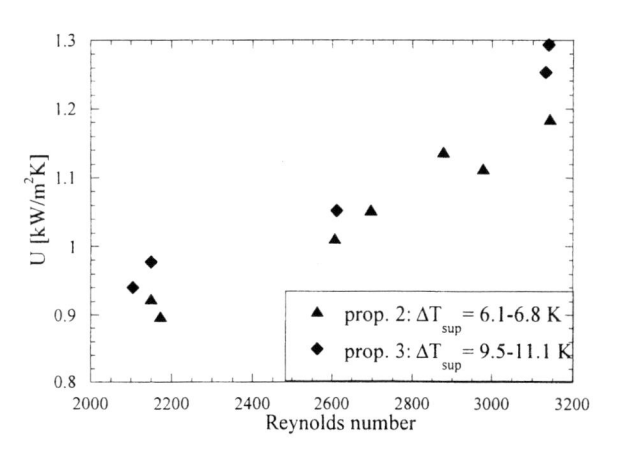

Fig. 6. Global overall heat transfer coefficient versus Reynolds number.

Figure 6 shows the global heat transfer coefficient (U) as a function of the Reynolds number calculated at the inlet of the evaporator, after distributors.

Although ΔT_{sup} values higher, the counter current flow involves U values greater than co-current case.

In Fig. 7 is shown the trend of average heat flux (q") versus Reynolds number. The comparison confirms the better performance of counter current case although the differences in the values are very small.

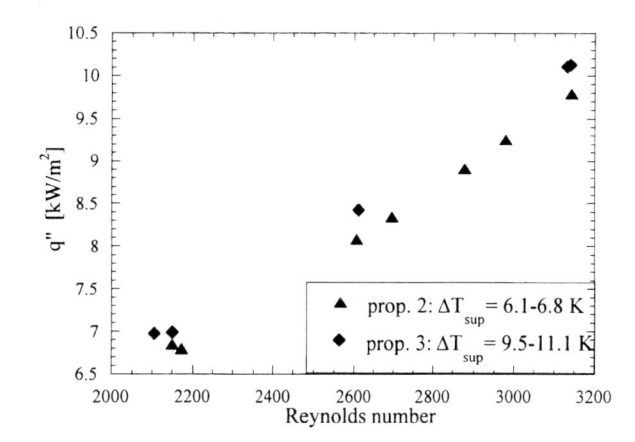

Fig. 7 Heat flux versus Reynolds number

Figure 8 shows q" as a function of pressure loss in the evaporator. The differences between two cases are little with lower pressure loss, at same q", for counter current data.

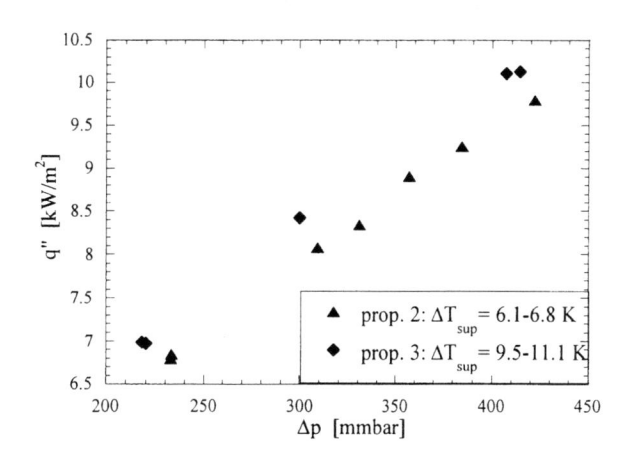

Fig. 8 Heat flux versus pressure loss.

To conclude, these data are interesting for examining the evaporator performance in the two-water flow rate and for evaluating the advantage of a design interview for changing the water flow rate in cooling condition.

Comparison of Propane Performance Respect on R407c, R410a, R134a and R22

To complete the evaluating of propane as substitute of R22 we have to compare it with other alternative fluids. For this aim we used experimental data obtained during a previous experimental campaign carried out on COMHETA plant (Boccardi et al., 2000). This research was aimed at getting information on the thermal efficiency of the compact heat exchangers, under reference commercial conditions, when using new ozone-friendly refrigerants to replace CFCs and HCFCs.

Specifically, the influence of some thermal-hydraulic parameters on the heat flux and the overall heat transfer coefficient were investigated for two evaporators, using R134a, R407C, R410A and R22 as reference.

Moreover, in Boccardi et al., 2000) a New Thermodynamical Method was applied with the purpose of defining the saturation temperature and investigating the other main parameters of zeotropic mixtures in the two phase region for the fluids R407C and R410A, that are blends of, respectively, three and two pure refrigerants. The use of this Method is very important for a correct comparison of data of R407C (glide almost 6-7 K) while for R410A, that has an almost azeotropic behaviour, it is practically irrelevant.

A series of tests was carried out with an evaporator of the same type of that used in PROHPHETA circuit; they are only different for the number of plate.

Some test parameters were different, in particular saturation temperature and water flow rate were not controlled and the evaporator works in counter current flow. ΔT_{sup} values were in the range 5-7 K. In the table 3 the working ranges of the some main parameters observed during these tests are shown.

Table 3: Evaporator test conditions for COMHETA tests

Fluid	G_r kg/s		W_r kW		X_{in} chan. %		T_{eva} °C	
	min	max	min	max	min	max	min	max
R134a	0.17	0.27	32.8	45.2	15.0	21.5	-3.5	0.4
R407C	0.20	0.31	37.9	55.8	22.5	25.5	0.3	3.6
R410A	0.19	0.32	32.5	54.5	23.8	27.5	0.9	3.4
R22	0.24	0.35	40.1	58.2	20.6	22.2	-1.4	1.1

For a qualitative comparison, we used propane_2 data that have ΔT_{sup} comparable; therefore in the evaluation we have to consider that these data were taken for co-current flow condition.

The global overall heat transfer coefficient as a function of the Reynolds number calculated at the inlet of the evaporator, after the distributors, is plotted in Fig. 9.

The propane experimental points have Reynolds values lower than R22, but the trend is similar. R410A confirms the best U values [..], but propane shows better than R407C and R134a. Moreover, while for R410A and R22, U increases lightly with Reynolds number, for R407C and R134a, the U values became practically constant when Reynolds number greater than 4000. This behaviour is examined in (Boccardi et al., 2000) and could depend on the working conditions.

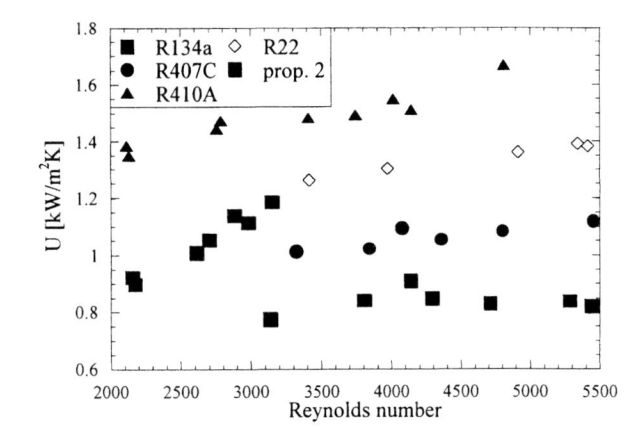

Fig. 9 Global overall heat transfer coefficient versus Reynolds number.

Figure 10 depicts q" versus Reynolds number for the five refrigerants. The data points of propane, R22 and R410A are very close while the other two fluids have worse performance.

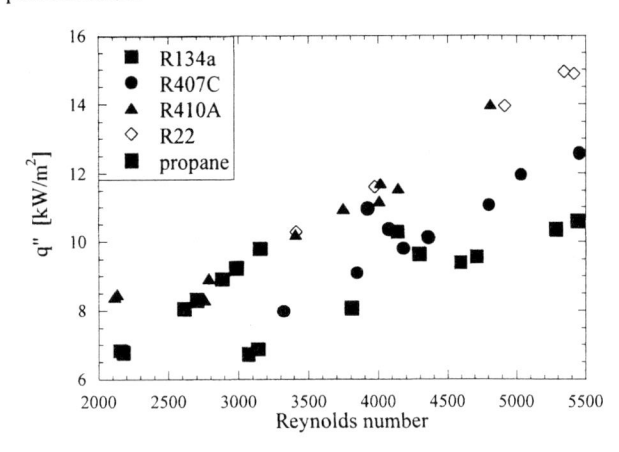

Fig. 10 Heat flux versus Reynolds number.

In Fig. 11 is shown the trend of U versus heat flux. U increases together q" except for R134a that has almost constant value. For the same Reynolds value, R410A has a higher q", followed by R22 together propane, R134a, and R407C, respectively.

Figure 12 shows q" as function of pressure loss in the evaporators. R22, propane and R410A have close values and, for the same q", their Δp are lowest. R134a has the worse behaviour, Δp values almost 600 mmbar greater for the same q", while R407C shows intermediate values.

Propane data enter coherently among the previous data, also in comparison with results of the first analysis step; only U values look a bit higher respect on R22 (although the first ones are taken in co-current flow) while the behaviour of propane respect R22 in Figs. 10 and 11 looks like that in Figs. 4 and 5.

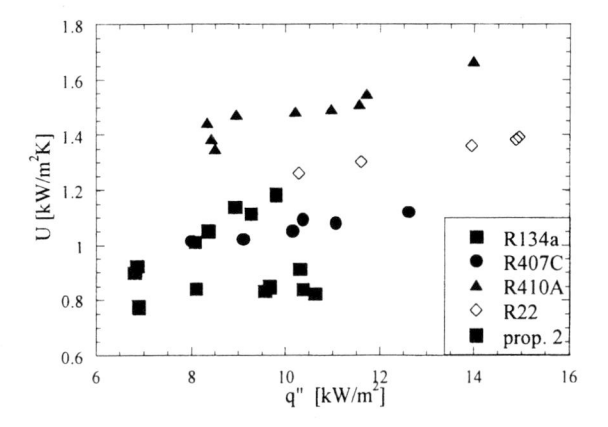

Fig. 11 Global overall heat transfer coefficient versus heat flux.

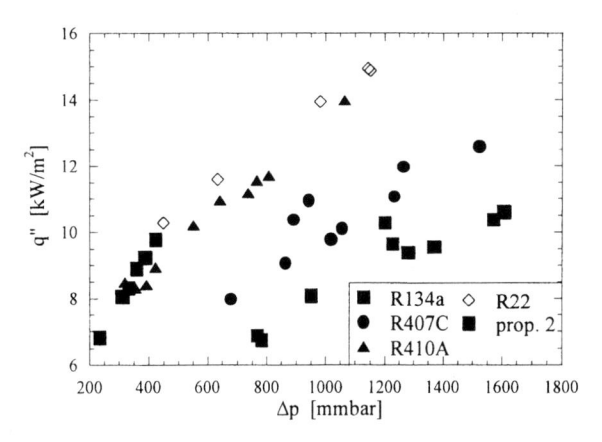

Fig. 12 Heat flux versus pressure loss.

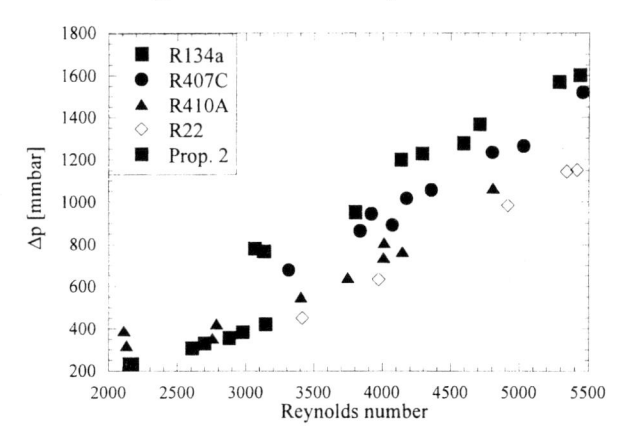

Fig. 13. Pressure loss versus Reynolds number.

We can think that the differences in U values, apart from different exchange conditions (inlet quality, P_{eva}, ΔT_{sup}), are due to the less number of plate in the first evaporator (almost the half of the second one) that supports a better flow distribution.

Figure 13 depicts Δp versus Reynolds number for the five refrigerants; the behaviours of propane, R22 and R410A are similar and corroborate suggestions obtained from previous figures.

To conclude, propane performance is comparable with R22 and R410A, and looks better than R134a and R407C.

CONCLUSIONS

The analysis of experimental data confirms the interesting characteristics of brazed plate heat exchangers for air conditioning application with propane as refrigerant (Granryd, 2001).

Propane performance in an evaporator (designed for R22) is comparable to those of R22 and R410A and is better than R407C and R134a.

Brazed plate heat exchangers, used in a heat pump, show an acceptable performance reduction changing the flow conditions; this is very useful for an optimization of heat pump design, allowing choosing smaller exchangers.

Moreover, propane has not zeotropic mixtures problems (R07C), does not need of polyester oil (R407C, R410A, R134a), is cheap and easy available.

The safety problem is the main handicap for using propane in retrofit of R22 circuit and in new refrigerant plants: the reduction of plant charge, the development of new components (in particular compressors), new plant layout are some of researches carry out to go on this obstacle.

Brazed plate heat exchangers thank to their less capacity, the absence of gaskets and the reduction of dimension of circuit, show very interesting prospects for using in plants with propane.

NOMENCLATURE

A	plate width, m
b	pressure depth, m
d_h	hydraulic diameter, m
F	void fraction
G	mass flow rate, kg/s
k	thermal conductivity, W/mK
P	Pressure, bar
Q	volumetric flow rate, m^3/s
q"	heat flux, kW/m^2
Re	Reynolds number
S	heat exchange surface area, m^2
T	Temperature, °C

U	overall heat-transfer coefficient, kW/m²K
u	Velocity, m/s
W	thermal power, kW
x	vapour quality
α	heat transfer coefficient, W/m² K
ΔT_{lm}	logarithmic mean temperature difference, K
ΔT_{lm}	logarithmic mean temperature difference, K
γ	corrugation angle, deg
μ	dynamic viscosity, Pa s
ρ	density, kg/m³

Subscripts

Ch	channel
dew	dew point
e	evaporator
i	inlet
l	liquid
o	outlet
r	refrigerant
sat	saturation
Tp	two phases
v	vapour
w	water

REFERENCES

Boccardi, G. and Celata, G.P. and Cumo, M. and Gerosa, A. and Marchesi-Donati, F. and Zorzin, A. 2000. R22 Replacement Aspects in Compact Heat Exchangers for Air Conditioning, *International Journal of Heat Exchangers*, vol. I, pp. 81-95.

Boccardi, G. and Celata, G.P. and Cumo, M. and Boccardi, G. and Celata, G.P. and Cumo, M. and Gerosa, A. and Giuliani, A. and Zorzin, A. 2000. The Use of New Refrigerants in Compact Exchangers for the Refrigeration Industry, *Heat Transfer Engineering*, vol. 21, n. 4, pp. 53-62.

G.Boccardi and G.P. Celata and A. Lattanzi, 2000, Influenza del fluido di carica e del surriscaldamento sul funzionamento di una valvola termostatica convenzionale con equalizzatore esterno, *Atti 18th UIT National Heat Transfer Conference*, vol. II, pp. 507-519.

Calm, J. M., Didion, D. A., 1997. Trade-offs in refrigerants selection: past, present and future, *Proceedings of the Ashrae-Nist Refrigerant Conference, "Refrigerants for the 21th Century"*, Gaithersburg, MD, USA, pp.6-12.

Cavallini, A., 1998. I Nuovi Fluidi Frigorigeni per la Climatizzazione, Centro Studi Galileo - *Industria e Formazione*, n° 217 pp. 26-32.

Chang, Y.S., Kim, M.S., Ro, S.T., 2000. Performance and heat transfer characteristics of hydrocarbon refrigerants in a heat pump system, *International Journal of Refrigeration*, n° 23, pp. 232-242.

J.M. Corberan and J. Urchueguia and J. Gonzalvez and T. Setaro and Gino Boccardi and Bjorn Palm, 2000. Two phase heat transfer in brazed plate heat exchangers, evaporators and condenser for r22 and propane, *Proceedings of 3rd European Thermal Sciences Conference*, pp. 1193-1198.

J.M. Corberan and S. Ortuno and F. Ferri and P. Fernandez de Cordoba and T. Setaro and Gino Boccardi, 2000. Modelling of tube and fin coils working as evaporators or condenser, *Proceedings of 3rd European Thermal Sciences Conference*, pp. 1199-1204.

Edwards, M.F. and Wilkinson, W.L., 1984. *Plate Heat Excahngers*, HTFS DR36.

McLinden, M.O., Klein, S.A., Lemmon, E.W., Peskin, A.P., 1998. Nist thermodynamic and transport properties of refrigerants and refrigerant mixtures-REFPROP, Version 6.0, NIST, Boulder, CO, USA.

Granryd, G., 2001, Hydrocarbons as refrigerants - an overview, *International Journal of Refrigeration* vol. 24, pp. 15-24.

CHARACTERISTICS OF HEAT TRANSFER AND PRESSURE DROP OF R-22 INSIDE AN EVAPORATING TUBE WITH SMALL DIAMETER HELICAL COIL

Ju-Won Kim[1], Jeung-Hoon Kim[2], Jong-Soo Kim[3]

[1] Graduate school, Pukyong National University, Pusan, 608-737, KOREA; E-Mail: juwkk@lycos.co.kr
[2] RCOID, Pukyong National University, Pusan, 608-737, KOREA; E-Mail: hoonkj@netian.com
[3] Pukyong National University, Pusan, 608-737, KOREA; E-Mail: kimjs@dolphin.pknu.ac.kr

ABSTRACT

There have been many studies conducted by engineers to enhance the efficiency of heat exchangers, but it has been found that it is very difficult to overcome various limitations in order to enhance performance. Lately, other methods of enhancing the performance of compact heat exchangers have been studied by engineers, including various ways of reducing the hydraulic diameter, and enlarging the heat transfer area per cubic meter. In spite of a lot of research on compact heat exchangers, there are still few reports on the characteristics of micro-tube heat exchangers, especially using refrigerants used widely in air conditioners and refrigerators.

To make a compact evaporator, experiments that show characteristics of evaporating heat transfer and pressure drop in a helically coiled small diameter tube were conducted in this research. The experiments were performed with HCFC-22 in a helically coiled small diameter tube; inner diameter= 1.0 mm, tube length=2.0 m, and curvature diameter=31, 34, 46.2 mm. The experiments were also carried out with the following test conditions; saturation pressure=0.588 MPa, mass velocity=150~500 kg/m²s, and heat flux=1~5 kW/m². An empirical correlation was presented based on the experimental results to predict the evaporating heat transfer coefficient in a helically coiled small diameter tube. It was found that dry-out occurred in a low quality region for evaporation heat transfer because of the breaking of the annular liquid film. The friction factor of the single-phase flow of the helically coiled tube corresponded to Prandtl's correlation. Finally, a correlation was proposed that can predict the friction factor of a two-phase flow in a helically coiled tube.

INTRODUCTION

As resources are becoming depleted, demands for high effective systems are increasing all over the world. The supply and demand have diversified and demand is increasing for energy saving features and reduction in size etc., of air conditioning and refrigeration systems.

Heat exchangers of fin and tube type have been mainly used as heat exchangers for air conditioning systems. Until now, to maximize the efficiency of heat exchangers, the performance of heat exchangers have been enhanced by manufacturing slit or louver to the fin surface and small grooves inside the heat transfer area. But, in making the fin, there are problems caused by the difficulties in the production process and remodeling.

It has been promoted that a small-sized heat transfer tube saves energy and overcomes space restriction through the high performance and compactness of the heat exchanger, which is a main part of air conditioning and refrigerant systems(Kim and Katsuta, 1995; Kureta et al., 1997; Moriyama et al., 1992; Wambsganss et al., 1991).

Observing the recent international trend, the diameter of the heat transfer tube decreased from 7 mm to 4 mm and consequently the volume of heat transfer decreased about 75% from the volume of a tube with an outer diameter of 9.52 mm which was widely used as the heat transfer tube.

This tendency will be dominant for the tubes with diameters of less than 1 mm to improve the performance by the use of HFC refrigerants and to make of a perfect counter flow to form the Lorentz cycle. With a small-sized inclination heat transfer tube, there are many merits such as the curtailment of the development period and costs needed to develop a fin as well as avoiding troublesome fin development as occurs in the conventional heat exchanger. The hydraulic diameter was less than 5mm in a compact heat exchanger using a capillary tube. Additionally, it was reported that surface tension, viscosity, and mass velocity influenced heat transfer characteristics in the capillary tube(Wang and Peng, 1994). A helically coiled tube was widely used in waste heat recovery and refrigeration systems of the chemical, pharmacy, food industries, and nuclear reactor. Its structure is very compact and it has high thermal characteristics.

The flow in helically coiled tubes is more complicated than that of straight tubes(Feng et al., 1997; Kazi et al., 1997). Taylor(1929) found that the secondary flow in helically coiled tubes depends on the curvature ratio. When the fluid flowed in the center of tube at high speed, the centrifugal force of fluid occurred and this was resisted by low speed flow relatively by influence of the viscosity at the tube wall. Thus, this kind of secondary flow happened (Louis, 1983). Until now, the analysis of helically coiled tubes(Han and Park, 1998) was limited to large heat exchangers with large diameter such as boilers and nuclear reactors. And research is lacking in the areas of evaporation heat transfer and pressure drop in forced convection of inside tubes according to the small-sized inclination of the tube diameter.

Therefore, in this study, to provide information on small-sized inclination of heat transfer tubes, these experiments were conducted on the evaporation heat transfer characteristic and pressure drop at liquid single-phase and two-phase flow in a helically coiled tube with inner diameter of 1 mm. Based on these experiments, basic design data will be obtained to make flexible heat exchangers of high performance by maximizing the heat transfer area.

EXPERIMENTAL APPARATUS AND METHODS

Experimental Apparatus

A schematic diagram of the experimental apparatus is shown in Fig. 1. The experimental setup was composed of a basic simple refrigeration cycle. After the refrigerant-oil mixture was compressed into a superheated vapor of high temperature and pressure in the compressor, it was passed through an oil separator and the oil was separated and sent to the compressor. The pure refrigerant was cooled by cold water in the condenser. After the subcooled liquid refrigerant was stored in the receiver, it was passed through a dryer-filter and strainer and its total flow rate was measured using a flow meter. Only a part of the refrigerant was flowed into the test section through an accurate refrigerant injector and after most of the refrigerant expanded isenthalpically to the evaporating pressure by expansion valve, it was evaporated at the sub evaporator and recirculated to the compressor.

When the mass velocity of the refrigerant was 300 kg/m^2, the mass flow rate of refrigerant was about 8 g/min. So, it is difficult to measure the flow rate by the general flow meter. Thus, the mass flow rate of refrigerant was measured using an accurate refrigerant injector(TAIATSU TECHNO, HPG-96).

The refrigerant which flowed into the accurate refrigerant injector expanded to the evaporation temperature 5℃ by adjusting the opening of the needle valve. It was passed through test section and the helically coiled tube was heated by the electric heater, and it was recirculated to the compressor.

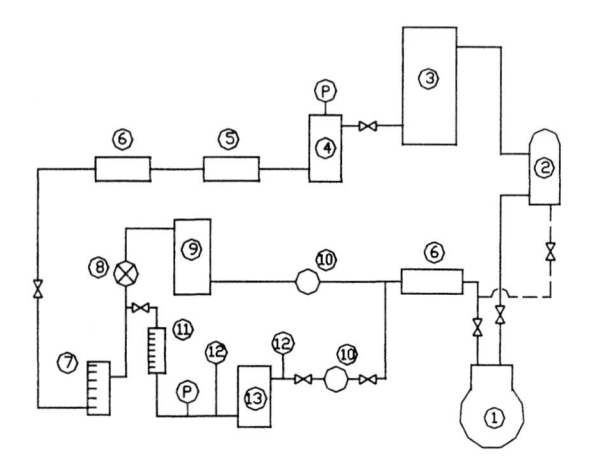

——— Refrigerant
----- Oil

1. Compressor 2. Oil separator
3. Condenser 4. Receiver
5. Dryer-filter 6. Strainer
7. Refrigerant flow meter 8. Expansion valve
9. Sub-evaporator 10. Sight glass
11. Accurate refrigerant injector
12. Digital pressure indicator
13. Test section

Fig. 1 Schematic diagram of experimental apparatus

Table 1. Dimension of test tubes

Tube types	Type A	Type B	Type C
Total tube length (mm)	2000	2000	2000
Number of helical turn	20	18	14
Pitch (mm)	9.1	7.5	13.8
Curvature diameter (mm)	31	34	46.2

Test Section and Experimental Methods

The heat transfer tube used in this study was made of copper, its inner diameter was 1 mm, outer diameter was 2.2 mm, and total length was 2 m. The specification of the helically coiled tube made to measure heat transfer coefficient and pressure drop are represented in Table 1.

The shape, pitch and curvature diameter of the helically coiled tube are represented in Fig. 2 and the details of the test section are represented in Fig. 3.

Sheath type(T type) thermocouples were installed at the inlet and outlet of the test section to measure the mixed average temperature of refrigerant. And, thermocouples (T type) of 6 pieces at the side of the test section were installed at constant intervals to measure the outside wall temperature. The inlet and outlet pressure of the test section were measured using a digital pressure gauge. Thermocouples

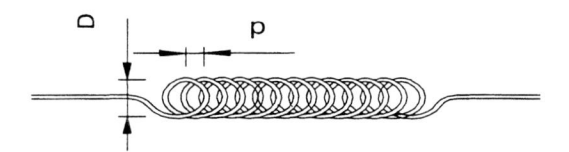

Fig. 2 Geometry of helically coiled tube

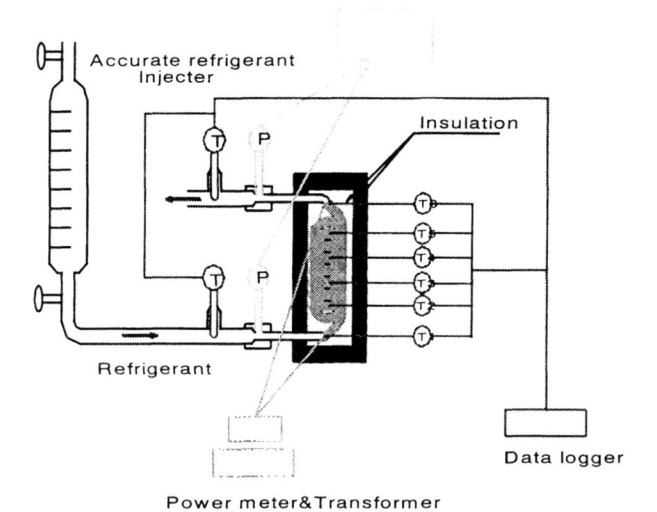

Fig. 3 Schematic diagram of test section

fixed to the test section were connected to the terminal of a data logger. All data was measured using a PC through RS-232C communication.

The copper wire was wound around the surface of the helically coiled tube. The test section was heated as a constant heat flux. The diameter, length, and resistance of the copper wire was 0.6 mm, 23 m, and 0.1 Ω/m, respectively. The test section was sufficiently insulated by using asbestos and styrofoam. It was confirmed that the heat balance of the test section was within 5%.

HCFC-22 was used as a working fluid and experimental conditions were as follows: evaporating temperature was 5°C, evaporation pressure was 0.588 MPa, mass velocity was 150~500 kg/m^{2}s and heat flux was 1~5 kW/m^2. All data were measured 10 times after it was maintained at a steady state and their average was used to data reduction.

EXPERIMENTAL RESULTS AND DISCUSSION

Heat Transfer of Single Liquid Phase Flow

Water was used as the working fluid to understand the heat transfer characteristics of forced convection for the liquid single-phase flow to the helically coiled tube. The heat transfer coefficient was calculated by Eq. (1). where, q'' is the average heat flux and $T_w - T_b$ is the superheat.

$$h_{fo} = \frac{q''}{T_w - T_b} \quad (1)$$

The non-dimensional liquid single-phase heat transfer coefficient for fully developed turbulent flow is represented by the Dittus-Boelter Eq. (2).

$$Nu = \frac{h_{fo} d_i}{k_f} = 0.023 \left[\frac{Gd_i}{\mu_f}\right]^{0.8} \left[\frac{c_{p_f} \mu_f}{k_f}\right]^{0.4} \quad (2)$$

But, for a helically coiled tube, the influence of the curvature ratio (δ) that is, the ratio of the inner diameter to the curvature diameter, has to be considered. Schmidt proposed the laminar flow heat transfer correlation to fully developed regions of a helically coiled tube as follows (Smith, 1997):

$$Nu_{fd} = \{3.65 + 0.08(1 + 0.8\delta^{-0.9})\} Pr^{1/3} Re^m \left(\frac{Pr}{Pr_w}\right)^{0.14}$$
$$m = 0.5 + 0.2903\delta^{-0.194}, \delta = \frac{D}{d_i} \quad (3)$$

The heat transfer coefficient was measured for single-phase flow using three kinds of test sections of A, B, and C represented in Table 1, which is shown in Fig. 4. According to the Dittus-Boelter equation, in the same conditions, the heat transfer coefficient for a helically coiled tube increases as the inner diameter decreases. It has been reported that heat transfer coefficient in a helically coiled tube has lower than the predicted value by the Dittus-Boelter equation(Wang and Peng, 1994). The following heat transfer correlation was proposed for a helically coiled tube at the laminar flow region based on experimental data.

$$Nu_{fd} = \{0.9 + 0.01(1 + 0.8\delta^{-0.9})\} Pr^{1/3} Re^m \left(\frac{Pr}{Pr_w}\right)^{0.14}$$
$$m = 0.5 + 0.2903\delta^{-0.194}, \delta = \frac{D}{d_i} \quad (4)$$

The heat transfer coefficient of a liquid single-phase flow measured in the helically coiled tube and the value proposed by Eq. (4) is compared in Fig. 5. It is well known that the predicted value agreed with experimental values within 15%.

Pressure Drop of Single Liquid Phase Flow

Before the calculation of the pressure drop for a two-phase flow, the pressure drop for a single-phase flow was measured in a straight tube and it was confirmed that the friction factors of laminar and turbulent flow agree well with

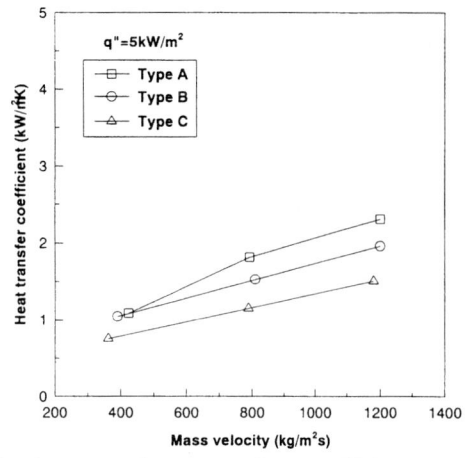

Fig. 4 Average heat transfer coefficient vs. mass velocity for water

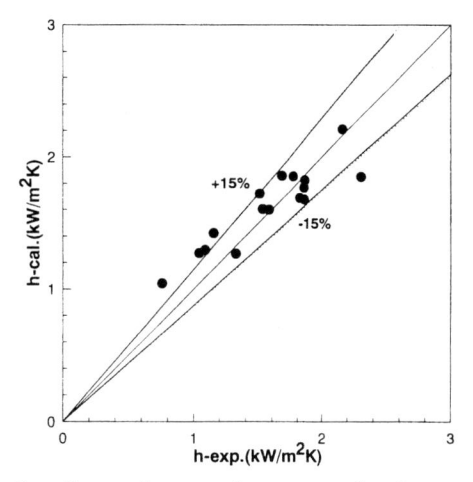

Fig. 5 Comparison of measured heat transfer coefficient and calculated data by Eq.(4)

those of Eqs. (6) and (7), respectively. Then, the pressure drop of the liquid single-phase flow was investigated with a variation on the curvature diameter D and pitch p of the helically coil tube. The friction factor $f_{straight}$ for straight tube was obtained by Eq. (5) after measuring the pressure drop.

$$f_{straight} = \frac{\Delta P}{4(L/d_i)\left(\frac{\rho u_m^2}{2}\right)} \quad (5)$$

Generally, the friction factors for laminar and turbulent flow are represented by the function of Re as shown in the Eqs. (6) and (7).

1) Laminar flow
   For $Re \leq 2300$

   $$f_{straight} = \frac{64}{Re} \quad (6)$$

2) Turbulent flow
   For $4{,}000 \leq Re \leq 10^5$

   $$f_{straight} = 0.3164 Re^{-0.25} \quad (7)$$

The friction factor of the helically coiled tube was proposed by a modification of the correlation proposed by Prandtl, White and Ito for laminar and turbulent flows as the follows(Louis, 1983).

1) Laminar flow
i) Prandtl

for $\frac{10^{1.6}}{2} < De < 500$

$$f_{helical} = f_{straight}\, 0.37(De)^{0.36} \quad (8)$$

2) Turbulent flow
i) White

$$f_{helical} = f_{straight}\left[1 + 0.075 Re^{1/4}\left(\frac{d_i}{D}\right)^{1/2}\right] \quad (9)$$

ii) Ito

for $Re\left(\frac{d_i}{D}\right)^2 > 6$

$$f_{helical} = f_{straight}\left[Re\left(\frac{d_i}{D}\right)^2\right]^{1/20} \quad (10)$$

where, Dean number De, the ratio of centrifugal force to inertial force, can be expressed as in Eq. (11).

$$De = Re\sqrt{\frac{d_i}{D}} \quad (11)$$

The friction factor, when the pitch was changed for the helically coiled tube of the same curvature diameter of 23.35 mm in the laminar flow region, is represented in Fig. 6. The friction factor for liquid single-phase flow agreed well with that the correlation of Prandtl. The friction factor decreased a little as the pitch increased. But, it is known that the friction factor in the helically coiled tube is not largely influenced by the variation of pitch.

The friction factor of the helically coiled tube as a function of curvature diameter is represented in Fig. 7. The measured value and predicted value by Prandtl's correlation were compared to each curvature diameter. From the view of Prandtl's correlation, the pressure drop in a helically coiled tube is a function of its curvature diameter. It was found that the pressure drop was inversed proportional to the curvature diameter at the same tube inner diameter. In experimental results, it has been also found that the pressure drop decreases as the curvature diameter increases and,

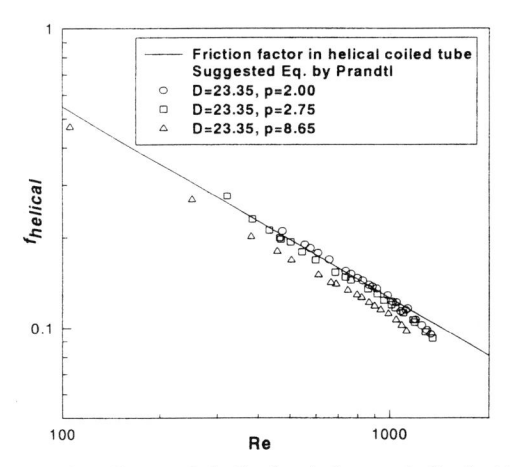

Fig. 6 The effect of helical pitch on helical friction factor.

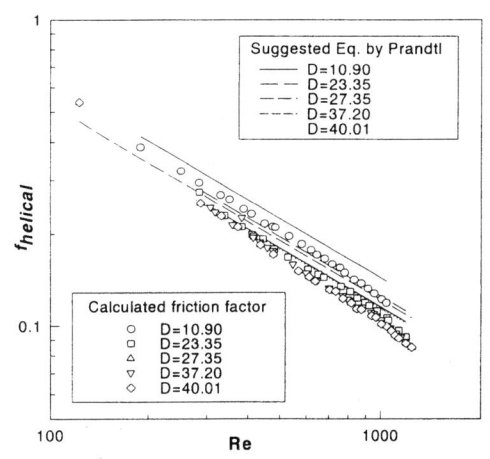

Fig. 7 The effect of helical curvature diameter on helical friction factor.

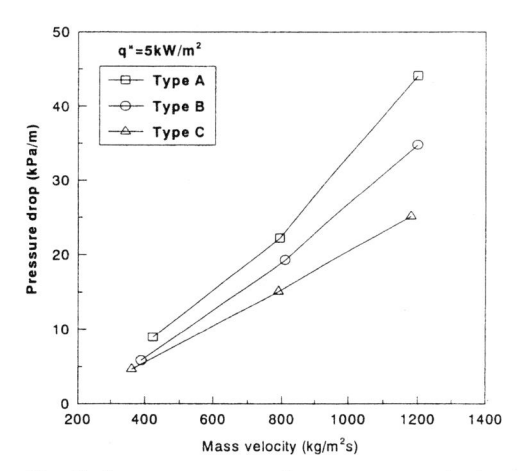

Fig. 8 Average pressure drop vs. mass velocity

generally, agrees well with the value of the correlation at the lower Reynolds number.

In Fig. 8, the pressure drop per unit length is represented as a function of mass velocity at the same heat flux condition. As the curvature diameter increased, pressure drop decreased. As the mass velocity increased, it was found that the influence of curvature diameter on pressure drop was gradually enlarged.

Evaporation Heat Transfer Characteristics

Experiments were conducted for evaporation heat transfer using A and B types of tubes as represented in Fig. 4 in which the heat transfer characteristics of liquid single-phase were good. For the forced convective boiling, nucleate boiling and two-phase forced convection are considered as two main heat transfer mechanisms.

Not like a straight tube, in a helically coiled tube, there is a large difference in the heat transfer coefficient at top, bottom, and the inside and outside of circumferential direction because secondary flow occurred by centrifugal force. Because a capillary tube of 1 mm inner diameter was used in this study, there was no large difference in heat transfer coefficient to the circumferential direction of the tube. Thus, average heat transfer coefficient at the wall was measured.

In Figs. 9 and 10, the local heat transfer coefficients at each part of the test section were represented according to heat flux. In the type A tube in which curvature diameter was smaller than that of the type B, the local heat transfer coefficient was higher and as the mass velocity and heat flux increased, local heat transfer coefficient increased.

As represented in Fig. 10(a), at the low heat flux condition, local heat transfer did not increases largely as the mass velocity increased. For high heat flux conditions as shown in Fig. 10(b) and (c), local heat transfer coefficient increased largely as the mass velocity increased. This was due to the following reasons: nucleate boiling occurred in the low quality region and, as heat flux increased, the mixing of liquid and vapor actively occurred by secondary flow in helically coiled tube. Annular flow occurred at the low quality region, as well.

From Figs. 9 and 10, it was ascertained that dry-out occurred in the lower quality region of the helically coiled tube compared with the straight tube. Vapor refrigerant of relatively rapid speed compared with liquid film of annular flow collided at the tube wall and the liquid film of annular flow collapsed. And then, the flow pattern changed to a misty flow. Therefore, dry-out occurred in the low quality region compared with the straight tube with large diameter.

Pressure Drop Characteristics

The pressure drop of the two-phase flow liquid and vapor in the heat transfer tube is a very important consideration when designing heat exchangers for refrigeration and air conditioning systems. Especially, centrifugal force caused by the curvature of the tube and simultaneously the buoyancy by the heating of the tube wall

(a) q" = 1.5 kW/m²   (b) q" = 2.1 kW/m²   (c) q" = 3.0 kW/m²

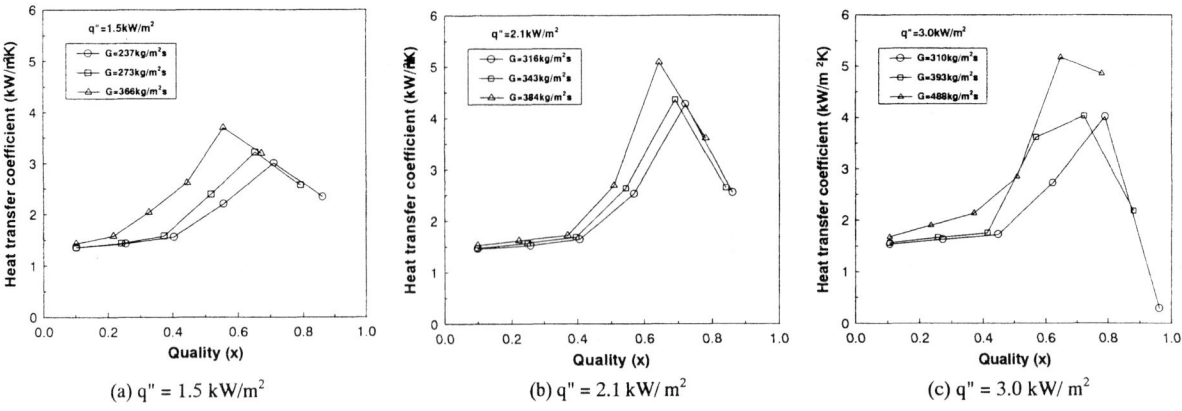

Fig. 9 Profile of local heat transfer coefficient according to heat flux and mass velocity for Type A

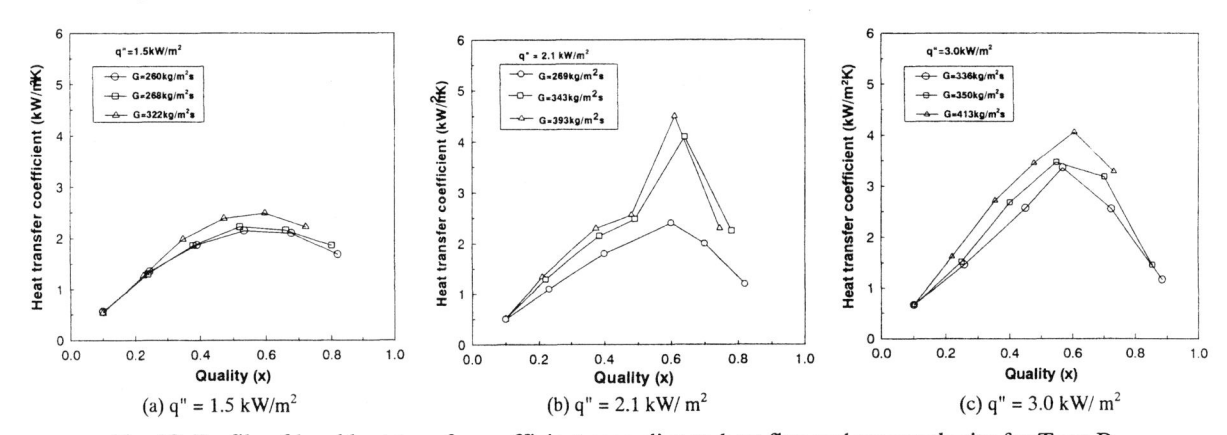

(a) q" = 1.5 kW/m²   (b) q" = 2.1 kW/m²   (c) q" = 3.0 kW/m²

Fig. 10 Profile of local heat transfer coefficient according to heat flux and mass velocity for Type B

acts in fluids flowing through a helically coiled tube. Generally, because these body forces act to another direction, the main velocity profile in viscous flow is represented by a shape different from that in a straight tube. Flow pattern is more complicated than the case without buoyancy(Sillekens, et al., 1997; Yang, et al., 1995). A secondary flow occurs when these two kinds of body forces are present and the pressure drop would be larger than that of a horizontal straight tube's secondary flow.

In Fig. 11(a) and (b), the pressure loss per unit length was represented as functions of mass velocity and heat flux. As the mass velocity and the heat flux increased, the pressure drop also increased. Generally, the pressure drop was largely influenced by mass velocity of the internal flow. But, the effects of heat flux on pressure drop were small compared with the influence of heat flux. Because the mixing of liquid and vapor was activated by secondary flow in the helically coiled tubes and annular flow occurred at the low quality region, heat transfer was enhanced and pressure drop increased as the heat flux increased at the same flow rate. So, the pressure drop in a helically coiled capillary tube was largely influenced by heat flux. To develop the correlation of the friction factor of a two-phase flow in a helically coiled tube, Blasius's correlation, which agreed well with the experimental results of the friction factor in a two-phase flow, was used.

$$f_{TP} = 0.079[Gd_i / \mu_{TP}]^{-1/4} \quad (12)$$

The viscosity equation proposed by Dukler was used to calculate the viscosity of a two-phase flow in Eq. (12)(John and John, 1994).

$$\mu_{TP} = \rho_{TP}[xv_g\mu_g + (1-x)v_l\mu_l] \quad (13)$$

Therefore, if the friction factor(f_{TP}) for a two-phase flow(Eq. (12)) were substituted to the friction factor($f_{straight}$) for straight tube(Eqs. (9) and (10)), the correlation of the friction factor for a two-phase flow in a helically coiled tube can be represented by Eqs. (14) and (15).

$$f_{helical.TP} = f_{TP}\left[1 + 0.075\,\text{Re}^{1/4}\left(\frac{d_i}{D}\right)^{1/2}\right] \quad (14)$$

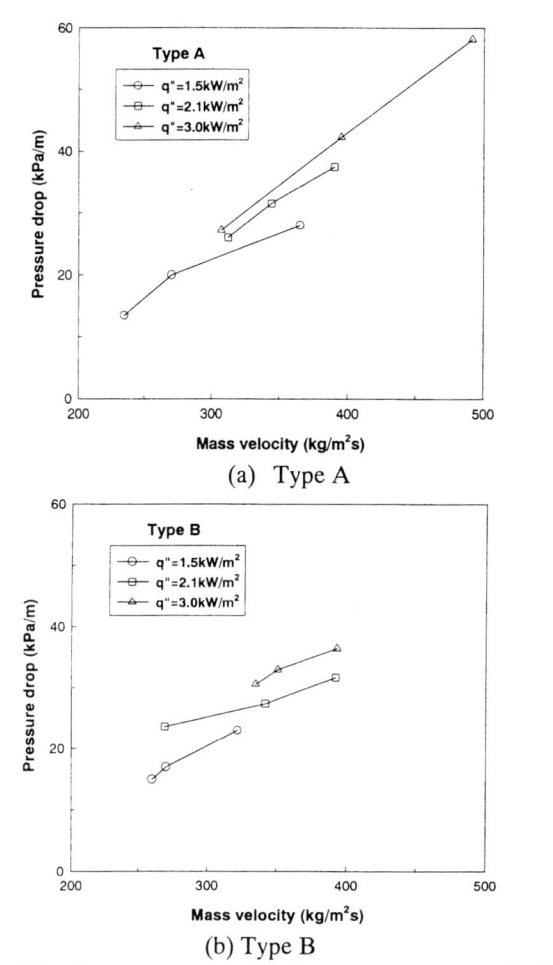

Fig. 11 Average pressure drop vs. mass velocity for two-phase flow

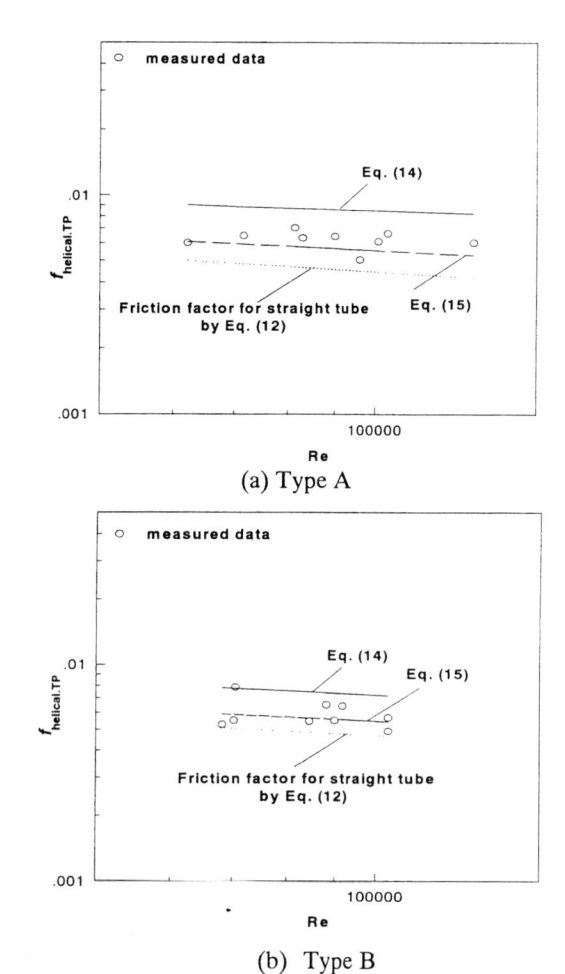

Fig. 12 Comparison of helical friction factor between friction correlations and measured data

$$f_{helical.TP} = f_{TP}\left[\text{Re}\left(\frac{d_i}{D}\right)^2\right]^{1/20} \quad (15)$$

$$\text{Re}\left(\frac{d_i}{D}\right)^2 > 6$$

The experimental results of the friction factor for a two-phase flow, in a straight tube and in a helically coiled tube were compared each other and represented in Fig. 12 (a) and (b). It was found that the friction factor of the helically coiled tube agreed well with that of Eq. (15) which was modified from Ito's correlation.

CONCLUSION

The following conclusions are made through experiments for evaporation heat transfer and pressure drop in single-phase and two-phase flow in helically coiled capillary tubes to provide basic data for making compact heat exchangers.

1. A heat transfer correlation was proposed to predict heat transfer coefficient of a liquid single-phase flow considering the curvature ratio.
2. The liquid film of annular flow collapsed by vapor refrigerant of rapid speed and the flow pattern rapidly changed from annular flow to misty flow. So, dry-out in the helically coiled capillary tube occurred in the lower quality region than that of the straight tube.
3. The friction factor of liquid single-phase flow agreed well with Prandtl's correlation using the Dean number. The influence of the curvature diameter was larger than the variation of pitch on friction factor.
4. A correlation was proposed to predict the friction factor of a two-phase flow in helically coiled tubes exactly.

NOMENCLATURE

c_p Isobaric specific heat, J/kg K
D Curvature diameter, m
d Diameter, m
De Dean number

f	Friction factor
G	Mass velocity, kg/m²s
h	Heat transfer coefficient, W/m²K
k	Thermal conductivity, W/m K
L	The length of heat transfer tube, m
Nu	Nusselt number
P	Pressure, Pa
p	Pitch, m
Pr	Prandtl number
q''	Heat flux, W/m²
Re	Reynolds number
T	Temperature, K
u	velocity, m/s
v	Specific volume, m³/kg
x	Quality
δ	Curvature ratio
μ	Viscosity, kg/s m
ρ	Density, kg/m³
ϕ	Diameter, m

Subscripts

b	Bulk
f	Fluid
fd	Fully developed
fo	Liquid single phase
g	Vapor
helical	Helically coiled tube
i	Inner
l	Liquid
m	Average
straight	Straight tube
TP	Two phase flow
w	Wall

ACKNOWLEDGEMENT

This work was supported by Korea Energy Management Corporation under contract number 1997-E-ID01-P-53.

REFERENCES

Feng, Z. P., Guo, L. J., and Chen, X. J., 1997, Forced Convection Boiling Heat Transfer in Helically Coiled Tubes with Various Helix Axial Angles, *Experimental Heat Transfer, Fluid Mechanics and Thermodynamics*, pp. 593-599.

Han, K. I., and Park, J. U., 1998, A study on the heat transfer phenomena in coiled tubes with variable curvature ratios, *KSME(B)*, Vol. 22, No. 11, pp. 1509-1520.

Ishigaki, H., 1996, Fundamental characteristics of mixed convective laminar flow in heated curved pipes, *Trans. JSME(Series B)*, Vol. 62, No. 600, pp. 2965-2971.

John, G. C., and John, R. T., 1994, *Convective Boiling and Condensation*, Oxford University Press, New York, pp. 41-80.

Kakac, C., Bergles, A. E., and Mayinger, F., 1981, *Heat Exchangers : Thermal-Hydraulic Fundamentals and Design*, McGraw-Hill Book Company, New York, pp. 12-114.

Kazi, M., Mori, K., Oishi, M., Nakanishi, S., and Sawai, T., 1997, Boiling Heat Transfer and Dry-Out Characteristics in Helically Coiled Tubes," *Experimental Heat Transfer, Fluid Mechanics and Thermodynamics*, pp. 649-656.

Kim, J. S., and Katsuta, M., 1995, Development of high performance heat exchanger for CFC alternative refrigerants(1st report : Condensing heat transfer and pressure drop of HFC-134a in multi-pass tubes), *J. Refrigeration Engineering & Air Conditioning*, Vol. 14, No. 5, pp. 273-283.

Kureta, M., Kobayashi, T., Mishima, K., and Nishihara, H., 1997, Pressure drop and heat transfer for flow-boiling of water in small-diameter tubes, *Philos. Mag.*, Vol. 63, No. 615, pp. 216-224.

Louis, C. B., 1983, *Convective Heat Transfer*, John Wiley & Sons, Toronto, pp. 497-499.

Moriyama, K., Inoue, A., and Ohira, H., 1992, The thermohydraulic characteristics of two-phase flow in extremely narrow channels(The fictional pressure drop and heat transfer of boiling two-phase flow, analytical model), *Trans. JSME(Series B)*, Vol. 58, No. 546, pp. 393-400.

Sillekens, J. J. M., Rindt, C. C. M., and Van Steenhoven, A. A., 1997, Developing Mixed Convection in a Coiled Heat Exchanger, *Int. J. Heat Mass Transfer*, Vol. 41, No. 1, pp. 61-72.

Smith, E. M., 1997, *Thermal Design of Heat Exchanger*, John Wiley & Sons, Toronto, pp. 120-153.

Taylor, G. I., 1929, The Criterion of Turbulence in Curved Pipes, *Proc. Roy. Soc. London, A124*, pp. 234-249.

Wambsganss, M. W., Jendrzejczyk, J. A., and France, D. M., 1991, Two-Phase Flow Patterns and Transitions in a Small, Horizontal, Rectangular Channel, *Int. J. Multiphase Flow*, Vol. 17, No. 3, pp. 327-342.

Wang, B. X., and Peng, X. F., 1994, Experimental Investigation on Liquid Forced Convection Heat Transfer through Microchannels, *Int. J. Heat & Mass Transfer*, Vol. 37, Suppl. 1, pp. 73-82.

Yang, G., Dong, Z. F., and Ebadian, M. A., 1995, Laminar Forced Convection in a Helical Coil Pipe with Finite Pitch, *Int. J. Heat & Mass Transfer*, Vol. 38, No. 5, pp. 853-862.

CONDENSATION OF PURE HYDROCARBONS AND THEIR MIXTURE IN A COMPACT WELDED HEAT EXCHANGER

J. Gitteau, B. Thonon and A. Bontemps*

CEA-Grenoble, GRETh, 17 rue des martyrs, 38054 Grenoble, France, e-mail: thonon@cea.fr
* Université J. Fourier, LEGI/GRETh, 17 rue des martyrs, 38054 Grenoble, France, e-mail: andre.bontemps@cea.fr

Abstract

This paper presents a study on heat transfer in condensation of pure and mixtures of hydrocarbons in a compact welded plate heat exchanger. Three pure fluids (pentane, butane and propane) and two mixtures (butane + propane) have been used. The operating pressure ranges from 1.5 to 18 bar.

For pure fluids, two heat transfer mechanisms have been identified. For low Reynolds numbers, condensation occurs almost filmwise and the heat transfer coefficient decreases with increasing Reynolds numbers. For higher values of the Reynolds number, the heat transfer coefficient increases gently. The transition, between the two regimes, is between $Re = 100$ and 1000 and depends on the operating conditions.

For mixtures, the behaviour is different. For low Reynolds numbers mass transfer affects heat transfer and reduces the heat transfer coefficient by a factor of up to 4.

Correlations for filmwise and intube condensation do not predict accurately the results and a specific correlation is proposed for pure fluid condensation. For mixtures, the condensation curve method does not allow mass transfer effects to be taken into account and more work is required to establish an accurate predictive model.

INTRODUCTION

Condensation occurs in many industrial processes, but rarely with pure fluids. The fluids encountered are mixtures and often non-condensable gases are present, and this makes the condensation process very complex. In the case of mixtures or in the presence of a non condensable gas, the vapour condensing must diffuse through the gas to the interface. This requires a partial pressure gradient towards the interface.

The partial pressure of the vapour falls from a constant value at a rather large distance from the phase interface to a lower value at the interface. Correspondingly the accompanying saturation temperature also falls towards the interface. Therefore during condensation the condensing vapour arrives by diffusion at the condensate surface and it is the thermal resistance in the vapour which limits the process. Hence in order to improve the heat transfer, one must reduce the thermal resistance on the vapour side.

Several factors can enhance the condensation process by reducing the vapour side resistance. During condensation of mixtures or of vapours that contain non-condensables the heat transfer on the vapour side can be improved by raising the vapour velocity. It has been shown that the heat transfer coefficient can be improved by approximately 30% by increasing the vapour velocity. The use of finely undulated surface can also achieve significant augmentations in heat transfer during condensation. It has been shown that corrugation can promote turbulent equilibrium between the phases and thus contribute to the increase in heat transfer.

Compact heat exchangers are characterised by small hydraulic diameters (1-10 mm) and there is no reliable design method to estimate heat transfer coefficients during condensation in such small passages. In the open literature, condensation of mixtures and of vapour in presence of non-condensables have been studied, but essentially for conventional geometries (plain tubes), and only few results have been published with fluids representative of actual process conditions (hydrocarbons).

HEAT TRANSFER IN CONDENSATION

Gravity controlled regime

For the gravity controlled regime, the Nusselt theory for a plain wall allows the local heat transfer coefficient to be calculated. The average heat transfer coefficient is given by:

$$\overline{\alpha}_g = Co\, 1.47\, Re_L^{-1/3} \qquad (1)$$

where Co is a physical property number:

$$Co = \lambda_L \left(\frac{\mu_L^2}{\rho_L (\rho_L - \rho_g) g} \right)^{-1/3} \qquad (2)$$

The film Reynolds number is calculated by:

$$Re_L = \frac{4\,\dot{m}_L\,Dh}{\mu_L\,P} \qquad (3)$$

whre P denotes the wetted perimeter (for flat channels twice the plate width).

The Nusselt law is valid only if the liquid film is smooth, but for film Reynolds number over 30, some waves appear at the interface and increase the heat transfer coefficient. Kutatelatze have proposed a modified Nusselt correlation. The average heat transfer coefficient is:

$$\overline{\alpha}_{gr} = Co\,\frac{Re_L}{1.08\,Re_L^{1.22} - 5.2} \qquad (4)$$

For Reynolds numbers above 1600, the liquid film becomes turbulent and the Labuntsov correlation can be applied. The average value is given by:

$$\overline{\alpha}_{gr} = Co\,\frac{Re_L}{8570 + 58\,Pr_L^{-0.5}\left(Re_L^{0.75} - 253\right)} \qquad (5)$$

Improvements of these basic correlations have been proposed to take into account subcooling of the condensate or the vapour shear effect (Rohsenow [1985]).

Forced convection regime

In the shear controlled regime, several theoretical or semi-empirical correlations exist for predicting the heat transfer coefficient. The most common ones are the Akers correlation modified by Cavillini and Zecchin [1993], the Boyko-Kruzhilin [1967] and the Shah [1979] correlations.

- Akers or Cavallini-Zecchin:

$$\alpha_{cv} = a\,Re_{eq}^{b}\,Pr_L^{0.33}\,\frac{\lambda_L}{D_h} \qquad (6)$$

where a and b are two constants determined by correlating with experimental data. Re_{eq} is an equivalent Reynolds number for the liquid + vapour mixture:

$$Re_{eq} = \frac{\dot{m}\left((1-x) + x\left(\frac{\rho_L}{\rho_G}\right)^{0.5}\right)D_h}{\mu_L} \qquad (7)$$

- Boyko and Kruzhilin or Shah:

$$\alpha_{cv} = \alpha_{LO}\,F \qquad (8)$$

where the heat transfer coefficient of the liquid phase (α_{LO}) is deduced from specific correlations for forced convection and F is the enhancement factor.

CONDENSATION IN PLATE HEAT EXCHANGERS

Steam condensation

Tovazhnyanski and Kapustenko [1984] have performed some tests on a plate heat exchanger with a corrugation angle of 60°. Their tests covered a large range of Reynolds numbers. For Reynolds numbers above 300, a shear controlled regime dominates and the heat transfer coefficients are compared to a modified Boyko and Kruzhilin correlation whose enhancement factor is:

$$F = \left(1 + x\left(\frac{\rho_L}{\rho_G} - 1\right)\right)^{0.5} \qquad (9)$$

The use of this correlation means that the enhancement of the heat transfer coefficients in condensation is comparable to the enhancement in single phase flows. For example, for a corrugation angle of 60° the heat transfer coefficient is 6 times larger than for a comparable plain channel. For lower Reynolds numbers ($Re_L < 300$), condensation seems to be gravity controlled. However, the Nusselt law underestimates the heat transfer coefficient. The herringbone pattern of the corrugation tends to drain the condensate to the periphery of the channel and keeps the condensate flim thin.

Luo Di-an et al. [1988] have studied steam condensation in two different geometries: a plain rectangular channel and a corrugated channel (corrugation angle of 45°). For Reynolds numbers between 70 and 500, the Nusselt theory predicts the heat transfer coefficients with a good accuracy. For the corrugated channels, their results clearly outline the two regimes of condensation, for Reynolds numbers below 350 a gravity controlled regime exists, and for Reynolds numbers above 350 a shear controlled regime dominates. The change of regime at a Reynolds number of 350 is explained by the change in the flow pattern. In corrugated channels (Thonon et al. [1995]), the transition from laminar to turbulent regimes occurs at lower Reynolds numbers than for a plain tube (about Re = 400). The intensification compared to the plain channel is 60% at a Reynolds number of 250 and of 220% for a Reynolds number of 900.

Zhong-zeng Wang [1993] studied steam condensation on a plate heat exchanger with a corrugation angle of 45°. Their results clearly indicate that the heat transfer coefficient is shear controlled: the heat transfer coefficient increases with the liquid film Reynolds number. The intensification of the heat transfer coefficient compared to the Nusselt theory varies from a factor 2 to 3 depending on the flow conditions.

Wang et al. [1999a and 1999b] have studied steam condensation in one plate and frame and two brazed heat exchangers. The effect of mass flux and pressure has been studied. The results have been compared to the Shah and Boyko-Kruzhilin correlations and large discrepancy has been observed. The density exponent of the Boyko-Kruzhilin correlation has been reduced to 0.45 and allows a better accuracy.

Condensation of refrigerants

Compact heat exchangers are often used in refrigerant systems. A review has been recently proposed by Palm and Thonon [1999].

For ammonia applications, Panchal [1993] has carried out two series of tests with high and low corrugation angle (60° and 30°). The results indicate that the plates with low corrugation angle give higher heat transfer coefficients than the plates with a corrugation angle of 60°. This is probably caused by an increase of the liquid hold-up in high corrugation angle plates.

Kumar [1992] has reported some results for condensation of R22 and ammonia in several types of plate heat exchangers. The results clearly indicate that the condensation is gravity controlled for low Reynolds numbers, and shear controlled for higher Reynolds number. The measured heat transfer coefficients are 1.5 to 4 times larger than those expected in a comparable plain tube.

Chopard [1992] have performed some tests with R22 at high Reynolds numbers in welded plate heat exchangers. Three different geometries were tested (plain rectangular, in-line or staggered studs). The results are presented in term of an intensification factor of the single-phase heat transfer coefficient. The heat transfer coefficients of in-line or staggered studs are significantly higher than those of the plain channel. Furthermore, even for the plain channel, the Shah correlation underestimates the heat transfer coefficients. Local measurements of the void fraction have shown that the liquid is drained to the periphery and that a thin film of liquid remains in the central part of the channel.

Arman and Rabas [1995] made a serious effort to develop a general computer program for heat transfer and pressure drop of single component and binary mixtures during condensation in plate heat exchangers. Starting with a review of the sparse literature in the area they conclude that the condensation heat transfer is shear-controlled. The computation scheme for heat transfer was based on a correlation by Tovazhnyanskiy and Kapustenko [1984] and for pressure drop on the Lockhart-Martinelli method. The condensation channel is divided into a number of steps and the local heat transfer coefficient and pressure drop of each step is calculated iteratively. A number of constants in the correlations were determined by comparison with experimental data from the literature. Results from new experiments with pure saturated ammonia were then compared to the predictions of the computer program and the deviations were shown to be small for the case of single component fluids.

Recently Yan et al. [1999] have presented results on condensation of R134a in a compact brazed heat exchanger having a 60° corrugation angle. The effect of mass flux, average heat flux, local vapour quality and operating pressure has been studied. The results indicate that the heat transfer coefficient increases with mass flux indicating forced convection heat transfer. To correlate the results a modified Akers correlation is proposed :

$$Nu = a\, Re_{eq}^{b}\, Pr_L^{1/3} \qquad (10)$$

As outlined by Srinivasan and Shah [1997] the heat transfer performances are closely linked to the flow pattern, Most of the conventional intube correlations fail into predicting the heat transfer coefficient in compact heat exchangers as the flow patterns are radically different from intube flows. Furthermore for very small channels, the surface tension has to be taken into account.

EXPERIMENTAL APPARATUS

General description

The test rig BUPRO is devoted to hydrocarbon applications. Fluids such as butane, propane or mixtures can be used. The heat exchangers tested can be either evaporators or condensers. The pressure can be varied up to 20 bars with temperature up to 80°C. The hydrocarbon flow rate is up to 600 kg/h with a maximal heat duty of 70 kW.

The test rig has three independent circuits (figure 1):
- the hot water circuit (100)
- the cold water circuit (200)
- the hydrocarbon circuit (300)

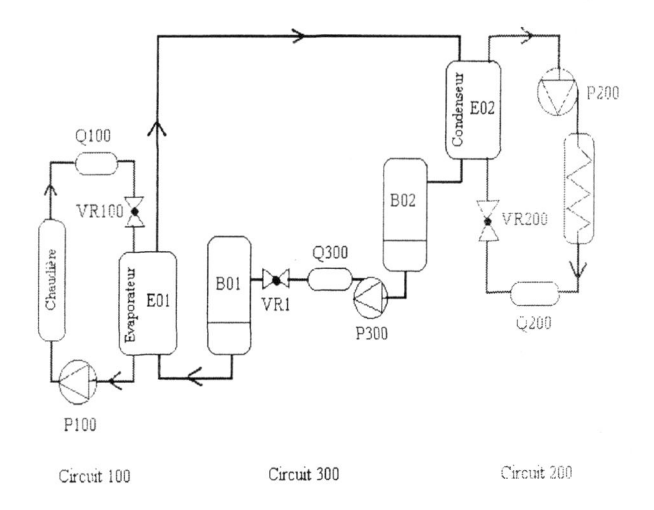

Fig. 1: The BUPRO test rig

The condenser is installed after the evaporator (figure 1) and is fed by saturated vapour. The saturation conditions are checked by measurement of the inlet pressure and temperature and by comparison with the saturation curve of the fluid ($T=f(p_{sat})$). At the heat exchanger outlet, the liquid flows directly to a storage tank. Consequently, there is no sub-cooling of the condensate and the outlet condition is saturated liquid.

Heat exchanger geometry

The plate arrangement of the tested heat exchanger is given on figure 2. The main characteristics are given below.

- Plate width   0.3 m
- Plate length   0.3 m
- Corrugation angle   45°
- Plate spacing   0.005 m
- Hydraulic diameter) 0.01 m

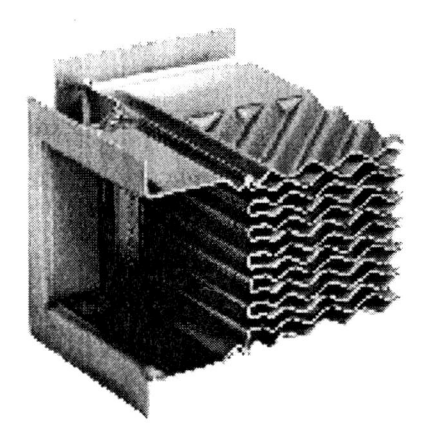

Fig. 2: Detail of the compact heat exchanger

Measurements

The temperatures and pressures are measured on both the hydrocarbon and the water side:

$T_{in\ 300}$:	inlet hydrocarbon temperature	[°C]
$T_{out\ 300}$:	outlet hydrocarbon temperature	[°C]
$P_{in\ 300}$:	inlet hydrocarbon pressure	[bar]
ΔP_{300}:	hydrocarbon pressure drop	[mbar]
$\dot{M}_{300}$:	hydrocarbon mass flow rate	[kg.s^{-1}]
$T_{in\ 200}$:	inlet water temperature	[°C]
$T_{out\ 200}$:	outlet water temperature	[°C]
$\dot{V}_{200}$:	water volumetric flow rate	[m^3.s^{-1}]

The accuracy of the temperature measurements (after calibration) is ± 0.2 °C. The flowmeters have an accuracy of ± 1%. The differential and absolute pressure taps are heated to avoid the presence of liquid in the connecting pipes and the accuracy is ± 1%.

RESULTS

Data reduction

Experimental differences have been observed between the heat duties calculated on the water and hydrocarbon sides. These differences are mainly due to uncertainties and fluctuations on the hydrocarbon mass flow rate. Therefore, the actual heat duty is assumed to be the one measured on the water side.

$$\dot{Q} = \dot{Q}_{water} = \dot{V}_{200}\ \rho_{200}\ Cp_{200}\ (T_{out,200} - T_{in,200}) \quad (11)$$

Then, from this value and assuming an exact heat balance, the hydrocarbon mass flow rate is recalculated from:

$$\dot{M}_{300} = \frac{\dot{Q}_{water}}{(h_{in} - h_{out})_{300}} \quad (12)$$

The heat transfer coefficient on the condensation side is deduced from the overall heat transfer coefficient U for which the heat duty is estimated from the water side.

$$U = \frac{\dot{Q}}{\Delta T_{ln} \cdot A} \quad (13)$$

The average heat transfer coefficient on the condensation side is then deduced from:

$$\frac{1}{U} = \frac{1}{\alpha_{exp}} + R_w + \frac{1}{\alpha_{CW}} \quad (14)$$

where α_{exp} and α_{CW} are the heat transfer coefficients on the hydrocarbon and water sides. The latter is calculated with the correlations presented by Thonon et al. [1995].

$$Nu = 0.347 \cdot Re^{0.653} \cdot Pr^{0.33} \quad (15)$$

The heat transfer coefficient is based on the total projected surface area (not including the surface extension).

The results, concerning the heat transfer coefficients are presented in function of the liquid Reynolds number Re_{LO} defined by assuming all the flow is liquid.

$$Re_{LO} = \frac{\dot{m}\ D_h}{\mu_L} \quad (16)$$

where $\dot{m}$ is the mass velocity. As average heat transfer coefficients are detrmined and total condensation occurs, Re_{LO} is the most appropriate Reynolds number.

Tests performed

Five sets of tests have been performed: three with pure fluids and two with mixtures. For each fluid, the absolute pressure and the mass flow rate on the condensation side were varied. The range covered is:
- Pentane (1.5 to 3 bar);
- Butane (4 to 8 bar);
- Propane (10 to 18 bar);
- 49% Butane-51% Propane in mass (8 to 14 bar);
- 28% Butane-72% Propane in mass (12 to 18 bar).

The pure fluids used are high quality and contain less than 0.1 % of other components. The physical properties have been calculated using the Prophy™ sofware developped by Prosim SA (France). For the two Butane+Propane mixtures, the mass concentration has been estimated by measuring the dew and bubble temperatures at different pressures and by comparing with the temperature predicted by the physical property software.

Pure fluid condensation

Three series of tests have been performed using Pentane, Butane and Propane and the overall thermal performances are presented in figures 3 to 5. The results clearly indicate that for low Reynolds number the heat transfer coefficient in condensation is similar to that of a laminar falling film and for higher Reynolds numbers turbulent effects tend to increase the heat transfer coefficient. The transition between these two mechanisms occurs for Reynolds numbers between 100 and 1000 depending on the pressure. The operating pressure has two effects. First, in the laminar zone, the heat transfer coefficients increase with pressure, which is essentially an effect of the physical properties. Secondly, the higher pressures give higher transition Reynolds numbers.

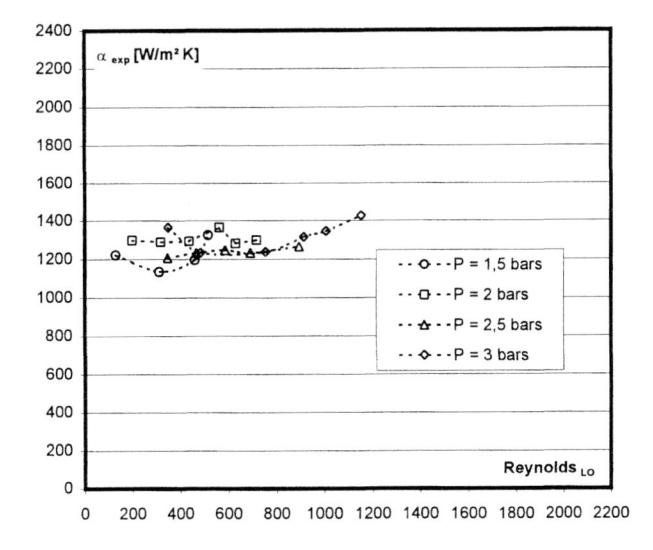

Fig. 3: Condensation of pure Pentane

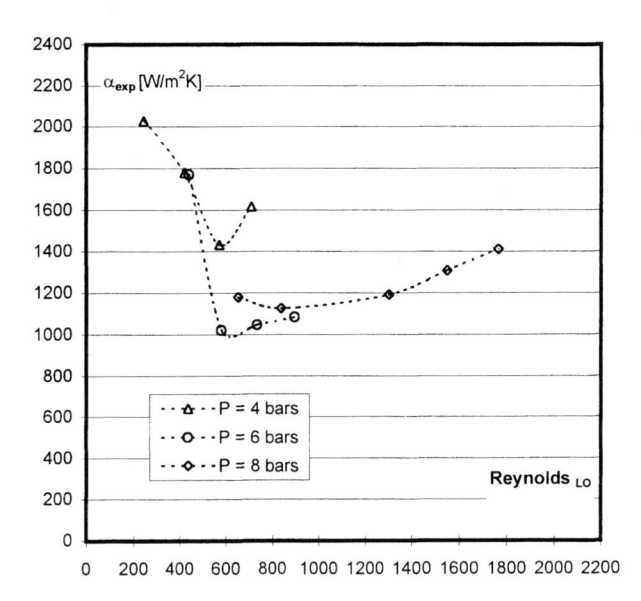

Fig. 4: Condensation of pure Butane

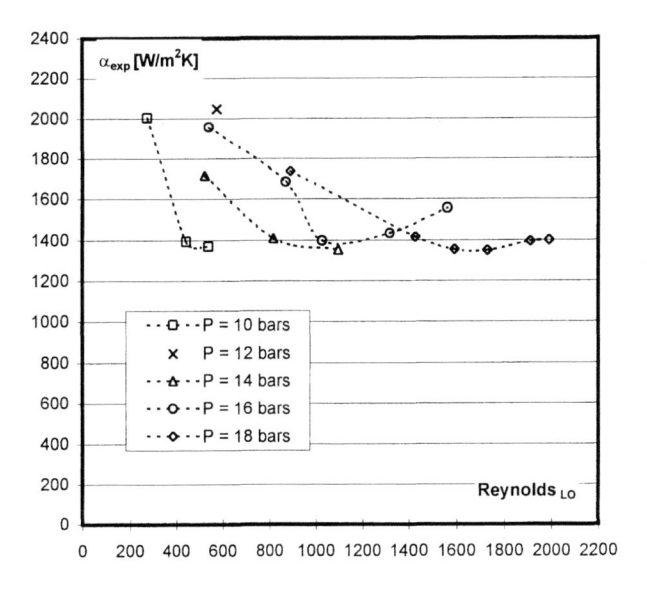

Fig. 5: Condensation of pure Propane

Mixture condensation

For the mixtures (figures 6 and 7), the behaviour is quite different. In the first zone, the heat transfer coefficient increases with the Reynolds number, then in the second zone the behaviour is similar to the one of pure fluids.

These observations outline the effect of mass transfer on the overall heat transfer coefficient. For low Reynolds numbers (laminar film) and for pure fluids the heat transfer resistance is mainly in the liquid film and this resistance increases with the Reynolds number (Nusselt theory). For higher Reynolds numbers, the liquid film becomes turbulent and with increasing Reynolds numbers the heat transfer coefficient increases.

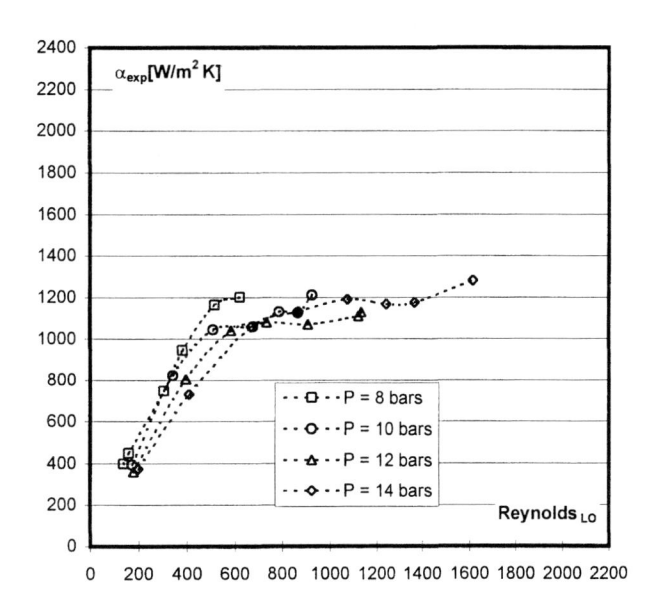

Fig. 6: Condensation of 49%-51% Butane-Propane mixture

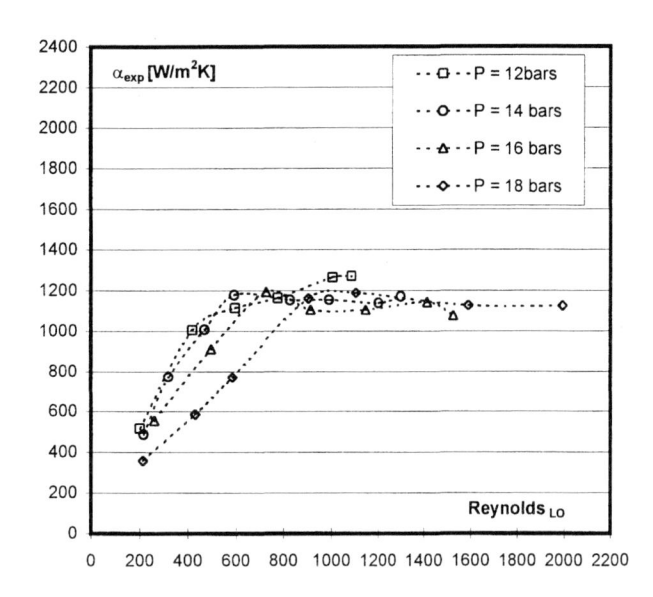

Fig. 7: Condensation of 28%-72% Butane-Propane mixture

For pure fluids there is no mass transfer effect during condensation. On the contrary with a mixture, there is a mass transfer resistance due to molecular diffusion in the gas phase. For low Reynolds numbers, the velocity is low and there is almost no mixing in the gas phase, consequently mass transfer effects will reduce the overall heat transfer coefficient. For higher Reynolds numbers, mixing occurs in the gas phase and enhances mass transfer; consequently the heat transfer resistance is mainly in the liquid film and the mixture behaves as a pure fluid.

DATA ANALYSIS

Pure fluids

The data have been compared to several correlations for filmwise and intube forced convection condensation, but all these correlations fail into predicting the heat transfer performances (figure 8).

In the laminar regime, condensate drainage may occur in the furrow of the corrugation and this will reduce the mean film thickness and consequently increase heat transfer. The vapour velocity at the heat exchanger inlet in relatively low and the shear effect should be negligible.

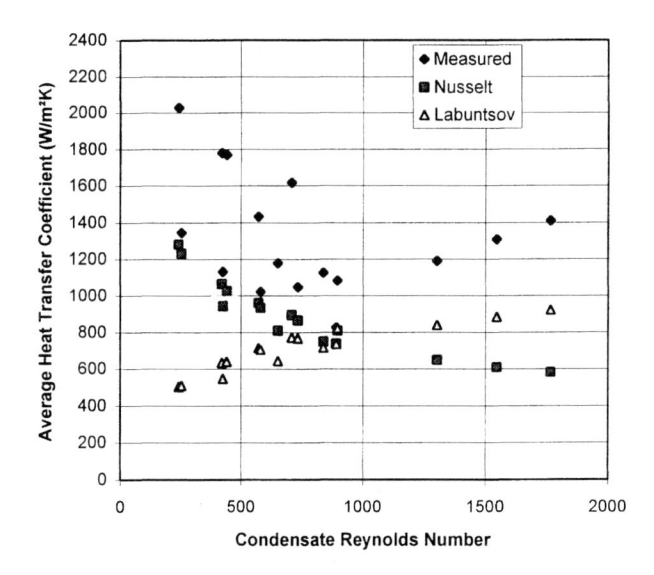

Fig. 8: Comparison of measured heat transfer coefficient using Butane with filmwise condensation models

To correlate the results, two approaches can be adopted. We can consider either falling film or forced convection correlations. For falling film models, the heat transfer coefficients are plotted introducing the physical property number Co. For forced convection, the enhancement factor F is used. These methods have been tested for correlating the results (figures 9 and 10), and using the enhancement factor F (eq. 8) plotted versus the equivalent Reynolds number (eq. 7) gives satisfactory results.

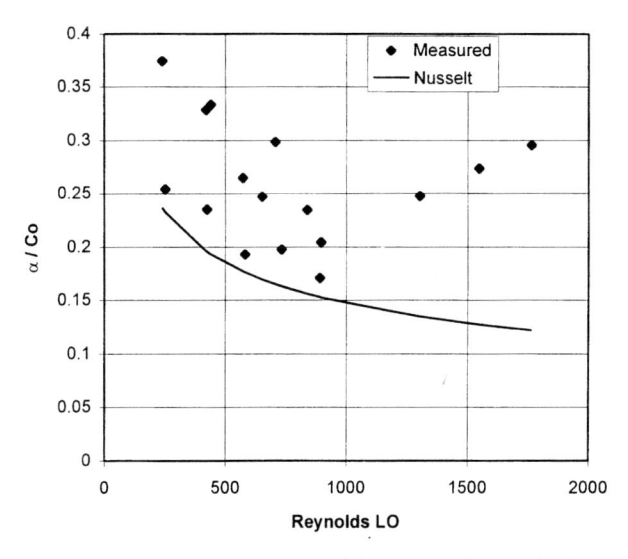

Fig. 9: Comparison of measured heat transfer coefficients using Butane introducing the physical property number Co with the Nusselt theory (eq. 1)

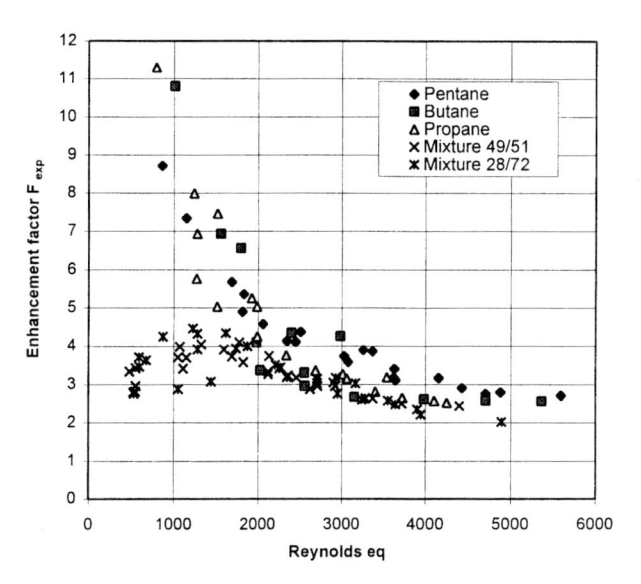

Fig. 11: Enhancement factors for pure fluids and mixtures

CONCLUSION

A small industrial compact condenser has been installed and instrumented on a hydrocarbon test rig at GRETh. Five sets of tests have been performed using hydrocarbons: three with pure fluids and two with mixtures. The pressure was varied between 1.5 and 18 bars for a range of Reynolds numbers between 100 and 2000. These conditions are representative for industrial cases.

For pure fluids (pentane, butane and propane), two regimes have been identified. For low Reynolds numbers, the heat transfer coefficient in condensation decreases, indicating a laminar regime. For higher Reynolds numbers, the heat transfer coefficient increases gently, indicating a transition to turbulent flows.

For mixtures (butane + propane), the behaviour is different. At low Reynolds numbers, the heat transfer coefficients are much lower than with pure fluids, and increases with the Reynolds number. At higher Reynolds numbers, the heat transfer coefficient remains almost constant or increases slightly and the values are close to those of pure fluids. Furthermore, a significant pressure effect is observed, low pressures give higher heat transfer coefficients. These observations indicate that mass transfer effects are significant and affect heat transfer during condensation.

A model for pure fluid condensation has been established and gives satisfactory results within the range of tested fluids. For mixtures, the condensation curve method does not allow estimating the heat transfer coefficient in the laminar regime. But for turbulent regimes, as the mass transfer effect is negligible, the pure fluid model allows predicting the condensation heat transfer coefficients of mixtures.

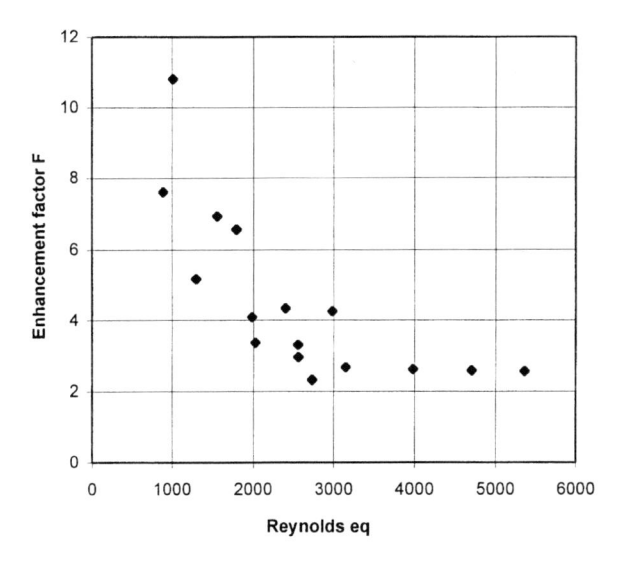

Fig. 10: Enhancement factor F versus Re_{eq} for Butane

Mixture

The experimental data (figure 11) suggest two heat transfer mechanisms: (1) a laminar regime where the mass transfer resistance is the dominant resistance; (2) a turbulent regime where heat transfer is controlled by the liquid film. The condensation curve method might be applied., with the mass transfer resistance assumed to be proportional to the heat transfer resistance in the vapour phase ($1/\alpha_g$).

NOMENCLATURE

A	heat transfer area, m²
Co	physical property number, W/m² K
Cp	heat capacity at constant pressure, J/kg K
Dh	hydraulic diameter, m
F	enhancement factor
g	gravity, m/s²
h	specific enthalpy, J/kg
$\dot{M}$	mass flow rate, kg/s
$\dot{m}$	mass velocity, kg/s m²
p	pressure, Pa
P	perimeter, m
Pr	Prandtl number
$\dot{Q}$	heat duty, W
Re	Reynolds number
Re_{eq}	equivalent Reynolds number
Rw	wall resistance, m² K/W
T	temperature, °C
ΔTln	log. mean temperature difference, K
U	overall heat transfer coefficient, W/m² K
$\dot{V}$	water volumetric flow rate, m³/s
x	vapour quality

Greek letters

α	heat transfer coefficient, W/m² K
$\bar{\alpha}$	average heat transfer coefficient, W/m² K
λ	thermal conductivity, W/m K
μ	dynamic viscosity, Pa.s
ρ	density, kg/m³

Subscript

CW	cooling water
g	vapour phase (fraction)
L	liquid phase (fraction)
LO	liquid phase only

ACKNOWLEDGEMENTS

This project JOE3-CT97-0062 has been partially supported by the EC, within the frame of the non-nuclear energy programme. The authors gratefully acknowledge Alfa-Laval Vicarb for supporting this study and coordinating the European project.

REFERENCES

Arman B. and Rabas T., Condensation analysis for plate and frame heat exchangers, HTD-Vol 314, *National Heat Transfer Conference* Volume 12, pp. 97-104, ASME 1995

Boyko L. and Kruzhilin G., Heat transfer and hydraulic resistance durig condensation of steam in an horizontal tube and in a bundle of tubes. *In. J. of Heat and Mass Transfer*, Vol 10, pp 361-373, 1967

Cavallini A., Longo G.A. and Rosetto L. Condensation heat transfer and pressure drop of refrigerants in tubes of finned tube heat exchangers. *Recent Development in Finned Tube Heat Exchangers : Theoretical and Practical Aspects*, DTI Energy Technology, Denmark, 1993

Chopard F., Marvillet C. and Pantaloni J. Assessment of heat transfer performance of rectangular channel geometries. *Proceedings of the First European Conference on Thermal Sciences*, IChemE, n° 129, volume 1, 1992

Di-an L. and Yongren L.. Steam condensation in a vertical corrugated duct. *Proc. of the National Heat Transfer Conference*, HTD-Vol 96, Volume 2, pp 389-393, 1988

Kumar H.. The design of plate heat exchangers for refrigerants. *Proc. Inst. Refrigeration*, 1991-92.5-1, 1992

Palm B. and Thonon B., Thermal and hydraulic performances of compact heat exchangers for refrigeration systems, *Compact Heat Exchangers and Enhancement Technology for the Process Industry*, pp. 455-462, Begell House, 1999

Panchal C.B. and Rabas T.J.. Thermal performance of advanced heat exchanger for ammonia refrigeration systems. *Heat Transfer Engineering*, Vol 14, n° 4, pp 42-57, 1993

Rohsenow W.M.. Condensation : Part 1. *Handbook of Heat Transfer Fundamentals 2nd Edition*, McGraw-Hill Publication, 1985

Shah R. A general correlation for heat transfer during film condensation in pipes. *Int. J. of Heat and Mass Transfer*, Vol 22, pp 547-556, 1979

Thonon B., Vidil R. and Marvillet C.. Recent research and developments in plate heat exchangers. *Journal of Enhanced Heat Transfer*, Vol 2, n°1-2, pp 149-155, 1995

Tovazhnyanski L.L. and Kapustenko P.A.. Intensification of heat and mass transfer in channels of plate condensers. *Chem Eng Communication*, Vol 31, pp 351-366, 1984

Wang L., Christensen R. and Sunden B., Analysis of steam condensation heat transfer in plate heat exchangers, *Two-Phase Flow Modelling and Experimentation*, G.P.Celata, P. Di Marco and R.K. Shah (Editors), Edizioni ETS, pp.375-380, 1999

Wang L., Christensen R. and Sunden B., Calculation procedure for steam condensation in plate heat exchangers, *Compact Heat Exchangers and Enhancement Technology for the Process Industries*, pp.479-484, Begell House, 1999

Yan Y-Y, Liao H-S and Lin T-F, Condensation heat transfer and pressure drop of refrigerant R-134a in a plate heat exchanger, *Int. J. of Heat and Mass Transfer*, Vol 42, pp. 993-1006, 1998

Zhong-Zheng Wang and Zhen-Nan Zhao. Analysis of performance of steam condensation heat transfer and pressure drop in plate condensers. *Heat Transfer Engineering*, Vol 14, n° 4, pp 32-41, 1993

THE SIMULATION OF MULTICOMPONENT MIXTURES CONDENSATION IN PLATE CONDENSERS

L.L.Tovazhnyansky, P.O.Kapustenko, O.G.Nagorna, O.Perevertaylenko

National Technical University "KhPI", 21, Frunze st.,
61002 Kharkiv, Ukraine; tel/fax +380 572 40 06 32, e-mail kap@kpi.kharkov.ua

ABSTRACT

The application of plate heat exchangers for condensation of multi-component mixtures requires reliable well-grounded methods of calculation. Numerical simulation using semi-empirical equations of heat and mass transfer performance along the surface of plate condenser was curried out for different multi-component mixtures with non-condensable component. The plates with cross-corrugated patterns for plate condensers were used. The simulation was done for four different types of corrugated plates of industrially manufactured plate heat exchangers.

The results of the simulation are in a good accordance with experimental data obtained during long-time experiments for pilot plant at the pharmaceutical factory in Kharkiv.

It is shown that enhancement of heat and mass transfer in plate condenser for case of four-component mixture gives the possibility in 1.8-2 times decreasing of necessary heat transfer surface area comparatively with shell-and-tube unit for the same process parameters.

INTRODUCTION

Last thirty years plate heat exchangers had been known as effective and flexible components of industrial heat exchangers networks. The corrugated pattern of modern plate heat exchangers is the most important part of these units because it determines the efficiency of the heat exchanger. The corrugated walls causes the very intensive redistribution of velocity components. The transition from laminar to turbulent flow occurs at Reynolds numbers much lower comparatively with straight tubes and channels with flat walls. One of the most significant achievements in implementation of plate heat exchangers is the using of such units for condensation of vapors and vapor-gas mixtures.

The investigations of heat and mass transfer for condensation in the channels of plate units carried out in Kharkiv polytechnical institute led to receiving of important results which had been presented by Tovazhnyansky and Kapustenko (1984). Basing on these results the methods of calculation of plate condensers had been developed and new types of plate condensers designed.

The extension of duty of plate condensers for case of multicomponent mixtures and mixtures with non-condensable gases requires the appropriate reliable and well-grounded methods of calculation.

THE SIMULATION OF THE PROCESS

The presence in mixture to be condensed of several components makes the simulation process more complicated because of possible simultaneous permeability of the vapor-condensate interface with all components (Tovazhnyansky, Kapustenko et. al, 1989).

In case of the heat transfer simulation for two-phase flows the total thermal resistance is the sum of thermal resistance of vapor phase, of condensate, film fouling, wall and coolant. For multi-component mixtures very significant are the diffusion processes. The total value of pressure of the mixture equalized in normal direction to flow so the increasing of concentration values on interface occurs because the vapor pressure here must be lower than in bulk of the flow. The relatively high concentration of non-condensable components creates the additional thermal resistance which decreases the heat transfer coefficients. The rate of heat transfer significantly depends on diffusion interactions which define the complicated concentrations profiles in normal direction to flow of mixture and significant redistribution along heat transfer surface. To simplify the description of heat and mass transfer in vapor (gaseous) and liquid phases we suppose the equalization of temperatures and concentrations in turbulent bulk because of intensive mixing. The presence of gradients of concentrations, temperatures and velocities is assumed to occur in thin boundary layer at the wall.

To simulate heat and mass transfer for real unit the basic equations of conservation in three-dimensional form have to be used. But the solving of problem following to such consideration is mathematically very difficult.

The basic assumption for creation of the model of condenser for multi-component mixtures is to use one-dimensional model of simultaneous heat and mass transfer with turbulent flows and taking into account the data corelations for condensation of pure vapor and binary vapor-non-condensable gas mixtures (Tovazhnyansky, Kapustenko, Nagornaya, 1990). The mathematical model is the system of differential equations based on mass and heat balances (equations (1)-(7))

$$\frac{dG_m}{dF} = -N_t; \qquad \frac{dL}{dF} = N_t; \qquad (1)$$

$$\frac{dy_i}{dF} = -\frac{1}{G_m}\left(N_i^y - N_t\, y_i\right); \qquad (2)$$

$$\frac{dx_i}{dF} = \frac{1}{L}\left(N_i^x - N_t\, x_i\right); \qquad (3)$$

$$\frac{d\tau}{dF} = \frac{1}{G_m \frac{\partial H}{\partial \tau}}\left[-\alpha^y(\tau - t_f) - \sum_{i=1}^{n} N_i^y (H_{if} - H_i)\right]; \qquad (4)$$

$$\frac{dt}{dF} = \frac{1}{L\frac{\partial h}{\partial t}}\left\{2\alpha^x\left[\frac{(t_f + t_w)}{1} - t\right] - \sum_{i=1}^{n-1} N_i^x (h_i - h_i)\right\}; \qquad (5)$$

$$\frac{dt_c}{dF} = \frac{1}{G_{cs} c_{cs}}\left(\frac{1}{\alpha} + \frac{\delta}{\lambda} + R\right)(t_w - t_c); \qquad (6)$$

$$\frac{dP}{dF} = -\frac{1}{M \Pi D}\left(1 + 2{,}9 X_{TT}^{0{,}46}\right) \zeta \frac{\rho^y (W^y)^2}{2} - \frac{d}{dF}\left[\frac{\rho^y (W^y)^2}{2}\right] - \frac{d}{dF}(\rho^y g\ F) \qquad (7)$$

where

$$X_{TT} = \left[\frac{(G_{V0} - G_V)(G_{V0} + G_G)}{G_V + G_G}\right]^{\left(1-\frac{m}{2}\right)} \left(\frac{\rho^y}{\rho^x}\right)^{0{,}5} \left(\frac{\mu^x}{\mu^y}\right)^{\frac{m}{2}};$$

m is the exponent at Reynolds number in the friction factor correlation for single-phase flow.

The system (1)-(7) is filled up with correlations for definition of heat and mass transfer coefficients and temperature of wall and interface between phases (Tovazhnyansky, Kapustenko, Nagornaya, 1990). The influence of cross mass flow on kinetic coefficients is considered with Stewart technique (1964) based on linearized theory of mass transfer. Non-ideality of mixture's components considered with coefficients of activity.

The system of equations (1)-(7) is solved with prediction-correction type methods, which requires the Jacobean calculation on each step of integration. The method foresee the automatic correction for step of integration.

The analysis of development of the multicomponent mixture condensation showed on complex heat and mass transfer either along the surface of condensation or in normal direction to main flow. So the investigation of the process is possible only with taking into account the change of local parameters.

The simulation carried out for the following multicomponent mixtures:

1 – water – ethanol (weight fractions respectively 0.45 and 0.55);
2 – water – ethanol – air (0.45; 0.48; .07);
3 – water – ethanol – air (0.45; 0.35; 0.2);
4 – ethyl – acetate – ethanol (0.88; 0.12);
5 – ethyl – acetate – ethanol – air (0.85; 0.12; .03);
6 – ethyl – acetate – ethanol – air (0.83; 0.12; .05).

As we can see the part of mixtures contains non-condensable component (air). The simulation carried out for corrugated patterns which were formed with industrially manufactured plates of different types. Some parameters of appropriate plates are presented in table 1.

Table 1. Parameters of corrugated patterns.

Parameter	Types of plates			
	0.2	0.3P	0.5	0.3
Surface area of one plate, sq. m	0.2	0.3	0.5	0.3
Average hydraulic diameter, m	.0088	.00346	.0096	.008
Cross area of the channel, sq. m	.00178	.0034	.003	.0011
Length of channel, m	0.518	0.7	0.86	1.12

RESULTS AND DISCUSSION

The distribution of local parameters along the surface of condensation was obtained.

The changes of temperatures of vapor-gas mixture and condensate τ and t respectively, temperature of interface t_f, temperature of wall from hot t_w' and from cold side t_w, temperature of cold stream t_c along the surface of

condensation are presented on Figure 1 for mixtures 1,2,3. As the content of inert (non-condensable) component increases the saturation temperature of vapor-gas mixture significantly decreases. It is necessary to notice, that the significant change of temperatures takes place during condensation process.

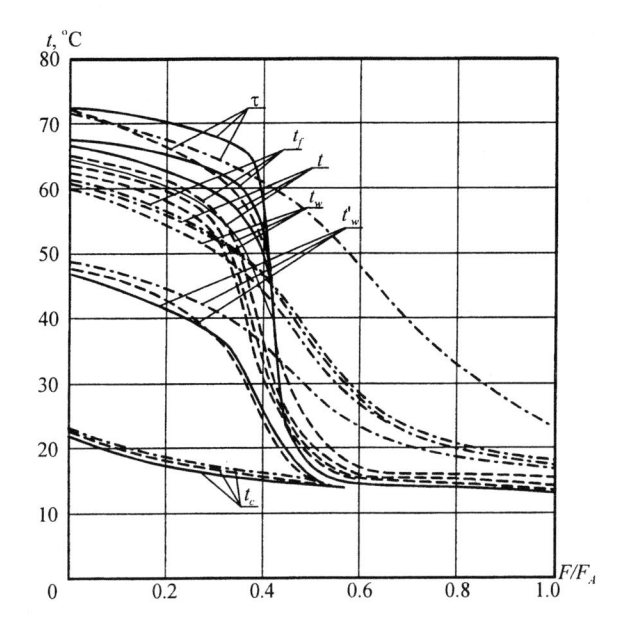

Fig.1 The profile of temperatures along the surface of condensation for mixtures 1,2,3 (plate type: 0.2 sq.m).

For case of multicomponent mixture condensation the distinctive feature is the significant change of components concentration during the condensation process. The change of component's concentrations of vapor phase in bulk and on interface along the surface of condensation is presented on Figure 2 for mixture 3.

The change of convective specific heat flux and specific heat flux of mixture condensed is presented on Figure 3 for mixtures 1,2,3 and corresponding change of non-condensable gas faction in vapor-gas mixtures. The specific heat flux caused by mass transfer through vapor-condensate interface along the surface of channel is much more than convective specific heat flux. It is not depended of air content in mixture on inlet.

For three-component mixtures 5 and 6 with dominating of one vapor component the distribution of specific mass fluxes is close to the similar for binary mixtures. This distribution for mixtures 4,5 and 6 is presented on Figure 4a. For condensation of mixtures 1,2,3 which are supplied slightly superheated and have approximately same weight fractions of vapor component the main part of heat removed compensate the decreasing of superheating on the entrance part of channel and the rate of condensation process reaches to the maximum approximately in middle part of the full condensation surface. The significant change of mass transfer rate for each component along the heat transfer surface takes place (Figure 4b).

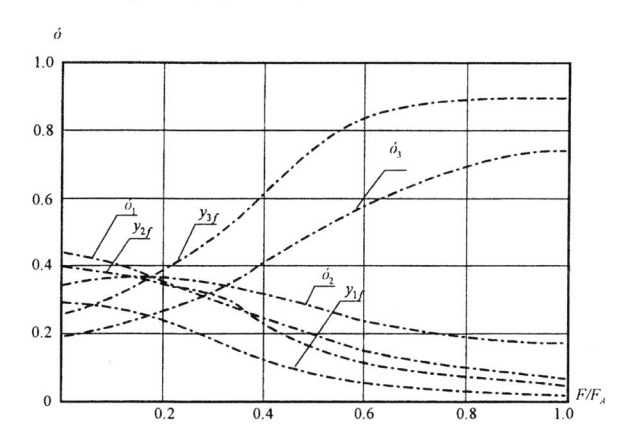

Fig. 2 The profile of concentrations of components along the surface of condensation in bulk and on interface. Case for mixture 3 condensation.

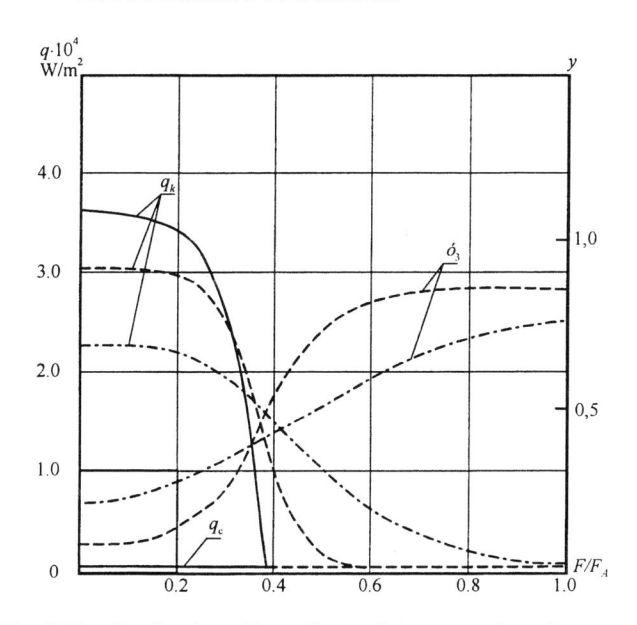

Fig. 3 The distribution of heat flux and concentration of non-condensable gas along the surface of condensation y.

——————— mixture 1
– – – – – – – mixture 2
– · – · – · – · mixture 3

The simulation of four-component vapor-gas mixture pentane-hexane-butane-air was carried out for plate heat exchanger. The results obtained were compared with data for shell-and-tube for same duty show that the plate unit has surface area 1.8 time less than tubular one, it is shown on Figure 5.

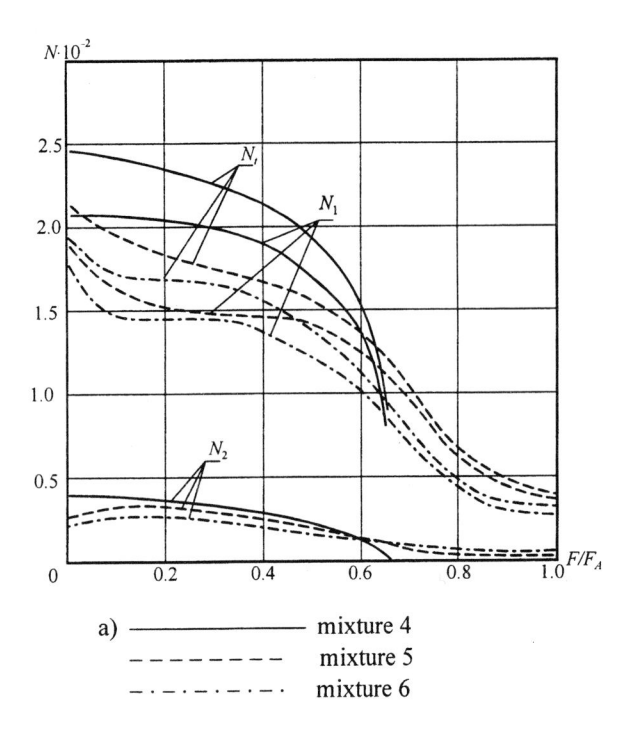

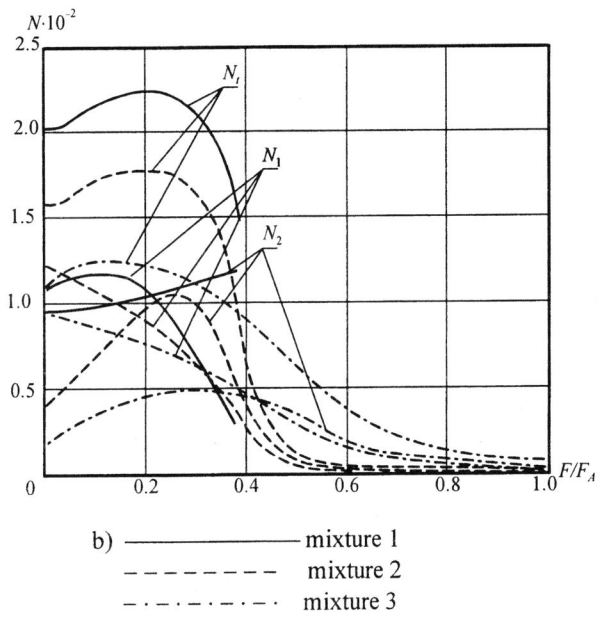

Fig. 4 The profile of components and total specific mass flux along the surface of condensation for mixtures 4,5,6 (a) and 1,2,3, (b).

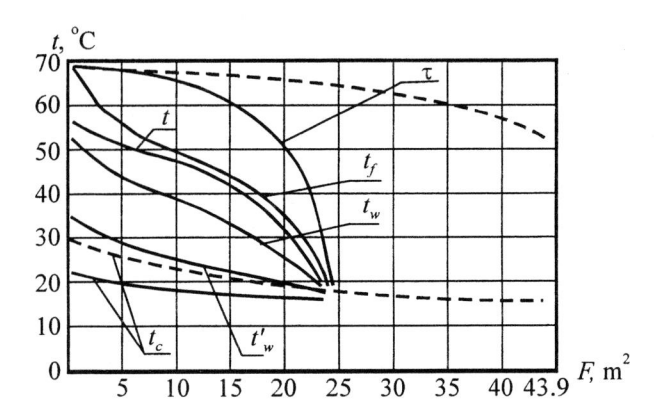

Fig. 5 The distribution of temperatures along the surface of condensation for plate and shell-and-tube condensers. Case of four-component mixture condensation.
———— plate unit
– – – – – – – shell-and-tube unit.

The condensation of mixtures containing vapors of water, ethanol, ethylacetate had been investigated for plate units with two types of corrugated plates. For small flowrates of mixture the process of condensation completed at the middle of channel with sharp decreasing of temperature along the very short distance and about a half of the unit works at low temperature differences. If flow rate increases about two times which corresponds to nominal process duty it was possible to see that the condensation takes place on more than 90% of heat transfer surface, only 5% corresponds to zone of supercooling of condensate. The experimental values of temperatures are in a good accordance with calculated data so it proves the accuracy of simulation results discussed above.

The using of plate condensers in pharmaceutical plant in Kharkiv lets to debottleneck some processes and to improve the temperature program.

Industrial tests

The accuracy of mathematical model and simulation results had been examined using test data on condensation of vapor-gas mixtures in processes of pharmaceutical industry.

CONCLUSIONS

The mathematical model of multi-component mixtures condensation in the channels of plate heat exchangers had been developed. A good accordance of results of simulation to industrial experiments lets to develop the design technique for plate units used for condensation of multicomponent mixtures for wide range of duties.

NOMENCLATURE

B coefficient
c specific heat capacity, J/(kg·°C)
D equivalent diameter of channel, m
F heat surface area, m^2
g specific gravity, m/s^2
G mass flow-rate, kg/s

H specific enthalpy of vapor-gas mixture, J/kg
h specific enthalpy of condensate, J/kg
L mass flow-rate of condensate, J/kg
M number of passages for vapor-gas mixture
m exponent
N specific mass flux, kg/(m^2·s)
P vapor-gas mixture pressure, Pa
q specific heat flux, w/m^2
R total thermal resistance of fouling, m^2·K/w
Re Reynolds number, $\frac{wD\rho}{\mu}$, dimensionless
t temperature, °C
w velocity of vapor-gas flow, m/s
x concentration in liquid phase, weight fraction
y concentration in vapor phase, weight fraction
α film heat transfer coefficients, w/(m^2·K)
δ thickness of heat transfer plate, m
ζ coefficient of pressure drop for single-phase flow in corrugated pattern, $B\operatorname{Re}^{-m}$, dimensionless
λ heat conductivity of plate, w/(m·K)
μ dynamic viscosity, Pa·s
Π perimeter of interplate passage, m
ρ density, kg/m^3
τ temperature of vapor-gas mixture, °C

Subscripts and superscripts

A relates to all plate unit
c convective flow
cs cold stream
G gas (non-condensable)
f interface vapor-condensate
k condensation process
v vapor
w wall
x liquid phase
y vapor phase
o inlet of vapor
t total mass flux of components

REFERENCES

Stewart W.E. Multicomponent Mass Transfer in Turbulent Flow. – AIChE Journal, 1973, v.19, N 2, pp. 398-400.

Tovazhnyanski L.L., Kapustenko P.A. Intensification of Heat and Mass Transfer in Channels of Plate Condensers. Chemical Engineering Commun., 1984, 31, N 6, p. 351.

Tovazhnyanski L.L., Kapustenko P.A., Kedrov M.S., Nagornaya E. The simulation of plate condensers for multicomponent mixtures. – Summary of report on III[d] Conference on cybernetic of chemical processing systems, 1989, Moscow, 1989 (In Russian).

Tovazhnyanski L.L., Kapustenko P.A., Nagornaya E. The simulation of condensation process for multi-component mixtures. – Chemical Industry (Khimicheskaya Promyshlennost), 1990, N 8, p. 13-15 (In Russian).

PHASE-CHANGE HEAT TRANSFER EXCHANGER DEVELOPMENT AND APPLICATIONS

USE OF PLATE HEAT EXCHANGERS IN REFINERY AND PETROCHEMICAL PLANTS

Jinn H. Wang

UOP LLC, 25 East Algonquin Road, Des Plaines, Illinois 60017, E-Mail: jhwang@uop.com

ABSTRACT

This paper reviews the application of plate type heat exchangers in refinery and petrochemical plants. The technology, applications, economics, limitations, and future challenges are reviewed and discussed.

INTRODUCTION

In the past decade, refinery and petrochemical industries have been confronted with eroding profits and a great deal of global competition. With each new project, engineers are pressured to improve profitability, reduce operating costs and enhance equipment reliability. Often, they are forced to look into alternative heat transfer equipment such as plate heat exchangers (PHE) to improve energy recovery and reduce capital investment costs.

The use of plate heat exchangers is not only driven by economics, but also driven by technical requirements. As new process plants demand larger throughput, higher energy efficiency, and tighter temperature approach, conventional shell-and-tubes (S&T) heat exchangers can no longer meet the requirements. Flow mal-distribution, poor thermal performance, high pressure drops, and vibration damages are just some of the common problems experienced by S&T exchangers. As a result, the search for alternative type heat exchangers becomes necessary.

Although plate heat exchangers have been used in food, HVAC, and utility industries since the 1960s (Sloan, 1998), the process industry did not start paying attention to plate exchangers for process applications until the mid 1980s. Since then, the use of plate heat exchangers has gradually widened from non-critical services such as water coolers into hydrocarbon process areas. They have been used as reactor feed effluent exchangers, vaporizers, condensers, stabilizer feed bottoms exchangers, gas coolers and even in slurry services. However, concerns of fouling, cleanability, and reliability continue to exist.

To ensure plate heat exchangers perform satisfactorily in hydrocarbon process services, not only should the equipment meet design requirements, but it also needs to accommodate operational changes including start-up, shutdown, turndown and upset conditions. Furthermore, process flows need to be evaluated to mitigate potential fouling due to contamination, oxidation, or salt formation. Proper protection must be provided to prevent exchangers from plugging. Lastly, on-line and off-line cleaning procedures should be developed to restore the exchanger's performance in the event of a process upset. With this comprehensive analysis, engineers will be able to weight the benefits and risks of applying plate heat exchanger in process units.

EVOLUTION OF PLATE HEAT EXCHANGERS FOR HYDROCARBON INDUSTRY

Plate heat exchangers typically consist of a series of thin corrugated plates staggered together to form a plate bundle. Each plate is precision formed into corrugated patterns. See Figure 1 for the typical construction of a plate bundle. Stainless steel or high alloy materials are required for ease of forming, high mechanical strength, and corrosion resistance.

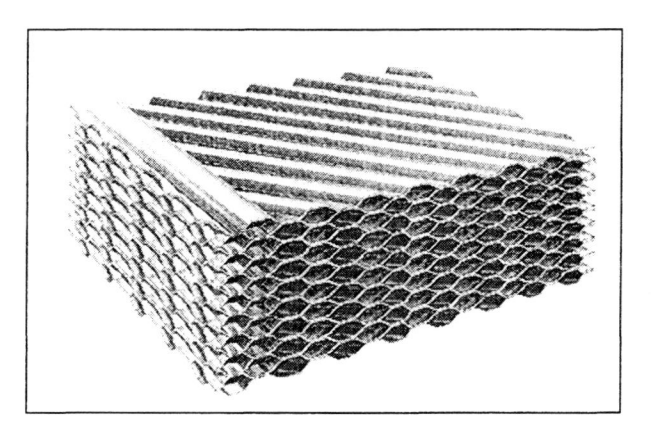

Figure 1. Typical plate heat exchanger bundle using corrugated plates

Early designs used an external frame to hold plates together, with gaskets around the periphery of the plates to form flow channels in the plate bundle. This type of plate and frame exchanger limited the design temperature to approximately 149°C (300°F) and pressure to approximately 21 Barg (300 psig). In addition, the gasket material had to be compatible with the process fluids. Their applications had been limited to non-hydrocarbon services such as cooling water to seawater exchangers.

To overcome the temperature and pressure limits, welded plate exchangers were developed. They eliminated gaskets and expanded temperature limit to over 568°C (1000°F) and pressure limit to over 69 Bar (1000 psig). Unlike the gasketed type, welded plate heat exchangers can not be dismantled for mechanical cleaning and have been proven to be more difficult to repair. They are used in light hydrocarbon services with low fouling applications.

The size of the plate heat exchanger has increased from approximately 50 m² to approximately 1000 m² per bundle. Modern welded plate type exchanger using explosion-forming technique can produce plates up to 15 meters in length and 1.4 meters in width with a total heat transfer surface area of 5000 m² per plate bundle.

Compared to S&T heat exchangers, plate heat exchangers are lightweight, high efficiency and potentially cost less. To meet process requirements, a large variety of plate heat exchangers have been introduced to the market including plate and frame type, semi-welded type, all welded plate type, spiral type, printed circuit type, and brazed aluminum type. The plate and frame heat exchangers are primarily used in the non-hydrocarbon services due to the large amount of gasketed areas within each plate. Semi-welded and all welded heat exchangers can be used for hydrocarbons services with proper precautions of fouling and cleaning. Printed circuit and brazed aluminum type exchangers have a very small flow passages and are primarily used for gas separation applications.

HEAT TRANSFER AND PRESSURE DROP PERFORMANCE

Plate heat exchangers use corrugated plates to form an array of small flow channels enabling high flow turbulence resulting in high heat transfer efficiency. The heat transfer and pressure drop characteristics depend on the width and depth of the flow channels, angles of corrugations, plate length, and the number of plates.

Figure 2 shows that the heat transfer coefficients (j factor) of a typical plate exchanger, with mild plate corrugation angle are about twice the coefficient of a tubular exchanger. Unlike S&T heat exchangers, plate heat exchangers have no stagnant flow regions. The counter current operation along with the absence of bypassing streams generates high thermal efficiency. Using the same allowable pressure drop, a plate heat exchanger can produce a heat transfer rate 1.5 to 2.5 times higher than the heat transfer value of a S&T heat exchanger.

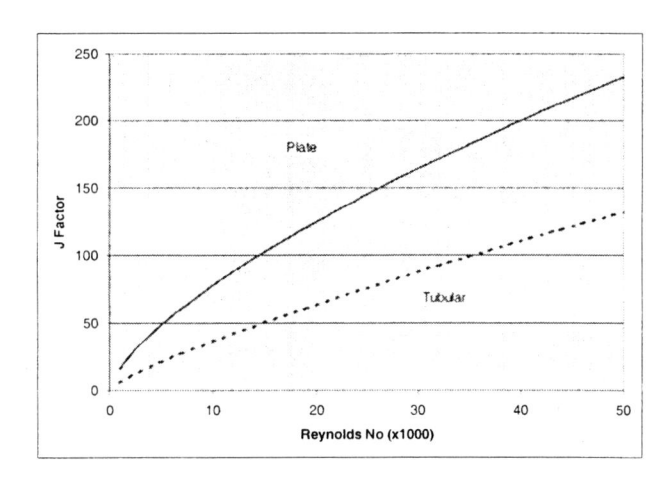

Figure 2. Heat Transfer vs. Reynolds Number

Due to the small flow passage and corrugated flow patterns, plate exchangers yield a higher frictional factor (f factor) than tubular exchangers. Figure 3 shows a comparison of frictional factor (f factor) versus Reynolds number. At the same Reynolds number, plate type exchangers give a frictional factor about 10 times higher than the frictional factor of tubular exchangers. Fortunately, the flow lengths are short and nominal velocities are low for plate heat exchangers. Therefore the pressure drop can be maintained within reasonable limits. In general, plate heat exchangers favor applications with higher allowable pressure drops.

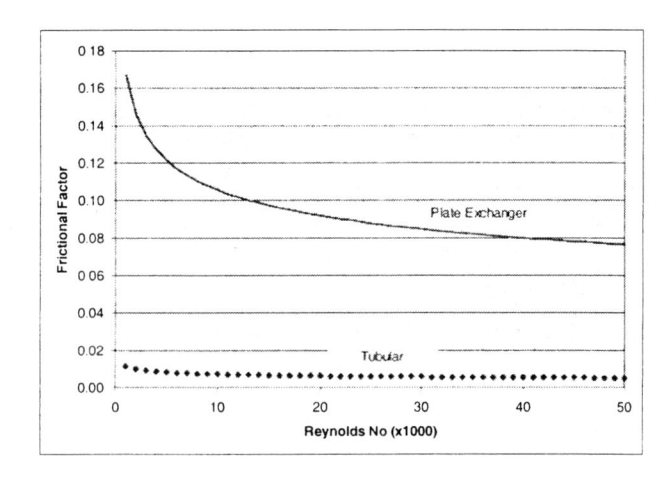

Figure 3. Frictional Factor vs. Reynolds Number

The pressure drops of plate heat exchangers are very sensitive to the angles of the corrugate patterns. As illustrated in Figure 4, a small change in corrugation angle can double the pressure drop but only improve heat transfer by a small fraction. It is very important for the exchanger designer to choose proper channel size, plate length and corrugation angles to optimize the exchanger design.

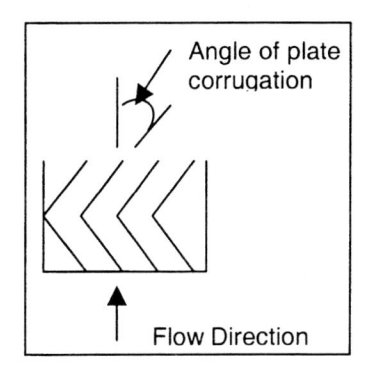

Figure 4a. Corrugation Angle Definition

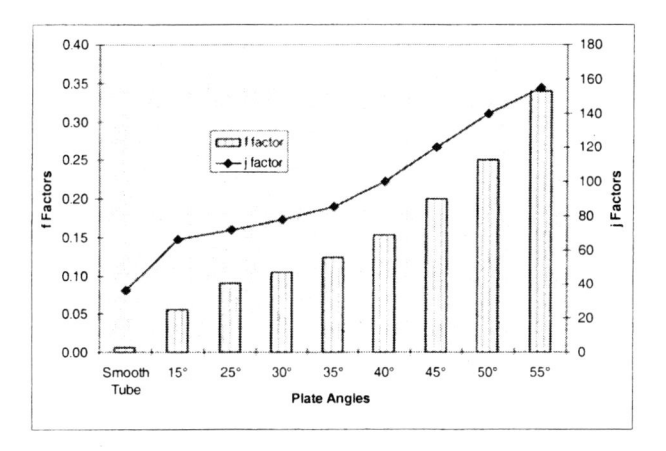

Figure 4b. Heat transfer and pressure drop vs. corrugated angle

APPLICATION OF PLATE HEAT EXCHANGERS FOR HYDROCARBON SERVICES

Plate heat exchangers are most conducive for applications with clean fluids, high allowable pressure drop and long temperature gradient services where multiple shells in series are required. As the technology for plate forming and welding continues to improve, plate heat exchangers will be able to handle larger capacity and more severe temperature and pressure conditions. As a result, the application of plate exchangers in the process industry will continue to expand. In refineries and petrochemical plants, plate heat exchangers have been applied to many hydrocarbon processes including catalytic reforming, desulfurization, isomerization, aromatics recoveries, sour water treatment, gas separation and many chemical processes. The following are examples of successful applications of plate heat exchangers for hydrocarbon process services.

1. Catalytic Reformer Process Unit

A large size welded plate heat exchanger was introduced in 1984 to replace the vertical single tube pass type heat exchanger for UOP Platforming combined feed vs. reactor effluent service (Wang, 2000). Since then, more than 80 large size plate type heat exchangers have been installed in the catalytic reformer process unit. A single plate heat exchanger can replace 2 or 4 vertical tubular exchangers in parallel. Advantages include more energy recovery, free of vibration damage problems, and lower installation cost.

This large plate type heat exchanger uses an explosion-forming technique to form large size plates (1.4 m in wide by 15 m in length). The plates are then welded together to avoid using gasket seals. The welded plate bundles are then placed within a pressurized vessel. The liquid feed is introduced into the exchanger through a feed distributor to assure even flow distribution among all plate channels. The recycle gas fills the containment vessel, pressurizes the plate bundle, and enters into plate channels via a gas distributor. The liquid feed and recycle gas mix at the plate entrance and flow upward. The corrugated plate pattern ensures proper mixing of the two-phase flow and even flow distribution across the entire plate bundle. Figure 5 shows typical UOP Platfoming process unit using plate heat exchanger for combined feed vs. reactor effluent service.

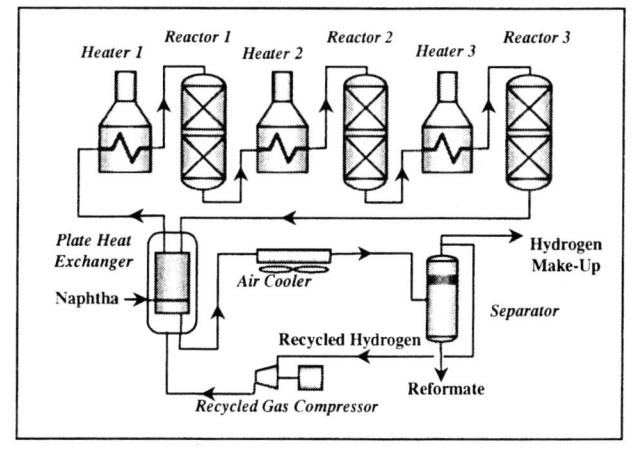

Figure 5. UOP Platforming Process Unit

2. Desulfurization Process Unit

Plate type heat exchangers are used in hydroprocessing process units in combined feed vs. reactor effluent exchanger services. A single vertical plate heat exchanger can replace S&T type heat exchangers with 4 to 6 shells in a series. It also reduces total pressure drop from approximately 3.5 Bars (50 psi) to 1 Bar (15 psi), thus reducing compressor power consumption.

Due to the frequent change in feed source and high volatility of feed compositions, the exchangers in the desulfurization unit are susceptible to fouling problems. Care must be exercised to avoid fouling and plugging caused by polymerization, iron sulfide, or ammonia salt formation. On line water injection is normally provided at the cold end of the exchanger for salt removal. The feed storage tank should be blanked with inert gas and a filter or strainer should be provided in the feed stream. The plate bundles should also be designed as free draining to avoid aquatic acid attack on the stainless steel during shutdown. Approximately 15 plate heat exchangers have been installed in desulfurization units for the combined feed vs. reactor effluent service.

Figure 6 shows typical flow diagrams of a desulfurization process unit. In this case, one plate type heat exchanger is able to replace 10 tubular exchanger bundles, configured 5 in series with 2 trains in parallel.

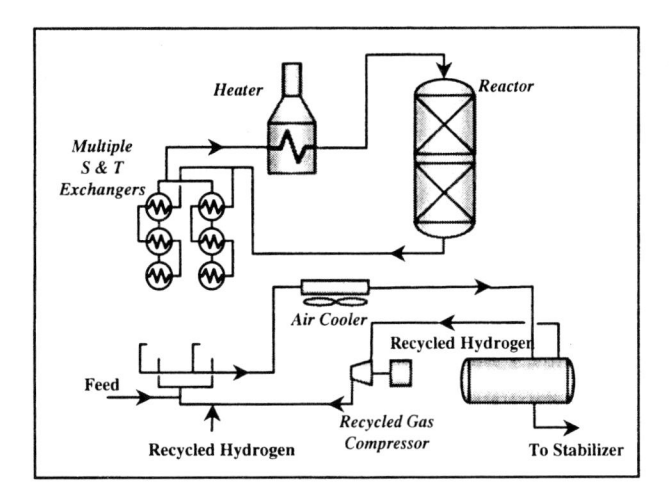

Figure 6a. Hydroprocessing process unit using multiple tubular exchangers

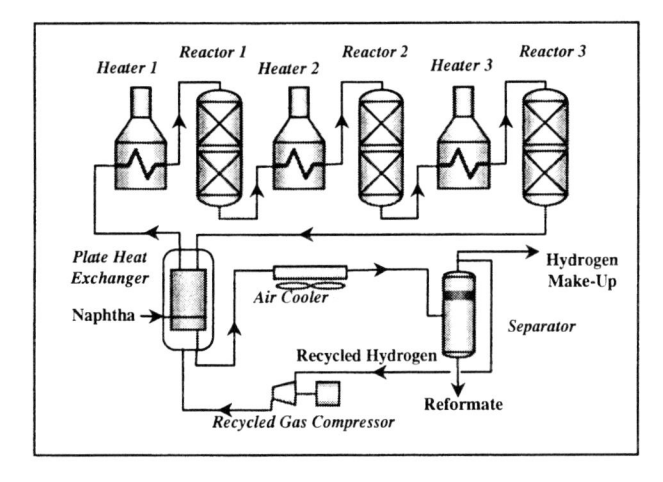

Figure 6b. Hydroprocessing process unit using plate heat exchanger

3. Isomerization Process Unit

A semi-welded type plate heat exchanger has been used in the reactor feed vs. effluent exchangers for light hydrocarbon services to debottleneck an existing unit with S&T heat exchangers and to fit into the existing plot space. Process streams are very clean with low fouling tendency. See figure 7 for a depiction of the semi-welded exchanger and Figure 8 for a typical Isomerization process flow diagram.

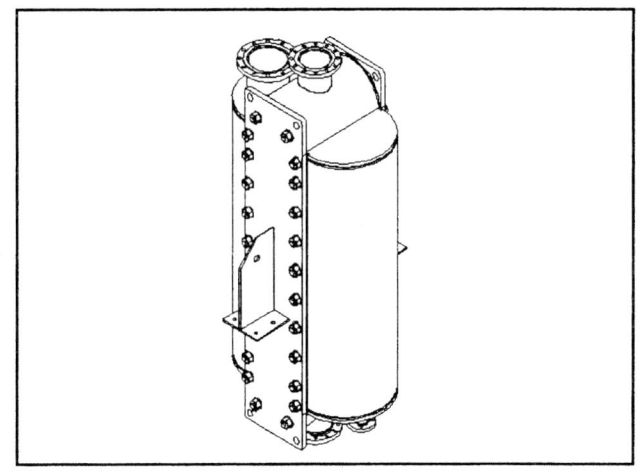

Figure 7. Semi-welded plate heat exchanger

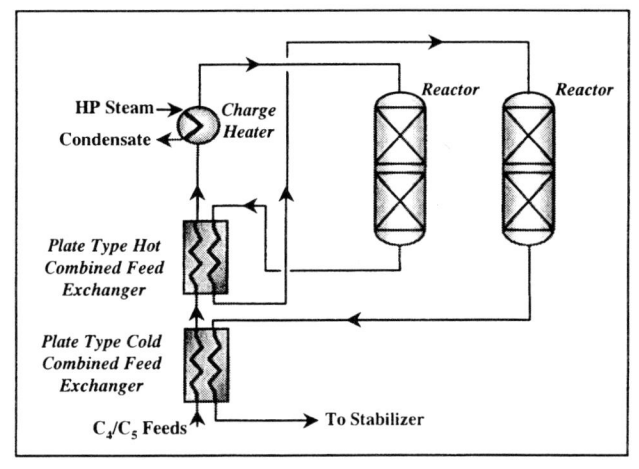

Figure 8. Isomerization Process Unit

4. Fluid Catalytic Cracking Process Unit

The FCC main column bottoms product cooler contains catalyst fines. This process stream is highly viscous and requires close temperature control. S&T type exchangers are not recommended for this service due to the concern of trapping catalyst fines in stagnant flow regions. Therefore, multiple sections of double pipe type heat exchangers have been used in order to avoid the accumulation of catalyst fines inside the tube bundle. This exchanger is normally equipped with 100 percent spare to allow for on-line cleaning. The poor heat transfer performance associated with a large heat transfer surface makes this exchanger very costly.

To reduce equipment cost and improve performance, a special spiral type plate heat exchanger, Figure 9, was introduced for this slurry service. The spiral flow path which has no stagnant flow regions keeps the catalyst fines in suspension while providing low fouling rates and high heat transfer performance. This type of exchanger has been used in the slurry services for the last 10 years with good success. The cost of the spiral heat exchanger is about one third of the double pipe exchanger, though it provides the same thermal and hydraulic performances. Figure 10 shows a simplified flow diagram of the FCC main column bottoms section.

Figure 9. Spiral Heat Exchanger

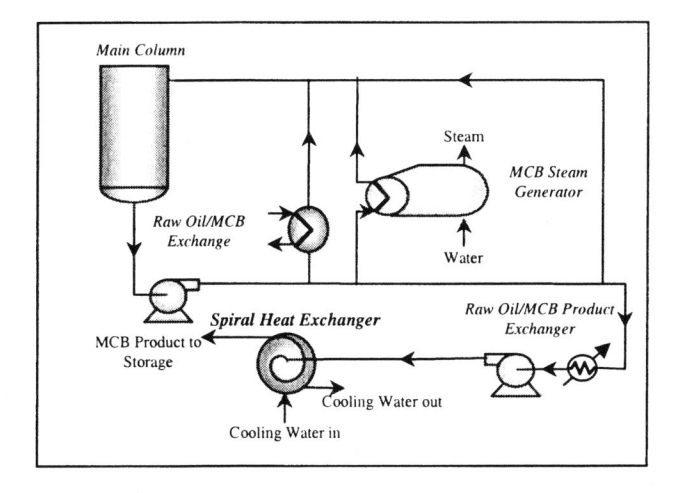

Figure 10. FCC Main Column Bottoms Flow Diagram

5. Gas Processing and Miscellaneous Services

For gas separation and cryogenic services, specialized plate type brazed aluminum and printed circuit heat exchangers are used to obtain tight temperature approach. The terminal temperature difference between the hot stream and cold stream can be as close as 2°C (3.6°F). In phenol and styrene plants, semi-welded heat exchangers are used for vent condensers. The exchangers can be mounted on the top of the column due to its light weight and compactness. Plate and frame heat exchangers are used in sour water cooling in crude and vacuum units for the ease of cleaning.

COST OPTIMIZATION

Plate heat exchangers are typically made of thin stainless steel alloy plates. Compared to S&T heat exchangers using carbon steel, the cost per unit area is typically higher. However, plate heat exchangers have a higher thermal efficiency, thus they require less heat transfer surface area. In addition, plate heat exchangers are compact, light weight, and require less plot space. A single plate heat exchanger can replace multiple S&T exchangers consequently reducing cost of piping, foundation and structure. In determining the economics of using plate heat exchangers, engineers must take into consideration total installation cost including the equipment, piping, foundation, support structure, plot space, controls and labors.

Figure 11 shows a case study comparing the cost of a S&T exchanger using carbon steel versus a welded plate exchanger using stainless steel. This study is based on a semi-welded plate heat exchanger for light hydrocarbon versus cooling water service. The high material cost is offset by the high thermal efficiency of the plate exchanger. Thus, the cost per thermal unit (UA) of the two exchanger types is about the same. The real cost savings of using a plate exchanger in this case has to come from the lower total installation cost.

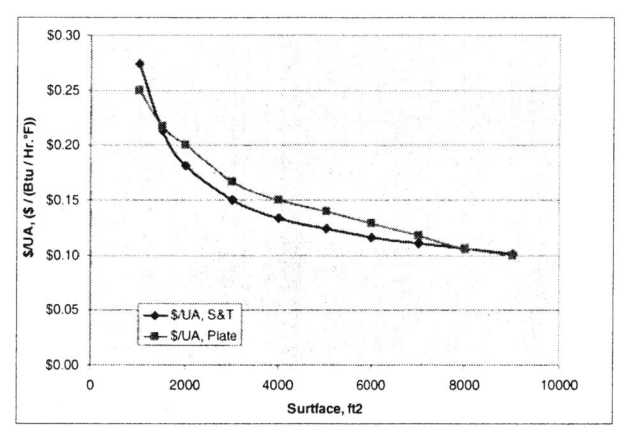

Figure 11. Cost comparison of plate exchanger vs. shell and tube exchanger

To optimize the size of a plate heat exchanger, design engineers consider the incremental equipment cost versus the value of energy savings. In addition, the cost impact on the upstream and downstream equipment such as heaters and condensers should also be taken into consideration. A larger plate heat exchanger costs more but will reduce the overall energy consumption. The final decision on the heat exchanger size shall be based on the equipment cost, energy saving, return of investment (ROI), and equipment reliability.

OPPORTUNITIES AND CHALLENGES

As refineries and petrochemical plants strive for lower operation cost, tighter product specifications, lower environmental impact, and higher profits, they will have to continue looking for higher efficiency and more reliable heat transfer equipment. Plate heat exchangers provide an alternative to gain better energy recovery, reduce fired heater or boiler emissions, reduce pressure drops, and increase profits. As potential users become more familiar with plate heat exchangers, and plate heat exchanger suppliers become more educated about their user's needs, plate heat exchangers could become an integral part of hydrocarbon process units.

Fouling and reliability are two major obstacles of using plate heat exchangers. Because plate exchangers use stainless steel and high alloy with high flow turbulence, their fouling tendency is low. However, the small flow passages make the plate heat exchangers susceptible to plugging. Once the plugging starts, the pressure drop can increase 3 to 4 fold within days and could cause premature unit shut down.

To address these concerns, plate heat exchangers need to be reviewed in the context of the process unit and address the corresponding problems accordingly. For example, operation procedures must be reviewed to avoid temperature or pressure excursion. Reactor screens should be provided when catalyst migration is a possibility. Filters or strainers may be required if the process stream contains iron scale or lose particles even if only at start-up. Intermediate feed storage tanks for streams with unsaturated hydrocarbons must be blanked with inert gas to reduce the formation of gumming components. Cleaning procedures must be developed for the provisions of on-line or off-line cleaning. Comprehensive process and equipment analyses are essential to the success of using plate heat exchangers.

CONCLUSIONS

Plate heat exchangers can be successfully employed in the process plants by taking a systematical approach. The main potential applications are in the following areas:
1. Where heat transfer area is large and expensive material is required. A single plate heat exchanger could replace multiple parallel trains of S&T bundles. Typical applications are reformer combined feed exchangers and hydrotreater combined feed exchangers.
2. Where close temperature approach is required and may be difficult to achieve by tubular exchangers. Typical applications are in gas processing and cryogenics applications or for revamp requirement.
3. Where high allowable pressure drops is available and can not be utilized by S&T heat exchangers.
4. Where compactness is important such as the low pressure vent condensers mounted on the top of separator columns.
5. Where streams contain suspended solids and conventional S&T heat exchanger can not prevent accumulation of the solids.
6. Where process streams are clean and plot space is limited especially for revamp projects.

REFERENCES

Sloan, Mark D., Designing and trouble shooting plate heat exchangers, *Chemical Engineering*, May 1998, pp. 78-83.

Wang, J.H., Heat exchanger evolution in UOP Platforming units, *AICHE Heat Exchanger Seminar at University of Illinois in Chicago, February 9, 2000.*

DEVELOPMENT AND APPLICATION OF SEMICONDUCTOR DEVICE COOLER USING HIGH PERFORMANCE PLATE FIN HEAT EXCHANGER

Kenji Ando

Sumitomo Precision Products co., ltd., 1-10, Fuso-Cho, Amagasaki, Hyogo, Japan; E-mail: ando-k@spp.co.jp

ABSTRACT

In recent years, development of power electronics technology which is represented by IGBT (Insulated Gate Bipolar Transistor) is remarkable, and control of power sources by an inverter is becoming a mainstream in various fields. The capacity of semiconductor devices, which is used for these power sources, has been progressing. At the same time the cooling technology of the devices is becoming very important. In the past, heat pipes were generally used to cool the semiconductor device, with a large capacity. However, the cooling equipment tends to be larger in this method, so further superior performance of the cooler is required. Therefore, we developed a compact and high-performance cooler (called SIPHOREX) using plate fin heat exchanger (hereinafter called PFHE) technology, which has been widely used for electric railway cars, etc.

This paper introduces the basic structure, process of

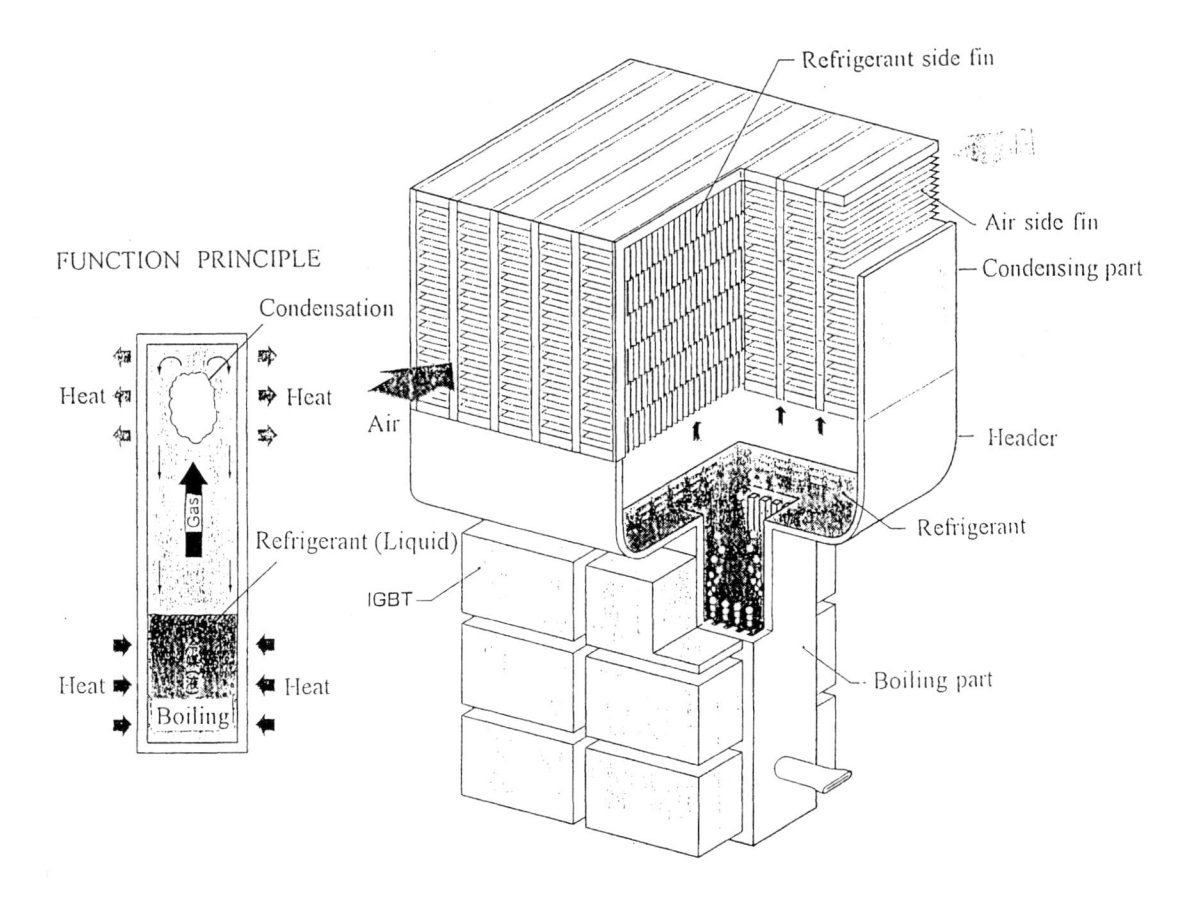

Fig. 1 Basic Structure and Function Principle of SIPHOREX

development, advantages, and application example of SIPHOREX.

BASIC STRUCTURE OF SIPHOREX

As shown in Fig.1, SIPHOREX is a combined heat exchanger made of aluminum alloy with a boiling part and a condensing part which contains internally refrigerant and hermetically sealed. The boiling part is a hollow container and has flat and smooth surfaces on outside to mount semiconductor devices (IGBT etc.). A high-performance PFHE is used for the condensing part. Refrigerant (liquid) in the boiling part boils by heat of the devices. Vaporized refrigerant gas enters the condensing part, and the gas is condensed by outside air. Condensed refrigerant liquid flows down by gravity, and return to the boiling part again and circulates. In this way, SIPHOREX dissipates the heat generated by IGBT efficiently by using vaporization and condensation of refrigerant.

Fundamental Construction of Plate Fin Heat Exchanger

PFHE is used at condensing part of SIPHOREX, and plays an important role for creating a high performance.

As shown in Fig. 2, parting sheets, corrugated fins, and side bars (which make partition between inside and outside) are stacked, and the stacked core is joined by vacuum brazing method. Each layer forms each passage, and heat exchange is made between fluids that flow through one passage and next passage. The parting sheet functions as a primary heat transfer surface. The corrugated fin brazed between the parting sheets functions as a secondary heat transfer surface and at the same time as a strength member against the internal pressure.

When the material is aluminum alloy, the filler metal is cladded onto and integrated with the parting sheet.

Brazing part is required for not only strength but also airtight and vibration-resistant. As these requirements must be satisfied constantly, it is important to control precisely the dimension (height) of the corrugated fin and heating condition at the brazing process.

The brazed core (condensing part) is joined to the header and the boiling part made of a hollow shape by welding.

EXAMINATION OF BOILING PERFORMANCE IMPROVEMENT

It was necessary to get performance of boiling part and to attempt superior performance in addition when this cooler is developed. The author thought three ways to improve the boiling performance, which are to make boiling surface rough by a blast method, to increase boiling heat transfer surface, and to combine the above methods. We selected to increase boiling transfer surface from these methods, because the blast process was costly.

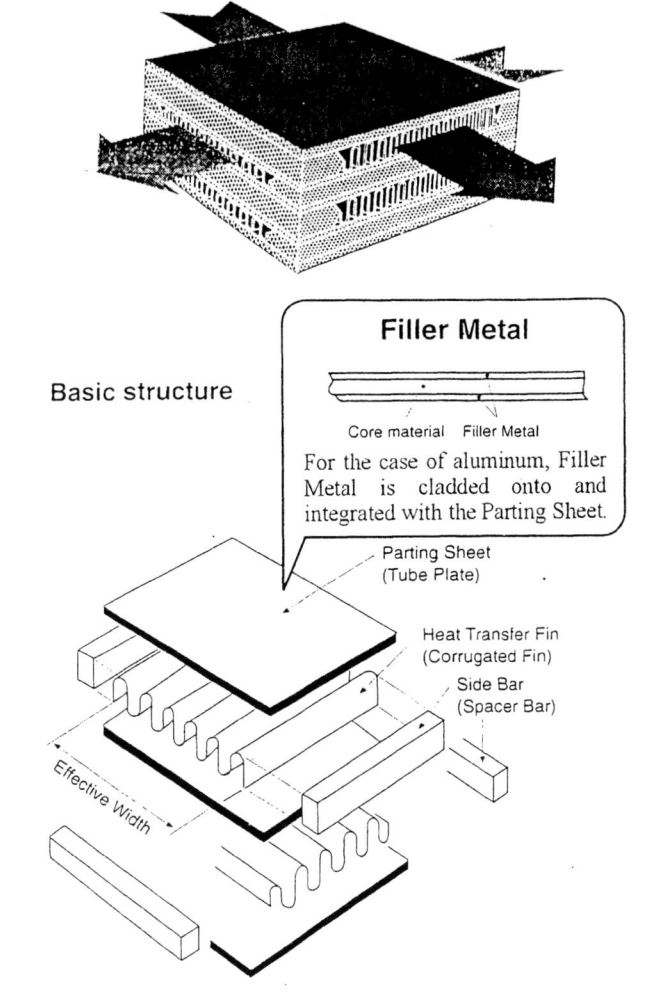

Fig. 2 Fundamental Construction of PFHE

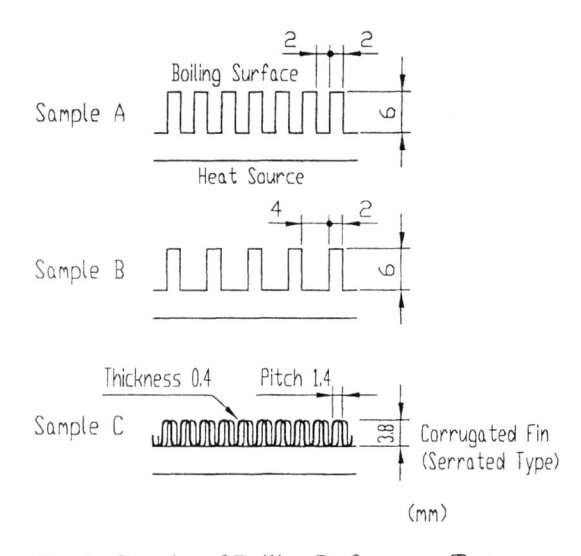

Fig. 3 Samples of Boiling Performance Tests

Condition of Boiling Performance Tests

The author conducted the experiment of boiling performance using three kinds of samples shown in Fig. 3. The sample A and B had projected fins which act as a boiling surface, and the pitch of sample A was smaller than sample B. Boiling surface of the sample C was brazed with a corrugated fin (serrated type) whose fin pitch was smaller than the sample A. The refrigerant used was perfluorohexan (PFC-51-14), and the temperature during the test was kept around 80 ℃. The position of boiling surface was either horizontal or vertical. (Sample B was tested at the vertical position only.)

The Test Results

Fig. 4 shows the test result of the vertical boiling surface. The transverse axis is the heat flux of installed heater, and the vertical axis is the boiling heat transfer coefficient that is calculated based on the installation area of the heater (not boiling heat transfer area).

In the region of heat flux less than about $1.2 \times 10^5 w/m^2$, sample C shows the highest heat transfer coefficient. It is considered that because sample C (which is brazed serrated fin with small pitch) has many local points where boiling starts easily, this sample had the highest heat transfer coefficient. But, in the region of heat flux greater than about $1.2 \times 10^5 w/m^2$, sample A shows higher values than C. It is considered that, in the case of corrugated fin whose pitch is small like sample C, since the bubble generated by boiling covers the boiling surface easily, the dry surface increases (boiling surface area decreases) in the region of high heat flux where generation of bubbles are large. Moreover, comparing the samples A and B which have projection fins, the heat transfer coefficients are almost same in the lowest heat flux ($2.0 \times 10^4 w/m^2$), and as the heat flux becomes high, and sample A shows higher heat transfer coefficient than B and the difference increases. It is considered that since starting points of boiling are few in the region of low heat flux, the performances are almost same although the heat transfer areas of these samples differ.

The test results at horizontal position are shown in Fig. 5. In this conditions, sample A shows higher heat transfer coefficient than C in the all region of heat flux. It was considered that because corrugated fins become the shape which obstruct the bubble to leave the boiling surface at horizontal position, sample A with a simple heat transfer surface shows a better performance than sample C.

Consideration

It has been understood from the above mentioned results that to increase boiling heat transfer, the fins with moderate pitch is effective, but when the pitch is too small, the efficiency decreases in the region of high heat flux. Moreover, because the surface with projected fins made by extrusion was more advantageous when thinking about the

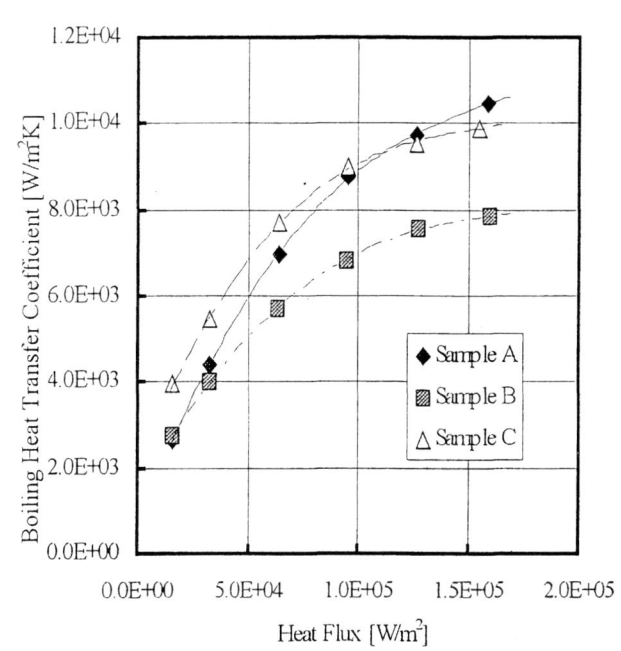

Fig. 4 Boiling Heat Transfer Coefficients .vs. Heat Flux (Vertical Boiling Surface)

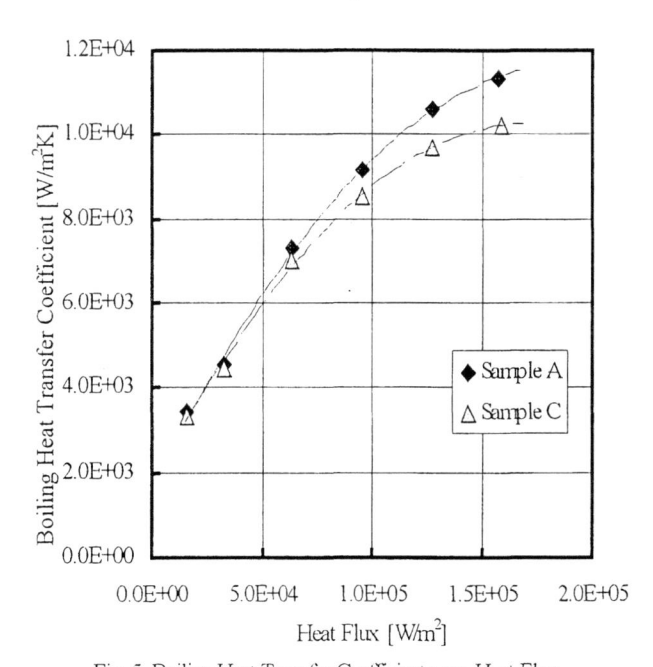

Fig. 5 Boiling Heat Transfer Coefficients .vs. Heat Flux (Horizontal Boiling Surface)

cost, we decided to apply the projected fin to the boiling surface for the product.

It was thought to improve the performance by making the pitch of fins smaller, but the fin pitch could not be made too small in hollow shape (made by extrusion) used for the mass production. As a result, we decided to adopt the pitch of sample A for the product.

Table 1 Comparison between each refrigerant

		Substitution refrigerant candidate		Present use
Name		HFC-43-10mee	HFE-7100	PFC-51-14
Chemical formula		$C_5H_2F_{10}$	$C_4F_9OCH_3$	C_6F_{14}
Liquid density (at 25°C)	kg/L	1.58	1.52	1.68
Specific heat (at 25°C)	kcal/kg°C	0.28	0.28	0.25
Latent heat	kcal/kg	31.5	30	21
Boiling point	°C	55	60	56
ODP (CFC11=1)		0	0	0
GWP (CO_2=1)		1300	320	7400

REFRIGERANT

The refrigerant enclosed in SIPHOREX should meet the following conditions.
- It dose not react to aluminum alloy.
- It dose not resolve in the operating conditions.
  (Saturated condition, Maximum temperature: about 90°C, Duration: about 20 years)
- The saturated vapor presser is low. (Because the working temperature is high)
- Ozone depleting potential (ODP) is very low.
- Global warming potential (GWP) is as low as possible.

The refrigerant, which we are using now, is PFC-51-14. Because GWP of PFC-51-14 is extremely large, this refrigerant is environmentally undesirable material, though the ODP is zero. Therefore, we are examining HFC-43-10mee and HFE-7100 as the candidate refrigerant. Table 1 shows the comparison of these refrigerants. GWP of the candidate refrigerants is greatly lower than PFC, and the ODP is zero of course. We have confirmed that the heat performance of either refrigerants are equal to or more than PFC-51-14. We want to change to the candidate refrigerant if stability in the operation condition can be confirmed in the future.

ADVANTAGES OF SIPHOREX

Advantages of SIPHOREX are shown in the following.

1. Possible to cool the devices with high heat density.
   Because the inside of the boiling part is filled with refrigerant (liquid), dryout dose not occurs easily. Therefore, it is possible to cool the devices with high heat flux (about 15w/cm^2 or less).
2. Temperature difference between the devices is very small.
   Because the inside of the boiling part is filled with refrigerant (liquid) of approximately uniform temperature, very small temperature difference between the devices is expected even if there are a lot of number of devices. Moreover, even if the heat load of each devices are imbalance, small temperature difference between the devices is expected. Fig. 6 and 7 shows examples of the distribution temperature rise in a uniform heat load and an imbalance heat load. The number of mounted devices is 12 (6 one side × 2), and the numerical value in this figure shows the temperature differences between the temperature of surface mounted devices and the inlet temperature of cooling air. The difference between the maximum value and the minimum value of the temperature on the surface mounted the devices is only about 1.0°C at uniform heat load. In the case of the imbalance (±40%) condition, the difference is only 6°C (±10%), and it is extremely smaller than the ratio of the difference of heat load.

Condensing Part

26.8	27.3
27.9	28.0
28.2	28.0

26.8	27.1
27.7	28.2
28.0	28.3

Boiling Part
200w / 1 device   °C   200w / 1 device

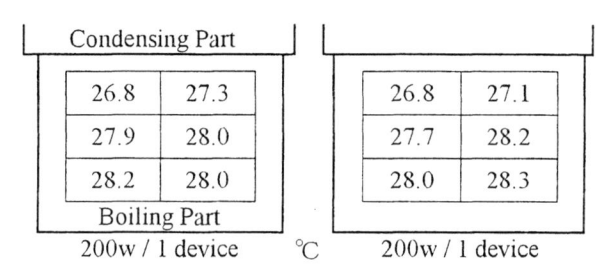

Fig. 6 Example of Temperature Rise Distribution at Uniform Heat load (Total: 2400w)

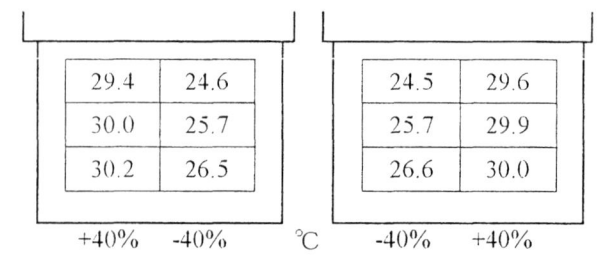

29.4	24.6
30.0	25.7
30.2	26.5

24.5	29.6
25.7	29.9
26.6	30.0

+40% -40%   °C   -40% +40%

Fig. 7 Example of Temperature Rise Distribution at Imbalance Heat load (Total: 2400w)

3. Compact and lightweight.

Because an efficient PHHE is used for the condensing part, it is possible to make the cooler compact and lightweight. Table 2 shows an example of comparison between SIPHOREX and heat pipe. In this condition, the size (volume) of SIPHOREX is about 50% of heat pipe and the mass is 60%.

Table 2 Comparison between SIPHOREX and Heat Pipe

	SIPHOREX	Heat Pipe
Volume (m^3)	0.008 (8 litters)	0.017 (17 litters)
Mass (kg)	8.0	14.5

Number of devices: 12(500A class), Heat load: 4kw, Air flow rate: 10m^3/min

4. The wide design flexibility for wide application.

If only the boiling part is arranged under the condensing part, mounting position of devices and direction of cooling air can be freely. The examples are shown in Fig. 8.

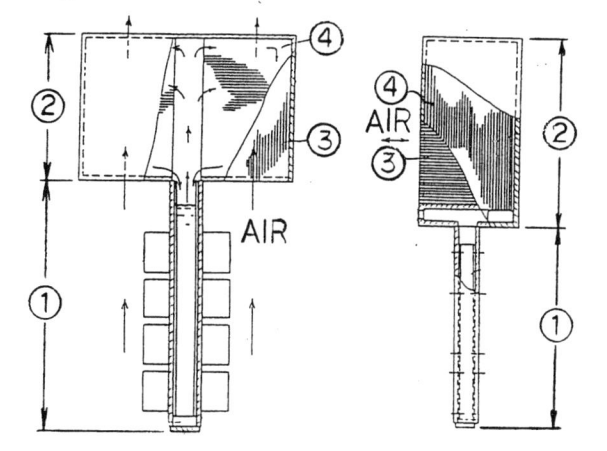

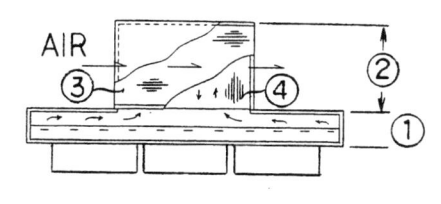

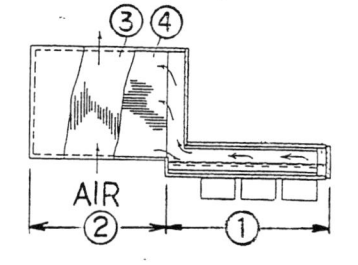

① Boiling part
② Condensing part
③ Air side fin
④ Refrigerant side fin

Fig. 8 Design Variations of SIPHOREX.

APPLICATION EXAMPLE OF SIPHOREX

SIPHOREX is being used for the following usage now, and is useful for making various power source equipments compact and lightweight.

1. For drive power sources of railway cars (Shinkansen etc.).

As for Shinkansen which is a super express train in Japan, several kinds of new cars have been introduced during these four years. IGBT with large capacity is used for these drive power source, and SIPHOREX is used for the almost all types shown in Table 3. The outward appearance of 700 Series Nozomi is shown in Fig. 9, and the outward appearance of SIPHOREX used for this train is shown in Fig. 10. IGBT is horizontally mounted on the underside of the boiling part, and the cooling air flows horizontally in this type. SIPHOREX is being rated highly, that it is possible to make the power source equipment compact and lightweight of course, and deal with the imbalance heat load of IGBT with large capacity, too.

Table 3 Types of Shinkansen using SIPHOREX

Type	Nickname	Railroad Company
E2	Nagano Sinkansen	JR East
E4	Max	JR East
700	700 series Nozomi	JR Tokai
700N	Rail Star	JR West

Fig. 9 Outward Appearance of 700 Series Nozomi which is a Super Express Train in Japan (Maximum operation speed: 285km/h)

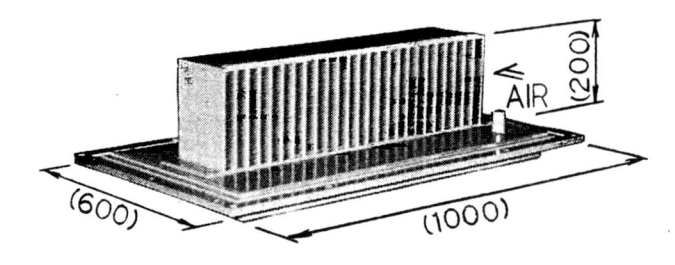

Number of devices: 12
Fig. 10 Outward Appearance of SIPHOREX used for 700 Series Nozomi

2. For the drive power sources in rolling plant etc.

IGBT has been used also for the power sources of drive equipment in such as the rolling plants and the paper manufacture plants, and SIPOHREX has been applied as the cooler. The outward appearance is shown in Fig. 11. IGBT is vertically mounted on side of the boiling part, and the cooling air flows vertically in this type. It is of good repute, because the temperature difference between each devices is very small, though a lot of IGBT is mounted and the heat load sometimes becomes imbalance.

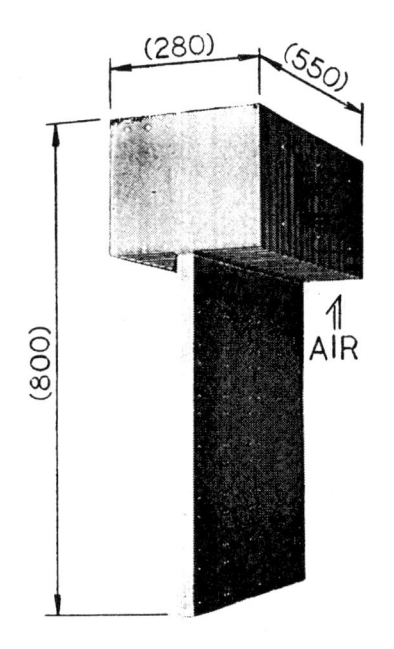

Total heat load: 10kw
Number of devices: 48
Cooling air flow rate: 25m^3/min

Fig. 11 Outward Appearance of SIPHOREX used for Drive Equipment.

3. Others

SIPHOREX is used for UPS (Uninterruptible Power Source), the drive power source of the elevator, etc. It is useful for the miniaturization of these equipments. An outward appearance of SIPHOREX for UPS is shown in Fig. 12. IGBT is mounted vertically and the cooling air flows horizontally in this type.

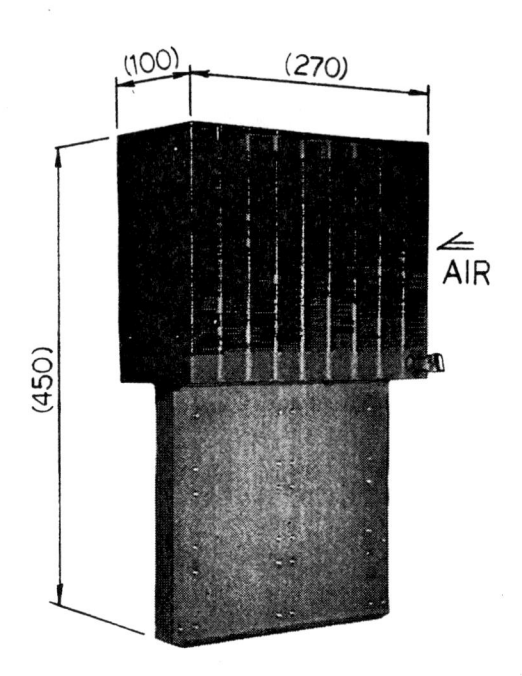

Total heat load: 4kw
Number of devices: 12
Cooling air flow rate: 10m^3/min

Fig. 12 Outward Appearance of SIPHOREX used for UPS

CONCLUSIONS

As mentioned above, the semiconductor device cooler that had a high efficiency and was compact and lightweight could be produced. Especially, this cooler is more effective for mounting a lot of devices and using high heat density devices, and it can cope with the development of power electronics technology in recent years.

In the future, we are going to advance the examination changing the refrigerant used for this cooler now into the other material which is more gentle to the earth environment.

PATENTS

Following patents are applied.
① One patent for International
② Three patents for domestic

BOILING REFRIGERANT TYPE COMPACT COOLING UNIT FOR COMPUTER CHIP

Hiroshi Tanaka, Takahide Ohara, Kiyoshi Kawaguchi, Tadayoshi Terao

DENSO CORPORATION, Kariya-shi Aichi-pref., Japan, E-mail: hiroshi_h_tanaka@notes.denso.co.jp

ABSTRACT

In recent years, CPU's (Central Processing Unit) calorific power has been increasing rapidly involved in the current computer trend for higher performance with larger frequency. So computer industries are requiring high performance cooling devices for CPU's large calorific power and high heat flux.

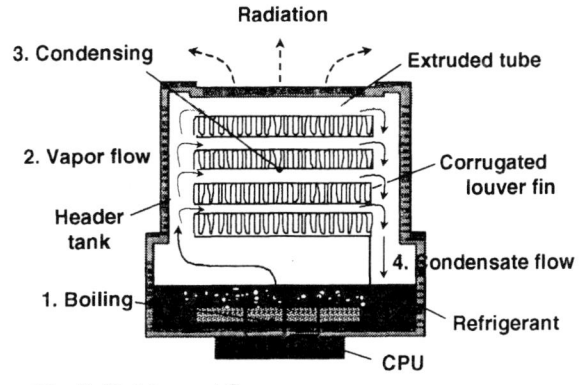

We have developed a new compact boiling refrigerant type cooling unit. [1),2),3)] The main feature of this cooling unit is high cooling performance with all aluminum lightweight and compact body, comparing to the conventional aluminum heat sink.

In the cooling unit we enclose refrigerant HFC-134a, alternative fluorocarbon utilized in the car air conditioning systems. Refrigerant transfers CPU's heat from the boiling area to the condensing area efficiently, changing phases gas to liquid, and then radiates heat into the cooling air. In the cooling unit, refrigerant circulates continuously with gravitation.[4)]

In this paper we report cooling performance of the boiling refrigerant type cooling unit comparing with the conventional aluminum heat sink.

STRUCTURE OF THE BOILING REFRIGERANT TYPE COOLING UNIT

This boiling refrigerant type cooling unit has high pressure resistant and airtight structure for enclosing refrigerant HFC-134a, alternative fluorocarbon utilized in the car air conditioning systems.

Fig. 1 shows photo and structure of the cooling unit. This cooling unit mainly consists of the refrigerant bath and the radiator core. All parts are made of aluminum and brazed at once.

Fig. 2 shows refrigerant flow in the cooling unit. In the refrigerant bath, refrigerant boils on the extended surface and circulates toward one header tank. This structure reduces heat resistance at the boiling area in the refrigerant bath.

In the radiator core, extruded tubes and corrugated louver fins structure utilized in automobile condensers, form refrigerant pass with the flat header tanks. Refrigerant gas from the refrigerant bath condenses through the tubes, radiating heat into the cooling air through the corrugated louver fins. Then condensate returns through the other header tank to the refrigerant bath. Thus, refrigerant changes phases from gas to liquid and circulates continuously with gravitation.

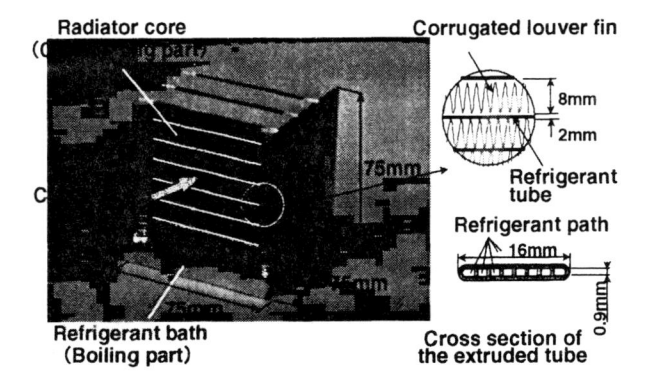

Fig. 1 Boiling refrigerant type cooling unit for CPU

Fig. 2 Refrigerant flow in the boiling refrigerant type cooling unit

Table 1 Prototype's specification of the boiling refrigerant type cooling unit

Items	Contents
Cooling performance (Heat resistance)	$\Delta T = 24$ K (0.2 K/W or less) ΔT= Contact surface temperature with CPU - Ambient temperature
Heat load	120 W
Cooling air velocity	1.6 m/s
Refrigerant	HFC-134a
Size	75(W) × 75(D) × 75(H)mm

Table 1 shows prototype's specification of the boiling refrigerant type cooling unit.

We defined the cooling performance as the temperature difference - ΔT - between the contact surface temperature on the cooling unit - Ts - and the inlet cooling air temperature -Ta - , and also the thermal resistance, dividing the difference between the contact surface temperature and the cooling air temperature by the heat load.

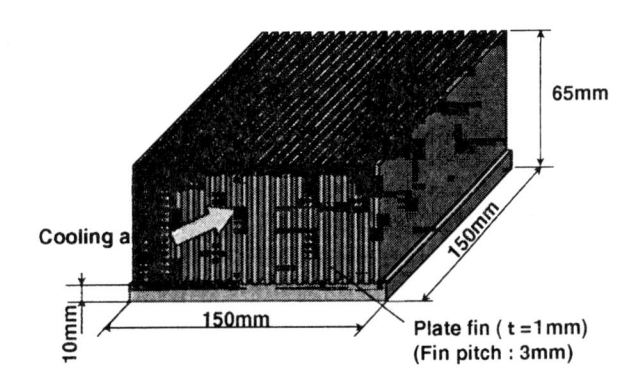

STRUCTURE OF THE ALUMINUM HEAT SINK

We prepared the conventional aluminum heat sink to compare the size at the same cooling performance.
Fig. 3 shows schematic diagram of the aluminum heat sink. Aluminum base plate and fins are all brazed to keep high heat conductivity.
Table 2 shows prototype's specification of the aluminum heat sink.

Table 2 Prototype's specification of the aluminum heat sink

Items	Contents
Cooling performance (Heat resistance)	$\Delta T = 24$ K (0.2 K/W or less) ΔT= Contact surface temperature with CPU - Ambient temperature
Heat load	120 W
Cooling air velocity	1.3 m/s
Size	150(W) × 150(D) × 65(H)mm

EXPERIMENTAL METHOD

Fig. 4 shows photo of the boiling refrigerant type cooling unit and the aluminum heat sink for the experiments.

Fig. 5 shows the experimental apparatus for these cooling units. We evaluate both cooling units attaching Cu heater block - contact surface area 25mm x 25mm - on the bottom of the cooling units instead of the CPU. We adjust applied voltage to control heat load onto the cooling units. The radiator core and fins of

Fig. 4 Prototype of the aluminum heat sink

Fig. 3 Prototype of the aluminum heat sink

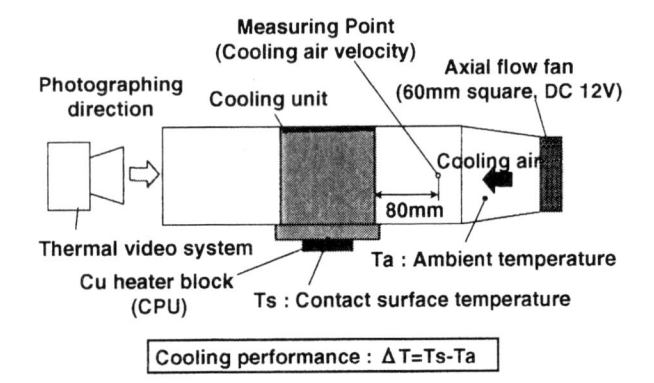

Fig. 5 Experimental apparatus

the cooling units are surrounded by the duct, and then the 60mm square axial flow fan - thickness 25mm, standard voltage 12V - provided cooling air. We measured the average cooling air velocity at the point of 80 mm upstream apart from the radiator core. We measured 1.6 m/s for the boiling refrigerant type cooling unit and 1.3 m/s for the aluminum heat sink. We defined the cooling performance as the temperature difference - ΔT - between the contact surface temperature - Ts - on the cooling unit and the inlet cooling air temperature -Ta - (1).

$$\Delta T = Ts - Ta \qquad (1)$$

We measured the contact surface temperature - Ts - and the inlet cooling air temperature -Ta - with the T type thermocouple. And we also analyzed the temperature distribution on the radiator core and fins with the thermal video system.

EXPERIMENTAL RESULTS

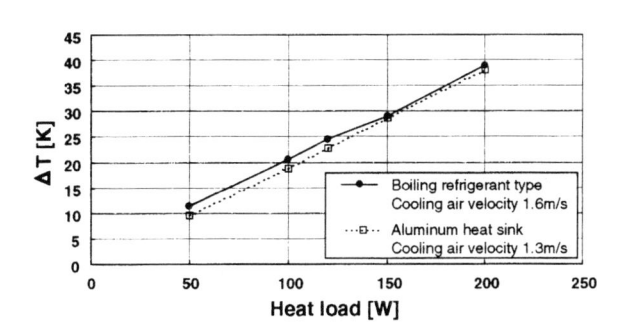

Fig. 6 shows comparison of the cooling performance - ΔT - against the heat load both the boiling refrigerant type cooling unit and the aluminum heat sink. At the 120W heat load with a 60mm square axial flow fan, the cooling performance - ΔT - of the boiling refrigerant type cooling unit is 25K, while that of the aluminum heat sink is 24K. Almost the same cooling performance has been obtained.

Fig. 7 shows comparison of the temperature distribution between the boiling refrigerant type cooling unit and the aluminum heat sink at the same cooling performance. The boiling refrigerant type cooling unit has uniform thermal distribution, while the aluminum heat sink has poor temperature l distribution.

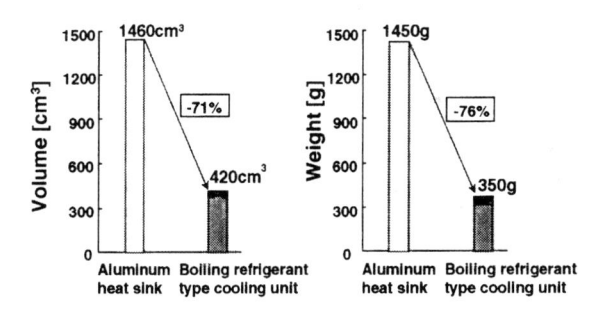

Fig. 8 shows comparison of the volume and weight between the boiling refrigerant type cooling unit and the aluminum heat sink at the same cooling performance. The boiling refrigerant type cooling unit decreases by 71% in volume and 76% in weight from the aluminum heat sink.

Fig. 9 shows the comparison of heat abstraction way of the boiling refrigerant type cooling unit and the aluminum heat sink.

In case of the aluminum heat sink, heat from CPU is transferred only by the aluminum thermal conductivity and then the temperature difference between the contact surface and the cooling fins results in 15.9K.

In case of the boiling refrigerant type cooling unit, the latent heat transfer with boiling and condensing and the thermal conductivity transfers heat from CPU.

Fig. 6 Comparison of the cooling performance

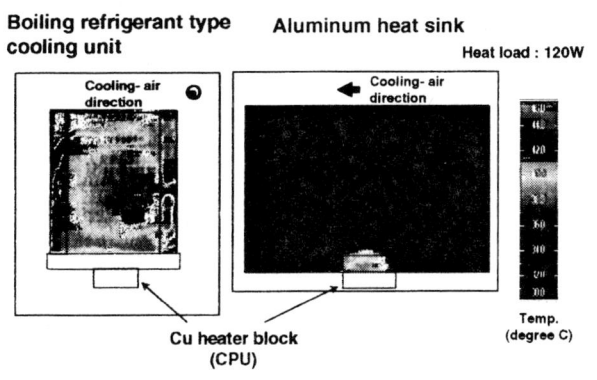

Fig. 7 Comparison of the thermal distribution

Fig. 8 Comparison of the volume and weight

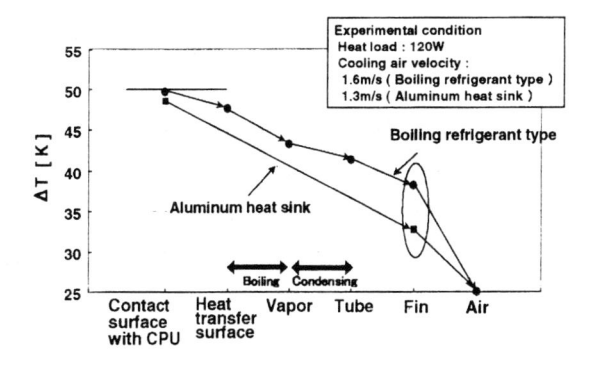

Fig. 9 Comparison of the heat abstraction way

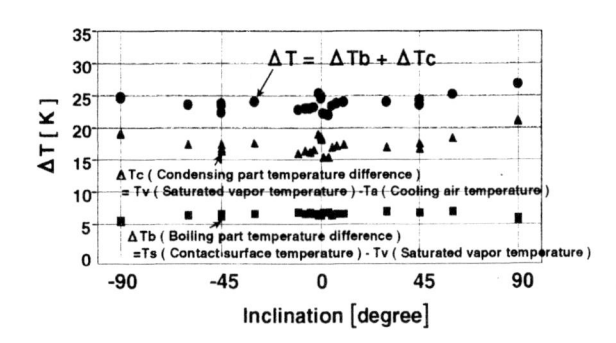

Fig. 10 Cooling performance with inclination

Fig. 10 shows the cooling performance of the boiling refrigerant type cooling unit in case of inclination. Around +10 degree or -10 degree inclination improves the cooling performance 10.8% - 2.7K -. This inclination could support refrigerant flow through the tubes in the radiator core and improve refrigerant circulation in the cooling unit.

We measured inner pressure to estimate the saturated vapor temperature - Tv - and defined the boiling part temperature difference - $\Delta T b$ - (2) and the condensing part temperature difference - $\Delta T c$ - (3).

$$\Delta T b = Ts - Tv \quad (2)$$
$$\Delta T c = Tv - Ts \quad (3)$$
$$\Delta T b + \Delta T c = Ts - Ta = \Delta T \quad (4)$$

When the inclination degree is getting larger, the effective condensing area in the radiator core becomes shorter and the condensing part temperature difference - $\Delta T c$ - increases. In case of 90 degree inclination the cooling performance declines only 7.6% - 1.9K -. We enclose adequate quantity of refrigerant into the cooling unit - approx. 60% of the internal volume - to receive CPU's heat to refrigerant constantly even in case of 90-degree inclination. When refrigerant quantity is too little, the boiling part temperature difference - $\Delta T b$ - would increase. Too much refrigerant decreases effective condensing area and too little refrigerant cannot receive CPU's heat, especially in case of larger inclination.

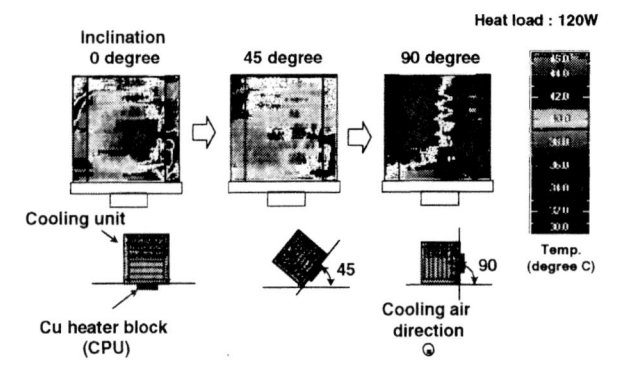

Fig. 11 Temperature distribution under inclination

Fig. 11 shows temperature distribution of the boiling refrigerant type cooling unit in case of inclination. This cooling unit works from +90-degree to -90-degree less than 11% decline in the cooling performance.

CONCLUSION

We have developed new compact boiling refrigerant type cooling units, in which refrigerant circulates continuously with gravitation. This cooling unit achieves a 71% smaller and a 76% lighter weight than the aluminum heat sink of the same cooling performance.

REFERENCE

1) K. Kawaguchi et al, 1997, Compact Cooling Unit for Power Modules Using Boiling heat transfer, ASME International Joint Power Generation Conference, pp. 453-460

2) M. Suzuki et al, Jun. 1998, Compact Thermosyphon using Multi-stacked Radiator Cores for Automobile, JSMA, Vol.64 - 622, pp1861 - 1866.

3) M. Suzuki et al, Jul. 1998, Compact Thermosyphon using Multi-stacked Radiator Cores for Automobile, JSMA, Vol.64 - 623, pp2244 - 2249.

4) T. Terao et al, Mar. 2000, Boiling Refrigerant Type Compact Cooling Unit for Computer Chip, JSMA, No.003-1, pp99 - 100.

REDUCTION OF MALDISTRIBUTION IN LARGE RISING FILM PLATE EVAPORATORS

Martin Holm, Inge Nilsson and Björn Wilhelmsson

Alfa Laval Thermal AB, Rudeboksvägen SE-22100 Lund, Sweden;
martin.holm@alfalaval.com; inge.nilsson@alfalaval.com; bjorn.wilhelmsson@alfalaval.com

ABSTRACT

Large plate heat exchangers working with two-phase fluids are often exposed to flow maldistribution due to the differences in density and velocity of the fluids in the heat exchanger. To overcome this maldistribution many types of methods have been applied to solve the problem. However, none of them works in a satisfying way. In this paper a novel method of improving the distribution is presented. The new distribution system has been tested in model scale in a rig working with water and air as two-phase media. From the experiments it is clear that the new distribution system works well and that the improvement in the distribution is significant for many types of two-phase flows.

BACKGROUND AND OBJECTIVES

Obtaining a correct distribution of two-phase fluid in a plate heat exchanger has always been difficult. The main problem with the two-phase fluid is that the two phases do not distribute evenly between the different individual channels of the heat exchanger. This means that some channels will have an excess of liquid and other channels will have a deficit of liquid. This phase maldistribution results in less than optimal performance and in lower product quality.

The intended use of the herein-described distribution system is in large evaporators used to concentrate industrial fluids, e.g. sugar, stillage and caustic. These are often installed in multi-stage evaporation plants that run in co-current mode. This means that steam and product both go from a higher to a lower pressure. The pressure differential between two consecutive effects is often 0.3 bar or higher. As the product leaves one effect it is at its boiling temperature and on its way to the next effect, its pressure will be reduced due to frictional and local pressure drops in the pipe runs. As a result, flash vapour will be generated; thus, the next effect has to handle a two-phase feed.

The evaporator can be of considerable length and can therefore suffer from large problems with uneven distribution of the two phases.

There are of course other applications that can benefit from this new distributor, e.g. refrigeration plate heat exchangers or any other plate heat exchanger working in two-phase applications.

Two-phase flow

Dividing a two-phase flow in two or more branches is generally difficult, as the physical properties of the phases are very. The problem of distributing two-phase flow in junctions has not been studied extensively. Examples from the literature are Peng et al. [1], Collier [2] and Azzopardi [3]. Much work on the topic has been performed within companies or research organisations; this work is generally not public. From the literature it is clear that the orientation of the branch have a strong influence on the phase distribution. The common problem at the junction is that the lighter fluids, e.g. steam or gas, moves more easily than the heavier component, e.g. water or sugar juice. This can result in completely different flow qualities in the two branches after the junction. The distribution of liquid and gas or steam in the cross section of a tube can be completely different depending on the flow rates of the two phases. Two examples of two-phase distributions are shown in fig. 1.

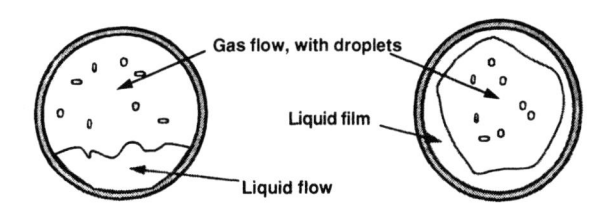

Figure 1. Examples of separated two-phase flow in a horizontal straight tube: stratified and annular flows.

Field observations and Present solutions

Observations from a large base of installed plate evaporator units indicate that there can be considerable maldistribution along a heat exchanger plate pack. It has been observed that below a certain maximum length, 1.5 to 2 m, no distribution problem occur. Today several methods exist to overcome the problem of maldistribution. Some of the methods that have been tested are shown in table 1.

Table 1. Methods of improving the flow distribution

Method	Positive	Negative
Multiple and smaller units	Effective	Very costly
Feed from two sides	Effective	Costly and inconvenient
Restrictions in inlet to each individual plate channel	Inexpensive, enhance boiling stability	Clogging in the restriction, moderate improvement of the distribution
Manifold pipe in the inlet	Effective	Inconvenient, costly

The method employed by Alfa Laval for large evaporators involved restrictions in the inlet of the individual heat exchanger channels. The optimal restriction size is a compromise between large holes to prevent clogging and small holes to improve distribution.

EXPERIMENTAL WORK

The general idea of the new distributor is that a vertical plate deflects part of the flow from the main stream. The amount of deflected flow divided by the full inlet flow is defined as the deflection ratio. To verify the suggested new method of improving the two-phase distribution, experiments had to be carried out. The heat exchangers targeted for this new distribution system are, however, very large rising film evaporators. In order to carry out the experiments the heat exchanger had to be scaled down to a more manageable size. Instead of sugar juice and vapour, water and compressed air were used as test fluids. To even more simplify the tests they were performed adiabatically. A single-phase test was also performed. To determine the performance of the distributor system, four different sets of experiments were carried out. The first tests were carried out with only a single flow distributor. Test no. 2, 3 and 4 were carried out in a special test heat exchanger in which different distributor combinations could be tested.

Experimental set-up, test series 1

The aim of the first set of tests was to determine how the deflection ratio was dependent on the two-phase regime and the position of the deflector plate, see fig 2. A whole range of two-phase flow types was tested. In the test rig a distributor was placed after the mixing device with 15 diameter of straight tube in between the mixer and the distributor. The mixing device sprays the water radially into the passing stream of air, see fig. 3. Only the water flow rate was measured. The distributor was made of a 72 mm stainless tube with a 15° slot cut into the tube. The deflecting plate was mounted onto the slot. The deflecting plate could be mounted in different positions by moving part of the distributor along the tube, see fig. 3. The flow-measuring device consisted of an open tank in which the not deflected part of the water was collected. This flow was measured with the bucket-and-stopwatch method. The water flow rate before the distributor was measured with an inductive volumetric flow meter, Process data, PD 340 flow-transmitter, and the air flow rate with a mass flow meter working on the principle of heat transfer, Bronkhorst HIGH-TECH B.V.

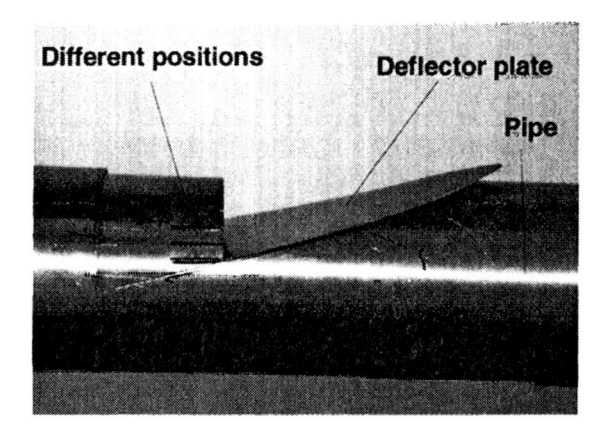

Figure 2. Picture of distributor used in the first test.

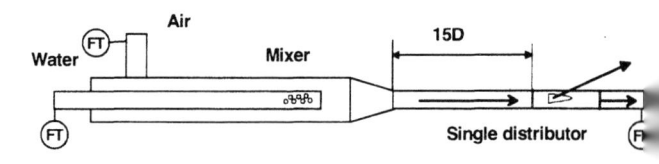

Figure 3. Set-up of the first experiment

Table 2. Tested flow rates for test series no. 1 with a single distributor

Water (kg/s)	0.5	0.5	0.5
Air (kg/s)	0.01	0.05	0.18
X (%)	2	10	36

Water (kg/s)	2	2	2
Air (kg/s)	0.01	0.05	0.18
X (%)	0.5	2.5	9

Water (kg/s)	5	5	5
Air (kg/s)	0.01	0.05	0.18
X (%)	0.2	1	3.6

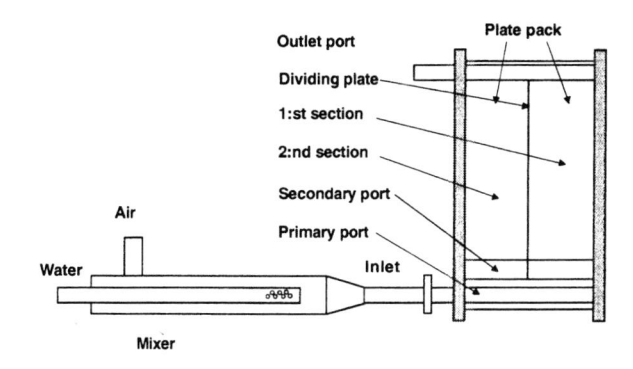

Figure 4. Set-up of test series 2, 3 and 4.

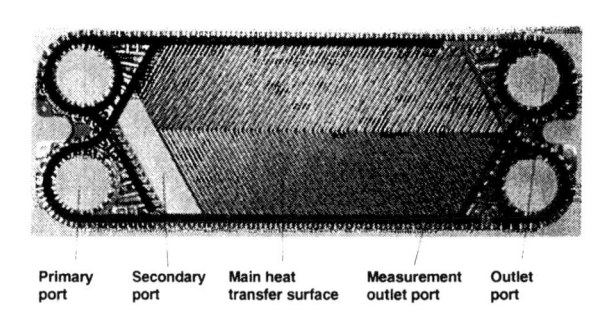

Figure 5. Picture of test plate used in the experiment no. 2, 3 and 4

Experimental set-up, test series 2, 3 and 4

The aim of these tests was to find out how well the distributor works inside a heat exchanger. The tests were conducted with a modified small-scale plate heat exchanger, an Alfa Laval P2 PHE, with 141 plates, see fig. 4. These channels form 70 active channels where the fluid flows. The plates in the test PHE have been modified so that a second port has been cut out from the plate, see fig. 5. It was then possible to place a distributor in the primary port. The inserted distributor had a 15° slot as in test no 1 and the same mixing device in front of the heat exchanger was also used.

The tests series performed were:
2 Heat exchanger without any distributor and full connection between the primary and secondary ports, fig. 6a.
3 Heat exchanger without any distributor and partial connection between the primary and secondary ports, fig. 6b.
4 Heat exchanger with distributor in the primary port in the first section and partial connection into the second section, fig. 6c.

The second test series was conducted with completely full connection between the primary and secondary ports. This test series resembles a normal heat exchanger with only one port. In the third test series there was only partial connection between the primary and secondary port at two separate locations. The number of open channels between the ports was 14 in the first section and 11 in the second section. This resembles an earlier test series conducted with another test object. In the fourth test series the distributor were tested with one distributor inserted in the primary port of the first section and an opening between the primary and secondary ports of the second chamber. The number of open channels was the same as in test series no. 3.

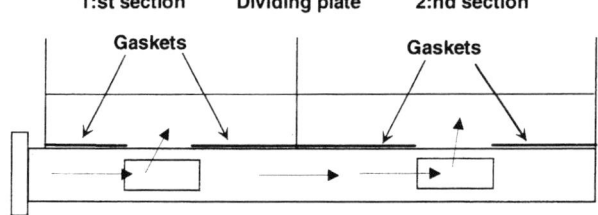

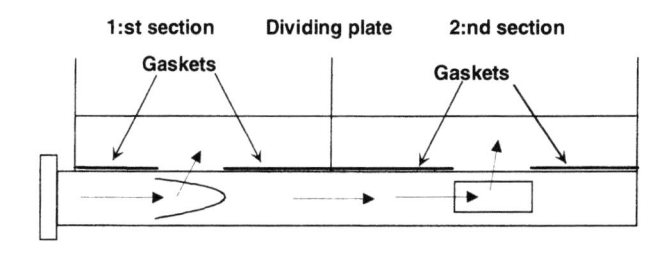

Figure 6. Set-up in the test using the small-scale heat exchanger, 2. 3 and 4. From above a, b and c.

In these tests the intention was to individually measure the flow rate in some specially prepared measuring channels. These channels were placed at equal distance between each other in the plate pack. The flow in these channels was blocked from the outlet port by gaskets. The flows instead emptied to the ambient through openings close to the outlet port and then through a pressure-regulating valve, fig. 7 and 8. These channels are not believed to significantly change the general appearance of the pressure in the outlet channel. Furthermore, only one measurement channel is opened at a time so only about 1/70 of the total flow rate is missing at the exit of the heat exchanger.

The test procedure consists of two moments. First the pressure profile is measured along the heat exchanger at the four measuring channels. During this process all the exit valves on the special measuring channels are closed. Then the pressure is measured at the outlet of one of the measuring channels at a time. Under identical feed conditions, the pressure is adjusted to the same value as in the first part of the test. If the channel pressure drop is the same in both cases, the flow rate should also be the same. The flow rate from the measuring channel was then measured with the same method as in test series no. 1.

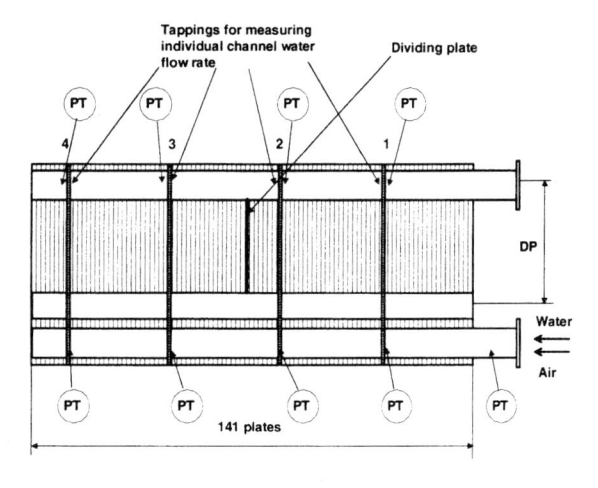

Figure 7. Set-up of P2 test PHE

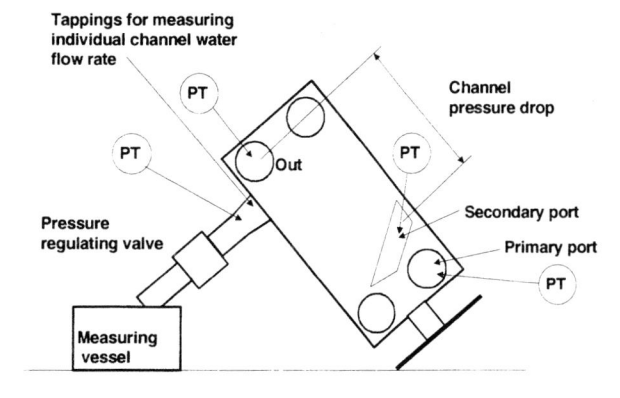

Figure 8 Set-up of P2 test PHE

Table 3. Tested flow rates for the P2 test PHE.

Water (kg/s)	5	4	3
Air (kg/s)	0.05	0.06	0.075
x (%)	1	1.5	2.5

Water (kg/s)	2.5	1.5	1
Air (kg/s)	0.125	0.15	0.18
x (%)	5	10	18

Scaling of experiment

There are some parameters necessary to consider when scaling the experiment. The most important is to have the same two-phase flow conditions for the model and the heat exchangers. This means that as long as the position on the flow regime map is the same in the experiments as in real duty, the two-phase distribution in the inlet of the primary port should be the same in both cases. Further into the primary port of the test heat exchanger this may not be the truth. The next factor to scale is the relation between the pressure drop in the outlet port to the pressure drop in the heat exchanger channel. Both these parameters were considered when scaling the test object.

RESULTS

Test series 1.

Visual observations during the test confirmed that the two-phase flow conditions correspond well to those in the Baker [4] chart, see fig. 9. In the flowchart the direction is indicated how the two-phase flow will change from dispersed to slug and stratified as one looks at different positions along the primary port. This assumes equal tapping of gas and liquid in each channel. Results from the experiments with the single distributor reveals an almost linear relationship between the deflected water flow rate and the total flow rate into the distributor. The spread in the data is considerable but this is reduced as the deflector opening increase, see fig. 10. There is no clear correlation between the two-phase regime entering the distributor and the deflection ratio. The relation obtained from these experiments was later used when designing the further experiments and in the design of the real heat exchanger. Below a certain minimum deflection ratio the spread in the data was very high and sensitive to the different two-phase profile entering the distributor. Thus giving a lowest recommended deflection rate of about 20%. The reason for this is probably found in the nature of the two-phase flow in the primary port, see fig. 1.

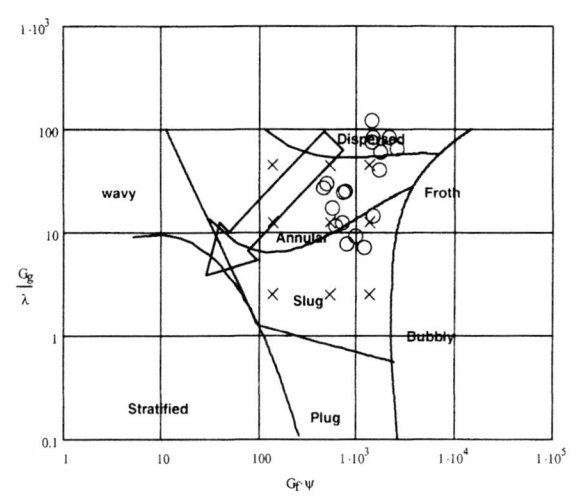

Figure 9. Baker two-phase flow chart for test no. 1. × indicate tested points O indicates real cases as often found in the sugar industry.

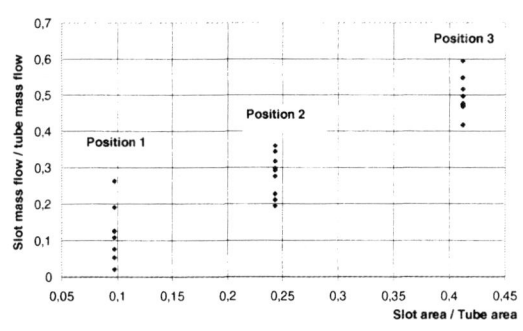

Figure 10. Result from test no. 1 with a single distributor, positions refer to fig. 2.

Test series 2, 3 and 4

Results from the measurements with the specially designed small-scale PHE reveals interesting data. In principle the distributor system works well but it will take some tuning of some of the parameters to prevent phase maldistribution. The tuning factors are the slot angle, slot opening and the size of the hole. The results from these experiments are displayed together for one flow mass quality, x, at a time for all three sets, see fig. 11. From the measurements it is clear that the combination of one distributor and partial connection works best for higher mass qualities. It is also clear that only partial connection between the primary and secondary ports is of little help to improve the distribution. The distributor system is also effective for one-phase flow. This confirms earlier experiments that were conducted with a completely different experimental set-up.

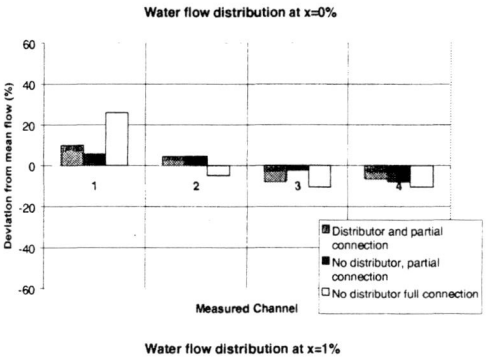

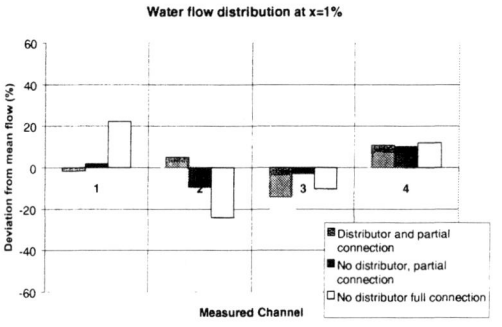

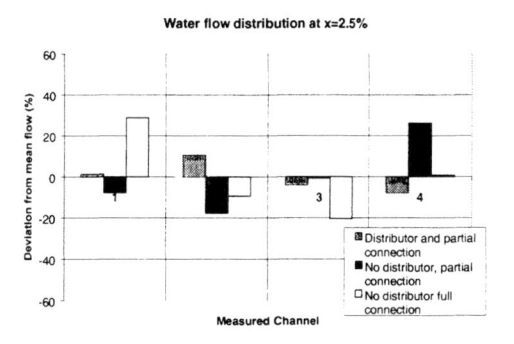

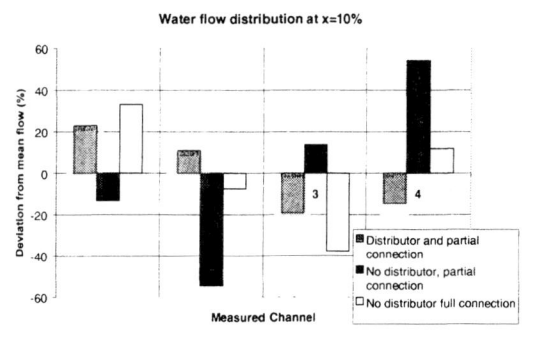

Figure 11. Result from different tests with the small-scale PHE model.

The cost of the distribution system is pressure drop. The pressure drop on the heat transferring surfaces for each two-phase case is measured to be constant. The pressure loss between the primary port and the secondary port divided by the total pressure drop has been measured, see table. 4.

Table 4 Relative pressure drop of the different distributors, refer to fig. 6.

Distributor system	Pressure loss / total pressure loss
None (a)	10%
Holes (b)	20%
Distributor and holes (c)	30%

Accuracy

Due to the relative simplicity of the set-up and the fact that the two-phase flow is extremely intermittent in its behaviour, the accuracy in the measurements is not too high. Fortunately, the flow rate is proportional to the difference in pressure to the square root and this tends to smooth out the fluctuations. Generally, the flow was pulsating thereby making the pressure drop measurements difficult. The stability of the air pressure was not satisfying as other events in the laboratory affected the experiment. One way of estimating the error is to compare the total flow rate per channel and the average flow rates of the measurement channels. The measured flows in the measuring channels are always lower than the total mean flow rate. The average deviation for all the tested qualities is 4%, 5% and 14% for test no. 2, 3 and 4.

CONCLUSIONS AND FUTURE WORK

It is clear that the distribution system works well in a model scale. The improvement in distribution is significant for all types of two-phase flow. The cost of the improved distribution is an increase in pressure drop of less than 30%.

This extra pressure drop is situated outside the heat transferring surfaces and thereby does not interfere with the boiling process. The pressure drop in the distributor results in flashing of the liquid and this reduces the liquid column that have to be lifted thus reducing the gravitational pressure drop. This is beneficial for the heat transfer and temperature difference between the two sides in the heat exchanger. Although the aim was to improve the two-phase distribution the one-phase distribution was also considerably improved. The test performed so far only covers one of many possible configurations that can be used. This means that new tests with other units may have to be performed. The next step is to test a longer unit with more than two sections and more than one distributor inserted in the primary port. For the new measurement an improvement of the measuring system is also needed.

REFERENCES

[1] Peng F., Shoukri M., Chan A. M. C.; "Effect of Branch Orientation on Annular Two-Phase Flow in T-Junctions"; Journal of Fluids Engineering, Transactions of the ASME, March 1996, Vol, 118 pp 166–171.

[2] Collier John G., Thome John R.; "Convective Boiling and Condensation"; 3:rd edition Oxford University Press, 1994.

[3] Azzopardi, B. J.; "The effect of a plate axial to the flow on the liquid distribution in annular two-phase flow"; Nuclear Engineering and Design, 152, 1994, 257–262.

[4] Baker O.; "Design of pipe lines for simultaneous flow of oil and gas"; Oil and gas J., July 26, 1954

Enhanced Reboilers for the Process Industry

B. Thonon

CEA-Grenoble, GRETh, 17 rue des martyrs, 38054 Grenoble, France, e-mail: thonon@cea.fr
Technip, La Défense 6, Cedex 23, 92090 Paris la Défense, France, e-mail: jjdelorme@technip.fr

Abstract

This paper presents the results of a European project dedicated to the improvement and the development of high efficient reboilers used in the oil and gas processing industries. Process intensification is achieved by reducing the size of the equipment and increasing heat transfer, which lead to higher heat recovery and lower energy consumption.

This project has clear industrial objectives:
- the establishment of thermal and hydraulic performances of innovative heat transfer surfaces under actual flow conditions (boiling of hydrocarbons);
- the development of a new design methodology, based on local and overall flow modelling;
- the construction and the tests of two prototypes of plate-fin reboilers;
- the construction and the test of a 5 MW pilot unit incorporating enhanced tubes.

The objectives have been reached with the active participation of 10 partners including 7 industrial companies (CIAT, CS-Informatique, Nordon, Shell, Technip, Targor and Wieland) from 4 countries.

In a first step, several innovative boiling structures have been developed and tested using hydrocarbons as working fluid. Upon the four geometries tested, the tube No.4 has the higher thermal performance especially for low and intermediate heat fluxes (10 to 30 kW/m²). With propane, heat transfer coefficients between 12 to 20 kW/m²K were measured for a saturation temperature of 293 K. This corresponds to a factor 3 increase compare to plain tubes and a factor 2 increase compared to low fin tubes. The best enhanced boiling tube has been selected, then tests on a bundle of 45 tubes with the same operating conditions have been performed. The data obtained have confirmed the performances measured on single tube experiments and it has been decided to build a 5 MW pilot unit incorporating theses new enhanced tubes. The selected case is a C3 splitter reboiler in a polypropylene plant.

INDUSTRIAL BACKGROUND

Shell side boiling in horizontal tube bundles is used in a variety of heat transfer applications in the chemical and petrochemical industries (Mc Carthy and Smith (1995), Sloley [1997] and Huchler [1999]), and the heat transfer and fluid flow processes are significantly different than what occurs in in-tube flows. Applications to bundle boiling of correlations or model obtained for boiling on single tube or in-tube flows is inappropriate and leads to inaccurately designed units. For single tube boiling of pure fluids, data and design methods are available in the open literature, but for an entire tube bundle with process fluids basic information and reliable predictive method are still not available. An enhanced structured surface tubes (re-entrant cavities, which promote nucleate boiling) have been studied for refrigeration application, with fluid such as CFC or HFC. But these fluids do not allow any extension to process fluids, as the structured tubes are sensible to thermo-physical properties such as surface tension, viscosity. These properties are significantly different between refrigerants and hydrocarbons (propane, propylene,..)

Process industries are very conservative regarding to innovative technologies and uses well established and proven technologies (plain tubes). The first reason is that the process industry deals with a variety of fluids and that the physical properties are not always specially known. As there is a lack of data and reliable design methods, especially with non-plain tubes, design engineers will select conventional solutions. Secondly, the capital costs of heat transfer equipments is rarely a significant fraction of the product costs, and any shut down of the plant will cost considerably more than the cost of the individual unit. This latter reason must be balanced when in the running costs the energy consumption represents a significant amount. In this case, the use of enhanced tube bundle is cost effective and an energy efficient alternative technology to plain tube units.

The application of enhanced tubes in shell-and-tube heat exchangers within the hydrocarbon processing industry has a long tradition. The first applications have been realized in situations of revamps or capacity enhancements. Presently a range of low-finned and double enhanced tubes are available and can be considered to be 'proven technology' in their specific applications fields (figure 1). An overview with typical case studies as well as realized references for olefin plants and refrigeration systems (Moore [1974], Kassem et [1995], Kenney [1979] and Webber [1960]) illustrates the state of this technology have been presented by Mertz et al [1999].

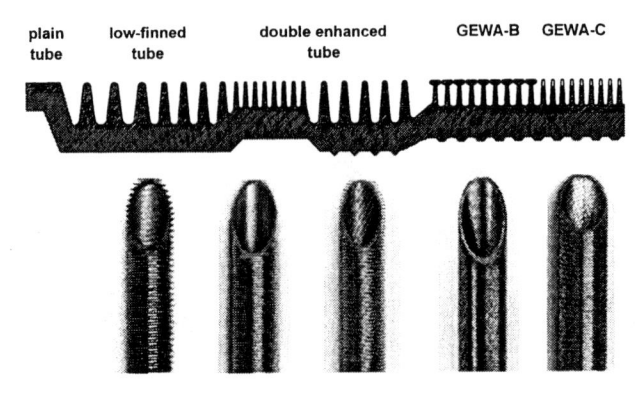

Fig. 1: Development of enhanced tubes

STATE OF THE ART

Extensive information is available on boiling of pure fluids on a single plain or enhanced tube, but for tube bundles the literature is scarce. Review of enhanced boiling on single and tube bundle has been carried out by Thome [1990], Mertz et al [1999] and Browe and Bansal [1999].

Three kinds of studies can be distinguished and except for the overall studies the test fluids were always refrigerants :
- overall measurements on actual heat exchangers Yilmaz and co-authors [1981] [1984]
- local measurements in pool boiling Hahne and co-authors [1983] [1994], Marto and Anderson [1992], Fujita et al [1986], Mertz and Groll [1998] and [1999]
- local measurements in forced convective boiling Hwang and Yao [1986], Jensen and Hsu [1988], Cornwell and Scoones [1988] Jensen et al [1992], Gupte and Webb [1994] and Roser et al [1999].

This brief literature survey of boiling in tube bundles have shown that:
- there is a bundle effect which increases the heat transfer coefficient of a tube bundle compared to a single tube. This bundle effect depends on the heat flux and on the bundle arrangement. For high heat flux, the bundle effect is negligible.
- Enhanced tube bundles have higher heat transfer coefficients than plain tubes and the bundle effect is less than for plain tubes. Tubes with re-entrant cavities have high heat transfer performances, and there is no significant bundle effect.

DEVELOPMENT OF A HIGH PERFORMANCE BOILING TUBE

Tube structure

The tested structured tubes with re-entrant cavities (outer diameter 19.05 mm) are new developments of Wieland Werke AG, Ulm. The main heat transfer area of these heat exchanger surfaces is located in the tube wall and consists of sub-surface channels and cavities (Figures 2 and 3). These internal structures are connected to the surrounding fluid by small openings, e.g. slits and pores. The dimensions of the sub-surface geometries are all in the sub-millimetre range.

Figure 2 : Magnified view of an enhanced evaporation surface

Figure 3 : Cross-sectional view of the sub-surface structures

Single and two-tubes experiments

Experiments have been conducted at IKE (Stuttgart University) with one smooth reference tube and four variations of enhanced tubes with structured surfaces (Mertz et al [2000] and [2001]). The tubes were made of carbon steel St35.8 and as working fluid the hydrocarbon propane was used. The experiments were carried out with single tubes and mini bundles (two tubes inline) at saturation conditions corresponding to temperatures between 253 K and 293 K. The employed heat fluxes ranged from about 2 kW/m^2 to 100 kW/m^2 for single tube experiments and from about 2 kW/m^2 to 70 kW/m^2 for mini bundle experiments. The obtained heat transfer coefficients of the enhanced surfaces were compared with the results of the smooth reference tube to calculate the respective improvement factors (table 1).

The enhanced tube variant 4 shows the best heat transfer coefficients for all tested saturation temperatures. This variant was chosen to be used for the bundle tests at GRETh and to be employed in the shell-and-tube heat exchanger prototype. Experiments with mini bundles, i.e. two tubes inline, were carried out at IKE to investigate the influence of the two-phase flow from the lower tube on the heat transfer of the upper tube. The tests were carried out with a mini bundle consisting of smooth reference surfaces and two mini bundles each employed with one type of enhanced tube, variant 1 and variant 4. In the mini bundle tests with smooth tubes the upper tubes have always higher heat transfer coefficients than the lower tubes. For low heat fluxes the upper tubes show higher heat transfer coefficients by a factor of about 1.4 and for high heat fluxes by a factor of about 1.2. In the mini bundle experiments with enhanced tubes variant 4, the results indicate that for low heat fluxes the upper tube shows higher heat transfer coefficients than the lower tube. With increasing heat flux the differences between the lower and the upper tube become smaller and for high heat fluxes almost no differences can be found. The results for low heat fluxes indicate a strong influence of the two-phase flow from the lower tube on the heat transfer of the upper tube. This influence decreases with increasing heat flux.

Fig. 4: View of the experimental test rig

Bundle experiments

Greth has performed tests on a small bundle of 45 tubes at (figure 4). The general configuration is representative of an industrial reboiler, but differs by the presence of two vertical walls confining the flow. In this manner, the internal shellside recirculating flow is avoided. Consequently, the mass flow rate through the bundle can be controlled. The loop has been designed to operate with flammable fluids (hydrocarbons) under low to moderate pressures (up to 16 bars). The operating fluid (hydrocarbon) is pumped as saturated liquid from the shell of the boiler, then subcooled and reentered to the test section in the tube bundle channel, as shown in figure 5. The subcooled liquid passes through a flow straightener, enters the tube bundle where evaporation takes place after the boiling point is reached, and finally passes the weir level (the end of the channel walls) where the vapour separates from the liquid. The vapour, which leaves the boiler by the top, is condensed and returned to the shell where it mixes with the liquid pouring over the weir. This configuration reduces the two-phase flow instabilities, allowing to work with very low mass fluxes which are representatives of the typical industrial reboilers operating conditions. The tests have been performed using propane as working fluids and the range of operating pressure was similar to the tests performed at IKE.

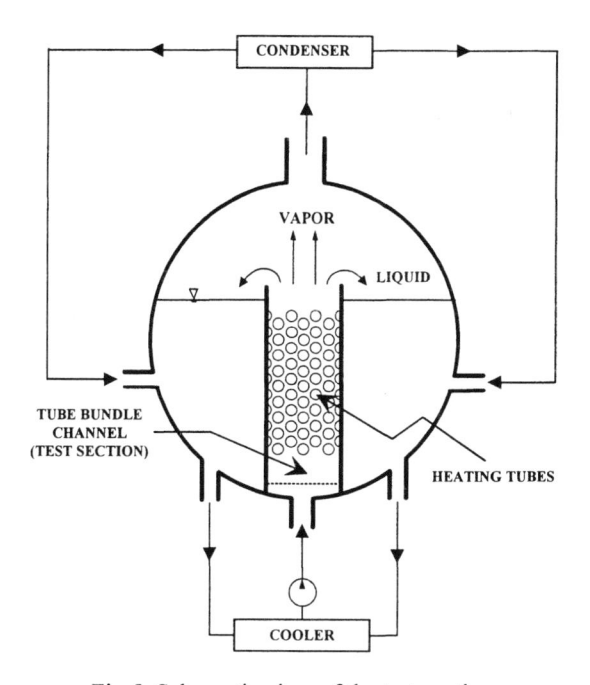

Fig.5: Schematic view of the test section

Table 1: Bundle geometry.

Tube outside diameter	D = 19.05 mm
Tube length	L = 500 mm
Pitch to diameter ratio	P/D = 1.33
Arrangement	staggered inverse equilateral
Number of columns	5
Number of rows	18
Number of tubes	45
Minimum cross section area	$A_{min} = 6*(P-D)*L =$ 0.01886 m^2

To evaluate the thermal performances, an overall heat transfer coefficient is given for the entire bundle. The local analysis has shown that there is no significant bundle effect and that the heat transfer coefficient seems almost constant along the bundle. Furthermore for a constant heat flux, increasing the mass flux does not affect the heat transfer coefficient (figure 6). This indicates that the heat transfer mechanism is nucleate boiling and that the convective effects are negligible.

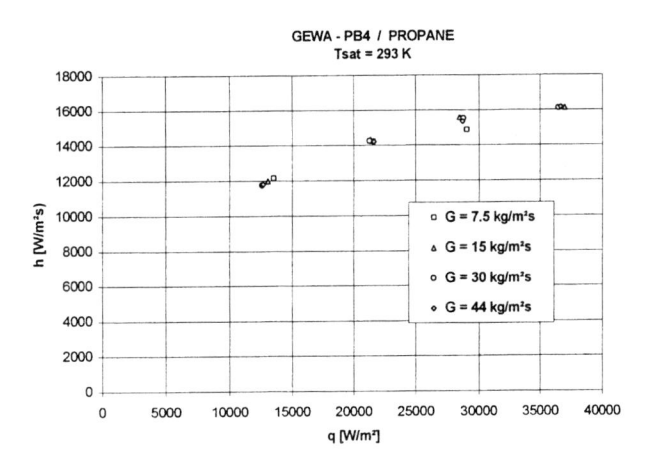

Fig. 6: Effect of the mass velocity on the heat transfer coefficient (bundle tests)

There is a relatively good agreement (up to 25% difference) between the data obtained at IKE on the two-tubes experiments and the data obtained at Greth (figure 7). Several reasons can explain these differences:
- the rising two-phase flow affects the heat transfer performance in the bundle experiments,
- the heating modes are different : constant heat flux at IKE and water heating at GREThn,
- impurities in the fluids.

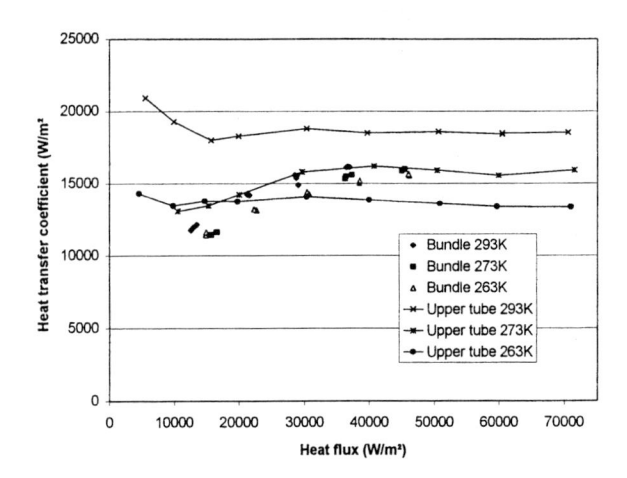

Fig.7: Comparison of Greth and IKE data

Based on the data generated at IKE and Greth is has been decided to manufacture a pilot unit incorporating the new enhanced tubes.

DEVELOPMENT OF A PILOT UNIT

Objectives

In the Knapsack Polypropylene plant near Cologne, Germany, with an annual capacity of 260 kt, the reboiler, named E711, of a propane/propylene – splitter column, was chosen for this project (figure 8). The C3 boiling takes place on the shell side according to the thermosiphon principle by operating conditions from about 13.4 barg and 40 °C. On the tube side the reboiler E 711 is part of a heat recovering system. Via a water circulation system, the heat of reaction in the polymerisation reactor is used for heating of the reboiler E 711. Remaining of the heat which could not be recovered in the reboiler E711 is rejected via a plate heat exchanger to the open re-circulating cooling water system and sent to the atmosphere (figure 9). In order to recover heat to the maximum extend, the full circulation water flow is pumped through the reboiler E 711. Depending on the reaction conditions, the flow rate and temperature of the circulation water are fluctuating. The flow rate is between 400 m³/h up to 455 m³/h and the water inlet temperature between 50 °C and 53 °C. To guarantee stable operating conditions in the splitter, a second steam operated liquid level controlled reboiler E 712 is installed. In this way also a constant liquid static head for the reboiler E 711 is provided.

Fig. 8 : Propane/propylene splitter (Targor plant)

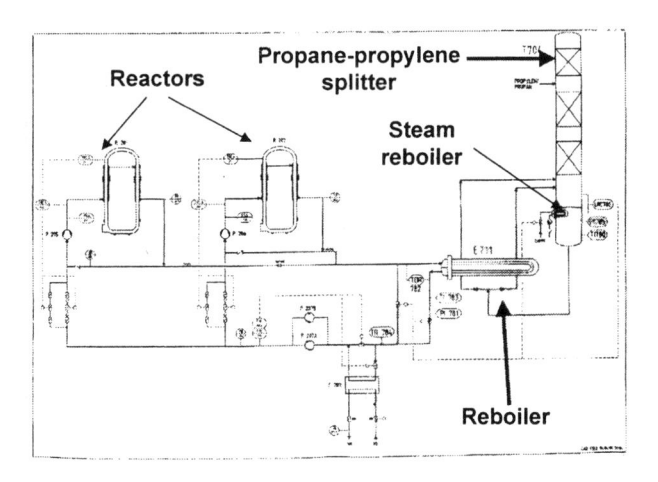

Fig.9: Process description

The objectives for Targor for taking part in this project result from an increase of the annual production rate from 180 kt up to 260 kt. The capacity of the splitter and the reboiler also had to be increased for that reason. As we had a temperature difference between circulation water outlet and boiling propane for about 4 up to 5 K, we decided to enlarge the reboiler E 711. In addition to this, there will be a saving in energy by the reduction of the steam consumption in the reboiler E 712. Due to the limited space in the plant, the new developed "high performances" tubes for boiling of C3 enable us to limit the size of the reboiler, so even the existing pipe couplings could be used.

Heat exchanger manufacturing

Concerning the supply of the enhanced tubes for the field test, Wieland has carried out screening tests using prototypes; these first tests have been quite successful. Thereby it is confirmed that the tube structure can be fabricated with the given tools and that the tube dimensions required for the field test can be achieved (10 m long U-bent tubes). The heat exchanger has been manufactured by Ciat according to the Targor specifications (figure 10).

The general characteristics of the prototype are :
- TEMA type BXU horizontal
- shell diameter : 1.022 m
- number of tubes : 1150
- tube length : 5.25 m
- tube diameter : 19.05 mm
- total heat transfer surface : 360 m²

Fig. 10: Bundle during construction (CIAT factory)

Results

To compare the predicted performances made during the design of the new reboiler (based on experimental curves and correlations) with the real performances observed on site, test-run data have been collected during a few days in study operating conditions. Global performances can be calculated as water information (flowrate, T out, ΔT) and propane information (boiling temperature) are available from the control room.

The main information are the following:
- C3 temperature: 39.7°C
- Water flowrate: 452 m³/h
- Water outlet temperature: 40.6 °C
- Water temperature difference : 9.6 °C

Complementary pressure drop measurement was made directly on the reboiler and lead to about 0.45 bar (water side). The exploitation of the recorded data gives the following results (table 2):

Table 2 : Measured performances

Duty	MW	5.0
Cold approach	°C	0.9
LMTD	°C	3.3
Global HTC (3)	W/m² °C(1)	3670
Waterside HTC	W/m² °C(2)	10800
Boiling side HTC	W/m² °C(1)	8600
Waterside pressure drop	bar	0.45

1. per m² of bare external surface
2. per m² of bare internal surface
3. total equivalent external bare surface = 377 m².

It should be noted that the precision of the temperature measurements has a big influence on the global heat transfer coefficient calculation. In a first estimation, the uncertainty of the global HTC is about ± 20 %. The recalculated boiling side HTC is in accordance with the GRETh curves, as the range of heat flux of the new reboiler is about 2 000 W/m² to 40 000 W/m².

Benefits

To quantify the benefits of PB4 tubes obtained on the C3 splitter reboiler application, we have simulated the performances of an equivalent bare tube reboiler. This simulation has been performed with conventional tools (commercial and Technip in-house programs). The results obtained are presented in table 3. The general data are propane temperature of 39.7°C, water flowrate of 452 t/h and water inlet temperature of 50.2°C

This comparison shows that the gain obtained on the global heat transfer coefficient (+ 90 %) is dispatched between a bigger duty (+ 25 %) and a smaller LMTD (- 35 %) (which leads to a smaller cold approach : 0.9 °C). It should be noted that the repartition of the gain between duty and LMTD is dependant of the basic data (C3 temperature, water inlet temperature, water flowrate).

The aim of the replacement of the existing reboiler was to improve the thermal performances, without any constraint of size reduction. But it could be interesting to imagine that the maximum length of the tube is limited ; let's start with the bare tube solution mentioned here above, which results in 4.0 MW. The table 4 compares the respective sizes of the bare tubes solution and a PB-4 tubes solution designed for the same thermal performances (4.0 MW). The tube length is reduced of 47 %. It should be noted that the water pressure drop reminds the same at equivalent thermal performances.

Table 3 : Data comparison

		Bare tube solution	PB4 tube solution	Ratio
Water T out	°C	42.6	40.6	
Duty	MW	4.0	5.0	
Cold approach	°C	2.0	0.9	
LMTD	°C	5.5	3.6	
Global HTC (1)	W/m² K	1940	3670	1.9
Waterside HTC (2)	W/m² K	6200	10800	1.75
Boiling side HTC (1)	W/m²°C	3800	8600	2.3
Waterside pressure drop	Bar	0.22	0.45	2.0

(1) per m² of external bare surface
(2) per m² of internal bare surface
HTC : Heat Transfer Coefficient

Table 4: Data comparison

Results (1)		Bare tube solution	PB4 tube solution
Water T° out	°C	42.6	42.6
Duty	MW	4.0	4.0
LMTD	°C	5.5	5.5
Global HTC	W/m²°C	1940	3670
Surface required	m²	377	199
Length required	m	5.25	2.8
Water pressure drop	bar	0.2	0.2

1. considering that the same number of tubes is kept.

HTC : Heat Transfer Coefficient

Energy and utility saving

To increase the duty of the C3 splitter reboiler without changing the water flowrate, it was imperative to decrease the cold approach between water and propane. This was observed during the test-run, with a typical cold approach of 0.9 °C. It's very important to notice that this approach would not have been possible with bare tubes, as it corresponds to heat flux values less than 1 000 W/m². At this level of heat flux, boiling phenomenon probably doesn't appear. Due to this smaller cold approach, water pumps have been retained without any increase of the power consumption, which would not have been possible with a new "bare tubes" reboiler. Before the installation of the new reboiler, TARGOR sometimes was using a second reboiler E712 (on LP steam at 4 barg – 152 °C) directly inserted inside the column bottom, to increase the duty of the reboiler E711. When looking at the typical test-run data described in § 2.3.1, the gain of 1 MW represents 1.7 t/h of LP steam, which can be estimated to 150 000 Euro/year.

CONCLUSIONS AND PERSPECTIVES

Within the frame of this project, a new high performance boiling tube has been developed. Through a complete set of testing ranging from single tube experiments, bundle experiments (1 m²) and full size unit (360 m²), the thermal performances have been checked. The performances are up to 3 times higher than plain tubes and twice that of low-finned tubes.

The major applications envisioned are for propane splitters and LNG applications. For short and medium term developments, propane splitters seem to be the most appropriate application, and within the frame of this project field tests have been carried out in a polypropylene plant. The first application of the new high performance boiling tube in a base-load LNG-plant will be a development that will take some time as only few new base-load plants are built world wide.

For the LNG application, the chilling train has been evaluated using conventional design and the technology developed is this project. Within some assumptions, the total length of the heat exchanger is reduced from 67 m to 37 m (figure 11) leading to a weight reduction from 433 tons to 269 tons. Furthermore the liquid inventory of the chillers is reduced from 241 to 172 tons of hydrocarbons, reducing the impact in case of troubleshooting and providing a better controllability. The economic impact has to be evaluated taking into account the optimisation of the process, the cost of the equipment, but also the civil infrastructure and piping.

ACKNOWLEDGEMENTS

This project JOE3-CT97-0061 has been partially supported by the EC, within the frame of the non-nuclear energy programme.

Greth was the coordinator of this project and its success is greatly due to the excellent collaboration between the partners. The authors specially acknowledge Jean-Jacques Delorme and Fabrice Martin from Technip, Aachim Portz from Targor, Thomas Lang from Wieland and Romuald Jurkowsky from Ciat for the work performed during the development and the testing of the pilot unit.

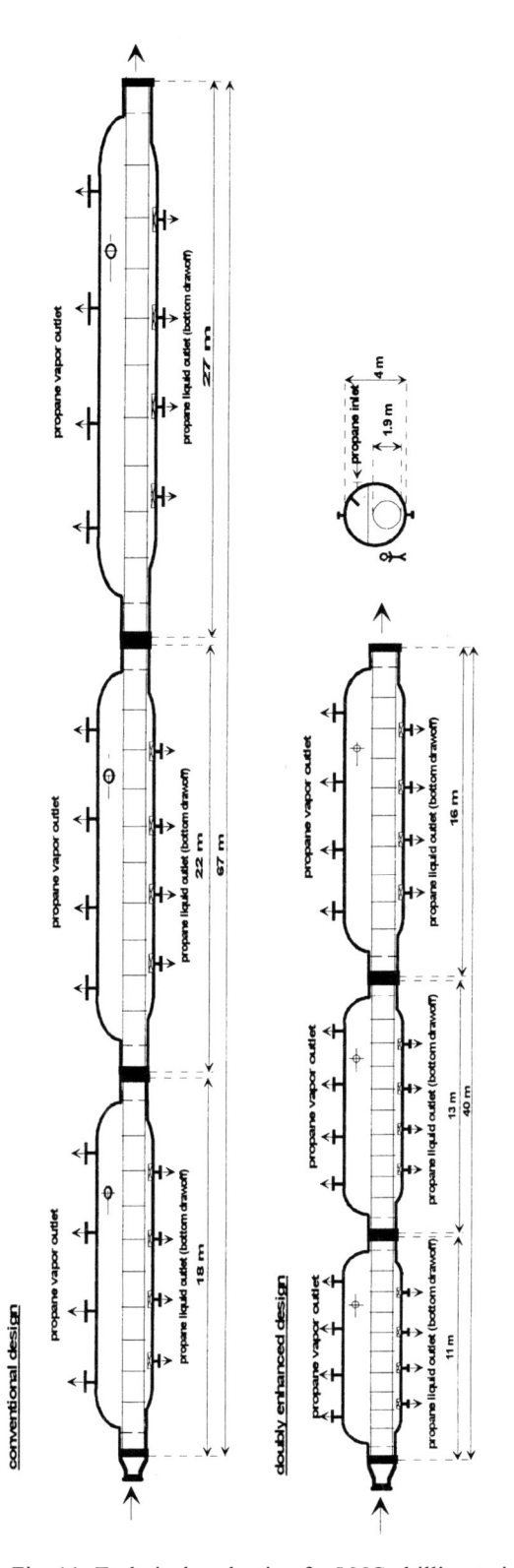

Fig. 11: Technical evaluation for LNG chilling train

REFERENCES

Cornwell K. and Scoones D.J, "Analysis of low quality boiling on plain and low-finned tube bundle", *Proceeding of the 2nd UK National Heat Transfer Conference*, Vol 1, pp 21-32, 1988.

Fujita Y., Ohta H., Hidaka S. and Nishikawa K., "Nucleate boiling heat transfer on horizontal tubes in bundles", *Proceedings of the 8th International Heat Transfer Conference*, vol 5, pp 2131-2136, San-Francisco, 1986.

Gupte N.S. and Webb R.L., "Convective vaporization of pure refrigerants in enhanced and integral-fin tube bank", *Enhanced Heat Transfer*, Vol 1, n°4, pp 351-364, 1994.

Hahne E. and Müller J., "Boiling on a finned tube and finned tube bundle", *International Journal of Heat and Mass Transfer*, Vol 26, n°6, pp 849-859, 1983.

Hahne E. and Windisch R., "Boiling heat transfer to bundle of plain tubes", *Proceedings of the 10th International Heat Transfer Conference*, vol 1, pp 117-128, Brighton, 1994.

Huchler L.A., "Improve reboiler operation and reliability", *Hydrocarbon Processing*, pp. 69-80, June 1999.

Hwang T. H. and Yao S.C., "Forced convective boiling in horizontal tube bundles", *International Journal of Heat and Mass Transfer*, Vol 29, n°5, pp 785-795, 1986

Jensen M.K. and Hsu J.T, A parametric study of boiling heat transfer in a horizontal tube bundle, *Transaction of the ASME : Journal of Heat Transfer*, Vol 110, pp 976-981, November., 1988.

Jensen M., Trewin R. and Bergles A., "Crosflow Boiling in Enhanced Tube Bundles", *Proc. of Pool and External Flow Boiling*, ASME, pp 373-379, 1992

Kassem E. A., Ali Mustansir, How to Design Shell and Tube Exchangers to Reduce Fouling and Enhance Performance, *Convective Flow Boiling Conf., Engineering Foundation, Banff*, Can, Paper III-2, 1995.

Kenney, W.F., Reducing the Energy Demand of Separation Processes, *CEP*, p. 68 - 71 March 1979.

Mc Carty A. and Smith B., "Reboilers sytem design : The tricks of the trade", *Chemical Engineering Progress*, pp. 34-47, May 1995

Martin F., Delorme J.J., Portz A. and Zoetemeijer L., Technical and Economic Case Studies of Intensified Reboilers, *Proc. of the EU seminar on Heat Equipments for the Process, Power and Refrigeration Industries*, Edited by GRETh, Grenoble, France, June 2000

Marto P.J. and Anderson C.L, "Nucleate boiling characteristics of R-113 in a small tube bundle", *Transaction of the ASME : Journal of Heat Transfer*, Vol 114, pp 425-433, May, 1992.

Mertz et al ,"Pool boiling from enhanced tubular heat transfer surfaces", Proc of the 11th Heat Transfer Conference, Vol. 2, pp 455-460, 1998

Mertz R., Groll M. and Thonon B., Tubular heat transfer elements for compact two-phase heat exchanger, *Compact Heat Exchangers and Enhancement Technology for the Process Industry*, pp.39-54, Begell House, 1999

Mertz, Kulenovic and Groll, Pool Boiling from Enhanced Tubular Surfaces, *Proc. of the 3rd European Thermal Conference*, Hahne, Heidemann and Spindler editors, Edizioni ETS, pp 791-796, 2000

Mertz R., Kulenovic R., Schäfer P. and Groll M.,boiling of hydrocarbons on tubes with subsurface structures, *Compact Heat Exchangers and Enhancement Technology for the Process Industry*, Davos, July 1-6, 2001

Moore, J.A., Fintubes Foil Fouling for Scaling Services, Chem Process. (7), pp. 8-10, 1974.

Roser R., Thonon B. and Mercier P., Experimental Investigations on Boiling of n-Pentane across an Horizontal Tube Bundle : Two-Phase Flow and Heat Transfer Characteristics", *Int. J. of Refrigeration*, Vol. 22, pp. 536-547, 1999

Sloley A.W., "Properly design themosyphon reboilers", *Chemical Engineering Progress*, pp.52-63, March 1997

Thome J.R., Heat Transfer Augmentation of Shell-and-Tube Heat Exchangers for the Chemical Processing Industry, 2nd European Thermal-Sciences and 14th UIT National Heat Transfer Conf., Rome, May 29-31, 1996.

Webb. R.L., Principles of Enhanced Heat Transfer, John Wiley & Sons, Inc., 1994.

Yilmaz S., Palen J.W. and Taborek J., "Enhanced boiling surfaces as single tubes and tube bundles", *ASME, HTD-18*, pp 123-129, 1981.

Yilmaz S. and Palen J.W., "Performances of finned tube reboilers in hydrocarbon service", *ASME paper*, 84-HT-91, 1984.

PROCESS INTENSIFICATION: PERFORMANCE STUDIES AND INTEGRATION OF MULTIFUNCTIONAL COMPACT CONDENSER IN DISTILLATION PROCESSES

B.D. Cailloux, J.M.G. Lee and R.J. Jachuck

Process Intensification & Innovation Centre (PIIC)
Dept. of Chemical and Process Engineering
Mertz Court
University of Newcastle upon Tyne
NE1 7RU, UK

ABSTRACT

The performance of shell and tube and cross flow plate heat exchangers has been compared for condensation duties. The comparison was made using HYSYS flow sheeting software for the condensation of pure chloroform and butane/propane mixtures. The heat transfer rate was higher and the pressure drop on the cooling water side was lower for the cross flow condenser. It has been found that it is more beneficial to use the increased performance of the compact condenser to reduce the cooling water requirement rather than using as a direct replacement for the existing condenser.

INTRODUCTION

Process intensification is a term used to describe the strategy of making major reductions in the size of a chemical plant needed to attain a given production rate. Intensified unit operations can be used to improve a process through better control and reduced utility requirement. The other benefits of Process Intensification are energy saving, reduced capital cost, improved product quality and intrinsic safety for the process plant.

The aim of this work was to compare the performance of two types of heat exchanger for condensing duty. One heat exchanger is a conventional shell & tube condenser, the other is a compact cross flow welded plate condenser manufactured by Alpha-Laval (Compablock®). This condenser will hereafter be referred to as the compact condenser. The comparison was made for the condensation of butane/propane system using a flow sheet simulation of the plant in a software package called HYSYS. A major petrochemical company provided the process flow sheet for the distillation unit (debutaniser) and the specification sheets for the component parts. The exchangers will be compared in terms of required heat transfer area, heat transfer coefficients and pressure drop. The benefits of replacing a shell and tube condenser with a compact one are discussed.

Design data for the compact condenser, provided by Alpha-Laval was incorporated in the simulation software in order to evaluate the performance of the compact unit. Laroche Industries provided performance data on a compact condenser that has been installed on their chlorinated methane plant. This data was used to validate the compact condenser model in HYSYS.

THEORY

Condensation heat transfer is a vital process in the chemical and power generation industries and has as a result been an area of research for over one hundred years. Over this period our understanding of the condensation process has gradually improved. Theories and models have become more accurate and are now applicable to a wide range of conditions (Whalley, 1987; Stephan, 1992).

Condensation on solid surfaces occurs by two methods, drop-wise and film-wise. On the one hand film-wise condensation occurs on a cooled surface that is easily wetted, whilst on the other drop-wise condensation occurs on non-wetted surfaces. In drop wise condensation vapour condenses in drops that grow by further condensation and coalescence as they move over the heat transfer surface. The normal mechanism for heat transfer in commercial condensers is film condensation. Drop wise condensation leads to higher heat fluxes but is unpredictable; and is not yet considered a practical proposition for general-purpose condensers. The experience and the knowledge gained in the past makes design of tubular condenser a relatively easy task. As shown in Appendix 1 correlations in terms of Reynolds and Prandtl numbers are available to calculate heat transfer coefficients and friction factors for the cooling fluid. For condensation expressions are also available which must be corrected by a factor chosen according to the mode of condensation (see Appendix 1).

Many duties can be carried out in both shell & tube and compact condensers. Whilst they are cheaper compact condensers are not often used for condensation duties. Correlations for the heat transfer coefficients are available for welded plate heat exchangers but tend to be specific to the type of plate and are generally given by the manufacturer. For condensation design correlations were obtained experimentally by Alpha-Laval.

EXPERIMENTAL SHELL & TUBE CONDENSER PERFORMANCE

For the shell & tube condenser the evaluation of its performance is straightforward. All the dimensions of the piece of equipment are input to the HYSYS software. In addition the inlet streams (the vapour from the column and the cooling water) are specified in terms of temperature, pressure, composition and flow rate. Consequently HYSYS uses the exchanger rating information to calculate the overall heat transfer coefficient and the pressure drops across both shell and tube side.

Compact Condenser Unit Performance

The evaluation of the performance of the compact condenser represents the core of the experimental work. HYSYS did not feature plate heat exchangers as a unit operation. To do this a feature of HYSYS, that allows the user to write his/her own unit operation was used. A user-defined unit operation can be integrated into HYSYS much like any other unit operation. The user defines its behaviour entirely with Visual Basic compatible code. The film heat transfer coefficients and pressure drop for the cooling water and the condensing vapour were calculated using the correlations based on experimental work.

It is envisaged that the compact condenser would replace the shell & tube condenser on a real plant. Therefore the flow rates of vapour and cooling water were known and the size and performance of the replacement plate condenser was determined. This was easily carried out using code inside the customised unit operation.

The overall heat transfer coefficient U was calculated using equation 1.

$$\frac{1}{U} = \frac{1}{\alpha_w} + \frac{1}{\alpha_c} + R_w + R_f \qquad (1)$$

In equation (1), α_w and α_c are the film heat transfer coefficients for the cooling water and the vapour. R_w is the wall resistance (0.00005 $m^2.K.W^{-1}$) and R_f is the fouling factor (assumed to be zero). The mean temperature difference for a cross flow plate heat exchanger with a large number of plates is given by equation (2).

$$\Delta T_{LM} = \frac{(T_{h,in} - T_{c,out}) - (T_{h,out} - T_{c,in})}{\ln((T_{h,in} - T_{c,out})/(T_{h,out} - T_{c,in}))} \qquad (2)$$

The mean temperature difference must be corrected because of the cross-flow pattern. According to Hewitt (1994) the value of F is generally close to 1, but can drop to 0.95. As a conservative estimate the value of F was kept constant and equal to 0.95. The temperature increase of the cooling water side was set and the flow-rate of cooling water obtained from an energy balance. The film heat transfer coefficients varies with the number of plates and passes. The heat transfer area and hence the number of plates for a given duty was calculated using the following standard procedure.

1. Choose an initial guess N for the number of plates.
2. Calculate the heat transfer area A, from the dimensions of the exchanger and N.
3. Calculate the two film heat transfer coefficients, h_s and h_t and the overall coefficient U.
4. Calculate ΔT_{LM}.
5. From U, ΔT_{LM} and Q, determine the heat transfer area.
6. If the guessed area and the calculated area do not match use a new N

The performance of the compact condenser and the shell and tube condenser were compared for condensation of saturated vapours without sub-cooling a. Using the inlet stream characteristics a compact condenser module was created within HYSYS with two options:

- Evaluate the performance of the compact condenser for a fixed heat transfer area.
- Design a compact condenser for a given duty.

The accuracy of the HYSYS model was tested by comparing the model predictiopns with data from a unit installed at a LaRoche Industries Inc. plant for the condensation of chloroform. For the same duty and with the data collected during normal operation of the condenser, the HYSYS model was used to predict the heat transfer area.

RESULTS

Model Validation

On plant the compablock exchanger installed by LaRoche for the condensation of chloroform (99.7%) has 300 plates (0.27m^2 per plate). The operating conditions are as follows: vapour 14700 kg/hr, water 43000 kg/hr, water inlet temperature 19 °C, water outlet temperature 40 °C, vapour temp. 51 °C, condensate temperature 50 °C.

The Hysys model predicts that 328 plates are required, compared to the 300 plates in the installed compact condenser. The 9% error on the predicted area comes from the fact that the condensing film coefficient for the chloroform has been calculated using the correlation for the butane/propane system.

Design with the Same Temperature Approach

For the gas plant, the first comparison was between condensers with the same flow-rate of cooling water and the same temperature approach. The two condensers were compared in terms of heat transfer area, local and overall heat transfer coefficients and pressure drop (for both sides). The operating conditions are shown in Table 1.

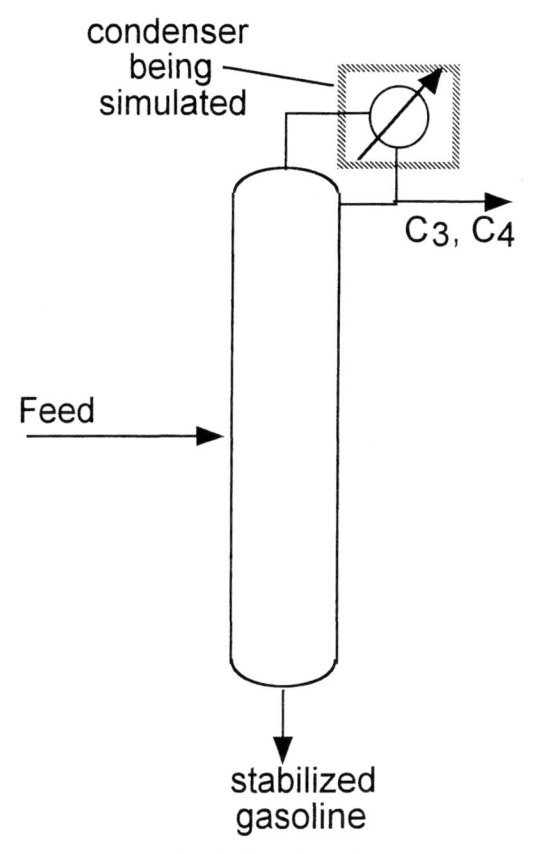

Fig. 1 Gas plant simulation.

The heat transfer coefficient for the condensing vapour is the limiting coefficient. It was observed that the heat transfer coefficient for condensation in the compact condenser is almost five times larger than the coefficient in the shell and tube condenser. Thus, for the compact condenser the overall heat transfer coefficient is roughly three times larger than the overall coefficient for the shell & tube, resulting in a very small heat transfer area for the compact condenser. The pressure drop in the shell & tube exchanger was larger than that for the plate and frame.

Design with a Different Temperature Approach

For this case the duty was unchanged but the flow-rate of coolant was decreased to 100,000 kg/hr and consequently the outlet temperature of the water is higher. The effect of reducing the cooling water flow on the size and performance of the compact condenser was examined. The stream data is given in Table 3 and the results of the simulation are given in Table 4.

When the temperature difference on the coolant side increases the required heat transfer area will increase and the overall heat transfer coefficient decrease due to the increased number of plates, the pressure drop for both streams is lower.

DISCUSSION

Two main advantages can be found for the replacing a shell & tube condenser with a compact one.

Table 1. Inlet Stream Characteristics

	Vapour	Cooling water
Flow (kg h^{-1})	35432	200000
Inlet temp. (°C)	66.4	10.0
Outlet temp. (°C)	54.6	23.6

Table 2. Performance Comparison

Heat exchanger	Shell & tube condenser	Compact condenser
Duty (MW)	32.3	32.3
A (m^2)	372	108
F ΔT_{lm} (°C)	43.1	39.3
U (W m^{-2} °C^{-1})	201	766
α_{wat} (W m^{-2} °C^{-1})	4269	2793
α_{vap} (W m^{-2} °C^{-1})	218	1123
ΔP_{wat} (Pa)	1783	180
ΔP_{vap} (Pa)	47390	5990

Table 3. Inlet Stream Characteristics

	Vapour	Cooling water
Flow (kg h^{-1})	35432	100000
Inlet temp. (°C)	66.4	10.0
Outlet temp. (°C)	54.6	37.3

Table 4. Compact Condenser Performance

Duty (MW)	32.3
A (m^2)	177
F ΔT_{lm} (°C)	32.7
U (W m^{-2} °C^{-1})	563
α_{wat} (W m^{-2} °C^{-1})	1286
α_{vap} (W m^{-2} °C^{-1})	1055
ΔP_{wat} (Pa)	140
ΔP_{vap} (Pa)	4200

Reduced Cooling Water Requirement

It has been observed that with a flow-rate of coolant half that of the shell & tube condenser (see Table 3), the compact condenser still achieves the same duty with a much better thermal effectiveness. This lower water requirement implies reduced utility consumption and reduced installation. The pressure drop through the compact unit are negligible is low (see Table 4). The flow-rate and the total head are smaller consequently, the required capacity of the centrifugal pump will be less and the savings on pumping cost will be large. To express the change in power requirement for each condenser equation 3 can be used (Coulson and Richardson, 1997)

$$\frac{P_{cb}}{P_{st}} = \frac{Q_{cb}\left(\Delta P_{cb} + CQ_{cb}^{1.75}\right)}{Q_{st}\left(\Delta P_{st} + CQ_{st}^{1.75}\right)}$$

(3)

$$C = \frac{0.2414\mu^{0.25}\rho^{0.75}L}{d^{4.75}}$$

The value of C depends on the distance the cooling water has to be pumped. If the shell and tube condenser is replaced by a compact condenser, whilst keeping the cooling water flow rate constant, the reduction in water pumping power requirement is 18.5%. Reducing the cooling water flow rate by 50%, leads to a reduction in pumping power requirement of 87.5%. The reduction in cooling water requirement alone represents a saving of 113000 Euro/annum. It is important to recognise that a reduction in the exchanger pressure drop does not lead to the largest energy savings. Pumping power reduction comes from a decrease in the pressure drop in the rest of the piping system. A shell and tube condenser could be designed to use a lower water flow but it would be larger than the existing unit where as the replacement compact condenser would be smaller.

Space Saving

A shell & tube exchanger is a robust piece of equipment requiring a large amount of space. Table 2 shows that the required heat transfer area for the same is duty is three times less for the compact condenser. The narrow passages and the corrugated surfaces produce a high overall heat transfer coefficient up to 4 times that of the equivalent shell & tube condenser. The volume of the compact condenser reduction is a tenth of that of the shell and tube condenser. As a result the plate heat exchanger takes up less floor space and imposes less floor loading. The compact equipment requires less heavy structures to support it. Due to its compactness and lighter weight, a cross flow plate condenser could also be placed closer to the top of the column in the production unit, resulting in a simplification of the piping system and a substantial savings in the piping costs.

CONCLUSIONS

At a time when equipment costs are rising and the available space is limited, compact heat exchangers are gaining a larger proportion of the heat exchanger market. Numerous types use special enhancement techniques in order to achieve the required heat transfer in smaller area and with less initial investment.

The present study shows that the use of cross flow welded plate exchanger (Compabloc®) for condensing duties, in particular as a reflux condenser for a distillation column would be judicious. Great savings can be realised by using this type of heat transfer equipment in terms of:

• Operating costs

• Capital costs

Furthermore a well-designed compact heat exchanger is safer as the liquid hold-up in a compact unit is smaller during operation owing to its size. Steady-sate simulations have proved the value of replacing the conventional condenser with an intensified one. In the future dynamic simulations will be used to show that the compact plate and frame condenser has an improved response to the variation of the process parameters and therefore making it an even more attractive alternative to shell and tube condensers.

NOMENCLATURE

c   specific heat capacity of water, J/kg °C

D_1 outside diameter of the tube, m

g	acceleration due to gravity, m/s^2
h$_{GL}$	latent heat, J/kg
L	pipe length in the cooling water system, m
P	power required to pump cooling water, W
ΔP	pressure drop on the cooling water side of the condenser, Pa
T$_{sat}$	saturation temperature, °C
L	pipe length in the cooling water system, m
P	power required to pump cooling water, W
ΔP	pressure drop on the cooling water side of the condenser, Pa
T$_{sat}$	saturation temperature, °C
T$_w$	wall temperature, °C
ρ$_L$	density of the condensing liquid, kg/m^3
U	overall heat transfer coefficient, W/m^2 °C
u	velocity of the cooling water, m/s
α	film heat transfer coefficient, W/m^2 °C
λ$_L$	thermal conductivity of the condensing liquid, W/m K
μ$_L$	viscosity of the condensing liquid, Pa s

REFERENCES

Baehr, H. D., 1998, *Heat and Mass Transfer*, Springer, Berlin.

Bennett, C.O., and Myers, J.E., 1974, *Momentum, Heat, and Mass Transfer*, 2nd Edition, McGraw-Hill, New York.

Collier, J.G., Thome, J.R., 1994, *Convective Boiling and Condensation*, 3rd Edition, Oxford Science Publications.

Coulson, J.M., Richardson, J.F., Backhurst, J.R. and Harker, J.H., 1997, *Chemical Engineering Vol. 1. Fluid flow, Heat Transfer and Mass Transfer*, 5th Edition, Butterworth-Heinemann, Oxford.

Hewitt, G.F., 1990, *Handbook of Heat Exchanger Design*, Hemisphere Pub. Corp., New York.

Hewitt, G.F., Shires, G.L., 1994, *Process Heat Transfer*, CRC Press, Table 3.14, p. 181.

Rohsenow, W.M., Choi, H.Y., 1961 *Heat, Mass and Momentum Transfer*, Prentice-Hall, Inc.

Stephan, K., 1992, *Heat Transfer in Condensation and Bboiling*, Springer-Verlag, New York.

Whalley, P.B., 1987, *Boiling, Condensation, and Gas-Liquid Flow*, Clarendon Press, Oxford.

APPENDIX 1

The following relationships were used to calculate the film coefficients and pressure drop for the shell and tube condenser. For turbulent flow of the cooling water in a smooth cylindrical pipe the Dittus-Boelter equation is used.

$$Nu = \frac{\alpha_w D_e}{\lambda_L} = 0.023 \times Re^{0.8} \times Pr^{0.4}$$

$$f = \frac{\Delta P}{4L/D(\rho u^2/2)} = 0.046 \times Re^{-1/5} \quad \text{for } Re \geq 2 \times 10^4$$

$$Re = \frac{uD\rho}{\mu_L} \qquad Pr = \frac{c\mu_L}{\lambda_L}$$

For film condensation when the flow in the film is laminar over a single horizontal tube, the mean film transfer coefficient is given by the Nusselt equation. For the condensation on a bank of horizontal tubes in the presence of a vapour flow, the situation is more complex. The flow of condensate from one tube to another thickens the condensate layer on the lower tubes and decreases the heat transfer coefficient. A correction factor must be used to obtain the average coefficient, α$_n$ for a bank of *n* tubes. The factor is chosen according to the mode of condensation flow on the shell side, for instance:

Nusselt idealised model,

$$\frac{\alpha_n}{\alpha_1} = n^{-1/4}$$

Ripples, splashing and turbulence,

$$\frac{\alpha_n}{\alpha_1} = n^{-1/6}$$

where α_1 is the heat transfer coefficient for a single tube:

$$\alpha_1 = 0.725 \left[\frac{\lambda^3 h_{GL} \rho_L^2 g}{D_1 \mu (T_{sat} - T_w)} \right]^{0.25}$$

A STUDY ON THE ADVANCED PERFORMANCE OF AN ABSORPTION HEATER/CHILLER USING WASTE GAS

E. Kim[1], J.I. Yoon[2], H.S. Lee[3], C.G. Moon[3] and S.H. Chun[4]

[1] Pukyong National University, Namgu, Pusan 608-737, Korea; E-mail: ekim@mail.pknu.ac.kr
[2] Pukyong National University, Namgu, Pusan 608-737, Korea; E-mail: yoonji@dolphin.pknu.ac.kr
[3] Graduate Student, Pukyong National University, Namgu, Pusan, 608-737, Korea; E-mail: mchg@mail1.pknu.ac.kr
[4] Korea Energy Management Corporation, Yongin, Kyonggido, 449-840, Korea; E-mail: shchun@kemco.or.kr

ABSTRACT

In this study an absorption heater/chiller is developed and experimented to save energy. Energy conservation is important for the environmental protection of the earth. A new method using the waste heat of the gas co-generation system is presented. The new method uses the waste heat below 250℃ for refrigeration. In the new method, the high temperature waste heat is fed into the weak solution line of the absorption heater/chiller, which is directly contacted with an auxiliary heat exchanger. To have the high COP of a refrigerator the solutions heated by a waste heat exchanger is connected with a low temperature generator and a high temperature exchanger. The results show that by supplying the waste gas at an auxiliary heat exchanger, the supplied gas to the high temperature generator for heating and cooling is reduced. At the partial load, the change of the circulating solution quantity is reduced. The development of the system for the use of waste gas reduces energy about 3 to 5%. Automatic operation of solution level change obtains the reliability for products, and the chilling capability is enhanced in a short time.

INTRODUCTION

Recently, refrigeration and air-conditioning applications have been widely applied due to the improvement of human life and used in economic activity. Especially the facilities of air-conditioner have been increased sharply in several decades and the rate of power consumption is increasing every year.

To maintain the reserved power according to the peak power requirement, it is necessary to construct new power plants every year. The construction of plants requires high cost and long time. The power requirements in spring and fall are much smaller than those of summer. The unbalance of energy requirements among seasons is occurred. Since the energy consumption of natural gas in summer reduces drastically to 1/5∼1/6 compared to the natural gas consumption in winter due to the decrease of heating requirement in summer. As a result, the significant unbalance between energy requirements and energy supply is occurred. (Yang, 1998)

The absorption cooling and heating applications have been developed to face the excessive peak power consumption during summer and reduce the environmental pollution problem. Gas driven absorption heating and cooling applications provide heating power during winter and cooling power during summer using an absorption chiller/heater. An absorption heater/chiller uses the environmentally safe fluids such as water or ammonia.

Studies on an absorption heater/chiller have been focused on the enhancement of system efficiency. Kim (1994) investigated an absorption heater/chiller, which uses waste heat as heat source. He showed the high efficiency on an absorption heat/chiller. Dong et al. (1991) reported that a single-effect absorption heat pump recovered about 40% of waste heat by using the 79℃ waste heat. Yoon et al. (1995) studied on a parallel absorption cycle using waste gas through a cycle simulation. They suggested the best driving conditions of the machine. Moon et al. (1998) did an experimental study on an absorption heat pump using the fuel cell thermal source. Kojima et al. (1997) performed a study on a two-stage absorption heat pump, which is driven by exhaust heat from a gas engine. Lee (1996) carried out the study on a double-effect parallel absorption chiller using sewage water as cooling water and evaluated the applicability of the system. Baek et al. (1998) evaluated dynamic characteristics of an absorption chiller, which is driven by solar heat.

The absorption heat pumps of previous studies

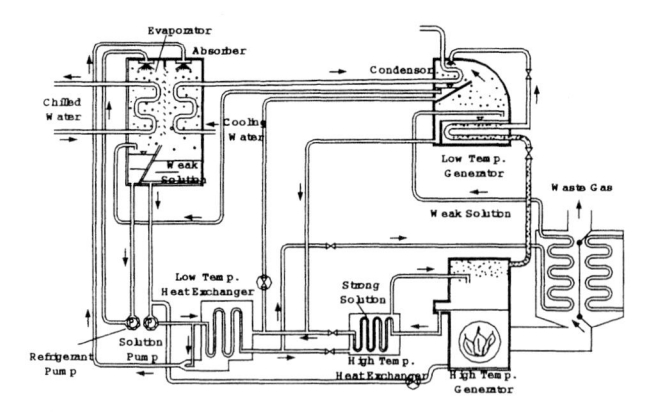

Fig. 1 Schematic diagram of an absorption heater/chiller using waste gas.

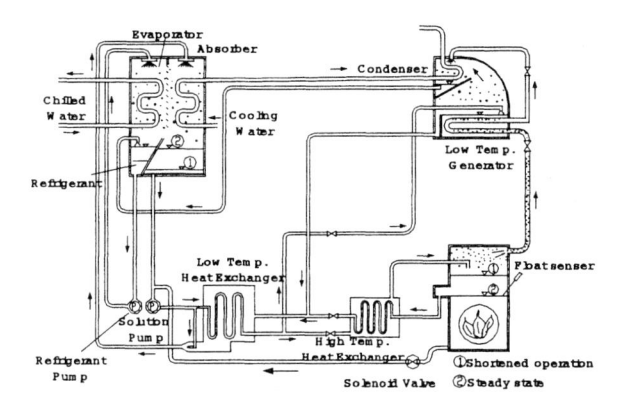

Fig. 2 Schematic diagram of a starting time shortened operation

Table 1 Specifications of a test plant.

Refrigeration capacity	125 [RT]
Cooling operation	
Inlet temperature of cooling water	32 [℃]
Outlet temperature of cooling water	37.0 [℃]
Flow rate of cooling water	135 [㎥/h]
Inlet temperature of chilled water	12.0 [℃]
Outlet temperature of chilled water	7 [℃]
Flow rate of chilled water	75.6 [㎥/h]
Gas consumption	338,000 [kcal/h]
Heating operation	
Inlet temperature of hot water	75 [℃]
Outlet temperature of hot water	80 [℃]
Flow rate of hot water	135 [㎥/h]
Gas consumption	392,080 [kcal/h]

EXPERIMENTAL APPARATUS

In a double-effect absorption heater/chiller, the exhaust heat circulates the tubes of the low temperature generator. This is an important factor in making the system more efficient. Figure 1 shows a conventional absorption cycle used outer and auxiliary heat sources or waste heat of other systems. In this study, the system does not use any auxiliary components to increase the efficiency. After a high temperature generator uses heating energy, the temperature of the exhausted gas is over 200℃. This energy is recovered by using an exhaust heat exchanger. Also, it is studied the cooling process to reduce the delay time at the starting operation. To enhance an efficiency of an absorption heater/chiller, heat exchangers in arrangement of the high temperature generator are installed to recover the waste heat. The overall performance of the system is investigated by recycling the exhaust gas from the high temperature generator. Additionally, the experiment is conducted to study the parameters of the load rate of cooling/heating.

with the heat recovery device for the exhaust gas. The heat exchange for the waste heat recovery is installed at the high temperature generator. The absorption solution of the weak concentration heated at the high temperature heat exchanger flows to the high temperature generator through a pipe. The weak solution heated at the low temperature heat exchanger flows into the low temperature generator. Between two separate pipes the heat exchanger is installed so that the weak solution flows into the high and low temperature generators to be reheated by the waste gas. Thus, the absorption solution becomes strong.

Figure 2 shows a schematic diagram of a starting time shortened operation mode. The heat exchanger for the waste heat recovery is installed between the low temperature solution heat exchanger and high temperature solution generator. In the experiment a double-effect parallel heater/chiller of a 125 RT (378Mcal/h) commercial product is used. COP level is measured after injection of the waste heat on the weak solution at the inlet of the low temperature generator.

Two improved components are as follows. First, at the exhaust gas position of the high temperature generator shown in Figure 1, a shell-and-tube typed heat exchanger for waste heat recovery is installed. A cooling system can be easily changed to a heating system by turning the damper in the heat recovery heat exchanger. Second, in the experiment for the starting time shortened operation mode, not only a float sensor is installed at the high temperature generator but also a solenoid valve is added for flow control. Then, pipes are connected to the exit of the absorber. Table 1 shows the specifications of the test plant, and Table 2 shows the specifications of the heat exchanger for waste heat recovery. The experiment is conducted to investigate the parameters of the load rate of cooling/heating and the inlet temperature of cooling water in the test plant. The experimental conditions are shown in Table 3. The cooling/heating load controlled by the quantity of gas injection is calculated by the temperature difference of cooling water.

Table 2 Specifications of a waste heat recovery heat exchanger.

Cooling operation	
Inlet temperature of weak solution	68 [°C]
Outlet temperature of weak solution	73.6 [°C]
Inlet temperature of waste gas	250 [°C]
Outlet temperature of waste gas	173 [°C]
Heating operation	
Inlet temperature of waste heat-hot water	75 [°C]
Outlet temperature of waste heat-hot water	77.71 [°C]
Flow rate of hot water	6.8 [m³/h]

Table 3 Conditions of the experiment

Parameters	Value
Chilled water flow rate	75.6 [m³/h]
Cooling water flow rate	135 [m³/h]
Chilled water outlet temperature	7 [°C]
Chilled water inlet temperature	12 [°C]
Cooling water inlet temperature	30, 32, 34 [°C]
Waste gas inlet temperature	250 [°C]
Cooling load, heating load	100, 66.7, 37.4 [%]
Cooling load, heating load	100, 81.5, 65.9, 50.6 [%]

Data measurement is controlled to satisfy the test conditions by operating the test plant. When the test plant is in a steady state, data are measured every 30 seconds, and average values during 30 minutes are selected as experimental values.

EXPERIMENTAL RESULTS AND DISCUSSION

Experiment of a cooling operation using waste gas.
Figure 3 shows the comparison of the experimental data with the waste heat recovery operation and the conventional operation, which does not use the waste gas. When the load is 100% and the inlet temperature of cooling water is 32°C, the COP levels of the waste gas mode and conventional mode are 1.24 and 1.21, respectively. The enhancement of the COP level between two cases is about 2.8%. When the inlet temperature of cooling water is low, the enhancement performance of the machine is increased. Figure 4 shows the variation of the COP with the inlet temperature of cooling water for two different operations. When the cooling load is low, the difference of the COP level between the waste gas operation and the conventional operation is large. When the cooling load increases, the difference of the COP gradually decreases. At 30°C of the inlet temperature of the cooling water the difference between two operations for 37.4% of the cooling load is about 7.4%, but the difference for 100% of the cooling load is about 4.1%. When the inlet temperature increases the difference decreases. Figure 5 shows a plot of the COP enhancement rate in term of cooling

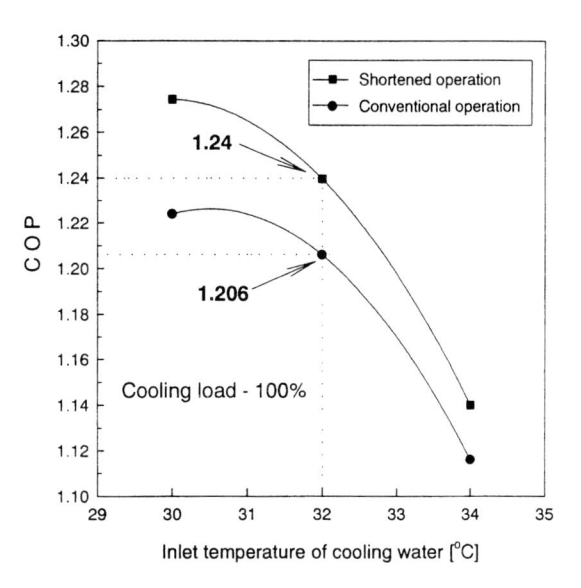

Fig. 3 Comparison with COP and inlet temperature of cooling water.

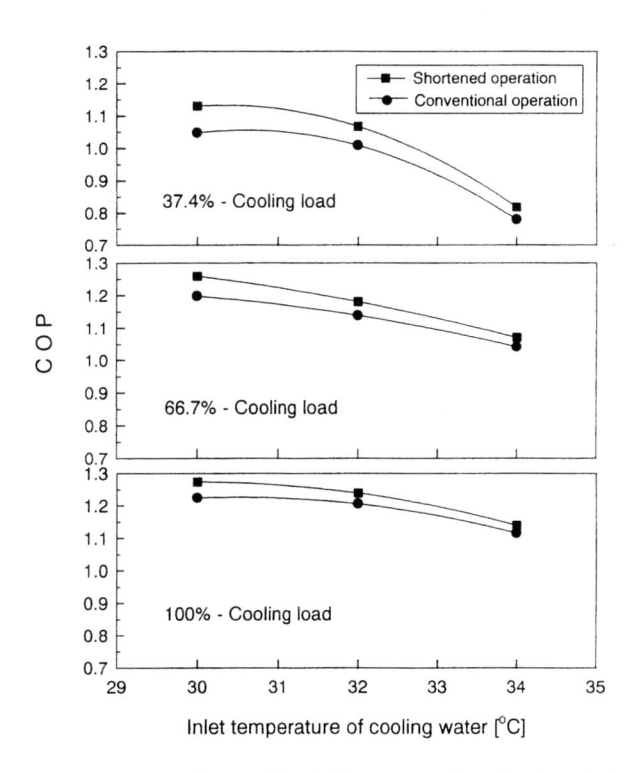

Fig. 4 Comparison with COP and cooling load on inlet temperature of cooling water.

load for different inlet temperatures of cooling water. When the inlet temperature of cooling water and the cooling load in the waste gas operation are low, the COP levels increase. The increase ratio of the COP enhancement rate at 30°C is larger than that at 34°C. It can be seen that as the temperature of cooling water is low, the absorption capacity of cooling increases.

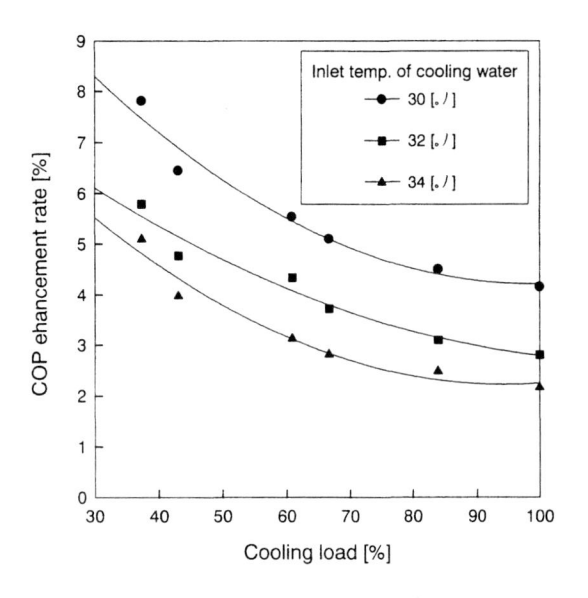

Fig. 5 Influence to COP by the inlet temperature of cooling water on cooling load.

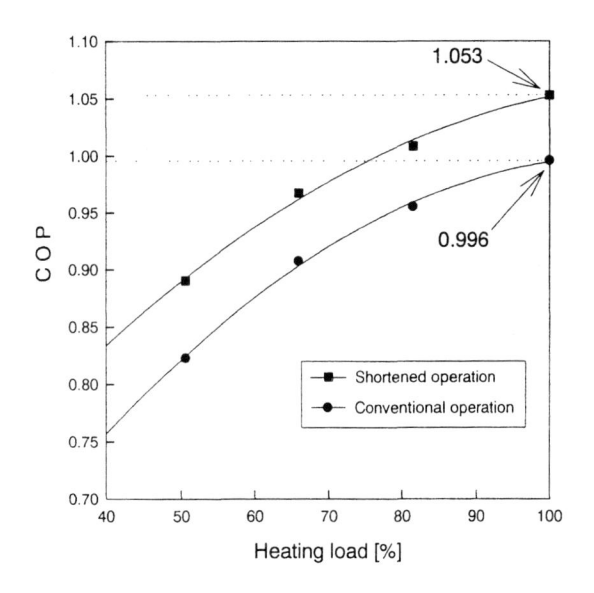

Fig. 6 Comparison with COP and heating load.

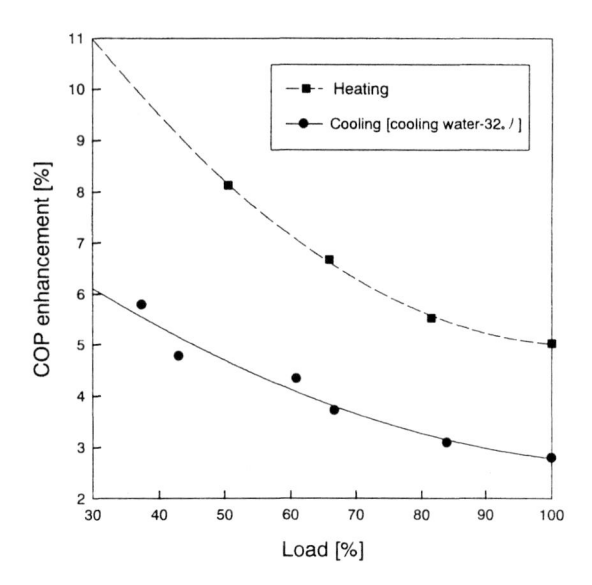

Fig. 7 Comparison with heating load and cooling load temperature generation

the COP level increases 5.1% at the heating operation and 2.8% at the cooling operation. This is due to the fact that at the heating operation mode using waste gas, the waste heat directly contacts the heating water whereas at the cooling operation the waste heat indirectly contacts the solution temperature.

Experiment of a starting time shortened operation.
When a double-effect absorption heater/chiller completes a whole cycle, the pressure of the high temperature generator becomes low. The absorption solution of the absorber and the low temperature generator flows into the high temperature generator by the pressure difference. As a result, the level of the solution at the high temperature generator becomes very high. If the chiller is operated again, the absorption solution, which is at the high level in the high temperature generator, moves to the high/low temperature heat exchangers only by vapor pressure. The weak solution is then sprayed to cooling water tubes by the spray equipment within the absorber. When time passes, the solution level becomes low and strong. About 40 minutes is required for the level of the absorption solution in the high temperature generator to be a normal condition. This operation delays a cooling process by taking a long heating time of the absorption solution in the high temperature generator until the level of the absorption solution is in a normal condition. In a conventional operation, it takes a long time to lower the solution level at the high temperature generator. It means during this time the cooling operation of the equipment is not properly working. To solve this problem, a solenoid valve will be installed to send the solution to the absorber and the low temperature generator separately until the operation works properly. By applying this technical method the initial operation time is investigated.

Experiment of heating operation using waste heat.
Figure 6 shows the comparison of the COP and the heating load for the waste heat recovery operation and the convention operation. When the heating load increases the difference of the COP of two operations is almost the same. This means that even though the heating load is low, the operation using waste gas has a good performance. At 100% of the heating load, the COP level of the conventional operation mode and waste gas operation mode is 0.996 and 1.056, respectively. Figure 7 compares the COP enhancement rates for two different operation modes: cooling and heating. The COP enhancement rate of the heating operation is higher than that of the cooling operation. As the load decreases, the heating operation shows better efficiency than the cooling operation. With 100% load,

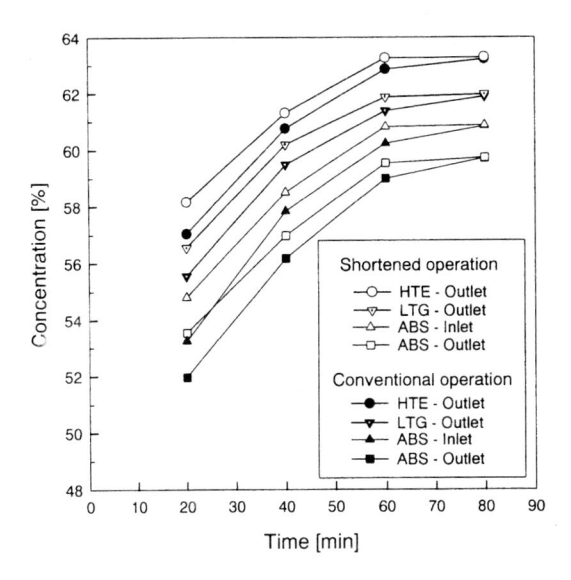

Fig. 8 Comparison with concentration and time

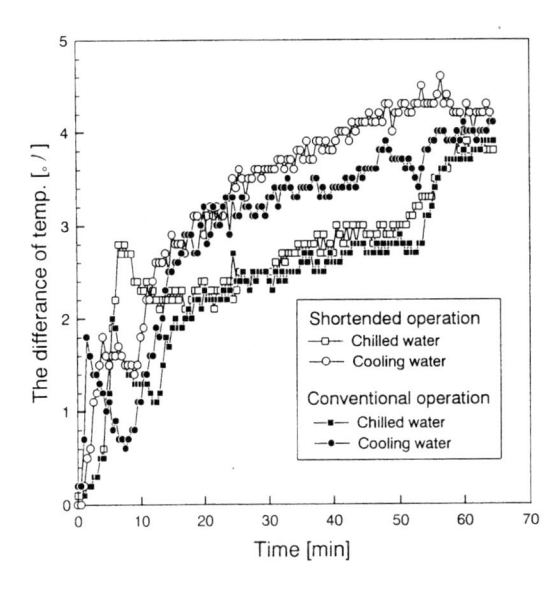

Fig. 9 Comparison with the difference of chilled water-temperature and time

Figure 8 shows the variation of the concentration in terms of time for two different operation modes. The time needed for the strong concentration is faster with the starting time shortened operation mode than with the conventional operation mode. The experimental values are shown after 20 minutes. It can be seen from the figure that the concentration has the highest value after the operation passes 60 minutes. After that time, the concentration becomes a constant value. The variation of the concentration at the inlet of the absorber is large at the starting time shortened operation. A solenoid valve driven by a float sensor is added to the high temperature generator. Consequently, the operation time to a steady state is reduced about 9 minutes, and the cooling capacity reaches to a steady state with very short time. Figure 9 shows comparison of the temperature difference in terms of time for two different modes with chilled water and cooling water. The temperature difference of a cooling water between the starting time shortened operation mode and the conventional operation mode is over 1.2□ □after 10 minutes is passed. After that time, the temperature difference becomes small. About temperature variation at the inlet/exit of the absorber, the starting time shortened operation mode shows a temperature difference after 5 minutes, and the temperature difference becomes large after 15 minutes. This is because the solution quantity of the high temperature generator decreases after 4 minutes. Consequently, the solution temperature in the high temperature generator increases and reaches to a steady state in a short time.

CONCLUSIONS

In this work, we discussed the energy efficiency of a double-effect absorption heater/chiller using waste gas. The conclusions are as follows:
(1) By supplying the waste gas at an auxiliary heat exchanger, the supplied gas to the high temperature generator for heating and cooling is reduced.
(2) The used quantity of waste gas increases when the inlet temperature of waste gas is high. The increase rate of the COP increases when the system operates at the partial load.
(3) At the partial load, the change of the circulating solution quantity reduces the gas supply.
(4) The development of the system for the use of waste gas reduces energy about 3 to 5%. Automatic operation by solution level changes obtains the reliability for products, and the reduction in the early operation time enhances the chilling capability in a short time.

ACKNOWLEDGEMENT

This work was sponsored partly the Korea Energy Management Corporation. The financial and technical support by the Samwon Mechanics Co. Ltd. is gratefully appreciated. The authors would like to thank Mr. J. W. Choi for his helpful assistance.

REFERENCES

Baek N.C., et al., 1998, "A Study on the Dynamic Performance of a Solar Absorption Cooling System", Solar Energy, Vol. 18, No. 3. pp. 81-87.

Dong T. L. et al., 1991, "Prototype Test of a Lithium Bromide-Water Absorption Heat Transformer using 78□ □ Waste Heat", Proceedings of Absorption Heat Pump Conference '91, Tokyo, pp. 357-361

Kim, K. J., 1994, "Absorption Chiller and Heat Pump Utilizing Waste Heat Source", Transactions of SAREK, Vol. 23, No. 4, pp. 299-305

Kojima, H., Akisawa, A., and Kashiwagi, T., 1997, "Characteristics of Two-Stage Absorption Heat Pump Cycle Driven by Waste Heat from Gas Engine", Trans. of the JSRAE, Vol. 14, No. 2, pp. 113-124.

Lee, Y.H., 1996, "Analysis of Absorption Refrigeration Cycles to Utilize Treated Sewage", Korean Journal of Air-Conditioning and Refrigeration Engineering, Vol. 8, No. 2, pp. 288-298

Moon, C. G., Kwon, O.K. and Yoon, J. I., 1998, "Characteristics Simulation of Absorption Cycles using the Fuel Cell Thermal , Proceeding of the 20th Anniversary Conference of KSES, pp.495 500.

Yang, Y.M., 1998, "Current Development Status of Gas Cooling Technologies and Measures for Sales Promotion of Gas Cooling Systems", Journal of Refrigeration Engineering and Air Conditioning, Vol. 17, No. 2, pp. 85-91.

Yoon, J. I., Oh, H. K. and Takao Kashiwagi, 1995, "Characteristic Simulation of the Waste Heat Utilization Absorption Cycles , Transactions of JAR. Vol. 12, No. 1, pp. 43 52.

The Experimental Analysis for the Application of High-Temperature Latent Heat Exchanger

Shr-Hau Huang[*] Hsiang-Hui Lin Bing-Chwen Yang

[*] Energy & Resources Laboratories, Industrial Technology Research Institute, Taiwan, ROC; E-mail: Calvin@itri.org.tw

ABSTRACT

The experimental analysis for the application of high-temperature latent heat exchanger is examined. The thermal energy storage materials with phase change that employed in these heat exchangers were NaCl-CaCl$_2$ composed of 32wt% NaCl and 68wt% CaCl$_2$ and LiOH with the fusion temperatures of 500 °C and 470 °C respectively. The analysis of these two phase change materials (PCMs) for the phase change temperature (T_m) and latent heat (ΔH) was also performed by the Differential Scanning Calorimeter (DSC) after 4 cycles of heating and cooling process. Besides, the experiments of two plate type heat exchangers modified as the latent heat exchanger with the thermal energy storage medium NaCl-CaCl$_2$ and LiOH were examined. The results showed that the modified plate type latent heat exchanger had increased the amount of thermal energy storage about 1629 kJ filled with NaCl-CaCl$_2$ and 2083kJ filled with LiOH by comparing with the same plate type heat exchanger without PCM. It was also shown that the contributions of thermal energy storage due to phase change materials were 56% and 62% respectively for these two different PCMs. The thermal energy storage rate was also increased more than 25% by comparing the heat exchanger without PCM. Moreover, it was found that the repeatability of NaCl-CaCl$_2$ and LiOH is acceptable, i.e., the amount of thermal energy storage in these two heat exchangers is not decreased significantly after long-term usage.

INTRODUCTION

The crisis of global warming effect made the reduction of carbon dioxide (CO$_2$) from the burning of fossil fuel necessary for the sustainability of our world. A high efficiency of energy usage and reduction of the consumption of fossil fuel are the important ways to achieve this purpose. Waste heat recovery from a boiler or any other industrial furnace is an easy way to increase the efficiency of energy usage (Yagi and Akiyama, 1995). There are several types of heat exchangers that can be used as a recuperator in the waste heat recovery system. Heat exchanger with thermal energy storage is a good candidate that is well suited for the waste heat recovery (Kang and Yabe, 1996). Traditionally, materials with high heat capacity are used as the material to storage thermal energy. These are easy to fabricate and have been used extensively in the boiler as a recuperator to preheat the supplied air. The disadvantages of this kind of heat exchanger are the large amount of energy storage material needed to meet the heat duty and the large thermal stress induced due to the large temperature difference during the operation cycles. In order to solve these two problems, a phase change material (PCM) was employed as the thermal energy storage medium (Lock, 1996 and Lane, 1989). In this study, two plate type heat exchangers were modified as the latent heat exchanger with the thermal energy storage medium NaCl-CaCl$_2$ and LiOH respectively. Some of the thermal and physical properties of NaCl-CaCl$_2$ and LiOH were examined experimentally in advance to understand its phenomena during phase change process. After that, a series of experiments were performed to study the total amount of thermal energy that can be charged and time required during each process.

EXPERIMENTAL ANALYSIS OF PCM

The criteria for the choice of phase change material that can be applied in the thermal energy storage system are thermophysical properties (i.e., latent heat, expansion rate, and stability during heating and cooling), safety, hazards, corrosiveness, toxicity, and cost. In this study, the Differential Scanning Calorimeter (DSC) was used to perform the quantitative measurement of the phase change temperature (T_m) and latent heat (ΔH). The operation temperature for this DSC is -65 °C to 725 °C and the rate of temperature increasing can be varied from 0.1 °C to 500 °C per minute. The accuracy of temperature is ±0.1 °C with the error of heat capacity measurement less than ±1%. The heating and cooling rate for these measurements was set in 10 °C per minute. The heating process was ended when the temperature is higher than the phase change temperature (i.e., a peak heat flux was observed in the experiment). The cooling process was started immediately after the heating process and until the room temperature. The typical curves of these two kinds of PCMs (eq. NaCl-CaCl$_2$ and LiOH) can be shown in Figure-1 and Figure-2. The upper line means the heating process and the lower is cooling process. For the heating process, the onset temperature (i.e., 503.7 °C for NaCl-CaCl$_2$ and 471.9 °C for LiOH) means the beginning of the phase change and peak temperature (i.e., 507.7 °C for NaCl-CaCl$_2$ and 480.2 °C for LiOH) was referred as phase change temperature (T_m) in this paper. Also, the integration of heat flux calculated from the onset temperature to the end temperature (i.e., 512.4 °C for NaCl-CaCl$_2$ and 481.7 °C for LiOH) is so-called latent heat (ΔH) in this paper. The analyses of these two PCMs for the phase change temperature (T_m) and latent heat (ΔH) were also performed by DSC after 4 cycles of heating and cooling process. The results of measurements and relevant material properties of PCMs were summarized in Table-1. The reasons for choosing NaCl-CaCl$_2$ composed of 32wt% NaCl and 68wt % CaCl$_2$ as the phase change material are no toxic gas released, the behavior similar to pure material, low expansion rate, easy availability, low corrosiveness, and low cost. The above advantages can compensate the disadvantage of low latent heat shown in Table-1. In terms of practicability, LiOH is not the most suitable material as the phase change material because of the toxicity and high cost, but it possesses the highest latent heat among the phase change materials in the range of 400 °C ~ 600 °C (Lin et al., 1998). If we can ensure no leakage of the toxic gas from the thermal storage unit and reduce the cost, it is feasible to attain the well performance of the thermal storage unit with LiOH as the thermal storage medium. From the above reasons, NaCl-CaCl$_2$ and LiOH are employed in this study with the fusion temperatures of 500 °C and 470 °C respectively.

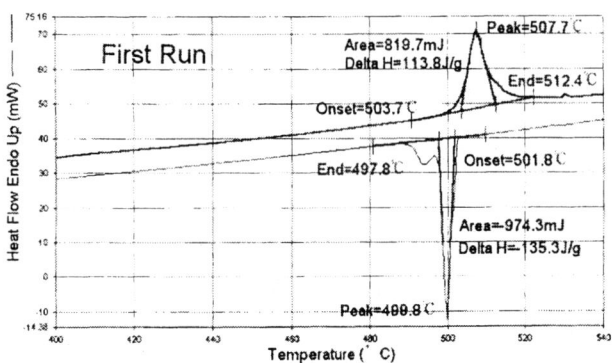

Figure-1 The variation of temperature with respect to heat flux for NaCl-CaCl$_2$

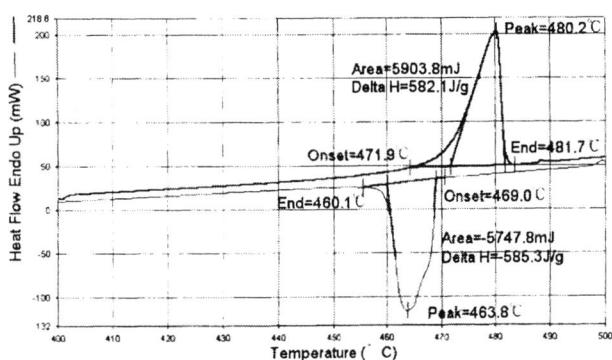

Figure-2 The variation of temperature with respect to heat flux for LiOH

Table-1 Summary of the material properties and the measured thermal properties of PCMs by DSC

Items	NaCl-CaCl$_2$		LiOH	
ρ (kg/dm)	2.15		1.43	
K (W/m·K)	K_ℓ=1.02		K_s=2.5 K_ℓ=0.48	
C_p (J/g·K)	C_{p_s}=0.8 C_{p_ℓ}=1.0		C_{p_s}=2.5 C_{p_ℓ}=3.9	
Hazard	nontoxic		toxic	
$NT/g	1.06		2.6	
Number of Cycles	T_m(°C)	ΔH(kJ/kg)	T_m(°C)	ΔH(kJ/kg)
1	503.7	113.8	470	582
2	504.0	132.6	471	576
3	503.9	142.1	470	580
4	503.9	142.0	472	579

TESTING OF THERMAL ENERGY STORAGE UNITS

A small plate heat exchanger (200mm by 200mm by 300mm) as shown in Figure-3 is made as the unit to store thermal energy. NaCl-CaCl$_2$ and LiOH are introduced as the thermal energy storage medium respectively. The plate type heat exchanger is made up of 20 stainless steel plate units filled with PCM inside each unit. Every plate unit is also completely soldered around 4 edges to ensure no leakage of PCM. The PCM is filled about 80% full inside each plate unit. This is to compensate the possible expansion of PCM resulting from heating. The specifications for this thermal storage unit can be summarized in Table-2.

Figure-3 The thermal energy storage unit with PCM used in this study

Table-2 The Specifications of Thermal Storage Unit

Items	Specification
Width of plate	20 cm
Length of plate	33 cm
Height of plate	0.6 cm
Number of plates	20
Amount of PCM for NaCl-CaCl$_2$	3kg
Amount of PCM for LiOH	2.6kg
Volume for thermal storage unit	12,000 cm^3
Front area of thermal storage unit	400 cm^2
Pressure drop of test section (mmAq)	2.3

The total amount of NaCl-CaCl$_2$ and LiOH used in these two units are 3kg and 2.6kg respectively. A test facility with testing area 20cm by 20cm by 30cm as shown in Figure-4 is used to study the behavior of this thermal storage unit during recycling heating and cooling. In this system a hot air generator with heating power up to 15kW and maximum temperature 650 °C is used to simulate the waste heat from industrial furnace. The PID temperature controller is also applied to control the procedures of heating and cooling processes in order to analyze the stability of these heat exchangers for the long-term operation. The performance (quantity of energy storage) of these two thermal energy storage units was studied with the airflow rate at 0.5 standard cubic meter per minute (SCMM). The inlet and outlet temperatures are also recorded. The total amount of energy storage can be calculated as:

$$Q_i = \int \dot{m} \times C_p \times (T_i - T_o) dt \quad (1)$$

$$Q_l = \int \bar{h} \times A_s \times (T_s - T_r) dt \quad (2)$$

$$\bar{h} = 1.89 \times (T_s - T_r)^{0.25} \quad (3)$$

$$Q_s = Q_i - Q_l \quad (4)$$

$$Q_s = Q_s - Q_{s(withoutPCM)} \quad (5)$$

The hot air generator was turned off as the outlet temperature reached 520 °C. The outlet temperature and the total amount of energy storage as a function of time for thermal storage unit with and without PCM were showed in figure-5. Also, the total amount of energy storage for each experiment was calculated by equations (1) to (4). The heating time was also recorded. The average results for this series of experiments can be summarized and shown in Table-3.

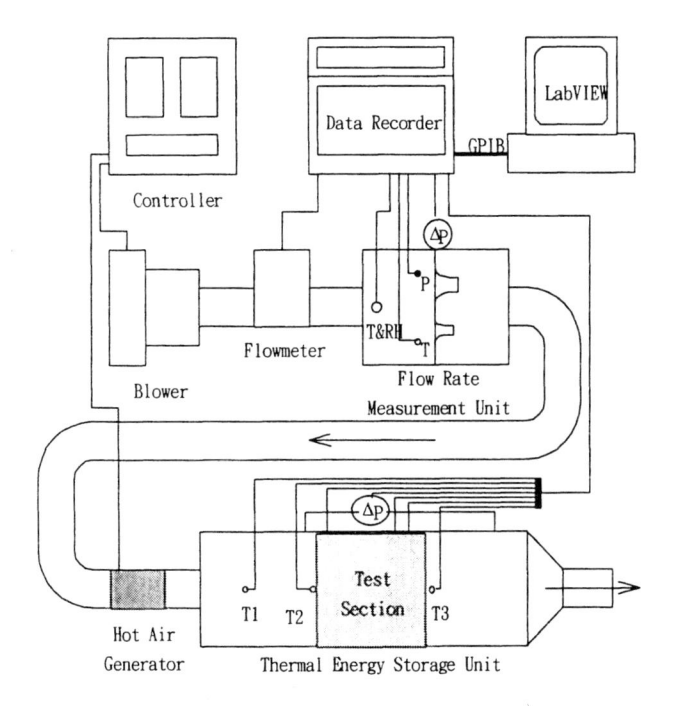

Figure-4 Experiment set up for the thermal energy storage system

Table-3 The results for different thermal energy storage units

Storage heat exchanger	Without PCM	With NaCl-CaCl$_2$	With LiOH
Flow Rate (SCMM)	0.5	0.5	0.5
Q_i (kJ)	1353	3119	3560
Q_s (kJ)	1307	2936	3390
Q_p (kJ)	0	1629	2083
Q_l (kJ)	46	183	170
Q_p/Q_s (%)	0	56	62
Q_l/Q_s (%)	85	80	76
Time (min)	18.1	32	34
Storage Rate (kJ/min)	72.2	91.8	99.7

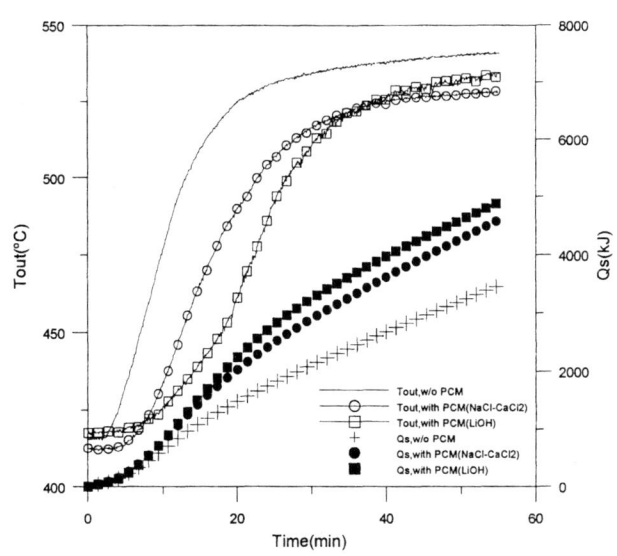

Figure-5 The variation of the outlet temperature and thermal energy storage for thermal storage unit with and without PCM

RESULTS AND DISCUSSION

The thermal properties of NaCl-CaCl$_2$ for the repeated heating and cooling were shown in Table-1. It is found that the latent heat for the second cycle is increased from 113 kJ/kg to 132 kJ/kg by comparing to the first cycle. However, the latent heat is maintained at the same value for the third and fourth cycles. This is due to the eutectic properties for the combination of NaCl and CaCl$_2$, which was more stable after two times of melting and freezing process. It is believed that the combination of NaCl and CaCl$_2$ was more homogenous after initial two heating and cooling cycles. This characteristic can ensure more exact design of thermal storage unit.

For LiOH, this phenomenon is not occurred, and the latent heat is almost maintained at constant value even after four cycles.

The experimental results in this study shown that the modified plate type latent heat exchanger have increased the amount of thermal energy storage about 1629 kJ filled with NaCl-CaCl$_2$ and 2083kJ filled with LiOH by comparing with the same plate type heat exchanger without PCM as shown in Table-3. It was also shown that the contributions of thermal energy storage due to phase change materials were 56% and 62% respectively for these two different PCMs. Moreover, the thermal energy storage rate was also increased more than 25% by comparing the heat exchanger without PCM. However, the percentage of thermal energy discharged for these two latent heat exchangers was decreased a little bit from 85% to 76%. It is probably due to the energy lost and the internal energy change of PCMs in the initial process. It should be level out after long-term operation. In addition, it is notable that the amount of LiOH used in the thermal storage unit is a little less than NaCl-CaCl$_2$, but the amount of thermal energy storage is higher than NaCl-CaCl$_2$ as shown in Figure-5. This is due to the latent heat for LiOH higher than NaCl-CaCl$_2$. The comparison of the outlet temperature for thermal storage unit with and without PCM was also showed in Figure-5. It is found that the outlet temperature increased rapidly up to 520 °C for thermal storage unit without PCM. However, the outlet temperature increased gradually due to the effect of latent heat thermal storage for the thermal storage unit with PCM.

In order to understand the thermal stability of PCMs for NaCl-CaCl$_2$ and LiOH during the long-term operation (60 times of heating and cooling processes), the samples of PCM taken from these two thermal storage units after every 10 cycle of process were measured by DSC as shown in Table-4. The results were shown that the phase change temperature and the latent heat for these two PCMs were almost maintained at constant. It is meant that the thermal properties are not affected significant by the cycles.

Table-4 Summary of the thermal properties of PCMs measured by DSC after every 10 cycles of experiment

Number of Cycles	NaCl-CaCl$_2$ T_m(°C)	NaCl-CaCl$_2$ ΔH(kJ/kg)	LiOH T_m(°C)	LiOH ΔH(kJ/kg)
10th	503	142	470	580
20th	504	143	471	581
30th	503	141	470	577
40th	503	140	472	579
50th	505	138	473	575
60th	502	136	471	573

CONCLUSIONS

The conclusions from this study can be summarized as following:

1. Adopting DSC to measure the thermophysical properties ($T_m, \Delta H, Cp$) of PCMs is very helpful in advance to understand its phenomena during phase change process. In this study, NaCl-CaCl$_2$ and LiOH are good candidates for the thermal energy storage materials of the applications of waste heat recovery operated at the temperature lower than 600 °C. However, it is still necessary to assess the effects of the high pressure, corrosion, leakage, toxicity and cost…etc..
2. In this study, the plate type heat exchangers modified as the latent heat exchanger with PCMs can substantially increased the percentage of the amount and rate of the thermal energy storage due to the contributions of the effect of phase change by comparing with the same plate type heat exchanger without PCMs. And it is repeatable and stable for these plate type heat exchangers after long-term usage.
3. Basically, most of the phase change materials are low thermal conductivity. The influence of low thermal conductivity of the PCMs is not examined in this study. However, it is the important parameter to design the well performance of the high-temperature latent heat exchanger. The enhanced heat transfer mechanism is required in order that all the phase change material can be subjected to phase change.
4. The choices of PCM and airflow rate are very important for the design of thermal energy storage unit. The higher latent heat of PCM can be resulted in more thermal storage efficiency, but more time is required for the inapposite airflow rate.
5. The economic and overall analysis for the waste heat recovery system with thermal energy storage that employed phase change material needs more study in order to set up the criteria for the application of phase change material.

NOMENCLATURE

A_s : Surface area of the test section;
C_p : Specific heat of the supply air or PCM;
ΔH : Latent heat of PCM;
$\bar{h}$: Mean natural convection heat transfer coefficient;
K : Conductivity of PCM;
$\dot{m}$: Mass flow rate of supply air;
Q_i : The total energy supply to the test section;
Q_l : The energy lost from test section to the environment;
Q_s : The energy absorbed by the energy storage unit;
Q_p : The energy charged by phase change material;
T_i : Temperature at the inlet of test section;
T_o : Temperature at the outlet of test section;
T_s : Temperature at the surface of test section;
T_r : Room temperature;
ρ : Density of PCM.

Subscripts

i : inlet, or total;
l : liquid, or loss;
m : melt;
o : outlet;
p : phase change material;
s : solid, or storage.

REFERENCES

Kang, Byung Ha and Yabe, Akira, 1996, *Performance Analysis of Metal-Hydride Heat Transformer for Waste Heat Recovery,* Applied Thermal Engineering, Vol. 16, P. 677 – 690.

Lane, George A., 1989, *Phase-Change Thermal Storage Materials,* Handbook of Applied Thermal Design, Chap. 1, pp. 6-1 – 6-11.

Lin H.H., Lin, G.T. and Yang, B.C., 1998, *"The Technologies for Energy Recovery and Application,* Final Report submitted to the Energy Commission, Taiwan, ROC.

Lock, G.S.H., 1996, *Latent Heat Transfer – An Introduction to Fundamentals,* Oxford University Press, pp. 1-52.

Yagi J. and Akiyama, T., 1995, *Storage of Thermal Energy for Effective Use of Waste Heat from Industries,* Journal of Material Processing Technology, v. 48, pp. 793-804.

FOULING IN HEAT EXCHANGERS

FOULING IN PLATE-AND-FRAME HEAT EXCHANGERS AND CLEANING STRATEGIES

Syed M. Zubair[1] and Ramesh K. Shah[2]

[1]King Fahd University of Petroleum & Minerals, Dhahran 31261, Saudi Arabia, e-mail: smzubair@kfupm.edu.sa
[2]Delphi Harrison Thermal Systems, Lockport, NY 14094-1896, USA, e-mail: rkshah@attglobal.net

ABSTRACT

Fouling of Plate-and-Frame Heat Exchangers (PHEs) is a complex phenomenon that is influenced by plate design, flow rates, and surface conditions. The understanding of basic fouling mechanisms is expected to guide the development of effective mitigation techniques, which are categorized as on-line and off-line techniques. It is shown through some case studies that properly designed PHEs are less prone to fouling than shell-and-tube heat exchangers (STHEs). The literature on fouling in PHEs having complex flow passages is somewhat limited; however, a reasonable progress has been made in understanding the crystallization and particulate deposition processes in PHEs. Both on-line and off-line cleaning methods are discussed along with probabilistic off-line cleaning schedule.

INTRODUCTION

Plate-and-frame or gasketed plate heat exchangers (PHEs) have many advantages over shell-and-tube heat exchangers (STHEs), which are considered as a workhorse of the process industry. These may be categorized as (i) higher heat transfer performance and exchanger efficiency, (ii) lower space requirements, (iii) easy accessibility to all areas, and (iv) lower overall capital and maintenance costs among others. Because of the much higher heat transfer coefficients for the PHEs, if the fouling resistances recommended by TEMA be used, the required additional heat transfer surface becomes excessive. Generally, it is assumed that PHEs foul less than comparable STHEs or double-pipe heat exchangers, because of a higher degree of turbulence in PHEs. For water-cooling applications, Cooper et al. (1980) report that the fouling resistance for the PHEs was only one-third that of the respective STHEs. It is, therefore, often recommended that the additional surface should not exceed 25% of that required for clean PHEs. Adding plates to a heat exchanger will reduce the fluid velocity; thereby resulting in higher fouling resistance than is anticipated. Long-term thermal-hydraulic performance under fouling conditions is still an important consideration that limits the PHE use and acceptance in the industry.

Due to the geometry configuration of PHEs, the flow characteristics are quite different from those in STHEs. Continuous change in the flow direction and cross-sectional area results in a flow regime that changes continuously with position throughout the PHEs. It is important to note that stagnation zones exist in the flow channels due to the typical location of the inlet and outlet ports. In addition, the flow velocities are low in the regions where the adjacent plates touch each other. Since the flow conditions affect the wall temperature and shear stresses, fouling varies across the heat exchanger plate. The flow characteristics within the heat exchanger are further changed by the deposit formation. Because the flow channels in PHEs are only a few millimeters deep, the accumulation of undesired matter can sometimes block the channels, resulting in higher local fluid velocities for the same mass flow rates. Because of all these flow-related complexities, it is very difficult to modify the fouling factors obtained for STHEs and use them for PHEs.

The objective of this paper is to discuss the results of several investigations on fouling in PHEs by various mechanisms, including crystallization, particulate, chemical reaction, and biological fouling. The operating conditions and mechanical design parameters that influence the fouling rate are highlighted. Both on-line and off-line cleaning methods are described. In this regard, a probabilistic (off-line) cleaning method is also discussed; it incorporates the risk factor p, representing the probability of heat exchanger plates being fouled to a critical level after which a cleaning is needed.

IMPACT OF FOULING

Fouling forms an essentially solid deposit on the heat transfer surface. If the thicknesses of the deposits are independent of time, then fouling simply adds another conduction resistance in series with the wall. In general, fouling varies with time. We, therefore, introduce a time-dependent fouling resistance in computing the overall heat transfer coefficient of the surface. Since the heat transfer resistances are in series, we can express the time-dependent, overall heat transfer coefficient in terms of the cleanliness factor, described as (Knudsen, 1998)

$$\frac{U(t)}{U_c} = \frac{1}{1 + U_c R_f(t)} \quad (1)$$

where U_c and $U(t)$ are the overall heat transfer coefficients of a clean and fouled surfaces, respectively; and $R_f(t)$ is the time-dependent fouling resistance which is influenced by the process under consideration.

Equation (1) is plotted in Fig. 1, where the cleanliness

factor U/U_c is shown as a function of fouling resistance R_f and overall clean heat transfer coefficient U_c, all being time dependent. It should be noted that larger the heat transfer coefficient U_c, the greater is the effect of fouling on the cleanliness factor $U(t)/U_c$. For example, for $R_f = 2.0 \times 10^{-4}$ m^{2}K/W, the cleanliness factor decreases by about 52% when U_c = 8,000 W/m^{2}K compared to the U_c = 500 W/m^{2}K case. This decrease in the cleanliness factor is reflected in additional materials, fabrication, installation and energy costs. As energy and materials costs increase, there will be a tendency to design heat transfer equipment (such as PHEs) with even higher overall heat transfer coefficients. Therefore, the control and assessment of fouling becomes more important.

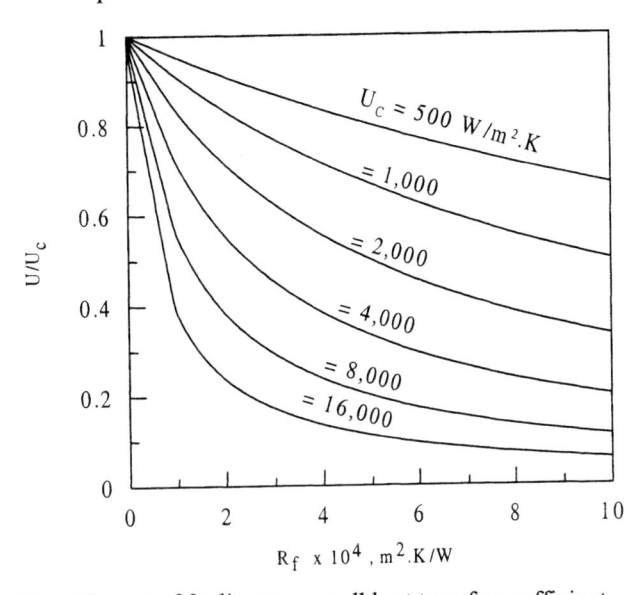

Fig. 1 Impact of fouling on overall heat transfer coefficient.

As discussed above, the PHEs have higher heat transfer coefficients than conventional heat exchangers; the fouling resistance value must not be overestimated. There is only a limited information available in the literature on fouling resistance values for PHEs. Mariott (1971) gives typical values, which are roughly 10 times lower than those recommended by the TEMA [see Chenoweth (1990)]. It must be noted that the fouling resistances for STHEs in the TEMA Standards are not experimental values, but rather consensus by knowledgeable experts. Cross (1979) recommends a limit of 25% additional heat transfer surface, and points out that adding plates to the heat exchanger will reduce the fluid velocity, and inevitably will result in increased fouling. A brief summary of recommended fouling resistance for PHEs versus those for STHEs (TEMA Standards) is shown in Table 1.

FOULING MECHANISMS IN PHEs

In the following paragraphs, we discuss various investigations on fouling in PHEs. They are subdivided according to the generally accepted mechanisms of fouling, namely crystallization, particulate, chemical reaction, corrosion, and biological.

Table 1. Recommended fouling resistance of PHEs versus TEMA values (Panchal and Rabas, 1999).

Process Fluid	R_f - PHEs m^{2}K/kW	R_f - TEMA m^{2}K/kW
Soft water	0.018	0.18 - 0.35
Cooling tower water	0.044	0.18 - 0.35
Sea water	0.026	0.18 - 0.35
River water	0.044	0.35 - 0.53
Lube oil	0.053	0.36
Organic solvents	0.018 - 0.053	0.36
Steam (Oil bearing)	0.009	0.18

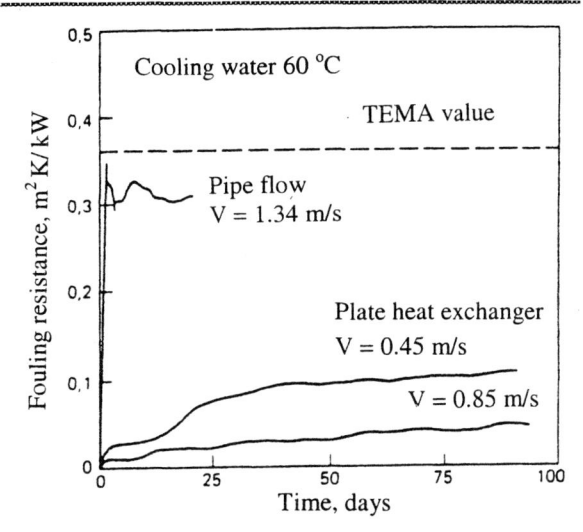

Fig. 2 Plate heat exchanger fouling curves (Cooper et al., 1980).

Crystallization Fouling

Cooper et al. (1980) investigated cooling water fouling in PHEs having stainless steel chevron plates with a corrugation angle β of 60°. Two relatively small STHEs were installed in parallel for comparison purpose. The water was chemically treated before entering the heat exchangers. Various stages of water treatment and characteristics of the cooling water are reported. Some of the important results are shown in Fig. 2. The fouling resistance in the PHEs is significantly lower than that of the STHEs, despite the low flow velocities. As expected, if the flow velocity is increased, the fouling resistance decreases significantly. The fouling resistance with increasing time seems to level off and approach asymptotic values which is about 15-30% that of the TEMA values. It should be noted that asymptotic fouling model is often observed in cooling water PHEs. In these heat exchangers, the conditions leading to the formation of a scale layer of a weak, less coherent structure are associated with (1) the simultaneous crystallization of salts of different crystal shapes, or (2) the presence of suspended particles embedded in the crystalline structure. Hasson (1999) has indicated that growth of such deposits is expected to create internal stresses in the scale layer, so that the removal processes become progressively more effective with the deposit thickness. Such considerations lead to an asymptotic scale thickness, at which the deposition is balanced by the scale removal mechanism.

The cooling water experiments described by Cooper et al. (1980) were for crystallization and particulate fouling and did not provide much insight into the deposition mechanisms. Bansal and Muller-Steinhagen (1993-1997) investigated crystallization fouling from $CaSO_4$ solutions in a commercial PHE and in a flat plate heat exchanger. They confirmed through X-ray diffraction analysis that gypsum (calcium sulphate dihydrate, $CaSO_4.2H_2O$) was the sole crystallizing matter. The rate of deposition increases with increasing wall temperature and bulk concentration and decreasing velocity as shown in Figs. 3-5. With increasing flow velocity, both the initial fouling rate as well as the fouling resistance decreases. The initial fouling rate decreases due to lower wall temperature at higher flow velocities whereas the fouling resistance decreases due to a higher removal rate in conjunction with a lower interface temperature. Bansal and Muller-Steinhagen (1997) found that the $CaSO_4$ concentration was only important in the initial stages of the fouling process. Later, the channel gap was reduced considerably by the deposit, thus causing other factors to dominate such as higher shear stresses and lower interface temperatures.

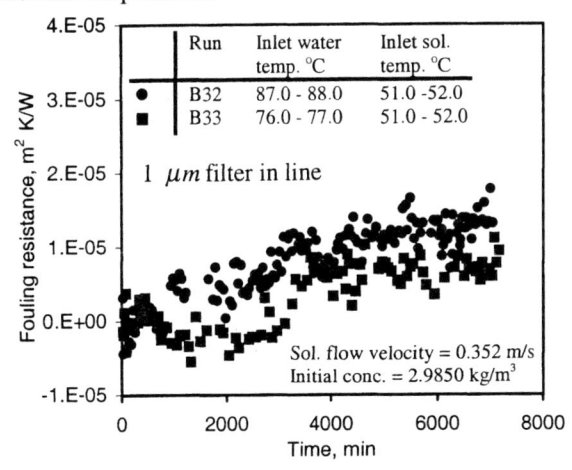

Fig. 3 Effect of wall temperature on fouling (Muller-Steinhagen, 1997).

It was demonstrated by Bansal et al. (1993-1997) that the diagonal flow (in the chevron plates) results in low velocity zones. In these areas, shear forces are at a minimum and the wall temperature is close to the temperature of the heating medium. Therefore, most deposition occurs in the top left-hand corner of the plates because of the low flow velocity and high fluid-surface temperature. The extent of the stagnant zones, and therefore the extent of the deposition, decreases with increasing flow velocity. The higher fluid flow velocity regions, for example, near the exit, usually remained free from fouling even though the local driving force for deposition was higher. It is obvious that flow distribution has a major effect on fouling and hence heat transfer rates of the exchanger. It is therefore expected that better design of corrugations and flow distribution section in PHEs may result in a higher heat transfer and lower pressure drop compared with the existing PHE units. However, under fouling conditions, the performance of these heat exchangers should be carefully monitored, because the formation of deposits with time may cause partial blockage of the flow channels that in turn may alter the flow patterns completely.

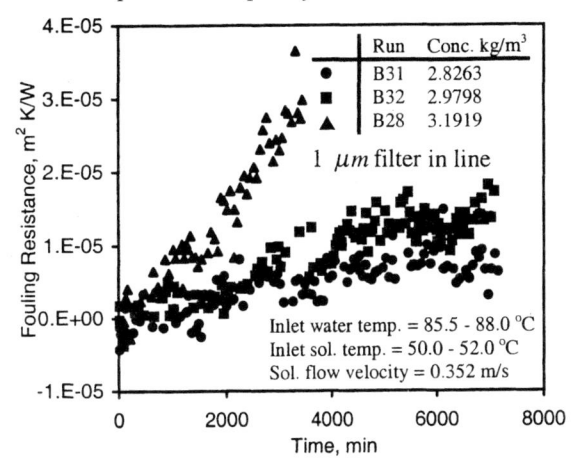

Fig. 4 Effect of $CaSO_4$ concentration on fouling (Muller-Steinhagen, 1997).

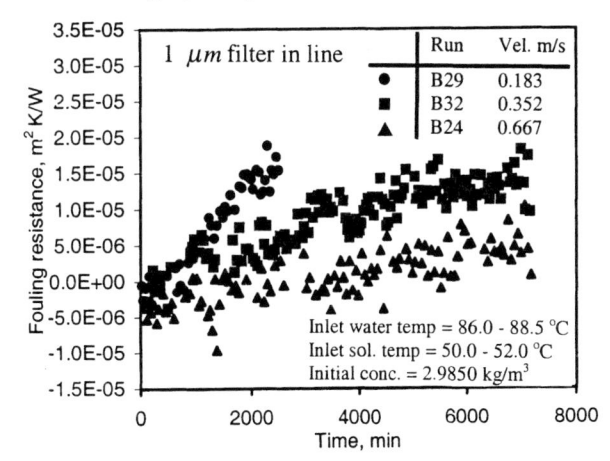

Fig. 5 Effect of flow velocity on fouling (Muller-Steinhagen, 1997).

Particulate Fouling

PHEs offer significant advantages over STHEs for operation with suspensions of finely divided particles such as sand, silt, etc, since the induced turbulence keeps solids in suspension. The smooth surface of the plates reduces the stick ability of the particles. Analysis of deposits formed during cooling water fouling in the experiments by Cooper (1980) showed that the amount of silica (sand) found at the shell side of a STHE was significantly higher than for the parallel plate heat exchanger. The high silica concentration was caused by the sedimentation in the stagnant flow zones at the corners between baffles and shell, whereas the high turbulence promoted by the corrugated plates in the PHE minimized sedimentation fouling.

Muller-Steinhagen and Middis (1989), using a suspension of Al_2O_3 particles in solvent X-2, investigated particulate fouling in a commercial PHE unit. The fouling resistance was found to increase rapidly without delay at the beginning of the experiment and approached an asymptotic value after some time. With increasing concentration, the rate of deposition and the value of the asymptotic fouling resistance increased. They noticed that the effect of particle concentration levels-off for higher particle concentrations. As expected for other fouling mechanisms, the asymptotic fouling resistance decreases with increasing fluid velocity due to the increase in wall shear stress.

Thonon et al. (1999) have shown with different types of particles that there is a great sensitivity of the fouling rate on the particle type and flow conditions. This is particularly true for (Al_2O_3 or TiO_2), where the solution pH has a strong influence. It is well known that solution pH affects the particle size distribution by modifying the particle zeta potential, in addition it changes the particle adhesion forces, and consequently the net fouling rate is also affected.

Some experiments, conducted by Bossan et al. (1995) and Grillot (1992) on particulate fouling in PHEs, have shown that the channel geometry and flow maldistribution between the channels have a major influence on the fouling rate.

Chemical Reaction Fouling

Chemical reaction fouling occurs in food industry, chemical industry and petroleum refineries. A critical variable is the surface temperature that determines the reaction rate. Shibuya et al. (1997) carried out fouling tests using pilot-scale PHE and STHE with untreated straight-run gas oils, including Heavy Gas Oil (HGO) and Light Cycle Oil (LCO). The duration of the tests was over 500 hours. Figure 6 shows fouling resistance of a PHE by Packinox and an STHE as well as the recommended values of TEMA by the crosshatched area. It is clearly demonstrated that fouling resistance of the PHE is almost negligible, and is much lower than that of the STHE after about 350 hours of operation.

Fouling in food industry is severe. It is important to note that in the petrochemical industry, it is common to clean only once or twice a year as compared to daily cleaning needed in food plants. Most fouling-related work has been reported on the milk systems (Sandu, 1989; Rene and Lalande, 1990; Tissier; 1992; Delplace et al., 1997; and Fryer et al., 1997). Fouling results from proteins, both β-lactoglobulin and caseins, and the decrease in solubility of milk salts with increasing temperature. The thermal response of milk is highly complex, particularly when investigating fouling effects on heat transfer and fluid flow. It is important to work closely with food chemists for understanding the fouling behavior of milk, but it must be combined with knowledge of flows and temperatures in PHEs. A number of authors have modeled fouling, combining kinetics of protein reactions with the fluid thermal-hydraulic conditions in PHEs. Paterson and Fryer (1988) showed that both bulk and surface reactions contribute to fouling. De Jong (1992, 1993) applied the kinetics of the β-lactoglobulin reaction to model fouling in PHEs up to $100°C$, and found that amounts of deposit could be correlated with reaction rates.

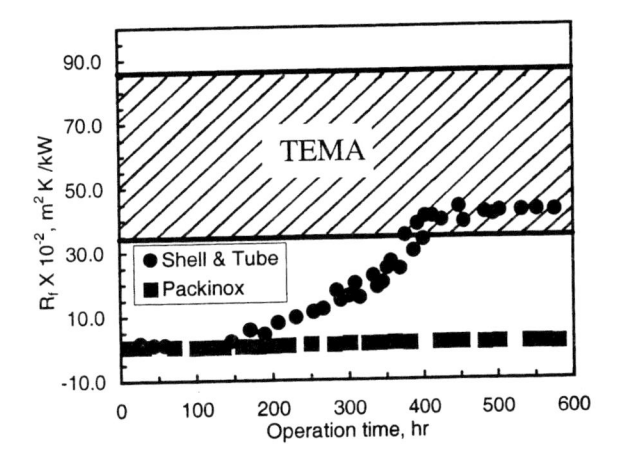

Fig. 6 A comparison of fouling resistances between PHE and STHE (Shibuya et al., 1997).

It is, however, important to note that the variation in chemical composition of real food fluids, coupled with complexity of food process plants, such as PHEs, makes it very difficult to apply such fouling models to real systems. There is no other possibility, but to monitor the PHEs undergoing fouling to determine the correct time to stop operation to clean the heat exchangers.

Corrosion Product Fouling

Corrosion product fouling is due to the chemical reaction of containment materials (including heat transfer surfaces of PHEs) with the circulating process streams. Sommerscales (1981) distinguishes between two types of corrosion related fouling. In one type, the fouling is a result of corrosion products formed at locations remote from the PHEs and transported to and deposited on the surfaces of PHEs. Lister (1981) has reviewed aspects of this type of ex situ corrosion fouling. The second type, in situ corrosion fouling in PHEs, is due to corrosion products forming and remaining at the heat transfer surface, because of nucleation promoting effect. It is important to mention that there are no systematic studies on corrosion-related fouling in PHEs since the effect of corrosion, in most cases, is very hard to separate from other fouling mechanisms.

There are numerous types of corrosion, such as general, crevice, pitting, stress, and microbiological observed in PHEs. Sloan (1998) indicated that operating parameters, which include the surface and bulk-fluid temperatures, the flow rate, solids content, pH value and oxygen level, influences each type on the corrosion of alloy materials that are used in PHEs. In general, chlorides attack stainless steel alloys, fluorides affect titanium, and fluids containing nitrogen attack copper. Chloride corrosion is usually directly proportional to the fluid's chloride concentration and temperature, and inversely proportional to its pH.

On stainless steel, chloride corrosion becomes more aggressive as fouling deposits such as calcium carbonate and silica scale accumulate. As the oxygen level under the deposits decreases, the pH value is also decreased. This lower pH results in a lower crevice temperature, thus increases the corrosion rates. Sloan (1998) mentioned that chloride corrosion pitting could typically be seen at the low-velocity crevice areas of PHEs such as contact points, or along the edge of the gasket grooves.

It should be noted that corrosion-related fouling deposits might break off if there are high concentrations of abrasive solids such as sand or crystal particles in the process fluid, or if the process fluid is moving with excessive velocity. Typically, this shows up as a smooth, shiny finish on the plate's inlet port and distribution area. Sloan (1998) mentioned that in improperly designed PHE units, the edges of the inlet port connection could erode away, causing external leakage in several cases.

Biological Fouling

PHEs are generally designed for heat transfer duties in which all kinds of natural water sources, such as sea, river, and lake water are used as a cooling medium. This type of water, with all its impurities, causes more-or-less heavy fouling of these units. Although different types of deposits may be formed, biological fouling is usually predominant up to temperature of 50°C. The presence of biological material in PHEs may promote other fouling mechanisms. It is not uncommon to find that scale formation, particulate deposition, and corrosion accompany biological growth. Bott (1995) mentioned that this is associated with metabolic activity involving chemical reactions and changed conditions, particularly pH beneath the biological mass. Microorganisms produce extracellular material that may be sticky and hence facilitates the adhesion of particulate material. The appearance of the foulant can be very varied depending on the organisms involved and the conditions under which the growth is taking place. It is therefore expected to have long filaments of bacteria or algae, and slime layers.

To study cooling water fouling in PHEs, Novak (1982) studied the fouling behavior of Rhine River water near Ludwigshafen, Germany, and of Oresund water in Sweden. For both waters, mainly biological fouling was noticed. Figure 7 shows the fouling resistance as a function of time for various flow velocities. In most cases, the fouling resistance increased almost linearly over the observed period. Figure 8 shows the influence of mean slime temperature on the dimensionless fouling rate due to biological deposits at the heat transfer surface. The fouling rate was normalized in terms of the fouling rate at 25°C for Rhine water. It reached a maximum at about 32°C. Novak (1982) also discussed the effect of wall shear stress and chlorination kinetics on the removal mechanism.

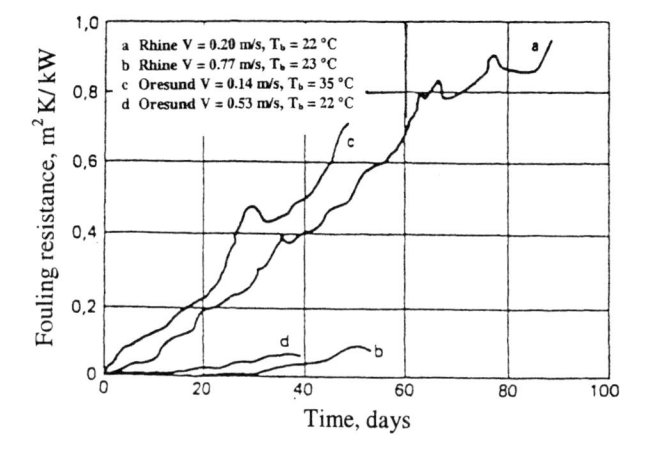

Fig. 7 Fouling resistance versus time (Novak, 1982).

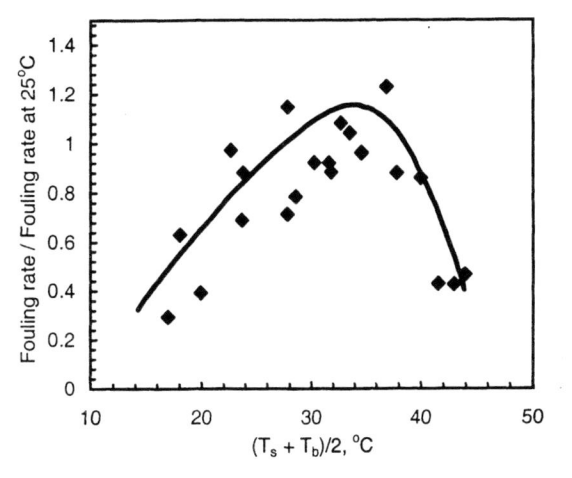

Fig. 8 Effect of mean slime temperature on bio-fouling in PHE (Novak, 1982).

FACTORS INFLUENCING FOULING IN PHEs

The parameters that influence or impact fouling on heat transfer or pressure drop performance in PHEs can be classified as: (a) wall shear stress, (b) surface and bulk temperatures, (c) mechanical design parameters, and (d) plate design.

Wall Shear Stress

The wall shear force due to fluid velocity is probably the most important parameter affecting the fouling behavior in PHEs. As discussed earlier, for most of the fouling mechanisms, the rate of fouling decreases as the velocity increases. It is expected that an increase in the velocity may promote the transport of foulants to the heat transfer surface. The associated increase of shear forces leads to an overall reduction of fouling, particularly for almost all-soft deposits. Experiments have shown that the fouling resistance in PHEs depends on the flow velocity in a similar way as the fouling resistance in STHEs. Thonon et al. (1999) have shown that for a 30° corrugation angle (refer to Fig. 9), the asymptotic fouling resistance is 10 times lower while the velocity is increased three fold. However, it should be noted that for the same flow velocity, fouling resistance would also depend on the plate design, which greatly influences the wall shear stress.

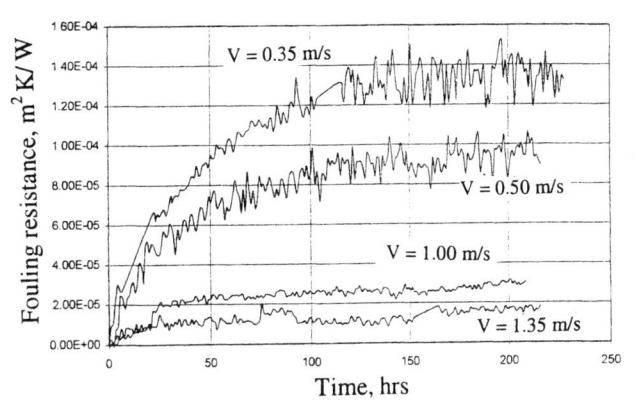

Fig. 9 Effect of the flow velocity on fouling resistance (Thonon et al., 1999).

It should be emphasized that the specification of a fouling resistance translates to installing additional heat transfer surface area for a heat exchanger. For PHEs with a given plate dimension, this is generally achieved by installing additional plates, which in turn leads to a reduced flow velocity in the channels between the plates. Since this will increase fouling, as discussed above, it is expected that deposition process may accelerate. It has been found (Muller-Steinhagen, 1997) that significantly over-designed PHEs often perform below design specification, after only a short period of operation. Figure 10 shows calculated heat duty of a relatively small commercial PHE as a function of the heat transfer surface area for constant flow rates. The inlet temperatures of hot water and solvent X-2 are 70°C and 30°C, respectively, and the concentration of particles suspended in the solvent is 300 ppm. As the heat transfer surface area is increased, the number of parallel plates is increased and the flow velocity of both fluids is reduced. We note that the heat duty for clean fluids increases continuously with increasing heat transfer surface area. However, a maximum heat duty is calculated for fouling conditions at about 0.8 m^2. We may say that adding plates to the heat exchanger is only beneficial to the left of this maximum.

The above results indicate longer plates (without affecting the fluid velocity in channels) rather than more number of plates (which would reduce the fluid velocity in channels) should provide that additional excess surface for higher heat transfer; but that option will also lead to higher fluid pressure drop. Another design option is to increase the flow velocity by connecting several flow channels in series. This approach increases the heat transfer coefficient but with a significant increase in the pressure drop, and it decreases the fouling resistance.

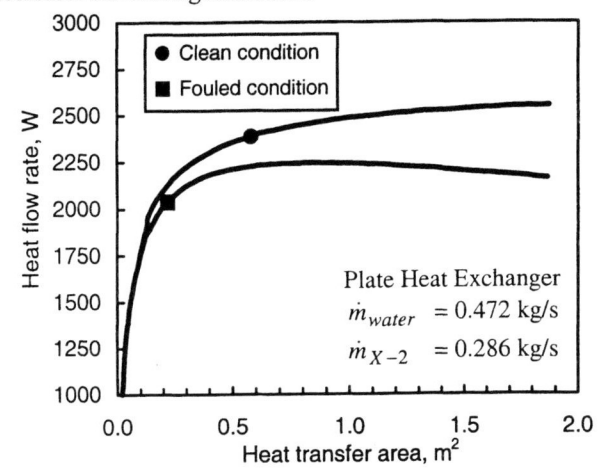

Fig. 10 Heat duty as a function of number of plates (Muller-Steinhagen and Middis, 1989).

Surface and Bulk Fluid Temperatures

The surface temperature of heat exchanger plates has a strong influence on deposition and hence the fouling rates through the chemical reaction rate and the solubility of dissolved substances. Salts with inverse solubility precipitate on heated surfaces. The driving concentration difference for this reaction will increase with increasing wall temperature. It should be noted that the reaction rate constant of chemical reactions increases exponentially with the surface temperature. In most cases, the effect on the reaction rate constant is more dominant than the effect on the solubility, causing increased deposition rates even for normal solubility salts as long as the solution is supersaturated.

For a given flow velocity, the wall temperature will increase with increasing bulk temperature, causing the fouling rates to accelerate, as discussed above. If nuclei for crystallization or reaction are present in the fluid, the fluid temperature increases the rate of polymerization, chemical reaction or crystallization in the bulk. Homogeneous bulk precipitation may occur at high temperatures. Particulate material formed in the fluid bulk will tend to deposit on the heat transfer surfaces, contributing to the overall fouling process.

Mechanical Design Parameters

The plate material has its strongest effect during the initiation stage of fouling. Once the surface is completely covered with deposits, the material effect will be less important. Heat exchanger plates are usually made from stainless steel, or from other corrosion resistant metals such as titanium for some specialized service. Therefore, corrosion fouling is not a significant problem in a well-designed PHE. In general, maximum adhesion occurs in systems undergoing a maximum decrease in surface energy, such as occurs when a low energy fluid spreads upon a high-energy surface (Rankin and Adamson, 1973). Therefore, one would expect that scale adherence is a function of the heat transfer surface material, and poorest scale adherence (or less fouling) should occur on materials that have low surface energies. Of the conventional heat transfer surface materials typically used in heat exchangers, Muller-Steinhagen (1997) indicated that the order of decreasing surface energy is: (1) Stainless steel, (2) Monel, (3) 70-30 Copper-Nickel, (4) 90-10 Copper-Nickel and (5) Titanium. Surface energies for these materials range from 1800 to 1300 ergs/cm^2. However, it should be noted that the effective surface energy might be altered by the surface crystallographic structure, the presence of small quantities of impurities in the metal, the roughness of the surface, and the presence of oxide on the surface. The scale formation can be greatly reduced if the metal surface energy is reduced by surface treatment such as Ion Assisted Reaction or by Plasma Polymerization Process. These investigations are presently expanded to include heat exchanger plates by some industrial organizations (Kim et al., 1999). The preliminary results of these investigations are promising.

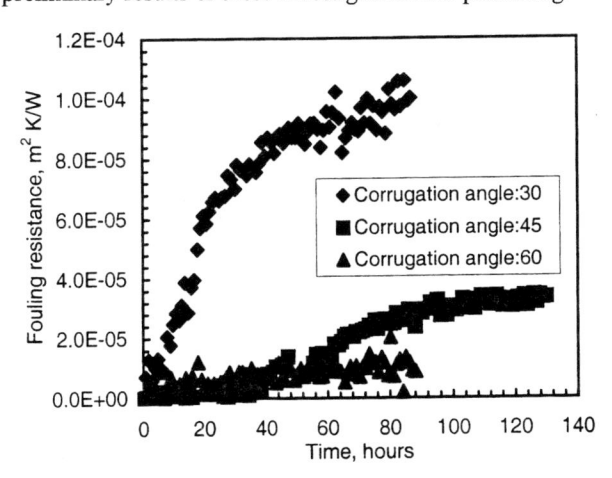

Fig. 11 Fouling resistance for chevron plates with three corrugation angles (Thonon et al., 1999).

Plate Design

To design for minimum fouling, the flow rates across the width of the plates in individual flow channels should be as uniform as possible. This offers a scope for improved plate design by optimizing flow distribution zone, height/width ratio and corrugation geometry. Thonon et al. (1999) studied three corrugated channels, differing only in the corrugation angles (β = 30, 45, and 60° from vertical axis). Figure 11 shows the fouling resistance as a function of time. It is clear from the figure that, for a given velocity, high corrugation angle β leads to low fouling resistance. This result is expected since the flow is highly three dimensional for high corrugation angles. Local wall shear stress measurements have shown that the average wall shear stress is higher than for a comparable straight channel.

MITIGATION OF FOULING

In this section, we discuss both on-line and off-line cleaning methods for PHEs.

On-line Methods

Two on-line cleaning methods are used for PHEs: Chemical Cleaning and Electronic Antifouling (EAF) technology. We will discuss these methods in the following paragraphs.

Various scale-inhibiting chemicals such as dispersing or chelating agents are used to prevent scales (Bott, 1995). Ion exchange and reverse osmosis are also used to reduce water hardness, alkalinity, and silica level; particularly in cooling water PHEs. However, these methods are expensive for PHEs applications in petrochemical and process industries and require heavy maintenance for proper operation. Once fouling occurs in heat exchangers, scales are removed by using acid chemicals, which shorten the life of heat exchanger plates by about 50%, thus necessitating premature replacement. When acid cleaning is not desirable, scraping, hydro blasting, sand blasting, metal or nylon brushes are used. These cleaning operations incur downtime and repair costs (Bott, 1995).

A new EAF technology has been developed by Cho et al. (1998) to mitigate primarily cooling water fouling in PHEs and STHEs. In this method, an oscillating electric field using time varying magnetic field generated in a solenoid is wrapped around a feed pipe carrying water. It is expected that the EAF technology can be used to reduce the maintenance efforts; for example, one may minimize or discontinue the use of scale-inhibiting or scale-removing chemicals, thus preserving a clean environment. The primary benefit of the EAF technology, if proven on a commercial scale, will be in maintaining the initial peak performance of a heat exchanger for a long period.

We know that as a PHE fouls, the pressure drop across the exchanger increases. Figure 12 shows the changes in the pressure drop due to fouling with and without the EAF technology with respect to time with circulation of 1000-ppm hard water at 1.5 gpm. Although the EAF technology did not completely prevent fouling inside the PHE, the results dramatically demonstrate the benefit of using the

EAF treatment for PHEs. It should be noted that the pressure drop results shown in Fig. 12 depict an interesting phenomenon on the rate of fouling. New 1000 ppm hard water was injected at t = 0 and 240 min in the PHE. The fouling rate is significantly greater at 240 min than at the startup (0 min). The former represents the rate of fouling from pristine clean plates, whereas the latter represents that from already fouled plates. Based on these results, one may conclude that the rate of fouling at a fouled plate is significantly greater than at a clean plate. However, the rate of fouling without the EAF treatment was consistently greater than that obtained with the EAF.

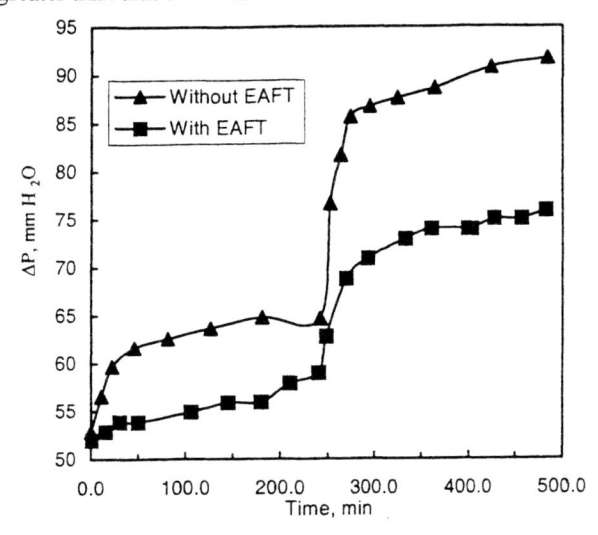

Fig. 12 Pressure drop across PHE with and without EAF treatment (Cho et al., 1998).

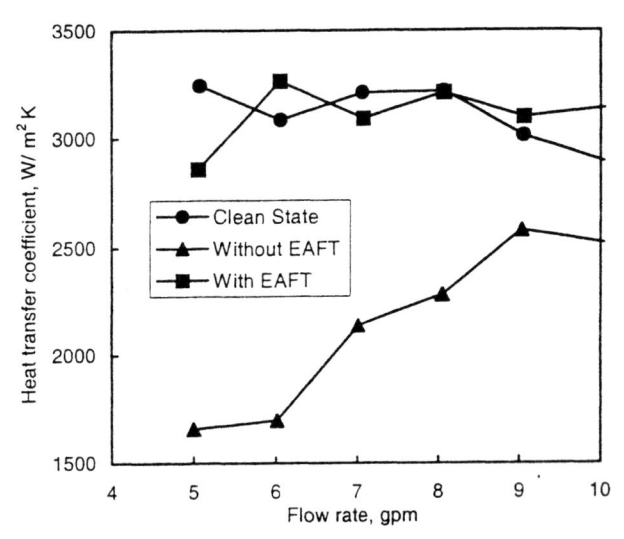

Fig. 13 Overall heat transfer coefficient of PHE with and without EAF treatment (Cho et al., 1998).

Figure 13 shows overall heat transfer coefficient versus flow rate measured after 8-hour of circulation of hard water at a flow rate of 1.5 gpm for three different cases: clean initial state, and tests with and without EAF treatment. The U values from clean initial state varied from 2900 to 3250 W/m^{2}K over a range of flow rate. When the PHE was used for 8-hour without the EAF treatment, the U values dropped from about 3250 to 1680 W/m^{2}K at 5 gpm (i.e., almost 50% drop from the initial U value). On the other hand, U values obtained for the case with the EAF treatment did not show any significant change in the U value. We may notice from the figure that the U values for the case without the EAF treatment increased with increasing flow rate. This was due to the fact that at high flow rates, for example, above 6 gpm, the PHE still had enough clean heat transfer surfaces available, resulting in relatively high U values, although the U values were still much smaller than those obtained at clean conditions.

Probabilistic Cleaning Method

Fouling is a time-dependent process traditionally analyzed in a deterministic framework of thinking. Comprehensive studies on these lines have been carried out and well documented in the literature (Bott, 1995). The complexity in controlling a large number of internal and external factors of a given process makes it very difficult to predict the fouling growth as a function of time using these deterministic models. The literature shows a substantial discrepancy between theoretical predictions and actual observations both in carefully controlled laboratory experiments and in industrial applications. These experimental uncertainties can best be treated by postulating that the fouling resistance is a time-dependent random process (Zubair et al., 1997a; and Sheikh et al., 2001), given by $R_f(t) = \varphi(\mathbf{A}, \mathbf{B}, t)$, where the cumulative effect of all the uncertainties is embedded in the parameters **A** and **B** of the underlying kinetic models. The intention here is to present a probabilistic approach to analysis of fouling models and its impact on heat exchangers cleaning.

Field investigations as well as replicate laboratory experiments (Sommerscales and Kassemi, 1987; Zubair et al., 1997a) in the study of fouling growth models suggest that there is (1) a considerable scatter in the values of R_f at time t, and (2) similarly for any fixed value of R_f, there is a corresponding scatter in the values of t. Both *trend* and *scatter* in R_f can be expressed by its probability distribution $f[R_f(t)]$; and main indicators of this distribution are its mean value $\mu[R_f(t)]$ and standard deviation $\sigma[R_f(t)]$, reflecting the *trend* and *scatter or uncertainty* of the process, respectively.

The evolution of $R_f(t)$ distribution with respect to t is represented by the random sample functions of the fouling resistance growth. Each sample function represents a realization of the process $R_f(t) = \varphi(\mathbf{A}, \mathbf{B}, t)$, where the parameters **A** and **B** as discussed above are the random process parameters. For example, consider a PHE that has many plates. The fouling resistance response of the individual plates will show a considerable scatter. This randomness of the process parameters is due to several reasons; some of these reasons are as follows (Zubair et al.,

1997b).

- Flow maldistribution in heat exchanger plates
- Variations and fluctuations in the velocity around the nominal value
- Variations and fluctuations in the pressure around the nominal value
- Variations and fluctuations in the surface temperature around the average value
- Perturbations in the foulant chemistry
- Fluctuations in environmental factors
- Plate material variability of metallurgical features
- Variability of surface finish
- Fluctuations in the initial quality characteristics of heat exchanger plates attributed to manufacturing and assembling process.

It is thus apparent that each heat exchanger plate will have its own fouling resistance growth curve. These curves will follow some type of fouling kinetic models such as linear, asymptotic or falling rate of the growth process. The ensemble of "m" such realizations for each of these curves is shown in Fig. 14. Mathematically, these functions represent fouling growth models, whose parameters should be treated as random due to a number of sources of randomness described above. These random functions represent fouling growth laws (Zubair et al., 1992). Here we will explain a linear random growth model and its implications on risk-based cleaning of PHEs.

A linear random fouling growth law, having no initial fouling and negligible induction time, can be expressed as $R_f(t) = Bt$. If B is normally distributed with mean $\mu(B)$ and standard deviation $\sigma(B)$, then the coefficient of variation $K(B)$ of the time rate of fouling resistance $[dR_f(t)/dt = B]$ is:

$$K(B) = \frac{\sigma(B)}{\mu(B)} \qquad (2)$$

To explain the cleaning strategy, we define $R_{f,max}$ as a limiting value of the fouling resistance). A preventive maintenance action needs to be taken periodically. In this regard, the time interval between cleaning for the kth preventive maintenance action taken at time t_k can be defined as

$$\theta_k = t_k - t_{k-1} \qquad (3)$$

where the time to the kth preventive maintenance action is measured from $t = 0$. The corresponding distribution of the interval between cleaning is given by (Zubair et al., 1992):

$$f(\theta_k) = \frac{M_k}{(2\pi\alpha)^{1/2} \theta_k^2} \exp\left[\frac{(1 - M_k/\theta_k)^2}{2\alpha}\right] \qquad (4)$$

Notice that fifty percent of the plates will reach $R_{f,max}$ before the median time M_k. The scatter in time θ_k for the kth preventive-maintenance action, $\alpha^{1/2}$, is approximately linked to the coefficient of variation of the fouling growth rate $K(B)$, i.e., $\alpha^{1/2} \approx K(B)$; in addition it is assumed that $\alpha_1 = \alpha_2 = \alpha_3 = \ldots = \alpha$.

Cleaning Scheme

Referring to Eq. (1), we note that at time $t = 0$, $U(0)/U_c = U_{max}/U_c = 1$. When $R_f(t)$ reaches a limiting value such as $R_{f,max}$, then $U(t)/U_c$ reaches $U_{f,min}/U_c$. At this stage, one or more of the following maintenance actions are needed:

- The PHE unit is hydro-tested to identify the damaged or leaking plate(s);
- Open the unit to clean and replace the damaged plate(s);
- All gaskets are thoroughly inspected for signs of aging, cracking, or chemical attack;
- Any plates that have cracks or show signs of corrosion, deformation from swollen gaskets, or over-tightening are replaced;
- Damaged plates are removed in pairs to maintain the critical plate pattern in the unit;
- After washing, the plates may be sent to a PHE service center for reconditioning (including chemical cleaning, if needed) and dye penetrant inspection, and a failure analysis on damaged plates or gaskets.

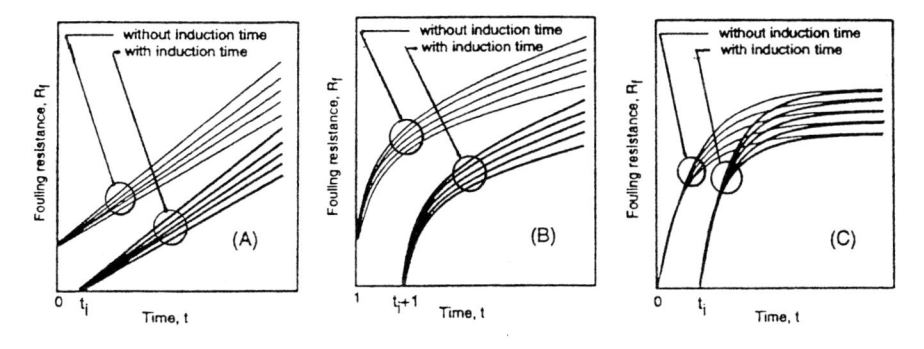

Fig. 14 Typical sample functions of fouling resistance (Zubair et al., 1997b).

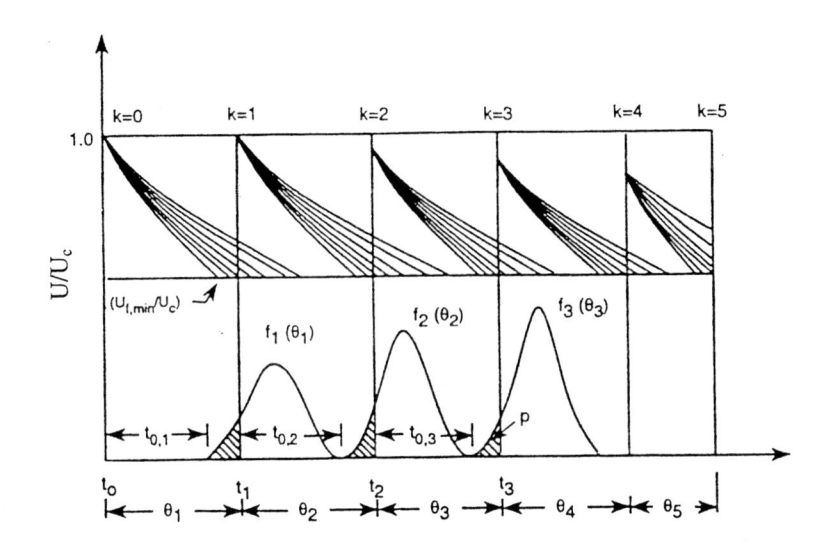

Fig. 15 Group of heat exchangers under same thermal-hydraulic conditions (Zubair et al., 1997b).

In a given life cycle of a PHE unit, there could be several cleaning cycles. Suppose t_k (k = 0, 1, 2, 3,..., j) represents the time locations where kth cleaning is done, where t_0 represents commissioning time of the heat exchanger. Therefore, after every successive cleaning cycle, $U_{max}(t_k)/U_c$ will either restore the heat exchanger to its original condition, or in some cases, the thermal performance of the unit will slightly decrease when specific opening and closing instructions of the plates are not followed. This means

$$\frac{U_{max}(t_0)}{U_c} = 1 > \frac{U_{max}(t_1)}{U_c} > ... > \frac{U_{max}(t_2)}{U_c} > ... \quad (5)$$

When $U_{max}(t_k)/U_c$ approaches U_{min}/U_c, then the interval between cleanings θ_k will start to decrease rapidly, resulting in a substantial increase in the operation and maintenance costs. As a result, we must either de-rate the system by further lowering U_{min}/U_c (which implies that lower heat exchanger performance is acceptable) or replace all the suspected plates. In the view of the probabilistic nature of $R_f(t)$, the reduced variable $U(t)/U_c$ will exhibit a stochastic pattern, i.e., the random variable "time between cleaning schedules" is distributed according to the probability model of Eq. (4). For a risk level p (i.e., 100p% of the plates have reached U_{min}/U_c or $U_{f,\ell}/U_c$ whichever is applicable), the time interval θ_{kp} is given by (Zubair et al., 1992):

$$\theta_{kp} = \frac{M_k}{[1 - \sqrt{\alpha}\, \Phi^{-1}(p)]} \quad (6)$$

where $\Phi^{-1}(p)$ is the inverse of Cumulative Normal Distribution function. The stochastic nature of Eq. (5) in a life cycle t_ℓ, with $t_{k-1} < t_\ell < t_k$, is shown in Fig. 15. In addition, the cleaning cycles $\theta_{1p} > \theta_{2p} > ... > \theta_{jp}$ are also shown in the figure, where p represents the associated risk level.

Reliability-based Cleaning Strategies

We now discuss various scenarios of the cleaning strategy (maintenance intervals) in terms of the scatter parameter ($\alpha^{1/2}$) and risk level (p). The scatter parameter represents the scatter in fouling resistance curves with respect to t, while the risk level represents probability of plates being fouled up to a critical level after which a cleaning is needed.

Maintenance restores the exchanger performance - Case I: Maintenance will fully restore the heat exchanger performance, i.e., a perfect maintenance scheme. Here, every preventive-maintenance cycle will be a replicate of the previous cycle, repeating at an equal interval. Thus, using Eq. (6), the reduced time interval θ_p is given by

$$\frac{\theta_p}{M} = \frac{1}{1 - \sqrt{\alpha}\, \Phi^{-1}(p)} \quad (7)$$

where $M_1 = M_2 = M_3 = ... = M_j$. It can be seen from the above equation that the reduced time will decrease considerably with the risk level p and scatter parameter $\alpha^{1/2}$. Zubair et al. (1997b) have shown the use of above equation with an example problem. It should be noted that this model may be considered as an appropriate model for cooling water PHEs.

Unequal maintenance intervals - Case II: There will be a fixed performance degradation "d" after each maintenance interval (or a block of equal maintenance intervals), resulting into a gradually decreasing preventive maintenance intervals. This imperfect maintenance scheme can be characterized by the following geometric sequence,

$$\frac{M_2}{M_1} = \frac{M_3}{M_2} = \ldots = \frac{M_k}{M_{k-1}} = (1-d) \quad (8)$$

Here $k = 1, 2, 3, \ldots$ may represent the successive blocks of maintenance cycles in which the risk level p is maintained fixed. In the limiting case, a block may have only one cleaning cycle. Thus, in general

$$M_k = (1-d)^{k-1} M_1 \quad (9)$$

and the reduced time between the cleanings as

$$\frac{\theta_{kp}}{M_1} = \frac{(1-d)^{k-1}}{[1-\sqrt{\alpha}\,\Phi^{-1}(p)]} \quad (10)$$

Maintenance at equal intervals - Case III: The user of heat transfer equipment will like to schedule maintenance at equal intervals for the case of an imperfect maintenance scheme described above. This will result in a gradual increase in the risk level p. In this situation, the risk level can be obtained from Eq. (10) to give

$$p_k = \Phi\left\{\frac{1-((1-d)^{k-1})}{(\theta_{kp}/M_1)\alpha^{1/2}}\right\} \quad (11)$$

where $\theta_{1p} = \theta_{2p} = \theta_{3p} = \ldots = \theta_{kp}$. It should be noted that operating a PHE at a critical risk level of a system or a component is important in some applications such as a PHE network in a refinery. In this situation, an acceptable level of heat exchanger overall heat transfer coefficient will primarily govern the maintenance strategy. However, in some situations, heat exchangers are neither in a network nor in a critical system; here maintaining the exchanger at a higher reliability level (or at a lower risk level p), implies more frequent maintenance intervals that can often result in an increasing operation and cleaning costs. Therefore, in situations, where the cost of operation and maintenance is an important factor along with the exchanger reliability (r = 1 - p), the decisions of maintenance can be optimized by developing cost as a function of reliability (or risk level). It is then possible to search for a minimum cost-based solution (Zubair et al., 1997b). This cost-optimized cleaning method will also result in an optimal level of heat-exchanger reliability (Zubair et al., 1997b, 1999).

Cost-Based Cleaning Strategies

We consider a specific heat exchanger whose fouling in plates is characterized by the α–distribution (or a linear random fouling model) with a median time M to reach a critical level of fouling $R_{f,c}$ and scatter parameter $\alpha^{1/2}$. It is possible to propose a cost-based operating and maintenance model of a heat exchanger as (Zubair et al., 1997b, 1999):

$$C = \frac{C_1}{(t/M)} + C_2(t/M) + C_3(t/M)^m \quad (12)$$

where C_1, C_2 and C_3 are various cost parameters representing anti-foulant (or on-line chemical cleaning), off-line cleaning and additional fuel consumption due to fouling, respectively expressed as \$/hr, and the exponent m is a constant reflecting the severity of financial penalty associated with the exchanger performance degradation due to fouling. It is normally greater than one, and is a function of the heat exchanger configuration. We note that the total cost function given by Eq. (12) is a convex function in every preventive-cleaning cycle. This characterization of operating and cleaning costs represents p = 0.50 and any value of $\alpha^{1/2}$. It can also be interpreted as a deterministic case ($\alpha^{1/2} = 0$). For a more general representation, corresponding to any risk level p and scatter parameter $\alpha^{1/2}$, the cost function can be expressed by incorporating the case of perfect cleaning represented by Eq. (7), or imperfect cleaning by Eq. (10). The resulting probabilistic cost functions will clearly demonstrate the relationship between the total operation and cleaning costs, risk level p and the scatter parameter $\alpha^{1/2}$. Zubair et al. (1992, 1997b) have discussed various scenarios of probabilistic cleaning cost models that are applicable to both a STHE and a PHE in a preheat train of a crude oil refining process.

CONCLUDING REMARKS

There is only limited design fouling resistances for PHEs that have been reported in the literature. These values indicate that the fouling resistances are about 15-30% those for the shell-and-tube heat exchangers for comparable conditions. Therefore, care should be taken to avoid unnecessary over design using higher values as a precaution. Over sizing of PHEs leads to lower fluid flow velocities, and hence, increased fouling should be expected if TEMA standard values are used. Operation at higher velocities and lower wall temperatures can significantly reduce the formation of deposits. This has to be balanced with thermoeconomic considerations, i.e. increase in pressure drop and required heat transfer area.

The plate corrugation angle β is also an important factor in assessing fouling. It is shown that, for a given velocity, a high corrugation angle that produces high turbulence maintains solids in suspension and allows low wall temperatures. This in turn leads to low fouling resistance than for a comparable straight channel. It is expected that the scale formation can be greatly reduced if the metal surface energy is reduced by surface treatment such as Ion Assisted Reaction or by Plasma Polymerization

Process (Kim et al., 1999).

The on-line cleaning method such as EAF technology appears to have promise in controlling (or minimizing) precipitation fouling in PHEs; however, additional tests are needed to demonstrate the benefits of this technology in other applications of PHEs. It should be noted that uncertainties associated with fouling could best be approached by characterizing fouling in a probabilistic framework. Its implications on off-line cleaning of PHEs are explained, particularly for systems where a target performance level is needed.

ACKNOWLEDGEMENTS

The first author acknowledges the support provided by KFUPM under the project SABIC/2000-08.

NOMENCLATURE

A	Random process parameter
B	Random process parameter
C	Total cost rate of operating and maintaining a heat exchanger, $/year
d	Performance degradation after heat exchanger cleaning ($0 \leq d \leq 1.0$)
f	Probability density function or PDF (Eq. (4)), 1/ hr
K	Coefficient of variation of time to reach a critical level of fouling
M	Median time for alpha distribution, hr or day
m	Constant reflecting severity of financial penalty due to fouling
PHEs	Plate-and-Frame Heat Exchangers
p	Risk level, defined as the probability of events that fouling resistance is less than or equal to the critical value, $p = P(R_f(t) \leq R_{f,c})$
R_f	Fouling resistance, $m^2 K/W$
R_f^*	Asymptotic fouling resistance, $m^2 K/W$
$R_{f,c}$	Critical fouling resistance, $m^2 K/W$
r	Reliability, $r = 1 - p$
STHEs	Shell-and-Tube Heat Exchangers
T_b	Water bulk temperature, °C
T_s	Wall temperature, °C
t	Time, days or hrs
U	Overall heat transfer coefficient, $W/m^2 K$
V	Flow velocity, m/s

Greek Symbols

β	Corrugation angle measured from the plate vertical (parallel to length) axis, degree
σ	Standard deviation, hr
μ	Mean value, hr
τ	Time constant, 1/ hr
Φ	Cumulative normal distribution function
φ	Function
$\alpha^{1/2}$	Scatter in time
θ	Time interval between cleaning, hr

Subscripts

C	Clean condition
F	Fouled condition or fouling
0	Commissioning time
k	kth cleaning cycle
kp	kth preventive cleaning cycle
ℓ	Life or limiting value
max	Maximum
min	Minimum

REFERENCES

Bansal, B. and Muller-Steinhagen, H., 1993, Crystallization Fouling in Plate Heat Exchangers, *ASME J. Heat Transfer*, Vol. 115, pp. 584-591.

Bansal, B., Muller-Steinhagen, H., and Deans, J., 1993, Formation of Deposits in a Plate and Frame Heat Exchanger, *AIChE Symp. Series 295*, Vol. 89, pp.359-364.

Bansal, B., 1995, Crystallization Fouling in Plate and Frame Heat Exchangers, Ph.D. Dissertation, Dept. Chem. Eng., University of Auckland, New Zealand.

Bansal, B., Muller-Steinhagen, H., and Chen, X. D., 1997, Effect of Suspended Particles on Crystallization Fouling in Plate Heat Exchangers, *ASME J. Heat Transfer*, Vol. 117, pp. 568-574.

Bossan, D., Grillot, J., Thonon, B., and Grandgeorge, S., 1995, Experimental Study of Particulate Fouling in an Industrial Plate Heat Exchanger, *J. Enhanced Heat Transfer*, Vol. 2, pp. 167-175.

Bott, T. R., 1995, *Fouling of Heat Exchangers*, Elsevier Science Publishers Ltd., Amsterdam.

Chenoweth, J. M., 1990, Final Report of the HTRI/TEMA Joint Committee to Review the Fouling Section of the TEMA Standards, *Heat Transfer Eng.*, Vol. 11, No. 1, pp. 73-107.

Cho, Y. I., Choi, B-G., and Drazner, B. J., 1998, Electronic Anti-Fouling Technology to Mitigate Precipitation Fouling in Plate-and-Frame Heat Exchangers, *Int. J. Heat Mass Transfer*, Vol. 41, pp. 2565-2571.

Cooper, A., Suitor, J., and Usher, J., 1980, Cooling Water Fouling in Plate Heat Exchangers, *Heat Transfer Eng.*, Vol. 1, No. 3, pp. 50-55.

Cross, P., 1979, Preventing Fouling in Plate Heat Exchangers, *Chem. Eng.*, pp. 87-90.

De Jong, P., Bouman, S., van der Linden, H.J.L.J, 1992, Fouling of Heat Treatment Equipment in Relation to the Denaturation of β-Lactoglobulin, *J. Soc. Dairy Tech.*, Vol. 45, pp. 3-8.

De Jong, P., Waalewijn, R., and van der Linden, H.J.L.J, 1993, Validity of a Kinetic Fouling Model for Heat Treatment of Whole Milk, *Lait.*, Vol. 73, pp. 293-302.

Delplace, F., Leuliet, J.C., and Bott, T.R., 1997, Influence of Plate Geometry on Fouling of Plate Heat Exchangers by Whey Proteins, in *Fouling Mitigation of Industrial Heat Exchangers,* eds. C. Panchal, T. Bott, E. Somerscales, and S. Toyama, Begell House, New York, NY, pp. 565-576.

Fryer, P.J., Robbins, P.T., Green, C., Schreier, P.J.R., Pritchard, A.M., Webb, P., Hasting, A.P.M., Royston, D.G., Davies, S.A., and Richardson, J.F., 1997, Fouling of a Plate Heat Exchangers by Whey Protein Solutions: Measurements and a Statistical Model, in *Fouling Mitigation of Industrial Heat Exchangers,* eds. C. Panchal, T. Bott, E. Somerscales, and S. Toyama, Begell House, New York, pp. 577-588.

Grillot, J.M., 1992, Gaseous Phase Particulate Fouling of Plate Heat Exchangers, in *Fouling Mechanisms: Theoretical and Practical Aspects,* eds. M. Bohnet, T. Bott, A. Karabelas, P. Pilavachi, R. Semeria, and R. Vidil, Eurotherm Seminar 23, pp. 219-230.

Hasson, D., 1999, Progress in Precipitation Fouling Research – A Review, in *Understanding Heat Exchanger Fouling and Its Mitigation,* eds. T. Bott, L. Melo, C. Panchal, E. Somerscales, Begell House, New York, NY, pp. 67-89.

Kim, K.H., Choi, S.C., Jung, H.J., Koh, S.K., Kim, C.H., Ha, S.C., and Choi, D.J., 1999, Effect of Plasma Polymerized Film on Fouling of Heat Exchangers, paper presented at Mitigation of Heat Exchanger Fouling and Its Economic and Environmental Implications, July 11-16, 1999, Banff, Canada.

Knudsen, J.G., 1998, Fouling in Heat Exchangers, in *Heat Exchanger Design Handbook,* ed. G.F. Hewitt, Section 3.17, Begell House, New York.

Lister, D.H., 1981, Corrosion Products in Power Generating Systems, in *Fouling of Heat Transfer Equipment,* eds. E. Sommerscales, and J. Knudsen, Hemisphere, Washington, DC, pp. 135-200.

Mariott, J., 1971, Where and How to Use Plate Heat Exchangers, *Chem. Eng.,* Vol. 78, pp. 156-162.

Muller-Steinhagen, H., and Middis, J., 1989, Particulate Fouling in Plate Heat Exchangers, *Heat Transfer Eng.,* Vol. 10, No. 4, pp. 30-36.

Muller-Steinhagen, H., 1997, Fouling in Plate and Frame Heat Exchangers, in *Fouling Mitigation of Industrial Heat Exchangers,* eds. C. Panchal, T. Bott, E. Somerscales, and S. Toyama, Begell House, New York, NY, pp. 101-112.

Novak, L., 1982, Comparison of the Rhine River and the Oresund Seawater Fouling and Its Removal by Chlorination, *ASME J. Heat Transfer,* Vol. 104, pp. 663-670.

Panchal, C.B., and Rabas, T.J., 1999, Fouling Characteristics of Compact Heat Exchangers and Enhanced Tubes, In *Compact Heat Exchangers and Enhancement Technology for the Process Industries,* ed. R.K. Shah, Begell House Inc., New York, pp. 497-502.

Paterson, W.R., and Fryer, P.J., 1988, A Reaction Engineering Approach to the Analysis of Fouling, *Chem. Eng. Sci.,* Vol. 43, pp. 1714-1717.

Rankin, B.H., and Adamson, W.L, 1973, Scale Formation as Related to Evaporation Surface Conditions, *Desalination,* Vol. 13, pp. 63-87.

Rene, G., and Lalande, H., 1990, Description and Measurement of the Fouling of Heat Exchangers in the Thermal Treatment of Milk, *Intl. Chem. Eng.,* Vol. 30, No. 4, p. 643.

Sandu, C., 1989, Physicomathematical Model for Milk Fouling in a Plate Heat Exchanger, Ph.D. Dissertation, Dept. of Food Science, University of Wisconsin, Madison, WI.

Sheikh, A.K, Zubair, S.M., Younas, M., and Budair, M.O., 2001, Statistical Aspects of Fouling Processes, accepted for publication in *proceedings of the IMechE, J. of Process Mech. Engng.*

Shibuya, H., Morohashi, M., Levy, W., and Costa, C., 1997, Fouling Tests Using Pilot-Scale Packinox Heat Exchangers With Untreated Straight-Run Gas Oils, in *Fouling Mitigation of Industrial Heat Exchangers,* eds. C. Panchal, T. Bott, E. Somerscales, and S. Toyama, Begell House, New York, pp. 525-536.

Sloan, M.D., 1998, Designing and Plate Heat, *Chem. Eng.,* May, pp. 78-83.

Somerscales, E.F.C., 1981, Corrosion Fouling, in *Fouling in Heat Exchange Equipment,* eds. J. Chenoweth, and M. Impagliazzo, ASME HTD Vol. 17, pp. 17-27.

Somerscales, E.F.C., and Kassemi, M., 1987, Fouling due to Corrosion Products Formed on a Heat Transfer Surface, *ASME J. Heat Transfer,* Vol. 109, pp. 262-271.

Tissier, J.P., 1992, Influence of Some Parameters on Milk Pasteurizer Soiling in a Plate Heat Exchanger, in *Fouling Mechanisms: Theoretical and Practical Aspects,* eds. M. Bohnet, T. Bott, A. Karabelas, P. Pilavachi, R. Semeria, and R. Vidil, Eurotherm Seminar 23, pp. 239-248.

Thonon, B., Grandgeorge, and Jallut, C., 1999, Effect of Geometry and Flow Conditions on Particulate Fouling in Plate Heat Exchangers, *Heat Transfer Eng.,* Vol. 20, No. 3, pp.12-24.

Zubair, S.M., Sheikh, A.K, and Shaik, M.N., 1992, A Probabilistic Approach to the Maintenance of Heat-Transfer Equipment Subject to Fouling, *Energy,* Vol. 17, pp. 769-776.

Zubair, S.M., Sheikh, A.K, Budair, M.O., Haq, M.U., Quddus, A., and Ashiru, O.A., 1997a, Statistical Aspects of $CaCO_3$ Fouling in AISI 316 Stainless Steel Tubes, *ASME J. Heat Transfer,* Vol. 119, pp. 581-588.

Zubair, S.M., Sheikh, A.K, Budair, M.A., and Badar, M.A., 1997b, A Maintenance Strategy for Heat-Transfer Equipment Subject to Fouling: A Probabilistic Approach, *ASME J. Heat Transfer,* Vol. 119, pp. 575-580.

Zubair, S.M., Sheikh, A.K, Budair, M.A., and Younas, M., 1999, A Maintenance Strategy for Heat-Transfer Equipment Subject to Fouling, Final report, project # ME/FOULING/176. King Fahd University of Petroleum & Minerals, Dhahran, Saudi Arabia, May 1999.

DIAGNOSING, CHARACTERIZING AND CONTROLLING A LIQUID TRANSFER SYSTEM CLEANING PROBLEM

Mark Fornalik[1], Dave Gruszczynski[2], Chris Puccini[3] and Margaret (Peggy) B. Wilcox[4]

[1]Eastman Kodak Company, Rochester, NY 14652-3712; E-mail:mark.fornalik@kodak.com
[2]Eastman Kodak Company, Rochester, NY 14652-3702; E-mail david.gruszczynski@kodak.com
[3]Eastman Kodak Company, Rochester, NY 14652-3702; E-mail christopher.puccini@kodak.com
[4]Eastman Kodak Company, Rochester, NY 14652-4869; E-mail margaret.wilcox@kodak.com

ABSTRACT

Contamination in liquid transport systems can generally be traced to problems with non-sanitary system components or design, insufficient purging/flushing of old product from the transfer line, interactions of product with system components, or formation of insoluble materials on the walls of the system (system fouling). Frequently, there are problems with all these. System design problems are straightforward (although expensive) to fix, as are product/component interaction problems. System purging/flushing requirements can be determined mathematically. Once all of these problems are fixed, though, system fouling can be insidious and difficult to remedy.

Insoluble films on solid surfaces in a transfer system as thin as several hundred Angstroms in thickness can be the cause of product quality problems. Worse, the chemistry of the wall fouling may be very different that the bulk solution flowing through the pipe, and cleaning of the system may not be intuitive. It is imperative to characterize the fouling in a system for chemistry and rate of formation. This knowledge can then drive appropriate cleaning technologies for the system.

In this paper, we describe the methods we used to characterize system fouling in a salt water transfer line. We also describe the lab modeling methods used, and the mechanical and chemical cleaning techniques used to control the fouling in the system.

INTRODUCTION

Fouling is the formation of insoluble residues on the surfaces of engineering materials in contact with flowing solutions. Few systems are immune from some level of fouling, and fouling is very dependent on system parameters (flow rates, solution chemistry, materials of construction, cleaning methods and frequency, etc.).

Entire volumes have been written about fouling and cleaning. Bott describes various forms of organic, inorganic, and biological fouling at surfaces of heat exchangers. Characklis and Marshall devote and an entire book to biological fouling on engineering materials. Costerton and Lappin-Scott and Baier have described biological fouling on the surfaces of medical devices and surfaces in contact with marine and oral environments. Belmar-Beiny and Fryer described fouling in the dairy processing area. Fouling may be considered ubiquitous whenever aqueous solutions are in contact with manmade engineering materials.

At Kodak, one of the principal materials used in the manufacture of silver halide products is salt. Various halide salts (NaCl, NaBr, KCl, etc.) are solubilized in water and transported throughout the plant in stainless steel pipelines. Great care has been taken to assure that incoming dry salt from vendors meets quality guidelines for various contaminants. But the liquid salt solution transfer lines have (until recently) not been examined as a potential source of contamination.

This study describes how liquid transfer system fouling was characterized in the manufacturing environment, how the fouling was modeled in the lab, and how designed experiments on mechanical and chemical cleaning have reduced the salt system fouling to manageable levels.

MATERIALS & METHODS

Manufacturing

Most parameters of the salt solutions and solution transfer system are proprietary. However, the temperature range of the salts and flushwater varied between 20°C and 60°C. The solutions used in the system varied between 0 molar and 3 molar salts. The subsequent fouling was resistant to this temperature and osmolarity range.

Visual inspection with a flashlight was the method for determining level of fouling in various pipelines. Lines were taken apart with the aid of a mechanic and assessed visually at various points throughout the distribution system.

Once several key locations of fouling had been identified, fouling cells (Figure 1) were used. A fouling cell is a modified pressure sensor housing with Tri-Clover™ sanitary connections. The pressure sensor itself has been removed, and replaced with witness plates in the form of plugs or discs of materials of interest. For this study, 316L stainless steel discs were used, polished to less than 2.54 micrometers surface roughness. Polishing was required for the analytical techniques, described below.

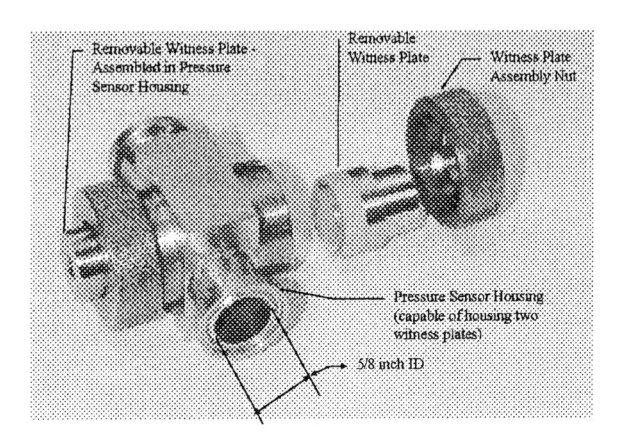

Fig. 1 Fouling Cell Used in Salt System Study.

Fouling cells were installed in series in the salt line. Generally, six fouling cell housings containing 2 stainless steel discs each were installed for each experiment. Housings could then be removed either one at a time – to produce a time exposure series of fouling – or removed all at once to produce numerous discs of simultaneously and equally fouled surfaces for cleaning experiments in the laboratory.

Fouling cell housings were cleaned prior to installation with detergent (Sparkleen, Fisher Scientific) and a bottlebrush in hot water. Housings were rinsed in Milli-Q™ ultrapure water and allowed to air dry. Fouling cell discs were cleaned with Sparkleen detergent using a soft, camel hair brush. The discs were then rinsed in hot tap water, then rinsed again with Milli-Q™ water. The discs were allowed to air dry in a position that encouraged water to drain from the surface. Once dry, the discs were placed in a plasma cleaner (Harrick Scientific) and plasma cleaned in air at low power for 30 seconds. Upon removal from the plasma cleaner, the discs were once again cleaned with Sparkleen detergent, rinsed with hot water and Milli-Q™ water, and allowed to air dry in a position that encouraged the water to drain from the surface.

The goal of this cleaning regimen was to first remove any adventitious contamination from exposure to air, handling artifacts, or other relatively thick film contamination. Once this thick layer had been removed, the plasma cleaner treatment was designed to remove very thin film contaminants. Upon plasma treatment, the discs were completely water wettable. The final cleaning with detergent was meant to ensure that all discs decayed to the same surface energy prior to installing the discs into a manufacturing piping system. This eliminated the effects of varying surface energy contaminants.

After exposure to the salt transfer system, the fouling cells were cleaned on-line by standard production system cleaning techniques – that is, the line was cleaned in the usual fashion with water, and the fouling cells were left in line for this cleaning. After the fouling cells were removed, they were transported back to the laboratory for analysis. Prior to analysis, the surfaces of the polished discs were flooded with Milli-Q™ water for 1-2 minutes (water was simply puddled up on the disc) to remove any soluble residual salts, then allowed to air dry. Once dry, the discs were ready for analysis.

Analysis

Analysis was accomplished by two primary techniques. Fourier Transform Infrared (FTIR) spectroscopy was used to characterize organic deposits on the surface, as well as to determine relative mass of the deposits by peak height analysis. Epifluorescence optical microscopy was used to examine the fouling for microorganisms.

Infrared analysis was performed with a Magna 550 FTIR (Nicolet Corporation). External specular reflectance from the polished stainless steel surface was accomplished with a "Seagull" optical accessory (Harrick Scientific). The Seagull sample holder was modified to accommodate fouling cell discs. Incident angle was set at 76°. All spectral manipulation (baseline corrections, water vapor subtraction, and peak height measurement) was carried out with the standard Omnic software package that came with the FTIR (Nicolet).

Epifluorescence optical microscopy was performed with an Olympus BH-60 optical microscope. Acridene orange and DAPI nucleic acid stains were used throughout the study for bacterial stains (Molecular Probes). Staining was accomplished with filtered, 10 ug/ml solutions of each stain, used at room temperature, for 20-30 minutes each, with a water flush between stains and after staining was completed. Blue and UV light were used for illumination on the microscope.

Lab Modeling

Two lab models were used in this work: glass bead reactors and a Rototorque™ annular reactor. Glass bead reactors were the first step in the modeling process. Glass columns (2.54 cm x 30.48 cm) were packed with 70 cc of glass beads (710 μ – 1,180μ, Sigma Chemical Co.). Glass beads were used to provide sufficient surface area for a biofilm to form. Each glass bead reactor was set up to mimic one factor in the experimental design. Design factors explored included aerobic/anaerobic conditions, presence/absence of light, presence/absence of flow, and steady state or alternating water/salt solutions.

After exposure, the glass bead reactors were disassembled. Samples of the exposed glass beads were vortexed to remove the fouling film from the glass surface, and the isolated fouling was placed on stainless steel plates for FTIR analysis. The goal of this work was to determine what environmental conditions produced films with infrared signatures similar to that found on the fouling cells in the manufacturing environment.

Once a match had been made with FTIR signatures, a Rototorque™ experiment was set up. The Rototorque™ is a commercially available benchtop bioreactor (Biosurface Technologies Corp.). 20 stainless steel slides were placed in slots around the internal drum in the reactor. Conditions in the reactor were set to mimic the successful conditions from the glass bead reactor trials. After sufficient exposure time, the stainless steel plates were removed from the reactor, rinsed with water and allowed to air dry in a vertical position. FTIR and epifluorescence were used to confirm that the fouling matched that found in production. The Rototorque™ was then used to produce identically fouled stainless steel plates for subsequent designed experiments in cleaning.

Cleaning Trials

Lab cleaning trials consisted of optimizing the current cleaning sequence and chemistry. FTIR peak height was used to determine which chemical cleaning agents were most successful at removing fouling from stainless steel plates. FTIR was used to judge the cleaning efficiency of chemical cleaner type, concentration, time and order of use. Cleaning agents evaluated included bleach, sodium hydroxide and phosphoric acid. All lab-cleaning trials were carried out at room temperature in glass beakers, with magnetic stir bars providing gentle mixing of the solutions. Upon removal from the cleaning solution, the stainless steel plates were rinsed in Milli-Q™ water and allowed to air dry in a vertical position before analysis by FTIR and epifluorescence optical microscopy.

Production cleaning trials evaluated mechanical cleaning with air/water powerflush (two-phase flow) as well as chemical cleaning. Results of both are presented in this study.

RESULTS

Production Fouling

The first set of fouling cells evaluated fouling formed during 2-week exposures of the fouling cells to the production salt system. Figure 2 presents an example of the fouling formed on stainless steel in two weeks in this system. Note that this fouling was resistant to the standard water flush cleaning then used in the system.

The fouling was comprised of hydrocarbon (peaks in the 2950 cm^{-1} region), ester (the 1740 cm^{-1} peak), protein (1650 cm^{-1}), and carbohydrate (1020 cm^{-1} and 1117 cm^{-1}). This material all resisted the cleaning efforts of the simple system water flush, thus this material was all water insoluble.

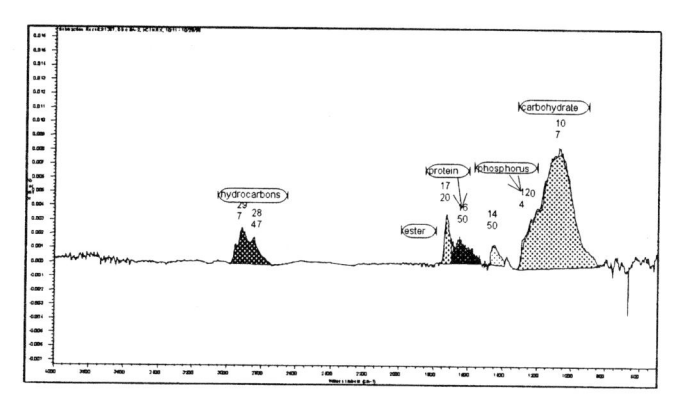

Fig. 2 Infrared Spectrum of Fouling on 316 Stainless Steel Disc After 2 Weeks Exposure to Liquid Salt Transfer System.

Figure 3 presents fouling cell results taken as a function of time in the salt system. The lower spectrum is the result of a 3-hour exposure to the salt solution system. Peaks for hydrocarbons and carbohydrates can already be seen even in this short exposure. The next spectrum up on this chart was the result of about 24 hours of exposure – note the increase in peak height for the hydrocarbon and carbohydrate peaks. The third spectrum up was from about 48 hours of exposure to the salt transfer system. Here, the ester band at 1740 cm^{-1} begins to appear. The top spectrum on the chart is from approximately 6 months of exposure. The fouling film has reached steady state. All spectra are plotted on a common scale for comparability.

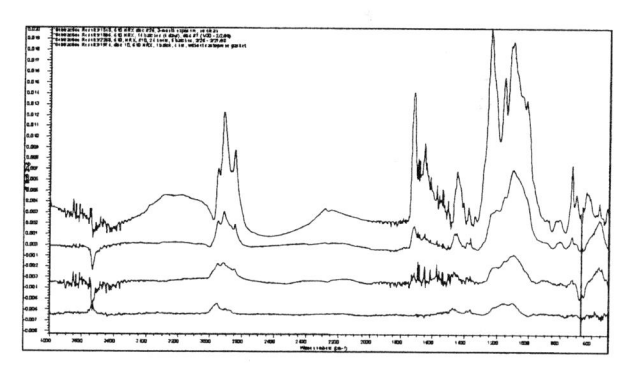

Fig. 3 Infrared Spectra of Salt System Fouling as a Function of Exposure Time on 316 Stainless Steel.

A typical epifluorescence optical micrograph of a 2-week fouling cell surface is shown in Figure 4. Amorphous clumps of material can be seen on the surface. There is no evidence of rod-shaped organisms, or traditional-appearing microorganisms on the surface. In fact, the material on the surface of the stainless steel plates fluoresced even without nucleic acid staining.

Due to the strong IR bands for hydrocarbon and ester from these samples, and the inconclusive microscopic examination, the fouling material was originally thought to be manmade in origin, or a synthetic organic chemical contaminant. However, in spite of numerous efforts to find a degraded or decomposed component from the system, the

Figure 4. Epifluorescence Optical Micrograph of 2-Week Fouling Cell from Liquid Salt Transfer System After Water Flush Cleaning

equipment in the system all appeared to be in good shape, with no obvious degradation of any component.

At this point, the fouling was thought to potentially be biological. Viable plate counting methods for culturing the organisms from the fouling failed; no viable growth was produced on R2A agar or special salt media in either aerobic or anaerobic conditions in the microbiology testing laboratory. However, samples of the transfer system fouling sludge were sent to Microbial Insights, Inc. (Rockford, Tenn.) for phospholipid fatty acid analysis. In this test, the phospholipids from cell walls were isolated from the sludge and both characterized and quantitated. The results were overwhelmingly positive – the fouling was definitely biological in nature, with an entire consortium of organisms living in the material in the salt system. Biofilms can be very difficult to grow in culture in the lab (Charaklis and Marshall, 1990), so alternate testing techniques can prove useful in determining the presence of organisms.

Manufacturing testing had demonstrated that liquid transfer system fouling was rapid (beginning within hours of exposure), water resistant (due to the hydrocarbons, esters, and long-chained carbohydrates), and comprised of microorganisms that formed a biofilm.

Lab Modeling of Fouling

It is difficult to run cleaning experiments on production equipment. The scale of the equipment, the windows of opportunity for experiments, the reproducibility of the fouling, and the ability to run dozens of experiments all become unwieldy. The solution is to mimic the fouling in the lab, to produce as many fouled surfaces as necessary for cleaning experiments in the lab.

Growing biofilms in the lab can be very difficult. The organisms found in the manufacturing system grow in those systems because of the local environment with respect to oxygen, nutrients, pH, ionic strength, flow conditions, etc. It can be very trying to determine which factors influence their growth, and reproducing those factors in the lab. Worse, even if the biofilm grows in the lab, it may not produce exopolymeric substances similar to that found in production.

Screening trials for all the major factors provide information that can lead to biofilm growth. Here, glass bead reactors were set up to determine which factors influenced growth in salt and water environments. Analyzing the biofilm deposit on the glass bead surface with FTIR and epifluorescence helped determine if the model system was behaving as the production system would.

The factors examined in this study were oxygen levels, light, flow conditions, and osmolarity. The IR signatures of glass bead reactor system, the FTIR signature of the biofilm from a glass bead reactor that did not simulate production growth, and the biofilm from a glass bead reactor that simulated well the growth found in production were demonstrated. The results of this study demonstrated that the biofilm required flow, anaerobic conditions, and alternating salt solution and water. Epifluorescence images supported these conclusions (results not shown here).

With the ability to form appropriate biofilms on stainless steel in the lab, the study now examined how best to clean these fouling layers, and at what frequency.

Mechanical Cleaning

The most rudimentary form of system cleaning is with water flush. One of the basic aims of a good water flush is to at least reach 1.5 m/s linear velocity of the flushwater through the system. Not all systems can attain this velocity, however. Older plants, larger-diameter pipelines, and varying demands on the water utilities system in a plant can make this goal unachievable or non-reproducible.

Two-phase flow cleaning has been described in the past. Here, air is added to the water flush to increase wall shear and dramatically increase turbulent flow, even with undersized water systems. Gruszczynski (US Patent 5,941,257 "Optimum Powerflush Ratio", August, 1999) has empirically determined the optimum ratio of air and water flows for systems ranging in diameter from 0.95 cm to 2.54 cm. Optimized two-phase flow cleaning was examined in the salt transfer system.

For this trial, fouling cells were installed in the salt system. Powerflush cleans with air/water two-phase flow were run every 3-4 hours (after each system use). Fouling cells were removed at 2-week intervals and examined by FTIR and epifluorescence optical microscopy.

Figure 5 presents the FTIR spectrum of 2-week fouling cells before and after implementing powerflush cleaning. The top spectrum presents the before cleaning case, and the bottom spectrum presents the after cleaning case. Both spectra are presented on a common scale for comparability. It is clear that powerflush cleaning reduced the carbohydrate peak significantly (about 90% reduction), and also reduced the hydrocarbon and ester peaks (about 50% each).

Figure 6 presents a chart of ester and carbohydrate peak heights for 2-week fouling cells before and after powerflush implementation at two system pressures. The set of bars for the before powerflush implementation data summarizes 3 2-week data sets. The set of bars for the after-powerflush data summarizes 12 2-week data sets. Error bars are 1 standard deviation from the mean of the FTIR data.

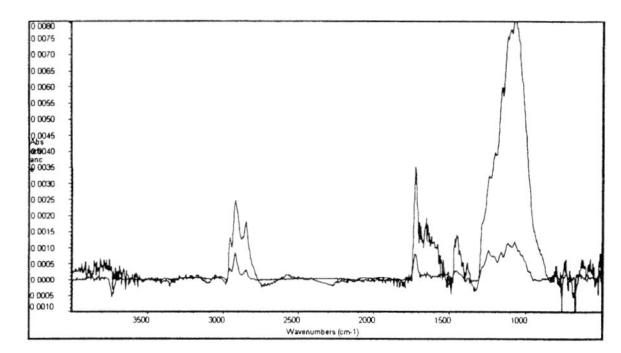

Fig. 5 Infrared Spectra of 2-Week Salt System Fouling on Stainless Steel Discs Before (Upper Spectrum) and After (Lower Spectrum) Powerflush Implementation.

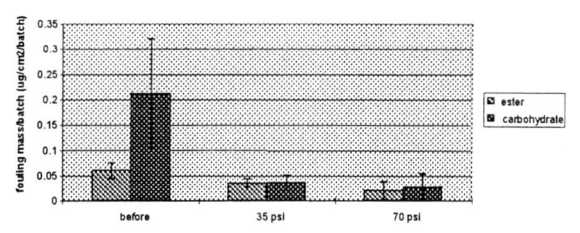

Figure 6. Summary of Infrared Peak Height Data Taken Before and After Powerflush Implementation for 2 Different Powerflush Conditions

Powerflushing at either system pressure clearly improved cleaning of the transfer system. However, FTIR indicated that there was still material remaining on the stainless steel witness plates after powerflush implementation. This material could act as nucleation sites for additional fouling. This material resisted the mechanical shear forces of air/water two-phase flow. A chemical cleaning treatment was required to eliminate this residual fouling material.

Chemical Cleaning

While two-phase flow cleaning was straightforward to try in production, chemical cleaning was not. There are many potential cleaning chemical solutions available, all of which can be made up into different concentrations, used in different sequences, and have different cleaning efficiencies. This part of the study was well suited to lab experiments, which were performed on lab-fouled stainless steel plates first, then in the lab on production-fouled fouling cell surfaces.

For this study, the list of cleaning chemicals was limited to those found easily in production – phosphoric acid, bleach and sodium hydroxide. Due to the design of the production salt transfer system, these chemicals were evaluated at room temperature, since heated solutions would be a safety hazard in the production environment. FTIR was used to determine cleaning efficiency. Peak height analysis for the ester and hydrocarbon peaks was monitored during the experiments; it was assumed the powerflush cleaning in production would control the carbohydrate peak.

This portion of the study was the subject of an extended series of designed experiments. Only the eventual best chemical cleaning treatment is presented here in Figure 7.

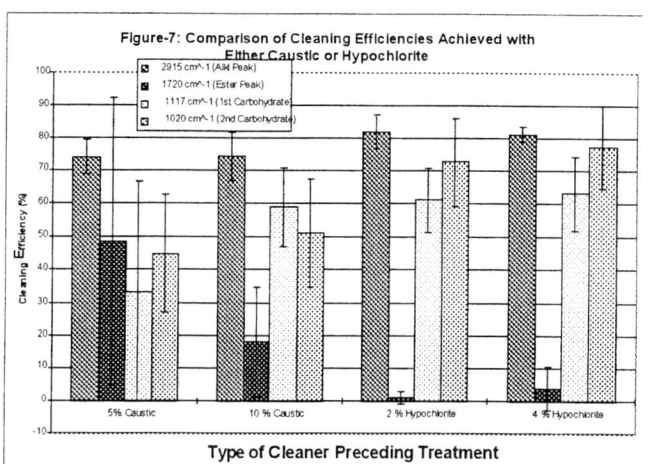

The FTIR peak height information of various combinations of the sodium hydroxide and hypochlorite cleaning agents (each cleaning agent was used sequentially, not as a mixture) with 65% phosphoric acid were measured. Clearly, 10% sodium hydroxide caustic was superior on efficiently cleaning the ester, while hypochlorite was ineffective and eliminated from the cleaning sequence.

CONCLUSIONS

The approach outlined in this study can be applied to other liquid transfer systems in industry:

- Characterization of system fouling with in-line witness plates (fouling cells)
- Determine the impact of improved mechanical cleaning
- If improved mechanical cleaning is not good enough, then chemical cleaning must be examined
- Model the fouling (if possible) in the laboratory
- Using lab-modeled fouling, run designed experiments in chemical cleaning in the lab
- Verify final chemical cleaning treatments in production

The combination of automated mechanical cleaning each product batch and periodic chemical cleaning eliminated production waste of $2M/year and enabled manufacture of a product that earned a $200M profit in 1999 for Eastman Kodak Company.

REFERENCES

Baier, R.E., 1982, Conditioning surfaces to Suit the Biomedical Environment: Recent Progress, *J. Biomech. Eng.*, Vol. 104, pp. 257-271.

Belmar-Beiny, M.T. and Fryer, P.J., 1993, Preliminary Stages of Fouling from Whey Protein Solutions, *J. Dairy.*, Vol. 60, pp. 467-483.

Bott, T.R., 1995, Fouling of Heat Exchangers, Chemical Engineering Monographs No. 26, Elsevier.

Charaklis, W.G. and Marshall, K.C., "Biofilms, 1990, A Basis for an Interdisciplinary Approach," in *Biofilms*, W.G. Charaklis and K.C. Marshall, Editors, John Wiley and Sons, New York.

Costerton, J.W. and Lappin-Scott, H.M., 1995, Introduction to Microbial Biofilms, *in Microbial Biofilms*, H.M. Lappin-Scott and J.W. Costerton, Editors, Cambridge University Press.

FOULING REDUCTION IN GELATIN-BASED PHOTOGRAPHIC EMULSIONS USING SPIRAL WOUND MEMBRANES

William Gately[1] and Margaret B. Wilcox[2]

[1]Eastman Kodak Company, Rochester, NY 14652-3701,USA; E-mail:wgately@kodak.com
[2]Eastman Kodak Company, Rochester, NY 14652-4869,USA; E-mail: mwilcox@kodak.com

ABSTRACT

Gelatin-based photographic emulsions of silver halides are desalted and concentrated using polymeric membranes. Because of the high cost of silver nitrate, yields of 99% ± 0.1% are desired in the photographic industry. Fouling of the clean membranes, especially when processing spherical feeds less than 0.3 microns in diameter, negatively affects waste and productivity.

Fouling of the membrane surface is a function of operating parameters as shown by correlation to scanning electron microscopy (SEM) and x-ray measurements of the membrane.

A critical flux is determined and a mass transfer-based model is used to control the process. A pilot scale yield-loss reduction of 90% and increase in production scale yield of 1.0% results from use of the optimized process control.

INTRODUCTION

Gelatin-based photographic emulsions have been produced commercially by a wide variety of processes. This three-step, batch process requires precipitation of silver nitrate and halide salts in a gelatin environment followed by chemical and/or spectral sensitization. Prior to chemical and/or spectral sensitization, the emulsion is desalted and concentrated. One production scheme to accomplish the desalting and dewatering step employs polymeric membranes. The concentration of the photographic emulsions is appropriate for the downstream thin film coating process following the chemical and/or spectral sensitization process. Because of the high cost of silver nitrate, yields of 99% ± 0.1% are desired in the photographic industry. Polymeric membranes in the ultrafiltration range used for the separation process are typically arranged either in the hollow fiber or spiral wound configuration.

Spiral wound ultrafiltration processes were chosen in this instance for their compact, cross-flow design and ability to concentrate the photographic emulsion. The process is pressure-driven at 6.9–8.6 bar inlet pressure set point through a polysulfone membrane with a 10,000 molecular weight cutoff. The photographic emulsions, processed at 40°C, range in size from 0.06 microns to 10 microns and can be cubic, rods, and/or platelets in morphology.

The biotechnology industry successfully demonstrated use of a control scheme controlled by maintaining a constant concentration at the wall (C_{wall}) of the ultrafiltration membrane rather than a constant transmembrane pressure. Use of C_{wall} process control in the biotechnology industry reduced membrane fouling, minimized yield losses, and improved process reproducibility and scale up.

When dewatering and desalting photographic emulsions using ultrafiltration, the flux, or permeate flowrate, is further increased with inlet pressure until an asymptomatic flux was reached, indicating ultrafiltration performance was limited by a concentration polarization at the wall of the membrane.

A model was sought to adequately describe the fouling. In addition, a process control scheme was needed to reduce the concentration polarization and improve yield of small-sized, cubic photographic emulsions.

EXPERIMENTS

A measurement system was required to determine yield losses on membrane surfaces. A pilot scale system (Fig. 1) was initially used measuring 1/100 the capacity of the production scale system. Experiments were also performed on production equipment. Membranes were freshly cleaned and various feeds processed using transmembrane pressure. Membranes were removed from the system and dissected. Representative samples were removed and imaged using scanning electron microscopy (SEM). Yield losses on the surface of the membrane could clearly be detected. Quantitative analysis was performed using x-ray fluorescence spectroscopy, or XRF. In Fig. 2, 18 L of an 0.18 micron spherical feed was processed using various transmembrane pressures with XRF used to measure the resultant yield losses on the membrane at the end of the process.

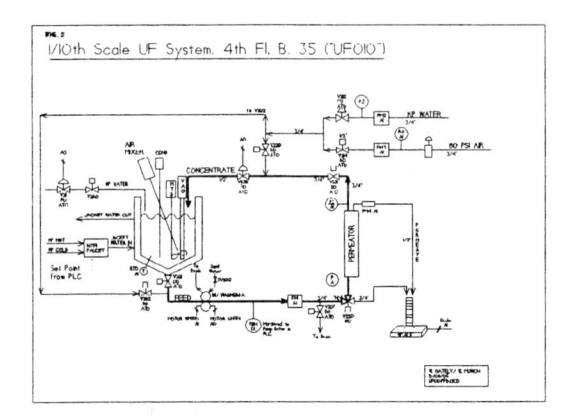

Fig. 1 Pilot scale UF plant (20 L capacity).

Fig. 2 Residual silver halide on membrane surface after processing 18 L of 0.18 micron photographic emulsion vs permeate flow rate (flux).

A critical flux was determined to occur using transmembrane pressure processing. Above the critical flux, further increase in operating pressure does not increase the flux through the system. In fact, flux begins to decrease as the concentration polarization builds up on the membrane surface as in Fig. 3.

Based on these observations, the current transmembrane pressure process appears to be operating above the critical flux. Although this regime is desirable for measuring mass transfer coefficients, it is generally undesirable for operation

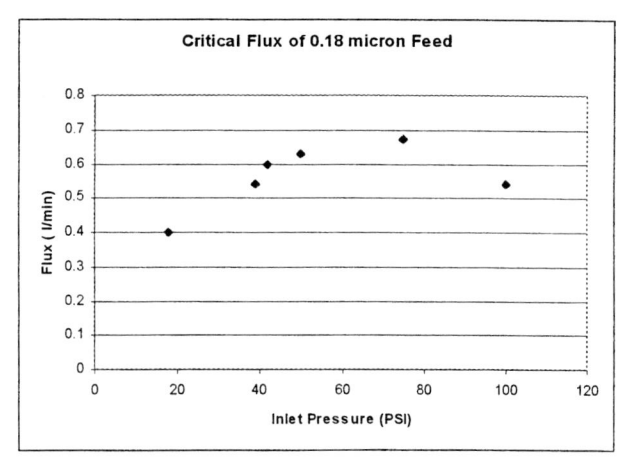

Fig. 3 Critical flux of 0.18 micron diameter spherical feedstream.

where minimizing adhesive losses to the membrane is desired.

A process model based on the Blatt stagnant film model adequately describes the observed variation in permeate flux with process variables such as pressure and crossflow velocity.

Furthermore, a novel process control algorithm for operation at optimal C_{wall}, also based on the Blatt stagnant film model was developed as follows:

$$PFR = k \, LN \, (C_{wall}/C_{bulk}) \qquad (1)$$

Pilot scale experiments were performed using transmembrane pressure processing at a constant pump speed of 330 rpm and an inlet pressure of 6.9 bar (100 psi). When operating above the cricital flux as with these parameters, measurement of the mass transfer coefficient is known to be reliable. The mass transfer coefficient and inherent concentration at the wall were measured from the permeate flowrate process data plotted against both silver halide concentration and gelatin concentration. Values of C_{wall} were calculated vastly exceeding 100% for silver halide, while values that were calculated for gelatin were at or below 100%, indicating that the mass transfer is primarily controlled by gelatin concentration. Fig. 4 demonstrates determination of a gelatin-based mass transfer coefficient using this method.

Pilot scale experiments at 1/100 scale were performed on photographic emulsion feeds of various grain sizes as measured by discreet wavelength turbidimetry. Spiral wound membranes were dissected and analyzed for adhesive losses on the membrane using x-ray fluorescence and SEM.

Inherent concentrations at the wall of the membrane were determined as well as the gelatin-based mass transfer coefficient. Using the maximum, or critical flux, an optimal concentration at the wall of the membrane was chosen using the following semi-empirical model (Fig. 5). Experiments using the destructive membrane test were

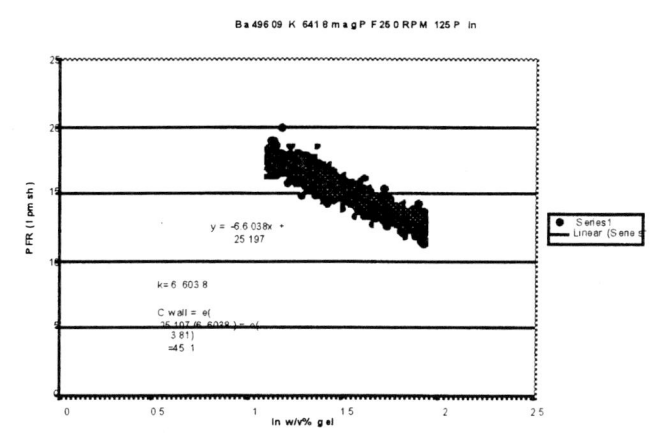

Fig. 4 Determination of a gelatin based mass transfer coefficient.

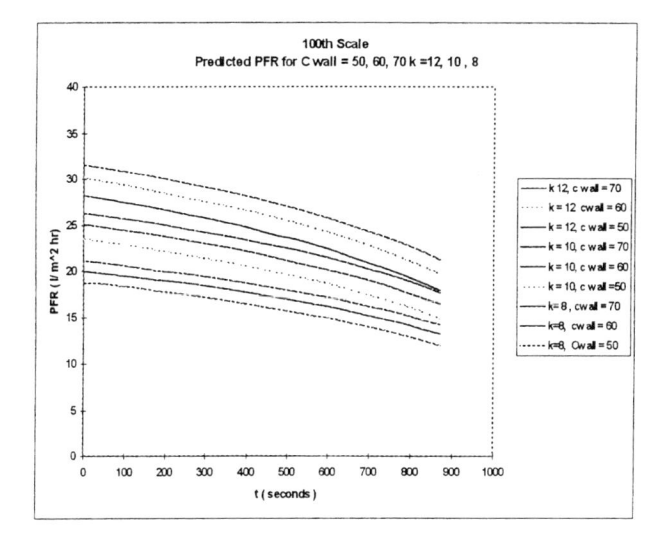

Fig. 5 Semi-empirical model for critical fluxes from 19–32 liters/meter² h.

performed to evaluate the reduction in adhesive losses. The reduction can also be translated to a theoretical yield improvement figure using the total area of membrane in the system.

Production scale experiments were carried out in an identical fashion. The production system is a master/slave system with 10 tubes of 2 membrane units each. While the pilot scale system concentrates, diafiltrates at constant volume, and concentrates again, the production scale system ends the process with an additional step of a reclaim of the feed from both banks of five tubes.

RESULTS

The C_{wall} setpoints are derived from the Blatt Stagnant film theory. Through experimentation, it was determined that the mass transfer coefficient is controlled largely by the gelatin in the silver halide emulsions and not by the silver halide grains. Yield losses caused by absorption of small-sized, fine grain cubic emulsions occur when using the current process where high operating pressures favor the formation of the concentration polarization layer. Yield losses due to absorption of small-sized, fine grain cubic emulsions are reduced when using the novel process control algorithm, referred to as "C_{wall}." C_{wall} operates at lower pressures but with higher crossflow, which constantly sweeps the membrane wall free of the concentrating feed and controls the process to a constant concentration of feed at the wall of the membrane.

This novel process control algorithm was shown to improve yields up to 1% with larger yield improvements occurring with smaller sized cubic emulsions. A pilot scale plant and several full-scale production plants were both retrofitted for the new process control scheme.

The pilot scale plant processes 20 L/h with a total area of 1.2 m². The production plant processes 1500 L/h with a total area of 170 m². The production scale plant included a reclamation operation that cannot be operated using the novel control algorithm because of the high-pressure constraints required to prevent membrane detachment from the membrane backing material. Because of this portion of the process not under C_{wall} control, the membrane losses for the production scale plant are more severe than for the pilot scale plant, but overall yields still improve.

CONCLUSIONS

The novel process control algorithm, called C_{wall}, reduces the adhesive losses on the membrane at pilot scale for a 0.12-micron photographic emulsion by 90%. The novel process control algorithm, called C_{wall}, increases the production yields of an 0.18 micron cubic photographic emulsion by 1% (reduction in adhesive losses of 30%). This represents a savings in yield and productivity for one photographic product family of $1M/year (Fig. 6).

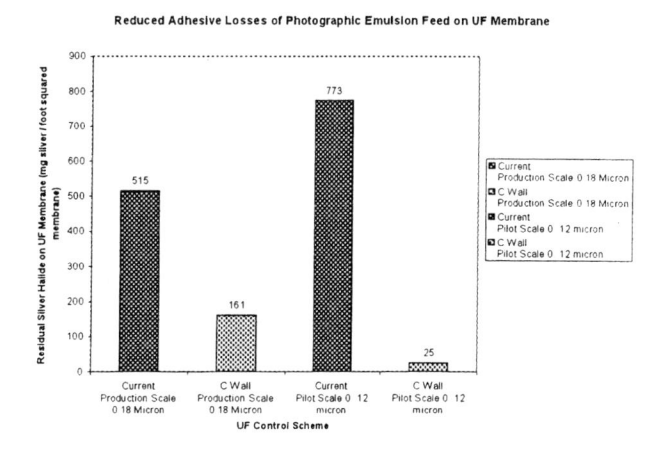

Fig. 6 Reduced adhesive losses of photographic emulsion on UF membrane.

The installed cost of the additional equipment (the permeate flowmeter) required to employ the novel process control algorithm is $15K per installation.

NOMENCLATURE

PFR   Permeate flow rate, L/m^2 h

k   mass transfer coefficient, L/m^2 h

C_{wall} Concentration of gel at wall of membrane, wt/vol % gel

C_{bulk} Concentration of gel in bulk solution, wt/vol % or $V_0 C_0/V$

V_0 volume at time = 0, L

V   volume at time = t, L

C_0 concentration of gel at time = 0, wt/vol %

REFERENCES

Blatt, W.F., 1976, Principles and Practice of Ultrafiltration, IN: P. Meares (Ed.), Membrane Separation Processes, Elsevier, Amsterdam.

Cheryan, M., 1986, Ultrafiltration Handbook, Technomic, Lancaster, Pa., 89-120.

Cheryan, M., Skula, R., Singh, N., Tandon, R., and Nguyen, M.H., 1999, Processing Starch and Its Hydrolysates Using Inorganic Membranes", 10th World Congress of Food Science and Technology, Sydney Australia.

Fane, T., Hoogland, M., October 1996 – November 1997, Personal Communication, UNESCO Center for Membrane Science and Technology, Department of Chemical Engineering, The University of New South Wales, Sydney, Australia.

Gately, W., Hall, R. and Wilcox, M. B., US Patent Docket #79732.

Mees, C.E.K. and James, T.H., 1971, The Theory of the Photographic Process, 31-44.

Munch, W., US Patent 5,164,092.

Munch, W., 1996, Industrial Ultrafiltration Concerns in the Photographic Industry, presented at North American Membrane Society, June 1996.

Van Reis, R., US Patent No. 5,256,294

Van Reis, R., US Patent No. 5,490,937.

Van Reis, R., Goodrich, E., Yson, C., Frautschy, L., Whiteley, R., and Zydney, A., L., 1996, Constant C Wall Ultrafiltration Process Control, *Journal of Membrane Science*.

Zehman, L.J., and Zydney, A.L., 1996, Microfiltration and Ultrafiltration: Principles and Applications, Marcel Dekker, New York, N.Y., 350-4 364.

Zydney, A. L., October 1997 – December 1998, Personal Communication, Department of Chemical Engineering, University of Delaware, USA.